Springer Monographs in Mathematics

A. Bonfiglioli • E. Lanconelli • F. Uguzzoni

Stratified Lie Groups and Potential Theory for their Sub-Laplacians

Springer

A. Bonfiglioli
Università Bologna, Dip.to Matematica
Piazza di Porta San Donato 5
40126 Bologna, Italy
e-mail: bonfigli@dm.unibo.it

F. Uguzzoni
Università Bologna, Dip.to Matematica
Piazza di Porta San Donato 5
40126 Bologna, Italy
e-mail: uguzzoni@dm.unibo.it

E. Lanconelli
Università Bologna, Dip.to Matematica
Piazza di Porta San Donato 5
40126 Bologna, Italy
e-mail: lanconel@dm.unibo.it

ISBN 978-3-642-09099-8 e-ISBN 978-3-540-71897-0

Mathematics Subject Classification (2000): 43A80, 35J70, 35H20, 35A08, 31C05, 31C15, 35B50, 22E60

ISSN 1439-7382

Springer is a part of Springer Science+Business Media
springer.com

Cover design: WMXDesign, Heidelberg, Germany

To Professor Bruno Pini

and to our Families

Preface

With this book we aim to present an introduction to the stratified Lie groups and to their Lie algebras of the left-invariant vector fields, starting from basic and elementary facts from linear algebra and differential calculus for functions of several real variables. The second aim of this book is to perform a potential theory analysis of the *sub-Laplacian* operators

$$\mathcal{L} = \sum_{j=1}^{m} X_j^2,$$

where the X_j's are vector fields, i.e. linear first order partial differential operators, generating the Lie algebra of a stratified Lie group.

In recent years, these operators have received considerable attention in literature, mainly due to their basic rôle in the theory of subelliptic second order partial differential equations with semidefinite characteristic form.

1. Some Historical Overviews

General second order partial differential equations with non-negative and degenerate characteristic form have appeared in literature since the early 1900s. They were first studied by M. Picone, who called them *elliptic-parabolic equations* and proved the celebrated weak maximum principle for their solutions [Pic13,Pic27].

The interest in this type of equations in application fields was originally found by A.D. Fokker, M. Planck and A.N. Kolmogorov. They discovered that partial differential equations with non-negative characteristic form arise in the mathematical modeling of theoretical physics and of diffusion processes [Fok14,Pla17,Kol34].

Since then, over the past half-century, this type of equations appeared in many other different research fields, both theoretical and applied, including geometric theory of several complex variables, Cauchy–Riemann geometry, partial differential equations, calculus of variations, quasiconformal mappings, minimal surfaces and convexity in sub-Riemannian settings, Brownian motion, kinetic theory of gases, mathematical models in finance and in human vision. We report a short list of references for these topics at the end of this preface.

A first systematic study of boundary value problems for wide classes of elliptic-parabolic operators was performed by G. Fichera. In 1956 [Fic56a,Fic56b], he proved existence theorems of *weak solutions* of the "Dirichlet problem" and found the right subset of the boundary on which the data have to be prescribed.

Some years later, several existence and regularity results for elliptic-parabolic operators were proved by O.A. Oleĭnik and E.V. Radkevič and by J.J. Kohn and L. Nirenberg (see the monograph [OR73] for a presentation and a wide survey on this subject). The methods used by these authors required particular assumptions on the *Fichera boundary set* and led to regularity results strongly depending on the regularity of the boundary data.

1.1. L. Hörmander's Theorem

The investigations of the *local* regularity properties of the solutions to elliptic-parabolic equations, that is, regularity properties *only depending on the given operator*, have produced more interesting results. The most beautiful ones have been obtained for elliptic-parabolic equations with underlying algebraic-geometric structures of sub-Riemannian type. The milestone of these research field is a celebrated theorem of L. Hörmander proved in 1967.

Theorem 1 (L. Hörmander, [Hor67]). *Let $X_1, \dots, X_m$ and Y be smooth vectors fields, i.e. linear first order partial differential operators with smooth coefficients in the open set $\Omega \subseteq \mathbb{R}^N$. Suppose*

$$\operatorname{rank}\big(\operatorname{Lie}\{X_1, \dots, X_m, Y\}(x)\big) = N \quad \forall\, x \in \Omega. \tag{P.1}$$

Then the operator

$$L = \sum_{j=1}^{m} X_j^2 + Y \tag{P.2}$$

is hypoelliptic in Ω, i.e. every distributional solution to $Lu = f$ is of class C^∞ whenever f is of class C^∞.

Condition (P.1) simply means that at any point of Ω one can find N *linearly independent* differential operators among $X_1, \dots, X_m, Y$ and all their commutators (the Lie algebra generated by $\{X_1, \dots, X_m, Y\}$).

Hörmander's work opened up a research field, the most remarkable contributions to which have been given by G.B. Folland, L.P. Rothschild and E.M. Stein. They developed and applied to (P.2) the singular integral theory in nilpotent Lie groups.[1]

[1] The application of this theory also occurs in the developments started from the works by J.J. Kohn on the $\bar{\partial}$-Neumann problem and the $\bar{\partial}_b$ complex.

By using these techniques, in 1975, G.B. Folland accomplished a functional analytic study of sub-Laplacians on stratified Lie groups [Fol75]. One year later L.P. Rothschild and E.M. Stein proved their celebrated *lifting theorem* (see [RS76]), enlightening the basic rôle played by the sub-Laplacians in the theory of second order partial differential equations which are sum of squares of vector fields. In force of this theorem, indeed, we can roughly say that:

Every operator $L = \sum_{j=1}^{m} X_j^2$ *satisfying the Hörmander rank condition* (P.1) *can be* lifted *to an operator* $\widehat{L}$ *"as close as we want" to a sub-Laplacian.*

1.2. The Rank Condition

The geometrical meaning of the rank condition (P.1) is clarified by the C. Carathéodory, W.L. Chow and P.K. Rashevsky theorem:

If (P.1) *is satisfied, then given two points* $x, y \in \Omega$, *sufficiently close, there exists a piecewise smooth curve, contained in* Ω *and connecting* x *and* y, *which is the sum of integral trajectories of the vector fields* $\pm X_1, \dots, \pm X_m, \pm Y$.

The appearance of (P.1) in Hörmander's theorem seems to be suggested by some deep properties of the Kolomogorov operators (see also the Introduction in [Hor67]), which we now aim to discuss.

In studying diffusion phenomena from a probabilistic point of view, A.N. Kolmogorov showed that the probability density of a system with $2n$ degrees of freedom satisfies an equation with non-negative characteristic form

$$Ku = 0 \quad \text{in } \mathbb{R}^{2n} \times \mathbb{R},$$

where $\mathbb{R}^{2n}$ is the phase-space of the system. A prototype for K is the following operator

$$K = \sum_{j=1}^{n} \partial_{x_j}^2 + \sum_{j=1}^{n} x_j \, \partial_{y_j} - \partial_t, \tag{P.3}$$

where $x = (x_1, \dots, x_n)$ and $y = (y_1, \dots, y_n)$ denote the velocity and the position vectors of the system, respectively. The operator K is "very degenerate": its second order part only contains derivatives with respect to the variables $x_1, \dots, x_n$. Nevertheless, as Kolmogorov showed, it has a *fundamental solution* Γ which is smooth out of its pole. This implies that K is hypoelliptic, that is, every distributional solution to $Ku = f$ is of class C^∞ whenever f is of class C^∞. The explicit expression of Γ is given by

$$\Gamma(z, t; \zeta, \tau) = \gamma\big(\zeta - E(t - \tau)z, t - \tau\big), \quad z = (x, y), \ \zeta = (\xi, \eta), \tag{P.4}$$

where $\gamma(z, t) = 0$ if $t \le 0$, and

$$\gamma(z, t) = \frac{(4\pi)^n}{\sqrt{\det C(t)}} \exp\left(-\frac{1}{4}\langle C^{-1} z, z\rangle\right) \quad \text{if } t > 0. \tag{P.5}$$

Here, $\langle \cdot, \cdot \rangle$ stands for the usual inner product in $\mathbb{R}^{2n}$; $E(t)$ and $C(t)$, respectively, denote the $2n \times 2n$ matrices

$$E(t) = \exp\left(-t\begin{pmatrix} 0 & 0 \\ \mathbb{I}_n & 0 \end{pmatrix}\right), \quad C(t) = \int_0^t E(s)A\,E(s)^T\,ds.$$

Moreover, $\mathbb{I}_n$ denotes the identity matrix of order n and $A = \begin{pmatrix} \mathbb{I}_n & 0 \\ 0 & 0 \end{pmatrix}$. We explicitly remark that

$$C(t) > 0 \quad \text{for every } t > 0. \tag{P.6}$$

This condition makes expression (P.5) meaningful and can be restated in geometrical–differential terms. Indeed, denoting

$$X_j = \partial_{x_j} \quad \text{and} \quad Y = \textstyle\sum_{k=1}^n x_k\,\partial_{y_k} - \partial_t,$$

it can be proved that (P.6) *is equivalent to the following rank condition*:

$$\operatorname{rank}\big(\operatorname{Lie}\{X_1, \ldots, X_n, Y\}(z,t)\big) = 2n + 1 \quad \forall\,(z,t) \in \mathbb{R}^{2n+1}. \tag{P.7}$$

It is also worthwhile to note that the Kolmogorov operator K can be written as

$$K = \sum_{j=1}^n X_j^2 + Y. \tag{P.8}$$

1.3. The Left Translation and Dilation Invariance

The structure (P.4) of Kolmogorov's fundamental solution suggests the relevance that a *Lie group theoretical approach* has in the analysis of Hörmander operators. Indeed, from the explicit expression of Γ one realizes that

$$\Gamma(z,t;\zeta,\tau) = \gamma\big((\zeta,\tau)^{-1} \circ (z,t)\big),$$

where $\circ$ is the following composition law making $\mathbb{K} := (\mathbb{R}^{2n} \times \mathbb{R}, \circ)$ a *non-commutative Lie group*

$$(z,t) \circ (z',t') := \big(z' + E(t')\,z, t + t'\big),$$

i.e. more explicitly,

$$(x,y,t) \circ (x',y',t') = \big(x + x', y + y' + t'\,x, t + t'\big).$$

In $\mathbb{K}$ one has $(\zeta,\tau)^{-1} = (-E(-t)\,\xi, -\tau)$. It is easy to check that K is invariant w.r.t. the left translations on $\mathbb{K}$ and commutes with the following *dilations*:

$$d_\lambda(z,t) := (\lambda x, \lambda^3 y, \lambda^2 t), \quad \lambda > 0.$$

For every $\lambda > 0$, d_λ is an automorphism of $\mathbb{K}$, so that $(\mathbb{R}^{2n} \times \mathbb{R}, \circ, d_\lambda)$ is a *homogeneous Lie group*. It can be seen that its *Lie algebra* is the one generated by the vector fields $X_j = \partial_{x_j}$ and $Y = \sum_{k=1}^n x_k\,\partial_{y_k} - \partial_t$ appearing in (P.8).

1.4. The Elliptic Counterpart: Stratified Groups and Sub-Laplacians

For a proper comprehension and appreciation of this type of "parabolic"-type operators such as the above Kolmogorov operator K, it is crucial to possess a deep knowledge of their "elliptic" counterpart. This seems unavoidable, also bearing in mind that the underlying algebraic–geometric structures of these two different classes of operators are almost identical. Let us go back again, for a moment, to the Kolmogorov operator (P.3). If in that operator we square the term $Y = \sum_{j=1}^n x_j\, \partial_{x_{j+n}} - \partial_t$, we obtain the following "sum of square"-operator (which we may refer to as the "elliptic counterpart" of K):

$$\mathcal{L} := \sum_{j=1}^n \partial_{x_j}^2 + \left(\sum_{j=1}^n x_j\, \partial_{x_{j+n}} - \partial_t \right)^2. \tag{P.9}$$

The characteristic form of $\mathcal{L}$ is a non-negative quadratic form with non-trivial kernel. Then $\mathcal{L}$ has to be considered as a *degenerate elliptic* operator. However, it is *hypoelliptic*: the Hörmander rank condition (P.7) does not distinguish between $\mathcal{L}$ and K! Moreover, $\mathcal{L}$ is left-invariant on $(\mathbb{R}^{2n} \times \mathbb{R}, \circ)$ (as we already know, so are the ∂_{x_j}'s and Y) but, this time, it commutes with the dilations

$$\delta_\lambda(z, t) = (\lambda x, \lambda^2 y, \lambda^2 t), \quad \lambda > 0.$$

Also these dilations are automorphisms of $(\mathbb{R}^{2n} \times \mathbb{R}, \circ)$, and $\mathbb{G} := (\mathbb{R}^{2n} \times \mathbb{R}, \circ, \delta_\lambda)$ becomes a *stratified Lie group* whose *generators* are the vector fields ∂_{x_j}'s and Y. Then, according to our general agreement, $\mathcal{L}$ is a *sub-Laplacian*[2] on $\mathbb{G}$.

1.5. The Heisenberg Group

In the lower-dimensional case $n = 1$, the operator (P.9) is

$$\partial_x^2 + (x\, \partial_y - \partial_t)^2, \quad (x, y, t) \in \mathbb{R}^3. \tag{P.10}$$

Up to a change and a relabeling of the variables, this can be written as follows:

$$(\partial_x^2 + 2\, y\, \partial_t)^2 + (\partial_y - 2\, x\, \partial_t)^2, \quad (x, y, t) \in \mathbb{R}^3,$$

which, in turn, is the lower-dimensional version of the celebrated sub-Laplacian on the *Heisenberg group*.

The Heisenberg group $\mathbb{H}^n$ is the stratified Lie group $(\mathbb{R}^{2n+1}, \circ)$ whose composition law is given by

$$(z, t) \circ (z', t') = \big(z + z', t + t + 2\,\mathrm{Im}\langle z, z'\rangle\big). \tag{P.11}$$

Here we identify $\mathbb{R}^{2n}$ with $\mathbb{C}^n$, and we use the notation

---

[2] All these notions will be properly introduced in Chapter 1.

$$(z, t) = (z_1, \dots, z_n, t) = (x_1, y_1, \dots, x_n, y_n, t)$$

for the points of $\mathbb{H}^n$. In (P.11), $\langle \cdot, \cdot \rangle$ stands for the usual Hermitian inner product in $\mathbb{C}^n$. The dilation

$$\delta_\lambda(z, t) = (\lambda z, \lambda^2 t) \tag{P.12}$$

is an automorphism of $\mathbb{H}^n$ and the vector fields

$$X_j = \partial_{x_j} + 2\, y_j\, \partial_t, \qquad Y_j = \partial_{y_j} - 2\, x_j\, \partial_t$$

are left-invariant on $(\mathbb{H}^n, \circ)$. One readily recognizes that the following commutation relations hold:

$$[X_j, Y_j] = -4\, \partial_t \tag{P.13a}$$

and

$$[X_j, X_k] = [Y_j, Y_k] = [X_j, Y_k] = 0 \quad \forall\, j \neq k. \tag{P.13b}$$

Identity (P.13a) is the canonical commutation relation between momentum and position in quantum mechanics. From (P.13a) it follows that

$$\operatorname{rank}\big(\operatorname{Lie}\{X_1, \dots, X_n, Y_1, \dots, Y_n, \partial_t\}\big) = 2n + 1$$

at any point of $\mathbb{R}^{2n+1}$. Then, by Hörmander's Theorem 1, the sub-Laplacian on $\mathbb{H}^n$

$$\Delta_{\mathbb{H}^n} := \sum_{j=1}^{n} (X_j^2 + Y_j^2)$$

is hypoelliptic. The Heisenberg group and its Lie algebra originally arose in the mathematical formalizations of quantum mechanics (see H. Weyl [Weyl31]). Today, they appear in many research fields such as several complex variables, CR geometry, Fourier analysis and partial differential equations of subelliptic type. The Heisenberg sub-Laplacian is undoubtedly the most important prototype of the sub-Laplacians on non-commutative stratified Lie groups.

1.4. The Lifting Theorem

Obviously, a generic Hörmander operator *sum of squares of vector fields* is not, in general, a sub-Laplacian on some stratified Lie group. Just consider, as an example,

$$M = \partial_x^2 + (x\, \partial_y)^2 \quad \text{in } \mathbb{R}^2.$$

This operator satisfies the Hörmander rank condition, hence it is hypoelliptic. However, there is no Lie group structure in $\mathbb{R}^2$ making M left-invariant on it. Nevertheless, adding the new variable t, M can be *lifted* to the operator in (P.10) which is the sub-Laplacian on (a group isomorphic to) the Heisenberg group $\mathbb{H}^1$. This is the source idea of the *lifting theorem* by L.P. Rothschild and E.M. Stein, which states, roughly speaking, that any Hörmander operator sum of squares of vector fields can be approximated by a sub-Laplacian on a stratified group. This result emphasizes the major rôle played by the sub-Laplacians in the theory of second order PDE's with non-negative and degenerate characteristic form.

1.5. Stratified Groups in Sub-Riemannian Geometry

Stratified groups also appear naturally in sub-Riemannian geometry (frequently referred to as "Carnot" geometry). Roughly speaking, stratified groups play a rôle, for sub-Riemannian manifolds, analogous to that played by Euclidean vector spaces for Riemannian manifolds.

More precisely, once it has been provided a suitable notion of *tangent space* at a point of a sub-Riemannian manifold, it turns out that (at a regular point) this tangent space is naturally endowed with a structure of nilpotent Lie group with dilations, a stratified Lie group (see J. Mitchell [Mit85] and A. Bellaïche in [BR96]).

Furthermore, the analysis of a left invariant sub-Laplacian on a connected nilpotent Lie group (or more generally on a Lie group *of polynomial growth*) and the geometry at infinity of this group is described by a canonically associated dilation-invariant sub-Laplacian on a stratified Lie group. See G. Alexopoulos [Ale92,Ale02], S. Ishiwata [Ish03] and N.Th. Varopoulos [Varo00].

2. The Contents of the Book: An Overview

A glance at the contents of the book and at our approach to the subjects is in order. The book is divided into three parts, and every part is, in its turn, subdivided in several chapters plus some appendices, if necessary.

2.1. Part I

The first four chapters of Part I are devoted to an elementary and self-contained introduction to the stratified Lie groups in $\mathbb{R}^N$. Our presentation does not require a specialized knowledge neither in algebra nor in differential geometry. The approach is completely elementary, "constructive" whenever possible, abundant in examples and intended to be understandable by readers with basic backgrounds only in linear algebra and differential calculus in $\mathbb{R}^N$. Subsequently, we present the formal and abstract approach to the stratified Lie groups commonly used in literature, and we prove the equivalence of the abstract notion of *stratified group* to the "constructive" notion of *homogeneous Carnot group*. This equivalence is also provided in Part I.

A very special emphasis is given to the examples. We introduce and discuss a wide range of explicit stratified Lie groups of arbitrarily large dimension and step. Some of them have been known in specialized literature for several years, such as the Heisenberg–Kaplan groups, the filiform groups and the Métivier groups. Many others have only appeared very recently, in particular what we shall call the Kolmogorov-type groups and the Bony-type groups. Other examples are completely new, some extracted from geometric control theory.

Our long list of examples is also intended to be appreciated by readers working in geometry and analysis on Carnot groups. It provides a valuable benchmark set to test new special properties of the groups, to exhibit explicit examples and counterexamples of the "pathologies" and the special features of Carnot groups.

It is also payed a special attention to the Lie algebras of the groups by stressing their links with second order partial differential operators of Hörmander type (sum of squares of vector fields). In particular, given such an operator, we show necessary and sufficient conditions for it to be a sub-Laplacian on a suitable homogeneous stratified Lie group, and we explicitly show how to construct the related composition law. As a byproduct, this enables the reader to build up another plenty of examples of stratified groups and sub-Laplacians.

Chapter 5 of Part I is dedicated to the analysis of the *fundamental solution* for the sub-Laplacians, a central topic of Part I. Here, the mainly used analytic tools are *integration by parts* and *coarea-formulas*. We start from the *hypoellipticity* of sub-Laplacians, easy consequence[3] of the Hörmander Theorem 1.

From this "assumption" on hypoellipticity, and with the aid of the *strong maximum principle*, whose proof is postponed to the Appendix of Chapter 5, we deduce the existence of a *gauge function* d for any given sub-Laplacian $\mathcal{L}$, i.e. the existence of a positive non-constant homogeneous function d such that d^{2-Q} is $\mathcal{L}$*-harmonic* away from the origin. Here Q stands for the homogeneous dimension of the group on which the sub-Laplacian lives (*we always assume that* $Q \geq 3$).

This property is one of the most striking analogies between $\mathcal{L}$ and the classical Laplace operator. We show that this leads to suitable *mean value formulas* on the *d-balls*, extending to this new setting the well-known Gauss theorem for classical harmonic functions. We then use these formulas (which will play a crucial rôle throughout the book) to prove *Liouville-type theorems*, *Harnack-type inequalities*, and a *Sobolev–Stein embedding theorem*. Furthermore, three sections are devoted to the following topics: some remarks on the analytic-hypoellipticity of sub-Laplacians, $\mathcal{L}$-harmonic approximations and an integral representation formula for the fundamental solution.

2.2. Part II

Part II of the book contains an exhaustive potential theory for the sub-Laplacians. Basically, our only starting point is the theorem by G.B. Folland asserting the *existence* for these operators of a *homogeneous* and *smooth* fundamental solution with pole at the origin. This key result allows us to perform a complete potential theory that parallel the one of the classical Laplace operator.

The lack of explicit Poisson integral formulas forces us to follow an abstract approach to this theory. For this reason, in Chapter 6 we present some topics from abstract harmonic space theory, mainly inspired by the ones developed by H. Bauer [Bau66] and C. Costantinescu and A. Cornea [CC72]. This chapter mainly involves Perron–Wiener–Brelot method for the Dirichlet problem, harmonic minorants and majorants and balayage theory.

[3] We do not go into the proof of this theorem, for it would require techniques very far from the ones developed in the book.

Next, in Chapter 7 we show that every sub-Laplacian equips $\mathbb{R}^N$ with a structure of harmonic space satisfying the axioms of the theory presented in Chapter 6. This is accomplished by using the Harnack-type theorem proved in Chapter 5, and then by showing the existence of a basis of the topology of $\mathbb{R}^N$ formed by $\mathcal{L}$-regular sets, i.e. by sets for which the Dirichlet problem for $\mathcal{L}$ is solvable in the usual classical sense (here we follow an idea by J.-M. Bony [Bon69]).

In the subsequent chapters of Part II, we use the full strength of the abstract theory, together with the remarkable properties of the fundamental solution for $\mathcal{L}$ to deal with the arguments listed below:

a) sub-mean characterizations of the $\mathcal{L}$-subharmonic functions, and applications to the notion of convexity in Carnot groups;
b) Green functions and Riesz representation theorems for $\mathcal{L}$-subharmonic functions, with applications (among which Bôcher-type theorems);
c) maximum principles on unbounded domains;
d) $\mathcal{L}$-capacity and $\mathcal{L}$-polar sets, with applications: the Poisson–Jensen formula and the so-called *fundamental convergence theorem*;
e) $\mathcal{L}$-thinness and $\mathcal{L}$-fine topology, with applications to the Dirichlet problem (and the derivation of Wiener's criterion);
f) the links between the Hausdorff measure naturally related to the gauge d and the capacity for $\mathcal{L}$.

In writing this part of the book we were also partially inspired by some monographs on potential theory for the classical Laplace operator—in particular the beautiful books by L.L Helms [Helm69], by W.K. Hayman and P.B. Kennedy [HK76] and by D.H. Armitage and S.J. Gardiner [AG01].

2.3. Part III

In Part III, we take up further topics on the algebraic and analytic theory of Carnot groups. In particular, this part of the book provides:

a) the study of *free Lie algebras*;
b) clear and complete proofs, in several contexts, of the fundamental and remarkable *Campbell–Hausdorff formula*[4];
c) the equivalence of sub-Laplacians under diffeomorphisms;
d) the *Rothschild–Stein lifting theorem* and *Folland's lifting theorem* (for stratified or homogeneous vector fields);
e) the study of the algebraic structure of *Heisenberg–Kaplan-type groups* (also providing an explicit characterization of them) with a special emphasis to the remarkable form of their fundamental solutions, discovered by G.B. Folland and A. Kaplan (we also present the *inversion* and the *Kelvin transform* in the H-type groups of *Iwasawa type*);

[4] In Chapter 15, we collect *four* theorems for the Campbell–Hausdorff formula: one for homogeneous vector fields, two for formal power series and one for general smooth vector fields.

f) the *Carathéodory–Chow–Rashevsky connectivity theorem* (for stratified vector fields) with applications;
g) *Taylor's formula* (with Lagrange and with integral remainder) on Carnot groups.

The difficulty of finding "easy" and complete proofs of some of the above mentioned results in the existing literature is well known. By working with stratified vector fields we are able to overcome some of the lengthy steps of the proofs, while maintaining a good amount of generality in the final results.

3. How to Read this Book

Besides Ph.D. students, the book is addressed to young and senior researchers. Indeed, one of the main efforts in presenting the material is to use an elementary approach and to reach, step by step, the level of current researches. Many parts of the book may be used for graduate courses and advanced lectures.

The first four chapters of Part I are addressed to non-specialists in Lie group and Lie algebra theory. The first two chapters can be skipped by the readers having familiarity with the basics of differential geometry and Lie group theory. The reader already acquainted with Carnot group theory can pass directly to Chapter 5. In any case, beginners and specialists in the theory of stratified groups can exploit the first four chapters as a source for examples.

Part II is the core of the monograph. The reader with some background in potential theory (and interested in the main case of sub-Laplacians) can pass directly to Chapter 7 and proceed throughout Part II, leaving Chapters 10 and 13 as a further reading.

Part III is thought of as a more specialized lecture. Nonetheless, a deep understanding of, e.g. the Campbell–Hausdorff formula or of Heisenberg-type groups are amongst the main goals of this monograph.

The book provides 21 illustrative figures, 250 exercises (each chapter has its own section of exercises) and an index of the basic notation. For the reading convenience, we furnish a synoptic diagram of the structure of the book on page XVII.

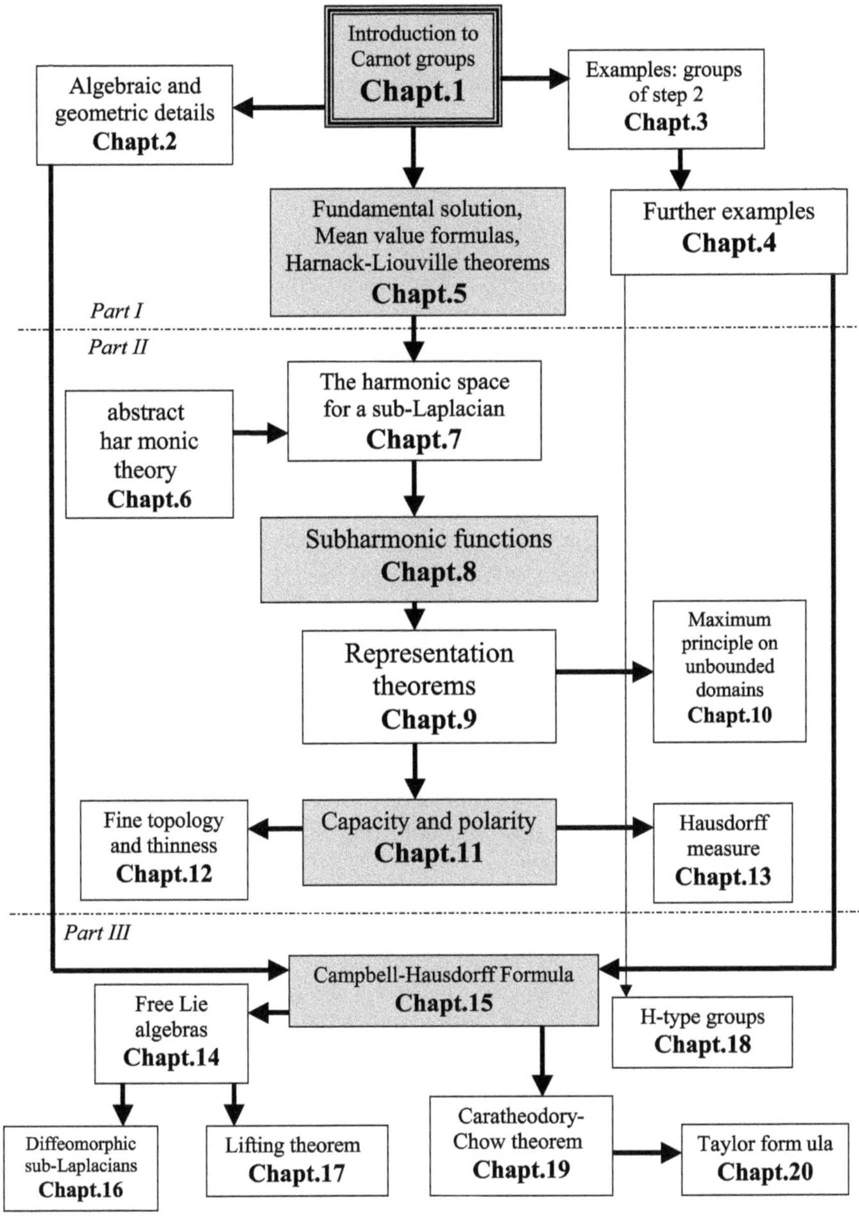

The synoptic diagram of the structure of the book.

4. Some References on Theoretical and Applied Related Topics

Here is a short list of references for related topics on analysis on stratified Lie groups and applications.[5]

Alexopoulos [Ale02], Altafini [Alt99], Bellaïche and Risler [BR96],
Birindelli, Capuzzo Dolcetta and Cutrì [BCC97], Bahri [Bah04,Bah03],
Barletta [Bar03], Barletta and Dragomir [BD04],
Barletta, Dragomir and Urakawa [BDU01],
Birindelli, Capuzzo Dolcetta and Cutrì [BCC98],
Brandolini, Rigoli and Setti [BRS98], Capogna [Cap99],
Capogna and Cowling [CC69], Capogna and Garofalo [CG98,CG03,CG06],
Capogna, Garofalo and Nhieu [CGN00,CGN02], Capuzzo Dolcetta [CD98],
Chandresekhar [Cha43], Citti [Cit98], Citti, Lanconelli and Montanari [CLM02],
Citti, Manfredini and Sarti [CMS04], Citti and Montanari [CM00],
Citti, Pascucci and Polidoro [CPP01], Citti and Sarti [CS06],
Citti and Tomassini [CT04] Cowling, De Mari, Korányi and Reimann [CDKR02],
Cowling and Reimann [CR03], Danielli, Garofalo, Nhieu and Tournier [DGNT04],
Danielli, Garofalo and Salsa [DGS03], Dragomir [Dra01],
Franchi, Gutiérrez and van Nguyen [FGvN05],
Franchi, Serapioni and Serra Cassano [FSS03a,FSS03b,FSS01],
Gamara [Gam01], Gamara and Yacoub [GY01],
Garofalo and Lanconelli [GL92], Garofalo and Tournier [GT06],
Garofalo and Vassilev [GV00], Golé and Karidi [GK95],
Gutiérrez and Lanconelli [GL03], Gutiérrez and Montanari [GM04a,GM04b],
Heinonen [Hei95b], Heinonen and Holopainen [HH97],
Heinonen and Koskela [HK98], Huisken and Klingenberg [HK99],
Jerison and Lee [JL87,JL88,JL89],
Juutinen, Lu, Manfredi and Stroffolini [JLMS07],
Korányi and Reimann [KR90,KR95],
Lanconelli [Lan03], Lanconelli, Pascucci and Polidoro [LPP02],
Lanconelli and Uguzzoni [LU00], Lu, Manfredi and Stroffolini [LMS04],
Lu and Wei [LW97], Malchiodi and Uguzzoni [MU02],
Manfredi and Stroffolini [MS02], Montanari [Mo01],
Montanari and Lascialfari [ML01], Montgomery [Mon02],
Montgomery, Shapiro and Stolin [MSS97], Monti and Morbidelli [MM05],
Monti and Rickly [MR05], Monti and Serra Cassano [MSC01],
Petitot and Tondut [PT99], Reimann [Rei01a,Rei01b],
Slodkowski and Tomassini [ST91], Stein [Ste81],
Uguzzoni [Ugu00], Varopoulos, Saloff-Coste and Coulhon [VSC92].

[5] The list is alphabetically ordered and the grouping of the references in different lines is only meant for typographical readability.

5. Acknowledgments

The authors would like to thank Chiara Cinti and Andrea Tommasoli for careful reading some chapters of the manuscript.

We would also like to express our gratitude to Italo Capuzzo Dolcetta, Cristian E. Gutiérrez, Guozhen Lu and Juan J. Manfredi for the kind encouraging appreciation of our work.

It is also a pleasure to thank the Springer-Verlag staff for the kind collaboration, in particular Dr. Catriona M. Byrne, Dr. Marina Reizakis and Dr. Susanne Denskus.

Some topics presented in this book have partially appeared in the following papers by joint collaborations of the authors (and of one of us with C. Cinti) [Bon04, BC04,BC05,BL01,BL02,BL03,BL07,BLU02,BLU03,BU04a,BU04b,BU05a].

Bologna, Italy
April 2007

Andrea Bonfiglioli
Ermanno Lanconelli
Francesco Uguzzoni

Contents

Part I Elements of Analysis of Stratified Groups

Part I

Elements of Analysis of Stratified Groups

1

Stratified Groups and Sub-Laplacians

In this first chapter, we introduce the main notation and the basic definitions concerning with vector fields in $\mathbb{R}^N$: algebras of vector fields, exponentials of smooth vector fields, Lie brackets. Then, we introduce the main geometric structure investigated throughout the book: the homogeneous Carnot groups.

To this end, we first study Lie groups $\mathbb{G}$ on $\mathbb{R}^N$ and the Lie algebra of their left-invariant vector fields. Subsequently, we equip $\mathbb{G}$ with a *homogeneous* structure by the datum of a well-behaved group of dilations $\{\delta_\lambda\}_{\lambda>0}$ on $\mathbb{G}$. The composition of $\mathbb{G}$ thus takes a transparent form, allowing us to study homogeneous Lie groups on $\mathbb{R}^N$ by very direct methods. In particular, it will be a simple exercise of calculus to verify that the relevant exponential and logarithmic maps are global polynomial diffeomorphisms, a result which will throughout account for a very useful tool. Finally, we introduce the notion of homogeneous Carnot group and of sub-Laplacian. A wide number of explicit examples of homogeneous Carnot groups will be given in Chapters 3 and 4, after (in Chapter 2) we have analyzed the due relationship between abstract Carnot groups and the homogeneous ones.

Indeed, despite our notions of Lie group on $\mathbb{R}^N$, homogeneous and homogeneous-Carnot group are dependent on a fixed system of coordinates for the group (thus being non-intrinsic notions), every abstract Carnot group is, as we shall see, isomorphic via the exponential map to its Lie algebra, which is a *homogeneous Carnot group*. This basic fact provides another motivation for this introductory chapter.

1.1 Vector Fields in $\mathbb{R}^N$. Exponential Maps. Lie Algebras of Vector Fields

1.1.1 Vector Fields in $\mathbb{R}^N$

Given $N \in \mathbb{N}$, we set, as usual, $\mathbb{R}^N = \{(x_1, \dots, x_N) : x_1, \dots, x_N \in \mathbb{R}\}$. We use any of the notation

$$\partial_j, \qquad \partial_{x_j}, \qquad \frac{\partial}{\partial x_j}, \qquad \partial/\partial x_j$$

to indicate the partial derivative operator with respect to the j-th coordinate of $\mathbb{R}^N$. Let $\Omega \subseteq \mathbb{R}^N$ be an open (and non-empty) set. Given an N-tuple of scalar functions $a_1, \dots, a_N$,

$$a_j : \Omega \to \mathbb{R}, \quad j \in \{1, \dots, N\},$$

the first order linear differential operator

$$X = \sum_{j=1}^{N} a_j \, \partial_j \tag{1.1}$$

will be called a *vector field on Ω with component functions* (or simply, *components*) $a_1, \dots, a_N$. If $O \subseteq \Omega$ is an open set and $f : O \to \mathbb{R}$ is a differentiable function, we denote by Xf the function on O defined by

$$Xf(x) = \sum_{j=1}^{N} a_j(x) \, \partial_j f(x), \qquad x \in O.$$

Occasionally, we shall also use the notation Xf when

$$f : O \to \mathbb{R}^m$$

is a vector-valued function, to mean the component-wise action of X. More precisely,[1]

$$\text{if } f(x) = \begin{pmatrix} f_1(x) \\ \vdots \\ f_m(x) \end{pmatrix}, \text{ we set } Xf(x) = \begin{pmatrix} Xf_1(x) \\ \vdots \\ Xf_m(x) \end{pmatrix}.$$

---

[1] We warn the reader that points in $\mathbb{R}^N$ will be usually denoted as N-tuples $x = (x_1, \dots, x_N)$. When this does not lead to confusion, the column-vector notation

$$x = \begin{pmatrix} x_1 \\ \vdots \\ x_N \end{pmatrix}$$

will also be allowed. For example, this last notation will be *sometimes* used (with no risk of misunderstanding) for vector-valued functions $f(x) = (f_1(x), \dots, f_N(x))^T$, e.g.

$$I(x) = (x_1, \dots, x_N)^T$$

will always denote (this time as a rule) the identity map in $\mathbb{R}^N$.

However, there are cases in which we shall keep the notation well distinguished: namely, if $f : \mathbb{R}^N \to \mathbb{R}$ is a differentiable function, its gradient $\nabla f = (\partial_1 f, \dots, \partial_N f)$ will *always be written as a row-vector*.

On the other side, we shall *always* use the column-vector notation *when vectors in $\mathbb{R}^N$ appear in matrix calculation*. For example, if $x, y \in \mathbb{R}^N$, $x^T y$ will denote the row$\times$column product

$$x^T y = (x_1, \dots, x_N) \begin{pmatrix} y_1 \\ \vdots \\ y_N \end{pmatrix} = \sum_{j=1}^{N} x_j y_j.$$

(Typically, this notation will be used when $f = I$ is the identity map of $\mathbb{R}^N$, i.e. $I(x) = (x_1, \dots, x_N)^T$; see (1.2) below.) Furthermore, given a differentiable function $f : O \to \mathbb{R}^m$, we shall denote by

$$\mathcal{J}_f(x), \quad x \in O$$

the Jacobian matrix of f at x.

Let $C^\infty(O, \mathbb{R})$ (for brevity, $C^\infty(O)$) be the set of smooth (i.e. infinitely-differentiable) real-valued functions. If the components a_j's of X are smooth, we shall call X a *smooth vector field* and we shall often consider X as an operator acting on smooth functions,

$$X : C^\infty(O) \to C^\infty(O), \qquad f \mapsto Xf.$$

We shall prevalently deal with smooth vector fields. We shall denote by

$$T(\mathbb{R}^N)$$

the set of all smooth vector fields in $\mathbb{R}^N$. Equipped with the natural operations, $T(\mathbb{R}^N)$ is a vector space over $\mathbb{R}$.

We adopt the following (non-conventional) notation: I will denote the identity map on $\mathbb{R}^N$ and, if X is the vector field in (1.1), then

$$XI := \begin{pmatrix} a_1 \\ \vdots \\ a_N \end{pmatrix} \tag{1.2}$$

will be the column vector of the components of X. This notation is obviously consistent with our definition of the action of X on a vector-valued function.[2] Thus, XI may also be regarded as a smooth map from $\mathbb{R}^N$ to itself (that is what some authors call a "vector field" on $\mathbb{R}^N$).

Often, many authors identify X and XI. Instead, in order to avoid any confusion between a smooth vector field *as a function* belonging to $C^\infty(\mathbb{R}^N, \mathbb{R}^N)$ and a smooth vector field *as a differential operator* from $C^\infty(\mathbb{R}^N)$ to itself,[3] we prefer to use the different notation XI and X as described in (1.2) and (1.1), respectively.

By consistency of notation, we may write

$$Xf = (\nabla f) \cdot XI,$$

where

$$\nabla = (\partial_1, \dots, \partial_N)$$

is the gradient operator in $\mathbb{R}^N$, f is any real-valued smooth function on $\mathbb{R}^N$ and $\cdot$ denotes the row$\times$column product. For example, for the following two vector fields on $\mathbb{R}^3$ (whose points are denoted by $x = (x_1, x_2, x_3)$)

$$X_1 = \partial_{x_1} + 2\,x_2\,\partial_{x_3}, \qquad X_2 = \partial_{x_2} - 2\,x_1\,\partial_{x_3}, \tag{1.3a}$$

we have

$$X_1 I(x) = \begin{pmatrix} 1 \\ 0 \\ 2\,x_2 \end{pmatrix}, \qquad X_2 I(x) = \begin{pmatrix} 0 \\ 1 \\ -2\,x_1 \end{pmatrix}. \tag{1.3b}$$

---

[2] Indeed, since $I = (I_1, \dots, I_N)$ with $I_j(x) = x_j$, we have $X(I_j) = a_j$.

[3] Or even from $C^\infty(\mathbb{R}^N, \mathbb{R}^N)$ to itself!

1.1.2 Integral Curves

A path $\gamma : \mathcal{D} \to \mathbb{R}^N$, $\mathcal{D}$ being an interval of $\mathbb{R}$, will be said an *integral curve* of the smooth vector field X if

$$\dot\gamma(t) = XI(\gamma(t)) \quad \text{for every } t \in \mathcal{D}.$$

If X is a smooth vector field, then, for every $x \in \mathbb{R}^N$, the Cauchy problem

$$\begin{cases} \dot\gamma = XI(\gamma), \\ \gamma(0) = x \end{cases} \tag{1.4}$$

has a unique solution

$$\gamma_X(\cdot, x) : \mathcal{D}(X, x) \to \mathbb{R}^N.$$

We agree to denote by $\mathcal{D}(X, x)$ the greatest open interval of $\mathbb{R}$ on which $\gamma_X(\cdot, x)$ exists. For example, if X_1 is as in (1.3a), the solution to the relevant problem (1.4) is obtained by solving the following system of ODE's (we write $\gamma = (\gamma_1, \gamma_2, \gamma_3)$)

$$\begin{cases} \dot\gamma_1(t) = 1, & \dot\gamma_2(t) = 0, & \dot\gamma_3(t) = 2\,\gamma_2(t), \\ \gamma_1(0) = x_1, & \gamma_2(0) = x_2, & \gamma_3(0) = x_3, \end{cases}$$

i.e. after simple computations, we obtain $\mathcal{D}(X_1, x) = \mathbb{R}$ and

$$\gamma_{X_1}(t, x) = (x_1 + t, x_2, x_3 + 2\,x_2\,t). \tag{1.5}$$

Since X is smooth, $t \mapsto \gamma_X(t, x)$ is a C^∞ function whose n-th Taylor expansion in a neighborhood of $t = 0$ is given by

$$\begin{aligned} \gamma_X(t, x) &= x + t\,X^{(1)}I(x) + \frac{t^2}{2!}\,X^{(2)}I(x) + \cdots + \frac{t^n}{n!}\,X^{(n)}I(x) \\ &\quad + \frac{1}{n!}\int_0^t (t-s)^n\,X^{(n+1)}I(\gamma_X(s, x))\,ds. \end{aligned} \tag{1.6}$$

Hereafter, for $k \in \mathbb{N}$, we denote by $X^{(k)}$ the *vector field*

$$X^{(k)} = \sum_{j=1}^N (X^{k-1}a_j)\partial_{x_j},$$

being $X^0 = I$ (the identity map) and X^h, $h \geq 1$, the h-th order iterated of X, i.e.

$$X^h := \underbrace{X \circ \cdots \circ X}_{h \text{ times}}.$$

In other words, we have

$$X^{(1)}I = XI = \begin{pmatrix} a_1 \\ \vdots \\ a_N \end{pmatrix}, X^{(2)}I = \begin{pmatrix} X(a_1) \\ \vdots \\ X(a_N) \end{pmatrix}, X^{(3)}I = \begin{pmatrix} X(X(a_1)) \\ \vdots \\ X(X(a_N)) \end{pmatrix}, \ldots.$$

We remark that X^h is a differential operator of order at most h, whereas $X^{(h)}$ is a differential operator of order at most 1. To check (1.6) we use (1.4). Writing $\gamma(t)$ instead of $\gamma_X(t,x)$, (1.4) gives

$$\gamma(0) = x, \qquad (\mathrm{d}/\mathrm{d}t)|_{t=0}\gamma(t) = XI(x) \quad \text{and}$$

$$\begin{aligned} \frac{\mathrm{d}^2}{\mathrm{d}t^2}\bigg|_{t=0}\gamma(t) &= \frac{\mathrm{d}}{\mathrm{d}t}\bigg|_{t=0}(XI)(\gamma(t)) = \mathcal{J}_{XI}(\gamma(0))\cdot\dot{\gamma}(0) = \mathcal{J}_{XI}(x)\cdot XI(x) \\ &= \begin{pmatrix} \nabla a_1(x)\cdot XI(x) \\ \vdots \\ \nabla a_N(x)\cdot XI(x) \end{pmatrix} = \begin{pmatrix} Xa_1(x) \\ \vdots \\ Xa_N(x) \end{pmatrix} = X^{(2)}I(x). \end{aligned}$$

By iterating this argument, we obtain

$$\gamma^{(k)}(0) := \frac{\mathrm{d}^k}{\mathrm{d}t^k}\bigg|_{t=0}\gamma(t) = X^{(k)}I(x), \quad k \geq 2.$$

Replacing this identity in the Taylor formula

$$\gamma(t) = x + \sum_{k=1}^{n}\frac{t^k}{k!}\gamma^{(k)}(0) + \frac{1}{n!}\int_0^t (t-s)^n\gamma^{(n+1)}(s)\,\mathrm{d}s,$$

we obtain (1.6). Since the identity map I is linear and since the first order part of X^h coincides with $X^{(h)}$, one has

$$X^{(h)}I \equiv X^h I.$$

Thus formula (1.6) can be rewritten as

$$\begin{aligned} \gamma_X(t,x) &= x + t\,XI(x) + \frac{t^2}{2!}\,X^2I(x) + \cdots + \frac{t^n}{n!}\,X^nI(x) \\ &\quad + \frac{1}{n!}\int_0^t (t-s)^n\,X^{n+1}I(\gamma_X(s,x))\,\mathrm{d}s. \end{aligned} \tag{1.7}$$

Example 1.1.1. For example, if X_1 is as in (1.3a), since

$$X_1^{(1)}I = \begin{pmatrix} 1 \\ 0 \\ 2x_2 \end{pmatrix}, \qquad X_1^{(2)}I = \begin{pmatrix} 0 \\ 0 \\ 0 \end{pmatrix} = X_1^{(k)}I \quad \forall k \geq 3,$$

we have

$$\gamma_{X_1}(t,x) = x + t\,X_1I(x) = \begin{pmatrix} x_1 \\ x_2 \\ x_3 \end{pmatrix} + t\begin{pmatrix} 1 \\ 0 \\ 2x_2 \end{pmatrix} = \begin{pmatrix} x_1 + t \\ x_2 \\ x_3 + 2x_2t \end{pmatrix},$$

as we directly found in (1.5). □

1.1.3 Exponentials of Vector Fields

The expansion in (1.7) suggests to use an "exponential-type" notation. For this (more than notational) reason, we give the following definition.

Definition 1.1.2 (Exponential of a vector field). *Let X be a smooth vector field on $\mathbb{R}^N$. Following all the above notation, we set*

$$\exp(tX)(x) := \gamma_X(t, x), \tag{1.8}$$

where $\gamma_X(\cdot, x)$ is the solution of (1.4).

This definition makes sense[4] *for every $X \in T(\mathbb{R}^N)$, for every $x \in \mathbb{R}^N$ and every $t \in \mathcal{D}(X, x)$.* (See Fig. 1.1.)

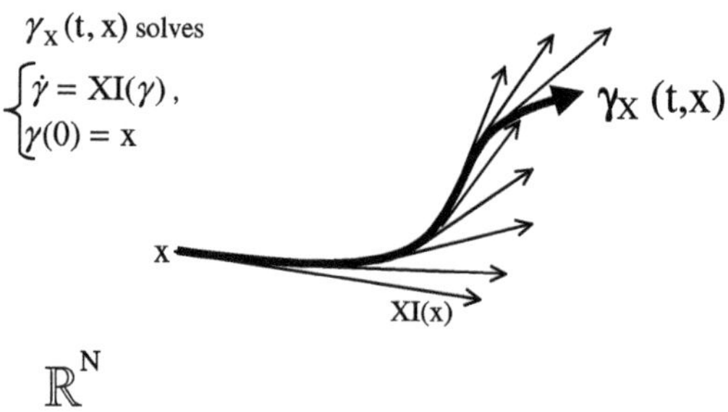

Fig. 1.1. Figure of Definition 1.1.2

Then, being X smooth, for every $n \in \mathbb{N}$, we have the expansion

$$\begin{aligned}\exp(tX)(x) &= \sum_{k=0}^{n} \frac{t^k}{k!} X^k I(x) \\ &\quad + \frac{1}{n!} \int_0^t (t-s)^n \, X^{n+1} I\big(\exp(sX)(x)\big) \, \mathrm{d}s.\end{aligned} \tag{1.9}$$

In particular, for $n = 1$,

$$\exp(tX)(x) = x + t \, XI(x) + \int_0^t (t-s) \, X^2 I\big(\exp(sX)(x)\big) \, \mathrm{d}s. \tag{1.10}$$

If we define

$$\mathcal{U} := \{(t, x) \in \mathbb{R} \times \mathbb{R}^N \mid x \in \mathbb{R}^N, \ t \in \mathcal{D}(X, x)\},$$

from the basic theory of ordinary differential equations (see, e.g. [Har82]) we know that $\mathcal{U}$ is open and the map

[4] Definition 1.1.2 is well-posed also when X has only Lipschitz-continuous component functions, but we shall prevalently deal, as already stated, with smooth vector fields.

$$\mathcal{U} \ni (t, x) \mapsto \exp(tX)(x) \in \mathbb{R}^N$$

is smooth. Moreover, from the unique solvability of the Cauchy problem related to smooth vector fields we get: $t \in \mathcal{D}(-X, x)$ iff $-t \in \mathcal{D}(X, x)$ and

$$\exp(-tX)(x) := \exp((-t)X)(x) = \exp(t(-X))(x), \tag{1.11a}$$
$$\exp(-tX)\big(\exp(tX)(x)\big) = x, \tag{1.11b}$$
$$\exp((t+\tau)X)(x) = \exp(tX)\big(\exp(\tau X)(x)\big), \tag{1.11c}$$
$$\exp((t\tau)X)(x) = \exp(t(\tau X))(x), \tag{1.11d}$$

when all the terms are defined. If $\mathcal{D}(X, x) = \mathbb{R}$, identities (1.11a)–(1.11d) hold true for every $t,\ \tau \in \mathbb{R}$. See also the note.[5]

Remark 1.1.3 (Pyramid-shaped vector fields). For our aims, the vector fields of the following type

$$X = \sum_{j=1}^{N} a_j(x_1, \dots, x_{j-1})\, \partial_{x_j} \tag{1.12}$$

will play a crucial rôle. In (1.12), the function a_j only depends on the variables $x_1, \dots, x_{j-1}$, and we agree to let $a_j(x_1, \dots, x_{j-1}) = \text{constant}$ when $j = 1$. Roughly speaking, such a remarkable kind of vector field is "pyramid"-shaped,

$$X = \begin{pmatrix} a_1 \\ a_2(x_1) \\ a_3(x_1, x_2) \\ a_4(x_1, x_2, x_3) \\ \vdots \\ a_N(x_1, \dots, x_{N-1}) \end{pmatrix}.$$

For example, the fields in (1.3b) have this form.

For any smooth vector field X of the form (1.12)*, the map*

$$(x, t) \mapsto \exp(tX)(x)$$

is well defined for every $x \in \mathbb{R}^N$ and $t \in \mathbb{R}$. Indeed, if $\gamma = (\gamma_1, \dots, \gamma_N)$ is the solution to the Cauchy problem

$$\begin{cases} \dot\gamma = XI(\gamma), \\ \gamma(0) = x, \qquad x = (x_1, \dots, x_N), \end{cases}$$

then $\dot\gamma_1 = a_1$ and $\dot\gamma_j = a_j(\gamma_1, \dots, \gamma_{j-1})$ for $j = 2, \dots, N$. As a consequence,

$$\gamma_1(x, t) = x_1 + t a_1, \quad \gamma_j(x, t) = x_j + \int_0^t a_j(\gamma_1(x, s), \dots, \gamma_{j-1}(x, s))\, ds,$$

[5] Strictly speaking, according to the very definition of $\exp(tX)(x) := \gamma_X(t, x)$, one may observe that t and X should be kept separate in notation. Though, we explicitly remark that identity (1.11d) justifies the notation "tX" in $\exp(tX)(x)$.

and $\gamma_j(x,t)$ is defined for every $x \in \mathbb{R}^N$ and $t \in \mathbb{R}$. Moreover, $\gamma_1(\cdot,t)$ only depends on x_1, whereas for $j = 2,\dots,N$, $\gamma_j(\cdot,t)$ only depends on $x_1,\dots,x_j$. Let us put $A_1(t) = ta_1$ and, for $j = 2,\dots,N$,

$$A_j(x,t) = A_j(x_1,\dots,x_{j-1},t) := \int_0^t a_j(\gamma_1(x,s),\dots,\gamma_{j-1}(x,s))\,ds.$$

Then, for every $x \in \mathbb{R}^N$, $t \in \mathbb{R}$,

$$\exp(tX)(x) = \begin{pmatrix} x_1 + A_1(t) \\ x_2 + A_2(x_1,t) \\ \vdots \\ x_N + A_N(x_1,\dots,x_{N-1},t) \end{pmatrix}, \tag{1.13}$$

and the map $x \mapsto \exp(tX)(x)$ is a *global diffeomorphism of* $\mathbb{R}^N$ *onto* $\mathbb{R}^N$ for every fixed $t \in \mathbb{R}$. Its inverse map $y \mapsto L(y,t)$ is given by

$$y \mapsto L(y,t) = \exp(-tX)(y). \tag{1.14}$$

This last statement follows from identity (1.11b). □

Remark 1.1.4. Let us consider a smooth function $u : \mathbb{R}^N \to \mathbb{R}$ and the vector field in (1.1). Then

$$Xu(x) = \lim_{t\to 0} \frac{u(\exp(tX)(x)) - u(x)}{t} \quad \forall x \in \mathbb{R}^N. \tag{1.15}$$

Indeed, since $\exp(tX)(x) = x + tXI(x) + \mathcal{O}(t^2)$, the limit on the right-hand side of (1.15) is equal to the following one:

$$\lim_{t\to 0} \frac{u(x + tXI(x)) - u(x)}{t} = \nabla u(x) \cdot XI(x) = Xu(x).$$

1.1.4 Lie Brackets of Vector Fields in $\mathbb{R}^N$

Given two smooth vector fields X and Y in $\mathbb{R}^N$, we define the Lie-bracket $[X,Y]$ as follows

$$[X,Y] := XY - YX.$$

This definition is only seemingly deceitful, for it writes $[X,Y]$ (which is a *first order* differential operator) as a difference of two second order differential operators. Indeed, if $X = \sum_{j=1}^N a_j\partial_j$ and $Y = \sum_{j=1}^N b_j\partial_j$, a direct computation shows that the Lie bracket $[X,Y]$ is the *vector field*

$$[X,Y] = \sum_{j=1}^N (Xb_j - Ya_j)\partial_j.$$

As a consequence,

$$[X, Y]I = \begin{pmatrix} Xb_1 \\ \vdots \\ Xb_N \end{pmatrix} - \begin{pmatrix} Ya_1 \\ \vdots \\ Ya_N \end{pmatrix} = \mathcal{J}_{YI} \cdot XI - \mathcal{J}_{XI} \cdot YI. \tag{1.16}$$

For example, if X_1, X_2 are as in (1.3b) (page 5), we have

$$[X_1, X_2] = (X_1(-2x_1) - X_2(2x_2))\, \partial_{x_3} = -4\, \partial_{x_3}.$$

It is quite trivial to check that $(X, Y) \mapsto [X, Y]$ is a bilinear map on the vector space $T(\mathbb{R}^N)$ satisfying the *Jacobi identity*

$$[X, [Y, Z]] + [Y, [Z, X]] + [Z, [X, Y]] = 0$$

for every $X, Y, Z \in T(\mathbb{R}^N)$.

We shall refer to $T(\mathbb{R}^N)$ (equipped with the above Lie-bracket) as the *Lie algebra* of the vector fields on $\mathbb{R}^N$. Any sub-algebra $\mathfrak{a}$ of $T(\mathbb{R}^N)$ will be called a *Lie algebra of vector fields*. More explicitly, $\mathfrak{a}$ is a Lie algebra of vector fields if $\mathfrak{a}$ is a *vector* subspace of $T(\mathbb{R}^N)$ closed with respect to $[\ ,\ ]$, i.e. $[X, Y] \in \mathfrak{a}$ for every $X, Y \in \mathfrak{a}$.

We now introduce some other notation on the algebras of vector fields. Given a set of vector fields $Z_1, \ldots, Z_m \in T(\mathbb{R}^N)$ and a multi-index

$$J = (j_1, \ldots, j_k) \in \{1, \ldots, m\}^k,$$

we set

$$Z_J := [Z_{j_1}, \ldots [Z_{j_{k-1}}, Z_{j_k}] \ldots].$$

We say that Z_J is a commutator of length (or height) k of $Z_1, \ldots, Z_m$. If $J = j_1$, we also say that $Z_J := Z_{j_1}$ is a commutator of length 1 of $Z_1, \ldots, Z_m$. A commutator of the form Z_J will also be called *nested*, in order to emphasize its difference from, e.g. a commutator of the form

$$[[Z_1, Z_2], [Z_3, Z_4]].$$

What is striking is that this last commutator is a linear combination of nested ones,[6] as we prove in Proposition 1.1.7.

First, we give a definition.

Definition 1.1.5 (The Lie algebra generated by a set). *If U is any subset of $T(\mathbb{R}^N)$, we denote by* Lie$\{U\}$ *the least sub-algebra of $T(\mathbb{R}^N)$ containing U, i.e.*

$$\mathrm{Lie}\{U\} := \bigcap \mathfrak{h}, \quad \text{where } \mathfrak{h} \text{ is a sub-algebra of } T(\mathbb{R}^N) \text{ with } U \subseteq \mathfrak{h}.$$

We define

$$\mathrm{rank}\big(\mathrm{Lie}\{U\}(x)\big) := \dim_{\mathbb{R}}\big\{ZI(x) \mid Z \in \mathrm{Lie}\{U\}\big\}.$$

---

[6] Namely,

$$[[Z_1, Z_2], [Z_3, Z_4]] = -[Z_3, [Z_4, [Z_1, Z_2]]] + [Z_4, [Z_3, [Z_1, Z_2]]].$$

Example 1.1.6. Let X_1 and X_2 be as in (1.3b) (page 5). Since $[X_1, X_2] = -4\partial_{x_3}$ and since any commutator involving X_1, X_2 more than twice is identically zero, then $\mathrm{Lie}\{X_1, X_2\} = \mathrm{span}\{X_1, X_2, [X_1, X_2]\}$, and

$$\mathrm{rank}(\mathrm{Lie}\{X_1, X_2\}(x)) = 3 \quad \text{for every } x \in \mathbb{R}^3$$

(the well-known *Hörmander condition*). □

The following result holds.

Proposition 1.1.7 (Nested commutators). *Let $U \subseteq T(\mathbb{R}^N)$ be any set of smooth vector fields on $\mathbb{R}^N$. We set*

$$U_1 := \mathrm{span}\{U\}, \qquad U_n := \mathrm{span}\{[u, v] \mid u \in U,\ v \in U_{n-1}\}, \quad n \geq 2.$$

Then we have

$$\mathrm{Lie}\{U\} = \mathrm{span}\{U_n \mid n \in \mathbb{N}\}.$$

Moreover,

$$[u, v] \in U_{i+j} \quad \textit{for every } u \in U_i,\, v \in U_j.$$

We explicitly remark that, from the very definition of U_n, the vector fields in U_n are linear combination of nested brackets, i.e. brackets of the type

$$[u_1[u_2[u_3[\cdots[u_{n-1}, u_n]\cdots]]]]$$

with $u_1, \ldots, u_n \in U$. The above proposition then states that *every element of Lie$\{U\}$ is a linear combination of nested brackets.*

To show the idea behind the proof, let us take $u_1, u_2, v_1, v_2 \in U$ and prove that $[[u_1, u_2], [v_1, v_2]]$ is a linear combination of nested brackets. By the Jacobi identity $[X, [Y, Z]] = -[Y, [Z, X]] - [Z, [X, Y]]$, one has

$$\begin{aligned}
[\underbrace{[u_1, u_2]}_{X}, [\underbrace{v_1}_{Y}, \underbrace{v_2}_{Z}]] &= -[v_1, [v_2, [u_1, u_2]]] - [v_2, [[u_1, u_2], v_1]] \\
&= -[v_1, [v_2, [u_1, u_2]]] + [v_2, [v_1, [u_1, u_2]]] \in U_4.
\end{aligned}$$

Proof (of Proposition 1.1.7). We set $U^* := \mathrm{span}\{U_n \mid n \in \mathbb{N}\}$. Obviously, U^* contains U and is contained in any algebra of vector fields which contains U. Hence, we are left to prove that U^* is closed under the bracket operation. Obviously, it is enough to show that, for any $i, j \in \mathbb{N}$ and for any $u_1, \ldots, u_i, v_1, \ldots, v_j \in U$, we have

$$\big[[u_1[u_2[\cdots[u_{i-1}, u_i]\cdots]]]; [v_1[v_2[\cdots[v_{j-1}, v_j]\cdots]]]\big] \in U_{i+j}.$$

We argue by induction on $k := i + j \geq 2$. For $k = 2$ and 3, the assertion is obvious. Let us now suppose the thesis holds for every $i + j \leq k$ with $k \geq 4$, and prove it also holds when $i + j = k + 1$. We can suppose, by skew-symmetry, $j \geq 3$. Exploiting repeatedly the induction hypothesis and the Jacobi identity, we have

$$\begin{aligned}
&\big[u; [v_1[v_2[\cdots[v_{j-1}, v_j]\cdots]]]\big] \\
&\quad = -[v_1, \underbrace{[[v_2, [v_3, \cdots]], u]}_{\text{length } k}] - [[v_2, [v_3, \cdots]], [u, v_1]] \\
&\quad = \{\text{element of } U_{k+1}\} - [[v_1, u], [v_2, [v_3, \cdots]]] \\
&\quad = \{\text{element of } U_{k+1}\} + [v_2, \underbrace{[[v_3, \cdots], [v_1, u]]}_{\text{length } k}] + [[v_3, \cdots], [[v_1, u]v_2]] \\
&\quad = \{\text{element of } U_{k+1}\} + [[v_2, [v_1, u]], [v_3, \cdots]] \\
&\quad \text{(after finitely many steps)} \\
&\quad = \{\text{element of } U_{k+1}\} + (-1)^{j-1}[[v_{j-i}, [v_{j-2}, \cdots [v_1, u]]], v_j] \\
&\quad = \{\text{element of } U_{k+1}\} + (-1)^{j}[v_j, [v_{j-i}, [v_{j-2}, \cdots [v_1, u]]]] \\
&\quad \in U_{k+1}.
\end{aligned}$$

This ends the proof. □

Corollary 1.1.8. *Let $Z_1, \ldots, Z_m \in T(\mathbb{R}^N)$ be fixed. Then*

$$\text{Lie}\{Z_1, \ldots, Z_m\} = \text{span}\big\{Z_J \mid \textit{with } J = (j_1, \ldots, j_k) \in \{1, \ldots, m\}^k, k \in \mathbb{N}\big\}.$$

This (non-trivial) fact (direct consequence of Proposition 1.1.7) will be used throughout the next sections, often without mention.

The following notation will be used when dealing with "stratified" (or "graded") Lie algebras. If V_1, V_2 are subsets of $T(\mathbb{R}^N)$, we denote

$$[V_1, V_2] := \text{span}\big\{[v_1, v_2] \mid v_i \in V_i,\ i = 1, \ , 2\big\}. \tag{1.17}$$

1.2 Lie Groups on $\mathbb{R}^N$

1.2.1 The Lie Algebra of a Lie Group on $\mathbb{R}^N$

We first recall a well-known definition.

Definition 1.2.1 (Lie group on $\mathbb{R}^N$). *Let $\circ$ be a given group law on $\mathbb{R}^N$, and suppose that the map*

$$\mathbb{R}^N \times \mathbb{R}^N \ni (x, y) \mapsto y^{-1} \circ x \in \mathbb{R}^N$$

is smooth. Then $\mathbb{G} := (\mathbb{R}^N, \circ)$ is called a Lie group on $\mathbb{R}^N$.

Convention. For the simplicity of notation, we shall assume that the origin 0 of $\mathbb{R}^N$ is the identity of $\mathbb{G}$. This assumption is not restrictive. Indeed, if $e \in \mathbb{G}$ is the identity of $\mathbb{G}$ (a Lie group on $\mathbb{R}^N$), we can consider new coordinates on $\mathbb{R}^N$ given by the C^∞-diffeomorphism defined by $T(x) = x - e$. Thus, we obtain a new Lie group on $\mathbb{R}^N$, $\widetilde{\mathbb{G}} = (\mathbb{R}^N, *)$, where

$$y * y' = (y + e) \circ (y' + e) - e,$$

with identity $T(e) = 0$. Obviously, $\mathbb{G}$ and $\widetilde{\mathbb{G}}$ are isomorphic Lie groups via the diffeomorphism T. Furthermore, we shall see that every homogeneous Lie group on $\mathbb{R}^N$ (see Section 1.3) has identity element equal to the origin of $\mathbb{R}^N$. Finally, the main subject of this book, i.e. Carnot groups, are naturally isomorphic to a homogeneous Lie group on $\mathbb{R}^N$. This justifies our convention and motivates the study of Lie groups on $\mathbb{R}^N$, i.e. roughly speaking, Lie groups with a *global* chart.

Fixed $\alpha \in \mathbb{G}$, we denote by $\tau_\alpha(x) := \alpha \circ x$ the left-translation by α on $\mathbb{G}$. A (smooth) vector field X on $\mathbb{R}^N$ is called *left-invariant* on $\mathbb{G}$ if

$$X(\varphi \circ \tau_\alpha) = (X\varphi) \circ \tau_\alpha \tag{1.18}$$

for every $\alpha \in \mathbb{G}$ and for every smooth function $\varphi : \mathbb{R}^N \to \mathbb{R}$. We denote by $\mathfrak{g}$ the set of the left-invariant vector fields on $\mathbb{G}$. It is quite obvious to recognize that

$$\begin{gathered}\text{for every } X, Y \in \mathfrak{g} \text{ and for every } \lambda, \mu \in \mathbb{R},\\ \text{we have } \lambda X + \mu Y \in \mathfrak{g} \text{ and } [X, Y] \in \mathfrak{g}.\end{gathered} \tag{1.19}$$

Then, $\mathfrak{g}$ is a Lie algebra of vector fields, sub-algebra of $T(\mathbb{R}^N)$. It will be called the *Lie algebra* of $\mathbb{G}$.

Example 1.2.2. The map

$$(x_1, x_2, x_3) \circ (y_1, y_2, y_3) = \big(x_1 + y_1, x_2 + y_2, x_3 + y_3 + 2\,(x_2\, y_1 - x_1\, y_2)\big)$$

endows $\mathbb{R}^3$ with a structure of Lie group. In several next examples, we shall refer to $\mathbb{H}^1 = (\mathbb{R}^3, \circ)$ as the Heisenberg–Weyl group on $\mathbb{R}^3$. It is a direct computation to show that the vector fields

$$X_1 = \partial_{x_1} + 2\,x_2\,\partial_{x_3}, \qquad X_2 = \partial_{x_2} - 2\,x_1\,\partial_{x_3} \tag{1.20}$$

are left invariant w.r.t. $\circ$. Consequently, $X_1, X_2, [X_1, X_2] \in \mathfrak{h}^1$, say, the Lie algebra of $\mathbb{H}^1$ (this notation is not standard). We shall show that, precisely, $\mathfrak{h}^1 = \operatorname{span}\{X_1, X_2, [X_1, X_2]\} = \operatorname{Lie}\{X_1, X_2\}$. □

From the theorem of differentiation of composite functions, we easily get the following characterization of left-invariant vector fields on $\mathbb{G}$.

Proposition 1.2.3 (Characterization of $\mathfrak{g}$. I). *Let $\mathbb{G}$ be a Lie group on $\mathbb{R}^N$, and let $\mathfrak{g}$ be the Lie algebra of $\mathbb{G}$. The (smooth) vector field X belongs to $\mathfrak{g}$ if and only if*

$$(XI)(\alpha \circ x) = \mathcal{J}_{\tau_\alpha}(x) \cdot (XI)(x) \quad \forall\, \alpha, x \in \mathbb{G}. \tag{1.21}$$

As usual, $\mathcal{J}_{\tau_\alpha}(x)$ denotes the Jacobian matrix at the point x of the map τ_α.

Proof. For every smooth function φ on $\mathbb{R}^N$, we have

$$(X(\varphi \circ \tau_\alpha))(x) = \nabla(\varphi \circ \tau_\alpha)(x) \cdot XI(x) = \big((\nabla\varphi)(\tau_\alpha(x)) \cdot \mathcal{J}_{\tau_\alpha}(x)\big) \cdot XI(x)$$

and

$$(X\varphi)(\tau_\alpha(x)) = (\nabla\varphi)(\tau_\alpha(x)) \cdot XI(\tau_\alpha(x)).$$

Then $X \in \mathfrak{g}$ if and only if

$$(\nabla\varphi)(\tau_\alpha(x)) \cdot \big(\mathcal{J}_{\tau_\alpha}(x) \cdot XI(x)\big) = (\nabla\varphi)(\tau_\alpha(x)) \cdot XI(\tau_\alpha(x)) \tag{1.22}$$

for every $\alpha, x \in \mathbb{R}^N$ and for every $\varphi \in C^\infty(C^\infty, \mathbb{R})$. By choosing $\varphi(x) = \sum_{j=1}^N h_j x_j$ with $h_j \in \mathbb{R}$ for $1 \le j \le N$, (1.22) gives

$$h^T \cdot \mathcal{J}_{\tau_\alpha}(x) \cdot XI(x) = h^T \cdot XI(\tau_\alpha(x))$$

for every $h \in \mathbb{R}^N$, which obviously implies (1.21). □

Interchanging α with x in (1.21), we obtain $(XI)(x \circ \alpha) = \mathcal{J}_{\tau_x}(\alpha) \cdot (XI)(\alpha)$ for all $\alpha,\ x \in \mathbb{G}$, so that, when $\alpha = 0$,

$$(XI)(x) = \mathcal{J}_{\tau_x}(0) \cdot (XI)(0) \quad \forall\, x \in \mathbb{G}. \tag{1.23}$$

This identity says that a left-invariant vector field on $\mathbb{G}$ is completely determined by its value at the origin (and by the Jacobian matrix at the origin of the left-translation). The following result shows that (1.23) characterizes the vector fields in $\mathfrak{g}$.

Proposition 1.2.4. *Let $\mathbb{G}$ be a Lie group on $\mathbb{R}^N$, and let $\mathfrak{g}$ be the Lie algebra of $\mathbb{G}$. Let η be a fixed vector of $\mathbb{R}^N$, and define the (component functions of the) vector field X as follows*

$$XI(x) := \mathcal{J}_{\tau_x}(0) \cdot \eta, \quad x \in \mathbb{R}^N. \tag{1.24}$$

Then $X \in \mathfrak{g}$.

Proof. Definition (1.24) gives

$$XI(\alpha \circ x) = \mathcal{J}_{\tau_{\alpha\circ x}}(0) \cdot \eta, \quad \alpha,\ x \in \mathbb{R}^N. \tag{1.25}$$

On the other hand, since the composition law on $\mathbb{G}$ is associative, we have $\tau_{\alpha\circ x} = \tau_\alpha \circ \tau_x$, so that

$$\mathcal{J}_{\tau_{\alpha\circ x}}(0) = \mathcal{J}_{\tau_\alpha}(x) \cdot \mathcal{J}_{\tau_x}(0).$$

Replacing this identity in (1.25), we get

$$XI(\alpha \circ x) = \mathcal{J}_{\tau_\alpha}(x) \cdot \mathcal{J}_{\tau_x}(0) \cdot \eta,$$

which implies, by (1.24), $XI(\alpha \circ x) = \mathcal{J}_{\tau_\alpha}(x) \cdot XI(x)$. Then, by Proposition 1.2.3, $X \in \mathfrak{g}$. This ends the proof. □

Corollary 1.2.5 (Characterization of $\mathfrak{g}$. II). *Let $\mathbb{G}$ be a Lie group on $\mathbb{R}^N$, and let $\mathfrak{g}$ be the Lie algebra of $\mathbb{G}$. The vector field X belongs to $\mathfrak{g}$ iff*

$$(XI)(x) = \mathcal{J}_{\tau_x}(0) \cdot (XI)(0) \quad \forall\, x \in \mathbb{G}. \tag{1.26}$$

Proof. If X satisfies (1.26), then, by Proposition 1.2.4 (setting $\eta := XI(0)$), we have $X \in \mathfrak{g}$. Vice versa, if $X \in \mathfrak{g}$, then we already showed that (1.26) follows from (1.21) of Proposition 1.2.3. □

Example 1.2.6. If $\mathbb{G} = \mathbb{H}^1$ (see Example 1.2.2), we have

$$\mathcal{J}_{\tau_x}(0) = \begin{pmatrix} 1 & 0 & 0 \\ 0 & 1 & 0 \\ 2x_2 & -2x_1 & 1 \end{pmatrix}.$$

For example, for $X_1 = \partial_{x_1} + 2\,x_2\,\partial_{x_3}$, we recognize that, for every $x \in \mathbb{H}^1$,

$$(X_1 I)(x) = \begin{pmatrix} 1 \\ 0 \\ 2\,x_2 \end{pmatrix} = \begin{pmatrix} 1 & 0 & 0 \\ 0 & 1 & 0 \\ 2x_2 & -2x_1 & 1 \end{pmatrix} \cdot \begin{pmatrix} 1 \\ 0 \\ 0 \end{pmatrix} = \mathcal{J}_{\tau_x}(0) \cdot (XI)(0).$$

The same obviously holds, e.g. for the fields $X_2 = \partial_{x_2} - 2\,x_1\,\partial_{x_3}$ and $[X_1, X_2] = -4\partial_{x_3}$.

From Proposition 1.2.3 and identity (1.23) it follows that $\mathfrak{g}$ is a vector space of dimension N. Indeed, the following proposition holds.

Proposition 1.2.7 (Characterization of $\mathfrak{g} \simeq \mathbb{R}^N$. III). *Let $\mathbb{G}$ be a Lie group on $\mathbb{R}^N$, and let $\mathfrak{g}$ be the Lie algebra of $\mathbb{G}$. The map*

$$J : \mathbb{R}^N \to \mathfrak{g}, \quad \eta \mapsto J(\eta)$$

with $J(\eta)$ defined by

$$J(\eta)I(x) = \mathcal{J}_{\tau_x}(0) \cdot \eta \tag{1.27}$$

is an isomorphism of vector spaces. In particular,

$$\dim \mathfrak{g} = N.$$

Proof. We first observe that J is well defined since, by Proposition 1.2.4, $J(\eta) \in \mathfrak{g}$ for every $\eta \in \mathbb{R}^N$. Moreover, by identity (1.23), $J(\mathbb{R}^N) = \mathfrak{g}$. The linearity of J is obvious. Then it remains to prove that J is injective.

Suppose $J(\eta) = 0$. This means that $\mathcal{J}_{\tau_x}(0)\cdot\eta = 0$ for every $x \in \mathbb{R}^N$. In particular $\mathcal{J}_{\tau_0}(0) \cdot \eta = 0$. On the other hand, since the left-translation τ_0 is the identity map, $\mathcal{J}_{\tau_0}(0) \cdot \eta = \eta$. Then $\eta = 0$, and J is one-to-one. □

Example 1.2.8. The Lie algebra $\mathfrak{h}^1$ of $\mathbb{G} = \mathbb{H}^1$ (see Example 1.2.2 for the notation) is given by span$\{X_1, X_2, [X_1, X_2]\}$. Indeed, $X_1, X_2, [X_1, X_2]$ are three linearly independent left-invariant vector fields and $\dim(\mathfrak{h}^1) = 3$, as stated in Proposition 1.2.7. Again using the same proposition, we could also argue as follows: $X_1, X_2, [X_1, X_2]$ are the vector fields obtained by multiplying $\mathcal{J}_{\tau_x}(0)$ respectively times the basis of $\mathbb{R}^3$

$$(1,0,0)^T, \qquad (0,1,0)^T, \qquad (0,0,-4)^T. \qquad \square$$

In what follows, the next remarks will be very useful.

Remark 1.2.9. Let $X \in \mathfrak{g}$, and denote by η the value of XI at $x = 0$, i.e. $\eta = XI(0)$. Then, by identity (1.23), $XI(x) = \mathcal{J}_{\tau_x}(0) \cdot \eta$. As a consequence, for every smooth function φ on $\mathbb{R}^N$,

$$\begin{aligned}\frac{\mathrm{d}}{\mathrm{d}t}\Big|_{t=0} \varphi(x \circ (t\eta)) &= \frac{\mathrm{d}}{\mathrm{d}t}\Big|_{t=0} \varphi(\tau_x(t\eta)) \\ &= \nabla\varphi(x) \cdot \mathcal{J}_{\tau_x}(0) \cdot \eta = \nabla\varphi(x) \cdot XI(x) = (X\varphi)(x).\end{aligned}$$

Then

$$(X\varphi)(x) = \frac{\mathrm{d}}{\mathrm{d}t}\Big|_{t=0} \varphi(x \circ (t\eta)), \qquad \eta = XI(0). \tag{1.28}$$

Identity (1.28) characterizes the left-invariant vector fields on $\mathbb{G}$. This follows from the next remark.

Remark 1.2.10. Let X be a vector field on $\mathbb{R}^N$. Assume that there exists $\eta \in \mathbb{R}^N$ such that, for every $\varphi \in C^\infty(\mathbb{R}^N, \mathbb{R})$,

$$(X\varphi)(x) = \frac{\mathrm{d}}{\mathrm{d}t}\Big|_{t=0} \varphi(x \circ (t\eta)) \quad \forall\, x \in \mathbb{R}^N. \tag{1.29}$$

Then $\eta = XI(0)$ and $X \in \mathfrak{g}$.

Indeed, by taking $\varphi(x) = x_j = I_j(x)$ and $x = 0$ in (1.29), one gets

$$(XI)_j(0) = \frac{\mathrm{d}}{\mathrm{d}t}\Big|_{t=0} (t\eta)_j = \eta_j,$$

i.e. $XI(0) = \eta$. Then (1.29) and the associativity of $\circ$ imply

$$\begin{aligned}(X\varphi)(\alpha \circ x) &= \frac{\mathrm{d}}{\mathrm{d}t}\Big|_{t=0} \varphi((\alpha \circ x) \circ (t\eta)) = \frac{\mathrm{d}}{\mathrm{d}t}\Big|_{t=0} (\varphi \circ \tau_\alpha)(x \circ (t\eta)) \\ &= X(\varphi \circ \tau_\alpha)(x)\end{aligned}$$

for every $\alpha, x \in \mathbb{G}$. Then X is left-invariant on $\mathbb{G}$. □

Collecting together the above remarks, we have proved the following result.

Proposition 1.2.11 (Characterization of $\mathfrak{g}$. IV). *Let $\mathbb{G}$ be a Lie group on $\mathbb{R}^N$, and let $\mathfrak{g}$ be the Lie algebra of $\mathbb{G}$. The vector field X belongs to $\mathfrak{g}$ iff there exists $\eta \in \mathbb{R}^N$ such that, for every $\varphi \in C^\infty(\mathbb{R}^N, \mathbb{R})$,*

$$(X\varphi)(x) = \frac{\mathrm{d}}{\mathrm{d}t}\Big|_{t=0} \varphi(x \circ (t\eta)) \quad \forall\, x \in \mathbb{R}^N. \tag{1.30}$$

In this case $\eta = XI(0)$.

Remark 1.2.12. For every $x \in \mathbb{R}^N$ and $X \in \mathfrak{g}$, the following expansion holds

$$\exp(tX)(x) = x \circ (t\eta) + o(t) \text{ as } t \to 0, \qquad \eta = XI(0). \tag{1.31}$$

Indeed, since $XI(x) = \mathcal{J}_{\tau_x}(0) \cdot \eta$,

$$x \circ t\eta = \tau_x(t\eta) = \tau_x(0) + t\mathcal{J}_{\tau_x}(0) \cdot \eta + o(t) = x + tXI(x) + o(t).$$

Then (1.31) follows from (1.10), see page 8.

We remark that two vector fields can be linearly independent in $T(\mathbb{R}^N)$ without being linearly independent at every point. Take, for example, ∂_{x_1} and $x_1\,\partial_{x_2}$ in $\mathbb{R}^2$. Moreover, two vector fields can be linearly dependent at every point without being linearly dependent in $T(\mathbb{R}^N)$. Take, for example, ∂_{x_1} and $x_1\,\partial_{x_1}$ in $\mathbb{R}^2$.

The following result shows that neither of the previous situations can occur for left-invariant vector fields on a Lie group $\mathbb{G}$ on $\mathbb{R}^N$. Indeed, given a family of vector fields $X_1, \ldots, X_m \in \mathfrak{g}$,

the rank of the subset of $\mathbb{R}^N$ spanned by
$\{X_1I(x), \ldots, X_mI(x)\}$ is independent of x.

More precisely, we have the following result.

Proposition 1.2.13 (Constant rank). *Let $\mathbb{G}$ be a Lie group on $\mathbb{R}^N$, and let $\mathfrak{g}$ be the Lie algebra of $\mathbb{G}$. Let $X_1, \ldots, X_m \in \mathfrak{g}$. Then the following statements are equivalent:*

(i) $X_1, \ldots, X_m$ *are linearly independent (in $\mathfrak{g}$);*
(ii) $X_1I(0), \ldots, X_mI(0)$ *are linearly independent (in $\mathbb{R}^N$);*
(iii) $\exists\, x_0 \in \mathbb{R}^N : X_1I(x_0), \ldots, X_mI(x_0)$ *are linearly independent (in $\mathbb{R}^N$);*
(iv) $X_1I(x), \ldots, X_mI(x)$ *are linearly independent (in $\mathbb{R}^N$) for all $x \in \mathbb{R}^N$.*

Proof. We first recall that, by identity (1.23),

$$X_jI(x) = \mathcal{J}_{\tau_x}(0) \cdot \eta_j, \qquad \text{with } \eta_j = X_jI(0),$$

for every $x \in \mathbb{R}^N$. On the other hand, since $\tau_{x^{-1}} \circ \tau_x = I$,

$$\mathcal{J}_{\tau_{x^{-1}}}(x) \cdot \mathcal{J}_{\tau_x}(0) = \mathbb{I}_N.$$

Hence $\mathcal{J}_{\tau_x}(0)$ is non-singular for every $x \in \mathbb{R}^N$. Then (ii), (iii) and (iv) are equivalent. The equivalence between (i) and (ii) follows from Proposition 1.2.7. Indeed, with the notation of that proposition, for every $j \in \{1, \ldots, m\}$, $X_j = J(\eta_j)$ with $\eta_j = X_jI(0)$, and J is an isomorphism of $\mathbb{R}^N$ onto $\mathfrak{g}$. □

Example 1.2.14. The vector fields on $\mathbb{R}^2$ defined by

$$X_1 = \partial_{x_1}, \qquad X_2 = x_1\partial_{x_2}$$

do satisfy the so-called Hörmander condition

$$\operatorname{rank}(\operatorname{Lie}\{X_1, X_2\}(x)) = 2 \qquad \text{for every } x \in \mathbb{R}^2.$$

However, since X_1 and X_2 are independent as vector fields but $X_1I(0)$, $X_2I(0)$ are dependent as vectors of $\mathbb{R}^2$, X_1 and X_2 are *not* left-invariant with respect to any group law on $\mathbb{R}^2$.

1.2.2 The Jacobian Basis

From Proposition 1.2.7 it follows that any basis of $\mathfrak{g}$ is the image via J of a basis of $\mathbb{R}^N$. A natural definition is thus in order.

Definition 1.2.15 (Jacobian basis). *Let $\mathbb{G}$ be a Lie group on $\mathbb{R}^N$, and let $\mathfrak{g}$ be the Lie algebra of $\mathbb{G}$. If $\{e_1, \dots, e_N\}$ is the canonical basis*[7] *of $\mathbb{R}^N$ and J is the map defined in Proposition 1.2.7, we call*

$$\{Z_1, \dots, Z_N\}, \qquad Z_j := J(e_j)$$

the Jacobian basis *of* $\mathfrak{g}$.

(*Note.* We warn the reader that the notion of Jacobian basis is strictly related to the fact that, presently, $\mathbb{G}$ is a Lie group on $\mathbb{R}^N$, and we are making reference to the fixed Cartesian coordinates on $\mathbb{R}^N$. Hence, the Jacobian basis is not well-posed on general Lie groups. Moreover, if we perform a change of coordinates on $\mathbb{R}^N$, even a linear one, the Jacobian basis changes. Despite this "non-invariant", non-coordinate-free nature of the Jacobian basis, the reader will soon recognize its usefulness.)

From the very definition of J we obtain

$$Z_j I(x) = \mathcal{J}_{\tau_x}(0) \cdot e_j = j\text{-th column of } \mathcal{J}_{\tau_x}(0) \quad \forall\, x \in \mathbb{R}^N, \tag{1.32}$$

so that, since $\mathcal{J}_{\tau_x}(0) = \mathbb{I}_N$,

$$Z_j I(0) = e_j.$$

From Remark 1.2.10 we also have

$$(Z_j\varphi)(x) = \left.\frac{\mathrm{d}}{\mathrm{d}t}\right|_{t=0} \varphi(x \circ te_j) = \left.\frac{\partial}{\partial y_j}\right|_{y=0} \varphi(x \circ y) \tag{1.33}$$

for every $\varphi \in C^\infty(\mathbb{R}^N)$ and every $x \in \mathbb{G}$.

Consequently, the Jacobian basis $\{Z_1, \dots, Z_N\}$ of $\mathfrak{g}$ is given by the N column of the Jacobian matrix

$$\mathcal{J}_{\tau_x}(0)$$

(whence the name). Moreover, $Z_j|_0 = \partial/\partial x_j|_0$ and

$$(Z_j\varphi)(x) = (\partial/\partial y_j)|_{y=0}\varphi(x \circ y) \quad \forall\, \varphi \in C^\infty(\mathbb{R}^N),\ x \in \mathbb{G}. \tag{1.34}$$

Summing up the above results, we have the following equivalent characterizations of the Jacobian basis.

---

[7] Id est, for every $j \in \{1, \dots, N\}$,

$$e_j = (0, \dots, \underbrace{1}_{j}, \dots, 0)^T.$$

Proposition 1.2.16 (Jacobian basis). *Let $\mathbb{G}$ be a Lie group on $\mathbb{R}^N$, and let $\mathfrak{g}$ be the Lie algebra of $\mathbb{G}$. Let $j \in \{1, \dots, N\}$ be fixed.*

Then there exists one and only one vector field in $\mathfrak{g}$, say Z_j, characterized by any of the following equivalent conditions:

1. $Z_j|_0 = (\partial/\partial x_j)|_0$, *i.e.*
$$(Z_j\varphi)(0) = \frac{\partial\varphi}{\partial x_j}(0) \quad \textit{for every } \varphi \in C^\infty(\mathbb{R}^N, \mathbb{R});$$
2. *for every $\varphi \in C^\infty(\mathbb{R}^N, \mathbb{R})$, it holds*
$$(Z_j\varphi)(x) = \frac{\partial}{\partial y_j}\bigg|_{y=0} \big(\varphi(x \circ y)\big) \quad \textit{for every } x \in \mathbb{G};$$
3. *if e_j denotes the j-th vector of the canonical basis of $\mathbb{R}^N$, then*
$$Z_j I(0) = e_j;$$
4. *the column vector of the component functions of Z_j is*
$$Z_j I(x) = \mathcal{J}_{\tau_x}(0) \cdot e_j = j\textit{-th column of } \mathcal{J}_{\tau_x}(0);$$
5. *for every $x \in \mathbb{G}$, we have*
$$(Z_j\varphi)(x) = \frac{\mathrm{d}}{\mathrm{d}t}\bigg|_{t=0} \varphi(x \circ (te_j)) \quad \textit{for every } \varphi \in C^\infty(\mathbb{R}^N, \mathbb{R}).$$

The system of vector fields $\mathcal{Z} := \{Z_1, \dots, Z_N\}$ is a basis of $\mathfrak{g}$, the Jacobian basis. The coordinates of $X \in \mathfrak{g}$ w.r.t. $\mathcal{Z}$ are, orderly, the entries of the column vector $XI(0)$.

(For the proof of the last statement, see Remark 1.2.20.)

In the sequel, when we need to endow $\mathfrak{g}$ with a differentiable structure, we shall consider the *vector space* structure of $\mathfrak{g}$, making $\mathfrak{g}$ (in a natural way) a differentiable manifold. Although the choice of a basis for $\mathfrak{g}$ is completely immaterial, we shall prevalently fix a system of coordinates on $\mathfrak{g}$ by choosing the Jacobian basis, then identifying $\mathfrak{g}$ with $\mathbb{R}^N$ in a fixed way.

Example 1.2.17 (The Jacobian basis of $\mathbb{H}^1$). The Jacobian basis for the Lie algebra of $\mathbb{H}^1$ (see Example 1.2.2) is given by
$$Z_1 = \partial_{x_1} + 2\,x_2\,\partial_{x_3}, \quad Z_2 = \partial_{x_2} - 2\,x_1\,\partial_{x_3}, \quad Z_3 = \partial_{x_3},$$
since, in this case, the Jacobian matrix at 0 of the left-translation is
$$\mathcal{J}_{\tau_x}(0) = \begin{pmatrix} 1 & 0 & 0 \\ 0 & 1 & 0 \\ 2x_2 & -2x_1 & 1 \end{pmatrix}. \qquad \square$$

Example 1.2.18 (A non-polynomial nilpotent Lie group on $\mathbb{R}^3$). It is a simple exercise to verify that the following operation

$$\begin{aligned}
&(x_1, x_2, x_3) \circ (y_1, y_2, y_3) \\
&\quad := (\operatorname{arcsinh}(\sinh(x_1) + \sinh(y_1)), x_2 + y_2 + \sinh(x_1)y_3, x_3 + y_3)
\end{aligned}$$

endows $\mathbb{R}^3$ with a Lie group structure $(\mathbb{G}, \circ)$. The Lie algebra $\mathfrak{g}$ of $\mathbb{G}$ is spanned by the vector fields, coefficient-vectors are given by the columns of the Jacobian matrix

$$\mathcal{J}_{\tau_x}(0) = \begin{pmatrix} \frac{1}{\cosh(x_1)} & 0 & 0 \\ 0 & 1 & \sinh(x_1) \\ 0 & 0 & 1 \end{pmatrix}$$

(the Jacobian basis), i.e. $\mathfrak{g} = \operatorname{span}\{Z_1, Z_2, Z_3\}$, where

$$Z_1 = \frac{1}{\cosh(x_1)}\,\partial_{x_1}$$
$$Z_2 = \partial_{x_2}$$
$$Z_3 = \partial_{x_3} + \sinh(x_1)\,\partial_{x_2}.$$

Note that $\mathbb{G}$ is nilpotent (of step two).

Example 1.2.19 (A non-polynomial non-nilpotent Lie group on $\mathbb{R}^2$). It is a simple exercise to verify that the following operation on $\mathbb{R}^2$

$$(x_1, x_2) \circ (y_1, y_2) = (x_1 + y_1, y_2 + x_2\, e^{y_1})$$

defines a Lie group structure, and the Jacobian basis is $Z_1 = \partial_{x_1} + x_2\,\partial_{x_2}$, $Z_2 = \partial_{x_2}$. Hence, the relevant Lie algebra *is not nilpotent*, for $[Z_2, Z_1] = Z_2$, so that, inductively,

$$[\cdots\underbrace{[[[Z_2, Z_1], Z_1], Z_1]\cdots Z_1]}_{k \text{ times}} = Z_2 \qquad \text{for all } k \in \mathbb{N}.$$

Remark 1.2.20. Let us consider the map

$$\pi : \mathfrak{g} \to \mathbb{R}^N, \qquad X \mapsto \pi(X) := XI(0). \tag{1.35}$$

From the very definition of π, the following fact follows: if $X \in \mathfrak{g}$ and we write $\eta := \pi(X)$, then we have

$$\mathcal{J}_{\tau_x}(0) \cdot \eta = \mathcal{J}_{\tau_x}(0) \cdot XI(0) = (XI)(x), \tag{1.36}$$

the last equality following from (1.23). Let now J be the map introduced in Proposition 1.2.7. Comparing (1.36) to (1.27), we recognize that

$$J(\pi(X)) = J(\eta) = X, \quad \pi(J(\eta)) = \eta \qquad \forall\, X \in \mathfrak{g},\ \eta \in \mathbb{R}^N.$$

Thus π is the inverse map of J.

We explicitly remark that π is the linear map which assigns, to every vector field X in $\mathfrak{g}$, the N-tuple η in $\mathbb{R}^N$ of the coordinates of X with respect to the Jacobian basis. The coordinate of X with respect to this basis is simply given by $XI(0)$ (see also Fig. 1.2).

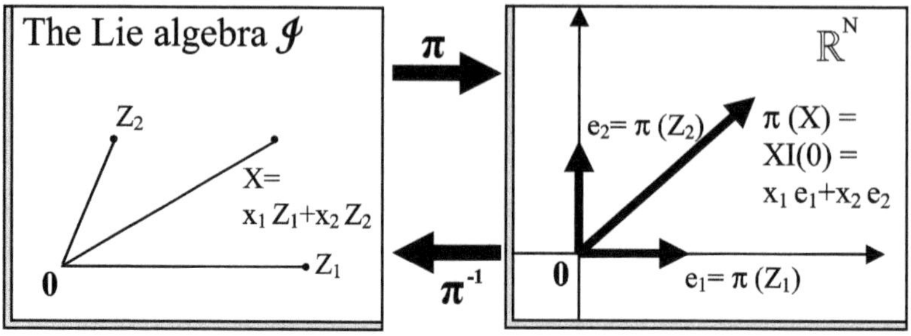

Fig. 1.2. Figure of Remark 1.2.20

1.2.3 The (Jacobian) Total Gradient

Let $\mathbb{G} = (\mathbb{R}^N, \circ)$ be a Lie group on $\mathbb{R}^N$, and let $Z_1, \ldots, Z_N$ be the Jacobian basis[8] of the Lie algebra $\mathfrak{g}$ of $\mathbb{G}$.

For any differentiable function u defined on an open set $\Omega \subseteq \mathbb{R}^N$, we consider a sort of "intrinsic" gradient of u given by $(Z_1 u, \ldots, Z_N u)$ (in the sequel, we shall call it *(Jacobian) total gradient*). Then it follows from (1.32) that

$$(Z_1 u(x), \ldots, Z_N u(x)) = \nabla u(x) \cdot \mathcal{J}_{\tau_x}(0) \quad \forall\, x \in \Omega. \tag{1.37}$$

On the other hand, since $\mathcal{J}_{\tau_x}(0)$ is non-singular and its inverse is given by $\mathcal{J}_{\tau_{x^{-1}}}(0)$, we can write the Euclidean gradient of u in terms of its total gradient in the following way

$$\nabla u(x) = (Z_1 u(x), \ldots, Z_N u(x)) \cdot \mathcal{J}_{\tau_{x^{-1}}}(0) \qquad \forall\, x \in \Omega. \tag{1.38}$$

From (1.38) we immediately obtain the following result. We shall follow the notation of Remark 1.2.3.

Proposition 1.2.21. *Let $\mathbb{G}$ be a Lie group on $\mathbb{R}^N$, and let $Z_1, \ldots, Z_N$ be the relevant Jacobian basis (or any basis for $\mathfrak{g}$). Let $\Omega \subseteq \mathbb{R}^N$ be an open and connected set. A function $u \in C^1(\Omega, \mathbb{R})$ is constant in Ω if and only if its total gradient $(Z_1 u, \ldots, Z_N u)$ vanishes identically on Ω.*

(*Note.* A significant improvement of this result will be available in the stratified setting. See, e.g. Proposition 1.5.6).

Proof. From (1.37) and (1.38) it follows that the total gradient of u vanishes at $x \in \Omega$ if and only if $\nabla u(x) = 0$. $\square$

---

[8] In this section, we consider the Jacobian basis for the sake of brevity. In fact, $Z_1, \ldots, Z_N$ can be replaced by any basis $X_1, \ldots, X_N$ of $\mathfrak{g}$. Indeed, note that in this case there exists a $N \times N$ non-singular constant matrix M such that

$$(X_1 I(x) \cdots X_N I(x)) = M \cdot (Z_1 I(x) \cdots Z_N I(x)).$$

Example 1.2.22. When $\mathbb{G} = \mathbb{H}^1$, it indeed holds

$$(Z_1 u, Z_2 u, Z_3 u) = (\partial_{x_1} u + 2x_2\, \partial_{x_3} u, \partial_{x_2} u - 2x_1\, \partial_{x_3} u, \partial_{x_3} u)$$
$$= \left(\partial_{x_1} u, \partial_{x_2} u, \partial_{x_3} u\right) \cdot \begin{pmatrix} 1 & 0 & 0 \\ 0 & 1 & 0 \\ 2x_2 & -2x_1 & 1 \end{pmatrix} = \nabla u \cdot \mathcal{J}_{\tau_x}(0),$$

and, vice versa,

$$(Z_1 u, Z_2 u, Z_3 u) \cdot \mathcal{J}_{\tau_{x^{-1}}}(0) = (Z_1 u, Z_2 u, Z_3 u) \cdot \begin{pmatrix} 1 & 0 & 0 \\ 0 & 1 & 0 \\ -2x_2 & 2x_1 & 1 \end{pmatrix} = \nabla u.$$

1.2.4 The Exponential Map of a Lie Group on $\mathbb{R}^N$

The next lemma will be useful to define the notion of *Exponential map* from $\mathfrak{g}$ to $\mathbb{G}$, one of the most important tools in the Lie group theory.

Lemma 1.2.23 (Completeness of $\mathfrak{g}$). *Let $(\mathbb{G}, \circ)$ be a Lie group on $\mathbb{R}^N$, and let $\mathfrak{g}$ be its Lie algebra. Let $X \in \mathfrak{g}$, and let $\gamma : [t_0, t_0 + T] \to \mathbb{R}^N$ be an integral curve of X. Then:*

(i) *$\alpha \circ \gamma$ is an integral curve of X for every $\alpha \in \mathbb{G}$.*
(ii) *γ can be continued to an integral curve of X on the interval $[t_0 - T, t_0 + 2T]$.*

Proof. (i) For every $t \in [t_0, t_0 + T]$, we have (by (1.21))

$$\frac{\mathrm{d}}{\mathrm{d}t}(\alpha \circ \gamma(t)) = \frac{\mathrm{d}}{\mathrm{d}t}(\tau_\alpha(\gamma(t))) = \mathcal{J}_{\tau_\alpha}(\gamma(t)) \cdot \dot{\gamma}(t)$$
$$= \mathcal{J}_{\tau_\alpha}(\gamma(t)) \cdot XI(\gamma(t)) = X(\alpha \circ \gamma(t)).$$

(ii) Define $\varGamma : [t_0 - T, t_0 + 2T] \to \mathbb{R}^N$ as follows:

$$\varGamma(t) := \begin{cases} \gamma(t_0) \circ (\gamma(t_0 + T))^{-1} \circ \gamma(t + T) & \text{if } t_0 - T \le t \le t_0, \\ \gamma(t), & \text{if } t_0 \le t \le t_0 + T, \\ \gamma(t_0 + T) \circ (\gamma(t_0))^{-1} \circ \gamma(t - T) & \text{if } t_0 + T \le t \le t_0 + 2T. \end{cases}$$

Then, by (i), $\varGamma$ is an integral curve of X and, obviously, $\varGamma|_{[t_0, t_0+T]} \equiv \gamma$. $\square$

From assertion (ii) of this Lemma we immediately obtain the following important statement: *For every $X \in \mathfrak{g}$, the map*

$$(x, t) \mapsto \exp(tX)(x)$$

is well-defined for every $x \in \mathbb{R}^N$ and every $t \in \mathbb{R}$.

The next corollary easily follows from the assertion (i) of Lemma 1.2.23.

Corollary 1.2.24. *Let* $(\mathbb{G}, \circ)$ *be a Lie group on* $\mathbb{R}^N$*, and let* $\mathfrak{g}$ *be its Lie algebra. Let* $X \in \mathfrak{g}$ *and* $x, y \in \mathbb{G}$*. Then*

$$x \circ \exp(tX)(y) = \exp(tX)(x \circ y) \tag{1.39}$$

for every $t \in \mathbb{R}$*. In particular, for* $y = 0$,

$$\exp(tX)(x) = x \circ \exp(tX)(0).$$

Proof. By Lemma 1.2.23-(i), $t \mapsto x \circ \exp(tX)(y)$ is an integral curve of X. Moreover,

$$(x \circ \exp(tX)(y))|_{t=0} = x \circ y.$$

Then (1.39) follows. □

Definition 1.2.25 (Exponential map). *Let* $\mathbb{G}$ *be a Lie group on* $\mathbb{R}^N$*, and let* $\mathfrak{g}$ *be its Lie algebra. The* exponential map *of the Lie group* $\mathbb{G}$ *is defined by*

$$\operatorname{Exp} : \mathfrak{g} \to \mathbb{G}, \quad \operatorname{Exp}(X) = \exp(1 \cdot X)(0).$$

More explicitly, $\operatorname{Exp}(X)$ *is the value at the time* $t = 1$ *of the path* $\gamma(t)$ *solution to* $\dot{\gamma}(t) = XI(\gamma(t))$, $\gamma(0) = 0$. (See Fig. 1.3.)

From Corollary 1.2.24 and identity (1.11b) (with $\tau = -t$) we get

$$\operatorname{Exp}(-X) \circ \operatorname{Exp}(X) = 0.$$

Indeed,

$$\begin{aligned}\operatorname{Exp}(-X) \circ \operatorname{Exp}(X) &= \operatorname{Exp}(-X) \circ \exp(X)(0) = \exp(X)(\operatorname{Exp}(-X)) \\ &= \exp(X)(\exp(-X)(0)) = 0.\end{aligned}$$

Then we have

$$(\operatorname{Exp}(X))^{-1} = \operatorname{Exp}(-X). \tag{1.40}$$

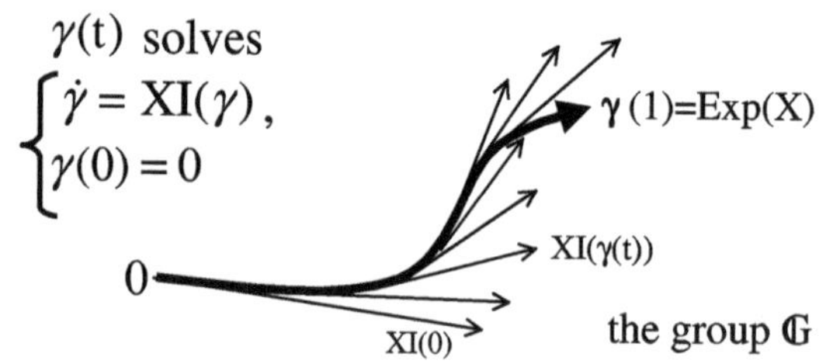

Fig. 1.3. Figure of Definition 1.2.25

We give an explicit example of exponential map.

Example 1.2.26 (The Exp *map on* $\mathbb{H}^1$*).* Let us consider once again the Heisenberg–Weyl group $\mathbb{H}^1$ on $\mathbb{R}^3$. In Example 1.2.8, we showed that a basis for its Lie algebra $\mathfrak{h}^1$ is given by X_1, X_2, X_3, where $X_1 = \partial_{x_1} + 2x_2\,\partial_{x_3}$, $X_2 = \partial_{x_2} - 2x_1\,\partial_{x_3}$ and $X_3 = [X_1, X_2] = -4\partial_{x_3}$. Let us construct the exponential map. We set, for $\xi \in \mathbb{R}^3$,

$$\begin{aligned} \xi \cdot X &:= \xi_1 X_1 + \xi_2 X_2 + \xi_3 X_3 \\ &= \xi_1 \begin{pmatrix} 1 \\ 0 \\ 2x_2 \end{pmatrix} + \xi_2 \begin{pmatrix} 0 \\ 1 \\ -2x_1 \end{pmatrix} + \xi_3 \begin{pmatrix} 0 \\ 0 \\ -4 \end{pmatrix} = \begin{pmatrix} \xi_1 \\ \xi_2 \\ -4\xi_3 + 2\xi_1 x_2 - 2\xi_2 x_1 \end{pmatrix}. \end{aligned}$$

By Definition 1.1.2, for fixed $x \in \mathbb{H}^1$, we have $\exp(\xi \cdot X)(x) = \gamma(1)$, where $\gamma(s) = (\gamma_1(s), \gamma_2(s), \gamma_3(s))$ is the solution to

$$\begin{cases} \dot\gamma(s) = (\xi \cdot X)I(\gamma(s)) = (\xi_1, \xi_2, -4\xi_3 + 2\xi_1\gamma_2(s) - 2\,\xi_2\gamma_1(s)), \\ \gamma(0) = x. \end{cases}$$

Solving the above system of ODE's, one gets

$$\exp(\xi \cdot X)(x) = \begin{pmatrix} x_1 + \xi_1 \\ x_2 + \xi_2 \\ x_3 - 4\,\xi_3 + 2\,\xi_1\,x_2 - 2\,\xi_2\,x_1 \end{pmatrix}.$$

As a consequence, by Definition 1.2.25, we obtain

$$\mathrm{Exp}\,(\xi \cdot X) = \exp(\xi \cdot W)(0) = \begin{pmatrix} \xi_1 \\ \xi_2 \\ -4\,\xi_3 \end{pmatrix},$$

so that Exp is globally invertible and its inverse map is given by

$$\mathrm{Log}\,(y) := (\mathrm{Exp}\,)^{-1}(y) = \begin{pmatrix} y_1 \\ y_2 \\ -\frac{1}{4}\,y_3 \end{pmatrix} \cdot X.$$

For example, we have

$$\mathrm{Exp}\,(-\xi \cdot X) = \begin{pmatrix} -\xi_1 \\ -\xi_2 \\ +4\,\xi_3 \end{pmatrix} = -\mathrm{Exp}\,(\xi \cdot X) = \big(\mathrm{Exp}\,(-\xi \cdot X)\big)^{-1},$$

since the inverse of x in $\mathbb{H}^1$ coincides with $-x$. This fact tests, in this simple example, the validity of (1.40). □

Remark 1.2.27 (Local invertibility of Exp*).* Let $\{X_1, \ldots, X_N\}$ be a basis of $\mathfrak{g}$. Then, for every $X \in \mathfrak{g}$,

$$X = \sum_{j=1}^{N} \xi_j X_j \quad \text{for a suitable} \quad \xi = (\xi_1, \ldots, \xi_N) \in \mathbb{R}^N,$$

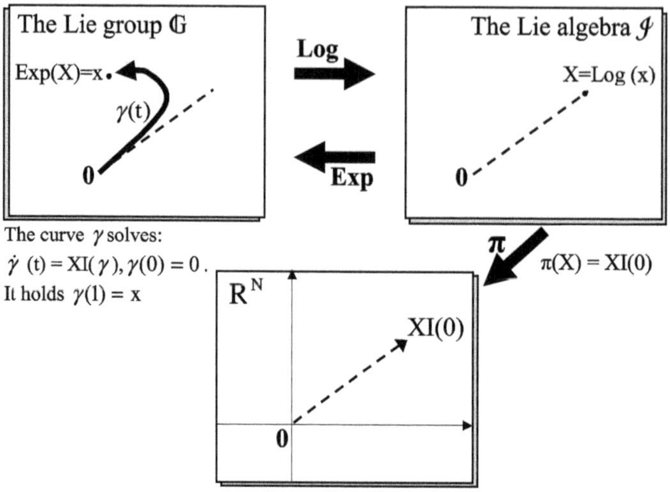

Fig. 1.4. The relation between the group $\mathbb{G}$, its algebra $\mathfrak{g}$ and $\mathbb{R}^N$

so that

$$\operatorname{Exp}(X) = \exp\left(\sum_{j=1}^N \xi_j X_j\right)(0).$$

From the classical theory of ODE's we know that the map

$$(\xi_1, \ldots, \xi_N) \mapsto \exp\left(\sum_{j=1}^N \xi_j X_j\right)(0)$$

is smooth. Then we can say that the map $\mathfrak{g} \ni X \mapsto \operatorname{Exp}(X) \in \mathbb{G}$ is smooth. From the Taylor expansion (1.10) (page 8) we get

$$\operatorname{Exp}(X) = \sum_{j=1}^N \xi_j \eta_j + \mathcal{O}(|\xi|^2), \quad \text{as } |\xi| \to 0,$$

where $\eta_j = X_j I(0)$.

Denote by E the matrix whose column vectors are $\eta_1, \ldots, \eta_N$. Then

$$\mathcal{J}_{\operatorname{Exp}}(0) = E \qquad \text{(we fix on } \mathfrak{g} \text{ coordinates related to the } X_j\text{'s).}$$

In particular, if $\{X_1, \ldots, X_N\} = \{Z_1, \ldots, Z_N\}$ is the Jacobian basis of $\mathfrak{g}$, then

$$\mathcal{J}_{\operatorname{Exp}}(0) = \mathbb{I}_N \qquad \text{(we fix on } \mathfrak{g} \text{ coordinates related to the } Z_j\text{'s).} \tag{1.41}$$

As a consequence, Exp is a diffeomorphism from a neighborhood of $0 \in \mathfrak{g}$ onto a neighborhood of $0 \in \mathbb{G}$. Where defined,[9] we denote by Log the inverse map of Exp.

Example 1.2.28. With the notation of Example 1.2.26, we recall that the Jacobian basis for $\mathfrak{h}^1$ is given by (see Example 1.2.17)

$$Z_1 = X_1, \qquad Z_2 = X_2, \qquad Z_3 = -\frac{1}{4} X_3.$$

Hence, by the computations in Example 1.2.26, we have

$$\mathrm{Exp}\,(\xi_1 Z_1 + \xi_2 Z_2 + \xi_3 Z_3) = \mathrm{Exp}\left(\xi_1 X_1 + \xi_2 X_2 - \frac{1}{4}\,\xi_3 X_3\right) = (\xi_1, \xi_2, \xi_3), \quad (1.42)$$

and (1.41) is readily verified. □

The next proposition is an easy consequence of Corollary 1.2.24 and shows an important link between the composition law in $\mathbb{G}$ and the exponential map.

Proposition 1.2.29 (Exponentiation and composition). *Let* $(\mathbb{G}, \circ)$ *be a Lie group on* $\mathbb{R}^N$*, and let* $\mathfrak{g}$ *be its Lie algebra. Let* $x, y \in \mathbb{G}$*. Assume* Log (y) *is defined. Then*

$$x \circ y = \exp(\mathrm{Log}\,(y))(x). \qquad (1.43)$$

Proof. Let $X = \mathrm{Log}\,(y)$. This means that

$$y = \mathrm{Exp}\,(X) = \exp(X)(0).$$

Then, by Corollary 1.2.24, we infer

$$x \circ y = x \circ \exp(X)(0) = \exp(X)(x).$$

This is precisely (1.43), and the proof is complete. □

By writing $y = \mathrm{Exp}\,(X)$ in (1.43), we obtain

$$x \circ \mathrm{Exp}\,(X) = \exp(X)(x) \quad \text{for every } X \in \mathfrak{g} \text{ and every } x \in \mathbb{G}. \qquad (1.44)$$

We give two examples of this proposition, the first is very simple, the second one a little more elaborated.

Example 1.2.30 (of (1.43) *for* $\mathbb{H}^1$*).* Let us consider once again the Heisenberg–Weyl group $\mathbb{H}^1$ on $\mathbb{R}^3$. Proceeding with the computations in Example 1.2.26, we have

---

[9] We shall see in Chapter 2 that, in many important situations, such as for connected and simply connected nilpotent Lie groups, Exp is globally invertible. Any Carnot group is a connected and simply connected nilpotent Lie group.

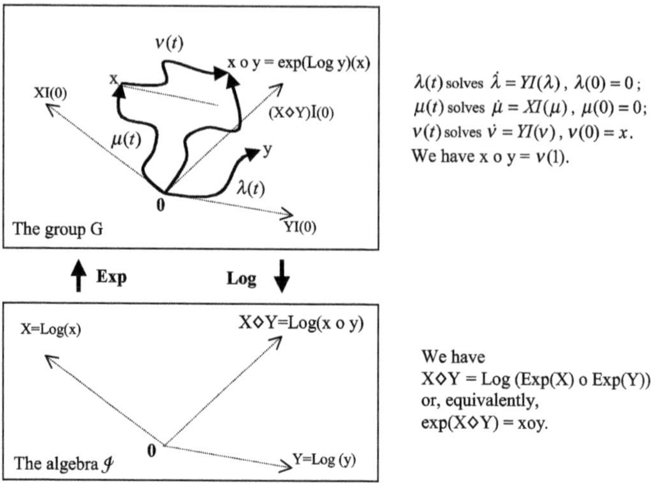

Fig. 1.5. Figure of Proposition 1.2.29

$$\exp(\mathrm{Log}\,(y))(x) = \exp\Big(\Big(y_1, y_2, -\frac{1}{4}\,y_3\Big)\cdot X\Big)(x)$$
$$= \begin{pmatrix} x_1 + y_1 \\ x_2 + y_2 \\ x_3 + y_3 + 2\,y_1\,x_2 - 2\,y_2\,x_1 \end{pmatrix} = x \circ y,$$

which tests, in this simple example, the validity of (1.43). □

Example 1.2.31. Let us consider once again the Lie group $\mathbb{G}$ introduced in Example 1.2.18. The relevant Jacobian basis is $\{Z_1, Z_2, Z_3\}$, where

$$Z_1 = \frac{1}{\cosh(x_1)}\,\partial_{x_1}, \quad Z_2 = \partial_{x_2}, \quad Z_3 = \partial_{x_3} + \sinh(x_1)\,\partial_{x_2}.$$

Given $\xi = (\xi_1, \xi_2, \xi_3) \in \mathbb{R}^3$ and $x \in \mathbb{G}$, we set $\xi \cdot Z = \sum_{i=1}^{3} \xi_i\, Z_i$, so that $\exp(t(\xi \cdot Z))(x) = \gamma(t)$ is the solution to

$$\dot\gamma(t) = \xi_1\, Z_1 I(\gamma(t)) + \xi_2\, Z_2 I(\gamma(t)) + \xi_3\, Z_3 I(\gamma(t)), \qquad \gamma(0) = x,$$

i.e. more explicitly (writing $\gamma(t) = (\gamma_1(t), \gamma_2(t), \gamma_3(t))$),

$$\begin{cases} \dot\gamma_1(t) = \xi_1\,(\cosh(\gamma_1(t)))^{-1}, & \gamma_1(0) = x_1, \\ \dot\gamma_2(t) = \xi_2 + \xi_3\,\sinh(\gamma_1(t)), & \gamma_2(0) = x_2, \\ \dot\gamma_3(t) = \xi_3, \quad \gamma_3(0) = x_3. \end{cases}$$

A direct computation gives

$$\begin{cases} \gamma_1(t) = \mathrm{arcsinh}(\sinh(x_1) + \xi_1 t), \\ \gamma_2(t) = x_2 + (\xi_2 + \xi_3 \sinh(x_1))t + \xi_1\xi_3 \frac{t^2}{2}, \\ \gamma_3(t) = x_3 + \xi_3 t. \end{cases}$$

In particular, $\operatorname{Exp}(\xi \cdot Z) = \exp(1\,(\xi \cdot Z))(0)$, i.e.

$$\operatorname{Exp}(\xi \cdot Z) = \left(\operatorname{arcsinh}(\xi_1), \xi_2 + \frac{1}{2}\,\xi_1\xi_3, \xi_3\right),$$

so that

$$\operatorname{Log}(y_1, y_2, y_3) = \left(\sinh(y_1),\ y_2 - \frac{1}{2}\sinh(y_1)\,y_3,\ y_3\right) \cdot Z.$$

Collecting the above facts together, we get

$$\begin{aligned}
\exp(\operatorname{Log}(y))(x) &= \exp\left(\left(\sinh(y_1),\ y_2 - \frac{1}{2}\sinh(y_1)\,y_3,\ y_3\right) \cdot Z\right)(x) \\
&= \begin{pmatrix} \operatorname{arcsinh}(\sinh(x_1) + \sinh(y_1)) \\ x_2 + (y_2 - \frac{1}{2}\sinh(y_1)\,y_3 + y_3\,\sinh(x_1)) + \sinh(y_1)\,y_3\,\frac{1}{2} \\ x_3 + y_3 \end{pmatrix} \\
&= \begin{pmatrix} \operatorname{arcsinh}(\sinh(x_1) + \sinh(y_1)) \\ x_2 + y_2 + y_3\,\sinh(x_1) \\ x_3 + y_3 \end{pmatrix} = x \circ y,
\end{aligned}$$

as stated in (1.43) of Proposition 1.2.29. □

Remark 1.2.32 (The composition of $\mathbb{G}$ induces a composition on $\mathfrak{g}$ via Exp: The Campbell–Hausdorff operation). Suppose that

$$\operatorname{Exp} : \mathfrak{g} \to \mathbb{G} \quad \text{and} \quad \operatorname{Log} : \mathbb{G} \to \mathfrak{g}$$

are globally defined C^∞ maps, inverse to each other. We then define on $\mathfrak{g}$ the operation

$$X \diamond Y := \operatorname{Log}(\operatorname{Exp}(X) \circ \operatorname{Exp}(Y)), \qquad X,\ Y \in \mathfrak{g}. \tag{1.45}$$

It is immediately seen that $\diamond$ *defines a Lie group structure on* $\mathfrak{g}$ *and*

$$\operatorname{Exp} : (\mathfrak{g}, \diamond) \to (\mathbb{G}, \circ)$$

is a Lie-group isomorphism. Indeed, this last fact is obvious from the very definition of $\diamond$, whereas the associativity of $\diamond$ on $\mathfrak{g}$ follows immediately from the associativity of $\circ$ on $\mathbb{G}$.

One of the most striking facts about Lie algebras and Lie groups is that (under suitable hypotheses) the operation $\diamond$ on $\mathfrak{g}$ is well-posed and can be expressed in a somewhat *"universal"* way as a sum of iterated Lie-brackets of X and Y (see (2.43) and Theorem 2.2.13, page 129). For example, the first few terms are

$$X \diamond Y = X + Y + \frac{1}{2}[X, Y] + \frac{1}{12}[X, [X, Y]] - \frac{1}{12}[Y, [X, Y]] + \cdots. \tag{1.46}$$

We shall deal extensively on the composition law $\diamond$ throughout the book. We give an example of $\diamond$ when $\mathbb{G}$ is the Heisenberg–Weyl group on $\mathbb{R}^3$.

Example 1.2.33 (The operation $\diamond$ for $\mathbb{H}^1$). The reader is invited to compare this example to the diagram in Fig. 1.6.

We use the notation and computations of Example 1.2.26. When $\mathbb{G} = \mathbb{H}^1$, we saw that Exp is globally invertible. Hence (1.45) makes sense. We fix $X \in \mathfrak{h}^1$. If Z_1, Z_2, Z_3 is the Jacobian basis for $\mathfrak{h}^1$, and we set, for brevity, $\xi := XI(0)$, we have (see Remark 1.2.20)

$$X = \xi_1 Z_1 + \xi_2 Z_2 + \xi_3 Z_3 =: \xi \cdot Z.$$

Analogously, if $Y \in \mathfrak{h}^1$, we set $\eta := YI(0)$, so that $Y = \eta \cdot Z$. Thus, we derive

$$\begin{aligned}
&\mathrm{Log}\,\big(\mathrm{Exp}\,(X) \circ \mathrm{Exp}\,(Y)\big)\\
&= \mathrm{Log}\,\big(\mathrm{Exp}\,(\xi \cdot Z) \circ \mathrm{Exp}\,(\eta \cdot Z)\big)\\
&= \text{(see (1.42))} \quad \mathrm{Log}\,(\xi \circ \eta)\\
&= \mathrm{Log}\,(\xi_1 + \eta_1, \xi_2 + \eta_2, \xi_3 + \eta_3 + 2\,\eta_1\,\xi_2 - 2\,\eta_2\,\xi_1)\\
&\text{(again from (1.42))}\\
&= (\xi_1 + \eta_1, \xi_2 + \eta_2, \xi_3 + \eta_3 + 2\,\eta_1\,\xi_2 - 2\,\eta_2\,\xi_1) \cdot Z\\
&= (\xi_1 + \eta_1)\, Z_1 + (\xi_2 + \eta_2)\, Z_2 + (\xi_3 + \eta_3 + 2\,\eta_1\,\xi_2 - 2\,\eta_2\,\xi_1)\, Z_3. \qquad (1.47)
\end{aligned}$$

On the other hand, we consider (1.46), truncated to the commutators of length two (sine $\mathfrak{h}^1$ is nilpotent of step two!), and we explicitly write down $X \diamond Y$ in our case, thus obtaining

$$\begin{aligned}
(\xi \cdot Z) \diamond (\eta \cdot Z) &= \xi \cdot Z + \eta \cdot Z + \frac{1}{2}\,[\xi \cdot Z, \eta \cdot Z]\\
&= \xi_1 Z_1 + \xi_2 Z_2 + \xi_3 Z_3 + \eta_1 Z_1 + \eta_2 Z_2 + \eta_3 Z_3\\
&\quad + \frac{1}{2}\,[\xi_1 Z_1 + \xi_2 Z_2 + \xi_3 Z_3, \eta_1 Z_1 + \eta_2 Z_2 + \eta_3 Z_3]\\
&\quad \text{(here we use } [Z_1, Z_2] = -4\, Z_3,\ [Z_1, Z_3] = [Z_2, Z_3] = 0)\\
&= (\xi_1 + \eta)\, Z_1 + (\xi_2 + \eta_2)\, Z_2 + (\xi_3 + \eta_3)\, Z_3\\
&\quad + \frac{1}{2}\big((-4\,\xi_1\,\eta_2 + 4\,\xi_2\,\eta_1)\, Z_3\big)\\
&= (\xi_1 + \eta)\, Z_1 + (\xi_2 + \eta_2)\, Z_2 + (\xi_3 + \eta_3 + 2\,\eta_1\,\xi_2 - 2\,\eta_2\,\xi_1)\, Z_3,
\end{aligned}$$

which equals the last term in (1.47). As a consequence, we have proved that in this case it holds

$$\mathrm{Log}\,\big(\mathrm{Exp}\,(X) \circ \mathrm{Exp}\,(Y)\big) = X + Y + \frac{1}{2}\,[X, Y].$$

For the group considered in Examples 1.2.18 and 1.2.31, the reader is invited to test a similar formula. □

1.3 Homogeneous Lie Groups on $\mathbb{R}^N$

We begin by giving the definition of homogeneous Lie group (see also E.M. Stein [Ste81]).

Definition 1.3.1 (Homogeneous Lie group (on $\mathbb{R}^N$)). *Let* $\mathbb{G} = (\mathbb{R}^N, \circ)$ *be a Lie group on* $\mathbb{R}^N$ *(according to Definition* 1.2.1*). We say that* $\mathbb{G}$ *is a* homogeneous (Lie) group (on $\mathbb{R}^N$) *if the following property holds:*

(H.1) *There exists an* N*-tuple of real numbers* $\sigma = (\sigma_1, \dots, \sigma_N)$, *with* $1 \le \sigma_1 \le \dots \le \sigma_N$, *such that the* "dilation"

$$\delta_\lambda : \mathbb{R}^N \to \mathbb{R}^N, \quad \delta_\lambda(x_1, \dots, x_N) := (\lambda^{\sigma_1} x_1, \dots, \lambda^{\sigma_N} x_N)$$

is an automorphism of the group $\mathbb{G}$ *for every* $\lambda > 0$.

We shall denote by $\mathbb{G} = (\mathbb{R}^N, \circ, \delta_\lambda)$ *the datum of a homogeneous Lie group on* $\mathbb{R}^N$ *with composition law* $\circ$ *and dilation group* $\{\delta_\lambda\}_{\lambda>0}$.

(*Note.* A note similar to that given after Definition 1.2.15 of Jacobian basis applies to the notion of homogeneous Lie group: the notion of *homogeneous* Lie group is not coordinate-free and strongly depends on the choice of a fixed system of coordinates on $\mathbb{R}^N$. Nonetheless, the reader will soon recognize the suitability of this notion.)

The family of dilations $\{\delta_\lambda\}_{\lambda>0}$ forms a one-parameter group of automorphisms of $\mathbb{G}$ whose identity is

$$\delta_1 = I,$$

the identity map of $\mathbb{R}^N$. Indeed, we have

$$\delta_{r\,s}(x) = \delta_r\big(\delta_s(x)\big) \qquad \forall\, x \in \mathbb{G},\ r, s > 0.$$

Moreover, $(\delta_\lambda)^{-1} = \delta_{\lambda^{-1}}$. In the sequel, $\{\delta_\lambda\}_{\lambda>0}$ will be referred to as the *dilation group* (or *group of dilations*) *of* $\mathbb{G}$.

From (H.1) it follows that

$$\delta_\lambda(x \circ y) = (\delta_\lambda\, x) \circ (\delta_\lambda\, y) \qquad \forall\, x, y \in \mathbb{G} \tag{1.48}$$

and, if e denotes the identity of $\mathbb{G}$, $\delta_\lambda(e) = e$ for every $\lambda > 0$. This obviously implies that $e = 0$. This is consistent with our previous assumption that the origin is the identity of $\mathbb{G}$.

For example, the Heisenberg–Weyl group $\mathbb{H}^1$ (see Example 1.2.2, page 14) is a homogeneous Lie group if $\mathbb{R}^3$ is equipped with the dilations $\delta_\lambda(x_1, x_2, x_3) = (\lambda x_1, \lambda x_2, \lambda^2 x_3)$.

Remark 1.3.2. Suppose $\mathbb{G} = (\mathbb{R}^N, \circ)$ is a Lie group on $\mathbb{R}^N$ such that there exists an N-tuple of *positive* real numbers $\sigma = (\sigma_1, \dots, \sigma_N)$ such that

$$\delta_\lambda : \mathbb{R}^N \to \mathbb{R}^N, \quad d_\lambda(x_1, \dots, x_N) := (\lambda^{\sigma_1} x_1, \dots, \lambda^{\sigma_N} x_N)$$

is an automorphism of the group $\mathbb{G}$ for every $\lambda > 0$. Then, modulo a permutation of the variables of $\mathbb{R}^N$, it is always not restrictive to suppose that $\sigma_1 \le \cdots \le \sigma_N$. Obviously, this permutation of the coordinates does not alter neither (the new permuted) $\mathbb{G}$ being a Lie group on $\mathbb{R}^N$ nor the (relevant permuted) dilation δ_λ satisfying (1.48). Moreover, there exists a group of dilations δ_λ on $\mathbb{G}$ such that

$$\delta_\lambda(x_1, \ldots, x_N) = (\lambda^{\widetilde{\sigma}_1} x_1, \ldots, \lambda^{\widetilde{\sigma}_N} x_N)$$

with $1 = \widetilde{\sigma}_1 \le \cdots \le \widetilde{\sigma}_N$. Indeed, it suffices to take (once the σ_j's have been ordered increasingly)

$$\widetilde{\sigma}_j := \sigma_j / \sigma_1 \quad \text{for every } j = 1, \ldots, N.$$

Indeed, with this choice, we have

$$\delta_\lambda \equiv d_{\lambda^{1/\sigma_1}},$$

and $\delta_\lambda(x \circ y) = \delta_\lambda(x) \circ \delta_\lambda(y)$ follows from (1.48), restated for d_λ, with λ replaced by λ^{1/σ_1}.

1.3.1 δ_λ-homogeneous Functions and Differential Operators

Before we continue the analysis of homogeneous Lie groups, we show some basic properties of homogeneous functions and homogeneous differential operators with respect to the family $\{\delta_\lambda\}_\lambda$.

In this subsection, no group law is required on $\mathbb{R}^N$. Here, we only suppose that it is given on $\mathbb{R}^N$ a family of maps δ_λ of the form

$$\delta_\lambda : \mathbb{R}^N \to \mathbb{R}^N, \quad \delta_\lambda(x_1, \ldots, x_N) := (\lambda^{\sigma_1} x_1, \ldots, \lambda^{\sigma_N} x_N), \tag{1.49}$$

with fixed *positive* real numbers $\sigma_1, \ldots, \sigma_N$. We set $\sigma := (\sigma_1, \ldots, \sigma_N)$.

A real function a defined on $\mathbb{R}^N$ is called *δ_λ-homogeneous* of degree $m \in \mathbb{R}$ if a does not vanish identically and, for every $x \in \mathbb{R}^N$ and $\lambda > 0$, it holds

$$a(\delta_\lambda(x)) = \lambda^m a(x).$$

A non-identically-vanishing linear differential operator X is called *δ_λ-homogeneous* of degree $m \in \mathbb{R}$ if, for every $\varphi \in C^\infty(\mathbb{R}^N)$, $x \in \mathbb{R}^N$ and $\lambda > 0$, it holds

$$X(\varphi(\delta_\lambda(x))) = \lambda^m (X\varphi)(\delta_\lambda(x)).$$

Let a be a smooth δ_λ-homogeneous function of degree $m \in \mathbb{R}$ and X be a linear differential operator δ_λ-homogeneous of degree $n \in \mathbb{R}$. Then Xa is a δ_λ-homogeneous function of degree $m - n$ (unless $Xa \equiv 0$). Indeed, for every $x \in \mathbb{R}^N$ and $\lambda > 0$, we have

$$\lambda^n (Xa)(\delta_\lambda(x)) = X(a(\delta_\lambda(x))) = X(\lambda^m a(x)) = \lambda^m (Xa)(x).$$

Given a multi-index $\alpha \in (\mathbb{N} \cup \{0\})^N$, $\alpha = (\alpha_1, \ldots, \alpha_N)$, we define the *$\delta_\lambda$-length* (or *$\delta_\lambda$-height*) of α as

$$|\alpha|_\sigma = \langle \alpha, \sigma \rangle = \sum_{i=1}^{N} \alpha_i \, \sigma_i. \tag{1.50}$$

Definition 1.3.3 ($\mathbb{G}$-length of a multi-index. $\mathbb{G}$-degree). *When $\mathbb{G} = (\mathbb{R}^N, \circ, \delta_\lambda)$ is a homogeneous Lie group on $\mathbb{R}^N$ with its given group of dilations $\{\delta_\lambda\}_\lambda$, we shall use the notation*

$$|\alpha|_{\mathbb{G}}$$

for the relevant δ_λ-length as defined in (1.50). *In this case, we shall refer to $|\alpha|_{\mathbb{G}}$ as the* $\mathbb{G}$-length (*or* $\mathbb{G}$-height) *of α. Moreover, if $p : \mathbb{G} \to \mathbb{R}$ is a polynomial function (the sum below is intended to be finite)*

$$p(x) = \sum_\alpha c_\alpha \, x^\alpha, \qquad c_\alpha \in \mathbb{R},$$

we say that

$$\deg_{\mathbb{G}}(p) := \max\{|\alpha|_{\mathbb{G}} : c_\alpha \neq 0\}$$

is the $\mathbb{G}$-degree *or* δ_λ-(homogeneous) degree *of p.*

Since $x \mapsto x_j$ and $\partial/\partial x_j$, $j \in \{1, \dots, N\}$, are obviously δ_λ-homogeneous of degree σ_j, the function $x \mapsto x^\alpha$ and the differential operator D^α are both δ_λ-homogeneous of degree $|\alpha|_\sigma$.

If a is a continuous function, δ_λ-homogeneous of degree m and $a(x_0) \neq 0$ for some $x_0 \in \mathbb{R}^N$, then $m \geq 0$. Indeed, from $a(\delta_\lambda(x_0)) = \lambda^m a(x_0)$ we get

$$\lim_{\lambda \to 0} \lambda^m = \lim_{\lambda \to 0} \frac{a(\delta_\lambda(x_0))}{a(x_0)} = \frac{a(0)}{a(x_0)}.$$

Moreover, the continuous and δ_λ-homogeneous of degree 0 functions are precisely the constant (non-zero) functions. Indeed,

$$a(x) = a(\delta_\lambda(x)) = \lim_{\lambda \to 0^+} a(\delta_\lambda(x)) = a(0).$$

Let us now consider a smooth and δ_λ-homogeneous of degree $m \in \mathbb{R}$ function a and a multi-index α. Assume that $D^\alpha a$ is not identically zero. Then, since $D^\alpha a$ is smooth and δ_λ-homogeneous of degree $m - |\alpha|_\sigma$, it has to be $m - |\alpha|_\sigma \geq 0$, i.e. $|\alpha|_\sigma \leq m$. This result can be restated as follows:

$$D^\alpha a \equiv 0 \quad \forall \, \alpha \text{ such that } |\alpha|_\sigma > m.$$

Thus a is a *polynomial function*. Let $a(x) = \sum_{\alpha \in \mathcal{A}} a_\alpha \, x^\alpha$, where $\mathcal{A}$ is a finite set of multi-indices and $a_\alpha \in \mathbb{R}$ for every $\alpha \in \mathcal{A}$. Since a is δ_λ-homogeneous of degree m, we have

$$\sum_{\alpha \in \mathcal{A}} \lambda^m \, a_\alpha \, x^\alpha = \lambda^m \, a(x) = a(\delta_\lambda(x)) = \sum_{\alpha \in \mathcal{A}} a_\alpha \, \lambda^{|\alpha|_\sigma} \, x^\alpha.$$

Hence $\lambda^m \, a_\alpha = \lambda^{|\alpha|_\sigma} \, a_\alpha$ for every $\lambda > 0$, so that $|\alpha|_\sigma = m$ if $a_\alpha \neq 0$. Then

$$a(x) = \sum_{|\alpha|_\sigma = m} a_\alpha \, x^\alpha . \tag{1.51}$$

It is quite obvious that every polynomial function of the form (1.51) is δ_λ-homogeneous of degree m. Thus, we have proved the following proposition.

Proposition 1.3.4 (Smooth δ_λ-homogeneous functions). *Let δ_λ be as in* (1.49). *Suppose that $a \in C^\infty(\mathbb{R}^N, \mathbb{R})$. Then a is δ_λ-homogeneous of degree $m \in \mathbb{R}$ if and only if a is a polynomial function of the form* (1.51) *with some $a_\alpha \neq 0$. As a consequence, the set of the degrees of the smooth δ_λ-homogeneous (non-vanishing) functions is precisely the set of the nonnegative real numbers*

$$\mathcal{A} = \{|\alpha|_\sigma : \alpha \in (\mathbb{N} \cup \{0\})^N\},$$

with $|\alpha|_\sigma = 0$ if and only if a is constant.

From the proposition above one easily obtains the following characterization of the smooth δ_λ-homogeneous vector fields.

Proposition 1.3.5 (Smooth δ_λ-homogeneous vector fields). *Let δ_λ be as in* (1.49). *Let X be a smooth non-vanishing vector field on $\mathbb{R}^N$,*

$$X = \sum_{j=1}^{N} a_j(x)\, \partial_{x_j}.$$

Then X is δ_λ-homogeneous of degree $n \in \mathbb{R}$ if and only if a_j is a polynomial function δ_λ-homogeneous of degree $\sigma_j - n$ (unless $a_j \equiv 0$). Hence, the degree of δ_λ-homogeneity of X belongs to the set of real (possibly negative) numbers

$$\mathcal{A}_j = \{\sigma_j - |\alpha|_\sigma \; : \; \alpha \in (\mathbb{N} \cup \{0\})^N\},$$

whenever j is such that a_j is not identically zero.

Proof. A direct computation shows the "if" part of the proposition. Vice versa, if $X(\varphi \circ \delta_\lambda) = \lambda^n (X\varphi) \circ \delta_\lambda$, the choice $\varphi(x) = x_j$ yields $\lambda^{\sigma_j} a_j(x) = \lambda^n a_j(\delta_\lambda(x))$, whence a_j is a (smooth) δ_λ-homogeneous function of degree $\sigma_j - n$. By Proposition 1.3.4, a_j is a polynomial function. □

For example, the differential operators

$$X_1 = \partial_{x_1} + 2x_2\, \partial_{x_3}, \qquad X_2 = \partial_{x_2} - 2x_1\, \partial_{x_3} \tag{1.52}$$

on $\mathbb{R}^3$ are δ_λ-homogeneous of degree one with respect to the dilation

$$\delta_\lambda(x_1, x_2, x_3) = (\lambda x_1, \lambda x_2, \lambda^2 x_3).$$

Also, the vector fields $x_1^3 X_1 = x_1^3 \partial_{x_1} + 2x_1^3 x_2\, \partial_{x_3}$ and $x_2 X_2 = x_2 \partial_{x_1} - 2x_1 x_2\, \partial_{x_3}$ are respectively δ_λ-homogeneous of degrees -2 and 0 w.r.t. the same dilation.

Corollary 1.3.6. *Let δ_λ be as in* (1.49). *Let X be a smooth non-vanishing vector field. Then X is δ_λ-homogeneous of degree $n \in \mathbb{R}$ iff*

$$\delta_\lambda\big(XI(x)\big) = \lambda^n\, XI(\delta_\lambda(x)).$$

Proof. Let $X = \sum_{j=1}^N a_j\, \partial_{x_j}$. By Proposition 1.3.5, X is δ_λ-homogeneous of degree n iff $a_j(\delta_\lambda(x)) = \lambda^{\sigma_j - n}\, a_j(x)$ for any $j \in \{1, \dots, N\}$. This is equivalent to

$$\begin{aligned}\delta_\lambda(XI(x)) &= \delta_\lambda\big(a_1(x), \dots, a_N(x)\big)^T = \big(\lambda^{\sigma_1} a_1(x), \dots, \lambda^{\sigma_N} a_N(x)\big)^T \\ &= \lambda^n \big(a_1(\delta_\lambda(x)), \dots, a_N(\delta_\lambda(x))\big)^T = \lambda^n\, XI(\delta_\lambda(x)).\end{aligned}$$

This ends the proof. □

As a straightforward consequence, we have the following simple fact.

Remark 1.3.7. Let δ_λ be as in (1.49). Let $X \neq 0$ be a smooth vector field on $\mathbb{R}^N$ of the form

$$X = \sum_{j=1}^N a_j(x)\, \partial_{x_j}.$$

If X is δ_λ-homogeneous of degree $n \in \mathbb{R}$, then, for every a_j non-identically zero, we must have $n \le \sigma_j$. As a consequence, it has to be $n \le \sigma_N$ (i.e. the set of the δ_λ-homogeneous degrees of the smooth vector fields is bounded above by the maximum exponent of the dilation). Hence, X has the form

$$X = \sum_{j \le N,\, \sigma_j \ge n} a_j(x)\, \partial/\partial x_j.$$

Suppose now $n > 0$. Since a_j is a polynomial function of degree $\sigma_j - n$ and $n > 0$, then a_j *does not depend on $x_j, \dots, x_N$,*

$$a_j(x) = a_j(x_1, \dots, x_{j-1})$$

(we agree to let $a_j(x_1, \dots, x_{j-1}) =$ constant when $j = 1$). We already highlighted in the previous section the importance of these "pyramid"-shaped vector fields (see Remark 1.1.3).

From this remark the next proposition straightforwardly follows.

Proposition 1.3.8. *Let δ_λ be as in* (1.49). *Let $X = \sum_{j=1}^N a_j(x)\, \partial_{x_j}$ be a smooth vector field δ_λ-homogeneous of positive degree. Then its adjoint operator $X^* = -\sum_{j=1}^N \partial_j(a_j\, \cdot)$ satisfies $X^* = -X$ and*

$$X^2 = \operatorname{div}(A \cdot \nabla^T), \tag{1.53}$$

where A is the square matrix $(a_i\, a_j)_{i,j \le N}$. Finally, X has null divergence.

Proof. By the previous remark, the coefficient a_j does not depend on x_j. Then, for every smooth function φ,

$$X^*\varphi = -\sum_{j=1}^{N} \partial_j(a_j\,\varphi) = -\sum_{j=1}^{N} a_j\,\partial_j\varphi = -X\varphi.$$

Moreover, it holds

$$X^2 = \sum_{i,j=1}^{N} a_i\partial_i(a_j\,\partial_j) = \sum_{i=1}^{N} \partial_i\left(\sum_{j=1}^{N} a_i\,a_j\,\partial_j\right) = \operatorname{div}(A\cdot\nabla^T),$$

where A is as in the statement of the proposition. Finally, $\operatorname{div}(XI) = \sum_{j=1}^{N} \partial_j(a_j) = 0$, since a_j is independent of x_j. □

Vector fields with different degree of δ_λ-homogeneity are linearly independent if they do not vanish at the origin. Indeed, the following proposition holds.

Proposition 1.3.9. *Let δ_λ be as in (1.49). Let $X_1, \ldots, X_k \in T(\mathbb{R}^N)$ be δ_λ-homogeneous vector fields of degree $n_1, \ldots, n_k$, respectively.*

If $n_i \neq n_j$ for $i \neq j$ and if $X_jI(0) \neq 0$ for every $j \in \{1, \ldots, k\}$, then $X_1, \ldots, X_k$ are linearly independent.

Proof. Let $c_1, \ldots, c_k \in \mathbb{R}$ be such that $\sum_{j=1}^{k} c_j\,X_j = 0$. Then, for every smooth function φ,

$$0 = \sum_{j=1}^{k} c_j\,X_j(\varphi(\delta_\lambda x)) = \sum_{j=1}^{k} c_j\,\lambda^{n_j}\,(X_j\varphi)(\delta_\lambda x) \qquad \forall\, x \in \mathbb{R}^N.$$

If we take $\varphi(x) = \langle h, x\rangle = \sum_{j=1}^{N} h_j\,x_j$, this identity at $x = 0$ gives

$$0 = \sum_{j=1}^{k} c_j\,\lambda^{n_j}\,\langle \eta_j, h\rangle \qquad \forall\, h \in \mathbb{R}^N, \quad \forall\, \lambda > 0,$$

where $\eta_j = X_jI(0)$. Equivalently,

$$0 = \left\langle \sum_{j=1}^{k} c_j\,\lambda^{n_j}\,\eta_j, h\right\rangle.$$

Due to the arbitrariness of $h \in \mathbb{R}^N$, this gives $\sum_{j=1}^{k} c_j\,\lambda^{n_j}\,\eta_j = 0$ for all $\lambda > 0$, so that (since $n_i \neq n_j$ if $i \neq j$)

$$c_j\,\eta_j = 0 \text{ for any } j \in \{1, \ldots, k\}.$$

This implies $c_j = 0$, since, for every $j = 1, \ldots, k$, $\eta_j \neq 0$ by the hypothesis. □

The following simple proposition will be useful in the sequel.

Proposition 1.3.10. *Let δ_λ be as in* (1.49). *Let X_1, X_2 be δ_λ-homogeneous vector fields of degree n_1, n_2, respectively. Then $[X_1, X_2]$ is δ_λ-homogeneous of degree $n_1 + n_2$ (unless X_1 and X_2 commute).*

As a consequence, if n_1, n_2 are both positive, then every commutator of X_1, X_2 containing k_1 times X_1 and k_2 times X_2 vanish identically whenever $k_1 n_1 + k_2 n_2 > \sigma_N$.

Proof. It suffices to note that, for every smooth function φ on $\mathbb{R}^N$, one has

$$(X_1X_2)(\varphi(\delta_\lambda(x))) = \lambda^{n_2} X_1((X_2\varphi)(\delta_\lambda(x))) = \lambda^{n_2+n_1} (X_1X_2)(\varphi(\delta_\lambda(x))).$$

This proves the first part of the assertion, since $[X_1, X_2] = X_1X_2 - X_2X_1$ (and $[X_1, X_2] \equiv 0$ iff X_1 and X_2 commute).

Finally, let X be a commutator of X_1, X_2 containing k_1 times X_1 and k_2 times X_2. By the first part of this proof, it follows inductively that X is δ_λ-homogeneous of degree $k_1 n_1 + k_2 n_2$ (unless $X \equiv 0$). By Remark 1.3.7, we know that if a smooth vector field is δ_λ-homogeneous of degree $n \in \mathbb{R}$, then $n \le \sigma_N$. This ends the proof. □

For example, the differential operators $X_1 = \partial_{x_1} + 2x_2\partial_{x_3}$, $X_2 = \partial_{x_2} - 2x_1\partial_{x_3}$ on the Heisenberg–Weyl group $\mathbb{H}^1$ are homogeneous of degree one with respect to the dilation $\delta_\lambda(x_1, x_2, x_3) = (\lambda x_1, \lambda x_2, \lambda^2 x_3)$, and $[X_1, X_2] = -4\partial_{x_3}$ is indeed δ_λ-homogeneous of degree two. Moreover, any commutator of X_1, X_2 of length ≥ 3 vanish identically, as stated in the last part of Proposition 1.3.10.

Corollary 1.3.11. *Let $\mathbb{G} = (\mathbb{R}^N, \circ, \delta_\lambda)$ be a homogeneous Lie group on $\mathbb{R}^N$, and let $\mathfrak{g}$ be the Lie algebra of $\mathbb{G}$. Let $X_1, \dots, X_k \in \mathfrak{g}$ be non-identically vanishing and δ_λ-homogeneous of degrees $n_1, \dots, n_k$, respectively. If $n_i \neq n_j$ for $i \neq j$, then $X_1, \dots, X_k$ are linearly independent.*

Proof. Since $X_jI(x) = \mathcal{J}_{\tau_x}(0) \cdot X_jI(0)$ for every $x \in \mathbb{R}^N$, and X_j is non-identically vanishing, then $X_jI(0) \neq 0$ for any $j \in \{1, \dots, k\}$. Hence the assertion follows from the previous proposition. □

Proposition 1.3.12 (Nilpotence of homogeneous Lie groups on $\mathbb{R}^N$). *Let $\mathbb{G} = (\mathbb{R}^N, \circ, \delta_\lambda)$ be a homogeneous Lie group on $\mathbb{R}^N$, and let $\mathfrak{g}$ be the Lie algebra of $\mathbb{G}$. Then $\mathbb{G}$ is nilpotent of step $\le \sigma_N$, i.e. every commutator of vector fields in $\mathfrak{g}$ containing more than σ_N terms vanishes identically.*

Moreover, if Z_j is the j-th element of the Jacobian basis of $\mathfrak{g}$, Z_j is δ_λ-homogeneous of degree σ_j.

Proof. Let Z_j be the j-th element of the Jacobian basis of $\mathfrak{g}$. By Proposition 1.2.16-(5), for every $\varphi \in C^\infty(\mathbb{R}^N, \mathbb{R})$, we have

$$\begin{aligned} Z_j(\varphi(\delta_\lambda(x))) &= \left.\frac{\mathrm{d}}{\mathrm{d}t}\right|_{t=0} (\varphi(\delta_\lambda(x \circ (te_j)))) = \left.\frac{\mathrm{d}}{\mathrm{d}t}\right|_{t=0} (\varphi(\delta_\lambda(x) \circ \delta_\lambda(te_j))) \\ &= \lambda^{\sigma_j} \left.\frac{\mathrm{d}}{\mathrm{d}r}\right|_{r=0} (\varphi(\delta_\lambda(x) \circ (r\, e_j))) = \lambda^{\sigma_j} (Z_j\varphi)(\delta_\lambda(x)), \end{aligned}$$

i.e. Z_j is δ_λ-homogeneous of degree σ_j. This proves the last part of the assertion.

Since $(Z_1, \dots, Z_N)$ is a linear basis for $\mathfrak{g}$, the above arguments show that every $X \in \mathfrak{g}$ is a linear combination of δ_λ-homogeneous smooth vector fields of degrees at least $\sigma_1 \geq 1$. The first part of the assertion hence follows from Proposition 1.3.10. $\square$

1.3.2 The Composition Law of a Homogeneous Lie Group

By using the elementary properties of the homogeneous functions showed in the previous section, we shall obtain a structure theorem for the composition law in a homogeneous Lie group $(\mathbb{R}^N, \circ, \delta_\lambda)$. We first prove two lemmas.

Lemma 1.3.13. *Let δ_λ be as in (1.49). Let $P : \mathbb{R}^N \times \mathbb{R}^N \to \mathbb{R}$ be a smooth non-vanishing function such that*

$$P(\delta_\lambda(x), \delta_\lambda(y)) = \lambda^{\sigma_j} P(x, y) \qquad \forall\, x, y \in \mathbb{R}^N,\ \forall\, \lambda > 0,$$

for some j such that $1 \leq j \leq N$. Assume also that

$$P(x, 0) = x_j, \quad P(0, y) = y_j. \tag{1.54}$$

Then $P(x, y) = x_1 + y_1$ if $j = 1$ and, if $j \geq 2$,

$$P(x, y) = x_j + y_j + \widetilde{P}(x_1, \dots, x_{j-1}, y_1, \dots, y_{j-1}),$$

where $\widetilde{P}$ is a polynomial, the sum of mixed monomials in $x_1, \dots, x_{j-1}, y_1, \dots, y_{j-1}$. Moreover, $\widetilde{P}(\delta_\lambda(x), \delta_\lambda(y)) = \lambda^{\sigma_j} \widetilde{P}(x, y)$. Finally, $P(x, y)$ only depends on the x_k's and y_k's with $\sigma_k < \sigma_j$.

Proof. By Proposition 1.3.4, P is a polynomial function of the following type:

$$P(x, y) = \sum_{|\alpha|_\sigma + |\beta|_\sigma = \sigma_j} c_{\alpha,\beta}\, x^\alpha\, y^\beta, \qquad c_{\alpha,\beta} \in \mathbb{R}.$$

On the other hand, by (1.54),

$$x_j = P(x, 0) = \sum_{|\alpha|_\sigma = \sigma_j} c_{\alpha,0}\, x^\alpha$$

and

$$y_j = P(0, y) = \sum_{|\beta|_\sigma = \sigma_j} c_{0,\beta}\, y^\alpha.$$

Then

$$P(x, y) = x_j + y_j + \sum_{|\alpha|_\sigma + |\beta|_\sigma = \sigma_j,\ \alpha,\beta \neq 0} c_{\alpha,\beta}\, x^\alpha\, y^\beta. \tag{1.55}$$

We can complete the proof by noticing that the condition $|\alpha|_\sigma + |\beta|_\sigma = \sigma_j, \alpha, \beta \neq 0$ is empty when $j = 1$, whereas it implies $\alpha = (\alpha_1, \dots, \alpha_{j-1}, 0, \dots, 0)$, $\beta = (\beta_1, \dots, \beta_{j-1}, 0, \dots, 0)$ when $j \geq 2$.

As for the last assertion of the lemma, being $\alpha, \beta \neq 0$ in the sum in the right-hand side of (1.55), the sum itself may depend only on the α's and β's with $|\alpha|_\sigma, |\beta|_\sigma < \sigma_j$, hence, on the x_k's and y_k's with $\sigma_k < \sigma_j$. $\square$

Lemma 1.3.14. *Let δ_λ be as in* (1.49). *Let $Q : \mathbb{R}^N \times \mathbb{R}^N \to \mathbb{R}$ be a smooth function such that*

$$Q(\delta_\lambda(x), \delta_\lambda(y)) = \lambda^m Q(x, y) \qquad \forall\, x, y \in \mathbb{R}^N,\ \forall\, \lambda > 0,$$

where $m \geq 0$. Then

$$x \mapsto \frac{\partial Q}{\partial y_j}(x, 0)$$

is δ_λ-homogeneous of degree $m - \sigma_j$ (unless it vanishes identically).

Proof. By Proposition 1.3.13, Q is a polynomial of the following type

$$Q(x, y) = \sum_{|\alpha|_\sigma + |\beta|_\sigma = m} c_{\alpha,\beta} x^\alpha y^\beta.$$

Then, denoting by e_j the j-th element of the canonical basis of $\mathbb{R}^N$, we have

$$\frac{\partial Q}{\partial y_j}(x, y) = \sum_{|\alpha|_\sigma + |\beta|_\sigma = m} c_{\alpha,\beta} \beta_j x^\alpha\, y^{\beta - e_j},$$

so that, since $|e_j|_\sigma = \sigma_j$,

$$\frac{\partial Q}{\partial y_j}(x, 0) = \sum_{|\alpha|_\sigma = m - \sigma_j, \beta = e_j} c_{\alpha,\beta} x^\alpha.$$

This completes the proof. □

Now, we are in the position to prove the previously mentioned structure theorem for the composition law of a homogeneous Lie group on $\mathbb{R}^N$.

Theorem 1.3.15 (Composition of a homogeneous Lie group on $\mathbb{R}^N$). *Let $(\mathbb{R}^N, \circ, \delta_\lambda)$ be a homogeneous Lie group on $\mathbb{R}^N$. Then $\circ$ has polynomial component functions. Furthermore, we have*

$$(x \circ y)_1 = x_1 + y_1, \quad (x \circ y)_j = x_j + y_j + Q_j(x, y), \quad 2 \leq j \leq N,$$

and the following facts hold:

1. *Q_j only depends on $x_1, \ldots, x_{j-1}$ and $y_1, \ldots, y_{j-1}$;*
2. *Q_j is a sum of mixed monomials in x, y;*
3. *$Q_j(\delta_\lambda x, \delta_\lambda y) = \lambda^{\sigma_j} Q_j(x, y)$.*

More precisely, $Q_j(x, y)$ only depends on the x_k's and y_k's with $\sigma_k < \sigma_j$.

Proof. Let $j \in \{1, \ldots, N\}$, and define

$$P_j : \mathbb{R}^N \times \mathbb{R}^N \to \mathbb{R}, \qquad P_j(x, y) = (x \circ y)_j.$$

Since δ_λ is an automorphism of $\mathbb{G}$, we have

$$P_j(\delta_\lambda(x), \delta_\lambda(y)) = (\delta_\lambda(x \circ y))_j = \lambda^{\sigma_j}(x \circ y)_j = \lambda^{\sigma_j} P_j(x, y).$$

Moreover, since $x \circ 0 = x$, $0 \circ y = y$, we have

$$P_j(x, 0) = x_j, \quad P_j(0, y) = y_j.$$

Then the proof follows from Lemma 1.3.13. □

For example, the Lie group $\mathbb{G} = (\mathbb{R}^3, \circ)$ considered in Example 1.2.18 is not homogeneous with respect to *any* dilation on $\mathbb{R}^3$, since its composition

$$x \circ y = (\operatorname{arcsinh}(\sinh(x_1) + \sinh(y_1)), x_2 + y_2 + \sinh(x_1)y_3, x_3 + y_3)$$

does not fulfill the requirements of Theorem 1.3.15.

Corollary 1.3.16 (Inversion of a homogeneous Lie group on $\mathbb{R}^N$). *Let $\mathbb{G} = (\mathbb{R}^N, \circ, \delta_\lambda)$ be a homogeneous Lie group on $\mathbb{R}^N$. Let $j \in \{1, \dots, N\}$. For every $y \in \mathbb{G}$, we have*

$$(y^{-1})_j = -y_j + q_j(y),$$

where $q_j(y)$ is a polynomial function in y, δ_λ-homogeneous of degree σ_j, only depending on the y_k's with $\sigma_k < \sigma_j$.

Proof. We let

$$p : \mathbb{R}^N \to \mathbb{R}, \quad p(y) = y^{-1}.$$

For every $y \in \mathbb{G}$, we have

$$0 = \delta_\lambda(y^{-1} \circ y) = \delta_\lambda(y^{-1}) \circ \delta_\lambda(y),$$

whence $\delta_\lambda(y^{-1}) = (\delta_\lambda(y))^{-1}$, i.e.

$$\delta_\lambda(p(y)) = p(\delta_\lambda(y)) \quad \forall\, y \in \mathbb{R}^N. \tag{1.56}$$

If $j \in \{1, \dots, N\}$ is fixed and p_j is the j-th component function of p, (1.56) means that p_j is δ_λ-homogeneous of degree σ_j.

As a consequence, since the inversion is a smooth map (by definition of Lie group), we can apply Proposition 1.3.4 and infer that p_j is a polynomial function, δ_λ-homogeneous of degree σ_j.

Now, we exploit the explicit form of the composition in Theorem 1.3.15. If $x \circ y = 0$, then we have

$$(\star) \qquad x_j = -y_j \quad \text{whenever } \sigma_j = 1.$$

Hence, if $\sigma_j = 2$, we have $x_j = -y_j + Q_j(x, y)$, where Q_j only depends on the x_k's and y_k's with $\sigma_k = 1$. As a consequence of $(\star)$, we infer

$$x_j = -y_j + q_j(y) \quad \text{whenever } \sigma_j = 2,$$

where q_j only depends on the y_k's with $\sigma_k = 1$. An inductive argument now proves that $x_j = -y_j + q_j(y)$, where q_j only depends on the y_k's with $\sigma_k < \sigma_j$, and the proof is complete. □

Example 1.3.17. Let us consider on $\mathbb{R}^4$ the composition law

$$x \circ y = \begin{pmatrix} x_1 + y_1 \\ x_2 + y_2 \\ x_3 + y_3 + y_1 x_2 \\ x_4 + y_4 + y_1 x_2^2 + 2\, x_2 y_3 \end{pmatrix}.$$

Then $\circ$ equips $\mathbb{R}^4$ with a homogeneous Lie group structure with dilations

$$\delta_\lambda(x_1, x_2, x_3, x_4) = (\lambda x_1, \lambda x_2, \lambda^2 x_3, \lambda^3 x_4).$$

Notice that the inversion on this group is given by

$$y^{-1} = \begin{pmatrix} -y_1 \\ -y_2 \\ -y_3 + y_1\, y_2 \\ -y_4 - y_1\, y_2^2 + 2\, y_2\, y_3 \end{pmatrix}.$$

Corollary 1.3.18. *Let $\mathbb{G} = (\mathbb{R}^N, \circ, \delta_\lambda)$ be a homogeneous Lie group on $\mathbb{R}^N$. Let $j \in \{1, \dots, N\}$. For every $x, y \in \mathbb{G}$, we have*

$$(y^{-1} \circ x)_j = x_j - y_j + \sum_{k:\sigma_k < \sigma_j} P_k^{(j)}(x, y)(x_k - y_k),$$

where $P_k^{(j)}(x, y)$ is a polynomial function in x and y only depending on the x_k's and y_k's with $\sigma_k < \sigma_j$.

Proof. We fix $j \in \{1, \dots, N\}$ and $y \in \mathbb{R}^N$, and we let

$$f_y : \mathbb{R}^N \to \mathbb{R}, \quad f_y(x) := (y^{-1} \circ x)_j.$$

By Theorem 1.3.15, f_y is a polynomial function on $\mathbb{R}^N$. Hence, Taylor's formula centered at $\xi \in \mathbb{R}^N$ gives

$$f_y(x) = \sum_\alpha \frac{(D^\alpha f_y)(\xi)}{\alpha!} (x - \xi)^\alpha, \tag{1.57}$$

where the sum is finite. Choose $\xi = y$ and note that the summand with $\alpha = 0$ in (1.57) is $f_y(\xi) = f_y(y) = (y^{-1} \circ y)_j = 0$. Consequently, (1.57) gives

$$(y^{-1} \circ x)_j = \sum_{\alpha \neq 0} \frac{(D^\alpha f_y)(y)}{\alpha!} (x - y)^\alpha. \tag{1.58}$$

Note that any summand in (1.58) contains at least one factor of the type $x_k - y_k$ (since $\alpha \neq 0$). On the other hand, by Theorem 1.3.15, we have

$$(y^{-1} \circ x)_j = (y^{-1})_j + x_j + Q_j(y^{-1}, x), \tag{1.59}$$

where $Q_j(x, y)$ only depends on the x_k's and y_k's with $\sigma_k < \sigma_j$. Then, by Corollary 1.3.16, (1.59) gives

$$(y^{-1} \circ x)_j = x_j - y_j + q_j(y) + Q_j(y^{-1}, x), \tag{1.60}$$

where $q_j(y)$ is a polynomial function in y, only depending on the y_k's with $\sigma_k < \sigma_j$. Note that $Q_j(y^{-1}, x)$ depends on the x_k's with $\sigma_k < \sigma_j$ and the $(y^{-1})_k$'s with $\sigma_k < \sigma_j$. Exploiting once again Corollary 1.3.16, this proves that $Q_j(y^{-1}, x)$ depends on the x_k's and the y_k's with $\sigma_k < \sigma_j$.

As a consequence, (1.60) proves that

$$(y^{-1} \circ x)_j = x_j - y_j + \widetilde{Q}_j(x, y), \tag{1.61}$$

where $\widetilde{Q}_j(x, y)$ is a polynomial function only depending on the x_k's and y_k's with $\sigma_k < \sigma_j$. By collecting together (1.58) and (1.61), we immediately get the assertion of the corollary. □

For example, on the Heisenberg–Weyl group $\mathbb{H}^1$, we have

$$\begin{aligned} y^{-1} \circ x &= (x_1 - y_1, x_2 - y_2, x_3 - y_3 + 2\,(-y_2\, x_1 + y_1\, x_2)) \\ &= (x_1 - y_1, x_2 - y_2, x_3 - y_3 - 2\, y_2(x_1 - y_1) + 2\, y_1(x_2 - y_2)). \end{aligned}$$

Also, the map

$$x \circ y = (x_1 + y_1, x_2 + y_2, x_3 + y_3 + x_1\, y_2)$$

endows $\mathbb{R}^3$ with a homogeneous Lie group structure. Since the inversion is

$$y^{-1} = (-y_1, -y_2, -y_3 + y_1\, y_2),$$

we have

$$\begin{aligned} y^{-1} \circ x &= (x_1 - y_1, x_2 - y_2, x_3 - y_3 + 2\,(-y_2\, x_1 + y_1\, x_2)) \\ &= (x_1 - y_1, x_2 - y_2, x_3 - y_3 - y_1(x_2 - y_2)). \end{aligned}$$

Note that $y^{-1} \circ x$ contains also non-mixed monomials of degree > 1,

$$y^{-1} \circ x = (x_1 - y_1, x_2 - y_2, x_3 - y_3 - y_1\, x_2 + \underline{\underline{y_1 y_2}}).$$

The following result describes in a very explicit way the Jacobian matrix at 0 of the left-translation τ_x on a homogeneous Lie group on $\mathbb{R}^N$.

Corollary 1.3.19 (The Jacobian basis of a homogeneous Lie group). *Let $\mathbb{G} = (\mathbb{R}^N, \circ, \delta_\lambda)$ be a homogeneous Lie group on $\mathbb{R}^N$. Then we have*

$$\mathcal{J}_{\tau_x}(0) = \begin{pmatrix} 1 & 0 & \cdots & 0 \\ a_2^{(1)} & 1 & \ddots & \vdots \\ \vdots & \ddots & \ddots & 0 \\ a_N^{(1)} & \cdots & a_N^{(N-1)} & 1 \end{pmatrix}, \tag{1.62}$$

where $a_i^{(j)}$ is a polynomial function δ_λ-homogeneous of degree $\sigma_i - \sigma_j$. As a consequence, if we let

$$Z_j = \partial_{x_j} + \sum_{i=j+1}^{N} a_i^{(j)}\, \partial_{x_i} \quad \textit{for } 1 \le j \le N-1 \textit{ and } \quad Z_N = \partial_{x_N},$$

then Z_j is a left-invariant vector field δ_λ-homogeneous of degree σ_j. Moreover,

$$\mathcal{J}_{\tau_x}(0) = (Z_1 I(x) \cdots Z_N I(x)).$$

In other words, the Jacobian basis $Z_1, \dots, Z_N$ for the Lie algebra $\mathfrak{g}$ of $\mathbb{G}$ is formed by δ_λ-homogeneous vector fields of degree $\sigma_1, \dots, \sigma_N$, respectively.

Proof. By Theorem 1.3.15, the Jacobian matrix $\mathcal{J}_{\tau_x}(0)$ takes the form (1.62) with

$$a_i^{(j)}(x) = \frac{\partial Q_i}{\partial y_j}(x, 0).$$

Then, by Lemma 1.3.14, $a_i^{(j)}(x)$ is a polynomial function, δ_λ-homogeneous of degree $\sigma_i - \sigma_j$. This proves the first part of the corollary. The second one follows from Proposition 1.3.12. □

Example 1.3.20. In Example 1.2.6, we showed that the Jacobian matrix of the left translation on $\mathbb{H}^1$ is

$$\mathcal{J}_{\tau_x}(0) = \begin{pmatrix} 1 & 0 & 0 \\ 0 & 1 & 0 \\ 2x_2 & -2x_1 & 1 \end{pmatrix}.$$

We recognize that the three columns of this matrix give raise to the Jacobian basis $Z_1 = \partial_{x_1} + 2x_2\, \partial_{x_3}$, $Z_2 = \partial_{x_2} - 2x_1\, \partial_{x_3}$ and $Z_3 = \partial_{x_3}$ and these vector fields are homogeneous of degree, respectively, 1, 1, 2 with respect to $\delta_\lambda(x_1, x_2, x_3) = (\lambda x_1, \lambda x_2, \lambda^2 x_3)$. □

The structure Theorem 1.3.15 of the composition law of $(\mathbb{R}^N, \circ, \delta_\lambda)$ implies that the Lebesgue measure on $\mathbb{R}^N$ is invariant under left and right translations on $\mathbb{G}$. Indeed, by Theorem 1.3.15, the Jacobian matrices of the functions $x \mapsto \alpha \circ x$ and $x \mapsto x \circ \alpha$ have the following lower triangular form

$$\begin{pmatrix} 1 & 0 & \cdots & 0 \\ \star & 1 & \ddots & \vdots \\ \vdots & \ddots & \ddots & 0 \\ \star & \cdots & \star & 1 \end{pmatrix}.$$

Then, we have proved the following proposition.

Proposition 1.3.21 (Haar measure on a homogeneous Lie group). *Let $\mathbb{G}$ be a homogeneous Lie group on $\mathbb{R}^N$. Then the Lebesgue measure on $\mathbb{R}^N$ is invariant with respect to the left and the right translations on $\mathbb{G}$.*

The above proposition is also restated as: *the Lebesgue measure on* $\mathbb{R}^N$ *is the Haar measure for* $\mathbb{G}$.

If we denote by $|E|$ the Lebesgue measure of a measurable set $E \subseteq \mathbb{R}^N$, we then have

$$|\alpha \circ E| = |E| = |E \circ \alpha| \quad \forall\, \alpha \in \mathbb{G}.$$

We also have that the Lebesgue measure is homogeneous with respect to the dilations $\{\delta_\lambda\}_{\lambda>0}$. More precisely, as a trivial computation shows,

$$|\delta_\lambda(E)| = \lambda^Q\, |E|,$$

where

$$Q = \sum_{j=1}^{N} \sigma_j. \tag{1.63}$$

The positive number Q is called the *homogeneous dimension* of the group $\mathbb{G} = (\mathbb{R}^N, \circ, \delta_\lambda)$.

For example, in the case of the Heisenberg–Weyl group $\mathbb{H}^1$, where τ_α is given by

$$\tau_\alpha(x) = (\alpha_1 + x_1, \alpha_2 + x_2, \alpha_3 + x_3 + 2\,(\alpha_2\, x_1 - \alpha_1\, x_2)),$$

and $\delta_\lambda(x_1, x_2, x_3) = (\lambda x_1, \lambda x_2, \lambda^2 x_3)$, we have

$$\mathcal{J}_{\tau_\alpha}(x) = \begin{pmatrix} 1 & 0 & 0 \\ 0 & 1 & 0 \\ 2\alpha_2 & -2\alpha_1 & 1 \end{pmatrix}, \qquad \mathcal{J}_{\delta_\lambda}(x) = \begin{pmatrix} \lambda & 0 & 0 \\ 0 & \lambda & 0 \\ 0 & 0 & \lambda^2 \end{pmatrix},$$

so that, for every $\alpha, x \in \mathbb{H}^1$ and every $\lambda > 0$, we have

$$\det \mathcal{J}_{\tau_\alpha}(x) = 1, \qquad \det \mathcal{J}_{\delta_\lambda}(x) = \lambda^4 = \lambda^Q,$$

since the homogeneous dimension of $\mathbb{H}^1$ is $Q = 1 + 1 + 2 = 4$.

1.3.3 The Lie Algebra of a Homogeneous Lie Group on $\mathbb{R}^N$

The following remark holds.

Remark 1.3.22. Let $\mathbb{G}$ be a homogeneous Lie group on $\mathbb{R}^N$ with Lie algebra $\mathfrak{g}$. From Corollary 1.3.19 we easily obtain the splitting of $\mathfrak{g}$ as a direct sum of linear spaces spanned by vector fields of constant degree of δ_λ-homogeneity.

More precisely, let us recall that the exponents σ_j's in the dilation δ_λ of $\mathbb{G}$ (see (H.1) in Definition 1.3.1) satisfy $\sigma_1 \le \cdots \le \sigma_N$ and can then be grouped together to produce real and natural numbers, respectively, say

$$n_1, \ldots, n_r \quad \text{and} \quad N_1, \ldots, N_r,$$

such that

$$n_1 < n_2 < \cdots < n_r, \qquad N_1 + N_2 + \cdots + N_r = N,$$

defined by

$$\begin{cases} n_1 = \sigma_j & \text{for } 1 \le j \le N_1, \\ n_2 = \sigma_j & \text{for } N_1 < j \le N_1 + N_2, \\ \quad\vdots & \\ n_r = \sigma_j & \text{for } N_1 + \cdots + N_{r-1} < j \le N_1 + \cdots + N_{r-1} + N_r. \end{cases}$$

Let $Z_1, \ldots, Z_N$ be the Jacobian basis of $\mathfrak{g}$. Define

$$\begin{aligned} \mathfrak{g}_1 &= \operatorname{span}\{Z_j \mid 1 \le j \le N_1\} \qquad \text{and, for } i = 2, \ldots, r, \\ \mathfrak{g}_i &= \operatorname{span}\{Z_j \mid N_1 + \cdots + N_{i-1} < j \le N_1 + \cdots + N_{i-1} + N_i\}. \end{aligned}$$

By Corollary 1.3.19, the generators Z_j's of $\mathfrak{g}_i$ are δ_λ-homogeneous vector fields of degree n_i, $1 \le i \le r$. Moreover, we obviously have

$$\mathfrak{g} = \mathfrak{g}_1 \oplus \cdots \oplus \mathfrak{g}_r.$$

We also explicitly notice that, by Proposition 1.3.9, a vector field $X \in \mathfrak{g}$ is δ_λ-homogeneous of degree n iff, for a suitable $i \in \{1, \ldots, r\}$, $n = n_i$ and $X \in \mathfrak{g}_i$.

In the next section, we shall deal with homogeneous groups in which $n_i = i$ for $1 \le i \le r$, and the *layer* (or *slice*) $\mathfrak{g}_i$, $i \in \{1, \ldots, r\}$, is precisely generated by the commutators of length i of the vector fields in $\mathfrak{g}_1$.

Example 1.3.23. The usual additive group $(\mathbb{R}^3, +)$ is a homogeneous Lie group if equipped with the dilation

$$\delta_\lambda(x_1, x_2, x_3) = (\lambda^2 x_1, \lambda^\pi x_2, \lambda^4 x_3).$$

The decomposition of the Lie algebra as in Remark 1.3.22 is

$$\operatorname{span}\{\partial_{x_1}\} \oplus \operatorname{span}\{\partial_{x_2}\} \oplus \operatorname{span}\{\partial_{x_3}\}.$$

Moreover, $\mathbb{R}^4$ is a homogeneous Lie group if equipped with the group law

$$x \circ y = \begin{pmatrix} x_1 + y_1 \\ x_2 + y_2 \\ x_3 + y_3 + 2\, y_1 x_2 - 2\, y_2 x_1 \\ x_4 + y_4 \end{pmatrix}$$

and the dilation

$$\delta_\lambda(x_1, x_2, x_3, x_4) = (\lambda x_1, \lambda x_2, \lambda^2 x_3, \lambda^2 x_4).$$

The decomposition of the Lie algebra as in Remark 1.3.22 is

$$\mathfrak{g}_1 \oplus \mathfrak{g}_2 = \operatorname{span}\{X_1, X_2\} \oplus \operatorname{span}\{\partial_{x_3}, \partial_{x_4}\},$$

where $X_1 = \partial_{x_1} + 2\, x_2\, \partial_{x_3}$, $X_2 = \partial_{x_2} - 2\, x_1\, \partial_{x_3}$. Note that (see the notation in (1.17))

$$[\mathfrak{g}_1, \mathfrak{g}_1] \subsetneq \mathfrak{g}_2.$$

Observe that the above $(\mathbb{R}^4, \circ)$ is isomorphic to the homogeneous Lie group $(\mathbb{R}^4, *)$ with the composition law

$$\xi * \eta = \begin{pmatrix} \xi_1 + \eta_1 \\ \xi_2 + \eta_2 \\ \xi_3 + \eta_3 \\ \xi_4 + \eta_4 + 2\,\eta_1\,\xi_2 - 2\,\eta_2\,\xi_1 \end{pmatrix}$$

and the new group of dilations

$$\delta_\lambda(\xi_1, \xi_2, \xi_3, \xi_4) = (\lambda\xi_1, \lambda\xi_2, \lambda\xi_3, \lambda^2 x_4).$$

The decomposition of the Lie algebra as in Remark 1.3.22 is

$$\mathfrak{g}_1 \oplus \mathfrak{g}_2 = \operatorname{span}\{Z_1, Z_2, \partial_{\xi_3}\} \oplus \operatorname{span}\{\partial_{\xi_4}\},$$

where $Z_1 = \partial_{\xi_1} + 2\,\xi_2\,\partial_{\xi_4}$, $Z_2 = \partial_{\xi_2} - 2\,\xi_1\,\partial_{\xi_4}$. Note that this time we have

$$[\mathfrak{g}_1, \mathfrak{g}_1] = \mathfrak{g}_2.$$

Definition 1.3.24 (Dilations on the Lie algebra of a homogeneous Lie group). *Let $\mathbb{G} = (\mathbb{R}^N, \circ, \delta_\lambda)$ be a homogeneous Lie group on $\mathbb{R}^N$ with Lie algebra $\mathfrak{g}$ and dilation*

$$\delta_\lambda(x_1, \ldots, x_N) = (\lambda^{\sigma_1} x_1, \ldots, \lambda^{\sigma_N} x_N).$$

We define a group of dilations on $\mathfrak{g}$ *(which we still denote by δ_λ) as follows: δ_λ is the (only) linear (auto)morphism of $\mathfrak{g}$ mapping the j-th element Z_j of the Jacobian basis for $\mathfrak{g}$ into $\lambda^{\sigma_j} Z_j$.*

In other words, if $X \in \mathfrak{g}$ is written w.r.t. the Jacobian basis $Z_1, \ldots, Z_N$ as

$$X = \sum_{j=1}^N c_j\, Z_j, \quad \text{we then have} \quad \delta_\lambda(X) = \sum_{j=1}^N c_j\, \lambda^{\sigma_j}\, Z_j.$$

We immediately recognize that, if $\pi : \mathfrak{g} \to \mathbb{R}^N$ is the map defined by $\pi(X) = XI(0)$ (see also Remark 1.2.20), it holds

$$\pi(\delta_\lambda(X)) = \delta_\lambda(\pi(X)) \quad \forall\, X \in \mathfrak{g}. \tag{1.64}$$

Indeed, we have

$$\begin{aligned} \delta_\lambda(\pi(X)) &= \delta_\lambda\left(\pi\left(\sum_{j=1}^N c_j\, Z_j\right)\right) = \delta_\lambda\left(\sum_{j=1}^N c_j\, \pi(Z_j)\right) \\ &= \delta_\lambda\left(\sum_{j=1}^N c_j\, (Z_j) I(0)\right) = \delta_\lambda(c_1, \ldots, c_N) \\ &= (\lambda^{\sigma_1}\, c_1, \ldots, \lambda^{\sigma_N}\, c_N) \end{aligned}$$

and, on the other hand,

$$\pi(\delta_\lambda(X)) = \pi\left(\delta_\lambda\left(\sum_{j=1}^N c_j\, Z_j\right)\right) = \pi\left(\sum_{j=1}^N c_j\, \lambda^{\sigma_j}\, Z_j\right)$$
$$= \sum_{j=1}^N c_j \lambda^{\sigma_j} \pi(Z_j) = (\lambda^{\sigma_1}\, c_1, \dots, \lambda^{\sigma_N}\, c_N).$$

The following simple and very useful fact holds.

Proposition 1.3.25. *Let $\mathbb{G}$ be a homogeneous Lie group on $\mathbb{R}^N$ with Lie algebra $\mathfrak{g}$. The dilation on $\mathfrak{g}$ introduced in Definition 1.3.24 is a Lie algebra automorphism of $\mathfrak{g}$, i.e.*

$$\delta_\lambda([X, Y]) = [\delta_\lambda(X), \delta_\lambda(Y)] \quad \forall\, X,\, Y \in \mathfrak{g}. \tag{1.65}$$

Proof. First we remark that, for every $i, j \in \{1, \dots, N\}$,

$$\delta_\lambda([Z_i, Z_j]) = \lambda^{\sigma_i + \sigma_j}\, [Z_i, Z_j]. \tag{1.66}$$

Indeed, since Z_i and Z_j are δ_λ-homogeneous of degrees σ_i and σ_j, respectively, then $[Z_i, Z_j]$ is a δ_λ-homogeneous vector field of degree $\sigma_i + \sigma_j$ (see Proposition 1.3.10). This implies that, if we express $[Z_i, Z_j]$ w.r.t. the Jacobian basis

$$[Z_i, Z_j] = \sum_{k=1}^N c_k\, Z_k,$$

then the sum runs over the k's such that

$$\sigma_k = \sigma_i + \sigma_j \tag{1.67}$$

(here we use, as a crucial tool, Corollary 1.3.11). Consequently,

$$\delta_\lambda([Z_i, Z_j]) = \delta_\lambda\left(\sum_{k=1}^N c_k\, Z_k\right) = \sum_{k=1}^N c_k \lambda^{\sigma_k} Z_k$$
$$\text{(by (1.67))} \quad = \lambda^{\sigma_i+\sigma_j} \sum_{k=1}^N c_k\, Z_k = \lambda^{\sigma_i+\sigma_j}\, [Z_i, Z_j].$$

This proves (1.66).

Let now $X = \sum_{i=1}^N x_i\, Z_i$ and $Y = \sum_{j=1}^N y_j\, Z_j$. Then we have

$$\delta_\lambda([X, Y]) = \delta_\lambda\left(\left[\sum_{i=1}^N x_i\, Z_i, \sum_{j=1}^N y_j\, Z_j\right]\right)$$
$$= \delta_\lambda\left(\sum_{i,j=1}^N x_i\, y_j\, [Z_i, Z_j]\right)$$

$$= \sum_{i,j=1}^{N} x_i\, y_j\, \delta_\lambda([Z_i, Z_j]) \quad \text{(see (1.66))}$$

$$= \sum_{i,j=1}^{N} x_i\, y_j\, \lambda^{\sigma_i+\sigma_j}([Z_i, Z_j])$$

$$= \left[\sum_{i=1}^{N} c_i\, \lambda^{\sigma_i} Z_i, \sum_{j=1}^{N} c_j\, \lambda^{\sigma_j} Z_j\right] = [\delta_\lambda(X), \delta_\lambda(Y)].$$

This completes the proof. □

1.3.4 The Exponential Map of a Homogeneous Lie Group

Let $\mathbb{G} = (\mathbb{R}^N, \circ, \delta_\lambda)$ be a homogeneous Lie group on $\mathbb{R}^N$ with Lie algebra $\mathfrak{g}$. The exponential map on $\mathfrak{g}$ has some remarkable properties, due to the homogeneous structure of $\mathbb{G}$. We prove such properties in what follows.

Let $Z_1, \dots, Z_N$ be the Jacobian basis of $\mathfrak{g}$. By Corollary 1.3.19, Z_j is δ_λ-homogeneous of degree σ_j and takes the form

$$Z_j = \sum_{k=j}^{N} a_k^{(j)}(x_1, \dots, x_{k-1})\partial_{x_k}, \tag{1.68}$$

where $a_k^{(j)}$ is a polynomial function δ_λ-homogeneous of degree $\sigma_k - \sigma_j$ and $a_j^{(j)} \equiv 1$. We now consider on $\mathfrak{g}$ the dilation group introduced in Definition 1.3.24, i.e. with abuse of notation (soon justified)

$$\delta_\lambda : \mathfrak{g} \longrightarrow \mathfrak{g}, \qquad \delta_\lambda\left(\sum_{j=1}^{N} \xi_j Z_j\right) := \sum_{j=1}^{N} \lambda^{\sigma_j} \xi_j Z_j. \tag{1.69}$$

Remark 1.3.26 (Consistency of the dilations on $\mathfrak{g}$ and $\mathbb{G}$). The dilation (1.69) is *consistent* with the one in $\mathbb{G}$. More precisely, if $Z \in \mathfrak{g}$ then, for every $\lambda > 0$, it holds

$$\delta_\lambda(ZI(x)) = (\delta_\lambda Z)I(\delta_\lambda(x)) \qquad \forall x \in \mathbb{G}. \tag{1.70}$$

We first check this identity in the case $Z = Z_j$, $j = 1, \dots, N$. Since Z_j is δ_λ-homogeneous of degree σ_j, by Corollary 1.3.6, we have $\delta_\lambda(Z_j I(x)) = \lambda^{\sigma_j}(Z_j I)(\delta_\lambda(x))$, so that (see (1.69))

$$\delta_\lambda(Z_j I(x)) = (\delta_\lambda Z_j)I(\delta_\lambda(x)).$$

Then, given $Z = \sum_{j=1}^{N} \xi_j Z_j \in \mathfrak{g}$, we have (since δ_λ is linear on $\mathfrak{g}$)

$$\delta_\lambda(ZI(x)) = \sum_{j=1}^{N} \xi_j\, \delta_\lambda(Z_j I(x)) = \sum_{j=1}^{N} \xi_j\, \big((\delta_\lambda Z_j)I(\delta_\lambda(x))\big)$$

$$= \left(\sum_{j=1}^{N} \xi_j\, (\delta_\lambda Z_j)\right) I(\delta_\lambda(x)) = (\delta_\lambda Z)I(\delta_\lambda(x)).$$

From the previous remark, we easily obtain the following lemma.

Lemma 1.3.27. *Let $\mathbb{G} = (\mathbb{R}^N, \circ, \delta_\lambda)$ be a homogeneous Lie group on $\mathbb{R}^N$ with Lie algebra $\mathfrak{g}$. Denote also by δ_λ the dilation* (1.69) *on $\mathfrak{g}$.*

Let $\gamma : [0, T] \to \mathbb{R}^N$ be an integral curve of Z with $Z \in \mathfrak{g}$. Then $\Gamma := \delta_\lambda(\gamma)$ is an integral curve of $\delta_\lambda(Z)$.

Proof. Identity (1.70) gives

$$\dot{\Gamma} = \delta_\lambda(\dot{\gamma}) = \delta_\lambda(ZI(\gamma)) = (\delta_\lambda Z)I(\delta_\lambda(\gamma)) = (\delta_\lambda Z)I(\Gamma).$$

This ends the proof. □

We are now in the position to prove the following important theorem.

Theorem 1.3.28 (Exponential map of a homogeneous Lie group). *Let $\mathbb{G} = (\mathbb{R}^N, \circ, \delta_\lambda)$ be a homogeneous Lie group with Lie algebra $\mathfrak{g}$. Then*

$$\mathrm{Exp} : \mathfrak{g} \to \mathbb{G} \quad \textit{and} \quad \mathrm{Log} : \mathbb{G} \to \mathfrak{g}$$

are globally defined diffeomorphisms with polynomial component functions (provided $\mathfrak{g}$ is equipped with its vector space structure and any fixed system of linear coordinates).

Moreover, denote also by δ_λ the dilation on $\mathfrak{g}$ defined in (1.69). *Then, for every $Z \in \mathfrak{g}$ and $x \in \mathbb{G}$, it holds*

$$\mathrm{Exp}\big(\delta_\lambda(Z)\big) = \delta_\lambda(\mathrm{Exp}\,(Z)) \quad \textit{and} \quad \mathrm{Log}\,(\delta_\lambda(x)) = \delta_\lambda(\mathrm{Log}\,(x)). \tag{1.71}$$

Proof. Let $Z \in \mathfrak{g}$, $Z = \sum_{j=1}^N \xi_j Z_j$. From (1.68) we obtain

$$Z = \sum_{k=1}^{N} \left(\sum_{j=1}^{k} \xi_j\, a_k^{(j)}(x_1, \ldots, x_{k-1}) \right) \partial_{x_k}. \tag{1.72}$$

Then the system of ODE's defining Exp (Z) is "pyramid"-shaped, and the first part of the theorem follows from Remark 1.1.3. In order to prove the first identity in (1.71), we consider the solution γ to the Cauchy problem

$$\dot{\gamma} = ZI(\gamma), \quad \gamma(0) = 0.$$

By the very definition of Exp (Z), we have $\gamma(1) = \mathrm{Exp}\,(Z)$. Let us put $\Gamma = \delta_\lambda(\gamma)$. By Lemma 1.3.27, Γ is an integral curve of $\delta_\lambda(Z)$. Moreover, $\Gamma(0) = \delta_\lambda(\gamma(0)) = \delta_\lambda(0) = 0$. Then $\Gamma(1) = \mathrm{Exp}\,(\delta_\lambda(Z))$, so that

$$\mathrm{Exp}\,(\delta_\lambda(Z)) = \Gamma(1) = \delta_\lambda(\gamma(1)) = \delta_\lambda(\mathrm{Exp}\,(Z)).$$

This proves the first identity in (1.71). The second one is trivially equivalent to the first one. □

The first part of Theorem 1.3.28 together with (1.40) and Proposition 1.2.29 (page 27) give the following corollary.

Corollary 1.3.29. *For every $x,\ y \in \mathbb{G}$, we have*

$$x \circ y = \exp(\mathrm{Log}\,(y))(x) \quad \textit{and} \quad x^{-1} = \mathrm{Exp}\,(-\mathrm{Log}\,(x)).$$

*Remark 1.3.30 (*Exp *and* Log *preserve the mass).* If Z is the vector field (1.72), then

$$ZI(x) = \left(\xi_1, \xi_2 + \xi_1\, a_2^{(1)}(x_1), \dots, \xi_N + \sum_{j=1}^{N-1} a_N^{(j)}(x_1, \dots, x_{N-1})\right).$$

This implies (see (1.13), page 10)

$$\mathrm{Exp}\,(Z) = \exp(Z)(0) = \big(\xi_1, \xi_2 + B_2(\xi_1), \dots, \xi_N + B_N(\xi_1, \dots, \xi_{N-1})\big), \quad (1.73)$$

where the B_j's are suitable polynomial functions. Then the Jacobian matrix of the map

$$\mathbb{R}^N \ni (\xi_1, \dots, \xi_N) \mapsto \mathrm{Exp}\,(\xi_1 Z_1 + \dots + \xi_N Z_N) \in \mathbb{R}^N$$

takes the following form

$$\begin{pmatrix} 1 & 0 & \cdots & 0 \\ \bigstar & 1 & \ddots & \vdots \\ \vdots & \ddots & \ddots & 0 \\ \bigstar & \cdots & \bigstar & 1 \end{pmatrix}. \quad (1.74)$$

Thus, with respect to the Jacobian basis of $\mathfrak{g}$ and the canonical basis of $\mathbb{G} \equiv \mathbb{R}^N$, Exp *preserves the Lebesgue measure. The same property holds for the map* Log, *since* $\mathrm{Log} = (\mathrm{Exp}\,)^{-1}$.

Equivalently, if $Z_1, \dots, Z_N$ is the Jacobian basis for $\mathfrak{g}$ and, as usual, $\xi \cdot Z = \sum_{j=1}^{N} \xi_j Z_j$, then Exp and Log have the following remarkable forms

$$\mathrm{Exp}\,(\xi \cdot Z) = \begin{pmatrix} \xi_1 \\ \xi_2 + B_2(\xi_1) \\ \vdots \\ \xi_N + B_N(\xi_1, \dots, \xi_{N-1}) \end{pmatrix} \quad (1.75a)$$

and

$$\mathrm{Log}\,(x) = \begin{pmatrix} x_1 \\ x_2 + C_2(x_1) \\ \vdots \\ x_N + B_N(x_1, \dots, x_{N-1}) \end{pmatrix} \cdot Z, \quad (1.75b)$$

where the B_i's and C_i's are polynomial functions (δ_λ-homogeneous of degree σ_i) completely determined by the composition law on $\mathbb{G}$.

For example, the above results are readily verified for $\mathbb{H}^1$, since in that case (as we proved in Example 1.2.28, page 27) Exp is represented by the identity matrix (if the algebra of $\mathbb{H}^1$ is equipped with the Jacobian basis). A little more elaborated example is given below.

Example 1.3.31. Let us consider on $\mathbb{R}^4$ the following composition law (we denote the points of $\mathbb{R}^4$ by $x = (x_1, x_2, x_3, x_4)$):

$$x \circ y = \begin{pmatrix} x_1 + y_1 \\ x_2 + y_2 \\ x_3 + y_3 + y_1 x_2 \\ x_4 + y_4 + y_1 x_2^2 + 2\, x_2 y_3 \end{pmatrix}.$$

It is readily verified that $\circ$ equips $\mathbb{R}^4$ with a homogeneous Lie group structure, provided that the dilation group is given by

$$\delta_\lambda(x_1, x_2, x_3, x_4) := (\lambda x_1, \lambda x_2, \lambda^2 x_3, \lambda^3 x_4).$$

Let us construct the exponential map. To begin with, the first two vector fields $X_1,\ X_2$ of the related Jacobian basis can be found as follows (see (1.33)): for every $\varphi \in C^\infty(\mathbb{R}^4)$ and every $x \in \mathbb{R}^4$, we have

$$(X_1\varphi)(x) = \frac{\partial}{\partial y_1}\bigg|_{y=0} \varphi(x \circ y), \qquad (X_2\varphi)(x) = \frac{\partial}{\partial y_2}\bigg|_{y=0} \varphi(x \circ y),$$

so that the chain rule straightforwardly gives

$$X_1 = \partial_{x_1} + x_2\, \partial_{x_3} + x_2^2\, \partial_{x_4}, \qquad X_2 = \partial_{x_2}.$$

A direct computation shows that

$$[X_1, X_2] = -\partial_3 - 2\, x_2\, \partial_4, \quad [X_1, [X_1, X_2]] = 0, \quad [X_2, [X_1, X_2]] = -2\, \partial_4,$$

whereas all commutators of length > 3 vanish. We now remark that

$$X_1, \quad X_2, \quad [X_1, X_2] \quad \text{and} \quad [X_2, [X_1, X_2]]$$

satisfy the following properties:

– they are left-invariant w.r.t. $\circ$ (as iterated commutators of left-invariant vector fields, see also (1.19));
– they are linearly independent vector fields[10];
– they form a basis of $\mathfrak{g}$ (since $\dim(\mathfrak{g}) = 4$, see Proposition 1.2.7).

[10] For example, we can notice that their respective evaluations at 0

$$\begin{pmatrix} 1 \\ 0 \\ 0 \\ 0 \end{pmatrix}, \quad \begin{pmatrix} 0 \\ 1 \\ 0 \\ 0 \end{pmatrix}, \quad \begin{pmatrix} 0 \\ 0 \\ -1 \\ 0 \end{pmatrix}, \quad \begin{pmatrix} 0 \\ 0 \\ 0 \\ -2 \end{pmatrix}$$

are linearly independent vectors of $\mathbb{R}^4$ (and use Proposition 1.2.13).

We now set $W_1 := X_1$, $W_2 := X_2$, $W_3 := [X_1, X_2]$, $W_4 := [X_2, [X_1, X_2]]$, and, for $\xi \in \mathbb{R}^4$, we also let $\xi \cdot W := \xi_1 W_1 + \xi_2 W_2 + \xi_3 W_3 + \xi_4 W_4$. By the definition in (1.8), we have $\exp(\xi \cdot W)(x) = \gamma(1)$, where $\gamma(s)$ solves

$$\begin{cases} \dot\gamma(s) = (\xi \cdot W)I(\gamma(s)) = (\xi_1, \xi_2, \xi_1\gamma_2 - \xi_3, \xi_1\gamma_2^2 - 2\,\xi_3\gamma_2 - 2\xi_4), \\ \gamma(0) = x. \end{cases}$$

Solving the above system of ODE's, one gets

$$\exp(\xi \cdot W)(x) = \begin{pmatrix} x_1 + \xi_1 \\ x_2 + \xi_2 \\ x_3 - \xi_3 + \frac{1}{2}\,\xi_1\xi_2 + \xi_1 x_2 \\ x_4 - 2\xi_4 + \frac{1}{3}\,\xi_1\xi_2^2 - \xi_2\xi_3 + \xi_1 x_2^2 + \xi_1\xi_2 x_2 - 2\xi_3 x_2 \end{pmatrix}.$$

As a consequence, by Definition 1.2.25, we obtain

$$\mathrm{Exp}\,(\xi \cdot W) = \exp(\xi \cdot W)(0) = \begin{pmatrix} \xi_1 \\ \xi_2 \\ -\xi_3 + \frac{1}{2}\,\xi_1\xi_2 \\ -2\xi_4 + \frac{1}{3}\,\xi_1\xi_2^2 - \xi_2\xi_3 \end{pmatrix},$$

so that the inverse map of Exp is given by

$$\mathrm{Log}\,(x) = \begin{pmatrix} x_1 \\ x_2 \\ -x_3 + \frac{1}{2}\,x_1x_2 \\ -\frac{1}{2}\,x_4 - \frac{1}{12}\,x_1x_2^2 + \frac{1}{2}\,x_2x_3 \end{pmatrix} \cdot W.$$

One can now directly check[11] the validity of Theorem 1.3.28, Corollary 1.3.29 and Remark 1.3.30. We explicitly remark the vectors W_i's do not form the Jacobian basis for $\mathfrak{g}$, which, instead, is given by

$$Z_1 = W_1, \quad Z_2 = W_2, \quad Z_3 = -W_3, \quad Z_4 = -\frac{1}{2}W_4.$$

Hence, if we write $\xi \cdot Z := \xi_1\, Z_1 + \xi_2\, Z_2 + \xi_3\, Z_3 + \xi_4\, Z_4$, we see that

$$(\xi, \xi_2, \xi_3, \xi_4) \cdot Z = \left(\xi, \xi_2, -\xi_3, -\frac{1}{2}\,\xi_4\right) \cdot Z,$$

[11] For example, we see that the inverse of $x \in \mathbb{G}$ is given by

$$x^{-1} = \mathrm{Exp}\left(-\mathrm{Log}\,(x)\right) = \begin{pmatrix} -x_1 \\ -x_2 \\ -x_3 + x_1x_2 \\ -x_4 + 2\,x_2x_3 - x_1x_2^2 \end{pmatrix}.$$

whence, with respect to the Jacobian coordinates, the exponential and the logarithmic maps are respectively given by

$$\mathrm{Exp}\,(\xi\cdot Z) = \begin{pmatrix} \xi_1 \\ \xi_2 \\ \xi_3 + \frac{1}{2}\,\xi_1\xi_2 \\ \xi_4 + \frac{1}{3}\,\xi_1\xi_2^2 + \xi_2\xi_3 \end{pmatrix},$$

$$\mathrm{Log}\,(x) = \begin{pmatrix} x_1 \\ x_2 \\ x_3 - \frac{1}{2}\,x_1x_2 \\ x_4 + \frac{1}{6}\,x_1x_2^2 - x_2x_3 \end{pmatrix}\cdot Z.$$

Compare this to (1.74). □

Theorem 1.3.28 has many important consequences. We collect some of them in the following remark.

Remark 1.3.32. From Theorem 1.3.28 we infer, in particular, that

$$\mathrm{Exp}\,:\mathfrak{g}\to\mathbb{G}\quad\text{and}\quad\mathrm{Log}\,:\mathbb{G}\to\mathfrak{g}$$

are globally defined C^∞ maps. Hence, by Remark 1.2.32, the operation on $\mathfrak{g}$

$$X\diamond Y := \mathrm{Log}\,(\mathrm{Exp}\,(X)\circ\mathrm{Exp}\,(Y)),\qquad X,\ Y\in\mathfrak{g},\tag{1.76}$$

defines a Lie group structure isomorphic to $(\mathbb{G},\circ)$. We consider on $\mathfrak{g}$ the dilation (still denoted by δ_λ) introduced in Definition 1.3.24. We claim that

δ_λ *is a Lie group automorphism of* $(\mathfrak{g},\diamond)$,

i.e.

$$\delta_\lambda(X\diamond Y) = (\delta_\lambda(X))\diamond(\delta_\lambda(Y))\quad\forall\,X,\ Y\in\mathfrak{g}.\tag{1.77}$$

Roughly speaking, $(\mathfrak{g},\diamond,\delta_\lambda)$ is a homogeneous Lie group too. To prove the claim, we notice that

$$\begin{aligned}
\delta_\lambda(X\diamond Y) &= \delta_\lambda\{\mathrm{Log}\,(\mathrm{Exp}\,(X)\circ\mathrm{Exp}\,(Y))\}\\
\text{(see (1.71))}\quad &= \mathrm{Log}\,\{\delta_\lambda(\mathrm{Exp}\,(X)\circ\mathrm{Exp}\,(Y))\}\\
&= \mathrm{Log}\,\{(\delta_\lambda(\mathrm{Exp}\,(X)))\circ(\delta_\lambda(\mathrm{Exp}\,(Y)))\}\\
\text{(see (1.71))}\quad &= \mathrm{Log}\,\{(\mathrm{Exp}\,(\delta_\lambda(X)))\circ(\mathrm{Exp}\,(\delta_\lambda(Y)))\}\\
&= (\delta_\lambda(X))\diamond(\delta_\lambda(Y)).
\end{aligned}$$

This proves our claim.[12]

[12] We explicitly remark that, if we already knew that $\diamond$ is defined by a "universal" composition of iterated Lie brackets, we could derive (1.77) from (1.65) in Proposition 1.3.25. Indeed,

$$\begin{aligned}
\delta_\lambda(X\diamond Y) &= \delta_\lambda\Big(X+Y+\frac{1}{2}\,[X,Y]+\frac{1}{12}\,[X,[X,Y]]+\cdots\Big)\\
&= \delta_\lambda(X)+\delta_\lambda(Y)+\frac{1}{2}\,[\delta_\lambda(X),\delta_\lambda(Y)]+\frac{1}{12}\,[\delta_\lambda(X),[\delta_\lambda(X),\delta_\lambda(Y)]]+\cdots\\
&= (\delta_\lambda(X))\diamond(\delta_\lambda(Y)),
\end{aligned}$$

identity following by a repeated application of (1.65).

We now identify $\mathfrak{g}$ with $\mathbb{R}^N$ taking coordinates with respect to the Jacobian basis. In other words, we consider the map (see Remark 1.2.20)

$$\pi : \mathfrak{g} \to \mathbb{R}^N, \qquad X \mapsto \pi(X) := XI(0).$$

Again, we transfer the Lie group structure of $(\mathfrak{g}, \diamond)$ into a Lie group $(\mathbb{R}^N, *)$ in the natural way, by setting

$$\xi * \eta := \pi(\pi^{-1}(\xi) \diamond \pi^{-1}(\eta)), \qquad \xi, \eta \in \mathbb{R}^N. \tag{1.78}$$

As a consequence, $(\mathbb{R}^N, *)$ is isomorphic to $(\mathfrak{g}, \diamond)$ and hence to $(\mathbb{G}, \circ)$. We finally consider on $\mathbb{R}^N$ the same dilation δ_λ defined on $\mathbb{G}$ (this makes sense, since the underlying manifold for $\mathbb{G}$ is $\mathbb{R}^N$ too).

We claim that

$$(\mathbb{R}^N, *, \delta_\lambda) \textit{ is a homogeneous Lie group.}$$

In other words, we have to show that δ_λ is a Lie group automorphism of $(\mathbb{R}^N, *)$. This follows from the argument below,

$$\begin{aligned}
\delta_\lambda(\xi * \eta) &= \delta_\lambda\{\pi(\pi^{-1}(\xi) \diamond \pi^{-1}(\eta))\} \\
\text{(see (1.64))} \quad &= \pi\{\delta_\lambda(\pi^{-1}(\xi) \diamond \pi^{-1}(\eta))\} \\
\text{(see (1.77))} \quad &= \pi\{(\delta_\lambda(\pi^{-1}(\xi))) \diamond (\delta_\lambda(\pi^{-1}(\eta)))\} \\
\text{(see (1.64))} \quad &= \pi\{(\pi^{-1}(\delta_\lambda(\xi))) \diamond (\pi^{-1}(\delta_\lambda(\eta)))\} = (\delta_\lambda(\xi)) * (\delta_\lambda(\eta)).
\end{aligned}$$

We can summarize the above remarked facts as follows.

Given a homogeneous Lie group $\mathbb{G} = (\mathbb{R}^N, \circ, \delta_\lambda)$, we can consider a somewhat "more canonical" homogeneous Lie group on $\mathbb{R}^N$

$$\text{C-H}(\mathbb{G}) := (\mathbb{R}^N, *, \delta_\lambda)$$

(which we may call *"of Campbell–Hausdorff type"*) obtained by the natural identification of the Lie algebra of $\mathbb{G}$ (equipped with the *Campbell–Hausdorff* composition law $\diamond$ in (1.76)) to $\mathbb{R}^N$ (via coordinates w.r.t. the Jacobian basis). (See also the Fig. 1.6.)

Example 1.3.33. Let us consider the homogeneous Lie group on $\mathbb{R}^3$ with the dilation

$$\delta_\lambda(x) = \delta_\lambda(x_1, x_2, x_3) = (\lambda x_1, \lambda x_2, \lambda^2 x_3)$$

and the composition law defined by

$$x \circ y = \begin{pmatrix} x_1 + y_1 \\ x_2 + y_2 \\ x_3 + y_3 + x_1 y_2 \end{pmatrix}.$$

The Jacobian basis for $\mathfrak{g}$ (the Lie algebra of $\mathbb{G}$) is

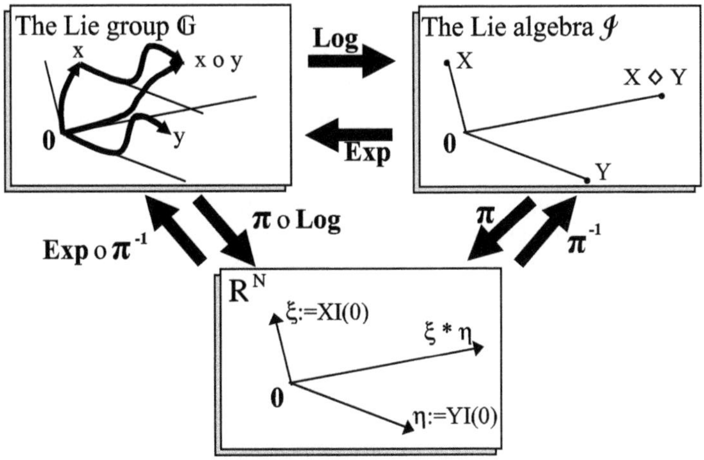

Fig. 1.6. Figure related to Remark 1.3.32

$$Z_1 = \partial_1, \quad Z_2 = \partial_2 + x_1\,\partial_3, \quad Z_3 = [Z_1, Z_2] = \partial_3.$$

For $\xi = (\xi_1, \xi_2, \xi_3) \in \mathbb{R}^3$, we fix the notation $\xi \cdot Z := \sum_{i=1}^3 \xi_i\, Z_i \in \mathfrak{g}$. It is easy to show that the exponential and logarithmic maps are given by

$$\operatorname{Exp}(\xi \cdot Z) = \begin{pmatrix} \xi_1 \\ \xi_2 \\ \xi_3 + \frac{\xi_1\xi_2}{2} \end{pmatrix}, \quad \operatorname{Log}(x) = \begin{pmatrix} x_1 \\ x_2 \\ x_3 - \frac{x_1x_2}{2} \end{pmatrix} \cdot Z.$$

Hence, the map $*$ considered in (1.78) is given by

$$\xi * \eta = \begin{pmatrix} \xi_1 + \eta_1 \\ \xi_2 + \eta_2 \\ \xi_3 + \eta_3 + \frac{1}{2}(\xi_1\eta_2 - \xi_2\eta_1) \end{pmatrix}.$$

The Jacobian basis related to C-H($\mathbb{G}$) = ($\mathbb{R}^3$, $*$, δ_λ) is

$$\widetilde{Z_1} = \partial_1 - \frac{\xi_2}{2}\,\partial_3, \quad \widetilde{Z_2} = \partial_2 + \frac{\xi_1}{2}\,\partial_3, \quad \widetilde{Z_3} = [\widetilde{Z_1}, \widetilde{Z_2}] = \partial_3.$$

Now, it is interesting to see what happens if we iterate this "C-G" process. It is easy to see that, if we consider once again the group obtained from C-H($\mathbb{G}$) in the same way (i.e. C-H(C-H($\mathbb{G}$))), we obtain nothing else than C-H($\mathbb{G}$) itself (a rigorous formulation of this fact will be given in Proposition 2.2.24).

We remark that $\mathbb{G}$ and C-H($\mathbb{G}$) are isomorphic and the canonical sub-Laplacian of $\mathbb{G}$

$$\Delta_{\mathbb{G}} = \{(\partial/\partial x_1)\}^2 + \{(\partial/\partial x_2) + x_1(\partial/\partial x_3)\}^2$$

is "equivalent" to the canonical sub-Laplacian of C-H($\mathbb{G}$)

$$\Delta_{\text{C-H}(\mathbb{G})} = \left\{(\partial/\partial\xi_1) - \frac{\xi_2}{2}(\partial/\partial x_3)\right\}^2 + \left\{(\partial/\partial\xi_2) + \frac{\xi_1}{2}(\partial/\partial x_3)\right\}^2.$$

1.4 Homogeneous Carnot Groups

We now enter into the core of the chapter by introducing the central definition of this book. Our definition here of Carnot group will be properly compared to the classical one in Section 2.2, page 121.

Definition 1.4.1 (Homogeneous Carnot group). *We say that a Lie group on* $\mathbb{R}^N$, $\mathbb{G} = (\mathbb{R}^N, \circ)$, *is a* (homogeneous) Carnot group *or a* homogeneous stratified group, *if the following properties hold*:

(C.1) $\mathbb{R}^N$ can be split as $\mathbb{R}^N = \mathbb{R}^{N_1} \times \cdots \times \mathbb{R}^{N_r}$, and the dilation $\delta_\lambda : \mathbb{R}^N \to \mathbb{R}^N$

$$\delta_\lambda(x) = \delta_\lambda(x^{(1)}, \dots, x^{(r)}) = (\lambda x^{(1)}, \lambda^2 x^{(2)}, \dots, \lambda^r x^{(r)}), \quad x^{(i)} \in \mathbb{R}^{N_i},$$

is an automorphism of the group $\mathbb{G}$ for every $\lambda > 0$.

Then $(\mathbb{R}^N, \circ, \delta_\lambda)$ *is a homogeneous Lie group on* $\mathbb{R}^N$, *according to Definition* 1.3.1. *Moreover, the following condition holds*:

(C.2) If N_1 is as above, let $Z_1, \dots, Z_{N_1}$ be the left invariant vector fields on $\mathbb{G}$ such that $Z_j(0) = \partial/\partial x_j|_0$ for $j = 1, \dots, N_1$. Then[13]

$$\text{rank}(\text{Lie}\{Z_1, \dots, Z_{N_1}\}(x)) = N \qquad \text{for every } x \in \mathbb{R}^N.$$

If (C.1) *and* (C.2) *are satisfied, we shall say that the triple* $\mathbb{G} = (\mathbb{R}^N, \circ, \delta_\lambda)$ *is a homogeneous Carnot group.*

We also say that $\mathbb{G}$ *has* step r *and* N_1 *generators. The vector fields* $Z_1, \dots, Z_{N_1}$ *will be called the* (Jacobian) generators *of* $\mathbb{G}$, *whereas any basis for* $\text{span}\{Z_1, \dots, Z_{N_1}\}$ *is called a system of* generators *of* $\mathbb{G}$.

(*Note.* As already remarked for Lie groups on $\mathbb{R}^N$ and homogeneous ones, the notion of homogeneous Carnot group is not coordinate-free. This fact will not distract us from recognizing its importance.)

In the sequel, we use the following notation to denote the points of $\mathbb{G}$

$$x = (x_1, \dots, x_N) = (x^{(1)}, \dots, x^{(r)}) \tag{1.79a}$$

with

$$x^{(i)} = (x_1^{(i)}, \dots, x_{N_i}^{(i)}) \in \mathbb{R}^{N_i}, \quad i = 1, \dots, r. \tag{1.79b}$$

Furthermore, we shall denote by $\mathfrak{g}$ the Lie algebra of $\mathbb{G}$.

Remark 1.4.2 (Equivalent definition of homogeneous Carnot group). An equivalent definition of homogeneous Carnot group can be given:

Suppose $\mathbb{G} = (\mathbb{R}^N, \circ)$ *is a Lie group on* $\mathbb{R}^N$, *and there exist positive real numbers* $\tau_1 \le \cdots \le \tau_N$ *such that* $d_\lambda(x) = (\lambda^{\tau_1} x_1, \dots, \lambda^{\tau_N} x_N)$ *is a Lie group morphism of* $\mathbb{G}$ *for every* $\lambda > 0$. *Let* $\mathfrak{g}$ *be the Lie algebra of* $\mathbb{G}$, *and let* $\mathfrak{g}_1$ *be the linear subspace*

[13] See the notation in Definition 1.1.5.

of $\mathfrak{g}$ *of the left-invariant vector fields which are* d_λ*-homogeneous of degree* τ_1. If $\mathfrak{g}_1$ Lie-generates[14] the whole $\mathfrak{g}$, *then* $\mathbb{G}$ *is a homogeneous Carnot group according to Definition* 1.4.1. *Precisely,* $\mathbb{G}$ *has step* $r := \tau_N/\tau_1$, *it has* $m := \dim(\mathfrak{g}_1)$ *generators, and it is a homogeneous Lie group with respect to the dilation*

$$\delta_\lambda = d_{\lambda^{1/\tau_1}}.$$

Also, set $\sigma_j := \tau_j/\tau_1$, *then* $\{\sigma_1, \sigma_2, \dots, \sigma_N\}$ *are consecutive integers starting from* 1 *up to* r. A sketch of the proof is in order.

As we observed in Remark 1.3.2, δ_λ is a morphism of $(\mathbb{G}, \circ)$, i.e. $\mathbb{G} = (\mathbb{R}^N, \circ, \delta_\lambda)$ is a homogeneous Lie group on $\mathbb{R}^N$. Obviously, $X \in \mathfrak{g}_1$ if and only if X is δ_λ-homogeneous of degree 1. Let ν be the maximum of the integers k's such that $\sigma_k = 1$. Let us denote by $\{Z_1, \dots, Z_N\}$ the Jacobian basis related to $\mathbb{G}$ and observe that (by Proposition 1.3.12), for every $j \le N$, Z_j is δ_λ-homogeneous of degree σ_j. We claim that

$$(\star) \qquad \nu = \dim(\mathfrak{g}_1) =: m, \text{ and } \{Z_1, \dots, Z_m\} \text{ is a basis for } \mathfrak{g}_1.$$

Indeed, let $X \in \mathfrak{g}_1$. Then $X = \xi_1 Z_1 + \dots + \xi_N Z_N$ for suitable scalars ξ_j's. Since X is δ_λ-homogeneous of degree 1, by Corollary 1.3.11 and the definition of ν, it holds $\xi_j = 0$ for every $j > \nu$. Hence, $\mathfrak{g}_1$ is spanned by $\{Z_1, \dots, Z_\nu\}$ whence (this system of vectors being linearly independent) the claimed $(\star)$ holds.

By the assumption $\mathrm{Lie}(\mathfrak{g}_1) = \mathfrak{g}$ and $(\star)$, it follows

$$(\star\star) \qquad \mathrm{Lie}(Z_1, \dots, Z_m) = \mathfrak{g}.$$

For every $j \in \mathbb{N}$, $j \ge 2$, let us set (see the notation in (1.17)) $\mathfrak{g}_j := [\mathfrak{g}_1, \mathfrak{g}_{j-1}]$. By Proposition 1.3.12, $\mathfrak{g}_j = \{0\}$ for every $j > r := \sigma_N$. Also, by Proposition 1.3.10, any $X \in \mathfrak{g}_j$ is δ_λ-homogeneous of degree j. Let now $j \in \{m+1, \dots, N\}$ be fixed. Then, by $(\star\star)$, Z_j is a linear combination of nested commutators of $Z_1, \dots, Z_m$. But any such commutator is δ_λ-homogeneous of an *integer* degree in $1, \dots, r$. This proves that σ_j (the δ_λ-homogeneous degree of Z_j) is integer and (again from Corollary 1.3.11) $\sigma_j \in \{1, \dots, r\}$. As a consequence, we have the splitting of $\mathbb{R}^N$, as requested in (C.1) of Definition 1.4.1, with $N_1 = m$.

Finally, let us prove that (C.2) holds too. This is obvious thanks to $(\star\star)$, since (see the notation in Definition 1.1.5)

$$\mathrm{rank}(\mathfrak{g}(x)) \ge \mathrm{rank}\big(Z_1 I(x), \dots, Z_N I(x)\big) = \mathrm{rank}\big(Z_1 I(0), \dots, Z_N I(0)\big) = N$$

for every $x \in \mathbb{G}$ (see Proposition 1.2.13). □

Example 1.4.3. The Heisenberg–Weyl group $\mathbb{H}^1$ is a Carnot group of step two and two generators. Indeed, it is a homogeneous Lie group (with dilations $\delta_\lambda(x_1, x_2, x_3) = (\lambda x_1, \lambda x_2, \lambda^2 x_3)$). Moreover (since the first two vector fields of the Jacobian basis are $Z_1 = \partial_{x_1} + 2x_2 \partial_{x_3}$ and $Z_2 = \partial_{x_2} - 2x_1 \partial_{x_3}$), we have

$$\mathrm{rank}(\mathrm{Lie}\{Z_1, Z_2\}(x)) = 3 \quad \text{for every } x \in \mathbb{R}^3,$$

[14] This means that $\mathrm{Lie}(\mathfrak{g}_1) = \mathfrak{g}$, see the notation in Proposition 1.1.7.

as we proved in Example 1.1.6. Thus, the above properties (C.1) and (C.2) are fulfilled.

We now give an example of a homogeneous Lie group which is *not* a Carnot group. Let us consider the following composition law on $\mathbb{R}^2$

$$(x_1, x_2) \circ (y_1, y_2) = (x_1 + y_1, x_2 + y_2 + x_1 y_1).$$

It can be readily verified that $\mathbb{G} = (\mathbb{R}^2, \circ)$ is a Lie group (here $(x_1, x_2)^{-1} = (-x_1, -x_2 + x_1^2)$). Moreover, $\mathbb{G}$ is a homogeneous group, if equipped with the dilation $\delta_\lambda(x_1, x_2) := (\lambda x_1, \lambda^2 x_2)$. Hence (C.1) is satisfied. However, (C.2) is not. Indeed, if $Z_1 = \partial_{x_1} + x_1 \partial_{x_2}$ is the first vector field of the Jacobian basis, we have

$$\operatorname{rank}(\operatorname{Lie}\{Z_1\}(x)) = 1 \neq 2 \quad \text{for every } x \in \mathbb{R}^2.$$

Hence $\mathbb{G}$ is not a homogeneous Carnot group.

Finally, let us remark that the triple $(\mathbb{R}^2, +, \delta_\lambda)$ is a homogeneous Carnot group if $\delta_\lambda(x_1, x_2) = (\lambda x_1, \lambda x_2)$, whereas if $\delta_\lambda(x_1, x_2) = (\lambda x_1, \lambda^2 x_2)$, $(\mathbb{R}^2, +, \delta_\lambda)$ is a homogeneous Lie group but not a Carnot one. □

From properties (C.1) and (C.2) of Definition 1.4.1 and the results on the homogeneous Lie groups showed in Section 1.3 we immediately get the assertions contained in the following remarks.

Remark 1.4.4. Let $(\mathbb{R}^N, \circ, \delta_\lambda)$ be a homogeneous Carnot group. Then $\circ$ has polynomial component functions. Moreover, following the notation in (1.79a) and denoting $x \circ y$ by $((x \circ y)^{(1)}, \dots, (x \circ y)^{(r)})$, we have

$$(x \circ y)^{(1)} = x^{(1)} + y^{(1)}, \quad (x \circ y)^{(i)} = x^{(i)} + y^{(i)} + Q^{(i)}(x, y), \quad 2 \le i \le r,$$

where

1. $Q^{(i)}$ only depends on $x^{(1)}, \dots, x^{(i-1)}$ and $y^{(1)}, \dots, y^{(i-1)}$;
2. the component functions of $Q^{(i)}$ are sums of mixed monomials in x, y;
3. $Q^{(i)}(\delta_\lambda x, \delta_\lambda y) = \lambda^i Q^{(i)}(x, y)$.

Remark 1.4.5. Let $(\mathbb{R}^N, \circ, \delta_\lambda)$ be a homogeneous Carnot group. Then we have

$$\mathcal{J}_{\tau_x}(0) = \begin{pmatrix} \mathbb{I}_{N_1} & 0 & \cdots & 0 \\ J_2^{(1)}(x) & \mathbb{I}_{N_2} & \ddots & \vdots \\ \vdots & \ddots & \ddots & 0 \\ J_r^{(1)}(x) & \cdots & J_r^{(r-1)}(x) & \mathbb{I}_{N_r} \end{pmatrix}, \tag{1.80}$$

where $\mathbb{I}_n$ is the $n \times n$ identity matrix, whereas $J_j^{(i)}(x)$ is a $N_j \times N_i$ matrix whose entries are δ_λ-homogeneous polynomials of degree $j - i$. In particular, if we let

$$\mathcal{J}_{\tau_x}(0) = \big(Z^{(1)}(x) \cdots Z^{(r)}(x)\big),$$

where $Z^{(i)}(x)$ is a $N \times N_i$ matrix, then the column vectors of $Z^{(i)}(x)$ define δ_λ-homogeneous vector fields of degree i: those of the relevant Jacobian basis.

Remark 1.4.6. Let $\mathbb{G} = (\mathbb{R}^N, \circ, \delta_\lambda)$ be a homogeneous Carnot group with Lie algebra $\mathfrak{g}$. Let $Z_1, \dots, Z_N$ be the Jacobian basis of $\mathfrak{g}$, i.e.

$$Z_j \in \mathfrak{g} \quad \text{and} \quad Z_j(0) = \partial_{x_j}|_0, \quad j = 1, \dots, N.$$

With a notation consistent with (1.79a) and (1.79b), we shall also denote the Jacobian basis by

$$Z_1^{(1)}, \dots, Z_{N_1}^{(1)}; \; \dots \; ; \; Z_1^{(r)}, \dots, Z_{N_r}^{(r)}.$$

Obviously, $Z_j^{(1)} = Z_j$ for $1 \le j \le N_1$. By Corollary 1.3.19, $Z_j^{(i)}$ is δ_λ-homogeneous of degree i and takes the form

$$Z_j^{(i)} = \partial/\partial x_j^{(i)} + \sum_{h=i+1}^{r} \sum_{k=1}^{N_h} a_{j,k}^{(i,h)}(x^{(1)}, \dots, x^{(h-i)})\, \partial/\partial x_k^{(h)}, \qquad (1.81)$$

where $a_{j,k}^{(i,h)}$ is a δ_λ-homogeneous polynomial function of degree $h - i$. In particular, the Jacobian generators of $\mathbb{G}$, i.e. the vector fields $Z_1^{(1)}, \dots, Z_{N_1}^{(1)}$ are δ_λ-homogeneous of degree 1. □

Remark 1.4.7. With the notation of the above remark, the Lie algebra $\mathfrak{g}$ is generated by $Z_1, \dots Z_{N_1}$,

$$\mathfrak{g} = \mathrm{Lie}\{Z_1, \dots Z_{N_1}\}. \qquad (1.82)$$

Indeed, the inclusion $\mathrm{Lie}\{Z_1, \dots Z_{N_1}\} \subseteq \mathfrak{g}$ is obvious. Since $\dim(\mathfrak{g}) = N$, in order to show the opposite inclusion, it is enough to prove that

$$\dim(\mathrm{Lie}\{Z_1, \dots Z_{N_1}\}) = N.$$

By condition (C.2), there exists $X_1, \dots, X_N \in \mathrm{Lie}\{Z_1, \dots Z_{N_1}\}$ such that $X_1 I(0)$, $\dots, X_N I(0)$ are linearly independent vectors in $\mathbb{R}^N$. Then (by Proposition 1.2.13) $X_1, \dots, X_N$ are linearly independent in $\mathfrak{g}$. Hence

$$N \ge \dim(\mathrm{Lie}\{Z_1, \dots Z_{N_1}\}) \ge N,$$

and this ends the proof. □

Remark 1.4.8 (Stratification of the algebra of a homogeneous Carnot group). Let the notation of Remark 1.4.6 be employed. Let us denote by $W^{(k)}$ the vector space spanned by the commutators of length k of $Z_1, \dots, Z_{N_1}$,

$$W^{(k)} := \mathrm{span}\big\{Z_J \mid J \in \{1, \dots, N_1\}^k\big\}.$$

Obviously, $W^{(k)} \subseteq \mathfrak{g}$, and every $Z \in W^{(k)}$ is δ_λ-homogeneous of degree k. Then, by Corollary 1.3.11 and Proposition 1.3.12, $W^{(k)} = \{0\}$ if $k > r$, while

$$W^{(k)} \subseteq \mathrm{span}\{Z_1^{(k)}, \dots, Z_{N_k}^{(k)}\} \qquad \text{if } 2 \le k \le r. \qquad (1.83)$$

Then, if we agree to let

$$W^{(1)} = \text{span}\{Z_1, \ldots, Z_{N_1}\} = \text{span}\{Z_1^{(1)}, \ldots, Z_{N_1}^{(1)}\},$$

we have

$$\dim(W^{(k)}) \leq N_k \qquad \text{for any } k \in \{1, \ldots, r\}. \tag{1.84}$$

On the other hand, by Proposition 1.1.7,

$$\text{span}\{W^{(1)}, \ldots, W^{(r)}\} - \text{Lie}\{Z_1^{(1)}, \ldots, Z_{N_1}^{(1)}\}.$$

Thus, by Remark 1.4.7,

$$\mathfrak{g} = \text{span}\{W^{(1)}, \ldots, W^{(r)}\},$$

so that, since $W^{(h)} \cap W^{(k)} = \{0\}$ if $h \neq k$ (see Corollary 1.3.11), we have

$$\mathfrak{g} = W^{(1)} \oplus W^{(2)} \oplus \cdots \oplus W^{(r)}.$$

As a consequence,

$$\dim(\mathfrak{g}) = \sum_{k=1}^{r} \dim(W^{(k)}).$$

On the other hand, $\dim(\mathfrak{g}) = N = \sum_{k=1}^{r} N_k$. Then, by (1.84),

$$\dim(W^{(k)}) = N_k \qquad \text{for any } k \in \{1, \ldots, r\},$$

and, by (1.83),

$$W^{(k)} = \text{span}\{Z_1^{(k)}, \ldots, Z_{N_k}^{(k)}\} \qquad \text{if } 1 \leq k \leq r.$$

We also have

$$[W^{(1)}, W^{(i-1)}] = W^{(i)} \qquad \text{for } 2 \leq k \leq r \tag{1.85a}$$

and

$$[W^{(1)}, W^{(r)}] = \{0\}. \tag{1.85b}$$

Indeed, let us put $V_1 := W^{(1)}$ and

$$V_i := [V_1, V_{i-1}] \qquad \text{for } i = 2, \ldots, r.$$

By the definition of $W^{(k)}$ and Proposition 1.1.7, $V_i \subseteq W^{(i)}$ for $i = 2, \ldots, r$. Then $\dim(V_i) \leq \dim(W^{(i)}) = N_i$. On the other hand, by Proposition 1.3.12, $[V_1, V_r] = \{0\}$, and, by Proposition 1.1.7,

$$\mathfrak{g} = \text{Lie}\{Z_1^{(1)}, \ldots, Z_{N_1}^{(1)}\} = \text{span}\{V_1, V_2, \ldots, V_r\}.$$

Then $N = \sum_{i=1}^{r} \dim(V_i) \leq \sum_{i=1}^{r} N_i = N$. This implies $\dim(V_i) = N_i$ for every $i \in \{1, \ldots, r\}$. As a consequence, $V_i = W^{(i)}$ for every $i \in \{1, \ldots, r\}$, and (1.85a) and (1.85b) hold.

Summing up, *we have proved the "stratification" of the Lie algebra* $\mathfrak{g}$, *i.e. the decomposition*

$$\mathfrak{g} = W^{(1)} \oplus W^{(2)} \oplus \cdots \oplus W^{(r)}$$

with

$$[W^{(1)}, W^{(i-1)}] = W^{(i)} \qquad \textit{for } 2 \le k \le r,$$
$$[W^{(1)}, W^{(r)}] = \{0\},$$

where

$$W^{(k)} = \text{span}\{Z_1^{(k)}, \dots, Z_{N_k}^{(k)}\} \qquad \textit{if } 1 \le k \le r.$$

Remark 1.4.9. Following all the notation and definitions in Remark 1.3.32, if $\mathbb{G} = (\mathbb{R}^N, \circ, \delta_\lambda)$ is a homogeneous Carnot group, then the Lie group C-H($\mathbb{G}$) := $(\mathbb{R}^N, *, \delta_\lambda)$ obtained by the natural identification of the algebra of $\mathbb{G}$ to $\mathbb{R}^N$ is a homogeneous *Carnot* group too. The proof of this fact is left to the reader as an exercise.

Remark 1.4.10 (Stratified change of basis on a homogeneous Carnot group). Let $(\mathbb{R}^N, \circ, \delta_\lambda)$ be a homogeneous Carnot group according to the Definition 1.4.1. As usual, we denote the points of $\mathbb{G}$ by

$$x = (x^{(1)}, \dots, x^{(r)}) \qquad \text{with } x^{(i)} \in \mathbb{R}^{N_i}$$

and the dilation group by $\delta_\lambda(x) = (\lambda x^{(1)}, \dots, \lambda^r x^{(r)})$. Let

$$C^{(1)}, \dots, C^{(r)}$$

be r fixed non-singular matrices with $C^{(i)}$ of dimension $N_i \times N_i$ for every $i = 1, \dots, r$. We denote by C the $N \times N$ matrix having $C^{(1)}, \dots, C^{(r)}$ as diagonal blocks and 0's elsewhere, i.e.

$$C = \begin{pmatrix} C^{(1)} & \cdots & 0 \\ \vdots & \ddots & \vdots \\ 0 & \cdots & C^{(r)} \end{pmatrix}.$$

Finally, we denote again by C the relevant linear change of basis on $\mathbb{R}^N$, i.e. the linear map

$$C : \mathbb{R}^N \to \mathbb{R}^N, \quad C(x) = C \cdot x.$$

We define on $\mathbb{R}^N$ a new composition law $*$ obtained by writing $\circ$ in the new coordinates defined by $\xi = C(x)$. More precisely, we have

$$\xi * \eta := C\big((C^{-1}(\xi)) \circ (C^{-1}(\eta))\big) \qquad \forall\, \xi, \eta \in \mathbb{R}^N. \tag{1.86}$$

We claim that $\mathbb{H} = (\mathbb{R}^N, *, \delta_\lambda)$ *is a homogeneous Carnot group isomorphic to* $\mathbb{G} = (\mathbb{R}^N, \circ, \delta_\lambda)$. The proof of this (not obvious) assertion is left as an exercise (see also Section 16.3 of Chapter 16, page 637).

1.5 The Sub-Laplacians on a Homogeneous Carnot Group

Definition 1.5.1 (Sub-Laplacian on a homogeneous Carnot group). *If $Z_1, \dots, Z_{N_1}$ are the Jacobian generators of the homogeneous Carnot group $\mathbb{G} = (\mathbb{R}^N, \circ, \delta_\lambda)$, the second order differential operator*

$$\Delta_{\mathbb{G}} = \sum_{j=1}^{N_1} Z_j^2 \tag{1.87}$$

is called the canonical sub-Laplacian *on $\mathbb{G}$. Any operator*

$$\mathcal{L} = \sum_{j=1}^{N_1} Y_j^2 \tag{1.88}$$

where $Y_1, \dots, Y_{N_1}$ is a basis of span$\{Z_1, \dots, Z_{N_1}\}$, *is simply called a* sub-Laplacian *on $\mathbb{G}$. The vector valued operator*

$$\nabla_{\mathbb{G}} = (Z_1, \dots, Z_{N_1}) \tag{1.89}$$

will be called the canonical (*or* horizontal) $\mathbb{G}$-gradient.

Finally, if $\mathcal{L}$ is as in (1.88), *the notation $\nabla_{\mathcal{L}} = (Y_1, \dots, Y_{N_1})$ will be used to denote the* $\mathcal{L}$-gradient (*or* horizontal $\mathcal{L}$-gradient).

Example 1.5.2. The canonical sub-Laplacian of the Heisenberg–Weyl group $\mathbb{H}^1$ is

$$\begin{aligned}\Delta_{\mathbb{H}^1} &= \{\partial_{x_1} + 2\,x_2\,\partial_{x_3}\}^2 + \{\partial_{x_2} - 2\,x_1\,\partial_{x_3}\}^2 \\ &= (\partial_{x_1})^2 + (\partial_{x_2})^2 + 4(x_1^2 + x_2^2)\,(\partial_{x_3})^2 + 4\,x_2\,\partial_{x_1,x_3} - 4\,x_1\,\partial_{x_2,x_3}.\end{aligned}$$

A (non-canonical) sub-Laplacian on $\mathbb{H}^1$ is, for example,

$$\begin{aligned}\mathcal{L} &= \{(\partial_{x_1} + 2\,x_2\,\partial_{x_3}) - (\partial_{x_2} - 2\,x_1\,\partial_{x_3})\}^2 + \{\partial_{x_2} - 2\,x_1\,\partial_{x_3}\}^2 \\ &= (\partial_{x_1})^2 + 2(\partial_{x_2})^2 + 4(x_1^2 + (x_1 + x_2)^2)\,(\partial_{x_3})^2 \\ &\quad - 2\partial_{x_1,x_2} + 4(x_1 + x_2)\,\partial_{x_1,x_3} - 4(x_1 + (x_1 + x_2))\,\partial_{x_2,x_3}.\end{aligned}$$

The following one is a (non-canonical!) sub-Laplacian on the classical additive group $(\mathbb{R}^2, +)$

$$\mathcal{L} = \{2\,\partial_{x_1} - 5\,\partial_{x_2}\}^2 + \{-\partial_{x_1} + 3\,\partial_{x_2}\}^2 = 5(\partial_{x_1})^2 + 34(\partial_{x_2})^2 - 26\,\partial_{x_1,x_2}.$$

It is not difficult to verify that $\mathbb{R}^4$ equipped with the operation

$$x \circ y = \begin{pmatrix} x_1 + y_1 \\ x_2 + y_2 \\ x_3 + y_3 + \frac{1}{2}\,(y_2 x_1 - y_1 x_2) \\ x_4 + y_4 + \frac{1}{2}\,(y_3 x_1 - y_1 x_3) + \frac{1}{12}\,(x_1 - y_1)(y_2 x_1 - y_1 x_2) \end{pmatrix}$$

and the dilation $\delta_\lambda(x_1, x_2, x_3, x_4) := (\lambda x_1, \lambda x_2, \lambda^2 x_3, \lambda^3 x_4)$ is a homogeneous Carnot group, say $\mathbb{G}$, whose canonical sub-Laplacian is

$$\begin{aligned}\Delta_{\mathbb{G}} &= \left\{\partial_1 - \frac{1}{2}x_2\partial_3 - \frac{1}{2}x_3\partial_4 - \frac{1}{12}x_1x_2\partial_4\right\}^2 \\ &\quad + \left\{\partial_2 + \frac{1}{2}x_1\partial_3 + \frac{1}{12}x_1^2\partial_4\right\}^2 \\ &= \partial_{11} + \partial_{22} + \frac{1}{4}(x_1^2 + x_2^2)\partial_{33} + x_1\partial_{23} - x_2\partial_{13} \\ &\quad + \frac{1}{4}\left(\left(\frac{x_1^2}{6}\right)^2 + \left(x_3 + \frac{x_1x_2}{6}\right)^2\right)\partial_{44} + \frac{x_1^2}{6}\partial_{24} - \left(x_3 + \frac{x_1x_2}{6}\right)\partial_{14} \\ &\quad + \frac{1}{2}\left(x_1\left(\frac{x_1^2}{6}\right) + x_2\left(x_3 + \frac{x_1x_2}{6}\right)\right)\partial_{34} + \underline{\frac{1}{6}x_2\,\partial_4}.\end{aligned}$$

We notice that $\Delta_{\mathbb{G}}$ also contains a first order (underlined) partial differential operator.[15] □

We would like to list some basic properties of the sub-Laplacians, straightforward consequences of the properties of the vector fields $Z_1^{(1)}, \dots, Z_{N_1}^{(1)}$. In what follows $\mathcal{L} = \sum_{j=1}^{N_1} Y_j^2$ will denote any sub-Laplacian on $\mathbb{G}$.

(A0) $\mathcal{L}$ is hypoelliptic, i.e. every distributional solution to $\mathcal{L}u = f$ is of class C^∞ whenever f is of class C^∞ (see Section 5.10, page 280, for further comments). This follows from the celebrated Hörmander hypoellipticity theorem [Hor67, Theorem 1.1], recalled in the Preface (Theorem 1, page VIII) and the fact that, if $\mathcal{L} = \sum_{j=1}^{N_1} Y_j^2$, then the following rank-condition holds

$$\operatorname{rank} \operatorname{Lie}\big(\{Y_1, \dots, Y_{N_1}\}(x)\big) = N \qquad \forall\, x \in \mathbb{R}^N.$$

This is an obvious consequence of hypothesis (C.2) in Definition 1.4.1, page 56.

(A1) $\mathcal{L}$ is invariant with respect to the left translations on $\mathbb{G}$, i.e. for every fixed $\alpha \in \mathbb{G}$,

$$\mathcal{L}(u(\alpha \circ x)) = (\mathcal{L}u)(\alpha \circ x) \qquad \text{for every } x \in \mathbb{G} \text{ and every } u \in C^\infty(\mathbb{R}^N).$$

This holds since the Y_j's are left-translation invariant on $\mathbb{G}$.

(A2) $\mathcal{L}$ is δ_λ-homogeneous of degree two, i.e. for every fixed $\lambda > 0$,

$$\mathcal{L}(u(\delta_\lambda(x))) = \lambda^2(\mathcal{L}u)(\delta_\lambda(x)) \quad \text{for every } x \in \mathbb{G} \text{ and every } u \in C^\infty(\mathbb{R}^N).$$

This holds since the Y_j's are δ_λ-homogeneous of degree one, see Remark 1.4.6.

---

[15] This cannot happen for the sub-Laplacians on groups of step two, *provided the inverse map on the group is* $-x$, i.e. any sub-Laplacian on a step-two homogeneous Carnot group (whose inverse map is $-x$) contains only second order coordinate partial derivatives (see Section 3.2).

(A3) $\mathcal{L}$ can be written as

$$\mathcal{L} = \operatorname{div}(A(x)\nabla^T), \tag{1.90a}$$

where div denotes the divergence operator in $\mathbb{R}^N$, $\nabla = (\partial_1, \ldots, \partial_N)$, A is the $N \times N$ symmetric matrix

$$A(x) = \sigma(x)\,\sigma(x)^T \tag{1.90b}$$

and $\sigma(x)$ is the $N \times N_1$ matrix whose columns are $Y_1 I(x), \ldots, Y_{N_1} I(x)$. In other words, $A(x)$ is the Gram matrix of the system of vectors $\{Y_1 I(x), \ldots, Y_{N_1} I(x)\}$; hence, since these vectors are linearly independent for every $x \in \mathbb{G}$, the rank of $A(x)$ is N_1 for every $x \in \mathbb{G}$. Now, (1.90a) is a consequence of the following computation

$$\begin{aligned}
\mathcal{L} &= \sum_{k=1}^{N_1} Y_k^2 = \sum_{k=1}^{N_1}\sum_{i=1}^{N} (Y_k I)_i(x)\ \partial_i \left(\sum_{j=1}^{N} (Y_k I)_j(x) \partial_j \right) \\
&= \sum_{i=1}^{N} \partial_i \left\{ \sum_{j=1}^{N} \left(\sum_{k=1}^{N_1} (Y_k I)_i(x)\,(Y_k I)_j(x) \right) \partial_j \right\},
\end{aligned}$$

since $(Y_k I)_i(x)$ does not depend on x_i. Hence, this proves that $\mathcal{L} = \operatorname{div}(A(x)\nabla^T)$ with

$$A(x) = \left(\sum_{k=1}^{N_1} (Y_k I)_i(x)\,(Y_k I)_j(x) \right)_{i,j=1,\ldots,N} = \sigma(x)\,\sigma(x)^T.$$

The matrix A takes the following block form

$$A = \begin{pmatrix} A_{1,1} & A_{1,2} \\ A_{2,1} & A_{2,2} \end{pmatrix}, \tag{1.91}$$

where $A_{i,j}$ stands for a $m_i \times m_j$ matrix with polynomial entries, with $m_1 = N_1$ and $m_2 = N - N_1$. Furthermore, $A_{1,1}$ is constant and non-singular. Indeed, for a suitable non-singular matrix $B = (b_{j,k})_{j,k=1}^{N_1}$, we have

$$Y_j = \sum_{k=1}^{N_1} b_{j,k}\, Z_k, \quad j = 1, \ldots, N_1. \tag{1.92}$$

On the other hand, see (1.81),

$$Z_k = \partial_k + \sum_{i=N_1+1}^{N} a_i^{(k)}\,\partial_i = \partial_k + \sum_{i=N_1+1}^{N} \partial_i (a_i^{(k)}\,\cdot), \tag{1.93}$$

where the $a_i^{(k)}$'s are suitable polynomial functions independent of x_i. Replacing (1.93) in (1.92) and squaring, we obtain

$$\mathcal{L} = \sum_{j=1}^{N_1} Y_j^2 = \sum_{h,k=1}^{N_1} a_{h,k}\, \partial_{h,k} + \sum_{h,k\le N,\ h\vee k>N_1} \partial_h (a_{h,k} \partial_k),$$

where

$$A_{1,1} = (a_{h,k})_{h,k\le N_1} = B^T\, B$$

is a constant $N_1 \times N_1$ matrix. Moreover, when $h \vee k := \max\{h,k\} > N_1$, then $a_{h,k}$ is a suitable polynomial function. We notice that if $\mathcal{L} = \Delta_{\mathbb{G}}$, then $B = \mathbb{I}_{N_1}$, so that $A_{1,1} = \mathbb{I}_{N_1}$. The above computations also give the expression of $\mathcal{L}$ with respect to the usual coordinate partial derivatives,

$$\mathcal{L} = \sum_{k=1}^{N_1} Y_k^2 = \sum_{i,j=1}^{N} a_{i,j}(x)\partial_{i,j} + \sum_{j=1}^{N} b_j(x)\, \partial_j,$$

where

$$a_{i,j}(x) = \sum_{k=1}^{N_1} (Y_k I)_i(x)\,(Y_k I)_j(x), \quad b_j(x) = \sum_{k=1}^{N_1} Y_k((Y_k I)_j(x)).$$

Analogous formulas hold for general sum of squares of vector fields (see Ex. 4 at the end of the chapter).

(A4) If $x \in \mathbb{G}$ is fixed and $A(x)$ is the matrix in (1.90a), then the quadratic form in $\xi \in \mathbb{R}^N$

$$q_{\mathcal{L}}(x,\xi) := \langle A(x)\xi, \xi\rangle$$

is called the *characteristic form* of $\mathcal{L}$. We have

$$q_{\mathcal{L}}(x,\xi) = \sum_{j=1}^{N_1} \langle Y_j I(x), \xi\rangle^2,$$

so that $q_{\mathcal{L}}(x,\cdot)$ is obtained by formally replacing in $\mathcal{L}$ the coordinate derivatives $\partial_1, \dots, \partial_N$ by $\xi_1, \dots, \xi_N$. This can be easily seen from (1.90b), for we have

$$\begin{aligned} q_{\mathcal{L}}(x,\xi) &= \langle A(x)\,\xi, \xi\rangle = \langle \sigma(x)\,\sigma(x)^T\,\xi, \xi\rangle \\ &= \langle \sigma(x)^T\,\xi, \sigma(x)^T\,\xi\rangle = |\sigma(x)^T\,\xi|^2 = \sum_{j=1}^{N_1} \langle Y_j I(x), \xi\rangle^2. \end{aligned}$$

Then $q_{\mathcal{L}}(x,\xi) \ge 0$ for every $x \in \mathbb{G}$ and every $\xi \in \mathbb{R}^N$, i.e. $A(x)$ is positive semi-definite for every $x \in \mathbb{G}$. Moreover, $q_{\mathcal{L}}(x,\xi) = 0$ iff $\langle Y_j I(x), \xi\rangle = 0$ for every $j \in \{1, \dots, N_1\}$.

Hence, for a fixed $x \in \mathbb{G}$, the set $N(x)$ of vectors ξ's which annihilate the quadratic form related to $A(x)$ is a linear space given by

$$N(x) := \{\xi \in \mathbb{R}^N \mid q_{\mathcal{L}}(x,\xi) = 0\} = \left(Y_1 I(x)\right)^{\perp} \cap \dots \cap \left(Y_{N_1} I(x)\right)^{\perp}. \quad (1.94)$$

We recall that, by Proposition 1.2.13, since $Y_1, \dots, Y_{N_1}$ are linearly independent in $\mathfrak{g}$, then $Y_1 I(x), \dots, Y_{N_1} I(x)$ are linearly independent in $\mathbb{R}^N$ for every fixed x. Thus, if $N_1 < N$, that is if $r \geq 2$, for every $x \in \mathbb{G}$ there exists $\xi \in \mathbb{R}^N \setminus \{0\}$ such that $q_{\mathcal{L}}(x, \xi) = 0$. More precisely, the set of isotropic vectors for the quadratic form related to $A(x)$ is a linear subspace of $\mathbb{R}^N$ of dimension $N - N_1$, equal to the kernel[16] of the matrix $A(x)$. This means that

if $r \geq 2$, then $\mathcal{L}$ is not elliptic at any point of $\mathbb{G}$.

On the other hand, if $N_1 = N$ (that is the step r of $\mathbb{G}$ is 1) the block $A_{1,1}$ in (1.91) has dimension $N \times N$, while the other blocks disappear. Then $\mathcal{L}$ is a constant coefficient operator of the form $\mathcal{L} = \sum_{i,j=1}^N a_{i,j}\, \partial_{i,j}$ with $A = (a_{i,j})_{i,j}$ symmetric and strictly positive definite. Thus, we can summarize these results as follows:

The sub-Laplacian $\mathcal{L}$ is a second order differential operator
in divergence form with polynomial coefficients.
The characteristic form of $\mathcal{L}$ is positive semi-definite.
If the step of $\mathbb{G}$ is ≥ 2, then $\mathcal{L}$ is not elliptic at any point of $\mathbb{G}$.
If the step of $\mathbb{G}$ is 1, then $\mathcal{L}$ is an elliptic operator
with constant coefficients.

Example 1.5.3. The canonical sub-Laplacian of the Heisenberg–Weyl group $\mathbb{H}^1$ has been written in Example 1.5.2: in that case the characteristic form is

$$\begin{aligned} q(x,\xi) &= (\xi_1)^2 + (\xi_2)^2 + 4(x_1^2 + x_2^2)(\xi_3)^2 + 4x_2\xi_1\xi_3 - 4x_1\xi_2\xi_3 \\ &= (\xi_1 + 2x_2\xi_3)^2 + (\xi_2 - 2x_1\xi_3)^2. \end{aligned}$$

Hence $q(x, \xi) = 0$ if and only if

$$\langle (1, 0, 2\,x_2), \xi\rangle = \langle (0, 1, -2\,x_1), \xi\rangle = 0,$$

[16] Indeed, given a $N \times N$ real symmetric matrix A, we denote by $\mathrm{Isotr}(A)$ the set of the isotropic vectors w.r.t. the quadratic form related to A, i.e.

$$\mathrm{Isotr}(A) := \{\xi \in \mathbb{R}^N \mid \langle A\xi, \xi\rangle = 0\}.$$

Obviously, it holds $\mathrm{Ker}(A) \subseteq \mathrm{Isotr}(A)$. In general, the reverse inclusion does not necessarily hold (as $\mathrm{Isotr}(A)$ is not necessarily a vector space!) as the following example shows: when

$$A := \begin{pmatrix} 1 & 0 \\ 0 & -1 \end{pmatrix},$$

we have $\mathrm{Ker}(A) = \{(0,0)\}$, whereas $\mathrm{Isotr}(A) = \mathrm{span}\{(1,1)\} \cup \mathrm{span}\{(1,-1)\}$. However, if A is positive semi-definite, then $\mathrm{Ker}(A) = \mathrm{Isotr}(A)$. Indeed, let R be a real, symmetric matrix such that $A = R^2$. If $\xi \in \mathrm{Isotr}(A)$, it holds

$$0 = \langle A\xi, \xi\rangle = \langle R^2\xi, \xi\rangle = \langle R^T R\xi, \xi\rangle = \langle R\xi, R\xi\rangle = \|R\,\xi\|^2.$$

Hence $R\,\xi = 0$, so that $A\,\xi = R^2\xi = RR\xi = 0$. This gives $\xi \in \mathrm{Ker}(A)$.

so that, for any fixed $x \in \mathbb{H}^1$, the set $N(x)$ of vectors ξ's which annihilate the relevant quadratic form is (see (1.94))

$$N(x) = (1, 0, 2x_2)^\perp \cap (0, 1, -2x_1)^\perp = \mathrm{span}\{(-2x_2, 2x_1, 1)\}.$$

Therefore, it is always one-dimensional. As we showed in (A4), $N(x)$ can also be found as $N(x) = \mathrm{Ker}(A(x))$, where

$$\begin{aligned} A(x) = \sigma(x)\sigma(x)^T &= \begin{pmatrix} 1 & 0 \\ 0 & 1 \\ 2x_2 & -2x_1 \end{pmatrix} \cdot \begin{pmatrix} 1 & 0 & 2x_2 \\ 0 & 1 & -2x_1 \end{pmatrix} \\ &= \begin{pmatrix} 1 & 0 & 2x_2 \\ 0 & 1 & -2x_1 \\ 2x_2 & -2x_1 & 4(x_1^2 + x_2^2) \end{pmatrix}. \end{aligned}$$

This kernel is one-dimensional, as $A(x)$ has rank 2 for every $x \in \mathbb{H}^1$.

(A5) The sub-Laplacian $\mathcal{L}$ is the second order partial differential operator related to the *Dirichlet form*

$$u \mapsto \int |\nabla_{\mathcal{L}} u|^2 \, \mathrm{d}x.$$

More precisely, let $\Omega \subseteq \mathbb{R}^N$ be an open set, and consider the functional

$$C^\infty(\Omega, \mathbb{R}) \ni u \mapsto J(u) = \frac{1}{2} \int_\Omega |\nabla_{\mathcal{L}} u|^2 \, \mathrm{d}x, \quad |\nabla_{\mathcal{L}} u|^2 = \sum_{j=1}^{N_1} (Y_j u)^2.$$

Denoting by $\langle\, , \rangle$ the inner product in $\mathbb{R}^{N_1}$, we have

$$J(u+h) - J(u) = \int_\Omega \langle \nabla_{\mathcal{L}} u, \nabla_{\mathcal{L}} h \rangle \, \mathrm{d}x + J(h)$$

for every $h \in C_0^\infty(\Omega, \mathbb{R})$. We call *critical point* of J any function $u \in C^\infty(\Omega, \mathbb{R})$ such that

$$\int_\Omega \langle \nabla_{\mathcal{L}} u, \nabla_{\mathcal{L}} h \rangle \, \mathrm{d}x = 0 \quad \forall\, h \in C_0^\infty(\Omega, \mathbb{R}).$$

Then, given $u \in C^\infty(\Omega, \mathbb{R})$, we have

u is a critical point of J if and only if $\mathcal{L}u = 0$ in Ω.

Indeed, since $Y_j^* = -Y_j$, an integration by parts gives

$$\begin{aligned} \int_\Omega \langle \nabla_{\mathcal{L}} u, \nabla_{\mathcal{L}} h \rangle \, \mathrm{d}x &= \sum_{j=1}^{N_1} \int_\Omega Y_j u Y_j h \, \mathrm{d}x = -\sum_{j=1}^{N_1} \int_\Omega (Y_j^2 u) h \, \mathrm{d}x \\ &= -\int_\Omega (\mathcal{L}u) h \, \mathrm{d}x \end{aligned}$$

for every $u \in C^\infty(\Omega, \mathbb{R})$ and $h \in C_0^\infty(\Omega, \mathbb{R})$.

1.5.1 The Horizontal $\mathcal{L}$-gradient

We end this section with some useful results on the horizontal $\mathcal{L}$-gradient.

Proposition 1.5.4. *Let $\mathcal{L} = \sum_{j=1}^{N_1} X_j^2$ be a sub-Laplacian on the homogeneous Carnot group $\mathbb{G}$. Let $u \in C^\infty(\mathbb{G}, \mathbb{R})$ be such that $X_j u$ is a polynomial function of $\mathbb{G}$-degree not exceeding m for every $j = 1, \dots, N_1$. Then u is a polynomial function of $\mathbb{G}$-degree not exceeding $m + 1$.*

Proof. Let $Z_1, \dots, Z_N$ be the Jacobian basis of $\mathfrak{g}$, the Lie algebra of $\mathbb{G}$. Since the X_j's have polynomial coefficients and $\mathfrak{g} = \mathrm{Lie}\{X_1, \dots, X_{N_1}\}$, $p_k := Z_k u$ is a polynomial function, $k = 1, \dots, N$. Moreover, if we denote

$$\delta_\lambda(x) = \delta_\lambda(x_1, \dots, x_N) := (\lambda^{\sigma_1} x_1, \dots, \lambda^{\sigma_N} x_N)$$

with $1 = \sigma_1 \le \dots \le \sigma_N = r$, keeping in mind the stratification of $\mathfrak{g}$, one easily recognizes that

$$\deg_{\mathbb{G}} p_k \le m + 1 - \sigma_k, \quad k = 1, \dots, N.$$

Now, by using (1.38) and (1.62) of Section 1.1, we see that

$$\partial_{x_j} u(x) = \sum_{k=j}^{N} p_k(x)\, a_k^{(j)}(x^{-1}), \quad j = 1, \dots, N,$$

where $y \mapsto a_k^{(j)}(y)$ is δ_λ-homogeneous of degree $\sigma_k - \sigma_j$. Then, since $\delta_\lambda(x^{-1}) = (\delta_\lambda(x))^{-1}$ and the map $x \mapsto x^{-1} = \mathrm{Exp}\,(-\mathrm{Log}\, x)$ has polynomial components (see Theorem 1.3.28 and Corollary 1.3.29), the function $x \mapsto a_k^{(j)}(x^{-1})$ is a polynomial δ_λ-homogeneous of degree $\sigma_k - \sigma_j$. It follows that

$$x \mapsto \langle \nabla u(x), x \rangle = \sum_{j=1}^{N} x_j\, \partial_{x_j} u(x)$$

is a polynomial function of $\mathbb{G}$-degree not exceeding $m + 1$. Therefore, also

$$u(x) - u(0) = \int_0^1 \frac{\mathrm{d}}{\mathrm{d}t}\big(u(tx)\big)\, \mathrm{d}t = \int_0^1 \langle \nabla u(tx), tx \rangle \frac{\mathrm{d}t}{t}$$

is a polynomial function of $\mathbb{G}$-degree not exceeding $m + 1$. □

In order to state the next corollary, we introduce a new notation. Let $\beta = (i_1, \dots, i_k)$ be a multi-index with components in the set $\{1, \dots, N_1\}$. We set

$$X^\beta := X_{i_1} \circ \dots \circ X_{i_k} \text{ and } |\beta| = k.$$

Corollary 1.5.5. *Let the hypotheses of Proposition 1.5.4 hold. Let $u \in C^\infty(\mathbb{G}, \mathbb{R})$ be such that*

$$X^\beta u = 0 \quad \forall\, \beta : |\beta| = m$$

for a suitable integer $m \ge 1$. Then u is a polynomial function on $\mathbb{G}$ of $\mathbb{G}$-degree not exceeding $m - 1$.

Proof. We argue by induction on m. By Proposition 1.5.6, the assertion holds if $m = 1$.

Suppose it holds for $m = p$, and let us prove that it holds for $m = p + 1$. Now, if $X^\beta u = 0$ for any multi-index β with $|\beta| = p + 1$, then

$$X^\gamma (X_j u) = 0 \quad \forall\, \gamma : |\gamma| = p$$

and for every $j \in \{1, \dots, N_1\}$. Hence, by the induction assumption, $X_j u$ is a polynomial function of $\mathbb{G}$-degree not exceeding $p - 1$ for every $j \in \{1, \dots, N_1\}$. From Proposition 1.5.4 it follows that u is a polynomial function of $\mathbb{G}$-degree not exceeding p. This completes the proof. □

Proposition 1.5.6. *Let Ω be an open and connected subset of the homogeneous Carnot group $\mathbb{G}$. Let $\mathcal{L}$ be any sub-Laplacian on $\mathbb{G}$.*

Then a function $u \in C^1(\Omega, \mathbb{R})$ is constant in Ω if and only if the relevant horizontal $\mathcal{L}$-gradient $\nabla_{\mathcal{L}} u$ vanishes identically on Ω.

Proof. It is obviously non-restrictive to suppose that $\mathcal{L} = \Delta_{\mathbb{G}}$.

Suppose $Z_1 u, \dots, Z_{N_1} u$ vanish identically on Ω. Since the Lie algebra of $\mathbb{G}$ is given by

$$\text{Lie}\{Z_1, \dots Z_{N_1}\}$$

(see (1.82)), then for every vector field Z_j of the Jacobian basis, we have $Z_j u \equiv 0$. We end by applying Proposition 1.2.21. □

Example 1.5.7. Our proof of Proposition 1.5.6, despite its simplicity, conceals a deep geometric argument, which can be applied in more general situations (namely, for vector fields satisfying Hörmander's condition). We describe the underlying geometric idea by an explicit example, leaving to the reader the task to generalize it in more general cases. (See also Chapter 19, where we study the so-called Carathéodory–Chow–Rashevsky connectivity theorem.)

Consider the Heisenberg–Weyl group $\mathbb{H}^1$ on $\mathbb{R}^3$. The Jacobian generators of its Lie algebra are the vector fields

$$X_1 = \partial_1 + 2x_2\partial_3, \qquad X_2 = \partial_2 - 2x_1\partial_3.$$

Proposition 1.5.6 then states that if a C^1-function u satisfies

$$X_1 u = 0 \quad \text{and} \quad X_2 u = 0 \quad \text{in a domain } \Omega \subseteq \mathbb{R}^3, \tag{1.95}$$

then u is constant in Ω. For the sake of brevity, rather than applying some (simple) connectedness argument, we may suppose $\Omega = \mathbb{R}^3$.

The basic idea is the following one. Suppose that two points $x, y \in \mathbb{R}^3$ can be joined by an integral curve γ of one of the fields $\pm X_1$ or $\pm X_2$, then $u(x) = u(y)$. Indeed, suppose that $\gamma : [0, T] \to \mathbb{R}^3$ is such that $\gamma(0) = x$, $\gamma(T) = y$ and

$$\dot{\gamma}(s) = XI(\gamma(s)) \quad \text{for all } s \in [0, T],$$

where X is one of the fields $\pm X_1$ or $\pm X_2$. Then, taking into account (1.95), we have

$$\begin{aligned} u(y) - u(x) &= \int_0^T \frac{\mathrm{d}}{\mathrm{d}s}(u(\gamma(s)))\,\mathrm{d}s = \int_0^T \langle (\nabla u)(\gamma(s)), \dot\gamma(s)\rangle\,\mathrm{d}s = \\ &= \int_0^T \big\langle (\nabla u)(\gamma(s)), XI(\gamma(s))\big\rangle\,\mathrm{d}s = \int_0^T (Xu)(\gamma(s))\,\mathrm{d}s = 0, \end{aligned}$$

whence $u(x) = u(y)$. As a consequence, we can prove that u is constant, if we show that

$$\text{\textit{any couple of points in } } \mathbb{H}^1 \text{ \textit{can be joined by a finite sequence of paths which are integral curves of} } \pm X_1 \text{ \textit{and} } \pm X_2. \tag{1.96}$$

We shall refer to (1.96) by saying that $\mathbb{H}^1$ is *(X_1, X_2)-connected.*

The (X_1, X_2)-connectedness has a deep motivation, namely the fact that

$$X_1, \quad X_2, \quad [X_1, X_2] \quad \text{\textit{are linearly independent.}}^{17} \tag{1.97}$$

In its turn, the fact that (1.97) implies the (X_1, X_2)-connectedness has a profound motivation too, mainly based on the so-called *Campbell–Hausdorff formula*, as we shall explain at the end of this section.

Before entering into the details of the Campbell–Hausdorff formula, we show how simple the argument is in the case of $\mathbb{H}^1$. Indeed, let us fix a point $P_0 := (x_1, x_2, x_3) \in \mathbb{H}^1$, and let us consider the path γ, integral curve of X_1 starting from this point. As we showed in (1.5), we have

$$\gamma(s) = (x_1 + s, x_2, x_3 + 2\,x_2\,s). \tag{1.98}$$

We denote this point by P_1. We then proceed along the integral curve of X_2 starting from P_1: at the time s, we arrive to the following point (as a simple calculation shows)

$$P_2 = (x_1 + s, x_2 + s, x_3 + 2x_2 s - 2(x_1 + s)s).$$

Moreover, we proceed along the integral curve of $-X_1$ starting from P_2: at the time s, we arrive to the following point (it is enough to have in mind (1.98) and replace s with $-s$)

$$P_3 = (x_1, x_2 + s, x_3 - 2(x_1 + s)s - 2s^2).$$

Finally, we proceed along the integral curve of $-X_2$ starting from P_3: at the time s, we arrive to the following point (again, it is enough to notice that the integral curve of $-X_2$ at time s coincides with the integral curve of X_2 at time $-s$)

[17] When X_1, X_2 are arbitrary smooth vector fields in $\mathbb{R}^3$, the result

$$(X_1,\ X_2,\ [X_1, X_2] \text{ \textit{linearly independent}}) \Longrightarrow (\mathbb{R}^3 \text{ is } [X_1, X_2]\text{-connected}),$$

is known as *Carathéodory's theorem*. For a more general version of this result in $\mathbb{R}^N$ (but under some further assumptions on the fields X_i's) we refer the reader to Chapter 19.

$$P_4 = (x_1, x_2, x_3 - 4s^2).$$

An analogous calculation shows that, if we start from (x_1, x_2, x_3) and proceed along the integral curves of, respectively, $X_2,\ X_1,\ -X_2,\ -X_1$, we arrive to

$$\widetilde{P}_4 = (x_1, x_2, x_3 + 4s^2).$$

Being s arbitrary in P_4 and $\widetilde{P}_4$, this shows that we can join any two points having the same x_1, x_2-coordinates and the third one arbitrarily given.

We now start from (x_1, x_2, x_3) and, along an integral curve of X_1, at the time s we arrive to $(x_1 + s, x_2, x_3 + 2x_2 s)$. By the preceding argument, keeping fixed the first two coordinates, we can vary the third one, in order to arrive to the point $(x_1 + s, x_2, x_3)$ (after finitely many integral curves of $\pm X_1, \pm X_2$). Being s arbitrary, this shows that we can join any two points having the same $x_2,\ x_3$ coordinates and the first one arbitrarily given. Finally, an obvious analogous argument shows that we can join any two points having the same $x_1,\ x_3$ coordinates and the second one arbitrarily given. All these facts together prove (1.96).

To end the section, we describe the reason why the validity of (1.97) implies the (X_1, X_2)-connectedness. For instance, let us denote by $\Gamma(s)$ the point we obtain if (as we did above) we follow paths of $X_1,\ X_2,\ -X_1,\ -X_2$ for a time s, this is the same as saying that we follow paths of

$$sX_1,\ sX_2,\ -sX_1,\ -sX_2$$

for unit time. More explicitly,

$$\Gamma(s) = \exp(-sX_2)(\exp(-sX_1)(\exp(sX_2)(\exp(sX_1)(x)))).$$

Very simple arguments (showed in the proof of Lemma 5.13.18, page 301) show that (see precisely (5.117), page 302)

$$\lim_{s\to 0} \frac{\Gamma(s) - x}{s^2} = [X_1, X_2]I(x). \tag{1.99}$$

Roughly speaking, (1.99) ensures that, by following *suitable* integral curves of $\pm X_1$, $\pm X_2$, we can arrive as close as we want to the endpoints of the integral curves of the commutator $[X_1, X_2]$. So, if $X_1, X_2, [X_1, X_2]$ span $\mathbb{R}^3$ at every point, it is intuitively evident that we can connect any two points by suitable integral curves of $\pm X_1, \pm X_2$.

This argument becomes completely apparent if we make use of the so-called Campbell–Hausdorff formula. For two vector fields X_1, X_2 generating an algebra nilpotent of step two, the Campbell–Hausdorff formula states that[18]

$$\exp(X_2)(\exp(X_1)(x)) = \exp\left(X_1 + X_2 + \frac{1}{2}[X_1, X_2]\right)(x). \tag{1.100}$$

---

[18] Further details on the Campbell–Hausdorff formula can be found in Definition 2.2.11, Theorem 2.2.13 (page 129), in Lemma 4.2.4 (page 194) and, mostly, in Theorem 15.1.1 (page 595).

We now apply three times formula (1.100), in order to simplify the above $\Gamma(s)$. The following calculation then applies:

$$\begin{aligned}
\Gamma(s) &= \exp(-sX_2)(\exp(-sX_1)(\exp(sX_2)(\exp(sX_1)(x)))) \\
&= \exp(-sX_2)\Big(\exp(-sX_1)\Big(\exp\Big(sX_1 + sX_2 + \frac{s^2}{2}[X_1, X_2]\Big)(x)\Big)\Big) \\
&= \exp(-sX_2)\Big(\exp\Big(sX_1 + sX_2 + \frac{s^2}{2}[X_1, X_2] - sX_1 - \frac{s^2}{2}[X_2, X_1]\Big)(x)\Big) \\
&= \exp(-sX_2)(\exp(sX_2 + s^2[X_1, X_2])(x)) \\
&= \exp(sX_2 + s^2[X_1, X_2] - sX_2)(x) = \exp(s^2[X_1, X_2])(x).
\end{aligned}$$

This says that, following suitable integral curves of $\pm X_1$, $\pm X_2$, we can arrive wherever the integral curves of the commutator $[X_1, X_2]$ do arrive! So, again, if $X_1, X_2, [X_1, X_2]$ span $\mathbb{R}^3$ at every point, we can obviously connect any two points by suitable integral curves of $\pm X_1$, $\pm X_2$.

We explicitly remark that the assumption that X_1, X_2 generate an algebra nilpotent of step two has made the above calculation very transparent. For general smooth vector fields, the Campbell–Hausdorff formula leads to a formula with a remainder term,

$$\begin{aligned}
&\exp(-sX_2)(\exp(-sX_1)(\exp(sX_2)(\exp(sX_1)(x)))) \\
&= \exp(s^2[X_1, X_2] + \mathcal{O}_x(s^3))(x).
\end{aligned}$$

Bibliographical Notes. "Carnot" groups seem to owe their name after an paper by C. Carathéodory [Car09] (related to a mathematical model of thermodynamics) dated 1909. The same denomination was then used in the school of M. Gromov [Gro96] and it is nowadays commonly used.

The definition of stratified group given in this chapter is seemingly different from the classical one by G.B. Folland [Fol75] and by G.B. Folland & E.M. Stein [FS82] (see also M. Gromov [Gro96], P. Pansu [Pan89]).

Indeed, here we focused on stratified groups having an underlying *homogeneous* structure. Nonetheless, we shall prove in Chapter 2 that any (abstract) stratified group is canonically isomorphic to a homogeneous one, so that our definition here is non-restrictive and seems to be more operative and easier to deal with.

A direct approach to homogeneous Carnot groups can also be found in E.M. Stein [Ste81] or in N.T. Varopoulos, L. Saloff-Coste, T. Coulhon [VSC92]; see also P. Hajlasz and P. Koskela [HK00].

1.6 Exercises of Chapter 1

Ex. 1) Verify that the following operation

$$
\begin{aligned}
&(x_1, x_2, x_3) \circ (y_1, y_2, y_3) \\
&\quad := (\operatorname{arcsinh}(\sinh(x_1) + \sinh(y_1)), x_2 + y_2 + \sinh(x_1) y_3, x_3 + y_3)
\end{aligned}
$$

endows $\mathbb{R}^3$ with a Lie group structure.

Ex. 2) **a)** Verify that, for every fixed $\alpha \in \mathbb{R}$, the following operation

$$
\begin{pmatrix}
x_1 + y_1, \\
x_2 + y_2, \\
x_3 + y_3 + \frac{1}{2}(x_1\, y_2 - x_2\, y_1), \\
x_4 + y_4 + \frac{1}{2}(x_1\, y_3 - x_3\, y_1) + \frac{\alpha}{2}(x_2\, y_3 - x_3\, y_2) \\
+\frac{1}{12}(x_1 - y_1)\,(x_1\, y_2 - x_2\, y_1) + \frac{\alpha}{12}(x_2 - y_2)\,(x_1\, y_2 - x_2\, y_1)
\end{pmatrix}
$$

defines on $\mathbb{R}^4$ a homogeneous Carnot group $\mathbb{G}$.

b) Verify that the Jacobian basis for the algebra $\mathfrak{g}$ of the above $\mathbb{G}$ is

$$
\begin{aligned}
Z_1 &= \partial_1 - \frac{1}{2} x_2\, \partial_3 - \left(\frac{1}{2} x_3 + \frac{1}{12} x_2 (x_1 + \alpha\, x_2)\right) \partial_4, \\
Z_2 &= \partial_2 + \frac{1}{2} x_1\, \partial_3 + \left(-\frac{1}{2} \alpha\, x_3 + \frac{1}{12} x_1 (x_1 + \alpha\, x_2)\right) \partial_4, \\
Z_3 &= [Z_1, Z_2] = \partial_3 + \left(\frac{1}{2} x_1 + \frac{1}{2} \alpha\, x_2\right) \partial_4, \\
Z_4 &= [Z_1, [Z_1, Z_2]] = [Z_1, Z_3] = \partial_4.
\end{aligned}
$$

Verify that the only other non-trivial commutator identity is

$$[Z_2, [Z_1, Z_2]] = \alpha\, [Z_1, [Z_1, Z_2]].$$

c) Verify that the exponential map for $\mathbb{G}$ (written w.r.t. the Jacobian basis) is the "identity map" in the following sense: if $\xi = (\xi_1, \xi_2, \xi_3, \xi_4) \in \mathbb{R}^4$, and $\xi \cdot Z \in \mathfrak{g}$ denotes the vector field $\sum_{i=1}^4 \xi_i\, Z_i$, then

$$\operatorname{Exp}(\xi \cdot Z) = \xi \in \mathbb{G}.$$

Ex. 3) With reference to what we proved in Example 1.2.33 (page 30), prove that, for the group considered in Examples 1.2.18 and 1.2.31 (pages 21 and 28) it holds

$$\operatorname{Log}\big(\operatorname{Exp}(X) \circ \operatorname{Exp}(Y)\big) = X + Y + \frac{1}{2}\,[X, Y].$$

Ex. 4) In this exercise, we provide some compact formulas for general sum of squares of vector fields. Suppose we are assigned m vector fields of class at least C^1 on $\mathbb{R}^N$,

$$X_k = \sum_{j=1}^{N} \alpha_{j,k}\, \partial_j, \quad \text{i.e. as usual,} \quad X_k I = \begin{pmatrix} \alpha_{1,k} \\ \alpha_{2,k} \\ \vdots \\ \alpha_{N,k} \end{pmatrix}, \quad k \le m.$$

Let S be the $N \times m$ matrix whose k-th column vector is given by $X_k I$,

$$S := \begin{pmatrix} \alpha_{1,1} & & \alpha_{1,k} & & \alpha_{1,m} \\ \vdots & \dots & \vdots & \dots & \vdots \\ \alpha_{N,1} & & \alpha_{N,k} & & \alpha_{N,m} \end{pmatrix} = (\alpha_{j,k})_{\substack{j=1,\dots,N \\ k=1,\dots,m}}.$$

With this notation, for every $u \in C^1(\mathbb{R}^N, \mathbb{R})$, we have

$$\nabla u \cdot S = (X_1 u, \dots, X_m u) =: \nabla_X u,$$

where $\nabla = (\partial_1, \dots, \partial_N)$ is the usual gradient operator, and

$$\nabla_X := (X_1, \dots, X_m)$$

is the "intrinsic gradient" related to the family $\{X_1, \dots, X_m\}$. We also define the $N \times N$ symmetric matrix

$$A := S \cdot S^T = \left(\sum_{k=1}^{m} \alpha_{i,k}\, \alpha_{j,k} \right)_{i,j=1,\dots,N} =: (a_{i,j})_{i,j \le N}.$$

For every $u \in C^1(\mathbb{R}^N, \mathbb{R})$, it holds

$$\begin{aligned} |\nabla_X u|^2 &:= \sum_{k=1}^{m} (X_k u)^2 = \langle \nabla_X u, \nabla_X u \rangle = \langle \nabla u \cdot S, \nabla u \cdot S \rangle \\ &= (\nabla u \cdot S) \cdot (\nabla u \cdot S)^T = \nabla u \cdot (S \cdot S^T) \cdot (\nabla u)^T = \nabla u \cdot A \cdot (\nabla u)^T, \end{aligned}$$

i.e.

$$|\nabla_X u|^2 = \sum_{i,j=1}^{N} a_{i,j}\, \partial_i u\, \partial_j u.$$

Show that, for every $k = 1, \dots, m$, it holds

$$X_k^2 = \sum_{i,j=1}^{N} \alpha_{i,k}\, \alpha_{j,k}\, \partial_{i,j}^2 + \sum_{j=1}^{N} \left(\sum_{i=1}^{N} \alpha_{i,k}\, (\partial_i \alpha_{j,k}) \right) \partial_j.$$

Derive that, in the coordinate form, the sum of squares related to the vector fields X_k's, $L := \sum_{k=1}^{m} X_k^2$, is given by

$$L = \sum_{i,j=1}^{N} a_{i,j}\, \partial_{i,j}^2 + \sum_{j=1}^{N} b_j\, \partial_j,$$

$$\text{with} \quad a_{i,j} = \sum_{k=1}^{m} \alpha_{i,k}\, \alpha_{j,k},$$

$$b_j = \sum_{k=1}^{m} \sum_{i=1}^{N} \alpha_{i,k}\, (\partial_i \alpha_{j,k}) = \sum_{k=1}^{m} X_k \alpha_{j,k}.$$

We note that b_j is a sort of "X-divergence" of the j-th row of the matrix S. Besides, show that

$$\begin{aligned} X_k^2 &= \sum_{i=1}^{N} \partial_i \left(\sum_{j=1}^{N} \alpha_{i,k}\, \alpha_{j,k}\, \partial_j \right) - \left(\sum_{i=1}^{N} \partial_i \alpha_{i,k} \right) \cdot \left(\sum_{j=1}^{N} \alpha_{j,k} \partial_j \right) \\ &= \sum_{i=1}^{N} \partial_i \left(\sum_{j=1}^{N} \alpha_{i,k}\, \alpha_{j,k}\, \partial_j \right) - \operatorname{div}(X_k I) \cdot X_k. \end{aligned}$$

Deduce that L can be written in the following equivalent ways:

$$\begin{aligned} L &= \sum_{i=1}^{N} \partial_i \left(\sum_{j=1}^{N} \sum_{k=1}^{m} \alpha_{i,k}\, \alpha_{j,k}\, \partial_j \right) - \sum_{j=1}^{N} \left(\sum_{k=1}^{m} \sum_{i=1}^{N} \alpha_{j,k}\, (\partial_i \alpha_{i,k}) \right) \partial_j \\ &= \sum_{i=1}^{N} \partial_i \left(\sum_{j=1}^{N} a_{i,j}\, \partial_j \right) - \sum_{j=1}^{N} \left(\sum_{k=1}^{m} \alpha_{j,k} \operatorname{div}(X_k I) \right) \partial_j \\ &= \operatorname{div}(A \cdot \nabla^T) - \sum_{k=1}^{m} \operatorname{div}(X_k I) \cdot X_k \\ &= \operatorname{div}(A \cdot \nabla^T) - \left\langle S \cdot \begin{pmatrix} \operatorname{div}(X_1 I) \\ \vdots \\ \operatorname{div}(X_m I) \end{pmatrix}, \nabla^T \right\rangle \\ &= \operatorname{div}(A \cdot \nabla^T) - \langle (\operatorname{div}(X_1 I), \dots, \operatorname{div}(X_m I)), \nabla_X \rangle. \end{aligned}$$

In particular, the sum of squares L is in divergence form if and only if

$$\forall\, j \le N, \qquad \sum_{k=1}^{m} \sum_{i=1}^{N} \alpha_{j,k}\, (\partial_i \alpha_{i,k}) = 0, \quad \text{i.e.} \quad S \cdot \begin{pmatrix} \operatorname{div}(X_1 I) \\ \vdots \\ \operatorname{div}(X_m I) \end{pmatrix} \equiv 0. \tag{1.101}$$

For example, re-derive that any sub-Laplacian $\mathcal{L}$ on a *homogeneous* Carnot group is in divergence form, because in this case it holds $\operatorname{div}(X_k I) = \sum_{i=1}^{N} \partial_i \alpha_{i,k} = 0$, for $\alpha_{i,k}$ does not depend on x_i. We explicitly remark that, in the case of *non*-homogeneous Carnot groups (which will be introduced in the next chapter), *this is not necessarily true*, as the following example shows: $\mathbb{R}^3$ equipped with the composition

$$x \circ y = (\operatorname{arcsinh}(\sinh(x_1) + \sinh(y_1)), x_2 + y_2 + \sinh(x_1)y_3, x_3 + y_3)$$

is a *non*-homogeneous Carnot group; the first two vector fields of the relevant Jacobian basis are

$$X_1 = (\cosh(x_1))^{-1}\,\partial_1, \quad X_2 = \partial_3 + \sinh(x_1)\,\partial_2,$$

so that the canonical sub-Laplacian is not a divergence form operator, for (1.101) is not satisfied, being, for $j = 1$,

$$\alpha_{1,1}\operatorname{div}(X_1 I) = -\sinh(x_1)/(\cosh(x_1))^3.$$

Ex. 5) Prove the assertions made in Remark 1.4.9 (page 61).

Ex. 6) Prove the following simple formulas of calculus for vector fields. Here, $u, f, g \in C^2(\mathbb{R}^N, \mathbb{R})$, $\alpha \in C^2(\mathbb{R}, \mathbb{R})$, $\varphi = (\varphi_1, \ldots, \varphi_N) \in C^2(\mathbb{R}^N, \mathbb{R}^N)$, $X = \nabla(\cdot) \cdot XI$ is a vector field on $\mathbb{R}^N$, $\mathcal{L} = \sum_{j=1}^m X_j^2$ is a sum of squares of vector fields on $\mathbb{R}^N$, $\nabla_{\mathcal{L}} = (X_1, \ldots, X_m)$ is the intrinsic gradient related to $\mathcal{L}$.

- $X(f\,g) = (Xf)\,g + f\,(Xg)$,
- $\nabla_{\mathcal{L}}(f\,g) = (\nabla_{\mathcal{L}}f)\,g + f\,(\nabla_{\mathcal{L}}g)$,
- $X^2(f\,g) = (X^2 f)\,g + 2\,(Xf)\,(Xg) + f\,(X^2 g)$,
- $\mathcal{L}(f\,g) = (\mathcal{L}f)\,g + 2\,\langle\nabla_{\mathcal{L}}f, \nabla_{\mathcal{L}}g\rangle + f\,(\mathcal{L}g)$,
- $X(\alpha(u)) = \alpha'(u)\,Xu$,
- $\nabla_{\mathcal{L}}(\alpha(u)) = \alpha'(u)\,\nabla_{\mathcal{L}}u$,
- $X^2(\alpha(u)) = \alpha''(u)\,(Xu)^2 + \alpha'(u)\,X^2 u$,
- $\mathcal{L}(\alpha(u)) = \alpha''(u)\,|\nabla_{\mathcal{L}}u|^2 + \alpha'(u)\,\mathcal{L}u$,
- $X(u(\varphi(x))) = \sum_{j=1}^N u_j(\varphi(x))\,X\varphi_j(x) = \langle(\nabla u)(\varphi(x)), X\varphi(x)\rangle$,
- $\nabla_{\mathcal{L}}(u(\varphi(x))) = \sum_{j=1}^N (\partial_j u)(\varphi(x))\,\nabla_{\mathcal{L}}\varphi_j(x)$; if we use the usual column-vector notation

$$\varphi(x) = \begin{pmatrix} \varphi_1(x) \\ \vdots \\ \varphi_N(x) \end{pmatrix}$$

for vectors in $\mathbb{R}^N$, the row-vector notation for the gradients

$$\nabla u = (\partial_1 u, \ldots, \partial_N u), \qquad \nabla_{\mathcal{L}}u = (X_1 u, \ldots, X_m u),$$

and we introduce the X-Jacobian matrix

$$\frac{\partial\varphi}{\partial X}(x) := \begin{pmatrix} X_1\varphi_1(x) & \cdots & X_m\varphi_1(x) \\ \vdots & \cdots & \vdots \\ X_1\varphi_N(x) & \cdots & X_m\varphi_N(x) \end{pmatrix} = \begin{pmatrix} \nabla_{\mathcal{L}}\varphi_1(x) \\ \vdots \\ \nabla_{\mathcal{L}}\varphi_N(x) \end{pmatrix}$$

(which is a $N \times m$ matrix) which can also be denoted by

$$\nabla_{\mathcal{L}}\varphi(x) = (X_1\varphi(x)\cdots X_m\varphi(x))$$
$$= \left(X_1 \begin{pmatrix} \varphi_1(x) \\ \vdots \\ \varphi_N(x) \end{pmatrix} \cdots X_m \begin{pmatrix} \varphi_1(x) \\ \vdots \\ \varphi_N(x) \end{pmatrix} \right),$$

then last formula can be rewritten as

$$\nabla_{\mathcal{L}}(u(\varphi(x))) = (\nabla u)(\varphi(x)) \cdot \frac{\partial \varphi}{\partial X}(x),$$

or equivalently

$$\nabla_{\mathcal{L}}(u(\varphi(x))) = (\nabla u)(\varphi(x)) \cdot \nabla_{\mathcal{L}}\varphi(x),$$

which is resemblant to the classical chain rule

$$\nabla(u(\varphi(x))) = (\nabla u)(\varphi(x)) \cdot \mathcal{J}_\varphi(x),$$

- prove that

$$X^2(u(\varphi(x))) = \sum_{i,j=1}^{N} (\partial_{i,j}u)(\varphi(x))X\varphi_i(x)\, X\varphi_j(x) + \sum_{j=1}^{N}(\partial_j u)(\varphi(x))X^2\varphi_j(x),$$

  or equivalently, following the previous notation (here $\text{Hess}_u(x)$ is the usual Hessian matrix of u at x),

$$X^2(u(\varphi(x))) = (X\varphi(x))^T \cdot \text{Hess}_u(\varphi(x)) \cdot (X\varphi(x)) + (\nabla u)(\varphi(x)) \cdot X^2\varphi(x),$$

- deduce from the previous formulas that

$$\mathcal{L}(u(\varphi(x))) = \sum_{i,j=1}^{N} (\partial_{i,j}u)(\varphi(x))\big\langle \nabla_{\mathcal{L}}\varphi_i(x), \nabla_{\mathcal{L}}\varphi_j(x)\big\rangle + \sum_{j=1}^{N}(\partial_j u)(\varphi(x))\mathcal{L}\varphi_j(x),$$

  which can also be rewritten as

$$\mathcal{L}(u(\varphi(x))) = \sum_{k=1}^{m}(X_k\varphi(x))^T \cdot \text{Hess}_u(\varphi(x)) \cdot (X_k\varphi(x)) + (\nabla u)(\varphi(x)) \cdot \mathcal{L}\varphi(x)$$

or, introducing the Gram matrix of the system of N vectors in $\mathbb{R}^m$

$$\{\nabla_{\mathcal{L}}\varphi_1(x), \ldots, \nabla_{\mathcal{L}}\varphi_N(x)\},$$

i.e. the $N \times N$ symmetric matrix

$$\begin{aligned}
G_\varphi(x) &:= \frac{\partial \varphi}{\partial X}(x) \cdot \left(\frac{\partial \varphi}{\partial X}(x)\right)^T \\
&= \begin{pmatrix} \nabla_{\mathcal{L}}\varphi_1(x) \\ \vdots \\ \nabla_{\mathcal{L}}\varphi_N(x) \end{pmatrix} \cdot \left((\nabla_{\mathcal{L}}\varphi_1(x))^T \cdots (\nabla_{\mathcal{L}}\varphi_N(x))^T \right) \\
&= \begin{pmatrix} \langle \nabla_{\mathcal{L}}\varphi_1(x), \nabla_{\mathcal{L}}\varphi_1(x)\rangle & \cdots & \langle \nabla_{\mathcal{L}}\varphi_1(x), \nabla_{\mathcal{L}}\varphi_N(x)\rangle \\ \vdots & \ddots & \vdots \\ \langle \nabla_{\mathcal{L}}\varphi_N(x), \nabla_{\mathcal{L}}\varphi_1(x)\rangle & \cdots & \langle \nabla_{\mathcal{L}}\varphi_N(x), \nabla_{\mathcal{L}}\varphi_N(x)\rangle \end{pmatrix},
\end{aligned}$$

the above formula can be rewritten as

$$\mathcal{L}(u(\varphi(x))) = \mathrm{trace}(\mathrm{Hess}_u(\varphi(x)) \cdot G_\varphi(x)) + (\nabla u)(\varphi(x)) \cdot \mathcal{L}\varphi(x).$$

Ex. 7) Let X be a smooth vector field on some open set $\Omega \subseteq \mathbb{R}^N$, and let $x \in \Omega$. Let γ be the solution to the system of ODE's

$$\dot\gamma(t) = XI(\gamma(t)), \quad \gamma(0) = x,$$

defined on the open (maximal) interval $\mathcal{D}(X, x) \subseteq \mathbb{R}$. Denote (momentarily) $\gamma(t)$ by $E(X, x, t)$. Let $\lambda \in \mathbb{R}$ be arbitrary. Prove the homogeneity relation

$$E(X, x, \lambda t) = E(\lambda X, x, t) \qquad \text{for all } t \text{ such that } \lambda t \in \mathcal{D}(X, x). \tag{1.102}$$

Let now $t_0 \in \mathcal{D}(X, x)$ be fixed. Take $\lambda = t_0$ and $t = 1$ in (1.102) and derive that (for the arbitrariness of t_0)

$$E(X, x, t) = E(t X, x, 1) \qquad \text{for all } t \in \mathcal{D}(X, x). \tag{1.103}$$

Throughout the following exercise, we shall adopt our usual notation $\exp(tX)(x)$ to denote $E(X, x, t)$. Equation (1.103) says that the notation is not improper.

Ex. 8) In this exercise, we give a generalization of formula (1.7) (page 7). Let $\Omega \subseteq \mathbb{R}^N$ be an open set, and let X be a smooth vector field on Ω. Let also f be a smooth function on Ω. Finally, we fix $x \in \Omega$. Then the function $t \mapsto f(\exp(tX)(x))$ is C^∞ near $t = 0$, and its Taylor expansion at 0 is given by

$$\begin{aligned}
f(\exp(tX)(x)) &= \sum_{k=0}^{n} \frac{t^k}{k!} (X^k f)(x) \\
&\quad + \frac{1}{n!} \int_0^t (t-s)^n X^{n+1} f(\exp(sX)(x)) \, ds.
\end{aligned} \tag{1.104}$$

Indeed, prove by induction that it holds

$$\frac{\mathrm{d}^k}{\mathrm{d}t^k}(f(\exp(tX)(x))) = (X^k f)(\exp(tX)(x)).$$

Derive from it the (very useful) formula

$$\left.\frac{\mathrm{d}^k}{\mathrm{d}t^k}\right|_{t=0}(f(\exp(tX)(x))) = (X^k f)(x).$$

Derive from (1.44) that if $\mathbb{G} = (\mathbb{R}^N, \circ)$ is a Lie group on $\mathbb{R}^N$ and X is left-invariant on $\mathbb{G}$, then

$$\left.\frac{\mathrm{d}^k}{\mathrm{d}t^k}\right|_{t=0}(f(x \circ \operatorname{Exp}(tX))) = (X^k f)(x).$$

Ex. 9) Let $\Omega \subseteq \mathbb{R}^N$ be an open set, and let X be a smooth vector field on Ω. Let us consider the (autonomous) equation $\dot{\gamma} = XI(\gamma)$. From general results on the existence and uniqueness of solution of ODE's (see, e.g. [Har82]) we know that, for every fixed compact set $K \subset \Omega$, there exists $\delta = \delta(X, K) > 0$ such that the solution $\gamma(t, x)$ to $\dot{\gamma} = XI(\gamma)$, $\gamma(0) = x$ exists for every $t \in]-\delta, \delta[$ and every initial value $x \in K$. From the uniqueness of the solution (and the autonomous nature of the equation) derive that

$$\gamma(t, \gamma(s, x)) = \gamma(t + s, x), \qquad |t| + |s| < \delta, \quad x \in K. \tag{1.105}$$

Deduce from this fact that, for every fixed $t_0 \in]-\delta, \delta[$, the mapping

$$K \ni x \mapsto \gamma(t_0, x)$$

is injective. (*Hint*: If $\gamma(t_0, x) = \gamma(t_0, y)$, then $\gamma(t+t_0, x) = \gamma(t, \gamma(t_0, x)) = \gamma(t, \gamma(t_0, y)) = \gamma(t + t_0, y)$, and evaluate at $t = -t_0$.)
If O is an open set such that $\overline{O} \subset \Omega$ is compact, prove that (whenever $|t_0| < \delta(X, \overline{O})$) the map $O \ni x \mapsto \gamma(t_0, x)$ is a C^∞-diffeomorphism onto its image. (*Hint*: Again thanks to (1.105), the inverse map is given by $y \mapsto \gamma(-t_0, x)$.)

Ex. 10) Let $Y_1, \ldots, Y_m$ be smooth vector fields on an open set $\Omega \subseteq \mathbb{R}^N$. Let $\xi = (\xi_1, \ldots, \xi_m) \in \mathbb{R}^m$. Since the dependence on ξ of $\xi_1 Y_1 I + \cdots + \xi_m Y_m I$ is smooth, by general results on ODE's (see, e.g. [Har82]), given a compact set $K \subset \Omega$, we infer the existence of $\varepsilon > 0$ such that the solution γ to

$$\dot{\gamma} = \sum_{j=1}^{m} \xi_j Y_j I(\gamma), \quad \gamma(0) = x$$

exists for every $x \in K$, every ξ satisfying $|\xi| < \varepsilon$ and every $t \in]-\varepsilon, \varepsilon[$ (here $|\xi|$ denotes any fixed norm on $\mathbb{R}^m$). Moreover, the dependence of γ on (x, ξ, t) is smooth. Using (1.103) (and the notation therein), observe that

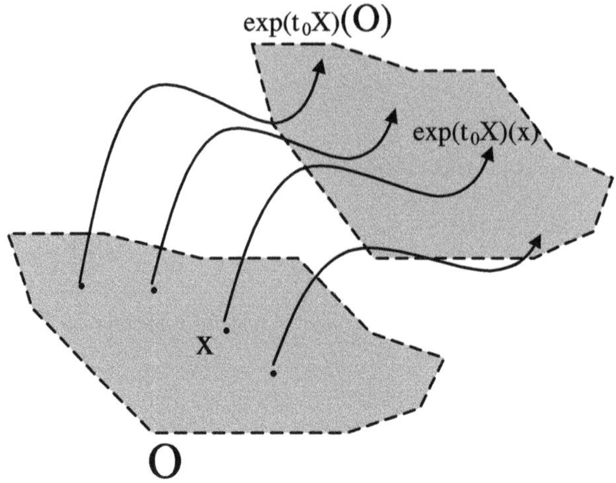

Fig. 1.7. The flow of the vector field X

$$E\Big(\sum_j \xi_j Y_j, x, t\Big) = E\Big(\sum_j t\xi_j Y_j, x, 1\Big) \qquad \text{for all } t \in]-\varepsilon, \varepsilon[$$

and derive that the map

$$(u_1, \dots, u_m) \mapsto \Theta_x(u) := \exp\Big(\sum_j u_j Y_j\Big)(x)$$

is well-defined (and smooth) on the open ball $B = \{u \in \mathbb{R}^m : |u| < \varepsilon^2\}$ (for any $x \in K$). Prove that the Jacobian matrix of Θ_x at $u = 0$ is the matrix whose j-th column vector is $Y_j I(x)$ or, equivalently, that the differential $\mathrm{d}_0\Theta_x$ sends the tangent vector $(\partial/\partial u_j)|_0 \in T_0(B)$ to $Y_j|_x \in T_x(\Omega)$. *Hint*: Use the following fact

$$\begin{aligned}\frac{\partial}{\partial u_j}\Big|_0 \exp\Big(\sum_j u_j Y_j\Big)(x) &= \frac{\partial}{\partial u_j}\Big|_0 \Big(I(x) + \sum_j u_j Y_j I(x) + \mathcal{O}_{u\to 0}(|u|^2)\Big)\\ &= Y_j I(x).\end{aligned}$$

As a corollary, deduce that if $m = N$ and $Y_1 I(x), \dots, Y_N I(x)$ are linearly independent, then Θ_x is a C^∞-diffeomorphism of a neighborhood of 0 onto a neighborhood of x. Finally, generalize (1.104) of Exercise 8 (we follow the notation therein) and obtain the following Taylor expansion at $u = 0$

$$f\Big(\exp\Big(\sum_{j=1}^m u_j Y_j\Big)(x)\Big) \sim \sum_{k=0}^\infty \frac{1}{k!}\Big(\sum_{j=1}^m u_j Y_j\Big)^k f(x). \tag{1.106}$$

Hint: Set

$$F(u) := f\left(\exp\left(\sum_{j=1}^{m} u_j Y_j\right)(x)\right) \quad \text{and} \quad G(t,u) := F(tu).$$

Hence $G(t,u) = f(\exp(tX)(x))$, where $X = \sum_{j=1}^{m} u_j Y_j$. Use Ex. 8 to derive that

$$\left.\frac{\partial^k}{\partial t^k}\right|_{t=0} G(t,u) = \left(\sum_{j=1}^{m} u_j Y_j\right)^k f(x),$$

and the Taylor expansion of F at $u = 0$ follows, as usual, from that of G at $t = 0$.

Ex. 11) (Rellich–Pohozaev identities). We introduce the following notation. If $\Omega \subset \mathbb{R}^N$ is a domain with boundary regular enough, we denote by ν the outer unit normal to $\partial\Omega$ and by $\mathrm{d}\sigma$ the Hausdorff $(N-1)$-dimensional measure on $\partial\Omega$. Provide the details for the following proposition.

Proposition 1.6.1. *Let $\mathbb{G}$ be an arbitrary homogeneous Carnot group, and let $\mathcal{L} = \sum_{i=1}^{m} X_i^2$ be a sub-Laplacian on $\mathbb{G}$. Let Ω be a bounded domain in $\mathbb{G}$, regular for the divergence theorem. Finally, let Z be a vector field of class $C^1(\mathbb{G})$. Then, for every $\varphi \in C^2(\overline{\Omega})$, we have*

$$\begin{aligned}
&2\int_{\partial\Omega} \sum_{i=1}^{m} X_i\varphi\, \langle X_i I, \nu\rangle\, Z\varphi \,\mathrm{d}\sigma - \int_{\partial\Omega} \langle ZI, \nu\rangle\, |\nabla_{\mathcal{L}}\varphi|^2 \,\mathrm{d}\sigma \\
&= 2\int_{\Omega} \sum_{i=1}^{m} X_i\varphi\, [X_i, Z]\varphi + 2\int_{\Omega} \mathcal{L}\varphi\, Z\varphi - \int_{\Omega} |\nabla_{\mathcal{L}}\varphi|^2 \operatorname{div}(ZI). \quad (1.107)
\end{aligned}$$

Proof. By applying two times the divergence theorem (and recalling that $X_i^* = -X_i$), we obtain

$$\begin{aligned}
\int_{\partial\Omega} |\nabla_{\mathcal{L}}\varphi|^2 \langle ZI, \nu\rangle \,\mathrm{d}\sigma &= \int_{\Omega} \operatorname{div}(|\nabla_{\mathcal{L}}\varphi|^2\, ZI) \\
&= \int_{\Omega} |\nabla_{\mathcal{L}}\varphi|^2 \operatorname{div}(ZI) + \int_{\Omega} Z(|\nabla_{\mathcal{L}}\varphi|^2) \\
&= \int_{\Omega} |\nabla_{\mathcal{L}}\varphi|^2 \operatorname{div}(ZI) + 2\int_{\Omega} \langle Z(\nabla_{\mathcal{L}}\varphi), \nabla_{\mathcal{L}}\varphi\rangle \\
&= \int_{\Omega} |\nabla_{\mathcal{L}}\varphi|^2 \operatorname{div}(ZI) + 2\int_{\Omega} \sum_{i=1}^{m} [ZI, X_i]\varphi\, X_i\varphi \\
&\quad + 2\int_{\Omega} \sum_{i=1}^{m} X_i(Z\varphi)\, X_i\varphi \\
&= \int_{\Omega} |\nabla_{\mathcal{L}}\varphi|^2 \operatorname{div}(ZI) + 2\int_{\Omega} \sum_{i=1}^{m} [ZI, X_i]\varphi\, X_i\varphi \\
&\quad + 2\int_{\partial\Omega} \sum_{i=1}^{m} Z\varphi\, X_i\varphi\, \langle X_i I, \nu\rangle \,\mathrm{d}\sigma - 2\int_{\Omega} Z\varphi\, \mathcal{L}\varphi.
\end{aligned}$$

This ends the proof. □

We now specify the integral identity (1.107) when Z is given by the so-called *generator of the translations*. Let $\circ$ be the group law on the Carnot group $\mathbb{G}$, and fix $z_0 \in \mathbb{G}$. We denote by Z^{z_0} the following vector field on $\mathbb{G}$

$$Z^{z_0} I(z) = \frac{\mathrm{d}}{\mathrm{d}h}\bigg|_{h=0} ((hz_0) \circ z). \tag{1.108}$$

Recalling that, for every $i = 1, \dots, m$,

$$X_i(z) = (\mathrm{d}/\mathrm{d}h)_{h=0}(z \circ (he_i))$$

(here e_i is the i-th versor of the canonical basis of $\mathbb{R}^N$), derive that *the bracket* $[X_i, Z^{z_0}]$ *vanishes identically. Hint*: $\circ$ is associative and

$$\begin{aligned}[X_i, Z^{z_0}]f(z) &= \frac{\mathrm{d}}{\mathrm{d}h}\bigg|_{h=0} \frac{\mathrm{d}}{\mathrm{d}s}\bigg|_{s=0} f((sz_0) \circ (z \circ (he_i))) \\ &\quad - \frac{\mathrm{d}}{\mathrm{d}s}\bigg|_{s=0} \frac{\mathrm{d}}{\mathrm{d}h}\bigg|_{h=0} f(((sz_0) \circ z) \circ (he_i)).\end{aligned}$$

Furthermore, *the divergence of the vector field* Z^{z_0} *vanishes identically*. Indeed, we recall that $\circ$ has the form $z \circ \zeta = ((z \circ \zeta)^{(1)}, \dots, (z \circ \zeta)^{(r)})$, where

$$(z \circ \zeta)^{(1)} = z^{(1)} + \zeta^{(1)}, \quad (z \circ \zeta)^{(j)} = z^{(j)} + \zeta^{(j)} + Q^{(j)}(z, \zeta), \quad 2 \le j \le r,$$

$Q^{(j)}$ being a function with values in $\mathbb{R}^{N_j}$ and whose components are mixed polynomials in z and ζ such that $Q^{(j)}(\delta_\lambda z, \delta_\lambda \zeta) = \lambda^j Q^{(j)}(z, \zeta)$. We then recognize that the components of $Z^{z_0}(z)$ in the j-th layer have the following form

$$(Z^{z_0}(z))^{(j)} = z_0^{(j)} + \langle z_0^{(1)}, q_1^{(j)}(z)\rangle + \cdots + \langle z_0^{(j-1)}, q_{j-1}^{(j)}(z)\rangle,$$

where $q_i^{(j)}$ is a $\mathbb{R}^{N_i}$-valued function whose components are polynomials δ_λ-homogeneous of degree $j - i$. In particular, $(Z^{z_0}(z))^{(j)}$ does not depend on z_j, whence $\operatorname{div}(Z^{z_0}) = 0$. From Proposition 1.6.1 and the above remarks, the next result immediately follows.

Proposition 1.6.2. *Let* $\mathbb{G}$ *be a homogeneous Carnot group, and let* $\mathcal{L} = \sum_{i=1}^m X_i^2$ *be a sub-Laplacian on* $\mathbb{G}$. *Let* Ω *be a bounded domain in* $\mathbb{G}$, *regular for the divergence theorem. Finally, for a fixed* $z_0 \in \mathbb{G}$, *let* Z^{z_0} *be the vector field defined in* (1.108). *Then, for every* $\varphi \in C^2(\overline{\Omega})$, *we have*

$$\begin{aligned}&2\int_{\partial\Omega} \sum_{i=1}^m X_i\varphi\, \langle X_i I, \nu\rangle\, Z^{z_0}\varphi \,\mathrm{d}\sigma - \int_{\partial\Omega} \langle Z^{z_0} I, \nu\rangle\, |\nabla_{\mathcal{L}}\varphi|^2 \,\mathrm{d}\sigma \\ &= 2\int_{\Omega} \mathcal{L}\varphi\, Z^{z_0}\varphi.\end{aligned}$$

Ex. 12) (Maps commuting with a sum of squares).

a) Consider a second order differential operator

$$L := \sum_{i,j=1}^{N} a_{i,j}(x)\, \partial^2_{i,j} + \sum_{i=1}^{N} b_i(x)\, \partial_i$$

with coefficients $a_{i,j}$ and b_i in $C^2(\mathbb{R}^N)$ such that $a_{i,j} = a_{j,i}$ for all i and j. Consider $\psi \in C^2(\mathbb{R}^N, \mathbb{R}^N)$. Prove that the necessary and sufficient conditions in order to have

$$L(u \circ \psi) = (L\, u) \circ \psi \qquad \forall\, u \in C^2(\mathbb{R}^N) \tag{1.109}$$

are the following ones:

$$\begin{cases} b_k(\psi) \;\; = L\psi_k & \forall\, k = 1, \dots, N, \\ a_{i,j}(\psi) = \displaystyle\sum_{r,s=1}^{N} a_{r,s}\, \partial_r \psi_i\, \partial_s \psi_j & \forall\, i,\, j = 1, \dots, N. \end{cases} \tag{1.110}$$

(*Hint*: Take $u(x) = x_k$ and then $u(x) := x_i\, x_j$.) If $\mathcal{J}_\psi$ denotes the usual Jacobian matrix of ψ and

$$A(x) := \big(a_{i,j}(x)\big)_{1 \le i,j \le N}, \qquad B(x) := \begin{pmatrix} b_1(x) \\ \vdots \\ b_N(x) \end{pmatrix},$$

so that

$$L = \mathrm{trace}\big(A(x) \cdot \mathrm{Hess}\big) + \langle \nabla, B(x) \rangle,$$

then system (1.110) can be rewritten as

$$\begin{cases} B(\psi(x)) = L\psi(x) \\ A(\psi(x)) = \mathcal{J}_\psi(x) \cdot A(x) \cdot (\mathcal{J}_\psi(x))^T. \end{cases} \tag{1.111}$$

b) Suppose furthermore that L is a sum of squares $L = \sum_{k=1}^{p} X_k^2$ with $X_k = \sum_{i=1}^{N} \sigma_i^{(k)}\, \partial_i$, so that

$$L = \sum_{i,j=1}^{N} \left(\sum_{k=1}^{p} \sigma_i^{(k)}\, \sigma_j^{(k)} \right) \partial^2_{i,j} + \sum_{i=1}^{N} \left(\sum_{k=1}^{p} X_k\, \sigma_i^{(k)} \right) \partial_i.$$

Prove that the necessary and sufficient conditions for (1.109) to hold are

$$\begin{cases} \displaystyle\sum_{k=1}^{p} (X_k\, \sigma_i^{(k)})(\psi) \quad = L\psi_i & \forall\, i = 1, \dots, N, \\ \displaystyle\sum_{k=1}^{p} \sigma_i^{(k)}(\psi)\, \sigma_j^{(k)}(\psi) = \langle \nabla_L \psi_i, \nabla_L \psi_j \rangle & \forall\, i,\, j = 1, \dots, N, \end{cases} \tag{1.112}$$

where $\nabla_L u := (X_1 u, \dots, X_p u)$. Set, as usual,

$$X_k I := \begin{pmatrix} \sigma_1^{(k)} \\ \vdots \\ \sigma_N^{(k)} \end{pmatrix}.$$

Considering the $N \times p$ matrix

$$S(x) := (\sigma_i^{(j)}(x))_{i \le N,\ j \le p} = (X_1 I(x) \cdots X_p I(x)),$$

we have

$$A(x) = \left(\sum_{k=1}^{p} \sigma_i^{(k)} \sigma_j^{(k)} \right)_{i,j \le N} = S(x) \cdot (S(x))^T,$$

so that

$$\mathcal{J}_\psi(x) \cdot A(x) \cdot (\mathcal{J}_\psi(x))^T = (\mathcal{J}_\psi(x) \cdot S(x)) \cdot (\mathcal{J}_\psi(x) \cdot S(x))^T.$$

Moreover,

$$\mathcal{J}_\psi(x) \cdot S(x) = (X_j \psi_i(x))_{i \le N,\ j \le p} = \begin{pmatrix} \nabla_{\mathcal{L}} \psi_1 \\ \vdots \\ \nabla_{\mathcal{L}} \psi_N \end{pmatrix}.$$

As a consequence,

$$(\mathcal{J}_\psi(x) \cdot S(x)) \cdot (\mathcal{J}_\psi(x) \cdot S(x))^T = (\langle \nabla_{\mathcal{L}} \psi_i(x), \nabla_{\mathcal{L}} \psi_j(x) \rangle)_{i,j \le N}.$$

Moreover,

$$B(x) = \left(\sum_{k=1}^{p} X_k \sigma_i^{(k)} \right)_{i \le N} = \begin{pmatrix} \sum_{k=1}^p X_k \sigma_1^{(k)} \\ \vdots \\ \sum_{k=1}^p X_k \sigma_N^{(k)} \end{pmatrix}.$$

Finally, (1.112) becomes

$$\begin{cases} B(\psi(x)) = \mathcal{L}\psi(x), \\ A(\psi(x)) = (\langle \nabla_{\mathcal{L}} \psi_i(x), \nabla_{\mathcal{L}} \psi_j(x) \rangle)_{i,j \le N}. \end{cases}$$

Ex. 13) Consider the Lie group (and the notation) in Example 1.2.19 (page 21). Show that, for every $\xi_1, \xi_2 \in \mathbb{R}$, the integral curve of $\xi_1 Z_1 + \xi_2 Z_2$ starting at $(0, 0)$ is

$$\gamma(t) = \begin{cases} (\xi_1 t, (\xi_2 e^{\xi_1 t} - \xi_2)/\xi_1) & \text{if } \xi_1 \ne 0, \\ (0, \xi_2 t) & \text{if } \xi_1 = 0. \end{cases}$$

Equivalently, considering the function

$$f : \mathbb{R} \to \mathbb{R}, \qquad f(z) := \frac{e^z - 1}{z} \text{ if } z \neq 0, \quad f(0) := 1,$$

we have

$$\gamma(t) = (\xi_1 \, t, \xi_2 \, t \, f(\xi_1 t)).$$

Derive that

$$\text{Exp} : \mathfrak{g} \to \mathbb{G}, \quad \exp(\xi_1 Z_1 + \xi_2 Z_2) = (\xi_1, \xi_2 \, f(\xi_1)).$$

Prove that Exp is smooth (indeed, real analytic!) and globally invertible. Find the Log function. (*Hint*: The function f is analytic and invertible.)

2

Abstract Lie Groups and Carnot Groups

The aim of this chapter is to prove that, up to a canonical isomorphism, the classical definition of stratified group (or Carnot group) (see Definition 2.2.3) coincides with our definition of homogeneous Carnot group, as given in Chapter 1. To this aim, we begin by recalling some basic facts about abstract Lie groups, providing all the terminology and the main results about manifolds, tangent vectors, left-invariant vector fields, Lie algebras, homomorphisms, the exponential map. Our exposition in Section 2.1 is self-contained and is intended to provide the topics from differential geometry and Lie group theory, which are strictly necessary to read this book.

In Section 2.2, we provide the cited equivalence between the two notions of Carnot group. This is accomplished by showing that the Lie algebra $\mathfrak{g}$ of the abstract Carnot group $\mathbb{G}$ possesses a natural structure of *homogeneous* Carnot group. Indeed, the group operation on $\mathfrak{g}$ is the one induced by that of $\mathbb{G}$ via the exponential map, and the dilations on $\mathfrak{g}$ are modeled on its stratification. A central rôle here will be played by the *Campbell–Hausdorff formula*, that we assume in this chapter by recalling (without proofs) some of its abstract and very general properties. Nonetheless, we shall devote Chapter 15 to provide a self-contained investigation of such a remarkable formula in the significant case of homogeneous stratified groups.

2.1 Abstract Lie Groups

2.1.1 Differentiable Manifolds

Let $N \in \mathbb{N}$, and let us define, for $i = 1, \dots, N$, the *coordinate projections* on $\mathbb{R}^N$ (whose points will be denoted by $\xi = (\xi_1, \dots, \xi_N) \in \mathbb{R}^N$ with $\xi_1, \dots, \xi_N \in \mathbb{R}$)

$$\pi_i : \mathbb{R}^N \longrightarrow \mathbb{R}, \qquad \pi_i(\xi) := \xi_i.$$

Definition 2.1.1 (N-dimensional locally Euclidean space). *An* N-dimensional locally Euclidean space M *is a Hausdorff topological space such that every point of* M

has a neighborhood in M homeomorphic to an open subset of $\mathbb{R}^N$. If φ is a homeomorphism between a connected open set $U \subseteq M$ and an open subset of $\mathbb{R}^N$, we say that $\varphi : U \to \mathbb{R}^N$ is a coordinate map,

$$x_i := \pi_i \circ \varphi : U \to \mathbb{R}$$

is a coordinate function, *and the pair (U, φ) (sometimes also denoted by $(U, x_1, \ldots, x_N)$) is a* coordinate system *or* a chart. *If $m \in U$ and $\varphi(m) = 0$, we say that the coordinate system is* centered *at m.*

Definition 2.1.2 (Differentiable manifold). *A C^∞* differentiable structure *$\mathcal{F}$ on a locally Euclidean space M is a collection of coordinate systems*

$$\{(U_\alpha, \varphi_\alpha) : \alpha \in A\}$$

with the following properties:

- $\bigcup_{\alpha \in A} U_\alpha = M$;
- *$\varphi_\alpha \circ \varphi_\beta^{-1}$ is C^∞ for every $\alpha, \beta \in A$ (whenever it is defined);*
- *$\mathcal{F}$ is maximal w.r.t. the second property in the sense that if (U, φ) is a coordinate system such that $\varphi \circ \varphi_\alpha^{-1}$ and $\varphi_\alpha \circ \varphi^{-1}$ are C^∞ for every $\alpha \in A$, then $(U, \varphi) \in \mathcal{F}$.*

An N-dimensional C^∞ differentiable manifold *is a couple $(M, \mathcal{F})$, where M is a second countable N-dimensional locally Euclidean space and $\mathcal{F}$ is a C^∞ differentiable structure.*

As usual, when we say "M is an N-dimensional C^∞ differentiable manifold", we leave unsaid that M is equipped with the fixed datum of a C^∞ differentiable structure $\mathcal{F}$ on M.

Let M and M' be differentiable manifolds of dimension N and N', respectively, and let $f : M \longrightarrow M'$. Then we say that f is C^∞ in $m \in M$ if, for every (or, equivalently, for at least one) coordinate system (U, φ) of M and for every (or, equivalently, for at least one) coordinate system (U', φ') of M' such that $m \in U$ and $f(m) \in U'$, the function

$$\varphi' \circ f \circ \varphi^{-1} : \varphi(U) \subseteq \mathbb{R}^N \longrightarrow \varphi'(U') \subseteq \mathbb{R}^{N'}$$

is C^∞ in a neighborhood of $\varphi(m)$. Let now $\mu(t)$ be a C^∞ function defined on a real interval and with values in M. We say that $\mu(t)$ is a *curve* passing through m at the time t_0 if $\mu(t_0) = m$. We say that a C^∞ real valued function f defined in a neighborhood of $m \in M$ is *horizontal* in m if

$$\left.\frac{\mathrm{d} f(\mu(t))}{\mathrm{d} t}\right|_{t=0} = 0$$

for every curve passing through m at the time $t = 0$.

Remark 2.1.3. The following characterizations hold:

(I). It is immediate to see that a real-valued function f is C^∞ in a neighborhood of $m \in M$ if and only if there exist a coordinate system (U, φ) with $m \in U$ and a (ordinary) C^∞ real-valued function $\widetilde{f}$ on (the open subset of $\mathbb{R}^N$) $\varphi(U)$ such that $f = \widetilde{f} \circ \varphi$ on U. (For example, any coordinate map $x_i = \pi_i \circ \varphi$ is C^∞ on U for π_i is smooth.)

(II). Analogously, μ is a curve passing through $m \in M$ at t_0 if and only if there exist a coordinate system (U, φ) with $m \in U$ and a (ordinary) C^∞ curve

$$\nu :]a, b[\subseteq \mathbb{R} \longrightarrow \varphi(U) \subseteq \mathbb{R}^N$$

with $t_0 \in]a, b[$, $\nu(t_0) = \varphi(m)$ and $\mu = \varphi^{-1} \circ \nu$ on $]a, b[$.

(III). Furthermore, we show that f is horizontal in $m \in M$ if and only if there exist a coordinate system (U, φ) with $m \in U$ and a C^∞ real-valued function $\widetilde{f}$ on $\varphi(U)$ such that $f = \widetilde{f} \circ \varphi$ on U and the ordinary gradient

$$(\nabla \widetilde{f})(\varphi(m))$$

is null.

Indeed, suppose this fact holds. Let μ be any curve passing through m at $t = 0$. By (II), there exist a coordinate system (V, ψ) with $m \in V$ and a C^∞ curve $\nu :]a, b[\longrightarrow \psi(U)$ with $0 \in]a, b[$, $\nu(0) = \psi(m)$ and $\mu = \psi^{-1} \circ \nu$ on $]a, b[$. Then, for $\varepsilon > 0$ sufficiently small (so that $\nu(t) \in \psi(U \cap V)$ for every $t \in]-\varepsilon, \varepsilon[$), the composition

$$f \circ \mu = \widetilde{f} \circ \varphi \circ \psi^{-1} \circ \nu$$

is well defined and smooth on $]-\varepsilon, \varepsilon[$ and, by the ordinary chain rule, it holds

$$\left.\frac{\mathrm{d}\, f(\mu(t))}{\mathrm{d} t}\right|_{t=0} = (\nabla \widetilde{f})\big(\varphi(\psi^{-1}(\nu(0)))\big) \cdot \mathcal{J}_{\varphi \circ \psi^{-1}}(\nu(0)) \cdot \dot{\nu}(0) = 0,$$

for $(\nabla \widetilde{f})(\varphi(\psi^{-1}(\nu(0)))) = (\nabla \widetilde{f})(\varphi(m)) = 0$. This proves that f is horizontal in m. Vice versa, suppose f is horizontal in m. By (I), there exist a coordinate system (U, φ) with $m \in U$ and a C^∞ real-valued function $\widetilde{f}$ on $\varphi(U)$ such that $f = \widetilde{f} \circ \varphi$ on U. We aim to prove that $(\nabla \widetilde{f})(\varphi(m)) = 0$. Let $h \in \mathbb{R}^N$ be any fixed vector. Let $\varepsilon > 0$ be small enough, so that, for every $|t| < \varepsilon$, we have

$$\nu(t) := \varphi(m) + t\, h \in \varphi(U).$$

Then, by (II), $\mu := \varphi^{-1} \circ \nu$ is a C^∞ curve passing through m at $t = 0$. Hence, being f horizontal in m, we have

$$\begin{aligned} 0 &= \left.\frac{\mathrm{d}\, f(\mu(t))}{\mathrm{d} t}\right|_{t=0} = \left.\frac{\mathrm{d}}{\mathrm{d} t}\right|_{t=0} \big(\widetilde{f} \circ \varphi \circ \varphi^{-1} \circ \nu\big)(t) \\ &= (\nabla \widetilde{f})(\nu(0)) \cdot \dot{\nu}(0) = \big\langle (\nabla \widetilde{f})(\varphi(m)), h \big\rangle. \end{aligned}$$

Consequently, due to the arbitrariness of h, this proves that $(\nabla \widetilde{f})(\varphi(m)) = 0$.

(IV). Finally, we prove that f is horizontal in $m \in M$ if and only if *for every* coordinate system (V, ψ) with $m \in V$, the C^∞ real-valued function $\widetilde{f} = f \circ \psi^{-1}$ on $\psi(V)$ satisfies

$$(\nabla \widetilde{f})(\psi(m)) = 0.$$

The "if" part follows by (III). Vice versa, suppose f is horizontal in $m \in M$. By (III), there exist a coordinate system (U, φ) with $m \in U$ and a C^∞ real-valued function g on $\varphi(U)$ such that $f = g \circ \varphi$ on U and $(\nabla g)(\varphi(m)) = 0$. As a consequence,

$$\begin{aligned}(\nabla \widetilde{f})(\psi(m)) &= (\nabla(f \circ \psi^{-1}))(\psi(m)) = (\nabla(g \circ \varphi \circ \psi^{-1}))(\psi(m)) \\ &= (\nabla g)(\varphi \circ \psi^{-1} \circ \psi(m)) \cdot \mathcal{J}_{\varphi \circ \psi^{-1}}(\psi(m)) \\ &= (\nabla g)(\varphi(m)) \cdot \mathcal{J}_{\varphi \circ \psi^{-1}}(\psi(m)) = 0,\end{aligned}$$

for $(\nabla g)(\varphi(m)) = 0$. □

Example 2.1.4. (i). Any constant function on M is trivially horizontal at every point of M. The same is true for a function vanishing near m.

(ii). Let $m \in M$, and fix any coordinate system (U, φ) with $m \in U$. Let $i, j \in \{1, \dots, N\}$ be chosen. Consider the relevant coordinate functions x_i, x_j, i.e. $x_i = \pi_i \circ \varphi$ and $x_j = \pi_j \circ \varphi$ on U. Denote $x^0 := \varphi(m)$. We show that

$$f := (x_i - x_i^0)\,(x_j - x_j^0)$$

is horizontal at m. By (III) of Remark 2.1.3, in order to prove that f is horizontal at m, it suffices to show that

$$\nabla(f \circ \varphi^{-1})(\varphi(m)) = 0.$$

But this is obvious since, for every $u \in \varphi(U)$,

$$\begin{aligned}(f \circ \varphi^{-1})(u) &= \big((\pi_i \circ \varphi - x_i^0)\,(\pi_j \circ \varphi - x_j^0)\big)(\varphi^{-1}(u)) \\ &= (u_i - x_i^0)\,(u_j - x_j^0),\end{aligned}$$

and trivially

$$\frac{\partial}{\partial u_k}\bigg|_{x^0} \big((u_i - x_i^0)(u_j - x_j^0)\big) = 0$$

for every $k = 1, \dots, N$.

(iii). If f is horizontal at $m \in M$, $f(m) = 0$ and g is any smooth function in a neighborhood of m, then $f\,g$ is horizontal at $m \in M$. Indeed, let (U, φ) be a coordinate system with $m \in M$. By (IV) of Remark 2.1.3, we need to show that $\nabla((f\,g) \circ (\varphi^{-1}))(\varphi(m)) = 0$. This is obvious since

$$\nabla\big((f\,g) \circ (\varphi^{-1})\big) = (g \circ \varphi^{-1})\,\nabla\big(f \circ (\varphi^{-1})\big) + (f \circ \varphi^{-1})\,\nabla\big(g \circ (\varphi^{-1})\big)$$

and recalling that $\nabla(f \circ \varphi^{-1})(\varphi(m)) = 0$ (for f is horizontal in m) and $f(m) = 0$ by hypothesis. □

As we shall see in the proof of Proposition 2.1.7, the functions in (i), (ii), (iii) of Example 2.1.4 are "infinitesimally" the only functions worth considering.

2.1.2 Tangent Vectors

Definition 2.1.5 (Tangent vector, space and bundle). *Let M be an N-dimensional C^∞ differentiable manifold. A* tangent vector $\mathbf{v}$ at $m \in M$ *is a linear functional, defined on the collection of the real-valued functions C^∞ in some neighborhood of m, such that*

$$\mathbf{v}(f) = 0$$

whenever f is horizontal in m.

We denote by M_m the set of the tangent vectors at $m \in M$, and we say that M_m is the tangent space *to M at m. We finally set*

$$T(M) := \bigcup_{m \in M} \{m\} \times M_m = \big\{(m, \mathbf{v}) : m \in M, \mathbf{v} \in \mathbf{M_m}\big\}. \tag{2.1}$$

$T(M)$ is called the tangent bundle *to M.*

Remark 2.1.6. Suppose f is a real-valued C^∞ function defined in a neighborhood of $m \in M$. Hence, there exists a coordinate system (U, φ) such that $m \in U$, and f is defined on U. Consider the open set $\varphi(U) \subseteq \mathbb{R}^N$. For $n \in \{1, 2, 3, 4\}$, denote by B_n the Euclidean ball centered at $\varphi(m)$ with radius $n\,\varepsilon$, where $\varepsilon > 0$ is small enough, so that B_4 is contained in $\varphi(U)$. Let χ be a smooth cut-off function on $\mathbb{R}^N$ such that $\chi \equiv 1$ on B_1 and $\chi \equiv 0$ on $\mathbb{R}^N \setminus B_2$. Then, consider the function on M

$$\overline{f}(n) := \begin{cases} \chi(\varphi(n))\, f(n) & \text{if } n \in \varphi^{-1}(B_3), \\ 0 & \text{otherwise,} \end{cases} \qquad n \in M.$$

It is easy to see that $\overline{f} \equiv f$ on the set $\varphi^{-1}(B_1)$, open neighborhood of n in M, and that φ is defined and smooth on M (note that $\chi \circ \varphi$ is C^∞ on U by Remark 2.1.3-(I)).

Since $f - \overline{f}$ vanishes in a neighborhood of m, by Example 2.1.4-(i), we have $\mathbf{v}(f - \overline{f}) = 0$ for every tangent vector $\mathbf{v}$ at m. Hence, by linearity, $\mathbf{v}(f) = \mathbf{v}(\overline{f})$.

This shows that in the definition of vector field it is not restrictive to suppose that a tangent vector at a point is a linear functional defined on the collection of the real-valued functions C^∞ in M.

We remark that M_m is a vector space with the usual operations of sum of functionals and multiplication of a functional times a scalar factor. Denoting by $\dim M$ the dimension N of the differentiable manifold M, we have the following result.

(*Note.* The reader will recognize in the proof below that a tangent vector at a point is indeed a linear "differential" operator of the first order defined on the C^∞ functions on M.)

Proposition 2.1.7. *Let M be an N-dimensional C^∞ differentiable manifold. Then* $\dim(M_m) = N = \dim M$.

Proof. Let $\mathbf{v} \in M_m$. Let (U, φ) be a coordinate system such that $m \in U$. Let $\widetilde{f}$ be an arbitrary C^∞ function defined in a neighborhood of $\varphi(m)$. We set

$$\widetilde{\mathbf{v}}(\widetilde{f}) := \mathbf{v}(f), \qquad \text{where } f := \widetilde{f} \circ \varphi.$$

By Remark 2.1.3-(I), $\widetilde{\mathbf{v}}$ is meaningful, and it is easily seen that $\widetilde{\mathbf{v}}$ is a linear functional defined on the functions C^∞ in a neighborhood of $x^0 := \varphi(m)$. In particular, $\widetilde{\mathbf{v}}$ (precisely, a suitable restriction of $\widetilde{\mathbf{v}}$) is linear on the vector space $\mathcal{V}$ of the real-valued linear functions $a(x)$ defined on $\mathbb{R}^N$ (where $N := \dim M$; note that $\mathcal{V} = (\mathbb{R}^N)^*$, the usual dual space of $\mathbb{R}^N$). Now, any linear functional on $\mathcal{V}$ is uniquely represented by a vector $\widetilde{\lambda} \in \mathbb{R}^N$, i.e.

$$\widetilde{\mathbf{v}}(a(u)) = a(\widetilde{\lambda}).$$

Let now $\widetilde{f} = \widetilde{f}(u)$ be any function C^∞ in a neighborhood of x^0. By Taylor expansion with the integral remainder, we have

$$\begin{aligned}
\widetilde{f}(u) &= \widetilde{f}(x^0) + \langle \nabla \widetilde{f}(x^0), u - x^0 \rangle \\
&\quad + \sum_{i,j=1}^{N} (u_i - x_i^0)(u_j - x_j^0) \int_0^1 (1-t) \frac{\partial^2 \widetilde{f}}{\partial u_i \partial u_j} \big(x^0 + t(u - x^0)\big)\, \mathrm{d}t \\
&=: \widetilde{f}(x^0) - a(x^0) + a(u) + \mathcal{R}(u)
\end{aligned}$$

for every u near x^0. By Example 2.1.4, we see that $\widetilde{f}(x^0) - a(x^0)$ and $\mathcal{R} \circ \varphi$ are horizontal in m. Indeed, the summand in the (i, j)-sum defining $\mathcal{R} \circ \varphi$ is the product of

$$(u_i - x_i^0)(u_j - x_j^0) \circ \varphi = (x_i - x_i^0)(x_j - x_j^0)$$

(where x_i, x_j are the relevant coordinate functions), which is horizontal and vanishing in m, times the smooth function near m

$$U \ni n \mapsto g(n) := \int_0^1 (1-t) \frac{\partial^2 \widetilde{f}}{\partial u_i \partial u_j} \big(x^0 + t(\varphi(n) - x^0)\big)\, \mathrm{d}t.$$

We now set $f = \widetilde{f} \circ \varphi$, so that

$$\begin{aligned}
f &= \widetilde{f}(x^0) - a(x^0) + a \circ \varphi + \mathcal{R} \circ \varphi \\
&= a \circ \varphi + \{\text{a horizontal function in } m\}.
\end{aligned}$$

As a consequence of the definition of tangent vector, we infer

$$\begin{aligned}
\mathbf{v}(f) &= \mathbf{v}(a \circ \varphi) = \widetilde{\mathbf{v}}\big((a \circ \varphi) \circ \varphi^{-1}\big) \\
&= \widetilde{\mathbf{v}}(a(u)) = \big\langle \nabla (f \circ \varphi^{-1})(x^0), \widetilde{\lambda} \big\rangle.
\end{aligned} \tag{2.2}$$

Now, it is easy (and is left as an exercise) to prove that the map

$$M_m \ni \mathbf{v} \mapsto \widetilde{\lambda} \in \mathbb{R}^N$$

is linear, injective and surjective. This completes the proof. □

Remark 2.1.8. It is easy to see that $\mathbf{v} \in M_m$ if and only if for every coordinate map (U, φ) with $m \in U$ (or, equivalently, for at least one such coordinate map) there exists $\lambda = \lambda(\varphi, m) \in \mathbb{R}^N$ such that

$$\mathbf{v}(f) = \left\langle \nabla(f \circ \varphi^{-1})(\varphi(m)), \lambda \right\rangle \qquad \text{for every } f \text{, real valued and } C^\infty \text{ near } m.$$

The "only if" part follows from the proof of Proposition 2.1.7 (see in particular (2.2)). Suppose now that there exists (U, φ) and λ as above. Let f be horizontal in $m \in M$. We need to show that

$$\left\langle \nabla(f \circ \varphi^{-1})(\varphi(m)), \lambda \right\rangle = 0.$$

By Remark 2.1.3, the function $\widetilde{f} = f \circ \varphi^{-1}$ satisfies $(\nabla \widetilde{f})(\varphi(m)) = 0$. This definitely suffices for what we needed to prove. □

Remark 2.1.9. Remark 2.1.8 furnishes another very useful characterization of tangent vectors (actually, an alternative definition frequently used in literature). If $m \in M$, then $\mathbf{v} \in M_m$ if and only if there exists a C^∞ curve on M passing through m at t_0 such that

$$\mathbf{v}(f) = \left.\frac{\mathrm{d}}{\mathrm{d}t}\right|_{t_0} (f(\mu(t))) \qquad \text{for every } f \text{, real valued and } C^\infty \text{ near } m.$$

Indeed, if $\mathbf{v}$ is defined in this way, we have, for any given coordinate map (U, φ), $m \in U$,

$$\begin{aligned} \mathbf{v}(f) &= \left.\frac{\mathrm{d}}{\mathrm{d}t}\right|_{t_0} (f(\mu(t))) = \left.\frac{\mathrm{d}}{\mathrm{d}t}\right|_{t_0} \big((f \circ \varphi^{-1}) \circ (\varphi \circ \mu)\big)(t) \\ &= \left\langle \nabla(f \circ \varphi^{-1})(\varphi(m)), \frac{\mathrm{d}}{\mathrm{d}t}(\varphi \circ \mu)(t_0) \right\rangle, \end{aligned}$$

whence $\mathbf{v}$ satisfies the condition in Remark 2.1.8 with

$$\lambda = \frac{\mathrm{d}}{\mathrm{d}t}(\varphi \circ \mu)(t_0).$$

Vice versa, suppose $\mathbf{v} \in M_m$. Fix any coordinate map (U, φ) with $m \in U$. Then, again by Remark 2.1.8, there exists $\lambda \in \mathbb{R}^N$ such that

$$(\star) \qquad \mathbf{v}(f) = \left\langle \nabla(f \circ \varphi^{-1})(\varphi(m)), \lambda \right\rangle$$

for every f, real valued and C^∞ near m. Consider the curve on M given by (see (II) in Remark 2.1.3)

$$\mu(t) := \varphi^{-1}(\varphi(m) + t\,\lambda), \quad |t| < \varepsilon,$$

with $\varepsilon > 0$ small enough. Obviously, μ passes through m at $t = 0$. Moreover,

$$\begin{aligned}\frac{\mathrm{d}}{\mathrm{d}t}\Big|_0 (f(\mu(t))) &= \frac{\mathrm{d}}{\mathrm{d}t}\Big|_0 \big((f\circ\varphi^{-1})\circ(\varphi\circ\mu)\big)(t)\\ &= \Big\langle \nabla(f\circ\varphi^{-1})(\varphi(m)), \frac{\mathrm{d}}{\mathrm{d}t}\Big|_0 \big(\varphi(m)+t\,\lambda\big)(0)\Big\rangle\\ &= \big\langle \nabla(f\circ\varphi^{-1})(\varphi(m)), \lambda\big\rangle.\end{aligned}$$

Comparing to ($\star$), we have proved our assertion. □

Definition 2.1.10 (Partial derivatives on M). *Let M be an N-dimensional C^∞ differentiable manifold. Let (U, φ) be a coordinate system with coordinate functions $x_1, \dots, x_N$ ($x_i := \pi_i \circ \varphi$), and let $m \in U$. For every $i \in \{1, \dots, N\}$ we define a tangent vector, denoted*

$$\frac{\partial}{\partial x_i}\Big|_m \in M_m,$$

by setting

$$\frac{\partial}{\partial x_i}\Big|_m (f) := \frac{\partial}{\partial \xi_i}\Big|_{\varphi(m)} (f\circ\varphi^{-1})(\xi) \tag{2.3}$$

for every C^∞ function f defined in a neighborhood of m.

We remark that $f \mapsto ((\partial/\partial x_i)|_m)(f)$ actually defines an element of M_m (see Remark 2.1.8 with $\lambda =$ the i-th element of the canonical basis of $\mathbb{R}^N$). Note that the definition of $(\partial/\partial x_i)|_m$ is not coordinate-free: despite the notation, forgetful of φ, *it depends on the coordinate map φ*, as x_i itself.

Example 2.1.11. With the notation of the above definition,

$$\frac{\partial}{\partial x_i}\Big|_m (x_j) = \delta_{i,j} \quad \text{(of Kronecker)}.$$

Indeed, $x_j \circ \varphi^{-1} = \pi_j$, so that (2.3) gives

$$\frac{\partial}{\partial x_i}\Big|_m (x_j) = \frac{\partial}{\partial \xi_i}\Big|_{\varphi(m)} (\xi_j) = \delta_{i,j}.$$

Remark 2.1.12. (i). If $\mathbf{v} \in M_m$, then (by collecting together Remark 2.1.8 and Example 2.1.11) we have the following suggesting formula:

$$\mathbf{v} = \sum_{i=1}^N \mathbf{v}(x_i)\cdot \frac{\partial}{\partial x_i}\Big|_m, \qquad x_i := \pi_i\circ\varphi, \tag{2.4}$$

and then

$$(\partial/\partial x_1)|_m, \dots, (\partial/\partial x_N)|_m$$

is a basis for M_m.

(ii). In particular, *two tangent vectors $\mathbf{v}$, $\mathbf{w} \in M_m$ coincide if and only if*

$$\mathbf{v}(\pi_i \circ \varphi) = \mathbf{w}(\pi_i \circ \varphi) \quad \forall\, i = 1, \dots, N,$$

for every (or, equivalently, for at least one) coordinate system (U, φ) *such that* $m \in U$.

(iii). The above formula (2.4) allows to represent a tangent vector in a very explicit way, once a coordinate system has been fixed. Indeed, if $\varphi : U \longrightarrow \mathbb{R}^N$ is a coordinate map, φ_i is the i-th component of φ, $(\xi_1, \dots, \xi_N)$ are the coordinates on $\mathbb{R}^N$, $m \in M$, f is a C^∞ function defined in U and $\mathbf{v} \in M_m$, then we have

$$\mathbf{v}(f) = \sum_{i=1}^{N} \mathbf{v}(\varphi_i) \cdot \frac{\partial (f \circ \varphi^{-1})}{\partial \xi_i}(\varphi(m)). \tag{2.5}$$

2.1.3 Differentials

Definition 2.1.13 (Differential at a point). *Let* $\psi : M \longrightarrow M'$ *be a* C^∞ *map between two differentiable manifolds, and let* $m \in M$. *The* differential of ψ at m *is the linear map*

$$\mathrm{d}_m\psi : M_m \longrightarrow M'_{\psi(m)}$$

defined as follows: if $\mathbf{v} \in M_m$, $\mathrm{d}_m\psi(\mathbf{v})$ *is the tangent vector in* $M'_{\psi(m)}$ *acting in the following way: if* f *is a* C^∞ *function in a neighborhood of* $\psi(m)$, *we set*

$$\big(\mathrm{d}_m\psi(\mathbf{v})\big)(f) := \mathbf{v}(f \circ \psi). \tag{2.6}$$

Even if many authors use to write $\mathrm{d}\psi$ instead of $\mathrm{d}_m\psi$, we shall keep the notation well distinguished, preserving "$\mathrm{d}\psi$" for a suitable further notion.

Remark 2.1.14. We verify that (2.6) actually defines a tangent vector at $\psi(m)$. Let f be a C^∞ function in a neighborhood of $\psi(m)$, and let (U, φ), (U', φ') be coordinate systems in M and M', respectively, with $m \in U$, $\psi(m) \in U'$. By Remark 2.1.8, there exists $\lambda \in \mathbb{R}^N$ such that

$$\mathbf{v}(g) = \big\langle \nabla(g \circ \varphi^{-1})(\varphi(m)), \lambda \big\rangle$$

for every g, C^∞ near m. Hence, we have (we denote λ as a column vector)

$$\begin{aligned}
\big(\mathrm{d}_m\psi(\mathbf{v})\big)(f) &= \mathbf{v}(f \circ \psi) = \mathbf{v}\big(f \circ (\varphi')^{-1} \circ \varphi' \circ \psi \circ \varphi^{-1} \circ \varphi\big) \\
&= \big\langle \nabla\big(f \circ (\varphi')^{-1} \circ \varphi' \circ \psi \circ \varphi^{-1}\big)(\varphi(m)), \lambda \big\rangle \\
&= \nabla\big(f \circ (\varphi')^{-1}\big)\big(\varphi'(\psi(m))\big) \cdot \mathcal{J}_{\varphi' \circ \psi \circ \varphi^{-1}}(\varphi(m)) \cdot \lambda \\
&= \big\langle \nabla\big(f \circ (\varphi')^{-1}\big)\big(\varphi'(\psi(m))\big), \mathcal{J}_{\varphi' \circ \psi \circ \varphi^{-1}}(\varphi(m)) \cdot \lambda \big\rangle.
\end{aligned}$$

This shows that

$$\big(\mathrm{d}_m\psi(\mathbf{v})\big)(f) = \big\langle \nabla\big(f \circ (\varphi')^{-1}\big)\big(\varphi'(\psi(m))\big), \lambda' \big\rangle, \tag{2.7a}$$

where

$$\lambda' = \mathcal{J}_{\varphi'\circ\psi\circ\varphi^{-1}}(\varphi(m)) \cdot \lambda. \tag{2.7b}$$

Again thanks to Remark 2.1.8, this proves that $d_m\psi(\mathbf{v})$ is a tangent vector at $\psi(m)$. □

Remark 2.1.15. We remark that, by (2.4), (2.7a) and (2.7b), if $(U, x_1, \ldots, x_N)$ and $(U', y_1, \ldots, y_{N'})$ are coordinate systems at $m \in M$ and $\psi(m) \in M'$, respectively, we have

$$d_m\psi\left(\frac{\partial}{\partial x_i}\bigg|_m\right) = \sum_{j=1}^{N'} \frac{\partial}{\partial x_i}\bigg|_m (y_j \circ \psi) \cdot \frac{\partial}{\partial y_j}\bigg|_{\psi(m)}. \tag{2.8}$$

Indeed, (2.8) follows from (2.7b) by taking $f = \pi_j \circ \varphi'$, $\lambda =$ the i-th element of the standard basis of $\mathbb{R}^N$ and by recalling that $y_j \circ \psi = \pi_j \circ \varphi' \circ \psi$ and

$$\frac{\partial}{\partial x_i}\bigg|_m (y_j \circ \psi) = \left(\partial_i(\pi_j \circ \varphi' \circ \psi \circ \varphi^{-1})\right)(\varphi(m)) = \left(\mathcal{J}_{\varphi'\circ\psi\circ\varphi^{-1}}(\varphi(m))\right)_{j,i}.$$

Remark 2.1.16 (Transformation of tangent vectors via a differential). In general, a vector field $\mathbf{v}$ at m transforms under the differential of ψ as follows:

$$\begin{aligned} &d_m\psi\left(\sum_{i=1}^{N} \mathbf{v}(\pi_i \circ \varphi) \cdot \frac{\partial}{\partial x_i}\bigg|_m\right) \\ &= \sum_{j=1}^{N'}\left(\sum_{i=1}^{N} \mathbf{v}(\pi_i \circ \varphi) \cdot \partial_i\left(\pi_j \circ \varphi' \circ \psi \circ \varphi^{-1}\right)\right)\frac{\partial}{\partial y_j}\bigg|_{\psi(m)} \\ &= \sum_{j=1}^{N'}\left(\mathcal{J}_{\varphi'\circ\psi\circ\varphi^{-1}}(\varphi(m)) \cdot \mathbf{v}(\varphi)\right)_j \frac{\partial}{\partial y_j}\bigg|_{\psi(m)}, \end{aligned} \tag{2.9}$$

where

$$\mathbf{v}(\varphi) = \begin{pmatrix} \mathbf{v}(\pi_1 \circ \varphi) \\ \vdots \\ \mathbf{v}(\pi_N \circ \varphi) \end{pmatrix}$$

and (U, φ), (U', φ') are, respectively, coordinate systems at $m \in M$, $\psi(m) \in M'$ (M, M' are differentiable manifolds of dimensions N, N', respectively).

Remark 2.1.17 (Differential of the composition and of the inverse). Let $\psi : M \longrightarrow M'$, $\phi : M' \longrightarrow M''$ be C^∞ maps between differentiable manifolds M, M', M''. Then it is easily seen that $\phi \circ \psi$ is C^∞, and we have

$$d_m(\phi \circ \psi) = d_{\psi(m)}\phi \circ d_m\psi \quad \forall\, m \in M. \tag{2.10}$$

Furthermore, let $\psi : M \longrightarrow M'$ be a C^∞ map between two differentiable manifolds. Suppose $\psi^{-1} : M' \longrightarrow M$ is a C^∞ map too. In this case, we say that ψ is a diffeomorphism.

It is immediate to observe that if ψ is a diffeomorphism and $m \in M$, then $\mathrm{d}_m\psi$ is also invertible (for every $m \in M$) and

$$\mathrm{d}_{\psi(m)}(\psi^{-1}) : M'_{\psi(m)} \longrightarrow M_m$$

is the inverse function of

$$\mathrm{d}_m\psi : M_m \longrightarrow M'_{\psi(m)}.$$

Definition 2.1.18 (dψ as a map on the tangent bundle). *Let $\psi : M \to M'$ be a C^∞ map between two differentiable manifolds M, M'. We set*

$$\mathrm{d}\psi : T(M) \to T(M'), \qquad \mathrm{d}\psi(m, \mathbf{v}) := (\psi(m), \mathrm{d}_m\psi(\mathbf{v})). \tag{2.11}$$

Note that, whereas $\mathrm{d}_m\psi$ is a map from M_m to $M'_{\psi(m)}$ (for any fixed $m \in M$), $\mathrm{d}\psi$ is a map from $T(M)$ to $T(M')$.

2.1.4 Vector Fields

The following definition is one of the most important in differential geometry.

Definition 2.1.19 (Vector field). *Let $\Omega \subseteq M$ be an open subset of a differentiable manifold M. A* vector field *X on Ω is an application*

$$X : \Omega \longrightarrow T(M)$$

such that,

$$X(m) = (m, \mathbf{v}(m)) \in T(M) \quad \forall\, m \in \Omega.$$

Equivalently, we have

$$X(m) = (m, \mathbf{v}(m)), \quad \textit{where } \mathbf{v}(m) \in M_m \textit{ for every } m \in \Omega.$$

In order to avoid any confusion, we make explicit the following conventional identification of a vector field with its projection onto the second argument. This identification is frequently tacitly employed in literature.

Convention–Notation. If $T(M)$ is the tangent bundle of a differentiable manifold M, and, for every $m \in M$, $\mathbf{v} \in M_m$, we set $\pi(m, \mathbf{v}) := \mathbf{v}$, then the following map is well posed on $T(M)$:

$$\pi : T(M) \to \bigcup_{m \in M} M_m, \quad (m, \mathbf{v}) \mapsto \mathbf{v}.$$

In the sequel, if X is a vector field on an open set $\Omega \subseteq M$, we shall use the notation $X(m)$ for the map

$$X : \Omega \to T(M), \quad m \mapsto X(m),$$

whereas X_m will denote the map

$$\Omega \to \bigcup_{m \in M} M_m, \quad m \mapsto X_m := (\pi \circ X)(m).$$

So, the above positions can be summarized as

$$X(m) = (m, X_m) \quad \text{for every } m \in M. \tag{2.12}$$

Finally, if f is a C^∞ function on Ω and X is a vector field on Ω, we shall denote (with an abuse of notation) by $X(f)$ or shortly Xf the function on Ω whose value at m is $X_m(f)$, i.e.

$$Xf : \Omega \to \mathbb{R}, \qquad (Xf)(m) := X_m(f). \tag{2.13}$$

Definition 2.1.20 (Smooth vector field). *Let X be a vector field defined on a manifold M. We say that X is C^∞ (or* smooth*) if, for every open set $\Omega \subseteq M$ and every smooth real-valued function f on Ω, the function Xf as defined in* (2.13) *is smooth on Ω.*

Remark 2.1.21. It is straightforward to verify that X is a smooth vector field on M if and only if, for every coordinate system $(U, x_1, \ldots, x_N)$, the functions $a_1, \ldots, a_N$ defined on U by

$$X_m = \sum_{i=1}^N a_i(m) \cdot \left.\frac{\partial}{\partial x_i}\right|_m$$

(see (2.12) and (2.4)), are C^∞ functions on U.

Following the above notation, we have

$$a_i(m) = X_m(x_i), \quad \text{where } x_i = \pi_i \circ \varphi.$$

Recalling (2.5), we see that, if (U, φ) is a coordinate system, a smooth vector field acts on a function $f \in C^\infty(U)$ in the following way:

$$\begin{aligned} X_m(f) &= \sum_{i=1}^N a_i(m) \cdot \left\{\partial_i(f \circ \varphi^{-1})\right\}(\varphi(m)) \\ &= \sum_{i=1}^N X_m(\pi_i \circ \varphi) \cdot \left\{\partial_i(f \circ \varphi^{-1})\right\}(\varphi(m)), \end{aligned} \tag{2.14}$$

where $a_1, \ldots, a_N \in C^\infty(U)$ are fixed functions depending on X and on the coordinate map. This shows that *a vector field X on M is smooth iff, for every coordinate system (U, φ), the functions*

$$m \mapsto X_m(\pi_1 \circ \varphi), \ldots, X_m(\pi_N \circ \varphi)$$

are C^∞ on U. □

Remark 2.1.22 (Smooth vector fields as operators on $C^\infty(M, \mathbb{R})$). Let X be a smooth vector field on a differentiable manifold M. Besides a map from M to $T(M)$, it is possible to identify X with the map

$$X : C^\infty(M, \mathbb{R}) \to C^\infty(M, \mathbb{R}), \quad f \mapsto Xf,$$

where (see (2.13))

$$Xf : M \to \mathbb{R}, \quad m \mapsto (Xf)(m) = X_m f.$$

We denote by $\mathcal{X}(M)$ the set of the smooth vector fields considered as linear operators (i.e. endomorphisms) on $C^\infty(M, \mathbb{R})$. Note that $\mathcal{X}(M)$ is a vector space over $\mathbb{R}$.

The correspondence between X as a vector field on M and as an endomorphism on $C^\infty(M, \mathbb{R})$ is faithful in the sense that, if X, Y are vector fields such that $Xf \equiv Yf$ for every $f \in C^\infty(M, \mathbb{R})$, then $X = Y$. (Indeed, take any $m \in M$, any coordinate system (U, φ) around m, any $i \in \{1, \dots, N\}$ and choose $f = \pi_i \circ \varphi$. Then apply Remark 2.1.12-(ii).)

Remark 2.1.23. Definition 2.1.13 defines $\mathrm{d}_m\psi$ as a map from M_m to $M'_{\psi(m)}$. Moreover, in Definition 2.1.18, we introduced a natural map denoted by $\mathrm{d}\psi$ between the tangent bundles $T(M)$ and $T(M')$. It may be thought that a third natural map can be defined between $\mathcal{X}(M)$ and $\mathcal{X}(M')$ by mapping $X \in \mathcal{X}(M)$ into the vector field Y such that $Y_{\psi(m)} = \mathrm{d}_m\psi(X_m)$. Unfortunately, in general, this defines a "vector field" only on the points of $\psi(M)$ and not on the whole M'. However, if ψ is a diffeomorphism, this can be done as we describe below.

Definition 2.1.24 (d$\boldsymbol{\psi}$ as a map on $\mathcal{X}(M)$). *Suppose $\psi : M \to M'$ is a C^∞-diffeomorphism of differentiable manifolds M, M'. We set*

$$\mathbf{d}\psi : \mathcal{X}(M) \longrightarrow \mathcal{X}(M'), \quad X \mapsto \mathbf{d}\psi(X),$$

where, for every $f' \in C^\infty(M', \mathbb{R})$,

$$\{\mathbf{d}\psi(X)\}_{m'}(f') = \mathrm{d}_{\psi^{-1}(m')}\psi(X_{\psi^{-1}(m')})(f') \quad \forall\, m' \in M'. \tag{2.15a}$$

Since ψ is onto, (2.15a) is equivalent to set

$$\{\mathbf{d}\psi(X)\}_{\psi(m)} = \mathrm{d}_m\psi(X_m) \text{ for every } m \in M. \tag{2.15b}$$

We say that $X \in \mathcal{X}(M)$ and the above defined $\mathbf{d}\psi(X) \in \mathcal{X}(M')$ are *ψ-related* (see also Definition 2.1.34). We leave to the reader the verification that $\mathbf{d}\psi(X)$ is indeed a smooth vector field on M' according to the definition of $\mathcal{X}(M')$ in Remark 2.1.22. The following example shows that the above mapping $X \mapsto \mathbf{d}\psi(X)$ appears naturally when a "change of variable" occurs.

Example 2.1.25 (Related vector fields via a diffeomorphism). We consider a simple but significant example. Let $T : \mathbb{R}^N \to \mathbb{R}^N$ be a C^∞-diffeomorphism (i.e. an invertible map, smooth together with its inverse function) from $\mathbb{R}^N$ onto itself. We consider on the domain of T a fixed system of Cartesian coordinates $(x_1, \dots, x_N)$, and we equip the image of T (which still coincides with $\mathbb{R}^N = T(\mathbb{R}^N)$) with the new system of coordinates y defined by $y = T(x)$. We use the following notation

$$T : (\mathbb{R}^N, x) \to (\mathbb{R}^N, y).$$

To every smooth vector field X on $(\mathbb{R}^N, x)$, there corresponds a vector field $\widetilde{X}$ on $(\mathbb{R}^N, y)$ in a natural way, roughly speaking, by expressing X in the new coordinates y. Namely, given any $f = f(y)$, $f \in C^\infty((\mathbb{R}^N, y), \mathbb{R})$, we set

$$\widetilde{X}_y(f) := X_{T^{-1}(y)}(f \circ T),$$

also written as

$$\widetilde{X} f(y) := X(f \circ T)(T^{-1}(y)),$$

or equivalently,

$$(\widetilde{X} f)(T(x)) = X(f \circ T)(x). \tag{2.16}$$

With reference to (2.16), we say that T *turns* X *into* $\widetilde{X}$, or that X and $\widetilde{X}$ are T*-related*. Roughly speaking, $\widetilde{X}$ *is the representation*[1] *of* X *in the new system of coordinates defined by* $y = T(x)$.

---

[1] A simple example is in order. Let $\mathbb{R}^2$ be equipped with coordinates $x = (x_1, x_2)$ and consider the linear change of coordinates given by

$$y = (y_1, y_2) = T(x_1, x_2) := (2x_1 - x_2, -5x_1 + 3x_2).$$

Following the above definition, the ordinary partial derivatives $X_1 := \partial_{x_1}$ and $X_2 := \partial_{x_2}$ are turned into the operators $\widetilde{X}_1$ and $\widetilde{X}_2$, respectively, where

$$\begin{aligned}
\widetilde{X}_1 f(y) &= \frac{\partial}{\partial x_1}\Big|_{x=T^{-1}(y)} f(2x_1 - x_2, -5x_1 + 3x_2) \\
&= 2\partial_{y_1} f(y_1, y_2) - 5\,\partial_{y_2} f(y_1, y_2), \\
\widetilde{X}_2 f(y) &= \frac{\partial}{\partial x_2}\Big|_{x=T^{-1}(y)} f(2x_1 - x_2, -5x_1 + 3x_2) \\
&= -\partial_{y_1} f(y_1, y_2) + 3\,\partial_{y_2} f(y_1, y_2).
\end{aligned}$$

Since the Jacobian matrix of T equals

$$\mathcal{J}_T(x_1, x_2) = \begin{pmatrix} 2 & -1 \\ -5 & 3 \end{pmatrix},$$

then $\widetilde{X}_1$ and $\widetilde{X}_2$ are the vector fields whose component functions at y are respectively given by

$$\begin{aligned}
\mathcal{J}_T(T^{-1}(y)) \cdot X_1 I(T^{-1}(y)) &= \begin{pmatrix} 2 & -1 \\ -5 & 3 \end{pmatrix} \cdot \begin{pmatrix} 1 \\ 0 \end{pmatrix}, \\
\mathcal{J}_T(T^{-1}(y)) \cdot X_2 I(T^{-1}(y)) &= \begin{pmatrix} 2 & -1 \\ -5 & 3 \end{pmatrix} \cdot \begin{pmatrix} 0 \\ 1 \end{pmatrix}.
\end{aligned}$$

Hence ∂_{x_1} is turned by T into $2\,\partial_{y_1} - 5\,\partial_{y_2}$ and ∂_{x_2} is turned by T into $-\,\partial_{y_1} + 3\,\partial_{y_2}$. Consequently, the ordinary Laplace operator $\Delta = (\partial_{x_1})^2 + (\partial_{x_2})^2$ is turned by T into the following second order constant coefficient differential operator of elliptic type (which is a sub-Laplacian on $(\mathbb{R}^2, +)$!)

If we write, as usual, $X = \nabla \cdot XI$, the chain rule then gives

$$\begin{aligned}\widetilde{X} f(y) &= \nabla(f \circ T)(T^{-1}(y)) \cdot XI(T^{-1}(y)) \\ &= (\nabla_y f)(y) \cdot \mathcal{J}_T(T^{-1}(y)) \cdot XI(T^{-1}(y)).\end{aligned}$$

In other words, for every $y \in (\mathbb{R}^N, y)$, one has

$$\widetilde{X}_y = \nabla_y \cdot (\widetilde{X} I)(y) \quad \text{with} \quad (\widetilde{X} I)(y) = \mathcal{J}_T(T^{-1}(y)) \cdot XI(T^{-1}(y)), \tag{2.17a}$$

or equivalently, for every $x \in (\mathbb{R}^N, x)$,

$$\widetilde{X}_{T(x)} = \nabla_{T(x)} \cdot (\widetilde{X} I)(T(x)) \quad \text{with} \quad (\widetilde{X} I)(T(x)) = \mathcal{J}_T(x) \cdot XI(x). \tag{2.17b}$$

Keeping in mind (2.9) of Remark 2.1.16, (2.17b) shows that

$$\widetilde{X}_{T(x)} = \mathrm{d}_x T(X_x), \tag{2.17c}$$

which gives a significant interpretation of the differential map (at least in the present case when T is a diffeomorphism):

Given a vector field X, $\mathrm{d}_x T(X_x)$ is the tangent vector at $T(x)$ which represents X_x with respect to the change of variable $y = T(x)$. □

In what follows, we introduce an important definition. The adjectives "regular" and "smooth" will always mean "of class C^∞".

Definition 2.1.26 (Tangent vector to a curve). *Let $\mu : [a, b] \to M$ be a regular curve. The* tangent vector to the curve μ at time t *is defined by*

$$\dot{\mu}(t) := \mathrm{d}_t \mu \left(\left. \frac{\mathrm{d}}{\mathrm{d}r} \right|_{r=t} \right) \in M_{\mu(t)}. \tag{2.18}$$

Hence, fixed $t \in [a, b]$, if f is C^∞ near $\mu(t)$, we have

$$\dot{\mu}(t)(f) = \left. \frac{\mathrm{d}}{\mathrm{d}r} \right|_{r=t} (f(\mu(r))).$$

Remark 2.1.27. Note that, as Remark 2.1.9 shows, any tangent vector at a point of M can be represented as the tangent vector to a certain curve at a suitable time.

Definition 2.1.28 (Integral curve). *Let X be a smooth vector field on the differentiable manifold M. A regular curve $\mu : [a, b] \longrightarrow M$ is called an* integral curve *of X if*

$$\dot{\mu}(t) = X_{\mu(t)} \quad \textit{for every } t \in [a, b]. \tag{2.19}$$

$$\widetilde{\Delta} = 5(\partial_{y_1})^2 + 34(\partial_{y_2})^2 - 26\, \partial_{y_1}\, \partial_{y_2}.$$

In other words, $\widetilde{\Delta}$ is simply the ordinary Laplace operator expressed in a new system of coordinates y in $\mathbb{R}^2$. In Chapter 16, we show that every second order constant coefficient differential operator of elliptic type in $\mathbb{R}^N$ is simply the ordinary Laplace operator expressed in a suitable new system of coordinates via a linear change of basis.

More explicitly, (2.19) means that

$$\frac{\mathrm{d}}{\mathrm{d}r}\bigg|_{r=t}(f(\mu(r))) = X(f)(\mu(t))$$

for every smooth function f on M and every $t \in [a,b]$.

We remark that if $X_m = \sum_{i=1}^{N} a_i(m)(\partial/\partial x_i)|_m$ and (U,φ) is a coordinate system such that $\mu(t) \in U$, then (2.19) is also equivalent to

$$\frac{\mathrm{d}}{\mathrm{d}r}\bigg|_{r=t}\{\varphi_i(\mu(r))\} = a_i(\mu(t)),$$

which becomes (setting $\gamma(t) := \varphi(\mu(t)) \in \mathbb{R}^N$)

$$\dot{\gamma}_i(t) = (a_i \circ \varphi^{-1})(\gamma(t)), \quad i = 1,\dots,N, \tag{2.20}$$

which is an ODE on $\mathbb{R}^N$. Equivalently, making more explicit the rôle of X, $\mu : [a,b] \to M$ is an integral curve of X if and only if, whenever (U,φ) is a coordinate system and $(t_0,t_1) \subset [a,b]$ is such that $\mu(t) \in U$ for every $t \in (t_0,t_1)$, it holds $\mu(t) = \varphi^{-1}(\gamma(t))$, where

$$\dot{\gamma}(t) = X_{\varphi^{-1}(\gamma(t))}(\varphi) \quad \forall\, t \in (t_0,t_1)$$

and $X_m(\varphi) = (X_m(\pi_1 \circ \varphi),\dots,X_m(\pi_N \circ \varphi))$.

Remark 2.1.29. By the above observation on the coordinate form of Definition 2.1.28 and by recalling the existence "in small" for smooth systems of ODE's, we infer that, given any smooth vector field X on M and fixed any $m \in M$, there exists one and only one integral curve of X passing through m at time 0. Hence the following definition is well-posed.

Definition 2.1.30 (Complete vector field). *Let X be a smooth vector field on the differentiable manifold M. We say that X is* complete *if, for every $m \in M$, the integral curve μ of X such that $\mu(0) = m$ is defined on the whole $\mathbb{R}$ (i.e. its* maximal *interval of definition is $\mathbb{R}$).*

2.1.5 Commutators. φ-relatedness

In the sequel, we denote by $C^\infty(M,\mathbb{R})$ or, shortly, $C^\infty(M)$ the set of the smooth real-valued functions defined on a differentiable manifold M. It is immediate to observe that if X is a smooth vector field on M and $f \in C^\infty(M,\mathbb{R})$, we have $Xf \in C^\infty(M,\mathbb{R})$. We explicitly recall that, here and in the sequel, we use the notation in (2.13):

$$Xf : M \to \mathbb{R}, \quad (Xf)(m) = X_m(f).$$

As a consequence, the following definition is well posed.

Definition 2.1.31 (Commutators). *Let X and Y be smooth vector fields on a differentiable manifold M. We define a vector field on M (called the* commutator *of X and Y) in the following way:*

$$[X, Y] : M \to T(M), \quad [X, Y](m) := (m, [X, Y]_m),$$

where

$$[X, Y]_m(f) := X_m(Yf) - Y_m(Xf) \tag{2.21}$$

for every $m \in M$ and every $f \in C^\infty(M, \mathbb{R})$.

Definition 2.1.31 is well posed as it follows from (i) in the proposition below.

Proposition 2.1.32. *If X, Y and Z are smooth vector fields on M, we have:*

(i) *$[X, Y]$ is a smooth vector field on M;*
(ii) *$[X, Y]_m = -[Y, X]_m$ for every $m \in M$;*
(iii) *$[[X, Y], Z]_m + [[Y, Z], X]_m + [[Z, X], Y]_m = 0$ for every $m \in M$.*

Proof (Sketch). The verification that these facts hold true, can be performed via a computation in a coordinate map. This reduces the above identities (ii) and (iii) to identities between usual differential operators in $\mathbb{R}^N$.

For instance, let us prove (i). Let $m \in M$ be fixed and choose any coordinate map (U, φ) such that $m \in M$. According to Definition 2.1.20, we have

$$X_m(f) = \sum_{i=1}^{N} X_m(\pi_i \circ \varphi)\big\{\partial_i (f \circ \varphi^{-1})\big\}(\varphi(m)).$$

As a consequence,

$$\begin{aligned} X_m(Yf) &= \sum_{i,j=1}^{N} X_m(\pi_i \circ \varphi)\left\{\frac{\partial}{\partial \xi_i}\big(Y_{\varphi^{-1}(\xi)}(\pi_j \circ \varphi)\, \partial_{\xi_j}(f \circ \varphi^{-1})(\xi)\big)\right\}(\varphi(m)) \\ &= \sum_{i,j=1}^{N} X_m(\pi_i \circ \varphi) \cdot \left\{\partial_{\xi_i}\big(Y_{\varphi^{-1}(\xi)}(\pi_j \circ \varphi)\big)\, \partial_{\xi_j}(f \circ \varphi^{-1})(\xi)\right. \\ &\qquad \left. + Y_{\varphi^{-1}(\xi)}(\pi_j \circ \varphi)\, \frac{\partial^2}{\partial \xi_i \xi_j}(f \circ \varphi^{-1})(\xi)\right\}_{\xi=\varphi(m)}. \end{aligned}$$

By reversing the rôles of X and Y and then subtracting, we obtain (note, before the second equality sign, *the cancellation of the second order terms*)

$$\begin{aligned} &X_m(Yf) - Y_m(Xf) \\ &= \sum_{i,j=1}^{N} X_m(\pi_i \circ \varphi)\,\big\{\partial_{\xi_i}\big(Y_{\varphi^{-1}(\xi)}(\pi_j \circ \varphi)\big)\, \partial_{\xi_j}(f \circ \varphi^{-1})(\xi)\big\}_{\xi=\varphi(m)} \end{aligned}$$

$$
\begin{aligned}
&- \sum_{i,j=1}^{N} Y_m(\pi_i \circ \varphi) \left\{ \partial_{\xi_i} \left(X_{\varphi^{-1}(\xi)}(\pi_j \circ \varphi) \right) \partial_{\xi_j}(f \circ \varphi^{-1})(\xi) \right\}_{\xi=\varphi(m)} \\
&+ \sum_{i,j=1}^{N} X_m(\pi_i \circ \varphi)\, Y_m(\pi_j \circ \varphi) \left(\frac{\partial^2}{\partial\, \xi_i \xi_j} (f \circ \varphi^{-1}) \right)(\varphi(m)) \\
&- \sum_{i,j=1}^{N} Y_m(\pi_i \circ \varphi)\, X_m(\pi_j \circ \varphi) \left(\frac{\partial^2}{\partial\, \xi_i \xi_j} (f \circ \varphi^{-1}) \right)(\varphi(m)) \\
= &\sum_{j=1}^{N} \left\{ \sum_{i=1}^{N} \left\{ X_m(\pi_i \circ \varphi)\, \partial_{\xi_i} \left(Y_{\varphi^{-1}(\xi)}(\pi_j \circ \varphi) \right)(\varphi(m)) \right. \right. \\
&\left. \left. - Y_m(\pi_i \circ \varphi)\, \partial_{\xi_i} \left(X_{\varphi^{-1}(\xi)}(\pi_j \circ \varphi) \right)(\varphi(m)) \right\} \right\} \partial_{\xi_j}(f \circ \varphi^{-1})(\varphi(m)).
\end{aligned}
$$

Thus, $[X, Y]_m$ is actually a tangent vector at m (see, for instance, Remark 2.1.8), and, following the notation in Definition 2.1.20, we have proved that, in a coordinate system (U, φ) around m,

$$
\begin{aligned}
&[X, Y]_m(f) = \sum_{j=1}^{N} c_j(m)\, \partial_j (f \circ \varphi^{-1})(\varphi(m)) \quad \text{with} \\
&c_j(m) = [X, Y]_m(\pi_j \circ \varphi) = X_m\big(Y(\pi_j \circ \varphi)\big) - Y_m\big(X(\pi_j \circ \varphi)\big) \quad \forall\, j \le N. \qquad (2.22)
\end{aligned}
$$

We see that the maps c_j's are smooth on U, so that, by Remark 2.1.21, $[X, Y]$ is a smooth vector field on M.

Now, (iii) in the assertion can be proved with an analogous coordinate-computation, whereas (ii) is trivial. □

Remark 2.1.33 (Commutators in $\mathcal{X}(M)$). Consider the alternative definition of smooth vector field as an element of $\mathcal{X}(M)$, see Remark 2.1.33. The commutator operation rewrites as an operation on $\mathcal{X}(M)$ in the following way: Given $X, Y \in \mathcal{X}(M)$, we consider the operator on $C^\infty(M, \mathbb{R})$ defined by

$$
[X, Y] : C^\infty(M, \mathbb{R}) \to C^\infty(M, \mathbb{R}), \qquad f \mapsto [X, Y] f,
$$

where

$$
([X, Y] f)(m) := [X, Y]_m f = X_m(Y f) - Y_m(X f).
$$

Then, obviously, $[X, Y] \in \mathcal{X}(M)$ is the operator on $C^\infty(M, \mathbb{R})$ related to the (usual) vector field $[X, Y]$.

With this meaning of the commutation, Proposition 2.1.32 rewrites as: *If X, Y and Z belong to $\mathcal{X}(M)$, we have:*

(i) $[X, Y] \in \mathcal{X}(M)$;
(ii) $[X, Y] = -[Y, X]$;
(iii) $[[X, Y], Z] + [[Y, Z], X] + [[Z, X], Y] = 0.$

Definition 2.1.34 (φ-relatedness). *Let $\varphi : M \to \widetilde{M}$ be a C^∞ function between two differentiable manifolds. The vector fields X on M and $\widetilde{X}$ on $\widetilde{M}$ are called* φ-related *if we have*

$$d\varphi \circ X = \widetilde{X} \circ \varphi. \tag{2.23a}$$

Here, $d\varphi$ is intended as a map from $T(M)$ to $T(\widetilde{M})$ (see Definition 2.1.18) and, as usual, the vector fields are maps from the underlying differentiable manifolds to the relevant tangent bundles. Hence, (2.23a) is intended as an equality of functions from M to $T(\widetilde{M})$. Hence, φ-relatedness is equivalent to saying that the following diagram is commutative

$$\begin{array}{ccc} M & \xrightarrow{\quad\varphi\quad} & \widetilde{M} \\ {\scriptstyle X}\downarrow & & \downarrow{\scriptstyle \widetilde{X}} \\ T(M) & \xrightarrow{\quad d\varphi\quad} & T(\widetilde{M}). \end{array}$$

Since, for every $m \in M$, we have

$$(d\varphi \circ X)(m) = d\varphi(m, X_m) = (\varphi(m), d_m\varphi(X_m)),$$

whereas

$$(\widetilde{X} \circ \varphi)(m) = \widetilde{X}(\varphi(m)) = (\varphi(m), \widetilde{X}_{\varphi(m)}),$$

then (2.23a) can be rewritten as the collection of identities

$$d_m\varphi(X_m) = \widetilde{X}_{\varphi(m)} \quad \forall\, m \in M \tag{2.23b}$$

between tangent vectors at $\varphi(m)$, i.e. between two functionals on $C^\infty(\widetilde{M}, \mathbb{R})$.

Finally, condition (2.23a) is equivalent to the following ones:

$$X_m(f \circ \varphi) = \widetilde{X}_{\varphi(m)}(f), \tag{2.23c}$$

$$X(f \circ \varphi) = (\widetilde{X} f) \circ \varphi \tag{2.23d}$$

for every $m \in M$ and for every f smooth in a neighborhood of $\varphi(m)$ in $\widetilde{M}$.

Remark 2.1.35. When φ is a diffeomorphism, for every $X \in \mathcal{X}(M)$, there always exists a (unique) $\widetilde{X} \in \mathcal{X}(\widetilde{M})$ which is φ-related to X (see Remark 2.1.23), namely, with the notation of Definition 2.1.24, $\widetilde{X} = \mathbf{d}\varphi(X)$.

With reference to (2.17c), we have an interpretation of φ-relatedness when $\varphi : \mathbb{R}^N \to \mathbb{R}^N$ is a diffeomorphism: X and $\widetilde{X}$ are φ-related if and only if $\widetilde{X}$ is the expression of X in the new coordinates defined by the new Cartesian coordinates $y = \varphi(x)$. □

Remark 2.1.36 (φ-relatedness in $\mathcal{X}(M)$). Consider the alternative definition of vector field as in Remark 2.1.22. Then, the notion of φ-relatedness of vector fields rewrites as follows: $X \in \mathcal{X}(M)$ and $\widetilde{X} \in \mathcal{X}(\widetilde{M})$ are φ-related if, for every $f \in C^\infty(\widetilde{M}, \mathbb{R})$, the functions $X(f \circ \varphi)$, $(\widetilde{X} f) \circ \varphi$ on M do coincide, i.e.

$$X_m(f \circ \varphi) = \widetilde{X}_{\varphi(m)} f \quad \text{for every } m \in M,$$

or, equivalently, (2.23b) holds.

The following simple result will be soon of a crucial importance.

Proposition 2.1.37 (φ-relatedness and commutators). *Let $\varphi : M \to \widetilde{M}$ be a C^∞ function between two differentiable manifolds. Let X, Y be smooth vector fields on M and $\widetilde{X}, \widetilde{Y}$ be smooth vector fields on $\widetilde{M}$. If X is φ-related to $\widetilde{X}$ and Y is φ-related to $\widetilde{Y}$, then $[X, Y]$ is φ-related to $[\widetilde{X}, \widetilde{Y}]$.*

Proof. We have to prove

$$\mathrm{d}\varphi \circ [X, Y] = [\widetilde{X}, \widetilde{Y}] \circ \varphi,$$

provided $\mathrm{d}\varphi \circ X = \widetilde{X} \circ \varphi$ and $\mathrm{d}\varphi \circ Y = \widetilde{Y} \circ \varphi$. To this end, let $m \in M$ and $f \in C^\infty(\widetilde{M})$. By the equivalent restatement (2.23c) of φ-relatedness, we have to show that

$$(\star) \qquad \mathrm{d}_m\varphi\big([X, Y]_m\big)(f) = [\widetilde{X}, \widetilde{Y}]_{\varphi(m)}(f).$$

By definition of $\mathrm{d}_m\varphi$, we have

$$\begin{aligned} \mathrm{d}_m\varphi\big([X, Y]_m\big)(f) &= [X, Y]_m(f \circ \varphi) \\ \text{(see (2.21))} &= X_m\big(Y(f \circ \varphi)\big) - Y_m\big(X(f \circ \varphi)\big) \\ \text{(see (2.23d))} &= X_m\big((\widetilde{Y} f) \circ \varphi\big) - Y_m\big((\widetilde{X} f) \circ \varphi\big) \\ &= \mathrm{d}_m\varphi(X_m)\big(\widetilde{Y}(f)\big) - \mathrm{d}_m\varphi(Y_m)\big(\widetilde{X}(f)\big) \\ \text{(see (2.23b))} &= \widetilde{X}_{\varphi(m)}\big(\widetilde{Y}(f)\big) - \widetilde{Y}_{\varphi(m)}\big(\widetilde{X}(f)\big) \\ \text{(see (2.21))} &= [\widetilde{X}, \widetilde{Y}]_{\varphi(m)}(f). \end{aligned}$$

This gives $(\star)$ thus ending the proof. $\square$

2.1.6 Abstract Lie Groups

Definition 2.1.38 (Lie group). *A Lie group $\mathbb{G}$ is a differentiable manifold $\mathbb{G}$ along with a group law $* : \mathbb{G} \times \mathbb{G} \longrightarrow \mathbb{G}$ such that the applications*

$$\mathbb{G} \times \mathbb{G} \ni (x, y) \mapsto x * y \in \mathbb{G}, \qquad \mathbb{G} \ni x \mapsto x^{-1} \in \mathbb{G}$$

are smooth.[2]

In the following, we shall always denote by e the identity of $(\mathbb{G}, *)$. Moreover, fixed $\sigma \in \mathbb{G}$, we denote by τ_σ the *left translation on $\mathbb{G}$ by σ*, i.e. the map

$$\mathbb{G} \ni x \mapsto \tau_\sigma(x) := \sigma * x \in \mathbb{G}.$$

In case, when more than only one composition law is involved, we may also write τ_σ^* instead of τ_σ.

[2] The notion of "smoothness" of a map from $\mathbb{G} \times \mathbb{G}$ should be properly defined. The product of two differentiable manifolds is indeed endowed with a differentiable structure in a natural way. Here, we just remark that a function $f : \mathbb{G} \times \mathbb{G} \to \mathbb{R}$ is smooth if, for every couple of coordinate systems (U, φ) and (V, ϕ) on $\mathbb{G}$, the function

$$\mathbb{R}^N \times \mathbb{R}^N \supseteq \varphi(U) \times \phi(V) \ni (u, v) \mapsto f\Big(\varphi^{-1}(u) * \phi^{-1}(v)\Big) \in \mathbb{R}$$

is smooth. For more details see, e.g. [War83].

Definition 2.1.39 (Lie algebra). *A* (real) Lie algebra *is a real vector space* $\mathfrak{g}$ *with a bilinear operation* $[\cdot,\cdot] : \mathfrak{g} \times \mathfrak{g} \longrightarrow \mathfrak{g}$ *(called* (Lie) bracket*) such that, for every* X, $Y, Z \in \mathfrak{g}$, *we have*:

1. (anti-commutativity) $[X, Y] = -[Y, X]$;
2. (Jacobi identity) $[[X, Y], Z] + [[Y, Z], X] + [[Z, X], Y] = 0$.

A very remarkable fact is that, given any Lie group, there exists a certain finite-dimensional Lie algebra such that the group properties are reflected into properties of the algebra. For instance (as we shall see later on), any connected and simply connected Lie group is completely determined (up to isomorphism) by its Lie algebra. Therefore, the study of a Lie group is often reduced to the study of its Lie algebra.

Remark 2.1.40. Straightforwardly adapting the proof of Proposition 1.1.7 in Section 1.1 (see page 12), we have: *if* $X_1, \ldots, X_m$ *are elements of an (abstract) Lie algebra, then a system of generators of* $Lie\{X_1, \ldots, X_m\}$ *is given by the commutators*

$$X_I := [X_{i_1}, [X_{i_2}, [X_{i_3}, \ldots [X_{i_{k-1}}, X_{i_k}] \ldots]]],$$

where $\{i_1, i_2, \ldots, i_k\} \subseteq \{1, \ldots, m\}$ *and* $I = (i_1, i_2, \ldots, i_k)$, $k \in \mathbb{N}$. Indeed, the proof of Proposition 1.1.7 is only based on anti-commutativity and the Jacobi identity.

2.1.7 Left Invariant Vector Fields and the Lie Algebra

Definition 2.1.41 (Left invariant vector fields). *Let* $\mathbb{G}$ *be a Lie group. A smooth vector field* X *on* $\mathbb{G}$ *is called* left invariant *if, for every* $\sigma \in \mathbb{G}$, X *is* τ_σ*-related to itself, i.e.*

$$\mathrm{d}\tau_\sigma \circ X = X \circ \tau_\sigma. \tag{2.24}$$

Here $\mathrm{d}\tau_\sigma$ is intended as a map from $T(\mathbb{G})$ to itself (as in Definition 2.1.18). As shown by the remarks after Definition 2.1.34, condition (2.24) is equivalent to the following one:

$$(\mathrm{d}_x \tau_\sigma)(X_x) = X_{\sigma * x} \qquad \forall\, x, \sigma \in \mathbb{G}. \tag{2.25a}$$

Applying (2.25a) at the identity e, it follows immediately that if X is a left invariant vector field, we have

$$\mathrm{d}_e \tau_\sigma (X_e) = X_\sigma \qquad \forall\, \sigma \in \mathbb{G}, \tag{2.25b}$$

which proves that a left invariant vector field is determined by its action at the origin. Equality (2.25b) is actually equivalent[3] to (2.25a). Moreover, (2.25a) can also be written as

---

[3] Indeed, by applying (2.25b) with σ replaced by $\sigma * x$, subsequently (2.10) and finally using again (2.25b) with σ replaced by x, we have

$$X_{\sigma * x} = \mathrm{d}_e \tau_{\sigma * x}(X_e) = \mathrm{d}_x \tau_\sigma \Big(\mathrm{d}_e \tau_x (X_e)\Big) = \mathrm{d}_x \tau_\sigma (X_x).$$

$$X_x(f \circ \tau_\sigma) = X_{\sigma * x}(f)$$
$$\text{for every } x,\ \sigma \in \mathbb{G} \text{ and every } f \in C^\infty(\mathbb{G}, \mathbb{R}), \tag{2.25c}$$

or again as (the most commonly used)

$$X_x\big(y \mapsto f(\sigma * y)\big) = (Xf)(\sigma * x), \tag{2.25d}$$

i.e. comparing it to (1.18) (page 14), the very analogue of the usual left-invariance.

Before giving the following central Definition 2.1.42, we pause a moment in order to recall the multiple ways a smooth vector field can be thought of. A smooth vector field on $\mathbb{G}$ is a map $X : \mathbb{G} \to T(\mathbb{G})$ such that, for every $x \in \mathbb{G}$, it holds $X(x) = (x, X_x)$, where $X_x \in \mathbb{G}_x$ for every $x \in \mathbb{G}$ and such that, for every $f \in C^\infty(\mathbb{G}, \mathbb{R})$, the function $x \mapsto X_x(f)$ is smooth on $\mathbb{G}$. A smooth vector field can be identified to the operator

$$\begin{aligned} X : C^\infty(\mathbb{G}, \mathbb{R}) &\longrightarrow C^\infty(\mathbb{G}, \mathbb{R}), \\ f &\mapsto Xf : \mathbb{G} \to \mathbb{R}, \\ &\qquad\qquad x \mapsto X_x f. \end{aligned}$$

The set of the vector fields, as the above described operators, is denoted by $\mathcal{X}(\mathbb{G})$. Obviously, the set of the left invariant operators on $\mathbb{G}$ gives rise to a relevant subset in $\mathcal{X}(\mathbb{G})$, following the above identification.

We are ready to give the following central definition.

Definition 2.1.42 (Algebra of a Lie group). *Let $\mathbb{G}$ be a Lie group. Then the subset of $\mathcal{X}(\mathbb{G})$ of the smooth left invariant vector fields on $\mathbb{G}$ is called the* (Lie) algebra of $\mathbb{G}$. *It will be denoted by $\mathfrak{g}$.*

More precisely, following Remark 2.1.22, *we henceforth identify a left invariant vector field X on $\mathbb{G}$ with the following operator*

$$X : C^\infty(\mathbb{G}, \mathbb{R}) \to C^\infty(\mathbb{G}, \mathbb{R})$$

such that, for every $f \in C^\infty(\mathbb{G}, \mathbb{R})$, the function Xf on $\mathbb{G}$ is defined by

$$(Xf)(x) := X_x f \quad \forall\, x \in \mathbb{G}.$$

Hence, $\mathfrak{g}$ is a (linear) set of endomorphisms on $C^\infty(\mathbb{G}, \mathbb{R})$,

$$\mathfrak{g} \subseteq \mathcal{X}(\mathbb{G}).$$

Note that, from the left invariance of $X \in \mathfrak{g}$, we have

$$(Xf)(x) = X(f \circ \tau_x)(e) \quad \forall\, x \in \mathbb{G}\ \forall\, f \in C^\infty(\mathbb{G}, \mathbb{R}).$$

Along with the above definition of the algebra of a Lie group, there is a wide commonly used identification of $\mathfrak{g}$ with $\mathbb{G}_e$ described in the following theorem.

Theorem 2.1.43 (The Lie algebra of a Lie group). *Let $\mathbb{G}$ be a Lie group and $\mathfrak{g}$ be its algebra. Then we have:*

(i) $\mathfrak{g}$ *is a vector space, and the map*

$$\alpha : \mathfrak{g} \longrightarrow \mathbb{G}_e, \quad X \mapsto \alpha(X) := X_e$$

is an isomorphism between $\mathfrak{g}$ *and the tangent space* $\mathbb{G}_e$ *to* $\mathbb{G}$ *at the identity* e *of* $\mathbb{G}$. *As a consequence,* $\dim \mathfrak{g} = \dim \mathbb{G}_e = \dim \mathbb{G}$;

(ii) *The commutator of smooth left invariant vector fields (see also Remark 2.1.33) is a smooth left invariant vector field;*

(iii) $\mathfrak{g}$ *with the commutation operation is a Lie algebra.*

Proof. (i). It is evident that $\mathfrak{g}$ (thought of as a subset of $\mathcal{X}(\mathbb{G})$) is a vector space and that α is linear.

Let us prove that α is injective. If $\alpha(X) = \alpha(Y)$, we have (see (2.25b))

$$X_\sigma = \mathrm{d}\tau_\sigma(X_e) = \mathrm{d}\tau_\sigma(\alpha(X)) = \mathrm{d}\tau_\sigma(\alpha(Y)) = \mathrm{d}\tau_\sigma(Y_e) = Y_\sigma,$$

i.e. $X = Y$.

Let us prove that α is surjective. If $\mathbf{v} \in \mathbb{G}_e$, we set

$$X_\sigma := (\mathrm{d}_e\tau_\sigma)(\mathbf{v}) \quad \text{for every } \sigma \in \mathbb{G}.$$

Here, $\mathrm{d}_e\tau_\sigma$ is the differential of the map τ_σ at the identity e of $\mathbb{G}$. By definition, we have $X_\sigma \in \mathbb{G}_\sigma$, i.e. $\sigma \mapsto (\sigma, X_\sigma)$ is a vector field on $\mathbb{G}$. Moreover, it is not difficult[4] to prove that X is smooth. On the other hand, X is left invariant since

$$X_{\sigma * x} = \mathrm{d}_e\tau_{\sigma * x}(\mathbf{v}) = (\mathrm{d}_x\tau_\sigma \circ \mathrm{d}_e\tau_x)(\mathbf{v}) = \mathrm{d}_x\tau_\sigma(X_x),$$

which gives (2.25a). Finally, we have

---

[4] Indeed, let us prove that, for every $f \in C^\infty(\mathbb{G}, \mathbb{R})$, $\mathbb{G} \ni \sigma \mapsto X_\sigma f$ is smooth. Fix $\sigma_0 \in \mathbb{G}$, and let (U, φ) be a coordinate system around σ_0. We have to prove that the map $\varphi(U) \ni u \mapsto (Xf)(\varphi^{-1}(u))$ is smooth. By definition, it holds

$$(Xf)(\varphi^{-1}(u)) = X_{\varphi^{-1}(u)}(f) = (\mathrm{d}_e\tau_{\varphi^{-1}(u)})(\mathbf{v})(f) = \mathbf{v}(f \circ \tau_{\varphi^{-1}(u)}) = (\bigstar).$$

By Remark 2.1.8, fixed a coordinate system (E, χ) around e, there exists $\lambda = \lambda(\chi, e, \mathbf{v}) \in \mathbb{R}^N$ such that

$$(\bigstar) = \left\langle \lambda, \nabla\left(f \circ \tau_{\varphi^{-1}(u)} \circ \chi^{-1}\right)(\chi(e))\right\rangle = \left\langle \lambda, \nabla_v\left(f(\varphi^{-1}(u) * \chi^{-1}(v))\right)(\chi(e))\right\rangle.$$

By the very definition of Lie group,

$$\varphi(U) \times \chi(E) \ni (u, v) \mapsto f(\varphi^{-1}(u) * \chi^{-1}(v))$$

is smooth, so that

$$\varphi(U) \ni u \mapsto \nabla_v\left(f(\varphi^{-1}(u) * \chi^{-1}(v))\right)(\chi(e))$$

is smooth, and this ends the proof that X is smooth.

$$\alpha(X) = X_e = \mathrm{d}_e \tau_e(\mathbf{v}) = \mathbf{v},$$

i.e. α is surjective. Since α is an isomorphism of vector spaces, we have $\dim \mathfrak{g} = \dim \mathbb{G}_e = \dim \mathbb{G}$ (see Proposition 2.1.7).

(ii). Let $\sigma \in \mathbb{G}$ be arbitrary. Let X and $Y \in \mathfrak{g}$, i.e. X is τ_σ-related to itself and Y is τ_σ-related to itself. From Proposition 2.1.37 it follows that $[X, Y]$ is τ_σ-related to itself, i.e. $[X, Y] \in \mathfrak{g}$.

(iii). The last statement of the theorem follows from Proposition 2.1.32 (see also Remark 2.1.33). □

Incidentally, in the above proof, we have explicitly written the inverse map of the natural identification

$$\alpha : \mathfrak{g} \longrightarrow \mathbb{G}_e, \qquad X \mapsto \alpha(X) := X_e.$$

Indeed, the inverse of α is given by

$$\alpha^{-1} : \mathbb{G}_e \longrightarrow \mathfrak{g}, \quad \mathbf{v} \mapsto X, \qquad \text{where } X_\sigma = (\mathrm{d}_e \tau_\sigma)(\mathbf{v}) \text{ for every } \sigma \in \mathbb{G}. \tag{2.26}$$

Example 2.1.44 (The Lie algebra of $(\mathbb{R}, +)$*).* It is obvious that the Lie algebra $\mathfrak{r}$ of the usual Euclidean Lie group $(\mathbb{R}, +)$ is

$$\mathrm{span}\left\{ \frac{\mathrm{d}}{\mathrm{d}r} \right\},$$

where

$$\frac{\mathrm{d}}{\mathrm{d}r} : C^\infty(\mathbb{R}, \mathbb{R}) \to C^\infty(\mathbb{R}, \mathbb{R}), \quad f \mapsto f'.$$

With the usual formalism X_x for vector fields, this rewrites as

$$\left.\frac{\mathrm{d}}{\mathrm{d}r}\right|_t f = f'(t) \quad \text{for every } t \in \mathbb{R}.$$

Remark 2.1.45 (Left invariance in coordinates). We now make explicit the left invariance condition in terms of the coefficients a_i of the coordinate form of a smooth vector field

$$X_\sigma = \sum_{i=1}^N a_i(\sigma) \cdot \left.\frac{\partial}{\partial x_i}\right|_\sigma .$$

We know that a smooth vector field X is left invariant on $\mathbb{G}$ if and only if (see (2.25b))

$$(\star) \qquad \mathrm{d}_e \tau_\sigma (X_e) = X_\sigma \qquad \forall\, \sigma \in \mathbb{G}.$$

In turn (see Remark 2.1.21), fixed a coordinate system (E, χ) around the identity e, $(\star)$ is equivalent to state that, for every $\sigma \in \mathbb{G}$ and for every (or, equivalently, for at least one) coordinate system (U, φ) around σ, it holds

$$\sum_{i=1}^{N} X_\sigma(\varphi_i)\cdot\frac{\partial}{\partial u_i}(f\circ\varphi^{-1})(\varphi(\sigma)) = \sum_{i=1}^{N} X_e(\chi_i)\cdot\frac{\partial}{\partial v_i}(f\circ\tau_\sigma\circ\chi^{-1})(\chi(e))$$

for every smooth function f around σ (or, equivalently, for the N smooth functions $\varphi_1,\dots,\varphi_N$). Writing $f = f\circ\varphi^{-1}\circ\varphi$, on the right-hand side the above becomes

$$\begin{aligned}
&\sum_{i=1}^{N} X_\sigma(\varphi_i)\cdot\frac{\partial}{\partial u_i}(f\circ\varphi^{-1})(\varphi(\sigma))\\
&\quad=\sum_{j=1}^{N}\left\{\sum_{i=1}^{N} X_e(\chi_i)\,\frac{\partial}{\partial v_i}(\varphi_j\circ\tau_\sigma\circ\chi^{-1})(\chi(e))\right\}\cdot\frac{\partial}{\partial u_j}(f\circ\varphi^{-1})(\varphi(\sigma))\\
&\quad=\sum_{j=1}^{N}\left\{\sum_{i=1}^{N}\Big(\mathcal{J}_{\varphi\circ\tau_\sigma\circ\chi^{-1}}(\chi(e))\Big)_{j,i} X_e(\chi_i)\right\}\cdot\frac{\partial}{\partial u_j}(f\circ\varphi^{-1})(\varphi(\sigma))\\
&\quad=\sum_{j=1}^{N}\Big(\mathcal{J}_{\varphi\circ\tau_\sigma\circ\chi^{-1}}(\chi(e))\cdot X_e(\chi)\Big)_j\cdot\frac{\partial}{\partial u_j}(f\circ\varphi^{-1})(\varphi(\sigma)).
\end{aligned}$$

Here, $X_e(\chi) = (X_e(\chi_1),\dots,X_e(\chi_N))^T$. Comparing the far left-hand and right-hand sides, we rewrite all as the matrix identity

$$\mathcal{J}_{\varphi\circ\tau_\sigma\circ\chi^{-1}}(\chi(e))\cdot X_e(\chi) = X_\sigma(\varphi). \tag{2.27}$$

Here, $X_e(\varphi) = (X_e(\varphi_1),\dots,X_e(\varphi_N))^T$.

As a consequence, we have proved that *a smooth vector field X on the Lie group $\mathbb{G}$ is left invariant if and only if, fixed a coordinate system (E,χ) around the identity e, for every $\sigma\in\mathbb{G}$, and for every (or, equivalently, for at least one) coordinate system (U,φ) around σ, it holds*

$$\begin{aligned}
&X_\sigma = \sum_{i=1}^{N} a_i(\sigma)\frac{\partial}{\partial x_i}\bigg|_\sigma,\\
&\textit{where}\ \begin{pmatrix} a_1(\sigma)\\ \vdots\\ a_N(\sigma)\end{pmatrix} = \mathcal{J}_{\varphi\circ\tau_\sigma\circ\chi^{-1}}(\chi(e))\cdot\begin{pmatrix} X_e(\chi_1)\\ \vdots\\ X_e(\chi_N)\end{pmatrix}.
\end{aligned}\tag{2.28}$$

Formula (2.28) is particularly useful for those Lie groups admitting a single global coordinate system defined on the whole group (such as Carnot groups or $\mathrm{GL}_N(\mathbb{R})$, to give some example).

Remark 2.1.46 (Bracket in the identity). Since the Lie algebra can be identified to $\mathbb{G}_e$, it is interesting to analyze what form takes the bracket as seen as an operation on $\mathbb{G}_e$.

Let (E,χ) be a fixed coordinate system around the identity $e\in\mathbb{G}$. We know that (see (2.14) and (2.22))

$$X_e f = \sum_{j=1}^{N} \lambda_j \, \partial_j|_{\chi(e)}(f \circ \chi^{-1}),$$

$$Y_e f = \sum_{j=1}^{N} \mu_j \, \partial_j|_{\chi(e)}(f \circ \chi^{-1}),$$

$$[X, Y]_e f = \sum_{j=1}^{N} \nu_j \, \partial_j|_{\chi(e)}(f \circ \chi^{-1}),$$

where $\lambda_j = X_e(\chi_j)$, $\mu_j = Y_e(\chi_j)$, and

$$\begin{aligned}\nu_j &= X_e(Y\chi_j) - Y_e(X\chi_j) \\ &= \sum_i \lambda_i \partial_{u_i}|_{\chi(e)}\big(Y_{\chi^{-1}(u)}\chi_j\big) - \sum_i \mu_i \partial_{u_i}|_{\chi(e)}\big(X_{\chi^{-1}(u)}\chi_j\big) \\ &\quad \text{(after a lengthy computation using also (2.28))} \\ &= \sum_{i,k=1}^{N} (\lambda_i \mu_k - \mu_i \lambda_k) \left.\frac{\partial}{\partial u_i}\right|_{\chi(e)} \left.\frac{\partial}{\partial v_k}\right|_{\chi(e)} \Big\{\chi_j\big(\chi^{-1}(u) * \chi^{-1}(v)\big)\Big\}.\end{aligned}$$

2.1.8 Homomorphisms

Definition 2.1.47 (Homomorphisms). *Let* $(\mathbb{G}, \bullet)$ *and* $(\mathbb{H}, *)$ *be Lie groups. A map* $\varphi : \mathbb{G} \longrightarrow \mathbb{H}$ *is a* homomorphism *of Lie groups if it is* C^∞ *and if*

$$\varphi(x \bullet y) = \varphi(x) * \varphi(y) \qquad \forall\ x,\ y \in \mathbb{G}.$$

A map φ *is an* isomorphism *of Lie groups if it is a homomorphism of Lie groups and a diffeomorphism of differentiable manifolds. An isomorphism of* $\mathbb{G}$ *onto itself is called an* automorphism *of* $\mathbb{G}$.

Let $(\mathfrak{g}, [\cdot,\cdot]_1)$ *and* $(\mathfrak{h}, [\cdot,\cdot]_2)$ *be Lie algebras. A map* $\varphi : \mathfrak{g} \longrightarrow \mathfrak{h}$ *is a* homomorphism *of Lie algebras if it is linear and if*

$$\varphi\big([X, Y]_1\big) = [\varphi(X), \varphi(Y)]_2 \qquad \forall\, X,\ Y \in \mathfrak{g}.$$

A map φ *is an* isomorphism *of Lie algebras if it is a bijective homomorphism of Lie algebras. An isomorphism of* $\mathfrak{g}$ *onto itself is called an* automorphism *of* $\mathfrak{g}$.

Example 2.1.48. Suppose $(\mathbb{G}, \bullet)$ is a Lie group and M is a differentiable manifold. Suppose $T : \mathbb{G} \to M$ is a C^∞-diffeomorphism. We can consider on M a Lie group structure naturally induced by $\bullet$ via T. Precisely, we equip M with the following composition law

$$M \times M \ni (x, y) \mapsto x * y := T\big(T^{-1}(x) \bullet T^{-1}(y)\big) \in M. \tag{2.29}$$

It is an easy exercise to verify that $*$ defines on M a Lie group structure, $T : (\mathbb{G}, \bullet) \to (M, *)$ is a Lie-group isomorphism, the identity of $(M, *)$ is $T(e_\mathbb{G})$ ($e_\mathbb{G}$ being the identity of $(\mathbb{G}, \bullet)$) and, finally, the inverse of x in M with respect to $*$ is given by

$$x^{-1} = T\Big(\big(T^{-1}(x)\big)^{-1}\Big).$$

Now, let X be a $\bullet$-left invariant vector field on $\mathbb{G}$. We showed in Example 2.1.25 how to relate to X a vector field $\widetilde{X}$ on M simply setting (see, for instance (2.17c))

$$\widetilde{X}_y = \mathrm{d}_{T^{-1}(y)}T(X_{T^{-1}(y)}) \qquad \text{for every } y \in M. \tag{2.30}$$

This obviously ensures that X and $\widetilde{X}$ are T-related according to Definition 2.1.34. Moreover, we claim that

$$\widetilde{X} \textit{ is } *\textit{-left invariant.} \tag{2.31}$$

This is equivalent to

$$\widetilde{X}_y(f(\sigma * \cdot)) = \widetilde{X}_{\sigma * y}(f) \qquad \forall\, y,\ \sigma \in M \quad \forall\, f \in C^\infty(M, \mathbb{R}).$$

Now, this last equality follows from the following calculation

$$\begin{aligned}
&\widetilde{X}_y(f(\sigma * \cdot)) = \\
\text{(see (2.30))} \quad &= \big\{\mathrm{d}_{T^{-1}(y)}T(X_{T^{-1}(y)})\big\}(f(\sigma * \cdot)) \\
&= (X_{T^{-1}(y)})\big(f(\sigma * T(\cdot))\big) \\
\text{(see (2.29))} \quad &= (X_{T^{-1}(y)})\big\{f\big(T(T^{-1}(\sigma) \bullet (\cdot))\big)\big\} \\
(X \text{ is } \bullet\text{-left inv.}) \quad &= (X_{T^{-1}(\sigma)\bullet T^{-1}(y)})\big\{f\big(T(\cdot)\big)\big\} \\
\text{(see (2.29))} \quad &= (X_{T^{-1}(\sigma * y)})\big\{f\big(T(\cdot)\big)\big\} \\
&= \big(\mathrm{d}_{T^{-1}(\sigma * y)}T(X_{T^{-1}(\sigma * y)})\big)(f) \\
\text{(see (2.30))} \quad &= \widetilde{X}_{\sigma * y}(f).
\end{aligned}$$

Roughly speaking, this gives the following fact.

Let $T : \mathbb{R}^N \to \mathbb{R}^N$ be a C^∞-diffeomorphism defining on $\mathbb{R}^N$ a change of variable $y = T(x)$. Suppose the domain $(\mathbb{R}^N, x)$ is equipped with a group law $\bullet$, and define on the image $(\mathbb{R}^N, y)$ the induced group law $*$ as in (2.29). Then, if X is a left-invariant vector field w.r.t. $\bullet$ on $(\mathbb{R}^N, x)$, the vector field $\widetilde{X}$ which expresses X in the y-coordinates is a left-invariant vector field w.r.t. $*$ on $(\mathbb{R}^N, y)$. □

Let $\varphi : \mathbb{G} \longrightarrow \mathbb{H}$ be a homomorphism of Lie groups. Since φ sends the identity of $\mathbb{G}$ to the identity of $\mathbb{H}$, the differential $\mathrm{d}_e\varphi$ of φ is a linear map from $\mathbb{G}_e$ to $\mathbb{H}_e$ (for brevity we denote by e the identity element both in $\mathbb{G}$ and in $\mathbb{H}$).

Now, by means of the natural identification between $\mathbb{G}_e$ and the algebra $\mathfrak{g}$ of $\mathbb{G}$ and between $\mathbb{H}_e$ and the algebra $\mathfrak{h}$ of $\mathbb{H}$ (see Theorem 2.1.43), $\mathrm{d}_e\varphi$ thus induces a natural linear map between $\mathfrak{g}$ and $\mathfrak{h}$, which we denote by $\mathrm{d}\varphi$.

In other words, the following definition naturally arises.

Definition 2.1.49 (Differential of a homomorphism). *If $\varphi : \mathbb{G} \longrightarrow \mathbb{H}$ is a homomorphism of Lie groups, we define*

$$\mathrm{d}\varphi : \mathfrak{g} \longrightarrow \mathfrak{h}$$

as the linear map such that, for every $X \in \mathfrak{g}$, $\mathrm{d}\varphi(X)$ *is the only element of* $\mathfrak{h}$ *satisfying*

$$\{\mathrm{d}\varphi(X)\}_{e_{\mathbb{H}}} = \mathrm{d}_{e_{\mathbb{G}}}\varphi(X_{e_{\mathbb{G}}}) \quad (\in \mathbb{H}_{e_{\mathbb{H}}}). \tag{2.32}$$

At this point, some confusion in the notation may occur. Indeed, Definition 2.1.49 defines the differential of a Lie-group homomorphism $\varphi : \mathbb{G} \to \mathbb{H}$ as a map from the Lie-algebra $\mathfrak{g}$ to the Lie-algebra $\mathfrak{h}$; but we already defined (see Definition 2.1.18) $\mathrm{d}\varphi$ as a map between the relevant tangent bundles, when φ is smooth, and $\mathrm{d}\varphi$ (see Definition 2.1.24) as a map between the smooth vector fields as operators, when φ is a diffeomorphism.

The following theorem shows that the three definitions are consistent (see also Remark 2.1.51).

Theorem 2.1.50. *Let* $\mathbb{G}$ *and* $\mathbb{H}$ *be Lie groups with associated algebras* $\mathfrak{g}$ *and* $\mathfrak{h}$, *respectively. Let* $\varphi : \mathbb{G} \longrightarrow \mathbb{H}$ *be a homomorphism of Lie groups. Then:*

(i) *for every* $X \in \mathfrak{g}$, *we have that* X *and* $\mathrm{d}\varphi(X)$ *are* φ*-related, i.e.*

$$\big\{\mathrm{d}\varphi(X)\big\}_{\varphi(x)} = \mathrm{d}_x\varphi(X_x) \qquad \forall\, x \in \mathbb{G}; \tag{2.33}$$

Condition (2.33) *characterizes* $\mathrm{d}\varphi(X)$, *i.e.* $\mathrm{d}\varphi(X)$ *is the only left invariant vector field* $Y \in \mathfrak{h}$ *such that* $Y_{\varphi(x)} = \mathrm{d}_x\varphi(X_x)$ *for every* $x \in \mathbb{G}$;

(ii) $\mathrm{d}\varphi : \mathfrak{g} \longrightarrow \mathfrak{h}$ *is a homomorphism of Lie algebras.*

Proof. Let $X \in \mathfrak{g}$ be arbitrary. Let $\widetilde{X} := \mathrm{d}\varphi(X)$. We have to prove that (see Remark 2.1.36 and (2.23b))

$$(\star) \qquad \widetilde{X}_{\varphi(\sigma)} = \mathrm{d}_\sigma\varphi(X_\sigma) \quad \forall\, \sigma \in \mathbb{G}.$$

Let $\sigma \in \mathbb{G}$. We shall use the notation $\tau^{\mathbb{G}}_\sigma$ and $\tau^{\mathbb{H}}_{\varphi(\sigma)}$ for the left translations on $\mathbb{G}$ and on $\mathbb{H}$ and the notation $e_{\mathbb{G}}$, $e_{\mathbb{H}}$ for the identity elements in $\mathbb{G}$ and $\mathbb{H}$, respectively. Moreover, the group laws on $\mathbb{G}$ and $\mathbb{H}$ will be respectively denoted by $\circ^{\mathbb{G}}$ and $\circ^{\mathbb{H}}$. Since φ is a Lie group homomorphism, we have

$$(\tau^{\mathbb{H}}_{\varphi(\sigma)} \circ \varphi)(\cdot) = \varphi(\sigma) \circ^{\mathbb{H}} \varphi(\cdot) = \varphi(\sigma \circ^{\mathbb{G}} \cdot) = (\varphi \circ \tau^{\mathbb{G}}_\sigma)(\cdot),$$

i.e. the following identity of maps on $\mathbb{G}$ holds

$$\tau^{\mathbb{H}}_{\varphi(\sigma)} \circ \varphi \equiv \varphi \circ \tau^{\mathbb{G}}_\sigma. \tag{2.34}$$

As a consequence, we have (also recall that $\widetilde{X}$ and X are left invariant)

$$\begin{aligned}
\widetilde{X}_{\varphi(\sigma)} &= \\
\text{(see (2.25b))} &= \mathrm{d}_{e_{\mathbb{H}}}\tau^{\mathbb{H}}_{\varphi(\sigma)}(\widetilde{X}_{e_{\mathbb{H}}}) \\
\text{(see (2.32))} &= \mathrm{d}_{e_{\mathbb{H}}}\tau^{\mathbb{H}}_{\varphi(\sigma)}(\mathrm{d}_{e_{\mathbb{G}}}\varphi(X_{e_{\mathbb{G}}})) = \mathrm{d}_{e_{\mathbb{G}}}(\tau^{\mathbb{H}}_{\varphi(\sigma)} \circ \varphi)(X_{e_{\mathbb{G}}}) \\
\text{(see (2.34))} &= \mathrm{d}_{e_{\mathbb{G}}}(\varphi \circ \tau^{\mathbb{G}}_\sigma)(X_{e_{\mathbb{G}}}) = \mathrm{d}_\sigma\varphi(\mathrm{d}_{e_{\mathbb{G}}}\tau^{\mathbb{G}}_\sigma(X_{e_{\mathbb{G}}})) \\
\text{(see (2.25b))} &= \mathrm{d}_\sigma\varphi(X_\sigma).
\end{aligned}$$

This is precisely $(\star)$, and (i) is proved.

Finally, we have to prove

$$d\varphi([X,Y]) = [d\varphi(X), d\varphi(Y)] \qquad \forall\, X,\ Y \in \mathfrak{g}.$$

By means of the first part (i) of the assertion, we have that X (resp. Y) is φ-related to $d\varphi(X)$ (resp. $d\varphi(Y)$). Thus, by Proposition 2.1.37, $[X, Y]$ is φ-related to $[d\varphi(X), d\varphi(Y)]$, i.e.

$$d_x\varphi([X,Y]_x) = [d\varphi(X), d\varphi(Y)]_{\varphi(x)} \qquad \forall\, x \in \mathbb{G}.$$

In particular, for $x = e_{\mathbb{G}}$, we have

$$d_{e_{\mathbb{G}}}\varphi([X,Y]_{e_{\mathbb{G}}}) = [d\varphi(X), d\varphi(Y)]_{e_{\mathbb{H}}}.$$

Now, by definition, $d\varphi([X, Y])$ is the unique vector field in $\mathfrak{h}$ whose value in $e_{\mathbb{H}}$ is $d_{e_{\mathbb{G}}}\varphi([X,Y]_{e_{\mathbb{G}}})$, then $d\varphi([X, Y]) = [d\varphi(X), d\varphi(Y)]$, and the theorem is completely proved. □

Remark 2.1.51 (Consistency of the notation "dφ"). Let $\mathbb{G}$ and $\mathbb{H}$ be Lie groups with associated algebras $\mathfrak{g}$ and $\mathfrak{h}$, respectively. Let $\varphi : \mathbb{G} \to \mathbb{H}$ be a homomorphism of Lie groups. In Definition 2.1.18, $d\varphi$ was defined as the map

$$d\varphi : T(\mathbb{G}) \to T(\mathbb{H}), \qquad d\varphi(x, X_x) = (\varphi(x), Y_{\varphi(x)}),$$
$$\text{where } Y_{\varphi(x)} = d_x\varphi(X_x),$$

Hence, by Theorem 2.1.50, $d\varphi$, as defined in Definition 2.1.49, is just the "projection" of the previously defined $d\varphi$ on the second component.

In case φ is also an isomorphism, we have defined $\mathbf{d}\varphi$ in Definition 2.1.24. A comparison of (2.15b) and (2.33) shows that $\mathbf{d}\varphi$ and $d\varphi$ of Definition 2.1.49 actually coincide. □

Remark 2.1.52. (i) From Theorem 2.1.50 it immediately follows that if $\varphi : \mathbb{G} \longrightarrow \mathbb{H}$ is a Lie group isomorphism then $d\varphi : \mathfrak{g} \longrightarrow \mathfrak{h}$ is a Lie algebra isomorphism.

(ii) Let us now suppose we are given two Lie groups on $\mathbb{R}^N$ as defined in Chapter 1: $(\mathbb{G}, \circ)$ and $(\mathbb{H}, *)$. Let the associated Lie-algebras be $\mathfrak{g}$ and $\mathfrak{h}$, respectively. Suppose it is given a Lie group isomorphism φ from $\mathbb{G}$ to $\mathbb{H}$. As usual, the identity elements in $\mathbb{G}$ and $\mathbb{H}$ are supposed to be the origin 0 of $\mathbb{R}^N$. Then $d\varphi$ sends the $\circ$-Jacobian basis of $\mathfrak{g}$ in a basis of $\mathfrak{h}$ that in 0 coincides with the column vectors of the matrix $\mathcal{J}_\varphi(0)$.

Indeed, if $Z_1, \dots, Z_N$ is the Jacobian basis related to $(\mathbb{G}, \circ)$ and if $f \in C^\infty(\mathbb{H}, \mathbb{R})$, we have

$$\begin{aligned}\{d\varphi(Z_k)\}_0(f) &= \{d_0\varphi((Z_k)_0)\}(f) = \{Z_k\}_0(f \circ \varphi)\\ &= \partial_{x_k}|_0(f \circ \varphi) = \sum_{j=1}^N (\partial_{\xi_j} f)(0) \cdot (\partial_{x_k}\varphi_j)(0).\end{aligned}$$

In particular, $d\varphi$ *sends the $\circ$-Jacobian basis of $\mathfrak{g}$ in the $*$-Jacobian basis of $\mathfrak{h}$ if and only if $\mathcal{J}_\varphi(0)$ is the identity matrix.*

(iii) Let $\mathbb{F}$, $\mathbb{G}$, $\mathbb{H}$ be three Lie groups with Lie algebras $\mathfrak{f}$, $\mathfrak{g}$, $\mathfrak{h}$. Let

$$\varphi : \mathbb{F} \to \mathbb{G}, \quad \psi : \mathbb{G} \to \mathbb{H}$$

be Lie-group homomorphisms. Then $\mathrm{d}(\psi \circ \varphi) = \mathrm{d}\psi \circ \mathrm{d}\varphi$. Indeed, for every $X \in \mathfrak{f}$, $\mathrm{d}(\psi \circ \varphi)(X)$ is the only element of $\mathfrak{h}$ such that

$$(\star) \qquad \big(\mathrm{d}(\psi \circ \varphi)(X)\big)_{e_{\mathbb{H}}} = \mathrm{d}_{e_{\mathbb{F}}}(\psi \circ \varphi)(X_{e_{\mathbb{F}}}).$$

But also $(\mathrm{d}\psi \circ \mathrm{d}\varphi)(X)$ when applied on $e_{\mathbb{H}}$ coincides with the right-hand side of $(\star)$. Indeed,

$$\begin{aligned}\big((\mathrm{d}\psi \circ \mathrm{d}\varphi)(X)\big)_{e_{\mathbb{H}}} &= \big(\mathrm{d}\psi\big(\mathrm{d}\varphi(X)\big)\big)_{e_{\mathbb{H}}} = \mathrm{d}_{e_{\mathbb{G}}}\psi\big(\big(\mathrm{d}\varphi(X)\big)_{e_{\mathbb{G}}}\big)\\ &= \mathrm{d}_{e_{\mathbb{G}}}\psi\big(\mathrm{d}_{e_{\mathbb{F}}}\varphi(X_{e_{\mathbb{F}}})\big) = \mathrm{d}_{e_{\mathbb{F}}}(\psi \circ \varphi)(X_{e_{\mathbb{F}}}). \qquad \square\end{aligned}$$

2.1.9 The Exponential Map

We begin with a simple but crucial result on Lie groups, following from an ODE's result.

We recall that, according to Definition 2.1.30, a smooth vector field X on a Lie group $\mathbb{G}$ is complete if, for every $x \in \mathbb{G}$, the integral curve μ of X such that $\mu(0) = x$ is defined on the whole $\mathbb{R}$.

Proposition 2.1.53 (Completeness of the left invariant vector fields). *The left invariant vector fields on a Lie group $\mathbb{G}$ are complete.*

Proof. Let $X \in \mathfrak{g}$ be fixed, $\mathfrak{g}$ being the Lie algebra of $\mathbb{G}$. By simple prolongation results for ODE's, it is enough to prove that there exists $\varepsilon = \varepsilon(X) > 0$ such that, for every $x \in \mathbb{G}$, the integral curve μ_x of X such that $\mu_x(0) = x$ is defined on $[-\varepsilon, \varepsilon]$.

Let e be the identity of $\mathbb{G}$. Let $\varepsilon > 0$ be such that the integral curve μ_e of X with $\mu_e(0) = e$ is defined on $[-\varepsilon, \varepsilon]$. Then, it is an immediate consequence of the left invariance of X to derive that the map

$$[-\varepsilon, \varepsilon] \ni t \mapsto (\tau_x \circ \mu_e)(t) =: \nu(t) \in \mathbb{G}$$

coincides with the above μ_x. Indeed, $\nu(0) = \tau_x(\mu_e(0)) = \tau_x(e) = x$ and (see the Definition 2.1.26 for $\dot{\nu}(t)$)

$$\begin{aligned}\dot{\nu}(t) &= \mathrm{d}_t \nu\left(\frac{\mathrm{d}}{\mathrm{d}r}\bigg|_{r=t}\right) = \mathrm{d}_t(\tau_x \circ \mu_e)\left(\frac{\mathrm{d}}{\mathrm{d}r}\bigg|_{r=t}\right)\\ &= \big(\mathrm{d}_{\mu_e(t)}\tau_x \circ \mathrm{d}_t \mu_e\big)\left(\frac{\mathrm{d}}{\mathrm{d}r}\bigg|_{r=t}\right)\\ &= \mathrm{d}_{\mu_e(t)}\tau_x(\dot{\mu}_e(t)) = \mathrm{d}_{\mu_e(t)}\tau_x(X_{\mu_e(t)})\\ \text{(see (2.24))} &= X_{(\tau_x \circ \mu_e)(t)} = X_{\nu(t)}.\end{aligned}$$

This ends the proof. $\square$

Definition 2.1.54 (The exponential curve $\exp_X(t)$). *Let $\mathbb{G}$ be a Lie group with Lie algebra $\mathfrak{g}$. Let $X \in \mathfrak{g}$ be fixed. By Proposition 2.1.53, the integral curve $\mu(t)$ of X passing through the identity of $\mathbb{G}$ when $t = 0$ is defined on the whole $\mathbb{R}$. We set*

$$\exp_X(t) := \mu(t).$$

By the very Definition 2.1.28 of integral curve, we have

$$\exp_X : \mathbb{R} \to \mathbb{G} \quad \text{with} \quad \begin{cases} \exp_X(0) = e_{\mathbb{G}}, \\ \mathrm{d}_t \exp_X \left(\left.\dfrac{\mathrm{d}}{\mathrm{d}r}\right|_{r=t} \right) = X_{\exp_X(t)} \quad \forall\, t \in \mathbb{R}. \end{cases} \tag{2.35}$$

In terms of functionals on $C^\infty(\mathbb{G}, \mathbb{R})$, (2.35) can be written more explicitly as

$$\left(\frac{\mathrm{d}}{\mathrm{d}r}\right)_{r=t} \big\{ f(\exp_X(r)) \big\} = X_{\exp_X(t)}(f) \qquad \forall\, f \in C^\infty(\mathbb{G}, \mathbb{R}). \tag{2.36a}$$

In particular, when $t = 0$,

$$\left(\frac{\mathrm{d}}{\mathrm{d}r}\right)_{r=0} \big\{ f(\exp_X(r)) \big\} = X_e(f) \qquad \forall\, f \in C^\infty(\mathbb{G}, \mathbb{R}). \tag{2.36b}$$

Again from (2.35) with $t = 0$ we infer

$$\mathrm{d}_0 \exp_X \left\{ \left(\frac{\mathrm{d}}{\mathrm{d}t}\right)_0 \right\} = X_e. \tag{2.36c}$$

Remark 2.1.55 (The exponential curve as a homomorphism). It is a simple exercise on ODE's (see also Ex. 9 at the end of Chapter 1) to verify that

$$\exp_X : (\mathbb{R}, +) \longrightarrow (\mathbb{G}, *)$$

is a Lie group homomorphism. In other words, it holds

$$\exp_X(r + s) = \exp_X(r) * \exp_X(s) \quad \text{for every } r,\ s \in \mathbb{R}.$$

Hence, Definition 2.1.49 can be applied. The differential of $\exp_X$ is the following map (here $\mathfrak{r}$ is the Lie algebra of $(\mathbb{R}, +)$)

$$\mathrm{d}\exp_X : \mathfrak{r} \longrightarrow \mathfrak{g} \qquad \text{is such that} \quad \lambda \frac{\mathrm{d}}{\mathrm{d}t} \mapsto \lambda X \quad \forall\, \lambda \in \mathbb{R},$$

i.e. it holds

$$\mathrm{d}\exp_X \left(\frac{\mathrm{d}}{\mathrm{d}t} \right) = X. \tag{2.37}$$

Indeed, by definition of the differential of a homomorphism, $\mathrm{d}\exp_X(\mathrm{d}/\mathrm{d}t)$ is the unique vector field of $\mathfrak{g}$ such that

$$\left\{ \mathrm{d}\exp_X \left(\frac{\mathrm{d}}{\mathrm{d}t} \right) \right\}_e = \mathrm{d}_0 \exp_X \left\{ \left(\frac{\mathrm{d}}{\mathrm{d}t} \right)_0 \right\}.$$

The above right-hand side equals X_e thanks to (2.36c), whence (2.37). □

For future reference, we collect some other useful formulas for $\exp_X(t)$, immediate consequence of the facts proved above.

Theorem 2.1.56. *Let* $(\mathbb{G}, *)$ *be a Lie group with algebra* $\mathfrak{g}$. *Let* $X \in \mathfrak{g}$. *Then*:

(i) $\exp_X(r+s) = \exp_X(r) * \exp_X(s)$ *for every* $r,\ s \in \mathbb{R}$;
(ii) $\exp_X(-t) = (\exp_X(t))^{-1}$ *for every* $t \in \mathbb{R}$;
(iii) $\exp_X(0) = e$;
(iv) $\mathbb{R} \ni t \mapsto \exp_X(t) \in \mathbb{G}$ *is a smooth curve*;
(v) $\exp_X(t)$ *is the unique integral curve of* X *passing through the identity at time zero, so that, for every* $x \in \mathbb{G}$,

$$t \mapsto x * (\exp_X(t))$$

is the unique integral curve of X *passing through* x *at time zero.*

Note that (i) in the above theorem jointly with the left invariance of X and (2.36c) gives back[5] (2.36b).

We are ready to give the fundamental definition.

Definition 2.1.57 (Exponential map). *Let* $(\mathbb{G}, *)$ *be a Lie group with Lie algebra* $\mathfrak{g}$. *Following the notation in Definition* 2.1.54, *we set*

$$\begin{aligned} \mathrm{Exp}: \ \mathfrak{g} &\longrightarrow \mathbb{G}, \\ X &\mapsto \mathrm{Exp}(X) := \exp_X(1). \end{aligned}$$

Exp *is called the* exponential map (*related to the Lie group* $\mathbb{G}$).

The following results hold.

Proposition 2.1.58. *Let* $(\mathbb{G}, *)$ *be a Lie group with Lie algebra* $\mathfrak{g}$. *For every* $X \in \mathfrak{g}$, *we have*

(i) $\mathrm{Exp}(t\,X) = \exp_X(t)$ *for every* $t \in \mathbb{R}$;
(ii) $\mathrm{Exp}((r+s)X) = \mathrm{Exp}(r\,X) * \mathrm{Exp}(sX)$ *for every* $r,\ s \in \mathbb{R}$;
(iii) $\mathrm{Exp}(-tX) = (\mathrm{Exp}(tX))^{-1}$, *for every* $t \in \mathbb{R}$.

[5] Indeed, a direct computation gives

$$\begin{aligned} &\left(\frac{\mathrm{d}}{\mathrm{d}r}\right)_{r=t} \Big\{ f(\exp_X(r)) \Big\} = \left(\frac{\mathrm{d}}{\mathrm{d}r}\right)_{r=0} \Big\{ f(\exp_X(r+t)) \Big\} \\ &= \left(\frac{\mathrm{d}}{\mathrm{d}r}\right)_{r=0} \Big\{ f(\exp_X(t) * \exp_X(r)) \Big\} = \left(\frac{\mathrm{d}}{\mathrm{d}r}\right)_{r=0} \Big\{ (f \circ \tau_{\exp_X(t)})(\exp_X(r)) \Big\} \\ &= \left\{ \mathrm{d}_0 \exp_X \left\{ \left(\frac{\mathrm{d}}{\mathrm{d}r}\right)_{r=0} \right\} \right\} (f \circ \tau_{\exp_X(t)}) = X_e(f \circ \tau_{\exp_X(t)}) = X_{\exp_X(t)}(f). \end{aligned}$$

Here we used (2.36c) and (2.25c).

Proof. Fix $t \in \mathbb{R}$ and consider the curve $s \mapsto \mu(s) := \exp_X(s\,t)$. We have $\mu(0) = \exp_X(0) = e$. By Theorem 2.1.56, we have

$$\dot{\mu}(s) = t\, X_{\exp_X(s\,t)} = t\, X_{\mu(s)}.$$

Thus μ is the integral curve of tX, and, again by Theorem 2.1.56, we have $\mu(s) = \exp_{tX}(s)$, i.e.

$$\exp_X(s\,t) = \exp_{tX}(s).$$

For $s = 1$, we have $\exp_X(t) = \exp_{tX}(1) = \operatorname{Exp}(t\,X)$.

Therefore, this yields

$$\operatorname{Exp}((r+s)\,X) = \exp_X(r+s) = \exp_X(r) * \exp_X(s) = \operatorname{Exp}(r\,X) * \operatorname{Exp}(s\,X),$$

and moreover,

$$\operatorname{Exp}(-t\,X) = \exp_X(-t) = \big(\exp_X(t)\big)^{-1} = \big(\operatorname{Exp}(t\,X)\big)^{-1}.$$

The proposition is thus completely proved. □

Theorem 2.1.59. *Let $\mathbb{G}$ and $\mathbb{H}$ be Lie groups with associated algebras $\mathfrak{g}$ and $\mathfrak{h}$. We denote by $\operatorname{Exp}_{\mathbb{G}}$ and $\operatorname{Exp}_{\mathbb{H}}$ the exponential maps related to $\mathbb{G}$ and to $\mathbb{H}$, respectively. Finally, let $\varphi : \mathbb{G} \longrightarrow \mathbb{H}$ be a Lie group homomorphism. Then the following diagram is commutative:*

$$\begin{array}{ccc} \mathbb{G} & \xrightarrow{\ \varphi\ } & \mathbb{H} \\ \uparrow{\scriptstyle \operatorname{Exp}_{\mathbb{G}}} & & \uparrow{\scriptstyle \operatorname{Exp}_{\mathbb{H}}} \\ \mathfrak{g} & \xrightarrow{\ \mathrm{d}\varphi\ } & \mathfrak{h}. \end{array}$$

Proof. Let $X \in \mathfrak{g}$. We have to prove that

$$\operatorname{Exp}_{\mathbb{H}}(\mathrm{d}\varphi(X)) = \varphi\big(\operatorname{Exp}_{\mathbb{G}}(X)\big). \tag{2.38}$$

We set for brevity $x := \operatorname{Exp}_{\mathbb{H}}(\mathrm{d}\varphi(X))$. By definition, we have

$$x = \exp^{\mathbb{H}}_{\mathrm{d}\varphi(X)}(1),$$

where $\exp^{\mathbb{H}}_{\mathrm{d}\varphi(X)}(t)$ is the unique integral curve of $\mathrm{d}\varphi(X)$ on $\mathbb{H}$ passing through the identity $e_{\mathbb{H}}$ of $\mathbb{H}$ at time zero (see Theorem 2.1.56). We set $y := \varphi(\operatorname{Exp}_{\mathbb{G}}(X))$, i.e.

$$y = \varphi(\exp^{\mathbb{G}}_X(1)),$$

where $\exp^{\mathbb{G}}_X(t)$ is the unique integral curve of X on $\mathbb{G}$ passing through the identity $e_{\mathbb{G}}$ of $\mathbb{G}$ at time zero. We want to show that

$$\exp^{\mathbb{H}}_{\mathrm{d}\varphi(X)}(t) = \varphi(\exp^{\mathbb{G}}_X(t)) \qquad \forall\, t \in \mathbb{R}. \tag{2.39}$$

When $t = 1$, this gives $x = y$, which is the claimed (2.38).

To this end, we show that $\mu(t) := \varphi(\exp_X^{\mathbb{G}}(t))$ is an integral curve of $\mathrm{d}\varphi(X)$ on $\mathbb{H}$ passing through $e_{\mathbb{H}}$ at time zero. By uniqueness, this will yield $\mu(t) = \exp^{\mathbb{H}}_{\mathrm{d}\varphi(X)}(t)$, i.e. (2.39). Since φ is a homomorphism, we have

$$\mu(0) = \varphi(\exp_X^{\mathbb{G}}(0)) = \varphi(e_{\mathbb{G}}) = e_{\mathbb{H}}.$$

Finally, since φ and $\exp_X^{\mathbb{G}}$ are Lie-group homomorphisms, μ is a homomorphism. Thus, we have

$$\begin{aligned}\dot{\mu}(t) &= \mathrm{d}_t\mu\left(\frac{\mathrm{d}}{\mathrm{d}r}\Big|_{r=t}\right) = \left(\mathrm{d}\mu\left(\frac{\mathrm{d}}{\mathrm{d}t}\right)\right)_{\mu(t)} = \left(\mathrm{d}(\varphi\circ\exp_X^{\mathbb{G}})\left(\frac{\mathrm{d}}{\mathrm{d}t}\right)\right)_{\mu(t)} \\ &= \left((\mathrm{d}\varphi\circ\mathrm{d}\exp_X^{\mathbb{G}})\left(\frac{\mathrm{d}}{\mathrm{d}t}\right)\right)_{\mu(t)} = \big(\mathrm{d}\varphi(X)\big)_{\mu(t)}.\end{aligned}$$

In the first equality, we used (2.18); in the second, (2.33); in the third, we exploited the very definition of μ; the fourth follows from Remark 2.1.52-(iii); in the fifth equality, we used (2.37). This ends the proof. □

Remark 2.1.60 (Some notable commutative diagrams). Theorem 2.1.59 has some important consequences: suppose that $\mathrm{Exp}_{\mathbb{G}}$ and $\mathrm{Exp}_{\mathbb{H}}$ are diffeomorphisms with inverse functions $\mathrm{Log}_{\mathbb{G}}$ and $\mathrm{Log}_{\mathbb{H}}$, respectively. Let us analyze the following commutative diagram:

$$\begin{CD} \mathbb{G} @>{\varphi}>> \mathbb{H} \\ @A{\mathrm{Exp}_{\mathbb{G}}}AA @VV{\mathrm{Log}_{\mathbb{H}}}V \\ \mathfrak{g} @>{\mathrm{d}\varphi}>> \mathfrak{h}. \end{CD}$$

We deduce that, given the Lie group homomorphism φ, the map

$$\mathrm{Log}_{\mathbb{H}}\circ\varphi\circ\mathrm{Exp}_{\mathbb{G}} : \mathfrak{g} \longrightarrow \mathfrak{h}$$

coincides with the differential of φ, which is a Lie algebra homomorphism (in particular, a linear map!).

Under the same hypotheses on $\mathrm{Exp}_{\mathbb{G}}$ and $\mathrm{Exp}_{\mathbb{H}}$, suppose we are given a Lie algebra homomorphism $\psi : \mathfrak{g} \longrightarrow \mathfrak{h}$. Let us consider the map

$$\mathrm{Exp}_{\mathbb{H}}\circ\psi\circ\mathrm{Log}_{\mathbb{G}} : \mathbb{G} \longrightarrow \mathbb{H}.$$

If such a map is a Lie group homomorphism,[6] then the differential of this map coincides with ψ. Indeed, the following diagram is commutative

$$\begin{CD} \mathbb{G} @>{\mathrm{Exp}_{\mathbb{H}}\circ\psi\circ\mathrm{Log}_{\mathbb{G}}}>> \mathbb{H} \\ @V{\mathrm{Log}_{\mathbb{G}}}VV @AA{\mathrm{Exp}_{\mathbb{H}}}A \\ \mathfrak{g} @>{\psi}>> \mathfrak{h}. \end{CD}$$

[6] We shall see that $\mathrm{Exp}_{\mathbb{H}}\circ\psi\circ\mathrm{Log}_{\mathbb{G}}$ is always a Lie group homomorphism whenever $\mathbb{G}$ and $\mathbb{H}$ are connected and simply connected nilpotent Lie groups.

Finally, given a Lie group homomorphism $\varphi : \mathbb{G} \longrightarrow \mathbb{H}$, by the commutative diagram

$$\begin{array}{ccc} \mathbb{G} & \xrightarrow{\quad\varphi\quad} & \mathbb{H} \\ \Big\downarrow \mathrm{Log}_{\mathbb{G}} & & \Big\uparrow \mathrm{Exp}_{\mathbb{H}} \\ \mathfrak{g} & \xrightarrow{\quad \mathrm{d}\varphi \quad} & \mathfrak{h} \end{array}$$

the map

$$\mathrm{Exp}_{\mathbb{H}} \circ \mathrm{d}\varphi \circ \mathrm{Log}_{\mathbb{G}} : \mathbb{G} \longrightarrow \mathbb{H}$$

coincides with φ and is thus a Lie group homomorphism.

(For other related topics[7] not employed in this book, see, e.g. [War83].)

2.2 Carnot Groups

We give two definitions of Carnot groups: the first one (which we already introduced in Chapter 1, Section 1.4, page 56) is the most convenient for our purposes and it seems very natural in an Analysis context; the second one is the classical one from Lie group theory. Then, we compare the two definitions showing that, up to isomorphism, they are equivalent. For reader's convenience, we recall the definition of homogeneous Carnot group.

Definition 2.2.1 (Homogeneous Carnot group). *Let $\mathbb{R}^N$ be equipped with a Lie group structure by the composition law $\circ$. Let also $\mathbb{R}^N$ be equipped with a homogeneous group structure by a family $\{\delta_\lambda\}_{\lambda>0}$ of automorphisms of $(\mathbb{R}^N, \circ)$ (called* dilations*) of the following form*

$$\delta_\lambda(x^{(1)}, x^{(2)}, \dots, x^{(r)}) = (\lambda x^{(1)}, \lambda^2 x^{(2)}, \dots, \lambda^r x^{(r)}). \tag{2.40}$$

Here $x^{(i)} \in \mathbb{R}^{N_i}$ for $i = 1, \dots, r$ and $N_1 + N_2 + \dots + N_r = N$. Let $\mathfrak{g}$ be the algebra of the group $(\mathbb{R}^N, \circ)$. For $i = 1, \dots, N_1$, let Z_i be the (unique) vector field of $\mathfrak{g}$ agreeing with $\partial/\partial x_i$ at the origin.

If the following assumption holds

(H1) *the Lie algebra generated by $Z_1, \dots, Z_{N_1}$ is the whole $\mathfrak{g}$,*

then $\mathbb{G} = (\mathbb{R}^N, \circ, \delta_\lambda)$ is called a homogeneous Carnot group.

---

[7] For example, the following results hold.

Theorem 2.1.61. *Let $\mathbb{G}$ and $\mathbb{H}$ be Lie groups with related algebras $\mathfrak{g}$ and $\mathfrak{h}$.*

1. *Suppose that $\mathbb{G}$ is connected. Let $\varphi, \psi : \mathbb{G} \longrightarrow \mathbb{H}$ be Lie group homomorphisms. Then φ and ψ coincide if and only if the Lie algebra homomorphisms $\mathrm{d}\varphi$ and $\mathrm{d}\psi$ from $\mathfrak{g}$ to $\mathfrak{h}$ coincide.*
2. *Suppose that $\mathbb{G}$ is simply connected. Let $\psi : \mathfrak{g} \longrightarrow \mathfrak{h}$ be a Lie algebra homomorphism. Then there exists a unique Lie group homomorphism $\varphi : \mathbb{G} \longrightarrow \mathbb{H}$ such that $\mathrm{d}\varphi = \psi$.*
3. *Suppose that $\mathbb{G}$ and $\mathbb{H}$ are simply connected. If $\mathfrak{g}$ and $\mathfrak{h}$ are isomorphic Lie algebras, $\mathbb{G}$ and $\mathbb{H}$ are isomorphic Lie groups.*

Given a Lie algebra $\mathfrak{h}$, for any two subsets V, W of $\mathfrak{h}$, we set

$$[V, W] = \mathrm{span}\big\{[v, w] \,\big|\, v \in V,\, w \in W\big\}.$$

Remark 2.2.2. Let $m \in \mathbb{N}$, and let $X_1, \ldots, X_m \in \mathfrak{h}$. Then, for every $k \in \mathbb{N}$, any Lie bracket of height k of $X_1, \ldots, X_m$ is a linear combination of *nested* brackets of the form (see Proposition 1.1.7, page 12)

$$[X_{i_1}, [X_{i_2}, \ldots [X_{i_{k-1}}, X_{i_k}] \ldots]], \quad i_1, i_2, \ldots, i_k \in \{1, \ldots, m\}.$$

Definition 2.2.3 (Stratified Lie group). *A* stratified group (*or* Carnot group) $\mathbb{H}$ *is a connected and simply connected Lie group whose Lie algebra* $\mathfrak{h}$ *admits a* stratification, *i.e. a direct sum decomposition*

$$\mathfrak{h} = V_1 \oplus V_2 \oplus \cdots \oplus V_r \quad \textit{such that} \quad \begin{cases} [V_1, V_{i-1}] = V_i & \textit{if } 2 \le i \le r, \\ [V_1, V_r] = \{0\}. \end{cases} \tag{2.41}$$

Convention. In the rest of the chapter, we shall use the term "stratified group" to refer to a Carnot group according to the above abstract, classical Definition 2.2.3. We shall instead use the prefix "homogeneous" for Carnot groups as in Definition 2.2.1. From Chapter 3 onwards, after we have proved that (see Theorem 2.2.18) every stratified group is isomorphic to a homogeneous Carnot group, we shall use the term "Carnot" group almost indifferently. Yet, whenever the homogeneous structure will have to be stressed, or whenever it will be important to distinguish stratified *versus* homogeneous Carnot groups, we shall re-invoke the prefix homogeneous.

Remark 2.2.4. Our Definition 2.2.1 is very natural to deal with in an analytic context. However, its only inconvenient property is that *it is not invariant under isomorphism of Lie groups*, since the homogeneity property heavily depends on the choice of coordinates on the underlying manifold $\mathbb{R}^N$. For example, consider the following group law in $\mathbb{R}^3$:

$$x \circ y = \big(\mathrm{arcsinh}(\sinh(x_1) + \sinh(y_1)),\, x_2 + y_2 + \sinh(x_1) y_3,\, x_3 + y_3\big).$$

It is an easy exercise to prove that $(\mathbb{R}^3, \circ)$ is a homogeneous Carnot group[8] according to Definition 2.2.3: indeed, the stratification is given by

$$\mathrm{span}\{(\cosh(x_1))^{-1}\, \partial_{x_1}, \partial_{x_3} + \sinh(x_1)\partial_{x_2}\} \oplus \mathrm{span}\{\partial_{x_2}\}.$$

It is also evident that $(\mathbb{R}^3, \circ)$ is not a homogeneous group (if it were so, the composition law should have polynomial component functions!, see Theorem 1.3.15). However, $(\mathbb{R}^3, \circ)$ is isomorphic to the *homogeneous* Carnot group $\mathbb{H}^1 = (\mathbb{R}^3, *, \delta_\lambda)$, being

[8] We remark that

$$\big((\cosh(x_1))^{-1}\partial_1\big)^2 + \big(\partial_3 + \sinh(x_1)\partial_2\big)^2$$

is the canonical sub-Laplacian on this group.

$$x * y = (x_1 + y_1, x_2 + y_2, x_3 + y_3 + 2x_2y_1 - 2x_1y_2),$$

and

$$\delta_\lambda(x) := (\lambda x_1, \lambda x_2, \lambda^2 x_3).$$

The main goal of this section is to prove that *in every equivalence class of isomorphic stratified groups there is one group which is homogeneous, according to Definition* 2.2.1.

Example 2.2.5. Consider the open subset of $\mathbb{R}^3$

$$\Omega := (0, \infty) \times \left(-\frac{\pi}{2}, \frac{\pi}{2}\right) \times \mathbb{R}.$$

Then Ω is equipped with a structure of (non-homogeneous) Carnot group by the composition ($\xi = (\xi_1, \xi_2, \xi_3) \in \Omega$ and analogously for $\eta \in \Omega$)

$$\xi * \eta = \begin{pmatrix} \xi_1 \eta_1 \\ \arctan(\xi_1 + \eta_1 + \tan \xi_2 + \tan \eta_2 - \xi_1 \eta_1) \\ \xi_3 + \eta_3 + 2(\xi_1 \ln \eta_1 + \tan \xi_2 \ln \eta_1 - \eta_1 \ln \xi_1 - \tan \eta_2 \ln \xi_1) \end{pmatrix}.$$

This can be seen, for example, by remarking that $\mathbb{G} = (\Omega, *)$ is isomorphic to the Heisenberg–Weyl group $\mathbb{H}^1$ (see Example 1.2.2) via the isomorphism

$$\varphi : \Omega \to \mathbb{H}^1, \qquad \varphi(\xi_1, \xi_2, \xi_3) = (\ln \xi_1, \xi_1 + \tan \xi_2, \xi_3).$$

Indeed, recalling that the composition law on $\mathbb{H}^1$ is given by

$$(x_1, x_2, x_3) \circ (y_1, y_2, y_3) = \big(x_1 + y_1, x_2 + y_2, x_3 + y_3 + 2\,(x_2\, y_1 - x_1\, y_2)\big),$$

and the inverse map of φ is

$$\varphi^{-1} : \mathbb{H}^1 \to \Omega, \qquad \varphi^{-1}(x_1, x_2, x_3) = \big(\exp(x_1), \arctan(x_2 - \exp(x_1)), x_3\big),$$

we notice that

$$\xi * \eta = \varphi^{-1}(\varphi(\xi) \circ \varphi(\eta)) \qquad \forall\, \xi,\ \eta \in \mathbb{G}.$$

This suffices to prove that $\mathbb{G}$ is a Carnot group. Equivalently, we show the stratification condition for $\mathfrak{g}$, the algebra of $\mathbb{G}$: first we find the $*$-left-invariant vector fields which coincide with the partial derivatives $\partial_{\xi_1}, \partial_{\xi_2}, \partial_{\xi_3}$ at the identity of $\mathbb{G}$, say

$$e_{\mathbb{G}} = \varphi^{-1}(0, 0, 0) = \left(1, -\frac{\pi}{4}, 0\right).$$

These vector fields Z_1, Z_2, Z_3 can be found via the following formula (here $f \in C^\infty(\Omega, \mathbb{R})$)

$$(Z_i f)(\xi) = \frac{\partial}{\partial \eta_i}\bigg|_{\eta = e_{\mathbb{G}}} (f(\xi * \eta)) \qquad \forall\, \xi \in \mathbb{G}.$$

A direct computation then shows that

$$
\begin{aligned}
Z_1 &= \xi_1\, \partial_{\xi_1} + \frac{1-\xi_1}{1+\tan^2(\xi_2)}\, \partial_{\xi_2} + (2\xi_1 + 2\tan\xi_2 - 2\ln\xi_1)\, \partial_{\xi_3},\\
Z_2 &= \frac{2}{1+\tan^2(\xi_2)}\, \partial_{\xi_2} - 4\,\ln\xi_1\, \partial_{\xi_3},\\
Z_3 &= \partial_{\xi_3}.
\end{aligned}
$$

Equivalently, the Z_i's are the vector fields whose column-vector of the coefficients is given by the columns of the Jacobian matrix of the map $\tau_\xi(\eta) = \xi * \eta$ at $e_{\mathbb{G}}$,

$$
\mathcal{J}_{\tau_\xi}(e_{\mathbb{G}}) = \begin{pmatrix} \xi_1 & 0 & 0 \\ \frac{1-\xi_1}{1+\tan^2(\xi_2)} & \frac{2}{1+\tan^2(\xi_2)} & 0 \\ 2\xi_1 + 2\tan\xi_2 - 2\ln\xi_1 & -4\,\ln\xi_1 & 1 \end{pmatrix}.
$$

As a consequence, the stratification of $\mathfrak{g}$ is given by

$$
\mathfrak{g} = \operatorname{span}\{Z_1, Z_2\} \oplus \operatorname{span}\{[Z_1, Z_2]\},
$$

and all commutators of Z_1, Z_2 with length > 2 vanish identically. Indeed, another direct computation gives

$$
[Z_1, Z_2] = -8\,\partial_{\xi_3}.
$$

We can now directly check the validity of Theorem 2.1.50-(2), i.e. that $\mathrm{d}\varphi$ is an algebra-isomorphism from $\mathfrak{g}$ to $\mathfrak{h}^1$, the algebra of $\mathbb{H}^1$. To this end, we remark that $\mathrm{d}\varphi$ is the map that to a vector field $Z \in \mathfrak{g}$ assigns the vector field in $\mathfrak{h}^1$ whose vector of the coefficients (at $x \in \mathbb{H}^1$) is given by

$$
(\mathrm{d}\varphi(Z))I(x) = \mathcal{J}_\varphi(\varphi^{-1}(x)) \cdot (ZI)(\varphi^{-1}(x)).
$$

Now, we have

$$
\begin{aligned}
\mathcal{J}_\varphi(\xi) &= \begin{pmatrix} \xi^{-1} & 0 & 0 \\ 1 & 1+\tan^2(\xi_2) & 0 \\ 0 & 0 & 1 \end{pmatrix},\\
\mathcal{J}_\varphi(\varphi^{-1}(x)) &= \begin{pmatrix} e^{-x_1} & 0 & 0 \\ 1 & 1+(x_2 - e^{x_1})^2 & 0 \\ 0 & 0 & 1 \end{pmatrix}.
\end{aligned}
$$

Consequently, it holds

$$
(\mathrm{d}\varphi(Z_1))I(x) = \begin{pmatrix} e^{-x_1} & 0 & 0 \\ 1 & 1+(x_2 - e^{x_1})^2 & 0 \\ 0 & 0 & 1 \end{pmatrix} \begin{pmatrix} e^{x_1} \\ \frac{1-e^{x_1}}{1+(x_2-e^{x_1})^2} \\ 2x_2 - 2x_1 \end{pmatrix} = \begin{pmatrix} 1 \\ 1 \\ 2x_2 - 2x_1 \end{pmatrix}.
$$

This means that

$$
\mathrm{d}\varphi(Z_1) = \partial_{x_1} + \partial_{x_2} + (2x_2 - 2x_1)\,\partial_{x_3} = X_1 + X_2,
$$

where, as usual, $X_1 = \partial_{x_1} + 2x_2\partial_{x_3}$, $X_2 = \partial_{x_2} - 2x_1\partial_{x_3}$ are the generators of the algebra of $\mathbb{H}^1$. Analogously, we have

$$(\mathrm{d}\varphi(Z_2))I(x) = \begin{pmatrix} e^{-x_1} & 0 & 0 \\ 1 & 1+(x_2-e^{x_1})^2 & 0 \\ 0 & 0 & 1 \end{pmatrix} \begin{pmatrix} 0 \\ \frac{2}{1+(x_2-e^{x_1})^2} \\ -4x_1 \end{pmatrix} = \begin{pmatrix} 0 \\ 2 \\ -4x_1 \end{pmatrix},$$

i.e.

$$\mathrm{d}\varphi(Z_2) = 2\partial_{x_2} - 4x_1\,\partial_{x_3} = 2X_2.$$

Moreover,

$$(\mathrm{d}\varphi([Z_1, Z_2]))I(x) = \begin{pmatrix} e^{-x_1} & 0 & 0 \\ 1 & 1+(x_2-e^{x_1})^2 & 0 \\ 0 & 0 & 1 \end{pmatrix} \begin{pmatrix} 0 \\ 0 \\ -8 \end{pmatrix} = \begin{pmatrix} 0 \\ 0 \\ -8 \end{pmatrix},$$

i.e.

$$\mathrm{d}\varphi([Z_1, Z_2]) = -8\,\partial_{x_3} = 2[X_1, X_2].$$

Finally, we can straightforwardly check the algebra-isomorphism condition for $\mathrm{d}\varphi$ (see Theorem 2.1.50-(2)), namely

$$\mathrm{d}\varphi([Z_1, Z_2]) = 2[X_1, X_2] = [X_1 + X_2, 2X_2] = [\mathrm{d}\varphi(Z_1), \mathrm{d}\varphi(Z_2)].$$

This completes our example. □

2.2.1 Some Properties of the Stratification of a Carnot Group

If $\mathbb{H}$ is a Carnot group, its Lie algebra admits at least a stratification, but it can have more than one. For example, if $\mathbb{H} = \mathbb{H}^1$ is the Heisenberg–Weyl group on $\mathbb{R}^3$ (see Example 1.2.2 and its notation), its Lie algebra admits the stratifications

$$\begin{aligned} &\mathrm{span}\{X_1, X_2\} \oplus \mathrm{span}\{[X_1, X_2]\}, \\ &\mathrm{span}\{X_1 - 3\,[X_1, X_2], X_2\} \oplus \mathrm{span}\{[X_1, X_2]\}, \\ &\mathrm{span}\{X_1 + X_2, 3X_1 + [X_1, X_2]\} \oplus \mathrm{span}\{[X_1, X_2]\}. \end{aligned}$$

Definition 2.2.6 (Basis adapted to the stratification). *Let $\mathbb{H}$ be a Carnot group. Let $\mathcal{V} = (V_1, \dots, V_r)$ be a fixed stratification of the Lie algebra $\mathfrak{h}$ of $\mathbb{H}$ as in* (2.41). *We say that a basis $\mathcal{B}$ of $\mathfrak{h}$ is* adapted *to $\mathcal{V}$ if*

$$\mathcal{B} = \left(E_1^{(1)}, \dots, E_{N_1}^{(1)};\ \dots\ ; E_1^{(r)}, \dots, E_{N_r}^{(r)}\right), \tag{2.42}$$

where, for $i = 1, \dots, r$, we have $N_i := \dim V_i$, and

$$\left(E_1^{(i)}, \dots, E_{N_i}^{(i)}\right) \text{ is a basis for } V_i.$$

Obviously, every Carnot group admits an adapted basis to any of its stratifications.

Definition 2.2.7 (Step and number of generators). *Let $\mathbb{H}$ be a stratified group. Let $(V_1, \dots, V_r)$ be any stratification of the algebra of $\mathbb{H}$, as in Definition 2.2.3. We say that $\mathbb{H}$ as* step (of nilpotency) r *and has* m generators, *where* $m := \dim(V_1)$.

The following proposition shows that the above definitions are well-posed, i.e. they do not depend on the particular stratification of $\mathbb{H}$.

Proposition 2.2.8. *Let $\mathbb{H}$ be a stratified group. Suppose that $(V_1, \dots, V_r)$ and $(\widetilde{V}_1, \dots, \widetilde{V}_{\tilde{r}})$ be any two stratifications of the algebra of $\mathbb{H}$, as in Definition 2.2.3. Then $r = \tilde{r}$ and $\dim(V_i) = \dim(\widetilde{V}_i)$ for every $i = 1, \dots, r$. Moreover, the algebra of $\mathbb{H}$ is a nilpotent Lie algebra of step r.*

Hence, the natural number

$$Q := \sum_{i=1}^{r} i \dim(V_i)$$

depends only on the stratified nature of $\mathbb{H}$ and not on the particular stratification. Q is called the homogeneous dimension *of $\mathbb{H}$.*

Proof. From the very Definition 2.2.3, we see that if $(V_1, \dots, V_r)$ is a stratification of $\mathfrak{h}$, the algebra of $\mathbb{H}$, then $\mathfrak{h}$ is a nilpotent[9] Lie algebra of step r. Hence r depends only on $\mathfrak{h}$ and not on the stratification.

The fact that $\dim(V_i) = \dim(\widetilde{V}_i)$ for every $i = 1, \dots, r$ is not trivial and follows from Lemma 2.2.9 below. □

Lemma 2.2.9 (The two-stratification lemma). *Let $\mathbb{H}$ be a stratified group with Lie algebra $\mathfrak{h}$. Suppose $V := (V_1, \dots, V_r)$ and $W := (W_1, \dots, W_r)$ are two stratifications of $\mathfrak{h}$.*

Then, for every couple of bases $\mathcal{V}$ and $\mathcal{W}$ of $\mathfrak{h}$ respectively adapted to the stratifications V and W, the transition matrix between the two bases is non-singular and has the block-triangular form

$$\begin{pmatrix} M^{(1)} & 0 & \cdots & 0 \\ \star & M^{(2)} & \ddots & \vdots \\ \vdots & \ddots & \ddots & 0 \\ \star & \cdots & \star & M^{(r)} \end{pmatrix},$$

[9] We recall that, given an abstract Lie algebra $(\mathfrak{g}, [\cdot, \cdot])$, $\mathfrak{g}$ is called nilpotent of step r, $r \in \mathbb{N}$, if for every $X_1, \dots, X_{r+1} \in \mathfrak{g}$,

$$[X_1, [X_2, \dots [X_r, X_{r+1}] \dots]] = 0,$$

and there exist $Y_1, \dots, Y_r \in \mathfrak{g}$ such that

$$[Y_1, [Y_2, \dots [Y_{r-1}, Y_r] \dots]] \neq 0.$$

where, for every $i = 1, \dots, r$, *the block* $M^{(i)}$ *is a* $N_i \times N_i$ *non-singular matrix* (N_i *being the common value of* $\dim(V_i) = \dim(W_i)$).

Proof. Let the notation in the assertion be fixed. We also set, for every $i = 1, \dots, r$, $N_i = \dim(V_i)$ and $M_i = \dim(W_i)$. From $V_1 \subseteq W_1 \oplus \cdots \oplus W_r$ and from the stratification condition for V and for W, we infer that

$$\begin{aligned} V_i &= \underbrace{[V_1, \cdots [V_1, V_1] \cdots]}_{i \text{ times}} \\ &\subseteq \Big[\{W_1 \oplus \cdots\}, \cdots \big[\{W_1 \oplus \cdots\}, \{W_1 \oplus \cdots\}\big] \cdots\Big] \\ &\subseteq W_i \oplus W_{i+1} \oplus \cdots \oplus W_r. \end{aligned}$$

Hence (the second column obtained by reversing the rôles of V and W),

$$\begin{cases} V_1 \subseteq W_1 \oplus W_2 \oplus \cdots \oplus W_r, & W_1 \subseteq V_1 \oplus V_2 \oplus \cdots \oplus V_r, \\ \vdots & \vdots \\ V_{r-1} \subseteq W_{r-1} \oplus W_r, & W_{r-1} \subseteq V_{r-1} \oplus V_r, \\ V_r \subseteq W_r, & W_r \subseteq V_r. \end{cases}$$

All this proves that

$$(\star) \qquad \begin{cases} W_1 \oplus W_2 \oplus \cdots \oplus W_r = V_1 \oplus V_2 \oplus \cdots \oplus V_r, \\ \vdots \\ W_{r-1} \oplus W_r = V_{r-1} \oplus V_r, \\ W_r = V_r. \end{cases}$$

In particular (recalling that $\dim(A \oplus B) = \dim(A) + \dim(B)$), this yields

$$\begin{cases} M_1 + M_2 + \cdots + M_r = N_1 + N_2 + \cdots + N_r, \\ \vdots \\ M_{r-1} + M_r = N_{r-1} + N_r, \\ M_r = N_r, \end{cases}$$

whence, $N_i = M_i$, i.e. $\dim(V_i) = \dim(W_i)$ for every $i = 1, \dots, r$.

From $(\star)$ it immediately follows the block-form for the matrix of the change of basis between $\mathcal{V}$ and $\mathcal{W}$. □

The following proposition shows that "to be a Carnot group" is an invariant under isomorphism of Lie groups.

Proposition 2.2.10. *Let* $\mathbb{H}$ *be a stratified group. Suppose* $\mathbb{G}$ *is a Lie group isomorphic to* $\mathbb{H}$. *Then* $\mathbb{G}$ *is a stratified group too. Moreover,* $\mathbb{H}$ *and* $\mathbb{G}$ *have the same step, the same number of generators and even the dimensions of the layers of the relevant* *stratifications are preserved. Also,* $\mathbb{H}$ *and* $\mathbb{G}$ *have the same homogeneous dimension* Q.

More precisely, suppose $\varphi : \mathbb{H} \to \mathbb{G}$ *is a Lie group isomorphism and that* $(V_1, \ldots, V_r)$ *is a stratification of* $\mathfrak{h}$*, the algebra of* $\mathbb{H}$*, as in Definition* 2.2.3*. Then, if* $\mathfrak{g}$ *is the algebra of* $\mathbb{G}$*, a stratification for* $\mathfrak{g}$ *is given by* $(\mathrm{d}\varphi(V_1), \ldots, \mathrm{d}\varphi(V_r))$*, where* $\mathrm{d}\varphi$ *is the differential of* φ *(see Definition* 2.1.49*) which is an isomorphism of Lie algebras (and of vector spaces).*

Proof. We follow the notation in the assertion. Set $W_i := \mathrm{d}\varphi(V_i)$ for every $i = 1, \ldots, r$. The proposition will be proved if we demonstrate that

$$\mathfrak{g} = W_1 \oplus W_2 \oplus \cdots \oplus W_r$$

and

$$\begin{cases} [W_1, W_{i-1}] = W_i & \text{if } 2 \le i \le r, \\ [W_1, W_r] = \{0\}. \end{cases}$$

Now, from the linearity and the invertibility of $\mathrm{d}\varphi$, it holds

$$\mathfrak{g} = \mathrm{d}\varphi(\mathfrak{h}) = \mathrm{d}\varphi(V_1 \oplus \cdots \oplus V_r) = \mathrm{d}\varphi(V_1) \oplus \cdots \oplus \mathrm{d}\varphi(V_r) = W_1 \oplus \cdots \oplus W_r.$$

Finally, $\mathrm{d}\varphi$ being a Lie algebra homomorphism (see Theorem 2.1.50-(ii)) it is easy to see that $[W_1, W_{i-1}]$ equals

$$[\mathrm{d}\varphi(V_1), \mathrm{d}\varphi(V_{i-1})] = \mathrm{d}\big([V_1, V_{i-1}]\big) = \begin{cases} \mathrm{d}\varphi(V_i) = W_i, \ 2 \le i \le r, \\ \mathrm{d}\varphi(\{0\}) = \{0\}. \end{cases}$$

This ends the proof. □

2.2.2 The Campbell–Hausdorff Formula and Some General Results on Nilpotent Lie Groups

Before proving that, up to isomorphism, Carnot groups and homogeneous Carnot groups provide equivalent notions, we recall some results about the so-called Campbell–Hausdorff formula.

Definition 2.2.11. *Let* $\mathfrak{h}$ *be a nilpotent Lie algebra. For* $X, Y \in \mathfrak{h}$*, we set*[10]

$$\begin{aligned} X \diamond Y := \sum_{n \ge 1} \frac{(-1)^{n+1}}{n} \sum_{\substack{p_i + q_i \ge 1 \\ 1 \le i \le n}} \frac{(\mathrm{ad}\, X)^{p_1} (\mathrm{ad}\, Y)^{q_1} \cdots (\mathrm{ad}\, X)^{p_n} (\mathrm{ad}\, Y)^{q_n - 1} Y}{\left(\sum_{j=1}^n (p_j + q_j)\right) \, p_1! q_1! \cdots p_n! q_n!} \\ = X + Y + \frac{1}{2}[X, Y] + \end{aligned}$$

---

[10] We use the notation $(\mathrm{ad}\, A)B = [A, B]$. Moreover, if $q_n = 0$, the term in the sum (2.43) is by convention $\cdots (\mathrm{ad}\, X)^{p_{n-1}} (\mathrm{ad}\, Y)^{q_{n-1}} (\mathrm{ad}\, X)^{p_n - 1} X$. Clearly, if $q_n > 1$, or $q_n = 0$ and $p_n > 1$, the term is zero.

$$
\begin{aligned}
&+\frac{1}{12}[X,[X,Y]]-\frac{1}{12}[Y,[X,Y]]\\
&-\frac{1}{48}[Y,[X,[X,Y]]]-\frac{1}{48}[X,[Y,[X,Y]]]\\
&+\{\textit{brackets of height} \geq 5\}. \qquad (2.43)
\end{aligned}
$$

The sum over n actually runs on $\{1,\dots,r\}$, *where* $r<\infty$ *is the step of nilpotency of* $\mathfrak{h}$. *The same is true for the sum over the* p_i*'s and* q_i*'s, for which it is left unsaid that* $\sum_{i=1}^n(p_i+q_i)\le r$. *We shall refer to the operation defined in* (2.43) *as the* Campbell–Hausdorff operation *on* $\mathfrak{h}$.

The inner sum in (2.43) can be reordered so that the brackets appear with increasing height (here $\mathbb{N}_0=\mathbb{N}\cup\{0\}$),

$$
X\diamond Y=\sum_{n=1}^{r}\frac{(-1)^{n+1}}{n}
\times\sum_{h=1}^{r}\sum_{\substack{(p_1,q_1),\dots,(p_n,q_n)\in\mathbb{N}_0\times\mathbb{N}_0\\(p_1,q_1),\dots,(p_n,q_n)\neq(0,0)\\(p_1+q_1)+\dots+(p_n+q_n)=h}}
\times\frac{(\mathrm{ad}X)^{p_1}(\mathrm{ad}\,Y)^{q_1}\cdots(\mathrm{ad}X)^{p_n}(\mathrm{ad}\,Y)^{q_n-1}Y}{(\sum_{j=1}^n(p_j+q_j))\;p_1!\,q_1!\cdots p_n!\,q_n!}.
$$

Remark 2.2.12. Since $\mathfrak{h}$ is nilpotent, (2.43) is a *finite* sum, and $\diamond$ defines a binary operation in $\mathfrak{h}$. A striking fact is that this operation is actually *associative*! Indeed, much more holds: $(\mathfrak{h},\diamond)$ is a Lie group. We explicitly remark that the inverse of $X\in\mathfrak{h}$ w.r.t. $\diamond$ is simply $-X$.

We shall prove this fact in Corollary 2.2.15 below, by making use of two abstract remarkable results whose proofs are out of our scope here (we refer the reader to the monographs [CG90] and [Var84] for more references).

First of all we recall the following result (see [CG90, p. 13]), which also gives the well known *Campbell–Hausdorff formula* (2.44).

Theorem 2.2.13 (Corwin and Greenleaf [CG90, Theorem 1.2.1]). *Let* $(\mathbb{H},*)$ *be a connected and simply connected Lie group. Suppose that the Lie algebra* $\mathfrak{h}$ *of* $\mathbb{H}$ *is nilpotent. Then* $\diamond$ *defines a Lie group structure on* $\mathfrak{h}$ ($\mathfrak{h}$ *being equipped with the manifold structure resulting from its finite-dimensional vector space structure) and* $\mathrm{Exp}:(\mathfrak{h},\diamond)\to(\mathbb{H},*)$ *is a group-isomorphism. In particular, we have*

$$
\mathrm{Exp}\,(X)*\mathrm{Exp}\,(Y)=\mathrm{Exp}\,(X\diamond Y)\qquad\forall\,X,\ Y\in\mathfrak{h}. \qquad (2.44)
$$

We now recall the third fundamental theorem of Lie (see [Var84, Theorem 3.15.1]). This is a very deep result in Lie group theory.

Theorem 2.2.14 (The third fundamental theorem of Lie). *Let $\mathfrak{h}$ be a finite-dimensional Lie algebra. Then there exists a connected and simply connected Lie group whose Lie algebra is isomorphic to $\mathfrak{h}$.*

Collecting the above two theorems, we obtain the following result.

Corollary 2.2.15. *Let $\mathfrak{h}$ be a finite-dimensional nilpotent Lie algebra. Then $\diamond$ defines a Lie group structure on $\mathfrak{h}$.*

Moreover, the Lie algebra associated to the Lie group $(\mathfrak{h}, \diamond)$ is isomorphic to the algebra $\mathfrak{h}$.

Proof. Since $\mathfrak{h}$ is a finite-dimensional Lie algebra, by Theorem 2.2.14 there exists a connected and simply connected Lie group (say, $\mathbb{G}$) whose Lie algebra $\mathfrak{g}$ is isomorphic to $\mathfrak{h}$. Let $\varphi : \mathfrak{g} \to \mathfrak{h}$ be a Lie algebra isomorphism.

Since $\mathfrak{h}$ is nilpotent by hypothesis, so is $\mathfrak{g}$ (since $\mathfrak{g}$ is isomorphic to $\mathfrak{h}$). Hence $\mathbb{G}$ satisfies the hypotheses of Theorem 2.2.13. As a consequence, $\mathfrak{g}$ equipped with the operation $\diamond$ defined in (2.43) is a Lie group.

We denote by $\diamond_{\mathfrak{g}}$ such a composition law $\diamond$ on the Lie algebra $\mathfrak{g}$. Analogously, we denote by $\diamond_{\mathfrak{h}}$ a similar operation on $\mathfrak{h}$. Now, the isomorphism φ between the two Lie algebras $\mathfrak{g}$ and $\mathfrak{h}$ transfers the operation $\diamond_{\mathfrak{g}}$ into $\diamond_{\mathfrak{h}}$. More precisely, for every $X, Y \in \mathfrak{g}$, we have

$$\begin{aligned}
&\varphi(X \diamond_{\mathfrak{g}} Y)\\
&\quad = \varphi\Big(X + Y + \frac{1}{2}[X, Y]_{\mathfrak{g}} + \frac{1}{12}[X, [X, Y]_{\mathfrak{g}}]_{\mathfrak{g}} + \cdots \Big)\\
&\quad \text{(since } \varphi \text{ is a Lie algebra homomorphism)}\\
&\quad = \varphi(X) + \varphi(Y) + \frac{1}{2}[\varphi(X), \varphi(Y)]_{\mathfrak{h}} + \frac{1}{12}[\varphi(X), [\varphi(X), \varphi(Y)]_{\mathfrak{h}}]_{\mathfrak{h}} + \cdots\\
&\quad = \varphi(X) \diamond_{\mathfrak{h}} \varphi(Y). \qquad (2.45)
\end{aligned}$$

We explicitly remark that in the last equality we used, as a crucial tool, the "universal" way in which $\diamond$ is defined on a Lie algebra. In particular, the associativity of $\diamond_{\mathfrak{g}}$ on $\mathfrak{g}$ directly implies the associativity of $\diamond_{\mathfrak{h}}$ on $\mathfrak{h}$. Thus, the first part of the assertion of the corollary is proved. The second part follows from Ex. 2 at the end of the chapter. □

Remark 2.2.16. If $\mathfrak{h}$ and $\mathfrak{g}$ are finite-dimensional nilpotent Lie algebras and $\varphi : \mathfrak{h} \to \mathfrak{g}$ is an algebra-homomorphism, then φ is also a group-homomorphism between $(\mathfrak{h}, \diamond)$ and $(\mathfrak{g}, \diamond)$. This directly follows from (2.45).

2.2.3 Abstract and Homogeneous Carnot Groups

We now aim to prove that, up to isomorphism, the definitions of classical and homogeneous Carnot group are equivalent. To begin with, we prove the following simple fact:

Proposition 2.2.17 ((Homogeneous ⇒ stratified) Carnot). *A homogeneous Carnot group is a stratified group.*

Proof. Let $\mathbb{G} = (\mathbb{R}^N, \circ, \delta_\lambda)$ be a homogeneous Carnot group. Clearly, $\mathbb{G}$ is connected and simply connected. Let $\mathfrak{g}$ be the algebra of $\mathbb{G}$.

For $i = 1, \dots, r$ and $j = 1, \dots, N_i$, let $Z_j^{(i)}$ be the vector field of $\mathfrak{g}$ agreeing with $\partial/\partial x_j^{(i)}$ at the origin. We set

$$V_i := \operatorname{span}\{Z_1^{(i)}, \dots, Z_{N_i}^{(i)}\}.$$

Remark 1.4.8 proves that $(V_1, \dots, V_r)$ is a stratification of $\mathfrak{g}$, as in Definition 2.2.3. This ends the proof. □

We are now in the position to prove the main result of this section. The proof is a detailed argument of what is commonly used (without comments) in literature, i.e. *the identification of the group with its algebra.*

Theorem 2.2.18 ((Stratified $\overset{\text{isom.}}{\Longrightarrow}$ homogeneous) Carnot). *Let $\mathbb{H}$ be a stratified group, according to Definition* 2.2.3. *Then there exists a homogeneous Carnot group $\mathbb{H}^*$ (according to our Definition* 2.2.1) *which is isomorphic to $\mathbb{H}$.*

We can choose as $\mathbb{H}^*$ the Lie algebra $\mathfrak{h}$ of $\mathbb{H}$ (identified to $\mathbb{R}^N$ by a suitable choice of an adapted basis of $\mathfrak{h}$) equipped with the composition law $\diamond$ defined by the Campbell–Hausdorff operation (2.43) *in Definition* 2.2.11. *In this case, a group-isomorphism from $\mathbb{H}^*$ to $\mathbb{H}$ is the exponential map*

$$\operatorname{Exp} : (\mathfrak{h}, \diamond) \to (\mathbb{H}, *).$$

Proof. Let $(\mathbb{H}, *)$ be as in Definition 2.2.3. Let $\mathfrak{h}$ be the algebra of $\mathbb{H}$. Let $\mathfrak{h} = V_1 \oplus \cdots \oplus V_r$ be a *fixed* stratification of $\mathfrak{h}$ as in (2.41). By Proposition 2.2.8, $\mathfrak{h}$ is nilpotent of step r.

Then Theorem 2.2.13 yields that

$$\operatorname{Exp} : (\mathfrak{h}, \diamond) \to (\mathbb{H}, *) \text{ is a Lie-group isomorphism,} \tag{2.46}$$

where $\diamond$ is as in (2.43). We now prove that (a coordinate version of) $(\mathfrak{h}, \diamond)$ is a homogeneous Carnot group according to Definition 2.2.1.

We fix a basis for $\mathfrak{h}$ adapted to its stratification (see Definition 2.2.6): for $i = 1, \dots, r$, set $N_i := \dim V_i$, and let $(E_1^{(i)}, \dots, E_{N_i}^{(i)})$ be a basis for V_i. Then consider the basis for $\mathfrak{h}$ given by

$$\mathcal{E} = \left(E_1^{(1)}, \dots, E_{N_1}^{(1)}; \dots; E_1^{(r)}, \dots, E_{N_r}^{(r)}\right).$$

By means of this basis, we fix a coordinate system on $\mathfrak{h}$, and we identify $\mathfrak{h}$ with $\mathbb{R}^N$, where $N := N_1 + \cdots + N_r$. More precisely, we consider the map[11]

[11] See also the map introduced in (1.35) of Remark 1.2.20, page 21. Incidentally, we notice that $(\mathfrak{h}, \pi_{\mathcal{E}})$ is a coordinate system for the whole $\mathfrak{h}$, determining its differentiable structure. Obviously, along with $\pi_{\mathcal{E}}$, we have other coordinate maps (for example those arising by a different choice of a linear basis of $\mathfrak{h}$, adapted or not) and the differentiable structure of $\mathfrak{h}$ does not depend on $\mathcal{E}$.

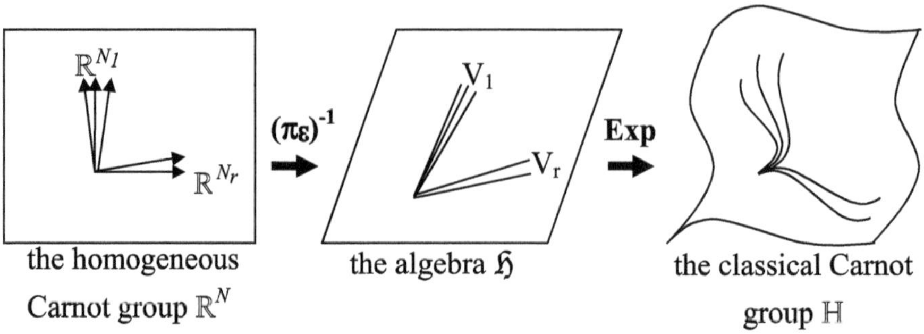

Fig. 2.1. Turning a stratified Carnot group into a homogeneous one

$$\pi_{\mathcal{E}} : \mathfrak{h} \to \mathbb{R}^N, \qquad E \cdot \xi := \sum_{i=1}^{r} \sum_{j=1}^{N_i} \xi_j^{(i)} E_j^{(i)} \mapsto (\xi^{(1)}, \ldots, \xi^{(r)}),$$

where $\xi^{(i)} = (\xi_1^{(i)}, \ldots, \xi_{N_i}^{(i)}) \in \mathbb{R}^{N_i}$ for every $i = 1, \ldots, r$. Next, we set

$$\Psi := \operatorname{Exp} \circ (\pi_{\mathcal{E}})^{-1} : \mathbb{R}^N \to \mathbb{H}, \quad \Psi(\xi) = (\operatorname{Exp}(E \cdot \xi)).$$

(See also Fig. 2.1.) Notice that, more explicitly,

$$\Psi(\xi) = \operatorname{Exp}\left(\sum_{i=1}^{r} \sum_{j=1}^{N_i} \xi_j^{(i)} E_j^{(i)} \right) \qquad \forall\, \xi \in \mathbb{R}^N. \tag{2.47}$$

Finally, we equip $\mathbb{R}^N$ with the composition law $\diamond_{\mathcal{E}}$ defined by

$$\xi \diamond_{\mathcal{E}} \eta := \Psi^{-1}\big(\Psi(\xi) * \Psi(\eta)\big), \qquad \xi, \eta \in \mathbb{R}^N, \tag{2.48}$$

We define a family of dilations $\{\Delta_\lambda\}_{\lambda>0}$ on the Lie algebra $\mathfrak{h}$ as follows:

$$\Delta_\lambda : \mathfrak{h} \to \mathfrak{h}, \qquad \Delta_\lambda\left(\sum_{i=1}^{r} X_i \right) := \sum_{i=1}^{r} \lambda^i X_i, \quad \text{where } X_i \in V_i. \tag{2.49a}$$

Obviously,

$$\Delta_\lambda \text{ is a vector-space automorphism of } \mathfrak{h}. \tag{2.49b}$$

(Note that N, the N_i's and the form of Δ_λ do not depend neither on the choice of the stratification of $\mathfrak{h}$ nor on the adapted basis; see, for instance, Lemma 2.2.9.) Obviously, Δ_λ turns into a family of dilations $\{\delta_\lambda\}_{\lambda>0}$ on $\mathbb{R}^N$ via Ψ by setting

$$\delta_\lambda := \pi_{\mathcal{E}} \circ \Delta_\lambda \circ \pi_{\mathcal{E}}^{-1}. \tag{2.49c}$$

We claim that
$\mathbb{H}^* := (\mathbb{R}^N, \diamond_{\mathcal{E}}, \delta_\lambda)$ *is a homogeneous Carnot group*
(of step r and N_1 generators)
isomorphic to $(\mathbb{H}, *)$ *via the Lie group isomorphism* Ψ.

To prove the claim, we split the proof in steps.

(I). By the very definition of $\diamond_{\mathcal{E}}$ and Ψ, we have

$$\Psi(\xi \diamond_{\mathcal{E}} \eta) = \Psi(\xi) * \Psi(\eta) \qquad \forall\, \xi,\, \eta \in \mathbb{R}^N, \tag{2.50a}$$

which, in turn, is equivalent to (exploit (2.46) and recall that $\Psi = \operatorname{Exp} \circ \pi_{\mathcal{E}}^{-1}$)

$$\pi_{\mathcal{E}}^{-1}(\xi \diamond_{\mathcal{E}} \eta) = \pi_{\mathcal{E}}^{-1}(\xi) \diamond \pi_{\mathcal{E}}^{-1}(\eta) \qquad \forall\, \xi,\, \eta \in \mathbb{R}^N, \tag{2.50b}$$

or, equivalently,

$$\pi_{\mathcal{E}}(X \diamond Y) = \pi_{\mathcal{E}}(X) \diamond_{\mathcal{E}} \pi_{\mathcal{E}}(Y) \qquad \forall\, X,\, Y \in \mathfrak{h}. \tag{2.50c}$$

Now, we recognize that (2.50a) and (2.50b) mean that

$$(\mathbb{R}^N, \diamond_{\mathcal{E}}), \quad (\mathfrak{h}, \diamond), \quad (\mathbb{H}, *)$$

are isomorphic Lie groups via the Lie-group isomorphisms

$$(\mathbb{R}^N, \diamond_{\mathcal{E}}) \overset{\pi_{\mathcal{E}}^{-1}}{\longrightarrow} (\mathfrak{h}, \diamond) \overset{\operatorname{Exp}}{\longrightarrow} (\mathbb{H}, *).$$

In particular,

$$\Psi = \operatorname{Exp} \circ \pi_{\mathcal{E}}^{-1} : (\mathbb{R}^N, \diamond_{\mathcal{E}}) \to (\mathbb{H}, *) \text{ is a Lie-group isomorphism.} \tag{2.51}$$

(II). We now investigate the dilation δ_λ. The stratified notation

$$\mathfrak{h} \ni E \cdot \xi = \sum_{i=1}^{r} \sum_{j=1}^{N_i} \xi_j^{(i)}\, E_j^{(i)}$$

for an arbitrary vector of $\mathfrak{h}$ and the fact that

$$\pi_{\mathcal{E}}(E \cdot \xi) = \xi \tag{2.52}$$

suggests the notation

$$\mathbb{R}^N \ni \xi = (\xi^{(1)}, \ldots, \xi^{(r)})$$

for the points in $\mathbb{R}^N$. We claim that, with the above notation, δ_λ introduced in (2.49c) has the form in (2.40), i.e.

$$\delta_\lambda(\xi^{(1)}, \xi^{(2)}, \ldots, \xi^{(r)}) = (\lambda \xi^{(1)}, \lambda^2 \xi^{(2)}, \ldots, \lambda^r \xi^{(r)}). \tag{2.53}$$

Indeed,

$$\begin{aligned}
\delta_\lambda(\xi) &= \text{(see (2.49c))} \quad \big(\pi_{\mathcal{E}} \circ \Delta_\lambda \circ \pi_{\mathcal{E}}^{-1}\big)(\xi) \\
\text{(see (2.52))} \quad &= \pi_{\mathcal{E}}\big(\Delta_\lambda(E \cdot \xi)\big) \\
&= \pi_{\mathcal{E}}\left(\Delta_\lambda\left(\sum_{i=1}^{r} \sum_{j=1}^{N_i} \xi_j^{(i)}\, E_j^{(i)}\right)\right)
\end{aligned}$$

$$
\begin{aligned}
\text{(see (2.49b))} \quad &= \pi_{\mathcal{E}}\left(\left(\sum_{i=1}^{r}\sum_{j=1}^{N_i} \xi_j^{(i)}\, \Delta_\lambda(E_j^{(i)})\right)\right)\\
\text{(see (2.49a))} \quad &= \pi_{\mathcal{E}}\left(\left(\sum_{i=1}^{r}\sum_{j=1}^{N_i} \xi_j^{(i)}\, \lambda^i\, E_j^{(i)}\right)\right)\\
&= \pi_{\mathcal{E}}\Big(E\cdot\big(\lambda\xi^{(1)},\dots,\lambda^r\xi^{(r)}\big)\Big)\\
\text{(see (2.52))} \quad &= \big(\lambda\xi^{(1)},\dots,\lambda^r\xi^{(r)}\big).
\end{aligned}
$$

Next, we proceed by showing that Δ_λ is an automorphism of the *Lie-group* $(\mathfrak{h}, \diamond)$, i.e.

$$\Delta_\lambda(X\diamond Y) = \Delta_\lambda(X)\diamond\Delta_\lambda(Y) \quad \forall\, X, Y\in\mathfrak{h}, \quad \forall\,\lambda>0. \tag{2.54}$$

Recalling Remark 2.2.16, it is enough to prove that

$$\Delta_\lambda\big([X,Y]\big) = [\Delta_\lambda(X), \Delta_\lambda(Y)] \quad \text{for every } X,\ Y\in\mathfrak{h}.$$

If $X=\sum_{i=1}^r X_i$ and $Y=\sum_{i=1}^r Y_i$, where $X_i, Y_i\in V_i$, we have $[X_i, Y_j]\in V_{i+j}$ (by the stratification condition), whence

$$
\begin{aligned}
\Delta_\lambda\big([X,Y]\big) &= \sum_{i,j=1}^{r}\Delta_\lambda\big([X_i,Y_j]\big) = \sum_{i,j=1}^{r}\lambda^{i+j}[X_i,Y_j]\\
&= \sum_{i,j=1}^{r}[\lambda^i X_i, \lambda^j Y_j] = \sum_{i,j=1}^{r}[\Delta_\lambda(X_i), \Delta_\lambda(Y_j)] = [\Delta_\lambda(X), \Delta_\lambda(Y)].
\end{aligned}
$$

Now, a joint application of (2.50b), (2.50c) and (2.54) prove that δ_λ is a Lie-group automorphism of $(\mathbb{R}^N, \diamond_{\mathcal{E}})$, i.e.

$$\delta_\lambda(\xi\diamond_{\mathcal{E}}\eta) = \delta_\lambda(\xi)\diamond_{\mathcal{E}}\delta_\lambda(\eta) \quad \forall\,\xi,\eta\in\mathbb{R}^N, \quad \forall\,\lambda>0.$$

(III). Thus, $\mathbb{H}^* := (\mathbb{R}^N, \diamond_{\mathcal{E}}, \delta_\lambda)$ is a homogeneous Lie group on $\mathbb{R}^N$, as defined in Definition 1.3.1, page 31. Let now $\mathfrak{h}^*$ be the Lie algebra of $\mathbb{H}^*$. Dealing with a Lie group on $\mathbb{R}^N$ (and the fixed Cartesian coordinates ξ's on $\mathbb{R}^N$), the Jacobian basis related to the composition $\diamond_{\mathcal{E}}$ is well-posed. We denote by

$$\mathcal{Z} = \left(Z_1^{(1)},\dots,Z_{N_1}^{(1)};\dots;Z_1^{(r)},\dots,Z_{N_r}^{(r)}\right)$$

this Jacobian basis, i.e. $Z_k^{(i)}$ is the vector field in $\mathfrak{h}^*$ agreeing at the origin with $\partial/\partial\xi_k^{(i)}$. The proof is complete if we show that the Lie algebra generated by $Z_1,\dots,Z_{N_1}$ coincides with the whole $\mathfrak{h}^*$.

To this end, we first observe that, thanks to (2.51), $\mathrm{d}\Psi:\mathfrak{h}^*\to\mathfrak{h}$ is an algebra-isomorphism (see Theorem 2.1.50). Furthermore, from Remark (2.1.52)-(iii), we have

$$\mathrm{d}\Psi = (\mathrm{d}\,\mathrm{Exp}) \circ \big(\mathrm{d}(\pi_{\mathcal{E}}^{-1})\big).$$

Moreover, since $E_1^{(1)}, \dots, E_{N_1}^{(1)}$ is a system of Lie-generators for $\mathfrak{h}$ (by the very definition of stratification!), it is enough to prove that

$$\mathrm{d}\,\Psi(Z_k^{(i)}) = E_k^{(i)} \quad \text{for every } i = 1, \dots, r \text{ and every } k = 1, \dots, N_i. \tag{2.55}$$

In order to prove (2.55), we recall that a left-invariant vector field is determined by its value at the identity. Hence, (2.55) will follow if we show that

$$\big(\mathrm{d}\,\Psi(Z_k^{(i)})\big)_e = (E_k^{(i)})_e.$$

For every $f \in C^\infty(\mathbb{H}, \mathbb{R})$, we have

$$\begin{aligned}
\big(\mathrm{d}\,\Psi(Z_k^{(i)})\big)_e(f) &= \big(\mathrm{d}_0\Psi(Z_k^{(i)})_0\big)(f) = (Z_k^{(i)})_0(f \circ \Psi) \\
&= (\partial/\partial\,\xi_k^{(i)})|_{\xi=0} f\left(\mathrm{Exp}\left(\sum_{i=1}^{r}\sum_{j=1}^{N_i} \xi_j^{(i)}\, E_j^{(i)}\right)\right) \\
&= \frac{\mathrm{d}}{\mathrm{d}t}\bigg|_{t=0} f\big(\mathrm{Exp}\,(t\, E_k^{(i)})\big) \\
&= \frac{\mathrm{d}}{\mathrm{d}t}\bigg|_{t=0} f\big(\exp_{E_k^{(i)}}(t)\big) \\
&= (E_k^{(i)})_e(f).
\end{aligned}$$

In the first equality, we used the very definition of the differential of a homomorphism (see Definition 2.1.49); in the second one, we used the definition of the differential at a point; in the third equality, we exploited the very definition of $Z_k^{(i)}$ and that of Ψ (see (2.47)); the fourth equality is a triviality from calculus; the fifth and the sixth equalities are the core of the computation: they follow, respectively, from Proposition 2.1.58-(1) and from (2.36b). The theorem is thus completely proved. □

Remark 2.2.19. From Theorem 2.2.18 and Remark 2.2.12, it follows that any stratified group is isomorphic to a homogeneous Carnot group in which the group inversion law is simply given by $x^{-1} = -x$. This is commonly assumed in great part of the literature involving Carnot groups, without further comments. See Proposition 2.2.22 for more details.

Remark 2.2.20 (Change of adapted basis). We give some more information on Theorem 2.2.18. Let $\mathbb{H}$ be a fixed stratified group with Lie algebra $\mathfrak{h}$. Let $\mathfrak{h} = V_1 \oplus \cdots \oplus V_r$ be a given stratification of $\mathfrak{h}$, as in (2.41). We know that r and every $N_i := \dim(V_i)$ (for $i = 1, \dots, r$) are invariants of the stratified group. So is $N := N_1 + \cdots + N_r$ ($= \dim(\mathfrak{h})$).

We fix *any arbitrary* basis $\mathcal{E}$ for $\mathfrak{h}$ adapted to its stratification. We use the notation in the proof of Theorem 2.2.18. We therein demonstrated that $\mathbb{R}^N$ is a *homogeneous Carnot group* isomorphic to $(\mathbb{H}, *)$ if $\mathbb{R}^N$ is equipped with the composition law $\diamond_{\mathcal{E}}$ defined by (2.48) and the dilation (2.53).

The map Ψ is a Lie-group isomorphism from $(\mathbb{R}^N, \diamond_{\mathcal{E}})$ to $(\mathbb{H}, *)$. Another way to write (2.48) is obviously

$$\operatorname{Exp}\big(E\cdot(\xi\diamond_{\mathcal{E}}\eta)\big) = (\operatorname{Exp}(E\cdot\xi)) * (\operatorname{Exp}(E\cdot\eta)). \tag{2.56a}$$

Suppose now that we choose *another* basis $\widetilde{\mathcal{E}}$ (adapted to the *same* stratification), with analogous notation as for $\mathcal{E}$. $\mathbb{R}^N$ can be equipped with an another composition $\diamond_{\widetilde{\mathcal{E}}}$ characterized by

$$\operatorname{Exp}\big(\widetilde{E}\cdot(\widetilde{\xi}\diamond_{\widetilde{\mathcal{E}}}\eta)\big) = (\operatorname{Exp}(\widetilde{E}\cdot\widetilde{\xi})) * (\operatorname{Exp}(\widetilde{E}\cdot\widetilde{\eta})). \tag{2.56b}$$

For every $i\in\{1,\dots,r\}$ and every $j\in\{1,\dots,N_i\}$, there exist scalars $c_{h,j}^{(i)}$'s such that

$$E_j^{(i)} = c_{1,j}^{(i)}\,\widetilde{E}_1^{(i)} + \dots + c_{N_i,j}^{(i)}\,\widetilde{E}_{N_i}^{(i)}.$$

Let us introduce the notation

$$C^{(i)} := \big(c_{h,j}^{(i)}\big)_{1\le h\le N_i,\,1\le j\le N_i}$$

and C to denote the $N\times N$ matrix whose diagonal blocks are $C^{(1)},\dots,C^{(r)}$,

$$C = \begin{pmatrix} C^{(1)} & \cdots & 0\\ \vdots & \ddots & \vdots\\ 0 & \cdots & C^{(r)} \end{pmatrix},$$

and, finally, let us use once again the notation C to denote the linear map

$$C:\mathbb{R}^N\to\mathbb{R}^N,\qquad x\mapsto Cx.$$

We note that if a vector field in $\mathfrak{h}$ has coordinates $\xi\in\mathbb{R}^N$ with respect to the basis $\mathcal{E}$, then[12] it has coordinates $C\xi$ with respect to the basis $\widetilde{\mathcal{E}}$, i.e.

$$E\cdot\xi = \widetilde{E}\cdot(C\xi)\qquad\text{for every } \xi\in\mathbb{R}^N. \tag{2.57}$$

[12] Indeed, it holds

$$\begin{aligned}
E\cdot\xi &= \sum_{i=1}^r\sum_{j=1}^{N_i}\xi_j^{(i)}\,E_j^{(i)}\\
&= \sum_{i=1}^r\sum_{j=1}^{N_i}\xi_j^{(i)}\sum_{h=1}^{N_i}c_{h,j}^{(i)}\,\widetilde{E}_h^{(i)} = \sum_{i=1}^r\sum_{h=1}^{N_i}\left(\sum_{j=1}^{N_i}c_{h,j}^{(i)}\,\xi_j^{(i)}\right)\widetilde{E}_h^{(i)}\\
&= \sum_{i=1}^r\sum_{h=1}^{N_i}\left(C^{(i)}\cdot\xi^{(i)}\right)_h\widetilde{E}_h^{(i)} = \sum_{i=1}^r\sum_{h=1}^{N_i}(C\cdot\xi)_h^{(i)}\,\widetilde{E}_h^{(i)}\\
&= \widetilde{E}\cdot(C\xi).
\end{aligned}$$

We now claim that the same happens for the relevant group structures

$$\mathbb{G} := (\mathbb{R}^N, \diamond_{\mathcal{E}}), \qquad \widetilde{\mathbb{G}} := (\mathbb{R}^N, \diamond_{\widetilde{\mathcal{E}}}),$$

i.e. *the linear map* $C : \mathbb{R}^N \to \mathbb{R}^N$ *is a group isomorphism from* $\mathbb{G}$ *to* $\widetilde{\mathbb{G}}$. Roughly speaking, the same change of basis in $\mathfrak{h}$ from the basis $\mathcal{E}$ to the basis $\widetilde{\mathcal{E}}$ acts as a Lie-group isomorphism from $(\mathbb{R}^N, \diamond_{\mathcal{E}})$ to $(\mathbb{R}^N, \diamond_{\widetilde{\mathcal{E}}})$.

Indeed, from (2.56a), (2.56b) and (2.57) we have

$$\begin{aligned}
\mathrm{Exp}\big(E \cdot (\xi \diamond_{\mathcal{E}} \eta)\big) &= (\mathrm{Exp}\,(E \cdot \xi)) * (\mathrm{Exp}\,(E \cdot \eta)) \\
&= \mathrm{Exp}\big(\widetilde{E} \cdot (C\xi)\big) * \mathrm{Exp}\big(\widetilde{E} \cdot (C\eta)\big) \\
&= \mathrm{Exp}\big(\widetilde{E} \cdot ((C\xi) \diamond_{\widetilde{\mathcal{E}}} (C\eta))\big) \\
&= \mathrm{Exp}\big(E \cdot \big(C^{-1}\{(C\xi) \diamond_{\widetilde{\mathcal{E}}} (C\eta)\}\big)\big),
\end{aligned}$$

which implies

$$\xi \diamond_{\mathcal{E}} \eta = C^{-1}\big\{(C\xi) \diamond_{\widetilde{\mathcal{E}}} (C\eta)\big\}, \tag{2.58a}$$

i.e.

$$C(\xi \diamond_{\mathcal{E}} \eta) = (C\xi) \diamond_{\widetilde{\mathcal{E}}} (C\eta) \quad \forall\, \xi, \eta \in \mathbb{R}^N. \tag{2.58b}$$

Now, (2.58b) is equivalent to say that

$$C : (\mathbb{R}^N, \diamond_{\mathcal{E}}) \to (\mathbb{R}^N, \diamond_{\widetilde{\mathcal{E}}}), \quad x \mapsto C\xi$$

is a Lie-group isomorphism. □

Example 2.2.21 (From "stratified" to "homogeneous"). We consider once again the Lie group introduced in Examples 1.2.18 and 1.2.31. To be consistent with the notation in Theorem 2.2.18, we denote this group by $\mathbb{H}$. We explicitly remark that $\mathbb{H}$ is a Carnot group which is not homogeneous (w.r.t. the coordinates initially assigned to it).

We find an explicit isomorphism of Lie groups turning $\mathbb{H}$ into a *homogeneous* Carnot group: the existence of such an isomorphism is indeed ensured by Theorem 2.2.18, whereas an explicit way to construct it comes from the very proof of Theorem 2.2.18 (see also Remark 2.2.20).

We have $\mathbb{H} = (\mathbb{R}^3, *)$, where

$$x * y = \Big(\mathrm{arcsinh}(\sinh(x_1) + \sinh(y_1)), x_2 + y_2 + \sinh(x_1)y_3, x_3 + y_3\Big).$$

A stratification of the algebra $\mathfrak{h}$ of $\mathbb{H}$ is $V_1 \oplus V_2$, where $V_1 = \mathrm{span}\{E_1^{(1)}, E_2^{(1)}\}$, $V_2 = \mathrm{span}\{E_1^{(2)}\}$, being

$$E_1^{(1)} = \frac{1}{\cosh(x_1)}\,\partial_{x_1}, \quad E_2^{(1)} = \partial_{x_3} + \sinh(x_1)\,\partial_{x_2}; \qquad E_1^{(2)} = \partial_{x_2}.$$

The relevant dilation on $\mathfrak{h}$ is given by

$$\delta_\lambda : \mathfrak{h} \longrightarrow \mathfrak{h},$$
$$\delta_\lambda(\xi_1\, E_1^{(1)} + \xi_2\, E_2^{(1)} + \xi_3\, E_1^{(2)}) := \xi_1\, \lambda\, E_1^{(1)} + \xi_2\, \lambda\, E_2^{(1)} + \xi_3\, \lambda^2\, E_1^{(2)}.$$

Since $\mathfrak{h}$ is a nilpotent algebra of step two, the Campbell–Hausdorff operation $\diamond$ on $\mathfrak{h}$ is

$$X \diamond Y = X + Y + \frac{1}{2}\,[X, Y].$$

Now, since $[E_1^{(1)}, E_2^{(1)}] = E_1^{(2)}$, a direct computation gives

$$(\xi_1\, E_1^{(1)} + \xi_2\, E_2^{(1)} + \xi_3\, E_1^{(2)}) \diamond (\eta_1\, E_1^{(1)} + \eta_2\, E_2^{(1)} + \eta_3\, E_1^{(2)})$$
$$= (\xi_1 + \eta_1)\, E_1^{(1)} + (\xi_2 + \eta_2)\, E_2^{(1)} + \left(\xi_3 + \eta_3 + \frac{1}{2}\,(\xi_1\, \eta_2 - \xi_2\, \eta_1)\right) E_1^{(2)}.$$

Via the choice of the basis $\mathcal{E} = \{E_1^{(1)}, E_2^{(1)}, E_1^{(2)}\}$, the Lie group $(\mathfrak{h}, \diamond)$ can be identified to $\mathbb{H}^* = (\mathbb{R}^3, \diamond_\mathcal{E})$, where

$$\xi \diamond_\mathcal{E} \eta = \left(\xi_1 + \eta_1, \xi_2 + \eta_2, \xi_3 + \eta_3 + \frac{1}{2}\,(\xi_1\, \eta_2 - \xi_2\, \eta_1)\right).$$

The Lie group isomorphism between $(\mathbb{H}^*, \diamond_\mathcal{E})$ and $(\mathbb{H}, *)$ is the "exponential-type map"

$$\mathbb{H}^* \ni (\xi_1, \xi_2, \xi_3) \mapsto \operatorname{Exp}(\xi_1\, E_1^{(1)} + \xi_2\, E_2^{(1)} + \xi_3\, E_1^{(2)}) \in \mathbb{H}.$$

Now, from the computation[13] in Example 1.2.31 (page 28) we derive that this map is given by

$$(\xi_1, \xi_2, \xi_3) \mapsto \Psi(\xi_1, \xi_2, \xi_3) = \left(\operatorname{arcsinh}(\xi_1), \xi_3 + \frac{1}{2}\,\xi_1\xi_2, \xi_2\right).$$

The map Ψ is the isomorphism of Lie groups turning the *homogeneous* Carnot group $\mathbb{H}^* = (\mathbb{R}^3, \diamond_\mathcal{E}, \delta_\lambda)$ (where $\delta_\lambda(\xi_1, \xi_2, \xi_3) = (\lambda\xi_1, \lambda\xi_2, \lambda^2\xi_3)$) into the *non-homogeneous* Carnot group $\mathbb{H}$. We recognize that

$$x * y = \Psi(\Psi^{-1}(x) \diamond_\mathcal{E} \Psi^{-1}(y)) \quad \forall\, x,\ y \in \mathbb{H}. \qquad \square$$

2.2.4 More Properties of the Lie Algebra of a Stratified Group

In the proof of Theorem 2.2.18, we described how to identify a stratified group with a homogeneous Carnot group on $\mathbb{R}^N$. More precisely, given a stratified group $\mathbb{H}$, the equivalence class of the Lie groups which are isomorphic to $\mathbb{H}$ contains at least one (in fact, infinite) homogeneous Carnot group $\mathbb{H}^*$ on $\mathbb{R}^N$, according to Definition 2.2.1. Namely, $\mathbb{H}^*$ is a "coordinate-copy" of $(\mathfrak{h}, \diamond)$, the Lie algebra of $\mathbb{H}$ equipped with the Campbell–Hausdorff operation.

[13] We warn the reader that a slight change of notation here is needed, if compared to the notation in Example 1.2.31, interchanging ξ_2 and ξ_3.

Since, in the specialized literature, one often meets with phrases such as "it is not restrictive to suppose that..." $\mathbb{H}$ has some distinguished properties, it is advisable to look more closely to the properties of $(\mathfrak{h}, \diamond)$. We furnish some of these properties in the following proposition which collect several already proved facts.

Proposition 2.2.22. *Let $\mathbb{H}$ be a stratified group with Lie algebra $\mathfrak{h}$ and exponential map* $\operatorname{Exp}_{\mathbb{H}} : \mathfrak{h} \to \mathbb{H}$. *Let also $\diamond$ be the Campbell–Hausdorff operation on $\mathfrak{h}$ defined in* (2.43).

Let $V_1 \oplus \cdots \oplus V_r$ be a stratification of $\mathfrak{h}$, as in (2.41). *Let $\mathcal{E}$ be any basis for $\mathfrak{h}$ adapted to the stratification, as in Definition* (2.2.6). *Set $N := \dim(\mathfrak{h})$, consider the map $\pi_{\mathcal{E}} : \mathfrak{h} \to \mathbb{R}^N$, where, for every $X \in \mathfrak{h}$, $\pi_{\mathcal{E}}(X)$ is the N-tuple of the coordinates of X w.r.t. $\mathcal{E}$.*

Then the binary operation on $\mathbb{R}^N$ defined by

$$x \diamond_{\mathcal{E}} y = \pi_{\mathcal{E}}\Big(\big(\pi_{\mathcal{E}}^{-1}(x)\big) \diamond \big(\pi_{\mathcal{E}}^{-1}(y)\big)\Big) \quad \forall\, x, y \in \mathbb{R}^N$$

has the following properties:

(1) *$\mathbb{G} := (\mathbb{R}^N, \diamond_{\mathcal{E}})$ is a Lie group on $\mathbb{R}^N$; $\mathbb{G}$ is isomorphic to $\mathbb{H}$ via the map $\Psi = \operatorname{Exp}_{\mathbb{H}} \circ \pi_{\mathcal{E}}^{-1}$ and to $(\mathfrak{h}, \diamond)$ via $\pi_{\mathcal{E}}$, whence $(\mathbb{G}, \diamond_{\mathcal{E}})$ and $(\mathfrak{h}, \diamond)$ are stratified groups.*

(2) *Let $\mathcal{Z} = \{Z_1, \ldots, Z_N\}$ be the Jacobian basis related to $\mathbb{G}$; then, denoting the adapted basis by $\mathcal{E} = \{E_1, \ldots, E_N\}$, we have*

$$\mathrm{d}\Psi(Z_i) = E_i \quad \text{for every } i = 1, \ldots, N,$$

or, equivalently,

$$Z_i(f \circ \Psi) \equiv E_i(f) \circ \Psi \text{ on } \mathbb{G}$$

for every $f \in C^\infty(\mathbb{H}, \mathbb{R})$. Moreover, if $\mathfrak{g}$ is the algebra of $\mathbb{G}$, the exponential map $\operatorname{Exp}_{\mathbb{G}} : \mathfrak{g} \to \mathbb{G}$ is a linear map and it sends Z_i in the i-th element of the standard basis of $\mathbb{G} \equiv \mathbb{R}^N$, whence

$$\operatorname{Exp}_{\mathbb{G}}\big((x_1, \ldots, x_N)_{\mathcal{Z}}\big) = (x_1, \ldots, x_N),$$

being $(x_1, \ldots, x_N)_{\mathcal{Z}} = x_1 Z_1 + \cdots + x_N Z_N$.

(3) *The inversion on $\mathbb{G}$ is the Euclidean inversion $-x$.*

(4) *For every $i \in \{1, \ldots, N\}$, we have*

$$(x \diamond_{\mathcal{E}} y)_i = x_i + y_i + \mathcal{R}_i(x, y),$$

where $\mathcal{R}_i(x, y)$ is a polynomial function depending on the x_k's and y_k's with $k < i$, and $\mathcal{R}_i(x, y)$ can be written as a sum of polynomials each containing a factor of the following type

$$x_h y_k - x_k y_h \qquad \text{with } h \neq k \text{ and } h, k < i.$$

(5) *Let Δ_λ be the linear map on $\mathfrak{h}$ such that, for every $i = 1, \ldots, r$,*

$$\Delta_\lambda(X) = \lambda^i X \quad \text{whenever } X \in V_i.$$

Let $\delta_\lambda := \pi_{\mathcal{E}} \circ \Delta_\lambda \circ \pi_{\mathcal{E}}^{-1}$. Then $(\mathbb{R}^N, \diamond_{\mathcal{E}}, \delta_\lambda)$ is a homogeneous Carnot group of the same step and number of generators as $\mathbb{H}$.

Proof. (1). The proof of (1) is contained in the proof of Theorem 2.2.18.

(2). In the proof of Theorem 2.2.18, we demonstrated that (see (2.55))

$$\begin{aligned} &\mathrm{d}\Psi(Z_k^{(i)}) = E_k^{(i)} \quad \forall\, i \in \{1,\ldots,r\},\ k \in \{1,\ldots,N_i\} \\ &\text{if}\quad \mathcal{E} = \left(E_1^{(1)},\ldots,E_{N_1}^{(1)};\ldots;E_1^{(r)},\ldots,E_{N_r}^{(r)}\right) \\ &\text{and}\quad \mathcal{Z} = \left(Z_1^{(1)},\ldots,Z_{N_1}^{(1)};\ldots;Z_1^{(r)},\ldots,Z_{N_r}^{(r)}\right) \end{aligned} \tag{2.59}$$

are, respectively, the usual notation for the adapted basis $\mathcal{E}$ and the notation for the Jacobian basis resulting from the coordinates induced by $\pi_{\mathcal{E}}$ on $\mathbb{R}^N$. This proves the first part of (2).

As to the first part, consider the following diagram ($\mathfrak{h}^*$ denotes the algebra of the Lie group $(\mathfrak{h},\diamond)$ and $\operatorname{Exp}_{\mathfrak{h}} : \mathfrak{h}^* \to \mathfrak{h}$ the related exponential map)

$$\begin{array}{ccccc} (\mathbb{G},\diamond_{\mathcal{E}}) & \xrightarrow{\ \pi_{\mathcal{E}}^{-1}\ } & (\mathfrak{h},\diamond) & \xrightarrow{\ \operatorname{Exp}_{\mathbb{H}}\ } & (\mathbb{H},*) \\ \uparrow{\scriptstyle \operatorname{Exp}_{\mathbb{G}}} & & \uparrow{\scriptstyle \mathrm{dExp}_{\mathfrak{h}}} & & \uparrow{\scriptstyle \operatorname{Exp}_{\mathbb{H}}} \\ \mathfrak{g} & \xrightarrow{\ \mathrm{d}\pi_{\mathcal{E}}^{-1}\ } & \mathfrak{h}^* & \xrightarrow{\ \mathrm{d}\operatorname{Exp}_{\mathbb{H}}\ } & \mathfrak{h}. \end{array}$$

From the commutativity result in Theorem 2.1.59 it holds

$$\begin{aligned} \operatorname{Exp}_{\mathbb{G}} &= \pi_{\mathcal{E}} \circ \operatorname{Exp}_{\mathfrak{h}} \circ \mathrm{d}\pi_{\mathcal{E}}^{-1} = \pi_{\mathcal{E}} \circ \left\{(\operatorname{Exp}_{\mathbb{H}})^{-1} \circ \operatorname{Exp}_{\mathbb{H}} \circ \mathrm{d}\operatorname{Exp}_{\mathbb{H}}\right\} \circ \mathrm{d}\pi_{\mathcal{E}}^{-1} \\ &= \pi_{\mathcal{E}} \circ \{\mathrm{d}\operatorname{Exp}_{\mathbb{H}}\} \circ \mathrm{d}\pi_{\mathcal{E}}^{-1}, \end{aligned}$$

whence

$$\operatorname{Exp}_{\mathbb{G}} = \pi_{\mathcal{E}} \circ \mathrm{d}\operatorname{Exp}_{\mathbb{H}} \circ \mathrm{d}\pi_{\mathcal{E}}^{-1}. \tag{2.60}$$

Hence, $\operatorname{Exp}_{\mathbb{G}}$ is a linear map since $\pi_{\mathcal{E}}$, $\mathrm{d}\operatorname{Exp}_{\mathbb{H}}$ and $\mathrm{d}\pi_{\mathcal{E}}^{-1}$ are linear.

We now prove that

$$\operatorname{Exp}_{\mathbb{G}}(Z_k^{(i)}) = \left(0^{(1)},\ldots,e_k^{(i)},\ldots,0^{(r)}\right) \quad \forall\, i \le r,\ k \le N_i, \tag{2.61}$$

where $e_k^{(i)}$ denotes the k-th element of the standard basis of $\mathbb{R}^{N_i}$ (for $k \in \{1,\ldots,N_i\}$). To this end, first notice that

$$\pi_{\mathcal{E}}(E_k^{(i)}) = \left(0^{(1)},\ldots,e_k^{(i)},\ldots,0^{(r)}\right) \quad \forall\, i \le r,\ k \le N_i. \tag{2.62}$$

Then, by applying (2.60), we infer

$$\begin{aligned} \operatorname{Exp}_{\mathbb{G}}(Z_k^{(i)}) &= \left(\pi_{\mathcal{E}} \circ \mathrm{d}\operatorname{Exp}_{\mathbb{H}} \circ \mathrm{d}\pi_{\mathcal{E}}^{-1}\right)(Z_k^{(i)}) = \pi_{\mathcal{E}}\left(\mathrm{d}\Psi(Z_k^{(i)})\right) \\ &= \pi_{\mathcal{E}}(E_k^{(i)}) = \left(0^{(1)},\ldots,e_k^{(i)},\ldots,0^{(r)}\right). \end{aligned}$$

In the second equality, we invoked the definition of Ψ and Remark (2.1.52)-(iii), whereas in the third we exploited (2.59); finally, the last equality is (2.62).

Now, the linearity of $\mathrm{Exp}_{\mathbb{G}}$ and (2.61) produce

$$\begin{aligned}
&\mathrm{Exp}_{\mathbb{G}}\big((x_1,\dots,x_N)_{\mathcal{Z}}\big)\\
&= \mathrm{Exp}_{\mathbb{G}}\left(\sum_{i=1}^{r}\sum_{j=1}^{N_i} x_j^{(i)}\, Z_j^{(i)}\right)\\
&= \sum_{i=1}^{r}\sum_{j=1}^{N_i} x_j^{(i)}\, \mathrm{Exp}_{\mathbb{G}}(Z_j^{(i)}) = \sum_{i=1}^{r}\sum_{j=1}^{N_i} x_j^{(i)}\, \big(0^{(1)},\dots,e_j^{(i)},\dots,0^{(r)}\big)\\
&= \sum_{i=1}^{r}\left(0^{(1)},\dots,\sum_{j=1}^{N_i} x_j^{(i)}\, e_j^{(i)},\dots,0^{(r)}\right) = \sum_{i=1}^{r}\big(0^{(1)},\dots,x^{(i)},\dots,0^{(r)}\big)\\
&= (x^{(1)},\dots,x^{(r)}).
\end{aligned}$$

(3). Since in (1) we proved that $(\mathbb{G},\circ_{\mathcal{E}})$ is isomorphic to $(\mathfrak{h},\diamond)$ and $\pi_{\mathcal{E}}$ provides a Lie-group isomorphism between them, which is also a linear map of vector spaces, we infer that, for any $x\in\mathbb{G}$,

$$(x)^{-1} = \pi_{\mathcal{E}}\Big(\big(\pi_{\mathcal{E}}^{-1}(x)\big)^{-1}\Big) = \pi_{\mathcal{E}}\big(-\pi_{\mathcal{E}}^{-1}(x)\big) = -\pi_{\mathcal{E}}\big(\pi_{\mathcal{E}}^{-1}(x)\big) = -x.$$

Here we used the fact that the inversion in $(\mathfrak{h},\diamond)$ is $X\mapsto -X$, for, by the Campbell–Hausdorff operation, we have

$$X\diamond(-X) = X - X + \frac{1}{2}[X,-X] + \frac{1}{12}[X,[X,-X]] + \cdots = 0.$$

(4). To begin with, we know that, by definition,

$$x\circ_{\mathcal{E}} y = \pi_{\mathcal{E}}\Big(\big(\pi_{\mathcal{E}}^{-1}(x)\big)\diamond\big(\pi_{\mathcal{E}}^{-1}(y)\big)\Big) \quad \forall\, x,y\in\mathbb{R}^N.$$

Writing $x=(x_1,\dots,x_N)$, we have $X:=\pi_{\mathcal{E}}^{-1}(x)=\sum_{i=1}^{N} x_i E_i$. Analogously, $Y:=\pi_{\mathcal{E}}^{-1}(y)=\sum_{j=1}^{N} y_j E_j$. Let us consider the Campbell–Hausdorff operation $X\diamond Y$. Considering the form of the operation in (2.43), we see that, besides the summands

$$X+Y=\sum_{i=1}^{N} x_i E_i + \sum_{j=1}^{N} y_j E_j = \sum_{i=1}^{N}(x_i+y_i)E_i,$$

any other summand has the following form: it is given by a rational number times

$$(\star)\qquad (\mathrm{ad}\,X)^{\alpha_1}(\mathrm{ad}\,Y)^{\beta_1}\cdots(\mathrm{ad}\,X)^{\alpha_n}(\mathrm{ad}\,Y)^{\beta_n}\big([X,Y]\big),$$

or an analogous term with $[X,Y]$ replaced by $[Y,X]$. Let us analyze the case $[X,Y]$ (the other being analogous). Replacing X and Y (but in the final $[X,Y]$) by $\sum_{i=1}^{N} x_i E_i$ and $\sum_{j=1}^{N} y_j E_j$, respectively, $(\star)$ is a finite sum of summands of the following type:

$$(\star\star) \qquad p(x,y)\,[E_{k_1}, \cdots [E_{k_M}, [X,Y]]],$$

where $p(x, y)$ is a polynomial, $M \in \mathbb{N}$ and $i_1, \ldots, i_M \in \{1, \ldots, N\}$. Now, $(\star\star)$ is equal to

$$\begin{aligned} & p(x,y)[E_{k_1}, \cdots [E_{k_M}, [X,Y]]] \\ & = \sum_{i,j=1}^{N} p(x,y)\, x_i\, y_j\, [E_{k_1}, \cdots [E_{k_M}, [E_i, E_j]]] \\ & = \sum_{i \neq j} \{\cdots\} = \sum_{i<j} \{\cdots\} + \sum_{i>j} \{\cdots\} \\ & = \sum_{1 \le h < k \le N} p(x,y)\,(x_h\, y_k - x_k\, y_h)\,[E_{k_1}, \cdots [E_{k_M}, [E_h, E_k]]]. \end{aligned}$$

In the last equality, we suitably renamed the dummy variables, and we exploited anti-commutativity $[E_i, E_j] = -[E_j, E_i]$. All these facts together prove that

$$\begin{aligned} x \diamond_{\mathcal{E}} y = \pi_{\mathcal{E}}(X \diamond Y) = \pi_{\mathcal{E}} \Bigg(& \sum_{i=1}^{N} (x_i + y_i) E_i + \Big\{ \text{finite sum of terms like} \\ & p(x,y)\,(x_h\, y_k - x_k\, y_h)\,[E_{k_1}, \cdots [E_{k_M}, [E_h, E_k]]] \Big\} \Bigg). \end{aligned}$$

Finally, the linearity of $\pi_{\mathcal{E}}$ proves that, for every $i = 1, \ldots, N$, it holds

$$(x \diamond_{\mathcal{E}} y)_i = x_i + y_i + \mathcal{R}_i(x,y),$$

where $\mathcal{R}_i(x, y)$ is the sum of polynomials each containing a factor of the type $x_h y_k - x_k y_h$ with $h \neq k$. The fact that $\mathcal{R}_i(x, y)$ does depend only on the x_k's and y_k's with $k < i$ easily follows from (5) of the assertion and the results on homogeneous groups from Section 1.3.2.

(5). The proof of (5) is contained in the proof of Theorem 2.2.18 (see also (1) and Proposition 2.2.8). □

A natural question arises from the proof of Theorem 2.2.18. In that proof, we handled with a Lie group $\mathbb{H}$, its Lie algebra $\mathfrak{h}$ which (equipped with the Campbell–Hausdorff operation $\diamond$) becomes a Lie group itself and, finally, we considered the Lie algebra $\mathfrak{h}^*$ of the Lie group $(\mathfrak{h}, \diamond)$ (roughly speaking, "the algebra of the algebra"). We now prove the very natural fact that $\mathfrak{h}^*$ is "essentially" $\mathfrak{h}$ itself. Moreover, the exponential map $\operatorname{Exp}_{\mathfrak{h}} : \mathfrak{h}^* \to \mathfrak{h}$ is essentially the "identity map".

Example 2.2.23. Before entering the details, we give an example. Let us consider the homogeneous Lie group $(\mathbb{H}, *)$ on $\mathbb{R}^3$ with the composition law

$$x * y = \begin{pmatrix} x_1 + y_1 \\ x_2 + y_2 \\ x_3 + y_3 + x_1\, y_2 \end{pmatrix}.$$

The Jacobian basis for $\mathfrak{h}$ (the algebra of $\mathbb{H}$) is

$$Z_1 = \partial_1, \quad Z_2 = \partial_2 + x_1\,\partial_3, \quad Z_3 = [Z_1, Z_2] = \partial_3.$$

Then a natural Lie group structure $(\mathfrak{h}, \diamond)$ is obtained by considering on $\mathfrak{h}$ the Campbell–Hausdorff law $\diamond$ and, at the same time, by identifying $\mathfrak{h}$ to $\mathbb{R}^3$ via coordinates w.r.t. the Jacobian basis. This gives

$$\xi \diamond \eta = \begin{pmatrix} \xi_1 + \eta_1 \\ \xi_2 + \eta_2 \\ \xi_3 + \eta_3 + \frac{1}{2}\,(\xi_1\eta_2 - \xi_2\eta_1) \end{pmatrix}.$$

The next Jacobian basis related to $(\mathfrak{h}, \diamond)$ is

$$\widetilde{Z_1} = \partial_1 - \frac{\xi_2}{2}\,\partial_3, \quad \widetilde{Z_2} = \partial_2 + \frac{\xi_1}{2}\,\partial_3, \quad \widetilde{Z_3} = [\widetilde{Z_1}, \widetilde{Z_2}] = \partial_3.$$

Finally, consider the Lie algebra $\mathfrak{h}^*$ of $(\mathfrak{h}, \diamond)$. It is easily checked that the exponential map

$$\mathrm{Exp}_{\mathfrak{h}} : \mathfrak{h}^* \to \mathfrak{h}$$

maps $\widetilde{Z_i}$ into Z_i for $i = 1, 2, 3$. In other words, by an abuse of language, $\mathfrak{h}^*$ coincides with $\mathfrak{h}$. $\square$

A general formulation of the above facts is given by the following proposition, which highlights once more the relevance of the Jacobian basis. Notice that no stratification condition is supposed in the first part of the following statement.

Proposition 2.2.24 (Algebra of the algebra). *Let* $(\mathbb{H}, *)$ *be a connected and simply connected nilpotent Lie group with Lie algebra* $\mathfrak{h}$. *Let* $\diamond$ *be the Campbell–Hausdorff operation on* $\mathfrak{h}$ *defined in Definition* 2.2.11, *and let* $(\mathfrak{h}, \diamond)$ *be equipped with its natural Lie group structure* (*see Theorem* 2.2.13). *Finally, let* $\mathfrak{h}^*$ *denote the Lie algebra of the Lie group* $(\mathfrak{h}, \diamond)$, *and let*

$$\mathrm{Exp}_{\mathfrak{h}} : \mathfrak{h}^* \to \mathfrak{h}$$

denote the related exponential map.

Then, $\mathrm{Exp}_{\mathfrak{h}}$ *is a linear map of vector spaces.*

For example, if $\mathbb{H}$ *is a homogeneous Carnot group, and we equip the above Lie algebras* $\mathfrak{h}$ *and* $\mathfrak{h}^*$ *with the relevant Jacobian bases, then* $\mathrm{Exp}_{\mathfrak{h}}$ *is the linear map related to the identity matrix.*

Proof. We follow the notation of the assertion, and we denote by $\mathrm{Exp}_{\mathbb{H}} : \mathfrak{h} \to \mathbb{H}$ the relevant exponential map. We know that the following diagram is commutative (see Theorem 2.1.59 jointly with Theorem 2.2.13):

$$\begin{array}{ccc} (\mathfrak{h}, \diamond) & \xrightarrow{\ \mathrm{Exp}_{\mathbb{H}}\ } & (\mathbb{H}, *) \\ \uparrow{\scriptstyle \mathrm{Exp}_{\mathfrak{h}}} & & \uparrow{\scriptstyle \mathrm{Exp}_{\mathbb{H}}} \\ \mathfrak{h}^* & \xrightarrow{\ d\,\mathrm{Exp}_{\mathbb{H}}\ } & \mathfrak{h}. \end{array}$$

As a consequence, we have

$$\operatorname{Exp}_{\mathfrak{h}} = (\operatorname{Exp}_{\mathbb{H}})^{-1} \circ \operatorname{Exp}_{\mathbb{H}} \circ (\mathrm{d}\operatorname{Exp}_{\mathbb{H}}) = \mathrm{d}\operatorname{Exp}_{\mathbb{H}},$$

so that $\operatorname{Exp}_{\mathfrak{h}}$ is a linear map (here, the vector space structures on $\mathfrak{h}$ and $\mathfrak{h}^*$ are considered).

Let now $(\mathbb{H}, *)$ be a homogeneous Carnot group on $\mathbb{R}^N$. We consider on $\mathfrak{h}$ the Jacobian basis $Z_1, \dots, Z_N$ (related to the operation $*$). We equip $(\mathfrak{h}, \diamond)$ with the natural system of coordinates given by

$$\mathfrak{h} \ni \sum_{j=1}^{N} \xi_j Z_j \mapsto (\xi_1, \dots, \xi_N) \in \mathbb{R}^N,$$

thus identifying $\mathfrak{h}$ to $\mathbb{R}^N$. On $\mathfrak{h} \equiv \mathbb{R}^N$, we have the Lie group law $\diamond$: we then consider the Jacobian basis $Z_1^*, \dots, Z_N^*$ of $\mathfrak{h}^*$ (related to the operation $\diamond$).

We have to show that $\mathrm{d}\operatorname{Exp}_{\mathbb{H}}$ sends Z_k^* into Z_k. To this end, given $f \in C^\infty(\mathbb{H}, \mathbb{R})$, we have

$$\begin{aligned}
&(\mathrm{d}\operatorname{Exp}_{\mathbb{H}}(Z_k^*))_0(f) = (Z_k^*)_0(f \circ \operatorname{Exp}_{\mathbb{H}}) \\
&\quad = (\partial/\partial\xi_k)|_{\xi=0} f\left(\operatorname{Exp}_{\mathbb{H}}\left(\sum_{j=1}^{N} \xi_j Z_j\right)\right) = \frac{\mathrm{d}}{\mathrm{d}t}\bigg|_0 f\big(\operatorname{Exp}(t Z_k)\big) = (Z_k)_0(f).
\end{aligned}$$

This ends the proof. □

2.2.5 Sub-Laplacians of a Stratified Group

We begin with a central definition.

Definition 2.2.25 (Sub-Laplacian of a stratified group). *Let $\mathbb{H}$ be a stratified group with Lie algebra $\mathfrak{h}$. The second order differential operator*

$$\mathcal{L} = \sum_{j=1}^{m} X_j^2 \tag{2.63}$$

is referred to as a sub-Laplacian on $\mathbb{H}$, *if there exists a stratification for* $\mathfrak{h}$

$$\mathfrak{h} = V_1 \oplus \cdots \oplus V_r,$$

as in Definition 2.2.3, such that

$$X_1, \dots, X_m \text{ is a (linear) basis of } V_1.$$

With the above notation, we also say that $\mathcal{L}$ is a sub-Laplacian on $\mathbb{H}$ with related *(or,* generating) stratification $(V_1, \dots, V_r)$.

We shall treat a sub-Laplacian $\mathcal{L}$ as an operator on smooth functions on the open subsets of $\mathbb{H}$. More precisely, if $\Omega \subseteq \mathbb{H}$ is open and $f \in C^\infty(\Omega, \mathbb{R})$, by $\mathcal{L}(f)$ we mean the function on Ω defined by

$$\Omega \ni x \mapsto \sum_{j=1}^{m} (X_j)_x \Big(\Omega \ni m \mapsto (X_j)_m(f)\Big).$$

Note that if $\mathcal{L}$ is as in (2.63), then $m = \dim(V_1)$ is the number of generators of $\mathbb{H}$ according to Definition 2.2.7. Observe also that if $\sum_{j=1}^m X_j^2$ is a sub-Laplacian on $\mathbb{H}$ with related stratification $(V_1, \dots, V_r)$ and if $\{Y_1, \dots, Y_m\}$ is another basis of V_1, then $\sum_{j=1}^m Y_j^2$ is a sub-Laplacian on $\mathbb{H}$.

Since stratified groups and homogeneous Carnot's are deeply related, a comparison of the above notion of sub-Laplacian and that of sub-Laplacian on a homogeneous Carnot group is in order.

Remark 2.2.26. Let $\mathbb{H}$ be a stratified group with an algebra $\mathfrak{h}$. Let $\mathcal{L} = \sum_{j=1}^m X_j^2$ be a sub-Laplacian on $\mathbb{H}$, and let $V = (V_1, \dots, V_r)$ be the stratification of $\mathfrak{h}$ related to $\mathcal{L}$ according to Definition 2.2.25. Let also $\mathcal{E}$ be a basis for $\mathfrak{h}$ adapted to the stratification V and such that $X_1, \dots, X_m$ are the first m elements of $\mathcal{E}$. (The existence of such a basis $\mathcal{E}$ is evident.)

By Proposition 2.2.22 (whose notation we presently follow), there exists a homogeneous Carnot group $\mathbb{G}$ on $\mathbb{R}^N$ which is isomorphic to $\mathbb{H}$. Let $\mathfrak{g}$ be the Lie algebra of $\mathbb{G}$, and let $Z_1, \dots, Z_N$ be the Jacobian basis of $\mathfrak{g}$ (the Jacobian basis is well-posed for $\mathbb{G}$ is a homogeneous Carnot group on $\mathbb{R}^N$). Let $\Psi : \mathbb{G} \to \mathbb{H}$ be the isomorphism, as in Proposition 2.2.22-(1). By Proposition 2.2.22-(2), we know, in particular, that

$$\mathrm{d}\Psi(Z_i) = X_i \quad \text{for every } i = 1, \dots, m, \tag{2.64}$$

or, equivalently, $Z_i(f \circ \Psi) = X_i(f) \circ \Psi$ on $\mathbb{G}$ for every $f \in C^\infty(\mathbb{H}, \mathbb{R})$. Set

$$\Delta_{\mathbb{G}} := \sum_{i=1}^{m} Z_i^2 \quad \text{(which is the canonical sub-Laplacian on } \mathbb{G}\text{)}.$$

We infer that, for every $f \in C^\infty(\mathbb{H}, \mathbb{R})$,

$$\Delta_{\mathbb{G}}(f \circ \Psi) \equiv \big(\mathcal{L}(f)\big) \circ \Psi, \text{ i.e. } \mathcal{L}(f) \equiv \big(\Delta_{\mathbb{G}}(f \circ \Psi)\big) \circ \Psi^{-1}. \tag{2.65a}$$

Equivalently, since we have (with obvious meaning of the notation)

$$C^\infty(\mathbb{H}, \mathbb{R}) \circ \Psi \equiv C^\infty(\mathbb{G}, \mathbb{R}), \quad C^\infty(\mathbb{G}, \mathbb{R}) \circ \Psi^{-1} \equiv C^\infty(\mathbb{H}, \mathbb{R}), \tag{2.65b}$$

for every given $g \in C^\infty(\mathbb{G}, \mathbb{R})$, if we set $f := g \circ \Psi^{-1}$, (2.65a) rewrites as

$$\Delta_{\mathbb{G}} g \equiv \big(\mathcal{L}(g \circ \Psi^{-1})\big) \circ \Psi \quad \forall\, g \in C^\infty(\mathbb{G}, \mathbb{R}). \tag{2.65c}$$

Roughly speaking, (2.65a) says that *the sub-Laplacian $\mathcal{L}$ of $\mathbb{H}$ is turned by Ψ^{-1} into the canonical sub-Laplacian $\Delta_{\mathbb{G}}$ of $\mathbb{G}$*, or, equivalently, *$\mathcal{L}$ and $\Delta_{\mathbb{G}}$ are Ψ-related.*

This fact ensures that the theory of sub-Laplacians on a stratified group can be recast in its homogeneous-Carnot-group dual version.

Moreover, let $\mathcal{L}_1$ and $\mathcal{L}_2$ be two sub-Laplacians on $\mathbb{H}$ with the same related stratification V. Suppose we perform the above identification of $\mathcal{L}_1$ with the canonical sub-Laplacian of $\mathbb{G}$. What is $\mathcal{L}_2$ Ψ-related to on $\mathbb{G}$? Obviously, to a (possibly non-canonical) sub-Laplacian on $\mathbb{G}$. Indeed, let

$$\mathcal{L}_1 = \sum_{i=1}^{m} X_i^2, \quad \mathcal{L}_2 = \sum_{i=1}^{m} Y_i^2.$$

We have that the X_i's and the Y_i's form two bases for V_1. Hence, there exists a non-singular $m \times m$ matrix $(a_{i,j})$ such that

$$Y_i = \sum_{j=1}^{m} a_{i,j}\, X_i \quad \text{for every } i = 1, \ldots, m.$$

Thus we have (see (2.64))

$$\widetilde{Y}_i := \mathrm{d}\Psi^{-1}(Y_i) = \sum_{j=1}^{m} a_{i,j}\, Z_i, \tag{2.66}$$

so that (being $\mathrm{d}\Psi(\widetilde{Y}_i) = Y_i$)

$$\left(\sum_{j=1}^{m} \widetilde{Y}_i^2\right)(f \circ \Psi) \equiv \left(\sum_{j=1}^{m} Y_i^2(f)\right) \circ \Psi$$

for every $f \in C^\infty(\mathbb{H}, \mathbb{R})$. This proves the claimed fact that $\mathcal{L}_2 = \sum_{j=1}^m Y_i^2$ is Ψ-related to $\sum_{j=1}^m \widetilde{Y}_i^2$. In turn, $\sum_{j=1}^m \widetilde{Y}_i^2$ is a sub-Laplacian on $\mathbb{G}$, in force of (2.66), which grants that $\widetilde{Y}_1, \ldots, \widetilde{Y}_m$ is a basis for $\mathrm{span}\{Z_1, \ldots, Z_m\}$. $\square$

We introduce another central definition.

Definition 2.2.27 ($\mathcal{L}$-harmonic function on a stratified group). *Let $\mathbb{H}$ be a stratified group, and let $\mathcal{L}$ be a fixed sub-Laplacian on $\mathbb{H}$, according to Definition 2.2.25. Let $\Omega \subseteq \mathbb{H}$ be an open set. A real-valued function f on Ω is called* $\mathcal{L}$-harmonic *if and only if it holds*

$$f \in C^\infty(\Omega, \mathbb{R}) \quad \text{and} \quad \mathcal{L}f = 0 \ \text{ on } \Omega. \tag{2.67}$$

We denote by $\mathcal{H}_{\mathcal{L}}(\Omega)$ the vector space of the $\mathcal{L}$-harmonic functions on Ω.

Remark 2.2.28. (i) Let $\mathbb{H}$ be a stratified group, and let $\mathcal{L} = \sum_{j=1}^m X_j^2$ be a sub-Laplacian on $\mathbb{H}$ with generating stratification $V = (V_1, \ldots, V_r)$. Let $\mathcal{E}$ be any basis of the algebra of $\mathbb{H}$ adapted to the stratification V (but *not necessarily* containing $X_1, \ldots, X_m$ as its first elements).

Let $\Psi = \operatorname{Exp}_{\mathbb{H}} \circ \pi_{\mathcal{E}}^{-1}$ be as in Proposition 2.2.22. Then, according to the results in Remark 2.2.26, $\mathcal{L}$ is Ψ-related to a sub-Laplacian on $\mathbb{G}$, say $\widetilde{\mathcal{L}}$. This means that

$$\mathcal{L}f = \big(\widetilde{\mathcal{L}}(f \circ \Psi)\big) \circ \Psi^{-1}, \qquad \widetilde{\mathcal{L}}\widetilde{f} = \big(\mathcal{L}(\widetilde{f} \circ \Psi^{-1})\big) \circ \Psi \tag{2.68}$$

for every f smooth on $\mathbb{H}$ and every $\widetilde{f}$ smooth on $\mathbb{G}$.

(ii) Let now $\Omega \neq \emptyset$ be an open subset of $\mathbb{H}$. Then $\widetilde{\Omega} := \Psi^{-1}(\Omega)$ is an open subset of $\mathbb{G}$. We claim that

$$\mathcal{H}_{\widetilde{\mathcal{L}}}(\widetilde{\Omega}) = \mathcal{H}_{\mathcal{L}}(\Omega) \circ \Psi, \qquad \mathcal{H}_{\mathcal{L}}(\Omega) = \mathcal{H}_{\widetilde{\mathcal{L}}}(\widetilde{\Omega}) \circ \Psi^{-1}. \tag{2.69}$$

More precisely,

$$\begin{aligned} f \in \mathcal{H}_{\mathcal{L}}(\Omega) \quad &\Longleftrightarrow \quad \widetilde{f} := f \circ \Psi \in \mathcal{H}_{\widetilde{\mathcal{L}}}(\widetilde{\Omega}), \\ \widetilde{f} \in \mathcal{H}_{\widetilde{\mathcal{L}}}(\widetilde{\Omega}) \quad &\Longleftrightarrow \quad f := \widetilde{f} \circ \Psi^{-1} \in \mathcal{H}_{\mathcal{L}}(\Omega). \end{aligned}$$

Indeed, (2.69) follows at once from (2.68).

All the above facts ensure that *the theory of the $\mathcal{L}$-harmonic functions on (the open subsets of) a stratified group can be recast in its homogeneous-Carnot-group dual version.*

This is what it will be done in the foregoing chapters.

Bibliographical Notes. In this chapter, we intended to provide only the topics of differential geometry and Lie group theory which are strictly necessary to read this book. The reader is referred to, e.g. M. Hausner, J.T. Schwartz [HS68], V.S. Varadarajan [Var84] and F.W. Warner [War83], for exhaustive introductions to the differential geometry and the theory of Lie groups and Lie algebras. See also L.J. Corwin and F.P. Greenleaf [CG90] for the theory of representation of *nilpotent* Lie groups.

2.3 Exercises of Chapter 2

Ex. 1) Following the lines of Example 2.2.21 (page 137), find an explicit isomorphism of Lie groups between a suitable *homogeneous* Carnot group and the (non-homogeneous) stratified group $\mathbb{G} = (\Omega, *)$, where

$$\Omega := \mathbb{R} \times \left(-\frac{\pi}{2}, \frac{\pi}{2}\right) \times (0, \infty)$$

and

$$\xi * \eta = \begin{pmatrix} \xi_1 + \eta_1 + 2(\xi_3 \ln \eta_3 + \tan \xi_2 \ln \eta_3 - \eta_3 \ln \xi_3 - \tan \eta_2 \ln \xi_3) \\ \arctan(\xi_3 + \eta_3 + \tan \xi_2 + \tan \eta_2 - \xi_3 \eta_3) \\ \xi_3 \eta_3 \end{pmatrix}.$$

Ex. 2) *Let $\mathfrak{h}$ be a nilpotent Lie algebra. Denote by $\diamond$ the Campbell–Hausdorff multiplication on $\mathfrak{h}$ (see Definition 2.2.11, page 128). Then $(\mathfrak{h}, \diamond)$ is a Lie group (see the first part of Corollary 2.2.15). Denote by $\mathfrak{h}^*$ the Lie algebra of this group. Prove that $\mathfrak{h}$ and $\mathfrak{h}^*$ are isomorphic as Lie algebras.*

Hint: Let $(\mathbb{G}, *)$ be a Lie group whose Lie algebra $\mathfrak{g}$ is isomorphic to $\mathfrak{h}$ (see the third fundamental theorem of Lie, Theorem 2.2.14). Let $\varphi : \mathfrak{h} \to \mathfrak{g}$ be a Lie-algebra isomorphism. Denote by $\operatorname{Exp}_{\mathbb{G}} : \mathfrak{g} \to \mathbb{G}$ the exponential map related to $\mathbb{G}$. Prove that $\operatorname{Exp}_{\mathbb{G}} \circ \varphi$ is a Lie-group isomorphism from $(\mathfrak{h}, \diamond)$ to $(\mathbb{G}, *)$, using the universality of the operation $\diamond$. Derive that $\mathrm{d}(\operatorname{Exp}_{\mathbb{G}} \circ \varphi) : \mathfrak{h}^* \to \mathfrak{g}$ is a Lie-algebra isomorphism. Consequently, $\varphi^{-1} \circ \mathrm{d}(\operatorname{Exp}_{\mathbb{G}} \circ \varphi)$ is a Lie-algebra isomorphism from $\mathfrak{h}^*$ to $\mathfrak{h}$. See also the following diagram

$$\begin{array}{ccccccc}
(\mathfrak{h}, \diamond) & \stackrel{\varphi}{\longrightarrow} & (\mathfrak{g}, \diamond) & \stackrel{\operatorname{Exp}_{\mathbb{G}}}{\longrightarrow} & (\mathbb{G}, *) & & \\
\uparrow & & & & \uparrow & & \\
\mathfrak{h}^* & \longrightarrow & \underset{\mathrm{d}(\operatorname{Exp}_{\mathbb{G}} \circ \varphi)}{\longrightarrow} & \longrightarrow & \mathfrak{g} & \stackrel{\varphi^{-1}}{\longrightarrow} & \mathfrak{h}.
\end{array}$$

Ex. 3) *Let $\mathfrak{h}$ be a nilpotent Lie algebra. Denote by $\diamond$ the Campbell–Hausdorff multiplication on $\mathfrak{h}$ (see Definition 2.2.11, page 128). Let $(\mathbb{G}, *)$ be a Lie group isomorphic to the Lie group $(\mathfrak{h}, \diamond)$. Prove that the Lie algebra $\mathfrak{g}$ of $\mathbb{G}$ is isomorphic to $\mathfrak{h}$ as Lie algebras.* (*Hint:* Use the preceding exercise: $\mathfrak{g}$ is isomorphic to $\mathfrak{h}^*$, which is isomorphic to $\mathfrak{h}$.)

Ex. 4) Prove that the following is another equivalent definition of stratified group:

(H$*$) *A stratified group is a connected and simply connected Lie group $\mathbb{G}$ whose Lie algebra $\mathfrak{g}$ admits a (vector space) decomposition of the type*

$$\mathfrak{g} = V_1 \oplus \cdots \oplus V_r,$$

where

$$\begin{cases} [V_i, V_j] \subseteq V_{i+j} & \forall\, i, j : i + j \le r, \\ [V_i, V_j] = 0 & \forall\, i, j : i + j > r, \end{cases} \tag{2.70}$$

and V_1 generates (by iterated commutators) all $\mathfrak{g}$.

In fact, in (2.70), it holds $[V_i, V_j] = V_{i+j}$ if $i + j \le r$. This is yet another equivalent definition of Carnot group, frequently adopted in literature.

Hint: Use the following facts: If $\mathbb{G}$ is a stratified group according to Definition 2.2.3, then (setting $V_i := \{0\}$ if $i > r$) it holds

$$V_i = \underbrace{[V_1, \cdots [V_1, V_1]]}_{i \text{ times}} \qquad \text{for every } i \in \mathbb{N}.$$

In particular, V_1 Lie-generates all the V_i's (whence it generates also $\mathfrak{g} = V_1 \oplus \cdots \oplus V_r$). This also gives (using Proposition 1.1.7, page 12)

$$[V_i, V_j] = \Big[\underbrace{[V_1, \cdots [V_1, V_1]]}_{i \text{ times}}, \underbrace{[V_1, \cdots [V_1, V_1]]}_{j \text{ times}}\Big]$$
$$\subseteq \underbrace{[V_1, \cdots [V_1, V_1]]}_{i+j \text{ times}} = V_{i+j}.$$

In particular, (2.70) holds. Vice versa, let $\mathbb{G}$ satisfy the above hypothesis (H$*$). Set $W_1 := V_1$ and

$$W_i := [W_1, W_{i-1}] = \underbrace{[V_1, \cdots [V_1, V_1]]}_{i \text{ times}} \qquad \text{for } i \geq 2.$$

Prove that condition (2.70) implies that $W_i \subseteq V_i$ for every $1 \leq i \leq r$ and $W_i = \{0\}$ for every $i > r$. Moreover, the second hypothesis in (H$*$) (i.e. V_1 Lie-generates $\mathfrak{g}$) ensures that $\mathfrak{g} = W_1 + \cdots + W_r$. Now, a simple linear algebra argument shows that the following conditions

$$W_1 + \cdots + W_r = \mathfrak{g} = V_1 \oplus \cdots \oplus V_r, \quad W_i \subseteq V_i \quad \forall\, i \leq r$$

are sufficient to derive that $W_i = V_i$ for every $1 \leq i \leq r$. As a consequence, we have $[V_1, V_j] = [W_1, W_j] = W_{j+1} = V_{j+1}$ whenever $1 + j \leq r$, and $[V_1, V_j] = [W_1, W_j] = \{0\}$ whenever $1+j > r$, so that $\mathbb{G}$ is a Carnot group according to Definition 2.2.3.

Ex. 5) Let $\mathfrak{g}$ be a Lie algebra. Consider the so-called lower central series of $\mathfrak{g}$, i.e. the sequence of subspaces defined by

$$\mathfrak{g}^{(1)} := \mathfrak{g}, \quad \mathfrak{g}^{(j)} := [\mathfrak{g}, \mathfrak{g}^{(j-1)}], \quad j \geq 1.$$

In other words, for every $j \geq 2$, we have

$$\mathfrak{g}^{(j)} = \underbrace{[\mathfrak{g}, [\mathfrak{g}, \cdots [\mathfrak{g}, \mathfrak{g}]]]}_{j \text{ times}}.$$

Prove the following facts:

- It holds $\mathfrak{g}^{(1)} \supseteq \mathfrak{g}^{(2)} \supseteq \cdots \supseteq \mathfrak{g}^{(j)} \supseteq \mathfrak{g}^{(j+1)} \supseteq \cdots$ for every $j \in \mathbb{N}$;
- if there exists $r \geq 1$ such that $\mathfrak{g}^{(r)} = \mathfrak{g}^{(r+1)}$, then $\mathfrak{g}^{(r)} = \mathfrak{g}^{(j)}$ for every $j \geq r$;
- $\mathfrak{g}$ is nilpotent of step r iff $\mathfrak{g}^{(r+1)} = \{0\}$, but $\mathfrak{g}^{(r)} \neq \{0\}$,
- $\mathfrak{g}$ is nilpotent of step r iff

$$\mathfrak{g}^{(1)} \supsetneq \mathfrak{g}^{(2)} \supsetneq \cdots \supsetneq \mathfrak{g}^{(r)} \supsetneq \mathfrak{g}^{(r+1)} = \{0\}.$$

Ex. 6) Prove the following fact:

A (finite dimensional) nilpotent Lie algebra $\mathfrak{g}$ *of step two is necessarily stratified.*

Indeed, let us set $V_2 = [\mathfrak{g}, \mathfrak{g}]$ and choose any V_1 such that $\mathfrak{g} = V_1 \oplus V_2$: then it also holds $[V_1, V_1] = V_2$ and $[V_1, V_2] = \{0\}$.
Hint: for every $X, Y \in \mathfrak{g}$, write $X = X_1 + X_2$ and $Y = Y_1 + Y_2$, where $X_i, Y_i \in V_i$, and observe that $[\mathfrak{g}, \mathfrak{g}] \ni [X, Y] = [X_1, Y_1] \in [V_1, V_1]$.

Ex. 7) Consider the Carnot group on $\mathbb{R}^4$ (whose points are denoted by (x, y) with $x \in \mathbb{R}$, $y = (y_1, y_2, y_3) \in \mathbb{R}^3$) with the composition law

$$(x, y) \circ (\xi, \eta) = \begin{pmatrix} x + \xi \\ y_1 + \eta_1 \\ y_2 + \eta_2 + \frac{1}{2}(x\eta_1 - \xi y_1) \\ y_3 + \eta_3 + \frac{1}{2}(x\eta_2 - \xi y_2) + \frac{1}{12}(x - \xi)(x\eta_1 - \xi y_1) \end{pmatrix}.$$

Its Lie algebra $\mathfrak{g}$ is spanned by

$$\begin{aligned} X &= \partial_x - \frac{1}{2} y_1 \, \partial_{y_2} - \frac{1}{2} y_2 \, \partial_{y_3} - \frac{1}{12} x y_1 \, \partial_{y_3}, \\ Y_1 &= \partial_{y_1} + \frac{1}{2} x \, \partial_{y_2} + \frac{1}{12} x^2 \, \partial_{y_3}, \\ Y_2 &= \partial_{y_2} + \frac{1}{2} x \, \partial_{y_3}, \\ Y_3 &= \partial_{y_3}, \end{aligned}$$

and the following commutator relations hold:

$$[X, Y_1] = Y_2, \quad [X, Y_2] = Y_3, \quad [X, Y_3] = 0, \quad [Y_i, Y_j] = 0,$$
$$i, j \in \{1, 2, 3\}.$$

Prove that the following are three different stratifications of $\mathfrak{g}$:

$$\begin{aligned} \mathfrak{g} &= \text{span}\{X, Y_1\} \oplus \text{span}\{Y_2\} \oplus \text{span}\{Y_3\}, \\ \mathfrak{g} &= \text{span}\{X, Y_1 + Y_2\} \oplus \text{span}\{Y_2 + Y_3\} \oplus \text{span}\{Y_3\}, \\ \mathfrak{g} &= \text{span}\{X, Y_1 + Y_2 + Y_3\} \oplus \text{span}\{Y_2 + Y_3\} \oplus \text{span}\{Y_3\}. \end{aligned}$$

Ex. 8) Let $\mathfrak{g}$ be a (finite dimensional) stratified Lie algebra, i.e. suppose we have $\mathfrak{g} = V^{(1)} \oplus V^{(2)} \oplus \cdots \oplus V^{(r)}$ with (set $V^{(i)} := \{0\}$ whenever $i > r$) $[V^{(1)}, V^{(i)}] = V^{(i+1)}$ for every $i \in \mathbb{N}$. Prove the following facts:

- $\mathfrak{g}$ is nilpotent of step r;
- $V^{(r)} = \mathfrak{g}^{(r)}$, the r-th element of the lower central series for $\mathfrak{g}$ (see Ex. 5); (*Hint*: for every $X_1, \cdots, X_r \in \mathfrak{g}$, write $X_i = X_i^{(1)} + \cdots + X_i^{(r)}$, where $X_i^{(1)} \in V^{(1)}, \cdots, X_i^{(r)} \in V^{(r)}$, and observe that $[X_1, \cdots [X_{r-1}, X_r]] = [X_1^{(1)}, \cdots [X_{r-1}^{(1)}, X_r^{(1)}]]$.)
- Suppose $V^{(1)} \oplus \cdots \oplus V^{(r)}$ and $W^{(1)} \oplus \cdots \oplus W^{(r)}$ are two stratifications of $\mathfrak{g}$. Prove that

$$W^{(r)} = V^{(r)},$$
$$W^{(r-1)} = V^{(r-1)} \oplus U^{(r)}_{r-1} \quad \text{(where } U^{(r)}_{r-1} \text{ is a subspace of } V^{(r)}\text{)},$$
$$W^{(r-2)} = V^{(r-2)} \oplus U^{(r-1)}_{r-2} \oplus U^{(r)}_{r-2}$$
(where $U^{(r-1)}_{r-2}$, $U^{(r)}_{r-2}$ are subspaces of $V^{(r-1)}$, $V^{(r)}$, respectively).

In other words, we have

$$\begin{aligned} W^{(r)} &= V^{(r)}, \\ W^{(r-1)} &= V^{(r-1)} \text{ (modulo } V^{(r)}\text{)}, \\ W^{(r-2)} &= V^{(r-2)} \text{ (modulo } V^{(r-1)} + V^{(r)}\text{)}, \\ &\vdots \\ W^{(1)} &= V^{(1)} \text{ (modulo } V^{(2)} + \cdots + V^{(r-1)} + V^{(r)}\text{)}. \end{aligned}$$

Precisely, prove the following facts:
(i) $\dim(V^{(i)}) = \dim(W^{(i)}) =: N_i$ for every $i = 1, \dots, r$. Choose two bases $\mathcal{V}$ and $\mathcal{W}$ of $\mathfrak{g}$ adapted respectively to the stratifications with the $V^{(i)}$'s and with the $W^{(i)}$'s. Set $\mathcal{V} = \{V_1, \dots, V_N\}$ and $\mathcal{W} = \{W_1, \dots, W_N\}$ and prove the existence of non-singular matrices $M^{(i)}$'s of order $N_i \times N_i$ such that (with clear meaning of the notation)

$$\begin{pmatrix} W_1 \\ \vdots \\ W_N \end{pmatrix} = \begin{pmatrix} M^{(1)} & \bigstar & \cdots & \bigstar \\ 0 & M^{(2)} & \ddots & \vdots \\ \vdots & \ddots & \ddots & \bigstar \\ 0 & \cdots & 0 & M^{(r)} \end{pmatrix} \begin{pmatrix} V_1 \\ \vdots \\ V_N \end{pmatrix}.$$

(*Hint*: By the preceding part of the exercise, we have $W^{(r)} = V^{(r)} = \mathfrak{g}^{(r)}$. Now consider the quotient $\mathfrak{g}_{/\mathfrak{g}^{(r)}}$. It holds

$$W^{(1)}_{/W^{(r)}} \oplus \cdots \oplus W^{(r-1)}_{/W^{(r)}} = \mathfrak{g}_{/\mathfrak{g}^{(r)}} = V^{(1)}_{/V^{(r)}} \oplus \cdots \oplus V^{(r-1)}_{/V^{(r)}},$$

and these are stratifications of the Lie algebra $\mathfrak{g}_{/\mathfrak{g}^{(r)}}$ too. Consequently, $W^{(r-1)}_{/W^{(r)}} = V^{(r-1)}_{/V^{(r)}}$. Proceed inductively. Alternatively, argue as in the two-stratification Lemma 2.2.9.)

Ex. 9) Let $\mathfrak{g}$ be a stratified Lie algebra, i.e. $\mathfrak{g} = V^{(1)} \oplus V^{(2)} \oplus \cdots \oplus V^{(r)}$ with (set $V^{(i)} := \{0\}$ whenever $i > r$) $[V^{(1)}, V^{(i)}] = V^{(i+1)}$ for every $i \in \mathbb{N}$. Prove that the lower central series of $\mathfrak{g}$ (see Ex. 5 above) is given by

$$\mathfrak{g}^{(k)} = \bigoplus_{i=k}^{r} V^{(i)} \quad \forall\, k = 1, \dots, r, \qquad g^{(k)} = \{0\} \quad \forall\, k > r.$$

(*Hint*: First prove that $\mathfrak{g}^{(k)} \subseteq \bigoplus_{i=k}^{r} V^{(i)}$ arguing inductively as follows: $\mathfrak{g}^{(2)} = [\mathfrak{g}, \mathfrak{g}] = [V^{(1)} + \cdots + V^{(r)}, V^{(1)} + \cdots + V^{(r)}] \subseteq \bigoplus_{i=2}^{r} V^{(i)}$;

$\mathfrak{g}^{(3)} = [\mathfrak{g}, [\mathfrak{g}, \mathfrak{g}]] \subseteq [\bigoplus_{i=1}^r V^{(i)}, \bigoplus_{i=2}^r V^{(i)}] \subseteq \bigoplus_{i=3}^r V^{(i)}$, etc. Vice versa, it holds

$$\begin{aligned} V^{(i)} &= \text{span}\{[X_1, \cdots [X_{i-1}, X_i] \cdots] : X_1, \ldots, X_i \in V^{(1)}\} \\ &\subseteq \underbrace{[[\mathfrak{g}, \cdots [\mathfrak{g}, \mathfrak{g}] \cdots]]}_{i \text{ times}} = \mathfrak{g}^{(i)}, \end{aligned}$$

then derive $\bigoplus_{i=k}^r V^{(i)} = V^{(k)} \oplus \cdots \oplus V^{(r)} \subseteq \mathfrak{g}^{(k)} \oplus \cdots \oplus \mathfrak{g}^{(r)} = \mathfrak{g}^{(k)}$.)

Ex. 10) The natural identification between the Lie algebra $\mathfrak{g}$ of Lie group $\mathbb{G}$ with the tangent space $\mathbb{G}_e$ to $\mathbb{G}$ at the identity must be "handled with care", as the following example shows. Suppose $\mathbb{G}$ and $\mathbb{H}$ are Lie groups with Lie algebras $\mathfrak{g}$, $\mathfrak{h}$, respectively. We denote by e both the identity of $\mathbb{G}$ and that of $\mathbb{H}$. Suppose $F : \mathbb{G} \to \mathbb{H}$ is a C^∞-map such that $F(e) = e$. Obviously, we have $\mathrm{d}_e F : \mathbb{G}_e \to \mathbb{H}_e$. But $\mathbb{G}_e$ is identified with $\mathfrak{g}$ and $\mathbb{H}_e$ is identified with $\mathfrak{h}$. Consequently, F induces a natural map, say f, between $\mathfrak{g}$ and $\mathfrak{h}$. The following question arises: as in the case when F is a group homomorphism, does f represent pointwise the differential of F? Id est (see (2.33)), does it hold (for every $X \in \mathfrak{g}$ and every $x \in \mathbb{G}$)

$$(f(X))_{F(x)} \overset{?}{=} \mathrm{d}_x F(X_x).$$

The answer is in general negative. Indeed, with the above notation (see also (2.26)) the map $f : \mathfrak{g} \to \mathfrak{h}$ is defined in the following way: for every $X \in \mathfrak{g}$ and every $y \in \mathbb{H}$, we have

$$(f(X))_y = \mathrm{d}_e \tau_y (\mathrm{d}_e F(X_e)),$$

i.e. $f(X) \in \mathfrak{h}$ is the vector field on $\mathbb{H}$ defined by

$$\mathbb{H} \ni y \mapsto \mathrm{d}_e(\tau_y \circ F)(X_e).$$

We ask whether

$$(*1) \qquad \mathrm{d}_x F(X_x) \overset{?}{=} \mathrm{d}_e(\tau_{F(x)} \circ F)(X_e) \quad \forall\, X \in \mathfrak{g},\ x \in \mathbb{G}.$$

Since X is left-invariant, it holds $X_x = \mathrm{d}_e \tau_x(X_e)$, so that $\mathrm{d}_x F(X_x) = \mathrm{d}_x F(\mathrm{d}_e \tau_x(X_e)) = \mathrm{d}_e(F \circ \tau_x) X_e$. As a consequence, $(*1)$ holds iff (recall that $\mathbb{G}_e = \{X_e : X \in \mathfrak{g}\}$ by Theorem 2.1.43-(1))

$$(*2) \qquad \mathrm{d}_e(F \circ \tau_x) \overset{?}{=} \mathrm{d}_e(\tau_{F(x)} \circ F) \qquad \forall\, x \in \mathbb{G}.$$

(Note that this certainly holds if F is a group homomorphism!) In turn, this is equivalent to

$$(*3) \qquad \mathrm{d}_x F \overset{?}{=} \mathrm{d}_x(\tau_{F(x)} \circ F \circ \tau_{x^{-1}}) \qquad \forall\, x \in \mathbb{G}.$$

Note that (∗3) is a "differential equation"

$$\begin{cases} \mathrm{d}_x F = \mathrm{d}_e \tau_{F(x)} \circ \mathrm{d}_e F \circ \mathrm{d}_x \tau_{x^{-1}} & \text{for every } x \in \mathbb{G}, \\ F(e) = e. \end{cases}$$

For example, prove that the above system holds for $\mathbb{G} = \mathbb{H} = (\mathbb{R}^N, +)$ iff F is a linear map, i.e. F is an automorphism of the Lie group $(\mathbb{R}^N, +)$. Consequently, the map $F : (\mathbb{R}, +) \to (\mathbb{R}, +)$, $F(x) = x^2$ furnishes a counterexample.

Ex. 11) Suppose that $\mathbb{G} = (\mathbb{R}^N, \circ)$ and $\mathbb{H} = (\mathbb{R}^n, *)$ are two Lie groups on $\mathbb{R}^N$ and $\mathbb{R}^n$, respectively. Let $\mathfrak{g}$ and $\mathfrak{h}$ denote the relevant Lie algebras. Finally, let $\varphi : \mathbb{G} \to \mathbb{H}$ be a homomorphism of Lie groups. If $\mathrm{d}\varphi : \mathfrak{g} \to \mathfrak{h}$ is the map defined in Definition 2.1.49, prove that it operates in the following way: for every $X \in \mathfrak{g}$, $\mathrm{d}\varphi(X)$ is the left invariant vector field on $\mathbb{H}$ whose column vector of the coefficients at the point $y \in \mathbb{R}^n$ is given by

$$\big(\mathrm{d}\varphi(X)\big)I(y) = \mathcal{J}_{\tau_y \circ \varphi}(0) \cdot XI(0),$$

or, equivalently,

$$\big(\mathrm{d}\varphi(X)\big)I(y) = \mathcal{J}_{\tau_y}(0) \cdot \mathcal{J}_{\varphi}(0) \cdot XI(0).$$

Observe that (since $\mathcal{J}_{\tau_0}(0)$ is the identity matrix, and since the coordinates of a vector field Y w.r.t. the Jacobian basis are given by $YI(0)$) the matrix representing the linear map $\mathrm{d}\varphi$ with respect to the relevant Jacobian bases on $\mathfrak{g}$ and $\mathfrak{h}$, respectively, is simply

$$\mathcal{J}_{\varphi}(0).$$

Ex. 12) Provide the ODE's details for what is stated in the first paragraph of the proof of Proposition 2.1.53, page 116.

Ex. 13) Suppose $\mathbb{G} = (\mathbb{R}^N, \circ)$ and $\mathbb{H} = (\mathbb{R}^n, *)$ are two Lie groups on $\mathbb{R}^N$ and $\mathbb{R}^n$, respectively (according to the definition given in Chapter 1). Let $\mathfrak{g}$ and $\mathfrak{h}$ be the relevant Lie algebras. Let $\varphi : \mathbb{G} \to \mathbb{H}$ be a homomorphism of Lie groups. Show that the differential of φ, as defined in Definition 2.1.49, is the map $\mathrm{d}\varphi : \mathfrak{g} \to \mathfrak{h}$ such that, for every $X \in \mathfrak{g}$, $\mathrm{d}\varphi(X)$ is the only vector field in $\mathfrak{h}$ such that

$$\big(\mathrm{d}\varphi(X)I\big)(\varphi(x)) = \mathcal{J}_{\varphi}(x)\,(XI)(x). \tag{2.71}$$

Ex. 14) The following is an example to the "change of adapted basis", described in Remark 2.2.20. We follow the therein notation.

Example 2.3.1 (Stratified groups of step two). Let $(\mathbb{H}, *)$ be a homogeneous Carnot group of step two on $\mathbb{R}^N$. We shall prove in Section 3.2, page 158, that if Z denotes the Jacobian basis related to $\mathbb{H}$, then it holds

$$(\xi,\tau)\circ_Z(\xi',\tau') = \begin{pmatrix} \xi_j+\xi'_j \quad (j=1,\ldots,m) \\ \tau_i+\tau'_i+\frac{1}{2}\langle B^{(i)}\xi,\xi'\rangle \quad (i=1,\ldots,n) \end{pmatrix},$$

where m is the number of generators of $\mathbb{H}$, $n=N-m$ and if

$$\{X_1,\ldots,X_m;\ T_1,\ldots,T_n\}$$

re-denotes the Jacobian basis, then $B^{(i)}$ is the $m\times m$ matrix

$$B^{(i)} = \big(b^{(i)}_{h,k}\big)_{1\le h,k\le m}$$

defined by

$$[X_k,X_h] = \textstyle\sum_{k=1}^n b^{(i)}_{h,k}T_i.$$

If $\widetilde{Z}$ is another basis of the algebra of $\mathfrak{g}$ adapted to its stratification, say

$$\widetilde{Z} = \{\widetilde{X}_1,\ldots,\widetilde{X}_m;\widetilde{T}_1,\ldots,\widetilde{T}_n\},$$

there exist two non-singular matrices U (of order $m\times m$) and V (of order $n\times n$) such that

$$\big(\widetilde{X}_1I\cdots\widetilde{X}_mI\big) = \big(X_1I\cdots X_mI\big)\cdot U,$$
$$\big(\widetilde{T}_1I\cdots\widetilde{T}_nI\big) = \big(T_1I\cdots T_nI\big)\cdot V.$$

A simple computation shows that the new composition law $\circ_{\widetilde{Z}}$ is given by (using, for example, (2.58a))

$$(\xi,\tau)\circ_{\widetilde{Z}}(\xi',\tau') = \begin{pmatrix} \xi_j+\xi'_j \quad (j=1,\ldots,m) \\ \tau_i+\tau'_i+\frac{1}{2}\langle \widetilde{B}^{(i)}\xi,\xi'\rangle \quad (i=1,\ldots,n) \end{pmatrix},$$

where, if $V^{-1}=(w_{i,j})_{i,j\le n}$,

$$\widetilde{B}^{(i)} = U^T\cdot\left(\sum_{j=1}^n w_{i,j}\,B^{(j)}\right)\cdot U.$$

The test of this fact is left to the reader. □

3

Carnot Groups of Step Two

The aim of this chapter is to collect some results and many explicit examples of Carnot groups of *step two*. Some examples are well known in literature, some are new. To begin with, we present the most studied (and by far one of the most important) among Carnot groups, the Heisenberg–Weyl group. Then, we turn our attention to general *homogeneous* Carnot groups of step two and m generators, $m \geq 2$. In particular, we show that they are naturally given with the data on $\mathbb{R}^{m+n}$ of n skew-symmetric matrices of order m.

The set of examples that we provide here contains the *free* step-two homogeneous groups, the prototype groups of *Heisenberg-type* (which will be widely studied in Chapter 18) and the H-groups *in the sense of Métivier.*

3.1 The Heisenberg–Weyl Group

Let us consider in $\mathbb{C}^n \times \mathbb{R}$ (whose points we denote by (z, t) with $t \in \mathbb{R}$ and $z = (z_1, \dots, z_n) \in \mathbb{C}^n$) the following composition law

$$(z, t) \circ (z', t') = (z + z', t + t' + 2\,\mathbb{I}\mathrm{m}(z \cdot \overline{z'})). \tag{3.1}$$

In (3.1), we have set (i obviously denotes the imaginary unit) $\mathbb{I}\mathrm{m}(x + iy) = y$ $(x, y \in \mathbb{R})$, whereas $z \cdot \overline{z'}$ denotes the usual Hermitian inner product in $\mathbb{C}^n$,

$$z \cdot \overline{z'} = \sum_{j=1}^{n} (x_j + iy_j)(x'_j - iy'_j).$$

Hereafter we agree to identify $\mathbb{C}^n$ with $\mathbb{R}^{2n}$ and to use the following notation[1] to denote the points of $\mathbb{C}^n \times \mathbb{R} \equiv \mathbb{R}^{2n+1}$:

[1] Someone may say that the notation $(z, t) \equiv (x_1, y_1, \dots, x_n, y_n, t)$ would be more appropriate, but the other notation is so deeply entrenched that we have no choice but to go along with it.

$$(z, t) \equiv (x, y, t) = (x_1, \dots, x_n, y_1, \dots, y_n, t)$$

with $z = (z_1, \dots, z_n)$, $z_j = x_j + i y_j$ and $x_j, y_j, t \in \mathbb{R}$. Then, the composition law $\circ$ can be explicitly written as

$$(x, y, t) \circ (x', y', t') = \big(x + x', y + y', t + t' + 2\langle y, x'\rangle - 2\langle x, y'\rangle\big), \qquad (3.2)$$

where $\langle \cdot, \cdot \rangle$ denotes the usual inner product in $\mathbb{R}^n$. It is quite easy to verify that $(\mathbb{R}^{2n+1}, \circ)$ is a Lie group whose identity is the origin and where the inverse is given by $(z, t)^{-1} = (-z, -t)$. Let us now consider the dilations

$$\delta_\lambda : \mathbb{R}^{2n+1} \to \mathbb{R}^{2n+1}, \quad \delta_\lambda(z, t) = (\lambda z, \lambda^2 t).$$

A trivial computation shows that δ_λ is an automorphism of $(\mathbb{R}^{2n+1}, \circ)$ for every $\lambda > 0$. Then $\mathbb{H}^n = (\mathbb{R}^{2n+1}, \circ, \delta_\lambda)$ is a homogeneous group. It is called the *Heisenberg–Weyl group* in $\mathbb{R}^{2n+1}$. For example, when $n = 1$, the Heisenberg–Weyl group $\mathbb{H}^1$ in $\mathbb{R}^3$ is equipped with the composition law

$$(x, y, t) \circ (x', y', t') = (x + x', y + y', t + t' + 2\,(y x' - x y')),$$

while, when $n = 2$, the Heisenberg–Weyl group $\mathbb{H}^2$ in $\mathbb{R}^5$ is equipped with the composition law

$$\begin{aligned}&(x_1, x_2, y_1, y_2, t) \circ (x_1', x_2', y_1', y_2', t') \\ &\quad = (x_1 + x_1', x_2 + x_2', y_1 + y_1', y_2 + y_2', t + t' + 2\,(y_1 x_1' + y_2 x_2' - x_1 y_1' - x_2 y_2')).\end{aligned}$$

The Jacobian matrix at the origin of the left translation $\tau_{(z,t)}$ is the following block matrix

$$\mathcal{J}_{\tau_{(z,t)}}(0, 0) = \begin{pmatrix} \mathbb{I}_n & 0 & 0 \\ 0 & \mathbb{I}_n & 0 \\ 2\,y^T & -2\,x^T & 1 \end{pmatrix},$$

where $\mathbb{I}_n$ denotes the $n \times n$ identity matrix, while $2\,y^T$ and $-2\,x^T$ stand for the $1 \times n$ matrices $(2y_1 \cdots 2y_n)$ and $(-2x_1 \cdots -2x_n)$, respectively. Then, the Jacobian basis of $\mathfrak{h}_n$, the Lie algebra of $\mathbb{H}^n$, is given by

$$X_j = \partial_{x_j} + 2 y_j\, \partial_t, \qquad Y_j = \partial_{y_j} - 2 x_j\, \partial_t, \quad j = 1, \dots, n, \quad T = \partial_t.$$

Since $[X_j, Y_j] = -4\,\partial_t$, we have

$$\begin{aligned}&\operatorname{rank}\big(\operatorname{Lie}\{X_1, \dots, X_n, Y_1, \dots, Y_n\}(0, 0)\big) \\ &\quad = \dim\big(\operatorname{span}\{\partial_{x_1}, \dots, \partial_{x_n}, \partial_{y_1}, \dots, \partial_{y_n}, -4\partial_t\}\big) = 2n + 1.\end{aligned}$$

This shows that $\mathbb{H}^n$ is a Carnot group with the following stratification[2]

$$\mathfrak{h}_n = \operatorname{span}\{X_1, \dots, X_n, Y_1, \dots, Y_n\} \oplus \operatorname{span}\{\partial_t\}. \qquad (3.3)$$

---

[2] Obviously, there exist other possible stratifications, but the above one is the most frequently adopted and the one we shall refer to.

The step of $(\mathbb{H}^n, \circ)$ is $r = 2$ and its Jacobian generators are the vector fields X_j, Y_j $(j = 1, \dots, n)$. The canonical sub-Laplacian on $\mathbb{H}^n$ (also referred to as *Kohn Laplacian*) is then given by

$$\Delta_{\mathbb{H}^n} = \sum_{j=1}^{n} \left(X_j^2 + Y_j^2\right).$$

An explicit formula for $\Delta_{\mathbb{H}^n}$ can be found in Ex. 1, at the end of this chapter. Finally, we exhibit the explicit form of the exponential map for $\mathbb{H}^n$. It is given by[3]

$$\operatorname{Exp}((\xi, \eta, \tau) \cdot Z) = (\xi, \eta, \tau). \tag{3.4}$$

Here we have set $(\xi, \eta, \tau) \cdot Z = \sum_{j=1}^{n} (\xi_j X_j + \eta_j Y_j) + \tau T$.

We now want to perform a change of variables in $\mathbb{H}^n$, inspired by the change of basis in $\mathfrak{h}_n$, which turns the Jacobian basis into the new basis

$$X_j^* := X_j, \qquad Y_j^* := Y_j \quad (j = 1, \dots, n), \qquad T^* := [X_1, Y_1] = -4\,T.$$

As above, we set $(\xi, \eta, \tau) \cdot Z^* = \sum_{j=1}^{n} (\xi_j X_j^* + \eta_j Y_j^*) + \tau T^*$. In other words, we have chosen *another* basis of $\mathfrak{h}_n$ adapted to the stratification (3.3).

Then, what we aim to do (see the proof of Theorem 2.2.18, page 131) is to equip $\mathfrak{h}_n$ with a Lie-group structure isomorphic (via Exp) to that of $(\mathbb{H}^n, \circ)$; we identify $\mathfrak{h}_n$ with $\mathbb{R}^{2n+1}$ via the cited basis Z^*: this amounts to equip $\mathbb{R}^{2n+1}$ with the composition $*$ such that

$$\begin{aligned} &\operatorname{Log}\big((\operatorname{Exp}((\xi, \eta, \tau) \cdot Z^*)) \circ (\operatorname{Exp}((\xi', \eta', \tau') \cdot Z^*))\big) \\ &\quad = \big((\xi, \eta, \tau) * (\xi', \eta', \tau')\big) \cdot Z^*. \end{aligned}$$

The explicit expression for $*$ is easily found,[4]

$$\begin{aligned} &(\xi, \eta, \tau) * (\xi', \eta', \tau') \\ &\quad = \left(\xi + \xi', \eta + \eta', \tau + \tau' - \frac{1}{2}\langle \eta, \xi' \rangle + \frac{1}{2}\langle \xi, \eta' \rangle\right), \end{aligned} \tag{3.5a}$$

or, equivalently,

$$(\zeta, \tau) * (\zeta', \tau') = \left(\zeta + \zeta', \tau + \tau' + \frac{1}{2}\langle B\zeta, \zeta' \rangle\right) \tag{3.5b}$$

with

$$B = \begin{pmatrix} 0 & -\mathbb{I}_n \\ \mathbb{I}_n & 0 \end{pmatrix}.$$

The natural isomorphism $\varphi : (\mathbb{R}^{2n+1}, *) \to (\mathbb{H}^n, \circ)$ is given by

$$\varphi(\xi, \eta, \tau) = \operatorname{Exp}((\xi, \eta, \tau) \cdot Z^*) = \operatorname{Exp}((\xi, \eta, -4\tau) \cdot Z) = (\xi, \eta, -4\tau).$$

---

[3] See (3.11a) in Section 3.5, page 167, and Remark 3.2.4, page 163.

[4] For all the details, see Section 4.3.2, page 200.

3.2 Homogeneous Carnot Groups of Step Two

Let $m, n \in \mathbb{N}$. Set $\mathbb{R}^N := \mathbb{R}^m \times \mathbb{R}^n$ and denote its points by $z = (x, t)$ with $x \in \mathbb{R}^m$ and $t \in \mathbb{R}^n$. Given an n-tuple $B^{(1)}, \dots, B^{(n)}$ of $m \times m$ matrices with real entries, let

$$(x, t) \circ (\xi, \tau) = \left(x + \xi, t + \tau + \frac{1}{2} \langle Bx, \xi \rangle \right). \tag{3.6}$$

Here $\langle Bx, \xi \rangle$ denotes the n-tuple

$$\left(\langle B^{(1)} x, \xi \rangle, \dots, \langle B^{(n)} x, \xi \rangle \right) \quad \left(\text{also written as} \sum_{i,j=1}^{m} B_{i,j} x_j \, \xi_i \right)$$

and $\langle \cdot, \cdot \rangle$ stands for the inner product in $\mathbb{R}^m$. One can easily verify that $(\mathbb{R}^N, \circ)$ is a Lie group whose identity is the origin and where the inverse is given by

$$(x, t)^{-1} = \left(-x, -t + \langle Bx, x \rangle \right).$$

We highlight that the inverse map is the usual $-(x, t)$ if and only if, for every $k = 1, \dots, n$, it holds

$$\langle B^{(k)} x, x \rangle = 0 \quad \forall\, x \in \mathbb{R}^m,$$

i.e. iff the matrices $B^{(k)}$'s are skew-symmetric. It is also quite easy to recognize that the dilation

$$\delta_\lambda : \mathbb{R}^N \to \mathbb{R}^N, \quad \delta_\lambda(x, t) = (\lambda x, \lambda^2 t) \tag{3.7}$$

is an automorphism of $(\mathbb{R}^N, \circ)$ for any $\lambda > 0$. Then $\mathbb{G} = (\mathbb{R}^N, \circ, \delta_\lambda)$ is a homogeneous Lie group.

We explicitly remark that the composition law of any Lie group in $\mathbb{R}^m \times \mathbb{R}^n$, homogeneous w.r.t. the dilations $\{\delta_\lambda\}_\lambda$ as in (3.7), takes the form (3.6) (see Theorem 1.3.15, page 39).

The Jacobian matrix at $(0, 0)$ of the left translation $\tau_{(x,t)}$ takes the following block form

$$\mathcal{J}_{\tau_{(x,t)}}(0, 0) = \left(\begin{array}{c|c} \mathbb{I}_m & 0 \\ \hline \frac{1}{2} B x & \mathbb{I}_n \end{array} \right),$$

where, if $B^{(k)} = (b^{(k)}_{i,j})_{i,j \le m}$ for $k = 1, \dots, n$, Bx denotes the matrix

$$\left(\sum_{j=1}^{m} b^{(k)}_{i,j} \, x_j \right)_{k \le n,\, i \le m}.$$

More explicitly, we have

$$\mathcal{J}_{\tau_{(x,t)}}(0, 0) = \left(\begin{array}{ccc|c} & \mathbb{I}_m & & 0_{m \times n} \\ \hline \frac{1}{2} \sum_{j=1}^{m} b^{(1)}_{1,j} x_j & \cdots & \frac{1}{2} \sum_{j=1}^{m} b^{(1)}_{m,j} x_j & \\ \vdots & \cdots & \vdots & \mathbb{I}_n \\ \frac{1}{2} \sum_{j=1}^{m} b^{(n)}_{1,j} x_j & \cdots & \frac{1}{2} \sum_{j=1}^{m} b^{(n)}_{m,j} x_j & \end{array} \right).$$

Then the Jacobian basis of $\mathfrak{g}$, the Lie algebra of $\mathbb{G}$, is given by

$$\begin{aligned} X_i &= (\partial/\partial x_i) + \frac{1}{2}\sum_{k=1}^{n}\left(\sum_{l=1}^{m} b_{i,l}^{(k)} x_l\right)(\partial/\partial t_k) \\ &= (\partial/\partial x_i) + \frac{1}{2}\langle (Bx)_i, \nabla_t\rangle, \quad i = 1,\dots,m, \\ T_k &= \partial/\partial t_k, \quad k = 1,\dots,n. \end{aligned} \tag{3.8}$$

Here, we briefly denoted by $(Bx)_i$ the vector of $\mathbb{R}^n$

$$\big((B^{(1)}x)_i, \dots, (B^{(n)}x)_i\big),$$

where $(B^{(k)}x)_i$ is the i-th component of $B^{(k)}x$. An easy computation shows that

$$[X_j, X_i] = \sum_{k=1}^{n} \frac{1}{2}\big(b_{i,j}^{(k)} - b_{j,i}^{(k)}\big)\,\partial_{t_k} =: \sum_{k=1}^{n} c_{i,j}^{(k)}\,\partial_{t_k}.$$

We have denoted by $C^{(k)} = (c_{i,j}^{(k)})_{i,j\le m}$ the skew-symmetric part of $B^{(k)}$, i.e.

$$C^{(k)} = \frac{1}{2}\big(B^{(k)} - (B^{(k)})^T\big).$$

Let us now assume that $C^{(1)}, \dots, C^{(n)}$ are linearly independent. This implies that the $m^2 \times n$ matrix

$$\begin{pmatrix} c_{1,1}^{(1)} & \cdots & c_{1,1}^{(n)} \\ c_{1,2}^{(1)} & \cdots & c_{1,2}^{(n)} \\ \vdots & \cdots & \vdots \\ c_{1,m}^{(1)} & \cdots & c_{1,m}^{(n)} \\ c_{2,1}^{(1)} & \cdots & c_{2,1}^{(n)} \\ \vdots & \cdots & \vdots \\ \text{(proceed} & \text{analogously} & \text{up to)} \\ \vdots & \cdots & \vdots \\ c_{m,m}^{(1)} & \cdots & c_{m,m}^{(n)} \end{pmatrix}$$

has rank equal to n. As a consequence,

$$\operatorname{span}\{[X_j, X_i] \,|\, i, j = 1,\dots,m\} = \operatorname{span}\{\partial_{t_1}, \dots, \partial_{t_n}\}.$$

Therefore,

$$\begin{aligned} &\operatorname{rank}\big(\operatorname{Lie}\{X_1,\dots,X_m\}(0,0)\big) \\ &\quad = \dim\big(\operatorname{span}\{\partial_{x_1},\dots,\partial_{x_m},\partial_{t_1},\dots,\partial_{t_n}\}\big) = m+n. \end{aligned}$$

This shows that $\mathbb{G}$ is a Carnot group of step two and Jacobian generators $X_1, \dots, X_m$.

We explicitly remark that the linear independence of the matrices

$$C^{(1)}, \dots, C^{(n)}$$

is also *necessary* for $\mathbb{G}$ to be a Carnot group. Then, we have proved the following proposition.

Proposition 3.2.1 (Characterization. I). *Every homogeneous Lie group $\mathbb{G}$ on $\mathbb{R}^N$, homogeneous with respect to the dilation*

$$\delta_\lambda : \mathbb{R}^N \to \mathbb{R}^N, \quad \delta_\lambda(x,t) = (\lambda x, \lambda^2 t)$$

(where $x \in \mathbb{R}^m$, $t \in \mathbb{R}^n$ and $N = m+n$), is equipped with the composition law

$$(x,t) \circ (\xi,\tau) = \left(x+\xi, t_1+\tau_1+\frac{1}{2}\langle B^{(1)}x,\xi\rangle, \dots, t_n+\tau_n+\frac{1}{2}\langle B^{(n)}x,\xi\rangle \right)$$

for n suitable $m \times m$ matrices $B^{(1)}, \dots, B^{(n)}$.

Moreover, a characterization of homogeneous *Carnot groups of step two and m generators is given by the above $\mathbb{G} = (\mathbb{R}^{m+n}, \circ, \delta_\lambda)$, where the skew-symmetric parts of the $B^{(k)}$'s are linearly independent.*

We remark that the above arguments show that there exist Carnot groups of any dimension $m \in \mathbb{N}$ of the first layer and any dimension

$$n \le m(m-1)/2$$

of the second layer: it suffices to choose n linearly independent matrices $B^{(1)}$, $\dots$, $B^{(n)}$ in the vector space of the skew-symmetric $m \times m$ matrices (which has dimension $m(m-1)/2$) and then define the composition law as in (3.6). Finally, by means of the general results on stratified groups in Chapter 2, we obtain the following theorem.

Theorem 3.2.2 (Characterization. II). *Every N-dimensional* (*not necessarily homogeneous*) *stratified group of step two and m generators is naturally isomorphic to a* homogeneous *Carnot group* $(\mathbb{R}^N, \circ, \delta_\lambda)$ *with Lie group law as in* (3.6) *for some $m \times m$ linearly independent* skew-symmetric *matrices $B^{(k)}$'s. The group of dilations is given by* (3.7), *and the inverse of x is $-x$.*

By (3.8), we can write explicitly the canonical sub-Laplacian of the Lie group $\mathbb{G} = (\mathbb{R}^N, \circ)$ with $\circ$ as in (3.6). It is given by

$$\begin{aligned} \Delta_{\mathbb{G}} = \Delta_x &+ \frac{1}{4}\sum_{h,k=1}^{n} \langle B^{(h)}x, B^{(k)}x\rangle\, \partial_{t_h t_k} \\ &+ \sum_{k=1}^{m} \langle B^{(k)}x, \nabla_x\rangle\, \partial_{t_k} + \frac{1}{2}\sum_{k=1}^{n} \operatorname{trace}(B^{(k)})\, \partial_{t_k}. \end{aligned} \tag{3.9}$$

Here, we denoted

$$\Delta_x = \sum_{i=1}^{m} \partial_{x_i,x_i} \quad \text{and} \quad \nabla_x = (\partial_{x_1}, \ldots, \partial_{x_m}).$$

We recognize that $\Delta_{\mathbb{G}}$ contains partial differential terms of *second order only* if $\mathrm{trace}(B^{(k)}) = 0$ for every $k = 1, \ldots, n$. This happens, for example, if the $B^{(k)}$'s are skew-symmetric, i.e. if the inverse map on $\mathbb{G}$ is $x \mapsto -x$.

Example 3.2.3. Following all the above notation, let us take $m = 3, n = 2$ and

$$B^{(1)} = \begin{pmatrix} 1 & 1 & 0 \\ -1 & 0 & 0 \\ 0 & 0 & 0 \end{pmatrix}, \qquad B^{(2)} = \begin{pmatrix} 0 & 0 & -1 \\ 0 & 1 & 0 \\ 1 & 0 & 0 \end{pmatrix}.$$

Then the composition law on $\mathbb{R}^5 = \mathbb{R}^3 \times \mathbb{R}^2$ as in (3.6) becomes (denoting $(x,t) = (x_1, x_2, x_3, t_1, t_2)$ and analogously for (ξ, τ))

$$(x,t) \circ (\xi, \tau) = \begin{pmatrix} x_1 + \xi_1 \\ x_2 + \xi_2 \\ x_3 + \xi_3 \\ t_1 + \tau_1 + \frac{1}{2}\left(x_1\xi_1 + \xi_1 x_2 - \xi_2 x_1\right) \\ t_2 + \tau_2 + \frac{1}{2}\left(x_2\xi_2 - \xi_1 x_3 - \xi_3 x_1\right) \end{pmatrix},$$

and the dilation is

$$\delta_\lambda(x_1, x_2, x_3, t_1, t_2) = (\lambda x_1, \lambda x_2, \lambda x_3, \lambda^2 t_1, \lambda^2 t_2).$$

Then $\mathbb{G} = (\mathbb{R}^5, \circ, \delta_\lambda)$ is a homogeneous Carnot group, for the skew-symmetric parts of $B^{(1)}$ and $B^{(2)}$ are linearly independent,

$$\frac{1}{2}\left(B^{(1)} - (B^{(1)})^T\right) = \begin{pmatrix} 0 & 1 & 0 \\ -1 & 0 & 0 \\ 0 & 0 & 0 \end{pmatrix}, \qquad \frac{1}{2}\left(B^{(2)} - (B^{(2)})^T\right) = \begin{pmatrix} 0 & 0 & -1 \\ 0 & 0 & 0 \\ 1 & 0 & 0 \end{pmatrix}.$$

In fact, we can compute the first three vector fields of the Jacobian basis and verify that they are Lie-generators for the whole Lie algebra,

$$\begin{aligned}
X_1 &= \partial_{x_1} + \frac{1}{2}(x_1 + x_2)\partial_{t_1} - \frac{1}{2}x_3\partial_{t_2}, \\
X_2 &= \partial_{x_2} - \frac{1}{2}x_1\partial_{t_1} + \frac{1}{2}x_2\partial_{t_2}, \\
X_3 &= \partial_{x_3} + \frac{1}{2}x_1\partial_{t_2}, \\
[X_1, X_2] &= -\partial_{t_1}, \\
[X_1, X_3] &= \partial_{t_2}, \\
[X_2, X_3] &= \frac{1}{2}\partial_{t_2}.
\end{aligned}$$

The related canonical sub-Laplacian is

$$\begin{aligned}\Delta_{\mathbb{G}} &= \partial_{x_1,x_1} + \partial_{x_2,x_2} + \partial_{x_3,x_3} \\ &\quad + \frac{1}{4}\Big\{\big\{(x_1+x_2)^2 + (-x_1)^2\big\}\,\partial_{t_1,t_1} + \big\{(-x_3)^2 + (x_2)^2 + (x_1)^2\big\}\,\partial_{t_2,t_2} \\ &\qquad + 2\big\{(x_1+x_2)(-x_3) + (-x_1)(x_2)\big\}\,\partial_{t_1,t_2}\Big\} \\ &\quad + \{(x_1+x_2)\,\partial_{x_1} - x_1\,\partial_{x_2}\}\,\partial_{t_1} + \{-x_3\,\partial_{x_1} + x_2\,\partial_{x_2} + x_1\,\partial_{x_3}\}\,\partial_{t_2} \\ &\quad + \frac{1}{2}\,\partial_{t_1} + \frac{1}{2}\,\partial_{t_2}.\end{aligned}$$

$\Delta_{\mathbb{G}}$ contains first order terms, for $\mathrm{trace}(B^{(1)}) \neq 0 \neq \mathrm{trace}(B^{(2)})$.

On the contrary, if

$$B^{(1)} = \begin{pmatrix} 1 & 1 & 0 \\ -1 & 0 & 0 \\ 0 & 0 & 0 \end{pmatrix}, \qquad B^{(2)} = \begin{pmatrix} 0 & -2 & 0 \\ 2 & 1 & 0 \\ 0 & 0 & 0 \end{pmatrix},$$

then the composition law on $\mathbb{R}^5$ given by

$$(x,t)\circ(\xi,\tau) = \begin{pmatrix} x_1+\xi_1 \\ x_2+\xi_2 \\ x_3+\xi_3 \\ t_1+\tau_1+\frac{1}{2}\,(x_1\xi_1+\xi_1x_2-\xi_2x_1) \\ t_2+\tau_2+\frac{1}{2}\,(x_2\xi_2-2\xi_1x_2+2\xi_2x_1) \end{pmatrix}$$

does not define a *homogeneous* Carnot group, because the skew-symmetric parts of $B^{(1)}$ and $B^{(2)}$ are linearly dependent,

$$\frac{1}{2}\big(B^{(1)} - (B^{(1)})^T\big) = \begin{pmatrix} 0 & 1 & 0 \\ -1 & 0 & 0 \\ 0 & 0 & 0 \end{pmatrix}, \qquad \frac{1}{2}\big(B^{(2)} - (B^{(2)})^T\big) = \begin{pmatrix} 0 & -2 & 0 \\ 2 & 0 & 0 \\ 0 & 0 & 0 \end{pmatrix}.$$

In fact, the only admissible dilation would be

$$\delta_\lambda(x,t) = (\lambda\,x_1, \lambda\,x_2, \lambda\,x_3, \lambda^2\,t_1, \lambda^2\,t_2),$$

but the first three vector fields of the related Jacobian basis are not Lie-generators for the whole Lie algebra, since

$$\begin{aligned} X_1 &= \partial_{x_1} + \frac{1}{2}(x_1+x_2)\partial_{t_1} - x_2\partial_{t_2}, \\ X_2 &= \partial_{x_2} - \frac{1}{2}x_1\partial_{t_1} + \Big(\frac{1}{2}x_2 + x_1\Big)\partial_{t_2}, \\ X_3 &= \partial_{x_3}, \\ [X_1,X_2] &= -\partial_{t_1} + 2\partial_{t_2}, \\ [X_1,X_3] &= [X_2,X_3] = 0. \end{aligned}$$

Remark 3.2.4. The composition law on the Heisenberg–Weyl group $\mathbb{H}^k$ on $\mathbb{R}^{2k+1}$, which is given by

$$(x, y, t) \circ (x', y', t') = \big(x + x', y + y', t + t' + 2\,(\langle y, x'\rangle - \langle x, y'\rangle)\big),$$

is of the form (3.6) with $m = 2k$, $n = 1$ and

$$B^{(1)} = 4 \begin{pmatrix} 0 & \mathbb{I}_k \\ -\mathbb{I}_k & 0 \end{pmatrix}.$$

3.3 Free Step-two Homogeneous Groups

In this section, following the notation of the previous one, we fix a particular set of matrices $B^{(k)}$'s and consider the relevant homogeneous Carnot group $(\mathbb{F}_{m,2}, \star)$, which will serve as prototype for what we shall call *free* Carnot group of step two and m generators. Throughout the section, $m \geq 2$ is a fixed integer.

Let $i, j \in \{1, \dots, m\}$ be fixed with $i > j$, and let $S^{(i,j)}$ be the $m \times m$ skew-symmetric matrix whose entries are -1 in the position (i, j), $+1$ in the position (j, i) and 0 elsewhere. For example, if $m = 3$, we have

$$S^{(2,1)} = \begin{pmatrix} 0 & 1 & 0 \\ -1 & 0 & 0 \\ 0 & 0 & 0 \end{pmatrix}, \qquad S^{(3,1)} = \begin{pmatrix} 0 & 0 & 1 \\ 0 & 0 & 0 \\ -1 & 0 & 0 \end{pmatrix},$$

$$S^{(3,2)} = \begin{pmatrix} 0 & 0 & 0 \\ 0 & 0 & 1 \\ 0 & -1 & 0 \end{pmatrix}.$$

Then, we agree to denote by $(\mathbb{F}_{m,2}, \star)$ the Carnot group on $\mathbb{R}^N$ associated to these $m(m-1)/2$ matrices according to (3.6) of the previous section. We set

$$n := m(m-1)/2, \qquad N = m + n = m(m+1)/2,$$
$$\mathcal{I} := \{(i, j) \mid 1 \leq j < i \leq m\}.$$

We observe that the set $\mathcal{I}$ has exactly n elements.

In the sequel of this section, we shall use the following notation, different from the one used in the previous section: instead of using the notation t for the coordinate in the "second layer" of the group, we denote the points of $\mathbb{F}_{m,2}$ by (x, γ), where $x = (x_1, \dots, x_m) \in \mathbb{R}^m$, $\gamma \in \mathbb{R}^n$, and the coordinates of γ are denoted by

$$\gamma_{i,j} \quad \text{where } (i, j) \in \mathcal{I}.$$

Here *we have ordered $\mathcal{I}$ in an arbitrary (henceforth) fixed way.* Then, the composition law $\star$ is given by

$$(x, \gamma) \star (x', \gamma') = \begin{pmatrix} x_h + x'_h, \quad h = 1, \dots, m \\ \gamma_{i,j} + \gamma'_{i,j} + \frac{1}{2}(x_i x'_j - x_j x'_i), \quad (i, j) \in \mathcal{I} \end{pmatrix}.$$

For example, when $m = 3$, we have

$$(x, \gamma) \circ (x', \gamma') = \begin{pmatrix} x_1 + x_1' \\ x_2 + x_2' \\ x_3 + x_3' \\ \gamma_{2,1} + \gamma_{2,1}' + \frac{1}{2}(x_2 x_1' - x_1 x_2') \\ \gamma_{3,1} + \gamma_{3,1}' + \frac{1}{2}(x_3 x_1' - x_1 x_3') \\ \gamma_{3,2} + \gamma_{3,2}' + \frac{1}{2}(x_3 x_2' - x_2 x_3') \end{pmatrix}.$$

By (3.8), we can compute the Jacobian basis

$$X_h, \quad h = 1, \dots, m, \qquad \Gamma_{i,j}, \quad (i, j) \in \mathcal{I},$$

of $\mathfrak{f}_{m,2}$, the Lie algebra of $\mathbb{F}_{m,2}$: it holds

$$\begin{aligned} X_h &= (\partial/\partial x_h) + \frac{1}{2} \sum_{1 \le j < i \le m} \left(\sum_{l=1}^{m} S_{h,l}^{(i,j)} x_l \right) (\partial/\partial \gamma_{i,j}) \\ &= \begin{cases} \partial_{x_1} + \frac{1}{2} \sum_{1 < i \le m} x_i \, \partial \gamma_{i,1} & \text{if } h = 1, \\ \partial_{x_h} + \frac{1}{2} \sum_{h < i \le m} x_i \, \partial \gamma_{i,h} - \frac{1}{2} \sum_{1 \le j < m} x_j \, \partial \gamma_{h,j} & \text{if } 1 < h < m, \\ \partial_{x_m} + \frac{1}{2} \sum_{1 \le j < m} x_j \, \partial \gamma_{m,j} & \text{if } h = m, \end{cases} \\ \Gamma_{i,j} &= \partial/\partial \gamma_{i,j}, \quad (i, j) \in \mathcal{I}. \end{aligned}$$

Moreover, for every $(i, j) \in \mathcal{I}$, we have the commutator identities

$$[X_j, X_i] = \sum_{1 \le k < h \le m} S_{i,j}^{(h,k)} \, \partial \gamma_{h,k} = \partial \gamma_{j,i} - \partial \gamma_{i,j},$$

whence we recognize that the algebra $\mathfrak{f}_{m,2}$ is "the most non-Abelian as possible" (as it is allowed for an algebra with m generators and step two).

This is the reason why we shall refer to (any algebra isomorphic to) $\mathfrak{f}_{m,2}$ as a *free Lie algebra with m generators and step two* (see Chapter 14 for more details on free Lie algebras). For example, when $m = 3$, we have

$$\begin{aligned} X_1 &= \partial x_1 + \frac{1}{2}(x_2 \, \partial \gamma_{2,1} + x_3 \, \partial \gamma_{3,1}), \\ X_2 &= \partial x_2 + \frac{1}{2}(x_3 \, \partial \gamma_{3,2} - x_1 \, \partial \gamma_{2,1}), \\ X_3 &= \partial x_3 - \frac{1}{2}(x_1 \, \partial \gamma_{3,1} - x_2 \, \partial \gamma_{3,2}) \\ \Gamma_{2,1} &= \partial \gamma_{2,1}, \quad \Gamma_{3,1} = \partial \gamma_{3,1}, \quad \Gamma_{3,2} = \partial \gamma_{3,2}. \end{aligned}$$

From (3.9), we derive the explicit expression for the canonical sub-Laplacian of $\mathbb{F}_{3,2}$,

$$\begin{aligned}
\Delta_{\mathbb{F}_{3,2}} &= (\partial x_1)^2 + (\partial x_2)^2 + (\partial x_3)^2 \\
&+ \frac{1}{4}\left\{(x_2^2 + x_1^2)(\partial \gamma_{2,1})^2 + (x_3^2 + x_1^2)(\partial \gamma_{3,1})^2 + (x_3^2 + x_2^2)(\partial \gamma_{3,2})^2\right\} \\
&+ \frac{1}{2} x_2 x_3\, (\partial \gamma_{2,1}\, \partial \gamma_{3,1}) - \frac{1}{2} x_1 x_3\, (\partial \gamma_{2,1}\, \partial \gamma_{3,2}) + \frac{1}{2} x_1 x_2\, (\partial \gamma_{3,1}\, \partial \gamma_{3,2}) \\
&+ (x_2 \partial x_1 - x_1 \partial x_2)\partial \gamma_{2,1} + (x_3 \partial x_1 - x_1 \partial x_3)\partial \gamma_{3,1} \\
&+ (x_3 \partial x_2 - x_2 \partial x_3)\partial \gamma_{3,2}. \qquad (3.10)
\end{aligned}$$

3.4 Change of Basis

In this section we consider "stratified" changes of basis on a homogeneous Carnot group of step two, according to the definition in Remark 1.4.10, page 61. We use the usual notation, as in the beginning of Section 3.2.

Let C, D be two fixed non-singular matrices, with C of dimension $m \times m$ and D of dimension $n \times n$. We denote by L the following $N \times N$ matrix

$$L = \begin{pmatrix} C & 0 \\ 0 & D \end{pmatrix}.$$

Finally, we denote again by L the relevant linear change of basis on $\mathbb{R}^N$, i.e., the linear map ($\cdot$ denotes row-times-column matrix product)

$$L : \mathbb{R}^N \to \mathbb{R}^N, \quad L(x, t) = L \cdot \begin{pmatrix} x \\ t \end{pmatrix} = \begin{pmatrix} C \cdot x \\ D \cdot t \end{pmatrix}.$$

We define on $\mathbb{R}^N$ a new composition law $*$ obtained by writing $\circ$ in the new coordinates defined by $(\xi, \tau) = L(x, t)$. More precisely, we have

$$(\xi, \tau) * (\xi', \tau') := L\big((L^{-1}(\xi, \tau)) \circ (L^{-1}(\xi', \tau'))\big).$$

An explicit computation gives (here $D = (d_{h,k})_{h,k \le n}$)

$$\begin{aligned}
&(\xi, \tau) * (\xi', \tau') \\
&= \begin{pmatrix} \xi_i + \xi_i', \quad i = 1, \dots, m \\ \tau_h + \tau_h' + \frac{1}{2}\big\langle (C^{-1})^T \cdot (\sum_{k=1}^n d_{h,k}\, B^{(k)}) \cdot C^{-1}\xi, \xi' \big\rangle, \quad h = 1, \dots, n \end{pmatrix},
\end{aligned}$$

i.e. setting

$$\widetilde{B}^{(h)} := (C^{-1})^T \cdot \left(\sum_{k=1}^n d_{h,k}\, B^{(k)} \right) \cdot C^{-1} \quad \text{for every } h = 1, \dots, n,$$

$*$ can be written as follows:

$$(\xi, \tau) * (\xi', \tau') = \begin{pmatrix} \xi_i + \xi_i', \quad i = 1, \dots, m \\ \tau_h + \tau_h' + \frac{1}{2}\langle \widetilde{B}^{(h)}\xi, \xi' \rangle, \quad h = 1, \dots, n \end{pmatrix}.$$

We explicitly remark that, if the matrices $B^{(h)}$'s are skew-symmetric, the same is true for the matrices $\widetilde{B}^{(h)}$'s. It can easily be proved that $\mathbb{H} = (\mathbb{R}^N, *, \delta_\lambda)$ is a homogeneous Carnot group isomorphic to $\mathbb{G} = (\mathbb{R}^N, \circ, \delta_\lambda)$ via L (see also Section 16.3, page 637).

It is immediately seen that the canonical basis of $\mathbb{H}$ is

$$(\partial/\partial\xi_i) + \frac{1}{2}\sum_{h=1}^{n}\left(\sum_{j=1}^{m}\widetilde{B}_{i,j}^{(h)}x_j\right)(\partial/\partial\tau_h), \quad i = 1, \dots, m,$$

$$\partial/\partial\tau_k, \quad k = 1, \dots, n.$$

Moreover, it is not difficult to see that the expression $\widetilde{X}_i$ of the i-th vector X_i of the canonical basis of $\mathbb{G}$ in the new coordinates (ξ, τ) on $\mathbb{H}$ (i.e. the image via $\mathrm{d}L$ of X_i) coincides at the origin with

$$L \cdot X_i I(0) = (C \cdot e_i, 0),$$

where $(e_1, \dots, e_m)$ denotes the canonical basis of $\mathbb{R}^m$. In particular, *given an arbitrary sub-Laplacian $\mathcal{L} = \sum_{j=1}^m Y_j^2$ on $\mathbb{G}$, it is always possible to perform a linear change of basis on $\mathbb{R}^N$ so that $\mathcal{L}$ is turned into the canonical sub-Laplacian on a homogeneous Carnot group isomorphic to $\mathbb{G}$.*

3.5 The Exponential Map of a Step-two Homogeneous Group

We now turn to compute the exponential map of a homogeneous group of step two. As usual, if $\xi \in \mathbb{R}^m$ and $\tau \in \mathbb{R}^n$, we write

$$(\xi, \tau) \cdot Z = \sum_{j=1}^{m} \xi_j X_j + \sum_{i=1}^{n} \tau_i T_i,$$

where $X_1, \dots, X_m, T_1, \dots, T_n$ is the Jacobian basis. More explicitly, by (3.8), we have

$$\big((\xi, \tau) \cdot Z\big)I = \begin{pmatrix} \xi \\ \frac{1}{2}\sum_{i,j=1}^{m} b_{i,j}^{(1)}\xi_i x_j \\ \vdots \\ \frac{1}{2}\sum_{i,j=1}^{m} b_{i,j}^{(n)}\xi_i x_j \end{pmatrix}.$$

By the definition of exponential map, we have $\mathrm{Exp}\,((\xi, \tau) \cdot Z) = (x(1), t(1))$, where $(x(s), t(s))$ solves

$$(\dot{x}(s), \dot{t}(s)) = ((\xi, \tau) \cdot Z)I(x(s), t(s)), \quad (x(0), t(0)) = (0, 0),$$

i.e. more explicitly,

$$\begin{cases} \dot{x}(s) = \xi, & x(0) = 0, \\ \dot{t}(s) = \tau + \frac{1}{2} \sum_{i,j=1}^{m} B_{i,j}\, \xi_i\, x_j(s), & t(0) = 0. \end{cases}$$

Here, we have set for brevity

$$B_{i,j} := \begin{pmatrix} b_{i,j}^{(1)} \\ \vdots \\ b_{i,j}^{(n)} \end{pmatrix}.$$

A straightforward computation gives

$$\operatorname{Exp}((\xi,\tau)\cdot Z) = \left(\xi, \tau + \frac{1}{4}\sum_{i,j=1}^{m} B_{i,j}\,\xi_i\xi_j\right), \tag{3.11a}$$

(here we again set $B_{i,j} = (B_{i,j}^{(1)}, \ldots, B_{i,j}^{(n)})$), whence

$$\operatorname{Log}(x,t) = \left(x, t - \frac{1}{4}\sum_{i,j=1}^{m} B_{i,j}\, x_i x_j\right)\cdot Z. \tag{3.11b}$$

We explicitly remark that, by (3.11a), the exponential map acts like a sort of "identity map" (provided $\mathfrak{g}$ is equipped with coordinates w.r.t. the Jacobian basis) if and only if the matrices $B^{(i)}$'s are skew-symmetric.

We now operate a change of coordinates in $\mathbb{R}^N$ inspired by the exponential map. In other words (see the proof of Theorem 2.2.18, page 131), we equip the algebra $\mathfrak{g}$ of $\mathbb{G}$ by a Lie-group structure isomorphic via Exp to that of $(\mathbb{G}, \circ)$, and we identify $\mathfrak{g}$ with $\mathbb{R}^N$ via the Jacobian basis.

This amounts to equip $\mathbb{R}^N = \mathbb{R}^m \times \mathbb{R}^n$ with the composition law $\diamond_Z$ (the reason for this notation will become clearer in further sections) such that

$$\operatorname{Log}\Big((\operatorname{Exp}((\xi,\tau)\cdot Z)) \circ (\operatorname{Exp}((\xi',\tau')\cdot Z))\Big) = \big((\xi,\tau)\diamond_Z(\xi',\tau')\big)\cdot Z.$$

We find an explicit expression for $\diamond_Z$. It holds

$$\begin{aligned} &\operatorname{Log}\Big((\operatorname{Exp}((\xi,\tau)\cdot Z)) \circ (\operatorname{Exp}((\xi',\tau')\cdot Z))\Big) \\ &= \operatorname{Log}\left(\left(\xi, \tau + \frac{1}{4}\sum_{i,j=1}^{m} B_{i,j}\,\xi_i\xi_j\right) \circ \left(\xi', \tau' + \frac{1}{4}\sum_{i,j=1}^{m} B_{i,j}\,\xi'_i\xi'_j\right)\right) \\ &= \operatorname{Log}\left(\xi+\xi', \tau+\tau' + \frac{1}{4}\sum_{i,j=1}^{m} B_{i,j}\,(\xi_i\xi_j + \xi'_i\xi'_j) + \frac{1}{2}\sum_{i,j=1}^{m} B_{i,j}\,\xi_j\xi'_i\right) \\ &= \Bigg(\xi+\xi', \tau+\tau' + \frac{1}{4}\sum_{i,j=1}^{m} B_{i,j}\,(\xi_i\xi_j + \xi'_i\xi'_j) + \frac{1}{2}\sum_{i,j=1}^{m} B_{i,j}\,\xi_j\xi'_i \end{aligned}$$

$$-\frac{1}{4}\sum_{i,j=1}^{m} B_{i,j}\,(\xi_i+\xi_i')(\xi_j+\xi_j')\Bigg)\cdot Z$$

$$=\left(\xi+\xi',\tau+\tau'+\frac{1}{4}\sum_{i,j=1}^{m} B_{i,j}\,\xi_j\xi_i'-\frac{1}{4}\sum_{i,j=1}^{m} B_{i,j}\,\xi_i\xi_j')\right)\cdot Z$$

$$=\left(\xi+\xi',\tau+\tau'+\frac{1}{2}\sum_{i,j=1}^{m}\frac{B_{i,j}-B_{j,i}}{2}\,\xi_j\xi_i'\right)\cdot Z.$$

As a consequence, the composition law $*$ on $\mathbb{R}^N$ has the form

$$(\xi,\tau)\diamond_Z(\xi',\tau')=\left(\xi+\xi',\tau+\tau'+\frac{1}{2}\sum_{i,j=1}^{m}\frac{B_{i,j}-B_{j,i}}{2}\,\xi_j\xi_i'\right), \tag{3.12}$$

i.e. it has the form as in (3.6) with the matrices $B^{(k)}$'s replaced by the *skew-symmetric* matrices

$$\widetilde{B}^{(k)}=\frac{B^{(k)}-(B^{(k)})^T}{2},\qquad k=1,\dots,n,$$

the skew-symmetric parts of the $B^{(k)}$'s. The inverse map on $(\mathbb{R}^N,\diamond_Z)$ is the usual $(-x,-t)$. The homogeneous Lie group $(\mathbb{R}^N,\diamond_Z,\delta_\lambda)$ is isomorphic to $(\mathbb{R}^N,\circ,\delta_\lambda)$ via the Lie-group isomorphism

$$\Psi:(\mathbb{R}^N,\diamond_Z)\to(\mathbb{R}^N,\circ),\quad \Psi(\xi,\tau):=\left(\xi,\tau+\frac{1}{4}\sum_{i,j=1}^{m} B_{i,j}\,\xi_i\xi_j\right),$$

which "essentially" equals the exponential map (written in the Jacobian coordinates). Since the differential of Ψ is associated to the matrix

$$\mathcal{J}_\Psi(\xi,\tau)=\left(\begin{array}{c|c}\mathbb{I}_m & 0\\\hline \left(\sum_{j=1}^m \frac{B^{(k)}_{i,j}+B^{(k)}_{j,i}}{2}\,\xi_j\right)_{k\le n,i\le m} & \mathbb{I}_n\end{array}\right),$$

we see that the Jacobian basis of $(\mathbb{R}^N,\diamond_Z)$ is turned into the Jacobian basis of $(\mathbb{R}^N,\circ)$ by the change of coordinates Ψ (see, for instance, (2.17b), page 101, or (2.71), page 153).

For future references, we summarize what we have just proved in the following proposition.

Proposition 3.5.1. *Let $\mathbb{R}^N=\mathbb{R}^m\times\mathbb{R}^n$ be equipped with a homogeneous Lie group structure by the composition law*

$$(x,t)\circ(\xi,\tau)=\left(x+\xi,t_1+\tau_1+\frac{1}{2}\langle B^{(1)}x,\xi\rangle,\dots,t_n+\tau_n+\frac{1}{2}\langle B^{(n)}x,\xi\rangle\right)$$

and the dilation $\delta_\lambda(x,t)=(\lambda x,\lambda^2 t)$, where $B^{(1)},\dots,B^{(n)}$ are fixed $m\times m$ matrices. Then $\mathbb{G}=(\mathbb{R}^N,\circ,\delta_\lambda)$ is isomorphic to the homogeneous Lie group $\widetilde{\mathbb{G}}=(\mathbb{R}^N,\diamond_Z,\delta_\lambda)$, where:

1) δ_λ *is the same dilation as above;*
2) $\diamond_Z$ *is defined by*

$$(x,t)\diamond_Z(\xi,\tau)=\left(x+\xi, t_1+\tau_1+\frac{1}{2}\langle\widetilde{B}^{(1)}x,\xi\rangle,\dots,t_n+\tau_n+\frac{1}{2}\langle\widetilde{B}^{(n)}x,\xi\rangle\right)$$

where $\widetilde{B}^{(i)}$ *is the skew-symmetric part of* $B^{(i)}$ *for every* $i=1,\dots,n$; *the inverse map on* $\widetilde{\mathbb{G}}$ *is* $(-x,-t)$;
3) *the Lie-group isomorphism is* $\Psi:\widetilde{\mathbb{G}}\to\mathbb{G}$ *with*

$$\Psi(\xi,\tau)=\left(\xi,\tau_1+\frac{1}{4}\langle B^{(1)}\xi,\xi\rangle,\dots,\tau_n+\frac{1}{4}\langle B^{(n)}\xi,\xi\rangle\right),$$

so that Ψ *is the identity map iff all the* $B^{(i)}$*'s are skew-symmetric;*
4) *the Jacobian basis of* $\widetilde{\mathbb{G}}$ *corresponds (via* $\mathrm{d}\Psi$*) to the Jacobian basis of* $\mathbb{G}$*, in particular the canonical sub-Laplacian of* $\widetilde{\mathbb{G}}$ *corresponds to that of* $\mathbb{G}$;
5) *the exponential map of the group* $\widetilde{\mathbb{G}}$ *is the linear map sending the Jacobian basis of the algebra of* $\widetilde{\mathbb{G}}$ *into the canonical basis of* $\mathbb{G}$;
6) *if* $\mathbb{G}$ *is a homogeneous Carnot group, then the same is true for* $\widetilde{\mathbb{G}}$.

3.6 Prototype Groups of Heisenberg Type

Definition 3.6.1 ((Prototype) Heisenberg-type group). *Consider the homogeneous Lie group*

$$\mathbb{H}=(\mathbb{R}^{m+n},\circ,\delta_\lambda)$$

with composition law as in (3.6) *(page* 158*), i.e.*

$$(x,t)\circ(\xi,\tau)=\left(x+\xi, t_1+\tau_1+\frac{1}{2}\langle B^{(1)}x,\xi\rangle,\dots,t_n+\tau_n+\frac{1}{2}\langle B^{(n)}x,\xi\rangle\right),$$

where $B^{(1)},\dots,B^{(n)}$ *are fixed* $m\times m$ *matrices, and dilations as in* (3.7). *Let us also assume that the matrices* $B^{(1)},\dots,B^{(n)}$ *have the following properties:*

1) $B^{(j)}$ *is an* $m\times m$ *skew-symmetric and orthogonal matrix for every* $j\le n$;
2) $B^{(i)}B^{(j)}=-B^{(j)}B^{(i)}$ *for every* $i,\ j\in\{1,\dots,n\}$ *with* $i\ne j$.

If all these conditions are satisfied, $\mathbb{H}$ *is called a (prototype) group of Heisenberg-type, in short, a (prototype) H-type group.*

A H-type group is a Carnot group, since conditions 1) and 2) imply the linear independence of $B^{(1)},\dots,B^{(n)}$.

Indeed, if $\alpha=(\alpha_1,\dots,\alpha_n)\in\mathbb{R}^n\setminus\{0\}$, then

$$\frac{1}{|\alpha|}\sum_{s=1}^{n}\alpha_s\,B^{(s)}$$

is orthogonal (hence non-vanishing), as the following computation shows,

$$\begin{aligned}
&\left(\frac{1}{|\alpha|}\sum_{s=1}^{n}\alpha_s B^{(s)}\right)\cdot\left(\frac{1}{|\alpha|}\sum_{s=1}^{n}\alpha_s B^{(s)}\right)^T \\
&= -\frac{1}{|\alpha|^2}\sum_{r,s\le n}\alpha_r\alpha_s\, B^{(r)}B^{(s)} \\
&= -\frac{1}{|\alpha|^2}\sum_{r\le n}\alpha_r^2\,(B^{(r)})^2 - \frac{1}{|\alpha|^2}\sum_{r,s\le n,\, r\ne s}\alpha_r\alpha_s\, B^{(r)}B^{(s)} \\
&= \mathbb{I}_m.
\end{aligned}$$

Here we used the following facts: $(B^{(r)})^2 = -\mathbb{I}_m$, since $B^{(r)}$ is skew-symmetric and orthogonal; $B^{(r)}B^{(s)} = -B^{(s)}B^{(r)}$ according to condition 2).

The generators of $\mathbb{H}$ are the vector fields (see (3.8), page 159)

$$X_i = \partial_{x_i} + \frac{1}{2}\sum_{k=1}^{n}\left(\sum_{l=1}^{m} b_{i,l}^{(k)} x_l\right)\partial_{t_k}, \quad i = 1,\dots,m. \tag{3.13a}$$

Moreover, if we set

$$T_k := \partial/\partial t_k, \quad k = 1,\dots,n, \tag{3.13b}$$

then (again from (3.8)) we know that

$$\{X_1,\dots,X_m;\ T_1,\dots,T_n\} \tag{3.13c}$$

is the Jacobian basis for $\mathbb{H}$.

A direct computation shows that the canonical sub-Laplacian $\Delta_{\mathbb{H}} = \sum_{i=1}^m X_i^2$ can be written as follows (see (3.9))

$$\begin{aligned}
\Delta_{\mathbb{H}} = \Delta_x &+ \frac{1}{4}\sum_{h,k=1}^{n}\langle B^{(h)}x, B^{(k)}x\rangle\,\partial_{t_h t_k} \\
&+ \sum_{k=1}^{m}\langle B^{(k)}x, \nabla_x\rangle\,\partial_{t_k} + \frac{1}{2}\sum_{k=1}^{n}\mathrm{trace}(B^{(k)})\,\partial_{t_k}.
\end{aligned}$$

On the other hand, by conditions 1) and 2),

$$\langle B^{(h)}x, B^{(h)}x\rangle = |x|^2,$$

while, for $h \ne k$,

$$\langle B^{(h)}x, B^{(k)}x\rangle = 0$$

since

$$\begin{aligned}
\langle B^{(h)}x, B^{(k)}x\rangle &= -\langle B^{(k)}\,B^{(h)}x, x\rangle = \langle B^{(h)}\,B^{(k)}x, x\rangle \\
&= -\langle B^{(k)}x, B^{(h)}x\rangle.
\end{aligned}$$

We also have $\text{trace}(B^{(k)}) = 0$, since $B^{(k)}$ is skew-symmetric. Then $\Delta_{\mathbb{H}}$ takes the very compact form

$$\Delta_{\mathbb{H}} = \Delta_x + \frac{1}{4}|x|^2 \Delta_t + \sum_{k=1}^{n} \langle B^{(k)} x, \nabla_x \rangle \, \partial_{t_k}. \tag{3.14}$$

By the computations made in Section 3.5 in finding the exponential map of a general homogeneous group of step two, we infer that (see precisely (3.11a), page 167) if $\{X_1, \dots, X_m; T_1, \dots, T_n\}$ is the Jacobian basis for $\mathbb{H}$ (where the X_i's and the T_k's are as in (3.13a), (3.13b)) then *the exponential map for the (prototype) H-type group* $\mathbb{H}$ *is*

$$\text{Exp} : \mathfrak{h} \to \mathbb{H}, \quad \text{Exp}\,(x_1 X_1 + \cdots + x_m X_m + t_1 T_1 + \cdots + t_n T_n) = (x, t). \tag{3.15}$$

Remark 3.6.2. From (3.8) one obtains

$$\begin{aligned} |\nabla_{\mathbb{H}} u|^2 = |\nabla_x u|^2 &+ \frac{1}{4} \sum_{i=1}^{m} \langle (Bx)_i, \nabla_t u \rangle^2 \\ &+ \sum_{i=1}^{m} \langle (Bx)_i, \nabla_t u \rangle \, \partial_{x_i} u, \quad u \in C^\infty. \end{aligned}$$

On the other hand,

$$\begin{aligned} \sum_{i=1}^{m} \langle (Bx)_i, \nabla_t u \rangle^2 &= \sum_{h,k=1}^{n} \langle B^{(h)} x, B^{(k)} x \rangle \partial_{t_h} u \, \partial_{t_k} u \\ &= |x|^2 \, |\nabla_t u|^2 \end{aligned}$$

and

$$\sum_{i=1}^{m} \langle (Bx)_i, \nabla_t u \rangle \, \partial_{x_i} u = \sum_{k=1}^{n} \langle B^{(k)} x, \nabla_x u \rangle \, \partial_{t_k} u.$$

Thus, for every smooth real-valued function u, it holds

$$|\nabla_{\mathbb{H}} u|^2 = |\nabla_x u|^2 + \frac{1}{4} |x|^2 \, |\nabla_t u|^2 + \sum_{k=1}^{n} \langle B^{(k)} x, \nabla_x u \rangle \, \partial_{t_k} u. \tag{3.16}$$

Remark 3.6.3. The first layer of a H-type group has even dimension m. Indeed, if B is a $m \times m$ skew-symmetric orthogonal matrix, we have $\mathbb{I}_m = B \cdot B^T = -B^2$, whence $1 = (-1)^m (\det B)^2$.

Remark 3.6.4. With the previous notation, if $\mathbb{H} = (\mathbb{R}^{m+n}, \circ, \delta_\lambda)$ is a H-type group, then

$$\mathfrak{z} = \{(0, t) \mid t \in \mathbb{R}^n\}$$

is the center of $\mathbb{H}$. Indeed, let $(y, t) \in \mathbb{H}$ be such that

$$(x,s)\circ(y,t)=(y,t)\circ(x,s)\quad \text{for every } (x,s)\in\mathbb{H}.$$

This holds iff

$$\langle B^{(k)}x, y\rangle = \langle B^{(k)}y, x\rangle$$

for any $x\in\mathbb{R}^m$ and any $k\in\{1,\dots,n\}$. Then, since $(B^{(k)})^T=-B^{(k)}$,

$$\langle B^{(k)}y, x\rangle = 0 \quad \forall\, x\in\mathbb{R}^m\ \forall\, k\in\{1,\dots,n\},$$

so that $y=0$ because $B^{(k)}$ is orthogonal (hence non-singular).

Remark 3.6.5. The classical Heisenberg–Weyl group $\mathbb{H}^k$ on $\mathbb{R}^{2k+1}$ is canonically isomorphic to a prototype H-type group. Precisely (see (3.5b), page 157), it is isomorphic to the prototype H-type group $(\mathbb{H},*)$ corresponding to the case $m=2k$, $n=1$ and

$$B^{(1)}=\begin{pmatrix} 0 & -\mathbb{I}_k \\ \mathbb{I}_k & 0\end{pmatrix}.$$

The isomorphism $\varphi:(\mathbb{R}^{2k+1},*)\to(\mathbb{H}^k,\circ)$ is given by

$$\varphi(\xi,\eta,\tau)=(\xi,\eta,-4\tau).$$

Moreover, $\mathbb{H}$ is in its turn isomorphic to the prototype H-type group with $m=2k$, $n=1$ and

$$\widetilde{B}^{(1)}=\text{diag}\left\{\begin{pmatrix} 0 & -1 \\ 1 & 0\end{pmatrix},\dots,\begin{pmatrix} 0 & -1 \\ 1 & 0\end{pmatrix}\right\},\quad \text{the block occurring } k \text{ times.}$$

This type of prototype Heisenberg-group is the only (up to isomorphism) H-type group with one-dimensional center (see Chapter 18).

Remark 3.6.6. Groups of Heisenberg type with center of dimension $n\geq 2$ do exist. For example, the following two matrices

$$B^{(1)}=\begin{pmatrix} 0 & -1 & 0 & 0 \\ 1 & 0 & 0 & 0 \\ 0 & 0 & 0 & -1 \\ 0 & 0 & 1 & 0\end{pmatrix},\qquad B^{(2)}=\begin{pmatrix} 0 & 0 & 1 & 0 \\ 0 & 0 & 0 & -1 \\ -1 & 0 & 0 & 0 \\ 0 & 1 & 0 & 0\end{pmatrix}$$

satisfy conditions 1)–2) and hence they define in $\mathbb{R}^6=\mathbb{R}^4\times\mathbb{R}^2$ a H-type group whose center has dimension 2. The composition law is

$$(x,t)\circ(\xi,\tau)=\begin{pmatrix} x_1+\xi_1 \\ x_2+\xi_2 \\ x_3+\xi_3 \\ x_4+\xi_4 \\ t_1+\tau_1+\frac{1}{2}(-x_2\xi_1+x_1\xi_2-x_4\xi_3+x_3\xi_4) \\ t_2+\tau_2+\frac{1}{2}(x_3\xi_1-x_4\xi_2-x_1\xi_3+x_2\xi_4)\end{pmatrix}.$$

The above matrices $B^{(1)}$ and $B^{(2)}$, together with

$$B^{(3)} = \begin{pmatrix} 0 & 0 & 0 & 1 \\ 0 & 0 & 1 & 0 \\ 0 & -1 & 0 & 0 \\ -1 & 0 & 0 & 0 \end{pmatrix},$$

define in $\mathbb{R}^7 = \mathbb{R}^4 \times \mathbb{R}^3$ a H-type group whose center has dimension 3. The composition law is

$$(x,t) \circ (\xi, \tau) = \begin{pmatrix} x_1 + \xi_1 \\ x_2 + \xi_2 \\ x_3 + \xi_3 \\ x_4 + \xi_4 \\ t_1 + \tau_1 + \frac{1}{2}(-x_2\xi_1 + x_1\xi_2 - x_4\xi_3 + x_3\xi_4) \\ t_2 + \tau_2 + \frac{1}{2}(x_3\xi_1 - x_4\xi_2 - x_1\xi_3 + x_2\xi_4) \\ t_3 + \tau_3 + \frac{1}{2}(x_4\xi_1 + x_3\xi_2 - x_2\xi_3 - x_1\xi_4) \end{pmatrix}.$$

Remark 3.6.7. The following result holds (see [Kap80, Corollary 1]). *Let m, n be two positive integers. Then there exists a H-type group of dimension $m + n$ whose center has dimension n if and only if it holds $n < \rho(m)$, where ρ is the so-called Hurwitz–Radon function, i.e.*

$$\rho : \mathbb{N} \to \mathbb{N}, \quad \rho(m) := 8p + q, \quad \text{where } m = (\text{odd}) \cdot 2^{4p+q}, \ 0 \le q \le 3.$$

We explicitly remark that if m is odd, then $\rho(m) = 0$, whence the first layer of any H-type group has even dimension (as we already proved in Remark 3.6.3).

Remark 3.6.8. The groups of Heisenberg-type were introduced by A. Kaplan in [Kap80]. Kaplan's definition of H-type groups is more abstract than the one given here. In Chapter 18, we shall show that, up to an isomorphism, the two definitions are equivalent.

3.7 H-groups (in the Sense of Métivier)

Following G. Métivier [Met80], we give the following definition.

Definition 3.7.1 (H-group (in the sense of Métivier)). *Let $\mathfrak{g}$ be a (finite-dimensional real) Lie algebra, and let us denote by $\mathfrak{z}$ its center. We say that $\mathfrak{g}$ is of* H-type (in the sense of Métivier) *if it admits a vector space decomposition*

$$\mathfrak{g} = \mathfrak{g}_1 \oplus \mathfrak{g}_2 \qquad \begin{cases} [\mathfrak{g}_1, \mathfrak{g}_1] \subseteq \mathfrak{g}_2, \\ \mathfrak{g}_2 \subseteq \mathfrak{z} \end{cases}$$

with the following additional property: for every $\eta \in \mathfrak{g}_2^*$ (the dual space of $\mathfrak{g}_2$), the skew-symmetric bilinear form on $\mathfrak{g}_1$ defined by

$$B_\eta : \mathfrak{g}_1 \times \mathfrak{g}_1 \to \mathbb{R}, \quad B_\eta(X, X') := \eta([X, X'])$$

is non-degenerate[5] *whenever $\eta \neq 0$.*

[5] We recall that a bilinear map on a finite-dimensional vector space is non-degenerate if any (or equivalently, if one) of the matrices representing it w.r.t. a fixed basis is non-singular.

We say that a Lie group is a H-group (in the sense of Métivier), *or a* HM-group *in short, if its Lie algebra is of H-type (in the sense of Métivier).*

Remark 3.7.2. In this remark, we follow the notation of Definition 3.7.1.

First, a (Métivier-)H-type algebra is obviously nilpotent of step two. Moreover, we have

$$[\mathfrak{g}, \mathfrak{g}] = [\mathfrak{g}_1 + \mathfrak{g}_2, \mathfrak{g}_1 + \mathfrak{g}_2] \subseteq [\mathfrak{g}_1, \mathfrak{g}_1] \quad (\text{since } \mathfrak{g}_2 \subseteq \mathfrak{z}),$$
$$[\mathfrak{g}_1, \mathfrak{g}_1] \subseteq [\mathfrak{g}, \mathfrak{g}] \quad (\text{since } \mathfrak{g}_1 \subseteq \mathfrak{g}).$$

Consequently, it holds

$$[\mathfrak{g}, \mathfrak{g}] = [\mathfrak{g}_1, \mathfrak{g}_1]. \tag{3.17a}$$

Finally, we claim that

$$\mathfrak{g}_2 = [\mathfrak{g}, \mathfrak{g}]. \tag{3.17b}$$

Indeed, from (3.17a) we first derive that $[\mathfrak{g}, \mathfrak{g}] = [\mathfrak{g}_1, \mathfrak{g}_1] \subseteq \mathfrak{g}_2$ (by the very definition of (Métivier-)H-type algebra). We are left to show that $\mathfrak{g}_2 \subseteq [\mathfrak{g}, \mathfrak{g}]$. Suppose to the contrary that there exists $Z \in \mathfrak{g}_2$ such that $Z \notin [\mathfrak{g}, \mathfrak{g}]$. This implies, in particular, that $Z \neq \{0\}$. Moreover, since both $Z \in \mathfrak{g}_2$ and $[\mathfrak{g}, \mathfrak{g}] \subseteq \mathfrak{g}_2$, there certainly exists $\eta \in \mathfrak{g}_2^*$ such that $\mathfrak{g}_2(Z) \neq 0$ (whence $\eta \neq 0$) and η vanishes identically on $[\mathfrak{g}, \mathfrak{g}]$ (here, we are using the fact that $Z \notin [\mathfrak{g}, \mathfrak{g}]$). But this implies that, for every $X, X' \in \mathfrak{g}_1$, we have

$$B_\eta(X, X') = \eta([X, X']) = 0,$$

for

$$[X, X'] \in [\mathfrak{g}_1, \mathfrak{g}_1] = [\mathfrak{g}, \mathfrak{g}] \quad \text{and} \quad \eta|_{[\mathfrak{g}, \mathfrak{g}]} \equiv 0.$$

This is in contradiction with the non-degeneracy of B_η.

Collecting together (3.17a) and (3.17b), we see that a (Métivier-)H-type algebra is stratified: indeed we have

$$\mathfrak{g} = \mathfrak{g}_1 \oplus \mathfrak{g}_2 \quad \text{with } [\mathfrak{g}_1, \mathfrak{g}_1] = \mathfrak{g}_2 \text{ and } [\mathfrak{g}_1, \mathfrak{g}_2] = \{0\}.$$

As a consequence, a HM-group is a Carnot group.

Moreover, if $\widetilde{\mathfrak{g}}_1$ is any other complement of $\mathfrak{g}_2 = [\mathfrak{g}, \mathfrak{g}]$ in $\mathfrak{g}$, it is easy to prove that also the decomposition $\mathfrak{g} = \widetilde{\mathfrak{g}}_1 \oplus \mathfrak{g}_2$ satisfies Definition 3.7.1.

Collecting the above results, we have proved the following proposition.

Proposition 3.7.3 (Characterization. I). *A HM-group is a Carnot group $\mathbb{G}$ of step two such that if*

$$\mathfrak{g} = \mathfrak{g}_1 \oplus \mathfrak{g}_2 \quad ([\mathfrak{g}_1, \mathfrak{g}_1] = \mathfrak{g}_2, \ [\mathfrak{g}_1, \mathfrak{g}_2] = \{0\})$$

is any stratification of the Lie algebra $\mathfrak{g}$ of $\mathbb{G}$, then the following property holds: For every non-vanishing linear map η from $\mathfrak{g}_2$ to $\mathbb{R}$, the (skew-symmetric) bilinear form B_η on $\mathfrak{g}_1$ defined by

---

Equivalently, a bilinear map B on the vector space V is non-degenerate if, for every $v \in V \setminus \{0\}$, there exists $w \in V$ such that $B(v, w) \neq 0$.

$$B_\eta(X, X') := \eta([X, X']), \quad X,\ X' \in \mathfrak{g}_1,$$

is non-degenerate.

When $\mathbb{G}$ is expressed in its logarithmic coordinates (making it a homogeneous Carnot group), the above definition is easily re-written as follows. We consider a homogeneous Lie group of step two $\mathbb{G} = (\mathbb{R}^{m+n}, \circ, \delta_\lambda)$ with the composition law as in (3.6) (page 158), i.e.

$$(x, t) \circ (\xi, \tau) = \left(x + \xi, t_1 + \tau_1 + \frac{1}{2} \langle B^{(1)}x, \xi\rangle, \ldots, t_n + \tau_n + \frac{1}{2} \langle B^{(n)}x, \xi\rangle\right),$$

where $B^{(1)}, \ldots, B^{(n)}$ are fixed $m \times m$ matrices, and the group of dilations is $\delta_\lambda(x, t) = (\lambda x, \lambda^2 t)$. For the sake of simplicity, we may also suppose that the matrices $B^{(k)}$'s are skew-symmetric (see, for instance, Proposition 3.5.1). We make use of the results and the notation of Section 3.2.

Now, if η is a linear map from $\mathfrak{g}_2$ to $\mathbb{R}$, there exist n scalars $\eta_1, \ldots, \eta_n \in \mathbb{R}$ such that

$$\eta : \mathfrak{g}_2 \to \mathbb{R}, \quad \eta(\partial_{t_i}) = \eta_i \quad \text{for all } i = 1, \ldots, n.$$

In particular, the map B_η, as in Proposition 3.7.3, can be explicitly written (after a simple calculation which we leave to the reader) as follows[6]

$$\text{if } X = \textstyle\sum_{i=1}^m v_i\, X_i \text{ and } X' = \sum_{i=1}^m v_i'\, X_i,$$

$$\text{then } B_\eta(X, X') = \sum_{i,j=1}^{n} \left(- \sum_{k=1}^{n} \eta_k\, B_{i,j}^{(k)} \right) v_i\, v_j'.$$

In other words, the matrix representing the (skew-symmetric) bilinear map B_η w.r.t. the basis $X_1, \ldots, X_m$ of $\mathfrak{g}_1$ is the matrix

$$\eta_1 B^{(1)} + \cdots + \eta_n B^{(n)}.$$

Hence, to ask for B_η to be non-degenerate (for every $\eta \neq 0$) is equivalent to ask that any linear combination of the matrices $B^{(k)}$'s is non-singular, unless it is the null matrix (recall that the $B^{(k)}$'s are linearly independent). We have thus proved the following proposition.

---

[6] Recall that, by (3.8), page 159, a basis for $\mathfrak{g}_1$ is $\{X_1, \ldots, X_m\}$ with

$$X_i = (\partial/\partial x_i) + \frac{1}{2} \sum_{k=1}^{n} \left(\sum_{l=1}^{m} B_{i,l}^{(k)} x_l \right) (\partial/\partial t_k),$$

where we have set

$$B^{(k)} = \left(B_{i,l}^{(k)}\right)_{i,l \le m}.$$

Proposition 3.7.4 (Characterization. II). *Let* $\mathbb{G} = (\mathbb{R}^{m+n}, \circ)$ *be a homogeneous Carnot group of step two, with the composition law*

$$(x,t) \circ (\xi,\tau) = \left(x+\xi, t_1+\tau_1+\frac{1}{2}\langle B^{(1)}x,\xi\rangle, \ldots, t_n+\tau_n+\frac{1}{2}\langle B^{(n)}x,\xi\rangle \right),$$

where $B^{(1)}, \ldots, B^{(n)}$ *are* $m \times m$ *skew-symmetric linearly independent matrices. Then* $\mathbb{G}$ *is a HM-group if and only if every non-vanishing linear combination of the matrices* $B^{(k)}$*'s is non-singular.*

In particular, if the above $\mathbb{G}$ is a HM-group, then the $B^{(k)}$'s are all non-singular $m \times m$ matrices, but since the $B^{(k)}$'s are also skew-symmetric, this implies that m is necessarily even.

Remark 3.7.5 (Any H-type group is a HM-group). Any prototype H-type group (according to Definition 3.6.1, page 169) is a HM-group. Indeed, as it can be seen from the computations on page 169, in a prototype H-type group, for every $\eta = (\eta_1, \ldots, \eta_n) \in \mathbb{R}^n$, $\eta \neq 0$, we proved that $\sum_{k=1}^n \eta_k B^{(k)}$ is $|\eta|$ times an orthogonal matrix, hence (in particular) $\sum_{k=1}^n \eta_k B^{(k)}$ is non-singular. The converse is not true. For example, consider the group on $\mathbb{R}^5$ (the points are denoted by (x,t), $x \in \mathbb{R}^4$, $t \in \mathbb{R}$) with the composition law

$$(x,t) \circ (\xi,\tau) = \left(x+\xi, t+\tau+\frac{1}{2}\langle Bx,\xi\rangle \right),$$

where

$$B = \begin{pmatrix} 0 & 1 & 0 & 0 \\ -1 & 0 & 0 & 0 \\ 0 & 0 & 0 & 2 \\ 0 & 0 & -2 & 0 \end{pmatrix}.$$

Then $\mathbb{G}$ is obviously a HM-group, for B is a non-singular skew-symmetric matrix. But $\mathbb{G}$ is not a prototype H-type group, for B is not orthogonal. But more is true: as we shall prove in Chapter 18 (see Remark 18.2.7 and Corollary 18.2.8, page 695) $\mathbb{G}$ is not even isomorphic to *any* prototype H-type group.

Bibliographical Notes. The explicitness of the composition law of a homogeneous Carnot group of step two allows to face a great variety of problems. See, e.g. some recent results concerning with step-two Carnot groups: [Dai00] for mappings with bounded distortion, [FSS03b] for geometric measure theory results, [Mil88] for a microlocal version of some results concerning hypoellipticity, [Ric06] and [SY06] for results on the convex functions, [Tha94] for theorems of Paley–Wiener type.

For an introduction to harmonic analysis on the Heisenberg group, see the monograph by S. Thangavelu [Tha98], and for a survey see R. Howe [How80].

3.8 Exercises of Chapter 3

Ex. 1) With reference to Section 3.1, recognize that $\Delta_{\mathbb{H}^1}$ equals

$$(\partial_{x_1})^2 + (\partial_{x_2})^2 + 4(x_1^2 + x_2^2)\,(\partial_{x_3})^2 + 4\,x_2\,\partial_{x_1,x_3} - 4\,x_1\,\partial_{x_2,x_3}.$$

Prove the analogous (general) formula for $\Delta_{\mathbb{H}^n}$,

$$\Delta_{\mathbb{H}^n} = \sum_{j=1}^{n}(\partial_{x_j}^2 + \partial_{y_j}^2) + 4\left(\sum_{j=1}^{N}(x_j^2 + y_j^2)\right)\partial_t^2 + 4\sum_{j=1}^{N}(y_j\,\partial_{x_j} - x_j\,\partial_{y_j})\,\partial_t.$$

Recognize that the quadratic form related to $\Delta_{\mathbb{H}^n}$ at (x, y, t) is the quadratic form q related to the symmetric matrix

$$A = \begin{pmatrix} \multicolumn{2}{c}{\mathbb{I}_n} & \begin{matrix} 2\,y \\ -2\,x \end{matrix} \\ 2\,y^T & -2\,x^T & 4\,(|x|^2 + |y|^2) \end{pmatrix}.$$

For every fixed $(x, y, t) \in \mathbb{H}^n$, find all vectors $(\xi, \eta, \tau) \in \mathbb{R}^{2n+1}$ such that

$$(\xi, \eta, \tau)\cdot A \cdot (\xi, \eta, \tau)^T = 0.$$

Compare this to what was shown in (A4), page 65.

Ex. 2) Prove (3.9), page 160.

Ex. 3) Making use of (3.9), derive an explicit expression for the canonical sub-Laplacian of $\mathbb{F}_{m,2}$ generalizing (3.10) (page 165).

Ex. 4) Consider the Lie group on $\mathbb{R}^4$ (the points are denoted by $x = (x_1, x_2, x_3, x_4)$) with the composition law given by

$$x \circ y = \begin{pmatrix} x_1 + y_1 \\ x_2 + y_2 \\ x_3 + y_3 + x_1 y_2 - x_2 y_1 \\ x_4 + y_3 + x_1 y_2 - x_2 y_1 \end{pmatrix}.$$

Prove that $(\mathbb{R}^4, \circ)$ is a Lie group nilpotent of step two. According to Proposition 3.2.1, this is not a *homogeneous* Carnot group (for any dilation!) since (we use the usual notation of (3.6), page 158) the relevant matrices

$$B^{(1)} = B^{(2)} = \begin{pmatrix} 0 & 1 \\ -1 & 0 \end{pmatrix}$$

have the same skew-symmetric parts (whence the latter are not linearly independent). However, according to Ex. 6 in Chapter 2, page 149, since $(\mathbb{R}^4, \circ)$ is nilpotent of step two, its algebra $\mathfrak{g}$ is necessarily *stratified*! Indeed, setting

$$\begin{aligned} X_1 &= \partial_{x_1} - x_2\,\partial_{x_3} - x_2\,\partial_{x_4}, \\ X_2 &= \partial_{x_2} + x_1\,\partial_{x_3} + x_1\,\partial_{x_4}, \\ X_3 &= \partial_{x_3}, \qquad X_4 = \partial_{x_4}, \end{aligned} \tag{3.18}$$

we have

$$[X_1, X_2] = 2\,\partial_{x_3} + 2\,\partial_{x_4} = 2(X_3 + X_4),$$

whence

$$\mathfrak{g} = \text{span}\{X_1, X_2, X_3\} \oplus \text{span}\{X_3 + X_4\},$$

and this is a stratification, as in Definition 2.2.3 (page 122). Consequently, this time thanks to Theorem 3.2.2, $(\mathbb{R}^4, \circ)$ is *isomorphic to a homogeneous Carnot group*. Indeed, retracing the arguments in the proof of Theorem 2.2.18 (page 131, see also the notation therein), prove that $(\mathbb{R}^4, \circ)$ is isomorphic to $(\mathbb{R}^4, \diamond, \delta_\lambda)$, where

$$\delta_\lambda(\xi_1, \xi_2, \xi_3, \xi_4) = (\lambda\xi_1, \lambda\xi_2, \lambda\xi_3, \lambda^2\xi_4),$$

and $\diamond$ is obtained by the Campbell–Hausdorff multiplication on $\mathfrak{g}$ equipped with the basis $\{X_1, X_2, X_3, X_4 + X_3\}$, which, in turn, is given by

$$\xi \circ \eta = \begin{pmatrix} \xi_1 + \eta_1 \\ \xi_2 + \eta_2 \\ \xi_3 + \eta_3 \\ \xi_4 + \eta_4 + \xi_1\eta_2 - \xi_2\eta_1 \end{pmatrix}.$$

A Lie group isomorphism between $(\mathbb{R}^4, \circ)$ and $(\mathbb{R}^4, \diamond)$ can be calculated noting that $(\mathbb{R}^4, \diamond)$ is essentially the Lie algebra of $(\mathbb{R}^4, \circ)$: we thus compute the exponential map via the usual system of ODE's as follows. Since

$$\big\{\xi_1\,X_1 + \xi_2\,X_2 + \xi_3\,X_3 + \xi_4\,(X_4 + X_3)\big\}I(x) = \begin{pmatrix} \xi_1 \\ \xi_2 \\ \xi_3 + \xi_4 - \xi_1\,x_2 + \xi_2\,x_1 \\ \xi_4 - \xi_1\,x_2 + \xi_2\,x_1 \end{pmatrix}$$

(denote this field by $YI(x)$), the solution γ to $\dot\gamma(t) = YI(\gamma(t))$, $\gamma(0) = 0$ is

$$\gamma(t) = \big(\xi_1\,t, \xi_2\,t, (\xi_3 + \xi_4)t, \xi_4\,t\big),$$

and we have $\gamma(1) = (\xi_1, \xi_2, \xi_3 + \xi_4, \xi_4)$. Define the map

$$\Phi : (\mathbb{R}^4, \diamond) \to (\mathbb{R}^4, \circ), \quad \Phi(\xi) := (\xi_1, \xi_2, \xi_3 + \xi_4, \xi_4)$$

and prove that it is a Lie group isomorphism, which turns the vector fields in (3.18), respectively, into

$$\widetilde{X}_1 = \partial_{\xi_1} - \xi_2\,\partial_{x_4}, \quad \widetilde{X}_2 = \partial_{\xi_2} + \xi_3\,\partial_{x_4},$$
$$\widetilde{X}_3 = \partial_{\xi_3}, \quad \widetilde{X}_4 = \partial_{\xi_4} - \partial_{\xi_3}.$$

Ex. 5) Prove in details the last statement in Remark 3.7.2, page 174.

Ex. 6) Consider the following homogeneous Carnot group of step two (which will be called *of Kolmogorov type*, see Section 4.1.4): this is $\mathbb{R}^5$ (the points are, by our usual notation, $(x,t) = (x_1, x_2, x_3, t_1, t_2)$) with the composition law

$$(x,t) \circ (\xi, \tau) = \begin{pmatrix} x_1 + \xi_1 \\ x_2 + \xi_2 \\ x_3 + \xi_3 \\ t_1 + \tau_1 - x_2\, \xi_1 \\ t_2 + \tau_2 - x_3\, \xi_1 \end{pmatrix}.$$

The relevant matrices are

$$B^{(1)} = \begin{pmatrix} 0 & -2 & 0 \\ 0 & 0 & 0 \\ 0 & 0 & 0 \end{pmatrix}, \qquad B^{(2)} = \begin{pmatrix} 0 & 0 & -2 \\ 0 & 0 & 0 \\ 0 & 0 & 0 \end{pmatrix}.$$

Prove that the related canonical sub-Laplacian is

$$\left(\partial_{x_1} - x_2\, \partial_{t_1} - x_3\, \partial_{t_2}\right)^2 + (\partial_{x_2})^2 + (\partial_{x_3})^2.$$

Now, with a slight change of notation, consider on $\mathbb{R}^{1+2n}$ (whose points are denoted by (t,x,y), $t \in \mathbb{R}$, $x, y \in \mathbb{R}^n$) the operator

$$\mathcal{L} := \Delta_x + \left(\partial - \langle x, \nabla_y \rangle\right)^2,$$

where $\Delta_x = \sum_{i=1}^n (\partial_{x_i})^2$ and $\nabla_y = (\partial_{y_1}, \ldots, \partial_{y_n})$. Find a homogeneous Carnot group structure on $\mathbb{R}^{1+2n}$ such that $\mathcal{L}$ is the related canonical sub-Laplacian.

Ex. 7) Operate a change of variables on $\mathbb{R}^5$ in such a way that the group $(\mathbb{R}^5, \circ)$ defined in the previous exercise is turned into a homogeneous Carnot group whose related matrices $B^{(k)}$'s are skew-symmetric. (Do the same for the general case of $\mathbb{R}^{1+2n}$.) Prove that the new composition law and the new canonical sub-Laplacian are, respectively,

$$(x,t) \circ (\xi, \tau) = \begin{pmatrix} x_1 + \xi_1 \\ x_2 + \xi_2 \\ x_3 + \xi_3 \\ t_1 + \tau_1 + \frac{1}{2}(x_1\, \xi_2 - x_2\, \xi_1) \\ t_2 + \tau_2 + \frac{1}{2}(x_1\, \xi_3 - x_3\, \xi_1) \end{pmatrix}$$

and

$$\left(\partial_{x_1} - \frac{1}{2} x_2\, \partial_{t_1} - \frac{1}{2} x_3\, \partial_{t_2}\right)^2 + \left(\partial_{x_2} + \frac{1}{2} x_1\, \partial_{t_1}\right)^2 + \left(\partial_{x_3} + \frac{1}{2} x_1\, \partial_{t_2}\right)^2.$$

Note that the relevant matrices are

$$\widetilde{B}^{(1)} = \frac{1}{2}\big(B^{(1)} - (B^{(1)})^T\big) = \begin{pmatrix} 0 & -1 & 0 \\ 1 & 0 & 0 \\ 0 & 0 & 0 \end{pmatrix},$$

$$\widetilde{B}^{(2)} = \frac{1}{2}\big(B^{(2)} - (B^{(2)})^T\big) = \begin{pmatrix} 0 & 0 & -1 \\ 0 & 0 & 0 \\ 1 & 0 & 0 \end{pmatrix}.$$

Ex. 8) (The polarized Heisenberg group). Let us denote by (x, y, t) the points of $\mathbb{R}^n \times \mathbb{R}^n \times \mathbb{R} \equiv \mathbb{R}^{2n+1}$. Consider the composition law

$$(x, y, t) \circ (x', y', t') := \big(x + x', y + y', t + t' + \langle x', y\rangle\big)$$

(where $\langle\cdot,\cdot\rangle$ denotes the Euclidean inner product) and the dilation

$$\delta_\lambda(x, y, t) := (\lambda x, \lambda y, \lambda^2 t).$$

Show that $\mathbb{H}_n^{\mathrm{pol}} = (\mathbb{R}^{2n+1}, \circ, \delta_\lambda)$ is a homogeneous Carnot group and write down its canonical sub-Laplacian.
Note. $\mathbb{H}_n^{\mathrm{pol}}$ is called the polarized Heisenberg group.

(Matrix representation of $\mathbb{H}_n^{\mathrm{pol}}$). Given $x, y \in \mathbb{R}^n$ and $t \in \mathbb{R}$, define the $(n+2) \times (n+2)$ matrix (x, y are being considered as column vectors)

$$m(x, y, t) := \mathbb{I}_{n+2} + \begin{pmatrix} 0 & y^T & t \\ 0 & 0 & x \\ 0 & 0 & 0 \end{pmatrix},$$

where $\mathbb{I}_{n+2}$ is the identity matrix of order $n + 2$. Show that

$$\mathbb{M}_n := \big\{ m(x, y, t) \mid (x, y, t) \in \mathbb{R}^{2n+1} \big\}$$

is a group with respect to the matrix multiplication. Show also that

$$m : \mathbb{H}_n^{\mathrm{pol}} \longrightarrow \mathbb{M}_n, \quad (x, y, t) \mapsto m(x, y, t)$$

is a group isomorphism.

(A unitary representation of $\mathbb{H}_n^{\mathrm{pol}}$). Given $x, y \in \mathbb{R}^n$ and $t \in \mathbb{R}$, define the operators

$$e(x),\ \tau(y),\ \chi(t)$$

from $L^2(\mathbb{R}^n, \mathbb{C})$ into itself as follows:

$$\begin{aligned} \big(e(x)f\big)(\xi) &:= e^{i\langle x,\xi\rangle} f(\xi), \quad \xi \in \mathbb{R}^n, \\ \big(\tau(y)f\big)(\xi) &:= f(\xi + y), \quad \xi \in \mathbb{R}^n, \\ \big(\chi(t)f\big)(\xi) &:= e^{i\,t} f(\xi), \quad \xi \in \mathbb{R}^n. \end{aligned}$$

(Again, $\langle\cdot,\cdot\rangle$ denotes the Euclidean inner product.) Show that the set of operators

$$\mathbb{U}_n := \left\{ e(x)\, \tau(y)\, \chi(t) \mid (x, y, t) \in \mathbb{R}^{2n+1} \right\}$$

is a group with respect to the composition of operators. Show also that

$$\pi : \mathbb{H}_n^{\text{pol}} \longrightarrow \mathbb{U}_n, \quad (x, y, t) \mapsto e(x)\, \tau(y)\, \chi(t)$$

is a group isomorphism.

Note. $e(x)$ and $\tau(y)$ are the unitary operators generated by the *position* and the *momentum* operators in quantum mechanics (see [Tha98] for more details).

4

Examples of Carnot Groups

The aim of this chapter is to provide a wide number of explicit examples of Carnot groups of step greater than two, thus enlarging the set of examples given in Chapter 3. Some are already known in literature (for instance, here we collect some results on the filiform groups), some are new. Among the latter, there are what we call the B-groups, the Kolmogorov-type groups (or K-type groups), the Bony-type groups.

Moreover, in Section 4.2, we furnish a criterion to recognize when a given operator sum of squares of vector fields in $\mathbb{R}^N$ is a sub-Laplacian on some *homogeneous* Carnot group. Precisely, given a set of linearly independent vector fields $X_1, \dots, X_m$, homogeneous of degree one with respect to a family of dilations $\{\delta_\lambda\}_{\lambda>0}$ in $\mathbb{R}^N$ and satisfying a suitable "rank-type" condition, we show how to explicitly define a composition law $\circ$ on $\mathbb{R}^N$ making $\mathbb{G} = (\mathbb{R}^N, \circ, \delta_\lambda)$ a Carnot group whose generators are $X_1, \dots, X_m$. Our construction rests on the solvability of relevant systems of ODE's.

In Section 4.3, we illustrate how this method can be applied to produce new examples of Carnot groups. In Chapter 17, Section 17.4, we shall use the results of the present chapter to produce "lifted" groups.

4.1 A Primer of Examples of Carnot Groups

4.1.1 Euclidean Group

The additive group $(\mathbb{R}^N, +)$ is a homogeneous group with respect to the dilations

$$\delta_\lambda(x) = \lambda\, x, \quad \lambda > 0.$$

We call $\mathbb{E} = (\mathbb{R}^N, +, \delta_\lambda)$ the *Euclidean group*. $\mathbb{E}$ is a Carnot group of step 1. Its generators are $\partial_{x_1}, \dots, \partial_{x_N}$. Thus, the canonical sub-Laplacian on $\mathbb{E}$ is the classical Laplace operator

$$\Delta = \sum_{j=1}^{N} \partial_{x_j}^2.$$

Moreover, any sub-Laplacian on $\mathbb{E}$ has the following form

$$\mathcal{L} = \sum_{j=1}^{N} Y_j^2 = \sum_{j=1}^{N} \left(\sum_{i=1}^{N} b_{j,i}\, \partial_{x_i} \right)^2,$$

where $B = (b_{i,j})_{i,j \le N}$ is a non-singular constant matrix. Hence, $\mathcal{L}$ can also be written as

$$\mathcal{L} = \sum_{i,k=1}^{N} \left(\sum_{j=1}^{N} b_{j,i}\, b_{j,k} \right) \partial_{x_i,x_k} = \sum_{i,k=1}^{N} (B^T \cdot B)_{i,k}\, \partial_{x_i,x_k},$$

so that $\mathcal{L}$ is a second order constant-coefficient strictly elliptic operator.

Vice versa, if $L = \sum_{i,k=1}^{N} a_{i,k}\, \partial_{x_i,x_k}$, where $A = (a_{i,k})_{i,k \le N}$ is a positive-definite symmetric matrix, then L is a sub-Laplacian on $\mathbb{E}$. Indeed, this immediately follows by writing $A = B^2$ (where B is a non-singular symmetric square-root of A), so that $L = \sum_{j=1}^{N} Y_j^2$, where $Y_j := \sum_{i=1}^{N} B_{j,i}\, \partial_{x_i}$.

We want to stress that

$\mathbb{E}$ is the only Carnot group
of step 1 *(and N generators).*

4.1.2 Carnot Groups with Homogeneous Dimension $Q \le 3$

Let $\mathbb{G} = (\mathbb{R}^N, \circ, \delta_\lambda)$ be a Carnot group with homogeneous dimension $Q \le 3$. With reference to the results and notation in Remark 1.4.8, we recall that $Q = \sum_{j=1}^{r} j\, N_j$, where r and $N_1, \dots, N_r$ are, respectively, the step of $\mathbb{G}$ and the dimensions of the layers $W^{(1)}, \dots, W^{(r)}$ of $\mathfrak{g}$.

Obviously, the group is not the Euclidean group in $\mathbb{R}^N$ iff $W^{(2)} \ne \{0\}$, i.e. $r \ge 2$. In this case, the first layer $W^{(1)}$ must be at least two-dimensional since $[W^{(1)}, W^{(1)}] = W^{(2)} \ne \{0\}$.

This shows that any non-Euclidean Carnot group has homogeneous dimension $Q \ge 4$. Thus, if $Q \le 3$, then $\mathbb{G}$ is the Euclidean group in $\mathbb{R}^N$, i.e. $\circ = +$ and $\delta_\lambda(x) = \lambda x$. The sub-Laplacians on $\mathbb{G}$ are the second order elliptic operators with constant coefficients. The canonical sub-Laplacians are $\mathrm{d}^2/\mathrm{d}x_1^2$ in $\mathbb{R}$ ($Q = N = 1$), $\partial_{x_1}^2 + \partial_{x_2}^2$ in $\mathbb{R}^2$ ($Q = N = 2$), and $\partial_{x_1}^2 + \partial_{x_2}^2 + \partial_{x_3}^2$ in $\mathbb{R}^3$ ($Q = N = 3$).

4.1.3 B-groups

Let us consider an $N \times N$ matrix B with real entries $b_{i,j}$ (with $i, j = 1, \dots, N$). Let us put

$$E(t) := \exp(t\, B), \quad t \in \mathbb{R}.$$

In $\mathbb{R}^{1+N}$, whose points will be denoted by $z = (t, x)$, $t \in \mathbb{R}$, $x \in \mathbb{R}^N$, let us introduce the following composition law

$$(t,x) \circ (t',x') = (t+t', x' + E(t')x).$$

One easily verifies that $\mathbb{B} = (\mathbb{R}^{1+N}, \circ)$ is a Lie group whose identity is the origin $(0,0)$, and where the inverse is given by

$$(t,x)^{-1} = (-t, -E(-t)x).$$

The Jacobian matrix at the origin of the left translation $\tau_{(t,x)}$ is the following block matrix

$$\mathcal{J}_{\tau_{(t,x)}}(0,0) = \begin{pmatrix} 1 & 0 \\ \mathbf{b} & \mathbb{I}_N \end{pmatrix},$$

where $\mathbf{b}$ stands for the $N \times 1$ column-matrix

$$\left.\frac{\mathrm{d}}{\mathrm{d}s}\right|_{s=0} E(s)x = BE(s)x\big|_{s=0} = Bx.$$

Then the Jacobian basis of $\mathfrak{b}$, the Lie algebra of $\mathbb{B}$, is given by

$$Y = \partial_t + \nabla_x \cdot Bx, \quad \partial_{x_1}, \dots, \partial_{x_N}. \tag{4.1}$$

We explicitly remark that, for a general matrix B, the group $(\mathbb{B}, \circ)$ may not be nilpotent. Indeed, an easy computation shows that, for any $j \in \{1, \dots, N\}$,

$$[[\cdots[\partial_{x_j}\underbrace{Y]\cdots]Y, Y]}_{k \text{ times}} = \sum_{i=1}^{N} (B^k)_{i,j}\, \partial_{x_i}.$$

Hence, $(\mathbb{B}, \circ)$ is a nilpotent group iff B is a nilpotent matrix. For example, if $N = 1$ and $B = (1)$, the composition law is

$$(t,x) \circ (t',x') = \big(t+t', x' + x\ \exp(t')\big), \quad (t,x),\ (t',x') \in \mathbb{R}^2,$$

and the Jacobian basis is $\partial_t + x\partial_x, \partial_x$. Then, $\mathbb{R}^2$ equipped with the above composition law is *not* a *stratified* group (least of all a homogeneous Carnot group!) since it is not nilpotent. In particular, the second order differential operator on $\mathbb{R}^2$ defined by

$$L = \partial_x^2 + (\partial_t + x\,\partial_x)^2 = (1+x^2)\,\partial_x^2 + \partial_t^2 + 2x\,\partial_x\partial_t + x\,\partial_x$$

is a sum of squares of left-invariant vector fields on $(\mathbb{R}^2, \circ)$, it satisfies Hörmander's hypoellipticity condition

$$\mathrm{rank}\big(\mathrm{Lie}\{\partial_x, \partial_t + x\partial_x\}(t,x)\big) = 2 \quad \forall\,(t,x) \in \mathbb{R}^2,$$

but L *is not* a sub-Laplacian on any Carnot group. Indeed, L is elliptic at any point (but $\mathbb{B}$ is not Euclidean!).

4.1.4 K-type Groups

Let us now suppose that the matrix B in the previous example takes the following special block-form

$$B = \begin{pmatrix} 0 & 0 & \cdots & 0 & 0 \\ B_1 & 0 & \cdots & 0 & 0 \\ 0 & B_2 & \ddots & \vdots & \vdots \\ \vdots & \vdots & \ddots & 0 & 0 \\ 0 & 0 & \cdots & B_r & 0 \end{pmatrix}, \tag{4.2a}$$

where B_j is a $p_j \times p_{j-1}$ block with rank equal to p_j, for every $j = 1, 2, \ldots, r$. Moreover, $p_0 \geq p_1 \geq \cdots \geq p_r$ and $p_0 + p_1 + \cdots + p_r = N$. Finally, the 0 blocks in (4.2a) are suitably chosen in such a way that B has dimension $N \times N$. We want to show that the group $\mathbb{B}$ related to this matrix is a Carnot group. It will be called a group of *Kolmogorov type* or, in short, a K-type group.

Let us split $\mathbb{R}^N$ as follows

$$\mathbb{R}^N = \mathbb{R}^{p_0} \times \mathbb{R}^{p_1} \times \cdots \times \mathbb{R}^{p_r}$$

and define, for every $\lambda > 0$,

$$D_\lambda x = D_\lambda(x^{(0)}, x^{(1)}, \ldots, x^{(r)}) = (\lambda x^{(0)}, \lambda^2 x^{(1)}, \ldots, \lambda^{r+1} x^{(r)}), \tag{4.2b}$$

where $x^{(i)} \in \mathbb{R}^{p_i}$ for $0 \leq i \leq r$. We also put $\delta_\lambda(t, x) = (\lambda t, D_\lambda x)$.

Claim 1. *For every $\lambda > 0$, the dilation δ_λ is an automorphism of $\mathbb{B}$.*

To prove this claim we need the following lemma.

Lemma 4.1.1. *For every $t \in \mathbb{R}$ and $\lambda > 0$, we have*

$$E(\lambda t)\, D_\lambda = D_\lambda\, E(t) \tag{4.2c}$$

where $E(t) = \exp(tB)$, B is as in (4.2a) *and D_λ is the dilation in* (4.2b).

Proof. Since $B^k = 0$ for every $k \geq r + 1$, one has

$$E(t) = \sum_{k=0}^{r} t^k\, B^k / k!,$$

and (4.2c) holds for every $t \in \mathbb{R}$ and $\lambda > 0$ iff

$$\lambda^k\, B^k\, D_\lambda = D_\lambda\, B^k \quad \forall\, k \geq 0,\ \forall\, \lambda > 0. \tag{4.2d}$$

This identity holds true when $k = 0$. An easy direct computation shows that it also holds true for $k = 1$. As a consequence,

$$\begin{aligned} \lambda^2 B^2 D_\lambda &= \lambda B(\lambda B D_\lambda) = \lambda B(D_\lambda B) \\ &= (\lambda B D_\lambda) B = (D_\lambda B) B = D_\lambda B^2. \end{aligned}$$

Then (4.2d) holds true for $k = 2$. An iteration of this argument shows (4.2d) for $k \geq 2$. □

From this lemma the Claim 1 easily follows. Indeed, for every $z=(t,x)$, $z'=(t',x')\in\mathbb{R}^{1+N}$, we have

$$\begin{aligned}(\delta_\lambda z)\circ(\delta_\lambda z') &= (\lambda t, D_\lambda x)\circ(\lambda t', D_\lambda x')\\ &= (\lambda t+\lambda t', D_\lambda x' + E(\lambda t')D_\lambda x)\\ \text{(by Lemma 4.1.1)} \quad &= (\lambda(t+t'), D_\lambda x' + D_\lambda E(t')x)\\ &= \delta_\lambda(t+t', x'+E(t')x)\\ &= \delta_\lambda(z\circ z'). \quad \square\end{aligned}$$

Thus, $\mathbb{B}=(\mathbb{R}^{1+N},\circ,\delta_\lambda)$ is a homogeneous group whose first layer is

$$\mathbb{R}\times\mathbb{R}^{p_0}=\{(t,x^{(0)})\,|\,t\in\mathbb{R},\ x^{(0)}\in\mathbb{R}^{p_0}\}.$$

Moreover, the vector fields in the Jacobian basis related to this first layer are given by

$$Y=\partial_t+\langle Bx,\nabla_x\rangle,\quad \partial_{x_1},\dots,\partial_{x_{p_0}}. \tag{4.2e}$$

Claim 2. *We have* rank(*Lie*$\{Y,\partial_{x_1},\dots,\partial_{x_{p_0}}\}(0,0))=1+N$.

Once this claim is proved, it will follow that $(\mathbb{R}^{1+N},\circ,\delta_\lambda)$ is a Carnot group of step $r+1$ with $1+p_0$ generators, which are the vector fields in (4.2e). Thus the related canonical sub-Laplacian is given by

$$\Delta_{\mathbb{B}}=Y^2+\Delta_{\mathbb{R}^{p_0}},\quad \text{where}\quad \Delta_{\mathbb{R}^{p_0}}=\sum_{j=1}^{p_0}\partial_{x_j}^2. \tag{4.2f}$$

This sub-Laplacian will be said of *Kolmogorov type*. To prove Claim 2, the following lemma will be useful.

Lemma 4.1.2. *In* $\mathbb{R}^p\times\mathbb{R}^q$, *let us consider the vector field*

$$Z=Ay\cdot(\nabla_z)^T, \text{ where } A \text{ is a } q\times p \text{ matrix, } y\in\mathbb{R}^p \text{ and } z\in\mathbb{R}^q.$$

Suppose $\operatorname{rank}(A)=q\le p$. *Then*

$$\operatorname{span}\{[\partial_{y_i},Z]\,|\,i=1,\dots,p\}=\operatorname{span}\{\partial_{z_1},\dots,\partial_{z_q}\}. \tag{4.2g}$$

Proof. Let $A=(a_{i,j})_{i\le q,\,j\le p}$. Then

$$[\partial_{y_i},Z]=\sum_{j=1}^q a_{j,i}\,\partial_{z_j},\quad i=1,\dots,p,$$

so that, since $\operatorname{rank}(A)=q$,

$$\dim\big(\operatorname{span}\{[\partial_{y_i},Z]\,|\,i=1,\dots,p\}\big)=q.$$

This implies (4.2g). $\square$

We now prove Claim 2. Since B has the form (4.2a), we can write

$$Y = \partial_t + \sum_{i=1}^{r} B_i x^{(i-1)} \cdot (\nabla_{x^{(i)}})^T.$$

Then, by applying Lemma 4.1.2, we get

$$\begin{aligned}&\operatorname{span}\big\{[\partial_{x_i}, Y] \,|\, i = 1, \dots, p_0\big\}\\&= \operatorname{span}\big\{[\partial_{x_i}, B_1 x^{(0)} \cdot (\nabla_{x^{(1)}})^T] \,|\, i = 1, \dots, p_0\big\}\\&= \operatorname{span}\big\{\partial_{x_i^{(1)}} \,|\, i = 1, \dots, p_1\big\}.\end{aligned}$$

Another application of Lemma 4.1.2 gives

$$\operatorname{span}\big\{[\partial_{x_i^{(1)}}, Y] \,|\, i = 1, \dots, p_1\big\} = \operatorname{span}\big\{\partial_{x_i^{(2)}} \,|\, i = 1, \dots, p_2\big\}.$$

Iterating this argument, we get

$$\operatorname{Lie}\{Y, \partial_{x_1}, \dots, \partial_{x_{p_0}}\} = \operatorname{Lie}\{Y, \partial_{x_1}, \dots, \partial_{x_N}\}.$$

This obviously proves the claim. □

Note. The groups of Kolmogorov type were introduced by E. Lanconelli and S. Polidoro [LP94] in studying a class of hypoelliptic ultraparabolic operators including the classical prototype operators of Kolmogorov–Fokker–Planck. The composition law in [LP94] was suggested by the structure of the fundamental solution of the operator $\partial^2_{x_1} + x_1\,\partial_{x_2} - \partial_{x_3}$ in $\mathbb{R}^3$ given by A.N. Kolmogorov in [Kol34].

Example 4.1.3. With the notation of Section 4.1.4, an example of K-type group is given by the choice $p_0 = p_1 = 1$, $B_1 = (1)$, whence

$$B = \begin{pmatrix} 0 & 0 \\ (1) & 0 \end{pmatrix}, \qquad N = p_0 + p_1 = 2, \qquad \exp(s\,B) = \begin{pmatrix} 1 & 0 \\ s & 1 \end{pmatrix}.$$

Hence, our K-type group $\mathbb{B}$ is $\mathbb{R}^3$ (whose points are denoted by (t, x_1, x_2)) equipped with the operation

$$(t, x_1, x_2) \circ (s, y_1, y_2) = (t + s, x_1 + y_1, x_2 + y_2 + s\,x_1)$$

and the dilation

$$\delta_\lambda(t, x_1, x_2) = (\lambda t, \lambda x_1, \lambda^2 x_2).$$

This is naturally isomorphic to the Heisenberg–Weyl group $\mathbb{H}^1$.

Again following the notation of Section 4.1.4, another example of K-type group is given by the choice $p_0 = p_1 = p_2 = 1$, $B_1 = B_2 = (1)$, whence

$$B = \begin{pmatrix} 0 & 0 & 0 \\ (1) & 0 & 0 \\ 0 & (1) & 0 \end{pmatrix}, \quad N = p_0 + p_1 + p_2 = 3, \quad \exp(s\,B) = \begin{pmatrix} 1 & 0 & 0 \\ s & 1 & 0 \\ \frac{s^2}{2} & s & 1 \end{pmatrix}.$$

Hence, our K-type group $\mathbb{B}$ is $\mathbb{R}^4$ (whose points are denoted by (t, x_1, x_2, x_3)) equipped with the operation

$$(t, x_1, x_2, x_3) \circ (s, y_1, y_2, y_3)$$
$$= \left(t + s, y_1 + x_1, y_2 + x_2 + s\, x_1, y_3 + x_3 + s x_2 + \frac{s^2}{2} x_1\right),$$

and the dilation

$$\delta_\lambda(t, x_1, x_2, x_3) = (\lambda t, \lambda x_1, \lambda^2 x_2, \lambda^3 x_3).$$

Alternatively, the same non-trivial block

$$\begin{pmatrix} (1) & 0 \\ 0 & (1) \end{pmatrix}$$

in the latter matrix B can also be realized by the choice

$$p_0 = p_1 = 2, \qquad B_1 = \begin{pmatrix} 1 & 0 \\ 0 & 1 \end{pmatrix},$$

whence we have $N = p_0 + p_1 = 4$ and

$$B = \begin{pmatrix} \mathbf{0} & \mathbf{0} \\ \begin{pmatrix} 1 & 0 \\ 0 & 1 \end{pmatrix} & \mathbf{0} \end{pmatrix}, \quad \text{where } \mathbf{0} = \begin{pmatrix} 0 & 0 \\ 0 & 0 \end{pmatrix}.$$

Hence, our K-type group $\mathbb{B}$ is $\mathbb{R}^5$ (whose points are denoted by $(t, x) = (t, x_1, x_2, x_3, x_4)$) equipped with the operation

$$(t, x) \circ (s, y) = (t + s, y_1 + x_1, y_2 + x_2, y_3 + x_3 + s\, x_1, y_4 + x_4 + s\, x_2),$$

and the dilation

$$\delta_\lambda(t, x_1, x_2, x_3, x_4) = (\lambda t, \lambda x_1, \lambda x_2, \lambda^2 x_3, \lambda^2 x_4).$$

Remark 4.1.4. Assume that the matrix B is as in (4.2a). If we define

$$d_\lambda : \mathbb{R}^{1+N} \to \mathbb{R}^{1+N},$$
$$d_\lambda(t, x^{(0)}, \dots, x^{(r)}) = (\lambda^2 t, \lambda x^{(0)}, \lambda^3 x^{(1)}, \dots, \lambda^{2r+1} x^{(r)}),$$

then $\{d_\lambda\}_{\lambda>0}$ is a group of automorphisms of $\mathbb{B}$. For a proof of this statement, we directly refer to [LP94]. This remark shows that $(\mathbb{R}^{1+N}, \circ, d_\lambda)$ is a homogeneous Lie group. It can be also easily proved that the *ultraparabolic operator*

$$L = \Delta_{p_0} + Y \tag{4.3}$$

is left-invariant (w.r.t. $\circ$) and homogeneous of degree two with respect to $\{d_\lambda\}_{\lambda>0}$. Operator (4.3) generalizes the prototypes of the ones introduced by Kolmogorov in [Kol34].

4.1.5 Sum of Carnot Groups

Suppose we are given two homogeneous Carnot groups $\mathbb{G}^{(1)} = (\mathbb{R}^N, \circ^{(1)})$, $\mathbb{G}^{(2)} = (\mathbb{R}^M, \circ^{(2)})$ with dilations

$$\delta_\lambda^{(1)}(x) = (\lambda\, x^{(1)}, \dots, \lambda^r x^{(r)}), \quad x \in \mathbb{G}^{(1)},$$
$$\delta_\lambda^{(2)}(y) = (\lambda\, y^{(1)}, \dots, \lambda^s y^{(s)}), \quad y \in \mathbb{G}^{(2)},$$

where

$$x^{(i)} \in \mathbb{R}^{N_i}, i \le r, \quad N_1 + \dots + N_r = N$$

and

$$y^{(i)} \in \mathbb{R}^{M_i}, i \le s, M_1 + \dots + M_s = M.$$

Let

$$\Delta_{\mathbb{G}^{(1)}} = \sum_{j=1}^{N_1} X_j^2 \quad \text{and} \quad \Delta_{\mathbb{G}^{(2)}} = \sum_{j=1}^{M_1} Y_j^2$$

be the canonical sub-Laplacians on $\mathbb{G}^{(1)}$ and $\mathbb{G}^{(2)}$, respectively. We define a homogeneous Carnot group $\mathbb{G}$ on $\mathbb{R}^{N+M}$ as follows. Suppose $r \le s$. If $(x, y) \in \mathbb{R}^N \times \mathbb{R}^M$, we consider the following permutation of the coordinates

$$R(x, y) = (x^{(1)}, y^{(1)}, \dots, x^{(r)}, y^{(r)}, y^{(r+1)}, \dots, y^{(s)}).$$

We then denote the points of $\mathbb{G} \equiv \mathbb{R}^{N+M}$ by $z = R(x, y)$. We finally define the group law $\circ$ and the dilation δ_λ on $\mathbb{G}$ as one can expect: for every $z = R(x, y)$, $\zeta = R(\xi, \eta) \in \mathbb{G}$, we set

$$z \circ \zeta = R(x \circ^{(1)} \xi, y \circ^{(2)} \eta), \qquad \delta_\lambda z = R(\delta_\lambda^{(1)} x, \delta_\lambda^{(2)} y).$$

It is then easily checked that $(\mathbb{G}, \circ, \delta_\lambda)$ is a homogeneous stratified group of step s and $N_1 + M_1$ generators. Moreover, the canonical sub-Laplacian on $\mathbb{G}$ is the sum of the sub-Laplacians on $\mathbb{G}^{(1)}$ and $\mathbb{G}^{(2)}$:

$$\Delta_{\mathbb{G}} = \Delta_{\mathbb{G}^{(1)}} + \Delta_{\mathbb{G}^{(2)}} = \sum_{j=1}^{N_1} X_j^2 + \sum_{j=1}^{M_1} Y_j^2.$$

For example, if $\mathbb{G}^{(1)}$ is the ordinary Euclidean group on $\mathbb{R}^2$ and $\mathbb{G}^{(2)}$ is the Heisenberg–Weyl group on $\mathbb{R}^3$, then the "sum" of $\mathbb{G}^{(1)}$ and $\mathbb{G}^{(2)}$ is the Carnot group on $\mathbb{R}^5$ (whose points are denoted $z = (x, y) = (x_1, x_2, y_1, y_2, y_3)$) with the composition law

$$(x, y) \circ (x', y') = \begin{pmatrix} x_1 + x_1', \quad x_2 + x_2', \\ y_1 + y_1', \quad y_2 + y_2', \quad y_3 + y_3' + 2(y_2\, y_1' - y_1\, y_2') \end{pmatrix},$$

and dilation $\delta_\lambda(x, y) = (\lambda x_1, \lambda x_2, \lambda y_1, \lambda y_2, \lambda^2 y_3)$.

4.2 From a Set of Vector Fields to a Stratified Group

In the analysis of PDE's, the following problem naturally arises: given a linear second order operator $\mathcal{L} = \sum_{j=1}^m X_j^2$, where the X_j's are smooth vector fields on $\mathbb{R}^N$, does there exist a Lie group on $\mathbb{R}^N$ with respect to which $\mathcal{L}$ is a sub-Laplacian? And, if the answer is affirmative, is the group law explicitly expressible? The aim of this section is to answer these questions.

First of all, we recall some notation. For every $k \in \mathbb{N}$, denote

$$W^{(k)} = \operatorname{span}\{X_J \mid J \in \{1, \dots, m\}^k\},$$

where, if $J = (j_1, \dots, j_k)$,

$$X_{(j_1,\dots,j_k)} = [X_{j_1}, \cdots [X_{j_{k-1}}, X_{j_k}] \cdots].$$

Assume the vector fields X_j's satisfy the following conditions:

(H0) $X_1, \dots, X_m$ are linearly independent and δ_λ-homogeneous of degree one with respect to a suitable family of dilations $\{\delta_\lambda\}_{\lambda>0}$ of the following type

$$\delta_\lambda : \mathbb{R}^N \to \mathbb{R}^N, \quad \delta_\lambda(x) = \delta_\lambda(x^{(1)}, \dots, x^{(r)}) := (\lambda x^{(1)}, \dots, \lambda^r x^{(r)}),$$

where $r \geq 1$ is an integer, $x^{(i)} \in \mathbb{R}^{N_i}$ for $i = 1, \dots, r$, $N_1 = m$ and $N_1 + \cdots + N_r = N$;

(H1) $\dim(W^{(k)}) = \dim\{XI(0) : X \in W^{(k)}\}$ for every $k = 1, \dots, r$;

(H2) $\dim(\operatorname{Lie}\{X_1, \dots, X_m\}I(0)) = N$.

By the results of the previous sections, conditions **(H0)**–**(H1)**–**(H2)** are necessary for $\sum_{j=1}^m X_j^2$ to be a sub-Laplacian on a suitable homogeneous Carnot group (see Proposition 1.2.13, Remark 1.4.8 and the very definitions of Carnot group and sub-Laplacian). Moreover, the hypotheses **(H0)**–**(H1)**–**(H2)** are independent, as the following examples show:

- The vector fields $\partial_{x_1}, \partial_{x_2}$ on $\mathbb{R}^3$ satisfy **(H0)** with respect to the dilation $(\lambda x_1, \lambda x_2, \lambda^2 x_3)$. Moreover, they satisfy **(H1)** but not **(H2)**;
- The vector fields $X_1 = \partial_{x_1} + x_2\partial_{x_4}$, $X_2 = \partial_{x_2}$, $X_3 = \partial_{x_3} + x_2\partial_{x_4} + x_2^2\partial_{x_5}$ in $\mathbb{R}^5$ satisfy **(H0)** with respect to the dilations $(\lambda x_1, \lambda x_2, \lambda x_3, \lambda^2 x_4, \lambda^3 x_5)$. Moreover, since $[X_1, X_2] = -\partial_{x_4}$, $[X_1, X_3] = 0$, $[X_2, X_3] = \partial_{x_4} + 2x_2\partial_{x_5}$ and $[X_2, [X_2, X_3]] = 2\partial_{x_5}$, the given vector fields satisfy **(H2)** but not **(H1)**;
- the vector fields $\partial_{x_1} + x_1\partial_{x_2}$, ∂_{x_2} on $\mathbb{R}^2$ satisfy **(H1)** and **(H2)** but do not satisfy **(H0)** with respect to *any* dilation $(\lambda^{\sigma_1} x_1, \lambda^{\sigma_2} x_2)$.

(See also Section 4.4.)

We are going to show that conditions **(H0)**–**(H1)**–**(H2)** are sufficient for the solvability of our problem.

To begin with, we notice that the vector fields in $W^{(k)}$ are δ_λ-homogeneous of degree k (see Proposition 1.3.10). Moreover, by Proposition 1.3.9 and hypothesis **(H1)**

$$W^{(i)} \cap W^{(j)} = \{0\} \quad \text{if } i \neq j.$$

By Remark 1.3.7, we also have $W^{(k)} = \{0\}$ for every $k \geq r+1$. Then, by Proposition 1.1.7,

$$\text{Lie}\{X_1, \ldots, X_m\} = W^{(1)} \oplus \cdots \oplus W^{(r)}. \tag{4.4}$$

Moreover, by hypotheses **(H0)** and **(H2)**,

$$\dim(W^{(1)}) = \dim(\text{span}\{X_1, \ldots, X_m\}) = m$$

and $\sum_{k=1}^{r} \dim W^{(k)} = N$. The following proposition shows a crucial link between the dimension of the $W^{(k)}$'s and the N_k's in **(H0)**.

Proposition 4.2.1. *If* $X_1, \ldots, X_m$ *satisfy hypotheses* **(H0)–(H1)–(H2)**, *then* $\dim(W^{(k)}) = N_k$ *for any* $k \in \{1, \ldots, r\}$.

Proof. Let us set $M_k := \dim(W^{(k)})$ and fix a basis $\{Z_1^{(k)}, \ldots, Z_{M_k}^{(k)}\}$ of $W^{(k)}$, $1 \leq k \leq r$. Then

$$\{Z_1^{(k)} I(0), \ldots, Z_{M_k}^{(k)} I(0)\}$$

span $W^{(k)} I(0)$, so that, by **(H1)**, it is a basis of

$$W^{(k)} I(0) := \{X I(0) : X \in W^{(k)}\}.$$

By (4.4), the set of vector fields

$$Z_1^{(1)}, \ldots, Z_{M_1}^{(1)}, \ \ldots \ , Z_1^{(r)}, \ldots, Z_{M_r}^{(r)} \tag{4.5}$$

is a basis of $\text{Lie}\{X_1, \ldots, X_m\}$. Then, by **(H2)**, the column vectors of the matrix

$$A := \left(Z_1^{(1)} I(0) \cdots Z_{M_1}^{(1)} I(0) \cdots Z_1^{(r)} I(0) \cdots Z_{M_r}^{(r)} I(0) \right)$$

span $\mathbb{R}^N$. We shall show that they are also linearly independent. By Proposition 1.3.5 and Remark 1.3.7, the vector fields $Z_j^{(k)}$ can be written as

$$Z_j^{(k)} = \sum_{s=k}^{r} \sum_{i=1}^{N_s} a_{s,i}^{(k,j)} \, (\partial / \partial x_i^{(s)}),$$

where $a_{s,i}^{(k,j)}$ is a polynomial function δ_λ-homogeneous of degree $s-k$. In particular, $a_{s,i}^{(k,j)}(0) = 0$ for every $k < s \leq r$. As a consequence, the matrix A takes the form

$$\begin{pmatrix} A^{(1)} & \cdots & 0 \\ \vdots & \ddots & \vdots \\ 0 & \cdots & A^{(r)} \end{pmatrix}, \qquad \text{where } A^{(k)} = \left(a_{k,i}^{(k,j)}(0) \right)_{1 \leq i \leq N_k, \ 1 \leq j \leq M_k}.$$

The block $A^{(k)}$ has dimension $N_k \times M_k$ and rank M_k since

$$\{Z_1^{(k)}I(0), \dots, Z_{M_k}^{(k)}I(0)\}$$

is a basis for $W^{(k)}I(0)$ and $\dim(W^{(k)}I(0)) = \dim(W^{(k)}) = M_k$. It follows that $N_k \geq M_k$ for $1 \leq k \leq r$. On the other hand, since the column vectors of A span $\mathbb{R}^N$,

$$\sum_{k=1}^{r} M_k \geq N = \sum_{k=1}^{r} N_k.$$

Then $M_k = N_k$ for any $k \in 1, \dots, r$. □

Following the previous proof, we infer that the matrix

$$\big(Z_1^{(1)}I(x) \cdots Z_{N_1}^{(1)}I(x) \cdots Z_1^{(r)}I(x) \cdots Z_{N_r}^{(r)}I(x)\big), \quad x \in \mathbb{R}^N,$$

takes the following form

$$\begin{pmatrix} A^{(1)} & 0 & \cdots & 0 \\ \star & A^{(2)} & \cdots & 0 \\ \vdots & \vdots & \ddots & \vdots \\ \star & \star & \cdots & A^{(r)} \end{pmatrix},$$

where $A^{(1)}, \dots, A^{(r)}$ are square constant non-singular matrices. As a consequence, if the vector fields $X_1, \dots, X_m$ satisfy hypotheses **(H0)**–**(H1)**–**(H2)**, then they also satisfy

(H1)* $\dim\big(W^{(k)}I(x)\big) = \dim\big(W^{(k)}\big) \quad \forall\, k \leq r,\ \forall\, x \in \mathbb{R}^N.$

(H2)* $\dim\big(\mathrm{Lie}\{X_1, \dots, X_m\}I(x)\big) = N \quad \forall\, x \in \mathbb{R}^N.$

Condition **(H2)*** is the well-known Hörmander's hypoellipticity condition for the partial differential operator $\sum_{j=1}^m X_j^2$.

Throughout the remaining part of this section, $X_1, \dots, X_m$ will be a given set of smooth vector fields satisfying hypotheses **(H0)**–**(H1)**–**(H2)**. The family $\{\delta_\lambda\}_{\lambda>0}$ will denote the family of dilations in **(H0)**. We let

$$\mathfrak{a} = \mathrm{Lie}\{X_1, \dots, X_m\}. \tag{4.6}$$

Finally, for every $k = 1, \dots, r$, $Z_1^{(k)}, \dots, Z_{N_k}^{(k)}$ will be a fixed basis for $W^{(k)}$. We know that

$$\{Z_1, \dots, Z_N\} := \{Z_1^{(1)}, \dots, Z_{N_1}^{(1)}, \dots, Z_1^{(r)}, \dots, Z_{N_r}^{(r)}\}$$

is a basis of $\mathfrak{a}$. In particular,

$$\dim(\mathfrak{a}) = N. \tag{4.7}$$

For every $\xi = (\xi_1, \dots, \xi_N) = (\xi^{(1)}, \dots, \xi^{(r)})$, we set

$$\xi \cdot Z = \sum_{j=1}^{N} \xi_j Z_j = \sum_{k=1}^{r} \sum_{j=1}^{N_k} \xi_j^{(k)} Z_j^{(k)}.$$

Obviously, $\mathfrak{a} = \{\xi \cdot Z : \xi \in \mathbb{R}^N\}$. Moreover, from the structure of $Z_j^{(k)}$ (see the matrix (4.6)) we get

$$\xi \cdot Z = \sum_{k=1}^{N} \left(\sum_{j=1}^{r} \xi_j \, a_j^{(k)}(x_1, \ldots, x_{k-1}) \right) \partial_{x_k},$$

where $a_j^{(k)}$ is a suitable polynomial function independent of $x_k, \ldots, x_N$. Then, by Remark 1.1.3, the map $(x,t) \mapsto \exp(t\,\xi \cdot Z)(x)$ is well defined for every $x \in \mathbb{R}^N$ and $t \in \mathbb{R}$. Furthermore,[1]

$$\mathrm{Exp} : \mathbb{R}^N \longrightarrow \mathbb{R}^N, \quad \mathrm{Exp}(\xi) := \exp(\xi \cdot Z)(0)$$

is a global diffeomorphism with polynomial component functions. Its inverse function, which we shall denote by Log, has polynomial components too.

We are now ready to define a composition law on $\mathbb{R}^N$, suggested by Corollary 1.3.29. The notation "exp" is introduced in Definition 1.1.2, page 8.

Definition 4.2.2. *If $X_1, \ldots, X_m$ satisfy hypotheses* **(H0)–(H1)–(H2)**, *we set*

$$x,\ y \in \mathbb{R}^N, \quad x \circ y := \exp(\mathrm{Log}(y) \cdot Z)(x). \tag{4.8}$$

Remark 4.2.3. We shall show that $\mathbb{G} := (\mathbb{R}^N, \circ, \delta_\lambda)$ is a Carnot group whose Lie algebra $\mathfrak{g}$ is $\mathfrak{a}$ in (4.7). Since *the group operation depends only on the Lie algebra itself*, it will follow, in particular, that the definition of $\circ$ is independent of the choice of the basis

$$\{Z_j^{(k)} \,|\, 1 \le k \le r,\ 1 \le j \le N_k\}$$

of $\mathfrak{a}$. The main task of the proof is to show that $\circ$ is *associative*. To this end, we shall use the following result, which is a consequence of the Campbell–Hausdorff–Dynkin formula.

Lemma 4.2.4 (Particular case of Campbell–Hausdorff formula). *With the hypotheses* **(H0)–(H1)–(H2)** *on $X_1, \ldots, X_m$ and set $\mathfrak{a} = \mathrm{Lie}\{X_1, \ldots, X_m\}$, the following result holds*:

for every $X, Y \in \mathfrak{a}$, there exists a unique $V \in \mathfrak{a}$ such that

$$\exp(Y)\big(\exp(X)(x)\big) = \exp(V)(x) \quad \textit{for every } x \in \mathbb{R}^N. \tag{4.9}$$

[1] We explicitly note that we are using the notation Exp to denote a map on $\mathbb{R}^N$ instead of on an algebra of vector fields.

We postpone the proof of this deep result to Part III of the book (see Chapter 15). We would like to stress that V in (4.9) depends only on X and Y, in particular, it is independent of $x \in \mathbb{R}^N$. Note that no mention of the word "associativity" is made in the above lemma, yet it will be the turning point in proving the associativity of $\circ$.

The statement of Lemma 4.2.4 can be rewritten as follows: for every $\xi, \eta \in \mathbb{R}^N$, there exists a unique $\zeta \in \mathbb{R}^N$, which we shall denote by $\xi * \eta$, such that

$$\exp(\eta \cdot Z)\big(\exp(\xi \cdot Z)(x)\big) = \exp\big((\xi * \eta) \cdot Z\big)(x) \tag{4.10}$$

for every $x \in \mathbb{R}^N$. Then, by Definition 4.2.2, for every $x, y \in \mathbb{R}^N$, we have

$$\begin{aligned} x \circ y &= \exp(\mathrm{Log}(y) \cdot Z)(x) = \exp(\mathrm{Log}(y) \cdot Z)\big(\exp(\mathrm{Log}(x) \cdot Z)(0)\big) \\ &= \exp\big((\mathrm{Log}(x) * \mathrm{Log}(y)) \cdot Z\big)(0) = \mathrm{Exp}\big(\mathrm{Log}(x) * \mathrm{Log}(y)\big). \end{aligned}$$

Then we have the identity

$$x \circ y = \mathrm{Exp}\big(\mathrm{Log}(x) * \mathrm{Log}(y)\big) \quad \forall\, x,\ y \in \mathbb{R}^N,$$

which is equivalent to the following one

$$\mathrm{Log}(x \circ y) = \mathrm{Log}(x) * \mathrm{Log}(y), \quad \forall\, x,\ y \in \mathbb{R}^N. \tag{4.11}$$

With this identity at hand, we easily get the proof of the following theorem.

Theorem 4.2.5. *Let $\circ$ be the composition law introduced in Definition* 4.2.2. *Then $(\mathbb{R}^N, \circ)$ is a Lie group.*

Proof. 1) IDENTITY ELEMENT. Since $\mathrm{Exp}(0) = 0 = \mathrm{Log}(0)$, for every $x \in \mathbb{R}^N$, we have

$$\begin{aligned} x \circ 0 &= \exp(\mathrm{Log}(0) \cdot Z)(x) = x, \\ 0 \circ x &= \exp(\mathrm{Log}(x) \cdot Z)(0) = \mathrm{Exp}(\mathrm{Log} x) = x. \end{aligned}$$

Then 0 is the identity of $(\mathbb{R}^N, \circ)$.

2) INVERSE ELEMENT. For any $x \in \mathbb{R}^N$, we have

$$\begin{aligned} x \circ \mathrm{Exp}(-\mathrm{Log}(x)) &= \exp(-\mathrm{Log}(x) \cdot Z)(x) \\ &= \exp(-\mathrm{Log}(x) \cdot Z)\big(\exp(\mathrm{Log}(x) \cdot Z)(0)\big) = 0. \end{aligned}$$

An analogous argument shows that

$$\mathrm{Exp}(-\mathrm{Log}(x)) \circ x = 0.$$

Then, for every $x \in \mathbb{R}^N$,

$$x^{-1} = \mathrm{Exp}(-\mathrm{Log}(x)).$$

3) ASSOCIATIVITY. For every $x, y, z \in \mathbb{R}^N$, we have

$$
\begin{aligned}
(x \circ y) \circ z &= \exp(\mathrm{Log}(z) \cdot Z)(x \circ y) \\
&= \exp(\mathrm{Log}(z) \cdot Z)\big(\exp(\mathrm{Log}(y) \cdot Z)(x)\big) \\
\text{(by (4.10))} &= \exp\big((\mathrm{Log}(y) * \mathrm{Log}(z)) \cdot Z\big)(x).
\end{aligned}
$$

On the other hand, by (4.11),

$$x \circ (y \circ z) = \exp(\mathrm{Log}(y \circ z) \cdot Z)(x) = \exp\big((\mathrm{Log}(y) * \mathrm{Log}(z)) \cdot Z\big)(x).$$

This, together with the previous identity, shows the associativity of $\circ$.

To complete the proof of the theorem, we only have to note that the maps

$$(x, y) \mapsto x \circ y = \exp(\mathrm{Log}(y) \cdot Z)(x), \quad x \mapsto x^{-1} = \mathrm{Exp}(-\mathrm{Log}(x))$$

are smooth. □

In order to prove that $\{\delta_\lambda\}_{\lambda>0}$ is a family of automorphisms of $(\mathbb{R}^N, \circ)$, we need the following lemma.

Lemma 4.2.6. *For every $x, \xi \in \mathbb{R}^N$, we have*

$$\delta_\lambda\big(\exp(\xi \cdot Z)(x)\big) = \exp((\delta_\lambda \xi) \cdot Z)(\delta_\lambda(x)) \quad \forall\, \lambda > 0. \tag{4.12}$$

Proof. The path $\mathbb{R} \ni t \mapsto \gamma(t) = \exp(t\xi \cdot Z)(x)$ is the solution to the Cauchy problem

$$\dot\gamma(t) = \sum_{j=1}^N \xi_j Z_j I(\gamma(t)), \quad \gamma(0) = x.$$

We have

$$
\begin{aligned}
\frac{\mathrm{d}}{\mathrm{d}t}(\delta_\lambda(\gamma(t))) = \delta_\lambda(\dot\gamma(t)) &= \sum_{j=1}^N \xi_j\, \delta_\lambda\big(Z_j I(\gamma(t))\big) \\
\text{(by Corollary 1.3.6, page 35)} &= \sum_{j=1}^N \xi_j\, \lambda^{\sigma_j} (Z_j I)(\delta_\lambda(\gamma(t))) \\
&= ((\delta_\lambda \xi) \cdot Z)(\delta_\lambda(\gamma(t))).
\end{aligned}
$$

Moreover, $\delta_\lambda(\gamma(0)) = \delta_\lambda(x)$. This shows that $\mu := \delta_\lambda(\gamma)$ solves the Cauchy problem

$$\dot\mu(t) = (\delta_\lambda(\xi) \cdot Z) I(\mu(t)), \quad \mu(0) = \delta_\lambda(x).$$

Thus

$$\delta_\lambda(\gamma(t)) = \exp(t(\delta_\lambda \xi) \cdot Z)(\delta_\lambda(x)).$$

By replacing $t = 1$ in this identity, we obtain (4.12). □

Theorem 4.2.7. $\mathbb{G} := (\mathbb{R}^N, \circ, \delta_\lambda)$ *is a homogeneous group.*

Proof. By Theorem 4.2.5, we only have to prove that δ_λ is an automorphism of $(\mathbb{R}^N, \circ)$. By Lemma 4.2.6, for every $\xi \in \mathbb{R}^N$, we have

$$\delta_\lambda(\mathrm{Exp}(\xi)) = \delta_\lambda(\exp(\xi \cdot Z)(0)) = \exp((\delta_\lambda \xi) \cdot Z)(0) = \mathrm{Exp}(\delta_\lambda(\xi)),$$

so that, since $\mathrm{Log} = \mathrm{Exp}^{-1}$,

$$\mathrm{Log}(\delta_\lambda(\xi)) = \delta_\lambda(\mathrm{Log}(\xi)).$$

Then, for every $x,\ y \in \mathbb{R}^N$,

$$\begin{aligned} \delta_\lambda(x \circ y) &= \delta_\lambda(\exp(\mathrm{Log}(y) \cdot Z)(x)) \\ \text{(by Lemma 4.2.6)} &= \exp(\delta_\lambda(\mathrm{Log}(y)) \cdot Z)(\delta_\lambda(x)) \\ &= \exp(\mathrm{Log}(\delta_\lambda(y)) \cdot Z)(\delta_\lambda(x)) = (\delta_\lambda(x)) \circ (\delta_\lambda(y)). \end{aligned}$$

This completes the proof. □

By using the associativity property of the composition law $\circ$, it is easy to show that the vector fields $Z_1, \ldots, Z_N$ are invariant with respect to the left translations on $(\mathbb{R}^N, \circ, \delta_\lambda)$.

Theorem 4.2.8. *The vector fields $Z_1, \ldots, Z_N$ are left-invariant on $\mathbb{G}$.*

Proof. If $e_j = (0, \ldots, 1, \ldots, 0)$ (1 being the j-th component), then $e_j \cdot Z = Z_j$ and

$$x \circ \mathrm{Exp}(t\, e_j \cdot Z) = \exp(t Z_j)(x)$$

for every $x \in \mathbb{R}^N$ and $t \in \mathbb{R}$. For any fixed $\alpha \in \mathbb{R}^N$ and $u \in C^\infty(\mathbb{R}^N)$, let us denote $u_\alpha(x) := u(\alpha \circ x)$. From (1.15) (page 10) and the associativity of $\circ$, we obtain

$$\begin{aligned} Z_j(u(\alpha \circ x)) = Z_j(u_\alpha(x)) &= \frac{\mathrm{d}}{\mathrm{d}t}\Big|_{t=0} \big\{u_\alpha\big(x \circ \mathrm{Exp}(t e_j \cdot Z)\big)\big\} \\ &= \frac{\mathrm{d}}{\mathrm{d}t}\Big|_{t=0} \big\{u\big((\alpha \circ x) \circ \mathrm{Exp}(t e_j \cdot Z)\big)\big\} = (Z_j u)(\alpha \circ x), \end{aligned}$$

for every $x \in \mathbb{R}^N$. This completes the proof. □

Corollary 4.2.9. *Let $\mathfrak{g}$ be the Lie algebra of $\mathbb{G}$. Then $\mathfrak{g} = \mathfrak{a}$, where $\mathfrak{a} = \mathrm{Lie}\{X_1, \ldots, X_m\}$.*

Proof. By Theorem 4.2.8, we have $\mathfrak{a} \subseteq \mathfrak{g}$. On the other hand, by (4.7) and Proposition 1.2.7, $\dim(\mathfrak{a}) = N = \dim(\mathfrak{g})$. Thus $\mathfrak{a} = \mathfrak{g}$. □

Let us now consider the vector fields $Y_1, \ldots, Y_{N_1} \in \mathfrak{g}$ such that $Y_j(0) = \partial_{x_j}$, $1 \le j \le N_1$. We know that $N_1 = m$. Since Y_j is δ_λ-homogeneous of degree one (see Corollary 1.3.19), by Corollaries 4.2.9 and 1.3.11 (the latter on page 37) and identity (4.4), we have $Y_j \in \mathrm{span}\{X_1, \ldots, X_m\}$, $1 \le j \le m$. Then, since $Y_1, \ldots, Y_m$ are linearly independent,

$$\text{span}\{Y_1, \dots, Y_m\} = \text{span}\{X_1, \dots, X_m\}.$$

This identity obviously implies

$$\text{Lie}\{Y_1, \dots, Y_m\} = \text{Lie}\{X_1, \dots, X_m\} = \mathfrak{a} = \mathfrak{g}.$$

Thus $Y_1, \dots, Y_m$ are the generators of $\mathfrak{g}$. Then $\mathbb{G} = (\mathbb{R}^N, \circ, \delta_\lambda)$ is a homogeneous Carnot group, $\Delta_{\mathbb{G}} = \sum_{j=1}^m Y_j^2$ is its canonical sub-Laplacian and $\mathcal{L} = \sum_{j=1}^m X_j^2$ is a sub-Laplacian on $\mathbb{G}$. □

We can summarize the results of this section by stating the following theorem.

Theorem 4.2.10. *Let $X_1, \dots, X_m$ be smooth vector fields in $\mathbb{R}^N$ satisfying hypotheses* **(H0)–(H1)–(H2)**. *Let $\{\delta_\lambda\}_{\lambda>0}$ be the family of dilations defined in* **(H0)**. *Finally, let $\circ$ be the composition law on $\mathbb{R}^N$ introduced in Definition* 4.2.2. *Then*

$$\mathbb{G} = (\mathbb{R}^N, \circ, \delta_\lambda)$$

is a homogeneous Carnot group of step r and with m generators whose Lie algebra $\mathfrak{g}$ is Lie-generated by $X_1, \dots, X_m$, i.e.

$$\mathfrak{g} = \text{Lie}\{X_1, \dots, X_m\}.$$

Moreover, the second order partial differential operator $\mathcal{L} = \sum_{j=1}^m X_j^2$ is a sub-Laplacian on $\mathbb{G}$.

4.3 Further Examples

In this section, we exhibit some non-trivial examples of homogeneous Carnot groups constructed starting from a set of vector fields satisfying hypotheses **(H0)–(H1)–(H2)** of the previous section. We introduce the notation: given $n \in \mathbb{N}$, $\mathbb{B}_n$ will denote the following $n \times n$ (nilpotent of step n) matrix

$$\mathbb{B}_n := \begin{pmatrix} 0 & 0 & \cdots & 0 \\ 1 & 0 & \cdots & 0 \\ \vdots & \ddots & \ddots & \vdots \\ 0 & \cdots & 1 & 0 \end{pmatrix}. \tag{4.13}$$

4.3.1 The Vector Fields ∂_1, $\partial_2 + x_1\partial_3$

Let us consider in $\mathbb{R}^3$ the vector fields

$$X_1 = \partial_{x_1}, \quad X_2 = \partial_{x_2} + x_1 \partial_{x_3}.$$

We denote by $x = (x_1, x_2, x_3)$ the points of $\mathbb{R}^3$. It is straightforwardly verified that $\{X_1, X_2\}$ satisfy conditions **(H0)–(H1)–(H2)** of the previous section with respect to the dilations

$$\delta_\lambda(x_1, x_2, x_3) = (\lambda x_1, \lambda x_2, \lambda^2 x_3), \qquad \lambda > 0.$$

Then, by Theorem 4.2.10, the operator

$$\mathcal{L} = \partial_{x_1}^2 + (\partial_{x_2} + x_1 \partial_{x_3})^2$$

is a sub-Laplacian (indeed, the canonical one) on a suitable homogeneous Carnot group $\mathbb{G} = (\mathbb{R}^3, \circ, \delta_\lambda)$. We now construct the composition law $\circ$ by using Definition 4.2.2. With the notation of the previous section, we have

$$W^{(1)} = \text{span}\{X_1, X_2\}, \qquad W^{(2)} = \text{span}\{X_3\},$$

where $X_3 = [X_1, X_2] = \partial_{x_3}$. For every $\xi = (\xi_1, \xi_2, \xi_3) \in \mathbb{R}^3$, we have

$$\xi \cdot X = \sum_{j=1}^{3} \xi_j X_j = \xi_1 \partial_{x_1} + \xi_2 \partial_{x_2} + (\xi_2 x_1 + \xi_3)\partial_{x_3},$$

so that $\exp(\xi \cdot X)(x) = \gamma(1)$, where $\gamma = (\gamma_1, \gamma_2, \gamma_3)$ and

$$\begin{cases} \dot\gamma_1(t) = \xi_1, & \gamma_1(0) = x_1, \\ \dot\gamma_2(t) = \xi_2, & \gamma_2(0) = x_2, \\ \dot\gamma_3(t) = \xi_2 \gamma_1(t) + \xi_3, & \gamma_3(0) = x_3. \end{cases}$$

An easy computation shows that

$$\exp(\xi \cdot X)(x) = \left(x_1 + \xi_1, x_2 + \xi_2, x_3 + \xi_3 + \xi_2 x_1 + \frac{1}{2}\xi_1\xi_2 \right).$$

As a consequence,

$$\text{Exp}(\xi) = \exp(\xi \cdot X)(0) = \left(\xi_1, \xi_2, \xi_3 + \frac{1}{2}\xi_1\xi_2 \right),$$

$$\text{Log}(\eta) = \text{Exp}^{-1}(\eta) = \left(\eta_1, \eta_2, \eta_3 - \frac{1}{2}\eta_1\eta_2 \right).$$

Then, by Definition 4.2.2, the composition law determined by X_1, X_2, X_3 is given by

$$\begin{aligned} x \circ y &= \exp(\text{Log}(y) \cdot X)(x) \\ &= \exp\left(y_1 X_1 + y_2 X_2 + \left(y_3 - \frac{1}{2} y_1 y_2 \right) X_3 \right)(x_1, x_2, x_3) \\ &= \left(x_1 + y_1, x_2 + y_2, x_3 + \left(y_3 - \frac{1}{2} y_1 y_2 \right) + y_2 x_1 + \frac{1}{2} y_1 y_2 \right), \end{aligned}$$

i.e.

$$x \circ y = (x_1 + y_1, x_2 + y_2, x_3 + y_3 + x_1 y_2).$$

4.3.2 Classical and Kohn Laplacians

In this section, we re-derive the classical Laplace operator and the Kohn Laplacian (already introduced in Sections 4.1.1 and 3.1, respectively) starting from the relevant vector fields.

• The only example of homogeneous Carnot group of step 1 is the usual additive group $(\mathbb{R}^N, +)$. If δ_λ denotes the usual dilation on $\mathbb{R}^N$

$$\delta_\lambda(x_1, \dots, x_N) = (\lambda x_1, \dots, \lambda x_N),$$

a family of fields satisfying hypotheses **(H0)–(H1)–(H2)** is necessarily of the form $\{X_j\}_{j\le N}$, where

$$X_j = \sum_{i=1}^{N} a_{i,j}\, \partial_i$$

and $A = (a_{i,j})_{i,j}$ is a non-singular $N\times N$ matrix so that $\sum_{j=1}^{N} X_j^2$ is a strictly-elliptic constant-coefficient operator. Given $\xi,\ x \in \mathbb{R}^N$, it holds

$$\exp(\xi\cdot X)(x) = \gamma(1) \qquad \left(\text{set}\, \xi\cdot X := \sum_{j=1}^{N} \xi_j X_j\right),$$

where $\dot\gamma(r) = A\cdot\xi$ and $\gamma(0) = x$. Hence

$$\exp(\xi\cdot X)(x) = x + A\cdot\xi, \quad \text{Exp}(\xi) = A\cdot\xi, \quad \text{Log}(y) = A^{-1}\cdot y.$$

We find out $x\circ y = \exp(\text{Log}(y)\cdot X)(x) = x + A\cdot\text{Log}(y) = x + A\cdot A^{-1}\cdot y = x + y$, the usual additive structure of $\mathbb{R}^N$. The canonical sub-Laplacian related to this group is the ordinary Laplace operator

$$\Delta = \sum_{j=1}^{N} (\partial_{x_j})^2.$$

• Consider now on $\mathbb{R}^{2N+1}$ (whose points are denoted by $z = (x, y, t)$, $x,\ y \in \mathbb{R}^N$, $t\in\mathbb{R}$) the $2N$ vector fields

$$X_j := \partial_{x_j} + 2y_j\,\partial_t, \qquad Y_j := \partial_{y_j} - 2x_j\,\partial_t, \quad j = 1, \dots, N.$$

If $\mathbb{R}^{2N+1}$ is equipped with the dilation $\delta_\lambda(z) = (\lambda x, \lambda y, \lambda^2 t)$, the former $2N$ vector fields satisfy hypotheses **(H0)–(H1)–(H2)**. We set

$$T := [X_j, Y_j] = -4\,\partial_t.$$

Given $\zeta = (\xi, \eta, \tau)$, $z = (x, y, t) \in \mathbb{R}^{2N+1}$, one has

$$\exp\left(\sum_{j=1}^{N} (\xi_j X_j + \eta_j Y_j) + \tau T\right)(z) = (\mu(1), \nu(1), \rho(1)),$$

where

$$(\dot{\mu}, \dot{\nu}, \dot{\rho})(r) = \big(\xi, \eta, -4\tau + 2\langle \nu(r), \xi\rangle - 2\langle \mu(r), \eta\rangle\big), \quad (\mu, \nu, \rho)(0) = (x, y, t).$$

This gives (set $\zeta \cdot Z := \sum_{j=1}^N (\xi_j X_j + \eta_j Y_j) + \tau T$)

$$\exp(\zeta \cdot Z)(z) = \big(x + \xi, y + \eta, t - 4\tau + 2\langle y, \xi\rangle - 2\langle x, \eta\rangle\big),$$

whence

$$\mathrm{Exp}(\zeta \cdot Z) = (\xi, \eta, -4\tau), \qquad \mathrm{Log}(z') = (x', y', -t'/4) \cdot Z. \tag{4.14}$$

Fixed $z,\ z' \in \mathbb{R}^{2N+1}$, we have

$$z \circ z' = \exp\big(\mathrm{Log}(z') \cdot Z\big)(z) = \big(x + x', y + y', t + t' + 2\langle y, x'\rangle - 2\langle x, y'\rangle\big). \tag{4.15}$$

We recognize the well-known Heisenberg–Weyl group $(\mathbb{H}^N, \circ)$ on $\mathbb{R}^{2N+1}$. Its canonical sub-Laplacian $\Delta_{\mathbb{H}^N} = \sum_{j=1}^N (X_j^2 + Y_j^2)$ is the Kohn Laplacian on $\mathbb{H}^N$. As we know from Chapter 1, $\circ$ induces a Lie group structure $*$ (of "Campbell–Hausdorff-type") on the Lie algebra $\mathfrak{h}^N$ of $\mathbb{H}^N$ in the way re-described hereafter. For $(\xi, \eta, \tau) \in \mathbb{R}^N \times \mathbb{R}^N \times \mathbb{R}$, we agree to set

$$(\xi, \eta, \tau)_Z := (\xi, \eta, \tau) \cdot Z = \sum_{j=1}^N (\xi_j X_j + \eta_j Y_j) + \tau T \in \mathfrak{h}^N.$$

Then the group law $*$ on $\mathfrak{h}^N$ is defined by

$$\begin{aligned}
(\xi_1, \eta_1, \tau_1)_Z * (\xi_2, \eta_2, \tau_2)_Z &= \mathrm{Log}\big(\mathrm{Exp}(\xi_1, \eta_1, \tau_1)_Z \circ \mathrm{Exp}(\xi_2, \eta_2, \tau_2)_Z\big)\\
&= \mathrm{Log}\big((\xi_1, \eta_1, -4\,\tau_1) \circ (\xi_2, \eta_2, -4\,\tau_2)\big)\\
&= \mathrm{Log}\big(\xi_1 + \xi_2, \eta_1 + \eta_2, -4\tau_1 - 4\tau_2 + 2\langle \eta_1, \xi_2\rangle - 2\,\langle \xi_1, \eta_2\rangle\big)\\
&= \left(\xi_1 + \xi_2, \eta_1 + \eta_2, \tau_1 + \tau_2 - \frac{1}{2}\langle \eta_1, \xi_2\rangle + \frac{1}{2}\,\langle \xi_1, \eta_2\rangle\right)_Z.
\end{aligned}$$

If we drop the notation $(\cdot)_Z$ and identify $\mathfrak{h}^N$ with $\mathbb{R}^{2N+1}$ via the basis $(X_1, \dots, X_N, Y_1, \dots, Y_N, T)$, we have somewhat "more intrinsic" group $(\mathbb{R}^{2N+1}, *)$ canonically related to $\mathbb{H}^N$, where

$$\begin{aligned}
&(\xi_1, \eta_1, \tau_1) * (\xi_2, \eta_2, \tau_2)\\
&\quad = \left(\xi_1 + \xi_2, \eta_1 + \eta_2, \tau_1 + \tau_2 - \frac{1}{2}\langle \eta_1, \xi_2\rangle + \frac{1}{2}\langle \xi_1, \eta_2\rangle\right).
\end{aligned} \tag{4.16}$$

This group is canonically related to $\mathbb{H}^N$ in the sense that we make precise below: if we consider the change of coordinate system on $\mathbb{H}^N$ induced by the exponential-type coordinates in (4.14), i.e.

$$(x, y, t) = \mathrm{Exp}\big((\xi, \eta, \tau)_Z\big) = (\xi, \eta, -4\tau),$$

then, with respect to this new coordinates, the vector fields X_j and Y_j are respectively turned into[2]

$$\widetilde{X}_j := \partial_{\xi_j} - \frac{1}{2}\,\eta_j\,\partial_\tau, \qquad \widetilde{Y}_j := \partial_{\eta_j} + \frac{1}{2}\,\xi_j\,\partial_\tau.$$

Starting from these new vector fields $\widetilde{X}_j$'s and $\widetilde{Y}_j$'s, we turn to construct a related Carnot group (as we did above): after simple computations we obtain the very same group law $*$ as in (4.16).

4.3.3 Bony-type Sub-Laplacians

Let us consider in $\mathbb{R}^{1+N}$ the operator

$$L = \left(\frac{\partial}{\partial t}\right)^2 + \left(t\frac{\partial}{\partial x_1} + t^2\frac{\partial}{\partial x_2} + \cdots + t^N\frac{\partial}{\partial x_N}\right)^2, \quad (t, x_1, \ldots, x_N) \in \mathbb{R}^{1+N},$$

quoted by J.-M. Bony in [Bon69, Rémarque 3.1] as an example of a sum of squares satisfying Hörmander condition but nevertheless with a "very degenerate" characteristic form. Clearly, L is not a sub-Laplacian on any Carnot group since the vector field $\sum_{j=1}^N t^j\,\partial/\partial x_j$ vanishes on the hyperplane $t = 0$. It is however sufficient to add a new coordinate in order to *lift* L to a sub-Laplacian. Indeed, consider on $\mathbb{R}^{2+N}$, whose points are denoted by (t, s, x), $t,\ s \in \mathbb{R}$, $x \in \mathbb{R}^N$, the following operator

$$\mathcal{L} := T^2 + S^2,$$

where

$$T := \partial_t, \qquad S := \partial_s + t\,\partial_{x_1} + \frac{t^2}{2!}\,\partial_{x_2} + \cdots + \frac{t^N}{N!}\,\partial_{x_N}.$$

It is readily verified that the pair $T,\ S$ satisfies hypothesis **(H0)** with respect to the family of dilations defined by

$$\delta_\lambda(t, s, x) := (\lambda\,t, \lambda\,s, \lambda^2 x_1, \lambda^3 x_2, \ldots, \lambda^{N+1} x_N).$$

For every $k = 1, \ldots, N$, we then consider the vector field

$$X_k := \underbrace{[T, [T, \cdots [T}_{k \text{ times}}, S] \cdots]] = \partial_{x_k} + t\,\partial_{x_{k+1}} + \cdots + \frac{t^{N-k}}{(N-k)!}\,\partial_{x_N}.$$

With the notation of Section 4.2, we have $W^{(1)} = \text{span}\{T, S\}$ and, for $k = 1, \ldots, N$, $W^{(k+1)} = \text{span}\{X_k\}$. It is easy to recognize that the hypothesis **(H1)** is satisfied. Finally, we have

---

[2] Indeed, for every $u \in C^\infty(\mathbb{H}^N, \mathbb{R})$, $u = u(x, y, t)$, we set $v := u \circ \text{Exp}$, i.e. $v = v(\xi, \eta, \tau) = u(\xi, \eta, -4\tau)$, so that it holds $\partial_{\xi_j} v = (\partial_{x_j} u) \circ \text{Exp}$, $\partial_{\eta_j} v = (\partial_{y_j} u) \circ \text{Exp}$, $\partial_\tau v = -4\,(\partial_t u) \circ \text{Exp}$. Consequently,

$$(X_j u) \circ \text{Exp} = \partial_{\xi_j} v - \frac{1}{2}\,\eta_j\,\partial_\tau v, \qquad (Y_j u) \circ \text{Exp} = \partial_{\eta_j} v + \frac{1}{2}\,\xi_j\,\partial_\tau v.$$

$$\dim\big(\mathrm{Lie}\{T, S\}I(0)\big) = 2 + N,$$

whence also the hypothesis **(H2)** holds. As a consequence, $\mathcal{L}$ is a sub-Laplacian on a suitable homogeneous Carnot group $(\mathbb{G}, \circ)$ on $\mathbb{R}^{2+N}$ with step $1 + N$ and with 2 generators. We now turn to construct the group multiplication $\circ$ on $\mathbb{G}$ by using Definition 4.2.2. Let $\alpha,\ \beta \in \mathbb{R}$ and $\xi \in \mathbb{R}^N$ be fixed. We have

$$\alpha\, T + \beta\, S + \sum_{k=1}^{N} \xi_k\, X_k = \left(\alpha, \beta, \left(\beta\, t^j/j + \sum_{k=1}^{j} \xi_k\, t^{j-k}/(j-k)!\right)_{j=1,\dots,N}\right).$$

This yields

$$\begin{aligned}&\exp[\alpha, \beta, \xi](t, s, x)\\ &\quad := \exp\left(\alpha T + \beta S + \sum_{k=1}^{N} \xi_k\, X_k\right)(t, s, x) = (\tau, \sigma, \gamma)(1),\end{aligned}$$

where (here j runs from 1 to N)

$$\begin{cases} \dot\tau(r) = \alpha, \quad \tau(0) = t, \quad \dot\sigma(r) = \beta, \quad \sigma(0) = s, \\ \dot\gamma_j(r) = \beta\, \tau^j(r)/j! + \sum_{k=1}^{j} \xi_k\, \tau^{j-k}(r)/(j-k)!, \quad \gamma_j(0) = x_j. \end{cases}$$

From a direct integration it follows that $\exp[\alpha, \beta, \xi](t, s, x)$ is given by

$$\left(\alpha + t, \beta + s, x_j + \beta\, \frac{(\alpha + t)^{j+1} - t^{j+1}}{(j+1)\alpha} + \sum_{k=1}^{j} \xi_k\, \frac{(\alpha + t)^{j-k+1} - t^{j-k+1}}{(j-k+1)\alpha}\right).$$

We agree to put $((\alpha + t)^{j+1} - t^{j+1})/\alpha = (j+1)t^j$ when $\alpha = 0$. We now define the following matrices

$$F(\alpha) := \begin{pmatrix} 1 & 0 & 0 & \cdots & 0 \\ \frac{\alpha}{2!} & 1 & 0 & \cdots & 0 \\ \frac{\alpha^2}{3!} & \frac{\alpha}{2!} & 1 & \ddots & \vdots \\ \vdots & \vdots & \ddots & \ddots & 0 \\ \frac{\alpha^{N-1}}{N!} & \frac{\alpha^{N-2}}{(N-1)!} & \cdots & \frac{\alpha}{2!} & 1 \end{pmatrix},$$

$$V(\alpha) := \begin{pmatrix} \frac{\alpha}{2!} \\ \frac{\alpha^2}{3!} \\ \frac{\alpha^3}{4!} \\ \vdots \\ \frac{\alpha^N}{(N+1)!} \end{pmatrix}, \qquad U(\alpha) := \begin{pmatrix} \alpha \\ \frac{\alpha^2}{2!} \\ \frac{\alpha^3}{3!} \\ \vdots \\ \frac{\alpha^N}{N!} \end{pmatrix}$$

and the functions

$$\widetilde{F}(\alpha, t) := \alpha^{-1}((\alpha + t)F(\alpha + t) - t\,F(t)),$$
$$\widetilde{V}(\alpha, t) := \alpha^{-1}((\alpha + t)V(\alpha + t) - t\,V(t)).$$

It then holds

$$\begin{aligned}\exp[\alpha, \beta, \xi](t, s, x) &= \big(\alpha + t, \beta + s, x + \widetilde{F}(\alpha, t) \cdot \xi + \beta\,\widetilde{V}(\alpha, t)\big),\\ \mathrm{Exp}(\alpha, \beta, \xi) &= \big(\alpha, \beta, F(\alpha) \cdot \xi + \beta\,V(\alpha)\big),\\ \mathrm{Log}(\tau, \sigma, y) &= \big(\tau, \sigma, F^{-1}(\tau) \cdot (y - \sigma\,V(\tau)\big).\end{aligned}$$

Let (t, s, x) and $(\tau, \sigma, y) \in \mathbb{R}^{2+N}$ be given. Then we have

$$\begin{aligned}&(t, s, x) \circ (\tau, \sigma, y)\\ &\quad = \big(\tau + t, \sigma + s, x + \widetilde{F}(\tau, t) \cdot F^{-1}(\tau) \cdot (y - \sigma\,V(\tau)) + \sigma\,\widetilde{V}(\tau, t)\big).\end{aligned}$$

If we now prove that the following identities hold (see also (4.13))

$$\widetilde{F}(\tau, t) = \exp(t\,\mathbb{B}_N) \cdot F(\tau), \qquad \widetilde{V}(\tau, t) - \exp(t\,\mathbb{B}_N) \cdot V(\tau) = U(t), \tag{4.17}$$

then the explicit form of the multiplication $\circ$ turns out to be

$$\begin{aligned}&(t, s, x) \circ (\tau, \sigma, y) = (\tau + t, \sigma + s, x + \exp(t\,\mathbb{B}_N) \cdot y + \sigma\,U(t))\\ &= \left(\tau + t, \sigma + s, x_1 + y_1 + \sigma\,t, \ldots, x_N + \sum_{k=1}^{N} y_k\,t^{N-k}\,\frac{1}{(N-k)!} + \sigma\,t^N \frac{1}{N!}\right).\end{aligned}$$

The first identity in (4.17) follows by proving that, for every $i, j \in \{1, \ldots, N\}$ with $i \geq j$, one has

$$\frac{(\tau + t)^{i-j+1} - t^{i-j+1}}{(i - j + 1)!\,\tau} = \sum_{k=j}^{i} \frac{t^{i-k}}{(i-k)!} \cdot \frac{\tau^{k-j}}{(k - j + 1)!},$$

which readily follows by applying Newton's binomial formula to the left-hand side. The second identity is equivalent to

$$\frac{t^i}{i!} = \frac{(\tau + t)^{i+1} - t^{i+1}}{(i + 1)!\,\tau} - \sum_{k=1}^{i} \frac{\tau^k}{(k+1)!} \cdot \frac{t^{i-k}}{(i-k)!}, \qquad i = 1, \ldots, N,$$

which can be proved analogously.

4.3.4 Kolmogorov-type Sub-Laplacians

We now reconsider the examples in Sections 4.1.3 and 4.1.4, and we show how to obtain the composition law of the groups of Kolmogorov type by using the results of Section 4.2. We consider in $\mathbb{R}^{1+N}$ (whose points are denoted by $(t, x^{(0)}, \ldots, x^{(r)})$)

the vector fields introduced in (4.2e). Then, by the results proved in Sections 4.1.3–4.1.4, they satisfy conditions **(H0)–(H1)–(H2)** in Section 4.2 with respect to the following family of dilations

$$\delta_\lambda(t, x^{(0)}, \dots, x^{(r)}) = (\lambda\, t, \lambda\, x^{(0)}, \dots, \lambda^{r+1} x^{(r)})$$

(we refer the reader directly to the notation in Sections 4.1.3–4.1.4). For $k = 0, \dots, r$ and $j = 1, \dots, p_k$, we set $Z_j^{(k)} := \partial/\partial\, x_j^{(k)}$. If $t,\ \tau \in \mathbb{R}$ and $x,\ \xi \in \mathbb{R}^N$ are fixed, we have

$$\exp[\tau, \xi](t, x) := \exp\left(\tau Y + \sum_{k=1}^{r+1} \sum_{j=1}^{p_k} \xi_j^{(k)} Z_j^{(k)}\right)(t, x) = (\mu(1), \gamma(1)),$$

where

$$\dot{\mu}(r) = \tau, \quad \mu(0) = t; \quad \dot{\gamma}(r) = \xi + \tau\, B \cdot \gamma(r), \quad \gamma(0) = x.$$

This yields

$$\begin{aligned} \exp[\tau, \xi](t, x) &= \left(\tau + t, \exp(\tau B) \cdot x + \int_0^1 \exp(\tau(1 - r)B) \cdot \xi \, \mathrm{d}r\right), \\ \mathrm{Exp}(\tau, \xi) &= \left(\tau, \int_0^1 \exp(\tau(1 - r)B) \cdot \xi \, \mathrm{d}r\right), \\ \mathrm{Log}(s, y) &= \left(s, \left(\int_0^1 \exp(s(1 - r)B) \, \mathrm{d}r\right)^{-1} \cdot y\right). \end{aligned}$$

As a consequence,

$$(t, x) \circ (s, y) = \big(t + s, y + \exp(sB)x\big).$$

We explicitly remark that this is the same group multiplication treated in Section 4.1.3.

4.3.5 Sub-Laplacians Arising in Control Theory

We here discuss an example of homogeneous Carnot group arising from control theory, and we refer to [Alt99] for a description of the relevance of this example in that context. In $\mathbb{R}^N$, we consider the following vector fields

$$X_1 := \partial_1 + x_2\, \partial_3 + x_3\, \partial_4 + \cdots + x_{N-1}\, \partial_N, \qquad X_2 := \partial_2.$$

For every $k = 3, \dots, N$, we have $X_k := [X_{k-1}, X_1] = \partial_k$, whence it is readily verified that X_1 and X_2 fulfill hypotheses **(H0)–(H1)–(H2)** with respect to the family of dilations

$$\delta_\lambda(x_1, x_2, x_3, \dots, x_N) := (\lambda\, x_1, \lambda\, x_2, \lambda^2 x_3, \dots, \lambda^{N-1} x_N).$$

As a consequence,

$$\mathcal{L} = X_1^2 + X_2^2$$

is a sub-Laplacian on a suitable homogeneous Carnot group $(\mathbb{G}, \circ)$ on $\mathbb{R}^N$ with step $N-1$ and with 2 generators. In [Alt99] it is given a representation of $\mathbb{G}$ by means of matrices of the following form

$$\begin{pmatrix} 1 & x_2 & x_3 & x_4 & \cdots & x_N \\ 0 & 1 & x_1 & \frac{x_1^2}{2!} & \cdots & \frac{x_1^{N-2}}{(N-2)!} \\ 0 & 0 & 1 & x_1 & \ddots & \vdots \\ 0 & 0 & 0 & 1 & \ddots & \frac{x_1^2}{2!} \\ \vdots & \vdots & \ddots & \ddots & \ddots & x_1 \\ 0 & 0 & \cdots & 0 & 0 & 1 \end{pmatrix} \equiv (x_1, x_2, \dots, x_N) \in \mathbb{G},$$

whereas the Lie group law is given by the matrix product. We hereafter show how to obtain the composition law following the lines described in Section 4.2. Let $\xi \in \mathbb{R}^N$ be fixed. We have

$$\sum_{k=1}^{N} \xi_k X_k = (\xi_1, \xi_2, \xi_3 + \xi_1 x_2, \dots, \xi_N + \xi_1 x_{N-1}) = \xi + \xi_1 Hx,$$

where H is the following $N \times N$ matrix (see also (4.13))

$$H := \begin{pmatrix} 0 & 0 \\ 0 & \mathbb{B}_{N-2} \end{pmatrix}.$$

This gives $\exp[\xi](x) := \exp(\sum_{k=1}^N \xi_k X_k)(x) = \gamma(1)$, where $\dot{\gamma}(r) = \xi + \xi_1 H\gamma(r)$, $\gamma(0) = x$, whence

$$\gamma(r) = \exp(\xi_1 r H)x + \int_0^r \exp(\xi_1 (r-t) H)\xi \, dt.$$

In particular,

$$\exp[\xi](x) = \exp(\xi_1 H)x + \int_0^1 \exp(\xi_1 (1-t) H)\xi \, dt,$$

$$\operatorname{Exp}(\xi) = \int_0^1 \exp(\xi_1 (1-t) H)\xi \, dt.$$

It is straightforward to recognize that, for every $\rho \in \mathbb{R}$, we have

$$\exp(\rho H) = \begin{pmatrix} 1 & 0 \\ 0 & \exp(\rho \mathbb{B}_{N-1}) \end{pmatrix}.$$

Given $y = (y_1, \widehat{y}) \in \mathbb{R}^N$, the equation $y = \operatorname{Exp}(\xi)$ is equivalent to the following system (setting $\xi = (\xi_1, \widehat{\xi}) \in \mathbb{R}^N$)

$$y_1 = \xi_1, \qquad \widehat{y} = \int_0^1 \exp\big(\xi_1(1-t)\mathbb{B}_{N-1}\big)\widehat{\xi}\,\mathrm{d}t.$$

As a consequence,

$$\mathrm{Log}(y) = \left(y_1, \left(\int_0^1 \exp(y_1(1-t)\mathbb{B}_{N-1})\,\mathrm{d}t\right)^{-1} \cdot \widehat{y}\right).$$

For any fixed $x, y \in \mathbb{R}^N$, this gives

$$\begin{aligned} x \circ y &= \exp(\mathrm{Log}(y))(x) = y + \exp(y_1 H)x = (y_1 + x_1, \widehat{y} + \exp(y_1\,\mathbb{B}_{N-1})\widehat{x}) \\ &= \left(y_1 + x_1, y_2 + x_2, y_3 + x_3 + y_1 x_2, \dots, y_N + \sum_{j=2}^{N} \frac{1}{(N-j)!}\, y_1^{N-j}\, x_j\right). \end{aligned}$$

4.3.6 Filiform Carnot Groups

In this section, we give the definition of *filiform* Carnot group. To this end, we recall the definition of filiform Lie algebra (see, e.g. [OV94, page 61]).

Definition 4.3.1 (Filiform Lie algebra). *Let* $\mathfrak{h}$ *be a Lie algebra of finite dimension* $n \geq 3$. *For every* $k \in \mathbb{N}$, *we recall that the terms of the* lower (or descending) central series *for* $\mathfrak{h}$ *are*

$$\mathfrak{h}_1 := \mathfrak{h}, \quad \mathfrak{h}_2 := [\mathfrak{h}, \mathfrak{h}], \quad \dots, \quad \mathfrak{h}_k := [\mathfrak{h}, \mathfrak{h}_{k-1}] = \underbrace{[\mathfrak{h}[\mathfrak{h}[\cdots[\mathfrak{h}, \mathfrak{h}]\cdots]]]}_{k \text{ times}}, \quad k \in \mathbb{N}.$$

Then, the Lie algebra $\mathfrak{h}$ *is called* filiform *if*

$$\mathrm{codim}_{\mathfrak{h}}(\mathfrak{h}_k) = k \quad \text{for every } k \text{ such that } 3 \leq k \leq n.$$

In other words, $\mathfrak{h}$ is filiform if and only if

$$\dim(\mathfrak{h}_k) = n - k \quad \text{for every } k\text{: } 3 \leq k \leq n \text{ (where } n = \dim(\mathfrak{h})). \tag{4.18}$$

We explicitly remark that for the lower central series we have

$$\mathfrak{h}_1 \supseteq \mathfrak{h}_2 \supseteq \mathfrak{h}_3 \supseteq \cdots \supseteq \mathfrak{h}_{k-1} \supseteq \mathfrak{h}_k \quad \forall\, k \in \mathbb{N}.$$

Remark 4.3.2. If $\dim(\mathfrak{h}) = 3$, (4.18) says that $\mathfrak{h}$ is filiform iff $\dim(\mathfrak{h}_3) = 0$. Hence, simple arguments show that the only filiform Lie algebras of dimension three are:

1) $\mathrm{Lie}\{X_1, X_2, X_3\}$ with $\{X_1, X_2, X_3\}$ linearly independent and

$$[X_2, X_1] = [X_3, X_1] = [X_3, X_2] = 0.$$

Any Lie group with such a Lie algebra is isomorphic to the classical $(\mathbb{R}^3, +)$.

2) $\mathrm{Lie}\{X_1, X_2\}$ with $\{X_1, X_2, [X_2, X_1]\}$ linearly independent and

$$[X_2, [X_2, X_1]] = [X_1, [X_2, X_1]] = 0.$$

Any Lie group with such a Lie algebra is isomorphic to the Heisenberg–Weyl group $\mathbb{H}^1$ on $\mathbb{R}^3$.

Let us now suppose that $n := \dim(\mathfrak{h}) \geq 4$. From (4.18) we have

$$\dim(\mathfrak{h}_{n-1}) = 1 \quad \text{and} \quad \dim(\mathfrak{h}_n) = 0,$$

whence *every filiform algebra is nilpotent*, and, precisely, (if $n \geq 4$) an n-dimensional filiform algebra is nilpotent of step $n-1$ (i.e. any commutator of length $\geq n$ vanishes, and there exists at least one non-vanishing commutator of length $= n-1$). A very useful fact is that the converse is also true, as the following proposition states.

Proposition 4.3.3 (Characterization). *Let $\mathfrak{h}$ be a Lie algebra of finite dimension $n \geq 3$.*

- *If $n \geq 4$, then $\mathfrak{h}$ is filiform if and only if it is nilpotent of step $n-1$.*
- *If $n = 3$, the same is true as in the previous case, except the case of the commutative $\mathbb{R}^3$.*

Note 4.3.4. Since the linear dimension, the step of nilpotency and being isomorphic to $\mathbb{R}^3$ are all invariants of isomorphic Lie algebras, by Proposition 4.3.3, we derive: *if $\mathfrak{g}$ is a Lie algebra isomorphic to a filiform one, then $\mathfrak{g}$ is filiform too.*

Proof (of Proposition 4.3.3). The "only if" part follows from Remark 4.3.2. We turn to the "if" part. It can be proved by an inductive argument.

For example, we give the proof when $n = 4$. Suppose that $\mathfrak{h}$ is nilpotent of step 3. We have to prove that $\mathfrak{h}$ fulfills (4.18), i.e.

$$\dim(\mathfrak{h}_3) = 1 \quad \text{and} \quad \dim(\mathfrak{h}_4) = 0.$$

The second equality follows from the step-3-nilpotence of $\mathfrak{h}$. By contradiction, suppose that $\dim(\mathfrak{h}_3) \geq 2$. Since $\mathfrak{h}_2 \supseteq \mathfrak{h}_3$ and $\mathfrak{h}_2$ must contain a commutator of length 2 which is not of length 3, then $\dim(\mathfrak{h}_2) \geq 3$. Finally, since there must exist at least two elements in $\mathfrak{h}_1 \setminus \mathfrak{h}_2$, then the inequality $\dim(\mathfrak{h}_2) \geq 3$ implies $\dim(\mathfrak{h}_1) \geq 5$, which contradicts the assumption $\dim(\mathfrak{h}_1) = \dim(\mathfrak{h}) = 4$. This completes the proof for $n = 4$. The proof in the general case follows these ideas and is left to the reader. □

Example 4.3.5. Let $\mathfrak{h}$ be a Lie algebra of finite dimension n satisfying the following commutator identities: if $\{X_1, \ldots, X_n\}$ is a basis (in the sense of vector spaces) for$\mathfrak{h}$, we have

$$X_3 = [X_1, X_2], \quad X_4 = [X_1, X_3], \quad X_5 = [X_1, X_4], \quad \ldots, \quad X_n = [X_1, X_{n-1}],$$

whereas all other commutators of the X_i's vanish identically. In other words, it holds

$$\begin{cases} [X_1, X_i] = X_{i+1} & \text{for every } i = 2, \dots, n-1, \\ [X_1, X_n] = 0, & \\ [X_i, X_j] = 0 & \text{for every } 1 < i < j \le n. \end{cases}$$

Since $\mathfrak{h}$ is evidently nilpotent of step $n-1$, then, by Proposition 4.3.3, $\mathfrak{h}$ is filiform. In Sections 4.3.3 and 4.3.5 we have furnished two explicit models for a similar algebra.

Vice versa, the following remarkable fact holds.

Theorem 4.3.6 (Bratzlavsky [Bra74]). *Let $\mathfrak{h}$ be a filiform Lie algebra of finite dimension n. Then there exists a basis $\{X_i\}_i$ for $\mathfrak{h}$ (in the sense of vector spaces) such that $X_{i+1} = [X_1, X_i]$ (for every $i = 2, \dots, n-1$) and $[X_1, X_n] = 0$.*

Hence any filiform Lie algebra has a basis of the following form:

$$X_1, X_2, \underbrace{[X_1, X_2]}_{=:X_3}, \underbrace{[X_1, [X_1, X_2]]}_{=:X_4}, \underbrace{[X_1, [X_1, [X_1, X_2]]]}_{=:X_5}, \dots, \underbrace{[\overbrace{X_1, \cdots [X_1}^{n-2 \text{ times}}, X_2] \cdots]}_{=:X_n}.$$

In general, nothing is said about the remaining commutators $[X_i, X_j]$ for $1 < i < j \le n$.

Definition 4.3.7 (Filiform Carnot group). *A Carnot group is said* filiform *if its Lie algebra is a filiform Lie algebra.*

(*Note.* If $\mathbb{G}$ and $\mathbb{H}$ are isomorphic Carnot groups and $\mathbb{G}$ is filiform, then $\mathbb{H}$ is filiform too. This follows from Note 4.3.4.)

By Proposition 4.3.3 and Remark 4.3.2, we have the following result.

Proposition 4.3.8. *A Carnot group on $\mathbb{R}^N$ (with $N \ge 3$) is filiform if and only if it is nilpotent of step $N-1$ (except the trivial case of the usual Euclidean $(\mathbb{R}^3, +)$).*

Remark 4.3.9. For example, among the Heisenberg–Weyl groups $\mathbb{H}^n$ the only filiform one is $\mathbb{H}^1$; analogously, the only filiform Carnot group of step two (hence nilpotent of step two) has necessarily dimension 3: up to isomorphism, this is $\mathbb{H}^1$. Moreover, the Carnot groups considered in Sections 4.3.3 and 4.3.5 are filiform.

Again from Proposition 4.3.3 it follows that a filiform Carnot group $\mathbb{G}$ on $\mathbb{R}^N$ (non-Euclidean) is characterized by a stratification of its algebra $\mathfrak{g}$ of the following type

$$\mathfrak{g} = V_1 \oplus \cdots \oplus V_{N-1} \quad \text{with } \dim(V_1) = 2, \dim(V_i) = 1 \; \forall\, i = 2, \dots, N-1.$$

In particular, a non-Euclidean filiform Carnot group has necessarily two generators. □

4.4 Fields not Satisfying One of the Hypotheses (H0), (H1), (H2)

In this section, we exhibit some examples of vector fields not satisfying one of the hypotheses **(H0)**, **(H1)** or **(H2)** on page 191: we formally try to construct a group for each situation and we show that what we obtain is *not* a Carnot group!

4.4.1 Fields not Satisfying Hypothesis (H0): The Group is not Well Posed

We consider on $\mathbb{R}^2$ the following two polynomial vector fields

$$X := (1+x^2)\,\partial_x, \quad Y := (1+y^2)\,\partial_y.$$

It is easily seen that they satisfy hypotheses **(H1)** and **(H2)** on page 191, but not **(H0)**. Indeed, we have

$$\dim\bigl(W^{(1)}\bigr) = 2 = \dim\bigl(\text{span}\,\{(1+x^2,0),(0,1+y^2)\}\bigr) \quad \forall\,(x,y)\in\mathbb{R}^2,$$

and all the vector spaces $W^{(k)}$'s for $k \geq 2$ reduce to $\{0\}$, for X and Y commute. However, *X and Y are not δ_λ-homogeneous with respect to any dilation on $\mathbb{R}^2$* (for their component functions contain zero degree terms), hence **(H0)** is not fulfilled.

This invalidates our construction of a Carnot group canonically related to X and Y, the construction being heavily dependent on the well-behaved properties of the fields and, in particular, dependent on the delicate Campbell–Hausdorff-type Lemma 4.2.4.

It is nonetheless important to remark that a Campbell–Hausdorff formula holds in a more general setting than the one we presented here. For instance, it holds[3] for vector fields satisfying the Hörmander condition (i.e. our hypothesis **(H2)**) and hence in the present situation too.

However, we remark that the present case is complicated by the lack of homogeneity of X and Y, which implies that the exponential series $\sum_{k\geq 0} X^k I/k!$ may fail to converge everywhere.

We formally try to construct a group in the present situation. As done in previous sections, we fix $\zeta := (\xi,\eta)\in\mathbb{R}^2$ and consider the vector field $\zeta\cdot Z := \xi\,X + \eta\,Y$. We formally let, as usual, $\text{Exp}(\zeta\cdot Z) := (x(1),y(1))$, where $(x(s),y(s))$ solves

$$\begin{cases} (\dot{x}(s),\dot{y}(s)) = (\zeta\cdot Z)I(x(s),y(s)) = \bigl(\xi\,(1+x^2(s)),\eta\,(1+y^2(s))\bigr), \\ (x(0),y(0)) = (0,0). \end{cases}$$

After a simple computation, we get

$$\text{Exp}(\zeta\cdot Z) = (\tan\xi,\tan\eta), \qquad \text{Log}(x,y) = (\arctan x,\arctan y)\cdot Z.$$

In particular, we explicitly remark that Exp is only defined for

[3] The reader is referred to Nagel–Stein–Wainger [NSW85].

$$(\xi, \eta) \in \left(-\frac{\pi}{2}, +\frac{\pi}{2}\right) \times \left(-\frac{\pi}{2}, +\frac{\pi}{2}\right).$$

We now fix $(x_1, y_1), (x_2, y_2) \in \mathbb{R}^2$, and, again, we formally set

$$(x_1, y_1) \circ (x_2, y_2) := (x(1), y(1)),$$

where $(x(s), y(s))$ solves

$$\begin{cases} (\dot{x}(s), \dot{y}(s)) = \big(\arctan x_2\,(1 + x^2(s)), \arctan y_2\,(1 + y^2(s))\big) \\ (x(0), y(0)) = (x_1, y_1). \end{cases}$$

After a simple computation, we derive

$$\begin{aligned} (x_1, y_1) \circ (x_2, y_2) &= (\tan(\arctan x_1 + \arctan x_2), \tan(\arctan y_1 + \arctan y_2)) \\ &= \left(\frac{x_1 + x_2}{1 - x_1 x_2}, \frac{y_1 + y_2}{1 - y_1 y_2}\right). \end{aligned}$$

Where defined, this operation is commutative, associative, $(0, 0)$ is the neutral element and has the inverse $(x, y)^{-1} = (-x, -y)$. However, the operation is defined only away from the subset of $\mathbb{R}^2 \times \mathbb{R}^2$

$$\big\{\big((x_1, y_1), (x_2, y_2)\big) : x_1 x_2 = 1, \ \text{or } y_1 y_2 = 1\big\}.$$

Moreover, *there does not exist any* $\varepsilon > 0$ *such that* $|x| < \varepsilon$, $|y| < \varepsilon$ *implies* $|(x + y)/(1 - xy)| < \varepsilon$. *This makes it impossible to define a group from* $\circ$.

Another formal argument is more successful in this case. Indeed, we formally compute the Campbell–Hausdorff-type operation on the "algebra": we identify $(x, y) \in \mathbb{R}^2$ with $x\,X + y\,Y$ and we set

$$\begin{aligned} (\xi_1, \eta_1) * (\xi_2, \eta_2) &\equiv \mathrm{Log}\big(\mathrm{Exp}(((\xi_1, \eta_1) \cdot Z) \circ (\mathrm{Exp}(\xi_2, \eta_2) \cdot Z))\big) \\ &= \mathrm{Log}\big((\tan \xi_1, \tan \eta_1) \circ (\tan \xi_2, \tan \eta_2)\big) \\ &= \mathrm{Log}\left(\frac{\tan \xi_1 + \tan \xi_2}{1 - \tan \xi_1 \tan \xi_2}, \frac{\tan \eta_1 + \tan \eta_2}{1 - \tan \eta_1 \tan \eta_2}\right) \\ &= \mathrm{Log}\big(\tan(\xi_1 + \xi_2), \tan(\eta_1 + \eta_2)\big) \\ &= (\xi_1 + \xi_2, \eta_1 + \eta_2) \cdot Z \equiv (\xi_1 + \xi_2, \eta_1 + \eta_2) \in \mathbb{R}^2. \end{aligned}$$

Thus, the operation on the "algebra" of the formal group related to $\circ$ is indeed a group law (the usual Euclidean structure on $\mathbb{R}^2$! reflecting the commutative nature of the algebra generated by the commuting vector fields $\{X, Y\}$).

Analogously, we consider the change of coordinate system induced by the Log map, i.e. we consider new coordinates defined by

$$(\xi, \eta) := (\arctan x, \arctan y).$$

With respect to these new coordinates, the vector fields X and Y are respectively turned into[4]

$$\widetilde{X} = \partial_\xi, \qquad \widetilde{Y} = \partial_\eta.$$

These fields *do* fulfill hypotheses **(H0)**, **(H1)** and **(H2)** of page 191, and the relevant homogeneous Carnot group is obviously the usual additive structure on $\mathbb{R}^2$, exactly as for the operation $*$ above.

For another example of polynomial vector fields satisfying hypotheses **(H1)** and **(H2)**, but not **(H0)**, see Ex. 18 at the end of this chapter.

4.4.2 Fields not Satisfying Hypothesis (H1): The Operation is not Associative

We consider on $\mathbb{R}^3$ (whose points are denoted by (x, y, t)) the vector fields

$$X := \partial_x + y^2\,\partial_t, \qquad Y := \partial_y.$$

It holds

$$[X, Y] = -2y\,\partial_t, \qquad [X, [X, Y]] = 0, \qquad [Y, [X, Y]] = -2\partial_t,$$

and the commutators of length > 3 vanish identically. It is immediately seen that hypothesis **(H2)** is fulfilled, whereas **(H1)** is not. Indeed, $W^{(2)} = \mathrm{span}\{-2y\,\partial_t\}$ is one-dimensional, whereas

$$\{ZI(0) : Z \in W^{(2)}\} = \{(0, 0, 0)\}$$

is zero dimensional. We remark that X and Y are δ_λ-homogeneous of degree 1 with respect to the (unusual) "dilation":

$$\delta_\lambda : \mathbb{R}^3 \longrightarrow \mathbb{R}^3, \qquad \delta_\lambda(x, y, t) := (\lambda x, \lambda y, \lambda^3 t).$$

We try formally to consider the relevant "exponential" map: we fix $\zeta := (\xi, \eta, \tau) \in \mathbb{R}^3$ and consider the vector field $\zeta \cdot Z := \xi\, X + \eta\, Y + \tau\,[X, Y]$. We formally let, as usual, $\mathrm{Exp}(\zeta \cdot Z) := (x(1), y(1), t(1))$, where $(x(s), y(s), t(s))$ solves

$$\begin{cases} (\dot{x}(s), \dot{y}(s), \dot{t}(s)) = \big(\xi, \eta, \xi\, y^2(s) - 2\tau\, y(s)\big) \\ (x(0), y(0), t(0)) = (0, 0, 0). \end{cases}$$

---

[4] Indeed, let $v = v(\xi, \eta) \in C^\infty(\mathbb{R}^2, \mathbb{R})$ and set $u = u(x, y) := v(\arctan x, \arctan y)$. We also set $\mathrm{Log}(x, y) := (\arctan x, \arctan y)$, so that $u = v \circ \mathrm{Log}$. We have

$$\partial_x u(x, y) = \frac{1}{1 + x^2}\,(\partial_\xi v) \circ \mathrm{Log}, \qquad \partial_y u(x, y) = \frac{1}{1 + y^2}\,(\partial_\eta v) \circ \mathrm{Log},$$

i.e.

$$\partial_x u(x, y) = \left(\frac{1}{1 + \tan^2(\xi)}\,(\partial_\xi v)\right) \circ \mathrm{Log}, \qquad \partial_y u(x, y) = \left(\frac{1}{1 + \tan^2(\eta)}\,(\partial_\eta v)\right) \circ \mathrm{Log}.$$

This immediately gives $Xu(x, y) = (\partial_\xi v) \circ \mathrm{Log}$, $Yu(x, y) = (\partial_\eta v) \circ \mathrm{Log}$.

After a simple computation, we get

$$\mathrm{Exp}((\xi,\eta,\tau)\cdot Z) = \left(\xi,\eta,\frac{1}{3}\,\xi\,\eta^2 - \tau\,\eta\right),$$

and this map is not *globally* invertible since

$$\mathrm{Exp}((0,0,0)\cdot Z) = \mathrm{Exp}((0,0,1)\cdot Z).$$

The singularity of this map reflects the fact that the system of vector fields $X, Y, [X,Y]$ is pointwise non-everywhere linearly independent! On the other hand, the everywhere well-posedness of the map is a consequence of the δ_λ-homogeneity of the vector fields X, Y.

Since the pointwise dependence of $X, Y, [X,Y]$ plays a negative rôle, a natural question arises. What if we considered the pointwise everywhere linearly independent vector fields X, Y, $[Y,[X,Y]]$? We proceed once again formally: let (ξ,η,τ), $(x,y,t)\in\mathbb{R}^3$ be fixed, and let us consider

$$\exp\big(\xi X + \eta Y + \tau[Y,[X,Y]]\big)(x,y,t) = (\gamma_1(1),\gamma_2(1),\gamma_3(1)),$$

where

$$\begin{cases} (\dot\gamma_1(s),\dot\gamma_2(s),\dot\gamma_3(s)) = (\xi,\eta,\xi\,\gamma_2^2(s) - 2\,\tau), \\ (\gamma_1(0),\gamma_2(0),\gamma_3(0)) = (x,y,t). \end{cases}$$

A simple computation now gives

$$(\gamma_1(1),\gamma_2(1),\gamma_3(1)) = \left(x+\xi,\, y+\eta,\, \frac{\xi\,\eta^2}{3} + \xi\,y^2 + \xi\,\eta\,y - 2\,\tau + t\right).$$

As a consequence, when $(x,y,t) = (0,0,0)$, it holds (by the slight abuse of notation $\mathrm{Exp}(\xi,\eta,\tau) := \mathrm{Exp}(\xi\,X + \eta\,Y + \tau\,[Y,[X,Y]])$)

$$\mathrm{Exp}(\xi,\eta,\tau) = \left(\xi,\eta,\frac{\xi\,\eta^2}{3} - 2\,\tau\right).$$

This map is now everywhere invertible, and its inverse function is (by an analogous abuse of notation)

$$\mathrm{Log}(x,y,t) = \left(x,y,\frac{x\,y^2}{6} - \frac{t}{2}\right).$$

Finally, given $(x_1,y_1,t_1), (x_2,y_2,t_2)\in\mathbb{R}^3$, we set (formally following the ideas in Definition 4.2.2)

$$\begin{aligned} &(x_1,y_1,t_1)\circ(x_2,y_2,t_2) \\ &\quad = \exp\big(x_2X + y_2Y + (x_2\,y_2^2/6 - t_2/2)[Y,[X,Y]]\big)(x_1,y_1,t_1) \\ &\quad = \big(x_1+x_2,\, y_1+y_2,\, t_1+t_2+x_2\,y_1^2 + x_2\,y_2\,y_1\big). \end{aligned} \tag{4.19}$$

This binary operation on $\mathbb{R}^3$ *is not associative*, for

$$((0,1,0)\circ(0,1,0))\circ(1,0,0) = (1,2,4) \neq (1,2,3)$$
$$= (0,1,0)\circ((0,1,0)\circ(1,0,0)).$$

Moreover, X and Y are not invariant under left "translations" with respect to $\circ$ (as we know, the left-invariance is closely linked to the associativity; see the proof of Theorem 4.2.8).

We explain more closely the fact that $\circ$ lacks to be associative by directly studying the Campbell–Hausdorff formula in the present situation. First, we fix $x = (x_1, x_2, x_3)$, $(\alpha_1, \alpha_2, \alpha_3)$ and $(\beta_1, \beta_2, \beta_3)$ in $\mathbb{R}^3$. Then, starting from x, we proceed along the integral path of the vector field

$$A := \alpha_1 X + \alpha_2 Y + \alpha_3 [Y, [X, Y]],$$

so that, at unit time, we arrive (by definition) in $\exp(A)(x)$. Then, starting from $\exp(A)(x)$ and following the integral curve of the field

$$B := \beta_1 X + \beta_2 Y + \beta_3 [Y, [X, Y]],$$

we arrive (at unit time) in $\exp(B)(\exp(A)(x))$. Now, the Campbell–Hausdorff formula says that[5] we get (at unit time) to this same final point even if we start from x and proceed along the integral path of the vector field

$$C = A \diamond B = A + B + \frac{1}{2}[A, B] + \frac{1}{12}[A, [A, B]] - \frac{1}{12}[B, [A, B]]. \tag{4.20}$$

A direct computation shows that it holds

$$C = (\alpha_1 + \beta_1)X + (\alpha_2 + \beta_2)Y + \frac{1}{2}(\alpha_1\beta_2 - \alpha_2\beta_1)[X, Y]$$
$$+ \left(\alpha_3 + \beta_3 + \frac{1}{12}(\alpha_2 - \beta_2)(\alpha_1\beta_2 - \alpha_2\beta_1)\right)[Y, [X, Y]].$$

As we explained after the statement of Lemma 4.2.4 (page 194), we could certainly construct the desired associative operation *on* $\mathbb{R}^3$, if C could be expressed as a *constant-coefficient* linear combination of $X,\ Y,\ [Y, [X, Y]]$. But here we have

$$C = (\alpha_1 + \beta_1)X + (\alpha_2 + \beta_2)Y$$
$$+ \left(\alpha_3 + \beta_3 + \frac{1}{2}(\alpha_1\beta_2 - \alpha_2\beta_1)\underline{\underline{x_2}} + \frac{1}{12}(\alpha_2 - \beta_2)(\alpha_1\beta_2 - \alpha_2\beta_1)\right)[Y, [X, Y]],$$

which shows that there do not exist three constants $\gamma_1,\ \gamma_2,\ \gamma_3$ such that

$$C = \gamma_1 X + \gamma_2 Y + \gamma_3 [Y, [X, Y]].$$

This can be seen as the very motivation for the lack of associativity of $\circ$ as defined in (4.19).

---

[5] We explicitly remark that the Campbell–Hausdorff formula holds in the present case too, see Chapter 15.

Remark 4.4.1. Nonetheless, we will be able to "lift" the above vector fields X, Y to suitable vector fields (on a larger vector space) satisfying hypotheses **(H0)**–**(H1)**–**(H2)**, only exploiting the homogeneity property of X and Y and the fact that they fulfil Hörmander's condition: this will be properly explained in Chapter 17 (see, in particular, Section 17.4, page 666), thus showing that the hypotheses of the present chapter can be somewhat weakened to produce, in case, "lifted" groups.

4.4.3 Fields not Satisfying Hypothesis (H2): The Group is Undefined

We consider on $\mathbb{R}^3$ the following two vector fields

$$X := \partial_{x_1}, \qquad X_2 := \partial_{x_2}.$$

It is easily seen that they satisfy hypothesis **(H0)** with dilation

$$\delta_\lambda(x_1, x_2, x_3) = (\lambda x_1, \lambda x_2, \lambda^2 x_3)$$

and hypothesis **(H1)**, but they do not satisfy hypothesis **(H2)** on page 191.

Obviously, the exponential map

$$(\xi_1, \xi_2, \xi_3) \mapsto \exp(\xi_1 X_1 + \xi_2 X_2) = (\xi_1, \xi_2, 0)$$

does not even define a bijection from $\mathbb{R}^3$ onto $\mathbb{R}^3$. Formally, the related "composition" would be

$$(x_1, x_2, x_3) \circ (y_1, y_2, y_3) = (x_1 + y_1, x_2 + y_2, 0)$$

which possess no inverse map!

Bibliographical Notes. For some topics on filiform Lie algebras, we followed A.L. Onishchik and E.B. Vinberg [OV94, page 61].

Some of the topics presented in this chapter also appear in [Bon04].

4.5 Exercises of Chapter 4

Ex. 1) Consider on $\mathbb{R}^4$ the vector fields

$$X_1 := \partial_1 - \frac{1}{2} x_2 \, \partial_3 - \frac{1}{2} x_3 \, \partial_4 - \frac{1}{12} x_1 x_2 \, \partial_4,$$
$$X_2 := \partial_2 + \frac{1}{2} x_1 \, \partial_3 + \frac{1}{12} x_1^2 \, \partial_4.$$

Prove that they fulfill hypotheses (**H0**)–(**H1**)–(**H2**) of page 191 with the dilation $\delta_\lambda(x_1, x_2, x_3, x_4) := (\lambda x_1, \lambda x_2, \lambda^2 x_3, \lambda^3 x_4)$. Set

$$Z_1^{(1)} := X_1, \quad Z_2^{(1)} := X_2, \quad Z_1^{(2)} := [X_1, X_2], \quad Z_1^{(3)} := [X_1, [X_1, X_2]].$$

Following the notation of Section 4.2, verify that $\exp(\xi \cdot Z)(x)$ equals

$$\begin{pmatrix} x_1 + \xi_1 \\ x_2 + \xi_2 \\ x_3 + \xi_3 + \frac{1}{2}(\xi_2 x_1 - \xi_1 x_2) \\ x_4 + \xi_4 + \frac{1}{2}(\xi_3 x_1 - \xi_1 x_3) + \frac{1}{12}(x_1 - \xi_1)(\xi_2 x_1 - \xi_1 x_2) \end{pmatrix},$$

whence $\mathrm{Exp}(\xi) = \xi$ and $\mathrm{Log}(x) = x$ (i.e. precisely, $\mathrm{Exp}(\xi \cdot Z) = \xi$ and $\mathrm{Log}(x) = x \cdot Z$). Finally, prove that the relevant composition law is

$$\begin{pmatrix} x_1 + y_1 \\ x_2 + y_2 \\ x_3 + y_3 + \frac{1}{2}(y_2 x_1 - y_1 x_2) \\ x_4 + y_4 + \frac{1}{2}(y_3 x_1 - y_1 x_3) + \frac{1}{12}(x_1 - y_1)(y_2 x_1 - y_1 x_2) \end{pmatrix}.$$

Ex. 2) Write down the canonical sub-Laplacian $\Delta_{\mathbb{G}}$ of the group obtained in Ex. 1 and observe that $\Delta_{\mathbb{G}}$ contains the first order differential term $\frac{1}{6}\, x_2\, \partial_4$.

Ex. 3) Consider on $\mathbb{R}^4$ the vector fields

$$X_1 := \partial_1 + x_2\, \partial_3 + x_2^2\, \partial_4, \quad X_2 := \partial_2.$$

Prove that they fulfill hypotheses (**H0**)–(**H1**)–(**H2**) of page 191 with the same dilation as in Ex. 1. Set

$$Z_1^{(1)} := X_1, \quad Z_2^{(1)} := X_2, \quad Z_1^{(2)} := [X_1, X_2], \quad Z_1^{(3)} := [X_2, [X_1, X_2]].$$

Following the notation of Section 4.2, verify that

$$\exp(\xi \cdot Z)(x) = \begin{pmatrix} x_1 + \xi_1 \\ x_2 + \xi_2 \\ x_3 - \xi_3 + \frac{1}{2}\xi_1 \xi_2 + \xi_1 x_2 \\ x_4 - 2\xi_4 + \frac{1}{3}\xi_1 \xi_2^2 - \xi_2 \xi_3 + \xi_1 x_2^2 + \xi_1 \xi_2 x_2 - 2\xi_3 x_2 \end{pmatrix}.$$

Deduce that

$$\mathrm{Exp}(\xi \cdot Z) = \begin{pmatrix} \xi_1 \\ \xi_2 \\ -\xi_3 + \frac{1}{2}\xi_1 \xi_2 \\ -2\xi_4 + \frac{1}{3}\xi_1 \xi_2^2 - \xi_2 \xi_3 \end{pmatrix}$$

and

$$\mathrm{Log}(x) = \begin{pmatrix} x_1 \\ x_2 \\ -x_3 + \frac{1}{2} x_1 x_2 \\ -\frac{1}{2} x_4 - \frac{1}{12} x_1 x_2^2 + \frac{1}{2} x_2 x_3 \end{pmatrix} \cdot Z.$$

Finally, prove that the relevant composition law is

$$x \circ y = \begin{pmatrix} x_1 + y_1 \\ x_2 + y_2 \\ x_3 + y_3 + y_1 x_2 \\ x_4 + y_4 + y_1 x_2^2 + 2\, x_2 y_3 \end{pmatrix}.$$

Ex. 4) Show that the inverse map for the Lie group in Ex. 3 is given by

$$x^{-1} = \begin{pmatrix} -x_1 \\ -x_2 \\ -x_3 + x_1 x_2 \\ -x_4 + 2\, x_2 x_3 - x_1 x_2^2 \end{pmatrix}.$$

Ex. 5) **a)** Consider the Lie group and the notation in Ex. 3. For every $\xi,\ \eta \in \mathbb{R}^N$, prove that $\mathrm{Log}(\mathrm{Exp}(\xi \cdot Z) \circ \mathrm{Exp}(\eta \cdot Z))$ equals

$$\begin{pmatrix} \xi_1 + \eta_1 \\ \xi_2 + \eta_2 \\ \xi_3 + \eta_3 + \frac{1}{2}\,(\xi_1 \eta_2 - \xi_2 \eta_1) \\ \xi_4 + \eta_4 + \frac{1}{2}\,(\xi_2 \eta_3 - \xi_3 \eta_2) + \frac{1}{12}\,(\xi_2 - \eta_2)\,(\xi_1 \eta_2 - \xi_2 \eta_1) \end{pmatrix}.$$

Denote the above composition law in $\mathbb{R}^4$ by $\xi * \eta$. Deduce that $\mathbb{F} = (\mathbb{R}^4, *)$ is a Carnot group isomorphic to $\mathbb{G} = (\mathbb{R}^4, \circ)$. Find the Lie-group isomorphism turning $\mathbb{F}$ into $\mathbb{G}$. Considering the natural identification $\mathbb{R}^4 \ni \xi \longleftrightarrow \xi \cdot Z \in \mathfrak{g}$ ($\mathfrak{g}$ being the Lie-algebra of $\mathbb{G}$) define a composition law on $\mathfrak{g}$ dual to $*$. Compare this operation to the Campbell–Hausdorff composition law $\diamond$ in (2.43), motivating your remarks.

b) Consider on $\mathbb{R}^4$ the change of coordinates modeled on the Log map, i.e. the new coordinates ξ on $\mathbb{R}^4$ defined by

$$\xi = L(x) := \begin{pmatrix} x_1 \\ x_2 \\ -x_3 + \frac{1}{2}\, x_1 x_2 \\ -\frac{1}{2}\, x_4 - \frac{1}{12}\, x_1 x_2^2 + \frac{1}{2}\, x_2 x_3 \end{pmatrix}.$$

Prove that X_1 and X_2 are respectively turned[6] by L into the following vector fields:

$$\widetilde{X}_1 := \partial_1 - \frac{1}{2}\,\xi_2 \partial_3 - \frac{1}{12}\,\xi_2^2\,\partial_4,$$
$$\widetilde{X}_2 := \partial_2 + \frac{1}{2}\,\xi_1 \partial_3 + \left(\frac{1}{12}\xi_1 \xi_2 - \frac{1}{2}\,\xi_3\right)\partial_4.$$

[6] I.e. if $v = v(\xi)$ and $u(x) = v(L(x))$, then $X_i u(x) = (\widetilde{X}_i v)(L(x))$.

c) Carry out an exercise similar to Ex. 3 for the above vector fields $\widetilde{X}_1$ and $\widetilde{X}_2$ on $\mathbb{R}^4$. Verify that the relevant composition law is the same as $*$ in Ex. 5-(a). Remark that the linear change of coordinates in $\mathbb{R}^4$ given by

$$(\xi_1, \xi_2, \xi_3, \xi_4) \mapsto (x_1, x_2, x_3, x_4) := (\xi_2, \xi_1, -\xi_3, -\xi_4)$$

turns $\widetilde{X}_1$ and $\widetilde{X}_2$ into X_2 and X_1, respectively, of Ex. 1. (Why?)

Ex. 6) Let $\alpha \in \mathbb{R}$ be fixed. Consider on $\mathbb{R}^4$ the vector fields

$$Z_1 = \partial_1 - \frac{1}{2}x_2\,\partial_3 - \left(\frac{1}{2}x_3 + \frac{1}{12}x_2(x_1 + \alpha\,x_2)\right)\partial_4,$$
$$Z_2 = \partial_2 + \frac{1}{2}x_1\,\partial_3 + \left(-\frac{1}{2}\alpha\,x_3 + \frac{1}{12}x_1(x_1 + \alpha\,x_2)\right)\partial_4.$$

Prove that they fulfill hypotheses **(H0)**–**(H1)**–**(H2)** of page 191 with a suitable dilation. Verify that the relevant composition law is the same as $\circ$ in Ex. 1-(a) of Chapter 1.

Ex. 7) For $(x_1, y_1), (x_2, y_2) \in \mathbb{R}^2 \setminus \{(x, y) \in \mathbb{R}^2 : xy = 1\}$, set

$$(x_1, y_1) \circ (x_2, y_2) := \left(\frac{x_1 + x_2}{1 - x_1\,x_2}, \frac{y_1 + y_2}{1 - y_1\,y_2}\right).$$

Prove that $\circ$ is commutative, associative, has $(0, 0)$ as neutral element and has the inverse $(x, y)^{-1} = (-x, -y)$. However, prove that there does not exist any $\varepsilon > 0$ such that $|x| < \varepsilon$, $|y| < \varepsilon$ implies $|(x + y)/(1 - xy)| < \varepsilon$. This makes it impossible to define a group from $\circ$.

Ex. 8) Consider on $\mathbb{R}^4$ the vector fields

$$X_1 := \partial_1 - \frac{1}{2}x_2\,\partial_3 + \frac{1}{6}x_2^2\,\partial_4,$$
$$X_2 := \partial_2 + \frac{1}{2}x_1\,\partial_3 + \left(x_3 - \frac{1}{6}x_1\,x_2\right)\partial_4.$$

Prove that they fulfill hypotheses **(H0)**–**(H1)**–**(H2)** of page 191 with the dilation $\delta_\lambda(x_1, x_2, x_3, x_4) := (\lambda x_1, \lambda x_2, \lambda^2 x_3, \lambda^3 x_4)$. Verify that the relevant composition law is

$$x \circ y = \begin{pmatrix} x_1 + y_1 \\ x_2 + y_2 \\ x_3 + y_3 + \frac{1}{2}(x_1 y_2 - x_2 y_1) \\ x_4 + y_4 + (x_3 y_2 - x_2 y_3) + \frac{1}{6}(y_2 - x_2)(x_1 y_2 - x_2 y_1) \end{pmatrix}.$$

Verify that the related Jacobian basis is

$$X_1, \quad X_2, \quad [X_1, X_2], \quad -\frac{1}{2}[X_2, [X_1, X_2]].$$

Ex. 9) Write the following second order constant-coefficient strictly elliptic operator L on $\mathbb{R}^2$ as a sum of squares of vector fields

$$L = 10\,(\partial_{x_1})^2 + 10\,\partial_{x_1,x_2} + 5\,(\partial_{x_2})^2.$$

Is L a sub-Laplacian on a suitable homogeneous Carnot group? Which one?

Ex. 10) Following the notation of Section 4.1.4, prove that an example of K-type group is given by the choice $p_0 = p_1 = p_2 = \cdots = p_r = 1$, $B_1 = B_2 = \cdots = B_r = (1)$, whence

$$B = \begin{pmatrix} 0 & 0 & \cdots & 0 & 0 \\ 1 & 0 & \cdots & 0 & 0 \\ 0 & 1 & \ddots & \vdots & \vdots \\ \vdots & \vdots & \ddots & 0 & 0 \\ 0 & 0 & \cdots & 1 & 0 \end{pmatrix}, \qquad N = p_0 + p_1 + \cdots + p_r = 1 + r,$$

so that the relevant K-type group $\mathbb{B}$ is $\mathbb{R}^{2+r}$ (whose points are denoted by $(t, x_1, x_2, x_3, \ldots, x_{r+1})$) equipped with the operation

$$\begin{pmatrix} t \\ x_1 \\ x_2 \\ x_3 \\ \vdots \\ x_{r+1} \end{pmatrix} \circ \begin{pmatrix} s \\ y_1 \\ y_2 \\ y_3 \\ \vdots \\ y_{r+1} \end{pmatrix} = \begin{pmatrix} s + t \\ y_1 + x_1 \\ y_2 + x_2 + s\,x_1 \\ y_3 + x_3 + s\,x_2 + \frac{s^2}{2}\,x_1 \\ \vdots \\ y_{r+1} + x_{r+1} + s\,x_r + \frac{s^2}{2}\,x_{r-1} + \cdots + \frac{s^r}{r!}\,x_1 \end{pmatrix}$$

and the dilation

$$\delta_\lambda(t, x_1, x_2, x_3, \ldots, x_{r+1}) = (\lambda t, \lambda x_1, \lambda^2 x_2, \lambda^3 x_3, \ldots, \lambda^{r+1} x_{r+1}).$$

Ex. 11) Following the notation of Section 4.1.4, find an explicit general expression for the composition law on the K-type group such that $r = 1$ and $B_1 = \mathbb{I}_{p_0}$.

Ex. 12) Consider on $\mathbb{R}^4$ the vector fields

$$Z_1 = \partial_{x_1} + x_1\,\partial_{x_4}, \quad Z_1 = \partial_{x_2}, \quad Z_1 = \partial_{x_3} + x_2\,\partial_{x_4}.$$

Prove that they fulfill hypotheses **(H0)–(H1)–(H2)** of page 191 with the dilation $\delta_\lambda(x_1, x_2, x_3, x_4) := (\lambda x_1, \lambda x_2, \lambda x_3, \lambda^2 x_4)$. Moreover, $[Z_1, Z_2] = 0 = [Z_1, Z_3]$, $[Z_2, Z_3] = \partial_{x_4}$. Set $Z_4 := [Z_2, Z_3]$. Following the notation of Section 4.2, verify that

$$\exp(\xi \cdot Z)(x) = \begin{pmatrix} x_1 + \xi_1 \\ x_2 + \xi_2 \\ x_3 + \xi_3 \\ x_4 + \xi_4 + x_1\xi_1 + \frac{1}{2}\,\xi_1^2 + \xi_3 x_2 + \frac{1}{2}\,\xi_3\xi_2 \end{pmatrix},$$

whence

$$\operatorname{Exp}(\xi \cdot Z) = \begin{pmatrix} \xi_1 \\ \xi_2 \\ \xi_3 \\ \xi_4 + \frac{1}{2}\,\xi_1^2 + \frac{1}{2}\,\xi_3\xi_2 \end{pmatrix},$$

$$\operatorname{Log}(y) = \begin{pmatrix} y_1 \\ y_2 \\ y_3 \\ y_4 - \frac{1}{2}\,y_1^2 - \frac{1}{2}\,y_3 y_2 \end{pmatrix} \cdot Z.$$

We explicitly remark that

$$\operatorname{Exp}(Z_1) = (1, 0, 0, 1/2) \neq (1, 0, 0, 0),$$

which shows that *the Jacobian basis of* $\mathfrak{g}$ *do not necessarily correspond to the canonical basis of* $\mathbb{G} \equiv \mathbb{R}^N$ *via the exponential map*. (But see also Proposition 2.2.22, page 139.)
Finally, prove that the relevant composition law is

$$\begin{pmatrix} x_1 + y_1 \\ x_2 + y_2 \\ x_3 + y_3 \\ x_4 + y_4 + y_3 x_2 + y_1 x_1 \end{pmatrix}.$$

Instead, consider the basis $\{X_1, \dots, X_4\}$ of $\mathfrak{g}$ corresponding to the canonical basis $\{e_1, \dots, e_4\}$ of $\mathbb{R}^4$ via the exponential map, i.e. $X_i := \operatorname{Log}(e_i) \cdot Z$. Verify that

$$X_1 = Z_1 - \frac{1}{2}\,Z_2, \quad X_2 = Z_2, \quad X_3 = Z_3, \quad X_4 = Z_4.$$

Now, equip $\mathbb{R}^4$ with the following composition law: given $\xi, \eta \in \mathbb{R}^4$, we define $\xi * \eta \in \mathbb{R}^4$ as the only vector such that

$$\operatorname{Log}(\operatorname{Exp}(\xi \cdot X) \circ \operatorname{Exp}(\eta \cdot X)) = (\xi * \eta) \cdot X.$$

Obviously, $(\mathbb{R}^4, *)$ is a Lie group isomorphic to $(\mathbb{R}^4, \circ)$. However, check out that

$$\xi * \eta = \begin{pmatrix} \xi_1 + \eta_1 \\ \xi_2 + \eta_2 \\ \xi_3 + \eta_3 \\ \xi_4 + \eta_4 - \frac{1}{2}\,\xi_1 - \frac{1}{2}\,\eta_1 - \frac{1}{2}\,\xi_3\eta_2 + \frac{1}{2}\,\xi_2\eta_3 \end{pmatrix},$$

so that $(\mathbb{R}^4, *)$ is not a *homogeneous* Carnot group, for the only "dilations" on $\mathbb{R}^4$ which are homomorphisms of $(\mathbb{R}^4, *)$ have the form

$$\delta_\lambda(x_1, x_2, x_3, x_4) = (\lambda^{\alpha_1} x_1, \lambda^{\alpha_2} x_2, \lambda^{\alpha_3} x_3, \lambda^{\alpha_4} x_4)$$

with $\alpha_4 = \alpha_1 = \alpha_2 + \alpha_3$.

Ex. 13) According to the definition given in Section 4.1.5, page 190, write the composition law of the sum of the Heisenberg–Weyl groups $\mathbb{H}^1$ and $\mathbb{H}^2$. Do the same for $\mathbb{H}^1$ and $\mathbb{H}^1$. The groups obtained are prototype H-type groups according to the definition given in Section 3.6, page 169? Why?

Ex. 14) Consider the fields on $\mathbb{R}^2$

$$X_1 := \partial_1, \qquad X_2 := x_1\, \partial_2.$$

Verify that $[X_1, X_2] = \partial_2$, whereas all commutators of length > 2 vanish. It holds

$$W^{(1)} I(x) = \mathrm{span}\{(1,0), (0,x_1)\}, \quad W^{(2)} I(x) = \mathrm{span}\{(0,1)\}.$$

Whence $\dim(W^{(1)} I(x)) = 2$ only when $x_1 \neq 0$, whereas

$$\dim\left(W^{(2)} I(x)\right) = 1$$

for every $x \in \mathbb{R}^2$, whence hypothesis **(H1)** is not satisfied, but **(H2)** is. Moreover, X_1 and X_2 are homogeneous of degree 1 w.r.t. the "dilation" $\delta_\lambda(x_1, x_2) := (\lambda x_1, \lambda^2 x_2)$. Try to construct formally the Exp, Log and $\circ$ maps as in Section 4.4.2, page 212, and verify that

$$\mathrm{Exp}(\xi_1, \xi_2) = \left(\xi_1, \frac{1}{2}\,\xi_1\xi_2\right),$$

$$\mathrm{Log}(x_1, x_2) = \left(x_1, \frac{2\,x_2}{x_1}\right), \qquad x \circ y = \left(x_1 + y_1, x_2 + y_2 + \frac{2\,x_1 y_2}{y_1}\right).$$

Observe that these maps are not everywhere defined, $\circ$ is not associative and X_1, X_2 are not invariant w.r.t. $\circ$.

Ex. 15) Let us consider on $\mathbb{R}^2$ the following polynomial vector fields

$$X_1 := \partial_{x_1}, \qquad X_2 := \partial_{x_2} + x_1\, \partial_{x_1}.$$

It is easily seen that they satisfy hypotheses **(H1)** and **(H2)** on page 191, but not[7] **(H0)**.
We explicitly remark that $W^{(k)} = \mathrm{span}\{X_1\}$ for every $k \in \mathbb{N}$, $k \geq 2$, whence $\mathrm{Lie}\{X_1, X_2\}$ is not a nilpotent algebra. Prove that the same holds for the group formally related to this case, according to the construction of Section 4.2. Following our usual notation, verify that the following facts hold.[8]

$$\gamma(t) = \left(x_1\, e^{t\xi_2} + \xi_1\, \frac{e^{t\xi_2} - 1}{\xi_2}, x_2 + t\xi_2\right)$$

[7] Precisely, the only "dilation" map for which X_1 and X_2 are homogeneous is $\delta_\lambda(x_1, x_2) = (\lambda^\alpha x_1, x_2)$ (for every α; X_1 is δ_λ-homogeneous of degree α and X_2 of degree 0).

[8] We agree to denote by $(e^\xi - 1)/\xi$ the (analytic) function equal to 1 when $\xi = 0$ and $(e^\xi - 1)/\xi$ when $\xi \neq 0$.

$$\text{solves}\begin{cases}\dot\gamma=(\xi_1X_1+\xi_2X_2)I(\gamma),\\ \gamma(0)=x,\end{cases}$$

$$\exp(\xi_1X_1+\xi_2X_2)(x_1,x_2)=\left(\xi_1\frac{e^{\xi_2}-1}{\xi_2},\xi_2\right),$$

$$\mathrm{Log}(y_1,y_2)=\left(y_1\frac{y_2}{e^{y_2}-1}\right)X_1+y_2X_2,$$

$$x\circ y=(x_1\,e^{y_2}+y_1,x_2+y_2).$$

Moreover, verify that $(\mathbb{R}^2,\circ)$ is a Lie group (not nilpotent) and that X_1 and X_2 Lie-generate the algebra of this group.

Ex. 16) Give a complete proof of Proposition 4.3.3.

Ex. 17) Let $n\in\mathbb{N}$ be fixed. Let us consider the Lie algebra $\mathfrak{k}_n$ with a basis $\{X,Y_1,Y_2,\dots,Y_n\}$ with commutator relations

$$\begin{aligned}&[Y_i,Y_j]=0, && 1\le i,j\le n,\\ &[X,Y_j]=Y_{j+1}, && 1\le j\le n-1,\\ &[X,Y_n]=0.\end{aligned}$$

Then $\mathfrak{k}_n$ is an $(n+1)$-dimensional Lie-algebra nilpotent of step n; also, $\mathfrak{k}_n$ is stratified with stratification

$$\mathfrak{k}_n=\mathrm{span}\{X,Y_1\}\oplus\mathrm{span}\{Y_2\}\oplus\mathrm{span}\{Y_2\}\oplus\cdots\oplus\mathrm{span}\{Y_n\}.$$

Prove that a model for $\mathfrak{k}_n$ is given by the algebra of vector fields on $\mathbb{R}^{1+n}$ (the points are denoted by (x,y) with $x\in\mathbb{R}$, $y=(y_1,\dots,y_n)\in\mathbb{R}^n$) spanned by

$$X=\partial_x,\qquad Y_j=\sum_{k=j}^{n}\frac{x^{k-j}}{(k-j)!}\partial_{y_k},\quad j=1,\dots,n.$$

Show that X, Y_1 fulfill hypotheses **(H0)–(H1)–(H2)** of page 191 with the dilation

$$\delta_\lambda(x,y_1,y_2,\dots,y_n)=(\lambda x,\lambda y_1,\lambda^2y_2,\dots,\lambda^ny_n).$$

Prove that the relevant composition law is

$$(x,y)\circ(\xi,\eta)$$
$$=\begin{pmatrix}x+\xi\\ y_1+\eta_1\\ y_2+\eta_2+\eta_1x\\ y_3+\eta_3+\eta_2x+\eta_1\frac{x^2}{2!}\\ y_4+\eta_4+\eta_3x+\eta_2\frac{x^2}{2!}+\eta_1\frac{x^3}{3!}\\ \vdots\\ y_n+\eta_n+\eta_{n-1}x+\eta_{n-2}\frac{x^2}{2!}+\cdots+\eta_2\frac{x^{n-2}}{(n-2)!}+\eta_1\frac{x^{n-1}}{(n-1)!}\end{pmatrix}.$$

Compare to the Bony-type sub-Laplacians in Section 4.3.3, page 202. Another realization of $\mathfrak{k}_n$ as a matrix algebra can be described as follows. Denote by K_n the set of the $(n+1) \times (n+1)$ matrices of the form

$$M(x,y) := \begin{pmatrix} 0 & x & 0 & \cdots & & 0 & y_n \\ 0 & 0 & x & \ddots & & 0 & y_{n-1} \\ \vdots & & 0 & \ddots & & \vdots & \vdots \\ & & & \ddots & x & 0 & y_3 \\ & & & & 0 & x & y_2 \\ & & & & & 0 & y_1 \\ 0 & & & & \cdots & 0 & 0 \end{pmatrix}.$$

Consider the map

$$\varphi : \mathfrak{k}_n \to K_n, \quad \varphi\left(xX + \sum_{j=1}^{n} y_j Y_j \right) = M(x,y).$$

Then φ is a Lie-algebra isomorphism. Now, consider the set

$$\mathbb{K}_n := \{\exp(M) : M \in K_n\},$$

where exp denotes the exponential of a square matrix. It is remarkable to observe that $(\mathbb{K}_n, \cdot)$ (where $\cdot$ denotes the usual product of matrices) is a Lie group with the Lie algebra isomorphic to $\mathfrak{k}_n$. In particular, $\mathbb{K}_n$ is nilpotent of step n and a Carnot group, for $\mathfrak{k}_n$ is stratified. Unfortunately, $\mathbb{K}_n$ is not a homogeneous group (but, obviously, it is isomorphic to $(\mathfrak{k}_n, \diamond)$, $\diamond$ being the Campbell–Hausdorff multiplication, which is a homogeneous Carnot group if we identify $\mathfrak{k}_n$ to $\mathbb{R}^{n+1}$ in the obvious way).
We sketch the verification that $\mathbb{K}_n$ is closed under the operation $\cdot$, for this involves the Campbell–Hausdorff formula. We aim to prove that if $A, A' \in \mathbb{K}_n$, then $A \cdot A' \in \mathbb{K}_n$. To this end, let $M, M' \in K_n$ be such that $A = \exp(M)$, $A' = \exp(M')$. Moreover, since φ is an isomorphism between $\mathfrak{k}_n$ and K_n, there exist $X, X' \in \mathfrak{k}_n$ such that $M = \varphi(X)$, $M' = \varphi(X')$. Then

$$\begin{aligned} A \cdot A' &= \exp(\varphi(X)) \cdot \exp(\varphi(X')) \\ &= \exp\left(\varphi(X) + \varphi(X') + \frac{1}{2}[\varphi(X), \varphi(X')] + \cdots \right) \\ &= \exp\left(\varphi(X + X' + \frac{1}{2}[X, X'] + \cdots) \right) \\ &= \exp\left(\varphi(X \diamond X') \right) \in \mathbb{K}_n, \qquad \text{since } X \diamond X' \in \mathfrak{k}_n. \end{aligned}$$

In the second equality, we have used the Campbell–Hausdorff formula for the exponential of matrices, which here involves a finite sum, since in K_n

we have strictly upper triangular matrices. In the third one, we used the fact that φ is a Lie-algebra morphism. Incidentally, we have also proved that

$$\exp \circ \varphi : (\mathfrak{k}_n, \diamond) \to (\mathbb{K}_n, \circ)$$

is a Lie-group isomorphism.

Ex. 18) With reference to the preceding exercise (we adopt the therein notation), we study the Lie algebra $\mathfrak{k}_3$. We have $\mathfrak{k}_3 = \operatorname{span}\{X, Y_1, Y_2, Y_3\}$ with

$$[X, Y_1] = Y_2, \quad [X, Y_2] = Y_3, \quad [X, Y_3] = 0, \quad [Y_i, Y_j] = 0, \quad i, j \in \{1, 2, 3\}.$$

Let us also set

$$K_3 = \left\{ M(x, y) := \begin{pmatrix} 0 & x & 0 & y_3 \\ 0 & 0 & x & y_2 \\ 0 & 0 & 0 & y_1 \\ 0 & 0 & 0 & 0 \end{pmatrix} : (x, y) = (x, y_1, y_2, y_3) \in \mathbb{R}^4 \right\}.$$

Verify that K_3 (equipped with the usual bracket of matrices) is a Lie algebra and that the map

$$\varphi : \mathfrak{k}_3 \to K_3, \quad \varphi\left(xX + \sum_{j=1}^{3} y_j Y_j \right) = M(x, y)$$

is a Lie-algebra isomorphism. Now, consider the set

$$\mathbb{K}_3 := \{\exp(M) : M \in K_3\},$$

where exp denotes the exponential of a square matrix. Verify that

$$\exp(M(x, y)) = \begin{pmatrix} 1 & x & \frac{x^2}{2!} & y_3 + \frac{1}{2!} x y_2 + \frac{1}{3!} x^2 y_1 \\ 0 & 1 & x & y_2 + \frac{1}{2!} x y_1 \\ 0 & 0 & 1 & y_1 \\ 0 & 0 & 0 & 1 \end{pmatrix}.$$

Verify that $(\mathbb{K}_3, \cdot)$ is a Lie group (here, $\cdot$ is the usual product of matrices), by first checking that

$$\exp(M(x, y)) \cdot \exp(M(\xi, \eta)) = \exp\big(M\big((x, y) \circ (\xi, \eta)\big)\big),$$

where

$$(x, y) \circ (\xi, \eta) = \begin{pmatrix} x + \xi \\ y_1 + \eta_1 \\ y_2 + \eta_2 + \frac{1}{2}(x\eta_1 - \xi y_1) \\ y_3 + \eta_3 + \frac{1}{2}(x\eta_2 - \xi y_2) + \frac{1}{12}(x - \xi)(x\eta_1 - \xi y_1) \end{pmatrix}.$$

Deduce that $(\mathbb{K}_3, \cdot)$ is isomorphic to the Lie group $(\mathfrak{k}_3, \diamond)$, where $\diamond$ is the Campbell–Hausdorff multiplication.

Ex. 19) Analogously to what we did in the preceding two exercises, we study the Lie algebra $\mathfrak{h}_1$ (related to the Heisenberg group $\mathbb{H}^1$). We have $\mathfrak{h}_1 = \operatorname{span}\{X, Y, Z\}$ with

$$[X, Y] = Z, \quad [X, Z] = [Y, Z] = 0.$$

Let us also set

$$H_1 = \left\{ M(x, y, z) := \begin{pmatrix} 0 & x & z \\ 0 & 0 & y \\ 0 & 0 & 0 \end{pmatrix} : (x, y, z) \in \mathbb{R}^3 \right\}.$$

Verify that H_1 (equipped with the usual bracket of matrices) is a Lie algebra and that the map

$$\varphi : \mathfrak{h}_1 \to H_1, \quad \varphi(xX + yY + zZ) = M(x, y, z)$$

is a Lie-algebra isomorphism. Now, consider the set

$$\mathbb{H}_1 := \{\exp(M) : M \in H_1\},$$

where exp denotes the exponential of a square matrix. Verify that

$$\exp(M(x, y, z)) = \begin{pmatrix} 1 & x & z + \frac{1}{2}xy \\ 0 & 1 & y \\ 0 & 0 & 1 \end{pmatrix}.$$

Verify that $(\mathbb{H}_1, \cdot)$ is a Lie group (here, $\cdot$ is the usual product of matrices), by first checking that

$$\exp(M(x, y, z)) \cdot \exp(M(\xi, \eta, \zeta)) \exp\big(M\big((x, y, z) \circ (\xi, \eta, \zeta)\big)\big),$$

where

$$(x, y, z) \circ (\xi, \eta, \zeta) = \begin{pmatrix} x + \xi \\ y + \eta \\ z + \zeta_2 + \frac{1}{2}(x\eta - \xi y) \end{pmatrix}.$$

Deduce that $(\mathbb{H}_1, \cdot)$ is isomorphic to the Lie group $(\mathfrak{h}_1, \diamond)$, where $\diamond$ is the Campbell–Hausdorff multiplication, which, in turn, is isomorphic to the usual Heisenberg–Weyl group $\mathbb{H}^1$ on $\mathbb{R}^3$.

Ex. 20) We here consider the Lie algebra $\mathfrak{n}_4$ of the strictly upper triangular matrices of dimension 4. We use the notation

$$\mathfrak{n}_4 = \left\{ M(x) := \begin{pmatrix} 0 & x_1 & x_4 & x_6 \\ 0 & 0 & x_2 & x_5 \\ 0 & 0 & 0 & x_3 \\ 0 & 0 & 0 & 0 \end{pmatrix} : x = (x_1, \dots, x_6) \in \mathbb{R}^6 \right\}.$$

Clearly, $\mathfrak{n}_4$ is a Lie algebra of dimension 6, nilpotent of step 3. Now, consider the set

$$\mathbb{N}_4 := \{\exp(M) : M \in N_4\},$$

where exp denotes the exponential of a square matrix. Verify that

$$\exp(M(x)) = \begin{pmatrix} 1 & x_1 & x_4 + \frac{1}{2}x_1x_2 & x_6 + \frac{1}{2!}(x_1x_5 + x_3x_4) + \frac{1}{3!}x_1x_2x_3 \\ 0 & 1 & x_2 & x_5 + \frac{1}{2}x_2x_3 \\ 0 & 0 & 1 & x_3 \\ 0 & 0 & 0 & 1 \end{pmatrix}.$$

Verify that $(\mathbb{N}_4, \cdot)$ is a Lie group (here, $\cdot$ is the usual product of matrices), by first checking that

$$\exp(M(x)) \cdot \exp(M(y)) = \exp(M(x \circ y)),$$

where

$$x \circ y = \begin{pmatrix} x_1 + y_1 \\ x_2 + y_2 \\ x_3 + y_3 \\ x_4 + y_4 + \frac{1}{2}(x_1y_2 - x_2y_1) \\ x_5 + y_5 + \frac{1}{2}(x_2y_3 - x_3y_2) \\ x_6 + y_6 + \frac{1}{2}(x_1y_5 - x_5y_1) + \frac{1}{2}(x_4y_3 - x_3y_4) \\ + \frac{1}{12}(x_3 - y_3)(x_2y_1 - x_1y_2) \\ + \frac{1}{12}(x_1 - y_1)(x_2y_3 - x_3y_2) \end{pmatrix}.$$

Deduce that $(\mathbb{N}_4, \cdot)$ is isomorphic to the Lie group $(\mathfrak{n}_4, \diamond)$, where $\diamond$ is the Campbell–Hausdorff multiplication.
Verify that the above $\circ$ defines on $\mathbb{R}^6$ a homogeneous Carnot group of step 3 with 3 generators and dilations

$$\delta_\lambda(x) = (\lambda x_1, \lambda x_2, \lambda x_3, \lambda^2 x_4, \lambda^2 x_5, \lambda^3 x_6).$$

The first three vector fields of the Jacobian basis are

$$X_1 = \partial_{x_1} - \frac{1}{2}x_2\partial_{x_4} + \left(-\frac{1}{2}x_5 + \frac{1}{12}x_2x_3\right)\partial_{x_6},$$
$$X_2 = \partial_{x_2} + \frac{1}{2}x_1\partial_{x_4} - \frac{1}{2}x_3\partial_{x_5} - \frac{1}{6}x_1x_3\partial_{x_6},$$
$$X_3 = \partial_{x_3} + \frac{1}{2}x_2\partial_{x_5} + \left(\frac{1}{2}x_4 + \frac{1}{12}x_1x_2\right)\partial_{x_6}.$$

The commutator relations are

$$[X_1, X_2] = \partial_{x_4} - \frac{1}{2}x_3\partial_{x_6}, \quad [X_2, X_3] = \partial_{x_5} + \frac{1}{2}x_1\partial_{x_6},$$
$$[X_1, [X_2, X_3]] = \partial_{x_6} = -[X_3, [X_1, X_2]],$$

whereas all other commutators are zero.

5

The Fundamental Solution for a Sub-Laplacian and Applications

In this chapter, we enter the core of the study of the sub-Laplacians $\mathcal{L}$ on the homogeneous Carnot groups (and hence on the stratified Lie groups) of homogeneous dimension $Q \geq 3$, by showing that they possess a *fundamental solution* Γ resemblant to the fundamental solution $c_N\,|x|^{2-N}$ of the usual Laplace operator Δ on $\mathbb{R}^N$, $N \geq 3$. This property is one of the most striking analogies between $\mathcal{L}$ and the classical Laplace operator. Indeed, we shall see that it holds

$$\Gamma = d^{2-Q},$$

where Q is the homogeneous dimension of $\mathbb{G}$ and d is a *symmetric homogeneous norm* on $\mathbb{G}$, smooth out of the origin (the relevant definitions will be given in Section 5.1). We shall also call d an *$\mathcal{L}$-gauge*.

To do this, we first fix some results on homogeneous norms and the Carnot–Carathéodory distance. Then, in Section 5.3, we define the fundamental solution Γ, whose existence follows from the hypoellipticity and the homogeneity properties of $\mathcal{L}$. We then collect many of its remarkable properties.

As a first application, we provide *mean value formulas* for $\mathcal{L}$, generalizing to the sub-Laplacian setting the Gauss theorem for classical harmonic functions. These formulas will play a central rôle throughout the book and are proved by only using integration by parts and the coarea theorem (see Theorems 5.5.4 and 5.6.1).

From the mean value formulas, we derive *Harnack-type inequalities* for $\mathcal{L}$ and the *Brelot convergence property* for monotone sequences of $\mathcal{L}$-harmonic functions (see Theorem 5.7.10). Furthermore, as an application of the Harnack theorem, in Section 5.8 we derive several *Liouville-type theorems* for $\mathcal{L}$. As another application of the properties of the fundamental solution of $\mathcal{L}$, we prove the *Sobolev–Stein embedding theorem* in the stratified group setting (see Section 5.9).

To end with the applications of Γ, we provide three sections devoted to the following topics: some remarks on the *analytic-hypoelliptic* sub-Laplacians, $\mathcal{L}$-harmonic approximations, and finally an integral representation formula for the fundamental solution by R. Beals, B. Gaveau and P. Greiner [BGG96].

Finally, three appendices close the chapter. The first is devoted to *the weak and the strong maximum principles* for $\mathcal{L}$. The second one provides an improved version of the pseudo-triangle inequality. In the third appendix, we prove in details the existence of geodesics on Carnot groups. As a direct application of the maximum principles, we give a decomposition theorem for $\mathcal{L}$-harmonic functions, resemblant to the decomposition of a holomorphic function on an annulus of $\mathbb{C}$ into the sum of the regular and singular parts from its Laurent expansion.

Convention. Throughout this chapter, we fix a stratified group $\mathbb{H}$ of step r and m generators. Q denotes the homogeneous dimension of $\mathbb{H}$.

Together with $\mathbb{H}$,
a stratification $V = (V_1, \dots, V_r)$ *of the algebra of* $\mathbb{H}$ *will be fixed.*

Moreover, $\mathcal{L}$ will be any sub-Laplacian on $\mathbb{H}$ related to the given stratification. We recall that any stratification of the algebra of $\mathbb{H}$ brings along a homogeneous Carnot group on $\mathbb{R}^N$ isomorphic to $\mathbb{H}$. Hence, together with the couple $(\mathbb{H}, V)$, we fix $\mathbb{G} = (\mathbb{R}^N, \circ, \delta_\lambda)$, a homogeneous Carnot group isomorphic to $\mathbb{H}$, as described in Proposition 2.2.22. We let $\Psi : \mathbb{G} \to \mathbb{H}$ be the Lie-group isomorphism, as in the cited proposition. We still denote by $\mathcal{L}$ the sub-Laplacian on $\mathbb{G}$ which is Ψ-related to the sub-Laplacian $\mathcal{L}$ on $\mathbb{H}$ (see (2.68), page 147).

Obviously, *the "homogeneous version"* $\mathbb{G}$ *of* $\mathbb{H}$ *depends upon the stratification* V *but not on the sub-Laplacian* $\mathcal{L}$.

Thus, any definition and result given henceforward for homogeneous Carnot groups has its counterpart (and is actually intended) for *any couple* $(\mathbb{H}, V)$, where $\mathbb{H}$ is an abstract stratified group and V is a stratification for $\mathbb{G}$.

Notation. We introduce the notation for the homogeneous Carnot group $\mathbb{G} = (\mathbb{R}^N, \circ, \delta_\lambda)$. Its dilations $\{\delta_\lambda\}_{\lambda>0}$ are denoted by

$$\delta_\lambda(x) = \delta_\lambda(x^{(1)}, \dots, x^{(r)}) = (\lambda x^{(1)}, \dots, \lambda^r x^{(r)}), \quad x^{(i)} \in \mathbb{R}^{N_i}, \quad 1 \le i \le r.$$

We denote by $m := N_1$ the number of generators of $\mathbb{G}$ and assume that the homogeneous dimension

$$Q = N_1 + 2\,N_2 + \cdots + r\,N_r \ge 3.$$

As we showed in Chapter 1, Section 1.4 (page 56), the sub-Laplacian $\mathcal{L}$ on $\mathbb{G}$ can be written as follows (see (1.90a))

$$\mathcal{L} = \sum_{j=1}^{m} X_j^2 = \operatorname{div}(A(x)\nabla^T),$$

where $\{X_1, \dots, X_m\}$ is a family of vector fields that form a linear basis of the first layer of $\mathfrak{g}$, the Lie algebra of $\mathbb{G}$. The matrix A is given by

$$A(x) = \begin{pmatrix} X_1 I(x) & \cdots & X_m I(x) \end{pmatrix} \cdot \begin{pmatrix} (X_1 I(x))^T \\ \vdots \\ (X_m I(x))^T \end{pmatrix}$$

and takes the following block form (see (1.91))

$$A = \begin{pmatrix} A_{1,1} & A_{1,2} \\ A_{2,1} & A_{2,2} \end{pmatrix},$$

where $A_{1,1}$ is a strictly positive definite constant $m \times m$ matrix.

The characteristic form of $\mathcal{L}$

$$q_{\mathcal{L}}(x,\xi) := \langle A(x)\xi, \xi\rangle = \sum_{j=1}^{N_1} \langle X_j I(x), \xi\rangle^2, \quad x,\ \xi \in \mathbb{R}^N, \tag{5.1a}$$

is non-negative definite and, for every fixed $x \in \mathbb{R}^N$, the set

$$\{\xi \in \mathbb{R}^N \mid q_{\mathcal{L}}(x,\xi) = 0\}$$

is a linear space of dimension $N - m$. The vector-valued operator

$$\nabla_{\mathcal{L}} := (X_1, \dots, X_m) \tag{5.1b}$$

is called the $\mathcal{L}$-gradient operator in $\mathbb{G}$. Due to identity (5.1a), we have

$$|\nabla_{\mathcal{L}} u|^2 = \sum_{j=1}^{m} |X_j u|^2 = \langle A\nabla^T u, \nabla^T u\rangle, \quad u \in C^1(\mathbb{R}^N, \mathbb{R}^N). \tag{5.1c}$$

5.1 Homogeneous Norms

Definition 5.1.1. *We call* homogeneous norm on (*the homogeneous Carnot group*) $\mathbb{G}$, *every* continuous[1] *function* $d : \mathbb{G} \to [0,\infty)$ *such that*:

1. $d(\delta_\lambda(x)) = \lambda\, d(x)$ *for every* $\lambda > 0$ *and* $x \in \mathbb{G}$;
2. $d(x) > 0$ *iff* $x \neq 0$.

   Moreover, we say that d is symmetric *if*
3. $d(x^{-1}) = d(x)$ *for every* $x \in \mathbb{G}$.

Example 5.1.2. Define

$$|x|_{\mathbb{G}} := \left(\sum_{j=1}^{r} |x^{(j)}|^{\frac{2r!}{j}} \right)^{\frac{1}{2r!}}, \quad x = (x^{(1)}, \dots, x^{(r)}) \in \mathbb{G}, \tag{5.2}$$

where $|x^{(j)}|$ denotes the Euclidean norm on $\mathbb{R}^{N_j}$. Then $|\cdot|_{\mathbb{G}}$ is a homogeneous norm on $\mathbb{G}$ smooth out of the origin. It is symmetric if $x^{-1} = -x$ for any $x \in \mathbb{G}$. In general, if $\mathbb{G}$ is any Carnot group (with inverse x^{-1} not necessarily equal to $-x$) the map $x \mapsto |\mathrm{Log}\,(x)|_{\mathbb{G}}$ is a symmetric homogeneous norm on $\mathbb{G}$ smooth out of the origin. This follows from the facts that $\mathrm{Log}\,(\delta_\lambda(x)) = \delta_\lambda(\mathrm{Log}\,(x))$ and $\mathrm{Log}\,(x^{-1}) = -\mathrm{Log}\,(x)$.

[1] With respect to the Euclidean topology.

Example 5.1.3 (Control norm). Let d be the control distance related to a system of generators of $\mathbb{G}$ (see Section 5.2). Define

$$d_0(x) := d(x, 0), \quad x \in \mathbb{G}.$$

Then (see Theorem 5.2.8 in Section 5.2) we shall see that d_0 is a symmetric homogeneous norm on $\mathbb{G}$.

From the next (elementary) proposition it will follow that the homogeneous norms on $\mathbb{G}$ are all equivalent.

Proposition 5.1.4 (Equivalence of the homogeneous norms). *Let d be a homogeneous norm on $\mathbb{G}$. Then there exists a constant $\mathbf{c} > 0$ such that*

$$\mathbf{c}^{-1} |x|_{\mathbb{G}} \le d(x) \le \mathbf{c}\, |x|_{\mathbb{G}} \quad \forall\, x \in \mathbb{G}, \tag{5.3}$$

where $|\cdot|_{\mathbb{G}}$ has been defined in (5.2).

Proof. Due to the δ_λ-homogeneity of d and $|\cdot|_{\mathbb{G}}$, inequalities (5.3) hold taking $\mathbf{c} := \max\{H, 1/h\}$, where

$$H := \sup\{d(x) : |x|_{\mathbb{G}} = 1\}, \quad h := \inf\{d(x) : |x|_{\mathbb{G}} = 1\}.$$

We explicitly remark that $H < \infty$ and $h > 0$, since the set

$$\{x : |x|_{\mathbb{G}} = 1\}$$

is a compact subset of $\mathbb{G}$ not containing the origin and d is a continuous function strictly positive in $\mathbb{G} \setminus \{0\}$. □

Corollary 5.1.5. *For every fixed (non-necessarily symmetric) homogeneous norm d on $\mathbb{G}$, there exists a constant $\mathbf{c} > 0$ such that*

$$\mathbf{c}^{-1} d(x) \le d(x^{-1}) \le \mathbf{c}\, d(x) \quad \forall\, x \in \mathbb{G}. \tag{5.4}$$

Proof. The function $x \mapsto d(x^{-1})$ is a homogeneous norm on $\mathbb{G}$. Indeed, recall that δ_λ is an automorphism of $\mathbb{G}$, whence $\delta_\lambda(x^{-1}) = (\delta_\lambda(x))^{-1}$. Then the assertion follows from Proposition 5.1.4. □

Any homogeneous norm turns out to be locally Hölder continuous with respect to the Euclidean metric in the following sense.

Proposition 5.1.6. *Let d be a homogeneous norm on $\mathbb{G}$. Then, for every compact set $K \subset \mathbb{R}^N$, there exists a constant $\mathbf{c}_K > 0$ such that*

$$d(y^{-1} \circ x) \le \mathbf{c}_K\, |x - y|^{1/r} \quad \forall\, x, y \in K, \tag{5.5}$$

where r is the step of $\mathbb{G}$.

(See also Proposition 5.15.1 (page 309) in Appendix C for an estimate from below of $d(y^{-1} \circ x)$.)

Proof. Let $K \subset \mathbb{R}^N$ be a compact set. It is easy to see that there exists a constant $\mathbf{c} = \mathbf{c}(K) > 0$ such that $|x|_{\mathbb{G}} \leq \mathbf{c}\,|x|^{1/r}$ for every $x \in K$, where $|\cdot|_{\mathbb{G}}$ has been defined in (5.2). We now use Proposition 5.1.4, and we obtain that there exists a constant $\mathbf{c} = \mathbf{c}(K) > 0$ such that $d(x) \leq \mathbf{c}\,|x|^{1/r}$ for every $x \in K$. Hence, (5.5) will follow if we prove that there exists a constant $\mathbf{c} = \mathbf{c}(K) > 0$ such that $|y^{-1} \circ x| \leq \mathbf{c}\,|x - y|$ for every $x,\ y \in K$. If we apply the mean value theorem to the function $F(x, y) := y^{-1} \circ x$, we obtain

$$\begin{aligned} |y^{-1} \circ x| &= |F(x, y) - F(x, x)| \\ &\leq \max_{t \in [0,1]} \|\mathcal{J}_F(x, t\,x + (1-t)y)\|\,|x - y| \leq \mathbf{c}\,|x - y|, \end{aligned}$$

and the assertion is proved. □

Any homogeneous norm satisfies a kind of *pseudo-triangle inequality.*

Proposition 5.1.7 (Pseudo-triangle inequalities. I). *Let d be a homogeneous norm on $\mathbb{G}$. Then there exists a constant $\mathbf{c} > 0$ such that:*

1) $d(x \circ y) \leq \mathbf{c}(d(x) + d(y))$,
2) $d(x \circ y) \geq \frac{1}{\mathbf{c}}\, d(x) - d(y^{-1})$,
3) $d(x \circ y) \geq \frac{1}{\mathbf{c}}\, d(x) - \mathbf{c}\, d(y)$

for every $x, y \in \mathbb{G}$.

Proof. Due to the δ_λ-homogeneity of d, inequality 1) is equivalent to the following one

$$d(x \circ y) \leq \mathbf{c} \quad \text{if } d(x) + d(y) = 1.$$

This inequality holds true taking

$$\mathbf{c} := \max\{d(x \circ y)\ :\ d(x) + d(y) = 1\}.$$

Obviously, $1 \leq \mathbf{c} < \infty$. From inequality 1) we now obtain

$$d(x) = d((x \circ y) \circ y^{-1}) \leq \mathbf{c}(d(x \circ y) + d(y^{-1})),$$

whence 2). Now, 3) follows from 2) and Corollary 5.1.5 with a suitable change of the constant $\mathbf{c}$. □

Given a homogeneous norm d_0 on $\mathbb{G}$, the function

$$\mathbb{G} \times \mathbb{G} \ni (x, y) \mapsto d(x, y) := d_0(y^{-1} \circ x)$$

is a *pseudometric* on $\mathbb{G}$. Indeed, we have the following proposition.

Proposition 5.1.8 (Pseudo-triangle inequalities. II). *With the above notation, there exists a positive constant $\mathbf{c} > 0$ such that:*

1) $d(x, y) \le \mathbf{c}\, d(y, x)$ *for every* $x, y \in \mathbb{G}$ *(here* $\mathbf{c}$ *can be taken* $= 1$ *iff* d_0 *is also symmetric),*
2) $d(x, y) \le \mathbf{c}(d(x, z) + d(z, y))$ *for every* $x, y, z \in \mathbb{G}$ (the pseudo-triangle inequality for d),
3) $d(x, y) = 0$ *iff* $x = y$.

Proof. It immediately follows from Corollary 5.1.5 and Proposition 5.1.7. □

5.2 Control Distances or Carnot–Carathéodory Distances

We begin with some important definitions.

Definition 5.2.1 (X-subunit path). *Let* $X = \{X_1, \dots, X_m\}$ *be any family of vector fields in* $\mathbb{R}^N$. *A piece-wise regular path* $\gamma : [0, T] \to \mathbb{R}^N$ *is said to be* X-subunit *if*

$$\langle \dot\gamma(t), \xi\rangle^2 \le \sum_{j=1}^{m} \langle X_j I(\gamma(t)), \xi\rangle^2 \quad \forall\, \xi \in \mathbb{R}^N,$$

almost everywhere in $[0, T]$. *We shall denote by* $\mathcal{S}(X)$ *the set of all* X*-subunit paths, and we put*

$$l(\gamma) = T$$

if $[0, T]$ *is the domain of* $\gamma \in \mathcal{S}(X)$.

We explicitly remark that every integral curve of $\pm X_j$ ($j \in \{1, \dots, m\}$) is X-subunit.

Convention. We assume $\mathbb{R}^N$ is X*-connected* in the following sense (a proof of this fact in the case of stratified vector fields will be given in Theorem 19.1.3 on page 716):

$$\text{For every } x, y \in \mathbb{R}^N, \text{ there exists}$$

$$\gamma \in \mathcal{S}(X), \quad \gamma : [0, T] \to \mathbb{R}^N \quad \text{such that } \gamma(0) = x \text{ and } \gamma(T) = y.$$

Then the following definition makes sense.

Definition 5.2.2 (X-Carnot–Carathéodory distance). *Suppose* $\mathbb{R}^N$ *is* X*-connected. Then, for every* $x, y \in \mathbb{R}^N$, *we set*

$$d_X(x, y) := \inf\bigl\{\, l(\gamma) : \ \gamma \in \mathcal{S}(X), \ \gamma(0) = x, \ \gamma(T) = y \,\bigr\}. \tag{5.6}$$

Under suitable hypotheses, the above inf is actually a minimum. For example, in Appendix C, we show that this occurs if $\{X_1, \dots, X_m\}$ are generators of the first layer of the stratified algebra of a homogeneous Carnot group (see also [HK00]).

Proposition 5.2.3 (d_X is a metric). *If* $\mathbb{R}^N$ *is* X*-connected, then the function* $(x, y) \mapsto d_X(x, y)$ *is a metric on* $\mathbb{R}^N$, *called the* X-control distance *or the* Carnot–Carathéodory distance *related to* X.

In what follows, when there is no risk of confusion, we shall simply write d instead of d_X.

Proof. It is quite easy to see that d is non-negative, symmetric and satisfies the triangle inequality. To prove positivity, i.e.

$$\big(d(x,y)=0\big) \quad \Longrightarrow \quad (x=y), \tag{5.7}$$

we compare d with the Euclidean metric (which has an interest in its own). Given $x \in \mathbb{R}^N$ and $r > 0$, define

$$M(x,r) := \sup\left\{ \sum_{j=1}^{n} |X_j I(z)| \; : \; |z-x| \le r \right\}, \tag{5.8a}$$

where $|\cdot|$ denotes the Euclidean norm. We next show the following inequality

$$M(x,|x-y|)\, d(x,y) \ge |x-y| \quad \forall\, x,y \in \mathbb{R}^N. \tag{5.8b}$$

This will obviously imply (5.7). By contradiction, assume (5.8b) is false for some x and y in $\mathbb{R}^N$. Then there exists $\gamma \in \mathcal{S}(X)$, $\gamma : [0,T] \to \mathbb{R}^N$, such that $\gamma(0) = x$, $\gamma(T) = y$ and $M(x,|x-y|)\, T < |x-y|$. As a consequence, if we put

$$t^* := \sup\big\{ t \in [0,T] \; : \; |\gamma(s)-x| < |x-y| \text{ for } 0 \le s \le t \big\},$$

we have

$$\begin{aligned} |\gamma(t^*)-x| &= \left| \int_0^{t^*} \dot{\gamma}(s)\, \mathrm{d}s \right| \le \sum_{j=1}^{m} \int_0^{t^*} |X_j I(\gamma(s))|\, \mathrm{d}s \\ &\le M(x,|x-y|)\, t^* \le M(x,|x-y|)\, T < |x-y|. \end{aligned}$$

It follows that $t^* = T$ and $|y-x| = |\gamma(T)-x| < |x-y|$. This contradiction proves (5.8b). □

If the vector fields $X_1, \dots, X_m$ are left invariant w.r.t. the translations on a Lie group $\mathbb{G} = (\mathbb{R}^N, \circ)$, then the $X = \{X_1, \dots, X_m\}$-control distance has the same property. Indeed, we have the following proposition.

Proposition 5.2.4 (Control distance of a left-invariant family). *Let $\mathbb{G} = (\mathbb{R}^N, \circ)$ be a Lie group on $\mathbb{R}^N$, and let d be the control distance related to a family of left invariant vector fields $X = \{X_1, \dots, X_m\}$ on $\mathbb{G}$. Then*

$$d(x,y) = d(y^{-1} \circ x, 0) \quad \forall\, x,y \in \mathbb{G}, \tag{5.9a}$$

and

$$d(x^{-1}, 0) = d(x,0) \quad \forall\, x \in \mathbb{G}. \tag{5.9b}$$

The proof of this proposition will easily follow from the next lemma.

Lemma 5.2.5. *In the hypotheses of Proposition 5.2.4, let $\gamma : [0,T] \to \mathbb{R}^N$ be a X-subunit curve. Then $\alpha \circ \gamma$ is X-subunit for every $\alpha \in \mathbb{G}$.*

Proof. If we denote by Γ the path $\alpha \circ \gamma$, we have

$$\dot{\Gamma}(s) = \mathcal{J}_{\tau_\alpha}(\gamma(s)) \cdot \dot{\gamma}(s) \text{ almost everywhere in } [0,T].$$

Then, for every $\xi \in \mathbb{R}^N$,

$$\begin{aligned}
\langle \dot{\Gamma}(s), \xi \rangle^2 &= \left\langle \dot{\gamma}(s), \left(\mathcal{J}_{\tau_\alpha}(\gamma(s))\right)^T \xi \right\rangle^2 \quad \text{(since } \gamma \text{ is } X\text{-subunit)} \\
&\le \sum_{j=1}^m \left\langle X_j I(\gamma(s)), \left(\mathcal{J}_{\tau_\alpha}(\gamma(s))\right)^T \xi \right\rangle^2 \\
&= \sum_{j=1}^m \left\langle \mathcal{J}_{\tau_\alpha}(\gamma(s)) \cdot X_j I(\gamma(s)), \xi \right\rangle^2 \quad \text{(Proposition 1.2.3, page 14)} \\
&= \sum_{j=1}^m \left\langle X_j I(\alpha \circ \gamma(s)), \xi \right\rangle^2 = \sum_{j=1}^m \left\langle X_j I(\Gamma(s)), \xi \right\rangle^2.
\end{aligned}$$

This proves that Γ is X-subunit. □

Proof (of Proposition 5.2.4). Let $x, y \in \mathbb{G}$, and let γ be a X-subunit path connecting x and y. For every $\alpha \in \mathbb{G}$, by the previous lemma, $\alpha \circ \gamma$ is a X-subunit path connecting $\alpha \circ x$ and $\alpha \circ y$. Then $d(\alpha \circ x, \alpha \circ y) \le d(x,y)$. Since x, y, α are arbitrary, this inequality obviously also implies $d(x,y) \le d(\alpha \circ x, \alpha \circ y)$. Thus, we have proved

$$d(\alpha \circ x, \alpha \circ y) = d(x,y) \quad \forall\, x,\ y,\ \alpha \in \mathbb{G}. \tag{5.10}$$

Choosing in (5.10) $\alpha = y^{-1}$, we obtain (5.9a).

Putting $x = 0$ in (5.9a), we obtain $d(y^{-1}, 0) = d(0, y) = d(y, 0)$, which is (5.9b). This completes the proof. □

The control distance related to homogeneous vector fields is homogeneous. More precisely, the following assertion holds.

Proposition 5.2.6 (Control distance of a homogeneous family). *Let d be the control distance related to a family $X = \{X_1, \dots, X_m\}$ of smooth vector fields in $\mathbb{R}^N$. Assume the X_j's are ϱ_λ-homogeneous of degree one with respect to the "dilations"*

$$\varrho_\lambda : \mathbb{R}^N \to \mathbb{R}^N, \quad \varrho_\lambda(x_1, \dots, x_N) = (\lambda^{\sigma_1} x_1, \dots, \lambda^{\sigma_N} x_N),$$

where $\sigma_1, \dots, \sigma_N$ are positive real numbers. Then

$$d(\varrho_\lambda(x), \varrho_\lambda(y)) = \lambda\, d(x,y) \quad \forall\, x, y \in \mathbb{R}^N\ \forall\, \lambda > 0. \tag{5.11}$$

For the proof of this proposition we need the following lemma.

Lemma 5.2.7. *In the hypotheses of Proposition* 5.2.6, *let* $\gamma : [0,T] \to \mathbb{R}^N$ *be a* X*-subunit curve. Then, for every* $\lambda > 0$, *the curve*

$$\Gamma : [0, \lambda T] \to \mathbb{R}^N, \quad \Gamma(s) := \varrho_\lambda(\gamma(s/\lambda))$$

is a X*-subunit path.*

Proof. For every $\xi \in \mathbb{R}^N$, we have (note that ϱ_λ is a linear and symmetric map for it is represented by a diagonal matrix)

$$\begin{aligned}
\langle \dot\Gamma(s), \xi\rangle^2 &= \lambda^{-2}\big\langle \varrho_\lambda(\dot\gamma(s/\lambda)), \xi\big\rangle^2 = \lambda^{-2}\big\langle \dot\gamma(s/\lambda), \varrho_\lambda(\xi)\big\rangle^2 \\
(\text{since } \gamma \text{ is } X\text{-subunit}) &\le \lambda^{-2} \sum_{j=1}^m \big\langle X_j I(\gamma(s/\lambda)), \varrho_\lambda(\xi)\big\rangle^2 \\
&= \sum_{j=1}^m \big\langle \lambda^{-1} \varrho_\lambda(X_j I(\gamma(s/\lambda))), \xi\big\rangle^2 \\
(\text{Corollary 1.3.6, page 35}) &= \sum_{j=1}^m \big\langle X_j I(\varrho_\lambda(\gamma(s/\lambda))), \xi\big\rangle^2.
\end{aligned}$$

Since $\varrho_\lambda(\gamma(s/\lambda)) = \Gamma(s)$, this proves the lemma. □

Proof (of Proposition 5.2.6). Let $x, y \in \mathbb{R}^N$, and let $\gamma : [0,T] \to \mathbb{R}^N$ be a X-subunit curve connecting x and y. By the previous lemma, $\Gamma(s) = \varrho_\lambda(\gamma(s/\lambda))$ $(0 \le s \le \lambda t)$ is a X-subunit curve, so that, since Γ connects $\varrho_\lambda(x)$ and $\varrho_\lambda(y)$,

$$d(\varrho_\lambda(x), \varrho_\lambda(y)) \le l(\Gamma) = \lambda T = \lambda\, l(\gamma).$$

Then, being γ an arbitrary X-subunit curve connecting x and y,

$$d(\varrho_\lambda(x), \varrho_\lambda(y)) \le \lambda\, d(x,y) \quad \forall\, x, y \in \mathbb{R}^N\ \forall\, \lambda > 0. \tag{5.12}$$

This inequality obviously implies (replace x and y with $\varrho_{1/\lambda}(\widetilde{x})$ and $\varrho_{1/\lambda}(\widetilde{y})$, respectively, and then λ with $1/\widetilde{\lambda}$; then remove "∼")

$$d(x,y) \le \frac{1}{\lambda} d(\varrho_\lambda(x), \varrho_\lambda(y)) \quad \forall\, x, y \in \mathbb{R}^N\ \forall\, \lambda > 0.$$

Then (5.12) holds with the equality sign and the proposition is proved. □

Theorem 5.2.8 (Control distance of a homogeneous Carnot group). *Let* $\mathbb{G}$ *be a homogeneous Carnot group on* $\mathbb{R}^N$, *and let* d *be the control distance related to any family of generators for* $\mathbb{G}$. *Then*

$$\mathbb{G} \ni x \mapsto d_0(x) := d(x, 0) \tag{5.13}$$

is a symmetric homogeneous norm on $\mathbb{G}$.

Proof. By means of Propositions 5.2.3, 5.2.4 and 5.2.6, we are only left to prove that d_0 is continuous. For the proof of this fact, we refer to Theorem 19.1.3 on page 716. □

Remark 5.2.9. The homogeneous norm (5.13), in general, is not smooth.

Corollary 5.2.10. *Let $\mathbb{G}$ be a Carnot group. Denote by d the control distance related to a family of generators for $\mathbb{G}$. Then, for every compact subset K of $\mathbb{G}$, there exists a positive constant $C(K)$ such that*

$$d(x, y) \leq C(K)\, |x - y|^{1/r} \quad \forall\, x, y \in K,$$

where r denotes the step of $\mathbb{G}$.

(See also Proposition 5.15.1 (page 309) in Appendix C for an estimate from below of $d(x, y)$.)

Proof. It follows from Propositions 5.1.6, 5.2.4 and Theorem 5.2.8. □

5.3 The Fundamental Solution

Throughout the sequel, we shall make use of some maximum principles for sub-Laplacians, which (for the reader's convenience) we postpone to Appendix A of the present chapter (see Section 5.13).

For our purposes, it is convenient to give the definition of fundamental solution of a sub-Laplacian $\mathcal{L}$ on a homogeneous Carnot group as follows.

Definition 5.3.1 (Fundamental solution). *Let $\mathbb{G}$ be a homogeneous Carnot group on $\mathbb{R}^N$. Let $\mathcal{L}$ be a sub-Laplacian on $\mathbb{G}$. A function $\Gamma : \mathbb{R}^N \setminus \{0\} \to \mathbb{R}$ is a* fundamental solution *for $\mathcal{L}$ if:*

(i) $\Gamma \in C^\infty(\mathbb{R}^N \setminus \{0\})$;
(ii) $\Gamma \in L^1_{\text{loc}}(\mathbb{R}^N)$ *and* $\Gamma(x) \longrightarrow 0$ *when x tends to infinity*;
(iii) $\mathcal{L}\Gamma = -\text{Dirac}_0$, *being* Dirac_0 *the Dirac measure supported at* $\{0\}$. *More explicitly (recall that $\mathcal{L}^* = \mathcal{L}$, being $\mathcal{L}^*$ the formal adjoint of $\mathcal{L}$),*

$$\int_{\mathbb{R}^N} \Gamma\, \mathcal{L}\varphi \,\mathrm{d}x = -\varphi(0) \qquad \forall\, \varphi \in C_0^\infty(\mathbb{R}^N). \tag{5.14}$$

Theorem 5.3.2 (Existence of the fundamental solution). *Let $\mathcal{L}$ be a sub-Laplacian on a homogeneous Carnot group $\mathbb{G}$ (whose homogeneous dimension Q is > 2). Then there exists a fundamental solution Γ for $\mathcal{L}$.*

(*Note*. Such a fundamental solution is indeed *unique*, as it will be proved in Proposition 5.3.10.)

Proof. The existence of such a fundamental solution may be proved by means of very general arguments from the theory of distributions, based on the hypoellipticity of $\mathcal{L}$ and of its formal adjoint $\mathcal{L}^*$ $(= \mathcal{L})$, jointly with the well-behaved homogeneity properties of $\mathcal{L}$.

Indeed, from the hypoellipticity of $\mathcal{L}$ (see property (A0), page 63) we infer the existence of a "local" fundamental solution satisfying $\mathcal{L}\Gamma = -\mathrm{Dirac}_0$ on a neighborhood of the origin (see F. Trèves [Tre67, Theorems 52.1, 52.2]). Moreover, by using the homogeneity properties of $\mathcal{L}$, a "local-to-global" argument can be performed. It is out of our scopes here to give the details. The complete proof is due to G.B. Folland and can be found in [Fol75, Theorem 2.1] (see also L. Gallardo [Gal82] for some further properties of Γ obtained via probabilistic techniques). An alternative proof can be found in [BLU02, Theorem 3.9]. □

From the integral identity (5.14) and condition (i) in the above Definition 5.3.1 we immediately get the $\mathcal{L}$-harmonicity of Γ out of the origin. Indeed, if we replace in (5.14) a test function φ with support in $\mathbb{R}^N \setminus \{0\}$, by the smoothness of Γ out of the origin, we can integrate by parts obtaining

$$\int_{\mathbb{R}^N} (\mathcal{L}\Gamma)\, \varphi \, \mathrm{d}x = 0 \qquad \forall\, \varphi \in C_0^\infty(\mathbb{R}^N \setminus \{0\}).$$

This obviously implies

$$\mathcal{L}\Gamma = 0 \text{ in } \mathbb{R}^N \setminus \{0\}. \tag{5.15}$$

A simple change of variable and the left-invariance of $\mathcal{L}$ w.r.t. the translations on $\mathbb{G}$ give the following theorem.

Theorem 5.3.3 (Γ left-inverts $\mathcal{L}$). *Let $\mathcal{L}$ be a sub-Laplacian on a homogeneous Carnot group $\mathbb{G}$. If Γ is a fundamental solution for $\mathcal{L}$, then*

$$\int_{\mathbb{R}^N} \Gamma(y^{-1} \circ x)\, \mathcal{L}\varphi(x)\, \mathrm{d}x = -\varphi(y) \qquad \forall\, \varphi \in C_0^\infty(\mathbb{R}^N) \tag{5.16}$$

and every $y \in \mathbb{R}^N$.

Proof. The change of variable $z = y^{-1} \circ x$ gives

$$\int_{\mathbb{R}^N} \Gamma(y^{-1} \circ x)\, \mathcal{L}\varphi(x)\, \mathrm{d}x = \int_{\mathbb{R}^N} \Gamma(z)\, (\mathcal{L}\varphi)(y \circ z)\, \mathrm{d}z. \tag{5.17}$$

On the other hand, since $\mathcal{L}$ is left-invariant on $\mathbb{G}$, then

$$(\mathcal{L}\varphi)(y \circ z) = \mathcal{L}(\varphi(y \circ z)).$$

Replacing this identity in (5.17) and using (5.14) with $\varphi(\cdot)$ replaced by $\varphi(y \circ \cdot)$, one gets the thesis. □

Remark 5.3.4. The integral identity (5.16) means that

$$\mathcal{L}(\Gamma(y^{-1} \circ \cdot)) = -\mathrm{Dirac}_y$$

in the weak sense of distributions. Here Dirac_y denotes the Dirac measure supported at $\{y\}$.

Due to identity (5.16), we can say that Γ is a left inverse of $\mathcal{L}$. We next prove that Γ is a right inverse too.

Theorem 5.3.5 (Γ right-inverts $\mathcal{L}$). *Let $\mathcal{L}$ be a sub-Laplacian on a homogeneous Carnot group $\mathbb{G}$. If Γ is a fundamental solution for $\mathcal{L}$, then, for every $\varphi \in C_0^\infty(\mathbb{R}^N)$, the function*

$$\mathbb{R}^N \ni y \mapsto u(y) := \int_{\mathbb{R}^N} \Gamma(y^{-1} \circ x)\, \varphi(x)\, dx \tag{5.18a}$$

is smooth and satisfies the equation

$$\mathcal{L}u = -\varphi. \tag{5.18b}$$

The proof of this theorem requires some prerequisites. First of all, we note that a change of variable in the integral at the right-hand side of (5.18a) gives

$$u(y) = \int_{\mathbb{R}^N} \Gamma(z)\, \varphi(y \circ z)\, dz.$$

Then we can differentiate under the integral sign and get the smoothness of u. Moreover, if $\mathrm{supp}(\varphi) \subseteq \{x \,:\, d(x) \le R\}$ (here d denotes a fixed homogeneous norm on $\mathbb{G}$), then

$$\begin{aligned} |u(y)| &\le \sup\{|\Gamma(z)| : d(y \circ z) \le R\} \int_{\mathbb{R}^N} |\varphi(z)|\, dz \\ &=: C(y) \int_{\mathbb{R}^N} |\varphi(z)|\, dz. \end{aligned} \tag{5.19}$$

On the other hand, by Corollary 5.1.5 and Proposition 5.1.7,

$$d(z) \ge \frac{1}{\mathbf{c}} d(y) - \mathbf{c} d(y \circ z),$$

for a suitable positive constant $\mathbf{c}$ independent of x, y, z. As a consequence,

$$C(y) \le \sup\left\{|\Gamma(z)| : d(z) \ge \frac{1}{\mathbf{c}} d(y) - \mathbf{c}R\right\},$$

so that, since $\Gamma(z)$ vanishes as z goes to infinity, inequality (5.19) implies

$$\lim_{y \to \infty} u(y) = 0. \tag{5.20}$$

We then show a crucial property of the $(\varepsilon, \mathbb{G})$-*mollifiers*. The relevant definition is the following one.

Definition 5.3.6 (Mollifiers). *Let $\mathbb{G} = (\mathbb{R}^N, \circ, \delta_\lambda)$ be a homogeneous Carnot group on $\mathbb{R}^N$. Let O be a fixed non-empty open neighborhood of the origin 0. Let also be given a function $J \in C_0^\infty(\mathbb{R}^N, \mathbb{R})$, $J \geq 0$, such that*

$$\mathrm{supp}(J) \subset O \text{ and } \int_{\mathbb{R}^N} J = 1.$$

For any $\varepsilon > 0$, we set $J_\varepsilon(x) := \varepsilon^{-Q} J(\delta_{1/\varepsilon}(x))$.

Let $u \in L^1_{\mathrm{loc}}(\mathbb{R}^N)$. We set, for every $x \in \mathbb{R}^N$,

$$\begin{aligned} u_\varepsilon(x) &:= (u *_\mathbb{G} J_\varepsilon)(x) := \int_{\mathbb{R}^N} u(y)\, J_\varepsilon(x \circ y^{-1})\, \mathrm{d}y \\ &= \int_{\delta_\varepsilon(O)} u(z^{-1} \circ x)\, J_\varepsilon(z)\, \mathrm{d}z. \end{aligned} \tag{5.21}$$

We call u_ε a mollifier of u *(or $(\varepsilon, \mathbb{G})$*-mollifier) related to the kernel J. Note that this mollifier depends only on $\mathbb{G} = (\mathbb{R}^N, \circ, \delta_\lambda)$ and J.

For the use of mollifiers in a context of subelliptic PDE's, see also [CDG97].

Example 5.3.7. Let $\mathbb{G} = (\mathbb{R}^N, \circ, \delta_\lambda)$ be a homogeneous Carnot group on $\mathbb{R}^N$. Let ϱ be a fixed homogeneous symmetric norm on $\mathbb{G}$. We set a notation which will be used throughout the book. For every $x \in \mathbb{G}$ and every $r > 0$, we set

$$B_\varrho(x, r) := \left\{ y \in \mathbb{G} : \varrho(x^{-1} \circ y) < r \right\}.$$

We say that $B_\varrho(x, r)$ is the *ϱ-ball with center x and radius r*. Also, fixed a point $x \in \mathbb{G}$ and a set $A \subset \mathbb{G}$, we let

$$\varrho\text{-dist}(x, A) := \inf \left\{ \varrho(x^{-1} \circ a) : a \in A \right\}.$$

We call ϱ-dist(x, A) the *ϱ-distance of x from A*. The notation $\mathrm{dist}_\varrho(x, A)$ will also be available.

Let now a function $J \in C_0^\infty(\mathbb{R}^N)$, $J \geq 0$ be given such that

$$\mathrm{supp}(J) \subset B_\varrho(0, 1) \text{ and } \int_{\mathbb{R}^N} J = 1.$$

For any $\varepsilon > 0$, we set $J_\varepsilon(x) := \varepsilon^{-Q} J(\delta_{1/\varepsilon}(x))$.

Let $u \in L^1_{\mathrm{loc}}(\Omega)$, $\Omega \subseteq \mathbb{R}^N$ open. For the open set

$$\Omega_\varepsilon := \{x \in \Omega : \varrho\text{-dist}(x, \partial\Omega) > \varepsilon\},$$

we define

$$\begin{aligned} u_\varepsilon(x) &:= (u *_\mathbb{G} J_\varepsilon)(x) := \int_{B_\varrho(x,\, \varepsilon)} u(y)\, J_\varepsilon(x \circ y^{-1})\, \mathrm{d}y \\ &= \int_{B_\varrho(0,\varepsilon)} u(y^{-1} \circ x)\, J_\varepsilon(y)\, \mathrm{d}y \end{aligned} \tag{5.22}$$

for every $x \in \Omega_\varepsilon$.

We call u_ε a *mollifier of* u (or $(\varepsilon, \mathbb{G})$-*mollifier*) *related to the homogeneous norm* ϱ. Note that this mollifier depends only on $\mathbb{G} = (\mathbb{R}^N, \circ, \delta_\lambda)$, J and ϱ.

Remark 5.3.8. Let the notation in Definition 5.3.6 be fixed. Let $u \in L^1_{\mathrm{loc}}(\mathbb{R}^N)$. Then the following fact holds:

$$(\star) \quad u_\varepsilon \to u \quad \text{as} \quad \varepsilon \to 0 \quad \text{in } L^1_{\mathrm{loc}}(\mathbb{R}^N).$$

Indeed, we have (perform the change of variable $y = \delta_{1/\varepsilon}(z)$)

$$u_\varepsilon(x) = \int_{\delta_\varepsilon(O)} u(z^{-1} \circ x)\, \varepsilon^{-Q} J(\delta_{1/\varepsilon}(z))\, \mathrm{d}z = \int_O u(\delta_\varepsilon(y^{-1}) \circ x) J(y)\, \mathrm{d}y.$$

As a consequence (recall that $\int_O J = 1$),

$$\begin{aligned}\int_{\mathbb{R}^N} |u_\varepsilon(x) - u(x)|\, \mathrm{d}x &= \int_{\mathbb{R}^N} \left| \int_O u(\delta_\varepsilon(y^{-1}) \circ x)\, J(y)\, \mathrm{d}y - u(x) \right| \mathrm{d}x \\ &= \int_{\mathbb{R}^N} \left| \int_O \big(u(\delta_\varepsilon(y^{-1}) \circ x) - u(x)\big)\, J(y)\, \mathrm{d}y \right| \mathrm{d}x \\ &\le \int_O \left\{ \int_{\mathbb{R}^N} \left| u(\delta_\varepsilon(y^{-1}) \circ x) - u(x) \right| \mathrm{d}x \right\} J(y)\, \mathrm{d}y.\end{aligned}$$

Given $\sigma > 0$, by well-known results, there exists $\overline{\varepsilon} = \overline{\varepsilon}(\sigma, \mathbb{G}, O) > 0$ such that if $0 < \varepsilon < \overline{\varepsilon}$, then the integral in braces is $\le \sigma$ for every fixed $y \in O$. This proves that

$$\int_{\mathbb{R}^N} |u_\varepsilon(x) - u(x)|\, \mathrm{d}x \le \sigma \int_O J(y)\, \mathrm{d}y = \sigma \quad \forall\, 0 < \varepsilon < \overline{\varepsilon},$$

i.e. $(\star)$ holds. $\square$

Proposition 5.3.9 ($\mathcal{L}$-harmonicity of the mollifier). *Let $\mathbb{G} = (\mathbb{R}^N, \circ, \delta_\lambda)$ be a homogeneous Carnot group on $\mathbb{R}^N$. Let $\mathcal{L}$ be a sub-Laplacian on $\mathbb{G}$. Let $u \in L^1_{\mathrm{loc}}(\mathbb{R}^N)$ be a weak solution to $\mathcal{L}u = 0$ in $\mathbb{R}^N$, i.e.*

$$\int_{\mathbb{R}^N} u\, \mathcal{L}\varphi\, \mathrm{d}y = 0 \qquad \forall\, \varphi \in C_0^\infty(\mathbb{R}^N). \tag{5.23}$$

Then, if u_ε denotes the mollification on $\mathbb{G}$ w.r.t. any kernel J (as in Definition 5.3.6), we have

$$\mathcal{L}u_\varepsilon = 0 \quad \textit{in } \mathbb{G} \quad \textit{for every } \varepsilon > 0. \tag{5.24}$$

Proof. First, note that, being $\mathrm{supp}(J_\varepsilon) \subset \delta_\varepsilon(O)$,

$$u_\varepsilon(x) = \int_{\mathbb{R}^N} u(y^{-1} \circ x)\, J_\varepsilon(y)\, \mathrm{d}y.$$

For every test function φ, we thus have (Fubini–Tonelli's theorem certainly applies)

$$
\begin{aligned}
\int_{\mathbb{R}^N} u_\varepsilon(x)\,\mathcal{L}\varphi(x)\,\mathrm{d}x &= \int_{\mathbb{R}^N} \mathcal{L}\varphi(x)\Big(\int_{\mathbb{R}^N} u(y^{-1}\circ x)\,J_\varepsilon(y)\,\mathrm{d}y\Big)\,\mathrm{d}x \\
&= \int_{\mathbb{R}^N} J_\varepsilon(y)\Big(\int_{\mathbb{R}^N} (\mathcal{L}\varphi)(x)u(y^{-1}\circ x)\,\mathrm{d}x\Big)\,\mathrm{d}y \\
&= \int_{\mathbb{R}^N} J_\varepsilon(y)\Big(\int_{\mathbb{R}^N} (\mathcal{L}\varphi)(y\circ z)\,u(z)\,\mathrm{d}z\Big)\,\mathrm{d}y \\
&= \int_{\mathbb{R}^N} J_\varepsilon(y)\Big(\int_{\mathbb{R}^N} \mathcal{L}\big(z\mapsto \varphi(y\circ z)\big)\,u(z)\,\mathrm{d}z\Big)\,\mathrm{d}y.
\end{aligned}
$$

The inner integral in the far right-hand side is equal to zero by the hypothesis (5.23). Then

$$\int_{\mathbb{R}^N} u_\varepsilon(x)\,\mathcal{L}\varphi(x)\,\mathrm{d}x = 0 \quad \forall\,\varphi\in C_0^\infty(\mathbb{R}^N),$$

so that the claimed (5.24) follows from the hypoellipticity of $\mathcal{L}$ and the fact that $\mathcal{L}^* = \mathcal{L}$. □

With Proposition 5.3.9 at hand, it is easy to prove the uniqueness of the fundamental solution.

Proposition 5.3.10 (Uniqueness of the fundamental solution). *Let $\mathcal{L}$ be a sub-Laplacian on a homogeneous Carnot group $\mathbb{G}$. The fundamental solution of $\mathcal{L}$ (whose existence is granted by Theorem 5.3.2) is unique.*

Proof. Let Γ and Γ' be fundamental solutions for $\mathcal{L}$. Then the function $u = \Gamma - \Gamma'$ has the following properties: $u\in L^1_{\mathrm{loc}}(\mathbb{R}^N)$, $u(x)\to 0$ as $x\to\infty$ and

$$\int_{\mathbb{R}^N} u\,\mathcal{L}\varphi\,\mathrm{d}y = 0 \qquad \forall\,\varphi\in C_0^\infty(\mathbb{R}^N).$$

As a consequence, by Proposition 5.3.9, $\mathcal{L}u_\varepsilon = 0$ in $\mathbb{R}^N$ for every $\varepsilon > 0$.

Thus, since $u_\varepsilon(x)\to 0$ as $x\to\infty$ (argue as in (5.20)), the maximum principle in Section 5.13 implies $u_\varepsilon = 0$ in $\mathbb{R}^N$. On the other hand, $u_\varepsilon \to u$ as $\varepsilon\to 0$ in $L^1_{\mathrm{loc}}(\mathbb{R}^N)$ (see Remark 5.3.8). Then $u = 0$ almost everywhere in $\mathbb{R}^N$, so that $\Gamma = \Gamma'$ in $\mathbb{R}^N\setminus\{0\}$. □

We now prove that Γ is "$\mathbb{G}$-symmetric" with respect to the origin. More precisely, the following assertion holds.

Proposition 5.3.11 (Symmetry of Γ). *Let $\mathcal{L}$ be a sub-Laplacian on a homogeneous Carnot group $\mathbb{G}$. Let Γ be the fundamental solution of $\mathcal{L}$. Then*

$$\Gamma(x^{-1}) = \Gamma(x) \quad \forall\,x\in\mathbb{G}\setminus\{0\}.$$

Proof. Given $\varphi\in C_0^\infty(\mathbb{R}^N)$, define

$$u(x) := -\int_{\mathbb{R}^N} \Gamma(y^{-1}\circ x)\,\mathcal{L}\varphi(y)\,\mathrm{d}y, \quad x\in\mathbb{G}.$$

The function u is smooth and vanishes at infinity (see (5.20)). Moreover, for every $\psi \in C_0^\infty(\mathbb{R}^N)$,

$$\begin{aligned}\int_{\mathbb{R}^N} \mathcal{L}u(x)\,\psi(x)\,\mathrm{d}x &= \int_{\mathbb{R}^N} u(x)\,\mathcal{L}\psi(x)\,\mathrm{d}x\\ &= -\int_{\mathbb{R}^N} \mathcal{L}\varphi(y)\Big(\int_{\mathbb{R}^N} \Gamma(y^{-1}\circ x)\,\mathcal{L}\psi(x)\,\mathrm{d}x\Big)\,\mathrm{d}y\\ &= \int_{\mathbb{R}^N} \mathcal{L}\varphi(x)\,\psi(x)\,\mathrm{d}x \quad \text{(by Theorem 5.3.3).}\end{aligned}$$

This proves that $\mathcal{L}(u-\varphi) = 0$ in $\mathbb{G}$. Thus, since $u - \varphi$ vanishes at infinity, by the maximum principle, $u = \varphi$ in $\mathbb{R}^N$. In particular,

$$\varphi(0) = u(0) = -\int_{\mathbb{R}^N} \Gamma(y^{-1}\circ x)\mathcal{L}\varphi(y)\,\mathrm{d}y \quad \forall\,\varphi \in C_0^\infty(\mathbb{R}^N),$$

so that $x \mapsto \Gamma(x^{-1})$ is a fundamental solution of $\mathcal{L}$ (see Definition 5.3.1). The uniqueness of the fundamental solution (Proposition 5.3.10) implies $\Gamma(x^{-1}) = \Gamma(x)$ for every $x \in \mathbb{G}\setminus\{0\}$. □

Finally, we are in the position to prove Theorem 5.3.5.

Proof (of Theorem 5.3.5). Let u be the function defined in (5.18a). Then $u \in C^\infty(\mathbb{R}^N)$ and, for any test function $\psi \in C_0^\infty(\mathbb{R}^N)$, one has

$$\begin{aligned}\int_{\mathbb{R}^N} (\mathcal{L}u)(y)\,\psi(y)\,\mathrm{d}y &= \int_{\mathbb{R}^N} u(y)\,\mathcal{L}\psi(y)\,\mathrm{d}y\\ &= \int_{\mathbb{R}^N} \mathcal{L}\psi(y)\Big(\int_{\mathbb{R}^N} \Gamma(y^{-1}\circ x)\,\varphi(x)\,\mathrm{d}x\Big)\,\mathrm{d}y\\ &= \int_{\mathbb{R}^N} \varphi(x)\Big(\int_{\mathbb{R}^N} \Gamma(y^{-1}\circ x)\mathcal{L}\psi(y)\,\mathrm{d}y\Big)\,\mathrm{d}x\\ &= \int_{\mathbb{R}^N} \varphi(x)\Big(\int_{\mathbb{R}^N} \Gamma(x^{-1}\circ y)\,\mathcal{L}\psi(y)\,\mathrm{d}y\Big)\,\mathrm{d}x.\end{aligned}$$

Here we used Proposition 5.3.11 ensuring that $\Gamma(y^{-1}\circ x) = \Gamma((x^{-1}\circ y)^{-1}) = \Gamma(x^{-1}\circ y)$. Now, by identity (5.16), the inner integral at the far right-hand side is equal to $-\psi(x)$. Then

$$\int_{\mathbb{R}^N} (\mathcal{L}u)(y)\,\psi(y)\,\mathrm{d}y = -\int_{\mathbb{R}^N} \varphi(x)\,\psi(x)\,\mathrm{d}x \quad \forall\,\psi \in C_0^\infty(\mathbb{R}^N).$$

This gives identity (5.18b). □

From the uniqueness of Γ we easily obtain its δ_λ-homogeneity.

Proposition 5.3.12 (δ_λ-homogeneity of Γ). *Let $\mathcal{L}$ be a sub-Laplacian on a homogeneous Carnot group $\mathbb{G}$. Let Γ be the fundamental solution of $\mathcal{L}$. Then Γ is δ_λ-homogeneous of degree $2-Q$, i.e.*

$$\Gamma(\delta_\lambda(x)) = \lambda^{2-Q}\Gamma(x) \qquad \forall\,x \in \mathbb{G}\setminus\{0\} \quad \forall\,\lambda > 0.$$

Proof. For any fixed $\lambda > 0$, define

$$\Gamma'(x) := \lambda^{Q-2}\Gamma(\delta_\lambda(x)) \quad \forall x \in \mathbb{G} \setminus \{0\}.$$

It is quite obvious that $\Gamma' \in C^\infty(\mathbb{R}^N \setminus \{0\}) \cap L^1_{\text{loc}}(\mathbb{R}^N)$ and $\Gamma'(x) \to 0$ as $x \to \infty$. Moreover, for every test function $\varphi \in C_0^\infty(\mathbb{R}^N)$,

$$\begin{aligned}\int_{\mathbb{R}^N} \Gamma'(x)\, \mathcal{L}\varphi(x)\, \mathrm{d}x &= \lambda^{Q-2} \int_{\mathbb{R}^N} \Gamma(\delta_\lambda(x))\, \mathcal{L}\varphi(x)\, \mathrm{d}x \\ &\quad \text{(by using the change of variable } y = \delta_\lambda(x)\text{)} \\ &= \lambda^{-2} \int_{\mathbb{R}^N} \Gamma(y)\, (\mathcal{L}\varphi)(\delta_{1/\lambda}(y))\, \mathrm{d}y \\ &\quad \text{(since } \mathcal{L} \text{ is } \delta_\lambda\text{-homogeneous of degree 2)} \\ &= \int_{\mathbb{R}^N} \Gamma(y)\mathcal{L}\big(\varphi(\delta_{1/\lambda}(y))\big)\, \mathrm{d}y = -\varphi(\delta_{1/\lambda}(0)) = -\varphi(0).\end{aligned}$$

This proves that Γ' is a fundamental solution of $\mathcal{L}$. Then, by Proposition 5.3.10, $\Gamma = \Gamma'$ and the assertion is proved. □

From the strong maximum principle in Theorem 5.13.8 we obtain another important property of Γ.

Proposition 5.3.13 (Positivity of Γ). *Let $\mathcal{L}$ be a sub-Laplacian on a homogeneous Carnot group $\mathbb{G}$. Let Γ be the fundamental solution of $\mathcal{L}$. Then*

$$\Gamma(x) > 0 \quad \forall x \in \mathbb{G} \setminus \{0\}.$$

Proof. Let $\varphi \in C_0^\infty(\mathbb{R}^N)$, $\varphi \geq 0$. Define

$$u(y) := \int_{\mathbb{R}^N} \Gamma(y^{-1} \circ x)\varphi(x)\, \mathrm{d}x, \quad y \in \mathbb{G}.$$

The function u is smooth, vanishes at infinity and satisfies the equation $\mathcal{L}u = -\varphi$ (see (5.20) and Theorem 5.3.5). Then

$$\mathcal{L}u \leq 0 \text{ in } \mathbb{G} \text{ and } \lim_{y \to \infty} u(y) = 0.$$

By the maximum principle in Corollary 5.13.6, it follows that $u \geq 0$ in $\mathbb{G}$. Hence

$$\int_{\mathbb{R}^N} \Gamma(y^{-1} \circ x)\, \varphi(x)\, \mathrm{d}x \geq 0 \qquad \forall \varphi \in C_0^\infty(\mathbb{R}^N),\ \varphi \geq 0.$$

Thus, $\Gamma \geq 0$, so that, since it is $\mathcal{L}$-harmonic in $\mathbb{G} \setminus \{0\}$, the strong maximum principle in Theorem 5.13.8 implies $\Gamma \equiv 0$ or $\Gamma(x) > 0$ for any $x \neq 0$. The first case would contradict identity (5.14). Then $\Gamma > 0$ at any point of $\mathbb{G} \setminus \{0\}$. This ends the proof. □

Corollary 5.3.14 (Pole of Γ). *Let $\mathcal{L}$ be a sub-Laplacian on a homogeneous Carnot group $\mathbb{G}$. The fundamental solution Γ of $\mathcal{L}$ has a pole at* 0, *i.e.*

$$\lim_{x \to 0} \Gamma(x) = \infty. \tag{5.25}$$

Proof. Since Γ is smooth and strictly positive out the origin, we have

$$h := \min\{\Gamma(x) : d(x) = 1\} > 0.$$

Here d denotes any fixed homogeneous norm on $\mathbb{G}$. Then, by Proposition 5.3.12,

$$\Gamma(x) = d(x)^{2-Q}\,\Gamma(\delta_{1/d(x)}(x)) \geq h\, d^{2-Q}(x).$$

From this inequality (5.25) immediately follows. □

5.3.1 The Fundamental Solution in the Abstract Setting

The aim of this brief section is to show how to derive a "fundamental solution" for a sub-Laplacian $\mathcal{L}$ on an abstract stratified group $\mathbb{H}$, *starting from* the fundamental solution of the related sub-Laplacian $\widetilde{\mathcal{L}}$ on a homogeneous-Carnot-group copy $\mathbb{G}$ of $\mathbb{H}$.

Many other alternative "more intrinsic" definitions may be certainly provided, for example, by making use of the integration on an abstract Lie group. We considered more "in the spirit" of our exposition to pass through the homogeneous group $\mathbb{G}$. Besides, this also makes unnecessary to furnish the (lengthy) theory of integration on manifolds.

Let $\mathbb{H}$ be a stratified group with an algebra $\mathfrak{h}$. Let $\mathcal{L} = \sum_{j=1}^{m} X_j^2$ be a sub-Laplacian on $\mathbb{H}$, and let $V = (V_1, \dots, V_r)$ be the stratification of $\mathfrak{h}$ related to $\mathcal{L}$, according to Definition 2.2.25, page 144. Let also $\mathcal{E}$ be a basis for $\mathfrak{h}$ adapted to the stratification V.

By Proposition 2.2.22 on page 139 (whose notation we presently follow), there exists a homogeneous Carnot group $\mathbb{G} = (\mathbb{R}^N, \diamond_{\mathcal{E}})$ which is isomorphic to $\mathbb{H}$. Let $\Psi : \mathbb{G} \to \mathbb{H}$ be the isomorphism as in Proposition 2.2.22-(1). With the therein notation, we have

$$\Psi = \operatorname{Exp} \circ \pi_{\mathcal{E}}^{-1},$$

where, for every $\xi \in \mathbb{R}^N$, $\pi_{\mathcal{E}}^{-1}(\xi)$ is the vector field in $\mathfrak{h}$ having ξ as N-tuple of the coordinates w.r.t. the basis $\mathcal{E}$.

Let $\widetilde{X}_1, \dots, \widetilde{X}_m$ be the vector fields in $\mathfrak{g}$ (the algebra of $\mathbb{G}$) which are Ψ-related to $X_1, \dots, X_m$, respectively, i.e.

$$\mathrm{d}\Psi(\widetilde{X}_j) = X_j \quad \text{for every } j = 1, \dots, m.$$

We set

$$\widetilde{\mathcal{L}} := \sum_{j=1}^{m} (\widetilde{X}_j)^2,$$

and (since $\widetilde{\mathcal{L}}$ is a sub-Laplacian on the homogeneous Carnot group $\mathbb{G}$) we let $\widetilde{\Gamma}$ denote its (unique) fundamental solution.

A possible definition of a fundamental solution for $\mathcal{L}$ is

$$\Gamma := \widetilde{\Gamma} \circ \Psi^{-1}. \tag{5.26}$$

Our task here is to show how Γ depends on the (arbitrary) choice of the above basis $\mathcal{E}$. We show that, roughly speaking "up to a multiplicative factor", Γ in (5.26) is intrinsic (see below for the precise statement).

To this aim, let $\mathcal{E}_1$ and $\mathcal{E}_2$ be two bases of $\mathfrak{h}$ adapted to the stratification V. With the above notation, let (for $i = 1, 2$) $\mathbb{G}_i = (\mathbb{R}^N, \diamond_{\mathcal{E}_i})$ and $\mathfrak{g}_i$ denotes the algebra of $\mathbb{G}_i$. Moreover, we set

$$\Psi_i := \operatorname{Exp} \circ \pi_{\mathcal{E}_i}^{-1}, \quad i = 1, 2. \tag{5.27}$$

Again for $i = 1, 2$, we also let

$$\widetilde{X}_1^{(i)}, \ldots, \widetilde{X}_m^{(i)}$$

be the vector fields in $\mathfrak{g}_i$ which are Ψ_i-related to $X_1, \ldots, X_m$, respectively, i.e.

$$\mathrm{d}\Psi_i(\widetilde{X}_j^{(i)}) = X_j \quad \text{for every } j = 1, \ldots, m. \tag{5.28}$$

We set

$$\widetilde{\mathcal{L}}_i := \sum_{j=1}^{m} (\widetilde{X}_j^{(i)})^2, \quad i = 1, 2.$$

Finally, $\widetilde{\Gamma}_i$ denotes the fundamental solution of $\widetilde{\mathcal{L}}_i$. We claim that

$$\widetilde{\Gamma}_2 = \mathbf{c}_{1,2} \cdot \widetilde{\Gamma}_1 \circ (\Psi_1^{-1} \circ \Psi_2), \quad \text{where} \quad \mathbf{c}_{1,2} := \left|\det\left(\pi_{\mathcal{E}_1} \circ \pi_{\mathcal{E}_2}^{-1}\right)\right|. \tag{5.29}$$

This will give

$$\widetilde{\Gamma}_2 \circ \Psi_2^{-1} = \mathbf{c}_{1,2} \cdot \widetilde{\Gamma}_1 \circ \Psi_1^{-1}, \tag{5.30}$$

which proves the "almost" intrinsic nature of the definition (5.26) of Γ. We explicitly remark that $\pi_{\mathcal{E}_1} \circ \pi_{\mathcal{E}_2}^{-1}$ is a linear automorphism of $\mathbb{R}^N$, so that its determinant is well-posed.

(*Note.* We explicitly remark that the uniqueness of the fundamental solution of $\mathcal{L}$ *up to a multiplicative factor* is a matter of fact. Indeed, even in the homogeneous Carnot setting, the fundamental solution *depends* on the measure within the integral in (5.14) of Definition 5.3.1. In order to have a unique fundamental solution, the choice of the Lebesgue measure on $\mathbb{R}^N$ was quite natural (though arbitrary). Instead, in the context of an abstract Lie group $\mathbb{H}$, the Haar measure is unique only *up to a multiplicative factor*, and there is no effective way to prefer a Haar measure instead of another. These remarks show that the constant in (5.30) is perfectly justified.)

We now turn to the claimed (5.29). By invoking Proposition 5.3.10, set

$$\widetilde{\Gamma}_{1,2} := \mathbf{c}_{1,2} \cdot \widetilde{\Gamma}_1 \circ (\Psi_1^{-1} \circ \Psi_2),$$

then (5.29) will follow if we show that $\widetilde{\Gamma}_{1,2}$ satisfies (i), (ii) and (iii) of Definition 5.3.1 w.r.t. the sub-Laplacian $\widetilde{\mathcal{L}}_2$.

To begin with, observe that (thanks to (5.27))

$$\Psi_1^{-1} \circ \Psi_2 = \left(\mathrm{Exp} \circ \pi_{\mathcal{E}_1}^{-1}\right)^{-1} \circ \left(\mathrm{Exp} \circ \pi_{\mathcal{E}_2}^{-1}\right) = \pi_{\mathcal{E}_1} \circ \pi_{\mathcal{E}_2}^{-1}, \tag{5.31}$$

and this last map is a linear isomorphism of $\mathbb{R}^N$. As a consequence, since $\widetilde{\Gamma}_1 \in C^\infty(\mathbb{R}^N \setminus \{0\})$, we immediately infer that the same holds for $\widetilde{\Gamma}_{1,2}$. This proves (i). Moreover, since $\widetilde{\Gamma}_1 \in L^1_{\mathrm{loc}}(\mathbb{R}^N)$ and $\widetilde{\Gamma}_1(x) \longrightarrow 0$ when x tends to infinity, the same holds for $\widetilde{\Gamma}_{1,2}$, again thanks to (5.31). This proves (ii).

Finally, we prove (iii). First, note that from (5.28) one gets

$$\mathrm{d}\Psi_1(\widetilde{X}_j^{(1)}) = X_j = \mathrm{d}\Psi_2(\widetilde{X}_j^{(2)}) \quad \text{for every } j = 1, \dots, m,$$

i.e. for every $j = 1, \dots, m$,

$$\widetilde{X}_j^{(1)} = \mathrm{d}(\Psi_1^{-1} \circ \Psi_2)(\widetilde{X}_j^{(2)}).$$

Set $\Psi_{1,2} := \Psi_1^{-1} \circ \Psi_2$. This gives

$$\widetilde{\mathcal{L}}_1 = \mathrm{d}\Psi_{1,2}(\widetilde{\mathcal{L}}_2),$$

i.e. it holds

$$(\widetilde{\mathcal{L}}_1 f) \circ \Psi_{1,2} = \widetilde{\mathcal{L}}_2(f \circ \Psi_{1,2}) \quad \forall\, f \in C^\infty(\mathbb{R}^N, \mathbb{R}). \tag{5.32}$$

Let now $\varphi \in C_0^\infty(\mathbb{R}^N)$. Then we have

$$\begin{aligned}
\int_{\mathbb{R}^N} \widetilde{\Gamma}_{1,2}(\widetilde{\mathcal{L}}_2\varphi) &= \mathbf{c}_{1,2} \cdot \int_{\mathbb{R}^N} \widetilde{\Gamma}_1(\Psi_{1,2}(x))(\widetilde{\mathcal{L}}_2\varphi)(x)\,\mathrm{d}x \\
&\quad \text{(by the linear change of variable } y = \Psi_{1,2}(x) = \pi_{\mathcal{E}_1} \circ \pi_{\mathcal{E}_2}^{-1}\text{, see (5.31))} \\
&= \mathbf{c}_{1,2} \cdot \int_{\mathbb{R}^N} \widetilde{\Gamma}_1(y)(\widetilde{\mathcal{L}}_2\varphi)\big(\Psi_{1,2}^{-1}(y)\big)\,\frac{\mathrm{d}y}{\mathbf{c}_{1,2}} \qquad \text{(see (5.32))} \\
&= \int_{\mathbb{R}^N} \widetilde{\Gamma}_1(y)\widetilde{\mathcal{L}}_1\big(\varphi \circ \Psi_{1,2}^{-1}\big)(y)\,\mathrm{d}y \\
&\quad (\widetilde{\Gamma}_1 \text{ is the fundamental solution of } \widetilde{\mathcal{L}}_1 \text{ and } \varphi \circ \Psi_{1,2}^{-1} \in C_0^\infty(\mathbb{R}^N)) \\
&= -\big(\varphi \circ \Psi_{1,2}^{-1}\big)(0) = -\varphi(0).
\end{aligned}$$

This proves (iii), and the proof is complete. □

5.4 $\mathcal{L}$-gauges and $\mathcal{L}$-radial Functions

On every homogeneous Carnot group, there exist distinguished smooth symmetric homogeneous norms playing a fundamental rôle for the sub-Laplacians. We call these norms *gauges*, according to the following definition.

Definition 5.4.1 ($\mathcal{L}$-gauge). *Let $\mathcal{L}$ be a sub-Laplacian on a homogeneous Carnot group $\mathbb{G}$. We call* $\mathcal{L}$-gauge on $\mathbb{G}$ *a homogeneous symmetric norm d* smooth *out of the origin and satisfying*

$$\mathcal{L}(d^{2-Q}) = 0 \quad \text{in } \mathbb{G} \setminus \{0\}. \tag{5.33}$$

An $\mathcal{L}$-radial function on $\mathbb{G}$ *is a function $u : \mathbb{G} \setminus \{0\} \to \mathbb{R}$ such that*

$$u(x) = f(d(x)) \qquad \forall\, x \in \mathbb{G} \setminus \{0\}$$

for a suitable $f : (0, \infty) \to \mathbb{R}$ and a given $\mathcal{L}$-gauge d on $\mathbb{G}$.

The $\mathcal{L}$-gauges are deeply related to the fundamental solution of $\mathcal{L}$.

Proposition 5.4.2. *Let $\mathcal{L}$ be a sub-Laplacian on a homogeneous Carnot group $\mathbb{G}$. Let Γ be the fundamental solution of $\mathcal{L}$. Then*

$$d(x) := \begin{cases} (\Gamma(x))^{1/(2-Q)} & \text{if } x \in \mathbb{G} \setminus \{0\}, \\ 0 & \text{if } x = 0 \end{cases}$$

is an $\mathcal{L}$-gauge on $\mathbb{G}$.

Proof. The assertion follows from condition (i) in Definition 5.3.1, (5.15), Propositions 5.3.11, 5.3.12, 5.3.13 and Corollary 5.3.14. □

In the next section, we shall show the reverse part of Proposition 5.4.2 (see Theorem 5.5.6): *if d is an $\mathcal{L}$-gauge on $\mathbb{G}$, then there exists a positive constant β_d such that $\Gamma = \beta_d\, d^{2-Q}$ is the fundamental solution of $\mathcal{L}$.* As a consequence, by Proposition 5.3.10, the $\mathcal{L}$-gauge is unique up to a multiplicative constant.

In Section 9.8, we shall also prove the following fact: *if d is a homogeneous norm on $\mathbb{G}$, smooth out of the origin and such that*

$$\mathcal{L}(d^{\alpha}) = 0 \text{ in } \mathbb{G} \setminus \{0\}$$

for a suitable $\alpha \in \mathbb{R}$, $\alpha \neq 0$, then $\alpha = 2 - Q$ and d is an $\mathcal{L}$-gauge on $\mathbb{G}$ (see Corollary 9.8.8, page 461, for the precise statement).

The sub-Laplacian of an $\mathcal{L}$-radial function takes a noteworthy form.

Proposition 5.4.3. *Let $\mathcal{L}$ be a sub-Laplacian on a homogeneous Carnot group $\mathbb{G}$. Let $f(d)$ be a smooth $\mathcal{L}$-radial function on $\mathbb{G} \setminus \{0\}$. Then*

$$\mathcal{L}(f(d)) = |\nabla_{\mathcal{L}} d|^2 \left(f''(d) + \frac{Q-1}{d}\, f'(d) \right), \tag{5.34}$$

where, if $\mathcal{L} = \sum_{j=1}^{m} X_j^2$, we set $\nabla_{\mathcal{L}} = (X_1, \ldots, X_m)$.

Proof. An easy computation gives (see also Ex. 6, Chapter 1)

$$\mathcal{L}(f(d)) = \sum_{j=1}^{m} X_j^2(f(d)) = f''(d) \sum_{j=1}^{m} (X_j d)^2 + f'(d) \sum_{j=1}^{m} X_j^2(d),$$

so that

$$\mathcal{L}(f(d)) = f''(d)\, |\nabla_{\mathcal{L}} d|^2 + f'(d)\, \mathcal{L}(d). \tag{5.35}$$

(Note that (5.35) holds with the only assumption that f and d are smooth functions on some open subsets of $\mathbb{G}$ and $\mathbb{R}$, respectively, and $f \circ d$ is defined.)

Applying this formula to the function $f(s) = s^{2-Q}$ and keeping in mind (5.33), we obtain

$$0 = (1 - Q)\, d^{-Q}\, |\nabla_{\mathcal{L}} d|^2 + d^{1-Q}\, \mathcal{L}(d),$$

hence

$$\mathcal{L}(d) = |\nabla_{\mathcal{L}} d|^2\, \frac{Q-1}{d}.$$

Identity (5.34) follows by replacing this last identity in (5.35). □

If ϱ is any (sufficiently smooth) homogeneous norm on $\mathbb{G}$, the integration of a ϱ-radial function over a "ϱ-radially-symmetric" domain reduces to an integration of a single variable function.

Proposition 5.4.4. *Let $\mathcal{L}$ be a sub-Laplacian on a homogeneous Carnot group $\mathbb{G}$. Let ϱ be any homogeneous norm on $\mathbb{G}$ smooth on $\mathbb{G} \setminus \{0\}$. Let $f(\varrho)$ be a function defined on the ϱ-ball $B_\varrho(0, r)$ of radius r centered at the origin,*

$$B_\varrho(0, r) := \{x \in \mathbb{G} : \varrho(x) < r\}.$$

Then, if $f(\varrho) \in L^1(B_\varrho(0, r))$, it holds

$$\int_{B_\varrho(0,r)} f(\varrho(x))\, \mathrm{d}x = Q\, \omega_\varrho \int_0^r s^{Q-1}\, f(s)\, \mathrm{d}s, \tag{5.36}$$

where ω_ϱ denotes the Lebesgue measure of $B_\varrho(0, 1)$,

$$\omega_\varrho := |B_\varrho(0, 1)|.$$

Proof. The coarea formula gives

$$\int_{B_\varrho(0,r)} f(\varrho(x))\, \mathrm{d}x = \int_0^r f(s) \left(\int_{\{\varrho=s\}} \frac{1}{|\nabla \varrho\,|}\, \mathrm{d}H^{N-1} \right) \mathrm{d}s. \tag{5.37}$$

On the other hand, by using the δ_λ-homogeneity of ϱ, we have

$$\int_0^r \left(\int_{\{\varrho=s\}} \frac{1}{|\nabla \varrho|}\, \mathrm{d}H^{N-1} \right) \mathrm{d}s = |B_\varrho(0, r)| = \omega_\varrho\, r^Q$$

for every $r > 0$. Differentiating this last identity with respect to r, we obtain

$$\int_{\{\varrho=r\}} \frac{1}{|\nabla \varrho|}\, \mathrm{d}H^{N-1} = Q \omega_\varrho\, r^{Q-1}. \tag{5.38}$$

By using this identity in (5.37), we immediately get (5.36). □

Corollary 5.4.5. *Let $\mathcal{L}$ be a sub-Laplacian on a homogeneous Carnot group $\mathbb{G}$. Let ϱ be any homogeneous norm on $\mathbb{G}$. The function ϱ^α is locally integrable in $\mathbb{R}^N$ if and only if $\alpha > -Q$.*

Proof. If ϱ is also smooth on $\mathbb{G} \setminus \{0\}$, by Proposition 5.4.4,

$$\int_{B_\varrho(0,r)} \varrho^\alpha(x)\,\mathrm{d}x = Q\omega_\varrho \int_0^r s^{\alpha+Q-1}\,\mathrm{d}s,$$

and the assertion trivially follows.

If ϱ is only continuous, we argue as follows. If $\alpha \geq 0$, the assertion is trivial. If $\alpha < 0$, we have

$$\begin{aligned}
\int_{B_\varrho(0,r)} \varrho^\alpha(x)\,\mathrm{d}x &= \sum_{k=0}^{N} \int_{\{r/2^{k+1} \leq \varrho < r/2^k\}} \varrho^\alpha(x)\,\mathrm{d}x \\
&\leq (r/2)^\alpha \sum_{k=0}^{N} \frac{1}{2^{k\alpha}} \int_{\{r/2^{k+1} \leq \varrho(x) < r/2^k\}} \mathrm{d}x \quad (\text{by } x = \delta_{r/2^k}(y)) \\
&= (r/2)^\alpha \sum_{k=0}^{N} \frac{1}{2^{k\alpha}} \left(\frac{r}{2^k}\right)^Q \int_{\{1/2 \leq \varrho(y) < 1\}} \mathrm{d}y \\
&= \mathbf{c}_\varrho\, r^Q\, (r/2)^\alpha \sum_{k=0}^{N} 2^{-k(\alpha+Q)},
\end{aligned}$$

hence, if $\alpha > -Q$, then ϱ^α is integrable on $B_\varrho(0,r)$. The reverse assertion follows by the same computation as above, by taking the obvious bound from below. □

A couple of examples of explicit $\mathcal{L}$-gauges are in order. See also Example 5.10.3 and Chapter 18.

Example 5.4.6 (Δ-gauge). The classical Laplace operator in $\mathbb{R}^N$, $N \geq 3$,

$$\Delta := \sum_{j=1}^{N} \partial_{x_j}^2$$

is the canonical (sub-)Laplacian on the Euclidean group

$$\mathbb{E} := (\mathbb{R}^N, +, \delta_\lambda)$$

with $\delta_\lambda x = \lambda x$. The homogeneous dimension of $\mathbb{E}$ is N and a Δ-gauge function is the Euclidean norm

$$x \mapsto |x| := \left(\sum_{j=1}^{N} x_j^2\right)^{1/2}, \qquad x = (x_1, \ldots, x_N).$$

Indeed, $|\cdot|$ is smooth and strictly positive out of the origin, δ_λ-homogeneous of degree 1 and, as it is well known,

$$\Delta(|x|^{2-N}) = 0 \quad \forall x \neq 0. \quad □$$

On every group of Heisenberg type, noteworthy *explicit* $\mathcal{L}$-gauges are known. They were discovered by A. Kaplan [Kap80].

Example 5.4.7 ($\Delta_{\mathbb{H}}$-gauges). Let $\mathbb{H} = (\mathbb{R}^{m+n}, \circ, \delta_\lambda)$ be a (prototype) group of Heisenberg type (see Section 3.6, page 169). Denoting by (x, t) the points of $\mathbb{H}$, $x \in \mathbb{R}^m$, $t \in \mathbb{R}^n$, we know that the canonical sub-Laplacian $\Delta_{\mathbb{H}}$ on $\mathbb{H}$ takes the form (see (3.14) on page 171)

$$\Delta_{\mathbb{H}} = \Delta_x + \frac{1}{4}\,|x|^2 \Delta_t + \sum_{k=1}^{n} \langle B^{(k)} x, \nabla_x \rangle\, \partial_{t_k}. \tag{5.39a}$$

Here, the $B^{(k)}$'s are n skew-symmetric $m \times m$ orthogonal matrices satisfying the relation

$$B^{(i)}\, B^{(j)} = -B^{(j)}\, B^{(i)} \text{ for every } i,\ j \in \{1, \ldots, n\} \text{ with } i \neq j.$$

The dilations $\{\delta_\lambda\}_{\lambda>0}$ are given by

$$\delta_\lambda(x, t) = (\lambda\, x, \lambda^2 t).$$

Then

$$Q = m + 2n \tag{5.39b}$$

is the homogeneous dimension of $\mathbb{H}$. We want to show that

$$d(x, t) := \left(|x|^4 + 16\,|t|^2\right)^{1/4} \tag{5.39c}$$

is a $\Delta_{\mathbb{H}}$-gauge. Here $|x|$ and $|t|$ denote respectively the Euclidean norm of $x \in \mathbb{R}^m$ and $t \in \mathbb{R}^n$. First of all, we remark that d is strictly positive and smooth out of the origin, it is δ_λ-homogeneous of degree one and symmetric, since $(x, t)^{-1} = (-x, -t)$. Now, we aim to compute $\Delta_{\mathbb{H}}(d^{2-Q})$. To this end, it is convenient to fix the following notation:

$$v(r, s) := r^4 + 16\, s^2,\ r = |x|,\ s = |t|; \quad \alpha = (2 - Q)/4.$$

Then

$$G(x, t) := (d(x, t))^{2-Q} = (r^4 + 16\, s^2)^\alpha = v^\alpha(r, s).$$

Since $B^{(k)}$ is skew-symmetric, we have

$$\langle B^{(k)} x, \nabla_x G(x, t)\rangle = 4\alpha\, v^{\alpha-1} r^2\, \langle B^{(k)} x, x\rangle = 0$$

for every $x \in \mathbb{R}^m$, $t \in \mathbb{R}^n$ and $1 \le k \le n$. Then, by using (5.39a),

$$\begin{aligned}
\Delta_{\mathbb{H}} G &= \Delta_x G + \frac{r^2}{4} \Delta_t G \\
&= \alpha v^{\alpha-1} \left(\Delta_x v + \frac{r^2}{4} \Delta_t v \right) \\
&\quad + \alpha(\alpha - 1)\, v^{\alpha-2} \left(|\nabla_x v|^2 + \frac{r^2}{4} |\nabla_t v|^2 \right).
\end{aligned} \tag{5.39d}$$

Now, thanks to the radial symmetry of v with respect to x and t, we easily obtain

$$\Delta_x v = (8+4m)r^2, \quad \frac{r^2}{4}\,\Delta_t v = 8nr^2 \tag{5.39e}$$

and

$$|\nabla_x v|^2 = 16r^6, \quad \frac{r^2}{4}\,|\nabla_t v|^2 = (16\,rs)^2. \tag{5.39f}$$

Replacing (5.39e) and (5.39f) in (5.39d), we get

$$\begin{aligned}\Delta_{\mathbb{H}} G &= 4\alpha\, v^{\alpha-1} r^2 (2+m+2n) + 16\alpha(\alpha-1)\, v^{\alpha-2} r^2 (r^4 + 16\, s^2)\\ &= 4\alpha\, v^{\alpha-1} r^2 (2+Q+4(\alpha-1))\\ &= 4\alpha\, v^{\alpha-1} r^2 (Q-2+4\alpha) = 0 \quad (\text{being } \alpha = (2-Q)/4).\end{aligned}$$

Then

$$\Delta_{\mathbb{H}}(d^{2-Q}) = 0 \text{ in } \mathbb{H}\setminus\{0\},$$

and d in (5.39c) is a $\Delta_{\mathbb{H}}$-gauge. □

Remark 5.4.8 (Cylindrically-symmetric functions on $\mathbb{H}$). In the notation of the above Example 5.4.7, we shall call *cylindrically-symmetric* any function u on $\mathbb{H}$ such that

$$u(x,t) = v(|x|,t)$$

for some function $v = v(r,t)$, $r \in \mathbb{R}$, $r > 0$, $t \in \mathbb{R}^n$. Assume v is smooth. Then

$$\langle B^{(k)}x, \nabla_x u(x,t)\rangle = \langle B^{(k)}x, x/|x|\rangle\, \partial_r v(|x|,t) = 0,$$

since $\langle B^{(k)}x, x\rangle = 0$ for every $x \in \mathbb{R}^m$. As a consequence, keeping in mind (3.14) and (3.16), we obtain the following form of $\Delta_{\mathbb{H}}$ and $|\nabla_{\mathbb{H}}|$ for cylindrically-symmetric functions $u(x,t) = v(|x|,t)$:

$$\begin{aligned}\Delta_{\mathbb{H}} u &= v_{rr} + \frac{m-1}{r}\, v_r + \frac{1}{4}\, r^2\, \Delta_t,\\ |\nabla_{\mathbb{H}} u|^2 &= v_r^2 + \frac{1}{4}\, r^2\, |\nabla_t v|^2.\end{aligned} \tag{5.39g}$$

Here, $r = |x|$ and $v_r = \delta_r v$, $v_{rr} = \delta_{rr} v$.

5.5 Gauge Functions and Surface Mean Value Theorem

Let $\mathcal{L}$ be a sub-Laplacian on the homogeneous Carnot group $\mathbb{G}$, and let d be an $\mathcal{L}$-gauge. The aim of this section is to prove a mean value theorem on the boundary of the d-balls for the $\mathcal{L}$-harmonic functions. When $\mathcal{L}$ is the classical Laplace operator, our result will give back the Gauss theorem for classical harmonic functions. We shall obtain the mean value theorem as a byproduct of a representation formula for general

C^2 functions. *These representation formulas will play a major rôle throughout the book.*

For any $x \in \mathbb{G}$ and $r > 0$, we recall that we have already defined the d-ball of center x and radius r as follows:

$$B_d(x,r) := \{y \in \mathbb{G} : d(x^{-1} \circ y) < r\}. \tag{5.40}$$

Then

$$B_d(x,r) = x \circ B_d(0,r).$$

By using the translation invariance of the Lebesgue measure and the δ_λ-homogeneity of d, one also easily recognizes that

$$|B_d(x,r)| = r^Q \, |B_d(0,1)| =: \omega_d \, r^Q. \tag{5.41}$$

We explicitly remark that

$$\partial B_d(x,r) := \{y \in \mathbb{G} : d(x^{-1} \circ y) = r\}$$

is a smooth manifold of dimension $N-1$. Indeed, by Sard's lemma, this holds true for almost every $r > 0$. The assertion then follows for *every* $r > 0$, since $\partial B_d(x,r)$ is diffeomorphic to $\partial B_d(x,1)$ via the dilation δ_r. (Note that, so far, d may be any homogeneous norm smooth out of the origin.)

Definition 5.5.1 (The kernels of the mean value formulas). *Let $\mathcal{L}$ be a sub-Laplacian on the homogeneous Carnot group $\mathbb{G}$, and let d be an $\mathcal{L}$-gauge. We set, for $x \in \mathbb{G} \setminus \{0\}$,*

$$\Psi_{\mathcal{L}}(x) := |\nabla_{\mathcal{L}} d|^2(x).$$

Moreover, for every $x, y \in \mathbb{G}$ with $x \neq y$, we define the functions

$$\Psi_{\mathcal{L}}(x,y) := \Psi_{\mathcal{L}}(x^{-1} \circ y) \quad \textit{and} \quad \mathcal{K}_{\mathcal{L}}(x,y) := \frac{|\nabla_{\mathcal{L}} d|^2(x^{-1} \circ y)}{|\nabla(d(x^{-1} \circ \cdot))|(y)}. \tag{5.42}$$

Remark 5.5.2. We explicitly remark that

$$\Psi_{\mathcal{L}} \text{ is } \delta_\lambda\text{-homogeneous of degree zero,}$$

a fact which will be used repeatedly. We would like to recall that $\Psi_{\mathcal{L}}$ appears in the "radial" form of $\mathcal{L}$, see (5.34).

We also explicitly remark that, while $\Psi_{\mathcal{L}}$ is translation-invariant (i.e. $\Psi_{\mathcal{L}}(\alpha \circ x, \alpha \circ y) = \Psi_{\mathcal{L}}(x,y)$), the function $\mathcal{K}_{\mathcal{L}}$ does not necessarily share the same property. Moreover, when $\mathcal{L} = \Delta$ is the classical Laplace operator, then $\Psi_{\mathcal{L}} = \mathcal{K}_{\mathcal{L}} = 1$. For a general sub-Laplacian $\mathcal{L}$, we shall prove the following fact (see Proposition 9.8.9, page 462):

The function $\Psi_{\mathcal{L}}$ is constant if and only if
$\mathbb{G}$ is the Euclidean group.

For instance, consider the following example.

Example 5.5.3. Let $\mathbb{H} = (\mathbb{R}^{m+n}, \circ, \delta_\lambda)$ be a group of Heisenberg type. Denoting by (x,t) the points of $\mathbb{H}$, $x \in \mathbb{R}^m$, $t \in \mathbb{R}^n$, we proved in Example 5.4.7 that the "Folland function"

$$d(x,t) := (|x|^4 + 16\,|t|^2)^{1/4}$$

is a $\Delta_{\mathbb{H}}$-gauge. Using (5.39g), we obtain

$$\begin{aligned}\Psi_{\mathbb{H}}(x,t) &= |\nabla_{\mathbb{H}} d(x,t)|^2 \\ &= \left(\frac{r}{d}\right)^6 + 16\left(\frac{r\,s}{d^3}\right)^2 = \left(\frac{r}{d}\right)^2\left(\frac{r^4}{d^4} + 16\,\frac{s^2}{d^4}\right),\end{aligned}$$

where $r = |x|$ and $s = |t|$. Then

$$\Psi_{\mathbb{H}}(x,t) = \frac{|x|^2}{\sqrt{|x|^4 + 16\,|t|^2}}, \qquad (x,t) \neq (0,0). \qquad \square$$

Let $\Omega \subseteq \mathbb{G}$ be an open set and $u, v \in C^2(\Omega)$. By using the divergence form of the sub-Laplacian $\mathcal{L}$ (see (1.90a) on page 64),

$$\mathcal{L} = \operatorname{div}(A(x)\cdot\nabla^T),$$

we easily get

$$v\,\mathcal{L}u - u\,\mathcal{L}v = \operatorname{div}(v\,A\cdot\nabla^T u) - \operatorname{div}(u\,A\cdot\nabla^T v). \tag{5.43a}$$

Let us now assume that Ω is bounded with boundary $\partial\Omega$ of class C^1 and exterior normal $\nu = \nu(y)$ at any point $y \in \partial\Omega$. Then, if u, v are of class C^2 in a neighborhood of $\overline{\Omega}$, integrating (5.43a) on Ω and using the divergence theorem, we obtain the Green identity[2]

$$\int_\Omega (v\,\mathcal{L}u - u\,\mathcal{L}v)\,\mathrm{d}H^N = \int_{\partial\Omega}\big(v\,\langle A\cdot\nabla^T u, \nu\rangle - u\,\langle A\cdot\nabla^T v, \nu\rangle\big)\,\mathrm{d}H^{N-1}. \tag{5.43b}$$

Hereafter, $\mathrm{d}H^N$ (respectively, $\mathrm{d}H^{N-1}$) stands for the N-dimensional (respectively, $(N-1)$-dimensional) Hausdorff measure in $\mathbb{R}^N$. If we choose $v \equiv 1$ in (5.43b), we

[2] Another way to write the Green identity is the following one: if $\mathcal{L} = \sum_{j=1}^m X_j^2$, we have

$$\int_\Omega (v\,\mathcal{L}u - u\,\mathcal{L}v)\,\mathrm{d}H^N = \int_{\partial\Omega}\left(v\sum_{j=1}^m X_j u\,\langle X_j I, \nu\rangle - u\sum_{j=1}^m X_j v\,\langle X_j I, \nu\rangle\right)\mathrm{d}H^{N-1}.$$

This follows from (5.43b), recalling that (see (1.90b), page 64) A is the $N \times N$ symmetric matrix $A(x) = \sigma(x)\,\sigma(x)^T$, where $\sigma(x)$ is the $N \times m$ matrix whose columns are $X_1 I(x), \dots, X_m I(x)$.

obtain[3]

$$\int_\Omega \mathcal{L}u \,\mathrm{d}H^N = \int_{\partial\Omega} \langle A\cdot\nabla^T u, \nu\rangle \,\mathrm{d}H^{N-1}, \tag{5.43c}$$

so that

$$\int_\Omega \mathcal{L}u \,\mathrm{d}H^N = 0 \quad \forall\, u \in C_0^\infty(\Omega). \tag{5.43d}$$

Let us now consider an arbitrary open set $O \subseteq \mathbb{R}^N$ such that $\overline{B_d(x,r)} \subset O$ for a suitable $r > 0$. For $0 < \varepsilon < r$, we define the "d-ring"

$$D_{\varepsilon,r} := B_d(x,r) \setminus \overline{B_d(x,\varepsilon)} = \{y \in \mathbb{R}^N : \varepsilon < d(x^{-1}\circ y) < r\}.$$

Given $u \in C^2(O)$, we apply the Green identity (5.43b) to the functions u and $v := d^{2-Q}(x^{-1}\circ\cdot)$ on the open set $D_{\varepsilon,r}$. Since v *is* $\mathcal{L}$*-harmonic in* $\mathbb{G}\setminus\{0\}$ (note that here we apply, for the first time and with crucial consequences, the fact d *is an* $\mathcal{L}$*-gauge*), we obtain

$$\int_{D_{\varepsilon,r}} v\,\mathcal{L}u = S_r(u) - S_\varepsilon(u) + T_\varepsilon(u) - T_r(u), \tag{5.43e}$$

where we have used the following notation

$$S_\rho(u) := \int_{\partial B_d(x,\rho)} v\,\langle A\cdot\nabla^T u, \nu\rangle \,\mathrm{d}H^{N-1},$$

$$T_\rho(u) := \int_{\partial B_d(x,\rho)} u\,\langle A\cdot\nabla^T v, \nu\rangle \,\mathrm{d}H^{N-1}.$$

Since v is constant on $\partial B_d(x,\rho)$, keeping in mind (5.43c), we have

$$S_\rho(u) = \rho^{2-Q}\int_{\partial B_d(x,\rho)} \langle A\cdot\nabla^T u, \nu\rangle \,\mathrm{d}H^{N-1} = \rho^{2-Q}\int_{B_d(x,\rho)} \mathcal{L}u\,\mathrm{d}H^N, \tag{5.43f}$$

so that, by means of (5.41),

$$S_\varepsilon(u) = \varepsilon^2\,\mathcal{O}(|\mathcal{L}u|) \longrightarrow 0, \quad \text{as } \varepsilon \to 0. \tag{5.43g}$$

To evaluate $T_\rho(u)$, we first remark that on $\partial B_d(x,\rho)$ one has

$$\nu = \frac{\nabla(d(x^{-1}\circ\cdot))}{|\nabla(d(x^{-1}\circ\cdot))|},$$

and

$$\langle A\cdot\nabla^T v, \nu\rangle = (2-Q)\,d^{1-Q}(x^{-1}\circ\cdot)\,\frac{\langle A\cdot\nabla^T(d(x^{-1}\circ\cdot)), \nabla^T(d(x^{-1}\circ\cdot))\rangle}{|\nabla(d(x^{-1}\circ\cdot))|}$$

$$\text{(see (5.1c) and (5.42))} = (2-Q)\,\rho^{1-Q}\,\frac{\Psi_{\mathcal{L}}(x^{-1}\circ\cdot)}{|\nabla(d(x^{-1}\circ\cdot))|}.$$

---

[3] Or, equivalently (see the previous note),

$$\int_\Omega \mathcal{L}u\,\mathrm{d}H^N = \int_{\partial\Omega} \sum_{j=1}^m X_j u\,\langle X_j I, \nu\rangle \,\mathrm{d}H^{N-1}.$$

Therefore, keeping in mind the second definition in (5.42),

$$T_\rho(u) = (2-Q)\,\rho^{1-Q}\int_{\partial B_d(x,\rho)} u(y)\,\mathcal{K}_{\mathcal{L}}(x,y)\,\mathrm{d}H^{N-1}(y), \tag{5.43h}$$

so that

$$T_\varepsilon(u) = (u(x)+o(1))\,T_\varepsilon(1), \quad \text{as } \varepsilon \to 0. \tag{5.43i}$$

To compute $T_\varepsilon(1)$, we observe that (5.43e) with $u = 1$ gives

$$T_\varepsilon(1) = T_r(1) \text{ for } 0 < \varepsilon < r < \infty. \tag{5.43j}$$

Hence, for every $\varepsilon > 0$,

$$T_\varepsilon(1) = T_1(1) = (2-Q)\int_{\partial B_d(x,1)} \mathcal{K}_{\mathcal{L}}(x,\cdot)\,\mathrm{d}H^{N-1}. \tag{5.43k}$$

Finally, since $v = d^{2-Q}(x,\cdot) \in L^1(B_d(x,r))$ (see Corollary 5.4.5)

$$\lim_{\varepsilon\to 0}\int_{D_{\varepsilon,r}} v\,\mathcal{L}u\,\mathrm{d}H^N = \int_{D_{0,r}} v\,\mathcal{L}u\,\mathrm{d}H^N.$$

Therefore, as $\varepsilon \to 0$, identity (5.43e) together with (5.43f)-(5.43k) give

$$\begin{aligned}\int_{B_d(x,r)} v\,\mathcal{L}u\,\mathrm{d}H^N &= r^{2-Q}\int_{B_d(x,r)} \mathcal{L}u\,\mathrm{d}H^N + T_1(1)\,u(x)\\ &\quad - (2-Q)\,r^{1-Q}\int_{\partial B_d(x,r)} u\,\mathcal{K}_{\mathcal{L}}(x,\cdot)\,\mathrm{d}H^{N-1}.\end{aligned} \tag{5.43l}$$

We now observe that

$$T_1(1) = (2-Q)\int_{\partial B_d(x,1)} \mathcal{K}_{\mathcal{L}}(x,\cdot)\,\mathrm{d}H^{N-1}$$

does not depend on x, i.e.

$$\begin{aligned}&(Q-2)\int_{\partial B_d(x,1)} \mathcal{K}_{\mathcal{L}}(x,\cdot)\,\mathrm{d}H^{N-1}\\ &\quad = (Q-2)\int_{\partial B_d(0,1)} \mathcal{K}_{\mathcal{L}}(0,\cdot)\,\mathrm{d}H^{N-1} =: (\beta_d)^{-1}.\end{aligned} \tag{5.43m}$$

Indeed, from (5.43j) we have

$$\begin{aligned}T_1(1)\,\frac{1}{Q} &= \int_0^1 T_r(1)\,r^{Q-1}\,\mathrm{d}r\\ &= (2-Q)\int_0^1\int_{\partial B_d(x,r)} \mathcal{K}_{\mathcal{L}}(x,y)\,\mathrm{d}H^{N-1}(y)\,\mathrm{d}r\\ &\qquad \text{(by the coarea formula)} = (2-Q)\int_{B_d(x,1)} \Psi_{\mathcal{L}}(x,y)\,\mathrm{d}H^{N-1}(y)\\ &= (2-Q)\int_{B_d(0,1)} \Psi_{\mathcal{L}}\,\mathrm{d}H^N.\end{aligned}$$

Here we used the very definition (5.42) of the functions $\Psi_{\mathcal{L}}$ and $\mathcal{K}_{\mathcal{L}}$ *and the left invariance of* $\Psi_{\mathcal{L}}$. Incidentally, we have also proved that

$$(\beta_d)^{-1} = Q(Q-2) \int_{B_d(0,1)} \Psi_{\mathcal{L}} \, \mathrm{d}H^N. \tag{5.43n}$$

From identity (5.43l), moving terms around, we obtain

$$u(x) = \frac{(Q-2)\,\beta_d}{r^{Q-1}} \int_{\partial B_d(x,r)} u\, \mathcal{K}_{\mathcal{L}}(x,\cdot)\, \mathrm{d}H^{N-1} - \beta_d \int_{B_d(x,r)} (d^{2-Q}(x^{-1}\circ\cdot) - r^{2-Q})\, \mathcal{L}u \, \mathrm{d}H^N. \tag{5.44}$$

We have thus proved the following fundamental result.

Theorem 5.5.4 (Surface mean value theorem). *Let $\mathcal{L}$ be a sub-Laplacian on the homogeneous Carnot group $\mathbb{G}$, and let d be an $\mathcal{L}$-gauge on $\mathbb{G}$. Let O be an open subset of $\mathbb{G}$, and let $u \in C^2(O, \mathbb{R})$.*

Then, for every $x \in O$ and $r > 0$ such that $\overline{B_d(x,r)} \subset O$, we have

$$u(x) = \mathcal{M}_r(u)(x) - \mathcal{N}_r(\mathcal{L}u)(x), \tag{5.45}$$

where

$$\begin{aligned} \mathcal{M}_r(u)(x) &= \frac{(Q-2)\,\beta_d}{r^{Q-1}} \int_{\partial B_d(x,r)} \mathcal{K}_{\mathcal{L}}(x,z)\, u(z)\, \mathrm{d}H^{N-1}(z), \\ \mathcal{N}_r(w)(x) &= \beta_d \int_{B_d(x,r)} (d^{2-Q}(x^{-1}\circ z) - r^{2-Q})\, w(z)\, \mathrm{d}H^N(z), \end{aligned} \tag{5.46}$$

and β_d and $\mathcal{K}_{\mathcal{L}}$ are defined, respectively, in (5.43m) *and* (5.42).

In particular, if $\mathcal{L}u = 0$, i.e. u is $\mathcal{L}$-harmonic in O, we have

$$u(x) = \mathcal{M}_r(u)(x). \tag{5.47}$$

Remark 5.5.5. When $\mathcal{L} = \Delta$ is the classical Laplace operator in $\mathbb{R}^N$, $N \geq 3$, the kernel $\mathcal{K}_{\mathcal{L}}$ is constant ($\mathcal{K}_{\mathcal{L}} \equiv 1$). Then, in this case,

$$\mathcal{M}_r(u)(x) = \frac{1}{\sigma_N\, r^{N-1}} \int_{|x-y|=r} u(y)\, \mathrm{d}H^{N-1}(y) =: \fint_{|x-y|=r} u(y)\, \mathrm{d}H^{N-1} y$$

(see (5.43m) and (5.46)) and (5.47) gives back the Gauss theorem for classical harmonic functions. □

From Theorem 5.5.4, we straightforwardly also obtain the following result (see also Corollary 9.8.8 on page 461 for yet another improvement).

Theorem 5.5.6 ("Uniqueness" of the $\mathcal{L}$-gauges. I). *Let $\mathcal{L}$ be a sub-Laplacian on the homogeneous Carnot group $\mathbb{G}$. Let d be an $\mathcal{L}$-gauge on $\mathbb{G}$, and let β_d be the positive constant defined in* (5.43m). *Then*

$$\Gamma = \beta_d\, d^{2-Q} \tag{5.48}$$

is the fundamental solution of $\mathcal{L}$.

Proof. Let $\varphi \in C_0^\infty(\Omega)$ and choose $r > 0$ such that $\operatorname{supp}(\varphi) \subset B_d(0,r)$. Then, by the mean value formula (5.45) (being $u \equiv 0$ on $\partial B_d(0,r)$),

$$\varphi(0) = -\beta_d \int_{B_d(0,r)} (d^{2-Q}(z) - r^{2-Q})\, \mathcal{L}\varphi(z)\, \mathrm{d}H^N(z).$$

On the other hand, by identity (5.43d),

$$\int_{B_d(0,r)} r^{2-Q}\, \mathcal{L}\varphi\, \mathrm{d}H^N = 0.$$

Then, if Γ is the function defined in (5.48), $\Gamma \in L^1_{\mathrm{loc}}(\mathbb{R}^N)$ thanks to Corollary 5.4.5 and

$$-\varphi(0) = \int_{\mathbb{R}^N} \Gamma(z)\, \mathcal{L}\varphi(z)\, \mathrm{d}H^N(z)$$

for every $\varphi \in C_0^\infty(\Omega)$. Moreover, Γ is smooth in $\mathbb{G} \setminus \{0\}$ and $\Gamma(z) \to 0$ as $z \to \infty$, since $Q - 2 > 0$. Thus, by Definition 5.3.1, Γ is the fundamental solution of $\mathcal{L}$. □

From Theorem 5.5.4 we obtain the following *asymptotic* formula for $\mathcal{L}$.

Theorem 5.5.7 (Asymptotic surface formula for $\mathcal{L}$). *Let $\mathcal{L}$ be a sub-Laplacian on the homogeneous Carnot group $\mathbb{G}$, and let d be an $\mathcal{L}$-gauge.*

Let $\Omega \subseteq \mathbb{G}$ be open, and let $u \in C^2(\Omega, \mathbb{R})$. Then, for every $x \in \Omega$, we have

$$\lim_{r \to 0^+} \frac{\mathcal{M}_r(u)(x) - u(x)}{r^2} = a_d\, \mathcal{L}u(x), \tag{5.49a}$$

where

$$a_d := \beta_d \int_{B_d(0,1)} (d^{2-Q}(y) - 1)\, \mathrm{d}y. \tag{5.49b}$$

Proof. The change of variable $z = x \circ \delta_r(y)$ in the integral defining $\mathcal{N}_r$ gives

$$\mathcal{N}_r(1)(x) = a_d\, r^2 \quad \text{for every } x \in \mathbb{G}.$$

As a consequence, if $u \in C^2(\Omega, \mathbb{R})$, from (5.45) we obtain

$$\begin{aligned}\mathcal{M}_r(u)(x) - u(x) &= \mathcal{N}_r(\mathcal{L}u)(x) = \mathcal{N}_r(\mathcal{L}u - \mathcal{L}u(x))(x) + \mathcal{N}_r(1)(x)\, \mathcal{L}u(x)\\ &= a_d\, r^2(\mathcal{L}u(x) + o(1)), \quad \text{as } r \to 0^+.\end{aligned}$$

This proves (5.49a). □

5.6 Superposition of Average Operators. Solid Mean Value Theorems. Koebe-type Theorems

As in the previous section, $\mathcal{L}$ and d will respectively denote a sub-Laplacian and an $\mathcal{L}$-gauge on the homogeneous Carnot group $\mathbb{G}$. We shall denote by $B_d(x,r)$ the

d-ball with center $x \in \mathbb{G}$ and radius $r \geq 0$ and by Q the homogeneous dimension of $\mathbb{G}$. We shall also assume, as usual, $Q \geq 3$.

Given an open set $O \subseteq \mathbb{R}^N$ and a real number $r > 0$, we let

$$O_r := \{x \in O \mid d\text{-dist}(x, \partial O) > r\},$$

where

$$d\text{-dist}(x, \partial O) := \inf_{y \in \partial O} d(y^{-1} \circ x).$$

Since the d-balls are connected, one easily recognizes that $B_d(x, \rho) \subseteq O$ for every $x \in O$ and $0 < \rho < d\text{-dist}(x, \partial O)$. It follows that

$$\overline{B_d(x, \rho)} \subseteq O \qquad \forall x \in O_r \text{ and } 0 < \rho \leq r.$$

Let $\varphi : \mathbb{R} \to \mathbb{R}$ be an L^1-function vanishing out of the interval $]0, 1[$ and such that $\int_{\mathbb{R}} \varphi = 1$. For $r > 0$, define

$$\varphi_r(t) := \frac{1}{r} \varphi\Big(\frac{t}{r}\Big), \quad t \in \mathbb{R}.$$

Let us now consider a function $u \in C^2(O)$. Then, if $x \in O_r$, from (5.45) we obtain

$$u(x) = \mathcal{M}_\rho(u)(x) - \mathcal{N}_\rho(\mathcal{L}u)(x) \qquad \text{for } 0 < \rho \leq r.$$

We now multiply both sides of this identity times $\varphi_r(\rho)$ and integrate with respect to ρ. We thus get

$$u(x) = \Phi_r(u)(x) - \Phi_r^*(\mathcal{L}u)(x), \quad x \in O_r, \tag{5.50a}$$

where

$$\Phi_r(u)(x) := \int_0^\infty \varphi_r(\rho)\, \mathcal{M}_\rho(u)(x)\, \mathrm{d}\rho \tag{5.50b}$$

and

$$\Phi_r^*(w)(x) := \int_0^\infty \varphi_r(\rho)\, \mathcal{N}_\rho(w)(x)\, \mathrm{d}\rho. \tag{5.50c}$$

By using the coarea formula, the average operator Φ_r can be written as follows (we agree to let $u = 0$ out of O)

$$\Phi_r(u)(x) = \int_{\mathbb{R}^N} u(z)\, \phi_r(x^{-1} \circ z)\, \mathrm{d}z \tag{5.50d}$$

with

$$\phi_r(z) := r^{-Q}\, \phi\big(\delta_{1/r}(z)\big)$$

and

$$\phi(z) := (Q-2)\beta_d\, \Psi_{\mathcal{L}}(z)\, \frac{\varphi(d(z))}{d(z)^{Q-1}}. \tag{5.50e}$$

We note that ϕ vanishes out of $B_d(0, 1)$ and

$$\int_{\mathbb{R}^N} \phi(z)\,\mathrm{d}z = (Q-2)\beta_d \int_0^\infty \frac{\varphi(\rho)}{\rho^{Q-1}} \left(\int_{d(z)=\rho} \frac{|\nabla_{\mathcal{L}} d(z)|^2}{|\nabla d(z)|}\,\mathrm{d}H^{N-1}(y) \right) \mathrm{d}\rho$$
$$= \int_0^\infty \varphi(\rho)\,\mathrm{d}\rho = 1.$$

Here we used (5.42), (5.43h), (5.43k) and (5.43m).

If the function φ is smooth and its support is contained in $]0, 1[$, then $\phi \in C_0^\infty(\mathbb{R}^N)$, $\operatorname{supp}(\phi) \subseteq B_d(0, 1)$ and

$$x \mapsto \Phi_r(u)(x)$$

is a smooth map in O_r, whenever u is just an $L^1_{\text{loc}}(O)$-function. We would like to explicitly remark that in the integral (5.50d) the function

$$z \mapsto \phi_r(x^{-1} \circ z) \text{ vanishes out of } B_d(x, r).$$

As a consequence, if $x \in O_r$, that integral is performed on a compact set contained in O, since $\overline{B_d(x, r)} \subseteq O$.

If we choose

$$\varphi(t) = \begin{cases} Q\,t^{Q-1} & \text{if } 0 < t < 1, \\ 0 & \text{otherwise,} \end{cases}$$

then

$$\Phi_r(u) = \mathrm{M}_r(u) \text{ and } \Phi_r^*(w) = \mathrm{N}_r(w),$$

where

$$\mathrm{M}_r(u)(x) := \frac{m_d}{r^Q} \int_{B_d(x,r)} \Psi_{\mathcal{L}}(x^{-1} \circ y)\, u(y)\,\mathrm{d}y, \tag{5.50f}$$

and

$$\mathrm{N}_r(w)(x) := \frac{n_d}{r^Q} \int_0^r \rho^{Q-1} \left(\int_{B_d(x,\rho)} \left(d^{2-Q}(x^{-1} \circ y) - \rho^{2-Q}\right) w(y)\,\mathrm{d}y \right) \mathrm{d}\rho, \tag{5.50g}$$

being

$$m_d := Q(Q-2)\beta_d \text{ and } n_d := Q\,\beta_d. \tag{5.50h}$$

Hence, from (5.50a), we obtain the following theorem.

Theorem 5.6.1 (Solid mean value theorem). *Let $\mathcal{L}$ be a sub-Laplacian on the homogeneous Carnot group $\mathbb{G}$, and let d be an $\mathcal{L}$-gauge on $\mathbb{G}$. Let O be an open subset of $\mathbb{G}$, and let $u \in C^2(O, \mathbb{R})$.*

Then, for every $x \in O$ and $r > 0$ such that $\overline{B_d(x, r)} \subset O$, we have

$$u(x) = \mathrm{M}_r(u)(x) - \mathrm{N}_r(\mathcal{L}u)(x), \tag{5.51}$$

where

$$\mathrm{M}_r(u)(x) = \frac{m_d}{r^Q}\int_{B_d(x,r)} \Psi_{\mathcal{L}}(x^{-1}\circ y)\, u(y)\,\mathrm{d}H^N(y),$$

$$\mathrm{N}_r(w)(x) = \frac{n_d}{r^Q}\int_0^r \rho^{Q-1}\left(\int_{B_d(x,\rho)} \left(d^{2-Q}(x^{-1}\circ y) - \rho^{2-Q}\right) w(y)\,\mathrm{d}y\right)\mathrm{d}\rho,$$

and m_d, n_d and $\Psi_{\mathcal{L}}$ are defined in (5.50h) *(see also* (5.43m)) *and* (5.42).

In particular, if $\mathcal{L}u = 0$, i.e. u is $\mathcal{L}$-harmonic in O, we have

$$u(x) = \mathrm{M}_r(u)(x). \tag{5.52}$$

Remark 5.6.2. When $\mathcal{L} = \Delta$ is the classical Laplace operator in $\mathbb{R}^N$, $N \geq 3$, one has $\Psi_{\mathcal{L}} \equiv 1$. Then, in this case,

$$\mathrm{M}_r(u)(x) = \frac{1}{\omega_N\, r^N}\int_{|x-y|<r} u(y)\,\mathrm{d}y =: \fint_{|x-y|<r} u(y)\,\mathrm{d}H^N(y)$$

and (5.52) gives back the “Solid” Gauss theorem for classical harmonic functions. □

The mean value theorems given by identities (5.47) and (5.52) characterize the $\mathcal{L}$-harmonic functions. Indeed, the following Gauss–Koebe–Levi–Tonelli type theorem holds.

Theorem 5.6.3 (Gauss–Koebe–Levi–Tonelli type theorem). *Let $\mathcal{L}$ be a sub-Laplacian on the homogeneous Carnot group $\mathbb{G}$, and let d be an $\mathcal{L}$-gauge.*

Let O be an open subset of $\mathbb{G}$, and let $u : O \to \mathbb{R}$ be a continuous function. Assume that one of the following conditions is satisfied:

(i) *$u(x) = \mathcal{M}_r(u)(x)$ for every $x \in O$ and $r > 0$ such that $\overline{B_d(x,r)} \subset O$,*
(ii) *$u(x) = \mathrm{M}_r(u)(x)$ for every $x \in O$ and $r > 0$ such that $\overline{B_d(x,r)} \subset O$.*

Then $u \in C^\infty(O)$ and

$$\mathcal{L}u = 0 \quad \textit{in } O.$$

Proof. Assume condition (i) is satisfied. Then

$$u(x) = \mathrm{M}_\rho(u)(x) \quad \text{for } 0 < \rho \leq r. \tag{5.53}$$

Let $\varphi \in C_0^\infty(]0,1[,\mathbb{R})$ be such that $\int_{\mathbb{R}}\varphi = 1$. Multiply both sides of (5.53) times $\varphi_r(\rho) = \varphi(\rho/r)/r$. An integration with respect to ρ gives

$$u(x) = \Phi_r(u)(x) \quad \forall x \in O_r,$$

where $\Phi_r(u)$ is the integral operator (5.50b). From (5.50b) we get

$$u(x) = \int_O u(z)\,\phi_r(x^{-1}\circ z)\,\mathrm{d}z,$$

where $\phi_r(z) = r^{-Q}\,\phi(\delta_{1/r}(z))$ and ϕ is the smooth function defined by (5.50e). It follows that $u \in C^\infty(O)$. Condition (i) and identity (5.45) now give $\mathcal{N}_r(\mathcal{L}u)(x) = 0$

for every $x \in O$ and $r > 0$ such that $\overline{B_d(x,r)} \subset O$. Since the kernel appearing in the integral operator $\mathcal{N}_r$ is strictly positive, this implies $\mathcal{L}u = 0$ in O. Thus, the theorem is proved if condition (i) is fulfilled.

Let us now assume (ii). Since $\rho \mapsto \mathcal{M}_\rho(u)(x)$ is continuous on $]0, r]$ and

$$\mathrm{M}_r(u)(x) = \frac{Q}{r^Q} \int_0^r \rho^{Q-1} \mathcal{M}_\rho(u)(x)\,\mathrm{d}\rho,$$

from (ii) we get

$$\begin{aligned} Qr^{Q-1}u(x) &= \frac{\mathrm{d}}{\mathrm{d}r}(r^Q u(x)) = Q\frac{\mathrm{d}}{\mathrm{d}r}\int_0^r \rho^{Q-1}\mathcal{M}_\rho(u)(x)\,\mathrm{d}\rho \\ &= Qr^{Q-1}\,\mathcal{M}_r(u)(x). \end{aligned}$$

Hence $u(x) = \mathcal{M}_r(u)(x)$ for every $x \in O$ and $r > 0$ such that $\overline{B_d(x,r)} \subset O$. Then u satisfies condition (i), so that $u \in C^\infty(O)$ and $\mathcal{L}u = 0$ in O. □

Remark 5.6.4 (Another Koebe-type result). In the hypotheses of Theorem 5.6.3, if $u : O \to \mathbb{R}$ is continuous and satisfies

$$u(x) = \Phi_r(u)(x) \qquad \forall\, r > 0\ \forall\, x \in O_r,$$

where Φ_r is the integral operator in (5.50d) related to a smooth function $\varphi \in C_0^\infty(]0,1[, \mathbb{R})$ (see also (5.50b)), then $u \in C^\infty(O)$. As a consequence, by identity (5.50a), $\Phi_r^*(\mathcal{L}u) = 0$ in O_r for every $r > 0$. It follows that

$$\mathcal{L}u = 0 \quad \text{in } O. \quad \square$$

From Theorem 5.6.1 we obtain another asymptotic formula for the sub-Laplacians.

Theorem 5.6.5 (Asymptotic solid formula for $\mathcal{L}$). *Let $\mathcal{L}$ be a sub-Laplacian on the homogeneous Carnot group $\mathbb{G}$, and let d be an $\mathcal{L}$-gauge.*

Let $\Omega \subseteq \mathbb{G}$ be open, and let $u \in C^2(\Omega, \mathbb{R})$. Then, for every $x \in \Omega$, we have

$$\lim_{r \to 0^+} \frac{\mathrm{M}_r(u)(x) - u(x)}{r^2} = \mathrm{a}_d\, \mathcal{L}u(x), \tag{5.54}$$

where $\mathrm{a}_d = Q\,a_d/(Q+2)$ (and a_d is as in (5.49b)).

Proof. From (5.50g) we obtain (by recalling the definition (5.49b) of a_d)

$$\mathrm{N}_r(1)(x) = \frac{Q\,\beta_d}{r^Q}\int_0^r \rho^{Q+1}\left(\int_{B_d(0,1)} \left(d^{2-Q}(y) - 1\right)\mathrm{d}y\right)\mathrm{d}\rho = r^2\,\mathrm{a}_d.$$

Then, by Theorem 5.6.1 (arguing as in the proof of Theorem 5.5.7), we get

$$\mathrm{M}_r(u)(x) - u(x) = \mathrm{a}_d\, r^2(\mathcal{L}u(x) + o(1)), \quad \text{as } r \to 0^+,$$

and (5.54) follows. □

5.7 Harnack Inequalities for Sub-Laplacians

In this section, we shall prove some type of Harnack inequalities for sub-Laplacians. Our main tool will be the solid average operator M_r in (5.50f).

Let d be an $\mathcal{L}$-gauge on $\mathbb{G}$, let $\mathbf{c}_d$ be the positive constant of the pseudo-triangle inequality (see Proposition 5.1.8)

$$d(x^{-1} \circ y) \le \mathbf{c}_d \left(d(x^{-1} \circ z) + d(z^{-1} \circ y) \right) \qquad \forall\, x,\, y,\, z \in \mathbb{G}, \tag{5.55}$$

and let $\Psi_{\mathcal{L}} = |\nabla_{\mathcal{L}} d|^2$ be the kernel appearing in the average operator M_r in (5.50f).

The following lemma will play a fundamental rôle in this section: it will allow us to compare the average on different d-balls of a given non-negative function.

Lemma 5.7.1. *Let $\mathcal{L}$ be a sub-Laplacian on the homogeneous Carnot group $\mathbb{G}$, and let d be an $\mathcal{L}$-gauge. For every $r > 0$, there exists a point $z_0 = z_0(r) \in \mathbb{G}$ satisfying the following conditions*:

(i) $d(z_0) = \lambda\, r$,
(ii) $\Psi_{\mathcal{L}}(x^{-1} \circ y) \ge \mu$ *and* $\Psi_{\mathcal{L}}(y^{-1} \circ x) \ge \mu$

for every $x \in B_d(z_0, r)$ and every $y \in B_d(0, 2\,\mathbf{c}_d\, r)$. Here $\lambda \ge 1$ and $\mu > 0$ are real constants independent of r (depending only on $\mathbb{G}$, d and $\mathcal{L}$).

Proof. We split the proof into three steps.

(I) The set

$$\{y \in \mathbb{G} \setminus \{0\} \mid \Psi_{\mathcal{L}}(y) = 0\}$$

has empty interior. Indeed, suppose by contradiction $\Psi_{\mathcal{L}}(y) = 0$ for every y in a neighborhood U of a suitable $y_0 \ne 0$. Then, since $\Psi_{\mathcal{L}} = |\nabla_{\mathcal{L}} d|^2$, this gives $\nabla_{\mathcal{L}} \Psi_{\mathcal{L}} \equiv 0$ on U, so that, by Proposition 1.5.6 (page 69), d is constant in an open set containing y_0. As a consequence, the function $r \mapsto d(\delta_r(y_0)) = r\, d(y_0)$ is constant near $r = 1$. This implies $d(y_0) = 0$, which is a contradiction with the assumption $y_0 \ne 0$.

(II) From step (I) and the homogeneity of $\Psi_{\mathcal{L}}$ we infer the existence of a point $y_0 \in \mathbb{G}$, $d(y_0) = 1$, such that

$$\Psi_{\mathcal{L}}(y_0) > 0 \quad \text{and} \quad \Psi_{\mathcal{L}}(y_0^{-1}) > 0.$$

Then, by the continuity of $\Psi_{\mathcal{L}}$ out of the origin, there exist two positive constants σ and μ (with $\sigma \le 1$) such that

$$\Psi_{\mathcal{L}}(\xi \circ y_0 \circ \eta) \ge \mu, \qquad \Psi_{\mathcal{L}}(\xi \circ y_0^{-1} \circ \eta) \ge \mu \tag{5.56}$$

for every $\xi, \eta \in \mathbb{G}$ satisfying the inequalities $d(\xi), d(\eta) \le 2\,\mathbf{c}_d\, \sigma$.

(III) For any fixed $r > 0$, we let $z_0 := \delta_{\lambda\, r}(y_0)$ with $\lambda = 1/\sigma$. Let $x \in B_d(z_0, r)$ and $y \in B_d(0, 2\,\mathbf{c}_d\, r)$. From the second inequality in (5.56) we obtain

$$\begin{aligned} \Psi_{\mathcal{L}}(x^{-1} \circ y) &= \Psi_{\mathcal{L}}\left((x^{-1} \circ z_0) \circ z_0^{-1} \circ y\right) \\ &\qquad \text{(by the homogeneity of } \Psi_{\mathcal{L}}) \\ &= \Psi_{\mathcal{L}}\left(\delta_{1/(\lambda\, r)}(x^{-1} \circ z_0) \circ y_0^{-1} \circ \delta_{1/(\lambda\, r)}(y)\right) \ge \mu, \end{aligned}$$

since $d(\delta_{1/(\lambda r)}(x^{-1} \circ z_0)) \le \sigma$ and $d(\delta_{1/(\lambda r)}(y)) \le 2\,\mathbf{c}_d\,\sigma$. Analogously, from the first inequality in (5.56) we obtain

$$\begin{aligned}\Psi_{\mathcal{L}}(y^{-1} \circ x) &= \Psi_{\mathcal{L}}\big(y^{-1} \circ z_0 \circ (z_0^{-1} \circ x)\big) \\ &= \Psi_{\mathcal{L}}\big(\delta_{1/(\lambda r)}(y^{-1}) \circ y_0 \circ \delta_{1/(\lambda r)}(z_0^{-1} \circ x)\big) \ge \mu,\end{aligned}$$

since $d(\delta_{1/(\lambda r)}(z_0^{-1} \circ x)) \le \sigma$ and $d(\delta_{1/(\lambda r)}(y^{-1})) \le 2\,\mathbf{c}_d\,\sigma$. □

To state the next theorem, it is convenient to introduce the following constants

$$\theta_0 := \mathbf{c}_d(1 + \mathbf{c}_d(1+\lambda)), \qquad \theta := \mathbf{c}_d(\lambda + \theta_0). \tag{5.57}$$

Theorem 5.7.2 (Non-homogeneous Harnack inequality). *Let $\mathcal{L}$ be a sub-Laplacian on the homogeneous Carnot group $\mathbb{G}$, and let d be an $\mathcal{L}$-gauge. Let $\Omega \subseteq \mathbb{G}$ be open. Finally, let $x_0 \in \Omega$ and $r > 0$ be such that $\overline{B_d(x_0, \theta\, r)} \subset \Omega$ (here θ is as in* (5.57)*, see also Lemma* 5.7.1 *and* (5.55)*).*

Then, for any $p \in (Q/2, \infty]$, we have

$$\sup_{B_d(x_0,r)} u \le \mathbf{c}\Big\{\inf_{B_d(x_0,r)} u + r^{2-Q/p} \,\|\, \mathcal{L}u \,\|_{L^p(B_d(x_0,\theta\, r))}\Big\} \tag{5.58}$$

for every non-negative smooth function $u : \Omega \to \mathbb{R}$. Here $\mathbf{c}$ is a positive constant depending only on $\mathbb{G}$, d, $\mathcal{L}$ and p and not depending on u, r, x_0 and Ω.

Proof. Since $\mathcal{L}$ is left invariant, we may assume $x_0 = 0$.

We split the proof in several steps. M_r will be the average operator in (5.50f) and $z_0 = z_0(r)$ the point of $\mathbb{G}$ given by Lemma 5.7.1.

(I) There exists an absolute constant[4] $\mathbf{c} > 0$ such that

$$\mathrm{M}_r(u)(x) \le \mathbf{c}\,\mathrm{M}_{\theta_0 r}(u)(z_0) \qquad \forall\, x \in B_d(0,r). \tag{5.59}$$

Indeed, for every $x \in B_d(0,r)$, we have

$$\overline{B_d(x,r)} \subseteq \overline{B_d(0, 2\,\mathbf{c}_d\, r)} \subseteq \overline{B_d(0, \theta_0\, r)} \subset \Omega,$$

whence (being u non-negative)

$$\begin{aligned}\mathrm{M}_r(u)(x) &\le \Big(\sup_{\mathbb{G}\setminus\{0\}} \Psi_{\mathcal{L}}\Big) \frac{m_d}{r^Q} \int_{B_d(x,r)} u(z)\, \mathrm{d}z \\ &\qquad \text{(by the first inequality in Lemma 5.7.1-(ii))} \\ &\le \frac{\mathbf{c}_1}{r^Q} \int_{B_d(x,r)} \Psi_{\mathcal{L}}(z_0^{-1} \circ z)\, u(z)\, \mathrm{d}z,\end{aligned}$$

where $\mathbf{c}_1 = m_d/\mu \, \sup_{\mathbb{G}\setminus\{0\}} \Psi_{\mathcal{L}}$. We remark that $\mathbf{c}_1 < \infty$, since $\Psi_{\mathcal{L}}$ is smooth out of the origin and δ_λ-homogeneous of degree zero. Then, since we also have

[4] Hereafter, we call *absolute constant* any positive real constant independent of u, r, x_0 and Ω.

$$\overline{B_d(x,r)} \subseteq \overline{B_d(z_0,\theta_0 r)} \subseteq \overline{B_d(0,\theta r)} \subset \Omega,$$

we get (again by the non-negativity of u)

$$\mathrm{M}_r(u)(x) \le \frac{\mathbf{c}_1}{r^Q} \int_{B_d(z_0,\theta_0 r)} \Psi_{\mathcal{L}}(z_0^{-1} \circ z)\, u(z)\, \mathrm{d}z = \frac{\mathbf{c}_1}{m_d} \mathrm{M}_{\theta_0 r}(u)(z_0).$$

This proves (5.59) with $\mathbf{c} = \mathbf{c}_1/m_d$.

(II) There exists an absolute constant $\mathbf{c} > 0$ such that

$$\mathrm{M}_r(u)(z_0) \le \mathbf{c}\, \mathrm{M}_{\theta_0 r}(u)(y) \qquad \forall\, y \in B_d(0,r). \tag{5.60}$$

This inequality can be proved just by proceeding as in the previous step, by using the second inequality in Lemma 5.7.1-(ii) and the inclusions

$$\overline{B_d(z_0,r)} \subseteq \overline{B_d(y,\theta_0 r)} \subseteq \overline{B_d(0,\theta r)} \subset \Omega.$$

(III) Let finally $x,\ y \in B_d(0,r)$. Then, by repeatedly using the solid mean value theorem 5.6.1, we have

$$\begin{aligned} u(x) &= \mathrm{M}_r(u)(x) - \mathrm{N}_r(\mathcal{L}u)(x) \qquad \text{(by (5.59))} \\ &\le \mathbf{c}\mathrm{M}_{\theta_0 r}(u)(z_0) - \mathrm{N}_r(\mathcal{L}u)(x) \\ &= \mathbf{c}\big(u(z_0) + \mathrm{N}_{\theta_0 r}(\mathcal{L}u)(z_0)\big) - \mathrm{N}_r(\mathcal{L}u)(x) \\ &= \mathbf{c}\big(\mathrm{M}_r(u)(z_0) - \mathrm{N}_r(\mathcal{L}u)(z_0) + \mathrm{N}_{\theta_0 r}(\mathcal{L}u)(z_0)\big) - N_r(\mathcal{L}u)(x). \end{aligned}$$

On the other hand, from (5.60),

$$\mathrm{M}_r(u)(z_0) \le \mathbf{c}\, \mathrm{M}_{\theta_0 r}(u)(y) = \mathbf{c}\, \big(u(y) - \mathrm{N}_{\theta_0 r}(\mathcal{L}u)(y)\big).$$

By using this last estimate in the previous one, we infer that $u(x)$ is bounded from above by a suitable absolute constant $\mathbf{c}$ times

$$u(y) + |\mathrm{N}_{\theta_0 r}(\mathcal{L}u)(y)| + |\mathrm{N}_r(\mathcal{L}u)(z_0)| + |\mathrm{N}_{\theta_0 r}(\mathcal{L}u)(z_0)| + |N_r(\mathcal{L}u)(x)|$$

for all $x,\ y \in B_d(0,r)$.

An elementary computation based on Hölder's inequality shows that

$$|\mathrm{N}_r(w)(x)| \le \mathbf{c}_p\, r^{2-Q/p}\, \|\, w\, \|_{L^{p'}(B_d(x,r))} \tag{5.61}$$

for $Q/2 < p \le \infty$, $p' = p/(p-1)$ and

$$\mathbf{c}_p := n_d \left(\int_{d(z)<1} (d(z)^{2-Q} - 1)\, \mathrm{d}z \right)^{1/p'}.$$

Thus, keeping in mind the inclusions

$$\overline{B_d(y,\theta_0 r)},\ \overline{B_d(z_0,r)},\ \overline{B_d(z_0,\theta_0 r)},\ \overline{B_d(x,r)} \subseteq \overline{B_d(0,\theta r)},$$

from the upper bound of $u(x)$ and from (5.61) we obtain (5.58) when $x_0 = 0$. This ends the proof. □

Theorem 5.7.2 contains the following “homogeneous” Harnack inequality.

Corollary 5.7.3 (Homogeneous invariant Harnack inequality). *Let $\mathcal{L}$ be a sub-Laplacian on the homogeneous Carnot group $\mathbb{G}$, and let d be an $\mathcal{L}$-gauge. Let Ω be an open subset of $\mathbb{G}$, and let $u : \Omega \to \mathbb{R}$ be a non-negative smooth solution to $\mathcal{L}u = 0$. Then*

$$\sup_{B_d(x_0,r)} u \leq \mathbf{c} \inf_{B_d(x_0,r)} u \tag{5.62}$$

for every $x_0 \in \Omega$ and $r > 0$ such that $\overline{B_d(x_0, \theta\, r)} \subset \Omega$. The constant $\mathbf{c}$ depends only on $\mathbb{G}$, $\mathcal{L}$ and d and does not depend on u, r, x_0 and Ω.

By using a covering argument, from the non-homogeneous Harnack inequality of Theorem 5.7.2 one obtains the following theorem.

Theorem 5.7.4 (Non-homogeneous, non-invariant Harnack inequality). *Let $\mathcal{L}$ be a sub-Laplacian on the homogeneous Carnot group $\mathbb{G}$, and let d be an $\mathcal{L}$-gauge. Let Ω be an open subset of $\mathbb{G}$, and let K and K_0 be compact and connected subsets of Ω such that $K \subset \mathrm{int}(K_0)$.*

Then, for every $p \in (Q/2, \infty]$, there exists a positive constant $\mathbf{c} = \mathbf{c}(K, K_0, \Omega, \mathcal{L}, d, p, Q)$ such that

$$\sup_K u \leq \mathbf{c}\left\{\inf_K u + \| \mathcal{L}u \|_{L^p(K_0)}\right\} \tag{5.63}$$

for every $u \in C^\infty(\Omega, \mathbb{R})$, $u \geq 0$.

Proof. Let $\{D_j \mid j = 1, \ldots, q\}$ be a finite family of d-balls $D_j = B_d(x_j, r_j)$ such that (here θ is as in (5.57)):

(i) $K \subset \bigcup_{j=1}^q D_j$,
(ii) $\theta\, D_j := B_d(x_j, \theta\, r_j) \subset K_0$ for any $j = 1, \ldots, q$,
(iii) $D_j \cap D_{j+1} \neq \emptyset$ for every $j \in \{1, \ldots, q-1\}$.

We explicitly remark that such a covering exists, since K is compact, connected and contained in the interior of K_0.

By Theorem 5.7.2, we have

$$\sup_{D_j} u \leq \mathbf{c}\left\{\inf_{D_j} u + \| \mathcal{L}u \|_{L^p(\theta\, D_j)}\right\}, \quad j = 1, \ldots, q,$$

for every $u \in C^\infty(\Omega, \mathbb{R})$, $u \geq 0$. The constant $\mathbf{c}$ is independent of u.

Then, inequality (5.63) will follow by a repeated application of the following elementary lemma. □

Lemma 5.7.5. *Let A_1 and A_2 be arbitrary sets such that $A_1 \cap A_2 \neq \emptyset$. Suppose $u : A_1 \cap A_2 \to \mathbb{R}$ is a non-negative function satisfying*

$$\sup_{A_i} u \leq \mathbf{c}\left\{\inf_{A_i} u + L_i\right\}, \quad i = 1, 2, \tag{5.64}$$

for suitable constants $\mathbf{c}\geq 1$ *and* $L_i \geq 0, i = 1, 2$. *Then*

$$\sup_{A_1\cup A_2} u \leq \mathbf{c}^2\Big\{\inf_{A_1\cup A_2} u + L_1 + L_2\Big\}.$$

Proof. We have to show that

$$u(x) \leq \mathbf{c}\{u(y) + L_1 + L_2\} \tag{5.65}$$

for every $x, y \in A_1 \cup A_2$. Now, if $x, y \in A_1$ or $x, y \in A_2$, then inequality (5.65) directly follows from (5.64). Suppose $x \in A_1$ and $y \in A_2$ and choose a point $z \in A_1 \cap A_2$. By hypothesis (5.64),

$$u(x) \leq \mathbf{c}\,\{u(z) + L_1\} \quad \text{and} \quad u(z) \leq \mathbf{c}\,\{u(y) + L_2\}.$$

Hence

$$u(x) \leq \mathbf{c}\,\{\mathbf{c}\,(u(y) + L_2) + L_1\},$$

and (5.65) follows. □

From Theorem 5.7.4 one obtains the following improvement of Theorem 5.7.2.

Corollary 5.7.6 (Non-homogeneous invariant Harnack inequality). *Let* $\mathcal{L}$ *be a sub-Laplacian on the homogeneous Carnot group* $\mathbb{G}$, *and let* d *be an* $\mathcal{L}$*-gauge. Let* $\Omega \subseteq \mathbb{G}$ *be open, and let* r, R *and* R_0 *be real constants such that* $0 < r < R < R_0$. *Assume*

$$\frac{r}{R}, \frac{R}{R_0} \leq \rho \quad \textit{and} \quad \overline{B_d(x_0, R_0)} \subset \Omega$$

for suitable $\rho < 1$ *and* $x_0 \in \Omega$.

Then, for every $p \in (Q/2, \infty]$, *there exists a constant* $\mathbf{c} > 0$ *such that*

$$\sup_{B_d(x_0,r)} u \leq \mathbf{c}\Big\{\inf_{B_d(x_0, r)} u + R^{2-Q/p} \parallel \mathcal{L}u \parallel_{L^p(B_d(x_0, R))}\Big\} \tag{5.66}$$

for every $u \in C^\infty(\Omega, \mathbb{R})$, $u \geq 0$. *The constant* $\mathbf{c}$ *depends only on* $\mathbb{G}$, $\mathcal{L}$, d, p *and* ρ *and does not depend on* u, r, R, R_0, Ω *and* x_0.

Proof. Since $\mathcal{L}$ is left invariant, we may assume $x_0 = 0$. Let us put

$$u_R(x) := u(\delta_R(x)), \quad x \in \delta_{1/R}(\Omega).$$

Then, by applying Theorem 5.7.4 to the function u_R, the compact sets $K = \overline{B_d(0, \rho)}$ and $K_0 = \overline{B_d(0, 1)}$ and the open set $B_d(0, 1/\rho)$ (which is contained in $B_d(0, R_0/R) \subseteq \delta_{1/R}(\Omega)$), we obtain

$$\begin{aligned}\sup_{B_d(0,\rho)} u_R &\leq \mathbf{c}\Big\{\inf_{B_d(0,\rho)} u_R + \parallel \mathcal{L}u_R \parallel_{L^p(B_d(0,1))}\Big\} \\ &= \mathbf{c}\Big\{\inf_{B_d(0,\rho)} u_R + R^{2-Q/p} \parallel \mathcal{L}u \parallel_{L^p(B_d(0,R))}\Big\},\end{aligned}$$

with $\mathbf{c} > 0$ depending only on the parameters in the assertion of the corollary. Note that we have also used the δ_λ-homogeneity of $\mathcal{L}$ (of degree 2). From this inequality (5.66) follows, since

$$\sup_{B_d(0,\rho)} u_R = \sup_{B_d(0,R\rho)} u \geq \sup_{B_d(0,r)} u \quad \text{and} \quad \inf_{B_d(0,\rho)} u_R \leq \inf_{B_d(0,r)} u.$$

This ends the proof. □

Theorem 5.7.4 and the δ_λ-homogeneity of $\mathcal{L}$ easily imply the following Harnack inequality on rings.

Corollary 5.7.7 (Harnack inequality on rings). *Let $\mathcal{L}$ be a sub-Laplacian on the homogeneous Carnot group $\mathbb{G}$, and let d be an $\mathcal{L}$-gauge. Let u be a smooth non-negative function on the ring*

$$A_{R,c} := \{x \in \mathbb{G} \,|\, cR < d(x) < R/c\},$$

where $R > 0$ and $0 < c < 1$. Let $0 < a < b < c$. Then, for every $p \in (Q/2, \infty]$, there exists a constant $\mathbf{c} > 0$ such that

$$\sup_{A_{R,a}} u \leq \mathbf{c}\Big\{\inf_{A_{R,a}} u + R^{2-Q/p} \,\|\, \mathcal{L}u \,\|_{L^p(A_{R,b})}\Big\}. \tag{5.67}$$

The constant $\mathbf{c}$ depends only on $\mathbb{G}$, $\mathcal{L}$, p, d, a, b and c and does not depend on u and R.

Proof. The change of variable $x \mapsto \delta_R(x)$ reduces (5.67) to the analogous inequality with $R = 1$. This last one follows from Theorem 5.7.4. □

Remark 5.7.8 (The Harnack inequality in the abstract setting). According the convention in the *incipit* of the chapter, given an abstract stratified group $\mathbb{H}$, via the isomorphism Ψ between $\mathbb{H}$ and a homogeneous Carnot group $\mathbb{G}$, all the Harnack inequalities of the present section do possess a counterpart in $\mathbb{H}$.

For example, it suffices to consider the results in Remark 2.2.28 in order to derive the following result from Theorem 5.7.4.

Theorem 5.7.9 (A Harnack inequality in the abstract setting). *Let $\mathbb{H}$ be an abstract stratified group, and let $\mathcal{L}$ be a sub-Laplacian on $\mathbb{H}$. Let Ω be an open subset of $\mathbb{H}$, and let K be a compact and connected subset of Ω.*

Then there exists a positive constant $\mathbf{c} = \mathbf{c}(\mathbb{H}, \mathcal{L}, \Omega, K)$ such that

$$\sup_K u \leq \mathbf{c} \inf_K u,$$

for every non-negative function $u \in C^\infty(\Omega, \mathbb{R})$ satisfying $\mathcal{L}u = 0$ in Ω.

We close this section by giving the following "monotone convergence" theorem.

Theorem 5.7.10 (The Brelot convergence property). *Let $\mathbb{H}$ be an abstract stratified group, and let $\mathcal{L}$ be a sub-Laplacian on $\mathbb{H}$. Let $\Omega \subseteq \mathbb{H}$ be open and connected. Let $\{u_n\}_{n\in\mathbb{N}}$ be a sequence of $\mathcal{L}$-harmonic functions in Ω, i.e.*

$$u_n \in C^\infty(\Omega, \mathbb{R}) \text{ and } \mathcal{L}u_n = 0 \text{ in } \Omega \quad \text{for every } n \in \mathbb{N}.$$

Assume the sequence $\{u_n\}_{n\in\mathbb{N}}$ is monotone increasing and

$$\sup_{n\in\mathbb{N}}\{u_n(x_0)\} < \infty \tag{5.68}$$

at some point $x_0 \in \Omega$. Then there exists an $\mathcal{L}$-harmonic function u in Ω such that $\{u_n\}_{n\in\mathbb{N}}$ is uniformly convergent on every compact subset of Ω to u.

Proof. By the results in Remark 2.2.28, it suffices[5] to consider the case when $\mathcal{L}$ is a sub-Laplacian on a homogeneous Carnot group $\mathbb{G}$.

Let K be a compact subset of Ω. Since Ω is connected, there exists a compact and connected set K^* such that

$$K \subseteq K^* \subset \Omega \text{ and } x_0 \in K_0.$$

Then, by Theorem 5.7.4,

$$\begin{aligned}\sup_K(u_n - u_m) &\le \sup_{K^*}(u_n - u_m) \le \mathbf{c}\inf_{K^*}(u_n - u_m)\\ &\le \mathbf{c}(u_n(x_0) - u_m(x_0)) \qquad \text{for every } n \ge m \ge 1.\end{aligned}$$

The constant $\mathbf{c}$ is independent of n and m. Then, by condition (5.68), $\{u_n\}_n$ is uniformly convergent on K. Since K is an arbitrary compact subset of Ω, $\{u_n\}_{n\in\mathbb{N}}$ is locally uniformly convergent to a continuous function $u : \Omega \to \mathbb{R}$. On the other hand, by the solid mean value Theorem 5.6.1, for every $x \in \Omega$ and $r > 0$ such that $\overline{B_d(x,r)} \subset \Omega$, we have

$$u_n(x) = \mathrm{M}_r(u_n)(x) \quad \forall n \in \mathbb{N}.$$

Letting n tend to infinity (by the uniform convergence $u_n \to u$), we get

$$u(x) = \mathrm{M}_r(u)(x) \qquad \forall\, x \in \Omega, r > 0 : \ \overline{B_d(x,r)} \subset \Omega,$$

and now the Koebe-type Theorem 5.6.3 implies

$$u \in C^\infty(\Omega, \mathbb{R}) \text{ and } \mathcal{L}u = 0 \text{ in } \Omega. \quad \square$$

[5] Indeed, following the notation in Remark 2.2.28 and in the assertion of the above theorem, the following facts hold: the (abstract) sub-Laplacian $\mathcal{L}$ is Ψ-related to the sub-Laplacian $\widetilde{\mathcal{L}}$ on $\mathbb{G}$; set $\widetilde{u}_n := u_n \circ \Psi$, $\widetilde{\Omega} := \Psi^{-1}(\Omega)$, $\widetilde{x}_0 := \Psi^{-1}(x_0)$, then $\widetilde{\Omega}$ is open and connected in $\mathbb{G}$ (recall that Ψ is a homeomorphism), the sequence $\{\widetilde{u}_n\}_{n\in\mathbb{N}}$ is monotone increasing on $\widetilde{\Omega}$, bounded in $\widetilde{x}_0$ and $\widetilde{\mathcal{L}}\widetilde{u}_n = 0$ on $\widetilde{\Omega}$. Finally, if K is a compact subset of $\mathbb{H}$, then $\widetilde{K} := \Psi^{-1}(K)$ is a compact subset of $\mathbb{G}$, and

$$\sup_K |u_n - u| = \sup_{\widetilde{K}} |\widetilde{u}_n - \widetilde{u}|.$$

The above Brelot convergence property implies the following strong minimum principle (see also Section 5.13 for a more exhaustive investigation of maximum–minimum principles).

Corollary 5.7.11 (Strong minimum principle). *Let $\mathcal{L}$ be a sub-Laplacian on an abstract stratified group $\mathbb{H}$. A non-negative solution to $\mathcal{L}u = 0$ on an open connected set $\Omega \subseteq \mathbb{H}$ vanishes identically iff it vanishes at a point.*

Proof. Apply the result of Theorem 5.7.10 to the sequence $\{n \cdot u \mid n \in \mathbb{N}\}$. □

5.8 Liouville-type Theorems

The classical Liouville theorem for entire harmonic functions also holds in the sub-Laplacian setting. Indeed, the Harnack inequality of Corollary 5.7.3 implies the following theorem:

Theorem 5.8.1 (Liouville theorem for sub-Laplacians). *Let $\mathbb{H}$ be an abstract stratified group. Let $\mathcal{L}$ be a sub-Laplacian on $\mathbb{H}$. Let $u \in C^\infty(\mathbb{H}, \mathbb{R})$ be a function satisfying*

$$u \geq 0 \text{ and } \mathcal{L}u = 0 \text{ in } \mathbb{H}.$$

Then u is constant.

Proof. By the results in Remark 2.2.28, it suffices[6] to consider the case when $\mathcal{L}$ is a sub-Laplacian on a homogeneous Carnot group $\mathbb{G}$. Define

$$m := \inf_{\mathbb{G}} u \quad \text{and} \quad v := u - m.$$

Then $v \geq 0$ and $\mathcal{L}v = 0$ in $\mathbb{G}$. From Harnack inequality (5.62) we obtain

$$\sup_{B_d(0,r)} v \leq \mathbf{c} \inf_{B_d(0,r)} v, \qquad \mathbf{c} \text{ independent of } r.$$

From this inequality, letting r tend to infinity, we obtain

$$0 \leq \sup_{\mathbb{G}} v \leq \mathbf{c} \inf_{\mathbb{G}} v = 0,$$

which implies $v \equiv 0$ and $u \equiv m$. □

Theorem 5.8.1 can be viewed as a particular case of the following stronger version of the Liouville property for $\mathcal{L}$.

[6] Indeed, the (abstract) sub-Laplacian $\mathcal{L}$ is Ψ-related to the sub-Laplacian $\widetilde{\mathcal{L}}$ on $\mathbb{G}$ and, set $\widetilde{u} := u \circ \Psi$, we have $\widetilde{u} \geq 0$ and $\widetilde{\mathcal{L}}\widetilde{u} = 0$ on $\mathbb{G}$. Once we have proved that $\widetilde{u}$ is constant on $\mathbb{G}$, the same follows for $u = \widetilde{u} \circ \Psi^{-1}$.

Theorem 5.8.2 (Liouville-type theorem-polynomial lower bound). *Let $\mathcal{L}$ be a sub-Laplacian on the homogeneous Carnot group $\mathbb{G}$. Let u be an entire solution to $\mathcal{L}u = 0$, i.e. a function $u \in C^\infty(\mathbb{G}, \mathbb{R})$ such that $\mathcal{L}u = 0$ in $\mathbb{G}$. Assume there exists a polynomial function p on $\mathbb{G}$ such that*

$$u \geq p \quad \text{in } \mathbb{G}.$$

Then[7] *u is a polynomial function and $\deg_{\mathbb{G}} u \leq \deg_{\mathbb{G}} p$.*

The inequality $\deg_{\mathbb{G}} u \leq \deg_{\mathbb{G}} p$ can be strict: take, for example, $u \equiv 0$, $p = -x_1^2$, so that $\deg_{\mathbb{G}} u = 0 < 2 = \deg_{\mathbb{G}} p$.

Remark 5.8.3. All the results in this section involving polynomial functions are obviously related to the fixed coordinates on a homogeneous Carnot group.

Abstract counterparts of these results are available in the obvious way. A polynomial function on the abstract stratified group $\mathbb{H}$ is a function $P : \mathbb{H} \to \mathbb{R}$ such that $p := P \circ \operatorname{Exp}$ is a polynomial function on the vector space $\mathfrak{h}$ (the Lie algebra of $\mathbb{H}$), i.e. p is a polynomial function when expressed in coordinates w.r.t. any (or equivalently, w.r.t. at least one) basis for $\mathfrak{h}$. All the details are left to the reader. □

Theorem 5.8.2 is an easy consequence of the following one.

Theorem 5.8.4 (Liouville-type theorem-polynomial sub-Laplacian). *Let $\mathcal{L}$ be a sub-Laplacian on the homogeneous Carnot group $\mathbb{G}$. Let u be a smooth function on $\mathbb{G}$ satisfying*

$$u \geq 0 \quad \text{and} \quad \mathcal{L}u = w \text{ in } \mathbb{G},$$

where w is a polynomial function. Then u is a polynomial function and

$$\deg_{\mathbb{G}} u \leq 2 + \deg_{\mathbb{G}} w.$$

More precisely,

$$\deg_{\mathbb{G}} u = \begin{cases} \deg_{\mathbb{G}} w & \text{if } \deg_{\mathbb{G}} u = 0, \\ 2 + \deg_{\mathbb{G}} w & \text{if } \deg_{\mathbb{G}} u \geq 2. \end{cases} \tag{5.69}$$

The case $\deg_{\mathbb{G}} u = 1$ cannot occur, since a polynomial of $\mathbb{G}$-degree 1 cannot be a non-negative function.

The proof of this theorem will follow from a representation formula that we shall deduce from identity (5.50a) in Section 5.6.

We first show how Theorem 5.8.2 can be obtained from Theorem 5.8.4.

Proof (of Theorem 5.8.2). We first recall that $\mathcal{L}$ is a differential operator, δ_λ-homogeneous of degree two, and its coefficients are polynomial functions. Then, if $u \geq p$ and $\mathcal{L}u = 0$, we have

$$u - p \geq 0 \text{ and } \mathcal{L}(u - p) = w,$$

---

[7] See Definition 1.3.3 on page 33 for the definition of the $\mathbb{G}$-degree $\deg_{\mathbb{G}}(p)$ of a polynomial p and the $\mathbb{G}$-length $|\alpha|_{\mathbb{G}}$ of the multi-index α.

where $w := -\mathcal{L}p$ is a polynomial whose $\mathbb{G}$-degree does not exceed

$$\max\{0, \deg_{\mathbb{G}} p - 2\}.$$

From Theorem 5.8.4 it follows that $u - p$ is a polynomial and that

$$\deg_{\mathbb{G}} (u-p) \le 2 + \deg_{\mathbb{G}} (w) \le 2 + \max\{0, \deg_{\mathbb{G}} p - 2\} = \max\{2, \deg_{\mathbb{G}} p\}.$$

Hence, $u = (u-p) + p$ is a polynomial function and its $\mathbb{G}$-degree does not exceed $\max\{2, \deg_{\mathbb{G}} p\}$.

If $\deg_{\mathbb{G}} p \ge 2$, this gives the assertion of Theorem 5.8.2. It remains to consider the cases $\deg_{\mathbb{G}} p = 0$ and $\deg_{\mathbb{G}} p = 1$ (indeed, in these cases the above argument only proves that $\deg_{\mathbb{G}} u \le 2$, which is weaker than $\deg_{\mathbb{G}} u \le \deg_{\mathbb{G}} p$). In both cases, $\mathcal{L}p \equiv 0$, so that the hypotheses of Theorem 5.8.2 rewrites as

$$u - p \ge 0, \quad \mathcal{L}(u-p) = 0 \qquad \text{in } \mathbb{G}.$$

Hence, by Liouville Theorem 5.8.1, $u - p$ is constant, whence

$$u = (u-p) + p = p + \text{constant},$$

so that u is a polynomial function of the same $\mathbb{G}$-degree as p. □

We next prove a representation formula having its own interest and useful for the proof of Theorem 5.8.4.

Proposition 5.8.5. *Let $\mathcal{L}$ be a sub-Laplacian on the homogeneous Carnot group $(\mathbb{G}, *)$, and let d be an $\mathcal{L}$-gauge on $\mathbb{G}$. Let $u \in C^\infty(\mathbb{G}, \mathbb{R})$ be such that*

$$\mathcal{L}u = w \quad \text{in } \mathbb{G}, \tag{5.70a}$$

where w is a polynomial function. Then

$$u(x) = \Phi_r(u)(x) - \sum_{|\alpha|_{\mathbb{G}} \le m} C_Q^{(\alpha)} \, w^{(\alpha)}(x) \, r^{2+|\alpha|_{\mathbb{G}}} \tag{5.70b}$$

for any $x \in \mathbb{G}$ and $r > 0$. Φ_r denotes the integral operator (5.50b) *related to a smooth function $\varphi \in C_0^\infty((0,1), \mathbb{R})$. Moreover, m is the $\mathbb{G}$-degree of w,*

$$w^{(\alpha)}(x) := D_y^\alpha|_{y=0} w(x * y)$$

and, for any α with $|\alpha|_{\mathbb{G}} \le m$, $C_Q^{(\alpha)}$ is a positive constant depending only on α, Q and d. In particular, $w^{(\alpha)}$ is a polynomial function with $\mathbb{G}$-degree not exceeding $m - |\alpha|_{\mathbb{G}}$.

Proof. Identity (5.50a) in Section 5.6 and hypothesis (5.70a) give

$$u(x) = \Phi_r(u)(x) - \Phi_r^*(w)(x),$$

where $\Phi_r^*(w)(x) = \int_0^\infty \varphi_r(\rho)\,\mathcal{N}_\rho(w)(x)\,\mathrm{d}\rho$, and

$$\begin{aligned}\mathcal{N}_\rho(w)(x) &= \beta_d \int_{B_d(x,\rho)} (d^{2-Q}(x^{-1} * y) - \rho^{2-Q})\, w(y)\,\mathrm{d}H^N(y)\\ &= \beta_d \int_{B_d(0,\rho)} (d^{2-Q}(y) - \rho^{2-Q})\, w(x * y)\,\mathrm{d}H^N(y).\end{aligned}$$

We now claim that

$$w(x * y) = \sum_{|\alpha|_{\mathbb{G}} \le m} \frac{w^{(\alpha)}(x)}{\alpha!}\, y^\alpha, \tag{5.71}$$

where $w^{(\alpha)} = D_y^\alpha|_{y=0} w(x * y)$ is a polynomial function of $\mathbb{G}$-degree $\le m - |\alpha|_{\mathbb{G}}$. Taking this claim for granted for a moment, we have

$$\begin{aligned}&\Phi_r^*(w)(x)\\ &\quad = \sum_{|\alpha|_{\mathbb{G}} \le m} \frac{w^{(\alpha)}(x)}{\alpha!}\, \beta_d \int_0^\infty \varphi\Big(\frac{\rho}{r}\Big)\Big(\int_{B_d(x,\rho)} (d^{2-Q}(y) - \rho^{2-Q})\, y^\alpha\,\mathrm{d}H^N(y)\Big)\frac{\mathrm{d}\rho}{r}\\ &\quad \text{(by using the change of variables } y = \delta_\rho(z) \text{ and } \rho = r\,\sigma)\\ &\quad = \sum_{|\alpha|_{\mathbb{G}} \le m} C_Q^{(\alpha)}\, w^{(\alpha)}(x)\, r^{2+|\alpha|_{\mathbb{G}}},\end{aligned}$$

where

$$C_Q^{(\alpha)} = \frac{\beta_d}{\alpha!} \int_0^\infty \varphi(\sigma)\Big(\int_{B_d(0,1)} (d^{2-Q}(z) - 1)\, z^\alpha\,\mathrm{d}H^N(z)\Big)\,\mathrm{d}\sigma.$$

Then, we are left with the proof of (5.71). Since w is a polynomial function and $(x, y) \mapsto x * y$ has polynomial components too, one has

$$w(x * y) = \sum_{|\alpha|_{\mathbb{G}} + |\beta|_{\mathbb{G}} \le n} c_{\alpha,\beta}\, x^\alpha\, y^\beta$$

for a suitable positive integer n and real constants $c_{\alpha,\beta}$. We have to prove only that $n \le m$. Now, since $\deg_{\mathbb{G}} w \le m$,

$$w(z) = \sum_{|\gamma|_{\mathbb{G}} \le m} c_\gamma\, z^\gamma, \quad c_\gamma \in \mathbb{R} \text{ for any } \gamma.$$

Then, since δ_λ is an automorphism of the group $\mathbb{G}$,

$$\begin{aligned}\sum_{|\alpha|_{\mathbb{G}} + |\beta|_{\mathbb{G}} \le n} c_{\alpha,\beta}\, \lambda^{|\alpha|_{\mathbb{G}} + |\beta|_{\mathbb{G}}}\, x^\alpha\, y^\beta &= w(\delta_\lambda(x) * \delta_\lambda(y))\\ &= w(\delta_\lambda(x * y)) = \sum_{|\gamma|_{\mathbb{G}} \le m} c_\gamma\, \lambda^{|\gamma|_{\mathbb{G}}}\, (x * y)^\gamma\end{aligned}$$

for every $x, y \in \mathbb{G}$ and $\lambda > 0$. As a consequence,

$$\sum_{m<|\alpha|_{\mathbb{G}}+|\beta|_{\mathbb{G}}\le n} c_{\alpha,\beta}\, x^\alpha\, y^\beta = 0 \quad \forall x,\ y \in \mathbb{G},$$

so that

$$w(x*y) = \sum_{|\alpha|_{\mathbb{G}}+|\beta|_{\mathbb{G}}\le m} c_{\alpha,\beta}\, x^\alpha\, y^\beta,$$

and the claim follows. This completes the proof. □

We are now in the position to prove Theorem 5.8.4.

Proof (of Theorem 5.8.4). In the Harnack inequality (5.58) of Theorem 5.7.2, take $x_0 = 0$, $r = 2\,|x|$ and $p = \infty$. We obtain

$$u(x) \le \mathbf{c}\left\{u(0) + |x|^{m+2}\right\}, \tag{5.72}$$

where $\mathbf{c}$ is a positive absolute constant.

Let us recall the notation used in Corollary 1.5.5 (page 68). If $\mathcal{L}$ is the sum of squares of the vector fields $X_1, \ldots, X_{N_1}$ and if $\beta = (i_1, \ldots, i_k)$ is a multi-index with components in $\{1, \ldots, N_1\}$, we set

$$X^\beta := X_{i_1} \circ \cdots \circ X_{i_k} \text{ and } |\beta| = k.$$

Let us now use the representation formula of Proposition 5.8.5. Since $w^{(\alpha)}$ in (5.70b) has $\mathbb{G}$-degree $\le m - |\alpha|$, for any non-negative multi-index β with $|\beta| > m$, we have

$$X^\beta u(x) = X^\beta\big(\Phi_r(u)(x)\big) \quad \forall x \in \mathbb{G}.$$

Then, since the X_j's are left-invariant on $(\mathbb{G}, *)$ and δ_λ-homogeneous of degree one, we have (see also (5.50d) and (5.50e))

$$\begin{aligned} X^\beta u(x) &= \int_{\mathbb{G}} u(y)\, X^\beta\big(\phi_r(x^{-1}*y)\big)\, dy \\ &= r^{-|\beta|} \int_{\mathbb{R}^N} u(y)\, \big(X^\beta\widetilde{\phi}\big)_r (y^{-1}*x)\, dy, \end{aligned}$$

where $\widetilde{\phi}(z) := \phi(z^{-1})$ and

$$\big(X^\beta\widetilde{\phi}\big)_r = r^{-Q}\,\big(X^\beta\widetilde{\phi}\big) * \delta_{1/r}.$$

Hence,

$$X^\beta u(x) = r^{-|\beta|} \int_{B_d(0,1)} u(x*\delta_r(z^{-1}))\, \big(X^\beta\widetilde{\phi}\big)(z)\, dz. \tag{5.73}$$

Using inequality (5.72) in (5.73), we obtain

$$|X^\beta u(x)| \le \mathbf{c}\, r^{-|\beta|}\,(1 + |x| + r)^{m+2}$$

for every $x \in \mathbb{G}$ and $r > 0$. The constant $\mathbf{c}$ depends on $u(0)$, but it is independent of x and r. Letting r tend to infinity, we obtain

$$X^\beta u(x) = 0 \qquad \forall x \in \mathbb{G}$$

and for every β with $|\beta| > m + 2$. By the cited Corollary 1.5.5, this implies that u is a polynomial function of $\mathbb{G}$-degree $\le m + 2$. □

Using the same argument as above, we can prove the following improvement of Theorem 5.8.4.

Theorem 5.8.6 (Liouville: polynomial sub-Laplacian and bound). *Let $\mathcal{L}$ be a sub-Laplacian on the homogeneous Carnot group $\mathbb{G}$. Let $u : \mathbb{G} \to \mathbb{R}$ be a smooth function satisfying*

$$u \geq p \quad \textit{and} \quad \mathcal{L}u = w \textit{ in } \mathbb{G},$$

where p and w are polynomial functions. Then u is a polynomial and

$$\deg_{\mathbb{G}} u \leq \max\{\deg_{\mathbb{G}} p, 2 + \deg_{\mathbb{G}} w\}.$$

Proof. Set $v = u - p$. We have

$$v \geq 0, \quad \mathcal{L}v = w - \mathcal{L}p \qquad \text{in } \mathbb{G}.$$

Since $\mathcal{L}$ has polynomial coefficients, we can apply Theorem 5.8.4 to derive

$$\begin{aligned}
\deg_{\mathbb{G}}(u-p) &\leq 2 + \deg_{\mathbb{G}}(w - \mathcal{L}p) \leq 2 + \max\{\deg_{\mathbb{G}} w, \deg_{\mathbb{G}} \mathcal{L}p\} \\
&\leq 2 + \begin{cases} \max\{\deg_{\mathbb{G}} w, \deg_{\mathbb{G}} p - 2\} & \text{if } \deg_{\mathbb{G}} p \geq 2, \\ \max\{\deg_{\mathbb{G}} w, 0\} & \text{if } \deg_{\mathbb{G}} p \leq 1 \end{cases} \\
&= \begin{cases} \max\{2 + \deg_{\mathbb{G}} w, \deg_{\mathbb{G}} p\} & \text{if } \deg_{\mathbb{G}} p \geq 2, \\ 2 + \deg_{\mathbb{G}} w & \text{if } \deg_{\mathbb{G}} p \leq 1 \end{cases} \\
&= \max\{\deg_{\mathbb{G}} p, 2 + \deg_{\mathbb{G}} w\}. \quad \square
\end{aligned}$$

Summing up all the above statements, we obtain the following Liouville theorem "of polynomial type".

Theorem 5.8.7 (Liouville-type theorem). *Let $\mathcal{L}$ be a sub-Laplacian on the homogeneous Carnot group $\mathbb{G}$. Let $u : \mathbb{G} \to \mathbb{R}$ be smooth and such that*

$$u \geq p \quad \textit{and} \quad \mathcal{L}u = w \quad \textit{in } \mathbb{G},$$

where p and w are polynomial functions. Then u is a polynomial, and

$$\deg_{\mathbb{G}} u \leq \begin{cases} \deg_{\mathbb{G}} p & \textit{if } w \equiv 0, \\ \max\{\deg_{\mathbb{G}} p, 2 + \deg_{\mathbb{G}} w\} & \textit{otherwise.} \end{cases}$$

5.8.1 Asymptotic Liouville-type Theorems

We close this section by giving some more "asymptotic" Liouville-type theorems, easy consequences of Theorem 5.8.2.

Theorem 5.8.8 (Asymptotic Liouville. I). *Let $\mathcal{L}$ be a sub-Laplacian on the homogeneous Carnot group $\mathbb{G}$. Let u be an entire solution to $\mathcal{L}u = 0$. Assume there exists a real number $m \geq 0$ such that*

$$u(x) = \mathcal{O}(\varrho^m(x)), \quad \textit{as } x \to \infty. \tag{5.74}$$

Then u is a polynomial function and

$$\deg_{\mathbb{G}} u \le [m]. \tag{5.75}$$

Here ϱ is (any fixed) homogeneous norm on $\mathbb{G}$ and $[m]$ denotes the integer part of m, i.e. $[m] \in \mathbb{Z}$ and $[m] \le m < [m] + 1$.

Proof. By condition (5.74) and Proposition 5.1.4, we get

$$u(x) \ge p(x) \quad \forall x \in \mathbb{G},$$

where

$$p(x) = -\mathbf{c}\left(1 + \sum_{j=1}^{r} |x^{(j)}|^{2r!/j}\right)^{[m]+1},$$

and $\mathbf{c}$ is a suitable positive constant. Then, by Theorem 5.8.2, u is a polynomial function. Let $n := \deg_{\mathbb{G}} u$. Assume, by contradiction, $n \ge [m] + 1$. Writing

$$u(x) = \sum_{k=0}^{n} u_k(x),$$

where u_k is δ_λ-homogeneous of degree k, from condition (5.74) we get

$$\sum_{k=0}^{n} \varrho(x)^{k-n}\, u_k\big(\delta_{1/\varrho(x)}(x)\big) = \frac{u(x)}{\varrho(x)^n} \longrightarrow 0, \quad \text{as } \varrho(x) \to \infty.$$

Hence, $u_n(y) = 0$ for every $y \in \mathbb{G}$ such that $\varrho(y) = 1$, which implies $\deg_{\mathbb{G}} u \le n-1$, a contradiction. Then $n \le [m]$ and the proof is complete. □

Theorem 5.8.8 together with Theorem 5.6.1 give the following corollary.

Corollary 5.8.9 (Asymptotic Liouville. II). *Let $\mathcal{L}$ be a sub-Laplacian on the homogeneous Carnot group $\mathbb{G}$, and let d be an $\mathcal{L}$-gauge.*

Let u be an entire solution to $\mathcal{L}u = 0$. Assume there exists $x_0 \in \mathbb{G}$ and a real number $m \ge 0$ such that

$$\fint_{B_d(x_0,r)} |u(y)|\,\mathrm{d}y = \mathcal{O}(r^m), \quad \text{as } r \to \infty. \tag{5.76}$$

Then u is a polynomial function and

$$\deg_{\mathbb{G}} u \le [m].$$

In (5.76) we have set

$$\fint_D := \frac{1}{|D|}\int_D.$$

Proof. Let $x \in \mathbb{G} \setminus \{0\}$ and put $r = \max\{d(x), d(x_0)\}$. By the pseudo-triangle inequality (5.55), $B_d(x, r) \subseteq B_d(x_0, 2\,\mathbf{c}_d\, r)$. Then, from the solid mean value Theorem 5.6.1, we obtain

$$|u(x)| = |\mathrm{M}_r(u)(x)| \le \mathrm{M}_r(|u|)(x) \\ \le \frac{m_d}{r^Q}\Big(\sup_{\mathbb{G}\setminus\{0\}} \Psi_{\mathcal{L}}\Big) \int_{B_d(x_0, 2\mathbf{c}_d\, r)} |u(y)|\, \mathrm{d}y,$$

so that

$$|u(x)| \le \mathbf{c} \fint_{B_d(x_0, 2\mathbf{c}_d\, r)} |u(y)|\, \mathrm{d}y,$$

being $\mathbf{c} > 0$ independent of x and r. This inequality and (5.76) give

$$u(x) = \mathcal{O}(r^m) = \mathcal{O}(d(x)^m), \quad \text{as } x \to \infty.$$

Then, by Theorem 5.8.8, u is a polynomial function with the $\mathbb{G}$-degree $\le [m]$. □

Remark 5.8.10. By using Proposition 5.1.4, one easily recognizes that condition (5.76) in the previous corollary is equivalent to the following one

$$\int_{B_r} |u(y)|\, \mathrm{d}y = \mathcal{O}(r^{m+Q}), \quad \text{as } r \to \infty, \tag{5.77}$$

where $B_r := \{y \in \mathbb{G} : \varrho(y) < r\}$ is the ball of radius r centered at the origin, with respect to a homogeneous norm ϱ on $\mathbb{G}$.

Corollary 5.8.11 (Asymptotic Liouville. III). *Let $\mathcal{L}$ be a sub-Laplacian on the homogeneous Carnot group $\mathbb{G}$. Let u be an entire solution to $\mathcal{L}u = 0$, and $1 \le p \le \infty$. Assume (with the notation of Remark* 5.8.10)

$$\| u \|_{L^p(B_r)} = \mathcal{O}(r^{m+Q/p}), \quad \textit{as } r \to \infty. \tag{5.78}$$

Then u is a polynomial function of the $\mathbb{G}$-degree not exceeding $[m]$.

Proof. The Hölder inequality gives

$$\int_{B_r} |u(y)|\, \mathrm{d}y \le \mathbf{c}\, r^{Q(1-1/p)} \, \| u \|_{L^p(B_r)} \,.$$

Then (5.78) implies (5.77), and the assertion follows from Remark 5.8.10 and Corollary 5.8.9. □

5.9 Some Results on $\mathbb{G}$-fractional Integrals and the Sobolev–Stein Embedding Inequality

Let $\mathcal{L}$ be a sub-Laplacian on the homogeneous Carnot group $\mathbb{G}$ of homogeneous dimension Q. Let $0 < \alpha < Q$. Given a function $f : \mathbb{G} \to \mathbb{R}$, we formally define

$$I_\alpha(f)(x) := \int_{\mathbb{G}} \frac{f(y)}{(d(x,y))^{Q-\alpha}}\, dy,$$

where $d(x, y)$ stands for $d(y^{-1} \circ x)$ and $z \mapsto d(z)$ is an $\mathcal{L}$-gauge function on $\mathbb{G}$. By analogy with the Euclidean setting, we shall call I_α the *$\mathbb{G}$-fractional integral of order α*.

The following theorem, when $\mathbb{G}$ is the usual Euclidean group $(\mathbb{R}^N, +)$, gives back a celebrated theorem by Hardy, Littlewood and Sobolev.

Theorem 5.9.1 (Hardy–Littlewood–Sobolev for sub-Laplacians). *Let $\mathcal{L}$ be a sub-Laplacian on the homogeneous Carnot group $\mathbb{G}$, and let d be an $\mathcal{L}$-gauge. Suppose $1 < \alpha < Q$ and $1 < p < \frac{Q}{\alpha}$. Let $q > p$ be defined by*

$$1/q = 1/p - \alpha/Q.$$

Then there exists a positive constant $C = C(\alpha, p, \mathbb{G}, \mathcal{L}, d)$ such that

$$\|I_\alpha(f)\|_q \le C \, \|f\|_p \quad \textit{for every } f \in L^p(\mathbb{G}).$$

Here, we use the notation $\|\cdot\|_r$ to denote the L^r norm in $\mathbb{G} \equiv \mathbb{R}^N$ with respect to the Lebesgue measure.

In the classical Euclidean setting, several proofs of this theorem are known. The simplest proof seems to be due to L.I. Hedberg [Hed72] and it makes use of the maximal function theorem by Hardy–Littlewood–Wiener. This last theorem holds in general metric spaces equipped with a *doubling measure*.[8] In particular, it holds in our context.

Fixed a sub-Laplacian $\mathcal{L}$ and an $\mathcal{L}$-gauge d, we define the *$\mathcal{L}$-maximal function* $M_{\mathcal{L}}(f)$ of a function $f \in L^p(\mathbb{G}, \mathbb{C})$, $1 < p < \infty$, as follows

$$M_{\mathcal{L}}(f)(x) := \sup_{r>0} \fint_{B_d(x,r)} |f(y)|\, dy, \qquad x \in \mathbb{G},$$

where we used the notation

$$\fint_D = \frac{1}{|D|} \int_D .$$

The function $x \mapsto M_{\mathcal{L}}(f)(x)$ is lower semicontinuous. Indeed, if $M_{\mathcal{L}}(f)(x) > \alpha$, there exists $r > 0$ such that $\fint_{B_d(x,r)} |f(y)| dy > \alpha$. Then, since

$$x \mapsto \fint_{B_d(x,r)} |f(y)|\, dy$$

is a continuous function (see Ex. 5 at the end of this chapter), there exists $\delta > 0$ such that

[8] A Radon measure μ on a quasi-metric space (X, d) is *doubling* if there exists a positive constant C_d such that

$$0 < \mu(B(x, 2r)) \le C_d \, \mu(B(x, r))$$

for every d-ball with center at x and radius r.

$$M_{\mathcal{L}}(f)(z) \geq \fint_{B_d(z,r)} |f(y)|\,\mathrm{d}y > \alpha \qquad \text{if } d(x^{-1}\circ z) < \delta.$$

The *$\mathcal{L}$-maximal function theorem* is the following one.

$\mathcal{L}$-Maximal Function Theorem. *Let $1 < p < \infty$. With the above notation, there exists a positive constant $C = C(p, \mathbb{G}, \mathcal{L}, d)$ such that*

$$\|M_{\mathcal{L}}(f)\|_p \leq C\,\|f\|_p \qquad \textit{for every } f \in L^p(\mathbb{G}, \mathbb{C}).$$

A proof of this theorem in general doubling metric spaces (which is out of our scopes here) can be found in the monograph [Ste81, Chapter 2], by E.M. Stein.

Starting from this result, we now prove Theorem 5.9.1 by using the idea in L.I. Hedberg's paper [Hed72].

Proof (of Theorem 5.9.1). For every fixed $t > 0$ and $x \in \mathbb{G}$, we have

$$\begin{aligned} I_\alpha(f)(x) &= \left(\int_{d(x,y)\leq t} + \int_{d(x,y)\geq t}\right) |f(y)|\,(d(x,y))^{\alpha-Q}\,\mathrm{d}y \\ &=: I_\alpha^{(t)}(f)(x) + E_\alpha^{(t)}(f)(x). \end{aligned}$$

The Hölder inequality gives

$$\begin{aligned} E_\alpha^{(t)}(f)(x) &\leq \|f\|_p \left(\int_{d(x,y)\leq t} (d(x,y))^{(\alpha-Q)p/(p-1)}\,\mathrm{d}y\right)^{1-1/p} \\ &= (\text{see } (5.36))\;\; C\,\|f\|_p \left(\int_t^\infty \tau^{Q-1+(\alpha-Q)p/(p-1)}\,\mathrm{d}\tau\right)^{1-1/p} \\ &= C'\,\|f\|_p\, t^{\alpha-Q/p}. \end{aligned}$$

On the other hand, one has

$$\begin{aligned} I_\alpha^{(t)}(f)(x) &= \sum_{k=0}^\infty \int_{\{2^{-k-1}<d(x,y)/t\leq 2^{-k}\}} |f(y)|\,(d(x,y))^{\alpha-Q}\,\mathrm{d}y \\ &\leq C \sum_{k=0}^\infty (t/2^k)^{\alpha-Q} \int_{\{d(x,y)\leq t2^{-k}\}} |f(y)|\,\mathrm{d}y \\ &\leq C' \sum_{k=0}^\infty (t/2^k)^{\alpha-Q}\,(2^{-k}\,t)^Q\, M_{\mathcal{L}}(f)(x) \\ &= C'\,t^\alpha\, M_{\mathcal{L}}(f)(x) \sum_{k=0}^\infty 2^{-k\alpha}. \end{aligned}$$

Then, summing up the above estimates, we derive

$$I_\alpha(f)(x) \leq C\left(t^{\alpha-Q/p}\,\|f\|_p + t^\alpha\, M_{\mathcal{L}}(f)(x)\right), \qquad x \in \mathbb{G},$$

for every $t > 0$. We now choose

$$t = \left(\frac{\|f\|_p}{M_{\mathcal{L}}(f)(x)}\right)^{p/Q},$$

so that

$$I_\alpha(f)(x) \le C\,\|f\|_p^{(p\,\alpha)/Q}(M_{\mathcal{L}}(f)(x))^{1-(p\,\alpha)/Q}.$$

Hence, being $q\,(1 - p\alpha/Q) = p$, we have

$$\begin{aligned}\|I_\alpha(f)\|_q^q &\le C\|f\|_p^{(pq\alpha)/Q}\int_{\mathbb{G}}(M_{\mathcal{L}}(f)(x))^p\,\mathrm{d}x \\ &\quad\text{(by the } \mathcal{L}\text{-maximal function theorem)} \\ &\le C\|f\|_p^{(p\,q\,\alpha)/Q}\|f\|_p^p = C\|f\|_p^q,\end{aligned}$$

since $(p\,q\,\alpha)/Q + p = q$. The theorem is proved. □

From this theorem and the representation formula (5.16), one easily obtains an inequality extending the classical Sobolev embedding theorem to the homogeneous Carnot groups.

Theorem 5.9.2 (Sobolev–Stein embedding). *Let $\mathcal{L}$ be a sub-Laplacian on the homogeneous Carnot group $\mathbb{G}$ of homogeneous dimension Q.*

Suppose $1 < p < Q$. Then there exists a positive constant $C = C(p, \mathbb{G}, \mathcal{L})$ such that

$$\|u\|_q \le C\,\|\nabla_{\mathcal{L}} u\|_p \qquad \textit{for every } u \in C_0^\infty(\mathbb{R}^N, \mathbb{R}),$$

where

$$1/q = 1/p - 1/Q \quad \left(\textit{i.e. } q = \frac{Qp}{Q-p}\right).$$

Proof. Let $u \in C_0^\infty(\mathbb{R}^N, \mathbb{R})$. Using the representation formula (5.16), we have

$$u(x) = -\int_{\mathbb{G}} \Gamma(x^{-1} \circ y)\,\mathcal{L}u(y)\,\mathrm{d}y.$$

Keeping in mind that $\mathcal{L} = \sum_{j=1}^m X_j^2$ and $X_j^* = -X_j$, by integrating by parts at the right-hand side, we obtain

$$u(x) = \int_{\mathbb{R}^N} (\nabla_{\mathcal{L}}\Gamma)(x^{-1} \circ y)\,\nabla_{\mathcal{L}} u(y)\,\mathrm{d}y. \tag{5.79}$$

On the other hand, out of the origin, we have

$$\nabla_{\mathcal{L}}\Gamma = \beta_d\,\nabla_{\mathcal{L}}(d^{2-Q}) = (2-Q)\,\beta_d\,d^{1-Q}\,\nabla_{\mathcal{L}} d,$$

so that, since $\nabla_{\mathcal{L}} d$ is smooth in $\mathbb{G} \setminus \{0\}$ and δ_λ-homogeneous of degree zero,

$$|\nabla_{\mathcal{L}}\Gamma| \le C\,d^{1-Q},$$

for a suitable constant $C > 0$ depending only on $\mathcal{L}$. Using this inequality in (5.79), we get

$$|u(x)| \leq C \int_{\mathbb{G}} |\nabla_{\mathcal{L}} u(y)| \, d(x,y)^{1-Q} \, dy = C \, I_1(|\nabla_{\mathcal{L}} u|)(x).$$

Then, by the Hardy–Littlewood–Sobolev Theorem 5.9.1,

$$\|u\|_q \leq C \, \|I_1(|\nabla_{\mathcal{L}} u|)\|_q \leq C \, \|\nabla_{\mathcal{L}} u\|_p,$$

where

$$1/q = 1/p - 1/Q \quad \left(\Leftrightarrow q = \frac{Qp}{Q-p} \right).$$

This ends the proof. □

5.10 Some Remarks on the Analytic Hypoellipticity of Sub-Laplacians

In this section, we collect some results on the hypoellipticity (especially in the analytic sense) of sub-Laplacians. It is far from our scopes here to give proofs of the results of this section, the interested reader will be properly referred to the existing literature.

First of all, we recall the relevant definition.

Definition 5.10.1 ((Analytic) hypoellipticity). *We say that a differential operator L defined on an open set $\Omega \subseteq \mathbb{R}^N$ is hypoelliptic* (*respectively, analytic hypoelliptic*) *in Ω if, for every open set $\Omega' \subseteq \Omega$ and every $f \in C^\infty(\Omega', \mathbb{R})$* (*respectively, f real analytic in Ω'*), *any solution u to the equation*

$$Lu = f \text{ on } \Omega' \text{ (in the weak sense of distributions)}$$

is of class $C^\infty(\Omega', \mathbb{R})$ (*respectively, is real analytic on Ω'*).

In the sequel, we shall write $u \in C^\omega(\Omega)$ to mean that u is real analytic on Ω. Moreover, we may also use the notation C^∞*-hypoelliptic* and C^ω*-hypoelliptic* to mean, respectively, hypoelliptic and analytic hypoelliptic.

In the very special case of a homogeneous differential operator L with *constant coefficients* in $\mathbb{R}^N$, the problem of hypoellipticity is completely solved by the following result (see, e.g. [Hor69]).

Let L be a homogeneous differential operator with constant coefficients in $\mathbb{R}^N$. Then the following statements are equivalent:

1) *L is C^∞-hypoelliptic in $\mathbb{R}^N$,*
2) *L is C^ω-hypoelliptic in $\mathbb{R}^N$,*
3) *L is elliptic in $\mathbb{R}^N$.*

Moreover, if L has constant coefficients (but it is not necessarily homogeneous), then the equivalence of (2) and (3) still holds true.

As a consequence of the above result, all sub-Laplacians on the Euclidean group $\mathbb{E} = (\mathbb{R}^N, +)$ (see Section 4.1.1) are C^∞ and C^ω-hypoelliptic.

We next focus our attention to the C^∞-hypoellipticity. Let $\{X_1, \dots, X_m\}$ be C^∞-vector fields on $\mathbb{R}^N$. We recall the well-known *rank* (or *bracket*) *condition* (also referred to as *Hörmander's hypoellipticity condition*):

$$\dim\big(\mathrm{Lie}\{X_1, \dots, X_m\}I(x)\big) = N \qquad \forall x \in \mathbb{R}^N,$$

i.e. for every $x \in \mathbb{R}^N$, there exists a set of N iterated brackets of the X_i's which are linearly independent at x. The following celebrated result holds (see Hörmander [Hor67]).

If $\{X_1, \dots, X_m\}$ are C^∞-vector fields on $\mathbb{R}^N$, then the rank condition is sufficient for the C^∞-hypoellipticity of the operator $L = \sum_{j=1}^n X_j^2$. Moreover, if the coefficients of the X_j's are analytic, then the rank condition is also necessary for the C^∞-hypoellipticity of L.

(See also Derridj [Der71], Helffer–Nourrigat [HN79], Kohn [Koh73], Oleĭnik–Radkevič [OR73], Rothschild–Stein [RS76].) See also Bony [Bon69,Bon70] for a partial converse of the rank condition, namely

If the sum of squares L of smooth vector fields is C^∞-hypoelliptic, then the rank condition holds on an open set, dense in $\mathbb{R}^N$.

(See also Fediĭ [Fed70], Kusuoka–Stroock [KuSt85], Bell–Mohammed [BM95] for examples of sums of squares which are C^∞-hypoelliptic but do not satisfy the rank condition everywhere.)

The problem of C^∞-hypoellipticity is thus completely solved for any sub-Laplacian $\mathcal{L}$ on any homogeneous Carnot group, since $\mathcal{L}$ is a sum of squares of polynomial vector fields satisfying the rank condition. It is also interesting to remark that, for any homogeneous left-invariant differential operator $\mathcal{L}$ on a stratified Lie group (hence in particular for our sub-Laplacians), the C^∞-hypoellipticity of $\mathcal{L}$ is *equivalent* to a Liouville-type property for $\mathcal{L}$, namely the property stating that *the only bounded functions u on $\mathbb{G}$ such that $\mathcal{L}u = 0$ are the constant functions*. (See [Rot83]; see also [HN79,Gel83].)

We finally turn our attention to the C^ω-hypoellipticity. The problem of analytic hypoellipticity is more involved and only partial results are known. To begin with, we consider the rank condition, which played a central rôle for C^∞-hypoellipticity. Unfortunately, if L is a sum of squares of analytic vector fields, then the rank-condition is *not sufficient* for analytic hypoellipticity. (See, for instance, Trèves [Tre78], Tartakoff [Tar80], Grigis–Sjöstrand [GS85]; see also explicit counterexamples in [BG72,Hel82,PR80,HH91,Chr91].) In the sequel of the section, we collect some of these results (in particular, several explicit negative ones) for our sub-Laplacians on Carnot groups. The first one is encouraging:

The canonical sub-Laplacian on the Heisenberg–Weyl group $\mathbb{H}^n$ is analytic hypoelliptic.

It is not difficult to prove this result making use of the real analyticity (out of the origin) of the fundamental solution Γ for the canonical sub-Laplacian on $\mathbb{H}^n$, for instance ($Q = 2n + 2$ denotes the homogeneous dimension of $\mathbb{H}^n$)

$$\Gamma(x,t) = c_Q \, \frac{1}{(|x|^4 + |t|^2)^{(Q-2)/4}}. \tag{5.80}$$

Since (see Kaplan [Kap80]; see also Example 5.4.7, page 250, and Chapter 18) the fundamental solution Γ for the canonical sub-Laplacian on every H-type group has exactly the same form as in (5.80), then *the canonical sub-Laplacian on any H-type group is analytic-hypoelliptic.*

This is only a partial result of what is true on (a class of operators containing the) Heisenberg-type groups, as we shall see below; but, unfortunately (as we shall also see in a moment), it must be soon realized that C^ω-hypoellipticity rarely occurs within the non-Euclidean setting of Carnot groups.

To this end, we cite a first "negative" result (see Helffer [Hel82]):

If $\mathbb{G}$ is a Carnot group of step two, and $\mathcal{L}$ is a sub-Laplacian on $\mathbb{G}$, then a necessary condition for $\mathcal{L}$ to be C^ω-hypoelliptic is that $\mathbb{G}$ is a HM-group. (See Definition 3.7.3, page 174, for the definition of HM-group.)

Hence, at least within the setting of step-two Carnot groups, in order to find a C^ω-hypoelliptic sub-Laplacian, we must restrict our attention to the sub-class of HM-groups. Fortunately, a complete answer on the C^ω-hypoellipticity for sub-Laplacians is available on HM-groups. This is given by the following result (see Métivier [Met81]).

If $\mathbb{G}$ is a HM-group and $L \in \mathfrak{U}_m(\mathfrak{g})$ (i.e. L is a homogeneous operator of degree m in the relevant enveloping algebra), then L is C^ω-hypoelliptic if and only if L is C^∞-hypoelliptic.

Consequently, since any sub-Laplacian $\mathcal{L}$ on $\mathbb{G}$ belongs to $\mathfrak{U}_2(\mathfrak{g})$ and $\mathcal{L}$ is C^∞-hypoelliptic by the rank condition, then $\mathcal{L}$ is also C^ω-hypoelliptic.

This result covers quite large classes of remarkable cases: for example, since the Heisenberg–Weyl groups, the Iwasawa-groups, and (more generally) the H-type groups all belong to the HM-group class, we now know that *all* their sub-Laplacians are C^ω-hypoelliptic. On the converse, it is easy to exhibit a step-two group where *no* sub-Laplacian is real analytic: take any Carnot group where the first layer has odd dimension (for, in that case, it cannot be a HM-group; see Example 5.10.2 below).

The problem finally rest on the investigation of Carnot groups of step *strictly greater than two*. Unfortunately, a complete answer to the analytic-hypoellipticity in that case is still an *open problem*. Quoting Rothschild [Rot84], it is reasonable to conjecture that if $\mathbb{G}$ is not a HM-group, then there is no $L \in \mathfrak{U}_m(\mathfrak{g})$ (hence, no sub-Laplacian) which is analytic-hypoelliptic. For example, from a result by M. Christ [Chr93, Theorem 1.5] we infer that if $\mathbb{G}$ is a filiform Carnot group of dimension ≥ 4, then no sub-Laplacian on $\mathbb{G}$ is analytic-hypoelliptic. In Examples 5.10.4, 5.10.5, 5.10.6 below, we exhibit some other explicit negative results.

To begin, we give two examples: the first (respectively, the second) is an example of a sub-Laplacian on a homogeneous Carnot group of step two which is not (respectively, which is) analytic-hypoelliptic; in the latter case, we explicitly write the (analytic) fundamental solution.

Example 5.10.2. Let $\mathbb{R}^4$ (the points are denoted by (x, y, z, t) with $x, y, z, t \in \mathbb{R}$) be equipped with the dilation $\delta_\lambda(x, y, z, t) = (\lambda x, \lambda y, \lambda z, \lambda^2 t)$ and the composition law

$$\begin{pmatrix} x \\ y \\ z \\ t \end{pmatrix} \circ \begin{pmatrix} \xi \\ \eta \\ \zeta \\ \tau \end{pmatrix} = \begin{pmatrix} x+\xi \\ y+\eta \\ z+\zeta \\ t+\tau+x\,\eta \end{pmatrix}.$$

Then $\mathbb{G} = (\mathbb{R}^4, \circ, \delta_\lambda)$ is a homogeneous Carnot group of step two, and

$$\Delta_{\mathbb{G}} = (\partial_x)^2 + (\partial_y + x\,\partial_t)^2 + (\partial_z)^2$$

is its canonical sub-Laplacian. It is easily seen that $\mathbb{G}$ is isomorphic to the sum (in the sense of Section 4.1.5) of the Heisenberg-Weyl group $\mathbb{H}^1$ on $\mathbb{R}^3$ and the usual Euclidean group $(\mathbb{R}, +)$ (note that the canonical sub-Laplacians on both these groups are analytic hypoelliptic!). *It can be proved that $\Delta_{\mathbb{G}}$ is not analytic hypoelliptic!* Indeed, by the cited result of Helffer [Hel82], this follows from the fact that $\mathbb{G}$ is not a HM-group, since the first layer of the stratification has odd dimension.

This example is a particular case of the family of *not analytic hypoelliptic* sub-Laplacians (see Rothschild [Rot84])

$$\mathcal{L} = \sum_{j=1}^{n} \left((\partial_{x_j})^2 + (\partial_{y_j} + x_j\,\partial_t)^2\right) + (\partial_z)^2, \tag{5.81}$$

which, in turn, are inspired by a famous counterexample by Baouendi–Goulaouic [BG72]. Indeed, in [BG72] it is proved that the operator

$$L := \sum_{j=1}^{n} \left((\partial_{x_j})^2 + (x_j\,\partial_t)^2\right) + (\partial_z)^2$$

on $\mathbb{R}^{n+2}$ (the points are denoted by (x, z, t), $x = (x_1, \dots, x_n) \in \mathbb{R}^n$, $z \in \mathbb{R}$, $t \in \mathbb{R}$) is not C^ω-hypoelliptic on $\mathbb{R}^{n+2}$. Now, we notice that the operator $\mathcal{L}$ in (5.81) is a "lifted" version of L, so that it is easy to prove that $\mathcal{L}$ is not C^ω-hypoelliptic on $\mathbb{R}^{2n+2}$ if L is not C^ω-hypoelliptic on $\mathbb{R}^{n+2}$.

Indeed, suppose to the contrary that $\mathcal{L}$ is C^ω-hypoelliptic on $\mathbb{R}^{2n+2}$. Then take a function $f = f(x, z, t)$ real analytic on an open set $\Omega \subseteq \mathbb{R}^{n+2}$ and a solution $u = u(x, z, t)$ to $Lu = f$ on Ω. If we set $\widetilde{f}(x, y, z, t) := f(x, z, t)$ and $\widetilde{u}(x, y, z, t) := u(x, z, t)$, then we notice that

$$\mathcal{L}\widetilde{u}(x, y, z, t) = Lu(x, z, t) = f(x, z, t) = \widetilde{f}(x, y, z, t) \tag{5.82}$$

on the open set $\widetilde{\Omega} := \{(x, y, z, t) \,:\, (x, z, t) \in \Omega,\ y \in \mathbb{R}\}$. Since $\widetilde{f}$ is clearly real analytic on $\widetilde{\Omega}$, then (5.82) and the supposed C^ω-hypoellipticity of $\mathcal{L}$ imply $\widetilde{u} \in C^\omega(\widetilde{\Omega})$. This obviously means that u is real analytic on Ω. Thus we have shown that L is C^ω-hypoelliptic on $\mathbb{R}^{n+2}$, contrarily to what is proved in [BG72]. □

Example 5.10.3. This example is taken from Balogh–Tyson [BT02]. Let us consider the group $\mathbb{G}$ on $\mathbb{R}^5$ (the points are denoted by $(x_1, x_2, x_3, x_4, t) \in \mathbb{G}$, $(x_1, x_2, x_3, x_4) \in \mathbb{R}^4$ corresponds to the first layer of the stratification, $t \in \mathbb{R}$ to the second one) with dilation $\delta_\lambda(x_1, x_2, x_3, x_4, t) = (\lambda x_1, \lambda x_2, \lambda x_3, \lambda x_4, \lambda^2 t)$ and the composition law

$$\begin{pmatrix} x_1 \\ x_2 \\ x_3 \\ x_4 \\ t \end{pmatrix} \circ \begin{pmatrix} \xi_1 \\ \xi_2 \\ \xi_3 \\ \xi_4 \\ \tau \end{pmatrix} = \begin{pmatrix} x_1 + \xi_1 \\ x_2 + \xi_2 \\ x_3 + \xi_3 \\ x_4 + \xi_4 \\ t + \tau + \frac{1}{2}(x_2\xi_1 - x_1\xi_2 + 2x_4\xi_3 - 2x_3\xi_4) \end{pmatrix}.$$

Following our conventional notation, $\mathbb{G}$ is the homogeneous Carnot group of step two (with $m = 4$ generators and $n = 1$) with relevant matrix

$$B^{(1)} = \begin{pmatrix} 0 & 1 & 0 & 0 \\ -1 & 0 & 0 & 0 \\ 0 & 0 & 0 & 2 \\ 0 & 0 & -2 & 0 \end{pmatrix}.$$

Then $\mathbb{G}$ is obviously a HM-group, for B is a non-singular skew-symmetric matrix. In particular, by the cited result of Métivier [Met81], the canonical sub-Laplacian $\mathcal{L}$ is C^ω-hypoelliptic. We can see this directly, for the relevant fundamental solution Γ has been explicitly written by Balogh–Tyson in [BT02] (making use of a remarkable formula by Beals–Gaveau–Greiner, see [BGG96]; we describe this formula closely in Section 5.12, page 291): it is apparent that Γ is analytic out of the origin! Indeed, it holds

$$\Gamma(x_1, x_2, x_3, x_4, t) = c\, d^{2-Q}(x_1, x_2, x_3, x_4, t),$$

where c is a suitable positive constant, $Q = 6$ is the homogeneous dimension of $\mathbb{G}$, and d is the homogeneous norm defined by

$$\begin{aligned} &d(x_1, x_2, x_3, x_4, t) \\ &= \left(\left(\frac{1}{2}x_1^2 + \frac{1}{2}x_2^2 + x_3^2 + x_4^2\right)^2 + t^2\right)^{1/8} \\ &\cdot \left(\frac{1}{2}x_1^2 + \frac{1}{2}x_2^2 + \sqrt{\left(\frac{1}{2}x_1^2 + \frac{1}{2}x_2^2 + x_3^2 + x_4^2\right)^2 + t^2}\right)^{3/8} \\ &\cdot \left(\frac{1}{2}x_1^2 + \frac{1}{2}x_2^2 + x_3^2 + x_4^2 + \sqrt{\left(\frac{1}{2}x_1^2 + \frac{1}{2}x_2^2 + x_3^2 + x_4^2\right)^2 + t^2}\right)^{-1/8}. \end{aligned} \tag{5.83}$$

This formula gives the remarkable example of an *explicit* fundamental solution of a group which is *not* a H-type group!

We then turn our attention to groups of step greater than two. Following the idea in the argument at the end of Example 5.10.2, we can give infinite examples of groups of arbitrarily high step with a sub-Laplacian which is not C^ω-hypoelliptic: it suffices to take the sum (in the sense of Section 4.1.5) of the group $\mathbb{G}$ of Example 5.10.2 with another Carnot group. Let us now quote some other examples taken or inspired by the existing literature.

Example 5.10.4. This example is taken from a paper by Christ [Chr95] (see also [Hel82,PR80]). Consider the Carnot group on $\mathbb{R}^4$ with the composition [9] low

$$\begin{pmatrix} x_1 \\ x_2 \\ x_3 \\ x_4 \end{pmatrix} \circ \begin{pmatrix} \xi_1 \\ \xi_2 \\ \xi_3 \\ \xi_4 \end{pmatrix} = \begin{pmatrix} x_1 + \xi_1 \\ x_2 + \xi_2 \\ x_3 + \xi_3 - \xi_2 x_1 \\ x_4 + \xi_4 + 2\xi_3 x_1 - \xi_2 x_1^2 \end{pmatrix}.$$

It is easily seen that $\mathbb{G}$ is a *filiform* homogeneous Carnot group of step three and two generators, with dilations

$$\delta_\lambda(x_1, x_2, x_3, x_4) = (\lambda x_1, \lambda x_2, \lambda^2 x_3, \lambda^3 x_4),$$

and such that the first two vector fields of the relevant Jacobian basis are

$$X_1 = \partial_{x_1}, \qquad X_2 = \partial_{x_2} - x_1 \partial_{x_3} - x_1^2 \partial_{x_4}.$$

Then, it can be proved that *the canonical sub-Laplacian on* $\mathbb{G}$

$$\Delta_{\mathbb{G}} = X_1^2 + X_2^2 = (\partial_{x_1})^2 + \left(\partial_{x_2} - x_1 \partial_{x_3} - x_1^2 \partial_{x_4}\right)^2$$

is not C^ω*-hypoelliptic.* (Compare the group in this example to the Bony-type sub-Laplacian with $N = 2$ in Section 4.3.3, page 202.) □

Example 5.10.5. It is known (see [Chr91,Chr93,HH91,PR80]) that in $\mathbb{R}^3$ (with coordinates (t, s, x)) the operator

$$L = (\partial_t)^2 + (\partial_s - t^m \partial_x)^2 \tag{5.84}$$

is not C^ω-hypoelliptic for any $m \in \mathbb{N}$, $m \geq 2$. In this example, fixed m as above, we give a suitable sub-Laplacian $\mathcal{L}$ "lifting" L: as a consequence (arguing as at the end of Example 5.10.2), $\mathcal{L}$ cannot be C^ω-hypoelliptic, since L does not possess this property.

Take $N \in \mathbb{N}$, $N \geq m \geq 2$, and consider the following Bony-type sub-Laplacian (see Section 4.3.3, page 202): we equip $\mathbb{R}^{2+N}$ (whose points are denoted by (t, s, x), $t, s \in \mathbb{R}$, $x \in \mathbb{R}^N$) by the composition law

$$(t, s, x_1, x_2, x_3, \ldots, x_N) \circ (\tau, \sigma, \xi_1, \xi_2, \xi_3, \ldots, \xi_N)$$
$$= \begin{pmatrix} t + \tau \\ s + \sigma \\ x_1 + \xi_1 + \sigma\, t \\ x_2 + \xi_2 + \xi_1\, t + \sigma\, \frac{t^2}{2!} \\ x_3 + \xi_3 + \xi_2\, t + \xi_1\, \frac{t^2}{2!} + \sigma\, \frac{t^3}{3!} \\ \vdots \\ x_N + \xi_N + \xi_{N-1}\, t + \xi_{N-2}\, \frac{t^2}{2!} + \cdots + \xi_1\, \frac{t^{N-1}}{(N-1)!} + \sigma\, \frac{t^N}{N!} \end{pmatrix}$$

---

[9] Compare to Ex. 3, Chapter 4, page 216.

and the group of dilations defined by

$$\delta_\lambda(t, s, x_1, x_2, x_3, \dots, x_N) = (\lambda t, \lambda s, \lambda^2 x_1, \lambda^3 x_2, \lambda^4 x_3, \dots, \lambda^{N+1} x_N).$$

Then, $\mathbb{G} = (\mathbb{R}^{2+N}, \circ, \delta_\lambda)$ is a filiform homogeneous Carnot group of step $N+1$ (note that, since $N \geq m \geq 2$, then the step $N+1$ is ≥ 3) and two generators, and

$$\Delta_{\mathbb{G}} = (\partial_t)^2 + \left(\partial_s + t\,\partial_{x_1} + \frac{t^2}{2!}\,\partial_{x_2} + \cdots + \frac{t^N}{N!}\,\partial_{x_N} \right)^2$$

is its canonical sub-Laplacian.

We now prove that $\Delta_{\mathbb{G}}$ is not C^ω-hypoelliptic (starting from the fact that L in (5.84) is not). Indeed, since L is not C^ω-hypoelliptic, there exists an open set $\Omega \subseteq \mathbb{R}^3$ and a function $u = u(t, s, x)$ on Ω such that $Lu \in C^\omega(\Omega)$ but $u \notin C^\omega(\Omega)$. Let us now consider the function (recall that $m \leq N$)

$$\widetilde{u} = \widetilde{u}(t, s, x_1, \dots, x_N) := u(t, s, -m! x_m)$$

defined on the open subset of $\mathbb{R}^{2+N}$

$$\widetilde{\Omega} := \big\{ (t, s, x_1, \dots, x_N) \,|\, (t, s, -m!\, x_m) \in \Omega \big\}.$$

As a consequence, it holds

$$\Delta_{\mathbb{G}}\big(\widetilde{u}(t, s, x_1, \dots, x_N)\big) = (Lu)(t, s, -m!\, x_m) \text{ on } \widetilde{\Omega},$$

whence $\Delta_{\mathbb{G}} \widetilde{u} \in C^\omega(\widetilde{\Omega})$ (for $Lu \in C^\omega(\Omega)$) and $\widetilde{u} \notin C^\omega(\widetilde{\Omega})$ (for $u \notin C^\omega(\Omega)$). This proves that $\Delta_{\mathbb{G}}$ is not analytic-hypoelliptic. □

Example 5.10.6. Let $m, k \in \mathbb{N}$ be such that $0 \leq m \leq k$. Consider the operator on $\mathbb{R}^3$ (whose points are denoted by (x_1, x_2, x_3))

$$L = (\partial_{x_1})^2 + (x_1^m\,\partial_{x_2})^2 + (x_1^k\,\partial_{x_3})^2. \tag{5.85}$$

O.A. Oleĭnik and E.V. Radkevič [OR72] (see also [Him98]) proved that L is C^ω-hypoelliptic if and only if $m = k$. Our aim in this example is to "lift" (in a suitable sense) the vector fields on $\mathbb{R}^3$ appearing in (5.85)

$$\partial_{x_1}, \qquad x_1^m\,\partial_{x_2}, \qquad x_1^k\,\partial_{x_3} \tag{5.86}$$

to three vector fields (in a larger space, namely $\mathbb{R}^{3+m+k}$) generating a homogeneous Carnot group. It will easily follow (arguing as in the last paragraph of Example 5.10.5) that the relevant sub-Laplacian is not C^ω-hypoelliptic if L is not.

When $m = k = 0$, then $L = \Delta_{\mathbb{R}^3}$ is the ordinary Laplace operator on $\mathbb{R}^3$; when $m = 0$, $k = 1$, we obtain a non-analytic-hypoelliptic operator already considered in Example 5.10.2 (see also Baouendi–Goulaouic [BG72]). Let us now suppose that $k > m \geq 1$. We equip $\mathbb{R}^{3+m+k}$ with the following coordinates (the semicolon will denote different layers in a suitable homogeneous Carnot group structure)

$$P = (x_1, y_1, z_1; y_2, z_2; y_3, z_3; \dots, y_m, z_m; \; x_2, z_{m+1}; z_{m+2}; z_{m+3}; \dots; z_k; x_3).$$

We define a group of dilations by setting (recall that $k \geq m+1$)

$$\begin{aligned} \delta_\lambda(P) = \Big(&\lambda x_1, \lambda y_1, \lambda z_1; \\ &\lambda^2 y_2, \lambda^2 z_2; \lambda^3 y_3, \lambda^3 z_3; \dots; \lambda^m y_m, \lambda^m z_m; \\ &\lambda^{m+1} x_2, \lambda^{m+1} z_{m+1}; \\ &\lambda^{m+2} z_{m+2}; \lambda^{m+3} z_{m+3}; \dots; \lambda^k z_k; \\ &\lambda^{k+1} x_3\Big). \end{aligned}$$

Let us also consider on $\mathbb{R}^{3+m+k}$ the following vector fields

$$\begin{aligned} X &= \partial_{x_1}, \\ Y &= \partial_{y_1} + x_1 \partial_{y_2} + x_1^2 \partial_{y_3} + \dots + x_1^{m-1} \partial_{y_m} + x_1^m \partial_{x_2}, \\ Z &= \partial_{z_1} + x_1 \partial_{z_2} + x_1^2 \partial_{z_3} + \dots + x_1^{k-1} \partial_{z_k} + x_1^k \partial_{x_3}. \end{aligned}$$

It is then not difficult to see that X, Y, Z are δ_λ-homogeneous of degree one and they fulfill hypotheses **(H0)–(H1)–(H2)** of page 191. Hence, by the results of Section 4.2 (page 191), we can define a suitable homogeneous Carnot group structure on $\mathbb{R}^{3+m+k}$ such that

$$\Delta_{\mathbb{G}} = X^2 + Y^2 + Z^2$$

is the relevant canonical sub-Laplacian. Now, since

$$\Delta_{\mathbb{G}}(u(x_1, x_2, x_3)) = (Lu)(x_1, x_2, x_3)$$

for any smooth function u on $\mathbb{R}^{3+m+k}$ depending only on x_1, x_2, x_3, we can argue as in the last paragraph of Example 5.10.5 to infer that $\Delta_{\mathbb{G}}$ is not C^ω-hypoelliptic (since L is not). $\square$

5.11 Harmonic Approximation

Let $\mathcal{L}$ be a sub-Laplacian on the homogeneous Carnot group $\mathbb{G}$. In this section, we give some conditions ensuring that a $\mathcal{L}$-harmonic function defined in a neighborhood of a compact set K contained in an open set Ω can be uniformly approximated on K by a sequence of $\mathcal{L}$-harmonic functions in Ω.

In some of the results of this section, we assume that $\mathcal{L}$ is *analytic-hypoelliptic* (see Section 5.10). $\Gamma = d^{2-Q}$, will denote its fundamental solution. To begin with, we prove the following lemma.

Lemma 5.11.1. *Let $\mathcal{L}$ be a sub-Laplacian on the homogeneous Carnot group $\mathbb{G} = (\mathbb{R}^N, \circ, \delta_\lambda)$, and let $\Gamma = d^{2-Q}$ be its fundamental solution. Suppose also that $\mathcal{L}$ is analytic-hypoelliptic.*

Let $K \subseteq \mathbb{G}$ be compact. Then there exists $R = R(K, \mathbb{G}, \mathcal{L}) > 0$ such that, for every $z \in \mathbb{G}$ with $d(z) > R$, it holds

$$\Gamma(z^{-1} \circ y) = \sum_{\alpha \in (\mathbb{N} \cup \{0\})^N} C_\alpha(z) \cdot y^\alpha, \qquad C_\alpha(z) = \frac{1}{\alpha!} D^\alpha \Big|_{y=0} \big(\Gamma(z^{-1} \circ y)\big),$$

the series converging uniformly on the d-disc $\{y \in \mathbb{G} : d(y) < d(K)\}$, where $d(K) := \sup_{z \in K} d(z)$.

Proof. Let $\zeta \in \mathbb{G}$ be such that $d(\zeta) = 1$. Since $\eta \mapsto \Gamma(\zeta^{-1} \circ \eta)$ is analytic close to $\eta = 0$ and $\partial B_d(0, 1) = \{\zeta \in \mathbb{G} : d(\zeta) = 1\}$ is compact, there exists $\rho > 0$, independent of ζ, such that

$$\Gamma(\zeta^{-1} \circ \eta) = \sum_{\alpha \in (\mathbb{N} \cup \{0\})^N} a_\alpha(\zeta) \cdot \eta^\alpha, \qquad \text{uniformly on } B_d(0, \rho),$$

where

$$a_\alpha(\zeta) = \frac{1}{\alpha!} D^\alpha \Big|_{\eta=0} \big(\Gamma(\zeta^{-1} \circ \eta)\big).$$

Let us now choose $R > 0$ such that $R > d(K)/\rho$. If $z \in \mathbb{G}$, $d(z) > R$, then

$$\begin{aligned}
\Gamma(z^{-1} \circ y) &= d^{2-Q}(z)\, \Gamma\big((\delta_{d^{-1}(z)} z^{-1}) \circ \delta_{d^{-1}(z)}(y)\big) \\
&= d^{2-Q}(z) \sum_{\alpha \in (\mathbb{N} \cup \{0\})^N} a_\alpha\big(\delta_{d^{-1}(z)} z^{-1}\big) \cdot \big(\delta_{d^{-1}(z)}(y)\big)^\alpha \\
&=: \sum_{\alpha \in (\mathbb{N} \cup \{0\})^N} C_\alpha(z) \cdot y^\alpha.
\end{aligned}$$

The series converges uniformly for $y \in \mathbb{G}$ such that

$$d\big(\delta_{d^{-1}(z)}(y)\big) = \frac{d(y)}{d(z)} < \rho.$$

In particular, this holds for

$$y \in B_d(0, R\,\rho) \supseteq B_d(0, d(K)). \qquad \square$$

The next lemma does not require the analytic-hypoellipticity of $\mathcal{L}$.

Lemma 5.11.2. *Let $\mathcal{L}$ be a sub-Laplacian on the homogeneous Carnot group $\mathbb{G} = (\mathbb{R}^N, \circ, \delta_\lambda)$, and let $\Gamma = d^{2-Q}$ be its fundamental solution. Let u be an $\mathcal{L}$-harmonic function on the ball $B_d(0, r)$. Suppose*

$$u(x) = \sum_{k=1}^{\infty} u_k(x), \qquad x \in B_d(0, r),$$

where u_k is a continuous δ_λ-homogeneous function in the whole $\mathbb{G}$ of δ_λ-degree m_k, $k \in \mathbb{N}$. If the series is uniformly convergent on the compact subsets of $B_d(0, r)$ and $m_k \neq m_h$ for every $k \neq h$, then every u_k is $\mathcal{L}$-harmonic in $\mathbb{G}$.

Proof. Let $\varphi \in C_0^\infty(\mathbb{R}^N, \mathbb{R})$ with $\operatorname{supp}\varphi \subseteq B_d(0,1)$. Since u is $\mathcal{L}$-harmonic in $B_d(0,r)$, we have

$$0 = \int_{B_d(0,r)} u(x)\, \mathcal{L}\big(\varphi(\delta_{\lambda^{-1}}(x))\big)\, dx \qquad \forall\, \lambda < r.$$

The homogeneity of $\mathcal{L}$ and the change of variable $y = \delta_{\lambda^{-1}}(x)$ give

$$\begin{aligned} 0 &= \lambda^{Q-2} \int_{B_d(0,r/\lambda)} u(\delta_\lambda(y))\, \mathcal{L}\varphi(y)\, dy \\ &= \lambda^{Q-2} \sum_{k=1}^{\infty} \int_{B_d(0,r/\lambda)} u_k(\delta_\lambda(y))\, \mathcal{L}\varphi(y)\, dy \\ &= \lambda^{Q-2} \sum_{k=1}^{\infty} \lambda^{m_k} \int_{B_d(0,r/\lambda)} u_k(y)\, \mathcal{L}\varphi(y)\, dy \\ &= \lambda^{Q-2} \sum_{k=1}^{\infty} \lambda^{m_k} \int_{B_d(0,1)} u_k(y)\, \mathcal{L}\varphi(y)\, dy. \end{aligned}$$

Note that, in the last equality, we were able to replace $B_d(0, r/\lambda)$ by $B_d(0,1)$, since $\lambda < r$ and φ is supported in a compact set in $B_d(0,1)$. Then

$$\int_{B_d(0,1)} u_k(y)\, \mathcal{L}\varphi(y)\, dy = 0 \qquad \forall\, \varphi \in C_0^\infty(B_d(0,1))\ \forall k \in \mathbb{N}.$$

This means that $\mathcal{L}u_k = 0$ in $B_d(0,1)$ in the weak sense of distributions. Since $\mathcal{L}$ is hypoelliptic, this implies the $\mathcal{L}$-harmonicity of u_k in $B_d(0,1)$, and so in $\mathbb{G}$, due to the δ_λ-homogeneity of u_k, $k \in \mathbb{N}$. □

We are now ready to prove the announced approximation theorem.

Theorem 5.11.3 (An approximation theorem). *Let $\mathcal{L}$ be a sub-Laplacian on the homogeneous Carnot group $\mathbb{G}$. Suppose that $\mathcal{L}$ is analytic-hypoelliptic.*

Let $\Omega \subseteq \mathbb{G}$ be an open set such that $\partial\Omega = \partial\overline{\Omega}$. Let $K \subseteq \Omega$ be compact and satisfy the following condition:

$$\text{if } \omega \text{ is a bounded connected component of } \Omega \setminus K, \text{ then } \partial\omega \subseteq \partial\Omega. \tag{5.87}$$

Then, for every function h which is $\mathcal{L}$-harmonic in a neighborhood of K, there exists a sequence $(h_n)_{n\in\mathbb{N}}$ of $\mathcal{L}$-harmonic functions in Ω such that

$$\lim_{n\to\infty} h_n = h, \quad \textit{uniformly on } K.$$

Proof. By general results from functional analysis, it is well known that it suffices to prove the following statement.[10]

[10] This is also known as "Caccioppoli's completeness method".

Let μ be a signed Radon measure supported on K and satisfying

$$\int_K h \,\mathrm{d}\mu = 0 \tag{5.88}$$

for every $\mathcal{L}$-harmonic function h in Ω. Then (5.88) holds for every function h which is $\mathcal{L}$-harmonic just in a neighborhood of K.

The crucial part of the proof is the following assertion.

Claim. If μ satisfies the previous hypotheses, then

$$u(x) := \int_K \Gamma(y^{-1} \circ x) \,\mathrm{d}\mu(y) = \int_K \Gamma(x^{-1} \circ y) \,\mathrm{d}\mu(y) \tag{5.89}$$

is identically zero in $\Omega \setminus K$.

We first show how to complete the proof of the theorem by using this claim. Let h be an $\mathcal{L}$-harmonic function in an open set $\Omega_0 \supseteq K$, $\Omega_0 \subseteq \Omega$. Choose a function $\phi \in C_0^\infty(\Omega_0)$ such that $\phi = 1$ in an open set $\Omega_1 \supseteq K$, $\overline{\Omega_1} \subseteq \Omega_0$. Then $\phi h \in C_0^\infty(\Omega_0)$ and $\phi h = h$ in Ω_1. By the representation formula (5.16), we have

$$\begin{aligned}(\phi h)(y) &= -\int_{\Omega_0} \Gamma(x^{-1} \circ y) \mathcal{L}(\phi h)(x) \,\mathrm{d}x \\ &= -\int_{\Omega_0 \setminus \overline{\Omega_1}} \Gamma(x^{-1} \circ y)\, \mathcal{L}(\phi h)(x) \,\mathrm{d}x, \qquad y \in \Omega_0.\end{aligned}$$

It follows that

$$\begin{aligned}\int_K h(y) \,\mathrm{d}\mu(y) &= \int_K (\phi h)(y) \,\mathrm{d}\mu(y) \\ &= -\int_K \left(\int_{\Omega_0 \setminus \overline{\Omega_1}} \Gamma(x^{-1} \circ y)\, \mathcal{L}(\phi h)(x) \,\mathrm{d}x \right) \mathrm{d}\mu(y) =: (\star).\end{aligned}$$

Thus, by interchanging the integrals and keeping in mind (5.89), we infer

$$(\star) = -\int_{\Omega_0 \setminus \overline{\Omega_1}} \mathcal{L}(\phi h)(x) u(x) \,\mathrm{d}x = 0,$$

since $u = 0$ in $\Omega \setminus K \supseteq \Omega_0 \setminus \overline{\Omega_1}$.

Thus, we are left with the proof of the *Claim.* Let ω be a connected component of $\Omega \setminus K$. We have to prove that $u \equiv 0$ in ω.

We first suppose that ω is bounded. Then $\emptyset \neq \partial\omega \subseteq \partial\Omega$. In particular, this implies that $\partial\Omega \neq \emptyset$, hence $\overline{\Omega} \neq \mathbb{G}$ for $\partial\overline{\Omega} = \partial\Omega$. For every $x_0 \in \mathbb{G} \setminus \overline{\Omega}$, the function $y \mapsto \Gamma(x_0^{-1} \circ y)$ is $\mathcal{L}$-harmonic in Ω. Therefore, by the assumption (5.88),

$$u(x_0) = \int_K \Gamma(y^{-1} \circ x_0) \,\mathrm{d}\mu(y) = 0.$$

This proves that $u \equiv 0$ in $\mathbb{G} \setminus \overline{\Omega}$. As a consequence,

$$D^\alpha u(x) = 0 \qquad \text{for every multi-index } \alpha \tag{5.90}$$

and for every $x \in \mathbb{G} \setminus \overline{\Omega}$. Since $u \in C^\infty(\mathbb{G} \setminus K)$, it follows that (5.90) also holds at any point $x \in \partial\overline{\Omega} = \partial\Omega$. In particular, since $\partial\omega \subseteq \partial\Omega$, (5.90) holds at some point $x \in \partial\omega$. Then $\omega \equiv 0$ in a neighborhood of x, since u is $\mathcal{L}$-harmonic, hence real analytic, close to x. The connectedness of ω and again the analyticity of u imply that $u \equiv 0$ in ω.

Let us now assume that ω is unbounded. By Lemma 5.11.1, for every $x \in \omega$ with $d(x)$ sufficiently large, we have

$$\Gamma(x^{-1} \circ y) = \sum_{\alpha \in (\mathbb{N} \cup \{0\})^N} C_\alpha(x) \cdot y^\alpha, \qquad \text{uniformly in } B_d(0, d(K)).$$

Then

$$u(x) = \int_K \Gamma(x^{-1} \circ y)\, \mathrm{d}\mu(y) = \sum_{m=0}^{\infty} \int_K u_m(x, y)\, \mathrm{d}\mu(y),$$

where

$$u_m(x, y) = \sum_{|\alpha|_G = k} C_\alpha(x) \cdot y^\alpha.$$

The function u_m is δ_λ-homogeneous of degree m and the series

$$\sum_{m=0}^{\infty} u_m(x, \cdot)$$

is uniformly convergent on $B_d(0, d(K))$ to the $\mathcal{L}$-harmonic function $y \mapsto \Gamma(x^{-1} \circ y)$. Then, by Lemma 5.11.2, $u_m(x, \cdot)$ is $\mathcal{L}$-harmonic in $\mathbb{G}$, so that, by the assumption (5.88),

$$\int_K u_m(x, y)\, \mathrm{d}\mu(y) = 0 \qquad \forall\, m \geq 0.$$

Thus we have proved that $u(x) = 0$ for every $x \in \omega$, with $d(x)$ sufficiently large. Since ω is connected and u is analytic in ω, this implies $u \equiv 0$ in ω and completes the proof of the theorem. □

5.12 An Integral Representation Formula for the Fundamental Solution on Step-two Carnot Groups

The aim of this section is to state a remarkable result in the paper [BGG96] by R. Beals, B. Gaveau and P. Greiner. This result provides a somewhat explicit integral representation formula of the fundamental solution of the canonical sub-Laplacian on a *general Carnot group of step two*.

We shall see in Section 16.3 (page 637 in Part III) that, given a Carnot group $\mathbb{G}_1$ and an arbitrary sub-Laplacian $\mathcal{L}$ on $\mathbb{G}_1$, there exists a Carnot group $\mathbb{G}_2$ isomorphic to $\mathbb{G}_1$ such that $\mathcal{L}$ corresponds (via the related isomorphism in the relevant Lie algebras) to the canonical sub-Laplacian $\Delta_{\mathbb{G}_2}$ on $\mathbb{G}_2$. Moreover, if $\mathbb{G}_2$ (whence $\mathbb{G}_1$) has step two, we saw in Proposition 3.5.1 (page 168) that we can perform another Lie-group isomorphism sending $\mathbb{G}_2$ into the homogeneous Carnot group $\mathbb{G}_3$ such that the composition law on $\mathbb{G}_3$ is given by (we follow our usual notation)

$$(x,t)\circ(\xi,\tau)=\left(x+\xi,t_1+\tau_1+\frac{1}{2}\langle B^{(1)}x,\xi\rangle,\dots,t_n+\tau_n+\frac{1}{2}\langle B^{(n)}x,\xi\rangle\right) \quad (5.91)$$

and the matrices $B^{(i)}$'s are skew-symmetric. This last isomorphism also sends the canonical sub-Laplacian of $\mathbb{G}_2$ into the canonical sub-Laplacian of $\mathbb{G}_3$. Now, the cited result in [BGG96] furnishes an integral formula for the *canonical* sub-Laplacian on a homogeneous Carnot group of step two whose composition law $\circ$ has precisely the above form and the matrices $B^{(i)}$'s are *skew-symmetric*. Our argument above shows that (and how) we can obtain a representation formula for the fundamental solution of any (not necessarily canonical) sub-Laplacian on any homogeneous Carnot group of step two.

We now state the remarkable result in [BGG96]. We explicitly remark that in [BGG96] the formalism of complex Hamiltonian mechanics is followed: we slightly change the therein notation.

Theorem 5.12.1 (Beals–Gaveau–Greiner, [BGG96]). *Let $\mathbb{G}=\mathbb{R}^{m+n}$ (whose points are denoted by (x,t), $x\in\mathbb{R}^m$, $t\in\mathbb{R}^n$) be equipped with a homogeneous Carnot group structure by the dilation $\delta_\lambda(x,t)=(\lambda x,\lambda^2 t)$ and the composition law in* (5.91), *where the $B^{(k)}$'s are n skew-symmetric linearly independent matrices of order $m\times m$. Consider the canonical sub-Laplacian $\Delta_{\mathbb{G}}=\sum_{i=1}^n X_i^2$, where*

$$X_i=\partial/\partial x_i+\frac{1}{2}\sum_{k=1}^n\left(\sum_{l=1}^m b_{i,l}^{(k)}x_l\right)\partial/\partial t_k,\qquad i=1,\dots,m$$

(here $b_{i,l}^{(k)}$ denotes the entry of position (i,l) of $B^{(k)}$). Then, for every (x,t) with $x\neq 0$, the fundamental solution Γ of $\Delta_{\mathbb{G}}$ is given by

$$\Gamma(x,t)=c_Q\int_{\mathbb{R}^n}\frac{\sqrt{\det(\mathbf{V}(B(\tau)))}}{\left(\frac{1}{2}\langle\mathbf{W}(B(\tau))\cdot x,x\rangle-\iota\,\langle t,\tau\rangle\right)^{Q/2-1}}\,d\tau, \quad (5.92)$$

where ι is the imaginary unit of $\mathbb{C}$ and c_Q is the dimensional constant

$$c_Q=\frac{\mathbf{\Gamma}(\frac{Q}{2}-1)}{2\,(2\pi)^{Q/2}}. \quad (5.93)$$

Here we used the following notation: $Q=m+2n$ is the homogeneous dimension of $\mathbb{G}$, $\mathbf{\Gamma}$ in (5.93) *is Euler's Gamma function, $\tau=(\tau_1,\dots,\tau_n)\in\mathbb{R}^n$,*

$$B(\tau)=\frac{1}{2}(\tau_1\,B^{(1)}+\dots+\tau_n\,B^{(n)}),$$

V *and* **W** *are the real-analytic functions prolonging* $z/\sin(z)$ *and* $z/\tan(z)$, *respectively, at* $z = 0$, *i.e.*

$$\mathbf{V}(z) = \sum_{j=0}^{\infty} \frac{(-1)^j}{(2j+1)!} z^{2j}, \qquad \mathbf{W}(z) = \sum_{j=0}^{\infty} \frac{(-1)^j\, 2^{2j}\, B_{2j}}{(2j)!} z^{2j}$$

(here the B_{2j}*'s are the Bernoulli numbers). Moreover, for every* $t \in \mathbb{R}^n \setminus \{0\}$, *we have*

$$\Gamma(0,t) = \lim_{0 \neq x \to 0} \Gamma(x,t).$$

We remark that a general integral formula for Γ is provided in [BGG96] comprising the case $x = 0$ too, by shifting the contour $\mathbb{R}^n$ into the complex domain $\mathbb{C}^n$ (see [BGG96, Theorem 3, page 315]).

The δ_λ-homogeneity of Γ in (5.92) (of degree $2 - Q$) should be noted.

As we saw in Example 5.10.3 (page 283), formula (5.92) can, in some cases, give explicit fundamental solutions. This can be done by using the fact that, if $\lambda_1(\tau), \ldots, \lambda_m(\tau)$ and $v_1(\tau), \ldots, v_m(\tau)$ denote the eigenvalues and corresponding eigenvectors of the matrix $B(\tau)$ (over the complex field), normalized in such a way that $|v_j(\tau)| = 1$ for $j = 1, \ldots, m$, we have

$$\det\big(\mathbf{V}(B(\tau))\big) = \prod_{j=1}^{m} \frac{\lambda_j(\tau)}{\sin(\lambda_j(\tau))},$$

$$\big\langle \mathbf{W}(B(\tau)) \cdot x, x \big\rangle = \sum_{j=1}^{m} \frac{\lambda_j(\tau)}{\tan(\lambda_j(\tau))} \left|\langle x, v_j(\tau)\rangle\right|^2.$$

(In the last formula, the inner product is, of course, that of $\mathbb{C}^m$.)

5.13 Appendix A. Maximum Principles

In this section, we shall prove some *weak* and *strong* maximum principles for $\mathcal{L}$, an arbitrary sub-Laplacian on a homogeneous Carnot group $\mathbb{G}$.

To begin with, we prove some elementary lemmas.

Lemma 5.13.1. *Let* $\Omega \subset \mathbb{R}^N$ *be a bounded open set and let* $u : \Omega \to \mathbb{R}$ *be an arbitrary function. Then there exists a point* $x_0 \in \overline{\Omega}$ *such that*

$$\limsup_{x \to x_0} u(x) = \sup_{\Omega} u. \tag{5.94}$$

Proof. We argue by contradiction and assume that (5.94) is false. Then, for every $x \in \overline{\Omega}$, there exists an open neighborhood V_x of x such that

$$\sup_{\Omega \cap V_x} u < \sup_{\Omega} u. \tag{5.95}$$

The family $\{V_x : x \in \overline{\Omega}\}$ is an open covering of $\overline{\Omega}$, so that, since $\overline{\Omega}$ is compact, we have

$$\overline{\Omega} \subseteq \bigcup_{j=1}^{p} V_{x_j}, \quad p \in \mathbb{N},$$

for suitable $x_1, \dots, x_p \in \overline{\Omega}$. Then

$$\sup_{\Omega} u = \max\left\{ \sup_{\Omega \cap V_{x_j}} u \,:\, j = 1, \dots, p \right\}. \tag{5.96}$$

On the other hand, by (5.95), the right-hand side of (5.96) is strictly less than $\sup_{\Omega} u$. This contradiction proves the lemma. □

Lemma 5.13.2. *Let A and B be $N \times N$ symmetric matrices with constant real entries. Assume $A \geq 0$ and $B \leq 0$. Then* $\mathrm{trace}(A \cdot B) \leq 0$.

Proof. Let $R := A^{1/2}$ be a symmetric square root of A. Then $\mathrm{trace}(A \cdot B) = \mathrm{trace}(R \cdot R \cdot B) = \mathrm{trace}(R \cdot B \cdot R) = \mathrm{trace}(R^T \cdot B \cdot R) \leq 0$, since $B \leq 0$. □

Lemma 5.13.3. *Let $\mathcal{L}$ be a sub-Laplacian on the homogeneous Carnot group $\mathbb{G}$. Let $\Omega \subseteq \mathbb{G}$ be an arbitrary open set, and let $u : \Omega \to \mathbb{R}$ be a C^2 real function. Assume that u has a local maximum at $x_0 \in \Omega$. Then*

$$\mathcal{L}u(x_0) \leq 0. \tag{5.97}$$

Proof. We know that $\mathcal{L} = \mathrm{div}(A \cdot \nabla^T)$, where A is a $N \times N$ symmetric matrix with polynomial entries and $A(x) \geq 0$ at any point $x \in \mathbb{R}^N$. Then

$$\mathcal{L} = \mathrm{trace}(A \cdot D^2 u) + \langle b, \nabla u \rangle, \tag{5.98}$$

where $D^2 u = (\partial_{x_i x_j})_{i,j \leq N}$ is the Hessian matrix of u and b is the vector-valued function whose j-th component is given by

$$b_j = \sum_{i=1}^{N} \partial_{x_i} a_{i,j}. \tag{5.99}$$

Since u has a local maximum at x_0, we have $\nabla u(x_0) = 0$ and $D^2 u(x_0) \leq 0$. Then, by Lemma 5.13.2,

$$\mathcal{L}u(x_0) = \mathrm{trace}(A(x_0) \cdot D^2 u(x_0)) \leq 0.$$

This ends the proof. □

We are now able to give a simple proof of the following weak maximum principle.

Theorem 5.13.4 (Weak maximum principle). *Let $\mathcal{L}$ be a sub-Laplacian on the homogeneous Carnot group $\mathbb{G}$. Let Ω be a bounded open subset of $\mathbb{G}$. Let $u : \Omega \to \mathbb{R}$ be a C^2 function such that*

$$\begin{cases} \mathcal{L}u \ge 0 & \text{in } \Omega, \\ \limsup_{x\to y} u(x) \le 0 & \text{for every } y \in \partial\Omega. \end{cases} \tag{5.100}$$

Then $u \le 0$ in Ω.

Proof. We know that the matrix A in (5.98) has the following block form (see also (1.91), page 64)

$$A = \begin{pmatrix} A_{1,1} & A_{1,2} \\ A_{2,1} & A_{2,2} \end{pmatrix},$$

where $A_{1,1} = (a_{i,j})_{i,j\le m}$ is a constant $m \times m$ symmetric matrix strictly positive definite. Then $a_{1,1} > 0$. Let b_1 be given by (5.99) with $j = 1$. Define

$$\lambda := 2 \sup_{x\in\Omega} \frac{b_1(x)}{a_{1,1}}, \qquad M := \sup_{(x_1,\dots,x_N)\in\Omega} \exp(\lambda\, x_1),$$

and

$$h(x) = h(x_1, \dots, x_N) := M - \exp(\lambda\, x_1).$$

A trivial computation shows that

$$h(x) \ge 0 \quad \text{and} \quad \mathcal{L}h(x) < 0 \text{ for every } x \in \Omega. \tag{5.101}$$

For an arbitrary $\varepsilon > 0$, let us now consider the function $u_\varepsilon := u - \varepsilon\, h$. Due to inequalities (5.101) and condition (5.100), we have

$$\mathcal{L}u_\varepsilon > 0 \text{ in } \Omega \quad \text{and} \quad \limsup_{x\to y} u_\varepsilon(x) \le 0 \text{ for every } y \in \partial\Omega. \tag{5.102}$$

By Lemma 5.13.1, there exists a point $x_0 \in \overline{\Omega}$ such that

$$\limsup_{x\to x_0} u_\varepsilon(x) = \sup_\Omega u_\varepsilon. \tag{5.103}$$

We want to show that $x_0 \in \partial\Omega$. Arguing by contradiction, we assume $x_0 \in \Omega$. Then, by the continuity of u in Ω,

$$u_\varepsilon(x_0) = \limsup_{x\to x_0} u_\varepsilon(x),$$

so that, by (5.103), $u_\varepsilon(x_0) = \max_\Omega u_\varepsilon$. As a consequence, by Lemma 5.13.3, $\mathcal{L}u_\varepsilon(x_0) \le 0$. This contradicts the first inequality in (5.102). Thus $x_0 \in \partial\Omega$. Then, by (5.103) and the second condition in (5.102),

$$\sup_\Omega u_\varepsilon = \limsup_{x\to x_0} u_\varepsilon(x) \le 0.$$

Hence, $u - \varepsilon\, h = u_\varepsilon \le 0$ in Ω for every $\varepsilon > 0$. Letting ε tend to zero, we obtain $u \le 0$ in Ω. The theorem is thus completely proved. □

Note 5.13.5. The previous proof can be applied also to *continuous* functions satisfying the inequality $\mathcal{L}u \geq 0$ in the asymptotic sense of Exercises 8 and 9 at the end of the chapter (see also Ex. 10).

Corollary 5.13.6. *Let $\mathcal{L}$ be a sub-Laplacian on the homogeneous Carnot group $\mathbb{G}$. Let Ω be an unbounded open subset of $\mathbb{G}$. Let $u : \Omega \to \mathbb{R}$ be a C^2 function such that*

$$\begin{cases} \mathcal{L}u \geq 0 & \textit{in } \Omega, \\ \limsup_{x \to y} u(x) \leq 0 & \textit{for every } y \in \partial\Omega, \\ \limsup_{|x| \to \infty} u(x) \leq 0. \end{cases} \tag{5.104}$$

Then $u \leq 0$ in Ω.

Proof. Let $\varepsilon > 0$ be arbitrary but fixed. The third condition in (5.104) implies the existence of a real positive constant R such that

$$u(x) - \varepsilon < 0 \quad \text{in } \Omega \setminus \Omega_R, \tag{5.105}$$

where $\Omega_R := \{x \in \Omega : |x| < R\}$. It follows that

$$\begin{cases} \mathcal{L}(u - \varepsilon) = \mathcal{L}u \geq 0 & \text{in } \Omega_R, \\ \limsup_{x \to y} u(x) \leq 0 & \text{for every } y \in \partial\Omega_R. \end{cases}$$

Then, by Theorem 5.13.4, $u - \varepsilon \leq 0$ in Ω_R. This inequality, together with (5.105), gives $u \leq \varepsilon$ in Ω for every $\varepsilon > 0$. Hence $u \leq 0$ in Ω. □

A particular case of Corollary 5.13.6 is the following one.

Corollary 5.13.7. *If $\mathcal{L}$ is as in Corollary* 5.13.6, *the only entire $\mathcal{L}$-harmonic function vanishing at infinity is the null function.*

Proof. Let $u : \mathbb{G} \to \mathbb{R}$ be an entire $\mathcal{L}$-harmonic function vanishing at infinity, i.e. $u \in C^\infty(\mathbb{G}, \mathbb{R})$ satisfies

$$\begin{cases} \mathcal{L}u = 0 & \text{in } \mathbb{G}, \\ \lim_{|x| \to \infty} u(x) = 0. \end{cases}$$

Then, by applying Corollary 5.13.6 both to u and $-u$, we get $u \equiv 0$. □

The rest of this section is devoted to the proof of the following strong maximum principle.

Theorem 5.13.8 (Strong maximum principle). *Let $\mathcal{L}$ be a sub-Laplacian on the homogeneous Carnot group $\mathbb{G}$. Let Ω be a connected open subset of $\mathbb{G}$. Let $u : \Omega \to \mathbb{R}$ be a C^2 function such that*

$$u \leq 0 \quad \textit{and} \quad \mathcal{L}u \geq 0 \quad \textit{in } \Omega. \tag{5.106}$$

Suppose there exists a point $x_0 \in \Omega$ such that $u(x_0) = 0$. Then $u(x) = 0$ for every $x \in \Omega$.

The proof of this theorem requires several preliminary results. In what follows, we shall denote by $|\cdot|$ the standard Euclidean norm and by $D(z,r)$ the ball

$$D(z,r) := \{x \in \mathbb{R}^N \,:\, |x-z| < r\}.$$

Definition 5.13.9. *Let F be a relatively closed subset of Ω. We say that a vector $\nu \in \mathbb{R}^N \setminus \{0\}$ is* orthogonal to F at a point $y \in \Omega \cap \partial F$ *if*

$$\overline{D(y+\nu, |\nu|)} \subseteq (\Omega \setminus F) \cup \{y\}. \tag{5.107}$$

If this inclusion holds, we shall write $\nu \perp F$ at y. We also put

$$F^* := \big\{ y \in \Omega \cap \partial F \,|\, \textit{there exists } \nu\text{: } \nu \perp F \textit{ at } y \big\}.$$

With the above notation, we explicitly remark that $F^* \neq \emptyset$ if $F \subset \Omega$, $F \neq \Omega$. Indeed, since Ω is connected, $\Omega \cap \partial F$ is not empty. Take a point $z \in \Omega \cap \partial F$, a ball $D(z,R) \subseteq \Omega$ and a point $x_0 \in D(z, R/2)$. Let $y \in \Omega \cap \partial F$ be such that

$$r := |x_0 - y| = \operatorname{dist}(x_0, \partial F).$$

Then $y \in F^*$ and $\nu := \frac{r}{2}\,(x_0 - y) \perp F$ at y.

The following Hopf-type lemma will be crucial for the proof of the strong maximum principle in Theorem 5.13.8.

Lemma 5.13.10 (A Hopf-type lemma for sub-Laplacians). *Let $\mathcal{L}$ be a sub-Laplacian on the homogeneous Carnot group $\mathbb{G}$. Let $\Omega \subseteq \mathbb{G}$ be open, and let $u : \Omega \to \mathbb{R}$ be a C^2 function satisfying the inequalities in* (5.106). *Let*

$$F := \{x \in \Omega : u(x) = 0\}. \tag{5.108}$$

Assume $\emptyset \neq F \neq \Omega$. Then, for every $y \in F^*$ and $\nu \perp F$ at y, we have

$$q_{\mathcal{L}}(y, \nu) = 0, \tag{5.109}$$

where $q_{\mathcal{L}}(x,\xi) := \langle A(x)\cdot \xi, \xi\rangle$ is the characteristic form of $\mathcal{L}$ defined in (5.1a).

Proof. Let $y \in F^*$ and $\nu \perp F$ at y. Then

$$\overline{D(y+\nu, |\nu|)} \subseteq (\Omega \setminus F) \cup \{y\}.$$

Since $F^* \subseteq F \cap \Omega$, y is a maximum point for u (see (5.108)). Then $\nabla u(y) = 0$. We now argue by contradiction assuming that (5.109) is false. Hence

$$q_{\mathcal{L}}(y, \nu) > 0.$$

Let us now consider the function

$$h(x) := \exp(-\lambda\,|x-z|^2) - \exp(-\lambda\, r^2),$$

where $z = y + \nu$ and $r = |\nu|$. The positive constant λ will be fixed later on. A direct and easy computation shows that

$$\mathcal{L}h(y) = 4\lambda^2 \exp(-\lambda r^2) \cdot \big(q_{\mathcal{L}}(y, \nu) + \mathcal{O}(1/\lambda)\big),$$

as $\lambda \to \infty$. Then, we can choose and fix $\lambda > 0$ in such a way that $\mathcal{L}h > 0$ in a suitable neighborhood V of y. Obviously, we may assume $\overline{V} \subset \Omega$. Let us now consider the bounded open set

$$U := V \cap D(z, r).$$

Note that $\partial U = \Gamma_1 \cup \Gamma_2$, where

$$\Gamma_1 = V \cap \partial D(z, r) \text{ and } \Gamma_2 = \overline{D(z, r)} \cap \partial V.$$

Since Γ_2 is a compact subset of $\Omega \setminus F$ and $u < 0$ in $\Omega \setminus F$, there exists $\varepsilon > 0$ such that $u + \varepsilon h < 0$ in Γ_2. On the other hand, being $h = 0$ on $\partial D(z, r)$ and $u \leq 0$ in Ω, we have $u + \varepsilon h \leq 0$ on Γ_1. Then, since $\mathcal{L}(u + \varepsilon h) \geq \varepsilon \mathcal{L}h \geq 0$ in U, from the maximum principle of Theorem 5.13.4, we obtain $u + \varepsilon h \leq 0$ in U. Since $u(y) = h(y) = 0$, this implies

$$\frac{u(y + t\nu) - u(y)}{t} \leq -\varepsilon \frac{h(y + t\nu) - h(y)}{t} \quad \text{for } 0 < t < 1. \tag{5.110}$$

Letting t tend to zero in this inequality, we get

$$\langle \nabla u(y), \nu \rangle \leq -\varepsilon \langle \nabla h(y), \nu \rangle = -2\varepsilon \exp(-\lambda r^2) r^2.$$

This contradicts the condition $\nabla u(y) = 0$ and completes the proof. □

Corollary 5.13.11. *Let the hypotheses and notation of the previous lemma hold. Let also* $\mathcal{L} = \sum_{j=1}^m X_j^2$. *Then we have*

$$\langle X_j I(y), \nu \rangle = 0 \quad \forall\, y \in F^* \ \forall\, \nu \perp F \text{ at } y$$

and for every $j = 1, \ldots, m$.

Proof. It follows from the previous lemma, by just noticing that

$$q_{\mathcal{L}}(x, \xi) = \sum_{j=1}^m \langle X_j I(y), \xi \rangle^2. \quad \square$$

Another crucial definition is the following one.

Definition 5.13.12 ((Positively) invariant set w.r.t. a vector field). *Let* $X \in T(\mathbb{R}^N)$ *be a smooth vector field in* $\mathbb{R}^N$, *and let* F *be a relatively closed subset of* Ω. *We say that* F *is* positively X-invariant *if, for any integral curve* γ *of* X, $\gamma : [0, T] \to \Omega$ *such that* $\gamma(0) \in F$, *we have* $\gamma(t) \in F$ *for every* $t \in [0, T]$. *We say that* F *is* X-invariant *if it is positively* X*-invariant with respect to both* X *and* $-X$.

It is easy to verify that the condition

$$\langle XI(y), \nu\rangle \le 0 \quad \forall\, y \in F^* \;\forall\, \nu \perp F \text{ at } y \tag{5.111}$$

is *necessary* for the positive X-invariance of F. Indeed, let $y \in F^*$, $\nu \perp F$ at y and $\gamma : [0, T] \to \Omega$ be an integral curve of X such that $\gamma(0) = y$. Let F be positively X-invariant. Since

$$\overline{D(y+\nu, |\nu|)} \subseteq (\Omega \setminus F) \cup \{y\},$$

we have

$$|\gamma(t) - (y+\nu)|^2 \ge |\nu|^2 \;\text{ and }\; |\gamma(0) - (y+\nu)|^2 = |\nu|^2$$

for every $t \in [0, T]$. This means that the real function $t \mapsto |\gamma(t) - (y+\nu)|^2$ has a minimum at $t = 0$. As a consequence,

$$0 \le \frac{\mathrm{d}}{\mathrm{d}t}\Big|_{t=0} |\gamma(t) - (y+\nu)|^2 = \langle \dot\gamma(0), \gamma(0) - (y+\nu)\rangle = \langle XI(y), -\nu\rangle.$$

Hence (5.111) holds. We will show that this condition is also *sufficient* for F to be positively X-invariant. To prove this statement, we need the following elementary lemma.

Lemma 5.13.13. *Let $g : [0, T] \to \mathbb{R}$ be a continuous function such that*

$$\limsup_{h\to 0^-} \frac{g(t+h) - g(t)}{h} \le M \quad \forall\, t \in (0, T], \tag{5.112}$$

for a suitable $M \in \mathbb{R}$. Then

$$g(t) \le g(0) + M\,t \quad \forall\, t \in [0, T].$$

Proof. Let $\varepsilon > 0$ be fixed. Condition (5.112) implies that the real function

$$t \mapsto g(t) - g(0) - (M+\varepsilon)t$$

has a maximum at $t = 0$. Indeed, suppose to the contrary that there exist $\varepsilon_0 > 0$ and $t_0 \in (0, T]$ such that

$$g(t) - g(0) - (M+\varepsilon)\,t \le g(t_0) - g(0) - (M+\varepsilon_0)\,t_0 \quad \forall\, t \in [0, T].$$

In particular, for $t = t_0 + h$ and $h < 0$ small enough, this gives

$$\frac{g(t_0+h) - g(t_0)}{h} \ge (M+\varepsilon_0),$$

which contradicts the hypothesis. Then, $g(t) - g(0) - (M+\varepsilon)t \le 0$ for every $t \in [0, T]$. Letting ε tend to zero, we obtain the assertion. □

Proposition 5.13.14 (Nagumo–Bony). *Let $X \in T(\mathbb{R}^N)$ be a smooth vector field in $\mathbb{R}^N$, and let F be a relatively closed subset of Ω. Then F is positively X-invariant if and only if*

$$\langle XI(y), \nu\rangle \le 0 \quad \forall\, y \in F^* \;\forall\, \nu \perp F \text{ at } y. \tag{5.113}$$

Proof. We only need to show the "if" part. Let $\gamma : [0, T] \to \Omega$ be an integral curve of X such that $x_0 := \gamma(0) \in F$. Define

$$\delta(t) := \text{dist}(\gamma(t), F), \quad 0 \le t \le T.$$

We have to prove that $\delta(t) = 0$ for every $t \in [0, T]$.

Let V be a bounded neighborhood of x_0 containing $\gamma([0, T])$, and let

$$L := \sup_{x,z \in V,\ x \ne z} \frac{|XI(x) - XI(z)|}{|x - z|} \tag{5.114}$$

be the Lipschitz constant of X on V. We may suppose $L\,T < 1/2$ and $V = D(x_0, r)$ with $D(x_0, 2r) \subseteq \Omega$. We claim that

$$L(t) := \limsup_{h \to 0^-} \frac{\delta(t+h) - \delta(t)}{h} \le L\,\delta(t) \quad \forall t \in (0, T]. \tag{5.115}$$

If $\delta(t) = 0$, inequality (5.115) is trivial, since $h < 0$ and $\delta(t + h) \ge 0$. Suppose $\delta(t) > 0$ and choose a sequence $h_n \uparrow 0$ such that

$$L(t) = \lim_{n \to \infty} \frac{\delta(t + h_n) - \delta(t)}{h_n}.$$

Let us now denote $x := \gamma(t)$ and $x_n := \gamma(t + h_n)$. Since $\gamma([0, T]) \subset D(x_0, r)$ and $D(x_0, 2r) \subseteq \Omega$, for every $n \in \mathbb{N}$ there exists a point $z_n \in F \cap \Omega$ such that

$$|x_n - z_n| = \text{dist}(x_n, F) = \delta(t + h_n).$$

Obviously, we may suppose that $z_n \to z \in F \cap \overline{D(x_0, r)}$, so that, since $x_n \to x$,

$$\begin{aligned} |x - z| &= \lim_{n \to \infty} |x_n - z| = \lim_{n \to \infty} \text{dist}(x_n, F) \\ &= \text{dist}(x, F) = \delta(t). \end{aligned}$$

Moreover,

$$\nu := \frac{1}{2}(x - z) \perp F \text{ at } z. \tag{5.116}$$

Then

$$\begin{aligned} \delta(t + h_n) - \delta(t) &= |x_n - z_n| - |x - z| \ge |x_n - z_n| - |x - z_n| \\ &\ge |x_n - x| \ge -\frac{\langle x_n - x, z_n - x \rangle}{|x - z_n|}. \end{aligned}$$

Hence

$$\begin{aligned} L(t) &\le \lim_{n \to \infty} \left\langle \frac{x - z_n}{|x - z_n|}, \frac{x_n - x}{h_n} \right\rangle = \left\langle \dot{\gamma}(t), \frac{x - z}{|x - z|} \right\rangle \\ &= \frac{2}{|x - z|} \langle XI(x), \nu \rangle \\ &= \frac{2}{|x - z|} \big(\langle XI(z) - XI(x), \nu \rangle + \langle XI(z), \nu \rangle \big). \end{aligned}$$

From (5.116) and (5.113), together with (5.114), we finally get

$$L(t) \leq L\,|x - z| = L\,\delta(t).$$

This completes the proof of (5.115). This inequality, combined with Lemma 5.13.13, gives

$$\delta(t) \leq \delta(t) - \delta(0) \leq L\,T \sup_{[0,T]} \delta,$$

so that $\sup_{[0,T]} \delta \leq 1/2 \cdot \sup_{[0,T]} \delta$. Hence $\delta \equiv 0$, and the proof is complete. □

Corollary 5.13.15. *The closed set F is X-invariant if and only if*

$$\langle XI(y), \nu \rangle = 0, \quad \forall\, y \in F^* \quad \forall\, \nu \perp F \text{ at } y.$$

Proof. It straightforwardly follows from Proposition 5.13.14 and Definition 5.13.12. □

This corollary, together with Corollary 5.13.11, immediately gives the following result.

Corollary 5.13.16. *Let $\mathcal{L} = \sum_{j=1}^m X_j^2$ be a sub-Laplacian on the homogeneous Carnot group $\mathbb{G}$. Let $u : \Omega \to \mathbb{R}$ be a C^2 function satisfying the inequalities* (5.106). *Let $F := \{x \in \Omega : u(x) = 0\}$. Assume $\emptyset \neq F \neq \Omega$. Then F is invariant with respect to $X_1, \ldots, X_m$.*

Proposition 5.13.17. *Let F be a relatively closed subset of the open set $\Omega \subseteq \mathbb{R}^N$. Assume $\emptyset \neq F \neq \Omega$. Then*

$$\mathfrak{a} := \{X \in T(\mathbb{R}^N) : F \text{ is } X\text{-invariant}\}$$

is a Lie algebra of vector fields.

Proof. Let $X, Y \in \mathfrak{a}$, and let λ, μ be real constants. By Corollary 5.13.15, we have $\langle XI(y), \nu \rangle = \langle YI(y), \nu \rangle = 0$ for every $y \in F^*$ and for every $\nu \perp F$ at y. Then

$$\langle \lambda\, XI(y) + \mu\, YI(y), \nu \rangle = 0 \quad \forall\, y \in F^*\, \forall\, \nu \perp F \text{ at } y.$$

Hence $\mathfrak{a}$ is a linear space. The next lemma will complete the proof of the proposition. □

Lemma 5.13.18. *In the notation of Proposition* 5.13.17, *if $X, Y \in \mathfrak{a}$, then $[X, Y] \in \mathfrak{a}$.*

Proof. Let $y \in F^*$ and $\nu \perp F$ at y. For every $t > 0$ define

$$\Gamma(t) := \big(\exp(-\sqrt{t}Y) \circ \exp(-\sqrt{t}X) \circ \exp(\sqrt{t}Y) \circ \exp(-\sqrt{t}X)\big)(y).$$

Here $\circ$ denotes the composition of maps, whereas "exp" denotes the exponential of a vector field as introduced in Definition 1.1.2, page 8.

Let $T > 0$ be such that $\Gamma(t) \in \Omega$ for $0 \le t \le T$. By using the Taylor expansion (1.7) of exp (on page 7) with $n = 2$, a direct computation gives

$$\begin{aligned}\Gamma(t) &= y + t\big(\mathcal{J}_{YI}(y) \cdot XI(y) - \mathcal{J}_{XI}(y) \cdot YI(y)\big) + o(t) \\ &= y + t[X, Y]I(y) + o(t), \quad \text{as } t \downarrow 0.\end{aligned}$$

Then

$$\lim_{t \downarrow 0} \frac{\Gamma(t) - y}{t} = [X, Y]I(y). \tag{5.117}$$

On the other hand, since F is X and Y invariant, $\Gamma(t) \in F$ for every $t \in [0, T]$. As a consequence, since $\overline{D(y + \nu, |\nu|)} \subseteq (\Omega \setminus F) \cup \{y\}$ and $\Gamma(0) = y$,

$$|\Gamma(t) - (y + \nu)|^2 \ge |\nu|^2 = |\Gamma(0) - (y + \nu)|^2.$$

Then, by using also (5.117), we have

$$0 \ge \frac{\mathrm{d}}{\mathrm{d}t}\bigg|_{t=0} |\Gamma(t) - (y + \nu)|^2 = 2\langle [X, Y]I(y), \nu\rangle.$$

Hence

$$\langle [X, Y]I(y), \nu\rangle \le 0 \quad \forall\, y \in F^* \ \forall\, \nu \bot F \text{ at } y.$$

By swapping X with Y, we also get $\langle [Y, X]I(y), \nu\rangle \le 0$, so that

$$\langle [X, Y]I(y), \nu\rangle = 0 \quad \forall\, y \in F^* \ \forall\, \nu \bot F \text{ at } y.$$

Then, by Corollary 5.13.15, $[X, Y] \in \mathfrak{a}$. □

Finally, we are in the position to give the proof of Theorem 5.13.8.

Proof (of Theorem 5.13.8). Let us define

$$F := \{x \in \Omega \, : \, u(x) = 0\}.$$

By hypothesis, $x_0 \in F$. Then F is a non-empty relatively closed subset of Ω. We have to prove that $F = \Omega$. By contradiction, assume $F \ne \Omega$. Then, since Ω is connected, $F^* \ne \emptyset$. Let $y \in F^*$, and let $\nu \bot F$ at y. By Proposition 5.13.17,

$$\langle ZI(y), \nu\rangle = 0 \quad \forall\, Z \in \mathfrak{g}.$$

Since $\dim(\mathfrak{g}) = N$, this obviously implies $\nu = 0$. On the other hand, by the very definition of a vector orthogonal to F, we have $\nu \in \mathbb{R}^N \setminus \{0\}$. This contradiction completes the proof. □

5.13.1 A Decomposition Theorem for $\mathcal{L}$-harmonic Functions

In this section, we give a decomposition theorem for $\mathcal{L}$-harmonic functions, resemblant to the decomposition of a holomorphic function on an annulus of $\mathbb{C}$ into the sum of the regular and singular parts from its Laurent expansion (for the classical case of the Laplace operator, see also S. Axler, P. Bourdon, W. Ramey [ABR92]). Our main tool is the maximum principle from the previous section (precisely, we use Corollary 5.13.7, page 296).

In the sequel, we assume $Q \geq 3$. Moreover, in the proof of Theorem 5.13.20, we adopt the following notation: $\mathbb{G} = (\mathbb{R}^N, \circ, \delta_\lambda)$ is a homogeneous Carnot group, $\mathcal{L}$ is a sub-Laplacian on $\mathbb{G}$, $\Gamma = d^{2-Q}$ is the fundamental solution for $\mathcal{L}$ (see Proposition 5.4.2). If d is the above $\mathcal{L}$-gauge, A is any subset of $\mathbb{G}$ and $\lambda > 0$, we agree to set

$$A_\lambda := \{x \in \mathbb{G} \mid d\text{-dist}(x, A) < \lambda\},$$

where

$$d\text{-dist}(x, A) := \inf\{d(x^{-1} \circ a) \mid a \in A\}.$$

Moreover, we use the following simple lemma.

Lemma 5.13.19. *Let K be a compact subset of $\mathbb{G}$, and let f be bounded on K. Then the function*

$$F : \mathbb{G} \to \mathbb{R}, \quad F(x) := \int_K \Gamma(y^{-1} \circ x)\, f(y)\, \mathrm{d}y$$

is $\mathcal{L}$-harmonic on $\mathbb{G} \setminus K$ and vanishes at infinity. Moreover, if μ is a Radon measure on $\mathbb{R}^N$ with compact support K, the same is true for $G(x) := \int_{\mathbb{R}^N} \Gamma(y^{-1} \circ x)\, \mathrm{d}\mu(y)$.

Proof. It easily follows by differentiation under the integral sign (recall also that Γ is locally integrable and vanishes at infinity). □

We are now ready to state and prove the following assertion.

Theorem 5.13.20 (The decomposition theorem). *Let the hypotheses in the incipit of this section hold. Let $K \subset \Omega \subseteq \mathbb{G}$, with K compact and Ω open. If u is $\mathcal{L}$-harmonic in $\Omega \setminus K$, then u has a decomposition of the form*

$$u = r + s,$$

where r is $\mathcal{L}$-harmonic in Ω and s is $\mathcal{L}$-harmonic in $\mathbb{G} \setminus K$. Furthermore, it can be assumed that s vanishes at infinity, and in this case the above decomposition is unique.

Proof. Suppose the theorem holds true whenever Ω is bounded. We show that it holds true even for an unbounded Ω. Indeed, let $u \in \mathcal{H}(\Omega \setminus K)$, where K is compact and Ω is an (unbounded) open set. Let $R > 0$ be such that $K \subset B_d(0, R)$. Set $\widetilde{\Omega} := \Omega \cap B_d(0, R)$. Then $K \subset \widetilde{\Omega}$ and, by our assumption, u can be decomposed as $u = \widetilde{r} + s$, where $\widetilde{r} \in \mathcal{H}(\widetilde{\Omega})$, $s \in \mathcal{H}(\mathbb{G} \setminus K)$ and $s \to 0$ at infinity. We consider the function $r := u - s$ on Ω. Then r is $\mathcal{L}$-harmonic in $\Omega \setminus K$ and extends $\mathcal{L}$-harmonically across K, since r coincides with $\widetilde{r}$ on $\widetilde{\Omega}$ (which is an open neighborhood of K). This ends the proof, since $u = r+s$, with $r \in \mathcal{H}(\Omega)$, $s \in \mathcal{H}(\mathbb{G}\setminus K)$ and $s \to 0$ at infinity.

We can then suppose that Ω is bounded. We fix the following notation (see also Fig. 5.1):

$$B(\lambda) := \text{closure of } (\partial\Omega)_\lambda, \quad C(\lambda) := \text{closure of } K_\lambda, \quad A(\lambda) := \Omega \setminus (B(\lambda) \cup C(\lambda)).$$

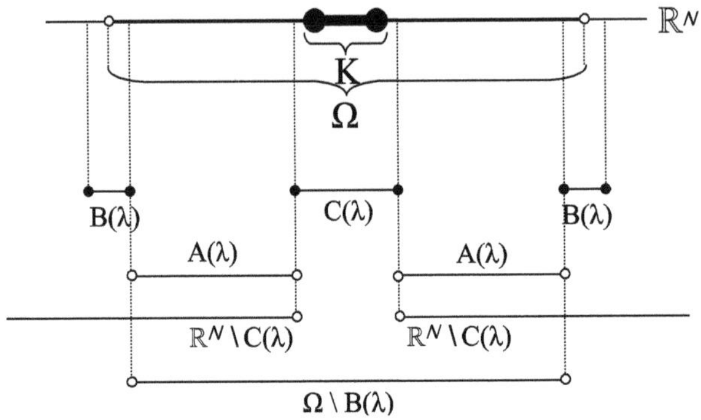

Fig. 5.1. Figure for the proof of Theorem 5.13.20

Since K is compact and $\partial\Omega$ is closed, we can choose $\lambda > 0$ small enough so that

$$B(\lambda) \cap C(\lambda) = \emptyset.$$

Since Ω is bounded, we can choose a cut-off function $\psi_\lambda \in C_0^\infty(\mathbb{R}^N)$ such that

$$\operatorname{supp}(\psi_\lambda) \subset \Omega \setminus K, \qquad \psi_\lambda \equiv 1 \text{ on } A(\lambda).$$

We consider the function $u\,\psi_\lambda$, and we agree to consider this function trivially prolonged on $\mathbb{R}^N$ to be zero. Hence this trivial prolongation belongs to $C_0^\infty(\mathbb{R}^N)$. By (5.16) in Theorem 5.3.3 (page 237) (being $\psi_\lambda \equiv 1$ on $A(\lambda)$),

$$\begin{aligned} u(x) = (u\,\psi_\lambda)(x) &= -\int_{\mathbb{R}^N} \Gamma(x^{-1} \circ y)\, \mathcal{L}(u\,\psi_\lambda)(y)\, dy \\ &= -\left(\int_{A(\lambda)} + \int_{B(\lambda)} + \int_{C(\lambda)} + \int_{\mathbb{R}^N \setminus (A(\lambda) \cup B(\lambda) \cup C(\lambda))} \right) \\ &= -\left(\int_{B(\lambda)} + \int_{C(\lambda)} \right) \Gamma(y^{-1} \circ x)\, \mathcal{L}(u\,\psi_\lambda)(y)\, dy \quad \forall\, x \in A(\lambda). \end{aligned} \tag{5.118}$$

In the last equality we used the fact that $\psi_\lambda \equiv 1$ on $A(\lambda)$ jointly with $\mathcal{L}u = 0$ on Ω, and the fact that ψ_λ is supported in Ω. We now set

$$r_\lambda(x) := -\int_{B(\lambda)} \Gamma(y^{-1} \circ x)\, \mathcal{L}(u\,\psi_\lambda)(y)\, dy \qquad \text{for } x \in \Omega \setminus B(\lambda),$$

$$s_\lambda(x) := -\int_{C(\lambda)} \Gamma(y^{-1} \circ x)\, \mathcal{L}(u\,\psi_\lambda)(y)\, dy \qquad \text{for } x \in \mathbb{G} \setminus C(\lambda).$$

Hence, (5.118) gives the decomposition

$$u(x) = r_\lambda(x) + s_\lambda(x) \qquad \forall\, x \in A(\lambda). \tag{5.119}$$

From Lemma 5.13.19, we infer that r_λ and s_λ are $\mathcal{L}$-harmonic on the respective sets of definition and that s_λ vanishes at infinity.

Let now $0 < \mu < \lambda$. Then obviously $A(\lambda) \subset A(\mu)$. From the decomposition in (5.119), we get an analogous decomposition

$$u(x) = r_\mu(x) + s_\mu(x) \qquad \forall\, x \in A(\mu). \tag{5.120}$$

We claim that the decompositions (5.119) and (5.120) are compatible, i.e.

$$\begin{aligned} r_\lambda(x) &= r_\mu(x) \qquad \forall\, x \in \Omega \setminus B(\lambda), \\ s_\lambda(x) &= s_\mu(x) \qquad \forall\, x \in \mathbb{G} \setminus C(\lambda). \end{aligned} \tag{5.121}$$

To prove the claim, we first remark that from (5.119) and (5.120) we obtain

$$r_\lambda(x) - r_\mu(x) = s_\mu(x) - s_\lambda(x) \qquad \forall\, x \in A(\lambda). \tag{5.122}$$

Now, let us consider the following function

$$S : \mathbb{G} \to \mathbb{R}, \qquad S(x) := \begin{cases} s_\mu(x) - s_\lambda(x) & \text{for every } x \in \mathbb{G} \setminus C(\lambda), \\ r_\lambda(x) - r_\mu(x) & \text{for every } x \in \Omega \setminus B(\lambda). \end{cases}$$

We claim that S has the following properties:

i) S is well-posed: indeed, thanks to (5.122), $s_\mu - s_\lambda$ coincides with $r_\lambda - r_\mu$ on the set
$$\{\mathbb{G} \setminus C(\lambda)\} \cap \{\Omega \setminus B(\lambda)\} = A(\lambda);$$
ii) S vanishes at infinity: indeed, this is true for s_μ and s_λ;
iii) S is $\mathcal{L}$-harmonic on $\mathbb{G}$: indeed, this is true for $s_\mu - s_\lambda$ on the open set $\mathbb{G} \setminus C(\lambda)$, and this is true for $r_\lambda - r_\mu$ on the open set $\Omega \setminus B(\lambda)$ (recall that $C(\mu) \subset C(\lambda)$ and $B(\mu) \subset B(\lambda)$).

Now, by the maximum principle (precisely, see Corollary 5.13.7, page 296) we infer that $S \equiv 0$, which is equivalent to the claimed (5.121). Now, let us fix a decreasing sequence of positive λ_n's such that $\lambda_n \to 0$ as $n \to \infty$. We set

$$\begin{aligned} r(x) &:= r_{\lambda_n}(x) \quad \forall\, x \in \Omega \quad (\text{where } n \in \mathbb{N} \text{ is such that } x \in \Omega \setminus B(\lambda_n)), \\ s(x) &:= s_{\lambda_n}(x) \quad \forall\, x \in \mathbb{G} \setminus K \quad (\text{where } n \in \mathbb{N} \text{ is such that } x \in \mathbb{G} \setminus C(\lambda_n)). \end{aligned}$$

Thanks to (5.121), the definition of $r(x)$ and $s(x)$ are unambiguous.[11] Now, the decomposition

$$u(x) = r(x) + s(x) \qquad \forall\, x \in \Omega \setminus K$$

[11] We are also using the trivial fact that $\Omega \setminus B(\lambda_n) \uparrow \Omega$ and $\mathbb{R}^N \setminus C(\lambda_n) \downarrow \mathbb{R}^N \setminus K$ as $n \to \infty$.

follows from the analogous decomposition (5.119) (and the fact that $A(\lambda) \uparrow \Omega \setminus K$ as $n \to \infty$). Moreover, it is easily seen that $r \in \mathcal{H}(\Omega)$, $s \in \mathcal{H}(\mathbb{G} \setminus K)$ and s vanishes at infinity. This gives the desired decomposition of u as in the assertion.

Finally, the uniqueness of the decomposition in the assertion (under the assumption that s vanishes at infinity) is another easy consequence of the same maximum principle quoted above. Indeed, suppose we are given two decompositions

$$r_2 + s_2 = u = r_1 + s_1 \qquad \text{on } \Omega \setminus K,$$

where

$$r_i \in \mathcal{H}(\Omega), \quad s_i \in \mathcal{H}(\mathbb{G} \setminus K), \quad \lim_{x \to \infty} s_i(x) = 0, \qquad i = 1, 2.$$

Then setting

$$S : \mathbb{G} \to \mathbb{R}, \qquad S(x) := \begin{cases} s_1(x) - s_2(x) & \text{for every } x \in \mathbb{G} \setminus K, \\ r_2(x) - r_1(x) & \text{for every } x \in \Omega, \end{cases}$$

we see that $S \in \mathcal{H}(\mathbb{G})$, S vanishes at infinity, and we end the proof following the same arguments as in the previous paragraph. □

5.14 Appendix B. The Improved Pseudo-triangle Inequality

Let $\mathbb{G} = (\mathbb{R}^N, \delta_\lambda, \circ)$ be a homogeneous Carnot group, and let d be a symmetric homogeneous norm on $\mathbb{G}$, *smooth* out of the origin.[12] For example, d could be an $\mathcal{L}$-gauge on $\mathbb{G}$ for some sub-Laplacian $\mathcal{L}$ on $\mathbb{G}$.

We know from Proposition 5.1.7 (page 231) that (even without assumptions on smoothness or symmetry of d) d satisfies the pseudo-triangle inequality

$$d(a \circ b) \le \mathbf{c}\,(d(a) + d(b)) \quad \text{for every } a,\ b \in \mathbb{G}.$$

Here $\mathbf{c} \ge 1$ is a constant depending on d and $\mathbb{G}$.

The aim of this brief appendix is to prove the following improvement of the pseudo-triangle inequality. For the following result, see also [DFGL05, (2.6)].

Proposition 5.14.1 (The improved pseudo-triangle inequality). *Let d be a symmetric homogeneous norm on the homogeneous Carnot group $\mathbb{G}$. Furthermore, suppose d is smooth out of the origin. Then there exists a constant $\beta \ge 1$ depending only on d and $\mathbb{G}$ such that*

$$d(y \circ x) \le \beta\, d(y) + d(x) \qquad \textit{for every } x,\ y \in \mathbb{G}. \tag{5.123}$$

[12] In other words (see the definition at the beginning of Section 5.1, page 229), the present d has the following properties:
$d \in C(\mathbb{G}, [0, \infty)) \cap C^\infty(\mathbb{G} \setminus \{0\})$; $d(\delta_\lambda(x)) = \lambda\, d(x)$ for every $\lambda > 0$ and $x \in \mathbb{G}$; $d(x) = 0$ iff $x = 0$; $d(x^{-1}) = d(x)$ for every $x \in \mathbb{G}$.

Proof. Since (5.123) holds when $y = 0$, we can suppose $y \neq 0$. Since d is δ_λ-homogeneous of degree one, (5.123) is equivalent to

$$d(\delta_{1/d(y)}(y) \circ \delta_{1/d(y)}(x)) \leq \beta + d(\delta_{1/d(y)}(x)). \tag{5.124a}$$

By using the symmetry of d and setting

$$\xi^{-1} = \delta_{1/d(y)}(y), \qquad \eta^{-1} = \delta_{1/d(y)}(x),$$

(5.124a) is equivalent to (note that $d(\delta_{1/d(y)}(y)) = 1$)

$$d(\eta \circ \xi) - d(\eta) \leq \beta \qquad \text{for every } \xi,\ \eta \in \mathbb{G}\text{: } d(\xi) = 1. \tag{5.124b}$$

By the usual[13] pseudo-triangle inequality for d, (5.124b) holds when $\eta \in B_d(0, M)$, i.e. $d(\eta) \leq M$ (where $M = M(d, \mathbb{G}) \gg 1$ will be chosen in the sequel). Indeed, if $\eta \in B_d(0, M)$, we have

$$d(\eta \circ \xi) - d(\eta) \leq \mathbf{c}\,(d(\eta) + d(\xi)) - d(\eta) = (\mathbf{c} - 1)d(\eta) + \mathbf{c} \leq (\mathbf{c} - 1)M + \mathbf{c} =: \beta.$$

We can hence suppose $\eta \notin B_d(0, M)$. Roughly speaking, we will show that we can drop η from (5.124b), by an argument of left-translation along curves which are supported away from zero, when M is large enough. Then, (5.124b) will follow from the classical mean value theorem.

We now make this precise. Set $Z := \operatorname{Log}(\xi) \in \mathfrak{g}$, where Log is the logarithmic map related to $\mathbb{G}$ and $\mathfrak{g}$ is the Lie algebra of $\mathbb{G}$. Consider the integral curve γ of Z starting from η, i.e. with our usual notation

$$\gamma(t) = \exp(tZ)(\eta) = \eta \circ \exp(tZ)(0) = \eta \circ \operatorname{Exp}(tZ).$$

Here we used Corollary 1.2.24 (page 24) and the definition of exponential map (see page 24). Obviously, we have $\gamma(0) = \eta$ and $\gamma(1) = \eta \circ \operatorname{Exp}(Z) = \eta \circ \operatorname{Exp}(\operatorname{Log}(\xi)) = \eta \circ \xi$. If we show that we can choose $M \gg 1$ such that

$$d(\gamma(t)) \geq 1 \qquad \text{for every } t \in [0, 1] \tag{5.124c}$$

(recall that γ depends on η, besides ξ), then $[0, 1] \ni t \mapsto u(t) := d(\gamma(t))$ is smooth (for d is smooth out of the origin by hypothesis) so that the classical Lagrange mean value theorem applies and gives

$$\begin{aligned} d(\eta \circ \xi) - d(\eta) = u(1) - u(0) &\leq \sup_{t\in[0,1]} |\dot u(t)| = \sup_{t\in[0,1]} \left|\langle (\nabla d)(\gamma(t)), \dot\gamma(t)\rangle\right| \\ &= \sup_{t\in[0,1]} \left|\langle (\nabla d)(\gamma(t)), (ZI)(\gamma(t))\rangle\right| \\ &= \sup_{t\in[0,1]} \left|(Zd)(\gamma(t))\right|. \end{aligned} \tag{5.124d}$$

Let $X_1, \ldots, X_N$ be the Jacobian basis for the Lie algebra $\mathfrak{g}$ of $\mathbb{G}$. Hence,

[13] See Proposition 5.1.7-1, page 231.

$$Z = \mathrm{Log}\,(\xi) = \sum_{j=1}^{N} p_j(\xi)\, Z_j, \tag{5.124e}$$

where the p_j's are polynomials, so that there exists a constant C such that

$$\sup_{d(\xi)=1} |p_j(\xi)| \le C_1 \quad \text{for all } j = 1, \dots, N, \tag{5.124f}$$

since $\{\xi \in \mathbb{G} : d(\xi) = 1\}$ is a compact set (see, e.g. Proposition 5.1.4, page 230). Moreover, Z_j is δ_λ-homogeneous of degree $\sigma_j \ge 1$ (see Corollary 1.3.19, page 42), so that $Z_j d$ is δ_λ-homogeneous of degree $1 - \sigma_j \le 0$. Consequently, it is bounded on $\mathbb{G} \setminus B_d(0, 1)$, say

$$\sup_{d(\zeta)\ge 1} |(Z_j d)(\zeta)| \le C_2 \quad \text{for all } j = 1, \dots, N. \tag{5.124g}$$

We now use again the claimed (5.124c) and, by collecting together (5.124f) to (5.124g), we derive that (5.124d) yields

$$\begin{aligned} d(\eta \circ \xi) - d(\eta) &\le \sup_{d(\zeta)\ge 1} \big|(Zd)(\zeta)\big| \\ &= \sup_{d(\zeta)\ge 1} \left| \sum_{j=1}^{N} p_j(\xi)\,(Z_j d)(\zeta) \right| \le N C_1 C_2. \end{aligned}$$

This proves (5.124b). We are then left to prove (5.124c). From the pseudo-triangle inequality for d (see Proposition 5.1.7-2, page 231) we have

$$\begin{aligned} d(\gamma(t)) = d(\eta \circ \mathrm{Exp}\,(tZ)) &\ge \frac{1}{\mathbf{c}}\, d(\eta) - d(\mathrm{Exp}\,(tZ)) \\ &\ge \frac{1}{\mathbf{c}}\, M - \sup_{t\in[0,1],\, d(\xi)=1} d(\mathrm{Exp}\,(t\,\mathrm{Log}\,(\xi))) =: \frac{1}{\mathbf{c}}\, M - m(d, G), \end{aligned}$$

whence (5.124c) follows by choosing $M = \mathbf{c}(1 + m(d, \mathbb{G}))$. The finiteness of $m(d, \mathbb{G})$ follows from

$$\begin{aligned} d(\mathrm{Exp}\,(t\,\mathrm{Log}\,(\xi))) &= d\left(\mathrm{Exp}\left(\sum_{j=1}^{N} t p_j(\xi)\, Z_j\right)\right) \\ &\le \sup_{|\zeta_j|\le C_1} d\left(\mathrm{Exp}\left(\sum_{j=1}^{N} \zeta_j\, Z_j\right)\right) < \infty, \end{aligned}$$

uniformly for $t \in [0, 1]$, $d(\xi) = 1$. Here we used (5.124f) and the fact that

$$q(\zeta) := \mathrm{Exp}\left(\sum_{j=1}^{N} \zeta_j\, Z_j\right)$$

has polynomial coefficient functions (see (1.75a), page 50). This completes the proof. □

Note that, β being the constant in Proposition 5.14.1, if we replace x, y in (5.123) by, respectively, $y^{-1} \circ z$ and $z^{-1} \circ x$, we get

$$d(y^{-1} \circ x) \leq \beta\, d(y^{-1} \circ z) + d(z^{-1} \circ x) \qquad \text{for every } x,\ y,\ z \in \mathbb{G}. \tag{5.125a}$$

Moreover, by interchanging y and z in the above inequality (and using the symmetry of d) one gets

$$|d(y^{-1} \circ x) - d(z^{-1} \circ x)| \leq \beta\, d(y^{-1} \circ z) \qquad \text{for every } x,\ y,\ z \in \mathbb{G}. \tag{5.125b}$$

5.15 Appendix C. Existence of Geodesics

Let $\mathbb{G} = (\mathbb{R}^N, \delta_\lambda, \circ)$ be a homogeneous Carnot group. Let

$$\mathfrak{g} = W^{(1)} \oplus W^{(2)} \oplus \cdots \oplus W^{(r)}$$

be a stratification of the Lie algebra of $\mathbb{G}$, as in Remark 1.4.8 (page 59). Let $X = \{X_1, \ldots, X_m\}$ be any basis of $W^{(1)}$. We consider the related Carnot–Carathéodory distance d_X. The aim of this section is to prove that the "inf" defining d_X in (5.6) on page 232 is actually a minimum.

In other words, fixed any $x, y \in \mathbb{G}$, we show the existence of a X-subunit curve $\gamma : [0, T] \to \mathbb{G}$ connecting x and y (i.e. $\gamma(0) = x$, $\gamma(T) = y$) such that $T = d_X(x, y)$. We shall call any such curve a X-*geodesic* (for x and y).

The existence of geodesics can be proved in many general cases (see [Bus55] and [HK00]). Our argument here will make crucial use of the δ_λ-homogeneity and left-invariant properties of the system X. The resulting proof will be quite simple (for a more general proof, see the note after Theorem 5.15.5).

First, we recall that, by Propositions 5.2.4, 5.2.6 and Theorem 5.2.8,

$$d_X(x, y) = d_0(y^{-1} \circ x) \quad \text{for every } x, y \in \mathbb{G}, \tag{5.126}$$

where

$$d_0(z) := d_X(z, 0), \quad z \in \mathbb{G}, \tag{5.127}$$

and d_0 is a homogeneous (symmetric) norm on $\mathbb{G}$. As usual, we denote the dilation of $\mathbb{G}$ by

$$\delta_\lambda(x) = \delta_\lambda(x_1, \ldots, x_N) = (\lambda^{\sigma_1} x_1, \ldots, \lambda^{\sigma_N} x_N), \quad \lambda > 0,\ x \in \mathbb{G},$$

where $1 = \sigma_1 \leq \cdots \leq \sigma_N = r$ are consecutive integers and r is the step of nilpotency of $\mathbb{G}$. We are ready to prove the following result.

Proposition 5.15.1. *Let $\mathbb{G}$ be a homogeneous Carnot group, and let d be any homogeneous norm on $\mathbb{G}$.*

For every compact set $K \subset \mathbb{G}$, there exists a constant $\mathbf{c}_K > 0$ such that

$$(\mathbf{c}_K)^{-1} |x - y| \leq d(y^{-1} \circ x) \leq \mathbf{c}_K\, |x - y|^{1/r} \quad \forall\, x,\ y \in K, \tag{5.128}$$

where r is the step of $\mathbb{G}$ and $|\cdot|$ is the Euclidean norm on $\mathbb{G} \equiv \mathbb{R}^N$.

In particular, (5.128) *holds true if* $d(y^{-1} \circ x)$ *is replaced by* $d_X(x, y)$, *where* d_X *is the control distance related to any basis of generators (of the first layer of the stratification of the Lie algebra) of* $\mathbb{G}$.

(*Note.* More generally, the estimates in (5.128) also hold when $d = d_X$ (with a suitable r), where X is a system of smooth vector fields satisfying the Hörmander condition, see [Lan83,NSW85,VSC92,Gro96]. The first equality in (5.128) holds for a general Carnot–Carathéodory distance $d = d_X$, see [HK00] and Ex. 25 at the end of the chapter.)

Proof. Once (5.128) has been proved, the last assertion of the proposition follows from (5.126), by taking $d = d_0$ (d_0 as in (5.127)).

We then turn to prove (5.128). If $|\cdot|$ denotes the absolute value on $\mathbb{R}$, set

$$\varrho : \mathbb{G} \to [0, \infty), \quad \varrho(x) = \sum_{j=1}^{N} |x_j|^{1/\sigma_j}. \tag{5.129}$$

Obviously, ϱ is a homogeneous (symmetric) norm on $\mathbb{G}$. By the equivalence of all homogeneous norms on $\mathbb{G}$ (see Proposition 5.1.4), there exists a constant $\mathbf{c} = \mathbf{c}(\varrho, d, \mathbb{G}) \geq 1$ such that

$$\mathbf{c}^{-1} \varrho(x) \leq d(x) \leq \mathbf{c}\, \varrho(x) \quad \forall\, x \in \mathbb{G}. \tag{5.130}$$

Thus (5.128) will follow if we show that, given a compact set $K \subset \mathbb{G}$,

$$(\mathbf{c}_K)^{-1} |x - y| \leq \varrho(y^{-1} \circ x) \leq \mathbf{c}_K\, |x - y|^{1/r} \quad \forall\, x,\ y \in K, \tag{5.131}$$

for a suitable constant $\mathbf{c}_K > 0$.

Now, we recall the result in Corollary 1.3.18 (page 41): for every $j \in \{1, \ldots, N\}$ and every $x, y \in \mathbb{G}$, we have

$$(y^{-1} \circ x)_j = x_j - y_j + \sum_{k:\, \sigma_k < \sigma_j} P_k^{(j)}(x, y)\,(x_k - y_k),$$

where $P_k^{(j)}(x, y)$ is a polynomial function. Thus, the very definition of ϱ gives

$$\varrho(y^{-1} \circ x) = \sum_{j=1}^{N} \left| x_j - y_j + \sum_{k:\, \sigma_k < \sigma_j} P_k^{(j)}(x, y)\,(x_k - y_k) \right|^{1/\sigma_j}. \tag{5.132}$$

Since any $P_k^{(j)}$ is a continuous function, we immediately get from (5.132) the estimate from above: for every $x, y \in K$

$$\begin{aligned} \varrho(y^{-1} \circ x) &\leq \sum_{j=1}^{N} \left(|x_j - y_j| + \mathbf{c} \sum\nolimits_{k:\, \sigma_k < \sigma_j} |x_k - y_k| \right)^{1/\sigma_j} \\ &\leq \mathbf{c}' \sum_{j=1}^{N} \Bigg(\sum_{k:\, \sigma_k \leq \sigma_j} |x_k - y_k| \Bigg)^{1/\sigma_j} \leq \mathbf{c}' \sum_{j=1}^{N} \sum_{k:\, \sigma_k \leq \sigma_j} |x_k - y_k|^{1/\sigma_j} \end{aligned}$$

$$\le \mathbf{c}'' \sum_{j=1}^{N} |x_j - y_j|^{1/\sigma_j} \le \mathbf{c}''' \sum_{j=1}^{N} |x_j - y_j|^{1/\sigma_N}$$
$$\le \mathbf{c}''' N \left(\sum_{j=1}^{N} |x_j - y_j| \right)^{1/\sigma_N} \le \mathbf{c}''' N^{1+1/\sigma_N} |x - y|^{1/\sigma_N}.$$

This gives the estimate from above in (5.131), since $\sigma_N = r$. (Yet another proof of the estimate from above can be obtained by the Lagrange mean value theorem.[14])

The estimate from below is just a little more delicate. We fix the notation: for every $j = 2, \dots, N$, let n_j be the cardinality of the set $\{k : \sigma_k < \sigma_j\}$. Let α be a vector in $\mathbb{R}^n$ with $n = n_2 + \cdots + n_N$, and let us denote α in the following way

$$\alpha = \left(\alpha^{(2)}, \dots, \alpha^{(N)}\right) \quad \text{with} \quad \alpha^{(j)} = \left(\alpha_1^{(j)}, \dots, \alpha_{n_j}^{(j)}\right), \; j = 2, \dots, N.$$

With the notation in (5.132), if K is a compact subset of $\mathbb{G}$, there exists a constant $M \ge 1$ such that (recall that the $P_k^{(j)}$'s are polynomial functions)

$$\sup_{x,y \in K} |P_k^{(j)}(x, y)| \le M \quad \forall\, j = 2, \dots, N \; \forall\, k : \sigma_k < \sigma_j.$$

As a consequence, (5.132) gives (here $\mathbf{M} = n\,M$)

$$\varrho(y^{-1} \circ x) = \sum_{j=1}^{N} \left| x_j - y_j + \sum_{k:\, \sigma_k < \sigma_j} P_k^{(j)}(x, y)\,(x_k - y_k) \right|^{1/\sigma_j}$$

---

[14] Indeed, fix any $y \in \mathbb{G}$ and set

$$f_y : \mathbb{R}^N \to \mathbb{R}, \quad f_y(x) := (y^{-1} \circ x)_j.$$

Given any $x, x_0 \in \mathbb{G}$, by the Lagrange mean value theorem, we have

$$(\star) \qquad |f_y(x) - f_y(x_0)| \le \sup_{|\xi - x_0| \le |x - x_0|} |\nabla f_y(\xi)|\, |x - x_0|.$$

Now, take $x_0 = y$ and observe that $f_y(x_0) = f_y(y) = 0$. Moreover, if x, y belong to a compact set K, and if $R_K \gg 1$ is such that $K \subseteq B(0, R_K)$, then

$$\sup_{|\xi - y| \le |x - y|} |\nabla f_y(\xi)| \le \sup_{|\xi| \le R_K + \mathrm{diam}(K)} |\nabla f_y(\xi)| =: M_j < \infty,$$

for $f_y(\xi)$ is a polynomial function in ξ and y. Thus, $(\star)$ gives (set $M = \max_{j \le N} M_j$)

$$|(y^{-1} \circ x)_j| \le M\,|x - y| \qquad \forall\, x, y \in K\; \forall\, j \le N.$$

Then, for every $x, y \in K$, we have

$$\varrho(y^{-1} \circ x) = \sum_{j=1}^{N} |(y^{-1} \circ x)_j|^{1/\sigma_j} \le \mathbf{c} \sum\nolimits_{j=1}^{N} |x - y|^{1/\sigma_j} \le \mathbf{c}' |x - y|^{1/r}.$$

$$\geq \inf_{|\alpha| \leq \mathbf{M}} \left\{ \sum_{j=1}^{N} \left| x_j - y_j + \sum_{k:\,\sigma_k < \sigma_j} \alpha_k^{(j)} (x_k - y_k) \right|^{1/\sigma_j} \right\} =: (\bigstar).$$

Being $x, y \in K$, we obviously have (whenever $\alpha \leq \mathbf{M}$)

$$\left| x_j - y_j + \sum_{k:\,\sigma_k < \sigma_j} \alpha_k^{(j)} (x_k - y_k) \right| \leq \mathbf{c}(K, \mathbf{M}) < \infty,$$

hence (notice that $1/\sigma_j \leq 1$)

$$\begin{aligned} (\bigstar) &\geq \frac{\mathbf{c}(K, \mathbf{M})^{1/r}}{\mathbf{c}(K, \mathbf{M})} \inf_{|\alpha| \leq \mathbf{M}} \left\{ \sum_{j=1}^{N} \left| x_j - y_j + \sum_{k:\,\sigma_k < \sigma_j} \alpha_k^{(j)} (x_k - y_k) \right| \right\} \\ &= \mathbf{c}'(K) \inf_{|\alpha| \leq \mathbf{M}} \left\{ f_\alpha(x - y) \right\} =: (\bigstar'), \end{aligned}$$

where

$$f_\alpha : \mathbb{R}^N \to \mathbb{R}, \quad f_\alpha(z) = \sum_{j=1}^{N} \left| z_j + \sum_{k:\,\sigma_k < \sigma_j} \alpha_k^{(j)} z_k \right|.$$

We explicitly remark that f_α is homogeneous of degree 1 with respect to the Euclidean dilations on $\mathbb{R}^N$. As a consequence (if $x \neq y$, otherwise there is nothing to prove),

$$f_\alpha(x - y) = |x - y|\, f_\alpha\left(\frac{x - y}{|x - y|} \right) \geq |x - y| \inf_{|\xi| = 1} f_\alpha(\xi).$$

This gives

$$(\bigstar') \geq |x - y|\, \mathbf{c}'(K) \inf_{|\alpha| \leq \mathbf{M},\, |\xi| = 1} f_\alpha(\xi) \geq |x - y|\, \mathbf{c}''(K).$$

Indeed, since $f_\alpha(\xi)$ is a continuous function in ξ and α, and (for every α) $f_\alpha(\cdot)$ is *positive outside the origin* (as a simple inductive argument shows), there exist α_0 and ξ_0 (with $|\alpha_0| \leq \mathbf{M}$ and $|\xi_0| = 1$) such that

$$\inf_{|\alpha| \leq \mathbf{M},\, |\xi| = 1} f_\alpha(\xi) = f_{\alpha_0}(\xi_0) > 0.$$

This gives the lower estimate in (5.131) and completes the proof. □

From Proposition 5.15.1 we obtain useful corollaries, as we show below. In the sequel, we write d_E to denote the usual Euclidean metric on $\mathbb{G} \equiv \mathbb{R}^N$.

Corollary 5.15.2. *Let $\mathbb{G}$ be a homogeneous Carnot group. Let d_X be the control distance related to any basis X of generators* (*of the first layer of the stratification of the Lie algebra*) *of $\mathbb{G}$.*

Then $A \subseteq \mathbb{G}$ is bounded in the metric space $(\mathbb{G}, d_X)$ if and only if it is bounded in the Euclidean metric space $(\mathbb{G}, d_E)$. More precisely, there exists a constant $\mathbf{c}\geq 1$ *such that (for every $R > 0$)*

$$B_{d_X}(0, R) \subseteq B_{d_E}(0, \mathbf{R}_1(R)) \quad \textit{and} \quad B_{d_E}(0, R) \subseteq B_{d_X}(0, \mathbf{R}_2(R)), \tag{5.133}$$

where

$$\mathbf{R}_1(R) = \mathbf{c}\max\{R, R^r\}, \quad \mathbf{R}_2(R) = \mathbf{c}\max\{R, R^{1/r}\}.$$

(*Note*. The fact that a bounded set in $(\mathbb{R}^N, d_E)$ is also bounded in $(\mathbb{R}^N, d_X)$ holds for a d_X related to a general system of vector fields satisfying the Hörmander condition, see the note after Proposition 5.15.1. The reverse assertion may be false even in the Hörmander case, see Ex. 27.)

Proof. By (5.130), we know that there exists a constant $\mathbf{c} = \mathbf{c}(\varrho, d_X, \mathbb{G}) \geq 1$ such that

$$\mathbf{c}^{-1}\varrho(x) \leq d_X(x, 0) \leq \mathbf{c}\,\varrho(x) \quad \forall\, x \in \mathbb{G}, \tag{5.134}$$

where ϱ is as in (5.129). Now, it is obvious that a set $A \subseteq \mathbb{G}$ is bounded in $(\mathbb{G}, d_E)$ iff there exists $R > 0$ such that $A \subseteq B_\varrho(0, R)$. This remark jointly with (5.134) proves the first assertion of the corollary.

Precisely (see the definition of ϱ), it holds (for every $R > 0$)

$$B_{d_E}(0, R) \subseteq B_\varrho\big(0, N\max\{R, R^{1/r}\}\big), \quad B_\varrho(0, R) \subseteq B_{d_E}\big(0, N\max\{R, R^r\}\big).$$

These inclusions, together with (5.134), prove (5.133). □

Obviously, (5.133) also gives the inclusions

$$B_{d_X}(0, \mathbf{R}_3(R)) \subseteq B_{d_E}(0, R) \quad \text{and} \quad B_{d_E}(0, \mathbf{R}_4(R)) \subseteq B_{d_X}(0, R), \tag{5.135}$$

where

$$\mathbf{R}_3(R) = \mathbf{c}^{-1}\min\{R, R^{1/r}\}, \quad \mathbf{R}_1(R) = \mathbf{c}^{-r}\min\{R, R^r\}.$$

Corollary 5.15.3. *Let the hypothesis of Corollary* 5.15.2 *hold.*

Then the map

$$\mathrm{id} : (\mathbb{G}, d_X) \to (\mathbb{G}, d_E), \quad \mathrm{id}(x) = x$$

is a homeomorphism. As a consequence, the topologies of the metric spaces $(\mathbb{G}, d_X)$, $(\mathbb{G}, d_E)$ coincide.

(*Note*. The continuity of $\mathrm{id} : (\mathbb{G}, d_X) \to (\mathbb{G}, d_E)$ holds for a general Carnot–Carathéodory distance d_X, see Ex. 25 at the end of the chapter. For a general d_X, the reverse continuity may be false, see Ex. 26 (see also [BR96]). Instead, this reverse continuity holds for all smooth vector fields satisfying the Hörmander condition, see the note after Proposition 5.15.1 and Ex. 25.)

Proof. Let $\{x_n\}_{n\in\mathbb{N}}$ be a sequence in $\mathbb{G}$. Suppose $x_n \to x$ in $(\mathbb{G}, d_E)$. Then there exists a compact set $K \subset \mathbb{R}^N$ such that $x, x_n \in K$ for every $n \in \mathbb{N}$. Hence, by the second inequality in (5.128), we derive

$$d_X(x, x_n) \leq \mathbf{c}_K\, |x - x_n| \quad \forall n \in \mathbb{N},$$

so that, letting $n \to \infty$, $x_n \to x$ in $(\mathbb{G}, d_X)$ too.

Vice versa, suppose $x_n \to x$ in $(\mathbb{G}, d_X)$. Then the sequence $\{x_n\}_n$ is bounded in $(\mathbb{G}, d_X)$. Hence, by Corollary 5.15.2, it is also bounded in $(\mathbb{G}, d_E)$. Consequently, there exists a compact subset K of $\mathbb{R}^N$ such that $x, x_n \in K$ for every $n \in \mathbb{N}$. Hence, by the first inequality in (5.128), we derive

$$|x - x_n| \leq \mathbf{c}_K\, d_X(x, x_n) \quad \forall n \in \mathbb{N},$$

so that, letting $n \to \infty$, $x_n \to x$ in $(\mathbb{G}, d_E)$ too.

We have thus proved that

$$x_n \xrightarrow{d_X} x \Rightarrow \mathrm{id}(x_n) \xrightarrow{d_E} x, \qquad x_n \xrightarrow{d_E} x \Rightarrow \mathrm{id}^{-1}(x_n) \xrightarrow{d_X} x.$$

As a consequence, by the well-known characterization of continuity (in a *metric* space) as *sequential*-continuity, we infer that $\mathrm{id} : (\mathbb{G}, d_X) \to (\mathbb{G}, d_E)$ is a homeomorphism. This implies that id and id^{-1} are open maps, i.e. $(\mathbb{G}, d_X)$ and $(\mathbb{G}, d_E)$ have the same open sets. □

By collecting together the above lemmas, we obtain the following result.

Proposition 5.15.4. *Let $\mathbb{G}$ be a homogeneous Carnot group. Let d_X be the control distance related to any basis X of generators (of the first layer of the stratification of the Lie algebra) of $\mathbb{G}$.*

Then, a subset of $\mathbb{G}$ is, respectively, open, closed, bounded or compact in the metric space $(\mathbb{G}, d_X)$ if and only if the same holds in the Euclidean metric space $\mathbb{G} \equiv \mathbb{R}^N$.

In particular, the compact subsets of $(\mathbb{G}, d_X)$ are precisely the closed and bounded subsets of $\mathbb{R}^N$ or, equivalently, the closed and bounded subsets of $(\mathbb{G}, d_X)$.

Proof. Let $A \subseteq \mathbb{G}$. Then, by Corollary 5.15.3, A is, respectively, open, closed or compact in the metric space $(\mathbb{G}, d_X)$ if and only if the same holds in $(\mathbb{G}, d_E)$.

By Corollary 5.15.2, A is bounded in $(\mathbb{G}, d_X)$ if and only if the same holds in $(\mathbb{G}, d_E)$. This also gives the last assertion of the proposition and completes the proof. □

We are in a position to prove the following result.

Theorem 5.15.5 (Existence of X-geodesics). *Let $\mathbb{G}$ be a homogeneous Carnot group. Let d_X be the control distance related to any basis X of generators (of the first layer of the stratification of the Lie algebra) of $\mathbb{G}$.*

*Then, for every $x, y \in \mathbb{G}$, there exists a X-*geodesic* connecting x and y, i.e. there exists a X-subunit path $\gamma : [0, T] \to \mathbb{G}$ such that $\gamma(0) = x$, $\gamma(T) = y$ and $T = d_X(x, y)$.*

(*Note*. Since we proved that the compact subsets of $(\mathbb{G}, d_X)$ are precisely the closed and bounded subsets of $(\mathbb{G}, d_X)$ (see Proposition 5.15.4) this result also follows from the general results in [Bus55, page 25].)

Proof. Fix $x, y \in \mathbb{G}$. By definition of d_X, there exists a sequence $\{\gamma_n\}_{n\in\mathbb{N}}$ of X-subunit paths

$$\gamma_n : [0, T_n] \to \mathbb{G}, \quad \gamma_n(0) = x, \; \gamma_n(T_n) = y, \quad T_n \searrow d_X(x, y).$$

By definition of X-subunit path, γ_n is an absolutely continuous curve such that, for every $n \in \mathbb{N}$, there exists $E_n \subseteq [0, T_n]$ such that $F_n := [0, T_n] \setminus E_n$ has vanishing Lebesgue measure and

$$\langle \dot\gamma_n(t), \xi\rangle^2 \le \sum_{j=1}^m \langle X_j I(\gamma_n(t)), \xi\rangle^2 \quad \forall\, \xi \in \mathbb{R}^N \;\forall\, t \in E_n. \tag{5.136}$$

Since $T_n \le T_{n+1}$ for all n, it is not restrictive to suppose that every γ_n is defined on $[0, T_1]$, by prolonging γ_n to be y on $(T_n, T_1]$. We still denote this prolongation by γ_n. Observe that the prolongation is still a X-subunit path (connecting x to y) because, for $t \in (T_n, T_1]$, γ_n is constant, whence the far left-hand side in (5.136) is 0. The resulting prolongation is also trivially absolutely continuous.

We claim that the family of functions $\mathcal{F} = \{\gamma_n\}_n$ on $[0, T_1]$ is uniformly bounded and equicontinuous.

• *$\mathcal{F}$ is equicontinuous*: Suppose we have proved that $\mathcal{F}$ is uniformly bounded, i.e.

$$\exists\, M > 0 : \quad \sup_{t\in[0,T_1]} |\gamma_n(t)| \le M \quad \forall\, n \in \mathbb{N}. \tag{5.137}$$

Then the equicontinuity of $\mathcal{F}$ follows. Indeed, for every $t, t' \in [0, T_1]$, it holds

$$\begin{aligned}
|\gamma_n(t) - \gamma_n(t')| &= \left|\int_{t'}^t \dot\gamma_n(s)\,\mathrm{d}s\right| \le \int_{t'}^t |\dot\gamma_n(s)|\,\mathrm{d}s \\
\text{(see the note}^{15}) &= \int_{t'}^t \sup_{\xi\in\mathbb{R}^N:\,|\xi|=1} \big|\langle \dot\gamma_n(s), \xi\rangle\big|\,\mathrm{d}s \\
\text{(by (5.136))} &\le \int_{t'}^t \sup_{\xi\in\mathbb{R}^N:\,|\xi|=1} \left(\sum_{j=1}^m \langle X_j I(\gamma_n(s)), \xi\rangle^2\right)^{1/2} \mathrm{d}s \\
\text{(Cauchy–Schwartz)} &\le \int_{t'}^t \sup_{\xi\in\mathbb{R}^N:\,|\xi|=1} \left(\sum_{j=1}^m |X_j I(\gamma_n(s))|^2\,|\xi|^2\right)^{1/2} \mathrm{d}s
\end{aligned}$$

---

[15] Here, we use a well-known fact: if $v \in \mathbb{R}^N$, then

$$|v| = \sup_{\xi\in\mathbb{R}^N:\,|\xi|=1} |\langle v, \xi\rangle|.$$

$$= \int_{t'}^{t} \big|(X_1 I, \dots, X_m I)(\gamma_n(s))\big| \, ds$$

$$\text{(by (5.137))} \quad \le |t - t'| \sup_{|\eta| \le M} \big|(X_1 I, \dots, X_m I)(\eta)\big| \le M' \, |t - t'|.$$

In the last inequality, we used the smoothness of the X_j's. Note that this proves more than the equicontinuity, namely, $\mathcal{F}$ is a uniform Lipschitz-continuous family.

• *$\mathcal{F}$ is uniformly bounded*: We are then left to prove (5.137). By the very definition of d_X, for every $t \in [0, T_1]$, we have

$$d_X(x, \gamma_n(t)) \le t \le T_1.$$

As a consequence, by the triangle-inequality for the distance d_X, we get

$$d_X(0, \gamma_n(t)) \le d_X(0, x) + d_X(x, \gamma_n(t)) \le d_X(0, x) + T_1 < \infty.$$

Hence, the set $\{\gamma_n(t) : n \in \mathbb{N}, \, t \in [0, T_1]\}$ is bounded in $(\mathbb{G}, d_X)$, so that, due to Corollary 5.15.2, the set is bounded in the Euclidean metric. This is precisely (5.137).

We are then entitled to apply the Arzelà–Ascoli theorem, ensuring that there exists a subsequence of $\{\gamma_n\}_{n \in \mathbb{N}}$ which converges uniformly on $[0, T_1]$, say to $\gamma : [0, T_1] \to \mathbb{G}$. For the sake of brevity, we still denote this uniformly convergent subsequence by $\{\gamma_n\}_{n \in \mathbb{N}}$.

Set $T := d_X(x, y)$. We aim to prove that the curve

$$\gamma^* : [0, T] \to \mathbb{G}, \quad \gamma^* := \gamma|_{[0,T]}$$

is X-subunit and γ connects x to y, which will end the proof since γ is defined on $[0, T]$.

First, $\gamma_n(0) = x$ for every $n \in \mathbb{N}$ yields $\gamma(0) = x$, by pointwise convergence. Second, since γ_n converges *uniformly* to γ on $[0, T_1]$, and $\{T_n\}_n$ is a sequence in $[0, T_1]$ converging to T, we infer

$$\gamma_n(T_n) \to \gamma(T),$$

so that

$$\gamma^*(T) = \gamma(T) = \lim_{n \to \infty} \gamma_n(T_n) = y,$$

because $\gamma_n(T_n) = y$ for every $n \in \mathbb{N}$. This proves that γ^* connects x to y.

Finally, we prove that γ is X-subunit. To begin with, we remark that γ *is absolutely continuous*. Indeed, in proving the equicontinuity of $\mathcal{F}$, we showed that there exists a constant M' such that

$$|\gamma_n(t) - \gamma_n(t')| \le M' \, |t - t'| \quad \text{for every } n \in \mathbb{N} \text{ and every } t, t' \in [0, T_1].$$

Letting $n \to \infty$, this shows that γ is Lipschitz continuous, whence it is absolutely continuous. Obviously, γ is X-subunit iff, for every fixed $\xi \in \mathbb{R}^N$, the functions

$$f^{\pm}(t) := \left(\sum_{j=1}^{m} \langle X_j I(\gamma(t)), \xi\rangle^2\right)^{1/2} \pm \langle \dot{\gamma}(t), \xi\rangle$$

are non-negative a.e. on $[0, T_1]$. To this end, it suffices to prove that (notice that $f_{\pm} \in L^1(0, T_1)$)

$$\int_0^{T_1} f^{\pm}(t)\,\varphi(t)\,\mathrm{d}t \geq 0 \quad \forall\, \varphi \in C_0^{\infty}(0, T_1),\ \varphi \geq 0. \tag{5.138}$$

First, being γ_n X-subunit, it holds

$$\begin{aligned} 0 &\leq \int_0^{T_1} \left\{ \left(\sum_{j=1}^{m} \langle X_j I(\gamma_n(t)), \xi\rangle^2\right)^{1/2} \pm \langle \dot{\gamma}_n(t), \xi\rangle \right\} \varphi(t)\,\mathrm{d}t \\ &= \int_0^{T_1} \left(\sum_{j=1}^{m} \langle X_j I(\gamma_n), \xi\rangle^2\right)^{1/2} \varphi \pm \int_0^{T_1} \langle \dot{\gamma}_n, \xi\rangle\,\varphi =: A_n \pm B_n \end{aligned}$$

for every $n \in \mathbb{N}$ and every $\varphi \in C_0^{\infty}(0, T_1)$, $\varphi \geq 0$ (exploit (5.136)), whence

$$0 \leq A_n \pm B_n \quad \text{for every } n \in \mathbb{N}. \tag{5.139}$$

Now, by dominated convergence (indeed, recall that X_j is smooth and $\{\gamma_n\}_{n\in\mathbb{N}}$ is uniformly bounded), the limit of A_n is easily obtained:

$$A_n = \int_0^{T_1} \left(\sum_{j=1}^{m} \langle X_j I(\gamma_n), \xi\rangle^2\right)^{1/2} \varphi \overset{n\to\infty}{\longrightarrow} \int_0^{T_1} \left(\sum_{j=1}^{m} \langle X_j I(\gamma), \xi\rangle^2\right)^{1/2} \varphi. \tag{5.140}$$

Furthermore, the limit of B_n can be obtained by recalling that absolutely continuous functions support integration by parts:

$$\begin{aligned} B_n &= \int_0^{T_1} \langle \dot{\gamma}_n, \xi\rangle\,\varphi = \sum_{i=1}^{N} \xi_i \int_0^{T_1} (\dot{\gamma}_n)_i\,\varphi = -\sum_{i=1}^{N} \xi_i \int_0^{T_1} (\gamma_n)_i\,\dot{\varphi} \\ &\overset{n\to\infty}{\longrightarrow} -\sum_{i=1}^{N} \xi_i \int_0^{T_1} \gamma_i\,\dot{\varphi} = \sum_{i=1}^{N} \xi_i \int_0^{T_1} \dot{\gamma}_i\,\varphi = \int_0^{T_1} \langle \dot{\gamma}, \xi\rangle\,\varphi. \end{aligned} \tag{5.141}$$

(Passing to the limit, we used another dominated convergence argument, recalling that $\gamma_n \to \gamma$ uniformly.) As a consequence, letting $n \to \infty$ in (5.139), from (5.140) and (5.141) we infer

$$0 \leq \int_0^{T_1} \left(\sum_{j=1}^{m} \langle X_j I(\gamma), \xi\rangle^2\right)^{1/2} \varphi \pm \int_0^{T_1} \langle \dot{\gamma}, \xi\rangle\,\varphi,$$

which is exactly (5.138) (by the very definition of $f^{\pm}$). This completes the proof. □

As a consequence of Theorem 5.15.5, we have the following *segment property* for the Carnot–Carathéodory distance on a homogeneous Carnot group. (See also B. Franchi and E. Lanconelli [FL83] for the segment property in a sub-elliptic context.)

Corollary 5.15.6 (The segment property). *Let $\mathbb{G}$ be a homogeneous Carnot group. Let d_X be the control distance related to any basis X of generators (of the first layer of the stratification of the Lie algebra) of $\mathbb{G}$.*

Then the metric space $(\mathbb{G}, d_X)$ has the segment property, *i.e. for every $x, y \in \mathbb{G}$, there exists a continuous curve $\gamma : [0, T] \to \mathbb{G}$ with $\gamma(0) = x$, $\gamma(T) = y$, and such that*

$$d_X(x, y) = d_X(x, \gamma(t)) + d_X(\gamma(t), y) \quad \textit{for every } t \in [0, T].$$

For instance, γ can be any X-geodesic connecting x and y, whose existence is granted by Theorem 5.15.5.

Proof. Let $x, y \in \mathbb{G}$ be fixed. Let γ be a X-geodesic connecting x and y: the existence of such a X-geodesic is granted by Theorem 5.15.5. We claim that

$$d_X(x, y) = d_X(x, \gamma(t)) + d_X(\gamma(t), y) \quad \text{for every } t \in [0, T].$$

Fix $t \in [0, T]$. First, by the very definition of d_X (being γ a X-subunit curve), we have

$$(\star) \qquad d_X(x, \gamma(t)) \le t.$$

(Actually, the equality will hold.) We next prove that

$$(\star\star) \qquad d_X(\gamma(t), y) \le T - t.$$

(Actually, the equality will hold.) Indeed, consider the curve

$$\widetilde{\gamma} : [0, T - t] \to \mathbb{G}, \quad \widetilde{\gamma}(s) := \gamma(s + t).$$

Obviously, $\widetilde{\gamma}$ is X-subunit and $\widetilde{\gamma}(0) = \gamma(t)$, $\widetilde{\gamma}(T - t) = \gamma(t) = y$. Hence, again by the definition of d_X, one has

$$d_X(\gamma(t), y) = d_X(\widetilde{\gamma}(0), \widetilde{\gamma}(T - t)) \le T - t,$$

i.e. $(\star\star)$ holds. Consequently, by $(\star)$ and $(\star\star)$ together with the triangle-inequality, we get

$$T = d_X(x, y) \le d_X(x, \gamma(t)) + d_X(\gamma(t), y) \le t + (T - t) = T.$$

Hence, these inequalities are in fact equalities, and the proof is complete. (Incidentally, this also proves that the equality holds in $(\star)$ and $(\star\star)$.) □

Bibliographical Notes. Some of the topics presented in this chapter also appear in [BL01].

For other equivalent definitions of the Carnot–Carathéodory distance, see, e.g. [JSC87,NSW85]. See [HK00] and the references therein for applications of Carnot–Carathéodory distances in PDE's problems. In particular, see the collection of papers [BR96] for an introduction to the geometry of C-C spaces. For explicit estimates of the Carnot–Carathéodory distance for "diagonal" vector fields, see [FL82]; for the segment property, see [FL83]. See also [GN98].

The gauge functions on the Heisenberg group and on H-type groups were discovered by G.B. Folland [Fol75] and A. Kaplan [Kap80], respectively.

For mean value formulas for the Hörmander sum of squares, see [Lan90,CGL93]; see [Gav77] for mean value formulas for the Heisenberg group; for a survey on mean value formulas in the classical setting of Laplace's operator, see [NV94] (see also the list of references therein for related results in a non-Riemannian setting).

The bibliography on Harnack-type inequalities for sub-elliptic operators is extremely vast: see, e.g. [FL83] for the first result on these topics. See also [BM95, Bon69,FGW94,FL82,GL03,LM97a,LK00,SaC90,SCS91,Varo87].

For Liouville-type theorems for homogeneous operators, see [Gel83,KoSt85, LM97a,Luo97]. For results on the maximum principles and propagation, see [Ama79, Bon69,Hil70,Hop52,PW67,Red71,Spe81,Tai88].

5.16 Exercises of Chapter 5

Ex. 1) Give a detailed proof of Lemma 5.13.19, page 303.

Ex. 2) The following operator

$$\mathcal{L} := \sum_{j=1}^{n} \left((\partial_{x_j})^2 + (\partial_{y_j} + x_j\, \partial_t)^2 \right) + (\partial_z)^2$$

in $\mathbb{R}^{2n+2}$ (the points are denoted by (x, y, z, t) with $x = (x_1, \ldots, x_n) \in \mathbb{R}^n$, $y = (y_1, \ldots, y_n) \in \mathbb{R}^n$, $z \in \mathbb{R}$, $t \in \mathbb{R}$) is not analytic-hypoelliptic (see [Rot84]). Find an explicit homogeneous Carnot group $\mathbb{G}$ such that $\mathcal{L}$ is the canonical sub-Laplacian of $\mathbb{G}$.

Ex. 3) Construct explicitly the Carnot group referred to in Example 5.10.6, page 286. (*Hint*: Model the needed composition law on the composition law in Example 5.10.5, page 285.)

Ex. 4) Consider the Kolmogorov-type group in Ex. 7 of Chapter 3, page 179. By means of formula (5.92), verify that the fundamental solution of its canonical sub-Laplacian is given by (when $x \neq 0$)

$$
\begin{aligned}
&\Gamma(x,t)\\
&= c\int_{\mathbb{R}^2} \mathrm{d}\tau_1 \mathrm{d}\tau_2\, \frac{|\tau|}{\sinh|\tau|}\\
&\quad\times\left(\frac{\{(\tau_2 x_2-\tau_1 x_3)^2+2\,\frac{|\tau|}{\tanh|\tau|}\,(|\tau|^2 x_1^2+(\tau_1 x_2+\tau_2 x_3)^2)\}}{\{\ldots\textit{the above braces}\ldots\}^2+(t_1\tau_1+t_2\tau_2)^2}\right)^{\frac{5}{2}}
\end{aligned}
$$

where c is the dimensional constant

$$c=\frac{3}{2^7\,\pi^3}.$$

Ex. 5) Prove that if $f\in L^1_{\mathrm{loc}}(\mathbb{R}^N)$, then

$$x\mapsto \fint_{B_d(x,r)} |f(y)|\,\mathrm{d}y$$

is a continuous function.

Ex. 6) Provide a detailed proof of (5.61) at page 264.

Ex. 7) Let $u\in C^2(B_d(0,R))\cap\mathcal{H}\big(B_d(0,R)\setminus\overline{B_d(0,r)}\big)$, $0<r<R$. Prove that there exist constants C_0, C_1 such that

$$\mathcal{M}_\rho(u)(0)=C_0+\frac{C_1}{\rho^{Q-2}}\quad\text{for every }\rho\in(r,R).$$

Ex. 8) **(The surface asymptotic sub-Laplacian).** Let $\Omega\subseteq\mathbb{G}$ be open, and let $u\in C(\Omega,\mathbb{R})$. Given $x\in\Omega$, we say that the *surface asymptotic sub-Laplacian* of u at x is non-negative, written $\mathcal{AL}u(x)\geq 0$, if

$$\liminf_{r\to 0^+}\frac{\mathcal{M}_r(u)(x)-u(x)}{r^2}\geq 0.$$

If $\mathcal{AL}(-u)(x)\geq 0$, we write $\mathcal{AL}u(x)\leq 0$. Prove the following statements:

(i) If $\mathcal{AL}u\geq 0$ and $\mathcal{AL}v\geq 0$, then $\mathcal{AL}(\lambda\,u+\mu\,v)\geq 0$ for every $\lambda,\mu\geq 0$.
(ii) If $\mathcal{AL}u\geq 0$ and $\lambda\leq 0$, then $\mathcal{AL}(\lambda\,u)\leq 0$.
(iii) If $u\in C^2(\Omega,\mathbb{R})$ and $\mathcal{L}u\geq 0$, then $\mathcal{AL}u(x)\geq 0$ for every $x\in\Omega$.
(iv) If $x_0\in\Omega$ is a local maximum point for u, then $\mathcal{AL}u(x_0)\leq 0$.

Ex. 9) **(The solid asymptotic sub-Laplacian).** Let $\Omega\subseteq\mathbb{G}$ be open, and let $u\in C(\Omega,\mathbb{R})$. Given $x\in\Omega$, we say that the *solid asymptotic sub-Laplacian* of u at x is non-negative, written $\mathrm{A}\mathcal{L}u(x)\geq 0$, if

$$\liminf_{r\to 0^+}\frac{\mathrm{M}_r(u)(x)-u(x)}{r^2}\geq 0.$$

If $\mathrm{A}\mathcal{L}(-u)(x)\geq 0$, we write $\mathrm{A}\mathcal{L}u(x)\leq 0$. Prove the statements of the previous exercise with $\mathcal{AL}$ replaced by $\mathrm{A}\mathcal{L}$.

Ex. 10) **(Maximum principle for $\mathcal{AL}$ and $\mathrm{A}\mathcal{L}$).** Let $\Omega \subseteq \mathbb{G}$ be open and bounded. Let $u \in C(\Omega, \mathbb{R})$ be such that

$$\begin{cases} \mathcal{AL}u \geq 0 & \text{in } \Omega, \\ \limsup_{y \to x} u(y) \leq 0 & \forall\, x \in \partial\Omega, \end{cases} \quad \text{or} \quad \begin{cases} \mathrm{A}\mathcal{L}u \geq 0 & \text{in } \Omega, \\ \limsup_{y \to x} u(y) \leq 0 & \forall\, x \in \partial\Omega. \end{cases}$$

Prove that $u \leq 0$ in Ω. (*Hint*: Proceed as in the proof of the weak maximum principle, Theorem 5.13.4.)

Ex. 11) Provide the detailed proof of the result in Remark 5.6.4.

Ex. 12) Let $\mathcal{L}$ be a sub-Laplacian on a homogeneous Carnot group $\mathbb{G}$ on $\mathbb{R}^N$. Suppose that $\mathcal{L}$ has no first order differential terms, i.e. $\mathcal{L}(x_k) \equiv 0$ for every $k = 1, \ldots, N$. Let $\psi \in C^2(\mathbb{G}, \mathbb{G})$ be such that

$$\mathcal{L}(u \circ \psi) = (\mathcal{L}\, u) \circ \psi \qquad \forall\, u \in C^2(\mathbb{G}). \tag{5.142}$$

Following the usual notation for the "stratification" of the variables of $\mathbb{G}$, we set

$$\psi = (\psi^{(1)}, \ldots, \psi^{(r)})$$

(here $\psi^{(i)} \in C^2(\mathbb{G}, \mathbb{R}^{N_i})$, $N_1 + \cdots + N_r = N$ and N_i is the dimension of the i-th layer in the stratification of the algebra of $\mathbb{G}$). Prove that, for every $1 \leq j \leq N_1$, it holds

$$\begin{cases} 0 = \mathcal{L}\, \psi_j^{(1)} \\ 1 = |\nabla_{\mathcal{L}}\, \psi_j^{(1)}|^2. \end{cases} \tag{5.143}$$

(*Hint*: Use (1.112) in Ex. 12, Chapter 1, page 83.)
Let now $u_j^{(1)} := (\psi_j^{(1)})^2$. Then $u_j^{(1)} \geq 0$, and from (5.143) we have $\mathcal{L}u_j^{(1)} = 2$. Using Liouville Theorem 5.8.4, derive that $u_j^{(1)}$ is a polynomial of $\mathbb{G}$-degree ≤ 2 and therefore $u_j^{(1)} = p_1(x^{(1)}) + p_2(x^{(2)})$, where p_1 and p_2 are polynomials of ordinary degrees 2 and 1, respectively. Deduce that $p_2 \equiv 0$. Since $|\psi_j^{(1)}| = \sqrt{u_j^{(1)}}$, using, for example, Theorem 5.8.8, derive that $\psi_j^{(1)}$ is a polynomial of $\mathbb{G}$-degree ≤ 1, depending at most on $x^{(1)}$.
Now suppose by induction that $\psi^{(1)}, \ldots, \psi^{(n)}$ have polynomial component functions of $\mathbb{G}$-degree respectively $1, \ldots, n$ at most. Write $\mathcal{L} = \sum_{k=1}^{p} X_k^2$ with $X_k = \sum_{i=1}^{N} \sigma_i^{(k)}\, \partial_i$. Prove that (using again (1.112)) if $i = n+1$ and $1 \leq j \leq N_{i+1}$, it holds

$$\begin{cases} 0 = \mathcal{L}\, \psi_j^{(n+1)} \\ \sum_{k=1}^{p} \left(\sigma_{n+1,\,j}^{(k)}\right)^2(\psi) = |\nabla_{\mathcal{L}}\, \psi_j^{(n+1)}|^2. \end{cases} \tag{5.144}$$

Let $u_j^{(n+1)} := (\psi_j^{(n+1)})^2$; then $u_j^{(n+1)} \geq 0$ and, from (5.144), derive

$$\mathcal{L}\, u_j^{(n+1)} = 2 \sum_{k=1}^{p} \left(\sigma_{n+1,\,j}^{(k)}\right)^2(\psi).$$

On the other hand, $\sigma^{(k)}_{n+1,\,j}$ is a polynomial of $\mathbb{G}$-degree n thus depending only on $x^{(1)}, \dots, x^{(n)}$. Consequently, by induction hypothesis,

$$\left(\sigma^{(k)}_{n+1,\,j}\right)^2(\psi) = \left(\sigma^{(k)}_{n+1,\,j}(\psi^{(1)}, \dots, \psi^{(n)})\right)^2$$

has $\mathbb{G}$-degree at most $2n$. Therefore (why?), $u_j^{(n+1)}$ is a polynomial of $\mathbb{G}$-degree at most $2 + 2n$, consequently $|\psi_j^{(n+1)}| = \sqrt{u_j^{(n+1)}} \le C(1 + |x|)^{n+1}$. This last estimate, together with $\mathcal{L}\psi_j^{(n+1)} = 0$ and Theorem 5.8.2, proves that $\psi_j^{(n+1)}$ is a polynomial of $\mathbb{G}$-degree at most $n + 1$. We have proved the following assertion.

Proposition 5.16.1. *Let ψ be a $C^2(\mathbb{G}, \mathbb{G})$ map such that*

$$\mathcal{L} \circ \psi = \psi \circ \mathcal{L},$$

where $\mathcal{L}$ is a sub-Laplacian without the first order differential terms. Following the usual notation of the coordinates on $\mathbb{G}$, we let $\psi = (\psi^{(1)}, \dots, \psi^{(r)})$, where $\psi^{(j)} \in C^2(\mathbb{G}, \mathbb{R}^{N_j})$ for every $j = 1, \dots, r$. Then each component of $\psi^{(j)}$ is a polynomial function of $\mathbb{G}$-degree less or equal to j.

Ex. 13) (Maps commuting with Laplace operator Δ). Consider the Laplace operator $\Delta = \sum_{i=1}^N \partial^2_{i,i}$ on $\mathbb{R}^N$. Let $\psi \in C^2(\mathbb{R}^N, \mathbb{R}^N)$ be such that

$$\Delta(u \circ \psi) = (\Delta u) \circ \psi \qquad \forall\, u \in C^2(\mathbb{R}^N). \tag{5.145}$$

With the aid of Ex. 12, Chapter 1, page 83, prove that (5.145) holds if and only if

$$\begin{cases} \Delta\psi_i = 0, & i = 1, \dots, N, \\ |\nabla\psi_i|^2 = 1, & i = 1, \dots, N, \\ \langle\nabla\psi_i, \nabla\psi_j\rangle = 0, & 1 \le i, j \le N, \quad i \ne j. \end{cases}$$

Derive that there exists an $N \times N$ *orthogonal* matrix M and a vector $C \in \mathbb{R}^N$ such that

$$\psi(x) = Mx + C.$$

This means that the set of the maps commuting with the Laplace operator coincides with the group of the isometries of $\mathbb{R}^N$.

Ex. 14) (Maps commuting with $\Delta_{\mathbb{H}^1}$). Let us consider the Kohn Laplacian $\Delta_{\mathbb{H}^1}$ on the Heisenberg-Weyl group $\mathbb{H}^1 \equiv \mathbb{R}^3$. We characterize the maps $\psi \in C^2(\mathbb{R}^3, \mathbb{R}^3)$ such that

$$\Delta_{\mathbb{H}^1}(u \circ \psi) = (\Delta_{\mathbb{H}^1} u) \circ \psi \qquad \forall\, u \in C^2(\mathbb{H}^1). \tag{5.146}$$

Set $\psi(x, y, t) = (\xi(x, y, t), \eta(x, y, t), \tau(x, y, t))$. With the aid of Ex. 12, Chapter 1, page 83, prove that (5.146) is equivalent to

$$\begin{cases} \text{(i)} & \begin{cases} 1 = |\nabla_{\mathbb{H}^1}\xi|^2, \\ 0 = \Delta_{\mathbb{H}^1}\xi, \end{cases} \\ \text{(ii)} & \begin{cases} 1 = |\nabla_{\mathbb{H}^1}\eta|^2, \\ 0 = \Delta_{\mathbb{H}^1}\eta, \end{cases} \\ \text{(iii)} & \begin{cases} 4\,(\xi^2+\eta^2) = |\nabla_{\mathbb{H}^1}\tau|^2, \\ 0 = \Delta_{\mathbb{H}^1}\tau, \end{cases} \\ \text{(iv)} & 2\,\eta = \langle \nabla_{\mathbb{H}^1}\xi, \nabla_{\mathbb{H}^1}\tau\rangle, \\ \text{(v)} & -\,2\,\xi = \langle \nabla_{\mathbb{H}^1}\eta, \nabla_{\mathbb{H}^1}\tau\rangle, \\ \text{(vi)} & 0 = \langle \nabla_{\mathbb{H}^1}\xi, \nabla_{\mathbb{H}^1}\eta\rangle. \end{cases}$$

Using Liouville Theorem 5.8.4, derive that ξ and η are polynomials of $\mathbb{H}^1$-degree ≤ 1, τ is a polynomial of $\mathbb{H}^1$-degree ≤ 2. Hence, there exist constants such that

$$\begin{aligned} \xi(x,y,t) &= c_0 + c_1\,x + c_2\,y, \\ \eta(x,y,t) &= d_0 + d_1\,x + d_2\,y, \\ \tau(x,y,t) &= e_0 + e_1\,x + e_2\,y + e_3\,t. \end{aligned}$$

From the first equation in (i) and the first equation in (ii) and from (vi) it follows that there exists $\theta \in [0, 2\pi[$ such that

$$(c_1, c_2) = (\cos\theta, \sin\theta), \qquad (d_1, d_2) = \pm(-\sin\theta, \cos\theta).$$

From (iv) and (v) one gets

$$\begin{pmatrix} e_1 + 2\,y\,e_3 \\ e_2 - 2\,x\,e_3 \end{pmatrix} = \begin{pmatrix} 2d_0c_1 - 2d_1c_0 \pm 2\,y \\ 2d_0c_2 - 2d_2c_0 \mp 2\,x \end{pmatrix},$$

i.e.

$$e_3 = \pm 1, \quad e_1 = 2d_0c_1 - 2d_1c_0, \quad e_2 = 2d_0c_2 - 2d_2c_0.$$

Recalling that on $\mathbb{H}^1$ it holds

$$(\alpha, \beta, \gamma) \circ (x, y, t) = (x+\alpha, y+\beta, t+\gamma+2x\beta-2y\alpha),$$

we have proved that *the only maps commuting with* $\Delta_{\mathbb{H}^1}$ *have the form*

$$\psi_+(x,y,t) = \begin{pmatrix} c_0 \\ d_0 \\ e_0 \end{pmatrix} \circ \begin{pmatrix} M_+ \cdot \binom{x}{y} \\ t \end{pmatrix},$$

$$\psi_-(x,y,t) = \begin{pmatrix} c_0 \\ d_0 \\ e_0 \end{pmatrix} \circ \begin{pmatrix} M_- \cdot \binom{x}{y} \\ -t \end{pmatrix},$$

where M_+ *and* M_- *are (respectively) isometries with the determinant equal to* $+1$ *(respectively,* -1*) and* (c_0, d_0, e_0) *is a fixed point in* $\mathbb{H}^1$.

Ex. 15) (Translation-formula in surface integrals over a d-sphere). Let $f : \mathbb{G} \to \mathbb{R}$ be non-negative (or with suitable summability properties). Let d be a smooth, homogeneous and symmetric norm on $\mathbb{G}$, and let $S(x_0, r)$ be the d-sphere with center x_0 and radius $r > 0$. Finally, let H^{N-1} denote the Hausdorff $(N-1)$-dimensional measure on $\mathbb{R}^N$. Prove that

$$\int_{S(x_0,r)} f(x)\,\mathrm{d}H^{N-1}(x) = \int_{S(0,r)} f(x_0 \circ y)\,K_d(x_0, y)\,\mathrm{d}H^{N-1}(y),$$

where

$$K_d(x_0, y) := \frac{|\nabla(d(x_0^{-1} \circ \cdot))|(x_0 \circ y)}{|\nabla d(y)|}.$$

Prove also that

$$K_d(x_0, y) = \frac{|\nabla d(y) \cdot \mathcal{J}_{\tau_{x_0^{-1}}}(x_0 \circ y)|}{|\nabla d(y)|} = \left| \frac{\nabla d(y)}{|\nabla d(y)|} \cdot \mathcal{J}_{\tau_{x_0^{-1}}}(x_0 \circ y) \right|.$$

(*Hint*: Consider the identity (why does it hold?)

$$\int_{S(x_0,r)} f(x)\,\mathrm{d}H^N(x) = \int_{S(0,r)} f(x_0 \circ y)\,\mathrm{d}H^N(y)$$

rewritten (by the coarea formula) as

$$\int_0^r \left(\int_{S(x_0,\lambda)} f(x) \frac{\mathrm{d}H^{N-1}(x)}{|\nabla d(x_0^{-1} \circ \cdot)|(x)} \right) \mathrm{d}r$$
$$= \int_0^r \left(\int_{S(0,\lambda)} f(x_0 \circ y) \frac{\mathrm{d}H^{N-1}(y)}{|\nabla d(y)|} \right) \mathrm{d}r$$

and differentiate w.r.t. r.)

Ex. 16) In the sequel, d is a symmetric homogeneous norm on $\mathbb{G}$. If $x \in \mathbb{G}$ and $A \subseteq \mathbb{G}$ is any set, we call the *d-distance of x to A*, the following real non-negative number

$$\mathrm{dist}_d(x, A) := \inf_{a \in A} d(x^{-1} \circ a).$$

Prove the following result.

Lemma 5.16.2. *Let $A \subset \mathbb{G}$ be any set. For every $x \in \mathbb{G}$, there exists $\overline{a}_x \in \overline{A}$ (the closure of A) such that*

$$\mathrm{dist}_d(x, A) = d(x^{-1} \circ \overline{a}_x).$$

(*Hint*: By definition, there exists $\{a_j\}_j$ in A such that $d(x^{-1} \circ a_j) \to \mathrm{dist}_d(x, A)$. Obviously, $\{a_j\}_j$ is bounded, otherwise $d(x^{-1} \circ a_j) \geq \frac{1}{\mathbf{c}} d(a_j) - d(x) \to \infty$. Then, extract a subsequence $a_{j_n} \to \overline{a}_x \in \overline{A}$, as $n \to \infty$, and use the continuity of d.)

Then prove the following result.

Proposition 5.16.3. *Let $A \subseteq \mathbb{G}$ be any set. The following assertions hold:*
(a) *For every $x \in \mathbb{G}$, we have $\mathrm{dist}_d(x, A) = \mathrm{dist}_d(x, \overline{A})$;*
(b) *The d-distance from A, i.e. the function*

$$\mathrm{dist}_d(\cdot, A) : \mathbb{G} \to [0, \infty), \qquad x \mapsto \mathrm{dist}_d(x, A),$$

is a continuous function.

Hint: (a). Let $\alpha \in \overline{A}$ be such that $\mathrm{dist}_d(x, \overline{A}) = d(x^{-1} \circ \alpha)$. Let $a_j \in A$ be such that $a_j \to \alpha$ as $j \to \infty$. Then consider the following inequalities:

$$\begin{aligned}\mathrm{dist}_d(x, A) &\geq \mathrm{dist}_d(x, \overline{A}) = d(x^{-1} \circ \alpha) \\ &= \lim_{j\to\infty} d(x^{-1} \circ a_j) \geq \mathrm{dist}_d(x, A).\end{aligned}$$

The last inequality follows from $d(x^{-1} \circ a_j) \geq \mathrm{dist}_d(x, A)$ for every $j \in A$, since $a_j \in A$, and by the very definition of $\mathrm{dist}_d(x, A)$.
(b). Provide the details of the following arguments: Let $x_0 \in \mathbb{G}$ be fixed. It suffices to show that from every sequence $\{x_j\}_{j\in\mathbb{N}}$ in $\mathbb{G}$ converging to x_0 we can extract a subsequence $\{x_{j_n}\}_{n\in\mathbb{N}}$ such that

$$\lim_{n\to\infty} \mathrm{dist}_d(x_{j_n}, A) = \mathrm{dist}_d(x_0, A). \tag{5.147}$$

It is not restrictive to suppose that A is closed. For every $j \in \mathbb{N}$, there exists $a_j \in A$ such that

$$\mathrm{dist}_d(x_j, A) = d(x_j^{-1} \circ a_j). \tag{5.148}$$

It is easy to see that $\{a_j\}_{j\in\mathbb{N}}$ is bounded. Hence, we can extract a converging subsequence from $\{a_j\}_{j\in\mathbb{N}}$, say $a_{j_n} \to a \in A$ as $n \to \infty$. From (5.148) we infer

$$\lim_{n\to\infty} \mathrm{dist}_d(x_{j_n}, A) = \lim_{n\to\infty} d(x_{j_n}^{-1} \circ a_{j_n}) = d(x_0^{-1} \circ a). \tag{5.149}$$

Thus, (5.147) will follow if we show that $d(x_0^{-1} \circ a) = \mathrm{dist}_d(x_0, A)$. Suppose to the contrary that we have $d(x_0^{-1} \circ a) \neq \mathrm{dist}_d(x_0, A)$. This may occur iff

$$\text{(i)} : d(x_0^{-1} \circ a) < \mathrm{dist}_d(x_0, A) \qquad \text{or} \quad \text{(ii)} : d(x_0^{-1} \circ a) > \mathrm{dist}_d(x_0, A).$$

Case (i) is impossible by the very definition of $\mathrm{dist}_d(x_0, A)$ (since $a \in A$). Case (ii) is absurd too, as we show below. Let $a_0 \in A$ be such that $\mathrm{dist}_d(x_0, A) = d(a_0^{-1} \circ x_0)$. Then, by (5.149), we derive

$$\begin{aligned}d(x_0^{-1} \circ a) &= \lim_{n\to\infty} \mathrm{dist}_d(x_{j_n}, A) \leq \lim_{n\to\infty} d(a_0^{-1} \circ x_{j_n}) \\ &= d(a_0^{-1} \circ x_0) = \mathrm{dist}_d(x_0, A).\end{aligned}$$

This contradicts (ii).

Ex. 17) Prove the following improvement of Proposition 5.16.3 above, when the extra hypothesis *d is smooth* holds.

Proposition 5.16.4. *Let d be a smooth homogeneous norm as in Proposition 5.14.1, and let $A \subseteq \mathbb{G}$. Prove that the function $\mathrm{dist}_d(x, A)$ is Lipschitz continuous with respect to d, i.e.*

$$|\mathrm{dist}_d(\xi, A) - \mathrm{dist}_d(\eta, A)| \le \beta\, d(\xi, \eta) \qquad \forall\, \xi, \eta \in \mathbb{G},$$

where β is the same constant as in Proposition 5.14.1.

(*Hint*: Let $a \in A$, and write (5.125a) with $x = a$, $y = \xi$, $z = \eta$,

$$d(\xi^{-1} \circ a) \le d(\eta^{-1} \circ a) + \beta\, d(\xi^{-1} \circ \eta).$$

Then take the infimum over $a \in A$, and derive $\mathrm{dist}_d(\xi, A) \le \mathrm{dist}_d(\eta, A) + \beta\, d(\xi, \eta)$. Then interchange ξ and η.)

Ex. 18) **(T: Another mean integral operator).** Let $\mathcal{L}$ be a sub-Laplacian on the homogeneous Carnot group $\mathbb{G}$. Let $\Omega \subseteq \mathbb{G}$ be a bounded open set. For every $x \in \Omega$, let us put

$$r_x := \frac{1}{2}\mathrm{dist}_d(x, \partial\Omega),$$

where d is an $\mathcal{L}$-gauge function. For every $u \in L^1_{\mathrm{loc}}(\Omega)$, define

$$T(u)(x) := \mathrm{M}_{r_x}(u)(x), \tag{5.150}$$

where M_r is the average operator defined in (5.50f).
Prove the following statements:

(i) $T(u)$ is continuous in Ω,
(ii) $u \le v \Rightarrow T(u) \le T(v)$,
(iii) $u \in L^\infty(\Omega) \Rightarrow \sup_\Omega |T(u)| \le \mathrm{ess\,sup}_\Omega |u|$.

Ex. 19) **(The strong maximum principle related to T).** Let T be as in the previous exercise, and let u be a continuous function in Ω such that $T(u) \le u$. If Ω is connected and u attains its minimum in Ω, then $u =$ constant in Ω.

Ex. 20) **(The weak maximum principle related to T).** Let T be as in the previous exercises, and let u be a continuous function in Ω such that

$$\begin{cases} T(u) \le u & \text{in } \Omega, \\ \liminf_{x \to y} u(x) \ge 0 & \forall\, y \in \partial\Omega. \end{cases}$$

Prove that $u \ge 0$ in Ω.

Ex. 21) Let u be an $\mathcal{L}$-harmonic function in an open set $\Omega \subseteq \mathbb{G}$, $\Omega \ne \mathbb{G}$. Assume $u \in L^p(\Omega)$, $1 \le p \le \infty$. Then

$$|u(x)| \le \frac{c}{\mathrm{dist}_d(x, \partial\Omega)}\, \|u\|_{L^p(\Omega)},$$

where c is independent of u and Ω, and d is an $\mathcal{L}$-gauge function.
(*Hint*: Formula (5.52) may prove useful.)

Ex. 22) **(Another Koebe-type result).** Let $\Omega \subseteq \mathbb{G}$ be open, and let $u \in C(\Omega, \mathbb{R})$. Suppose that one of the following conditions is satisfied:

(i) $\mathcal{M}_\rho(u)(x) = \mathcal{M}_r(u)(x)$ for every $\rho, r > 0$ such that $0 < \rho \le r$ and $\overline{B_d(x,r)} \subset \Omega$,

(ii) $\mathrm{M}_\rho(u)(x) = \mathrm{M}_r(u)(x)$ for every $\rho, r > 0$ such that $0 < \rho \le r$ and $\overline{B_d(x,r)} \subset \Omega$.

Show that $u \in C^\infty(\Omega, \mathbb{R})$ and $\mathcal{L}u = 0$ in Ω.

Ex. 23) Prove the following *Harnack inequality on d-spheres.*

Theorem 5.16.5. *There exists a constant $C > 1$ such that*

$$\sup_{\partial B_d(0,r)} h \le C \inf_{\partial B_d(0,r)} h \tag{5.151}$$

for every $0 < r \le 1/2$ and every $\mathcal{L}$-harmonic non-negative function h on $B_d(0,1) \setminus \{0\}$.

Ex. 24) Let d be any homogeneous norm on $\mathbb{G}$, smooth on $\mathbb{G} \setminus \{0\}$. Prove that

$$\int_{d(x)=r} \frac{\mathrm{d}H^{N-1}(x)}{|\nabla d(x)|} = c_d\, r^{Q-1} \quad \text{for every } r > 0,$$

where $c_d = Q\, H^N(B_d(0,1))$. (*Hint*: By the coarea formula and the δ_λ-homogeneity of d, we have

$$\int_0^r \left(\int_{\{d(x)=\rho\}} \frac{\mathrm{d}H^{N-1}(x)}{|\nabla d(x)|} \right) \mathrm{d}\rho = \int_{\{d(x)<r\}} \mathrm{d}H^N = r^Q\, H^N(B_d(0,1)).$$

Then, take the derivative w.r.t. r on both sides.)

Ex. 25) In this exercise, we follow the arguments from [HK00] in Lemma 11.1 and Proposition 11.2, jointly with some of our results in Section 5.2.

Lemma 5.16.6. *Let $X = \{X_1, \dots, X_m\}$ be a system of locally Lipschitz-continuous vector fields on $\mathbb{R}^N$.*
Let $B_E(x,R) = \{z \in \mathbb{R}^N : |z-x| < R\}$ denote the Euclidean ball centered at x with radius R. With reference to (5.8a) (page 233) set

$$M(x,R) := \sup_{z \in B_E(x,R)} \left\{ \sum_{j=1}^n |X_j I(z)| \right\}.$$

Suppose $\gamma : [0,T] \to \mathbb{R}^N$ is X-subunit, $\gamma(0) = x$ and $T < R/M(x,R)$. Then $\gamma([0,T]) \subseteq B_E(x,R)$.

(*Hint*: Suppose to the contrary that $\gamma([0,T]) \not\subseteq B_E(x,R)$. Hence, there exists a least $t \in (0,T]$, such that $y := \gamma(t) \notin B_E(x,R)$. From the minimality of t and the continuity of γ, we certainly have $y \in \partial B_E(x,R)$, i.e.

$|y-x| = R$. Notice also that $d_X(x, y) \le t \le T$. Hence, (5.8b) on page 233 gives

$$R = |x - y| \le M(x, |x - y|)\, d_X(x, y) \le M(x, R)T < R.$$

This is absurd.)
Deduce the following result.

Corollary 5.16.7. *Let $X = \{X_1, \dots, X_m\}$ be a system of locally Lipschitz-continuous vector fields on $\mathbb{R}^N$.*
Then, for every bounded set $D \subset \mathbb{R}^N$, there exist positive numbers $R_0 \ll 1$, $R_1 \gg 1$ (depending only on D and X) such that $B_{d_X}(x, R) \subseteq B_E(0, R_1)$ for every $x \in D$ and every $R \in [0, R_0]$. In particular, for every $x_0 \in \mathbb{R}^N$, every Carnot–Carathéodory ball $B_{d_X}(x_0, R)$ with small radius (dependently on x_0) is bounded in the Euclidean metric.

(*Hint*: With the notation of Lemma 5.16.6, let

$$\mathbf{M} := \sup\{M(x, 1) : x \in \overline{D}\}.$$

Obviously, $\mathbf{M} < \infty$, for the X_j's are continuous. Set $R_0 := (4\,\mathbf{M})^{-1}$. Let $x \in D$ and $R \le R_0$. (Note that $\mathbf{M} \ge M(x, 1)$, since $x \in D$.) Let also $y \in B_{d_X}(x, R)$, i.e. $d_X(x, y) < R$. Then there exists a X-subunit curve $\gamma : [0, T] \to \mathbb{R}^N$ such that $\gamma(0) = x$, $\gamma(T) = y$ and

$$T < d_X(x, y) + (4\mathbf{M})^{-1} < 2\, R_0 = (2\mathbf{M})^{-1} < 1/M(x, 1).$$

Then, we can apply Lemma 5.16.6 with $R = 1$ and derive that $\gamma([0, T]) \subseteq B_E(x, 1)$. In particular, this gives $|y - x| = |\gamma(T) - x| < 1$, i.e. $y \in B_E(x, 1)$. Due to the arbitrariness of $y \in B_{d_X}(x, R)$, this proves

$$B_{d_X}(x, R) \subseteq B_E(x, 1) \subseteq D_1 := \bigcup_{x \in D} B_E(x, 1) \Subset \mathbb{R}^N.$$

This ends the proof, by finding a suitable $R_1 \gg 1$ such that $D_1 \subseteq B_E(0, R_1)$.)

Proposition 5.16.8. *Let $X = \{X_1, \dots, X_m\}$ be a system of locally Lipschitz-continuous vector fields on $\mathbb{R}^N$.*
Let K be a compact subset of $\mathbb{R}^N$. Then there exists a constant $\mathbf{c} = \mathbf{c}(K, X) > 0$ such that

$$d_X(x, y) \ge \mathbf{c}\, |x - y| \quad \textit{for every } x, y \in K.$$

Proof. If $d_X(x, y) = \infty$, there is nothing to prove. Hence, we can suppose $d_X(x, y) < \infty$. Let $\gamma : [0, T] \to \mathbb{R}^N$ be any X-subunit curve such that $\gamma(0) = x$, $\gamma(T) = y$. Let $R := |x - y|$. Set

$$M := \sup\left\{ \sum_{j=1}^{n} |X_j I(z)| \ :\ z \in B_E(x, 1 + \operatorname{diam}(K)),\ x \in K \right\}.$$

Note that $M = M(K, X) < \infty$, since K is compact and the X_j's are continuous. Obviously, we do not have $\gamma([0, T]) \subseteq B_E(x, R)$, because $|\gamma(T) - x| = |y - x| = R$ is not $< R$. Hence, by Lemma 5.16.6,

$$T \geq R/M(x, R) \geq R/M =: \mathbf{c}\, R = \mathbf{c}|x - y|.$$

(Indeed, $M(x, R) \leq M$.) Passing to the infimum over the above γ's, we get $d_X(x, y) \geq \mathbf{c}\, |x - y|$. □

Derive the following assertion from Corollary 5.16.7 and Proposition 5.16.8.

Proposition 5.16.9. *Let $X = \{X_1, \dots, X_m\}$ be locally Lipschitz-continuous vector fields on $\mathbb{R}^N$. Suppose $\mathbb{R}^N$ is X-connected. Let d_X be the relevant Carnot–Carathéodory distance. Then the map*

$$\mathrm{id} : (\mathbb{G}, d_X) \to (\mathbb{G}, d_E)$$

is continuous.

(*Hint*: Dealing with metric spaces, we can prove sequential continuity. Let $x_n \to x_0$ in $(\mathbb{G}, d_X)$. We aim to prove that $x_n \to x_0$ in $(\mathbb{G}, d_E)$. By Corollary 5.16.7, there exists $\varepsilon = \varepsilon(x_0) > 0$ such that $B_{d_X}(x_0, \varepsilon)$ is bounded in $(\mathbb{G}, d_E)$. By definition of limit, there exists $\overline{n} = \overline{n}(\varepsilon) \in \mathbb{N}$ such that $x_n \in B_{d_X}(x_0, \varepsilon)$ whenever $n \geq \overline{n}$. Hence the set $\{x_n : n \geq \overline{n}\} \cup \{x_0\}$ is contained in a compact subset of $\mathbb{R}^N$. We can thus apply Proposition 5.16.8 to derive that, for a suitable $\mathbf{c} > 0$,

$$\mathbf{c}\, |x_n - x_0| \leq d_X(x_n, x_0) \quad \text{for all } n \geq \overline{n}.$$

Letting $n \to \infty$, we get $x_n \to x_0$ in $(\mathbb{G}, d_E)$.)

Now, consider a system X of Hörmander vector fields in $\mathbb{R}^N$ and the relevant d_X. Recall that, by the Carathéodory–Chow–Rashevsky theorem (see Chapter 19 for suitable references), $\mathbb{R}^N$ is X-connected. Moreover, by known results (see, e.g. [NSW85]) the inequalities in Proposition 5.15.1 hold (for a suitable $r > 0$) when $d(y^{-1} \circ x)$ is replaced by $d_X(x, y)$. In particular, given a compact subset K of $\mathbb{R}^N$, there exist $r, \mathbf{c} > 0$ (depending on K, X) such that

$$d_X(x, y) \leq \mathbf{c}\, |x - y|^{1/r} \quad \forall\, x,\ y \in K.$$

Arguing as in the first paragraph of the proof of Corollary 5.15.3, derive the following assertion:

Given a system X of Hörmander vector fields in $\mathbb{R}^N$, the map $\mathrm{id} : (\mathbb{G}, d_E) \to (\mathbb{G}, d_X)$ *is continuous* (indeed, it is a homeomorphism, see Proposition 5.16.9).

Ex. 26) Consider the vector fields in $\mathbb{R}^2$ defined by

$$X_1 = \partial_{x_1}, \qquad X_2 = \max\{0, x_1\}\,\partial_{x_2}.$$

Consider the relevant d_X, where $X = \{X_1, X_2\}$. Is $\mathbb{R}^2$ X-connected? Prove that the identity map $\mathrm{id} : (\mathbb{R}^2, d_E) \to (\mathbb{R}^2, d_X)$ is not continuous.
(*Hint*: It may be useful to notice that the integral curves of X_1 are the lines parallel to the x_1 axis; the integral curve of X_2 through (x_0, y_0) is

$$t \mapsto \begin{cases} (x_0, y_0) & \text{if } x_0 \le 0, \\ (x_0, y_0 + x_0\, t) & \text{if } x_0 > 0, \end{cases}$$

i.e. the single point (x_0, y_0) if $x_0 \le 0$ or the lines parallel to the x_2 axis if $x_0 > 0$.)

Ex. 27) Consider the vector fields in $\mathbb{R}^2$ defined by

$$X_1 = \partial_{x_1}, \qquad X_2 = (1 + x_2^2)\,\partial_{x_2}.$$

Verify that the system $X = \{X_1, X_2\}$ satisfies the Hörmander condition. Consider the relevant d_X. Is $\mathbb{R}^2$ X-connected? Prove that there exists a set $\Omega \subset \mathbb{R}^2$ which is bounded w.r.t. d_X but unbounded in the Euclidean metric. *Hint*: Observe that the integral curve of X_2 (hence, a X-subunit path!) starting from (x_0, y_0) is

$$t \mapsto \big(x_0, \tan(t + \arctan(y_0))\big).$$

Hence, for example,

$$d_X\left((0,0), \left(0, \tan\left(\frac{\pi}{2} - \frac{1}{n}\right)\right)\right) \le \pi/2 - 1/n < \pi/2,$$

but

$$d_E\left((0,0), \left(0, \tan\left(\frac{\pi}{2} - \frac{1}{n}\right)\right)\right) \longrightarrow \infty \quad \text{as } n \to \infty.$$

Ex. 28) Prove the following linear algebra result.

Lemma 5.16.10. *Let $\mathbf{v}_1, \dots, \mathbf{v}_m$ be vectors in $\mathbb{R}^N$. Suppose $\mathbf{w} \in \mathbb{R}^N$ is such that*

$$(\star) \qquad \langle \mathbf{w}, \mathbf{x}\rangle^2 \le \sum_{j=1}^m \langle \mathbf{v}_j, \mathbf{x}\rangle^2 \quad \forall\, \mathbf{x} \in \mathbb{R}^N.$$

Then there exist scalars $\alpha_1, \dots, \alpha_m$ such that

$$\mathbf{w} = \sum_{j=1}^m \alpha_j\, \mathbf{v}_j \quad \text{and} \quad \sum_{j=1}^m \alpha_j^2 \le 1.$$

Proof. Let us first prove that $\mathbf{w} \in \text{span}\{\mathbf{v}_1, \dots, \mathbf{v}_m\} =: V$. From the decomposition $\mathbb{R}^N = V \oplus V^\perp$ it follows

$$\mathbf{w} = \mathbf{a} + \mathbf{b}, \qquad \mathbf{a} \in V, \quad \mathbf{b} \in V^\perp.$$

If we choose $\mathbf{x} := \mathbf{b}$ in $(\star)$, we get

$$\|\mathbf{b}\|^4 = \langle \mathbf{a} + \mathbf{b}, \mathbf{b} \rangle^2 \le \sum_{j=1}^m \langle \mathbf{v}_j, \mathbf{b} \rangle^2 = 0,$$

whence $\mathbf{b} = 0$, i.e. $\mathbf{w} = \mathbf{a} \in V$.
Up to a permutation of the $\mathbf{v}_j$'s, it is not restrictive to suppose that $(\mathbf{v}_1, \dots, \mathbf{v}_q)$ is a basis of V. Then we have

$$\mathbf{v}_j = \sum_{k=1}^q \beta_{j,k}\, \mathbf{v}_k \quad \forall\, j = 1, \dots, m.$$

The $m \times q$ matrix B whose (j,k)-th entry is $\beta_{j,k}$ has the block form

$$B = \begin{pmatrix} \mathbb{I}_q \\ \widehat{B} \end{pmatrix},$$

where $\mathbb{I}_q$ is the identity matrix of order q. Let $\gamma_1, \dots, \gamma_q \in \mathbb{R}$ be such that $\mathbf{w} = \sum_{k=1}^q \gamma_k\, \mathbf{v}_k$. We characterize *all* the scalars $\alpha_1, \dots, \alpha_m \in \mathbb{R}$ such that $\mathbf{w} = \sum_{j=1}^m \alpha_j\, \mathbf{v}_j$. From the identity

$$\sum_{k=1}^q \gamma_k\, \mathbf{v}_k = \mathbf{w} = \sum_{k=1}^q \left(\sum_{j=1}^m \alpha_j\, \beta_{j,k} \right) \mathbf{v}_k$$

and the linear independence of $\mathbf{v}_1, \dots, \mathbf{v}_q$ we infer

$$(\bullet) \qquad \gamma_k = \sum_{j=1}^m \alpha_j\, \beta_{j,k} \quad \forall\, k = 1, \dots, q.$$

Let us suppose that, besides $(\bullet)$, there also exists a solution $\bar{\mathbf{x}} \in \mathbb{R}^N$ to the m-equation system

$$\text{(SL)} \qquad \begin{cases} \langle \mathbf{v}_j, \mathbf{x} \rangle = \alpha_j, \\ j = 1, \dots, m. \end{cases}$$

This will give

$$\left(\sum_{j=1}^m \alpha_j^2 \right)^2 = \left(\sum_{j=1}^m \alpha_j \,\langle \mathbf{v}_j, \bar{\mathbf{x}} \rangle \right)^2 = \left(\left\langle \sum_{j=1}^m \alpha_j\, \mathbf{v}_j, \bar{\mathbf{x}} \right\rangle \right)^2$$
$$= \langle \mathbf{w}, \bar{\mathbf{x}} \rangle^2 \le \sum_{j=1}^m \langle \mathbf{v}_j, \bar{\mathbf{x}} \rangle^2 = \sum_{j=1}^m \alpha_j^2,$$

whence $\sum_{j=1}^m \alpha_j^2 \le 1$, and the proof is complete.

We remark that, in order to (SL) to be solvable, it is necessary and sufficient that the rank of the coefficient-matrix of (SL) (i.e. q) equals the rank of the complete-matrix of (SL): thanks to the dependence of $\mathbf{v}_{q+1}, \dots, \mathbf{v}_m$ w.r.t. $\mathbf{v}_1, \dots, \mathbf{v}_q$, this is equivalent to

$$(\bullet\bullet) \qquad \alpha_j = \sum_{k=1}^{q} \beta_{j,k}\, \alpha_k \quad \forall\, j = q+1, \dots, m.$$

Hence, the proof is complete if we show that there exists $(\alpha_1, \dots, \alpha_m) \in \mathbb{R}^m$ satisfying $(\bullet)$ and $(\bullet\bullet)$, i.e. a solution to the m-equation and m-indeterminate system $\alpha_1, \dots, \alpha_m$

$$\text{(SL)'} \qquad \begin{cases} \displaystyle\sum_{j=1}^{m} \alpha_j\, \beta_{j,k} = \gamma_k, & k = 1, \dots, q, \\ \displaystyle\sum_{k=1}^{q} \beta_{j,k}\, \alpha_k - \alpha_j = 0, & j = q+1, \dots, m. \end{cases}$$

The coefficient-matrix of (SL)' is

$$\begin{pmatrix} \mathbb{I}_q & {}^t\widehat{B} \\ \widehat{B} & -\mathbb{I}_{m-q} \end{pmatrix}.$$

We show that this matrix is invertible, whence (SL)' is solvable.
In general, if A is a real $q \times (m-q)$ matrix, the block matrix

$$P := \begin{pmatrix} \mathbb{I}_q & A \\ {}^tA & -\mathbb{I}_{m-q} \end{pmatrix} \quad \text{satisfies} \quad P^2 = \mathbb{I}_m + \begin{pmatrix} A & 0 \\ 0 & {}^tA \end{pmatrix} \cdot \begin{pmatrix} {}^tA & 0 \\ 0 & A \end{pmatrix}.$$

Observe that P^2 is the sum of a positive-definite matrix plus a positive semi-definite one. In particular, P^2 is positive-definite, hence it is not singular. This gives $|\det P| = \sqrt{\det P^2} > 0$, so that P is not singular too. This ends the proof. □

By means of Lemma 5.16.10, derive the following equivalent characterization of X-subunit curve.

Proposition 5.16.11. *Let $X = \{X_1, \dots, X_m\}$ be a system of locally Lipschitz-continuous vector fields on an open set $\Omega \subseteq \mathbb{R}^N$.*
Let $\gamma : [0, T] \to \Omega$ be an absolutely continuous curve. Then γ is X-subunit if and only if there exist measurable functions $c_j : [0, T] \to \mathbb{R}$, $j = 1, \dots, m$, such that, almost everywhere on $[0, T]$,

$$\dot\gamma(t) = \sum_{j=1}^{m} c_j(t)\, X_j I(\gamma(t)), \quad \text{and} \quad \sum_{j=1}^{m} \big(c_j(t)\big)^2 \le 1. \tag{5.152}$$

Proof. Suppose $\gamma : [0, T] \to \Omega$, satisfies (5.152). The Cauchy-Schwartz inequality in $\mathbb{R}^m$ immediately yields

$$\langle \dot{\gamma}(t), \xi \rangle^2 \le \left(\sum_{j=1}^{m} (c_j(t))^2 \right) \cdot \left(\sum_{j=1}^{m} \langle X_j(\gamma(t)), \xi \rangle^2 \right) \le \sum_{j=1}^{m} \langle X_j(\gamma(t)), \xi \rangle^2,$$

i.e. γ is a X-subunit curve. Vice versa, let γ be X-subunit. If we apply Lemma 5.16.10 with the choice

$$\mathbf{v}_j := X_j(\gamma(t)), \quad j = 1, \dots, m, \quad \mathbf{w} := \dot{\gamma}(t),$$

then (5.152) is satisfied for suitable scalar functions c_j's defined a.e. on $[0, T]$. It is easy to prove[16] the measurability of the c_j's. $\square$

Ex. 29) Let u be an $\mathcal{L}$-harmonic function in an open set $\Omega \subseteq \mathbb{G}$. Let $\overline{B_d(x_0, r)} \subset \Omega$. Then, for every multi-index α, there exists a constant $C_\alpha > 0$ (independent of u, x_0 and r) such that

$$\left| X^\alpha u(x_0) \right| \le C_\alpha \, r^{-|\alpha|_{\mathbb{G}}} \sup_{B_d(x_0, r)} |u|.$$

(*Hint*: Since u is $\mathcal{L}$-harmonic, the representation formula (5.50a) (see also (5.50d)) gives

$$u(x_0) = \int_{\mathbb{R}^N} u(z)\, \phi_r(x^{-1} \circ z)\, \mathrm{d}z = \int_{\mathbb{R}^N} u(z)\, \widehat{\phi_r}(x^{-1} \circ z)\, \mathrm{d}z,$$

where $\widehat{\phi_r}(\zeta) = \phi_r(\zeta^{-1})$. Now, derive the integral.)

[16] Indeed, notice that $\dot{\gamma}$ is measurable (since γ is a absolutely continuous), $t \mapsto X_j(\gamma(t))$ is continuous and the operations which provide the components of a vector w.r.t. a system of vectors are continuous. We leave the details to the reader.

Part II

Elements of Potential Theory for Sub-Laplacians

6

Abstract Harmonic Spaces

In this chapter, we present some topics from the theory of abstract harmonic spaces. Although this theory is extremely vast, we only focus on the results which are crucial for the scopes of the next chapters.

In this abstract setting, we shall only deal with the Perron–Wiener–Brelot method for the Dirichlet problem, the study of harmonic minorants and majorants and balayage theory.

Our aim here is to furnish the background material for an exhaustive potential theory for the sub-Laplacians $\mathcal{L}$, which will be developed through the rest of Part II. *The lack of explicit Poisson integral formulas for $\mathcal{L}$ forced us to follow the abstract approach to this theory.*

In the subsequent chapters, we shall use this basic results, together with the particular structure of the fundamental solutions, to develop the deepest part of the potential theory for $\mathcal{L}$.

The following scheme describes our approach to the abstract potential theory for sub-Laplacians. Notation and definitions will be explained in due course.

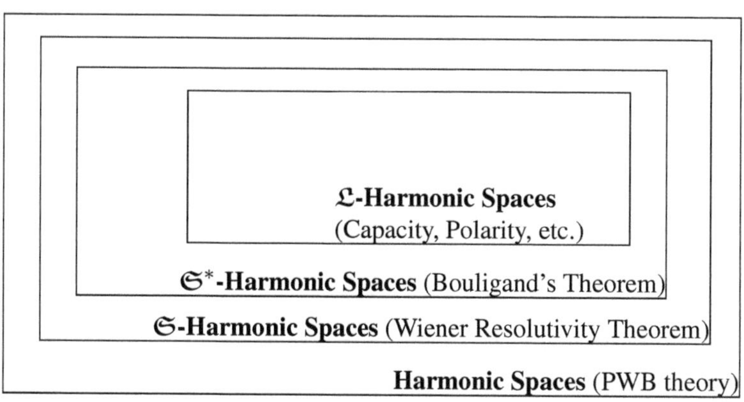

Convention. Throughout this chapter $(E, \mathcal{T})$ will denote a topological Hausdorff space, locally connected and locally compact. We also assume that the topology $\mathcal{T}$ has a *countable basis*.

6.1 Preliminaries

Let $A \subseteq E$ and $u : A \to [-\infty, \infty]$. If x is a point of $\overline{A}$, we define

$$\liminf_{y \to x} u(y) := \sup_{V \in \mathcal{U}_x} \Big(\inf_{V \cap A} u \Big),$$

$$\limsup_{y \to x} u(y) := \inf_{V \in \mathcal{U}_x} \Big(\sup_{V \cap A} u \Big),$$

where $\mathcal{U}_x$ denotes the *family of the neighborhoods* of x. (Note the slight difference w.r.t. the usual definition of lim sup and lim inf in that, for example, in the lim inf definition, the inner $\inf_{V \cap A}$ is usually replaced by $\inf_{(V \setminus \{x\}) \cap A}$. Our choice will be soon motivated.)

The function u is called *lower semicontinuous* (l.s.c., in short) at $x \in A$ if $u(x) = \liminf_{y \to x} u(y)$. If $u(x) = \limsup_{y \to x} u(y)$, then u will be called *upper semicontinuous* (u.s.c., in short) at x. If u is l.s.c. (u.s.c.) at any point of A, then u will be said l.s.c. (u.s.c.) on A.

We now list some elementary properties of semicontinuous functions, whose proofs will be left as exercises.

(P1) Suppose that f, g are l.s.c. (u.s.c.) on A. Then $f+g$, λf with $\lambda \geq 0$, $\max\{f, g\}$, $\min\{f, g\}$ are l.s.c. (u.s.c.) on A. Moreover, λf is u.s.c. if f is l.s.c. and $\lambda < 0$.

(P2) $f : A \to [-\infty, \infty]$ is l.s.c. on A if and only if

$$\{f > t\} := \{x \in A : f(x) > t\}$$

is open for every $t \in \mathbb{R}$. Similarly, f is u.s.c. on A if and only if

$$\{f < t\} := \{x \in A : f(x) < t\}$$

is open for every $t \in \mathbb{R}$.

(P3) Let $(u_\alpha)_{\alpha \in \mathcal{A}}$ be a family of l.s.c. (u.s.c.) functions on A. Then $\sup_{\alpha \in \mathcal{A}} u_\alpha$ ($\inf_{\alpha \in \mathcal{A}} u_\alpha$, respectively) is l.s.c. (u.s.c., respectively).

(P4) If A is compact and $u : A \to]-\infty, \infty]$ ($u : A \to [-\infty, \infty[$, respectively) is l.s.c. (u.s.c., respectively), then u attains its minimum (maximum, respectively) on A.

(P5) If A is compact and $u : A \to]-\infty, \infty]$ ($u : A \to [-\infty, \infty[$, respectively) is l.s.c. (u.s.c., respectively), then there exists a sequence (f_j) of real continuous functions on A such that

$$f_j \leq f_{j+1} \quad \text{and} \quad f_j \to u \text{ pointwise on } A$$
$$(f_j \geq f_{j+1} \quad \text{and} \quad f_j \to u \text{ pointwise on } A\text{, respectively}).$$

If $u : A \to [-\infty, \infty]$, one defines the *lower regularization* $\widehat{u}$ of u as follows

$$\widehat{u} : A \to [-\infty, \infty], \quad \widehat{u}(x) = \liminf_{y \to x} u(y).$$

Similarly, one defines the *upper regularization* $\check{u}$ of u as follows

$$\check{u} : A \to [-\infty, \infty], \quad \check{u}(x) = \limsup_{y \to x} u(y).$$

The functions $\widehat{u}$ and $\check{u}$ are l.s.c. and u.s.c., respectively, and

$$\widehat{u} \le u \le \check{u}.$$

We now state and prove some propositions on the envelopes of families of functions which will play a crucial rôle in our subsequent exposition. Our proofs are simple adaptation to an abstract setting of Lemmas 3.7.1 and 3.7.4 in [AG01].

Proposition 6.1.1 (Envelopes I). *Let $(u_\alpha)_{\alpha \in \mathcal{A}}$ be a family of l.s.c. functions from A $(\subseteq E)$ to $]-\infty, \infty]$. Then there exists a countable set $\mathcal{B} \subseteq \mathcal{A}$ such that*

$$\sup_{\alpha \in \mathcal{B}} u_\alpha = \sup_{\alpha \in \mathcal{A}} u_\alpha.$$

Proof. Let $u := \sup_{\alpha \in \mathcal{A}} u_\alpha$. Then u is l.s.c. (see (P3)) and

$$\{u > t\} = \bigcup_{\alpha \in \mathcal{A}} \{u_\alpha > t\}.$$

Since $(E, \mathcal{T})$ is locally compact and endowed with a countable basis of open sets, there exists a countable set $\mathcal{A}_t \subseteq \mathcal{A}$ such that

$$\{u > t\} = \bigcup_{\alpha \in \mathcal{A}_t} \{u_\alpha > t\}. \tag{6.1}$$

Define $\mathcal{B} := \bigcup_{t \in \mathbb{Q}} \mathcal{A}_t$ and $v := \sup_{\alpha \in \mathcal{B}} u_\alpha$. Let us prove that $v = u$. Assume by contradiction $v \ne u$. Since $v \le u$, this implies the existence of a point $y \in A$ such that $v(y) < u(y)$. Let $\tau \in \mathbb{Q}$ satisfy $v(y) \le \tau < u(y)$. By using (6.1), we can find $\alpha_\tau \in \mathcal{A}_\tau$ such that $\tau < u_{\alpha_\tau}(y)$. Then

$$v(y) = \sup_{\alpha \in \mathcal{B}} u_\alpha(y) \ge u_{\alpha_\tau}(y) > \tau \ge v(y).$$

This contradiction completes the proof. □

Proposition 6.1.2 (Envelopes II: Choquet). *Let $(u_\alpha)_{\alpha \in \mathcal{A}}$ be a family of functions from A $(\subseteq E)$ to $[-\infty, \infty]$, and let $u := \inf_{\alpha \in \mathcal{A}} u_\alpha$. Then there exists a countable set $\mathcal{B} \subseteq \mathcal{A}$ such that*

$$\widehat{u} = \widehat{v}, \quad \text{where} \quad v = \inf_{\alpha \in \mathcal{B}} u_\alpha.$$

Proof. By replacing u_α with $\arctan u_\alpha$, we may assume that u is bounded from below. Let $(B_n)_{n\in\mathbb{N}}$ be a sequence of open sets such that $\{B_n : n \in \mathbb{N}\}$ forms a basis of the topology $\mathcal{T}$ and such that, for every $m \in \mathbb{N}$,

$$\mathbb{N}_m := \{n \in \mathbb{N} : B_n = B_m\}$$

is infinite. For every $n \in \mathbb{N}$, we now choose $x_n \in B_n$ and $\alpha_n \in \mathcal{A}$ such that

$$u(x_n) < \inf_{B_n\cap A} u + \frac{1}{n} \quad \text{and} \quad u_{\alpha_n}(x_n) < u(x_n) + \frac{1}{n}.$$

Let us define $\mathcal{B} := \{\alpha_n : n \in \mathbb{N}\}$ and $v := \inf_{\alpha\in\mathcal{B}} u_\alpha$. Then, for every $m \in \mathbb{N}$ and $n \in \mathbb{N}_m$, we have

$$\inf_{B_m\cap A} v = \inf_{B_n\cap A} v \le u_{\alpha_n}(x_n) < u(x_n) + \frac{1}{n} < \inf_{B_n\cap A} u + \frac{2}{n} = \inf_{B_m\cap A} u + \frac{2}{n},$$

hence, being $\mathbb{N}_m$ infinite,

$$\inf_{B_m\cap A} v \le \inf_{B_m\cap A} u \quad \text{for every } m \in \mathbb{N}.$$

As a consequence, for every $x \in A$,

$$\widehat{v}(x) = \sup_{\{B_m : x\in B_m\}} \Big(\inf_{B_m\cap A} v\Big) \le \sup_{\{B_m : x\in B_m\}} \Big(\inf_{B_m\cap A} u\Big) = \widehat{u}(x),$$

so that $\widehat{v} \le \widehat{u}$. Since the reverse inequality is obvious, we are done. □

6.2 Sheafs of Functions. Harmonic Sheafs

We begin with the main definition of this section.

Definition 6.2.1 (Sheaf of functions. Harmonic sheaf). *Suppose we are given, for every open set $V \in \mathcal{T}$, a family $\mathcal{F}(V)$ of extended real-valued functions $u : V \to [-\infty, \infty]$. We say that the map*

$$\mathcal{F} : V \mapsto \mathcal{F}(V)$$

is a sheaf of functions *on E if the following properties hold:*

$$\text{if } V_1, V_2 \in \mathcal{T} \text{ with } V_1 \subseteq V_2 \text{ and } u \in \mathcal{F}(V_2), \text{ then } u|_{V_1} \in \mathcal{F}(V_1), \tag{6.2a}$$

$$\begin{aligned} &\text{if } (V_\alpha)_{\alpha\in\mathcal{A}} \subseteq \mathcal{T} \text{ and } u : \textstyle\bigcup_{\alpha\in\mathcal{A}} V_\alpha \to [-\infty, \infty] \\ &\text{is such that } u|_{V_\alpha} \in \mathcal{F}(V_\alpha) \text{ for every } \alpha \in \mathcal{A}, \text{ then } u \in \mathcal{F}(\textstyle\bigcup_{\alpha\in\mathcal{A}} V_\alpha). \end{aligned} \tag{6.2b}$$

A sheaf of functions $\mathcal{F}$ on E is called harmonic *if $\mathcal{F}(V)$ is a linear subspace of $C(V, \mathbb{R})$, the vector space of the real continuous functions defined on V. When $\mathcal{F}$ is a harmonic sheaf on E and V is an open set in $\mathcal{T}$, a function $u \in \mathcal{F}(V)$ will be called $\mathcal{F}$-harmonic.*

6.2.1 Regular Open Sets. Harmonic Measures. Hyperharmonic Functions

Definition 6.2.2 ($\mathcal{H}$-regular set). *Let $\mathcal{H}$ be a harmonic sheaf on E. We say that an open set $V \in \mathcal{T}$ is* $\mathcal{H}$-regular *if the following conditions are satisfied:*

(R1) *$\overline{V}$ is compact and $\partial V \neq \emptyset$;*
(R2) *for every continuous function $f : \partial V \to \mathbb{R}$, there exists a unique $\mathcal{H}$-harmonic function in V, denoted by H_f^V, such that*

$$\lim_{x \to y} H_f^V(x) = f(y) \qquad \textit{for every } y \in \partial V;$$

(R3) *if $f \geq 0$, then $H_f^V \geq 0$.*

When V is $\mathcal{H}$-regular, from the linearity of $\mathcal{F}(V)$ and the uniqueness assumption in condition (R2) it follows that

$$H_{f+g}^V = H_f^V + H_g^V, \qquad H_{\lambda f}^V = \lambda H_f^V$$

for every $f, g \in C(\partial V, \mathbb{R})$ and for every $\lambda \in \mathbb{R}$. Then, also keeping in mind (R3), for every $\mathcal{H}$-regular open set V and for every $x \in V$, the map

$$C(\partial V, \mathbb{R}) \ni f \mapsto H_f^V(x) \in \mathbb{R}$$

is *linear* and *positive*. Hence, the following definition is well posed.

Definition 6.2.3 ($\mathcal{H}$-harmonic measure. I). *Let $\mathcal{H}$ be a harmonic sheaf on E. Let $V \in \mathcal{T}$ be an $\mathcal{H}$-regular set. Then there exists a Radon measure μ_x^V on $C(V, \mathbb{R})$ such that*

$$H_f^V(x) = \int_{\partial V} f(y)\, \mathrm{d}\mu_x^V(y) \qquad \forall\, f \in C(\partial V, \mathbb{R}).$$

*The measure μ_x^V is called the $\mathcal{H}$-*harmonic measure *related to V and x.*

We provide a definition which will be used throughout the sequel.

Definition 6.2.4 ($\mathcal{H}$-hyperharmonic function). *Let $\mathcal{H}$ be a harmonic sheaf on $(E, \mathcal{T})$. Let $\Omega \in \mathcal{T}$. A function $u : \Omega \to]-\infty, \infty]$ is called $\mathcal{H}$-*hyperharmonic *in Ω if:*

(i) *u is lower semi-continuous;*
(ii) *for every $\mathcal{H}$-regular open set $V \subset \overline{V} \subseteq \Omega$, one has*

$$u(x) \geq \int_{\partial V} u(y)\, \mathrm{d}\mu_x^V(y) \qquad \forall\, x \in V.$$

We shall denote by $\mathcal{H}^*(\Omega)$ the set of the $\mathcal{H}$-hyperharmonic functions in Ω.

Since

$$\int_{\partial V} u\, \mathrm{d}\mu_x^V = \sup\left\{ \int_{\partial V} \varphi\, \mathrm{d}\mu_x^V \mid \varphi \in C(\partial V, \mathbb{R}),\ \varphi \leq u \right\},$$

condition (ii) can be rewritten as follows

(ii)' for every $\mathcal{H}$-regular open set $V \subseteq \overline{V} \subseteq \Omega$ and for every $\varphi \in C(\partial V, \mathbb{R})$ such that $\varphi \leq u|_{\partial V}$, one has

$$H_\varphi^V \leq u|_V.$$

A function $v : \Omega \to [-\infty, \infty[$ will be called $\mathcal{H}$-*hypoharmonic* if $-v \in \mathcal{H}^*(\Omega)$. We denote by

$$\mathcal{H}_*(\Omega) := -\mathcal{H}^*(\Omega)$$

the family of the $\mathcal{H}$-hypoharmonic functions in Ω.

Exercise 6.2.5. Let $\Omega \subseteq E$ be open, and let $\mathcal{F} \subseteq \mathcal{H}^*(\Omega)$. Suppose that

$$u_0 := \inf \mathcal{F}$$

is l.s.c. in Ω. Then $u_0 \in \mathcal{H}^*(\Omega)$. (*Hint*: if $V \subseteq \overline{V} \subseteq \Omega$ is $\mathcal{H}$-regular, then $u(x) \geq \int_{\partial V} u \, d\mu_x^V \geq \int_{\partial V} u_0 \, d\mu_x^V$ for every $x \in V$ and $u \in \mathcal{F}$.)

We will see later that $\mathcal{H}^*(\Omega)$ is a sheaf of functions. Now, we state a proposition whose easy proof will be left as an exercise.

Proposition 6.2.6. *Let $\Omega \subset E$ be open, and let $u,\ v \in \mathcal{H}^*(\Omega)$. Then:*

(i) $u + v \in \mathcal{H}^*(\Omega)$,
(ii) $\lambda\, u \in \mathcal{H}^*(\Omega)$ *for every* $\lambda \geq 0$,
(iii) $\min\{u, v\} \in \mathcal{H}^*(\Omega)$.

An analogous proposition also holds, obviously, for $\mathcal{H}_*(\Omega)$.

6.2.2 Directed Families of Functions

In this section we shall show some results on directed families of functions, extending well-known theorems related to monotone sequences of functions. The results we are going to show will be applied in the subsequent section to families of harmonic and hyperharmonic functions.

Let $\mathcal{F}$ be a family of functions from $A \subseteq E$ to $[-\infty, \infty]$.

Definition 6.2.7 (Up (down) directed family). *We say that $\mathcal{F}$ is* up directed, *and we shall write*

$$\mathcal{F} \uparrow,$$

if, for every $u, v \in \mathcal{F}$, there exists $w \in \mathcal{F}$ such that

$$u \leq w \quad \textit{and} \quad v \leq w.$$

Analogously, if, for every $u, v \in \mathcal{F}$, there exists $w \in \mathcal{F}$ such that $u \geq w,\ v \geq w$, we say that $\mathcal{F}$ is down directed, *and we shall write $\mathcal{F} \downarrow$.*

Theorem 6.2.8 (Dini–Cartan). *Let $A \subseteq E$ be compact, and let $\mathcal{F}$ be a* down directed *family of upper semicontinuous functions. Assume*

$$\inf \mathcal{F} := \inf_{u \in \mathcal{F}} u = 0.$$

Then $\mathcal{F}$ is uniformly convergent to zero, that is, for every $\varepsilon > 0$, there exists $u \in \mathcal{F}$ such that $u(x) < \varepsilon$ for every $x \in A$.

Proof. Let $\varepsilon > 0$. Since $\inf \mathcal{F} = 0$, for every $x \in A$, there exists $u_x \in \mathcal{F}$ such that $u_x(x) < \varepsilon$. The upper semicontinuity of u_x implies the existence of a neighborhood V_x of x such that $u_x(y) < \varepsilon$ for every $y \in V_x$. Since A is compact, we can find

$$x_1, \dots, x_p \in A \quad \text{such that} \quad A \subseteq \bigcup_{j=1}^{p} V_{x_j}.$$

On the other hand, since $\mathcal{F}$ is down directed, there exists $u \in \mathcal{F}$ such that $u \leq u_{x_j}$ for every $j \in \{1, \dots, p\}$. As a consequence, given any point $y \in A$, there exists $j \in \{1, \dots, p\}$ such that $y \in V_{x_j}$, so that $u(y) \leq u_{x_j}(y) < \varepsilon$, and the assertion follows. □

Corollary 6.2.9. *Let $\mathcal{F}$ be an up directed (respectively, down directed) family of l.s.c. (respectively, u.s.c.) functions on a compact set $A \subseteq E$. Assume that*

$$v := \sup \mathcal{F} \quad (\textit{respectively, } v := \inf \mathcal{F})$$

is continuous. Then $\mathcal{F}$ is uniformly convergent to u, that is, for every $\varepsilon > 0$, there exists $u \in \mathcal{F}$ such that $v - u < \varepsilon$ (respectively, $u - v < \varepsilon$).

Proof. Apply the previous theorem to the family

$$\mathcal{F}_v := \{v - u \ : \ u \in \mathcal{F}\} \quad (\text{respectively } \mathcal{F}_v := \{u - v \ : \ u \in \mathcal{F}\}). \quad \square$$

Now, we prove one of the main results of this section.

Theorem 6.2.10 (Levi–Cartan). *Let $A \subseteq E$ be compact, and let μ be a Radon measure in A. Suppose we are given an up directed family $\mathcal{F}$ of l.s.c. functions. Then*

$$\sup_{u \in \mathcal{F}} \int_A u \, \mathrm{d}\mu = \int_A \Big(\sup_{u \in \mathcal{F}} u \Big) \, \mathrm{d}\mu. \tag{6.3}$$

Analogously, if $\mathcal{F}$ is a down directed family of u.s.c. functions, then

$$\inf_{u \in \mathcal{F}} \int_A u \, \mathrm{d}\mu = \int_A \Big(\inf_{u \in \mathcal{F}} u \Big) \, \mathrm{d}\mu.$$

Proof. We only prove the first part of the theorem. The second one will follow by applying (6.3) to the family $-\mathcal{F} := \{-u \ : \ u \in \mathcal{F}\}$. The function

$$v := \sup_{u \in \mathcal{F}} u$$

is l.s.c. since the functions in $\mathcal{F}$ are l.s.c. Therefore,

$$\int_A v \, d\mu = \sup \left\{ \int_A \phi \, d\mu \ : \ \phi \in C(A, \mathbb{R}), \phi \le v \right\}. \tag{6.4}$$

Since $u \le v$ for any $u \in \mathcal{F}$, the inequality

$$\sup_{u \in \mathcal{F}} \int_A u \, d\mu \le \int_A v \, d\mu \tag{6.5}$$

readily follows. To prove the reverse one, let us pick $\phi \in C(A, \mathbb{R})$ such that $\phi \le v$ and consider

$$\mathcal{F}_\phi := \{\min\{u, \phi\} \ : \ u \in \mathcal{F}\}.$$

This family is up directed and

$$\sup \mathcal{F}_\phi = \min\{v, \phi\} = \phi.$$

Then, we can apply Corollary 6.2.9: for every $\varepsilon > 0$, there exists $u_0 \in \mathcal{F}$ such that $\phi - \min\{u_0, \phi\} < \varepsilon$. As a consequence,

$$\begin{aligned} \int_A \phi \, d\mu &\le \varepsilon\mu(A) + \int_A \min\{u_0, \phi\} \, d\mu \le \varepsilon\mu(A) + \int_A u_0 \, d\mu \\ &\le \varepsilon\mu(A) + \sup_{u \in \mathcal{F}} \int_A u \, d\mu. \end{aligned}$$

Noticing that $\mu(A) < \infty$, since A is compact, and letting ε tend to zero, we get

$$\int_A \phi \, d\mu \le \sup_{u \in \mathcal{F}} \int_A u \, d\mu.$$

Thanks to (6.4) this inequality implies

$$\int_A v \, d\mu \le \sup_{u \in \mathcal{F}} \int_A u \, d\mu.$$

This equality, together with (6.5), completes the proof. □

We close the section by proving the following theorem, an easy consequence of Proposition 6.1.1.

Theorem 6.2.11 (Cornea). *Let $\mathcal{F}$ be an up directed family of continuous functions in an open set $\Omega \in \mathcal{T}$. Suppose $\mathcal{F}$ has the following property:*

$$\begin{aligned} &\textit{for every increasing sequence } (f_n)_n \textit{ in } \mathcal{F}, \\ &\qquad \textit{one has} \quad \sup\nolimits_{n \in \mathbb{N}} f_n \in C(\Omega, \mathbb{R}). \end{aligned} \tag{6.6}$$

Then the function

$$v := \sup_{u \in \mathcal{F}} u$$

is continuous in Ω.

Proof. Proposition 6.1.1 implies the existence of a sequence (g_n) in $\mathcal{F}$ such that $v = \sup_n g_n$. Since $\mathcal{F}$ is up directed, we can construct a sequence (f_n) in $\mathcal{F}$ such that

$$f_1 = g_1,\ f_2 \geq \max\{g_2, f_1\}, \ldots,\ f_{n+1} \geq \max\{g_{n+1}, f_n\}, \ldots.$$

Then $\mathcal{F} \ni f_n \uparrow v$. Assumption (6.6) implies that v is continuous. □

6.3 Harmonic Spaces

Definition 6.3.1 (Harmonic space). *Let $\mathcal{H}$ be a harmonic sheaf on E. We say that $(E, \mathcal{H})$ is a* harmonic space *if the following axioms are satisfied:*

(A1) (Positivity). *For every relatively compact open set $\Omega \subseteq E$, there exist $h_0 \in \mathcal{H}^*(\Omega)$ and $k_0 \in \mathcal{H}_*(\Omega)$ satisfying*[1]*:*

$$\inf_{\Omega} h_0,\ \inf_{\Omega} k_0 > 0 \qquad \text{and } h_0(x) < \infty \quad \forall\, x \in \Omega.$$

(A2) (Convergence). *If $\{u_n\}_{n\in\mathbb{N}}$ is a monotone increasing sequence of $\mathcal{H}$-harmonic functions on an open set $\Omega \in \mathcal{T}$ such that*

$$\left\{x \in \Omega \ :\ \sup_{n\in\mathbb{N}} u_n(x) < \infty\right\}$$

is dense in Ω, then

$$u := \lim_{n\to\infty} u_n$$

is $\mathcal{H}$-harmonic in Ω.

(A3) (Regularity). *The family $\mathcal{T}_r$ of the $\mathcal{H}$-regular open sets is a basis for the topology $\mathcal{T}$.*

(A4) (Separation). *For every $x,\ y \in E$ with $x \neq y$, there exist $u, v \in \mathcal{H}^*(E)$ such that*

$$u(x)\, v(y) \neq u(y)\, v(x).$$

The following remark is often useful in applying the abstract harmonic theory to the linear partial differential equations without zero order terms.

Remark 6.3.2. If the constant functions are $\mathcal{H}$-harmonic in E, then the positivity axiom is satisfied. Moreover, the separation assumption (A4) can be restated as follows.

(A4)' For every $x,\ y \in E$ with $x \neq y$, there exists $u \in \mathcal{H}^*(E)$ such that

$$u(x) \neq u(y).$$

Indeed, (A4)' imply (A4) by taking in the latter $v \equiv 1$.

---

[1] An $\mathcal{H}$-hyperharmonic function taking real values in a dense subset of its domain will be called $\mathcal{H}$-superharmonic, see Section 6.5. Then we can say that $h_0, -k_0$ are $\mathcal{H}$-superharmonic.

If $\Omega \in \mathcal{T}$, $V \subseteq \overline{V} \subseteq \Omega$ is $\mathcal{H}$-regular and $u \in \mathcal{H}(\Omega)$, it readily follows from the property (R2) of the regular sets that

$$u|_V = H^V_{u|_{\partial V}}.$$

Hence

$$u(x) = \int_{\partial V} u \, d\mu^V_x \qquad \forall\, x \in V.$$

An important consequence of the *regularity axiom* is that this property characterizes the $\mathcal{H}$-harmonic functions.

Proposition 6.3.3 (Characterization of $\mathcal{H}$-harmonicity). *Let $(E, \mathcal{H})$ be a harmonic space. Let $\Omega \in \mathcal{T}$ and $u \in C(\Omega, \mathbb{R})$ be such that*

$$u(x) = \int_{\partial V} u \, d\mu^V_x \quad \text{for every } x \in V \tag{6.7}$$

and for every $\mathcal{H}$-regular open set $V \subseteq \overline{V} \subseteq \Omega$. Then $u \in \mathcal{H}(\Omega)$.

Proof. Let us define

$$\mathcal{V} := \{V \in \mathcal{T}_r \mid \overline{V} \subseteq \Omega\}.$$

We claim that

$$\Omega = \bigcup_{V \in \mathcal{V}} V. \tag{6.8}$$

Taking this claim for granted for a moment, from (6.7) we get

$$u|_V = H^V_{u|_{\partial V}} \qquad \forall\, V \in \mathcal{V}.$$

Then u is $\mathcal{H}$-harmonic in V for every $V \in \mathcal{V}$. Since $\mathcal{H}$ is a sheaf, (6.8) now implies that $u \in \mathcal{H}(\Omega)$.

So we are left with the proof of (6.8). For every $x \in \Omega$, there exists a regular open set W contained in Ω and containing x (by (A3)). Since ∂W is compact, there exists $V \in \mathcal{T}_r$ such that

$$V \subseteq W, \quad x \in V, \quad \overline{V} \cap \partial W = \emptyset$$

(we are again using (A3), together with the Hausdorff separation property of the topology $\mathcal{T}$). Consequently, $x \in V \subseteq \overline{V} \subseteq W \subseteq \Omega$. This shows (6.8) and completes the proof of the proposition. □

Exercise 6.3.4. For every open set $\Omega \subseteq E$, one has

$$\mathcal{H}_*(\Omega) \cap \mathcal{H}^*(\Omega) = \mathcal{H}(\Omega).$$

Corollary 6.3.5. *Let $(E, \mathcal{H})$ be a harmonic space. Let $\{u_n\}_{n \in \mathbb{N}}$ be a sequence in $\mathcal{H}(\Omega)$, $\Omega \in \mathcal{T}$, such that*

$$u_n \to u, \qquad \text{as } n \to \infty,$$

uniformly on every compact subset of Ω. Then $u \in \mathcal{H}(\Omega)$.

Proof. First of all, u is continuous in Ω, since it is the limit of a sequence of continuous functions, locally uniformly convergent. Moreover, for every regular open set $V \subseteq \overline{V} \subseteq \Omega$ and for every $x \in V$, we have

$$\begin{aligned} u(x) &= \lim_{n\to\infty} u_n(x) = \lim_{n\to\infty} \int_{\partial V} u_n \, \mathrm{d}\mu_x^V \qquad (u_n \in \mathcal{H}(\Omega)) \\ &= \int_{\partial V} \lim_{n\to\infty} u_n \, \mathrm{d}\mu_x^V \qquad (u_n \text{ is uniformly convergent on } \overline{V}) \\ &= \int_{\partial V} u \, \mathrm{d}\mu_x^V. \end{aligned}$$

Then, by Proposition 6.3.3, $u \in \mathcal{H}(\Omega)$. □

6.3.1 Directed Families of Harmonic and Hyperharmonic Functions

The results proved in the previous section are the ad hoc tools to show the following two theorems, which are consequences of the convergence axiom (A2) in Section 6.3.

Theorem 6.3.6 (Up directed families in $\mathcal{H}(\Omega)$). *Let $(E, \mathcal{H})$ be a harmonic space, and let Ω be an open subset of E. Suppose we are given a family $\mathcal{F} \subseteq \mathcal{H}(\Omega)$ such that:*

(i) *$\mathcal{F}$ is up directed,*
(ii) *$v := \sup_{u\in\mathcal{F}} u$ is finite in a dense subset of Ω.*

Then v is $\mathcal{H}$-harmonic in Ω.

Proof. First, we prove that v is continuous in Ω by using Cornea's Theorem 6.2.11. Let (f_n) be a monotone increasing sequence in $\mathcal{F}$. Obviously,

$$f := \sup_{n\in\mathbb{N}} f_n \le \sup_{u\in\mathcal{F}} u,$$

so that, by hypothesis (ii), $f < \infty$ in a dense subset of Ω. The convergence axiom (A2) implies that $f \in \mathcal{H}(\Omega)$, hence it is continuous. This shows that $\mathcal{F}$ satisfies the assumptions of Theorem 6.2.11, so that $v \in C(\Omega, \mathbb{R})$. Let us now take an $\mathcal{H}$-regular open set $V \subseteq \overline{V} \subseteq \Omega$ and a point $x \in \Omega$. Since ∂V is compact and μ_x^V is a Radon measure in ∂V, by Levi–Cartan's Theorem 6.2.10 we obtain

$$v(x) = \sup_{u\in\mathcal{F}} u(x) = \sup_{u\in\mathcal{F}} \int_{\partial V} u \, \mathrm{d}\mu_x^V = \int_{\partial V} \Big(\sup_{u\in\mathcal{F}} u \Big) \mathrm{d}\mu_x^V = \int_{\partial V} v \, \mathrm{d}\mu_x^V.$$

Then, by Proposition 6.3.3, $u \in \mathcal{H}(\Omega)$. □

Corollary 6.3.7. *In the hypotheses of the previous theorem, let $\mathcal{F} \subseteq \mathcal{H}(\Omega)$ be such that $\mathcal{F} \downarrow$ and $\inf \mathcal{F} > -\infty$ in a dense subset of Ω. Then $\inf \mathcal{F} \in \mathcal{H}(\Omega)$.*

Proof. Apply Theorem 6.3.6 to the family

$$-\mathcal{F} := \{-u \ : \ u \in \mathcal{F}\}. \quad \square$$

Theorem 6.3.8 (Up directed families in $\mathcal{H}^*(\Omega)$). *Let $(E, \mathcal{H})$ be a harmonic space, and let $\Omega \subseteq E$ be open. If $\mathcal{F} \subseteq \mathcal{H}^*(\Omega)$ and $\mathcal{F} \uparrow$, then*

$$v := \sup_{u \in \mathcal{F}} u \in \mathcal{H}^*(\Omega).$$

Proof. The function v is l.s.c. Moreover, for every $\mathcal{H}$-regular open set $V \subseteq \overline{V} \subseteq \Omega$ and for every $x \in V$, by using Theorem 6.2.10, we obtain

$$v(x) = \sup_{u \in \mathcal{F}} u(x) \geq \sup_{u \in \mathcal{F}} \int_{\partial V} u \, \mathrm{d}\mu_x^V = \int_{\partial V} \Big(\sup_{u \in \mathcal{F}} u \Big) \mathrm{d}\mu_x^V,$$

so that $v(x) \geq \int_{\partial V} v \, \mathrm{d}\mu_x^V$. This proves the theorem. □

Exercise 6.3.9. Let $V \subseteq E$ be an $\mathcal{H}$-regular open set, and let $f : \partial V \to \mathbb{R}$ be l.s.c. and such that

$$\int_{\partial V} f \, \mathrm{d}\mu_x^V < \infty \qquad \forall \, x \in V.$$

Then $x \mapsto \int_{\partial V} f \, \mathrm{d}\mu_x^V$ is $\mathcal{H}$-harmonic in V.
(*Hint*: $\int_{\partial V} f \, \mathrm{d}\mu_x^V = \sup\{H_\varphi^V(x) \ : \ \varphi \in C(\partial V, \mathbb{R}), \ \varphi \leq f\}$.)

Exercise 6.3.10. Let $V \subseteq E$ be an $\mathcal{H}$-regular open set, and let

$$f : \partial V \to \mathbb{R}$$

be μ_x^V-summable for every $x \in V$. Then $x \mapsto \int_{\partial V} f \, \mathrm{d}\mu_x^V$ is $\mathcal{H}$-harmonic in V. (*Hint*: Use Ex. 6.3.9 and Corollary 6.3.7.)

6.4 $\mathcal{B}$-hyperharmonic Functions. Minimum Principle

Definition 6.4.1 ($\mathcal{B}$-hyperharmonic function). *Let $(E, \mathcal{H})$ be a harmonic space and let Ω be an open subset of E. Assume that, for every $x \in \Omega$, we are given a basis $\mathcal{B}(x)$ of $\mathcal{H}$-regular neighborhoods of x with closure contained in Ω. A l.s.c. function*

$$u : \Omega \to \,]-\infty, \infty]$$

will be called $\mathcal{B}$-hyperharmonic *in Ω if*

$$u(x) \geq \int_{\partial V} u \, \mathrm{d}\mu_x^V$$

for every $x \in \Omega$ and every $V \in \mathcal{B}(x)$. We shall denote by $\mathcal{B}$-$\mathcal{H}^*(\Omega)$ the family of the $\mathcal{B}$-hyperharmonic functions in Ω.

Obviously, $\mathcal{H}^*(\Omega) \subseteq \mathcal{B}$-$\mathcal{H}^*(\Omega)$. We will see in Corollary 6.4.9 that $\mathcal{H}^*(\Omega) = \mathcal{B}$-$\mathcal{H}^*(\Omega)$. An application of the Levi–Cartan Theorem 6.2.10 gives the following lemma.

Lemma 6.4.2. *Let $K \subseteq \Omega$ be compact, and let μ be a measure supported inside K. Suppose we are given a family $\mathcal{K}$ of compact subset of K such that:*

(i) $\mathcal{K} \downarrow$, *that is, for every $A, B \in \mathcal{K}$, there exists $C \in \mathcal{K}$ contained in $A \cap B$;*
(ii) $\operatorname{supp}(\mu) \subseteq A$ *for every $A \in \mathcal{K}$.*

Then

$$\operatorname{supp}(\mu) \subseteq \bigcap_{A \in \mathcal{K}} A.$$

Proof. Let us put $B := \bigcap_{A \in \mathcal{K}} A$. We have to show that $\mu(\Omega \setminus A) = 0$. Obviously, since $\mu(\Omega \setminus K) = 0$ and $B \subseteq K$, this is equivalent to prove that $\mu(B) = \mu(K)$. Observing that $\mathcal{F} := \{\chi_A : A \in \mathcal{K}\}$ is a down directed family of u.s.c functions, we may apply the Levi–Cartan Theorem 6.2.10 to obtain

$$\mu(B) = \int_K \chi_B \, \mathrm{d}\mu = \int_K \Big(\inf_{A \in \mathcal{K}} \chi_A \Big) \mathrm{d}\mu = \inf_{A \in \mathcal{K}} \int_K \chi_A \, \mathrm{d}\mu = \inf_{A \in \mathcal{K}} \mu(A) = \mu(K).$$

The last equality follows from the assumptions $\operatorname{supp}(\mu) \subseteq A \subseteq K$ for every $A \in \mathcal{K}$. This ends the proof. □

Definition 6.4.3 ($\mathcal{B}$-invariant set). *Let $A \subseteq \Omega$ be relatively closed in Ω, that is $A = \Omega \cap A_0$ with A_0 closed in E. We say that A is* $\mathcal{B}$-invariant *if*

for every $y \in A$ and every $V \in \mathcal{B}(y)$, one has $\operatorname{supp}(\mu_y^V) \subseteq A$.

We now show several lemmas needed for proving a minimum principle for $\mathcal{B}$-hyperharmonic functions, which will be the main achievement of this section. The first one is a trivial corollary of Lemma 6.4.2.

Lemma 6.4.4. *Let $K \subseteq \Omega$ be compact, and let $\mathcal{K}$ be a family of closed $\mathcal{B}$-invariant subset of K. Assume $\mathcal{K} \downarrow$. Then*

$$B = \bigcap_{A \in \mathcal{K}} A$$

is $\mathcal{B}$-invariant.

The second lemma shows that the zero set of a non-negative $\mathcal{B}$-hyperharmonic function is a $\mathcal{B}$-invariant set.

Lemma 6.4.5. *Let $u \in \mathcal{B}\text{-}\mathcal{H}^*(\Omega)$, $u \geq 0$. Then*

$$u^{-1}(0) := \{x \in \Omega \, : \, u(x) = 0\}$$

is $\mathcal{B}$-invariant.

Proof. First of all, since u is l.s.c. and

$$u^{-1}(0) = \{x \in \Omega \, : \, u(x) \leq 0\},$$

the set $u^{-1}(0)$ is relatively closed in Ω. Moreover, if $y \in u^{-1}(0)$ and $V \in \mathcal{B}(y)$, we have

$$\begin{aligned} 0 = u(y) &\geq \int_V u \, \mathrm{d}\mu_y^V \qquad (\text{for } u \in \mathcal{H}^*(\Omega)) \\ &\geq 0 \qquad (\text{recall that } u \geq 0). \end{aligned}$$

Then $u = 0$ μ_y^V-almost everywhere on ∂V. That is, since $\mathrm{supp}(\mu_y^V) \subseteq \partial V \subseteq \Omega$, we have

$$0 = \mu_y^V(\partial V \setminus u^{-1}(0)) = \mu_y^V(\Omega \setminus u^{-1}(0)).$$

This means that $\mathrm{supp}(\mu_y^V) \subseteq u^{-1}(0)$, and the lemma is proved. □

The argument used in this proof can be used for proving the following assertion.

Lemma 6.4.6. *Let $A \subseteq \Omega$ be $\mathcal{B}$-invariant, and let $u \in \mathcal{B}$-$\mathcal{H}^*(\Omega)$ such that $u \geq 0$ in A. Then*

$$A_0 := \{x \in A \,:\, u(x) = 0\}$$

is $\mathcal{B}$-invariant.

The next lemma shows that every non-empty compact $\mathcal{B}$-invariant set contains a non-empty *minimal* $\mathcal{B}$-invariant set. Indeed, we have the following result.

Lemma 6.4.7 (Minimal $\mathcal{B}$-invariant set). *Let $K \subseteq \Omega$ be a non-empty compact $\mathcal{B}$-invariant set. Then there exists $A_0 \subseteq K$ such that:*

(i) *A_0 is non-empty and $\mathcal{B}$-invariant,*
(ii) *A_0 is minimal, that is,*

$$A \subseteq A_0, \ A \neq \emptyset, \ A \ \mathcal{B}\text{-invariant} \Rightarrow A = A_0.$$

Proof. Define

$$\mathcal{E} := \{A \subseteq K \,:\, A \text{ closed}, A \neq \emptyset, A\ \mathcal{B}\text{-invariant}\}.$$

We have to prove that $(\mathcal{E}, \subseteq)$ has a minimal element. We will use Zorn's lemma stating the existence of such an element if every linearly ordered family of elements of $\mathcal{E}$ has a minorant in $\mathcal{E}$. Then, let $\mathcal{E}_0 \subseteq \mathcal{E}$ be linearly ordered with respect to the inclusion $\subseteq$. Define

$$B := \bigcap_{A \in \mathcal{E}_0} A.$$

B is compact and non-empty, since every finite intersection of elements of $\mathcal{E}_0$ is non-empty, for $(\mathcal{E}_0, \subseteq)$ is linearly ordered. Moreover, by Lemma 6.4.4, B is invariant. Then $B \in \mathcal{E}_0$ and, obviously, $B \subseteq A$ for every $A \in \mathcal{E}_0$. Thus $\mathcal{E}_0$ has a minorant in $\mathcal{E}$. The lemma is proved. □

Finally, we are ready to prove our minimum principle.

Theorem 6.4.8 (Minimum principle for $\mathcal{B}$-$\mathcal{H}^*$ functions). *Let Ω be a relatively compact open set in a harmonic space $(E, \mathcal{H})$, and let $u \in \mathcal{B}\text{-}\mathcal{H}^*(\Omega)$. Assume*

$$\liminf_{x \to y} u(x) \geq 0 \qquad \forall\ y \in \partial\Omega. \tag{6.9}$$

Then $u(x) \geq 0$ for every $x \in \Omega$.

Proof. Argue by contradiction and assume the existence of a point $\overline{x} \in \Omega$ such that $u(\overline{x}) < 0$. Let $h_0 \in \mathcal{H}^*(\Omega)$ be such that $\inf_\Omega h_0 > 0$ and $h_0 < \infty$ at any point of Ω (see the positivity axiom (A1)). Let us prove the following claim: there exists a strictly positive real constant α such that

$$u + \alpha h_0 \geq 0 \text{ in } \Omega \text{ and } K := \{x \in \Omega \ : \ u(x) + \alpha h_0(x) = 0\} \text{ is compact.}$$

Consider the function

$$h := \frac{1}{h_0} \max\{0, -u\}.$$

Condition (6.9), together with the upper semicontinuity of h, the compactness of $\overline{\Omega}$, and the condition $u(\overline{x}) < 0$, implies the existence of $\max_\Omega h =: \alpha$, with $0 < \alpha < \infty$. Moreover:

(i) $u + \alpha h_0 \geq u + h\, h_0 = u + \max\{0, -u\} \geq 0$ in Ω.
(ii) $K := \{x \in \Omega \ : \ u(x) + \alpha h_0(x) = 0\}$ is non-empty, since the points of K are the maximum points of h.
(iii) $K = \overline{K}$. Indeed, first notice that K is relatively closed in Ω being the null set of a non-negative l.s.c. function. Then $K = \Omega \cap \overline{K}$. If we prove that $\overline{K} \subseteq \Omega$, we are done. For this it is enough to show that if $y \in \partial\Omega$, then $y \notin K$. Now, condition (6.9) implies

$$\liminf_{x \to y}(u(x) + \alpha h_0) \geq \alpha \inf_\Omega h_0 > 0,$$

so that $u + \alpha h_0 > 0$ in a suitable neighborhood of y. This, obviously, implies that $y \notin K$.
(iv) K is compact, since $K = \overline{K} \subseteq \overline{\Omega}$ and $\overline{\Omega}$ is compact.

By Lemma 6.4.5, K is $\mathcal{B}$-invariant and, by Lemma 6.4.6, there exists $A_0 \subseteq K$, compact, non-empty, $\mathcal{B}$-invariant and minimal.

We claim that

$$A_0 = \{x_0\} \tag{6.10}$$

for a suitable $x_0 \in \Omega$. Assuming this claim for true for a moment, we show how to reach a contradiction, thus proving the theorem. Let $V \in \mathcal{B}(x_0)$. Since A_0 is $\mathcal{B}$-invariant, it has to be $\operatorname{supp}(\mu_{x_0}^V) \subseteq A_0 = \{x_0\}$. On the other hand, again by the positivity axiom, we can find $k_0 \in \mathcal{H}_*(\Omega)$ such that $k_0(x_0) > 0$. It follows that

$$0 < k_0(x_0) \leq \int_{\partial V} k_0 \, \mathrm{d}\mu_{x_0}^\Omega,$$

so that $\mu_{x_0}^V(\partial V) > 0$. This is impossible, since $\text{supp}(\mu_{x_0}^V) \subseteq \{x_0\}$ and $x_0 \notin \partial V$. Then, we are left with the proof of (6.10). By contradiction, assume it is false. Then, since $\mathcal{H}^*(E)$ separates different points of E, there exists $w \in \mathcal{H}^*(E)$ such that $w \not\equiv \infty$. Fix such a hyperharmonic function and choose a real $\beta > 0$ such that $\beta u + w < 0$ somewhere in A_0 (recall that if $u + \alpha h_0 \equiv 0$ in A_0 and $\alpha > 0$, then $u < 0$ in A_0). Letting

$$-\gamma := \min_{A_0} \frac{\beta u + w}{h_0},$$

we have $\gamma > 0$, $\beta u + w + \gamma h_0 \geq 0$ in A_0 and

$$A_1 := \{x \in A_0 \ : \ (\beta u + w + \gamma h_0)(x) = 0\} \neq \emptyset.$$

By Lemma 6.4.6, A_1 is $\mathcal{B}$-invariant. Then, since $A_1 \subseteq A_0$ and A_0 is minimal, it has to be $A_1 = A_0$, so that

$$\beta u + w + \gamma h_0 \equiv 0 \text{ in } A_0.$$

On the other hand, being $A_0 \subseteq K$, $u + \alpha h_0 \equiv 0$ in A_0. It follows that

$$w = (\beta\alpha - \gamma)h_0 := \lambda h_0 \text{ in } A_0.$$

Summing up: for every $w \in \mathcal{H}^*(E)$, $w \not\equiv \infty$ in A_0, there exists $\lambda \in \mathbb{R}$ such that $w = \lambda h_0$. This implies that $\mathcal{H}^*(E)$ cannot separate different points of A_0, in contradiction with the separation axiom (A4). Then A_0 has to be a singleton. This proves (6.10) and completes the proof of the theorem. □

We now show some consequences of the previous theorem. First of all, we show that the $\mathcal{B}$-hyperharmonicity is equivalent to the hyperharmonicity.

Corollary 6.4.9. *Let Ω be an arbitrary open set in E, and let $u \in \mathcal{B}\text{-}\mathcal{H}^*(\Omega)$ for a suitable $\mathcal{B}$. Then $u \in \mathcal{H}^*(\Omega)$.*

Proof. Let V be an $\mathcal{H}$-regular set with the closure contained in Ω. We have to prove that

$$u(x) \geq \int_{\partial V} u \, \mathrm{d}\mu_x^V \qquad \forall \, x \in V. \tag{6.11}$$

For every $z \in V$, define

$$\mathcal{B}_V(z) = \{W \in \mathcal{B}(z) \ : \ \overline{W} \subseteq V\}.$$

Let us now choose $\phi \in C(\partial V, \mathbb{R})$, $\phi \leq u|_{\partial V}$. The function $u - H_\phi^V$ is $\mathcal{B}_V$-hyperharmonic in V. Moreover,

$$\liminf_{V \ni x \to y} (u - H_\phi^V)(x) \geq u(y) - \phi(y) \geq 0 \qquad \forall \, y \in \partial V.$$

Then, by Theorem 6.4.8, $u - H_\phi^V \geq 0$ in V, that is,

$$u(x) \geq \int_{\partial V} \phi \, \mathrm{d}\mu_x^V \qquad \forall \, x \in V \tag{6.12}$$

and for every $\phi \in C(\partial V, \mathbb{R})$, $\phi \leq u|_{\partial V}$. Taking the supremum at the right-hand side with respect to these functions ϕ's, we obtain (6.11). □

Corollary 6.4.10. *Let $(E, \mathcal{H})$ be a harmonic space, and let $\Omega \subseteq E$ be open. The map*

$$\Omega \mapsto \mathcal{H}^*(\Omega)$$

is a sheaf of functions.

Proof. Property (6.2a) (page 340) is trivial. Let us prove (6.2b). Let $(\Omega_\alpha)_{\alpha \in \mathcal{A}}$ be a family of open subsets of E, and let

$$u : \Omega \to [-\infty, \infty], \ \Omega := \bigcup_{\alpha \in \mathcal{A}} \Omega_\alpha,$$

be a function such that $u|_{\Omega_\alpha} \in \mathcal{H}^*(\Omega_\alpha)$ for every $\alpha \in \mathcal{A}$. Then u is l.s.c and $u > -\infty$ at any point. For every $x \in \Omega$, define

$$\mathcal{B}(x) := \{V \in \mathcal{T}_r \ : \ x \in V \subseteq \overline{V} \subseteq \Omega_\alpha \text{ for a suitable } \alpha \in \mathcal{A}\}.$$

One readily verifies that $u \in \mathcal{B}\text{-}\mathcal{H}^*(\Omega)$. Then, by the previous corollary, we infer that $u \in \mathcal{H}^*(\Omega)$. □

Exercise 6.4.11. Let $(E, \mathcal{H})$ be a harmonic space. Let $\Omega \subseteq E$ be open, and let $x_0 \in \Omega$. If a function $u : \Omega \to]-\infty, \infty]$ satisfies:

(i) $u \in \mathcal{H}^*(\Omega \setminus \{x_0\})$,
(ii) $\limsup_{x \to x_0} u(x) = \infty$,

then $u \in \mathcal{H}^*(\Omega)$.

6.5 Subharmonic and Superharmonic Functions. Perron Families

Definition 6.5.1 ($\mathcal{H}$-super- and $\mathcal{H}$-sub-harmonic function). *Let $(E, \mathcal{H})$ be a harmonic space, and let $\Omega \subseteq E$ be open. A function $u \in \mathcal{H}^*(\Omega)$ will be said* $\mathcal{H}$-superharmonic *if, for every $\mathcal{H}$-regular open set $V \subseteq \overline{V} \subseteq \Omega$, the function*

$$V \ni x \mapsto \int_{\partial V} u \, d\mu_x^V$$

is $\mathcal{H}$-harmonic in V. The set of the $\mathcal{H}$-superharmonic functions in Ω will be denoted by

$$\overline{\mathcal{S}}(\Omega).$$

A function $v : \Omega \to [-\infty, \infty[$ will be said $\mathcal{H}$-subharmonic *in Ω if $-v \in \overline{\mathcal{S}}(\Omega)$. We shall denote by*

$$\underline{\mathcal{S}}(\Omega) := -\overline{\mathcal{S}}(\Omega)$$

the set of the $\mathcal{H}$-subharmonic functions in Ω.

From Theorem 6.3.6 we readily obtain the following characterization of the $\mathcal{H}$-superharmonic functions.

Theorem 6.5.2 (Characterization of $\overline{\mathcal{S}}(\Omega)$). *Let $u \in \mathcal{H}^*(\Omega)$. Then $u \in \overline{\mathcal{S}}(\Omega)$ if and only if*

$$D := \{x \in \Omega \ : \ u(x) < \infty\}$$

is dense in Ω, that is, $\overline{D} \supseteq \Omega$.

Proof. By contradiction, suppose D is not dense in Ω. Then there exists an $\mathcal{H}$-regular open set V such that

$$\overline{V} \subseteq \{x \in \Omega \ : \ u(x) = \infty\}. \tag{6.13}$$

Invoking the positivity axiom, we may assume $\int_{\partial V} \mathrm{d}\mu_x^V > 0$ for every $x \in V$. Indeed, by this axiom, one can suppose the existence of a strictly positive $\mathcal{H}_*$ function k_0 in an open set containing $\overline{V}$. This implies

$$\int_{\partial V} k_0 \, \mathrm{d}\mu_x^V \geq k_0(x) > 0,$$

hence $\mu_x^V(\partial V) > 0$ for every $x \in \Omega$. Then, from (6.13) we get

$$\int_{\partial V} u \, \mathrm{d}\mu_x^V = \infty \qquad \forall \, x \in V,$$

showing that $x \mapsto \int_{\partial V} u \, \mathrm{d}\mu_x^V$ is not $\mathcal{H}$-harmonic in V. Vice versa, suppose $\overline{D} \supseteq \Omega$, and let $V \subseteq \overline{V} \subseteq \Omega$ be an $\mathcal{H}$-regular set. Consider

$$\mathcal{F} := \{H_\phi^V \ : \ \phi \in C(\partial V, \mathbb{R}), \ \phi \leq u|_{\partial V}\}.$$

$\mathcal{F}$ is an up directed family of $\mathcal{H}$-harmonic functions in V such that

$$\sup \mathcal{F} \leq u < \infty \text{ in } D,$$

since $u \in \mathcal{H}^*(\Omega)$. Theorem 6.3.6 implies that $v := \sup \mathcal{F}$ is $\mathcal{H}$-harmonic in V. This completes the proof, since

$$v(x) = \int_{\partial V} u \, \mathrm{d}\mu_x^V, \quad x \in V. \quad \square$$

A useful criterion for subharmonicity is given by the next corollary.

Corollary 6.5.3. *Let $(E, \mathcal{H})$ be a harmonic space and let $\Omega \subseteq E$ be open. Let $u \in \mathcal{H}^*(\Omega)$. Assume there exists $v \in \overline{\mathcal{S}}(\Omega)$ such that $u \leq v$. Then $u \in \overline{\mathcal{S}}(\Omega)$.*

Proof. Notice that

$$\overline{\{x \in \Omega \ : \ u(x) < \infty\}} \supseteq \overline{\{x \in \Omega \ : \ v(x) < \infty\}}$$

and apply Theorem 6.5.2. $\square$

We also have the following result.

Proposition 6.5.4. *Let* $u, v \in \overline{\mathcal{S}}(\Omega)$. *Then*:

(i) $\lambda u \in \overline{\mathcal{S}}(\Omega)$ *for every* $\lambda \geq 0$,
(ii) $u + v \in \overline{\mathcal{S}}(\Omega)$,
(iii) $\min\{u, v\} \in \overline{\mathcal{S}}(\Omega)$.

Proof. Exercise. □

Let us now introduce the following crucial definition.

Definition 6.5.5 (Perron-regularization). *Let* $(E, \mathcal{H})$ *be a harmonic space, and let* $\Omega \subseteq E$ *be open. Given* $u \in \mathcal{H}^*(\Omega)$ *and an* $\mathcal{H}$*-regular set* $V \subseteq \overline{V} \subseteq \Omega$, *define*

$$u_V : \Omega \to]-\infty, \infty], \quad u_V(x) := \begin{cases} u(x), & x \notin V, \\ \int_{\partial V} u \, \mathrm{d}\mu_x^V, & x \in V. \end{cases}$$

The function u_V *will be called the Perron-regularization of* u *related to* V.

The main properties of the Perron-regularization are given in by the following theorem.

Theorem 6.5.6 (Properties of the Perron-regularization). *Suppose that* $u \in \mathcal{H}^*(\Omega)$ *and* $V \subseteq \overline{V} \subseteq \Omega$ *is an open* $\mathcal{H}$*-regular set, then*:

(i) $u_V \leq u$,
(ii) $u_V \in \mathcal{H}^*(\Omega)$,
(iii) $u_V \leq v_V$ *if* $u, v \in \mathcal{H}^*(\Omega)$ *and* $u \leq v$.

Moreover, if $u \in \overline{\mathcal{S}}(\Omega)$, *then*

(iv) $u_V \in \overline{\mathcal{S}}(\Omega)$ *and* $u|_V \in \mathcal{H}(V)$.

Proof. (i) It simply follows from the very definition of hyperharmonic function.
(ii) We split the proof of this property in two steps.

(ii-a) u_V is lower semicontinuous. For this, we have to prove that

$$\liminf_{x \to y} u_V(x) \geq u_V(y) \qquad \text{for every } y \in \Omega.$$

The statement holds true if $y \notin \overline{V}$, since, in this case, $u_V = u$ in a neighborhood of y. On the other hand, for $y \in V$, we have

$$u_V(y) = \int_{\partial V} u \, \mathrm{d}\mu_y^V = \sup\{H_\varphi^V \mid \varphi \in C(\partial V, \mathbb{R}), \varphi \leq u|_{\partial V}\}. \tag{6.14}$$

Then u_V is $\mathcal{H}$-hyperharmonic in V (see Theorem 6.3.8), hence lower semicontinuous at any point $y \in V$.

Let us now suppose $y \in \partial V$. In this case $u_V(y) = u(y)$, so that

$$\liminf_{V \not\ni x \to y} u_V(x) = \liminf_{V \not\ni x \to y} u(x) \geq u(y) = u_V(y).$$

Then we only have to prove that

$$\liminf_{V \ni x \to y} u_V(x) \geq u(y). \tag{6.15}$$

Let $\varphi \in C(\partial V, \mathbb{R})$, $\varphi \leq u|_{\partial V}$. Then $u_V \geq H_\varphi^V$ in V, so that

$$\liminf_{V \ni x \to y} u_V(x) \geq \liminf_{V \ni x \to y} H_\varphi^V(x) = \varphi(y).$$

Since $u|_{\partial V} = \sup\{\varphi \in C(\partial V, \mathbb{R}) : \varphi \leq u|_{\partial V}\}$, (6.15) follows from the previous inequality.

(ii-b) $u_V \in \mathcal{B}\text{-}\mathcal{H}^*(\Omega)$ with $\mathcal{B}$ defined as follows

$$\mathcal{B}(x) = \begin{cases} \{B : B \ \mathcal{H}\text{-regular open set}, x \in B \subseteq \overline{B} \subseteq V\} & \text{if } x \in V, \\ \{B : B \ \mathcal{H}\text{-regular open set}, x \in B \subseteq \overline{B} \subseteq \Omega \setminus \overline{V}\} & \text{if } x \notin \overline{V}, \\ \{B : B \ \mathcal{H}\text{-regular open set}, x \in B \subseteq \overline{B} \subseteq \Omega\} & \text{if } x \in \partial V. \end{cases}$$

Since u_V is $\mathcal{H}$-hyperharmonic in $\Omega \setminus \overline{V}$ and in V (see (6.14)), we only have to prove that

$$u_V(y) \geq \int_{\partial B} u_V \, d\mu_y^B \qquad \forall y \in \partial B, \ \forall B \in \mathcal{B}(y).$$

We have

$$\begin{aligned} u_V(y) &= u(y) && (\text{since } y \in \partial V) \\ &\geq \int_{\partial B} u \, d\mu_y^B && (u \in \mathcal{H}^*(\Omega)) \\ &\geq \int_{\partial B} u_V \, d\mu_y^B && (u \geq u_V). \end{aligned}$$

This shows that $u_V \in \mathcal{B}\text{-}\mathcal{H}^*(\Omega)$. Thus, by Corollary 6.4.9, $u \in \mathcal{H}^*(\Omega)$.

(iii) This is quite obvious.

(iv) Since $u_V \in \mathcal{H}^*(\Omega)$ and $u_V \leq u$, if $u \in \overline{\mathcal{S}}(\Omega)$ then $u_V \in \overline{\mathcal{S}}(\Omega)$ (see Corollary 6.5.3). The harmonicity of u_V in V follows from the very definition of subharmonic function. □

We are now ready to give another crucial definition.

Definition 6.5.7 (Perron family). *Let $(E, \mathcal{H})$ be a harmonic space, and let $\Omega \subseteq E$ be open. A family of functions $\mathcal{F} \subseteq \mathcal{H}^*(\Omega)$ will be said a* Perron family *if the following conditions are satisfied:*

(i) *$\mathcal{F} \downarrow$, that is, $\mathcal{F}$ is down directed,*
(ii) *$\mathcal{F}$ has an $\mathcal{H}$-subharmonic minorant, i.e. there exists $v_0 \in \underline{\mathcal{S}}(\Omega)$ such that*

$$v_0 \leq u \quad \textit{for every } u \in \mathcal{F},$$

(iii) $u_V \in \mathcal{F}$ *for every* $u \in \mathcal{F}$ *and for every* $\mathcal{H}$*-regular open set* $V \subseteq \overline{V} \subseteq \Omega$,
(iv) $\mathcal{F}$ *contains at least one* $\mathcal{H}$*-superharmonic function, i.e. there exists* $u_0 \in \mathcal{F}$ *such that* $u_0 \in \overline{\mathcal{S}}(\Omega)$.

The following important result holds.

Theorem 6.5.8 (The fundamental theorem on Perron families). *Let* $(E, \mathcal{H})$ *be a harmonic space, and let* $\mathcal{F}$ *be a Perron family on an open set* $\Omega \subseteq E$. *Then*

$$\underline{u} := \inf \mathcal{F} \in \mathcal{H}(\Omega).$$

Proof. Let $V \subseteq \overline{V} \subseteq \Omega$ be an arbitrary fixed $\mathcal{H}$-regular set. It is enough to prove that $\underline{u}$ is $\mathcal{H}$-harmonic in V. Define

$$\mathcal{F}_V = \{u_V|_V \ : \ u \in \mathcal{F}, \ u \in \overline{\mathcal{S}}(\Omega)\}$$

and prove the following list of claims:

(C1) $\mathcal{F}_V \neq \emptyset$. This follows from assumption (iv) in Definition 6.5.7.
(C2) $\mathcal{F}_V \subseteq \mathcal{H}(V)$. This is granted by Theorem 6.5.6-(iv).
(C3) $\mathcal{F}_V \downarrow$. Indeed, let $u, v \in \mathcal{F} \cap \overline{\mathcal{S}}(\Omega)$. Since $\mathcal{F}$ is down directed, there exists $w \in \mathcal{F}$ such that $w \le \min\{u, v\}$. The function w is $\mathcal{H}$-superharmonic, for it is bounded from above by $\min\{u, v\}$ which is $\mathcal{H}$-superharmonic (see Corollary 6.5.3). It follows that $w_V \in \mathcal{F}_V$ and $w_V \le \min\{u_V, v_V\}$, proving that $\mathcal{F}_V \downarrow$.
(C4) $\mathcal{F}_V$ has an $\mathcal{H}$-subharmonic minorant, so that $\inf \mathcal{F}_V > -\infty$ in a dense subset of Ω. This directly follows from assumptions (ii) and (iii) in Definition 6.5.7.
(C5) $\underline{u}|_V = \inf \mathcal{F}_V$. Indeed, the inequality

$$\underline{u}|_V \le \inf \mathcal{F}_V$$

is trivial for $\mathcal{F}_V \subseteq \{u|_V \ : \ u \in \mathcal{F}\}$. On the other hand, let $u_0 \in \mathcal{F} \cap \overline{\mathcal{S}}(\Omega)$ be the function given by assumption (iv) in Definition 6.5.7. For every $u \in \mathcal{F}$, take $v \in \mathcal{F}$ satisfying $v \le \min\{u, u_0\}$ ($\mathcal{F} \downarrow$). Since $u_0 \in \overline{\mathcal{S}}(\Omega)$, also $v \in \overline{\mathcal{S}}(\Omega)$. It follows that $v|_V \in \mathcal{F}_V$ and that

$$\inf \mathcal{F}_V \le v_V \le v|_V \le u|_V.$$

Therefore,

$$\inf \mathcal{F}_V \le u|_V \qquad \forall u \in \mathcal{F},$$

and the inequality $\inf \mathcal{F}_V \le \underline{u}|_V$ follows.

Summing up: (C1)–(C4) and Corollary 6.3.7 imply $\inf \mathcal{F}_V \in \mathcal{H}(V)$. Hence, from (C5) we get $\underline{u}|_V \in \mathcal{H}(V)$. The proof is complete. □

6.6 Harmonic Majorants and Minorants

From the fundamental theorem on Perron families, one easily obtains conditions for the existence of least *harmonic majorants* of subharmonic functions and of greatest harmonic minorants of superharmonic functions.

Theorem 6.6.1 (The least $\mathcal{H}$-superharmonic majorant). *Let $(E, \mathcal{H})$ be a harmonic space, and let $\Omega \subseteq E$ be open. Let $v \in \underline{S}(\Omega)$ be such that*

$$v \le u_0$$

for a suitable $u_0 \in \overline{S}(\Omega)$. Then v has a least $\mathcal{H}$-superharmonic majorant h on Ω, and $h \in \mathcal{H}(\Omega)$.

Proof. Consider the family

$$\mathcal{F} := \{u \in \overline{S}(\Omega) \ : \ v \le u \le u_0\}.$$

If $u \in \mathcal{F}$ and V is an $\mathcal{H}$-regular open set with $\overline{V} \subseteq \Omega$, we have $v \le v_V \le u_V \le u \le u_0$. Hence $u_V \in \mathcal{F}$. It follows that $\mathcal{F}$ is a Perron family, with the $\mathcal{H}$-subharmonic minorant v and the $\mathcal{H}$-superharmonic majorant u_0. Then

$$h := \inf \mathcal{F} \in \mathcal{H}(\Omega),$$

by Theorem 6.5.8, and $h \ge v$ in Ω. Hence h is the least element of $\mathcal{F}$ and h is $\mathcal{H}$-harmonic in Ω. □

Corollary 6.6.2. *In a harmonic space $(E, \mathcal{H})$, every $\mathcal{H}$-superharmonic function with an $\mathcal{H}$-subharmonic minorant in an open set Ω has a greatest $\mathcal{H}$-subharmonic minorant which is $\mathcal{H}$-harmonic in Ω.*

The following proposition holds.

Proposition 6.6.3. *Let $\Omega \subseteq E$ be open, and let $v_1, v_2 \in \underline{S}(\Omega)$. Assume v_1, v_2 have an $\mathcal{H}$-superharmonic majorant. Then $v_1 + v_2$ has a least $\mathcal{H}$-superharmonic majorant given by*

$$h_1 + h_2,$$

where h_i is the least $\mathcal{H}$-(super)harmonic majorant of v_i.

Proof. Let h be the least $\mathcal{H}$-(super)harmonic majorant of $v_1 + v_2$. Obviously, $h \le h_1 + h_2$. On the other hand, being $h \ge v_1 + v_2$, the function $h - v_1$ is an $\mathcal{H}$-superharmonic majorant of v_2. Hence

$$h - v_1 \ge h_2.$$

It follows that $h - h_2 \ge v_1$ and $h - h_2$ is $\mathcal{H}$-harmonic in Ω. Then $h - h_2 \ge h_1$, that is $h \ge h_1 + h_2$. This completes the proof. □

6.7 The Perron–Wiener–Brelot Operator

Throughout this section, Ω will denote an open set, with compact closure and non-empty boundary, in a harmonic space $(E, \mathcal{H})$. We shall construct the Perron–Wiener–Brelot operator

$$f \mapsto H_f^\Omega$$

from a suitable linear set of real extended functions $f : \partial\Omega \to [-\infty, \infty]$ to the linear space $\mathcal{H}(\Omega)$ of the $\mathcal{H}$-harmonic functions in Ω. When the *$\mathcal{H}$-Dirichlet problem*

$$(\mathcal{H}\text{-D}) \qquad \begin{cases} u \in \mathcal{H}(\Omega), \\ \lim_{x\to y} u(x) = f(y) \quad \forall\, y \in \partial\Omega, \end{cases}$$

has a solution u, it will turn out that $u = H_f^\Omega$. For this reason, sometimes we shall call H_f^Ω the *generalized solution* to ($\mathcal{H}$-D) in the sense of Perron–Wiener–Brelot. Our construction, in the case of classic harmonic functions, i.e. of the solutions to the "usual" Laplace equation $\Delta u = 0$, goes back to O. Perron, N. Wiener and M. Brelot.

Definition 6.7.1 (Upper and lower functions). *Let $(E, \mathcal{H})$ be a harmonic space, and let $\Omega \subseteq E$ be open and such that $\overline{\Omega}$ is compact and $\partial\Omega \neq \emptyset$. Given a function $f : \partial\Omega \to [-\infty, \infty]$, we set*[2]

$$\overline{\mathcal{U}}_f^\Omega := \left\{ u \in \mathcal{H}^*(\Omega) \ : \ \liminf_{\partial\Omega} u \geq f, \ \inf u > -\infty \right\}$$

and

$$\underline{\mathcal{U}}_f^\Omega := \left\{ v \in -\mathcal{H}^*(\Omega) \ : \ \limsup_{\partial\Omega} v \leq f, \ \sup v < \infty \right\}.$$

The families $\overline{\mathcal{U}}_f^\Omega$ and $\underline{\mathcal{U}}_f^\Omega$ will be called, respectively, the family of the upper functions *and of the* lower functions *related to f and Ω.*

The function $u \equiv \infty$ ($v \equiv -\infty$, respectively) is an upper function (lower function, respectively). Therefore, $\overline{\mathcal{U}}_f^\Omega \neq \emptyset$ ($\underline{\mathcal{U}}_f^\Omega \neq \emptyset$, respectively).

Definition 6.7.2 (Upper and lower solutions). *With the hypotheses and notation of the previous definition, the real extended functions*

$$\overline{H}_f^\Omega := \inf \overline{\mathcal{U}}_f^\Omega, \qquad \underline{H}_f^\Omega := \sup \underline{\mathcal{U}}_f^\Omega$$

will be called the upper solution *and the* lower solution, *respectively, to the problem* ($\mathcal{H}$-D).

---

[2] The notation

$$\liminf_{\partial\Omega} u \geq f$$

means

$$\liminf_{x\to y} u(x) \geq f(y) \qquad \forall\, y \in \partial\Omega.$$

Analogous meaning for $\limsup_{\partial\Omega} u \leq f$.

It has to be noticed that

$$\underline{H}_f^{\Omega} = -\overline{H}_{-f}^{\Omega},$$

since $\underline{\mathcal{U}}_f^{\Omega} = -\overline{\mathcal{U}}_{-f}^{\Omega}$. We also have the following proposition, an easy consequence of the minimum principle for the $\mathcal{H}$-hyperharmonic functions.

Proposition 6.7.3. *In the hypotheses of Definition* 6.7.1, *for every* $f : \partial\Omega \to [-\infty, \infty]$, *one has*

$$\underline{H}_f^{\Omega} \le \overline{H}_f^{\Omega}. \tag{6.16}$$

Proof. For any $u \in \overline{\mathcal{U}}_f^{\Omega}$ and $v \in \underline{\mathcal{U}}_f^{\Omega}$, one has $u - v \in \mathcal{H}^*(\Omega)$ and

$$\liminf_{\partial\Omega}(u - v) \ge 0.$$

Then, from Theorem 6.4.8 (the minimum principle), $u - v \ge 0$ in Ω. Thus

$$u \ge v \qquad \forall\, u \in \overline{\mathcal{U}}_f^{\Omega}, \qquad \forall\, v \in \underline{\mathcal{U}}_f^{\Omega},$$

and (6.16) readily follows. □

The next proposition is a straightforward consequence of the definitions of upper and lower solution.

Proposition 6.7.4. *Let* $f, g : \partial\Omega \to [-\infty, \infty]$, $\alpha \in \mathbb{R}$, $\alpha > 0$. *Then*:

(i) $f \le g \;\Rightarrow\; \overline{H}_f^{\Omega} \le \overline{H}_g^{\Omega},\ \underline{H}_f^{\Omega} \le \underline{H}_g^{\Omega}$,
(ii) $\overline{H}_{f+g}^{\Omega} \le \overline{H}_f^{\Omega} + \overline{H}_g^{\Omega},\ \underline{H}_{f+g}^{\Omega} \ge \underline{H}_f^{\Omega} + \underline{H}_g^{\Omega}$, *whenever the sums are defined.*
(iii) $\overline{H}_{\alpha f}^{\Omega} = \alpha\overline{H}_f^{\Omega},\ \underline{H}_{\alpha f}^{\Omega} = \alpha\underline{H}_f^{\Omega},\ \overline{H}_{-\alpha f}^{\Omega} = -\alpha\underline{H}_f^{\Omega}$,
(iv) $f \ge 0 \;\Rightarrow\; \overline{H}_f^{\Omega} \ge 0,\ \underline{H}_f^{\Omega} \ge 0.$

Proof. It is left as an exercise. (*Hint*: to prove (iv) use the minimum principle and the fact that $u \equiv 0$ is $\mathcal{H}$-harmonic, hence both $\mathcal{H}$-hyperharmonic and $\mathcal{H}$-hypoharmo nic.) □

In Section 6.9, we will use the following Beppo Levi-type property of the map $f \mapsto \overline{H}_f^{\Omega}$, which we leave here as an exercise.

Exercise 6.7.5. Let $\Omega \subseteq E$ be a relatively compact open set, and let (f_n) be a sequence of real extended functions $f_n : \partial\Omega \to [-\infty, \infty]$ such that

$$f_n \nearrow f \quad \text{and} \quad \overline{H}_{f_n}^{\Omega} \in \mathcal{H}(\Omega) \quad \forall n \in \mathbb{N}.$$

Then

$$\overline{H}_{f_n}^{\Omega} \nearrow \overline{H}_f^{\Omega}.$$

Hint: $h = \sup_n \overline{H}_{f_n}^{\Omega} \in \mathcal{H}^*(\Omega)$, see Theorem 6.3.8. For any fixed $x \in \Omega$ and $\varepsilon > 0$ choose $u_n \in \overline{\mathcal{U}}_{f_n}^{\Omega}$ such that $u_n(x) < \overline{H}_{f_n}^{\Omega}(x) + \frac{\varepsilon}{2^n}$. Then

$$u = h + \sum_{n=1}^{\infty}(u_n - \overline{H}_{f_n}^{\Omega}) \in \overline{\mathcal{U}}_f^{\Omega}.$$

Definition 6.7.6 (Resolutive function). *Let $(E, \mathcal{H})$ be a harmonic space, and let $\Omega \subseteq E$ be an open set with compact closure and non-empty boundary. A real extended function $f : \partial\Omega \to [-\infty, \infty]$ will be said* resolutive *if:*

(i) $\overline{H}_f^{\Omega} = \underline{H}_f^{\Omega}$,
(ii) $\overline{H}_f^{\Omega} \in \mathcal{H}(\Omega)$.

In this case, we set

$$H_f^{\Omega} := \overline{H}_f^{\Omega} \ (= \underline{H}_f^{\Omega}),$$

and we say that H_f^{Ω} is the generalized solution, in the sense of Perron–Wiener–Brelot, to the problem ($\mathcal{H}$-D). We also call H_f^{Ω} the PWB function *related to Ω and f. The set of the resolutive functions $f : \partial\Omega \to [-\infty, \infty]$ will be denoted by $\mathcal{R}(\partial\Omega)$,*

$$\mathcal{R}(\partial\Omega) := \{f : \partial\Omega \to [-\infty, \infty] \mid f \textit{ is resolutive}\}. \tag{6.17}$$

The connection between H_f^{Ω} and the $\mathcal{H}$-Dirichlet problem ($\mathcal{H}$-D) is showed by the following proposition.

Proposition 6.7.7. *Let the hypotheses of Definition 6.7.6 hold. Let $f : \partial\Omega \to [-\infty, \infty]$ be a bounded function. Then the following statements are equivalent:*

(i) *f is resolutive and $\lim_{x\to y} H_f^{\Omega}(x) = f(y)$ for every $y \in \partial\Omega$.*
(ii) *There exists $u \in \mathcal{H}(\Omega)$ such that $\lim_{x\to y} u(x) = f(y)$ for every $y \in \partial\Omega$.*

In this latter case, $u = H_f^{\Omega}$.

Proof. (i) ⇒ (ii). This is trivial.

(ii) ⇒ (i). Just remark that u is bounded from above and below, and that $u \in \overline{\mathcal{U}}_f^{\Omega} \cap \underline{\mathcal{U}}_f^{\Omega}$. □

We would like to explicitly remark that $\mathcal{R}(\partial\Omega)$ is not empty, since it contains the null function. From Proposition 6.7.4 it follows that:

(i) if $f, g \in \mathcal{R}(\partial\Omega)$ and $f + g$ is defined, then $f + g \in \mathcal{R}(\partial\Omega)$ and $H_{f+g}^{\Omega} = H_f^{\Omega} + H_g^{\Omega}$,
(ii) if $f \in \mathcal{R}(\partial\Omega)$ and $c \in \mathbb{R}$, then $c f \in \mathcal{R}(\partial\Omega)$ and $H_{cf}^{\Omega} = cH_f^{\Omega}$.

In particular,

$$\mathcal{R}^{\infty}(\partial\Omega) := \mathcal{R}(\partial\Omega) \cap L^{\infty}(\partial\Omega)$$

is a real vector space, and the map

$$\mathcal{R}^{\infty}(\partial\Omega) \ni f \mapsto H_f^{\Omega} \in C(\Omega, \mathbb{R})$$

is *linear* and *monotone*.

We want to show that $\mathcal{R}(\partial\Omega)$ contains the restrictions to the boundary of Ω of the continuous superharmonic functions. We first show a proposition having an independent interest, an easy consequence of the fundamental theorem on the Perron families (Theorem 6.5.8, page 357).

Proposition 6.7.8. *Let $\Omega \subseteq E$ be an open set with compact closure and non-empty boundary. Let $f : \partial\Omega \to [-\infty, \infty]$. Assume that there exist*

$$u_0 \in \overline{\mathcal{U}}_f^{\Omega} \cap \overline{\mathcal{S}}(\Omega) \quad \textit{and} \quad v_0 \in \underline{\mathcal{U}}_f^{\Omega} \cap \underline{\mathcal{S}}(\Omega). \tag{6.18}$$

Then

$$\overline{H}_f^{\Omega}, \underline{H}_f^{\Omega} \in \mathcal{H}(\Omega).$$

Proof. Using (6.18), we recognize that $\overline{\mathcal{U}}_f^{\Omega}$ is a Perron family. Hence, by Theorem 6.5.8,

$$\overline{H}_f^{\Omega} \in \mathcal{H}(\Omega).$$

From (6.18) it also follows that $\overline{\mathcal{U}}_{-f}^{\Omega}$ is a Perron family, thus

$$\underline{H}_f^{\Omega} = \sup \underline{\mathcal{U}}_f^{\Omega} = -\inf \overline{\mathcal{U}}_{-f}^{\Omega} \in \mathcal{H}(\Omega).$$

This ends the proof. □

Theorem 6.7.9 (Resolutivity. I). *Let $(E, \mathcal{H})$ be a harmonic space, and let $\Omega \subseteq E$ be an open set with compact closure and non-empty boundary.*

If $u \in C(\overline{\Omega}, \mathbb{R})$ and

$$u|_{\Omega} \in \overline{\mathcal{S}}(\Omega),$$

then

$$f := u|_{\partial\Omega} \in \mathcal{R}(\partial\Omega).$$

Proof. By the positivity axiom (A1), there exists $h_0 \in \overline{\mathcal{S}}(\Omega)$ such that $\inf_{\Omega} h_0 > 0$. Denote

$$\lambda := \frac{\max\{0, -u\}}{\inf_{\Omega} h_0}.$$

Then $u \geq -\lambda h_0$ in Ω and $-\lambda h_0 \in \underline{\mathcal{S}}(\Omega) \cap \underline{\mathcal{U}}_f^{\Omega}$. On the other hand,

$$u|_{\Omega} \in \underline{\mathcal{S}}(\Omega) \cap \overline{\mathcal{U}}_f^{\Omega}.$$

By Proposition 6.7.8,

$$\overline{H}_f^{\Omega}, \underline{H}_f^{\Omega} \in \mathcal{H}(\Omega).$$

Moreover, $\overline{H}_f^{\Omega} \leq u|_{\Omega}$ so that, since u is continuous up to $\partial\Omega$,

$$\overline{H}_f^{\Omega} \in \underline{\mathcal{U}}_f^{\Omega}.$$

Thus $\overline{H}_f^{\Omega} \leq \underline{H}_f^{\Omega}$. The opposite inequality always being true, this implies $\overline{H}_f^{\Omega} = \underline{H}_f^{\Omega}$. Summing up,

$$\overline{H}_f^{\Omega} = \underline{H}_f^{\Omega} \in \mathcal{H}(\Omega),$$

that is, f is resolutive. □

This last theorem would achieve a full strength if we knew that the family of continuous subharmonic functions is sufficiently rich. This happens in the $\mathfrak{S}$-harmonic spaces introduced in the following section.

6.8 $\mathfrak{S}$-harmonic Spaces: Wiener Resolutivity Theorem

Definition 6.8.1 ($\mathfrak{S}$-harmonic space). *A harmonic space $(E, \mathcal{H})$ will be said* $\mathfrak{S}$-harmonic *if the family*

$$\overline{\mathcal{S}}_c^+(E) := \{u \in \overline{\mathcal{S}}(E) \cap C(E, \mathbb{R}) \ : \ u \geq 0\}$$

separates the points of E, that is,

$$\begin{aligned}&\textit{for every } x, y \in E, \ x \neq y, \ \textit{there exists}\\ &u, v \in \overline{\mathcal{S}}_c^+(E) \ \textit{such that } u(x)v(y) \neq u(y)v(x).\end{aligned} \tag{6.19}$$

In a $\mathfrak{S}$-harmonic space, the positivity property of the axiom (A1) takes a stronger form.

Proposition 6.8.2 (Positivity axiom in a $\mathfrak{S}$-harmonic space). *Let $(E, \mathcal{H})$ be a $\mathfrak{S}$-harmonic space. For every compact set $K \subseteq E$, there exists $w \in \overline{\mathcal{S}}_c^+(E)$ such that*

$$\inf_K w > 0.$$

Proof. Since $\overline{\mathcal{S}}_c^+(E)$ separates the points of E, for every $x \in K$, there exists $u_x \in \overline{\mathcal{S}}_c^+(E)$ such that $u_x(x) > 0$. By the lower continuity of u, we can find an open neighborhood V_x of x such that $\inf_V u_x > 0$. Since K is compact, there exist $x_1, \ldots, x_p \in K$ such that

$$K \subseteq \bigcup_{j=1}^{p} V_{x_j}.$$

Then, if we define

$$w := \sum_{j=1}^{p} u_{x_j},$$

we have $w \in \overline{\mathcal{S}}_c^+(E)$ and $\inf_K w > 0$. □

A crucial property of the $\mathfrak{S}$-harmonic spaces is given by the next proposition.

Proposition 6.8.3. *Let $(E, \mathcal{H})$ be a $\mathfrak{S}$-harmonic space. For every compact set $K \subseteq E$, for every $f \in C(K, \mathbb{R})$ and for every $\varepsilon > 0$, there exist $u, v \in \overline{\mathcal{S}}_c^+(E)$ such that*

$$\sup_K |f - (u - v)| < \varepsilon.$$

Proof. Define

$$\mathcal{A} := \{u - v \ : \ u, v \in \overline{\mathcal{S}}_c^+(E)\}.$$

It is immediate to recognize that $\mathcal{A}$ is a linear subspace of $C(E, \mathbb{R})$. Moreover, since $(E, \mathcal{H})$ is $\mathfrak{S}$-harmonic, $\mathcal{A}$ separates the points of E. Finally, if $p = u - v \in \mathcal{A}$, then also $p^+ := \max\{p, 0\}$ belongs to $\mathcal{A}$, since

$$p^+ = u - \min\{u, v\}.$$

Then, by the Stone–Weierstrass theorem,[3] for every $f \in C(K, \mathbb{R})$ and $\varepsilon > 0$, there exists $p \in \mathcal{A}$ such that

$$\sup_K |f - p| < \varepsilon.$$

This ends the proof. □

We are now ready to prove the resolutivity of any continuous function.

Theorem 6.8.4 (Wiener resolutivity theorem). *Let $(E, \mathcal{H})$ be a $\mathfrak{S}$-harmonic space, and let $\Omega \subseteq E$ be an open set with compact closure and non-empty boundary. Every continuous function $f : \partial\Omega \to \mathbb{R}$ is resolutive.*

Proof. By Theorem 6.7.9, if $u \in \overline{\mathcal{S}}_c^+(E)$, then $u|_{\partial\Omega}$ is resolutive. As a consequence, every function

$$p = (u - v)|_{\partial\Omega}, \quad u, v \in \overline{\mathcal{S}}_c^+(E),$$

is resolutive. By Proposition 6.8.2, for every $\varepsilon > 0$, there exist $u_\varepsilon, v_\varepsilon \in \overline{\mathcal{S}}_c^+(E)$ such that

$$|f - p| < \varepsilon \quad \text{on } \partial\Omega, \qquad p = p_\varepsilon := (u_\varepsilon - v_\varepsilon)|_{\partial\Omega}. \tag{6.20a}$$

By Proposition 6.8.2, there exists $w \in \overline{\mathcal{S}}_c^+(E)$ such that $\inf_{\partial\Omega} w > 1$. Then, from (6.20a) we have

$$p - \varepsilon w < f < p + \varepsilon w \quad \text{on } \partial\Omega,$$

so that, letting $v := w|_{\partial\Omega}$, we have

$$\underline{H}^\Omega_{p-\varepsilon v} \le \underline{H}^\Omega_f \le \overline{H}^\Omega_f \le \overline{H}^\Omega_{p+\varepsilon v}. \tag{6.20b}$$

On the other hand, we know that $p - \varepsilon v, \ p + \varepsilon v \in \mathcal{R}(\partial\Omega)$. It follows that

[3] See the Appendix to this section.

$$\underline{H}^{\Omega}_{p-\varepsilon v} = H^{\Omega}_{p-\varepsilon v} = H^{\Omega}_{p} - \varepsilon H^{\Omega}_{v},$$
$$\overline{H}^{\Omega}_{p+\varepsilon v} = H^{\Omega}_{p+\varepsilon v} = H^{\Omega}_{p} + \varepsilon H^{\Omega}_{v}.$$

Using these identities in (6.20b), we obtain

$$0 \le \overline{H}^{\Omega}_{f} - \underline{H}^{\Omega}_{f} \le 2\varepsilon H^{\Omega}_{v} \le 2\varepsilon w, \tag{6.20c}$$

where, in the last inequality, we used the fact that w is an upper function related to $w|_{\partial\Omega}$ and Ω. Letting ε tend to zero in (6.20c), we get

$$\overline{H}^{\Omega}_{f} = \underline{H}^{\Omega}_{f}. \tag{6.20d}$$

Inequalities (6.20b) also give,

$$0 \le \overline{H}^{\Omega}_{f} - H^{\Omega}_{p-\varepsilon v} \le H^{\Omega}_{p+\varepsilon v} - H^{\Omega}_{p-\varepsilon v} = 2\varepsilon H^{\Omega}_{v} \le 2\varepsilon w \le 2\varepsilon \sup_{\overline{\Omega}} w.$$

Then, as $\varepsilon \to 0$,

$$H^{\Omega}_{p-\varepsilon v} = H^{\Omega}_{p_\varepsilon-\varepsilon v} \to \overline{H}^{\Omega}_{f} \qquad \text{uniformly on } \Omega.$$

Since $H^{\Omega}_{p-\varepsilon v} \in \mathcal{H}(\Omega)$, from this limit and Corollary 6.3.5 we obtain

$$\overline{H}^{\Omega}_{f} \in \mathcal{H}(\Omega).$$

Together with (6.20d), this implies the resolutivity of f. □

Exercise 6.8.5. Let $\Omega \subseteq E$ be open with $\overline{\Omega}$ compact and $\partial\Omega \neq \emptyset$. Let $w \in \overline{\mathcal{S}}^{+}_{c}(E)$ be such that $\inf_{\overline{\Omega}} w =: \lambda > 0$. Then, for every bounded function $f : \partial\Omega \to \mathbb{R}$, we have

$$-\lambda \sup_{\partial\Omega} |f| \le \underline{H}^{\Omega}_{f} \le \overline{H}^{\Omega}_{f} \le \lambda \sup_{\partial\Omega} |f|.$$

In particular, if f is resolutive,

$$\sup_{\Omega} |H^{\Omega}_{f}| \le \lambda \sup_{\partial\Omega} |f|.$$

(*Hint*: Note that $w \sup_{\Omega} |f| \in \overline{\mathcal{U}}^{\Omega}_{f}$.)

Exercise 6.8.6 ($\mathfrak{R}(\Omega) = \overline{\mathcal{R}(\Omega)}$). Let $\Omega \subseteq E$ be an open set as in the previous exercise. Let $(f_n)_{n\in\mathbb{N}}$ be a sequence of bounded resolutive functions such that

$$\sup_{\partial\Omega} |f_n - f| \to 0 \qquad \text{as } n \to \infty.$$

Then f is resolutive.

Exercise 6.8.7. Let $\Omega \subseteq E$ be an open set as in the previous exercise, and let $f : \partial\Omega \to \mathbb{R}$ be a bounded function. Then

$$\overline{H}^{\Omega}_{f}, \underline{H}^{\Omega}_{f} \in \mathcal{H}(\Omega).$$

6.8.1 Appendix: The Stone–Weierstrass Theorem

The aim of this section is to prove the following well-known density result.

Theorem 6.8.8 (The Stone–Weierstrass theorem). *Let $(Y, \mathcal{T})$ be a compact topological space, and let $\mathcal{A} \subseteq C(Y, \mathbb{R})$ satisfy the following conditions:*

(i) *$\mathcal{A}$ is a real vector space,*
(ii) *if $p \in \mathcal{A}$, then $p^+ := \max\{0, p\} \in \mathcal{A}$,*
(iii) *$\mathcal{A}$ separates the points of E, that is,*

$$\text{for every } x, y \in Y,\ x \neq y,\ \text{there exists } p, q \in \mathcal{A} \text{ such that } p(x)q(y) \neq p(y)q(x).$$

Then, for every $f \in C(Y, \mathbb{R})$ and for every $\varepsilon > 0$, there exists $p \in \mathcal{A}$ such that

$$\sup_Y |f - p| < \varepsilon.$$

Proof. Assumptions (i) and (ii) imply

$$|u| = u^+ + (-u)^+ \in \mathcal{A} \quad \text{for every } u \in \mathcal{A}.$$

As a consequence, $\max\{u, v\} = u + v + |u - v|$ and $\min\{u, v\} = u + v - |u - v|$ belong to $\mathcal{A}$ for any $u, v \in \mathcal{A}$. Let us now fix $u \in C(Y, \mathbb{R})$. For every $x, y \in Y$, due to assumption (iii), there exist $\alpha, \beta \in \mathbb{R}$ and $u, v \in \mathcal{A}$ such that

$$\begin{cases} \alpha u(x) + \beta v(x) = f(x), \\ \alpha u(y) + \beta v(y) = f(y). \end{cases} \tag{6.21}$$

Obviously, $u_{x,y} := \alpha u + \beta v \in \mathcal{A}$, and, for every $\varepsilon > 0$, there exists a neighborhood V_y of y such that

$$u_{x,y}(z) < f(z) + \varepsilon \qquad \forall z \in V_y.$$

Since Y is compact, we can choose $y_1, \ldots, y_p \in Y$ such that $\bigcup_{j=1}^p V_{y_j} = Y$. Letting $u_x := \min\{u_{x,y_1}, \ldots, u_{x,y_p}\}$, we have

$$u_x \in \mathcal{A} \text{ and } u_x(z) < f(z) + \varepsilon \quad \forall z \in Y. \tag{6.22}$$

On the other hand, by the first identity in (6.21), $u_x(x) = f(x)$ for every $x \in Y$. Then, since u_x is continuous, for every $x \in Y$, there exists a neighborhood W_x of x such that

$$u_x(z) > f(z) - \varepsilon \quad \forall z \in W_x.$$

Since Y is compact, there exist $W_{x_1}, \ldots, W_{x_q}$ such that $\bigcup_{j=1}^q W_{x_j} = Y$. As a consequence, letting $u := \max\{u_{x_1}, \ldots, u_{x_q}\}$, we get $u \in \mathcal{A}$ and $u(z) > f(z) - \varepsilon$ for every $y \in Y$. On the other hand, keeping in mind (6.22), we also have

$$u = \max\{u_{x_1}, \ldots, u_{x_q}\} < f + \varepsilon \quad \text{in } Y.$$

This completes the proof. □

6.9 $\mathcal{H}$-harmonic Measures for General Relatively Compact Open Sets. A Characterization of the Resolutive Functions

As in the previous section, Ω will denote a relatively compact (i.e. with compact closure in E) open set with non-empty boundary, contained in a $\mathfrak{S}$-harmonic space $(E, \mathcal{H})$.

Let us fix a point $x \in \Omega$. By the Wiener resolutivity theorem (Theorem 6.8.4) and Proposition 6.7.4, the map

$$C(\partial\Omega, \mathbb{R}) \ni f \mapsto H_f^\Omega(x) \in \mathbb{R}$$

is well defined, *linear and positive*. Then by the Riesz representation theorem, the following definition is well posed.

Definition 6.9.1 ($\mathcal{H}$-harmonic measure. II). *Let Ω be a relatively compact open set with non-empty boundary, contained in the $\mathfrak{S}$-harmonic space $(E, \mathcal{H})$. There exists a* unique Radon measure, *denoted by μ_x^Ω, such that*

$$H_f^\Omega(x) = \int_{\partial\Omega} f \, \mathrm{d}\mu_x^\Omega \qquad \textit{for every } f \in C(\partial\Omega, \mathbb{R}).$$

*We shall call μ_x^Ω the $\mathcal{H}$-*harmonic measure *related to Ω and x.*

Exercise 6.9.2. For every relatively compact open set $\Omega \subseteq E$, one has $\mu_x^\Omega(\partial\Omega) < \infty$ for every $x \in \Omega$. (*Hint*: Note that $\mu_x^\Omega(\partial\Omega) = H_1^\Omega(x)$ and use the a priori estimate of Exercise 6.8.5.)

When Ω is an $\mathcal{H}$-regular open set, this definition gives back the one given in Definition 6.2.3. Our aim here is to show the following theorem.

Theorem 6.9.3 (Characterization of resolutivity). *Let $(E, \mathcal{H})$ be a $\mathfrak{S}$-harmonic space and let $\Omega \subseteq E$ be a relatively compact open set with non-empty boundary. Given $f : \partial\Omega \to [-\infty, \infty]$, the following statements are equivalent:*

(i) $f \in \mathcal{R}(\partial\Omega)$,
(ii) $f \in L^1(\partial\Omega, \mu_x^\Omega)$ *for every $x \in \Omega$.*

In this case

$$H_f^\Omega(x) = \int_{\partial\Omega} f \, \mathrm{d}\mu_x^\Omega \qquad \textit{for every } x \in \Omega.$$

The proof of this theorem requires some preliminary results having an independent interest.

Notation. Given $u : \Omega \to]-\infty, \infty]$, lower semicontinuous and bounded below, we set

$$u^* : \overline{\Omega} \to]-\infty, \infty], \quad u^*(x) := \begin{cases} u(x) & \text{if } x \in \Omega, \\ \liminf_{y \to x} u(y) & \text{if } x \in \partial\Omega. \end{cases}$$

It is quite easy to recognize that u^* is lower semicontinuous.

Similarly, given $u : \Omega \to [-\infty, \infty[$, upper semicontinuous and bounded above, we set

$$u_* : \overline{\Omega} \to [-\infty, \infty[, \quad u_*(x) := \begin{cases} u(x) & \text{if } x \in \Omega, \\ \limsup_{y \to x} u(y) & \text{if } x \in \partial\Omega. \end{cases}$$

The function u_* is upper semicontinuous.

Lemma 6.9.4. *With the previous notation, if $u \in \mathcal{H}^*(\Omega)$ and $\inf_\Omega u > -\infty$, then:*

(i) $u(x) \geq \int_{\partial\Omega} u^* \, d\mu_x^\Omega > -\infty$ *for every $x \in \Omega$,*
(ii) $x \mapsto \int_{\partial\Omega} u^* \, d\mu_x^\Omega$ *belongs to $\mathcal{H}^*(\Omega)$.*

An analogous result also holds if $u \in \mathcal{H}_*(\Omega)$ and $\sup_\Omega u < \infty$.

Proof. For every $\phi \in C(\partial\Omega, \mathbb{R})$, $\phi \leq u^*|_{\partial\Omega}$, we have

$$H_\phi^\Omega \leq u,$$

since $u \in \overline{\mathcal{U}}_\phi^\Omega$. Then

$$u(x) \geq \sup_\phi H_\phi^\Omega(x) = \sup_\phi \int_{\partial\Omega} \phi \, d\mu_x^\Omega = \int_{\partial\Omega} u^* \, d\mu_x^\Omega.$$

(The suprema are taken with respect to $\phi \in C(\partial\Omega, \mathbb{R})$, $\phi \leq u^*|_{\partial\Omega}$.) On the other hand, since $\inf_{\partial\Omega} u^* > -\infty$ and $\mu_x(\Omega) < \infty$ for every $x \in \Omega$ (see Exercise 12 at the end of the chapter), we have

$$\int_{\partial\Omega} u^* \, d\mu_x^\Omega > -\infty.$$

This proves (i). To obtain (ii), we just have to use Theorem 6.3.8 keeping in mind that $H_\phi^\Omega \in \mathcal{H}(\Omega)$. □

Lemma 6.9.5. *Let $f : \partial\Omega \to]-\infty, \infty]$ be a l.s.c. function. Then:*

(i) *for every $x \in \Omega$, we have*

$$\overline{H}_f^\Omega(x) = \int_{\partial\Omega} f \, d\mu_x^\Omega,$$

(ii) $\overline{H}_f^\Omega \in \mathcal{H}(\Omega)$ *if and only if*

$$\overline{H}_f^\Omega(x) < \infty \qquad \textit{for every } x \in \Omega.$$

Proof. (i) Let (f_n) be a sequence of continuous functions on $\partial\Omega$ such that $f_n \nearrow f$. Since $\overline{H}^{\Omega}_{f_n} = H^{\Omega}_{f_n} \in \mathcal{H}(\Omega)$ for every $n \in \mathbb{N}$, we have (see Exercise 6.7.5, page 360)

$$H^{\Omega}_{f_n} \nearrow \overline{H}^{\Omega}_{f}, \tag{6.23}$$

so that, by the Beppo Levi monotone convergence theorem,

$$\overline{H}^{\Omega}_{f}(x) = \lim_{n\to\infty} H^{\Omega}_{f_n}(x) = \lim_{n\to\infty} \int_{\partial\Omega} f_n \,\mathrm{d}\mu^{\Omega}_x = \int_{\partial\Omega} f \,\mathrm{d}\mu^{\Omega}_x$$

for every $x \in \Omega$.

(ii) The only if part is trivial. The if part follows from (6.23) and Theorem 6.2.8, page 343. □

Proof (of Theorem 6.9.3). (i) ⇒ (ii). Let $f \in \mathcal{R}(\partial\Omega)$. Then, for every $x \in \Omega$ and $\varepsilon > 0$, there exist $u \in \overline{\mathcal{U}}^{\Omega}_f$ and $v \in \underline{\mathcal{U}}^{\Omega}_f$ such that

$$u(x) - \varepsilon < H^{\Omega}_f(x) < v(x) + \varepsilon.$$

By Lemma 6.9.4,

$$-\infty < \int_{\partial\Omega} u^* \,\mathrm{d}\mu^{\Omega}_x - \varepsilon < H^{\Omega}_f(x) < \int_{\partial\Omega} v_* \,\mathrm{d}\mu^{\Omega}_x + \varepsilon < \infty,$$

and we have

$$\int_{\partial\Omega} (u^* - v_*) \,\mathrm{d}\mu^{\Omega}_x < 2\varepsilon,$$

so that, since $v_* \le f \le u^*$, u^* is l.s.c., v_* is u.s.c., and ε is an arbitrary positive number,

$$f \in L^1(\partial\Omega, \mathrm{d}\mu^{\Omega}_x) \qquad \forall\, x \in \Omega$$

and

$$H^{\Omega}_f(x) = \int_{\partial\Omega} f \,\mathrm{d}\mu^{\Omega}_x \qquad \text{for every } x \in \Omega.$$

(ii) ⇒ (i). It is not restrictive to assume $f \ge 0$. Define

$$\underline{\mathcal{F}} := \{g : \partial\Omega \to [-\infty, \infty[\; \mid g \text{ u.s.c.}, 0 \le g \le f\}$$

and

$$\overline{\mathcal{F}} := \{h : \partial\Omega \to \,]-\infty, \infty] \mid h \text{ l.s.c.}, h \ge f\}.$$

Let us fix $x \in \Omega$ and $\varepsilon > 0$. By Vitali–Carathéodory's theorem, there exist $g \in \underline{\mathcal{F}}$ and $h \in \overline{\mathcal{F}}$ such that

$$\int_{\partial\Omega} (h - g) \,\mathrm{d}\mu^{\Omega}_x < \varepsilon.$$

Then $h, g \in L^1(\partial\Omega, \mu^{\Omega}_x)$ and, by Lemma 6.9.5-(i),

$$-\infty < \int_{\partial\Omega} g \,\mathrm{d}\mu_x^\Omega = \underline{H}_g^\Omega(x) \le \underline{H}_f^\Omega(x) \le \overline{H}_f^\Omega(x)$$
$$= \int_{\partial\Omega} h \,\mathrm{d}\mu_x^\Omega \le \int_{\partial\Omega} g \,\mathrm{d}\mu_x^\Omega + \varepsilon < \infty.$$

This proves that $-\infty < \underline{H}_f^\Omega(x) \le \overline{H}_f^\Omega(x) < \infty$ for every $x \in \Omega$ and that

$$\sup_{g\in\underline{\mathcal{F}}} \underline{H}_g^\Omega = \underline{H}_f^\Omega = \overline{H}_f^\Omega = \inf_{h\in\overline{\mathcal{F}}} \overline{H}_h^\Omega. \tag{6.24}$$

On the other hand, by Lemma 6.9.5-(ii),

$$\underline{H}_g^\Omega \in \mathcal{H}(\Omega) \quad \text{for every } g \in \underline{\mathcal{F}},$$

since $g \ge 0$. Hence $\underline{H}_g^\Omega > -\infty$ at any point of Ω. Thus, from (6.24) and Theorem 6.3.6 we finally obtain

$$\underline{H}_f^\Omega = \overline{H}_f^\Omega \in \mathcal{H}(\Omega).$$

This completes the proof. □

Exercise 6.9.6. Prove that if $f : \partial\Omega \to \mathbb{R}$ is l.s.c. and bounded from above, then f is resolutive.

6.10 $\mathfrak{S}^*$-harmonic Spaces: Bouligand's Theorem

We now give a definition introducing a new property which is not usually assumed in the abstract potential theory. However, as we will see in Chapter 7, this assumption does not affect at all the possibility to apply our theory to the sub-Laplacians.

Definition 6.10.1 ($\mathfrak{S}^*$-harmonic space). *A $\mathfrak{S}$-harmonic space $(E, \mathcal{H})$ will be said $\mathfrak{S}^*$-harmonic if the following property holds:*

(A*) *For every $x_0 \in E$ there exists $s_{x_0} \in \underline{S}_c^+(E)$ such that $s_{x_0}(x_0) = 0$ and*

$$\inf_{E\setminus V} s_{x_0} > 0$$

for every neighborhood V of x_0. (See Definition 6.8.1 for the definition of $\underline{S}_c^+(E)$.)

This property will make transparent the rôle of the *barrier functions* in studying the continuity up to the boundary of the Perron–Wiener–Brelot functions.

Convention. Throughout this section, Ω will denote an open set in a $\mathfrak{S}^*$-harmonic space $(E, \mathcal{H})$. We will always assume, without any further comments, that $\overline{\Omega}$ is compact and $\partial\Omega \ne \emptyset$.

We know from the Wiener resolutivity theorem that every continuous function $f : \partial\Omega \to \mathbb{R}$ is resolutive, so that the Perron–Wiener–Brelot function H_f^Ω is $\mathcal{H}$-harmonic in Ω. However, in general, we cannot expect a "good" behavior of H_f^Ω at the boundary points of Ω.

Definition 6.10.2 ($\mathcal{H}$-regular point). *A point $y \in \partial\Omega$ will be called* $\mathcal{H}$-regular *if*

$$\lim_{\Omega \ni x \to y} H_f^\Omega(x) = f(y) \qquad \forall\ f \in C(\partial\Omega, \mathbb{R}).$$

Obviously, the function H_f^Ω is the solution (the unique solution, thanks to the minimum principle!) of the *$\mathcal{H}$-Dirichlet problem*

$$(\mathcal{H}\text{-D}) \qquad \begin{cases} u \in \mathcal{H}(\Omega), \\ \lim_{x\to y} u(x) = f(y) \quad \forall\ y \in \partial\Omega, \end{cases}$$

for every $f \in C(\partial\Omega, \mathbb{R})$, if and only if all the boundary points of Ω are $\mathcal{H}$-regular points.

Unfortunately, as we said before, we have to expect that, in general, $\partial\Omega$ contains boundary points which are not $\mathcal{H}$-regular.[4]

The notion of $\mathcal{H}$-barrier function will allow us to give a necessary and sufficient condition for the $\mathcal{H}$-regularity.

Definition 6.10.3 ($\mathcal{H}$-barrier function). *Let $y \in \partial\Omega$. A $\mathcal{H}$-barrier function for Ω at y is a function w defined in $\Omega \cap V$, being V a suitable open neighborhood of y, such that* (see also Fig. 6.1):

(i) $w \in \overline{\mathcal{S}}(\Omega \cap V)$,
(ii) $w(x) > 0$ *for every* $x \in \Omega \cap V$,
(iii) $\lim_{x\to y} w(x) = 0$.

The link between $\mathcal{H}$-regularity and $\mathcal{H}$-barrier functions is given by the following theorem.

Theorem 6.10.4 (Bouligand's theorem). *Let $(E, \mathcal{H})$ be a $\mathfrak{S}^*$-harmonic space, and let $\Omega \subseteq E$ be a relatively compact open set with non-empty boundary. A point $x_0 \in \partial\Omega$ is $\mathcal{H}$-regular for Ω if and only if there exists an $\mathcal{H}$-barrier function for Ω at x_0.*

The proof of this theorem is quite long. It is convenient to premise some lemmas.

Lemma 6.10.5. *Let $V \subseteq E$ be a regular open set, and let U be a relatively open subset of ∂V. For every $x_0 \in V$ and $\varepsilon > 0$, there exists a compact set $K \subseteq U$ and a non-negative function $h \in \mathcal{H}(V)$ such that:*

---

[4] See, for instance, Exercise 6.10.7 (page 375) related to the classical S. Zaremba's example of non-regular boundary points.

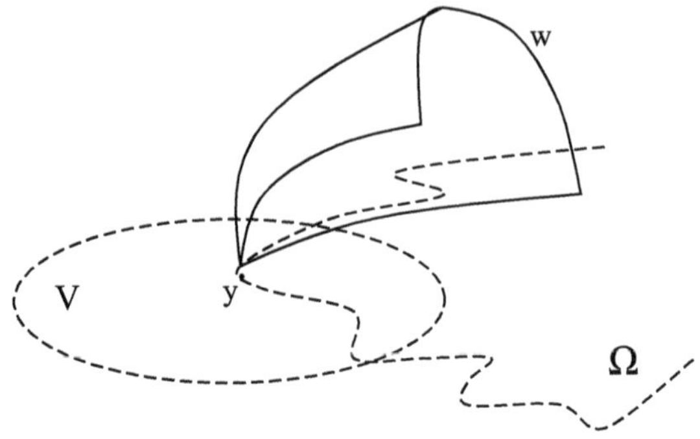

Fig. 6.1. A $\mathcal{H}$-barrier function

(i) $h(x_0) < \varepsilon$,
(ii) $\liminf_{V \ni x \to y} h(x) \geq 1$ *for every* $y \in U \setminus K$.

Proof. Let $K \subseteq U$ be such that $\mu_{x_0}^V(U \setminus K) < \varepsilon$, and let $f : \partial V \to \mathbb{R}$ be the characteristic function of $U \setminus K$. Since f is bounded and l.s.c., it is resolutive (see Ex. 6.9.6, page 370). Let us put $h := H_f^V$. Then $h \in \mathcal{H}(V)$, $h \geq 0$, since $f \geq 0$ and

$$h(x_0) = \int_{\partial V} f \, d\mu_{x_0}^V = \mu_{x_0}^V(U \setminus K) < \varepsilon.$$

We now choose a sequence (f_n) of continuous functions on ∂V such that $0 \leq f_n \leq f$ and $f_n \nearrow f$. Since V is regular, for every $y \in U \setminus K$ and $n \in \mathbb{N}$, we have

$$\liminf_{x \to y} h(x) \geq \lim_{x \to y} H_{f_n}^V(x) = f_n(y).$$

By taking the supremum with respect to n at the last right-hand side, we get

$$\liminf_{x \to y} h(x) \geq 1 \qquad \text{for every } y \in U \setminus K.$$

This completes the proof. □

Lemma 6.10.6. *Let* $\Omega \subseteq E$ *be a relatively compact open set, and let* $x_0 \in \partial\Omega$. *Define* $f : \partial\Omega \to \mathbb{R}$, $f(x) = s_{x_0}(x)$, *where* s_{x_0} *is the continuous subharmonic function of assumption* (A*). *Define*

$$s_{x_0}^{\Omega} := H_f^{\Omega}.$$

Then:

(i) $s_{x_0}^{\Omega}$ *is harmonic in* Ω,
(ii) $\inf_{\Omega \setminus U} s_{x_0}^{\Omega} > 0$ *for every neighborhood* U *of* x_0,
(iii) $\lim_{\Omega \ni x \to x_0} s_{x_0}^{\Omega}(x) = 0$ *if* x_0 *is* $\mathcal{H}$*-regular for* Ω,
(iv) $\lim_{\Omega \ni x \to x_0} s_{x_0}^{\Omega}(x) = 0$ *if there exists a barrier function for* Ω *at* x_0.

Proof. (i) This is obvious, since f is continuous, hence resolutive.

(ii) Since $s_{x_0}|_\Omega \in \underline{\mathcal{U}}_f^\Omega$, one has $s_{x_0}|_\Omega \leq H_f^\Omega$. Thus, the assertion follows from the positivity property of s_{x_0}.

(iii) Since $s_{x_0}^\Omega = H_f^\Omega$, if x_0 is $\mathcal{H}$-regular for Ω, then one has

$$\lim_{x \to x_0} s_{x_0}^\Omega(x) = f(x_0) = 0.$$

(iv) Suppose that there exists a barrier function b at x_0. Then there exists a neighborhood B of x_0 such that $b \in \overline{\mathcal{S}}(\Omega \cap B)$ and $b(x) \to 0$ as $x \to x_0$.

Let $w \in \overline{\mathcal{S}}_c^+(E)$ be such that $w \geq 1$ in $\overline{\Omega}$ (see Proposition 6.8.2). The maximum principle of Ex. 6.8.5 implies the existence of a positive constant M such that

$$0 \leq s_{x_0}^\Omega \leq M \qquad \text{in } \Omega.$$

Let $\varepsilon > 0$ be fixed. Choose a regular neighborhood V of x_0 such that

$$\overline{V} \subseteq B \qquad \text{and} \qquad \sup_{V \cap \partial\Omega} f \leq \varepsilon.$$

Let $K \subseteq \Omega \cap \partial V$ be compact, and let $h \in \mathcal{H}(V)$ be such that $h \geq 0$,

$$h(x_0) \leq \frac{\varepsilon}{M} \qquad \text{and} \qquad \liminf_{x \to y} h(x) \geq 1 \quad \forall y \in \Omega \cap \partial V \setminus K.$$

(see the previous lemma). Since $b > 0$ in $B \cap \Omega$ and $B \supseteq \overline{V}$, we have $m := \inf_K b > 0$. For a given function $u \in \underline{\mathcal{U}}_f^\Omega$, define

$$u_0 := u - \varepsilon w - M\left(\frac{b}{m} + h\right)$$

in $\Omega \cap V$. Then $u_0 \in \underline{\mathcal{S}}(\Omega)$, and $\limsup_{\partial(\Omega\cap V)} u_0 \leq 0$. Indeed, since $u \leq H_f^\Omega \leq M$, we have

$$y \in K \Rightarrow \limsup_{x \to y} u_0(x) \leq \sup_\Omega u - \frac{M}{m} \inf_K b \leq M - M = 0, \tag{6.25a}$$

$$y \in \Omega \cap \partial V \setminus K \Rightarrow$$
$$\limsup_{x \to y} u_0(x) \leq \sup_\Omega u - M \liminf_{x \to y} h(x) \leq M - M = 0, \tag{6.25b}$$

$$y \in V \cap \partial\Omega \Rightarrow$$
$$\limsup_{x \to y} u_0(x) \leq \limsup_{x \to y} u(x) - \varepsilon w(y) \leq f(y) - \varepsilon \leq 0. \tag{6.25c}$$

The maximum principle for subharmonic functions implies $u_0 \leq 0$ in $\Omega \cap V$, that is,

$$u \leq \varepsilon w + M\left(\frac{b}{m} + h\right) \qquad \text{in } \Omega \cap V.$$

This inequality holds for every $u \in \overline{\mathcal{U}}_f^\Omega$, so that

$$s_{x_0}^{\Omega} \equiv H_f^{\Omega} \le \varepsilon w + M\left(\frac{b}{m} + h\right) \qquad \text{in } \Omega \cap V.$$

Then, since $b(x) \to 0$ as $x \to x_0$, and $Mh(x_0) \le \varepsilon$,

$$0 \le \limsup_{x \to x_0} s_{x_0}^{\Omega}(x) \le \varepsilon w(x_0) + \varepsilon \qquad \text{for every } \varepsilon > 0.$$

This completes the proof. □

We are now ready to prove Bouligand's theorem.

Proof (of Theorem 6.10.4). If x_0 is $\mathcal{H}$-regular for Ω, then $s_{x_0}^{\Omega}$ is a barrier function for Ω at x_0, see Lemma 6.10.6-(i)–(iii). Vice versa, suppose that there exists a barrier function for Ω at x_0. Let us prove that

$$\lim_{\Omega \ni x \to x_0} H_{\varphi}^{\Omega}(x) = \varphi(x_0) \tag{6.26}$$

for every $\varphi \in C(\partial\Omega, \mathbb{R})$. Then $s_{x_0}^{\Omega}$ has properties (i)–(ii) and (iv) in Lemma 6.10.6. Let $w \in \mathcal{S}_c^+(\Omega)$ be such that $\inf_{\Omega} w \ge 1$ (see Proposition 6.8.2), and let V be a neighborhood of x_0. Define

$$\begin{aligned} \omega_{\varphi}(V) &:= \sup\left\{|\varphi(z)\, w(x_0) - \varphi(x_0)\, w(z)| \;:\; z \in V \cap \partial\Omega\right\}, \\ M_{\varphi} &:= \sup\left\{|\varphi(z)\, w(x_0) - \varphi(x_0)\, w(z)| \;:\; z \in \partial\Omega\right\}, \\ m &:= \inf\left\{s_{x_0}^{\Omega}(x) \;:\; x \in \Omega \setminus V\right\}, \end{aligned}$$

and

$$u := \left(\frac{\varphi(x_0)}{w(x_0)} + \omega_{\varphi}(V)\right) w + M_{\varphi}\, \frac{s_{x_0}^{\Omega}}{m}.$$

Then $u \in \overline{\mathcal{S}}(\Omega)$. Moreover,

$$\liminf_{\Omega \ni x \to z} u(x) \ge \varphi(z) \quad \forall\, z \in \partial\Omega.$$

Indeed, if $z \in V \cap \partial\Omega$, the very definition of $\omega_{\varphi}(V)$ and the inequality $w \ge 1$ in $\overline{\Omega}$ imply

$$\liminf_{x \to z} u(x) \ge \left(\frac{\varphi(x_0)}{w(x_0)} + \omega_{\varphi}(V)\right) w(z) \ge \varphi(z).$$

On the other hand, if $z \in \partial\Omega \setminus V$, keeping in mind the definitions of m and M_{φ}, one has

$$\liminf_{x \to z} u(x) \ge \frac{\varphi(x_0)}{w(x_0)}\, w(z) + M_{\varphi} \ge \varphi(z).$$

We also have $\inf_{\Omega} u > -\infty$. Thus, $u \in \overline{\mathcal{U}}_{\varphi}^{\Omega}$. Hence $H_{\varphi}^{\Omega} \le u$ and

$$\limsup_{\Omega \ni x \to x_0} H_{\varphi}^{\Omega}(x) \le \lim_{\Omega \ni x \to x_0} u(x) = \varphi(x_0) + \omega_{\varphi}(V) w(x_0),$$

since $s^{\Omega}_{x_0}(x) \to 0$ as $x \to x_0$. By taking the infimum with respect to V at the last right-hand side, we get

$$\limsup_{\Omega \ni x \to x_0} H^{\Omega}_{\varphi}(x) \le \varphi(x_0) \tag{6.27}$$

for every $\varphi \in C(\partial\Omega, \mathbb{R})$. As a consequence,

$$\liminf_{x \to x_0} H^{\Omega}_{\varphi}(x) = -\limsup_{\Omega \ni x \to x_0} H^{\Omega}_{-\varphi}(x) \ge \varphi(x_0).$$

This inequality, together with (6.27), implies (6.26). □

Exercise 6.10.7 (S. Zaremba's counterexample). Let $x_0 \in \Omega$. Consider the set $\widehat{\Omega} = \Omega \setminus \{x_0\}$. Assume there exists $u \in \overline{\mathcal{S}}_+(\Omega)$ such that

$$u(x_0) = \infty, \quad 0 \le u(x) < \infty \quad \forall\, x \in \Omega.$$

Define

$$f : \partial\widehat{\Omega} \to \mathbb{R}, \qquad f(x) = \begin{cases} 0 & \text{if } x \in \partial\Omega, \\ 1 & \text{if } x = x_0. \end{cases}$$

Then $H^{\widehat{\Omega}}_f \equiv 0$. In particular, x_0 is not $\mathcal{H}$-regular for $\widehat{\Omega}$. (*Hint*: the function $\varepsilon\, u$ belongs to $\overline{\mathcal{U}}^{\widehat{\Omega}}_f$ for all $\varepsilon > 0$.)

6.11 Reduced Functions and Balayage

The notions of reduced function and balayage are crucial in the potential theory. Before giving the basic definitions, we need a general preliminary result.

Theorem 6.11.1 (Envelopes in $\overline{\mathcal{S}}(\Omega)$). *Let $(E, \mathcal{H})$ be a harmonic space and let $\Omega \subseteq E$ be open. Given $\mathcal{F} \subseteq \overline{\mathcal{S}}(\Omega)$, assume that*

$$u_0 := \inf \mathcal{F}$$

is locally bounded from below. Then

$$\widehat{u_0} \in \overline{\mathcal{S}}(\Omega).$$

Remark 6.11.2. In the setting of the harmonic spaces related to the sub-Laplacians, the statement of Theorem 6.11.1 can be strengthened (see Theorem 9.5.6, page 449).

Proof. The function $\widehat{u_0}$ is l.s.c. in Ω and $> -\infty$ at any point. Moreover, for a given $\mathcal{H}$-regular set $V \subseteq \overline{V} \subseteq \Omega$, we have

$$\int_{\partial V} \widehat{u_0}\, \mathrm{d}\mu^V_x \le \int_{\partial V} u\, \mathrm{d}\mu^V_x \le u(x) \tag{6.28}$$

at any point $x \in V$ for every $u \in \mathcal{F}$. Then

$$\int_{\partial V} \widehat{u_0}\, d\mu_x^V \le u_0(x) \qquad \forall\, x \in V. \tag{6.29}$$

On the other hand, since $u \in \overline{\mathcal{S}}(\Omega)$, $x \mapsto \int_{\partial V} u\, d\mu_x^V$ is $\mathcal{H}$-harmonic in Ω, and the first inequality in (6.28) implies that also

$$x \mapsto \int_{\partial V} \widehat{u_0}\, d\mu_x^V$$

is $\mathcal{H}$-harmonic, hence continuous, in V (see Exercise 6.3.9). Therefore, from inequality (6.29) we obtain

$$\int_{\partial V} \widehat{u_0}\, d\mu_x^V \le \widehat{u_0}(x),$$

so that $\widehat{u_0} \in \overline{\mathcal{S}}(\Omega)$. □

Exercise 6.11.3. Let $\mathcal{F} \subseteq \mathcal{H}^*(\Omega)$, for an open set $\Omega \subseteq E$. Assume that

$$u_0 := \inf \mathcal{F}$$

is locally bounded from below. Then $\widehat{u_0} \in \mathcal{H}^*(\Omega)$.
(*Hint*: Apply the previous theorem to the families

$$\mathcal{F}_n := \{\min\{u, n\} \,:\, u \in \mathcal{F}\}, \qquad n \in \mathbb{N}.$$

Then use Theorem 6.3.8.)

Definition 6.11.4 ($\mathcal{H}$-reduced function and $\mathcal{H}$-balayage). *Let $f : E \to [0, \infty]$ be a non-negative real extended function, and let $A \subseteq E$. Define*

$$\mathbf{R}_A^f := \inf\left\{v \in \mathcal{H}^*(E) \,:\, v \ge 0 \text{ in } E, v \ge f \text{ in } A\right\}$$

and

$$\widehat{\mathbf{R}}_A^f := \widehat{(\mathbf{R}_A^f)}.$$

*$\mathbf{R}_A^f$ and $\widehat{\mathbf{R}}_A^f$ are called, respectively, the $\mathcal{H}$-*reduced function *and the* regularized $\mathcal{H}$-reduced function *of f on A.*

*$\widehat{\mathbf{R}}_A^f$ is also called the $\mathcal{H}$-*balayage *of f on A.*

We, obviously, have

$$0 \le \widehat{\mathbf{R}}_A^f \le \mathbf{R}_A^f.$$

Moreover, by Exercise 6.11.3,

$$\widehat{\mathbf{R}}_A^f \in \mathcal{H}^*(E).$$

If the function $f \in \mathcal{H}^*(E)$, $f \ge 0$, from the very definition of reduced function we get $\mathbf{R}_A^f \le f$. So that, in this case,

$$0 \le \widehat{\mathbf{R}}_A^f \le \mathbf{R}_A^f \le f,$$

and $\widehat{\mathbf{R}}_A^f$ is $\mathcal{H}$-superharmonic if f is $\mathcal{H}$-superharmonic. Moreover,

$$\widehat{\mathbf{R}}_A^f(x) = \mathbf{R}_A^f(x) = f(x) \quad \text{for every } x \in \operatorname{Int}(A). \tag{6.30}$$

Exercise 6.11.5. Prove the claimed (6.30). Prove also that if $f \in \overline{\mathcal{S}}(E)$, $f \geq 0$, then

$$\mathbf{R}_E^f = \inf\{v \in \overline{\mathcal{S}}(E) \,:\, v \geq 0 \text{ in } E, v \geq f \text{ in } A\},$$

and $\mathbf{R}_E^f = f$ in A. □

An application of the fundamental theorem on Perron families gives the following result.

Theorem 6.11.6 (Properties of $\mathcal{H}$-reduction and $\mathcal{H}$-balayage. I). *Let $(E, \mathcal{H})$ be a harmonic space. Let $f \in \overline{\mathcal{S}}(E)$, $f \geq 0$, and let $A \subseteq E$. Then*

(i) $\widehat{\mathbf{R}}_A^f$ *is $\mathcal{H}$-superharmonic in E,*
(ii) $\mathbf{R}_A^f = f$ *in A and* $\widehat{\mathbf{R}}_A^f = f$ *in* $\mathrm{int}(A)$,
(iii) $\mathbf{R}_A^f$ *and* $\widehat{\mathbf{R}}_A^f$ *are equal and $\mathcal{H}$-harmonic in* $\Omega := E \setminus \overline{A}$.

An improvement of (iii) will be given in the setting of sub-Laplacians.

Proof. We already know that (i) and (ii) hold true.
To prove (iii), consider the family

$$\mathcal{F} := \{v|_\Omega \,:\, v \in \overline{\mathcal{S}}(E),\ v \geq 0 \text{ in } E, v \geq f \text{ in } A\}.$$

It is easy to see that $\mathcal{F}$ is a Perron family in Ω and that f and $g \equiv 0$ are, respectively, an $\mathcal{H}$-superharmonic majorant and an $\mathcal{H}$-subharmonic minorant of $\mathcal{F}$. Then, by Theorem 6.5.8,

$$(\mathbf{R}_A^f)|_\Omega = \inf \mathcal{F} \in \mathcal{H}(\Omega).$$

In particular, $\mathbf{R}_A^f$ is continuous in Ω, hence

$$\widehat{\mathbf{R}}_A^f = \mathbf{R}_A^f \qquad \text{in } \Omega.$$

Then (iii) follows. □

The following theorem states some properties of the reduced function and the balayage. Their easy proofs are left as an exercise.

Theorem 6.11.7 (Properties of $\mathcal{H}$-reduction and $\mathcal{H}$-balayage. II). *Let $(E, \mathcal{H})$ be a harmonic space, and let $A, B \subseteq E$ and $f, g \in \overline{\mathcal{S}}(E)$, $f, g \geq 0$. Then*

$$\widehat{\mathbf{R}}_A^f \leq \widehat{\mathbf{R}}_B^f \text{ if } A \subseteq B, \tag{6.31a}$$
$$\widehat{\mathbf{R}}_A^f \leq \widehat{\mathbf{R}}_A^g \text{ if } f \leq g, \tag{6.31b}$$
$$\widehat{\mathbf{R}}_A^{\lambda f} = \lambda \widehat{\mathbf{R}}_A^f \text{ if } \lambda \geq 0. \tag{6.31c}$$

The same assertions hold for the operator $\mathbf{R}$.

Some other crucial properties of the reduced function and the balayage will be proved in the context of the harmonic spaces associated to the sub-Laplacians. The use of the average operators $\mathcal{M}_r$ and M_r will make the proofs much easier than in the general abstract setting.

Bibliographical Notes. In this chapter, we presented some topics from the theory of abstract harmonic spaces mainly inspired by the ones developed by H. Bauer in [Bau66] and C. Costantinescu and A. Cornea in [CC72]. See also J.L. Doob [Doo01].

6.12 Exercises of Chapter 6

Unlike the other chapters, several exercises of Chapter 6 are distributed throughout the text. For the reading convenience, we gather below the list of these exercises.

Ex. 1) Prove property (P1) in Section 6.1.
Ex. 2) Prove property (P2) in Section 6.1 by recognizing that $u : A \to [-\infty, \infty]$ is l.s.c. if and only if it is continuous from the topological space $(A, \mathcal{T}|_A)$ and $([-\infty, \infty], \mathcal{I})$, where $\mathcal{I}$ is the topology on $[-\infty, \infty]$ generated by the intervals $]t, \infty]$, $t \in \mathbb{R}$.
Ex. 3) Prove property (P3) in Section 6.1.
(*Hint*: Use (P2).)
Ex. 4) Prove property (P4) in Section 6.1.
(*Hint*: Use Ex. 2, first recognizing that $K \subseteq]-\infty, \infty]$ is compact w.r.t. the topology $\mathcal{I}$ if and only if K has real minimum.)
Ex. 5) Prove property (P5) in Section 6.1.
(*Hint*: For every $a \in A$ and $\varepsilon > 0$, there exists $V_{a,\varepsilon} \in \mathcal{U}_a$ such that $u(a) - \varepsilon < u(x)$ for every $x \in V_{a,\varepsilon}$. Let $m := \min_A u$. Urysohn's lemma implies the existence of a continuous function $f_{a,\varepsilon} : A \to \mathbb{R}$ such that, $m - \varepsilon \leq f \leq u(a) - \varepsilon$, $f(a) = u(a) - \varepsilon$ and $f(x) = m - \varepsilon$ for every $x \in A \setminus V_{a,\varepsilon}$. Since $\{f_{a,\varepsilon} : a \in A, \varepsilon > 0\} \subseteq \mathcal{F} := \{f \in C(A, \mathbb{R}) : f \leq u\}$, one has $\sup_{f \in \mathcal{F}} f = u$. To conclude the proof, use Proposition 6.1.1.)
Ex. 6) Solve Ex. 6.3.4, page 346, and prove Proposition 6.2.6, page 342.
Ex. 7) Solve Ex. 6.2.5, page 342.
Ex. 8) Solve Exercises 6.3.9 and 6.3.1, page 348.
Ex. 9) Solve Ex. 6.4.11, page 353.
Ex. 10) Prove Proposition 6.5.4, page 355.
Ex. 11) Prove Proposition 6.7.4, page 360.
Ex. 12) Solve Exercises 6.8.5, 6.8.6, 6.8.7, page 365.
Ex. 13) Solve Ex. 6.10.7, page 375.
Ex. 14) Solve Ex. 6.11.3, page 376.

Ex. 15) Solve Ex. 6.11.5, page 377.

Ex. 16) Prove Theorem 6.11.7.

Ex. 17) Prove directly that if f is any function, then $\widehat{f}$ is l.s.c. Moreover, if g is l.s.c. and $g \le f$, then $g \le \widehat{f}$.

Ex. 18) Let Ω be a relatively compact open set in a harmonic space $(E, \mathcal{H})$. Let $u \in \overline{S}(U)$, where U is an open set containing $\overline{\Omega}$. Show that the function

$$x \mapsto \int_{\partial\Omega} u \, \mathrm{d}\mu_x^{\Omega} =: h(x)$$

is $\mathcal{H}$-harmonic in Ω.
(*Hint*: $\mathcal{F} = \{H_{\varphi}^{\Omega} : \varphi \in C(\partial\Omega, \mathbb{R}),\ \varphi \le u|_{\partial\Omega}\}$ is an up directed family of harmonic functions such that $\sup \mathcal{F} = h$ and $\sup \mathcal{F} \le u|_{\Omega}$.)

Ex. 19) Let $(E, \mathcal{H})$ be a $\mathfrak{S}^*$-harmonic space, and let $\Omega \subseteq E$ be open and relatively compact. Let $z \in \partial\Omega$, and let U be an open neighborhood of z. Prove that

$$z \text{ is } \mathcal{H}\text{-regular for } \Omega \iff z \text{ is } \mathcal{H}\text{-regular for } \Omega \cap U.$$

Ex. 20) Let $(E, \mathcal{H})$ be a $\mathfrak{S}^*$-harmonic space, and let $\Omega_1 \subseteq \Omega_2 \subseteq E$ be open and relatively compact. Let $z \in \partial\Omega_1 \cap \partial\Omega_2$. Assume z is $\mathcal{H}$-regular for Ω_2. Prove that z is $\mathcal{H}$-regular for Ω_1.

Ex. 21) Let Ω be a relatively compact open set in a $\mathfrak{S}$-harmonic space $(E, \mathcal{H})$. Let (f_n) be a sequence of resolutive functions from $\partial\Omega$ into $[-\infty, \infty]$. Suppose that

$$\sup_n \overline{H}_{|f_n|}^{\Omega}(x) < \infty \qquad \text{for every } x \in \Omega.$$

Assume (f_n) is pointwise convergent to $f : \partial\Omega \to [-\infty, \infty]$. Prove that f is resolutive.

Ex. 22) Let Ω be a relatively compact open set in a $\mathfrak{S}$-harmonic space $(E, \mathcal{H})$. Let $f : \partial\Omega \to [-\infty, \infty[$ be bounded above. Prove that if $z \in \partial\Omega$ is $\mathcal{H}$-regular, one has

$$\limsup_{\Omega \ni x \to z} \overline{H}_f^{\Omega}(x) \le \limsup_{\partial\Omega \ni y \to z} f(y).$$

(*Hint*: For every fixed $\beta \in \mathbb{R}$, $\beta > \limsup_{\partial\Omega \ni y \to z} f(y)$, choose a function $\varphi \in C(\partial\Omega, \mathbb{R})$ such that $\varphi \ge f$ and $\varphi(z) = \beta$. Then $\overline{H}_f^{\Omega} \le \overline{H}_{\varphi}^{\Omega} = H_{\varphi}^{\Omega}$.)

Ex. 23) Let $(E, \mathcal{H})$ be a harmonic space such that the constant functions are $\mathcal{H}$-harmonic. Let $\varphi : \mathbb{R} \to \mathbb{R}$ be a convex function. Prove the following statements:

(i) if $\Omega \subseteq E$ is open and $u \in \mathcal{H}(\Omega)$, then $\varphi(u) \in \underline{S}(\Omega)$,

(ii) if we also assume φ monotone increasing, then $\varphi(u) \in \underline{S}(\Omega)$ whenever $u \in \underline{S}(\Omega)$.

(*Hint*: Since $1 \in \mathcal{H}(E)$, $\mu_x^V(\partial V) = 1$ for every $\mathcal{H}$-regular set V and every $x \in V$. Use Jensen's inequality.)

Ex. 24) Let Ω_1 and Ω be open subsets of E with $\Omega_1 \subseteq \Omega$. Given $u \in \mathcal{H}^*(\Omega)$ and $v \in \mathcal{H}^*(\Omega_1)$ define

$$w : \Omega \to]-\infty, \infty], \quad w = \begin{cases} u & \text{in } \Omega \setminus \Omega_1, \\ \min\{u, v\} & \text{in } \Omega_1. \end{cases}$$

Prove that, if w is l.s.c., then $w \in \mathcal{H}^*(\Omega)$. Moreover, $w \in \overline{\mathcal{S}}(\Omega)$ if $u \in \overline{\mathcal{S}}(\Omega)$.

Ex. 25) Let Ω_1 and Ω be open subsets of E with $\Omega_1 \subseteq \Omega$. Given $u \in \overline{\mathcal{S}}(\Omega)$ and $v \in \overline{\mathcal{S}}(\Omega_1)$ define w as in the previous exercise. Assume that

$$\liminf_{\Omega_1 \ni x \to y} v(x) \geq u(y) \quad \text{for every } y \in \Omega \cap \partial\Omega_1.$$

Then $w \in \overline{\mathcal{S}}(\Omega_1)$.

Ex. 26) Let Ω be a relatively compact open set in a $\mathfrak{S}^*$-harmonic space $(E, \mathcal{H})$ and let $x_0 \in \partial\Omega$. Prove that x_0 is $\mathcal{H}$-regular for Ω iff x_0 is $\mathcal{H}$-regular for every connected component O of Ω such that $x_0 \in \partial O$.

7

The $\mathcal{L}$-harmonic Space

The aim of this short chapter is to prove that the theory of abstract harmonic spaces developed in Chapter 6 can be applied to the setting of (homogeneous) Carnot groups $\mathbb{G}$. More precisely, fixed a sub-Laplacian $\mathcal{L}$ on $\mathbb{G}$, the $\mathcal{L}$-harmonic functions provide a harmonic sheaf ${}^{\mathcal{L}}\mathcal{H}$ on $\mathbb{G}$. We shall prove that $(\mathbb{G}, {}^{\mathcal{L}}\mathcal{H})$ turns out to be a $\mathfrak{S}^*$-harmonic space.

The ingredients we use are the structure of the fundamental solution for $\mathcal{L}$ and the Harnack-type theorem established in Chapter 5. The more involved task is to show the existence of a basis of the topology of $\mathbb{R}^N$ formed by $\mathcal{L}$-regular sets, i.e. by sets for which the Dirichlet problem for $\mathcal{L}$ is solvable in the usual classical sense.

For reading convenience, in the second part of the chapter, we recall some basic definitions and results from the abstract theory of Chapter 6, just to show how they read in the sub-Laplacian setting. Moreover, we prove a criterion of $\mathcal{L}$-subharmonicity for C^2-functions, we show that the gauge balls are $\mathcal{L}$-regular sets and we write an explicit formula for the density of the $\mathcal{L}$-harmonic measures of the gauge balls at their center.

Convention. The same convention as in the *incipit* of Chapter 5 applies. In other words, the potential theory developed in this chapter can be straightforwardly generalized and rephrased in the context of the *abstract* stratified groups. It suffices to consider the notion of $\mathcal{L}$-harmonic function on a stratified group in Definition 2.2.27 on page 146 and to keep in mind Remark 2.2.28. That remark ensures that, if $\mathbb{H}$ is an abstract stratified group, $\mathbb{G}$ is a "homogeneous copy" of $\mathbb{H}$ and $\Psi : \mathbb{G} \to \mathbb{H}$ the Lie group isomorphism in Remark 2.2.26 (page 145) the harmonic (hence the hyperharmonic, subharmonic, etc.) functions in $\mathbb{H}$ simply "pull back" via Ψ to the harmonic (hyperharmonic, subharmonic, etc.) functions in $\mathbb{G}$.

7.1 The $\mathcal{L}$-harmonic Space

Let $\mathbb{G} = (\mathbb{R}^N, \circ, \delta_\lambda)$ be a homogeneous Carnot group in $\mathbb{R}^N$ and let $\mathcal{L}$ be one of its sub-Laplacians. We know that

$$\mathcal{L} = \operatorname{div}(A(x)\nabla) = \sum_{i,j=1}^{N} a_{i,j}(x)\, \partial_{x_i,x_j} + \sum_{i=1}^{N} b_i(x)\, \partial_{x_i},$$

where $A(x) = (a_{i,j})_{i,j \le N}$ is a symmetric non-negative matrix with smooth entries and constant $a_{1,1} > 0$.

For every open set $\Omega \subseteq \mathbb{G}$ denote

$${}^{\mathcal{L}}\mathcal{H}(\Omega) := \big\{u \in C^\infty(\Omega, \mathbb{R}) \;:\; \mathcal{L}u = 0\big\}.$$

A function $u \in {}^{\mathcal{L}}\mathcal{H}(\Omega)$ will be called $\mathcal{L}$-harmonic in Ω. The map

$${}^{\mathcal{L}}\mathcal{H} : \Omega \mapsto {}^{\mathcal{L}}\mathcal{H}(\Omega)$$

is a harmonic sheaf. The aim of this section is to prove that

$$(\mathbb{R}^N, {}^{\mathcal{L}}\mathcal{H}) \textit{ is a } \mathfrak{S}^*\textit{-harmonic space.}$$

We split the proof in four steps, corresponding to the axioms we have to verify.

(A1) **(Positivity).** This axiom is trivially satisfied for the constant functions are $\mathcal{L}$-harmonic.

(A2) **(Convergence).** This readily follows from the Harnack theorem proved in Theorem 5.7.10, page 268.

We postpone to the end of the section the proof of (A3), which is our main task. Now we proceed with (A4) and (A*).

(A4) **(Separation).** We shall show that $\mathcal{S}_c^+(\mathbb{G})$, the family of continuous non-negative ${}^{\mathcal{L}}\mathcal{H}$-superharmonic functions, separates the points of $\mathbb{G}$. Since $\mathcal{L}$ is left translation invariant and the constant functions are $\mathcal{L}$-harmonic, it is enough to prove the following statement:

$$\text{for every } x_0 \neq 0, \text{ there exists } u \in \mathcal{S}_c^+(\mathbb{G}) : u(x_0) \neq u(0).$$

Let Γ be the fundamental solution of $\mathcal{L}$ with pole at the origin. Since Γ is $\mathcal{L}$-harmonic in $\mathbb{G} \setminus \{0\}$ and

$$\lim_{x \to 0} \Gamma(x) = \infty =: \Gamma(0),$$

one has $\Gamma \in \overline{\mathcal{S}}(\mathbb{G})$ (see Exercise 7.3). For $k \in \mathbb{N}$, define

$$\Gamma_k = \min\{\Gamma, k\}.$$

Obviously, $\Gamma_k \in \overline{\mathcal{S}}_c^+(\mathbb{G})$. Moreover, $\Gamma_k(0) \nearrow \infty$ as $k \nearrow \infty$ while, as $x_0 \neq 0$, $\Gamma_k(x_0) \nearrow \Gamma(x_0) \in]0, \infty[$. Then there exists $k \in \mathbb{N}$ such that

$$\Gamma_k(x_0) \neq \Gamma_k(0),$$

and we are done.

(A*) (The function s_{x_0}). For every fixed $x_0 \in \mathbb{G}$, let us define

$$s_{x_0}(x) = d(x_0^{-1} \circ x), \quad x \in \mathbb{G},$$

where d is an $\mathcal{L}$-gauge function (see Definition 5.4.1). Then s_{x_0} is continuous in $\mathbb{G}$, $s_{x_0}(x_0) = 0$ and

$$\inf_{\mathbb{G}\setminus V} s_{x_0} > 0$$

for every neighborhood V of x_0. Moreover, in $\mathbb{G} \setminus \{x_0\}$,

$$\mathcal{L}s_{x_0} = \frac{(Q-1)}{d} |\nabla_{\mathcal{L}} d|^2 \geq 0$$

(see (5.34)). Then $s_{x_0} \in \underline{S}(\mathbb{G} \setminus \{x_0\})$. Since $s_{x_0} \geq 0$ and $s_{x_0}(x_0) = 0$, it follows that $s_{x_0} \in \underline{S}(\mathbb{G})$ (see Ex. 7.3.4).

(A3) (Regularity). This is the most difficult part of our task. We shall use some results from the theory of linear second order partial differential equations of elliptic type.

From now on, for a fixed $\varepsilon > 0$, we shall denote

$$\mathcal{L}_\varepsilon := \mathcal{L} + \varepsilon \Delta, \tag{7.1}$$

where $\Delta = \sum_{j=1}^N \partial_{x_j}^2$ is the Laplace operator.

The characteristic form of $\mathcal{L}_\varepsilon$

$$q_{\mathcal{L}_\varepsilon}(x, \xi) := \langle A(x)\xi, \xi \rangle + \varepsilon |\xi|^2, \quad \xi \in \mathbb{R}^N,$$

is strictly positive definite at any point of $\mathbb{G}$. Moreover, for every compact set $K \subseteq \mathbb{G}$ there exists $\lambda = \lambda(K) > 0$ such that

$$\lambda^{-1} |\xi|^2 \leq q_{\mathcal{L}_\varepsilon}(x, \xi) \leq \lambda |\xi|^2 \quad \forall\, x \in K,\ \xi \in \mathbb{R}^N.$$

This means that $\mathcal{L}_\varepsilon$ is *uniformly elliptic* on every bounded open set Ω, so that if $\partial\Omega$ also satisfies some mild regularity condition, the Dirichlet problem

$$\begin{cases} \mathcal{L}_\varepsilon u = 0 & \text{in } \Omega, \\ u|_{\partial\Omega} = \varphi, & \varphi \in C(\partial\Omega, \mathbb{R}), \end{cases} \tag{7.2}$$

has a solution[1]

$$u_\varepsilon \in C^\infty(\Omega) \cap C(\overline{\Omega})$$

such that $u_\varepsilon|_{\partial\Omega} = \varphi$.

A condition ensuring this existence result is the following one[2]:

(R) For every $x_0 \in \partial\Omega$, there exists a function $h(x_0, \cdot)$ of class C^2 in an open set containing $\overline{\Omega}$ such that:

---

[1] $\mathcal{L}_\varepsilon$ is hypoelliptic for it is elliptic and its coefficients are smooth. Then $u \in C^\infty$ if $\mathcal{L}_\varepsilon u = 0$.

[2] See the monograph by D. Gilbarg and N.S. Trudinger [GT77, Chapter 6].

(a) $h(x_0, \cdot) > 0$ in $\overline{\Omega} \setminus \{x_0\}$, $h(x_0, x_0) = 0$.
(b) $\mathcal{L}_\varepsilon h(x_0, \cdot) \leq 0$ in Ω.

A geometric condition ensuring (R) is a particular form of the Poincaré exterior ball condition.

Definition 7.1.1 (Non-characteristic exterior ball condition). *We say that $\partial\Omega$ satisfies the non-characteristic exterior ball condition at a point $x_0 \in \partial\Omega$ if there exists $z \in \mathbb{G} \setminus \overline{\Omega}$ such that*

(i) $\overline{D(z, r)} \cap \overline{\Omega} = \{x_0\}$, $r = |x_0 - z|$,
(ii) $q_\mathcal{L}(x, z - x) := \langle A(x)(z - x), z - x\rangle > 0 \ \forall\, x \in \overline{\Omega}$.

Here $D(z, r)$ denotes the standard Euclidean ball with center z and radius r. The standard Euclidean norm is denoted by $|\cdot|$.

Note 7.1.2. If (i) and (ii) hold, then the boundary of Ω is non-characteristic at x_0, since $\nu := z - x_0$ is orthogonal to $\partial\Omega$ at x_0 and

$$q_\mathcal{L}(x_0, \nu) = \big\langle A(x_0)\, \nu, \nu\big\rangle > 0.$$

Lemma 7.1.3. *Let $\Omega \subseteq \mathbb{G}$ be a bonded open set satisfying the non-characteristic exterior ball condition at a point $x_0 \in \partial\Omega$. Define*

$$h(x_0, x) = \exp(-\mu\, r^2) - \exp(-\mu\, |z - x|^2), \quad x \in \mathbb{G},$$

where

$$\mu = \max_{x\in\Omega}\left\{\frac{|\langle b(x), z - x\rangle|}{q_\mathcal{L}(x, z - x)} : x \in \overline{\Omega}\right\}$$

(as before, $|\cdot|$ stands for the usual Euclidean norm). Then $h \in C^\infty(\mathbb{G}, \mathbb{R})$ and satisfies (a) *and* (b) *of condition* (R).

Proof. The smoothness of h is obvious. Condition (R)-(a) readily follows from (i) in Definition 7.1.1. Moreover, by (ii) in Definition 7.1.1 and the very definition of μ, $\mathcal{L}_\varepsilon h(x_0, x)$ equals

$$-\exp(-\mu\, |x - x_0|^2)(4\mu^2 q_\mathcal{L}(x, z - x) + 2\mu\langle b(x), z - x\rangle + 4\varepsilon N),$$

which is non-negative in $\overline{\Omega}$. □

We now prove another crucial lemma for our purposes.

Lemma 7.1.4. *Let Ω and h be as in Lemma* 7.1.3, *and let u_ε be the solution of the Dirichlet problem* (7.2). *Then*

$$|u_\varepsilon(x) - \varphi(x_0)| \leq 2\, h_\varphi(x_0, x) \quad \forall\, x \in \Omega. \tag{7.3}$$

The function h_φ is defined as follows. For $t > 0$, we let

$$\omega_\varphi(x_0, t) := t + \sup\{|\varphi(x) - \varphi(x_0)| : x \in \partial\Omega, |x - x_0| \le t\},$$

and

$$m(x_0, t) := \inf\{h(x_0, x) : |x - x_0| \ge t,\ |x - z| \ge r\}.$$

The function $t \mapsto \omega_\varphi(x_0, t) m(x_0, t)$ is continuous and strictly increasing. Denote by $T_\varphi(x_0, \cdot)$ its inverse function. We define

$$h_\varphi(x_0, x) := T_\varphi\Big(x_0, \sup_{\partial\Omega} |\varphi| h(x_0, x)\Big).$$

It should be noted that h_φ *is independent of* ε. Moreover,

$$\lim_{x \to x_0} h_\varphi(x, x_0) = 0.$$

Proof (of Lemma 7.1.4*).* Let $t > 0$ be fixed. Consider the function

$$w(x) := \varphi(x_0) + \omega_\varphi(x_0, t) + \sup_{\partial\Omega} |\varphi| \frac{h(x_0, x)}{m(x_0, t)}.$$

One readily verifies that

$$\lim_{x \to y} w(x) \ge \varphi(y) = \lim_{x \to y} u_\varepsilon(x) \quad \forall\, y \in \partial\Omega.$$

Then, since $\mathcal{L}_e w \le 0$, by the maximum principle we get $u_\varepsilon \le w$ in Ω, that is

$$u_\varepsilon(x) - \varphi(x_0) \le \omega_\varphi(x_0, t) + \sup_{\partial\Omega} |\varphi| \frac{h(x_0, x)}{m(x_0, t)}$$

for every $x \in \Omega$ and $t > 0$. The same inequality holds by replacing in it u_ε with $-u_\varepsilon$ and φ with $-\varphi$. Then

$$|u(x) - \varphi(x_0)| \le \omega_\varphi(x_0, t) + \sup_{\partial\Omega} |\varphi| \frac{h(x_0, x)}{m(t)}, \qquad x \in \Omega, \tag{7.4}$$

for every $t > 0$. Now we choose $t > 0$ such that

$$\omega(x_0, t) = \sup_{\partial\Omega} |\varphi| \frac{h(x_0, x)}{m(t)},$$

that is $t = T(x_0, \sup_{\partial\Omega} |\varphi| h(x_0, x))$. With this choice of t, from (7.4) we obtain (7.3). This completes the proof. □

From Lemma 7.1.3 and Lemma 7.1.4 we obtain a resolutivity result for $\mathcal{L}$.

Proposition 7.1.5. *Let $\Omega \subseteq \mathbb{G}$ be a bounded open set satisfying the non-characteristic exterior ball condition at any point of $\partial\Omega$. Then Ω is ${}^{\mathcal{L}}\mathcal{H}$-regular, that is: for every $\varphi \in C(\partial\Omega, \mathbb{R})$ there exists a unique function $u \in C^\infty(\Omega, \mathbb{R})$ such that*

$$\begin{cases} \mathcal{L}u = 0 & \text{in } \Omega, \\ \lim_{x \to y} u(x) = \varphi(y) & \forall\, y \in \partial\Omega. \end{cases}$$

Moreover, $u \geq 0$ if $\varphi \geq 0$.

Proof. First of all we prove the existence. Let $\varphi \in C(\partial\Omega, \mathbb{R})$ and let u_ε, $0 < \varepsilon < 1$, be the solution of the Dirichlet problem (7.2). By the maximum principle

$$\sup_{\Omega} |u_\varepsilon| \leq \sup_{\partial\Omega} |\varphi|,$$

so that

$$\sup_{0<\varepsilon<1} \int_{\Omega} |u_\varepsilon|^2 \, dx < \infty.$$

Then there exists $u \in L^2(\Omega)$ and a sequence $\varepsilon_j \searrow 0$ such that

$$\lim_{j\to\infty} u_{\varepsilon_j} = u$$

in the weak topology of $L^2(\Omega)$. As a consequence, for every test function $\psi \in C_0^\infty(\Omega)$,

$$\begin{aligned}
\int_{\Omega} u\mathcal{L}^*\psi \, dx &= \lim_{j\to\infty} \int_{\Omega} u_{\varepsilon_j} \mathcal{L}^*\psi \, dx \\
&= \lim_{j\to\infty} \left(\int_{\Omega} u_{\varepsilon_j} \mathcal{L}_\varepsilon^*\psi \, dx - \varepsilon_j \int_{\Omega} u_{\varepsilon_j} \Delta\psi \, dx \right) \\
&= \lim_{j\to\infty} \int_{\Omega} \psi \mathcal{L}_\varepsilon u_{\varepsilon_j} \, dx = 0.
\end{aligned}$$

Therefore $\mathcal{L}u = 0$ in the weak sense of distributions. Since $\mathcal{L}$ is hypoelliptic, we may assume

$$u \in C^\infty(\Omega, \mathbb{R}) \text{ and } \mathcal{L}u = 0 \text{ in } \Omega.$$

On the other hand, by Lemma 7.1.4, for every $x_0 \in \partial\Omega$ we have

$$\varphi(x_0) - 2\,h_\varphi(x_0, x) \leq u_{\varepsilon_j}(x) \leq \varphi(x_0) + 2\,h_\varphi(x_0, x) \quad \forall\, x \in \partial\Omega \tag{7.5}$$

and for every $j \in \mathbb{N}$. Then, since the weak limits in $L^2(\Omega)$ of non-negative functions are non-negative, inequalities (7.5) imply

$$|u_\varepsilon(x) - \varphi(x_0)| \leq h_\varphi(x_0, x) \quad \forall\, x \in \Omega,\ x_0 \in \partial\Omega.$$

Then, keeping in mind that $h_\varphi(x_0, x) \to 0$ as $x \to x_0$,

$$\lim_{x\to x_0} u(x) = \varphi(x_0) \quad \forall\, x_0 \in \partial\Omega.$$

This completes the proof of the existence. The uniqueness and the positivity of u if $\varphi \geq 0$ straightforwardly follow from the weak maximum principle of Theorem 5.13.4 (page 295). □

Finally, we are ready to complete the proof of (A3). We will construct a basis $(V_\lambda)_{\lambda>0}$ of $\mathcal{L}$-regular neighborhoods of the origin. It will follow that, for every $x_0 \in \mathbb{G}$,

$$\{x_0 \circ V_\lambda\}_{\lambda>0}$$

is a basis of $\mathcal{L}$-regular open sets containing x_0.

Let $e_1 = (1, 0, \dots, 0)$ be the first element of the canonical basis of $\mathbb{G}$. Then

$$q_{\mathcal{L}}(0, e_1) = \langle A(0)\, e_1, e_1 \rangle = a_{1,1} > 0.$$

As a consequence, there exists $\rho > 0$ such that $q_{\mathcal{L}}(x, \nu) > 0$ for every $x, \nu \in \mathbb{G}$, $|x| < \rho$, $|\nu - e_1| < \rho$.

Therefore, for $R > 0$ sufficiently large and $\varepsilon > 0$ sufficiently small,

$$V_1 := D(R\, e_1, R + \varepsilon) \cap D(-R\, e_1, R + \varepsilon) \tag{7.6}$$

satisfies the *non-characteristic exterior ball condition at any point of its boundary.* (See also Fig. 7.1.)

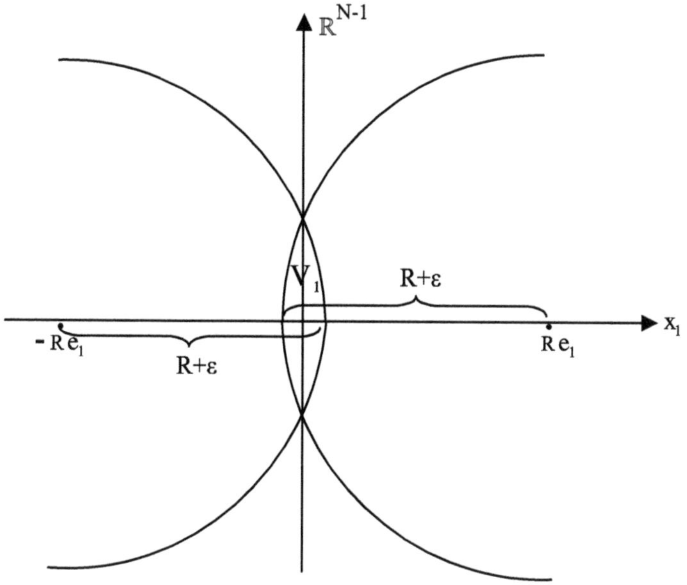

Fig. 7.1. Figure of (7.6)

Then, by Proposition 7.1.5, V_1 is ${}^{\mathcal{L}}\mathcal{H}$-regular. Since $\mathcal{L}$ commutes with the dilations δ_λ, $\lambda > 0$, if we put

$$V_\lambda := \delta_\lambda \circ V,$$

the family $\{V_\lambda\}_{\lambda>0}$ is a basis of ${}^{\mathcal{L}}\mathcal{H}$-regular neighborhoods of the origin. This completes the proof of (A3). $\square$

7.2 Some Basic Definitions and Selecta of Properties

Throughout this section, Ω will be an open subset of a homogeneous Carnot group $\mathbb{G}$.

Notation. *For the sake of brevity, we shall simply denote by $\mathcal{H}$ the harmonic sheaf ${}^{\mathcal{L}}\mathcal{H}$ introduced in the previous section.*

In Section 7.1, we have shown that

$$(\mathbb{G}, \mathcal{H})$$

is a $\mathfrak{S}^*$-*harmonic space*. Hence the abstract potential theory developed throughout Chapter 6 applies to the present setting. Some of the basic definitions and results from that theory will be recalled along this section, for reading convenience.

A bounded open set $V \subset \mathbb{G}$ will be called $\mathcal{L}$-*regular* if the boundary value problem

$$\begin{cases} \mathcal{L}u = 0 & \text{in } V, \\ u|_{\partial V} = \varphi \end{cases} \tag{7.7}$$

has a (unique) solution $u := H_\varphi^V$ for every continuous function $\varphi : \partial V \to \mathbb{R}$. We say that u solves (7.7) if u is $\mathcal{L}$-harmonic in V and

$$\lim_{x \to y} u(x) = \varphi(y) \quad \forall\, y \in \partial V.$$

The weak maximum principle of Theorem 5.13.4 implies the uniqueness of the solution to the boundary value problem (7.7). Moreover,

$$H_\varphi^V \geq 0 \text{ in } V \text{ whenever } \varphi \geq 0 \text{ on } \partial V.$$

Then, if V is $\mathcal{L}$-regular, for every fixed $x \in V$ the map

$$C(\partial V, \mathbb{R}) \ni \varphi \mapsto H_\varphi^V(x) \in \mathbb{R}$$

defines a linear positive functional on $C(\partial V, \mathbb{R})$. As a consequence (by the classical Riesz representation theorem for linear positive functionals), there exists a Radon measure μ_x^V supported in ∂V such that

$$H_\varphi^V(x) = \int_{\partial V} \varphi(y)\, \mathrm{d}\mu_x^V(y) \quad \forall\, \varphi \in C(\partial V, \mathbb{R}).$$

We shall call μ_x^V the $\mathcal{L}$-*harmonic measure* related to V and x.

The family of $\mathcal{L}$-regular sets is not empty. Indeed, in Section 7.1 we have proved the following proposition.

Proposition 7.2.1. *The family of the $\mathcal{L}$-regular open sets is a basis of the Euclidean topology of $\mathbb{G}$.*

The following definition of $\mathcal{L}$-subharmonic function is equivalent to the abstract one given in Definition 6.5.1 (page 353), in force of Theorem 6.5.2 and the definition of $\mathcal{H}$-hyperharmonic function given in Definition 6.2.4 (page 341).

Definition 7.2.2 ($\mathcal{L}$-subharmonic function). *Let Ω be an open subset of $\mathbb{G}$. A function $u : \Omega \to [-\infty, \infty[$ will be called* $\mathcal{L}$-subharmonic in Ω *if:*

(i) *u is upper semi-continuous and $u > -\infty$ in a dense subset of Ω,*
(ii) *for every $\mathcal{L}$-regular open set V with closure $\overline{V} \subset \Omega$ and for every $x \in V$,*

$$u(x) \leq \int_{\partial V} u(y)\, \mathrm{d}\mu_x^V(y). \tag{7.8}$$

The family of the $\mathcal{L}$-subharmonic functions in Ω will be denoted by

$$\underline{S}(\Omega).$$

Remark 7.2.3. As we have already noticed in the abstract harmonic spaces (see Section 6.2.1), condition (ii) in Definition 7.2.2 is equivalent to the following one:

$$\text{(ii)'} \quad H_\varphi^V \geq u|_V \quad \forall\, \varphi \in C(\partial V, \mathbb{R})\colon \varphi \geq u|_{\partial V},$$

for every $\mathcal{L}$-regular open set V such that $\overline{V} \subset \Omega$.

We shall denote by $\overline{S}(\Omega)$ the family of the $\mathcal{L}$-*superharmonic* functions, i.e. the set of functions $u : \Omega \to]-\infty, \infty]$ such that $-u \in \underline{S}(\Omega)$. Then, keeping in mind the previous definition, $u \in \overline{S}(\Omega)$ if and only if

(i) u is lower semi-continuous and $u < \infty$ in a dense subset of Ω,
(ii) for every $\mathcal{L}$-regular open set V with closure $\overline{V} \subset \Omega$ and for every $x \in V$,

$$u(x) \geq \int_{\partial V} u(y)\, \mathrm{d}\mu_x^V(y). \tag{7.9}$$

Proposition 7.2.4. *A function $u : \Omega \to \mathbb{R}$ is $\mathcal{L}$-harmonic in Ω if and only if $u \in \underline{S}(\Omega) \cap \overline{S}(\Omega)$. In other words,*

$$\mathcal{H}(\Omega) = \underline{S}(\Omega) \cap \overline{S}(\Omega).$$

Proof. See Exercise 6.3.4 on page 346. □

By using the weak maximum principle of Theorem 5.13.4 (page 295), we obtain the following criterion of $\mathcal{L}$-subharmonicity for functions of class C^2.

Proposition 7.2.5 (Smooth $\mathcal{L}$-subharmonic functions). *Let u be a function in $C^2(\Omega, \mathbb{R})$. Then u is $\mathcal{L}$-subharmonic if and only if*

$$\mathcal{L}u \geq 0 \quad \text{in } \Omega.$$

Proof. Let $V \subset \overline{V} \subset \Omega$ be an $\mathcal{L}$-regular open set and let $\varphi \in C(\partial V, \mathbb{R})$, $\varphi \geq u$. One has

$$\begin{cases} \mathcal{L}(u - H_\varphi^V) \geq 0 & \text{in } V, \\ \limsup_{x \to y}(u(x) - H_\varphi^V(x)) \leq u(y) - \varphi(y) \leq 0 & \text{for every } y \in \partial V. \end{cases}$$

Then, by the maximum principle, $u - H_\varphi^V \leq 0$ in V, so that u is $\mathcal{L}$-subharmonic. Vice versa, let $u \in \underline{\mathcal{S}}(\Omega)$ and assume, by contradiction, $\mathcal{L}u(x_0) < 0$ at some point $x_0 \in \Omega$. It follows that $\mathcal{L}u < 0$ in a neighborhood Ω_0 of x_0. Then, by the first part of the proof, $u \in \overline{\mathcal{S}}(\Omega_0)$. On the other hand, $u \in \underline{\mathcal{S}}(\Omega_0)$ since $u \in \underline{\mathcal{S}}(\Omega)$. By Proposition 7.2.4, it follows that $u \in \mathcal{H}(\Omega_0)$, i.e. $\mathcal{L}u = 0$ in Ω_0, a contradiction. □

Remark 7.2.6. The previous proof can be easily adapted to prove the following statement. Let $\Omega \subseteq \mathbb{G}$ be open and let $x_0 \in \Omega$. Let $u \in C^2(\Omega \setminus \{x_0\}, \mathbb{R})$ be such that

$$\mathcal{L}u \geq 0 \quad \text{in } \Omega \setminus \{x_0\}, \qquad \liminf_{x \to x_0} u(x) = -\infty.$$

Then, if we extend u at x_0 letting $u(x_0) := -\infty$, we have $u \in \underline{\mathcal{S}}(\Omega)$.

Indeed, let $V \subset \overline{V} \subset \Omega$ be a $\mathcal{L}$-regular open set and let $\varphi \in C(\partial V, \mathbb{R})$, $\varphi \geq u|_{\partial V}$. Proceeding as in the first part of the previous proof, we see that

$$u|_V \leq (H_\varphi^V)|_V \quad \text{if } x_0 \notin V$$

and

$$u|_{V \setminus \{x_0\}} \leq (H_\varphi^V)|_{V \setminus \{x_0\}} \quad \text{if } x_0 \in V.$$

Then, in any case, $u|_V \leq (H_\varphi^V)|_V$, so that $u \in \underline{\mathcal{S}}(\Omega)$. □

Example 7.2.7. Let d be an $\mathcal{L}$-gauge and let $x_0 \in \mathbb{G}$ be fixed. Define

$$\gamma_{x_0}(x) := \begin{cases} (d(x_0^{-1} \circ x))^{2-Q} & \text{if } x \neq x_0, \\ \infty & \text{if } x = x_0. \end{cases}$$

Then, γ_{x_0} is $\mathcal{L}$-superharmonic in $\mathbb{G}$. Indeed, since d is an $\mathcal{L}$-gauge, γ_{x_0} is $\mathcal{L}$-harmonic (hence, $\mathcal{L}$-superharmonic) in $\mathbb{G} \setminus \{x_0\}$. □

Let $\Omega \subset \mathbb{G}$ be a bounded open set. By the Wiener resolutivity Theorem 6.8.4 in $\mathfrak{S}$-harmonic spaces (page 364), we know that any function $\varphi \in C(\partial\Omega, \mathbb{R})$ is resolutive. Then there exists the *Perron–Wiener–Brelot generalized solution* $H_\varphi^\Omega \in \mathcal{H}(\Omega)$ to the boundary value problem

$$\begin{cases} \mathcal{L}u = 0 & \text{in } \Omega, \\ u|_{\partial\Omega} = \varphi. \end{cases} \tag{7.10}$$

As it is well-known, even in the classical case of Laplace operator, we cannot expect that

$$\lim_{x \to y} H_\varphi^\Omega(x) = \varphi(y) \tag{7.11}$$

for every $y \in \partial\Omega$. If (7.11) holds *for every* $\varphi \in C(\partial\Omega, \mathbb{R})$, we say that y is a *$\mathcal{L}$-regular point* of $\partial\Omega$.

Obviously, *problem* (7.10) *is solvable for every continuous boundary datum* φ *(i.e.* Ω *is an* $\mathcal{L}$*-regular set) if and only if every point of* $\partial\Omega$ *is* $\mathcal{L}$*-regular.* Bouligand's Theorem 6.10.4 in $\mathfrak{S}^*$-harmonic spaces (page 371) implies that

> *a point* $y \in \partial\Omega$ *is* $\mathcal{L}$*-regular if there exists an* $\mathcal{L}$*-barrier function at* y *in* Ω*, i.e. a function* $w \in \overline{\mathcal{S}}(\Omega)$ *such that*
>
> $$w(x) > 0 \text{ for every } x \in \Omega \text{ and } w(x) \to 0 \text{ as } x \to y.$$

By using this result, one easily proves the following proposition.

Proposition 7.2.8 (The d-balls are $\mathcal{L}$-regular). *Let* d *be an* $\mathcal{L}$*-gauge. Then, the* d*-balls*

$$B_d(x_0, r), \quad x_0 \in \mathbb{G},\ r > 0$$

are $\mathcal{L}$*-regular open sets.*

Proof. The function

$$v(x) := \begin{cases} (d(x_0^{-1} \circ x))^{2-Q} - r^{2-Q} & \text{if } x \neq x_0, \\ \infty & \text{if } x = x_0 \end{cases}$$

is $\mathcal{L}$-superharmonic in $B_d(x_0, r)$ (see Example 7.2.7). Moreover, $v > 0$ in $B_d(x_0, r)$ and $v(x) \to 0$ as $x \to y$ for every $y \in \partial B_d(x_0, r)$. Then v is an $\mathcal{L}$-barrier function for $B_d(x_0, r)$ at any point of its boundary. Bouligand's Theorem implies that $B_d(x_0, r)$ is $\mathcal{L}$-regular. □

The surface mean value formula (5.45) in Theorem 5.5.4 (page 256) provides an explicit "density" of the $\mathcal{L}$-harmonic measures related to $B_d(x_0, r)$ and x_0:

Theorem 7.2.9 (Density). *Let* d *be an* $\mathcal{L}$*-gauge. Given* $x_0 \in \mathbb{G}$ *and* $r > 0$*, we have*

$$d\mu_{x_0}^{B_d(x_0,r)}(y) = \frac{\beta_d\,(Q-2)}{r^{Q-1}}\,\mathcal{K}_{\mathcal{L}}(x_0, y) \cdot d\sigma(y), \tag{7.12}$$

where

$$\mathcal{K}_{\mathcal{L}} = \frac{|\nabla_{\mathcal{L}} d|^2(x^{-1} \circ y)}{|\nabla(d(x^{-1} \circ \cdot))|(y)}$$

and $d\sigma$ *denotes the Hausdorff* $(N-1)$*-dimensional measure on* $\partial B_d(x_0, r)$.

Proof. Let $0 < \rho < r$ and $\varphi \in C(\partial B_d(x_0, r), \mathbb{R})$. Since

$$h := H_{\varphi}^{B_d(x_0,r)}$$

is $\mathcal{L}$-harmonic in $B_d(x_0, r)$, by using the surface mean value formula (5.45) (page 256), we have

$$\int_{\partial B_d(x_0,r)} \varphi(y)\, d\mu_{x_0}^{B_d(x_0,r)}(y) = h(x_0) = \mathcal{M}_\rho(h)(x_0).$$

On the other hand, since $h(y) \to \varphi(z)$ as $y \to z$ for every $z \in \partial B_d(x_0, r)$,

$$\lim_{\rho \to r} \mathcal{M}_\rho(h)(x_0) = \mathcal{M}_r(h)(x_0)$$
$$= \frac{\beta_d\,(Q-2)}{r^Q} \int_{\partial B_d(x_0,r)} \varphi(y)\, \mathcal{K}_{\mathcal{L}}(x, y)\, \mathrm{d}\sigma(y).$$

Then

$$\int_{\partial B_d(x_0,r)} \varphi(y)\, \mathrm{d}\mu_{x_0}^{B_d(x_0,r)}(y) = \frac{\beta_d\,(Q-2)}{r^Q} \int_{\partial B_d(x_0,r)} \varphi(y)\, \mathcal{K}_{\mathcal{L}}(x_0, y)\, \mathrm{d}\sigma(y)$$

for every $\varphi \in C(\partial B_d(x_0, r), \mathbb{R})$ and the assertion follows. □

We refer the interested reader to the papers [UL97,UL02] for some estimates of $\mathcal{L}$-Poisson kernels on more general domains of Carnot groups.

Bibliographical Notes. In showing the existence of a basis of the topology of $\mathbb{G}$ formed by $\mathcal{L}$-regular sets, we followed an idea by J.-M. Bony [Bon69].

For other presentations of potential theory, linear and non-linear and in sub-Riemannian settings, see [BT03,Bir95,HKM93,Kil94,TW02b,VM96].

For some topics in potential theory in an elliptic degenerate context and estimates of the Poisson kernel, see [BAKS84,CGN02,HH87,UL97,UL02].

7.3 Exercises of Chapter 7

Ex. 1) Prove that $\Gamma \in \overline{\mathcal{S}}(\mathbb{G})$.
(*Hint*: Use Ex. 4.)

Ex. 2) Complete the proof of the following result, which is a consequence of Harnack's Theorem for sub-Laplacians.

Theorem 7.3.1. *Let $\Omega \subseteq \mathbb{G}$ be a bounded, open and connected set such that $0 \in \Omega$. For all $x \in \Omega$, there exists $h_x \in L^1(\partial\Omega, \mu_0^\Omega)$ such that*

$$\mathrm{d}\mu_x^\Omega(\xi) = h_x(\xi)\, \mathrm{d}\mu_0^\Omega(\xi).$$

Moreover, if K is a compact subset of Ω, there exists a positive constant $\mathbf{c}(K)$ depending only on K such that

$$\mathbf{c}(K)^{-1} \le \|h_x; L^\infty(\partial\Omega, \mu_0^\Omega)\| \le \mathbf{c}(K) \quad \textit{for every } x \in K.$$

Proof (Sketch). In this proof, for brevity, for any given $x \in \Omega$ we agree to set $\mu_x := \mu_x^\Omega$. Let $E \subseteq \partial\Omega$ be closed. Then χ_E is u.s.c., and we have (why?)

$$\mu_x(E) = \int_{\partial\Omega} \chi_E\, \mathrm{d}\mu_x = H_{\chi_E}^\Omega(x).$$

Now, $H^{\Omega}_{\chi_E}$ is a non-negative $\mathcal{L}$-harmonic function on the connected open set Ω, so that (why?) for every compact set $K \subset \Omega$ there exists $\mathbf{c} = \mathbf{c}(K) > 0$ such that

$$H^{\Omega}_{\chi_E}(x) \le \mathbf{c} H^{\Omega}_{\chi_E}(y) \quad \text{for every } x, y \in K.$$

Hence, $\mu_x(E) \le \mathbf{c}\, \mu_y(E)$ for every closed $E \subseteq \partial\Omega$. Consequently, we obtain (why?)

$$\mathbf{c}^{-1} \mu_y(E) \le \mu_x(E) \le \mathbf{c}\, \mu_y(E) \quad \text{for every Borel set } E \subseteq \partial\Omega.$$

This proves that $\mu_y \ll \mu_x \ll \mu_y$ for every couple of points $x, y \in \Omega$. Hence, by Lebesgue decomposition theorem, we infer the existence of $h_{x,y} \in L^1(\partial\Omega, \mu_y)$ as in the assertion. We claim that

$$\mathbf{c}^{-1} \le \|h_{x,y}(\cdot); L^\infty(\partial\Omega, \mu_y)\| \le \mathbf{c}.$$

Indeed, suppose first to the contrary that there exists a Borel set $E \subseteq \partial\Omega$ such that $\mu_y(E) > 0$ and $h_{x,y}(\xi) > \mathbf{c}$, μ_y-almost-everywhere on E. This would give the contradiction

$$\mathbf{c} \ge \frac{\mu_x(E)}{\mu_y(E)} = \frac{1}{\mu_y(E)} \int_E h_{x,y}(\xi)\, \mathrm{d}\mu_y(\xi) \gneqq \mathbf{c}\, \frac{\int_E \mathrm{d}\mu_y(\xi)}{\mu_y(E)} = \mathbf{c}.$$

Then take $y = 0$ and set $h_x := h_{x,0}$. □

Ex. 3) Let $f \in L^p(\partial B_d(0,1), \mu)$, $1 \le p < \infty$, where

$$\mu := \mu_0^{B_d(0,1)}$$

is the $\mathcal{L}$-harmonic measure of $B_d(0,1)$ at $x_0 = 0$. Then:
(i) f is resolutive,
(ii) letting $u := H_f^{B_d(0,1)}$, we have

$$\sup_{0 \le \lambda \le 1} \int_{\partial B_d(0,1)} |u(\delta_\lambda(x))|^p \, \mathrm{d}\mu(x) = \|f\|_{L^p(B_d(0,1))}.$$

Ex. 4) Let $\Omega \subseteq \mathbb{G}$ be open and let $x_0 \in \Omega$. Suppose we are given a function

$$u \in C(\Omega, \mathbb{R}) \cap \underline{S}(\Omega \setminus \{x_0\})$$

such that $u \ge 0$ and $u(x_0) = 0$. Then

$$u \in \underline{S}(\Omega).$$

(*Hint*: For every $y \in \Omega$, there exists a basis of regular neighborhoods of y such that

$$u(y) \le \int_{\partial V} u \, \mathrm{d}\mu_y^V.$$

Then use Corollary 6.4.9.)

Ex. 5) (Asymptotic Koebe Theorem). Let $\Omega \subseteq \mathbb{G}$ be open and let $u \in C(\Omega, \mathbb{R})$. Assume that

$$\mathcal{A}\mathcal{L}u = 0 \quad \text{in } \Omega \quad \text{or} \quad A\mathcal{L}u = 0 \quad \text{in } \Omega.$$

Prove that $u \in C^\infty(\Omega, \mathbb{R})$ and $\mathcal{L}u = 0$ in Ω. $\mathcal{A}\mathcal{L}$ and $A\mathcal{L}$ denote the asymptotic $\mathcal{L}$ Laplacians introduced in Ex. 8 of Chapter 5.
(*Hint*: Use the maximum principles of Ex. 10 of Chapter 5 to show that $u|_V = H^V_{u|\partial V}$ for every $\mathcal{L}$-regular open set $V \subseteq \overline{V} \subseteq \Omega$.)

Ex. 6) (M. Picone's Maximum Principle). Consider the following differential operator

$$L = \sum_{i,j=1}^N a_{i,j}(x)\, \partial_{x_i,x_j} + \sum_{j=1}^N b_j(x)\, \partial_{x_j},$$

where $a_{i,j} = a_{j,i}$, b_j are real functions in a bounded open subset Ω of $\mathbb{G}$. Assume the following facts hold:
(i) $q_L(x,\xi) := \sum_{i,j=1}^N a_{i,j}(x)\xi_i\xi_j \geq 0 \; \forall\, x \in \Omega,\ \forall\, \xi \in \mathbb{R}^N$,
(ii) there exists $h \in C^2(\Omega, \mathbb{R})$ such that $Lh < 0$ and $h > 0$ in Ω.
Then L satisfies the following weak maximum principle:
if $u \in C^2(\Omega, \mathbb{R})$ verifies

$$\begin{cases} Lu \leq 0 & \text{in } \Omega, \\ \limsup_{x\to y} u(x) \leq 0 & \forall\, y \in \partial\Omega \end{cases}$$

then $u \leq 0$ in Ω.
(*Hint*: See the proof of Theorem 5.13.4.)

Ex. 7) Let L be the differential operator introduced in Ex. 6. Together with the hypotheses (i) and (ii) assumed in that exercise, suppose that
(iii) L is hypoelliptic in Ω,
(iv) for every $x_0 \in \Omega$, there exists $\xi_0 \in \mathbb{R}^N$ such that $q_L(x_0, \xi_0) > 0$,
(v) for every $x_0 \in \Omega$, there exists a positive real function γ_{x_0} of class C^2 in $\Omega \setminus \{x_0\}$) such that

$$L\gamma_{x_0} = 0 \quad \text{in } \Omega \setminus \{x_0\}, \qquad \lim_{x\to x_0} \gamma_{x_0}(x) = \infty.$$

For every open set $U \subseteq \Omega$ define

$$^L\mathcal{H}(U) = \big\{u \in C^\infty(U, \mathbb{R}) \,:\, Lu = 0 \text{ in } U\big\}.$$

Prove that $(\Omega, {}^L\mathcal{H})$ is a $\mathfrak{S}^*$-harmonic space.
(*Hint*: Just follow the lines of Section 7.1. Take as s_{x_0} the function $1/\gamma_{x_0}$.)

Ex. 8) Let $\mathcal{L}$ be a sub-Laplacian on a homogeneous Carnot group $\mathbb{G}$. Let $\Omega \subseteq \mathbb{G}$ be an $\mathcal{L}$-regular connected open set. Show that

$$\operatorname{supp} \mu_x^\Omega = \partial\Omega$$

for every $x \in \Omega$
(*Hint*: Use the strong maximum principle.)
Note: This result shows that the $\mathcal{L}$-harmonic space on $\mathbb{G}$ is an "elliptic space in the sense of Brelot" (see [CC72]).

Ex. 9) (The Exterior $\mathcal{L}$-Ball Regularity Condition). Let $\mathbb{G}$ be a homogeneous Carnot group and let d be a $\mathcal{L}$-gauge, $\mathcal{L}$ being a sub-Laplacian on $\mathbb{G}$. In the following statement, we say that Ω has at y the property of the exterior $\mathcal{L}$-ball if there exists a d-ball $B_d(z, \rho)$ such that

$$\mathbb{G} \setminus \Omega \supseteq B_d(z, \rho) \quad \text{and} \quad y \in \partial B_d(z, \rho).$$

Prove the following result:

Let $\Omega \subseteq \mathbb{G}$ be a bounded open set and let $y \in \partial\Omega$. Assume Ω has at y the property of the exterior $\mathcal{L}$-gauge ball. Then y is an $\mathcal{L}$-regular point for Ω.

8

$\mathcal{L}$-subharmonic Functions

Throughout the chapter, $\mathbb{G} = (\mathbb{R}^N, \circ, \delta_\lambda)$, $\mathcal{L} = \sum_{j=1}^{N_1} X_j^2$ and d will denote, respectively, a homogeneous Carnot group, a sub-Laplacian on $\mathbb{G}$ and an $\mathcal{L}$-gauge on $\mathbb{G}$. Our main task is to provide some characterizations of the $\mathcal{L}$-subharmonic functions u in terms of suitable sub-mean properties w.r.t. the mean value operator $\mathcal{M}_r u$ and $\mathrm{M}_r u$ (introduced in Chapter 5), in terms of the monotonicity w.r.t. the radius of these operators and in terms of the sign of $\mathcal{L}u$ in the weak sense of distributions.

We also provide some maximum principles for $\mathcal{L}$-subharmonic functions and smoothing approximation theorems. Finally, in Section 8.3, we furnish a brief investigation on the continuous convex functions on $\mathbb{G}$.

Throughout the chapter $\underline{S}(\Omega)$ and $\overline{S}(\Omega)$ will denote the sheafs of the $\mathcal{L}$-subharmonic and the $\mathcal{L}$-superharmonic functions, respectively, in the open set $\Omega \subseteq \mathbb{G}$.

8.1 Sub-mean Functions

In this section, we shall denote by $\mathcal{M}_r$ and M_r the average operators defined in (5.46) and (5.50f), respectively (pages 256 and 259).

Definition 8.1.1 (Solid and surface sub-mean function). *If $\Omega \subseteq \mathbb{G}$ is an open set, we say that an u.s.c. function $u : \Omega \to [-\infty, \infty[$ satisfies the* local surface (local solid) sub-mean property *if, for every $x \in \Omega$, there exists $r_x > 0$ such that*

$$u(x) \le \mathcal{M}_r(u)(x) \quad \big(u(x) \le \mathrm{M}_r(u)(x)\big) \quad \textit{for } 0 < r < r_x. \tag{8.1}$$

If (8.1) holds for any $r > 0$ such that $\overline{B_d(x,r)} \subset \Omega$, we shall say that u satisfies the global surface (global solid) sub-mean property.

The next theorem shows that solid sub-mean functions satisfy weak and strong maximum principles.

Theorem 8.1.2 (Maximum principles for sub-mean functions). *Let $\Omega \subseteq \mathbb{G}$ be an open set. Let $u : \Omega \to [-\infty, \infty[$ be an u.s.c. function satisfying the local solid sub-mean property. The following statements hold:*

(i) *if Ω is connected and there exists $x_0 \in \Omega$ such that $u(x_0) = \max_\Omega u$, then $u \equiv u(x_0)$ in Ω,*
(ii) *if Ω is bounded and $\limsup_{\Omega \ni y \to x} u(y) \le 0$ for every $x \in \partial\Omega$, then $u(y) \le 0$ in Ω.*

Proof. (i) Let $x_0 \in \Omega$ be such that $u(x_0)$ is the maximum of u in Ω. We may suppose that $u(x_0) \neq -\infty$. Since u is $\mathcal{L}$-sub-mean, we have

$$0 \le \frac{m_d}{r^Q} \int_{B_d(x_0,r)} \Psi_{\mathcal{L}}(x_0^{-1} \circ y) \big(u(y) - u(x_0)\big)\, \mathrm{d}y \tag{8.2}$$

for some $r = r_{x_0} > 0$. Thus, since $\Psi_{\mathcal{L}} \ge 0$ and $u(y) \le u(x_0)$,

$$\Psi_{\mathcal{L}}(x_0^{-1} \circ y) \big(u(y) - u(x_0)\big) = 0 \quad \text{almost everywhere in } B_d(x_0, r).$$

On the other hand,

$$\Psi_{\mathcal{L}} > 0 \text{ in a dense open subset of } B_d(x_0, r)$$

(see the first step in the proof of Lemma 5.7.1, page 262) and u is u.s.c. This implies $u = u(x_0)$ on the whole $B_d(x_0, r)$. The assertion follows from a connectedness argument.

(ii) Let $x_0 \in \overline{\Omega}$ be such that

$$\sup_{B_d(x_0,r)\cap\Omega} u = \sup_\Omega u \quad \forall\, r > 0.$$

If $x_0 \in \partial\Omega$, by the hypothesis and the boundary behavior of u, we have

$$0 \ge \limsup_{\Omega \ni y \to x_0} u = \inf_{r>0} \sup_{B_d(x_0,r)\cap\Omega} u = \sup_\Omega u,$$

whence $u \le 0$ on Ω. If $x_0 \in \Omega$, by the upper semicontinuity of u, we have

$$u(x_0) \ge \inf_{r>0} \sup_{B_d(x_0,r)\cap\Omega} u = \inf_{r>0} \sup_\Omega u = \sup_\Omega u.$$

Hence $u(x_0) = \max_\Omega u$. From (i) it follows that $u \equiv u(x_0)$ on the connected component of Ω containing x_0, so that, for some $x \in \partial\Omega$,

$$0 \ge \limsup_{y \to x} u(y) \ge u(x_0) = \max_\Omega u.$$

The assertion is proved. □

Theorems 8.1.2 and 7.2.9 (page 391) allow us to show that the solid and surface sub-mean properties are equivalent.

Theorem 8.1.3 (Equivalence of sub-mean properties). *Let $u : \Omega \to [-\infty, \infty[$ be an u.s.c. function. Then the following statements are equivalent:*

(1) *u satisfies the local solid sub-mean property,*
(2) *u satisfies the global solid sub-mean property,*
(3) *u satisfies the local surface sub-mean property,*
(4) *u satisfies the global surface sub-mean property.*

Proof. (1) $\Rightarrow$ (4). Let $\overline{B_d}(x, r) \subset \Omega$ and $\varphi \in C(\partial B_d(x, r), \mathbb{R})$ be such that $u \leq \varphi$ in $\partial B_d(x, r)$. Since $u - H_\varphi^{B_d(x,r)}$ is a solid sub-mean function, by the maximum principle in Theorem 8.1.2, we have $u - H_\varphi^{B_d(x,r)} \leq 0$ in $B_d(x, r)$. In particular,

$$u(x) \leq H_\varphi^{B_d(x,r)}(x) = \int_{\partial B_d(x,r)} \varphi(y)\, \mathrm{d}\mu_x^{B_d(x,r)}(y)$$
$$= \text{(by Theorem 7.2.9)}\ \frac{\beta_d(Q-2)}{r^{Q-1}} \int_{\partial B_d(x,r)} \varphi(y)\, \mathcal{K}_{\mathcal{L}}(x, y)\, \mathrm{d}\sigma(y).$$

Then, taking the infimum with respect to the continuous functions $\varphi \geq u|_{\partial\Omega}$, we get

$$u(x) \leq \frac{\beta_d(Q-2)}{r^{Q-1}} \int_{\partial B_d(x,r)} u(y)\, \mathcal{K}_{\mathcal{L}}(x, y)\, \mathrm{d}\sigma(y) = \mathcal{M}_r(u)(x),$$

and (4) is proved.

(4) $\Rightarrow$ (3). This is trivial.

(3) $\Rightarrow$ (1). Suppose $u(x) \leq \mathcal{M}_r(u)(x)$ for $0 < r < r_x$. Then

$$u(x) \leq \frac{Q}{r^Q} \int_0^r \rho^{Q-1}\, \mathcal{M}_\rho(u)(x)\, \mathrm{d}\rho = \mathrm{M}_r(u)(x).$$

(4) $\Rightarrow$ (2). The same proof as the previous one.
(2) $\Rightarrow$ (1). This is trivial. □

From now on, we shall call *sub-mean* any function satisfying one of the properties (1)–(4) in Theorem 8.1.3.

Theorem 8.1.4 (L^1_{loc} of sub-mean functions). *Let Ω be a connected open subset of $\mathbb{G}$, and let $u : \Omega \to [-\infty, \infty[$ be a sub-mean u.s.c. function.*

Suppose that $u(x_0) > -\infty$ for some $x_0 \in \Omega$. Then $u \in L^1_{\mathrm{loc}}(\Omega)$.

Proof. We first prove that $u > -\infty$ in a dense subset of Ω. Since Ω is connected and there exists $x_0 \in \Omega$ such that $u(x_0) > -\infty$, it is enough to show that $u > -\infty$ in a dense subset of $B_d(x, r)$ for any $\overline{B_d(x, r)} \subset \Omega$ such that $u(x) > -\infty$. This last statement follows from

$$-\infty < u(x) \leq \mathrm{M}_r(u)(x) = \frac{m_d}{r^Q} \int_{B_d(x,r)} \Psi_{\mathcal{L}}(x^{-1} \circ y)\, u(y)\, \mathrm{d}y$$

and the property $\Psi_{\mathcal{L}} > 0$ in an open dense set.

Let us now fix $z_0 \in \Omega$ and $R > 0$ such that $\overline{B_d(z_0, R)} \subset \Omega$. We shall prove the existence of a constant $\lambda > 0$ such that

$$\int_{B_d(z_0, \lambda R)} u(z)\, dz > -\infty.$$

From this statement our theorem will follow.

Let us choose three positive constants λ_0, λ, Λ such that

$$\mathbf{c}\,(\lambda_0 + \lambda) \le \Lambda, \quad \mathbf{c}\,(\lambda_0 + \Lambda) \le 1, \tag{8.3}$$

where $\mathbf{c}$ is the constant appearing in the pseudo-triangle inequality for the gauge d (see (5.55), page 262). Since the set

$$\{x \in \Omega \mid \Psi_{\mathcal{L}}(x^{-1} \circ z_0) > 0\}$$

is an open and dense subset of Ω and

$$\{x \in \Omega \mid u(x) > -\infty\}$$

is dense in Ω too, there exists $x_0 \in B_d(z_0, \lambda_0 R)$ such that

$$x_0 \ne z_0,\ u(x_0) > -\infty \text{ and } \Psi_{\mathcal{L}}(x_0^{-1} \circ z_0) > 0.$$

Due to the smoothness of $\Psi_{\mathcal{L}}$ out of the origin, we may refine the choice of λ in such a way that, for a suitable $m_0 > 0$, we get $\Psi_{\mathcal{L}}(x_0^{-1} \circ z) \ge m_0$ for every $z \in B_d(z_0, \lambda R)$. On the other hand, inequalities (8.3) imply the inclusions

$$B_d(z_0, \lambda R) \subseteq B_d(x_0, \Lambda R) \subseteq B_d(z_0, R).$$

Then, if we put

$$U := \max_{\overline{B_d(z_0, R)}} u \ (\in \mathbb{R}),$$

we have (denoting by meas the Lebesgue measure in $\mathbb{G}$)

$$\begin{aligned}
&\int_{B_d(z_0, \lambda R)} u(z)\, dz \\
&= \int_{B_d(z_0, \lambda R)} \big(u(z) - U\big)\, dz + U\, \text{meas}\big(B_d(z_0, \lambda R)\big) \\
&\ge \frac{1}{m_0} \int_{B_d(z_0, \lambda R)} \Psi_{\mathcal{L}}(x_0^{-1} \circ z)\, \big(u(z) - U\big)\, dz + U\, \text{meas}\big(B_d(z_0, \lambda R)\big) \\
&\ge \frac{1}{m_0} \int_{B_d(x_0, \Lambda R)} \Psi_{\mathcal{L}}(x_0^{-1} \circ z)\, \big(u(z) - U\big)\, dz + U\, \text{meas}\big(B_d(z_0, \lambda R)\big) \\
&= \frac{(\Lambda R)^Q}{m_d\, m_0} \mathrm{M}_{\Lambda R}(u - U)(x_0) + U\, \text{meas}\big(B_d(z_0, \lambda R)\big) \\
&\ge \frac{(\Lambda R)^Q}{m_d\, m_0} \big(u(x_0) - U\big) + U\, \text{meas}\big(B_d(z_0, \lambda R)\big) > -\infty.
\end{aligned}$$

This completes the proof. □

We close this section by proving that every $(\varepsilon, \mathbb{G})$-mollifier preserves the sub-mean properties (see Definition 5.3.6 on page 239).

Theorem 8.1.5 (Mollifier of a sub-mean function). *Let $u : \Omega \to [-\infty, \infty[$ be u.s.c. and $u \in L^1_{\text{loc}}(\Omega)$. Then, if u is sub-mean in Ω, u_ε is sub-mean in Ω_ε too.*

Proof. For $x \in \Omega_\varepsilon$ and $r > 0$ such that $B_d(x,r) \subseteq \Omega_\varepsilon$, we have

$$\begin{aligned} \mathrm{M}_r(u_\varepsilon)(x) &= \int_{B_d(0,\varepsilon)} J_\varepsilon(z) \left(\frac{m_d}{r^Q} \int_{B_d(x,r)} \Psi_{\mathcal{L}}(x^{-1} \circ y)\, u(z^{-1} \circ y)\, \mathrm{d}y \right) \mathrm{d}z \\ &= \int_{B_d(0,\varepsilon)} J_\varepsilon(z)\, \mathrm{M}_r(u)(z^{-1} \circ x)\, \mathrm{d}z \\ &\geq \int_{B_d(0,\varepsilon)} J_\varepsilon(z)\, u(z^{-1} \circ x)\, \mathrm{d}z = u_\varepsilon(x), \end{aligned}$$

and the assertion follows. □

8.2 Some Characterizations of $\mathcal{L}$-subharmonic Functions

We begin by proving the following theorem.

Theorem 8.2.1 ($\mathcal{L}$-hypoharmonicity and sub-mean property). *Let Ω be an open subset of $\mathbb{G}$ and $u : \Omega \to [-\infty, \infty[$ be an u.s.c. function. Then u is $\mathcal{L}$-hypoharmonic if and only if u is sub-mean.*

Proof. Let u be a sub-mean function. By using the maximum principle in Theorem 8.1.2 and arguing as in the first part of the proof of Theorem 8.1.3, it is easy to prove that u is $\mathcal{L}$-hypoharmonic.

Vice versa, let us suppose that u is $\mathcal{L}$-hypoharmonic and prove that u satisfies the global surface sub-mean property. Let $\overline{B_d(x,r)} \subset \Omega$. Since $B_d(x,r)$ is an $\mathcal{L}$-regular open set, for all $\varphi \in C(\partial B_d(x,r), \mathbb{R})$ such that $\varphi \geq u$ on $\partial B_d(x,r)$, we have

$$u(x) \leq H_\varphi^{B_d(x,r)}(x) = \int_{\partial B_d(x,r)} \varphi(y)\, \mathrm{d}\mu_x^{B_d(x,r)}(y).$$

On the other hand, from Theorem 7.2.9 (page 391) we have

$$\int_{\partial B_d(x,r)} \varphi(y)\, \mathrm{d}\mu_x^{B_d(x,r)}(y) = \frac{\beta_d(Q-2)}{r^{Q-1}} \int_{\partial B_d(x,r)} \varphi(y)\, \mathcal{K}_{\mathcal{L}}(x,y)\, \mathrm{d}\sigma(y).$$

Hence

$$\frac{\beta_d(Q-2)}{r^{Q-1}} \int_{\partial B_d(x,r)} \varphi(y)\, \mathcal{K}_{\mathcal{L}}(x,y)\, \mathrm{d}\sigma(y) \geq u(x) \qquad \forall\, \varphi \in C(\partial B_d(x,r), \mathbb{R}),$$

so that

$$\mathcal{M}_r(u)(x) = \frac{\beta_d(Q-2)}{r^{Q-1}} \int_{\partial B_d(x,r)} u(y)\, \mathcal{K}_{\mathcal{L}}(x,y)\, \mathrm{d}\sigma(y) \geq u(x).$$

Thus, u has the surface sub-mean property. □

Corollary 8.2.2. *Let Ω be an open subset of $\mathbb{G}$, and let $u : \Omega \to [-\infty, \infty[$ be an u.s.c. function, finite in a dense subset of Ω. Then $u \in \underline{S}(\Omega)$ if and only if u is sub-mean.*

The following result will be used very often.

Corollary 8.2.3 (Characterization of $\underline{S}(\Omega)$). *Let Ω be a connected open subset of $\mathbb{G}$, and let $u : \Omega \to [-\infty, \infty[$ be an u.s.c. function.*

Then $u \in \underline{S}(\Omega)$ if and only if u is sub-mean and $u > -\infty$ in at least one point of Ω.

Proof. If $u \in \underline{S}(\Omega)$, then $u > -\infty$ in a dense subset of Ω, and u is sub-mean by Theorem 8.2.1. Vice versa, if u is sub-mean and $u > -\infty$ in at least one point of Ω, by Theorem 8.1.4, $u \in L^1_{\text{loc}}(\Omega)$. In particular, $u > -\infty$ in a dense subset of Ω. Then, by Theorem 8.2.1, $u \in \underline{S}(\Omega)$. □

Corollary 8.2.4. *If $u \in \underline{S}(\Omega)$, then $u \in L^1_{\text{loc}}(\Omega)$.*

Corollary 8.2.5 (A Brelot-type convergence result). *Let $\{u_n\}_{n\in\mathbb{N}}$ be a sequence of $\mathcal{L}$-subharmonic functions in a connected open set $\Omega \subseteq \mathbb{G}$. Assume $\{u_n\}_{n\in\mathbb{N}}$ is monotone decreasing.*

Then, if we set $u := \lim_{n\to\infty} u_n$, we have $u \in \underline{S}(\Omega)$ provided there exists $x_0 \in \Omega$ such that $u(x_0) > -\infty$.

Proof. The function u is u.s.c. and sub-mean, since so are the u_n's, and $u_n \geq u_{n+1}$ for every $n \in \mathbb{N}$. Then, the assertion follows from Corollary 8.2.3. □

Corollary 8.2.6. *Let $\Omega \subseteq \mathbb{G}$ be open, and let $x_0 \in \Omega$. Let $u \in \underline{S}(\Omega \setminus \{x_0\})$ be such that*

$$\lim_{x\to x_0} u(x) = -\infty.$$

Then, if we continue u at x_0 by letting $u(x_0) = -\infty$, we have $u \in \underline{S}(\Omega)$.

Proof. The continued function u is u.s.c. and finite in a dense subset of Ω. Moreover, for every $x \in \Omega \setminus \{x_0\}$ and $r > 0$ such that $\overline{B_d(x, r)} \subset \Omega \setminus \{x_0\}$, we have

$$u(x) \leq \mathrm{M}_r(u)(x),$$

since u is $\mathcal{L}$-subharmonic (hence sub-mean) in $\Omega \setminus \{x_0\}$. On the other hand,

$$u(x_0) = -\infty \leq \mathrm{M}_r(u)(x_0) \quad \forall\, r > 0 : \ \overline{B_d(x_0, r)} \subset \Omega.$$

Thus, u is locally sub-mean in Ω, hence $u \in \underline{S}(\Omega)$. □

From Corollary 8.2.3 and the abstract results on directed families of hyperharmonic functions in Section 6.3.1 (page 347), we obtain the following assertion.

Theorem 8.2.7 (Down directed families in $\underline{S}(\Omega)$). *Let $\mathcal{F}$ be a down directed family of $\mathcal{L}$-subharmonic functions in an open connected set $\Omega \subseteq \mathbb{G}$. The function*

$$u := \inf \mathcal{F}$$

is $\mathcal{L}$-subharmonic in Ω if and only if there exists a point $x_0 \in \Omega$ such that

$$u(x_0) > -\infty.$$

Proof. By Theorem 6.3.8 (page 348), u is $\mathcal{L}$-hypoharmonic in Ω. In particular, u is u.s.c. and sub-mean in Ω. Then, by Corollary 8.2.3, u is $\mathcal{L}$-subharmonic if and only if there exists a point $x_0 \in \Omega$ such that $u(x_0) > -\infty$. □

Corollary 8.2.8. *Let $\mathcal{F}$ be a down directed family of $\mathcal{L}$-subharmonic functions in an open set $\Omega \subseteq \mathbb{G}$. Then, on every connected component of Ω, the function*

$$u := \inf \mathcal{F}$$

is $\mathcal{L}$-subharmonic or identically equal to $-\infty$.

The $\mathcal{L}$-subharmonicity can be characterized in terms of the monotonicity with respect to r of the averaging operators $\mathcal{M}_r$ and M_r. We first prove the following lemma.

Lemma 8.2.9. *Let $m : [0, \infty[\to \mathbb{R}$ be a strictly increasing function. Let $f : [0, \infty[\to \mathbb{R}$ be an increasing function, integrable w.r.t. m, in the sense of Riemann-Stieltjes. If $r \in]0, \infty[$, we set*

$$M(r) := \int_0^r \mathrm{d}m(\rho), \qquad F(r) := \frac{1}{M(r)} \int_0^r f(\rho)\,\mathrm{d}m(\rho).$$

Then $F \le f$ and F is non-decreasing.

Proof. From the monotonicity of f we have

$$F(r) = \frac{1}{M(r)} \int_0^r f(\rho)\,\mathrm{d}m(\rho) \le f(r) \frac{1}{M(r)} \int_0^r \mathrm{d}m(\rho) = f(r).$$

Finally, if $0 < r_1 < r_2$,

$$\begin{aligned} &F(r_2) - F(r_1) \\ &= \left(\frac{1}{M(r_2)} - \frac{1}{M(r_1)} \right) \int_0^{r_1} f(\rho)\,\mathrm{d}m(\rho) + \frac{1}{M(r_2)} \int_{r_1}^{r_2} f(\rho)\,\mathrm{d}m(\rho) \\ &\ge \left(\frac{1}{M(r_2)} - \frac{1}{M(r_1)} \right) M(r_1)\, f(r_1) + \frac{1}{M(r_2)} \big(M(r_2) - M(r_1)\big)\, f(r_1) = 0. \end{aligned}$$

This completes the proof. □

In order to state the next theorem, it is convenient to introduce the following definition: if $\Omega \subseteq \mathbb{G}$ is an open set and $x \in \Omega$, we set

$$R(x) := d(x, \Omega) \quad (= \sup\{r > 0 \,:\, B_d(x,r) \subseteq \Omega\}).$$

Theorem 8.2.10 ($\underline{\mathcal{S}}(\Omega)$ and the monotonicity of $\mathcal{M}_r$ and M_r). *Let Ω be an open subset of $\mathbb{G}$, and let $u : \Omega \to [-\infty, \infty[$ be an u.s.c. function finite in a dense subset of Ω. Then the following statements are equivalent:*

(i) $u \in \underline{\mathcal{S}}(\Omega)$,
(ii) $r \mapsto \mathcal{M}_r(u)(x)$ *is monotone increasing for* $0 < r < R(x)$ *and*

$$u(x) = \lim_{r \to 0^+} \mathcal{M}_r(u)(x),$$

(iii) $r \mapsto \mathrm{M}_r(u)(x)$ *is monotone increasing for* $0 < r < R(x)$ *and*

$$u(x) = \lim_{r \to 0^+} \mathrm{M}_r(u)(x).$$

Proof. (i) $\Rightarrow$ (ii). Let $\overline{B_d(x,s)} \subset \Omega$ and $0 < r < s$. Define

$$\mathfrak{F} := \{\varphi \in C(\partial B_d(x,s), \mathbb{R}) \,:\, \varphi \geq u \text{ on } \partial B_d(x,s)\}.$$

Since $B_d(x,s)$ is an $\mathcal{L}$-regular open set, we have

$$H_\varphi^{B_d(x,s)}(y) \geq u(y) \qquad \forall\, y \in B_d(x,s),$$

so that, by the surface mean value Theorem 5.5.4,

$$\begin{aligned} \mathcal{M}_r(u)(x) &\leq \mathcal{M}_r(H_\varphi^{B_d(x,s)})(x) = H_\varphi^{B_d(x,s)}(x) = \mathcal{M}_s(H_\varphi^{B_d(x,s)})(x) \\ &= \mathcal{M}_s(\varphi)(x) = \frac{\beta_d(Q-2)}{r^{Q-1}} \int_{\partial B_d(x,s)} \mathcal{K}_{\mathcal{L}}(x,y)\, \varphi(y)\, \mathrm{d}\sigma(y). \end{aligned}$$

Hence

$$\begin{aligned} \mathcal{M}_r(u)(x) &\leq \inf_{\varphi \in \mathfrak{F}} \frac{\beta_d(Q-2)}{r^{Q-1}} \int_{\partial B_d(x,s)} \mathcal{K}_{\mathcal{L}}(x,y)\, \varphi(y)\, \mathrm{d}\sigma(y) \\ &= \frac{\beta_d(Q-2)}{r^{Q-1}} \int_{\partial B_d(x,s)} \mathcal{K}_{\mathcal{L}}(x,y)\, u(y)\, \mathrm{d}\sigma(y) = \mathcal{M}_s(u)(x). \end{aligned}$$

This proves the first part of the statement. In order to prove the second one, we just have to note that, by the upper semicontinuity of u, for every $\lambda > u(x)$ there exists $\overline{r} > 0$ such that

$$\mathcal{M}_r(u)(x) < \lambda \text{ for } 0 < r < \overline{r}.$$

On the other hand, $u(x) \leq \mathcal{M}_r(u)(x)$ for every $r < R(x)$. Then

$$u(x) \leq \mathcal{M}_r(u)(x) < \lambda \quad \forall\, r \in]0, \overline{r}[.$$

(ii) $\Rightarrow$ (iii). Since

$$\mathrm{M}_r(u)(x) = \frac{Q}{r^Q} \int_0^r \rho^{Q-1} \mathcal{M}_\rho(u)(x)\, \mathrm{d}\rho \qquad \forall\, r \in]0, R(x)[$$

and $\rho \mapsto \mathcal{M}_\rho(u)(x)$ is monotone increasing, by Lemma 8.2.9, $r \mapsto \mathrm{M}_r(u)(x)$ is monotone increasing too. Moreover, by arguing as in the previous step, $u(x) = \lim_{r\to 0^+} \mathrm{M}_r(u)(x)$.

(iii) $\Rightarrow$ (i). Since $r \mapsto \mathrm{M}_r(u)(x)$ is a monotone increasing function and $u(x) = \lim_{r\to 0^+} \mathrm{M}_r(u)(x)$, we have

$$u(x) \le \mathrm{M}_r(u)(x) \quad \text{if } 0 < r < R(x).$$

Then u satisfies the global solid sub-mean property, so that, by Theorem 8.2.1, $u \in \underline{\mathcal{S}}(\Omega)$. □

We know that a smooth function u is $\mathcal{L}$-subharmonic if and only if $\mathcal{L}u \ge 0$. A similar characterization holds for non-smooth $\mathcal{L}$-subharmonic functions. More precisely, we have the following result.

Theorem 8.2.11 ($\underline{\mathcal{S}}(\Omega)$ and the inequality $\mathcal{L}u \ge 0$ in $\mathcal{D}'(\Omega)$. I). *Let Ω be an open subset of $\mathbb{G}$, and let $u : \Omega \to [-\infty, \infty[$ be an u.s.c. function. Then the following statements are equivalent*:

(i) $u \in \underline{\mathcal{S}}(\Omega)$,
(ii) $u \in L^1_{\mathrm{loc}}(\Omega)$, $u(x) = \lim_{r\to 0^+} \mathrm{M}_r(u)(x)$ *for every* $x \in \Omega$, *and*

$$\mathcal{L}u \ge 0 \text{ in } \Omega$$

in the weak sense of distributions.[1]

Remark 8.2.12. We remark that in condition (ii) of Theorem 8.2.11 the hypothesis $u(x) = \lim_{r\to 0^+} \mathrm{M}_r(u)(x)$ cannot be removed. Consider, for example, the function $u = \chi_{\{0\}}$ (the characteristic function of the set $\{0\}$). Then, it is immediately seen that u is u.s.c., $u \in L^1_{\mathrm{loc}}(\mathbb{G})$, $\mathcal{L}u = 0$ in $\mathbb{G}$ (in the weak sense of distributions), but $u \notin \underline{\mathcal{S}}(\mathbb{G})$ for it is not sub-mean at the origin. On the other hand, if, in addition, u is continuous then $u(x) = \lim_{r\to 0^+} \mathrm{M}_r(u)(x)$ always holds true. □

In order to prove the theorem, we need the following real analysis lemma.

Lemma 8.2.13. *Let $f :]0, \infty[\to \mathbb{R}$ be a L^1_{loc} function such that*

$$\int_0^\infty f(t)\, h'(t)\, \mathrm{d}t \le 0 \quad \forall\, h \in C^1_0(]0, \infty[), \; h \ge 0. \tag{8.4}$$

[1] The inequality $\mathcal{L}u \ge 0$ in the weak sense of distributions means

$$\int_\Omega u\, \mathcal{L}\phi \ge 0 \qquad \forall\, \phi \in C^\infty_0(\Omega), \; \phi \ge 0.$$

Let $\alpha > 1$. Suppose that

$$F(t) := \frac{\alpha}{t^\alpha} \int_0^t s^{\alpha-1} f(s)\, \mathrm{d}s$$

is finite for $0 < t < T$. Then $t \mapsto F(t)$ is monotone increasing in $]0, T[$.

Proof. Let t_1, t_2 be Lebesgue points of f such that $0 < t_1 < t_2$. For a fixed $\varepsilon > 0$, $\varepsilon < \frac{t_2 - t_1}{2}$, we choose a sequence $\{h_n\}_n$ of positive $C_0^1(]0, \infty[)$ functions such that $\operatorname{supp} h_n \subseteq [t_1, t_2]$, $\sup_n \sup_t |h_n'(t)| < \infty$, and, as $n \to \infty$,

$$h_n'(t) \to \begin{cases} 1/\varepsilon & \text{if } t_1 < t < t_1 + \varepsilon, \\ -1/\varepsilon & \text{if } t_2 - \varepsilon < t < t_2, \\ 0 & \text{if } t \notin [t_1, t_1 + \varepsilon] \cup [t_2 - \varepsilon, t_2]. \end{cases}$$

Replacing h_n in (8.4) and letting n tend to infinity, we get

$$\frac{1}{\varepsilon} \int_{t_1}^{t_1+\varepsilon} f(t)\, \mathrm{d}t - \frac{1}{\varepsilon} \int_{t_2-\varepsilon}^{t_2} f(t)\, \mathrm{d}t \le 0.$$

From this inequality, letting $\varepsilon \to 0^+$, we obtain $f(t_1) \le f(t_2)$, since t_1 and t_2 are Lebesgue points for f.

Since almost every point of $]0, \infty[$ is a Lebesgue point for f, we can use the same argument as in the proof of Lemma 8.2.9 in order to show that $F(t_1) \le F(t_2)$ for every pair of Lebesgue points $t_1, t_2 \in]0, T[$, with $t_1 \le t_2$. This result, together with the continuity of F, proves the lemma. □

We are now in the position to prove Theorem 8.2.11.

Proof (of Theorem 8.2.11). (i) ⇒ (ii). If $u \in \underline{S}(\Omega)$, then, by Corollary 8.2.4, $u \in L^1_{\mathrm{loc}}(\Omega)$. Moreover, by Theorem 8.2.10, $u(x) = \lim_{r \to 0^+} \mathrm{M}_r(u)(x)$. By Theorem 8.2.1, we also have that u is sub-mean in Ω, so that, by Theorem 8.1.5, u_ε is sub-mean in Ω_ε. As a consequence, by Theorem 8.2.1, $u_\varepsilon \in \underline{S}(\Omega_\varepsilon)$, so that, since $u_\varepsilon \in C^\infty(\Omega_\varepsilon)$, $\mathcal{L}u_\varepsilon \ge 0$ in Ω_ε. Since $u_\varepsilon \to u$ in $L^1_{\mathrm{loc}}(\Omega)$, as $\varepsilon \searrow 0$, we get $\mathcal{L}u \ge 0$ in the weak sense of distributions.

(ii) ⇒ (i). The inequality $\mathcal{L}u \ge 0$ in the weak sense of distributions means

$$\int_\Omega u\, \mathcal{L}\phi \ge 0 \qquad \forall\, \phi \in C_0^\infty(\Omega),\ \phi \ge 0. \tag{8.5}$$

We now choose a suitable test function ϕ. Let us fix a d-ball $B_d(x_0, R) \subseteq \Omega$ and a non-negative function $g \in C^\infty([0, R[, \mathbb{R})$, constant in a right neighborhood of 0. Define

$$\phi(x) := g(d(x_0^{-1} \circ x)).$$

Then $\phi \in C_0^\infty(B_d(x_0, R))$ and, by Proposition 5.4.3 (page 247),

$$\mathcal{L}\phi = |\nabla_{\mathcal{L}} d|^2 \left(g''(d) + \frac{Q-1}{d}\, g'(d) \right).$$

Replacing the right-hand side in (8.5), using Federer's coarea formula, and keeping in mind the definition of the average operator $\mathcal{M}_t$, we get

$$0 \le \int_0^\infty \left(g''(t) + \frac{Q-1}{t} g'(t) \right) \left(\int_{\{d=t\}} u \, \mathcal{K}_{\mathcal{L}} \, d\sigma \right) dt$$

$$= \frac{1}{\beta_d (Q-2)} \int_0^\infty \left(t^{Q-1} g'(t) \right)' \mathcal{M}_t(u) \, dt.$$

Then, if we choose

$$g(t) := \int_t^\infty \frac{h(s)}{s^{Q-1}} \, ds,$$

where h is any non-negative $C_0^\infty(]0, R[; \mathbb{R})$ function, the previous inequality can be written as

$$\int_0^\infty h'(t) \, \mathcal{M}_t(u)(x_0) \, dt \le 0.$$

Then, by Lemma 8.2.13, $t \mapsto \mathrm{M}_t(u)(x_0)$ is monotone increasing in $]0, R[$. On the other hand, by hypothesis, $\lim_{r \to 0^+} \mathrm{M}_r(u)(x) = u(x)$. Thus, by Theorem 8.2.10, $u \in \underline{S}(\Omega)$. □

Remark 8.2.14. From the previous proof, we have that

$$t \mapsto \mathrm{M}_t(u)(x), \quad 0 \le t < R(x), \ x \in \Omega,$$

is monotone increasing whenever $u \in L^1_{\mathrm{loc}}(\Omega)$ and $\mathcal{L}u \ge 0$ in the weak sense of distributions. Here, we denoted

$$R(x) := \sup\{r > 0 : \overline{B_d(x,r)} \subset \Omega\}.$$

On the other hand, if u is just a $L^1_{\mathrm{loc}}(\Omega)$ function, very standard arguments show that:

(i) $(x,t) \mapsto \mathrm{M}_t(u)(x)$ is continuous in $\{(x,t) : x \in \Omega, 0 \le t < R(x)\}$,
(ii) $\lim_{j \to \infty} \mathrm{M}_{t_j}(u)(x) = u(x)$ almost everywhere in Ω, for a suitable sequence $t_j \downarrow 0$ as $j \uparrow \infty$. □

The above remark, together with Theorems 8.2.10 and 8.2.11, leads to the next theorem.

Theorem 8.2.15 ($\underline{S}(\Omega)$ and the inequality $\mathcal{L}u \ge 0$ in $\mathcal{D}'(\Omega)$. II). *Let $u \in L^1_{\mathrm{loc}}(\Omega)$, with open $\Omega \subseteq \mathbb{G}$. Then the following statements are equivalent:*

(1) *$\mathcal{L}u \ge 0$ in Ω, in the weak sense of distributions,*
(2) *$t \mapsto \mathrm{M}_t(u)(x)$ is monotone increasing on $]0, R(x)[$ for every $x \in \Omega$,*
(3) *there exists a function $\widetilde{u} \in \underline{S}(\Omega)$ such that $u(x) = \widetilde{u}(x)$ almost everywhere in Ω.*

The function in (3) *is unique. It is given by*

$$\widetilde{u}(x) = \lim_{t \downarrow 0} \mathrm{M}_t(u)(x), \quad x \in \Omega.$$

Proof. (1) $\Rightarrow$ (2). This follows from the previous remark.

(2) $\Rightarrow$ (3). The monotonicity of $t \mapsto \mathrm{M}_t(u)(x)$ implies the existence of

$$\widetilde{u}(x) := \lim_{t\downarrow 0} \mathrm{M}_t(u)(x) \quad \forall\, x \in \Omega.$$

Remark 8.2.14-(ii) implies $u(x) = \widetilde{u}(x)$ almost everywhere in Ω. Then, if $0 < t < \tau$, we have

$$\mathrm{M}_t(\widetilde{u})(x) = \mathrm{M}_t(u)(x) \le \mathrm{M}_\tau(u)(x) = \mathrm{M}_\tau(\widetilde{u})(x).$$

Hence, $t \mapsto \mathrm{M}_t(\widetilde{u})(x)$ is monotone increasing. Let us now show that $\widetilde{u}$ is u.s.c. Since

$$\widetilde{u}(x) = \lim_{t\downarrow 0} \mathrm{M}_t(u)(x) = \lim_{t\downarrow 0} \mathrm{M}_t(\widetilde{u})(x),$$

by Theorem 8.2.10, the assertion will follow.

Let $\lambda \in \mathbb{R}$ and $x \in \Omega$ be such that $\widetilde{u}(x) < \lambda$. Then, by the very definition of $\widetilde{u}$, we have $\mathrm{M}_t(u)(x) < \lambda$ for every t small enough. The continuity of $(t, y) \mapsto \mathrm{M}_t(u)(y)$ now implies

$$\mathrm{M}_t(u)(y) < \lambda$$

for every t sufficiently small and for any y close to x. As a consequence, being $\widetilde{u} \le \mathrm{M}_t(u)$,

$$\widetilde{u}(y) < \lambda$$

for every y in a neighborhood of x. Thus, $\widetilde{u}$ is u.s.c., and this part of the proof is complete.

(3) $\Rightarrow$ (1). By Theorem 8.2.11, if $\widetilde{u} \in \underline{\mathcal{S}}(\Omega)$, then $\widetilde{u} \in L^1_{\mathrm{loc}}(\Omega)$ and $\mathcal{L}\widetilde{u} \ge 0$ in the weak sense of distributions. Since $\widetilde{u} = u$ almost everywhere in Ω, these properties hold for u too. $\square$

Theorem 8.2.16 (Caccioppoli–Weyl's lemma for sub-Laplacians). *Let $u \in L^1_{\mathrm{loc}}(\Omega)$ be a weak solution to*

$$\mathcal{L}u = 0 \quad \textit{in } \Omega.$$

Then there exists a smooth $\mathcal{L}$-harmonic $\widetilde{u}$ in Ω such that

$$u(x) = \widetilde{u}(x) \textit{ almost everywhere in } \Omega.$$

Proof. By the previous theorem, there exists $\widetilde{u}_1 \in \underline{\mathcal{S}}(\Omega)$ and $\widetilde{u}_2 \in \overline{\mathcal{S}}(\Omega)$ such that $\widetilde{u}_1(x) = u(x) = \widetilde{u}_2(x)$ almost everywhere in Ω. Moreover,

$$\widetilde{u}_1(x) = \lim_{t\downarrow 0} \mathrm{M}_t(u)(x) = \widetilde{u}_2(x).$$

Hence, $\widetilde{u} := \widetilde{u}_1 = \widetilde{u}_2 \in \underline{\mathcal{S}}(\Omega) \cap \overline{\mathcal{S}}(\Omega)$, so that, by Proposition 7.2.4, $\widetilde{u}$ is $\mathcal{L}$-harmonic in Ω. We also obviously have $u(x) = \widetilde{u}(x)$ almost everywhere in Ω. This completes the proof. $\square$

From Theorem 8.2.11 and Theorem 8.2.15 we also obtain the following characterization of $\mathcal{L}$-subharmonic functions.

Theorem 8.2.17 ($\underline{\mathcal{S}}(\boldsymbol{\Omega})$ and the inequality $\mathcal{L}u \geq 0$ in $\mathcal{D}'(\boldsymbol{\Omega})$. III). *Let Ω be an open subset of $\mathbb{G}$. Then the following statements are equivalent:*

(i) $u \in \underline{\mathcal{S}}(\Omega)$,
(ii) $u \in L^1_{\text{loc}}(\Omega)$, $\mathcal{L}u \geq 0$ *in Ω in the weak sense of distributions, and*[2]

$$u(x) = \operatorname{ess\,lim\,sup}_{y \to x} u(y) \qquad \forall\, x \in \Omega. \tag{8.6}$$

We first prove the following lemma.

Lemma 8.2.18. *Let Ω be an open subset of $\mathbb{G}$. If $u \in \underline{\mathcal{S}}(\Omega)$, then* (8.6) *holds.*

Proof. Since u is u.s.c., we have $\overline{u} := \operatorname{ess\,lim\,sup} u \leq \limsup u \leq u$. Assume by contradiction that $u \neq \overline{u}$. Then there exists $x \in \Omega$ such that $\overline{u}(x) < \lambda < u(x)$ (for some $\lambda \in \mathbb{R}$). In particular $u \leq \lambda$ a.e. in a neighborhood of x, so that $\mathrm{M}_r(u)(x) \leq \lambda$ for any small $r > 0$. Recall that, by Theorem 8.2.11, we have $u(x) = \lim_{r \to 0^+} \mathrm{M}_r(u)(x)$. Hence we obtain $u(x) \leq \lambda$, which gives a contradiction. □

Proof (of Theorem 8.2.17). (i) $\Rightarrow$ (ii) follows from Theorem 8.2.11 and Lemma 8.2.18.

(ii) $\Rightarrow$ (i). According to Theorem 8.2.15, there exists $v \in \underline{\mathcal{S}}(\Omega)$ coinciding almost everywhere with u. From Lemma 8.2.18 we infer that $v = \operatorname{ess\,lim\,sup} v$. In order to obtain $u = v$, we now only need to observe that $v = u$ almost everywhere implies $\operatorname{ess\,lim\,sup} v = \operatorname{ess\,lim\,sup} u$ and to recall that $\operatorname{ess\,lim\,sup} u = u$ by hypothesis. □

Further, we explicit state some maximum principles for $\mathcal{L}$-subharmonic functions, which immediately follow from Theorems 8.1.2 and 8.2.1.

Theorem 8.2.19 (The strong maximum principle for $\underline{\mathcal{S}}(\boldsymbol{\Omega})$). *Let $\Omega \subseteq \mathbb{G}$ be an open set, and let $u \in \underline{\mathcal{S}}(\Omega)$. The following statements hold:*

(i) *if Ω is connected and there exists $x_0 \in \Omega$ such that $u(x_0) = \max_\Omega u$, then $u \equiv u(x_0)$ in Ω,*
(ii) *if* $\limsup_{\Omega \ni y \to x} u(y) \leq 0$ *for every $x \in \partial\Omega$, and* $\limsup_{\Omega \ni y \to \infty} u(y) \leq 0$ *if Ω is unbounded, then $u(y) \leq 0$ in Ω.*

---

[2] Here we use the following notation

$$\operatorname{ess\,lim\,sup}_{y \to x} u(y) = \inf\Big\{\operatorname{ess\,sup}_V u \,:\, V \text{ is a neighborhood of } x\Big\},$$

where

$$\operatorname{ess\,sup}_V u = \inf\{m \in \mathbb{R} \cup \{\infty\} \,:\, m \geq u(x) \text{ for almost every } x \in V\}.$$

Proof. Due to Theorems 8.1.2 and 8.2.1, we only have to prove (ii) in the case when Ω is unbounded. From the conditions at the boundary of Ω and at infinity we obtain: for every $\varepsilon > 0$, there exists $R > 0$ such that

$$\limsup_{\Omega \cap B_d(0,R) \ni x \to y} u(x) \le \varepsilon \quad \forall\, y \in \partial(\Omega \cap B_d(0, R))$$

and

$$u(x) < \varepsilon \quad \forall\, x \in \Omega \setminus B_d(0, R).$$

Then, by the maximum principle on bounded domains, $u(x) < \varepsilon$ for every $x \in \Omega$. Letting ε vanish, we get $u \le 0$ in Ω. □

We close the section by giving a theorem on superposition of superharmonic functions.

Theorem 8.2.20. *Let $\Lambda \subseteq \mathbb{R}^\nu$ and $\Omega \subseteq \mathbb{G}$ be connected open sets. Let*

$$u : \Lambda \times \Omega \to]-\infty, \infty]$$

be a l.s.c. function. Finally, let μ be a Radon measure with compact support contained in Λ. Assume that $u(z, \cdot) \in \overline{\mathcal{S}}(\Omega)$ for every fixed $z \in \Lambda$. Then

$$U(x) := \int_\Lambda u(z, x)\, \mathrm{d}\mu(z), \quad x \in \Omega,$$

is $\mathcal{L}$-superharmonic in Ω if there exists $x_0 \in \Omega$ such that

$$U(x_0) < \infty. \tag{8.7}$$

Proof. Since u is l.s.c., we have

$$\begin{aligned}\liminf_{x \to y} U(x) &\ge \int_\Lambda \liminf_{x \to y} u(z, x)\, \mathrm{d}\mu(z)\\ &\ge \int_\Lambda u(z, y)\, \mathrm{d}\mu(z) \quad \forall\, y \in \Omega.\end{aligned}$$

Thus, U is l.s.c. in Ω. Let $x \in \Omega$ and $r > 0$ be such that $\overline{B_d(x, r)} \subset \Omega$. By Fubini–Tonelli's theorem, we obtain

$$\begin{aligned}\mathcal{M}_r(U)(x) &= \int_\Lambda \mathcal{M}_r(u(z, \cdot))(x)\, \mathrm{d}\mu(z)\\ &\quad \text{(since } u(z, \cdot) \text{ is super-mean)}\\ &\ge \int_\Lambda u(z, x)\, \mathrm{d}\mu(z) = U(x).\end{aligned} \tag{8.8}$$

Thus, U is super-mean. From Corollary 8.2.3 and condition (8.7) it follows that $U \in \overline{\mathcal{S}}(\Omega)$. This completes the proof. □

Remark 8.2.21. The previous theorem still holds if we replace the condition

$$\text{supp}(\mu) \textit{ compact and contained in } \Lambda$$

with the following one:

$$\begin{gathered} \textit{for every compact set } K \subset \Omega, \\ \textit{there exists a function } \varphi_K \in L^1(\Lambda, \mathrm{d}\mu) \textit{ such that} \\ u(z,x) \geq \varphi_K(z) \quad \forall\, z \in \Lambda\, \forall\, x \in K. \end{gathered} \tag{8.9}$$

Indeed, if this condition is satisfied, we can still apply Fubini–Tonelli's theorem in (8.8). For example, we explicitly remark that (8.9) holds with $\varphi_K \equiv 0$ if u is non-negative.

8.3 Continuous Convex Functions on $\mathbb{G}$

Throughout this section, we shall follow two different approaches in defining convex functions in Carnot groups. Our main references will be G. Lu, J.J. Manfredi and B. Stroffolini [LMS04] (see also P. Juutinen and the same authors, [JLMS07]) and D. Danielli, N. Garofalo and D.-M. Nhieu [DGN03].

If an extra hypothesis is added (namely, *continuity*) these two notions are equivalent. Since the aim of this section is only to give an overview on convexity, we decided to restrict our attention to *continuous convex functions* on Carnot groups, since the dropping of the continuity hypothesis would lead us beyond our scopes. The reader will be referred to suitable references for more general results (see Note 8.3.18).

A convex function in $\mathbb{R}^N$ is subharmonic with respect to every constant coefficient elliptic operator, and vice versa (see, e.g. [Hor94]). This makes quite natural the following (first) extension of the notion of convexity to Carnot groups. To this end, we introduce the following notation. We shall denote by $\mathbb{S}^m_+$ the cone of the real $m \times m$ symmetric and strictly positive definite matrices.

Let $\mathbb{G}$ be an abstract stratified Lie group with the Lie algebra $\mathfrak{g}$. Let also $V_1 \oplus V_2 \oplus \cdots \oplus V_r$ be a stratification of $\mathfrak{g}$ according to Definition 2.2.3 (page 122). We set $V = (V_1, \ldots, V_r)$. Let finally $\mathcal{X} = \{X_1, \ldots, X_m\}$ be a basis of the first layer V_1 of the stratification (the so-called *horizontal layer*). Given a matrix $A = (a_{j,k})_{j,k} \in \mathbb{S}^m_+$, we denote by $\mathcal{L}_A$ the sub-Laplacian[3]

$$\mathcal{L}_A = \sum_{j,k=1}^{m} a_{j,k} X_j X_k. \tag{8.10}$$

Definition 8.3.1 (v-convex function). *Let $\mathbb{G}$ be a stratified group, and let $V = (V_1, \ldots, V_r)$ be a given stratification of the algebra of $\mathbb{G}$. Let $\{X_1, \ldots, X_m\}$ be a basis of the first layer V_1.*

A continuous real-valued function u in an open set $\Omega \subseteq \mathbb{G}$ is v-convex *if it is $\mathcal{L}_A$-subharmonic in Ω for every $A \in \mathbb{S}^m_+$, being $\mathcal{L}_A$ as in* (8.10).

[3] See Remark 8.3.3 below.

Note 8.3.2. Obviously, the notion of v-convexity is in general well-given for upper semi-continuous functions. The above definition is one of the equivalent forms of the notion of v-convexity in [LMS04], convexity in the *viscosity* sense. Our choice to add the extra hypothesis of continuity has already been justified by the sake of simplicity.

We explicitly remark the following fact: the notion of v-convexity depends on the stratification $V = (V_1, \dots, V_r)$ of $\mathfrak{g}$ or, more precisely, by the horizontal layer V_1 (since V_1 determines all the V_i's). Hence, another possible notation should be "H-convex function" (where "H" stands for "horizontal").

Remark 8.3.3. As we shall see in Proposition 16.1.1 in Chapter 16 (page 623), any sub-Laplacian on $\mathbb{G}$ is of the form (8.10) for a suitable symmetric positive definite matrix A, and vice versa, any operator $\mathcal{L}_A$ in that form is a sub-Laplacian $\mathcal{L}_A = \sum_{k=1}^m Y_k^2$ with $Y_k = \sum_{j=1}^m (A^{1/2})_{k,j} X_j$. This immediately proves the following proposition.

Proposition 8.3.4 (Equivalent definition of v-convexity). *A continuous real-valued function u in an open set $\Omega \subseteq \mathbb{G}$ is v-convex if and only if it is $\mathcal{L}$-subharmonic in Ω for every sub-Laplacian $\mathcal{L}$ (related to V).*

More explicitly, denoted by V_1 the first layer of the stratification V, a continuous function u is v-convex in Ω if and only if it is $\mathcal{L}$-subharmonic in Ω w.r.t. every $\mathcal{L} = \sum_{k=1}^m Y_k^2$, being $(Y_1, \dots, Y_m)$ any basis of V_1.

Note 8.3.5. Observe that the above Proposition 8.3.4 highlights the fact that the notion of v-convexity in Definition 8.3.1 does not depend on the particular basis $(X_1, \dots, X_m)$ of V_1.

Remark 8.3.6. We know from Section 8.2 that the $\mathcal{L}_A$-subharmonicity of u is equivalent to the inequality

$$\mathcal{L}_A u \geq 0 \qquad \text{in } \mathcal{D}'(\Omega), \tag{8.11}$$

i.e. in the weak sense of distributions. It follows that the v-convexity is invariant with respect to left translations and dilations on $\mathbb{G}$, and that locally uniform limits of v-convex functions are v-convex. Indeed, such properties hold for the solutions to the inequality (8.11). □

For a function u of class C^2, the v-convexity can be characterized in terms of its *horizontal Hessian* related to the family $\mathcal{X}$, i.e. of the matrix

$$\mathcal{X}^2 u(x) := \left(\frac{X_j X_k u(x) + X_k X_j u(x)}{2} \right)_{j,k=1,\dots,m}.$$

Proposition 8.3.7 (v-convexity and the horizontal Hessian). *With all the above notation, a function $u \in C^2(\Omega, \mathbb{R})$ is v-convex if and only if*

$$\mathcal{X}^2 u(x) \geq 0 \qquad \textit{for every } x \in \Omega.$$

Proof. Since $u \in C^2$, the distributional inequality (8.11) is equivalent to

$$\mathcal{L}_A u(x) \geq 0$$

for every $x \in \Omega$, so that u is v-convex if and only if

$$\sum_{j,k=1}^{m} a_{j,k}\, X_j X_k u(x) \geq 0 \quad \text{for every } A = (a_{j,k})_{j,k} \text{ in } \mathbb{S}_+^m \text{ and every } x \in \Omega.$$

Then the assertion follows from the following linear algebra lemma. □

Lemma 8.3.8. *Let $B = (b_{i,j})_{i,j}$ be an $m \times m$ real matrix. Then*

$$\sum_{j,k=1}^{m} a_{j,k}\, b_{j,k} \geq 0 \tag{8.12}$$

for every $A = (a_{j,k})_{j,k}$ in $\mathbb{S}_+^m$ if and only if the symmetric part of B is non-negative definite, i.e.

$$B^{\text{sym}} := \left(\frac{b_{j,k} + b_{k,j}}{2}\right)_{j,k \leq m} \geq 0.$$

Proof. Let $\xi = (\xi_1, \ldots, \xi_m)^T \in \mathbb{R}^m$ and $\varepsilon > 0$. Consider the matrix

$$A_\varepsilon := \xi \cdot \xi^T + \varepsilon\, \mathbb{I}_m = (\xi_j \xi_k)_{j,k \leq m} + \varepsilon\, \mathbb{I}_m,$$

where $\mathbb{I}_m$ denotes the $m \times m$ identity matrix. Clearly, $A_\varepsilon \in \mathbb{S}_+^m$ since it is symmetric and

$$\langle A_\varepsilon \eta, \eta \rangle = \eta^T \cdot \xi \cdot \xi^T \cdot \eta + \varepsilon\, |\eta|^2 = (\eta^T \cdot \xi)(\xi^T \cdot \eta) + \varepsilon\, |\eta|^2 = \langle \xi, \eta \rangle^2 + \varepsilon\, |\eta|^2,$$

for every (column-vector) $\eta \in \mathbb{R}^m$. Then, we can use A_ε in (8.12) obtaining

$$0 \leq \sum_{j,k=1}^{m} b_{j,k}\, \xi_j\, \xi_k + \varepsilon \sum_{j=1}^{m} b_{j,j}.$$

Letting $\varepsilon \to 0$, we get

$$0 \leq \sum_{j,k=1}^{m} b_{j,k}\, \xi_j\, \xi_k = \sum_{j,k=1}^{m} \left(\frac{b_{j,k} + b_{k,j}}{2}\right) \xi_j\, \xi_k \qquad \forall\, \xi \in \mathbb{R}^m.$$

This proves the "only if" part of the lemma.

To prove the "if" part, we first remark that inequality (8.12) rewrites as

$$\operatorname{trace}(A \cdot B^{\text{sym}}) \geq 0.$$

Then, for given $A \in \mathbb{S}_+^m$ taking a matrix $C \in \mathbb{S}_+^m$ such that $C^2 = A$, we have

$$\operatorname{trace}(A \cdot B^{\text{sym}}) = \operatorname{trace}(C^2 \cdot B^{\text{sym}}) = \operatorname{trace}(C \cdot B^{\text{sym}} \cdot C) \geq 0,$$

since the matrix $C \cdot B^{\text{sym}} \cdot C = C \cdot B^{\text{sym}} \cdot C^T$ is non-negative definite if B^{sym} is. This completes the proof. □

The following result holds.

Corollary 8.3.9 (v-convexity for C^2-functions). *A $C^2(\Omega, \mathbb{R})$ function u is v-convex if and only if one of the following equivalent statements hold:*

(1) $\mathcal{X}^2 u \geq 0$ *on* Ω *for one basis* $\mathcal{X} = (X_1, \dots, X_m)$ *of* V_1;
(2) $\mathcal{X}^2 u \geq 0$ *on* Ω *for every basis* $\mathcal{X} = (X_1, \dots, X_m)$ *of* V_1;
(3) $\mathcal{L} u \geq 0$ *on* Ω *for every sub-Laplacian* $\mathcal{L}$ *(related to the stratification* V*).*

Proof. (1) is a restatement of Proposition 8.3.7.

(2) follows from Proposition 8.3.7 and the independence of the notion of v-convexity w.r.t. the basis $\mathcal{X}$ of V_1 (see Note 8.3.5).

(3) follows from Proposition 8.3.4 and the characterization of $\mathcal{L}$-subharmonicity for C^2-functions. □

The smooth v-convex functions are dense in the set of the continuous v-convex functions. Indeed, the following result holds (see the definition of mollifier in Definition 5.3.6, page 239; precisely, see the particular case in Example 5.3.7 on page 239).

Theorem 8.3.10 (Smoothing of a v-convex function). *Let $\mathbb{G}$ be a homogeneous Carnot group. Let V be a fixed stratification of the algebra of $\mathbb{G}$. Let also d be any fixed homogeneous norm on $\mathbb{G}$.*

Suppose $u \in C(\Omega, \mathbb{R})$ is a v-convex function. For $\varepsilon > 0$, let Ω_ε be the set $\{y \in \mathbb{G} : B_d(y, \varepsilon) \subseteq \Omega\}$. For $x \in \Omega_\varepsilon$, define as usual

$$u_\varepsilon(x) = \int_\Omega u(y)\, J_\varepsilon(x \circ y^{-1})\, dy, \tag{8.13}$$

where $J \in C_0^\infty(B_d(0,1))$ and $J_\varepsilon(z) = \varepsilon^{-Q} J(\delta_{1/\varepsilon}(z))$. Then:

(i) $u_\varepsilon \in C^\infty(\Omega_\varepsilon)$ *and* $u_\varepsilon \to u$ *as* $\varepsilon \to 0$ *uniformly on the compact sets in* Ω,
(ii) u_ε *is v-convex in* Ω_ε.

Proof. (i) is a standard property of the mollifiers.

(ii). Since u is v-convex, u is $\mathcal{L}_A$-subharmonic for every $A \in \mathbb{S}^m_+$. It follows from Theorem 8.1.5 that u_ε is $\mathcal{L}_A$-subharmonic in its domain. Since this holds for every $A \in \mathbb{S}^m_+$, we infer that u_ε is v-convex in Ω_ε. □

The characterization of v-convexity in terms of the horizontal Hessian implies the convexity in the usual sense along the V-*horizontal segments* and vice versa. To make this statement precise, we first introduce the notion of V-horizontal segment and of V-horizontal subspace.

Definition 8.3.11 (V-horizontal segment and subspace). *Let $(\mathbb{G}, \circ)$ be a stratified group with the Lie algebra $\mathfrak{g}$. Let also $V = (V_1, \dots, V_r)$ be a given stratification of $\mathfrak{g}$, and let* Exp $: \mathfrak{g} \to \mathbb{G}$ *be the usual exponential map. We set*

$$\mathbb{V} := \operatorname{Exp}(V_1),$$

and we say that $\mathbb{V}$ is the V-horizontal subspace *of $\mathbb{G}$.*

Let $x \in \mathbb{G}$ be fixed. If h is any element of $\mathbb{V}$, we say that

$$[x \circ h^{-1}, x \circ h] := \big\{x \circ \operatorname{Exp}(t \operatorname{Log}(h)) \;:\; -1 \le t \le 1\big\}$$

is a V-horizontal segment (along h and) centered at x.

Finally, for a fixed $x \in \mathbb{G}$, we say that

$$\mathbb{V}_x := x \circ \mathbb{V} = \{x \circ \operatorname{Exp}(X) \;:\; X \in V_1\}$$

is the V-horizontal subspace through x.

Remark 8.3.12. Notice that $\mathbb{V} = \mathbb{V}_e$, being e the identity of $\mathbb{G}$. We explicitly remark that $\mathbb{V}$ is a subset of $\mathbb{G}$ closed under inversion (for $(\operatorname{Exp}(X))^{-1} = \operatorname{Exp}(-X)$), it contains the identity of $\mathbb{G}$, but $\mathbb{V}$ is not in general a subgroup of $\mathbb{G}$. Indeed

$$\operatorname{Exp}(X) \circ \operatorname{Exp}(Y) = \operatorname{Exp}(X \diamond Y) = \operatorname{Exp}\left(X + Y + \frac{1}{2}[X, Y] + \cdots\right),$$

and, if $X, Y \in V_1$, it does not generally hold $X \diamond Y \in V_1$. We remark that $\mathbb{V}$ is a submanifold of $\mathbb{G}$ of dimension $m = \dim(V_1)$. □

Theorem 8.3.13 (v-convexity and the horizontal segments). *Let $u \in C^2(\Omega, \mathbb{R})$ with open $\Omega \subseteq \mathbb{G}$. Then u is v-convex in Ω if and only if, for every $x \in \Omega$ and every $h \in \mathbb{V}$ such that $[x \circ h^{-1}, x \circ h] \subset \Omega$, the function*

$$(-1, 1) \ni t \mapsto u\big(x \circ \operatorname{Exp}(t\, X)\big) \quad (\textit{being } X = \operatorname{Log}(h))$$

is convex in the classical sense.

Proof. We split the proof in two parts.

Sufficiency. Let u be v-convex. Take any $h \in \mathbb{V}$ such that the segment along h centered at x is contained in Ω. Let $X := \operatorname{Log}(h)$. Hence, the function

$$(-1, 1) \ni t \mapsto f(t) := u\big(x \circ \operatorname{Exp}(t\, X)\big)$$

is well posed and, thanks to the C^2-regularity assumption on u, $f \in C^2(-1, 1)$. By (2.36a), page 117, we have

$$f''(t) = (X^2 u)\big(x \circ \operatorname{Exp}(t\, X)\big). \tag{8.14}$$

Recalling that $X \in V_1$ by definition of $\mathbb{V}$, we can suppose that X is the first element of a basis of V_1, say $\mathcal{X}$. By Proposition 8.3.7, $\mathcal{X}^2 u(z) \ge 0$ for every $z \in \Omega$. Since $z := x \circ \operatorname{Exp}(t\, X) \in \Omega$, for $[x \circ h^{-1}, x \circ h] \subset \Omega$, this gives

$$(X^2 u)(z) = \langle \mathcal{X}^2 u(z)\, e_1, e_1 \rangle \ge 0 \quad (\text{being } e_1 = (1, 0, \ldots, 0) \in \mathbb{R}^m).$$

Consequently, (8.14) gives $f''(t) \ge 0$. Being $t \in (-1, 1)$ arbitrary, this proves that f is convex.

Necessity. By the C^2-regularity of u and thanks to Corollary 8.3.9-(3), we have to prove that, given a basis $\mathcal{X} = (X_1, \dots, X_m)$ for V_1, it holds $\sum_{j=1}^m X_j^2 u(x) \geq 0$ for every $x \in \Omega$. Fix any $j \in \{1, \dots, m\}$ and any $x \in \Omega$. Since Ω is open, there exists $\varepsilon > 0$ so small that

$$x \circ \operatorname{Exp}(t\,\varepsilon X_j) \in \Omega \quad \text{for every } t \in [-1, 1].$$

This means that setting $h := \operatorname{Exp}(\varepsilon\, X_j)$, we have $[x \circ h^{-1}, x \circ h] \subset \Omega$. Since h clearly belongs to $\mathbb{V}$, the necessity assumption ensures that the function

$$(-1, 1) \ni t \mapsto f(t) := u\big(x \circ \operatorname{Exp}(t\,\varepsilon\, X_j)\big)$$

is a convex function in the classical sense. In particular, $f''(0) \geq 0$. On the other hand, again by (2.36a), page 117, we have

$$0 \leq f''(0) = \varepsilon^2 (X_j^2 u)(x),$$

whence $(X_j^2 u)(x) \geq 0$. The arbitrariness of j and x proves that $\sum_{j=1}^m X_j^2 u$ is nonnegative on Ω, and the proof is complete. □

We explicitly remark that, though Definition 8.3.1 of v-convexity needs some regularity assumption on u to be well posed (for example, upper semicontinuity or a L^1_{loc} assumption), the characterization of v-convexity as stated in Theorem 8.3.13 suggests another notion of convexity, which is free from any *a priori* assumption on u. We hence give the following definition.

Definition 8.3.14 (Horizontally convex function). *Let* $(\mathbb{G}, \circ)$ *be a stratified group with a fixed stratification* $V = (V_1, \dots, V_r)$. *Let* $\Omega \subseteq \mathbb{G}$ *be open. A function* $u : \Omega \to \mathbb{R}$ *will be called* horizontally convex *(or, in short,* H-convex*) in* Ω *if the function*

$$[-1, 1] \ni t \mapsto u(x \circ \operatorname{Exp}(t\,X))$$

is convex (in the classical sense) for every $X \in V_1$ *such that (set* $h = \operatorname{Exp}(X)$*) the horizontal segment* $[x \circ h^{-1}, x \circ h]$ *is contained in* Ω.

Note 8.3.15. The above notion of H-convex function is equivalent (see Proposition 8.3.17) to that of "weakly H-convex" (*weakly horizontally convex*) function introduced by D. Danielli, N. Garofalo and D.-M. Nhieu [DGN03]. The same notion is also referred to as "CC-convexity". L.A. Caffarelli's 1997 NSF Proposal is cited in [DGN03] as seemingly the first reference for a notion of convexity in the Heisenberg groups. The simpler name "H-convex" is now commonly used.

We now focus our attention on *homogeneous* Carnot groups. See Remark 8.3.20 for the details on the link between the notions of convexity in the abstract and in the homogeneous setting.

It turns out that for *continuous* functions v-convexity and horizontal convexity are equivalent. This will be easily seen first noticing the following facts:

(i) Pointwise limits of horizontally convex functions are horizontally convex;
(ii) If $u \in C(\Omega, \mathbb{R})$ is horizontally convex, then so is its mollifier u_ε defined in (8.13). Indeed, the change of variable $x \circ y^{-1} = z$ in the integral at the right-hand side of (8.13) gives

$$u_\varepsilon(x) = \int_\Omega u(z^{-1} \circ x) J_\varepsilon(z)\, dz.$$

Then, for every $X \in V_1$,

$$t \mapsto u_\varepsilon(x \circ \operatorname{Exp}(t X)) = \int_\Omega u\big(z^{-1} \circ x \circ \operatorname{Exp}(t X)\big)\, J_\varepsilon(z)\, dz$$

is convex since $t \mapsto u((z^{-1} \circ x) \circ \operatorname{Exp}(t X))$ is convex for every $z \in B_d(0, \varepsilon)$ with $\varepsilon > 0$ small enough.

We are now in the position to easily prove the equivalence between v- and H- convexity for continuous functions on homogeneous Carnot groups.

Theorem 8.3.16 (v- and H-convexity for continuous functions). *Let $\mathbb{G}$ be a homogeneous Carnot group. Let $u \in C(\Omega, \mathbb{R})$ with open $\Omega \subseteq \mathbb{G}$. Then u is v-convex in Ω if and only if it is horizontally convex in Ω.*

Proof. If u is v-convex, then by Theorem 8.3.10, the function u_ε in (8.13) is smooth and v-convex in Ω_ε. Then, by Theorem 8.3.13, u_ε is horizontally convex. On the other hand, u_ε converges to u as $\varepsilon \to 0$, uniformly on the compact subsets of Ω. The previous remark (i) implies that u is horizontally convex.

Vice versa, let us assume that u is horizontally convex. The previous remark (ii) tells us that u_ε has the same property, and Theorem 8.3.13 ensures that u_ε is v-convex. As $\varepsilon \to 0$, we get the v-convexity of u (see Remark 8.3.6). This completes the proof. □

In the following result, we compare Definition 8.3.14 to the notion of H-convex function in the recent literature. For example, 4) in the proposition below is [DGN03, Definition 5.5], whereas 5) is [Mag06, Definition 3.4]. For the sake of simplicity, we consider the case $\Omega = \mathbb{G}$. For the general case, see V. Magnani [Mag06, Proposition 3.9].

Proposition 8.3.17 (Characterizations of H-convexity). *Let $(\mathbb{G}, *)$ be a stratified group, and let $V = (V_1, \dots, V_r)$ be a given stratification of $\mathfrak{g}$, the Lie algebra of $\mathbb{G}$. For any $\lambda > 0$, let $\Delta_\lambda : \mathfrak{g} \to \mathfrak{g}$ be the linear map such that (for every $i = 1, \dots, r$) Δ_λ acts on V_i as the multiplication times λ^i. Let $\delta_\lambda := \operatorname{Exp} \circ \Delta_\lambda \circ \operatorname{Log}$. Finally, let $u : \mathbb{G} \to \mathbb{R}$ be a function.*

Then the following statements are equivalent:

(1) *u is horizontally convex on $\mathbb{G}$;*
(2) *for every $x \in \mathbb{G}$, $X \in V_1$ and $t \in (0, 1)$, it holds*

$$u\big(x * \operatorname{Exp}(t X)\big) \le (1 - t)\, u(x) + t\, u\big(x * \operatorname{Exp}(X)\big); \tag{8.15a}$$

(3) *for every* $x \in \mathbb{G}$, $h \in \mathbb{V}$ *and* $\lambda \in (0,1)$, *it holds*

$$u\big(x * \delta_\lambda(h)\big) \le (1-\lambda)\,u(x) + \lambda\,u(x*h); \tag{8.15b}$$

(4) *for every* $x \in \mathbb{G}$, $x' \in \mathbb{V}_x$ *and* $\lambda \in (0,1)$, *it holds*

$$u\big(x * \delta_\lambda(x^{-1} * x')\big) \le (1-\lambda)\,u(x) + \lambda\,u(x'); \tag{8.15c}$$

(5) *for every* $x,\ x' \in \mathbb{G}$ *satisfying the geometrical constraint* $x^{-1} * x' \in \mathbb{V}$ *and every* $\lambda \in (0,1)$, *it holds* (8.15c).

Proof. Obviously, (4) and (5) are equivalent, since the condition $x^{-1} * x' \in \mathbb{V}$ is equivalent to $x' \in x * \mathbb{V} = \mathbb{V}_x$.

Moreover, (3) is clearly equivalent to (4), by setting $h = x^{-1} * x'$.

We immediately notice that (2) is a restatement of (3), since the following facts hold: $\mathbb{V} = \mathrm{Exp}\,(V_1)$; $\delta_\lambda \circ \mathrm{Exp} = \mathrm{Exp} \circ \varDelta_\lambda$; for every $X \in V_1$, we have

$$\delta_\lambda(\mathrm{Exp}\,(X)) = \mathrm{Exp}\,(\varDelta_\lambda(X)) = \mathrm{Exp}\,(\lambda\,X).$$

We are then left with the proof of the equivalence of (1) and (2). Suppose first that (1) holds. Then the function

$$(\bigstar) \qquad [-1,1] \ni t \mapsto f(t) := u(x * \mathrm{Exp}\,(t\,X))$$

is convex for every $x \in \mathbb{G}$ and every $X \in V_1$. In particular, we have

$$\begin{aligned} u(x * \mathrm{Exp}\,(t\,X)) &= f(t) = f\big(t\cdot 1 + (1-t)\cdot 0\big) \le t\,f(1) + (1-t)\,f(0) \\ &= t\,u(x * \mathrm{Exp}\,(X)) + (1-t)\,u(x), \end{aligned}$$

and (8.15a) follows. Vice versa, suppose that (2) holds. We have to prove that the function f in ($\bigstar$) is convex. To this end, fix $t_1, t_2 \in [-1,1]$ and $\theta \in [0,1]$. We have to demonstrate that $f(\theta\,t_2 + (1-\theta)t_1) \le \theta\,f(t_2) + (1-\theta)\,f(t_1)$, i.e.

$$\begin{aligned} &u\big(x * \mathrm{Exp}\,(\theta\,t_2\,X + (1-\theta)t_1\,X)\big) \\ &\quad \le \theta\,u(x * \mathrm{Exp}\,(t_2\,X)) + (1-\theta)\,u(x * \mathrm{Exp}\,(t_1\,X)). \end{aligned} \tag{8.16}$$

Observe that

$$\begin{aligned} x * \mathrm{Exp}\,\big(\theta\,t_2\,X + (1-\theta)t_1\,X\big) &= x * \mathrm{Exp}\,\big(t_1\,X + \theta\,(t_2 - t_1)\,X\big) \\ &= x * \mathrm{Exp}\,\big((t_1\,X) \diamond (\theta\,(t_2 - t_1)\,X)\big) \\ &= x * \mathrm{Exp}\,(t_1\,X) * \mathrm{Exp}\,\big(\theta\,(t_2 - t_1)\,X\big). \end{aligned} \tag{8.17}$$

Here we used the Campbell–Hausdorff formula $\mathrm{Exp}\,(A \diamond B) = \mathrm{Exp}\,(A) * \mathrm{Exp}\,(B)$ (for every $A, B \in \mathfrak{g}$) jointly with the obvious fact $(\lambda\,X) \diamond (\mu\,X) = \lambda\,X + \mu\,X$ (for every $\lambda, \mu \in \mathbb{R}$), since it holds

$$(\lambda X) \diamond (\mu X) = \lambda X + \mu X + \frac{1}{2} [\lambda X, \mu X] + \cdots = \lambda X + \mu X.$$

Now, taking into account (8.17) and applying (8.15a) with x, t and X respectively replaced by $x * \mathrm{Exp}\,(t_1 X)$, θ and $(t_2 - t_1) X$ (which clearly belongs to V_1), we infer

$$\begin{aligned} & u\big(x * \mathrm{Exp}\,(\theta\, t_2\, X + (1-\theta) t_1\, X)\big) \\ & \quad \le (1-\theta)\; u(x * \mathrm{Exp}\,(t_1 X)) + \theta\, u\big(x * \mathrm{Exp}\,(t_1 X) * \mathrm{Exp}\,\big((t_2 - t_1) X\big)\big) \\ & \quad = (1-\theta)\; u(x * \mathrm{Exp}\,(t_1 X)) + \theta\, u(x * \mathrm{Exp}\,(t_2 X)), \end{aligned}$$

for arguing as above,

$$\begin{aligned} \mathrm{Exp}\,(t_1 X) * \mathrm{Exp}\,\big((t_2 - t_1) X\big) &= \mathrm{Exp}\,\big((t_1 X) \diamond \big((t_2 - t_1) X\big)\big) \\ &= \mathrm{Exp}\,\big((t_1 X) + (t_2 - t_1) X\big) \\ &= \mathrm{Exp}\,(t_2 X). \end{aligned}$$

This ends the proof. □

Note 8.3.18 (On the continuity of convex functions on Carnot groups). We record some references from the recent literature.

D. Danielli, N. Garofalo and D.-M. Nhieu [DGN03] proved that locally bounded H-convex functions on the Heisenberg groups are locally Lipschitz-continuous w.r.t. the quasi-distance generated by a homogeneous norm.

Z.M. Balogh and M. Rickly [BR02] have improved this result, removing the hypothesis of local bound.

For general Carnot groups, V. Magnani [Mag06] has proved that the H-convex functions which are locally bounded *from above* are locally Lipschitz-continuous (in the above sense).

Finally, M. Sun and X. Yang [SY06] have recently proved that, in general Carnot groups *of step two*, the H-convex functions are locally bounded, so that (by the cited result in [Mag06]) they are locally Lipschitz-continuous.

Note 8.3.19 (Some other references). We collect some recent references on convexity for Carnot groups. Besides the already mentioned D. Danielli, N. Garofalo, D.-M. Nhieu [DGN03] and G. Lu, J.J. Manfredi and B. Stroffolini (and P. Juutinen) [LMS04,JLMS07] and the references in the previous Note 8.3.18, the reader is also referred to C.E. Gutiérrez and A. Montanari [GM04a,GM04b], M. Rickly [Ric06], C. Wang [Wan05].

Remark 8.3.20 (Convexity in the homogeneous Carnot group setting). By Proposition 2.2.22 on page 139, given any abstract stratified group $(\mathbb{H}, *)$ and any of its stratifications $V = (V_1, \dots, V_r)$, there exists a homogeneous Carnot group $(\mathbb{G}, \bullet)$ with the following properties:

(1) $(\mathbb{G}, \bullet)$ is isomorphic to $(\mathbb{H}, *)$ via a Lie-group isomorphism $\Psi : \mathbb{G} \to \mathbb{H}$, whence (denoted by $\mathfrak{g}$ and $\mathfrak{h}$ the relevant Lie algebras) $d\Psi : \mathfrak{g} \to \mathfrak{h}$ is a Lie algebra iso-

morphism; Ψ is completely determined by a fixed choice of $\mathcal{E} = (E_1, \dots, E_N)$, a basis of $\mathfrak{h}$ adapted to the stratification V;

(2) Set $\widetilde{V}_i := (\mathrm{d}\Psi)^{-1}(V_i)$ for every $i = 1, \dots, r$, then $\widetilde{V} = (\widetilde{V}_1, \dots, \widetilde{V}_r)$ is a stratification of $\mathfrak{g}$; more explicitly, $\widetilde{V}_1$ is formed by the first m elements of the Jacobian basis of $\mathfrak{g}$, where $m = \dim(V_1)$;

(3) Denoting by $\mathrm{Exp}_{\mathbb{H}} : \mathfrak{h} \to \mathbb{H}$, $\mathrm{Exp}_{\mathbb{G}} : \mathfrak{g} \to \mathbb{G}$ the relevant exponential maps and setting $\mathbb{V} = \mathrm{Exp}_{\mathbb{H}}(V_1)$, $\widetilde{\mathbb{V}} = \mathrm{Exp}_{\mathbb{G}}(\widetilde{V}_1)$, we have $\Psi(\widetilde{\mathbb{V}}) = \mathbb{V}$.

(4) With the above notation, $\widetilde{\mathbb{V}}$ has a very explicit expression w.r.t. the usual coordinates on $\mathbb{G}$,

$$\widetilde{\mathbb{V}} = \big\{(h_1, \dots, h_m, 0, \dots, 0) : \ h_1, \dots, h_m \in \mathbb{R}\big\}.$$

The statements in (3) and (4) deserve explicit proofs:

(3). By Theorem 2.1.59 (page 119), we know that

$$(\star) \qquad \mathrm{Exp}_{\mathbb{H}} \circ \mathrm{d}\Psi = \Psi \circ \mathrm{Exp}_{\mathbb{G}},$$

whence

$$\Psi(\widetilde{\mathbb{V}}) = \Psi(\mathrm{Exp}_{\mathbb{G}}(\widetilde{V}_1)) = \Psi\big(\mathrm{Exp}_{\mathbb{G}}((\mathrm{d}\Psi)^{-1}(V_1))\big) = \mathrm{Exp}_{\mathbb{H}}(V_1) = \mathbb{V}.$$

(4). Let us denote by e_i ($i = 1, \dots, N$) the i-th element of the standard basis of $\mathbb{R}^N \equiv \mathbb{G}$. Also, let us denote by $Z_1, \dots, Z_N$ the Jacobian basis of $\mathfrak{g}$. By Proposition 2.2.22-(2), we know that $\mathrm{Exp}_{\mathbb{G}}$ is a linear map and that

$$\mathrm{Exp}_{\mathbb{G}}(Z_i) = e_i, \quad \mathrm{d}\Psi(Z_i) = E_i, \qquad i = 1, \dots, N.$$

As a consequence, it holds (use $(\star)$ and notice that $V_1 = \mathrm{span}\{E_1, \dots, E_m\}$)

$$\begin{aligned}
\widetilde{\mathbb{V}} &= \Psi^{-1}(\mathbb{V}) = \Psi^{-1}\big(\mathrm{Exp}_{\mathbb{H}}(V_1)\big) = \big(\mathrm{Exp}_{\mathbb{G}} \circ (\mathrm{d}\Psi)^{-1}\big)(V_1) \\
&\text{(notice that } \mathrm{Exp}_{\mathbb{G}} \circ (\mathrm{d}\Psi)^{-1} \text{ is linear)} \\
&= \mathrm{span}\big\{\big(\mathrm{Exp}_{\mathbb{G}} \circ (\mathrm{d}\Psi)^{-1}\big)(E_i) \ : \ i = 1, \dots, m\big\} \\
&= \mathrm{span}\big\{\big(\mathrm{Exp}_{\mathbb{G}}(Z_i) = e_i \ : \ i = 1, \dots, m\big\} \\
&= \big\{(h_1, \dots, h_m, 0, \dots, 0) : \ h_1, \dots, h_m \in \mathbb{R}\big\}. \quad \square
\end{aligned}$$

Let Ω be an open subset of $\mathbb{H}$, and let us denote by

$$V\text{-Conv}_{\mathrm{v}}(\Omega), \quad V\text{-Conv}_{\mathrm{H}}(\Omega),$$

respectively, the v-convex functions and the horizontally convex functions on Ω (w.r.t. the stratification V). Analogously, if $\widetilde{\Omega}$ is an open subset of $\mathbb{G}$, we use the notation

$$\widetilde{V}\text{-Conv}_{\mathrm{v}}(\widetilde{\Omega}), \quad \widetilde{V}\text{-Conv}_{\mathrm{H}}(\widetilde{\Omega})$$

(this time w.r.t. the stratification $\widetilde{V}$). We aim to prove that the following equalities hold:

$$\begin{aligned}
V\text{-Conv}_{\mathrm{v}}(\Omega) &= \{\widetilde{u}\circ\Psi^{-1} \;:\; \widetilde{u}\in\widetilde{V}\text{-Conv}_{\mathrm{v}}(\Psi^{-1}(\Omega))\},\\
V\text{-Conv}_{\mathrm{H}}(\Omega) &= \{\widetilde{u}\circ\Psi^{-1} \;:\; \widetilde{u}\in\widetilde{V}\text{-Conv}_{\mathrm{H}}(\Psi^{-1}(\Omega))\},\\
\widetilde{V}\text{-Conv}_{\mathrm{v}}(\widetilde{\Omega}) &= \{u\circ\Psi \;:\; u\in V\text{-Conv}_{\mathrm{v}}(\Psi(\widetilde{\Omega}))\},\\
\widetilde{V}\text{-Conv}_{\mathrm{H}}(\widetilde{\Omega}) &= \{u\circ\Psi \;:\; u\in V\text{-Conv}_{\mathrm{H}}(\Psi(\widetilde{\Omega}))\}.
\end{aligned}$$

The first (hence the third) equality holds true, since the notion of v-convexity only involve sub-harmonicity w.r.t. the relevant sub-Laplacians (and we know that the sub-Laplacians related to V and to those related to $\widetilde{V}$ are Ψ-related). The fourth equality (hence the second) follows by the arguments below.

Let $\widetilde{\Omega}\subseteq\mathbb{G}$ be open, let $x\in\widetilde{\Omega}$ and $\widetilde{X}\in\widetilde{V}_1$ be such that

$$x\bullet\operatorname{Exp}_{\mathbb{G}}(t\,\widetilde{X})\in\widetilde{\Omega}\quad\text{for every } t\in[-1,1].$$

This is equivalent to $\Psi(x\bullet\operatorname{Exp}_{\mathbb{G}}(t\,\widetilde{X}))\in\Psi(\widetilde{\Omega})$ for every $t\in[-1,1]$. But we have (see ($\star$) above)

$$\Psi\big(x\bullet\operatorname{Exp}_{\mathbb{G}}(t\,\widetilde{X})\big)=\Psi(x)*\Psi\big(\operatorname{Exp}_{\mathbb{G}}(t\,\widetilde{X})\big)=\Psi(x)*\operatorname{Exp}_{\mathbb{H}}\big(t\,\mathrm{d}\Psi(\widetilde{X})\big).$$

So, if $u\in V\text{-Conv}_{\mathrm{H}}(\Psi(\widetilde{\Omega}))$, by definition of H-convexity, we have that

$$[-1,1]\ni t\mapsto f(t):=u\big(\Psi(x)*\operatorname{Exp}_{\mathbb{H}}(t\,X)\big)$$

is convex (here we let $X:=\mathrm{d}\Psi(\widetilde{X})$ and we noticed that, by definition of $\widetilde{V}_1=(\mathrm{d}\Psi)^{-1}(V_1)$, it holds $X\in V_1$). On the other hand, by the above arguments,

$$f(t)=(u\circ\Psi)\big(x\bullet\operatorname{Exp}_{\mathbb{G}}(t\,\widetilde{X})\big),$$

and the convexity of f (jointly with the arbitrariness of $\widetilde{X}\in\widetilde{V}_1$) proves that $u\circ\Psi\in\widetilde{V}\text{-Conv}_{\mathrm{H}}(\widetilde{\Omega})$.

The above remarks demonstrate that *the study of v- and H- convexity can be entirely carried out in the homogeneous setting, without lack of generality.*

The (more) manageable structure of homogeneous Carnot groups (see, e.g. property (4) above) may be useful in studying convexity in a simpler way. □

Bibliographical Notes. A comprehensive introduction to the classical potential theory and to the study of sub-harmonicity can be found, e.g. in the following monographs: D.H. Armitage and S.J. Gardiner [AG01] and in W.K. Hayman and P.B. Kennedy [HK76]. See also H. Aikawa and M. Essén [AE96], S. Axler, P. Bourdon, W. Ramey [ABR92], N. du Plessis [duP70], L.L. Helms [Helm69], N.S. Landkof [Lan72].

For a bibliography concerning with the notions of convexity in Carnot groups, see the references within Section 8.3.

Some of the topics presented in this chapter also appear in [BL03].

8.4 Exercises of Chapter 8

Ex. 1) Prove that if u is a continuous function which coincides with an $\mathcal{L}$-subharmonic function almost everywhere, then u is $\mathcal{L}$-subharmonic. (*Hint*: Use Theorem 8.2.11.)

Ex. 2) Prove that two $\mathcal{L}$-superharmonic functions on an open set Ω which are equal a.e. in Ω coincide everywhere. (*Hint*: Use Theorem 8.2.11.)

Ex. 3) Prove that $\mathcal{L}_A$ in (8.10) is indeed a sub-Laplacian on $\mathbb{G}$.

Ex. 4) Let Ω be a connected bounded open set in $\mathbb{G}$ and let $u : \Omega \to [-\infty, \infty[$ be an $\mathcal{L}$-hypoharmonic function. Then u is $\mathcal{L}$-subharmonic in Ω if and only if there exists a point $x_0 \in \Omega$ such that

$$u(x_0) > -\infty.$$

Ex. 5) Show that if $u \in \underline{S}(\Omega)$, then

$$u(x) = \limsup_{\Omega \ni y \to x} u(y).$$

Ex. 6) Let $\mathcal{F}$ be a family of $\mathcal{L}$-subharmonic functions in an open set $\Omega \subseteq \mathbb{G}$. Assume that

$$u(x) := \sup\{v(x) \ : \ v \in \mathcal{F}\}$$

is u.s.c. and belongs to $L^1_{\mathrm{loc}}(\Omega)$. Then

$$u \in \underline{S}(\Omega).$$

Ex. 7) Let Γ be the fundamental solution of $\mathcal{L}$, and let A be a closed subset of $\mathbb{G}$. Define

$$u(x) := \sup\{\Gamma(y^{-1} \circ x) \ : \ y \in A\}.$$

Then $u \in \underline{S}(\mathbb{G} \setminus A)$ and it is locally Lipschitz-continuous with respect to d.

Ex. 8) There exists a function $u \in \underline{S}(\mathbb{G}) \cap C(\mathbb{G})$ such that $U_t := \{x \ : \ u(x) \le t\}$ is compact for every $t > 0$ and

$$\bigcup_{t>0} U_t = \mathbb{G}$$

(*Hint*: Use suitable powers of a gauge function.)

Ex. 9) Let Ω be any open set in $\mathbb{G}$. Then there exists a function $u \in \underline{S}(\Omega) \cap C(\Omega)$ such that $U_t := \{x \ : \ u(x) \le t\}$ is compact for every $t > 0$ and

$$\bigcup_{t>0} U_t = \Omega.$$

(*Hint*: Use the previous exercises.)

Ex. 10) Let u be an $\mathcal{L}$-superharmonic function in an open set $\Omega \subseteq \mathbb{G}$. Show that

$$\mathcal{M}_r(u)(x) < \infty$$

for every $r > 0$ such that $\overline{B_d(x,r)} \subseteq \Omega$. (*Hint*: $M_r(u)(x) < \infty$ since $u \in L^1_{\text{loc}}$. Then use the integral representation of M_r in terms of $\mathcal{M}_\rho$, and the monotonicity of $\mathcal{M}_\rho$, $0 < \rho \le r$.)

Ex. 11) Let $\Omega \subseteq \mathbb{G}$ be open, $u : \Omega \to]-\infty, \infty]$ be $\mathcal{L}$-superharmonic and $\alpha \in \mathbb{G}$. Show that

$$x \mapsto u(\alpha \circ x)$$

is $\mathcal{L}$-superharmonic in $\alpha^{-1} \circ \Omega$.

Ex. 12) Let Ω be a connected bounded open set in $\mathbb{G}$, and let $f : \partial\Omega \to [-\infty, \infty]$. Assume $\overline{H}_f^\Omega \not\equiv \infty$ and $\underline{H}_f^\Omega \not\equiv -\infty$. Then $\overline{H}_f^\Omega$ and $\underline{H}_f^\Omega$ are $\mathcal{L}$-harmonic in Ω. (*Hint*: If $\overline{H}_f^\Omega \not\equiv \infty$, then $\overline{\mathcal{U}}_f^\Omega$ contains an $\mathcal{L}$-superharmonic function. Similarly, if $\underline{H}_f^\Omega \not\equiv -\infty$, then $\underline{\mathcal{U}}_f^\Omega$ contains an $\mathcal{L}$-subharmonic function.)

Ex. 13) Let T be as in (5.150) of Ex. 18, Chapter 5 (page 326), and let $u \in \overline{\mathcal{S}}(\Omega)$. Prove the following statements:
(i) $T(u) \le u$,
(ii) $T^{(k)}(u) \ge T^{(k+1)}(u)$ for every $k \in \mathbb{N}$, where

$$T^{(k)} = \underbrace{T \circ \cdots \circ T}_{k \text{ times}}.$$

Ex. 14) Let T be as in the previous exercise, and let $u \in \overline{\mathcal{S}}(\Omega)$ and $v \in \underline{\mathcal{S}}(\Omega)$ be such that $v \le u$. Then:
(i) $v \le T^{(k)}(v) \le T^{(k)}(u) \le u$ for every $k \in \mathbb{N}$,
(ii) $T^{(k)}(v) \uparrow v^*$, $T^{(k)}(u) \downarrow u^*$ for suitable $v^*, u^* \in L^1_{\text{loc}}(\Omega)$. Moreover,

$$T(v^*) = v^* \le u^* = T(u^*).$$

Ex. 15) Let $\Omega \subseteq \mathbb{G}$ be a bounded open set, and let T be as in the previous exercises. Let $u \in \overline{\mathcal{S}}(\Omega) \cap C(\overline{\Omega})$ and $f = u|_{\partial\Omega}$. Then $(T^{(k)}(u))_{k\ge 1}$ is monotone decreasing and

$$\lim_{k\to\infty} T^{(k)}(u) = H_f^\Omega.$$

(*Hint*: If $h := \lim_{k\to\infty} T^{(k)}(u)$, then $h = T(h)$. It follows that $v \le h \le w$ for every $v \in \underline{\mathcal{U}}_f^\Omega$ and $w \in \overline{\mathcal{U}}_f^\Omega$.)

Ex. 16) We keep the notation of the previous exercise. Let $F \in C(\overline{\Omega})$ and $f = F|_{\partial\Omega}$. Prove that the sequence $(T^{(k)}(F))_{k\ge 1}$ is convergent and

$$\lim_{k\to\infty} T^{(k)}(F) = H_f^\Omega.$$

(*Hint*: For every $n \in \mathbb{N}$, there exists $u_n \in \overline{\mathcal{S}}(\mathbb{G}) \cap C(\mathbb{G})$ and $v_n \in \underline{\mathcal{S}}(\mathbb{G}) \cap C(\mathbb{G})$ such that $v_n - \frac{1}{n} < F < u_n + \frac{1}{n}$ in $\overline{\Omega}$. Then use the previous exercises.)

Ex. 17) Let $\mathcal{L}$ be a sub-Laplacian, and let d be an $\mathcal{L}$-gauge. Prove that

$$d^{\alpha} \in \underline{\mathcal{S}}(\mathbb{G}) \quad \Longleftrightarrow \quad \alpha \geq 0.$$

Prove also that

$$d^{\alpha} \in \overline{\mathcal{S}}(\mathbb{G}) \quad \Longleftrightarrow \quad 2 - Q \leq \alpha \leq 0.$$

(*Hint*: Compute $\mathcal{L}(d^{\alpha})$ in $\mathbb{G} \setminus \{0\}$. Remind that if $d^{\alpha} \in \underline{\mathcal{S}}(\mathbb{G})$, then $d^{\alpha} < \infty$ at any point.)

Ex. 18) Let $\Omega \subseteq \mathbb{G}$ be open, and let $u \in C(\Omega, \mathbb{R})$. Prove the equivalence of the following statements:

- $u \in \underline{\mathcal{S}}(\Omega)$,
- $\mathcal{AL}u \geq 0$ in Ω,
- $\mathrm{A}\mathcal{L}u \geq 0$ in Ω.

Here $\mathcal{AL}$ and $\mathrm{A}\mathcal{L}$ denote the asymptotic $\mathcal{L}$ sub-Laplacians introduced in Exercise 8 of Chapter 5.

9

Representation Theorems

The aim of this chapter is mainly to establish some representation theorems of Riesz-type for $\mathcal{L}$-subharmonic functions, being $\mathcal{L}$ a sub-Laplacian on a homogeneous Carnot group $\mathbb{G}$.

We start by introducing the $\mathcal{L}$-Green function G_Ω, first for an $\mathcal{L}$-regular domain Ω, then for general open sets. In order to prove the symmetry of G_Ω in the latter case, we show that any open set can be approximated by $\mathcal{L}$-regular domains. We then come to the core of the chapter, by introducing the $\mathcal{L}$-Green potentials and proving several Riesz-type representation theorems for $\mathcal{L}$-subharmonic functions.

In the rest of the chapter, we give some applications of the above results. In Section 9.5, we prove the Poisson–Jensen formula for $\mathcal{L}$-regular domains. We shall turn back to this formula in Chapter 11, where we shall prove the Poisson–Jensen formula for arbitrary domains (see Theorem 11.7.6 on page 518), by using the theory of $\mathcal{L}$-polar sets.

In Section 9.7, we prove a monotone approximation theorem for $\mathcal{L}$-subharmonic functions via smooth $\mathcal{L}$-subharmonic functions, by using the smoothing operators constructed in Section 5.6 (page 257) by superposition of surface mean operators.

In Section 9.8, we study isolated singularities for $\mathcal{L}$-harmonic functions and we prove some Bôcher-type theorems. As an application, we prove that the kernel appearing in the solid mean value formula of Theorem 5.6.1 (page 259) is constant on $\mathbb{G} \setminus \{0\}$ if and only if $\mathbb{G}$ is the Euclidean group.

9.1 $\mathcal{L}$-Green Function for $\mathcal{L}$-regular Domains

Definition 9.1.1 ($\mathcal{L}$-Green function for an $\mathcal{L}$-regular domain). *Let Ω be a bounded $\mathcal{L}$-regular open subset of $\mathbb{G}$. We call* $\mathcal{L}$-Green function of Ω with pole at $x \in \Omega$, *the function $G_\Omega(x, \cdot) : \Omega \to]-\infty, \infty]$ defined as follows*

$$G_\Omega(x, y) := \Gamma(x^{-1} \circ y) - h_x(y),$$

where Γ is the fundamental solution for $\mathcal{L}$, and h_x denotes the solution to the boundary value problem

$$\begin{cases} \mathcal{L}h = 0 & \textit{in } \Omega, \\ h(z) = \Gamma(x^{-1} \circ z) & \textit{for every } z \in \partial\Omega. \end{cases}$$

With the above definition, we have

$$G_\Omega(x, \cdot) \text{ is } \mathcal{L}\text{-harmonic in } \Omega \setminus \{x\}, \tag{9.1a}$$

$$G_\Omega(x, y) \longrightarrow 0 \text{ as } y \to z, \text{ for every } z \in \partial\Omega, \tag{9.1b}$$

and

$$G_\Omega(x, y) = \Gamma(x^{-1} \circ y) - \int_{\partial\Omega} \Gamma(x^{-1} \circ z)\, \mathrm{d}\mu_y^\Omega(z), \quad x, y \in \Omega. \tag{9.1c}$$

We recall that μ_y^Ω denotes the $\mathcal{L}$-harmonic measure related to (the $\mathcal{L}$-regular open set) Ω and the point y (see Section 7.2, page 388).

The following theorem states some other important properties of the $\mathcal{L}$-Green function.

Theorem 9.1.2. *For every $x, y \in \Omega$, $x \neq y$, we have:*

(i) $G_\Omega(x, y) \geq 0$,
(ii) $G_\Omega(x, y) > 0$ *iff x and y belong to the same connected component of Ω,*
(iii) $G_\Omega(x, y) = G_\Omega(y, x)$.

Proof. (i). Since $\Gamma(x^{-1} \circ z) \to \infty$ as $z \to x$ and h_x is smooth in Ω, there exists $r > 0$ such that $G_\Omega(x, z) > 0$ for every $z \in \overline{B_d(x, r)}$. Moreover,

$$\begin{cases} \mathcal{L}G_\Omega(x, \cdot) = 0 & \text{in } \Omega \setminus \overline{B_d(x, r)}, \\ \lim_{z \to \zeta} G_\Omega(x, z) \geq 0 & \text{for every } \zeta \in \partial(\Omega \setminus \overline{B_d(x, r)}), \end{cases}$$

so that, by the maximum principle, $G_\Omega(x, z) \geq 0$ in $\Omega \setminus \overline{B_d(x, r)}$. Thus, $G_\Omega(x, z) \geq 0$ for any $z \in \Omega$. In particular, $G_\Omega(x, y) \geq 0$.

(ii). Suppose $x, y \in \Omega_0$, with $\Omega_0 \subseteq \Omega$ open and connected. Assume by contradiction $G_\Omega(x, y) = 0$. Then, since $G_\Omega(x, \cdot)$ is non-negative and $\mathcal{L}$-harmonic in $\Omega_0 \setminus \{x\}$, by the strong maximum principle of Theorem 5.13.8 (page 296),

$$G_\Omega(x, z) = 0 \quad \text{for every } z \in \Omega_0 \setminus \{x\}.$$

This is impossible, because $G_\Omega(x, z) \to \infty$ as $z \to x$. Let us now suppose $y \in \Omega_1$, being Ω_1 a connected component of Ω not containing x. Then $z \mapsto \Gamma(x^{-1} \circ z)$ is $\mathcal{L}$-harmonic in an open set containing $\overline{\Omega_1}$, so that $h_x(z) = \Gamma(x^{-1} \circ z)$ for every $z \in \Omega_1$. It follows that $G_\Omega(x, \cdot) = 0$ in Ω_1. In particular, $G_\Omega(x, y) = 0$.

(iii). Let $y \in \Omega$ be fixed. Denote by g_y the Γ-potential of μ_y^Ω, i.e.

$$g_y(z) := \int_{\partial\Omega} \Gamma(\zeta^{-1} \circ z)\, \mathrm{d}\mu_y^\Omega(\zeta) = \int_{\partial\Omega} \Gamma(z^{-1} \circ \zeta)\, \mathrm{d}\mu_y^\Omega(\zeta), \quad z \in \mathbb{G}.$$

The function g_y is $\mathcal{L}$-harmonic in Ω since μ_y^Ω is supported on $\partial\Omega$ (see Theorem 9.3.5, page 433). On the other hand, (9.1c) together with the positivity of G_Ω, gives

$$g_y(z) \le \Gamma(z^{-1} \circ y) \quad \forall z \in \Omega.$$

It follows that $\limsup_{z\to\zeta} g_y(z) \le \Gamma(\zeta^{-1} \circ y)$ for every $\zeta \in \partial\Omega$. The definition of h_y and the maximum principle imply $g_y(z) \le h_y(z)$ for every $z \in \Omega$. In particular, $g_y(x) \le h_y(x)$. Then

$$\Gamma(x^{-1} \circ y) - G_\Omega(x, y) = g_y(x) \le h_y(x) = \Gamma(y^{-1} \circ x) - G_\Omega(y, x),$$

so that, since $\Gamma(x^{-1} \circ y) = \Gamma(y^{-1} \circ x)$,

$$G_\Omega(x, y) \ge G_\Omega(y, x).$$

By interchanging the rôles of x and y, we also get $G_\Omega(y, x) \ge G_\Omega(x, y)$. Hence, $G_\Omega(x, y) = G_\Omega(y, x)$. □

Example 9.1.3. We know that the d-ball $B_d(x, r)$ is an $\mathcal{L}$-regular domain. Since $\Gamma(x^{-1} \circ y) = \Gamma(r)$ if $y \in \partial B_d(x, r)$, we have $h_x \equiv \Gamma(r)$. Then

$$G_{B_d(x,r)}(x, y) = \Gamma(x^{-1} \circ y) - \Gamma(r). \tag{9.2}$$

9.2 $\mathcal{L}$-Green Function for General Domains

In this section, we extend the notion of $\mathcal{L}$-Green function to general open sets.

Definition 9.2.1 ($\mathcal{L}$-Green function for a general domain). *Let $\Omega \subseteq \mathbb{G}$ be open, and let $x \in \Omega$. The function $y \mapsto \Gamma(x^{-1} \circ y)$ is $\mathcal{L}$-superharmonic and non-negative in Ω. Then it has a greatest $\mathcal{L}$-harmonic minorant in Ω. Let us denote it by h_x. The function*

$$\Omega \times \Omega \ni (x, y) \mapsto G_\Omega(x, y) := \Gamma(x^{-1} \circ y) - h_x(y) \in [0, \infty]$$

is the $\mathcal{L}$-Green function *for Ω.*

We explicitly remark that h_x and $G_\Omega(x, \cdot)$ are $\mathcal{L}$-harmonic, respectively, in Ω and in $\Omega \setminus \{x\}$. Moreover, $G_\Omega(x, \cdot)$ is $\mathcal{L}$-superharmonic in Ω and

$$h_x = \sup\{v \in \underline{S}(\Omega) | v \le \Gamma(x^{-1} \circ \cdot)\}.$$

As a consequence, $0 \le h_x \le \Gamma(x^{-1} \circ \cdot)$ and $G_\Omega \ge 0$. For future references it is worth stating the following proposition.

Proposition 9.2.2. *Let $x \in \Omega$, and let $v \in \underline{S}(\Omega)$ be such that*

$$v \le G_\Omega(x, \cdot) \qquad \text{in } \Omega.$$

Then $v \le 0$. Hence, the null function is the greatest $\mathcal{L}$-harmonic minorant of the function $\Omega \ni x \mapsto G_\Omega(x, \cdot)$.

Proof. The hypothesis implies $v + h_x \le \Gamma(x^{-1} \circ \cdot)$. Then, since $v + h_x \in \underline{\mathcal{S}}(\Omega)$, we infer $v + h_x \le h_x$, that is $v \le 0$. The second part of the proposition trivially follows from the first one. □

Example 9.2.3. The $\mathcal{L}$-Green function for $\mathbb{G}$ is

$$G_{\mathbb{G}}(x, y) = \Gamma(x^{-1} \circ y), \qquad x, y \in \mathbb{G}.$$

Indeed, since $0 \le h_x \le \Gamma(x^{-1} \circ \cdot)$ and $\Gamma(x^{-1} \circ y) \to 0$ as $y \to \infty$, $h_x \equiv 0$ and $G_{\mathbb{G}}(x, \cdot) = \Gamma(x^{-1} \circ \cdot)$.

When Ω is bounded, G_Ω can be expressed in terms of the PWB operator. Indeed, the following theorem holds.

Theorem 9.2.4. *Let $\Omega \subseteq \mathbb{G}$ be open and bounded. Then, for every $x \in \Omega$, the greatest $\mathcal{L}$-harmonic minorant in Ω of the map $x \mapsto \Gamma(x^{-1} \circ y)$ is the PWB solution to the Dirichlet problem*

$$\begin{cases} \mathcal{L}h = 0 & \text{in } \Omega, \\ h|_{\partial\Omega} = \Gamma(x^{-1} \circ \cdot). \end{cases}$$

Remark 9.2.5. From this theorem it follows that

$$G_\Omega(x, y) = \Gamma(x^{-1} \circ y) - \int_{\partial\Omega} \Gamma\big(x^{-1} \circ z\big)\, d\mu_y^\Omega(z), \qquad x, y \in \Omega, \tag{9.3}$$

where, as usual, μ_y^Ω denotes the $\mathcal{L}$-harmonic measure related to Ω and y. When Ω is $\mathcal{L}$-regular, this formula gives back (9.1c) of Section 9.1.

Proof (of Theorem 9.2.4). Let $x \in \Omega$ be fixed, and let $\varphi := \Gamma(x^{-1} \circ \cdot)|_{\partial\Omega}$. Let $v \in \underline{\mathcal{S}}(\Omega)$. From the maximum principle in Theorem 8.2.19 (page 409) we obtain

$$\limsup_{\partial\Omega} v \le \varphi \quad \text{iff } v \le \Gamma(x^{-1} \circ \cdot) \quad \text{in } \Omega.$$

Then, since φ is resolutive,

$$h_x = \sup\{v \in \underline{\mathcal{S}}(\Omega) \,:\, v \le \Gamma(x^{-1} \circ \cdot)\} = \sup \underline{\mathcal{U}}_\varphi^\Omega = \underline{H}_\varphi^\Omega = H_\varphi^\Omega,$$

as we aimed to prove. □

The $\mathcal{L}$-Green function G_Ω is non-negative in $\Omega \times \Omega$ and such that, for every fixed $x \in \Omega$, $G_\Omega(x, \cdot)$ is the sum of $\Gamma(x^{-1} \circ \cdot)$ plus an $\mathcal{L}$-harmonic function on Ω. We now show that $\Gamma(x^{-1} \circ \cdot)$ does not exceed any other function sharing the same property.

Proposition 9.2.6. *Let $x \in \Omega$, and let $u \in \overline{\mathcal{S}}(\Omega)$, $u \ge 0$, be such that*

$$u = \Gamma(x^{-1} \circ \cdot) + v$$

with $v \in \overline{\mathcal{S}}(\Omega)$. Then

$$u \ge G_\Omega(x, \cdot).$$

Proof. The condition $u \geq 0$ implies $-v \leq \Gamma(x^{-1} \circ \cdot)$, so that (since $-v \in \underline{\mathcal{S}}(\Omega)$) $-v \leq h_x$. Hence $\Gamma(x^{-1} \circ \cdot) - u \leq h_x$, and the assertion follows. □

Corollary 9.2.7. *Let $\Omega_1 \subseteq \Omega_2 \subseteq \mathbb{G}$. Then*

$$G_{\Omega_1} \leq G_{\Omega_2} \qquad \text{in } \Omega_1 \times \Omega_1.$$

Proof. Let $x \in \Omega_1$. The function $G_{\Omega_2}(x, \cdot)|_{\Omega_1}$ is $\mathcal{L}$-superharmonic and non-negative in Ω_1 and is the sum of $\Gamma(x^{-1} \circ \cdot)$ plus an $\mathcal{L}$-harmonic function. Proposition 9.2.6 implies $G_{\Omega_1}(x, \cdot) \leq G_{\Omega_2}(x, \cdot)|_{\Omega_1}$. □

The following approximation theorem also holds.

Theorem 9.2.8. *Let $(\Omega_n)_{n\in\mathbb{N}}$ be a monotone increasing sequence of open sets, and let*

$$\Omega := \bigcup_{n\in\mathbb{N}} \Omega_n.$$

Then[1]

$$\lim_{n\to\infty} G_{\Omega_n} = G_{\Omega}. \tag{9.4}$$

Proof. Since $\Omega_n \subseteq \Omega_{n+1} \subseteq \Omega$, Corollary 9.2.7 gives

$$G_{\Omega_n} \leq G_{\Omega_{n+1}} \leq G_{\Omega}.$$

Hence the limit in (9.4) exists and is $\leq G_{\Omega}$. To prove the opposite inequality, we fix $x \in \Omega$ and consider $n \in \mathbb{N}$ such that $\Omega_n \ni x$. Then

$$\Gamma(x^{-1} \circ \cdot) - G_{\Omega_n}(x, \cdot) = h_n, \qquad h_n := h_{\Omega_n, x}$$

and

$$h_n \geq h_{n+1} \geq 0 \qquad \text{in } \Omega_n.$$

By the Harnack convergence Theorem 5.7.10, page 268, there exists a $\mathcal{L}$-harmonic function $h \in \mathcal{H}(\Omega)$ such that $h_n \downarrow h$. It follows that

$$U := \Gamma(x^{-1} \circ \cdot) - h$$

is non-negative in Ω since $\Gamma(x^{-1} \circ \cdot) - h_n \geq 0$ in Ω_n. Then, by Proposition 9.2.6,

$$G_{\Omega}(x, \cdot) \leq U = \lim_{n\to\infty} G_{\Omega_n}(x, \cdot).$$

This completes the proof. □

---

[1] We agree that $\lim_{n\to\infty} G_{\Omega_n} = G_{\Omega}$ means

$$x, y \in \Omega_p \Rightarrow \lim_{n\to\infty} G_{\Omega_{p+n}}(x, y) = G_{\Omega}(x, y).$$

In order to prove the symmetry of the $\mathcal{L}$-Green function for general domains, we need a result which is of independent interest.

Lemma 9.2.9 (Approximation from the inside by $\mathcal{L}$-regular sets). *Given an open set $\Omega \subseteq \mathbb{G}$, there exists a monotone increasing sequence of bounded $\mathcal{L}$-regular open sets $(\Omega_n)_{n\in\mathbb{N}}$ such that*

$$\bigcup_{n\in\mathbb{N}} \Omega_n = \Omega.$$

Proof. We first assume that Ω is bounded. For every $n \in \mathbb{N}$, let us cover $\partial\Omega$ by a finite family of $\mathcal{L}$-gauge balls

$$\left\{B_d(x_j^n, r_j^n)\right\}_{j=1,\dots,p_n} \quad \text{with } 0 < r_j^n < 1/n.$$

We choose the balls in a such a way that

$$\overline{B_d(x_j^{n+1}, r_j^{n+1})} \subseteq \bigcup_{i=1}^{p_n} B_d(x_i^n, r_i^n).$$

Then, defining

$$\Omega_n := \Omega \setminus \bigcup_{i=1}^{p_n} \overline{B_d(x_i^n, r_i^n)},$$

we have $\Omega_n \subseteq \Omega_{n+1}$ and $\bigcup_n \Omega_n = \Omega$. The open sets Ω_n are $\mathcal{L}$-regular. Indeed, if $z_0 \in \partial\Omega_n$, there exists $j \in \{1, \dots, p_n\}$ such that $z_0 \in \partial B_d(x_j^n, r_j^n)$. The function

$$w = \Gamma(r_j^n) - \Gamma\big((x_j^n)^{-1} \circ \cdot\big)$$

is $\mathcal{L}$-harmonic and strictly positive in the complement of $B_d(x_j^n, r_j^n)$, hence in Ω_n. Moreover, $\lim_{z\to z_0} w(z) = 0$. Thus w is an $\mathcal{L}$-barrier for Ω_n at z_0, i.e. z_0 is $\mathcal{L}$-regular for Ω_n.

Let us now suppose that Ω is unbounded and put

$$O_n := \Omega \cap B_d(0, n), \quad n \in \mathbb{N}.$$

Since O_n is bounded, we can find an increasing sequence of bounded $\mathcal{L}$-regular open sets $(O_n^k)_{k\in\mathbb{N}}$ such that $\bigcup_{k\in\mathbb{N}} O_n^k = O_n$. Using the first part of the proof, we can choose O_n^k such that $O_n^k \subseteq O_{n+1}^k$ for every n and for every k. It follows that $O_n^k \subseteq O_m^m$ if $k, n \leq m$. Let us now put (see also Fig. 9.1)

$$\Omega_n = O_n^n.$$

Then Ω_n is bounded and $\mathcal{L}$-regular. Moreover,

$$\Omega = \bigcup_n O_n = \bigcup_n \Big(\bigcup_k O_n^k\Big) \subseteq \bigcup_n O_n^n \subseteq \Omega.$$

Hence $\Omega = \bigcup_n \Omega_n$, and the proof is complete. □

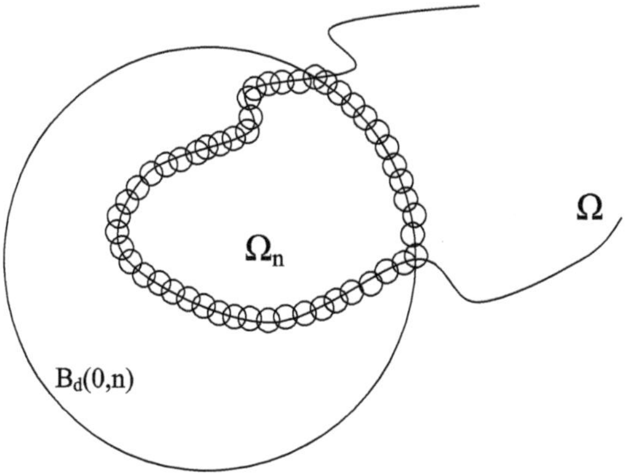

Fig. 9.1. Approximation of an open set by $\mathcal{L}$-regular open sets

This lemma, together with Theorem 9.1.2, page 426, immediately proves the following proposition.

Proposition 9.2.10 (Symmetry of the $\mathcal{L}$-Green function). *Let $\Omega \subseteq \mathbb{G}$ be an open set. Then*

$$G_\Omega(x, y) = G_\Omega(y, x) \qquad \text{for every } x, y \in \Omega.$$

In particular, for every fixed $y \in \Omega$, the function

$$x \mapsto G_\Omega(x, y)$$

is $\mathcal{L}$-harmonic in $\Omega \setminus \{y\}$.

Remark 9.2.11. From (9.3) and the above Proposition 9.2.10 we immediately get the following formula:

$$G_\Omega(x, y) = \Gamma(y^{-1} \circ x) - \int_{\partial\Omega} \Gamma(y^{-1} \circ \eta)\, \mathrm{d}\mu_x^\Omega(\eta) \qquad \forall\, x, y \in \Omega.$$

By collecting together some of the results of this section, we have the following characterizations of the $\mathcal{L}$-Green function, which we state for future reference and for the reading convenience.

Proposition 9.2.12. *Let $\Omega \subseteq \mathbb{G}$ be open, and let $x \in \Omega$ be fixed. Let us denote by h_x the greatest $\mathcal{L}$-harmonic minorant of $\Gamma(x^{-1} \circ \cdot)$ in Ω. Then the following facts hold:*

(i) *The $\mathcal{L}$-Green function $G_\Omega(x, y) = \Gamma(x^{-1} \circ y) - h_x(y)$ is a symmetric function, i.e. $G_\Omega(x, y) = G_\Omega(y, x)$ for all $x,\ y \in \Omega$. Moreover, G_Ω is continuous (in the extended sense) on $\Omega \times \Omega$, and the greatest $\mathcal{L}$-harmonic minorant of $G_\Omega(x, \cdot)$ in Ω is the null function.*

(ii) *It holds*

$$h_x = \sup \left\{ u \in \underline{\mathcal{S}}(\Omega) \,:\, u \le \Gamma(x^{-1} \circ \cdot) \ \text{on } \Omega \right\}.$$

(iii) *If Ω is a bounded domain, then $h_x = H^{\Omega}_{\Gamma(x^{-1}\circ\cdot)}$, in the sense of Perron-Wiener-Brelot, or equivalently*

$$h_x = \sup\Big\{u \in \underline{S}(\Omega) \,:\, \limsup_{z\to\zeta} u(z) \le \Gamma(x^{-1}\circ\zeta) \text{ for all } \zeta \in \partial\Omega\Big\}.$$

(iv) *If Ω is a $\mathcal{L}$-regular domain, then h_x is the solution (in the classical sense) to*

$$\mathcal{L}u = 0 \text{ in } \Omega \quad \text{and} \quad u = \Gamma(x^{-1}\circ\cdot) \text{ on } \partial\Omega. \tag{9.5}$$

(v) *An equivalent definition of the $\mathcal{L}$-Green function is the following one: G_Ω is a non-negative function on $\Omega\times\Omega$ such that (for every $x\in\Omega$) the function $G_\Omega(x,\cdot)$ is the sum of $\Gamma(x^{-1}\circ\cdot)$ plus a $\mathcal{L}$-harmonic function on Ω and, moreover, $G_\Omega(x,\cdot)$ does not exceed any other non-negative $\mathcal{L}$-superharmonic function on Ω which is the sum of $\Gamma(x^{-1}\circ\cdot)$ plus a $\mathcal{L}$-superharmonic function on Ω.*

9.3 Potentials of Radon Measures

Definition 9.3.1 (G_Ω-potential of a Radon measure). *Let $\Omega\subseteq\mathbb{G}$ be open and let G_Ω be its $\mathcal{L}$-Green function. Let μ be a Radon measure in Ω. The function*

$$G_\Omega * \mu : \Omega \to [0,\infty], \quad (G_\Omega*\mu)(x) := \int_\Omega G_\Omega(x,y)\,\mathrm{d}\mu(y)$$

is well defined and l.s.c. It is called the G_Ω-potential of μ.

The following theorem holds.

Theorem 9.3.2 (Subharmonicity of a G_Ω-potential). *Suppose Ω is connected. Then $G_\Omega*\mu \in \overline{S}(\Omega)$ if and only if there exists $x_0\in\Omega$ such that $(G_\Omega*\mu)(x_0) < \infty$.*

Proof. The "only if" part is trivial: actually, if $G_\Omega*\mu\in\overline{S}(\Omega)$, then

$$G_\Omega*\mu < \infty \text{ in a dense subset of } \Omega.$$

To prove the "if" part, by Corollary 8.2.3 (page 402), it is enough to check that $G_\Omega*\mu$ is super-mean. Given $x\in\Omega$ and $r>0$ such that $\overline{B_d(x,r)}\subseteq\Omega$, we have

$$\begin{aligned}
\mathrm{M}_r(G_\Omega*\mu)(x) &= \frac{m_d}{r^Q}\int_{B_d(x,r)} \Psi_{\mathcal{L}}(x^{-1}\circ y)\,(G_\Omega*\mu)(y)\,\mathrm{d}y\\
&= \frac{m_d}{r^Q}\int_{B_d(x,r)} \Psi_{\mathcal{L}}(x^{-1}\circ y)\Big(\int_\Omega G_\Omega(y,z)\,\mathrm{d}\mu(z)\Big)\,\mathrm{d}y\\
&= \frac{m_d}{r^Q}\int_\Omega\Big(\int_{B_d(x,r)} \Psi_{\mathcal{L}}(x^{-1}\circ y)\,G_\Omega(y,z)\,\mathrm{d}y\Big)\,\mathrm{d}\mu(z)\\
&\quad \text{(since } y\mapsto G_\Omega(y,z) \text{ is } \mathcal{L}\text{-superharmonic,}\\
&\quad \text{hence super-mean, see Example 7.2.7, page 390)}\\
&\le \int_\Omega G_\Omega(x,z)\,\mathrm{d}\mu(z) = (G_\Omega*\mu)(x).
\end{aligned}$$

Then $G_\Omega*\mu$ is super-mean, and the proof is complete. □

Corollary 9.3.3. *Let μ be a Radon measure in Ω such that*

$$\mu(\Omega) < \infty.$$

Then $G_\Omega * \mu \in \overline{\mathcal{S}}(\Omega)$.

Proof. Let $\overline{B_d(x_0, r)} \subseteq \Omega$. Since

$$\begin{aligned}\int_{B_d(x_0,r)} (G_\Omega * \mu)(x)\,\mathrm{d}x &= \int_\Omega \left(\int_{B_d(x_0,r)} G_\Omega(x, y)\,\mathrm{d}x \right) \mathrm{d}\mu(y) \\ &\le \mu(\Omega) \sup_{y\in\Omega} \int_{B_d(x_0,r)} G_\Omega(x, y)\,\mathrm{d}x < \infty,\end{aligned}$$

then $G_\Omega * \mu < \infty$ almost everywhere in $B_d(x_0, r)$. Then the assertion follows from the previous theorem. □

By using the Harnack inequality, Corollary 9.3.3 can be improved as follows.

Corollary 9.3.4. *Let $\Omega \subseteq \mathbb{G}$ be open and connected, and let μ be a Radon measure in Ω such that $K := \operatorname{supp}(\mu)$ is a compact subset of Ω. For every fixed $y_0 \in K$ and every bounded and connected open set U such that*

$$K \subseteq U, \quad \overline{U} \subseteq \Omega,$$

there exists a positive constant $C = C(U, y_0)$ such that

$$(G_\Omega * \mu)(x) \le C\, G_\Omega(x, y_0) \qquad \forall\, x \in \Omega \setminus U.$$

Proof. For every $x \in \Omega \setminus U$, the function $y \mapsto G_\Omega(x, y)$ is $\mathcal{L}$-harmonic and nonnegative in U. Then, by the Harnack inequality (Corollary 5.7.3, page 265), there exists a positive constant C independent of x, such that $G_\Omega(x, y) \le C\, G_\Omega(x, y_0)$ for every $y \in K$. As a consequence,

$$(G_\Omega * \mu)(x) = \int_K G_\Omega(x, y)\,\mathrm{d}\mu(y) \le C\, \mu(K)\, G_\Omega(x, y_0)$$

for every $x \in \Omega \setminus U$. □

Theorem 9.3.5 (Sub-Laplacian in $\mathcal{D}'$ of a G_Ω-potential). *Under the hypothesis of Theorem* 9.3.2, *we have*

$$\mathcal{L}(G_\Omega * \mu) = -\mu \quad \textit{in the weak sense of distributions.}$$

In particular, $G_\Omega * \mu$ is $\mathcal{L}$-harmonic in $\Omega \setminus \operatorname{supp}(\mu)$.

Proof. By Theorem 9.3.2 and Corollary 8.2.4 (page 402), $G_\Omega * \mu \in L^1_{\mathrm{loc}}(\Omega)$. Moreover, for every $\varphi \in C_0^\infty(\Omega)$,

$$\begin{aligned}\int_\Omega (G_\Omega * \mu)(x)\,\mathcal{L}\varphi(x)\,\mathrm{d}x &= \int_\Omega \left(\int_\Omega (\Gamma(y^{-1} \circ x) - h_y(x))\,\mathcal{L}\varphi(x)\,\mathrm{d}x \right) \mathrm{d}\mu(y) \\ (\text{since } \mathcal{L}h_y = 0) &= \int_\Omega \left(\int_\Omega \Gamma(y^{-1} \circ x)\,\mathcal{L}\varphi(x)\,\mathrm{d}x \right) \mathrm{d}\mu(y) \\ (\text{by Theorem 5.3.3}) &= -\int_\Omega \varphi(y)\,\mathrm{d}\mu(y).\end{aligned}$$

This proves the first part of the theorem. The second one follows from the hypoellipticity of $\mathcal{L}$. □

From this theorem and Corollary 9.3.4 we obtain a corollary that will be used very soon.

Corollary 9.3.6. *Let μ be a compactly supported Radon measure in Ω. Then*:

(i) $G_\Omega * \mu \in \overline{\mathcal{S}}(\Omega)$,
(ii) $G_\Omega * \mu$ *is $\mathcal{L}$-harmonic in* $\Omega \setminus \mathrm{supp}(\mu)$,
(iii) *if $v \in \underline{\mathcal{S}}(\Omega)$ and $v \le G_\Omega * \mu$, then $v \le 0$.*

Proof. (i) and (ii) directly follow from Corollary 9.3.3 and Theorem 9.3.5. To prove (iii), we assume that Ω is connected (this is not restrictive) and use Corollary 9.3.4. First of all, if $v \in \underline{\mathcal{S}}(\Omega)$ and $v \le G_\Omega * \mu$, there exists $h \in \mathcal{H}(\Omega)$ such that $v \le h \le G_\Omega * \mu$ (h is the greatest $\mathcal{L}$-harmonic minorant of $G_\Omega * \mu$). For a fixed $y_0 \in K := \mathrm{supp}(\mu)$ and a connected bounded open set $U \supseteq K, \overline{U} \subseteq \Omega$, we have

$$h(x) \le (G_\Omega * \mu)(x) \le G_\Omega(x, y_0) \qquad \forall\, x \in \Omega \setminus U.$$

Since $G_\Omega(\cdot, y_0)$ is $\mathcal{L}$-harmonic, hence continuous, in $\Omega \setminus \{y_0\}$, we have

$$\lim_{U \ni x \to \xi} (G_\Omega(x, y_0) - h(x)) = G_\Omega(\xi, y_0) - h(\xi) \ge 0$$

for every $\xi \in \partial U$. Moreover,

$$\lim_{U \setminus \{y_0\} \ni x \to y_0} (G_\Omega(x, y_0) - h(x)) = \infty.$$

Then, by the maximum principle, $G_\Omega(\cdot, y_0) \ge h$ in U. Summing up:

$$h \le G_\Omega(\cdot, y_0) \qquad \text{in } \Omega$$

so that, by the very definition of G_Ω, we get $h \le 0$. Hence $v \le 0$, and the proof is complete. □

If a G_Ω-potential is $\mathcal{L}$-superharmonic, then its $\mathcal{L}$-harmonic minorants are nonpositive constant functions. Indeed, the following theorem holds.

Theorem 9.3.7 ($\mathcal{L}$-harmonic minorants of a G_Ω-potential). *Let μ be a Radon measure in an open and connected set Ω such that $(G_\Omega * \mu)(x_0) < \infty$ for some $x_0 \in \Omega$. Let h be a $\mathcal{L}$-harmonic function in Ω such that*

$$h(x) \le (G_\Omega * \mu)(x) \quad \forall\, x \in \Omega. \tag{9.6}$$

Then $h \le 0$.

Proof. Let $\{K_n\}$ be a sequence of compact subsets of Ω such that

$$K_n \subseteq K_{n+1}, \qquad \bigcup_n K_n = \Omega.$$

For every $n \in \mathbb{N}$, we have

$$h \le G_\Omega * \mu = \mathbb{G}_\Omega * (\mu|_{K_n}) + \mathbb{G}_\Omega * (\mu|_{\Omega \setminus K_n}) =: v_n + w_n.$$

The functions v_n and w_n are non-negative and $\mathcal{L}$-superharmonic in Ω (see Theorem 9.3.2). Moreover, the greatest $\mathcal{L}$-harmonic minorant of v_n is the zero function (see Corollary 9.3.6). Then, by Proposition 6.6.3 (page 358), h is less than the greatest $\mathcal{L}$-harmonic minorant of w_n. In particular,

$$h \le w_n \qquad \text{in } \Omega \quad \forall\, n \in \mathbb{N}. \tag{9.7}$$

On the other hand, by the monotone convergence theorem, we infer

$$v_n \uparrow G_\Omega * \mu,$$

so that

$$w_n = G_\Omega * \mu - v_n \downarrow 0, \qquad \text{as } n \to \infty.$$

This, together with (9.7), implies $h \le 0$ and completes the proof. □

9.3.1 The Potentials Related to the Average Operators for $\mathcal{L}$

This section provides detailed computations of the $\mathcal{L}$-potentials of the measures naturally related to the mean integral operators $\mathcal{M}_r$ and M_r introduced in Sections 5.5 and 5.6 (pages 251 and 257, respectively, see in particular Theorems 5.5.4 and 5.6.1). For the sake of brevity, we rewrite our mean value formulas from those sections with a different notation.

Indeed, in order to *normalize* the mean value formulas, we make the following choice: fixed the sub-Laplacian $\mathcal{L}$ and its fundamental solution Γ, we take $d := \Gamma^{1/(2-Q)}$. Then, d is an $\mathcal{L}$-gauge (see Proposition 5.4.2, page 247), and by Theorem 5.5.6, page 256, the constant β_d appearing in the surface mean value formula (see (5.43n), page 256) equals 1. Moreover, it is immediately seen that with this choice of d, for every $y \in \partial B_d(x, r)$, we have

$$\frac{(Q-2)\,\beta_d}{r^{Q-1}}\, \mathcal{K}_{\mathcal{L}}(x, y) = \frac{|\nabla_{\mathcal{L}} \Gamma|^2(x^{-1} \circ y)}{\big|\nabla \Gamma(x^{-1} \circ \cdot)\big|(y)}.$$

As a consequence, the mean value formulas rewrite as follows:

If $u \in C^2(\Omega)$, $d := \Gamma^{1/(2-Q)}$ (where Γ is the fundamental solution for $\mathcal{L}$) and $\overline{B}_d(x, r) \subset \Omega$, it holds

$$u(x) = m_r[u](x) - \int_{B_d(x,r)} \left\{\Gamma(x^{-1} \circ y) - r^{2-Q}\right\} \mathcal{L}u(y)\, \mathrm{d}H^N(y),$$
$$u(x) = M_r[u](x) - \frac{Q}{r^Q} \int_0^r \rho^{Q-1} \left\{ \int_{B_d(x,\rho)} \left\{\Gamma(x^{-1} \circ y) - \rho^{2-Q}\right\} \mathcal{L}u(y)\, \mathrm{d}H^N(y) \right\} \mathrm{d}\rho,$$

where $\mathbf{k}(x, y) := |\nabla_{\mathcal{L}}\Gamma|^2(x^{-1} \circ y)/|\nabla \Gamma(x^{-1} \circ \cdot)|(y)$, $\mathbf{K} := |\nabla_{\mathcal{L}} d|^2$, *and*

$$\begin{aligned} m_r[u](x) &= \int_{\partial B_d(x,r)} u(y)\, \mathbf{k}(x, y)\, \mathrm{d}H^{N-1}(y), \\ M_r[u](x) &= \frac{Q(Q-2)}{r^Q} \int_{B_d(x,r)} u(y)\, \mathbf{K}(x^{-1} \circ y)\, \mathrm{d}H^N(y). \end{aligned} \tag{9.8}$$

We next give the notation which will be used in this section. If $x \in \mathbb{G}$ and $r > 0$ are fixed, we introduce the measures $\lambda_{x,r}$ and $\Lambda_{x,r}$ defined by

$$\begin{aligned} \mathrm{d}\lambda_{x,r}(y) &:= \chi_{\partial B_d(x,r)}(y)\, \mathbf{k}(x, y)\, \mathrm{d}H^{N-1}(y), \\ \mathrm{d}\Lambda_{x,r}(y) &:= \frac{Q(Q-2)}{r^Q} \chi_{B_d(x,r)}(y)\, \mathbf{K}(x, y)\, \mathrm{d}H^N(y), \end{aligned} \tag{9.9}$$

with $\mathbf{k}$ and $\mathbf{K}$ as above. We explicitly point out that, with this notations,

$$m_r[u](x) = \int u\, \mathrm{d}\lambda_{x,r} \quad \text{and} \quad M_r[u](x) = \int u\, \mathrm{d}\Lambda_{x,r}.$$

We are interested in computing the Γ-potentials $\Gamma * \lambda_{x,r}$ and $\Gamma * \Lambda_{x,r}$. We first need two lemmas.

Lemma 9.3.8. *With the above notation,* $\lambda_{x,r}$ *is equal to* $\mu_x^{B_d(x,r)}$ *(the* $\mathcal{L}$*-harmonic measure of* $B_d(x, r)$ *at* x*) prolonged to zero outside* $\partial B_d(x, r)$*.*

Proof. This is just Theorem 7.2.9, page 391. □

Lemma 9.3.9. *Let* Ω *be* $\mathcal{L}$*-regular and such that* $\partial\Omega$ *has vanishing Lebesgue measure. Then, for every* $x \in \Omega$*, we have*

$$(\Gamma * \mu_x^{\Omega})(z) = \begin{cases} \Gamma(z^{-1} \circ x) & \text{if } z \notin \Omega, \\ H^{\Omega}_{\Gamma(z^{-1}\circ \cdot)}(x) & \text{if } z \in \Omega, \end{cases}$$

where μ_x^{Ω} *denotes the measure obtained by prolonging* μ_x^{Ω} *to* 0 *outside* $\partial\Omega$*.*

The same assertion holds for a general $\mathcal{L}$-regular set (without the assumption that $\partial\Omega$ has vanishing Lebesgue measure). A proof in this case may be obtained by arguing as in (9.19) in the proof of Theorem 9.5.1 (page 445).

Proof. Let $x \in \Omega$ be fixed. With the notation in the assertion, for any $z \in \mathbb{G}$ we aim to consider

$$(\Gamma * \mu_x^\Omega)(z) = \int_{\partial\Omega} \Gamma(z^{-1} \circ y)\,\mathrm{d}\mu_x^\Omega(y).$$

If $z \notin \overline{\Omega}$, the map $y \mapsto \Gamma(z^{-1} \circ y)$ is $\mathcal{L}$-harmonic about $\overline{\Omega}$. Hence, by the definition of $\mathcal{L}$-harmonic measure, we have

$$\Gamma(z^{-1} \circ x) = \int_{\partial\Omega} \Gamma(z^{-1} \circ y)\,\mathrm{d}\mu_x^\Omega(y).$$

Let now $z \in \Omega$. Then the map $y \mapsto \Gamma(z^{-1} \circ y)$ is continuous on $\partial\Omega$, and the equality of

$$\int_{\partial\Omega} \Gamma(z^{-1} \circ y)\,\mathrm{d}\mu_x^\Omega(y) = H^\Omega_{\Gamma(z^{-1}\circ\cdot)}(x)$$

is the very definition of the latter.

We end by considering a fixed $z_0 \in \partial\Omega$. We set $J = \Gamma * \mu_x^\Omega$. Since $J \in \overline{\mathcal{S}}(\mathbb{G})$, by Theorem 8.2.10 (page 404) we have

$$M_r[J](z_0) \longrightarrow J(z_0) \quad \text{as } r \to 0.$$

Since the function

$$G_\Omega(z, x) = \Gamma(z^{-1} \circ x) - H^\Omega_{\Gamma(z^{-1}\circ\cdot)}(x)$$

and Γ are symmetric functions, we have

$$H^\Omega_{\Gamma(z^{-1}\circ\cdot)}(x) = H^\Omega_{\Gamma(x^{-1}\circ\cdot)}(z), \quad \text{whenever } x,\, z \in \Omega.$$

Then, since $B_d(z_0, r) \cap \partial\Omega$ has zero H^N-measure, we have (by applying what we proved above)

$$\begin{aligned} M_r[J](z_0) &= M_r[J\chi_\Omega](z_0) + M_r[J\chi_{\mathbb{G}\setminus\overline{\Omega}}](z_0) = M_r\big[H^\Omega_{\Gamma(x^{-1}\circ\cdot)}\,\chi_\Omega\big](z_0) \\ &\quad + M_r[\Gamma(x^{-1} \circ \cdot)\,\chi_{\mathbb{G}\setminus\overline{\Omega}}](z_0) \\ &= M_r[\Gamma(x^{-1} \circ \cdot)](z_0) - M_r[G_\Omega(x, \cdot)\chi_\Omega](z_0). \end{aligned}$$

Clearly, we have

$$M_r[\Gamma(x^{-1} \circ \cdot)](z_0) \longrightarrow \Gamma(x^{-1} \circ z_0) = \Gamma(z_0^{-1} \circ x) \quad \text{as } r \to 0.$$

On the other hand, since $G_\Omega(x, z)$ vanishes with continuity as $\Omega \ni z \to z_0 \in \partial\Omega$, and $M_r[1] \equiv 1$, we obtain

$$M_r[G_\Omega(x, \cdot)\chi_\Omega](z_0) \longrightarrow 0 \quad \text{as } r \to 0.$$

This completes the proof. □

Theorem 9.3.10. *For every $x \in \mathbb{G}$ and $r > 0$, we have* (see Fig. 9.2)

$$(\Gamma * \lambda_{x,r})(z) = m_r[\Gamma(z^{-1} \circ \cdot)](x) = \begin{cases} \Gamma(z^{-1} \circ x) & \text{if } z \notin B_d(x,r), \\ r^{2-Q} & \text{if } z \in B_d(x,r). \end{cases} \tag{9.10a}$$

$$(\Gamma * \Lambda_{x,r})(z) = M_r[\Gamma(z^{-1} \circ \cdot)](x) = \begin{cases} \Gamma(z^{-1} \circ x) & \text{if } z \notin B_d(x,r), \\ \dfrac{2-Q}{2}\dfrac{d^2(x,z)}{r^Q} + \dfrac{Q}{2} r^{2-Q} & \text{if } z \in B_d(x,r). \end{cases} \tag{9.10b}$$

In particular, the Γ-potentials $\Gamma * \lambda_{x,r}$, $\Gamma * \Lambda_{x,r}$ are continuous on $\mathbb{G}$. Moreover,

$$(\Gamma * \lambda_{x,r})(z),\ m_r[\Gamma(z^{-1} \circ \cdot)](x), \quad (\Gamma * \Lambda_{x,r})(z),\ M_r[\Gamma(z^{-1} \circ \cdot)](x)$$

are symmetric functions in x and z.

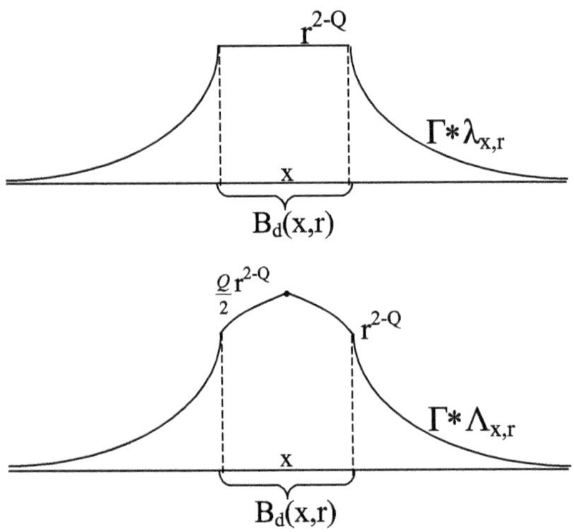

Fig. 9.2. The functions in Theorem 9.3.10

Proof. By Lemma 9.3.8, $\lambda_{x,r}$ is the trivial prolongation of $\mu_x^{B_d(x,r)}$ outside $\partial B_d(x,r)$. Then, we can apply Lemma 9.3.9 for $\Omega = B_d(x,r)$ in order to compute $\Gamma * \lambda_{x,r}$. We have to prove that

$$H^{B_d(x,r)}_{\Gamma(z^{-1}\circ\cdot)}(x) = r^{2-Q} \quad \text{for every } z \in B_d(x,r).$$

By the symmetry of the $\mathcal{L}$-Green function for $B_d(x,r)$, we have

$$H^{B_d(x,r)}_{\Gamma(z^{-1}\circ\cdot)}(x) = H^{B_d(x,r)}_{\Gamma(x^{-1}\circ\cdot)}(z).$$

Now, the function of z in the above right-hand side is the $\mathcal{L}$-harmonic function on $B_d(x,r)$ which equals $\Gamma(x^{-1}\circ\cdot)$ on $\partial B_d(x,r)$. Since $\Gamma(x^{-1}\circ\cdot)\equiv r^{2-Q}$ on $\partial B_d(x,r)$ (and the constants are $\mathcal{L}$-harmonic), (9.10a) follows.

Finally, we have

$$(\Gamma * \Lambda_{x,r})(z) = M_r[\Gamma(z^{-1}\circ\cdot)](x) = \frac{Q}{r^Q}\int_0^r \rho^{Q-1}\, m_\rho[\Gamma(z^{-1}\circ\cdot)](x)\,\mathrm{d}\rho.$$

If $z\notin B_d(x,r)$, then $z\notin B_d(x,\rho)$ also for any $0<\rho<r$, whence, by (9.10a),

$$(\Gamma * \Lambda_{x,r})(z) = \frac{Q}{r^Q}\int_0^r \rho^{Q-1}\,\Gamma(z^{-1}\circ x)\,\mathrm{d}\rho = \Gamma(z^{-1}\circ x).$$

On the other hand, suppose $z\in B_d(x,r)$. Arguing as above, we have

$$\begin{aligned}
&\frac{Q}{r^Q}\int_0^r \rho^{Q-1}\, m_\rho[\Gamma(z^{-1}\circ\cdot)](x)\,\mathrm{d}\rho\\
&= \frac{Q}{r^Q}\int_0^{d(x,z)} \rho^{Q-1}\,\Gamma(z^{-1}\circ x)\,\mathrm{d}\rho + \frac{Q}{r^Q}\int_{d(x,z)}^r \rho^{Q-1}\,\rho^{2-Q}\,\mathrm{d}\rho\\
&= \Gamma(z^{-1}\circ x)\,\frac{d^Q(x,z)}{r^Q} + \frac{Q}{2\,r^Q}(r^2-d^2(x,z)).
\end{aligned}$$

Thus, (9.10b) is proved. The last assertion of the theorem now follows from (9.10a) and (9.10b), since $\Gamma(z^{-1}\circ x)=\Gamma(x^{-1}\circ z)$ and $d(x,z)=d(z,x)$. □

For an application of Theorem 9.3.10, see Exercise 5 at the end of the chapter.

The next result gives another consequence of Theorem 9.3.10. Indeed, we prove that the mean integrals of a Γ-potential are $\mathcal{L}$-superharmonic continuous functions.

Theorem 9.3.11. *Let μ be a Radon measure in $\mathbb{G}$ and $r>0$. We suppose that $\Gamma*\mu\not\equiv\infty$. Then the map $x\mapsto m_r[\Gamma*\mu](x)$ belongs to $\overline{\mathcal{S}}(\mathbb{G})\cap C(\mathbb{G},\mathbb{R})$. Moreover, the map*

$$(x,r)\mapsto m_r[\Gamma*\mu](x)$$

is lower semicontinuous. The same assertion holds for $M_r[\Gamma*\mu]$.

Proof. (i). First we prove that $m_r[u]$ and $M_r[u]$ are finite-valued and continuous for every $u\in\overline{\mathcal{S}}(\mathbb{G})$. We fix $x\in\mathbb{G}$. Then

$$|M_r[u](x)| \le \|\mathbf{K}\|_\infty \int_{B_d(x,r)} |u(y)|\,\mathrm{d}H^N(y) < \infty,$$

since $u\in L^1_{\mathrm{loc}}(\mathbb{G})$. Let $R>r$ and set $\Omega=B_d(x,R)$. Set also (see Definition 9.4.1 in the next section)

$$\mu := (\mu[u])|_{\overline{\Omega}}.$$

Then, by Theorem 9.3.5, we have $\mathcal{L}(u - \Gamma * \mu) = 0$ in the distributional sense in Ω, so that[2] (by the hypoellipticity of $\mathcal{L}$) $u = h + \Gamma * \mu$ on Ω for a suitable $h \in \mathcal{H}(\Omega)$. Hence (being $\partial B_d(x,r) \subset B_d(x,R)$)

$$m_r[u](x) = m_r[h](x) + m_r[\Gamma * \mu](x) = h(x) + \int_\Omega m_r[\Gamma(z^{-1} \circ \cdot)](x)\, \mathrm{d}\mu(z)$$

$$\text{(by (9.10a))} = h(x) + \int_\Omega \min\{\Gamma(z^{-1} \circ x), r^{2-Q}\}\, \mathrm{d}\mu(z),$$

which is finite, since μ is a Radon measure and Ω is bounded. Moreover, the above representation of $m_r[u](x)$ also proves that $m_r[u]$ is continuous. Another argument of dominated convergence then shows that $M_r[u](x) = \frac{Q}{r^Q} \int_0^r \rho^{Q-1} m_\rho[u](x)\, \mathrm{d}\rho$ is continuous at x.

(ii). We now observe a simple fact (see also Theorem 8.2.20, page 410): let (A, λ) be an arbitrary measure space, and let $\{f_x\}_{x \in \mathbb{G}}$ be a family of non-negative λ-measurable functions on A such that, for every fixed $z \in A$, the function $x \mapsto f_x(z)$ is in $\overline{\mathcal{S}}(\mathbb{G})$. Then the integral function

$$F(x) = \int_A f_x(z)\, \mathrm{d}\lambda(z)$$

is in $\overline{\mathcal{S}}(\mathbb{G})$ too (unless it is $\equiv \infty$). Indeed, by Theorem 8.1.3, it is enough to prove that F is $\mathcal{L}$-supermean. By Tonelli's theorem we have

$$m_r[F](x) = \int_A m_r[y \mapsto f_y(z)](x)\, \mathrm{d}\lambda(z) \le \int_A f_x(z)\, \mathrm{d}\lambda(z) = F(x)$$

(since $y \mapsto f_y(z)$ is $\mathcal{L}$-supermean), i.e. F is $\mathcal{L}$-supermean.

(iii). Tonelli's theorem also ensures that

$$m_r[\Gamma * \mu](x) = \int_{\mathbb{G}} m_r[\Gamma(z^{-1} \circ \cdot)](x)\, \mathrm{d}\mu(z)$$

and $M_r[\Gamma * \mu](x) = \int_{\mathbb{G}} M_r[\Gamma(z^{-1} \circ \cdot)](x)\, \mathrm{d}\mu(z)$. Consequently, by (ii), the theorem is proved if we show that the maps

$$x \mapsto m_r[\Gamma(z^{-1} \circ \cdot)](x), \quad M_r[\Gamma(z^{-1} \circ \cdot)](x)$$

are in $\overline{\mathcal{S}}(\mathbb{G})$ for any $z \in \mathbb{G}$.

(iv). Now, the $\mathcal{L}$-superharmonicity of $m_r[\Gamma(z^{-1} \circ \cdot)]$ follows from (9.10a). Finally, we have

$$M_r[\Gamma(z^{-1} \circ \cdot)](x) = \frac{Q}{r^Q} \int_0^r \rho^{Q-1}\, m_\rho[\Gamma(z^{-1} \circ \cdot)](x)\, \mathrm{d}\rho.$$

Consequently, the $\mathcal{L}$-superharmonicity of $M_r[\Gamma(z^{-1} \circ \cdot)]$ follows by an application of (ii) for $A = [0, r]$ and $\mathrm{d}\lambda(\rho) = \frac{Q}{r^Q} \rho^{Q-1}\, \mathrm{d}\rho$, and making use of what we proved at the beginning of step (iv). The semicontinuity of $(x, r) \mapsto m_r[\Gamma * \mu](x)$ is a standard consequence of Fatou's lemma. □

[2] The reader has certainly realized that we have just proved a Riesz representation type result: we investigate the topic in details in Section 9.4.

Corollary 9.3.12. *Let $u \in \overline{\mathcal{S}}(\mathbb{G})$ and $r > 0$. Then $m_r[u]$, $M_r[u]$ belong to $\overline{\mathcal{S}}(\mathbb{G}) \cap C(\mathbb{G}, \mathbb{R})$.*

Proof. See (i) in the proof of Theorem 9.3.11. □

9.4 Riesz Representation Theorems for $\mathcal{L}$-subharmonic Functions

Definition 9.4.1 ($\mathcal{L}$-Riesz measure). *Let u be an $\mathcal{L}$-subharmonic function in an open set $\Omega \subseteq \mathbb{G}$. By Theorem* 8.2.11 *(page* 405*), $u \in L^1_{\mathrm{loc}}(\Omega)$ and $\mathcal{L}u \geq 0$ in the weak sense of distributions. Then*[3] *there exists a Radon measure μ in Ω such that*

$$\mathcal{L}u = \mu \tag{9.11}$$

in the weak sense of distributions. The measure μ will be called the $\mathcal{L}$-Riesz measure of u. *If u is $\mathcal{L}$-superharmonic in Ω, the $\mathcal{L}$-Riesz measure related to $-u$ will be referred to as the $\mathcal{L}$*-Riesz measure of u. *In this case, it holds $\mathcal{L}u = -\mu$, in the weak sense of distributions.*

With reference to the above definition, we shall sometimes also write $\mu[u]$ or μ_u instead of μ.

Example 9.4.2. If $u = \Gamma$, then $\mu[\Gamma] = \mathrm{Dirac}_0$, the Dirac mass supported at $\{0\}$. Indeed, $\mathcal{L}(-\Gamma) = \mathrm{Dirac}_0$, for (see (5.14), page 236)

$$\int_{\mathbb{G}} (-\Gamma)\, \mathcal{L}\varphi = \varphi(0) = \int_{\mathbb{G}} \mathrm{Dirac}_0\, \varphi \qquad \forall \varphi \in C_0^\infty(\mathbb{G}).$$

When $\mu[u]$ is compactly supported, a representation theorem easily follows.

Theorem 9.4.3 (Riesz representation. I). *Let $\Omega \subseteq \mathbb{G}$ be open, and let $u \in \underline{\mathcal{S}}(\Omega)$. Let μ be the $\mathcal{L}$-Riesz measure of u. Assume that* $\mathrm{supp}(\mu)$ *is a compact subset of Ω. Then there exists an $\mathcal{L}$-harmonic function h in Ω satisfying the identity*

$$u = h - G_\Omega * \mu \quad \text{in } \Omega. \tag{9.12}$$

Proof. By Corollary 9.3.3 and Theorem 9.3.5, $v := G_\Omega * \mu$ is $\mathcal{L}$-superharmonic in Ω, and $\mathcal{L}v = -\mu$ in the weak sense of distributions. It follows that $\mathcal{L}(u + v) = 0$ in Ω in the weak sense of distributions. Since $\mathcal{L}$ is hypoelliptic, there exists a function h, $\mathcal{L}$-harmonic in Ω, such that $h(x) = u(x) + v(x)$ almost everywhere in Ω. As a consequence, for every $x \in \Omega$ and for every $r < \mathrm{dist}_d(x, \partial\Omega)$

[3] Here we used the following result. *Given a linear map $L : C_0^\infty(\Omega) \to \mathbb{R}$ such that $L(\varphi) \geq 0$ whenever $\varphi \geq 0$, there exists a Radon measure μ on Ω such that $L(\varphi) = \int \varphi \, \mathrm{d}\mu$ for every $\varphi \in C_0^\infty(\Omega)$.* For a proof of this result, it suffices to rerun the proof of the classical Riesz representation theorem of positive functionals on C_0 as presented, e.g. in [Rud87].

$$\mathrm{M}_r(u)(x) = -\mathrm{M}_r(v)(x) + h(x),$$

where M_r is the solid average operator (5.50f) (page 259). Here we used the $\mathcal{L}$-harmonicity of h to write h in place of $\mathrm{M}_r(h)$. Letting r tend to zero in the last identity and using Theorem 8.2.11-(ii) (page 405), we get

$$u(x) = -v(x) + h(x) \quad \forall x \in \Omega.$$

This completes the proof. □

For the future reference, we explicitly show the following theorem, which can be proved as Theorem 9.4.3.

Theorem 9.4.4 (Riesz representation. II). *Let $\Omega \subseteq \mathbb{G}$ be open, and let $u \in \underline{S}(\Omega)$. Let μ be the $\mathcal{L}$-Riesz measure of u. Then, for every bounded open set Ω_1 such that $\overline{\Omega_1} \subset \Omega$, there exists a function h, $\mathcal{L}$-harmonic in Ω_1, satisfying the identity*

$$u(x) = -\int_{\overline{\Omega_1}} \Gamma(y^{-1} \circ x)\, \mathrm{d}\mu(y) + h(x) \quad \forall x \in \Omega_1. \tag{9.13}$$

Proof. The function

$$v(x) := -\int_{\overline{\Omega_1}} \Gamma(y^{-1} \circ x)\, \mathrm{d}\mu(y), \quad x \in \mathbb{G},$$

is $\mathcal{L}$-subharmonic in $\mathbb{G}$ and satisfies $\mathcal{L}v = \mu|_{\overline{\Omega_1}}$ in the weak sense of distributions. Therefore, $\mathcal{L}(u - v) = 0$ in $\mathcal{D}'(\Omega_1)$. Then, just proceeding as in the proof of the previous theorem, we show the existence of an $\mathcal{L}$-harmonic function in Ω_1 such that $u = v + h$ in Ω_1. □

Lemma 9.4.5. *Let $u \in \overline{S}(\Omega)$, let μ be its $\mathcal{L}$-Riesz measure and let $K \subseteq \Omega$ be compact. There exists $w \in \overline{S}(\Omega)$ such that*

$$u = G_\Omega * (\mu|_K) + w \qquad \textit{in } \Omega.$$

Proof. Let $H \subseteq \Omega$ be compact and such that $K \subseteq \mathrm{Int}(H)$. Let $k \in \mathcal{H}(\mathrm{Int}(K))$ and $h \in \mathcal{H}(\mathrm{Int}(H))$ be such that

$$u = h + G_\Omega * (\mu|_H) \quad \text{in } \mathrm{Int}(H).$$

Define

$$w : \Omega \to]-\infty, \infty], \ w := \begin{cases} h + G_\Omega * (\mu|_{H \setminus K}) & \text{in } \mathrm{Int}(H), \\ u - G_\Omega * (\mu|_K) & \text{in } \Omega \setminus K. \end{cases}$$

This definition is well-posed since the function $G_\Omega * (\mu|_K)$ is $\mathcal{L}$-harmonic in $\mathrm{Int}(H) \setminus K$, hence real-valued, and

$$h + G_\Omega * (\mu|_{H\setminus K}) = h + G_\Omega * (\mu|_H) - G_\Omega * (\mu|_K) = u - G_\Omega * (\mu|_K).$$

Moreover, w is $\mathcal{L}$-superharmonic in Ω since it is $\mathcal{L}$-superharmonic in $\mathrm{Int}(H)$ and in $\Omega \setminus K$, and $\Omega = \mathrm{Int}(H) \cup (\Omega \setminus K)$. □

Theorem 9.4.6 (Superharmonicity of $G_\Omega * \mathcal{L}u$). *Let $u \in \overline{\mathcal{S}}(\Omega)$, and let μ be its $\mathcal{L}$-Riesz measure. Then*

$$G_\Omega * \mu \in \overline{\mathcal{S}}(\Omega) \tag{9.14}$$

if and only if there exists $v \in \underline{\mathcal{S}}(\Omega)$ such that

$$v \le u.$$

Proof. We first prove the "if" part and assume $u \ge v$ in Ω with $v \in \underline{\mathcal{S}}(\Omega)$. Let $(K_n)_{n\in\mathbb{N}}$ be an increasing sequence of compact sets such that $\cup_n K_n = \Omega$. Let us put $\mu_n = \mu|_{K_n}$. By Lemma 9.4.5, there exists $w_n \in \overline{\mathcal{S}}(\Omega)$ such that $u = w_n + G_\Omega * \mu_n$. The hypothesis implies $0 \le u - v = (w_n - v) + G_\Omega * \mu_n$ in Ω, so that

$$G_\Omega * \mu_n \ge v - w_n.$$

By Corollary 9.3.6, we have $v - w_n \le 0$ in Ω. Hence

$$u - v \ge G_\Omega * \mu_n \qquad \text{in } \Omega \quad \forall\, n \in \mathbb{N}.$$

Letting n tend to infinity, we get $u - v \ge G_\Omega * \mu$. Since $u - v \in \overline{\mathcal{S}}(\Omega)$, this implies (9.14). Vice versa, assume (9.14) is true. Then, by Theorem 9.3.5,

$$\mathcal{L}(G_\Omega * \mu - u) = 0 \text{ in } \Omega, \text{ in the weak sense of distributions.}$$

Since $\mathcal{L}$ is hypoelliptic, there exists a function $h \in \mathcal{H}(\Omega)$ such that $G_\Omega * \mu = u + h$ a.e. in Ω, hence everywhere (see Exercise 2 of Chapter 8, page 422). Since $G_\Omega * \mu \ge 0$, we have $u \ge -h$, and the proof is complete. □

Theorem 9.4.7 (The Riesz representation). *Let $u \in \overline{\mathcal{S}}(\Omega)$, and let μ be the $\mathcal{L}$-Riesz measure of u. The following statements are equivalent*:

(i) *there exists $h \in \mathcal{H}(\Omega)$ such that*

$$u = G_\Omega * \mu + h \qquad \text{in } \Omega, \tag{9.15}$$

(ii) *there exists $v \in \underline{\mathcal{S}}(\Omega)$ such that $v \le u$ in Ω,*
(iii) *every connected component of Ω contains a point x_0 such that*

$$(G_\Omega * \mu)(x_0) < \infty.$$

Moreover, (9.15) *holds with $h \in \mathcal{H}(\Omega)$ if and only if h is the greatest $\mathcal{L}$-harmonic minorant of u in Ω.*

Proof. (i) ⇒ (ii). If (9.15) holds, then h is an $\mathcal{L}$-harmonic minorant of u (hence an $\mathcal{L}$-subharmonic function), since $G_\Omega * \mu \ge 0$.

(ii) ⇔ (iii). This follows from Theorems 9.4.6 and 9.3.2.

(iii) ⇒ (i). It is not restrictive to assume that Ω is connected. Let $x_0 \in \Omega$ be such that $(G_\Omega * \mu)(x_0) < \infty$, and let $\{\Omega_n\}_{n\in\mathbb{N}}$ be a sequence of bounded open sets such that

$$\overline{\Omega_n} \subset \Omega_{n+1}, \quad \overline{\Omega_{n+1}} \subset \Omega, \quad \bigcup_{n \in \mathbb{N}} \Omega_n = \Omega.$$

Define

$$\mu_n := \mu|_{\overline{\Omega_n}}, \qquad n \in \mathbb{N}.$$

Then, by the representation Theorem 9.4.3, there exists an $\mathcal{L}$-harmonic function h_n such that

$$u(x) = (G_\Omega * \mu_n)(x) + h_n(x) \quad \forall\, x \in \Omega_n, \quad \forall\, n \in \mathbb{N}. \tag{9.16}$$

Since $G_\Omega * \mu_n \nearrow G_\Omega * \mu$, we have

$$h_n(x) \geq h_{n+k}(x) \geq u(x) - (G_\Omega * \mu)(x) \quad \forall\, x \in \Omega_n, \quad \forall\, n,\, k \in \mathbb{N}.$$

On the other hand, by Theorem 9.3.2, $G_\Omega * \mu \in \overline{\mathcal{S}}(\Omega)$. Then, keeping in mind that $u(x) > -\infty$ for every $x \in \Omega$, for any $n \in \mathbb{N}$, we have

$$\inf_k h_{n+k} > -\infty \quad \text{in a dense subset of } \Omega_n.$$

By using Theorem 5.7.10 (page 268), we infer the existence of a function $h : \Omega \to \mathbb{R}$, $\mathcal{L}$-harmonic in Ω, such that

$$h(x) = \lim_{k \to \infty} h_{n+k}(x) \quad \forall\, x \in \Omega_n, \quad \forall\, n \in \mathbb{N}.$$

Then (9.15) follows from (9.16).

We are left with the proof of the second part of the theorem. Assume (9.15) holds with $h \in \mathcal{H}(\Omega)$. Then, if $k \in \mathcal{H}(\Omega)$ and $k \leq u$, we have $k - h \leq G_\Omega * \mu$, so that, by Theorem 9.3.7, we have $k - h \leq 0$, i.e. $k \leq h$. Vice versa, assume h is the greatest $\mathcal{L}$-harmonic minorant of u. Then, by (ii), there exists $k \in \mathcal{H}(\Omega)$ such that $u = G_\Omega * \mu + k$. This implies that k is the greatest $\mathcal{L}$-harmonic minorant of u, i.e. $k = h$. Thus $u = G_\Omega * \mu + h$. The proof is complete. □

Corollary 9.4.8 (Riesz representation in space). *Let $u \in \underline{\mathcal{S}}(\mathbb{G})$ be such that $U := \sup u < \infty$. Then*

$$u = -\Gamma * \mu + U,$$

where μ is the $\mathcal{L}$-Riesz measure of u.

Proof. Since U is an $\mathcal{L}$-harmonic majorant of u, by the previous theorem we have $u = -\Gamma * \mu + h$, where h is the least $\mathcal{L}$-harmonic majorant of u. Then $h \leq U$ so that, by the Liouville Theorem 5.8.1 (page 269) $h \equiv U_0$, a real constant. Hence $u = -\Gamma * \mu + U_0$. As a consequence $U = \sup u \leq U_0 \leq U$, that is, $U = U_0$. The proof is complete. □

We remark that a stronger version of the previous corollary will be provided in Theorem 9.6.1, page 451.

Corollary 9.4.9 (Riesz representation. III). *Let $u \in \underline{\mathcal{S}}(\Omega)$ be such that $\mathcal{L}u = 0$ outside a compact set $K \subset \Omega$. Then there exists an $\mathcal{L}$-harmonic function h in Ω such that*

$$u = -G_\Omega * \mu + h \quad \text{in } \Omega.$$

Proof. Since $\operatorname{supp}(\mu) \subseteq K$, we have $G_\Omega * \mu \in \overline{S}(\Omega)$. Then $G_\Omega * \mu < \infty$ in a dense subset of Ω, and the assertion follows from Theorem 9.4.7. □

For the future references, we explicitly write the following consequence of Theorems 9.3.2 and 9.3.5.

Corollary 9.4.10 (Riesz representation. IV). *Let $u \in \underline{S}(\Omega)$, and let μ be the $\mathcal{L}$-Riesz measure of u. Assume $(\Gamma * \mu)(x_0) < \infty$ at some point $x_0 \in \mathbb{G}$. Then there exists an $\mathcal{L}$-harmonic function h in Ω such that*

$$u(x) = -(\Gamma * \mu)(x) + h(x) \quad \forall x \in \Omega. \tag{9.17}$$

Proof. It is sufficient to note that $\Gamma * \mu \in \overline{S}\mathbb{G}$ and that $\mathcal{L}(u + \Gamma * \mu) = 0$ in Ω. □

9.5 The Poisson–Jensen Formula

The next theorem, when $\mathcal{L} = \Delta$ is the classical Laplace operator, will give back the classical Poisson–Jensen formula (see, e.g. [HK76, Theorem 3.14]). An improved version of Theorem 9.5.1 (removing the hypothesis of $\mathcal{L}$-regularity of Ω) will be given in Section 11 (see Theorem 11.7.6, page 518).

Theorem 9.5.1 (Poisson–Jensen's formula). *Let U, Ω be open subsets of $\mathbb{G}$, $\overline{\Omega} \subset U$ and Ω be $\mathcal{L}$-regular. Let $u \in \underline{S}(U)$ and $\mu = \mathcal{L}u$ be its $\mathcal{L}$-Riesz measure. Then*

$$u(x) = \int_{\partial\Omega} u(y)\, \mathrm{d}\mu_x^\Omega(y) - \int_\Omega G_\Omega(y, x)\, \mathrm{d}\mu(y), \quad x \in \Omega. \tag{9.18}$$

Here G_Ω is the $\mathcal{L}$-Green function of Ω and μ_x^Ω is the $\mathcal{L}$-harmonic measure related to Ω.

Proof. Let O be a bounded open set such that $\overline{\Omega} \subset O \subset \overline{O} \subset U$. By the Riesz representation Theorem 9.4.4, there exists an $\mathcal{L}$-harmonic function h in O such that

$$u(x) = -\int_{\overline{O}} \Gamma(y^{-1} \circ x)\, \mathrm{d}\mu(y) + h(x) =: v(x) + h(x) \qquad \forall x \in O.$$

We have $v \in \underline{S}(\mathbb{G})$ and $\mathcal{L}v = \mu|_{\overline{O}}$. Then, since

$$h(x) = \int_{\partial\Omega} h(y)\, \mathrm{d}\mu_x^\Omega(y) \quad \forall x \in \Omega,$$

it suffices to prove (9.18) with u replaced by v. We can also suppose that $v(x) > -\infty$. Indeed, if $v(x) = -\infty$, then $u(x) = -\infty$ and

$$\int_{\Omega} G_{\Omega}(y,x)\,d\mu(y) \geq \int_{\Omega} \Gamma(y^{-1}\circ x)\,d\mu(y) = -u(x) - h(x) = \infty.$$

Moreover, since $u \in \underline{S}(U)$, the function $x \mapsto \int_{\partial\Omega} u(y)\,d\mu_x^{\Omega}(y)$ is $\mathcal{L}$-harmonic, hence real-valued, in Ω (see Exercise 18 of Chapter 6). Thus, in this case, (9.18) trivially holds.

Let us fix $x \in \Omega$. We have

$$\int_{\partial\Omega} v(y)\,d\mu_x^{\Omega}(y) = -\int_{\overline{O}}\left(\int_{\partial\Omega} \Gamma(z^{-1}\circ y)\,d\mu_x^{\Omega}(y)\right)d\mu(z).$$

The crucial part of the proof is to show that

$$\int_{\partial\Omega} \Gamma(z^{-1}\circ y)\,d\mu_x^{\Omega}(y) = \begin{cases} \Gamma(z^{-1}\circ x), & z \in \overline{O}\setminus\Omega, \\ \Gamma(z^{-1}\circ x) - G_{\Omega}(z,x), & z \in \Omega. \end{cases} \tag{9.19}$$

With (9.19) at hand, and keeping in mind the assumption $v(x) = -\infty$, which implies that $z \mapsto \Gamma(z^{-1}\circ x)$ is μ-summable, we get the assertion. Indeed,

$$\begin{aligned} \int_{\partial\Omega} v(y)\,d\mu_x^{\Omega}(y) &= -\int_{\overline{O}} \Gamma(z^{-1}\circ x)\,d\mu(z) + \int_{\Omega} G_{\Omega}(z,x)\,d\mu(z) \\ &= v(x) + \int_{\Omega} G_{\Omega}(z,x)\,d\mu(z). \end{aligned}$$

Then, it remains to prove (9.19). If $z \in \overline{O}\setminus\overline{\Omega}$, the function $\Gamma(z^{-1}\circ\cdot)$ is harmonic in $O\setminus\{z\}$, so that

$$\Gamma(z^{-1}\circ x) = \int_{\partial\Omega} \Gamma(z^{-1}\circ y)\,d\mu_x^{\Omega}(y).$$

If $z \in \Omega$, the function $\Gamma(z^{-1}\circ\cdot)$ is continuous in $\partial\Omega$, hence the solution h_z to the Dirichlet problem

$$\begin{cases} \mathcal{L}h(x) = 0, & x \in \Omega, \\ h(y) = \Gamma(z^{-1}\circ y), & y \in \partial\Omega \end{cases}$$

is given by

$$h_z(x) := \int_{\partial\Omega} \Gamma(z^{-1}\circ y)\,d\mu_x^{\Omega}(y).$$

Then, by the definition of the $\mathcal{L}$-Green function,

$$\int_{\partial\Omega} \Gamma(z^{-1}\circ y)\,d\mu_x^{\Omega}(y) = h_z(x) = \Gamma(z^{-1}\circ x) - G_{\Omega}(z,x).$$

Finally, we fix $z_0 \in \partial\Omega$. Let us prove that

$$\int_{\partial\Omega} \Gamma(y^{-1}\circ z_0)\,d\mu_x^{\Omega}(y) = \Gamma(x^{-1}\circ z_0). \tag{9.20}$$

Since $\Gamma(x^{-1} \circ z_0) = \Gamma(z_0^{-1} \circ x)$, this will give (9.19) in the case $z_0 \in \partial\Omega$. Let us define

$$w(z) := \Gamma(x^{-1} \circ z) - \int_{\partial\Omega} \Gamma(y^{-1} \circ z)\, \mathrm{d}\mu_x^{\Omega}(y), \quad z \in \mathbb{G} \setminus \{x\}.$$

The function w is $\mathcal{L}$-subharmonic in $\mathbb{G} \setminus \{x\}$ and

$$\limsup_{z \notin \partial\Omega,\, z \to \zeta} w(z) = 0 \quad \forall \zeta \in \partial\Omega. \tag{9.21}$$

Indeed, (9.19) holds in $O \setminus \partial\Omega$ and $G_\Omega(z, x) \to 0$ as $z \to \zeta$ from inside of Ω, since Ω is $\mathcal{L}$-regular and G_Ω is symmetric. In order to prove (9.20), we have to show that $w \equiv 0$ on $\partial\Omega$. First of all, we observe that

$$w(\zeta) \geq \limsup_{z \to \zeta} w(z) \geq \limsup_{z \notin \partial\Omega,\, z \to \zeta} w(z) = 0 \quad \forall \zeta \in \partial\Omega.$$

Suppose, by contradiction, that $w > 0$ somewhere in $\partial\Omega$. Then we have $\max_{\partial\Omega} w > 0$. From (9.21) it follows that there exists an open set $V \subseteq O$ such that $\partial\Omega \subset V$, $x \notin V$ and $\max_V w = \max_{\partial\Omega} w$. Let $\zeta_0 \in \partial\Omega$ be such that $w(\zeta_0) = \max_V w$. Since w is $\mathcal{L}$-subharmonic in V, by Theorems 8.1.2 and 8.2.1 (pages 398 and 401), $w \equiv w(\zeta_0)$ in the connected component of V containing ζ_0. Thus,

$$\limsup_{z \notin \partial\Omega,\, z \to \zeta_0} w(z) = w(\zeta_0) = \max_V w > 0,$$

in contradiction with (9.21). This completes the proof of (9.19). □

If, in the previous theorem, we take $\Omega = B_d(x, r)$, we obtain an extension of the mean value formulas (5.45) and (5.51) (pages 256 and 259) to the $\mathcal{L}$-subharmonic functions.

Theorem 9.5.2 (Mean value formulas for $\mathcal{L}$-subharmonic functions). *Let $\Omega \subseteq \mathbb{G}$ be open, and let $u \in \underline{S}(\Omega)$. Then, for every $x \in \Omega$ and $r > 0$ such that*

$$\overline{B_d(x, r)} \subset \Omega,$$

we have

$$u(x) = \mathcal{M}_r(u)(x) - \int_{B_d(x,r)} \big(\Gamma(x^{-1} \circ y) - \Gamma(r)\big)\, \mathrm{d}\mu(y) \tag{9.22}$$

and

$$u(x) = \mathrm{M}_r(u)(x) - \frac{Q}{r^Q} \int_0^r \rho^{Q-1} \left(\int_{B_d(x,\rho)} (\Gamma(x^{-1} \circ y) - \Gamma(\rho))\, \mathrm{d}\mu(y) \right) \mathrm{d}\rho, \tag{9.23}$$

where $\mu := \mathcal{L}u$ is the $\mathcal{L}$-Riesz measure of u and $\Gamma(\rho) := \beta_d\, \rho^{2-Q}$.

Proof. Take $\Omega = B_d(x,r)$ in Poisson–Jensen's Theorem 9.5.1. By Theorem 7.2.9 (page 391), we get

$$\int_{\partial B_d(x,r)} u(y)\,\mathrm{d}\mu_x^{B_d(x,r)} = \mathcal{M}_r(u)(x).$$

Moreover, if G_Ω denotes the $\mathcal{L}$-Green function of $B_d(x,r)$, then

$$G_\Omega(x,y) = \Gamma(x^{-1}\circ y) - \Gamma(r). \tag{9.24}$$

Then (9.22) follows from (9.18). Identity (9.23) follows from (9.22) keeping in mind that

$$\mathrm{M}_r(u)(x) = \frac{Q}{r^Q}\int_0^r \rho^{Q-1}\,\mathcal{M}_\rho(u)(x)\,\mathrm{d}\rho.$$

This completes the proof. □

Remark 9.5.3. In Chapter 11, Theorem 9.5.1 will be extended to arbitrary bounded open sets Ω.

From Theorem 9.5.2 it is easy to derive the following left-continuity result of the surface average.

Proposition 9.5.4 (Left-continuity of the surface operator). *Let* $v \in \underline{S}(\Omega)$, *and let* $\overline{B_d(x,r)} \subseteq \Omega$. *Then*

$$\lim_{\rho\uparrow r} \mathcal{M}_\rho(u)(x) = \mathcal{M}_r(u)(x). \tag{9.25}$$

Proof. We first assume that $v(x) > -\infty$, and we write the Poisson–Jensen formula (9.22) as follows

$$\mathcal{M}_\rho(v)(x) = v(x) + \int_{B_d(x,\rho)} \big(\Gamma(x^{-1}\circ y) - \Gamma(\rho)\big)\,\mathrm{d}\mu(y), \quad 0<\rho\le r. \tag{9.26}$$

Since $B_d(x,\rho)\uparrow B_d(x,r)$ and, as $\rho\uparrow r$,

$$\big(\Gamma(x^{-1}\circ y) - \Gamma(\rho)\big) \uparrow \big(\Gamma(x^{-1}\circ y) - \Gamma(r)\big),$$

one obtains (9.25) from (9.26) by using the monotone convergence theorem.

If $v(x) = -\infty$, we replace v by its Perron-regularized $w := v_{B_d(x,r/2)}$, and we get

$$w \in \underline{S}(\Omega), \quad w(x) > -\infty$$

and

$$\mathcal{M}_\rho(u)(x) = \mathcal{M}_\rho(w)(x)$$

for $\frac{r}{2} < \rho \le r$. Then, from what we have already proved,

$$\lim_{\rho\nearrow r} \mathcal{M}_\rho(u)(x) = \lim_{\rho\nearrow r} \mathcal{M}_\rho(w)(x) = \mathcal{M}_r(w)(x) = \mathcal{M}_r(u)(x). \quad \square$$

The previous proposition is the needed tool to prove the following result.

Proposition 9.5.5. *Let $v \in \underline{S}(\Omega)$, and let $\overline{B_d(x_0, r)} \subseteq \Omega$. Denote $B := B_d(x_0, r)$. Then the function*

$$h : B \to \mathbb{R}, \quad h(x) = \int_{\partial B} v \, \mathrm{d}\mu_x^B$$

is the least harmonic majorant of v on B.

Proof. Since v is bounded from above on $\overline{B}$, by Theorem 6.6.1, v has a least harmonic majorant h_0 on B. Then $h_0 \leq h$, since h is harmonic on B and $h \geq v|_B$. On the other hand, for $0 < \rho < r$,

$$h_0(x_0) = \mathcal{M}_\rho(h_0)(x_0) \geq \mathcal{M}_\rho(u)(x_0).$$

As $\rho \uparrow r$, by using the previous proposition and Theorem 7.2.9, we get

$$h_0(x_0) \geq \mathcal{M}_r(v)(x_0) = h(x_0).$$

Therefore, $h_0 - h \leq 0$ in B and $(h_0 - h)(x_0) = 0$. The strong maximum principle of Theorem 8.2.19 implies $h_0 \equiv h$. □

We are now in the position to show that, in the present setting, Theorem 6.11.1 on page 375 can be improved.

Theorem 9.5.6 (Envelopes in $\overline{\mathcal{S}}(\Omega)$). *Let $\Omega \subseteq \mathbb{G}$ be open, and let $\mathcal{F} \subseteq \overline{\mathcal{S}}(\Omega)$. Assume $u_0 := \inf \mathcal{F}$ is locally bounded from below. Then $\widehat{u_0} \in \overline{\mathcal{S}}(\Omega)$ and*

$$u_0 = \widehat{u_0} \qquad \textit{a.e. in } \Omega.$$

Proof. We already know, from Theorem 6.11.1, that $\widehat{u_0} \in \overline{\mathcal{S}}(\Omega)$. Then, we only have to prove that $u_0 = \widehat{u_0}$ a.e. in Ω. By Choquet's lemma (Proposition 6.1.2), there exists a sequence (v_n) in $\mathcal{F}$ such that $\widehat{v} = \widehat{u_0}$, where $v = \inf_{n \in \mathbb{N}} v_n$. Defining $w_n := \min\{v_1, \ldots, v_n\}$, $n \in \mathbb{N}$, we obtain a decreasing sequence $\{w_n\}_{n \in \mathbb{N}}$ of $\mathcal{L}$-subharmonic functions in Ω such that $w_n \downarrow v$. For every $\mathcal{L}$-gauge ball B with $\overline{B} \subseteq \Omega$, we have

$$\int_{\partial B} \widehat{u_0} \, \mathrm{d}\mu_y^B \leq \int_{\partial B} w_n \, \mathrm{d}\mu_y^B \leq w_n(y) \qquad \forall\, y \in B,$$

so that, letting n tend to infinity,

$$\int_{\partial B} \widehat{u_0} \, \mathrm{d}\mu_y^B \leq \int_{\partial B} v \, \mathrm{d}\mu_y^B \leq v(y) \qquad \forall\, y \in B,$$

and $y \mapsto \int_{\partial B} v \, \mathrm{d}\mu_y^B$ is $\mathcal{L}$-harmonic in B (it is the limit of a decreasing sequence of $\mathcal{L}$-harmonic functions bounded from below by an $\mathcal{L}$-harmonic function). Then the previous inequalities extends to the following ones

$$\int_{\partial B} \widehat{u_0} \, \mathrm{d}\mu_y^B \leq \int_{\partial B} v \, \mathrm{d}\mu_y^B \leq \widehat{v}(y) = \widehat{u_0}(y) \qquad \forall\, y \in B. \tag{9.27}$$

On the other hand, by Proposition 9.5.5, $y \mapsto \int_{\partial B} \widehat{u}_0 \, \mathrm{d}\mu_y^B$ is the greatest harmonic minorant of $\widehat{u}_0$ in B. Thus, the first inequality in (9.27) actually is an equality, i.e.

$$\int_{\partial B} \widehat{u}_0 \, \mathrm{d}\mu_y^B = \int_{\partial B} v \, \mathrm{d}\mu_y^B \qquad \forall \, y \in B.$$

In particular, if $B = B_d(x_0, r)$, we have

$$\mathcal{M}_r(\widehat{u}_0)(x_0) = \mathcal{M}_r(v)(x_0).$$

Since this identity holds for every gauge ball $B_d(x_0, r)$ with $\overline{B_d(x_0, r)} \subseteq \Omega$, we also have

$$\mathrm{M}_r(\widehat{u}_0)(x_0) = \mathrm{M}_r(v)(x_0)$$

for every $x_0 \in \Omega$ and $r > 0$ such that $\overline{B_d(x_0, r)} \subseteq \Omega$. Keeping in mind that $\widehat{u}_0 \leq u_0 \leq v$, the assertion follows from the next lemma. □

Lemma 9.5.7. *Let $\Omega \subseteq \mathbb{G}$ be open, and let $f : \Omega \to [0, \infty]$ be a Lebesgue measurable function such that*

$$\mathrm{M}_r(f)(x_0) = 0 \tag{9.28}$$

for every $x_0 \in \Omega$ and $r > 0$ such that $\overline{B_d(x_0, r)} \subseteq \Omega$. Then $f = 0$ a.e. in Ω.

Proof. Let K be the kernel appearing in the average operator M_r (see (5.50f), page 259). Hypothesis (9.28) implies that $f = 0$ a.e. in

$$B_d^+(x_0, r) := \left\{ x \in B_d(x_0, r) \, : \, K(x_0^{-1} \circ x) > 0 \right\}$$

for every $x_0 \in \Omega$ and $r > 0$ such that $\overline{B_d(x_0, r)} \subseteq \Omega$.

Then, since $B_d^+(x_0, r)$ is open, it is enough to show that $\Omega = \bigcup B_d^+(x_0, r)$, where the union is taken over the family of the $\mathcal{L}$-gauge balls with the closure contained in Ω. Let $x \in \Omega$. Since K is smooth out of the origin and δ_λ-homogeneous of degree zero, there exists $z \in \mathbb{G}$ such that $K(z) > 0$ and

$$d(z) < r := \frac{1}{2} \, \mathrm{dist}_d(x, \partial\Omega).$$

Then, setting $x_0 = x \circ z^{-1}$, we have $x_0^{-1} \circ x = z$, so that $x \in B_d(x_0, r)$ and $K(x_0^{-1} \circ x) = K(z) > 0$. This means that $x \in B_d^+(x_0, r)$, and we are done, since $\overline{B_d(x_0, r)} \subseteq \Omega$. □

Corollary 9.5.8. *Let $\mathcal{F}$ and u_0 be as in Theorem 9.5.6. Then*

$$\widehat{u}_0(x) = \lim_{r \to 0} \mathrm{M}_r(u_0)(x) \qquad \text{for every } x \in \Omega.$$

Proof. From Theorem 8.2.11 we have

$$\widehat{u}_0(x) = \lim_{r \to 0} \mathrm{M}_r(\widehat{u}_0)(x) \qquad \text{for every } x \in \Omega.$$

The assertion follows, since $\widehat{u}_0 = u_0$ a.e. in Ω. □

9.6 Bounded-above $\mathcal{L}$-subharmonic Functions in $\mathbb{G}$

This section is devoted to the proof of the following sharp Riesz representation theorem for $\mathcal{L}$-subharmonic functions bounded from above in $\mathbb{G}$. We recall that we are always assuming that $\mathbb{G}$ is equipped with a homogeneous Carnot group structure $\mathbb{G}$ such that the homogeneous dimension Q of $\mathbb{G}$ is strictly greater than 2.

Theorem 9.6.1 (The $\mathcal{L}$-Riesz measure of a bounded-above $u \in \underline{\mathcal{S}}(\mathbb{G})$). *Let μ be a Radon measure in $\mathbb{G}$, and let $x_0 \in \mathbb{G}$. Then μ is the $\mathcal{L}$-Riesz measure of a bounded-above $\mathcal{L}$-subharmonic function u in $\mathbb{G}$ with $u(x_0) > -\infty$ if and only if the following condition holds:*

$$\int_0^\infty \frac{\mu(B_d(x_0,t))}{t^{Q-1}}\,\mathrm{d}t < \infty. \tag{9.29}$$

If this condition is satisfied, then there is a unique $\mathcal{L}$-subharmonic function u in $\mathbb{G}$ having the $\mathcal{L}$-Riesz measure μ, the least upper bound $U < \infty$ and such that $u(x_0) > -\infty$. It is given by $U - \Gamma * \mu$, i.e.

$$u(x) = U - \int_{\mathbb{G}} \Gamma(y^{-1} \circ x)\,\mathrm{d}\mu(y), \quad x \in \mathbb{G}. \tag{9.30}$$

Proof. It is not restrictive to assume $x_0 = 0$. For the sake of brevity, let us put $n(t) := \mu(B_d(0,t))$. The proof is split into the "sufficiency" and the "necessity" parts.

• *Sufficiency.* Let μ be a Radon measure satisfying (9.29) with $x_0 = 0$, and let $U \in \mathbb{R}$. Consider the function

$$u(x) := U - \int_{\mathbb{G}} \Gamma(y^{-1} \circ x)\,\mathrm{d}\mu(y), \quad x \in \mathbb{G}.$$

We shall show that

$$u(0) > -\infty, \tag{9.31a}$$

$$\sup_{\mathbb{G}} u = U, \tag{9.31b}$$

$$\left(v \in \underline{\mathcal{S}}(\mathbb{G}) : \ \mathcal{L}v = \mu, \ \sup_{\mathbb{G}} v = U\right) \implies v = u. \tag{9.31c}$$

Since (9.31a) implies $u \in \underline{\mathcal{S}}(\mathbb{G})$ and $\mathcal{L}u = \mu$ (see Theorems 9.3.2 and 9.3.5), this will prove the sufficiency part.

Proof (of (9.31a)). We put $\Gamma(t) := \beta_d\, t^{2-Q}$. Since $n(t)$ is monotone increasing and $Q > 2$, condition (9.29) implies

$$\mu(\{0\}) = \lim_{t \downarrow 0} n(t) = 0.$$

Then

$$u(0) - U = -\int_{\mathbb{G}\setminus\{0\}} \Gamma(y^{-1})\,\mathrm{d}\mu(y) = -\lim_{\lambda \downarrow 0} \int_\lambda^{1/\lambda} \Gamma(t)\,\mathrm{d}n(t). \tag{9.32}$$

On the other hand, we have

$$\begin{aligned}\int_\lambda^{1/\lambda} \Gamma(t)\,\mathrm{d}n(t) &= \big(\Gamma(1/\lambda)\,n(1/\lambda) - \Gamma(\lambda)\,n(\lambda)\big) - \int_\lambda^{1/\lambda} \Gamma'(t)\,n(t)\,\mathrm{d}t \\ &\le \Gamma(1/\lambda)\,n(1/\lambda) + \beta_d\,(Q-2)\int_\lambda^\infty \frac{n(t)}{t^{Q-1}}\,\mathrm{d}t \\ &\le 2\,\beta_d\,(Q-2)\int_\lambda^\infty \frac{n(t)}{t^{Q-1}}\,\mathrm{d}t. \end{aligned} \tag{9.33}$$

Here, being $n(\lambda)$ monotone increasing and $\Gamma' \le 0$, we have used the fact that

$$\begin{aligned}\Gamma(\lambda)\,n(\lambda) &= -n(\lambda)\int_\lambda^\infty \Gamma'(t)\,\mathrm{d}t \le -\int_\lambda^\infty \Gamma'(t)\,n(t)\,\mathrm{d}t \\ &= \beta_d\,(Q-2)\int_\lambda^\infty \frac{n(t)}{t^{Q-1}}\,\mathrm{d}t \quad \forall\,\lambda > 0.\end{aligned}$$

Identity (9.32), together with (9.33) and condition (9.29), implies (9.31a). □

Proof (of (9.31b)*).* For a fixed $R > 0$, let us split $u - U$ as follows

$$u - U = u_R + u_R^\infty, \tag{9.34}$$

where

$$u_R(x) := -\int_{B_d(0,R)} \Gamma(y^{-1}\circ x)\,\mathrm{d}\mu(y).$$

Then

$$\lim_{x\to\infty} u_R(x) = 0. \tag{9.35}$$

On the other hand, since u_R^∞ is the Γ-potential of the measure

$$\mu_{|\mathbb{G}\setminus B_d(0,R)}$$

and

$$u_R^\infty(0) \ge u(0) - U > -\infty,$$

by Theorem 9.3.2, u_R^∞ is $\mathcal{L}$-subharmonic in $\mathbb{G}$. Hence, u_R^∞ is sub-mean. Thus, for every $r > 0$, keeping in mind (9.33), we have

$$\begin{aligned}\mathcal{M}_r(u_R^\infty)(0) \ge u_R^\infty(0) &\ge -\int_R^\infty \Gamma(t)\,\mathrm{d}n(t) \\ &\ge -2\,\beta_d\,(Q-2)\int_R^\infty \frac{n(t)}{t^{Q-1}}\,\mathrm{d}t.\end{aligned}$$

This implies the existence of at least one point $y(r,R) \in \partial B_d(0,r)$ such that

$$u_R^\infty(y(r,R)) \ge -2\,\beta_d\,(Q-2)\int_R^\infty \frac{n(t)}{t^{Q-1}}\,\mathrm{d}t. \tag{9.36}$$

Since $d(y(r,R)) = r$ for every $R > 0$, (9.31b) follows from (9.34)–(9.36) and condition (9.29). □

Proof (of (9.31c)*).* Let $v \in \underline{S}(\mathbb{G})$ be such that $\mathcal{L}v = \mu$ and $\sup v = U$. Since v is bounded above by the $\mathcal{L}$-harmonic function U, by Theorem 9.4.7 there exists an $\mathcal{L}$-harmonic function h in $\mathbb{G}$ such that $v = h - \Gamma * \mu$. Since $v \le U$, we have $h - U \le \Gamma * \mu$ in $\mathbb{G}$. Theorem 9.3.7 and Liouville Theorem 5.8.1 (page 269) imply $h - U \equiv c$ with $c \in \mathbb{R}$, $c \le 0$. As a consequence, $v = U + c - \Gamma * \mu = u + c$. But v and u have the same upper bound. Hence $c = 0$ and $v \equiv u$. □

• *Necessity.* Let $u \in \underline{S}(\mathbb{G})$ be such that

$$u(0) > -\infty, \quad \sup_{\mathbb{G}} = U < \infty.$$

Since u is sub-mean, we have

$$-\infty < u(0) \le \mathcal{M}_r(u)(0) \le U,$$

so that, by the Poisson–Jensen formula (9.22),

$$\int_{B_d(0,r)} (\Gamma(y) - \Gamma(r))\, \mathrm{d}\mu(y) \le U - u(0) \quad \forall r > 0. \tag{9.37}$$

On the other hand, since

$$\Gamma(y) - \Gamma(r) \ge \left(1 - (1/2)^{Q-2}\right) \Gamma(y) \quad \text{if } 0 < d(y) \le t < r/2,$$

we have

$$\begin{aligned} n(t)\, \Gamma(t) = \mu(B_d(0,t))\, \Gamma(t) &\le \int_{B_d(0,t)} \Gamma(y)\, \mathrm{d}\mu(y) \\ &\le \mathbf{c} \int_{B_d(0,t)} (\Gamma(y) - \Gamma(r))\, \mathrm{d}\mu(y) \\ &\le \mathbf{c} \int_{B_d(0,r)} (\Gamma(y) - \Gamma(r))\, \mathrm{d}\mu(y), \qquad \mathbf{c} = \frac{1}{1 - (\frac{1}{2})^{Q-2}}. \end{aligned}$$

Then, $n(t)$ is bounded for $0 < t < r/2$. In particular,

$$\mu(\{0\}) = n(0^+) = \lim_{t \downarrow 0} n(t) = 0.$$

Therefore, for $0 < \varepsilon < r$, we have

$$\begin{aligned} U - u(0) &\ge \int_{\{\varepsilon < d < r\}} (\Gamma(y) - \Gamma(r))\, \mathrm{d}\mu(y) \\ &\ge \int_\varepsilon^r (\Gamma(t) - \Gamma(r))\, \mathrm{d}n(t) \\ &= -(\Gamma(\varepsilon) - \Gamma(r))\, n(\varepsilon) - \int_\varepsilon^r \Gamma'(t)\, n(t)\, \mathrm{d}t \\ &\ge -\Gamma(\varepsilon)\, n(\varepsilon) + \beta_d\, (Q-2) \int_\varepsilon^r \frac{n(t)}{t^{Q-1}}\, \mathrm{d}t. \end{aligned}$$

Using the boundedness of $\{\Gamma(\varepsilon)\, n(\varepsilon) \ : \ 0 < \varepsilon < r/2\}$ and letting ε tend to zero and r tend to ∞, we get

$$\beta_d\,(Q-2)\int_0^\infty \frac{n(t)}{t^{Q-1}}\,\mathrm{d}t \le U - u(0) < \infty.$$

The proof is complete. □

Example 9.6.2. The function $u = -\Gamma$ is $\mathcal{L}$-subharmonic and bounded from above in $\mathbb{G}$, for $U := \sup u = 0$. The $\mathcal{L}$-Riesz measure of u is $\mu :=$ Dirac$_0$, the Dirac mass supported at $\{0\}$ (see Example 9.4.2, page 441). For every $x_0 \in \mathbb{G}$ we have

$$\mu(B_d(x_0,t)) = \begin{cases} 1 & \text{if } d(x_0) < t, \\ 0 & \text{if } d(x_0) \ge t, \end{cases}$$

so that

$$\int_0^\infty \frac{\mu(B_d(x_0,t))}{t^{Q-1}}\,\mathrm{d}t = \int_{d(x_0)}^\infty \frac{1}{t^{Q-1}}\,\mathrm{d}t.$$

Since $Q \ge 3$, this integral is finite if and only if $d(x_0) > 0$, i.e. if and only if $x_0 \ne 0$. The representation formula (9.30) obviously holds true for

$$U - \Gamma * \mu = -\Gamma * \mathrm{Dirac}_0 = -\Gamma = u. \quad \square$$

In Theorem 9.6.1, we can remove the explicit mention of the point x_0. Precisely, the following theorem holds.

Theorem 9.6.3 ($\mathcal{L}$-Riesz measure of a bounded-above $u \in \underline{\mathcal{S}}(\mathbb{G})$. II). *A Radon measure μ in $\mathbb{G}$ is the $\mathcal{L}$-Riesz measure of a bounded-above $\mathcal{L}$-superharmonic function in $\mathbb{G}$ if and only if*

$$\int_1^\infty \frac{\mu(B_d(0,t))}{t^{Q-1}}\,\mathrm{d}t < \infty. \tag{9.38}$$

If this condition is satisfied and $U \in \mathbb{R}$, the function

$$u(x) := U - \int_{\mathbb{R}^N} \Gamma(y^{-1} \circ x)\,\mathrm{d}\mu(y), \quad x \in \mathbb{G},$$

is the unique $\mathcal{L}$-subharmonic function whose $\mathcal{L}$-Riesz measure is μ and the least upper bound is U.

Proof. Since every $\mathcal{L}$-subharmonic function is finite in a dense subset of its domain, by Theorem 9.6.1 it remains only to prove the first part of the theorem.

Let $u \in \underline{\mathcal{S}}(\mathbb{G})$. Denote by $\mu := \mathcal{L}u$ its $\mathcal{L}$-Riesz measure. Let

$$v := u_B, \quad \text{where } B := u_{B_d(0,1/2)}$$

is the Perron-regularization of u related to the d-ball B. Then $v \in \underline{\mathcal{S}}(\mathbb{G})$, $v(0) > -\infty$ and $v = u$ in $\mathbb{G} \setminus B_d(0,1/2)$. Let $\mu_v := \mathcal{L}v$ be the $\mathcal{L}$-Riesz measure of v. Then, by Theorem 9.6.1,

$$\int_0^\infty \frac{\mu_v(B_d(0,t))}{t^{Q-1}}\,dt < \infty. \tag{9.39}$$

Moreover, for every $t > 1$,

$$\mu(B_d(0,t)) = \mu(\overline{B_d(0,1)}) - \mu_v(\overline{B_d(0,1)}) + \mu_v(B_d(0,t)).$$

By using this identity in (9.39), we obtain (9.38). Vice versa, assume (9.38) is satisfied and define

$$\mu_1 := \mu|_{\overline{B_d(0,2)}} \quad \text{and} \quad \mu_2 := \mu|_{\mathbb{G}\setminus\overline{B_d(0,2)}}.$$

Since

$$\int_0^\infty \frac{\mu_2(B_d(0,t))}{t^{Q-1}}\,dt = \int_1^\infty \frac{\mu_2(B_d(0,t))}{t^{Q-1}}\,dt \le \int_1^\infty \frac{\mu(B_d(0,t))}{t^{Q-1}}\,dt < \infty,$$

by Theorem 9.6.1 there exists a function $u_2 \in \underline{S}(\mathbb{G})$ bounded from above and such that $\mathcal{L}u_2 = \mu_2$. Then, letting

$$u = -\Gamma * \mu_1 + u_2,$$

we get $u \in \underline{S}(\mathbb{G})$, $\sup u \le \sup u_2 < \infty$ and $\mathcal{L}u = \mu_1 + \mu_2 = \mu$. □

9.7 $\mathcal{L}$-subharmonic Smoothing of $\mathcal{L}$-subharmonic Functions

Let ϱ be a homogeneous norm on $\mathbb{G}$ (not necessarily an $\mathcal{L}$-gauge for some sub-Laplacian $\mathcal{L}$). Given a L^1_{loc}-function $u : \Omega \to [-\infty, \infty]$, $\Omega \subseteq \mathbb{G}$ open, we know that the $(\varepsilon, \mathbb{G})$-mollifier of u (see Example 5.3.7, page 239)

$$u_\varepsilon(x) = (u *_{\mathbb{G}} J_\varepsilon)(x) := \int_{B_\varrho(0,\varepsilon)} u(y^{-1} \circ x)\, J_\varepsilon(y)\, dy$$

is well defined and smooth in

$$\Omega_\varepsilon := \{x \in \Omega : \operatorname{dist}_\varrho(x, \partial\Omega) > \varepsilon\}.$$

We recall that $J_\varepsilon(x) = \varepsilon^{-Q} J(\delta_{1/\varepsilon}(x))$ and J is a smooth function supported in $B_d(0,1)$ with $\int_{\mathbb{G}} J = 1$.

We also know (see Theorem 8.1.5, page 401) that the mapping

$$u \mapsto u *_{\mathbb{G}} J_\varepsilon$$

preserves the $\mathcal{L}$-sub-mean property *for every* sub-Laplacian $\mathcal{L}$ on $\mathbb{G}$. As a consequence, by Theorem 8.2.1 (page 401),

the $(\varepsilon, \mathbb{G})$-mollifiers *preserve the $\mathcal{L}$-subharmonicity*
for every sub-Laplacian $\mathcal{L}$ on $\mathbb{G}$.

Note that the notion of $(\varepsilon, \mathbb{G})$-mollification depends only on the structure of $(\mathbb{G}, \circ, \delta_\lambda)$ and on the homogeneous norm ϱ and does not depend on the sub-Laplacian $\mathcal{L}$.

The following theorem holds.

Theorem 9.7.1 (Smoothing of $\mathcal{L}$-subharmonic functions). *Let Ω be an open subset of $\mathbb{G}$, let $\mathcal{L}$ be any sub-Laplacian on $\mathbb{G}$ and let $u : \Omega \to [-\infty, \infty[$ be an $\mathcal{L}$-subharmonic function. Then:*

(i) $u_\varepsilon \in C^\infty(\Omega_\varepsilon, \mathbb{R})$,
(ii) $u_\varepsilon \longrightarrow u$ *in* $L^1_{\mathrm{loc}}(\Omega)$ *as* $\varepsilon \to 0$,
(iii) u_ε *is $\mathcal{L}$-subharmonic in* Ω_ε.

By using the smoothing operators constructed by superposition of the surface mean operators (see Section 5.6, page 257), one can find *monotone* sequences of $\mathcal{L}$-smooth and $\mathcal{L}$-subharmonic functions approximating a given $\mathcal{L}$-subharmonic function. However, the smooth approximating functions constructed via this method are $\mathcal{L}$-subharmonic only with respect to the *given* sub-Laplacian $\mathcal{L}$.

The starting point of this construction is the following crucial lemma.

Lemma 9.7.2. *Let Γ be the fundamental solution for $\mathcal{L}$ and, for any fixed $z \in \mathbb{G}$, let us set*

$$\Gamma_z(x) := \Gamma(z^{-1} \circ x).$$

Then, for every $r > 0$,

$$\mathcal{M}_r(\Gamma_z)(x) = \min\{\Gamma(z^{-1} \circ x), \Gamma(r)\}, \quad x \in \mathbb{G}. \tag{9.40}$$

As a consequence, the function $x \mapsto \mathcal{M}_r(\Gamma_z)(x)$ is $\mathcal{L}$-superharmonic in $\mathbb{G}$. Moreover, the map

$$(x, z, r) \mapsto \mathcal{M}_r(\Gamma_z)(x)$$

is lower semicontinuous.

Proof. (9.40) is (9.10a), page 438. The $\mathcal{L}$-superharmonicity of $\mathcal{M}_r(\Gamma_z)$ also follows from Proposition 6.5.4-(iii) (page 355). The last part of the assertion is a standard consequence of Fatou's lemma.

An alternative proof which makes use of the Poisson–Jensen formula is the following one. Since $\Gamma_z \in \overline{\mathcal{S}}(\mathbb{G})$ and $\mathcal{L}\Gamma_z = -\mathrm{Dirac}_z$ (the Dirac mass supported at $\{z\}$) the Poisson–Jensen formula (9.22) gives

$$\Gamma(z^{-1} \circ x) = \Gamma_z(x) = \mathcal{M}_r(\Gamma_z)(x) + w_z(x), \tag{9.41}$$

where

$$w_z(x) = \begin{cases} 0 & \text{if } d(z^{-1} \circ x) \geq r, \\ \Gamma(x^{-1} \circ z) - \Gamma(r) & \text{if } d(z^{-1} \circ x) < r. \end{cases}$$

Then

$$w_z(x) = \max\{\Gamma(x^{-1} \circ z) - \Gamma(r), 0\}$$

and (9.40) follows from (9.41) and the symmetry of Γ. □

From Theorem 9.3.11 (page 439), one can easily obtain the approximation Theorem 9.7.3 below. For the reading convenience, we recall that in (5.50b) on page 258, we introduced the following integral operator Φ_r:

$$\Phi_r(u)(x) := \int_0^\infty \varphi_r(\rho)\,\mathcal{M}_\rho(u)(x)\,\mathrm{d}\rho,$$
$$\text{where}\quad \varphi_r(t) := \frac{1}{r}\varphi\left(\frac{t}{r}\right), \quad \varphi \in C_0^\infty(]0,1[,\mathbb{R}) \quad\text{and}\quad \int_0^1 \varphi = 1. \quad (9.42)$$

Theorem 9.7.3 (Approximation I. Γ-potentials). *Let $u \in \overline{\mathcal{S}}(\mathbb{G})$ be the Γ-potential of a Radon measure μ on $\mathbb{G}$. Then, for every $r > 0$,*

$$\Phi_r(u) \in \overline{\mathcal{S}}(\mathbb{G}) \cap C^\infty(\mathbb{G}), \tag{9.43}$$

where Φ_r is as in (9.42).

Moreover, the following assertions hold:

$$u(x) \geq \Phi_\rho(u)(x) \geq \Phi_r(u)(x) \quad \forall\, x \in \mathbb{G},\ 0 < \rho \leq r, \tag{9.44a}$$
$$\lim_{r\to 0} \Phi_r(u)(x) = u(x) \quad \forall\, x \in \mathbb{G}. \tag{9.44b}$$

Proof. Property (9.43) follows from (9.42), Theorem 9.3.11 (page 439) and Theorem 8.2.20, page 410 (in this last remark, we take $\Lambda = (0,\infty)$, $z = \rho$ and $\mathrm{d}\mu(\rho) = \varphi_r(\rho)\,\mathrm{d}\rho$).

To prove (9.44a), let us assume $0 < \rho \leq r$. By Theorem 8.2.10 (page 404), $\rho \mapsto \mathcal{M}_\rho(u)$ is monotone decreasing, so that

$$\begin{aligned}\Phi_\rho(u)(x) &= \int_0^\infty \varphi(t)\,\mathcal{M}_{\rho t}(u)(x)\,\mathrm{d}t \\ &\geq \int_0^\infty \varphi(t)\,\mathcal{M}_{r t}(u)(x)\,\mathrm{d}t = \Phi_r(u)(x).\end{aligned}$$

Moreover, since u is super-mean,

$$\Phi_\rho(u)(x) = \int_0^\infty \varphi(t)\,\mathcal{M}_{\rho t}(u)(x)\,\mathrm{d}t \leq \left(\int_0^\infty \varphi(t)\,\mathrm{d}t\right) u(x) = u(x).$$

Thus, (9.44a) holds. Finally, (9.44b) follows from the inequality $u(x) \geq \Phi_r(u)(x)$ and the lower semicontinuity of u. □

If we consider the integral operator Φ_r related to a non-smooth function φ, Theorem 9.7.3 still holds with (9.43) replaced by

$$\Phi_r(u) \in \overline{\mathcal{S}}(\mathbb{G}). \tag{9.43$'$}$$

In particular, by taking

$$\varphi(t) = \begin{cases} Q\,t^{Q-1} & \text{if } 0 < t < 1, \\ 0 & \text{otherwise,}\end{cases}$$

we obtain the following theorem.

Theorem 9.7.4 (Approximation II. Γ-potentials). *Let $u \in \overline{S}(\mathbb{G})$ be the Γ-potential of a Radon measure μ in $\mathbb{G}$. Then:*

(i) $x \mapsto \mathrm{M}_r(u)(x)$ *is $\mathcal{L}$-superharmonic in $\mathbb{G}$ for every* $r > 0$,
(ii) $\mathrm{M}_r(u)(x) \le \mathrm{M}_\rho(u)(x) \le u(x)$ *for every* $x \in \mathbb{G}$ *and* $0 < \rho \le r < \infty$,
(iii) $\lim_{r\downarrow 0} \mathrm{M}_r(u)(x) = u(x)$ *for every* $x \in \mathbb{G}$.

M_r *is the solid average operator defined in* (5.50f), *page 259.*

By using the Riesz representation Theorem 9.4.4, Theorem 9.7.3 extends to $\mathcal{L}$-subharmonic functions on arbitrary open subsets of $\mathbb{G}$:

Theorem 9.7.5 ($\mathcal{L}$-subharmonic-monotone-smooth approximation). *Let $u \in \underline{S}(\Omega)$ and $\Omega \subseteq \mathbb{G}$ be open. For every $r > 0$, let us define*

$$u_r(x) := \Phi_r(u)(x), \quad x \in \Omega_r := \{y \in \Omega \ : \ \mathrm{dist}_d(y, \partial\Omega) > r\},$$

where Φ_r is the smoothing operator of Theorem 9.7.3. Then:

(i) $u_r \in \underline{S}(\Omega_r) \cap C^\infty(\Omega_r)$ *for every* $r > 0$,
(ii) $u \le u_\rho \le u_r$ *for every* $0 < \rho \le r < \infty$,
(iii) $\lim_{r\downarrow 0} u_r(x) = u(x)$ *for every* $x \in \Omega$.

Proof. Let $x \in \Omega_r$, and let O be a bounded open set such that

$$\overline{B_d(y,r)} \subset O \subset \overline{O} \subset \Omega$$

for every $y \in V$, where V is a suitable neighborhood of x. By the Riesz-type representation Theorem 9.4.4, there exists a Radon measure μ supported in $\overline{O}$ and a function h, $\mathcal{L}$-harmonic in O, such that

$$u(y) = h(y) - (\Gamma * \mu)(y) \quad \forall\, y \in O.$$

As a consequence,

$$u_r(y) = h_r(y) - (\Gamma * \mu)_r(y) = h(y) - (\Gamma * \mu)_r(y) \quad \forall\, y \in V. \tag{9.45}$$

Then, by (9.43),

$$(u_r)|_V \in \underline{S}(V) \cap C^\infty(V),$$

and (i) is proved. Now, (ii) and (iii) directly follow from (9.45) with $y = x$ and, respectively, (9.44a) and (9.44b). □

9.8 Isolated Singularities—Bôcher-type Theorems

Given an open set $\Omega \subseteq \mathbb{G}$, a point $x_0 \in \Omega$ and an $\mathcal{L}$-harmonic function u in $\Omega \setminus \{x_0\}$, we say that x_0 is a *removable singularity of u* if there exists an $\mathcal{L}$-harmonic function v in Ω such that

$$u(x) = v(x) \quad \forall\, x \in \Omega \setminus \{x_0\}.$$

A first result on removable singularities will easily follow from the next lemma, in which we shall use the notation

$$\dot{B}_d(x_0, r) := B_d(x_0, r) \setminus \{x_0\}.$$

We recall that Q (the homogeneous dimension of $\mathbb{G}$) is always supposed > 2.

Lemma 9.8.1. *Let $h : \dot{B}_d(x_0, r) \to \mathbb{R}$ be an $\mathcal{L}$-harmonic function such that:*

(i) $\lim_{x \to y} h(x) = 0$ *for every* $y \in \partial B_d(x_0, r)$,
(ii) $\limsup_{x \to x_0} h(x) \left(d(x_0^{-1} \circ x)\right)^{Q-2} = 0.$

Then $h \equiv 0$.

Proof. Let $\varepsilon > 0$. Define

$$h_\varepsilon(x) := \pm h(x) - \varepsilon \left(d(x_0^{-1} \circ x)\right)^{2-Q}, \quad x \in \dot{B}_d(x_0, r).$$

Since d is an $\mathcal{L}$-gauge and h_ε is $\mathcal{L}$-harmonic, h_e is $\mathcal{L}$-harmonic in $\dot{B}_d(x_0, r)$. Moreover, conditions (i) and (ii) imply

$$\limsup_{x \to y} h_\varepsilon(x) \leq 0 \quad \forall\, y \in \partial \dot{B}_d(x_0, r).$$

Then, by the maximum principle, $h_\varepsilon \leq 0$ in $\dot{B}_d(x_0, r)$. Letting ε tend to zero, we obtain $h \leq 0$. The same inequality holds for $-h$. Thus $h = 0$. □

Theorem 9.8.2 (Characterization of removable singularities). *Let Ω be an open subset of $\mathbb{G}$. A point $x_0 \in \Omega$ is a removable singularity of the $\mathcal{L}$-harmonic function $u : \Omega \setminus \{x_0\} \to \mathbb{R}$ if and only if*

$$\limsup_{x \to x_0} u(x) \left(d(x_0^{-1} \circ x)\right)^{Q-2} = 0. \tag{9.46}$$

Proof. The "only if" part is trivial. We turn to the "if" part. Assume that (9.46) is satisfied and $r > 0$ is such that $\overline{B_d(x_0, r)} \subset \Omega$. Let w be the solution to the boundary value problem

$$\begin{cases} \mathcal{L}w = 0 & \text{in } B_d(x_0, r), \\ w|_{\partial B_d(x_0, r)} = u|_{\partial B_d(x_0, r)}. \end{cases}$$

Then $h := u - w$ satisfies the hypothesis of Lemma 9.8.1. It follows that $h = 0$ in $\dot{B}_d(x_0, r)$, i.e. $u = w$ in $\dot{B}_d(x_0, r)$. Thus, the function

$$v : \Omega \to \mathbb{R}, \quad v(x) := \begin{cases} u(x) & \text{in } \Omega \setminus B_d(x_0, r), \\ w(x) & \text{in } B_d(x_0, r) \end{cases}$$

is $\mathcal{L}$-harmonic in Ω and equals u in $\Omega \setminus \{x_0\}$. This completes the proof. □

Remark 9.8.3. The function

$$x \mapsto \Gamma(x_0^{-1} \circ x) = \left(d(x_0^{-1} \circ x)\right)^{2-Q}$$

is $\mathcal{L}$-harmonic in $\mathbb{G} \setminus \{x_0\}$ with a non-removable singularity at $x = x_0$. □

Non-negative $\mathcal{L}$-harmonic functions in $\mathbb{G} \setminus \{x_0\}$ can be characterized in terms of the function $x \mapsto \Gamma(x_0^{-1} \circ x)$.

We first show a local version of this statement, which extends the classical Bôcher theorem to the sub-Laplacian setting.

Theorem 9.8.4 (Bôcher-type theorem on punctured balls). *Let u be $\mathcal{L}$-harmonic and non-negative in the punctured ball $\dot{B}_d(x_0, R)$. Then there exist a non-negative constant a and an $\mathcal{L}$-harmonic function h in $B_d(x_0, R)$ such that*

$$u = a\,\Gamma + h \quad \text{in } \dot{B}_d(x_0, R). \tag{9.47}$$

Proof. It is not restrictive to assume $x_0 = 0$. If u were bounded near zero, then (9.47) would follow, with $a = 0$, from Theorem 9.8.2. Assume that

$$\limsup_{x \to 0} u(x) = \infty. \tag{9.48}$$

For $0 < r \le R/2$, define

$$S(r) := \sup_{d(x)=r} u(x) \quad \text{and} \quad s(r) := \inf_{d(x)=r} u(x).$$

The Harnack inequality on rings of Corollary 5.7.7 (page 267) implies

$$S(r) \le \mathbf{c}\, s(r), \qquad 0 < r \le R/2, \tag{9.49}$$

with a positive constant $\mathbf{c}$ independent of r. On the other hand, by the maximum principle,

$$\max\{S(r), S(R/2)\} = \max_{r \le d(x) \le R/2} u(x), \qquad 0 < r \le R/2,$$

so that, by (9.48),

$$\lim_{r \downarrow 0} S(r) = \infty. \tag{9.50}$$

Using (9.50) and (9.49), we get $\liminf_{x \to 0} u(x) = \infty$. Hence

$$\lim_{x \to 0} u(x) = \infty.$$

If we extend u at 0 letting $u(0) = \infty$, by Corollary 8.2.6 (page 402), $u \in \overline{\mathcal{S}}(B_d(0, R))$. Let us now observe that the $\mathcal{L}$-Riesz measure of u

$$\mu := -\mathcal{L}u$$

is supported at $\{0\}$, since u is $\mathcal{L}$-harmonic in $\dot{B}_d(0, r)$. Therefore,

$$\mu = a\,\mathrm{Dirac}_0,$$

for a suitable $a > 0$ (here Dirac_0 is the Dirac mass supported at the origin). Now, the Riesz representation result of Corollary 9.4.9 implies

$$u = \Gamma * \mu + h = a\,\Gamma + h,$$

where h is $\mathcal{L}$-harmonic in $B_d(0, R)$. The proof is complete. □

Theorem 9.8.5 (Bôcher-type theorem for $\mathcal{L}$). *Let $\Omega \subseteq \mathbb{G}$ be open, and let $x_0 \in \Omega$. Let $u : \Omega \setminus \{x_0\} \to \mathbb{R}$ be $\mathcal{L}$-harmonic and non-negative. Then there exist a non-negative constant a and an $\mathcal{L}$-harmonic function v in Ω such that*

$$u = a\,\Gamma + v \quad \text{in } \Omega \setminus \{x_0\}. \tag{9.51}$$

Proof. Let $R > 0$ be such that $\overline{B_d(x_0, R)} \subset \Omega$. By Theorem 9.8.4, there exist a constant $a \geq 0$ and an $\mathcal{L}$-harmonic function h in $B_d(x_0, R)$ such that

$$u = a\,\Gamma + h \text{ in } \dot{B}_d(x_0, R).$$

Thus, the function

$$v : \Omega \setminus \{x_0\} \to \mathbb{R}, \qquad v := \begin{cases} h & \text{in } B_d(x_0, R), \\ u - a\,\Gamma & \text{in } \Omega \setminus B_d(x_0, R) \end{cases}$$

is an $\mathcal{L}$-harmonic function in Ω satisfying (9.51). □

This theorem, together with Liouville Theorem 5.8.1 (page 269), implies the following corollary.

Corollary 9.8.6 (Non-negative $\mathcal{L}$-harmonic functions in $\mathbb{G} \setminus \{0\}$). *Let $u : \mathbb{G} \setminus \{0\} \to \mathbb{R}$ be a non-negative $\mathcal{L}$-harmonic function. Then there exist two non-negative constants a and b such that*

$$u = a\,\Gamma + b \quad \text{in } \mathbb{G} \setminus \{0\}.$$

Proof. By Theorem 9.8.5,

$$u = a\,\Gamma + v \quad \text{in } \mathbb{G} \setminus \{0\},$$

where a is a non-negative constant and v is $\mathcal{L}$-harmonic in $\mathbb{G}$. Since $u \geq 0$ and $\Gamma(x) \to 0$ as $x \to \infty$, the function v is bounded from below and $\liminf_{x \to \infty} v(x) \geq 0$. Then, by Liouville Theorem 5.8.1 (page 269), v is constant, and this constant is non-negative. □

With this corollary at hand, we can trivially characterize the δ_λ-homogeneous non-negative $\mathcal{L}$-harmonic functions in $\mathbb{G} \setminus \{0\}$.

Corollary 9.8.7 (Non-negative δ_λ-homogeneous functions in $\mathcal{H}(\dot{\mathbb{G}})$). *Suppose $u : \mathbb{G} \setminus \{0\} \to \mathbb{R}$ is non-negative, $\mathcal{L}$-harmonic and δ_λ-homogeneous of degree m. If u is not identically zero, then $m = 0$ or $m = 2 - Q$. Moreover:*

(i) *$u =$ constant if $m = 0$,*
(ii) *$u = a\,\Gamma$ for a suitable constant $a > 0$ if $m = 2 - Q$.*

As a consequence, we immediately obtain the following improvement of Theorem 5.5.6 on page 256.

Corollary 9.8.8 ("Uniqueness" of the $\mathcal{L}$-gauges. II). *Let $\mathcal{L}$ be a sub-Laplacian on the homogeneous Carnot group $\mathbb{G}$. Let d be a homogeneous norm on $\mathbb{G}$, smooth out of the origin and such that*

$$\mathcal{L}(d^\alpha) = 0 \ \text{ in } \mathbb{G} \setminus \{0\},$$

for a suitable $\alpha \in \mathbb{R}$, $\alpha \neq 0$. Then $\alpha = 2 - Q$ and d is an $\mathcal{L}$-gauge on $\mathbb{G}$.

9.8.1 An Application of Bôcher's Theorem

The aim of this section is to prove the following Proposition 9.8.9, as a consequence of Bôcher's Theorem 9.8.5 above and of the Liouville-type Theorem 5.8.4 (page 270). We recall that if d is an $\mathcal{L}$-gauge on $\mathbb{G}$, then

$$\Psi_{\mathcal{L}} = |\nabla_{\mathcal{L}} d|^2$$

is the kernel appearing in the relevant solid mean value formula of Theorem 5.6.1 (page 259).

Proposition 9.8.9. *Suppose $\mathcal{L}$ is a sub-Laplacian on the (homogeneous) Carnot group $\mathbb{G}$ and d is an $\mathcal{L}$-gauge on $\mathbb{G}$ such that $|\nabla_{\mathcal{L}} d|^2$ is constant on $\mathbb{G} \setminus \{0\}$. Then $\mathbb{G}$ has step* 1, *i.e.* $\mathbb{G} = (\mathbb{G}, +)$ *is the Euclidean group.*

Proof. Let us set, for brevity, $\dot{\mathbb{G}} := \mathbb{G} \setminus \{0\}$. Let d be an $\mathcal{L}$-gauge on $\mathbb{G}$. Then, by Definition 5.4.1 (page 247),

$$d \in C^\infty(\dot{\mathbb{G}}) \cap C(\mathbb{G})$$

and d^{2-Q} is $\mathcal{L}$-harmonic on $\dot{\mathbb{G}}$ (here Q is the homogeneous dimension of $\mathbb{G}$). From the formula

$$\mathcal{L}(\alpha(u)) = \alpha''(u)\, |\nabla_{\mathcal{L}} u|^2 + \alpha'(u)\, \mathcal{L}u$$

(for suitably regular functions $u : \mathbb{G} \to \mathbb{R}$, $\alpha : \mathbb{R} \to \mathbb{R}$), we have

$$0 = \mathcal{L}(d^{2-Q}) = (2-Q)(1-Q)\, d^{-Q} |\nabla_{\mathcal{L}} d|^2 + (2-Q) d^{1-Q}\, \mathcal{L}d$$

on $\dot{\mathbb{G}}$, i.e.

$$d\, \mathcal{L}d = (Q-1)\, |\nabla_{\mathcal{L}} d|^2 \quad \text{on } \dot{\mathbb{G}}. \tag{9.52}$$

Analogously, we have $\mathcal{L}(d^2) = 2\, |\nabla_{\mathcal{L}} d|^2 + 2\, d\, \mathcal{L}u$. Hence, by (9.52),

$$\mathcal{L}(d^2) = 2\, Q\, |\nabla_{\mathcal{L}} d|^2 \quad \text{on } \dot{\mathbb{G}}. \tag{9.53}$$

Consequently, (9.53) proves that

$$\big(|\nabla_{\mathcal{L}} d|^2 = \text{constant on } \dot{\mathbb{G}}\big) \quad \Longrightarrow \quad \big(\mathcal{L}(d^2) = \text{constant on } \dot{\mathbb{G}}\big). \tag{9.54}$$

We claim that

$$\big(|\nabla_{\mathcal{L}} d|^2 = \text{constant on } \dot{\mathbb{G}}\big) \quad \Longrightarrow \quad \big(d^2 \in C^\infty(\mathbb{G})\big). \tag{9.55}$$

From this claim the proposition follows. Indeed, by continuity, the claimed (9.55) and (9.54) prove that $\mathcal{L}(d^2) = $ constant on $\mathbb{G}$. Then, thanks to the Liouville-type Theorem 5.8.4 (page 270), d^2 is a polynomial function of $\mathbb{G}$-degree at most 2. Denoting, as usual, by $x = (x^{(1)}, x^{(2)}, \dots, x^{(r)})$ the "stratification" of the variables of the (homogeneous) group $\mathbb{G}$, we have

$$d^2(x) = p_2(x^{(1)}) + p_1(x^{(2)}),$$

where p_i is a polynomial function of *ordinary* degree i ($i = 1, 2$). Due to the non-negativity of d^2, this gives $p_1 \equiv 0$, so that $d^2(x) = p_2(x^{(1)})$. But since d is positive outside 0 (see the definition of homogeneous norm in Section 5.1, page 229), if $\mathbb{G}$ had step $r \geq 2$, fixed $x^{(2)} \neq 0$, we would have the contradiction

$$0 < d^2(0^{(1)}, x^{(2)}, 0^{(3)}, \dots, 0^{(r)}) = p_2(0^{(1)}) = 0.$$

(Recall that p_2 is a homogeneous polynomial of degree 2.)

It remains to prove (9.55). Suppose $|\nabla_{\mathcal{L}} d|^2 = c \in \mathbb{R}$ and take a *non-negative* cut-off function $\varphi \in C_0^\infty(\mathbb{G})$ such that $\varphi \equiv 1$ on a neighborhood U of the origin, say $U = B_d(0, \varepsilon)$ with ε small. Denote by Γ the fundamental solution of $\mathcal{L}$. From Theorem 5.3.5 (page 238) the function

$$u(y) := \int_{\mathbb{G}} \Gamma(y^{-1} \circ x)\, \varphi(x)\, dx$$

is non-negative, smooth on $\mathbb{G}$ and satisfies $\mathcal{L}u = -\varphi$ on $\mathbb{G}$. In particular,

$$\mathcal{L}u = -1 \quad \text{on } U. \tag{9.56}$$

By (9.53) and (9.56), we have

$$\mathcal{L}(d^2 + 2\, Q\, c\, u) = 0 \quad \text{on } U \setminus \{0\}.$$

Hence, $d^2 + 2\, Q\, c\, u$ is $\mathcal{L}$-harmonic and non-negative on $U \setminus \{0\}$. Furthermore, it is bounded there, for d and u are continuous on $\mathbb{G}$. By Theorem 9.8.5, there exists $h \in \mathcal{H}(U)$ such that $d^2 + 2\, Q\, c\, u = h$ on $U \setminus \{0\}$. By continuity, this holds on the whole U, thus proving (9.55). □

Bibliographical Notes. The topics developed in this chapter within the classical theory of Laplace's operator can be found in all the main monographs devoted to the potential theory; see the references in the Bibliographical Notes of Chapter 8. For Bôcher-type theorems, see [TW02b].

Some of the topics presented in this chapter also appear in [BL03,BL07].

9.9 Exercises of Chapter 9

Ex. 1) Prove that if $u > 0$ is $\mathcal{L}$-harmonic in an open set Ω, then $\log(u)$ is $\mathcal{L}$-superharmonic in Ω.

Ex. 2) Prove that any $\mathcal{L}$-superharmonic function u in an open set Ω belongs to $L^p_{\mathrm{loc}}(\Omega)$ for every

$$1 \le p < \frac{Q}{Q-2}.$$

(*Hint*: Use Theorem 9.4.4, in order to restrict the proof to the case where $u = \Gamma * \mu$ with μ compactly supported. Then use the Hölder inequality and the summability properties of Γ.)

Ex. 3) Prove that there exists an $\mathcal{L}$-superharmonic function u in $\mathbb{G}$ such that u does not belong to $L^p_{\mathrm{loc}}(\Omega)$ with $p = \frac{Q}{Q-2}$. (*Hint*: $u = \Gamma$.)

Ex. 4) Prove that if u is an $\mathcal{L}$-superharmonic function in an open set Ω, then $\nabla_{\mathcal{L}} u$ is well-defined in the weak sense of distributions, and it holds

$$|\nabla_{\mathcal{L}} u| \in L^p_{\mathrm{loc}}(\Omega)$$

for every

$$1 \le p < \frac{Q}{Q-1}.$$

(*Hint*: Use Theorem 9.4.4, to restrict to the case where $u = \Gamma * \mu$ with μ compactly supported. Then use the summability properties of $\nabla_{\mathcal{L}} \Gamma$.)

Ex. 5) Complete the sketched proof of the following proposition.

Proposition 9.9.1. *Let $u \in \overline{S}(\mathbb{G})$ be non-negative. Then the following assertions hold:*

a) *u is a Γ-potential iff $\lim_{R\to\infty} m_R[u](0) = 0$ or iff $\inf_{\mathbb{G}} u = 0$.*
b) *u is a Γ-potential iff u is bounded from above by a Γ-potential.*

Proof (Sketch). Let $u = \Gamma * \mu$, where μ is a Radon measure in $\mathbb{G}$. Then (why?)

$$\begin{aligned} m_R[u](0) &= \int_{\mathbb{G}} \left(\int_{\partial B_d(0,R)} \mathbf{k}(0,y)\Gamma(z^{-1} \circ y)\, \mathrm{d}H^{N-1}(y) \right) \mathrm{d}\mu(z) \\ &= \int_{\mathbb{G}} (\Gamma * \lambda_{0,R})(z)\, \mathrm{d}\mu(z) = \int_{\mathbb{G}} \min\{\Gamma(z), R^{2-Q}\}\, \mathrm{d}\mu(z). \end{aligned}$$

By monotone convergence derive $\lim_{R\to\infty} m_R[u](0) = 0$. Vice versa (why?), $u = \inf_{\mathbb{G}} u + \Gamma * (\mu_u)$. Hence

$$m_R[u](0) = \inf_{\mathbb{G}} u + m_R[\Gamma * (\mu_u)](0).$$

Then, if $\lim_{R\to\infty} m_R[u](0) = 0$, we obtain (why?) $\inf_{\mathbb{G}} u = 0$, i.e. $u = \Gamma * (\mu_u)$. This argument also proves the second "iff" in (1). Finally, derive (2) from (1). □

Ex. 6) Prove the assertion in Example 9.1.3, page 427.

Ex. 7) Let u be an $\mathcal{L}$-harmonic function in the punctured ball

$$\dot{B}_d(0,1) := B_d(0,1) \setminus \{0\}.$$

Assume that

$$\liminf_{|x|\to 0} u(x)\, d(x)^{2-Q} > -\infty.$$

Then there exists $b \in \mathbb{R}$ and v, $\mathcal{L}$-harmonic in $B_d(0,1)$, such that

$$u = b\Gamma + v \qquad \text{in } \dot{B}_d(0,1).$$

Ex. 8) Let Ω be an open set of $\mathbb{G}$. Denote

$$b^p(\Omega) := \left\{ u \ \mathcal{L}\text{-harmonic in } \Omega \ : \ \int_\Omega |u|^p < \infty \right\},$$

for $1 \le p < \infty$. Prove that $b^p(\Omega)$ is a closed subspace of $L^p(\Omega)$.

Ex. 9) Let $b^p(\Omega)$ be as in the previous exercise. Let $\Omega = B_d(0,1) \setminus \{0\}$, and let $u \in b^p(\Omega)$, $p \ge Q/(Q-2)$. Prove that 0 is a removable singularity of u.

Ex. 10) Let $1 \le p < Q/(Q-2)$. Prove the existence of a function

$$u \in b^p(B_d(0,1) \setminus \{0\})$$

such that 0 is not a removable singularity of u.

Ex. 11) Let $u : \mathbb{G} \setminus \{0\} \to \mathbb{R}$ be an $\mathcal{L}$-harmonic function. Assume that, for some $p \in [1, \infty[$,

$$\int_{\mathbb{G}\setminus\{0\}} |u|^p < \infty.$$

Prove that:

(i)

$$|u(x)|^p \le c|x|^{-Q} \int_{\mathbb{G}\setminus\{0\}} |u|^p \qquad \forall x \ne 0,$$

where $c > 0$ is independent of u and x,

(ii) $u \equiv 0$ in $\mathbb{G}$ provided that $p \ge Q/(Q-2)$.

Ex. 12) Prove the following result, as a consequence of Corollary 9.8.6.

Corollary 9.9.2. *Let ϱ be any homogeneous norm on $\mathbb{G}$. Suppose there exists a non-constant function $\alpha : (0,\infty) \to (0,\infty)$ such that $\alpha \circ \varrho$ is $\mathcal{L}$-harmonic in $\dot{\mathbb{G}}$. Then ϱ is an $\mathcal{L}$-gauge function and $\alpha(t) = a\, t^{2-Q} + b$, for some constants c, $b \ge 0$.*

(*Hint*: $\alpha \circ \varrho$ is a non-negative $\mathcal{L}$-harmonic function on $\dot{\mathbb{G}}$. Then, by Corollary 9.8.6, we have (here d is the $\mathcal{L}$-gauge function such that d^{2-Q} is the fundamental solution for $\mathcal{L}$)

$$\alpha(\varrho(x)) = a\, d^{2-Q}(x) + b \quad \text{for every } x \in \dot{\mathbb{G}}.$$

Since d and ϱ are both δ_λ-homogeneous of degree 1, this gives, for every $\xi \in \mathbb{G}$ such that $\varrho(\xi) = 1$,

$$\begin{aligned}\alpha(t) &= \alpha(t\,\varrho(\xi)) = \alpha(\varrho(\delta_t\,\xi)) = a\,d^{2-Q}((\delta_t\,\xi)) + b \\ &= a\,d^{2-Q}(\xi)\,t^{2-Q} + b.\end{aligned} \tag{9.57}$$

When $t = 1$, this proves that d is constant (say, $\mathbf{a}$) on the level set of ϱ $\{\xi : \varrho(\xi) = 1\}$, so that

$$d(x) = d\big(\delta_{\varrho(x)}\,\delta_{1/\varrho(x)}(x)\big) = \varrho(x)\,d\big(\delta_{1/\varrho(x)}(x)\big) = \varrho(x)\,\mathbf{a},$$

whence $\varrho = \mathbf{a}^{-1}\,d$ is an $\mathcal{L}$-gauge function. Finally, (9.57) also shows that $\alpha(t) = \widetilde{a}\,t^{2-Q} + b$, with $\widetilde{a} = a\,\mathbf{a}^{2-Q}$.)

Ex. 13) Let d be a homogeneous norm on $\mathbb{G}$, smooth in $\dot{\mathbb{G}}$. We recall: if there exists a real constant $\gamma \neq 0$ such that

$$d^\gamma \text{ is } \mathcal{L}\text{-harmonic in } \dot{\mathbb{G}}, \tag{9.58}$$

then we say that d is an $\mathcal{L}$-gauge. As we know, a remarkable homogeneous norm on $\mathbb{G}$ is given by

$$d_C(x) := d_X(x, 0), \quad x \in \mathbb{G}, \tag{9.59}$$

where d_X denotes the Carnot–Carathéodory distance related to the family of vector fields $X = \{X_1, \ldots, X_m\}$ (see Proposition 5.2.8). The norm d_C in (9.59) will be referred to as the $\mathcal{L}_C$-*norm* on $\mathbb{G}$. In studying surface measures in Carnot–Carathéodory spaces, R. Monti and F. Serra-Cassano proved that d_C satisfies the $\mathcal{L}$-Eikonal equation, i.e.

$$|\nabla_{\mathcal{L}} d_C| = 1 \quad \text{almost everywhere in } \mathbb{G}$$

(see [MSC01, Theorem 3.1]). Prove the following assertion.

Corollary 9.9.3. *Let $\mathbb{G}$ be a Carnot group different from an Euclidean group, and let $\mathcal{L}$ be a sub-Laplacian on $\mathbb{G}$. Then the $\mathcal{L}_C$-norm* (9.59) *is* not *an $\mathcal{L}$-gauge. More explicitly, for every $\gamma \in \mathbb{R} \setminus \{0\}$,*

$$d_C^\gamma \textit{ is } \text{not } \mathcal{L}\textit{-harmonic in } \dot{\mathbb{G}}.$$

Moreover, suppose there exists a function $\alpha : (0, \infty) \to (0, \infty)$ such that $\alpha \circ d_C$ is $\mathcal{L}$-harmonic in $\dot{\mathbb{G}}$. Then $t \mapsto \alpha(t)$ is a constant function.

(*Hint*: Use the cited result in [MSC01, Theorem 3.1] and our Theorem 9.8.9. Then use Corollary 9.9.2.)
By using Proposition 9.8.9, prove also the following assertion.
*If $\mathcal{L}$ is a sub-Laplacian on the (homogeneous) Carnot group $\mathbb{G}$ and d is an $\mathcal{L}$-gauge on $\mathbb{G}$ satisfying the $\mathcal{L}$-*Eikonal equation

$$|\nabla_{\mathcal{L}} d| = 1 \quad \textit{on } \mathbb{G} \setminus \{0\},$$

then $\mathbb{G}$ has step 1, i.e. $\mathbb{G} = (\mathbb{G}, +)$ is the Euclidean group.

Ex. 14) Prove the following result, by making use of the proof of Theorem 9.8.9.

Proposition 9.9.4. *Let $u : \mathbb{G} \to \mathbb{R}$ be a (smooth) solution to*

$$\mathcal{L}u = c \quad \text{in } \mathbb{G},$$

where c is a real constant. Assume u is coercive, i.e.

$$u(x) \longrightarrow \infty \quad \text{as } x \to \infty.$$

Then $\mathbb{G}$ is the Euclidean group.

Ex. 15) When d is an $\mathcal{L}$-gauge function, we already know that the kernel $\Psi_{\mathcal{L}} = |\nabla_{\mathcal{L}} d|^2$ appears in the (solid) average formula of Theorem 5.6.1 (see also (5.42) on page 252). For the sake of brevity, we introduce a new notation:

$$K := \Psi_{\mathcal{L}}, \quad \mathcal{K} := \frac{\Psi_{\mathcal{L}}}{|\nabla d|}, \quad m_r(u) := \int_{\partial B_d(0,r)} u\, \mathcal{K}\, \mathrm{d}H^{N-1} \tag{9.60}$$

(here, ∇ is the standard gradient vector in $\mathbb{R}^N$, and H^{N-1} denotes the $(N-1)$-Hausdorff measure in $\mathbb{G}$). Compare to our previous notation (see Theorem 5.5.4 and (5.42)) and notice that

$$\mathcal{K} = \mathcal{K}_{\mathcal{L}}(0, \cdot), \quad m_r(u) = \frac{1}{(Q-2)\beta_d}\, r^{Q-1}\, \mathcal{M}_r(u)(0).$$

Prove the following result.

Theorem 9.9.5. *Let d be an $\mathcal{L}$-gauge, and let γ be the related exponent as in (9.58). If u is $\mathcal{L}$-harmonic in the unit punctured ball $\dot{B} = B_d(0,1) \setminus \{0\}$, then there exist real constants a, b such that*

$$r^{\gamma-1} m_r(u) = a\, r^{\gamma} + b \quad \text{for every } r \in\,]0,1[. \tag{9.61}$$

(*Hint*: We have proved in Section 5.5 (page 251) that if the sub-Laplacian $\mathcal{L}$ is written in the divergence form $\mathcal{L} = \mathrm{div}(A(x)\,\nabla)$, then we have

$$\int_{A_{r,R}} (g\,\mathcal{L}f - f\,\mathcal{L}g) = \int_{\partial A_{r,R}} \big(g\,\langle A\nabla f, \nu\rangle - f\,\langle A\nabla g, \nu\rangle\big)\, \mathrm{d}H^{N-1} \tag{9.62}$$

for each $f, g \in C^2(\dot{B})$ and $0 < r < R < 1$, where $A_{r,R} := \{x : r < d(x) < R\}$ and ν is the normal outer vector on $\partial A_{r,R}$. If $f = u$ and $g \equiv 1$, then

$$\int_{\partial B_d(0,R)} \langle A\nabla u, \nu\rangle\, \mathrm{d}H^{N-1} = \int_{\partial B_d(0,r)} \langle A\nabla u, \nu\rangle\, \mathrm{d}H^{N-1}. \tag{9.63}$$

Hence, $\int_{\partial B_d(0,R)} \langle A\nabla u, \nu\rangle\, \mathrm{d}H^{N-1}$ is constant w.r.t. R, say α. We now choose $f = u$ and $g = d^{\gamma}$ in (9.62). By noticing that on $\partial B_d(0,\rho)$ we have

$\nu = \nabla d/|\nabla d|$, $g = \rho^\gamma$ and $\langle A\nabla g, \nu\rangle = \gamma\,\rho^{\gamma-1}\,|\nabla_{\mathcal{L}} d|^2/|\nabla d|$. Then (9.62) becomes (see the definition of m_r in (9.60) and use (9.63))

$$0 = \alpha\,R^\gamma - \gamma\,R^{\gamma-1} m_R(u) - \alpha\,r^\gamma + \gamma\,r^{\gamma-1} m_r(u).$$

In other words, the function $]0,1[\ni r \mapsto \alpha\,r^\gamma - \gamma\,r^{\gamma-1} m_r(u)$ is constant, say β. This is exactly (9.61) with $a = \alpha/\gamma$ and $b = -\beta/\gamma$.)

Ex. 16) Re-derive Theorem 5.5.6 on page 256 as a consequence of the above Theorem 9.9.5. In other words, prove the following assertion.

Corollary 9.9.6. *Let d be an $\mathcal{L}$-gauge function, and let γ be the related exponent as in* (9.58). *Then*

$$\gamma = 2 - Q, \quad \textit{and} \quad \Gamma = \beta_d\, d^{2-Q} \tag{9.64}$$

is the fundamental solution (with pole at the origin) for $\mathcal{L}$. Here Q is the homogeneous dimension of $\mathbb{G}$ and

$$\beta_d^{-1} = Q(Q-2)\int_{B_d(0,1)} K. \tag{9.65}$$

Hint: By means of the coarea formula, we have

$$\int_0^r \int_{d(x)=\rho} K(x)\frac{\mathrm{d}H^{N-1}(x)}{|\nabla d(x)|} = \int_{d(x)<r} K(x)\,\mathrm{d}x$$
$$= \int_{d(\xi)<1} K(\delta_r(\xi))\,r^Q\,\mathrm{d}\xi = r^Q \int_{B_d(0,1)} K(\xi)\,\mathrm{d}\xi =: r^Q\,\omega_d.$$

By differentiating both sides w.r.t. r, we get

$$\int_{d(x)=r} K(x)\frac{\mathrm{d}H^{N-1}(x)}{|\nabla d(x)|} = r^{Q-1}\,Q\,\omega_d \quad \text{for all } r>0. \tag{9.66}$$

Now, recalling (9.60), the above identity rewrites as

$$m_r(1) = r^{Q-1}\,Q\,\omega_d. \tag{9.67}$$

On the other hand, by Theorem 9.9.5, there exist constants a, b such that $m_r(1) = a\,r + b\,r^{1-\gamma}$, whence (9.67) gives $Q\omega_d\,r^{Q-2} = a + b\,r^{-\gamma}$ which is possible iff (recall that $Q>2$) $a = 0$ and $\gamma = 2-Q$.

Finally, let us set $\Gamma(x) := \beta_d\, d^{2-Q}(x)$ (β_d as in (9.65)), and let us show that Γ is the fundamental solution for $\mathcal{L}$. First, show that

$$\int_{B_d(0,1)} \Gamma(x)\,\mathrm{d}x = \beta_d\,H^N(B_d(0,1))\,\frac{Q}{2} < \infty.$$

Hence $\Gamma \in L^1_{\mathrm{loc}}(\mathbb{G})$. Moreover, for $2-Q<0$, $\Gamma(x)$ vanishes when $x\to\infty$. Finally, let us check that $\mathcal{L}\Gamma = -\mathrm{Dirac}_0$ in the weak sense of distributions. Let $\varphi \in C_0^\infty(\mathbb{G})$ be fixed. By the divergence theorem, show that

$$\begin{aligned}
\int \Gamma\, \mathcal{L}\varphi &= \lim_{\varepsilon\to 0} \int_{d>\varepsilon} d^{2-Q}\, \mathcal{L}\varphi \\
&= \lim_{\varepsilon\to 0}\bigg(- \int_{d(x)=\varepsilon} d^{2-Q}\, \langle \nabla_{\mathcal{L}}\varphi, \nabla_{\mathcal{L}} d\rangle\, \frac{\mathrm{d}H^{N-1}}{|\nabla d|} \\
&\quad - \int_{d>\varepsilon} \langle \nabla_{\mathcal{L}}(d^{2-Q}), \nabla_{\mathcal{L}}\varphi\rangle \bigg) \\
&= \lim_{\varepsilon\to 0}\bigg(-I_1(\varepsilon) + \int_{d=\varepsilon} \varphi\,(2-Q)\, d^{1-Q}\, K\, \frac{\mathrm{d}H^{N-1}}{|\nabla d|} \\
&\quad + \int_{d>\varepsilon} \mathcal{L}(d^{2-Q})\,\varphi \bigg) \\
&= \lim_{\varepsilon\to 0}(-I_1(\varepsilon) + I_2(\varepsilon) + 0) = -\varphi(0)\,\beta_d^{-1}.
\end{aligned}$$

Indeed, prove that

$$|I_1(\varepsilon)| \le \varepsilon^{2-Q}\, \sup\big\{ |\nabla_{\mathcal{L}}\varphi|\, |\nabla_{\mathcal{L}} d| \big\} \int_{d(x)=\varepsilon} \frac{\mathrm{d}H^{N-1}}{|\nabla d|} = c\,\varepsilon.$$

On the other hand, $\lim_{\varepsilon\to 0} I_2(\varepsilon) = -\varphi(0)\,\beta_d^{-1}$, since (recall (9.67))

$$I_2(\varepsilon) = -\beta_d^{-1}\varphi(0) + I_3(\varepsilon),$$

and (exploiting again (9.67))

$$|I_3(\varepsilon)| \le Q(Q-2)\omega_d \sup_{d(x)=\varepsilon} |\varphi(x) - \varphi(0)| \to 0 \quad \text{as } \varepsilon \to 0.$$

Ex. 17) Let m_r be as in (9.60). Consider our usual surface average operator $\mathcal{M}_r(u)(0)$, briefly denoted by $\mathcal{M}_r(u)$. Following an idea exploited by Axler, Bourdon and Ramey [ABR92] in the classical harmonic setting, given a smooth function $u : \dot{B} \to \mathbb{R}$, we define

$$\mathcal{S}(u)(x) := \mathcal{M}_{d(x)}(u), \quad x \in \dot{B}. \tag{9.68}$$

If u is $\mathcal{L}$-harmonic in $\dot{B}$ and d is an $\mathcal{L}$-gauge, by Theorem 9.9.5 and Corollary 9.9.6, we have

$$\mathcal{S}(u)(x) = a\, d^{2-Q}(x) + b, \quad x \in \dot{B},$$

for suitable $a, b \in \mathbb{R}$. Thus, $\mathcal{S}(u)$ is $\mathcal{L}$-harmonic in $\dot{B}$ and *radially symmetric* with respect to d, i.e. $\mathcal{S}(u)(x) = \mathcal{S}(u)(y)$ if $d(x) = d(y)$. Moreover, since $\mathcal{M}_r(1) = 1$ and d is constant on $\partial B_d(0, 1)$, we have

$$\mathcal{S}(\mathcal{S}(u)) = \mathcal{S}(a\, d^{2-Q} + b) = a\, d^{2-Q} + b = \mathcal{S}(u). \tag{9.69}$$

Prove the following result.

Lemma 9.9.7. *If u is $\mathcal{L}$-harmonic on $\dot{B}$ and d is an $\mathcal{L}$-gauge, there exist $a, b \in \mathbb{R}$ such that*

$$\begin{aligned} \mathcal{M}_r(u)(0) &= a\, r^{2-Q} + b \quad \text{for every } r \in]0,1[, \\ \mathcal{S}(u)(x) &= a\, d^{2-Q}(x) + b \quad \text{for every } x \in \dot{B}. \end{aligned} \tag{9.70}$$

Moreover, for every $n \in \mathbb{N}$, we have $\mathcal{S}^n(u) = \mathcal{S}(u)$, where $\mathcal{S}^n$ denotes the n-th iterate of the operator $\mathcal{S}$.

(*Hint*: Use Theorem 9.9.5, Corollary 9.9.6 and $\mathcal{M}_r(1) = 1$.)

The operator $\mathcal{S}$ can be used to prove the following *radial-symmetry* result for non-negative $\mathcal{L}$-harmonic functions in a punctured $\mathcal{L}$-gauge ball, vanishing on the boundary. Denote $B = B_d(0,1)$.

Theorem 9.9.8. *Let d be an $\mathcal{L}$-gauge, and let w be a non-negative $\mathcal{L}$-harmonic function in $\dot{B}$. Assume that w is continuous up to the boundary of B and $w(x) = 0$ for every $x \in \partial B$. Then w is d-radially symmetric. More precisely, $w(x) = \mathcal{S}(w)(x)$ for every $x \in \dot{B}$, so that, for a suitable real constant $a > 0$,*

$$w(x) = a\big(d^{2-Q}(x) - 1\big) \quad \text{for every } x \in \dot{B}.$$

Proof. Following an idea in [ABR92], we consider the Harnack inequality on d-spheres in Theorem 5.16.5 (page 327). Rewrite it to derive the existence of a constant $\mathbf{c} \in (0,1)$ such that

$$\mathbf{c}\, h(z) \le h(x) \quad \text{whenever } 0 < d(z) = d(x) \le \tfrac{1}{2} \text{ and } h \in \mathcal{H}(\dot{D}),\ h > 0. \tag{9.71}$$

We claim that

$$\begin{cases} h \in \mathcal{H}(\dot{B}) \cap C(\dot{\overline{B}}), \\ h \equiv 0 \text{ on } \partial B \text{ and } h \ge 0 \text{ on } \dot{B} \end{cases} \Longrightarrow \quad h - \mathbf{c}\, \mathcal{S}(h) \ge 0 \text{ on } \dot{B}, \tag{9.72}$$

where $\dot{\overline{B}} = \overline{B_d(0,1)} \setminus \{0\}$. Suppose that $w \in \mathcal{H}(\dot{B}) \cap C(\overline{B} \setminus \{0\})$ is non-negative on $\dot{B}$ and null on ∂B. Set

$$\mathbf{c}_n := 1 - (1 - \mathbf{c})^n, \quad n \in \mathbb{N} \cup \{0\}. \tag{9.73}$$

Let us prove by induction that

$$w - \mathbf{c}_n\, \mathcal{S}(w) \ge 0 \quad \text{on } \dot{B} \quad \text{for all } n \in \mathbb{N} \cup \{0\}. \tag{9.74}$$

The case $n = 0$ is obvious. Suppose (9.74) holds, and let us prove it for $n+1$. The function $h := w - \mathbf{c}_n\, \mathcal{S}(w)$ satisfies the hypothesis of (9.72). Indeed, $w|_{\partial B} = 0$ implies $\mathcal{S}(w)|_{\partial B} = 0$. Moreover, both w and $\mathcal{S}(w)$ belong to $\mathcal{H}(\dot{B}) \cap C(\dot{\overline{B}})$, since (see also Lemma 9.9.7, $b = -a$)

$$\mathcal{S}(w)(x) = \mathcal{M}_{d(x)}(w)(0) = a\, d^{2-Q}(x) - a. \tag{9.75}$$

Consequently, by (9.72), we have on $\dot{B}$

$$0 \leq h - \mathbf{c}\,\mathcal{S}(h) = w - \mathbf{c}_n\,\mathcal{S}(w) - \mathbf{c}\,\mathcal{S}\big(w - \mathbf{c}_n\,\mathcal{S}(w)\big) = w - \mathbf{c}_{n+1}\,\mathcal{S}(w).$$

(Use $\mathbf{c}_n + \mathbf{c} - \mathbf{c}\,\mathbf{c}_n = \mathbf{c}_{n+1}$ and Lemma 9.9.7.) Thus (9.74) is proved by induction. Letting $n \to \infty$ in it, we infer that $w \geq \mathcal{S}(w)$ on $\dot{B}$. On the other hand, the inequality $w(\overline{x}) > \mathcal{S}(w)(\overline{x})$ for some $\overline{x} \in \dot{B}$ is impossible, otherwise $\mathcal{S}(w)(\overline{x}) > \mathcal{S}(w)(\overline{x})$. Consequently, by using (9.75), $w(x) = \mathcal{S}(w)(x) = a(d^{2-Q}(x) - 1)$ for $x \in \dot{B}$.
It remains to prove (9.72). Let h be as in (9.72). From Lemma 9.9.7 we have $\mathcal{S}(h)(x) = a\, d^{2-Q}(x) + b$ on $\dot{B}$, so that

$$H := h - \mathbf{c}\,\mathcal{S}(h) \in \mathcal{H}(\dot{B}) \cap C(\overline{\dot{B}}).$$

Furthermore, $H = 0$ on ∂B. If we multiply the inequality in (9.71) times $(Q-2)\beta_d\, d^{1-Q}(x)\,\mathcal{K}(z)$ and then integrate w.r.t. $z \in B_d(0, d(x))$, we get $H(x) \geq 0$ for every $x \in B_d(0, 1/2)$. Then the weak maximum principle ensures that $H \geq 0$ on $\dot{B}$. □

Ex. 18) Derive Theorem 9.8.5 from Theorem 9.9.8.
(*Hint*: Let $B_d(x_0, \varepsilon) \subset \Omega$. Let h be the solution to the Dirichlet problem $\mathcal{L}h = 0$ in $B_d(x_0, \varepsilon)$, $h = u$ on $\partial B_d(x_0, \varepsilon)$. Consider the function $w(x) := (u-h)(x_0 * \delta_\varepsilon(x)) + \Gamma(x) - 1$. Prove, by the weak maximum principle, that $w \geq 0$ on $\dot{B}$. Hence, by Theorem 9.9.8, there exists $c \in \mathbb{R}$ such that $u = h + c\,\Gamma(x_0^{-1} * \cdot)$ for $x \in \dot{B}_d(x_0, \varepsilon)$. Then $c \geq 0$, otherwise $\lim_{x \to x_0} u(x) = -\infty$. Thus, $v := u - \Gamma(x_0^{-1} * \cdot)$ extends $\mathcal{L}$-harmonically through x_0.)

Ex. 19) *Let $x_0 \in \Omega \subseteq \mathbb{G}$, where Ω is open, and let $\dot{\Omega} = \Omega \setminus \{x_0\}$. Suppose $w : \dot{\Omega} \to \mathbb{R}$ satisfies the following conditions:*
(i) *$w \in C^\infty(\dot{\Omega})$ and w, $\nabla_{\mathcal{L}} w$ are bounded on $\dot{\Omega}$,*
(ii) *$\mathcal{L}w = f$ in $\dot{\Omega}$, where $f \in C^\infty(\Omega)$.*
Then w extends to a C^∞ function on the whole Ω.
(*Hint*: Suppose $x_0 = 0$. It is enough to prove that $\mathcal{L}w = f$ in Ω, in the weak sense of distributions. Let $\varphi \in C_0^\infty(\Omega)$, and let d be a homogeneous norm on $\mathbb{G}$, smooth on $\dot{\mathbb{G}}$. Argue as in the proof of Corollary 9.9.6, to show that

$$\begin{aligned}
\int_\Omega w\, \mathcal{L}\varphi &= \lim_{\varepsilon \to 0}\left(-\int_{d(x)=\varepsilon} w\, \langle \nabla_{\mathcal{L}}\varphi, \nabla_{\mathcal{L}} d \rangle \frac{\mathrm{d}H^{N-1}}{|\nabla d|} - \int_{d>\varepsilon} \langle \nabla_{\mathcal{L}} w, \nabla_{\mathcal{L}}\varphi \rangle \right) \\
&= \lim_{\varepsilon \to 0}\left(-I_1(\varepsilon) + \int_{d=\varepsilon} \varphi\, \langle \nabla_{\mathcal{L}} w, \nabla_{\mathcal{L}} d \rangle \frac{\mathrm{d}H^{N-1}}{|\nabla d|} + \int_{d>\varepsilon} \varphi\, \mathcal{L}w \right) \\
&= \lim_{\varepsilon \to 0}\left(-I_1(\varepsilon) + I_2(\varepsilon) + \int_{d>\varepsilon} f\,\varphi \right) = \int_\Omega f\,\varphi.
\end{aligned}$$

Indeed (see also Ex. 24, Chapter 5), $|I_1(\varepsilon)|, |I_2(\varepsilon)| \leq c\,\varepsilon^{Q-1}$.)

Ex. 20) Let $\Omega \subseteq \mathbb{G}$ be open, and let $u \in \underline{\mathcal{S}}(\Omega)$. Show that, for every $x \in \Omega$ and $r > 0$ such that $\overline{B_d(x,r)} \subseteq \Omega$, one has

$$u(x) = \mathcal{M}_r(u)(x) - \beta_d(Q-2)\int_0^r \frac{\mu(B_d(x,t))}{t^{Q-1}}\,\mathrm{d}t,$$

where $\mu := \mathcal{L}u$ is the $\mathcal{L}$-Riesz measure of u.
(*Hint*: Write

$$\Gamma(d) - \Gamma(r) = \int_d^r \Gamma'(t)\,\mathrm{d}t$$

and use Fubini's theorem in the Poisson–Jensen formula (9.22).)

Ex. 21) Assume the same hypotheses as in the previous exercise. Show that

$$\mathcal{M}_r(u)(x) - \mathcal{M}_\rho(u)(x) = \beta_d(Q-2)\int_\rho^r \frac{\mu(B_d(x,t))}{t^{Q-1}}\,\mathrm{d}t,$$

for $0 < \rho < r$.
(*Hint*: First, assume $u(x) > -\infty$ and use the Poisson–Jensen formula (9.22). Then complete the proof by replacing u with its Perron-regularized $u_{B_d(x,\rho/2)}$.)

Ex. 22) Let $\Omega \subseteq \mathbb{G}$ be open, and let $u \in \overline{\mathcal{S}}(\Omega)$. Show that, for every $x \in \Omega$,

$$\mu(\{x\}) = \frac{1}{\beta_d}\lim_{\rho\searrow 0} \rho^{Q-2}\mathcal{M}_\rho(u)(x),$$

where $\mu := -\mathcal{L}u$ is the $\mathcal{L}$-Riesz measure of u.
(*Hint*: Use the previous exercise.)

Ex. 23) Assume the same hypotheses as in the previous exercise. Show that

$$\mu(\{x\}) = 0 \quad \text{if } u(x) < \infty.$$

(*Hint*: Use the previous exercise.)

Ex. 24) Let $f : \mathbb{G} \to \mathbb{R} \cup \{\infty\}$ be defined as follows

$$f(x) = \begin{cases} (d(x))^{-2} & \text{if } 0 < d(x) < 1, \\ \infty & \text{if } x = 0, \\ 0 & \text{if } d(x) \geq 1. \end{cases}$$

Show that:
(i) $f \in L^1(\mathbb{G})$,
(ii) $x \mapsto u(x) := \int_{\mathbb{R}^N} \Gamma(y^{-1} \circ x) f(y)\,\mathrm{d}y$ is $\mathcal{L}$-superharmonic in $\mathbb{G}$,
(iii) $u(0) = \infty$,
(iv) $\mu(\{0\}) = 0$, where μ is the $\mathcal{L}$-Riesz measure of u.

10

Maximum Principle on Unbounded Domains

The aim of this chapter is to study the following version of the maximum principle for sub-Laplacians $\mathcal{L}$ on Carnot groups $\mathbb{G}$. An open set $\Omega \subseteq \mathbb{G}$ is called a *maximum principle set for* $\mathcal{L}$ (MP set, in short) if every *bounded from above* $\mathcal{L}$-subharmonic function u is ≤ 0 in Ω whenever

$$\limsup_{x \to y} u(x) \le 0 \qquad \text{for every } y \in \partial\Omega.$$

We already know that bounded domains are MP sets. Here, we introduce the notion of $\mathcal{L}$-thin set at infinity, and we prove that Ω is an MP set if and only if $\mathbb{G} \setminus \Omega$ is *not* $\mathcal{L}$-thin at infinity. Then, we show some geometrical conditions for non-$\mathcal{L}$-thinness at infinity, based on the notion of q-set. We prove that a sufficient condition for F to be not $\mathcal{L}$-thin at infinity is that F is not a q-set, for some $q > Q - 2$. As usual, Q is the homogeneous dimension of $\mathbb{G}$. To this end, a crucial tool will be Theorem 9.6.1 (page 451), the representation theorem for bounded from above $\mathcal{L}$-subharmonic functions on $\mathbb{G}$.

It will follow, in particular, that open sets Ω satisfying an exterior $\mathbb{G}$-cone condition are MP sets, since $\mathbb{G} \setminus \Omega$ is not a Q-set, hence not $\mathcal{L}$-thin at infinity. Some other explicit criteria are given.

As usual, $\mathbb{G} = (\mathbb{R}^N, \circ, \delta_\lambda)$ will denote a homogeneous Carnot group and $\mathcal{L}$ a fixed sub-Laplacian on $\mathbb{G}$.

10.1 MP Sets and $\mathcal{L}$-thinness at Infinity

The maximum principle of Theorem 5.13.4 (page 295) does not hold, in general, if the open set Ω is unbounded.[1] The aim of this section is to provide sufficient conditions on unbounded Ω's for an analogue of the maximum principle to hold for a given sub-Laplacian $\mathcal{L}$.

[1] Consider, for example, the classical case of the ordinary Laplace operator on $\mathbb{R}^2$, $\Omega = \{(x, y) \in \mathbb{R}^2 : y > 0\}$, $u(x, y) = \exp(x)\sin(y)$. Then u satisfies $\Delta u(x, y) = 0$ in Ω, $u = 0$ on $\partial\Omega$, but u is not identically zero.

To begin with, we introduce some notation and definitions. If $\Omega \subseteq \mathbb{G}$ is open, $\underline{S}_b(\Omega)$ will denote the (cone of the) *bounded from above* $\mathcal{L}$-subharmonic functions in Ω. For example, $-\Gamma \in \underline{S}_b(\mathbb{G})$.

Definition 10.1.1 (MP set for $\mathcal{L}$). *We say that Ω is a* maximum principle set for $\mathcal{L}$ *(MP set for $\mathcal{L}$, in short) if the following assertion holds:*

$$\begin{cases} u \in \underline{S}_b(\Omega) \\ \limsup_{x\to y} u(x) \le 0 \quad \forall\, y \in \partial\Omega \end{cases} \quad \Longrightarrow \quad u \le 0 \ \textit{in}\ \Omega. \tag{10.1}$$

By Theorem 8.2.19-(ii), page 409, it follows that

every bounded open set is an MP set for $\mathcal{L}$.

We would like to remark that an u.s.c. function $u : \Omega \to [-\infty, \infty[$ satisfying the boundary condition in (10.1) is bounded from above if Ω is bounded.[2] Then, if Ω is bounded, in (10.1), $\underline{S}_b(\Omega)$ may be replaced by $\underline{S}(\Omega)$.

The maximum principle is deeply related to the notion of $\mathcal{L}$-thinness at infinity, which we now introduce.

Definition 10.1.2 ($\mathcal{L}$-thinness at infinity). *A subset F of $\mathbb{G}$ will be said* $\mathcal{L}$-thin at infinity *if there exists a bounded from above $\mathcal{L}$-subharmonic function u in $\mathbb{G}$ such that*[3]

$$\limsup_{x\to\infty,\ x\in F} u(x) \lneqq \limsup_{x\to\infty,\ x\in\mathbb{G}} u(x). \tag{10.2}$$

From this definition it follows that a set $F \subseteq \mathbb{G}$ is *not* $\mathcal{L}$-thin at infinity iff for every $u \in \underline{S}_b(\mathbb{G})$

$$\limsup_{x\to\infty,\ x\in F} u(x) = \limsup_{x\to\infty,\ x\in\mathbb{G}} u(x).$$

The following theorem holds (see also Fig. 10.1).

Theorem 10.1.3 (Characterization of MP set). *An open set $\Omega \subseteq \mathbb{G}$ is an MP set for $\mathcal{L}$ if and only if $\mathbb{G} \setminus \Omega$ is not $\mathcal{L}$-thin at infinity.*

The proof of this result will easily follow from the next two lemmas.

Lemma 10.1.4. *Let $u : \Omega \to [-\infty, \infty[$ be an $\mathcal{L}$-subharmonic function in Ω satisfying*

$$\limsup_{x\to y} u(x) \le 0 \quad \forall\, y \in \partial\Omega. \tag{10.3}$$

Then the function

$$v : \mathbb{G} \to [-\infty, \infty[, \qquad v = \begin{cases} \max\{u, 0\} & \textit{in}\ \Omega, \\ 0 & \textit{in}\ \mathbb{G}\setminus\Omega \end{cases}$$

is $\mathcal{L}$-subharmonic in $\mathbb{G}$.

---

[2] See Exercise 1 at the end of the Chapter.

[3] We agree to let

$$\limsup_{x\to\infty,\ x\in F} u(x) = -\infty,$$

if F is bounded.

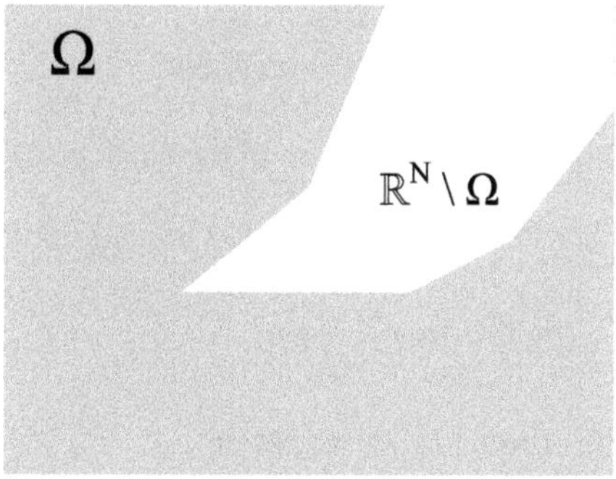

Fig. 10.1. A MP set Ω

Proof. Condition (10.3) implies that v is upper semicontinuous in $\mathbb{G}$. Moreover, since $\max\{u, 0\} \in \underline{S}(\Omega)$, v is locally sub-mean in Ω. On the other hand, since $v \geq 0$,

$$v(x) = 0 \leq \mathrm{M}_r(v)(x) \qquad \text{for every } x \in \mathbb{G} \setminus \Omega.$$

Then, v is everywhere locally sub-mean, hence $\mathcal{L}$-subharmonic in $\mathbb{G}$. □

Lemma 10.1.5. *Let $F \subseteq \mathbb{G}$ and let $u \in \underline{S}_b(\mathbb{G})$. Then*

$$\limsup_{x\to\infty,\ x\in F} u(x) = \limsup_{x\to\infty,\ x\in\mathbb{G}} u(x) \tag{10.4a}$$

if and only if

$$\sup_F u = \sup_{\mathbb{G}} u. \tag{10.4b}$$

Proof. We may assume that u is not constant. Then, by the maximum principle of Theorem 8.2.19 (page 409), we have

$$u(x) < \sup_{\mathbb{G}} u \qquad \forall\, x \in \mathbb{G}. \tag{10.5}$$

As a consequence,

$$\sup_{\mathbb{G}} u = \limsup_{x\to\infty,\ x\in\mathbb{G}} u(x). \tag{10.6}$$

Then

$$\limsup_{x\to\infty,\ x\in F} u(x) \leq \sup_F u \leq \sup_{\mathbb{G}} u = \limsup_{x\to\infty,\ x\in\mathbb{G}} u(x).$$

This shows that (10.4a) implies (10.4b).

Let us now assume (10.4b). Then

$$\sup_{F\cap\overline{B_d(0,R)}} u < \sup_{F\setminus\overline{B_d(0,R)}} u \quad \forall\, R > 0. \tag{10.7}$$

Indeed, if (10.7) were false, we would have

$$\sup_{F\cap\overline{B_d(0,R)}} u \geq \sup_{F\setminus\overline{B_d(0,R)}} u,$$

for a certain $R > 0$, so that, for a suitable x_0 in the closure if $F \cap \overline{B_d(0, R)}$,

$$\begin{aligned} u(x_0) &= \sup_{F\cap\overline{B_d(0,R)}} u = \sup_F u \\ \text{(by (10.4b))} &= \sup_{\mathbb{G}} u, \end{aligned}$$

thus contradicting (10.5). From (10.7), we obtain

$$\begin{aligned} \limsup_{x\to\infty,\ x\in F} u(x) &= \sup_F u \\ \text{(by (10.4b))} &= \sup_{\mathbb{G}} u \\ \text{(by (10.6))} &= \limsup_{x\to\infty,\ x\in\mathbb{G}} u(x), \end{aligned}$$

and (10.4a) follows. □

From Lemma 10.1.5 and the very definition of $\mathcal{L}$-thin set, we have the following assertion.

Corollary 10.1.6 (Characterization of $\mathcal{L}$-thinness at infinity). *Let $F \subseteq \mathbb{G}$. Then F is $\mathcal{L}$-thin at infinity iff there exists $u \in \underline{S}_b(\mathbb{G})$ such that $\sup_F u < \sup_{\mathbb{G}} u$. Vice versa, F is not $\mathcal{L}$-thin at infinity iff $\sup_F u = \sup_{\mathbb{G}} u$ for all $u \in \underline{S}_b(\mathbb{G})$.*

We are now ready to prove Theorem 10.1.3.

Proof (of Theorem 10.1.3). Assuming that Ω is an MP set for $\mathcal{L}$, we shall prove that $\mathbb{G} \setminus \Omega$ is not $\mathcal{L}$-thin at infinity. Let $u \in \underline{S}_b(\mathbb{G})$. Put

$$U := \sup_{\mathbb{G}\setminus\Omega} u. \tag{10.8}$$

Then

$$\limsup_{\Omega\ni x\to y} u(x) \leq u(y) \leq U \quad \forall\ y \in \partial\Omega,$$

so that, since Ω is an MP set, $u = U$ in Ω. Definition (10.8) now implies

$$\sup_{\mathbb{G}\setminus\Omega} u = \sup_{\mathbb{G}} u,$$

hence, by Lemma 10.1.5, $\mathbb{G} \setminus \Omega$ is not $\mathcal{L}$-thin at infinity.

Vice versa, assuming that $\mathbb{G} \setminus \Omega$ is not $\mathcal{L}$-thin at infinity, we shall prove that Ω is an MP set. Let $u \in \underline{S}_b(\Omega)$ be such that

$$\limsup_{x\to y} u(x) \leq 0 \quad \forall\, y \in \partial\Omega.$$

By Lemma 10.1.4, the function

$$v : \mathbb{G} \to [-\infty, \infty[, \qquad v = \begin{cases} \max\{u, 0\} & \text{in } \Omega, \\ 0 & \text{in } \mathbb{G} \setminus \Omega \end{cases}$$

is $\mathcal{L}$-subharmonic in $\mathbb{G}$. Then, being $\mathbb{G} \setminus \Omega$ not $\mathcal{L}$-thin at infinity,

$$\sup_{\mathbb{G}} v = \sup_{\mathbb{G}\setminus\Omega} v = 0.$$

In particular, $v \leq 0$ in Ω. Hence, $u \leq 0$ in Ω, and Ω is an MP set. □

From Theorem 10.1.3 and Corollary 10.1.6 we obtain the following equivalent characterization of MP sets.

Corollary 10.1.7 (Characterization of MP set. II). *An open set $\Omega \subseteq \mathbb{G}$ is an MP set for $\mathcal{L}$ if and only if $\sup_{\mathbb{G}\setminus\Omega} u = \sup_{\mathbb{G}} u$ for all $u \in \underline{S}_b(\mathbb{G})$.*

10.2 q-sets and the Maximum Principle

This section is devoted to some geometrical conditions for non-thinness (w.r.t. $\mathcal{L}$) at infinity. Our criteria will be based on the notion of q-set.

Throughout the sequel, $\mathcal{L}$ will denote a fixed sub-Laplacian on $\mathbb{G}$, Γ its fundamental solution and d any $\mathcal{L}$-gauge for $\mathcal{L}$. We adopt the usual notation

$$\Gamma = \beta_d \, d^{2-Q}$$

(see Theorem 5.5.6, page 256). We recall that Q denotes the homogeneous dimension of $\mathbb{G}$, and β_d is a suitable positive constant.

Definition 10.2.1 (q-set for a sub-Laplacian). *Let q be a real positive number. A subset F of $\mathbb{G}$ will be called a q-set (w.r.t. $\mathcal{L}$) if there exists a finite or countable family of d-balls $\{B_d(x_j, r_j)\}_{j\in J}$ such that:*

(i) $F \subseteq \bigcup_{j\in J} \overline{B_d(x_j, r_j)}$,
(ii) $\sum_{j\in J} (\frac{r_j}{d(x_j)})^q < \infty$.

(See also Fig. 10.2.) The next theorem will play a crucial rôle in what follows.

Theorem 10.2.2 ($\mathcal{L}$-thinness at infinity and q-sets). *Let $F \subseteq \mathbb{G}$ be $\mathcal{L}$-thin at infinity. Then F is a q-set for every $q > Q - 2$.*

The proof of Theorem 10.2.2 rests on the following deep lemma.

Lemma 10.2.3. *Let $u \in \underline{S}_b(\mathbb{G})$. Then, for every $q > Q - 2$, there exists a q-set $F \subseteq \mathbb{G}$ such that*

$$\lim_{x\to\infty,\ x\notin F} u = \sup_{\mathbb{G}} u.$$

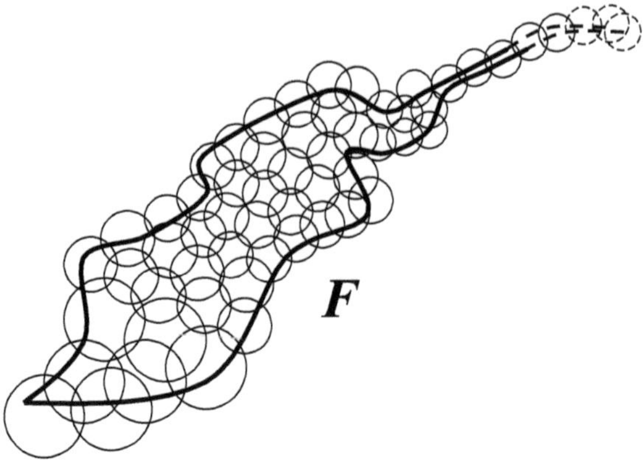

Fig. 10.2. A q-set F

We prove this result in the Appendix of this chapter. Note that the assertion of the above lemma may fail to be true for $q = Q - 2$ (see Ex. 7 at the end of the chapter).

We are now in the position to prove Theorem 10.2.2.

Proof (of Theorem 10.2.2). Assume by contradiction that F is not a q_0-set, being $q_0 > Q - 2$. Fixed $u \in \mathcal{S}_b(\mathbb{G})$, by Lemma 10.2.3, there exists a q_0-set $F_0 \subseteq \mathbb{G}$ such that

$$\lim_{x \to \infty,\ x \notin F_0} u = \sup_{\mathbb{G}} u. \tag{10.9}$$

On the other hand, $F \setminus F_0$ is not a q_0-set, since F is not. In particular, $F \setminus F_0$ is non-empty and unbounded. Then, by (10.9),

$$\lim_{x \to \infty,\ x \in F \setminus F_0} u = \sup_{\mathbb{G}} u,$$

so that

$$\limsup_{x \to \infty,\ x \in F} u(x) = \sup_{\mathbb{G}} u = \limsup_{x \to \infty} u(x).$$

This proves that F is not $\mathcal{L}$-thin at infinity, in contradiction with the hypothesis. □

For the future references, it is convenient to rephrase the previous theorem as follows.

Theorem 10.2.4. *Let $F \subseteq \mathbb{G}$. Assume there exists $q > Q - 2$ such that F is not a q-set. Then F is not $\mathcal{L}$-thin at infinity.*

Our more explicit geometric condition for non-$\mathcal{L}$-thinness at infinity will follow from the next proposition.

Proposition 10.2.5. *Let $F \subseteq \mathbb{G}$. Assume there exists a sequence of d-balls $\{B_d(z_j, R_j)\}_{j \in \mathbb{N}}$ contained in F and such that*

$$d(z_j) \to \infty, \qquad \liminf_{j\to\infty} \frac{R_j}{d(z_j)} > 0.$$

Then F is not a Q-set (hence, it is not $\mathcal{L}$-thin at infinity).

Proof. Let $\mathbf{c}$ be the constant in the pseudo-triangle inequality (see Proposition 5.1.7, page 231). Let us put

$$M := 4\mathbf{c}^2$$

and choose a subsequence $\{z_{k_j}\}_j$ of $\{z_j\}_j$ such that, for each $j \in \mathbb{N}$,

$$d(z_{k_{j+1}}) \geq M^2 \, d(z_{k_j}), \qquad R_{k_j} \geq \delta \, d(z_{k_j}), \tag{10.10}$$

with a suitable $\delta \in]0, 1/M[$. Let us now define

$$y_j := z_{k_j}, \quad \rho_j := \delta \, d(y_j), \quad B_j := B_d(y_j, \rho_j), \quad F_0 := \bigcup_{j\in\mathbb{N}} B_j.$$

Since $\rho_j \leq R_{k_j}$, we have $B_j \subseteq B_d(z_{k_j}, R_{k_j}) \subseteq F$ for every $j \in \mathbb{N}$. Then, in order to show that F is not a Q-set, it is enough to prove that F_0 is not a Q-set.

We argue by contradiction and assume the existence of a family of d-balls $\{B_d(x_k, r_k)\}_{k\in\mathbb{N}}$ covering F_0 and such that

$$\sum_{k=1}^{\infty} \left(\frac{r_k}{d(x_k)}\right)^Q < \infty.$$

Then, if we fix $\varepsilon \in]0, 1/M[$, there exists $\overline{k} \in \mathbb{N}$ such that

$$r_k \leq \varepsilon \, d(x_k) \quad \forall k \geq \overline{k}.$$

Moreover, there exists $\overline{j} \in \mathbb{N}$ such that the family

$$\{D_k\}_{k\geq\overline{k}}, \quad D_k := B_d(x_k, r_k),$$

is a covering of

$$\bigcup_{j\geq\overline{j}} B_j.$$

For every $j \in \mathbb{N}$, $j \geq \overline{j}$, we finally define

$$K_j := \{k \in \mathbb{N} \,|\, k \geq \overline{k}, \; B_j \cap D_k \neq \emptyset\}.$$

We claim that

$$\frac{1}{M} \leq \frac{d(x_k)}{d(y_j)} \leq M \qquad \forall k \in K_j, \qquad \forall j \geq \overline{j}. \tag{10.11}$$

These inequalities, together with the first one in (10.10), imply

$$K_i \cap K_j = \emptyset \text{ if } i \neq j.$$

By means of this last fact, we infer that

$$\begin{aligned}
\infty > \sum_{k \geq \overline{k}} \left(\frac{r_k}{d(x_k)}\right)^Q &\geq \sum_{j \geq \overline{j}} \sum_{k \in K_j} \left(\frac{r_k}{d(x_k)}\right)^Q \\
\text{(by (10.11))} \geq \left(\frac{1}{M}\right)^Q &\sum_{j \geq \overline{j}} \left(\frac{1}{d(y_j)}\right)^Q \sum_{k \in K_j} (r_k)^Q \\
&= \frac{1}{\omega_d\, M^Q} \sum_{j \geq \overline{j}} \left(\frac{1}{d(y_j)}\right)^Q \sum_{k \in K_j} \text{meas}(D_k) \\
&\text{(since } \{D_k\}_{k \in K_j} \text{ is a covering of } B_j) \\
&\geq \frac{1}{\omega_d\, M^Q} \sum_{j \geq \overline{j}} \left(\frac{1}{d(y_j)}\right)^Q \text{meas}(B_j) \\
&= \left(\frac{1}{M}\right)^Q \sum_{j \geq \overline{j}} \left(\frac{\rho_j}{d(y_j)}\right)^Q = \infty,
\end{aligned}$$

since $\rho_j = \delta\, d(y_j)$ for every $j \in \mathbb{N}$. This contradiction shows that F_0 cannot be a Q-set.

We are then left with the proof of (10.11). If $k \in K_j$, there exists $x' \in B_j \cap D_k$. Hence by the pseudo-triangle inequality,

$$\begin{aligned}
d(x_k) &\leq \mathbf{c}\,(d(y_j) + d(y_j, x_k)) \leq \mathbf{c}^2\big(d(y_j) + d(y_j, x') + d(x', x_k)\big) \\
&\leq \mathbf{c}^2\,(d(y_j) + \rho_j + r_k) \leq \mathbf{c}^2\big((1+\delta)\, d(y_j) + \varepsilon\, d(x_k)\big) \\
&\leq 2\,\mathbf{c}^2\, d(y_j) + \frac{1}{2}\, d(x_k).
\end{aligned}$$

Then

$$\frac{1}{2}\, d(x_k) \leq 2\,\mathbf{c}^2\, d(y_j),$$

and the second inequality in (10.11) follows. With the same argument, one can prove the first one too. □

Examples of subsets of $\mathbb{G}$ satisfying the hypothesis of Proposition 10.2.5 are the $\mathbb{G}$-cones. The relevant definition is the following one.

Definition 10.2.6 ($\mathbb{G}$-cone). *A subset $\mathcal{C}$ of $\mathbb{G}$ will be said a* $\mathbb{G}$-cone with vertex at the origin *if* (see also Fig. 10.3)

$$\delta_\lambda(x) \in \mathcal{C} \qquad \forall\, x \in \mathcal{C}, \quad \forall\, \lambda > 0.$$

If $\mathcal{C}$ is such a $\mathbb{G}$-cone, we shall call

$$z_0 \circ \mathcal{C} = \{z_0 \circ x \mid x \in \mathcal{C}\}$$

a $\mathbb{G}$-cone with vertex at z_0.

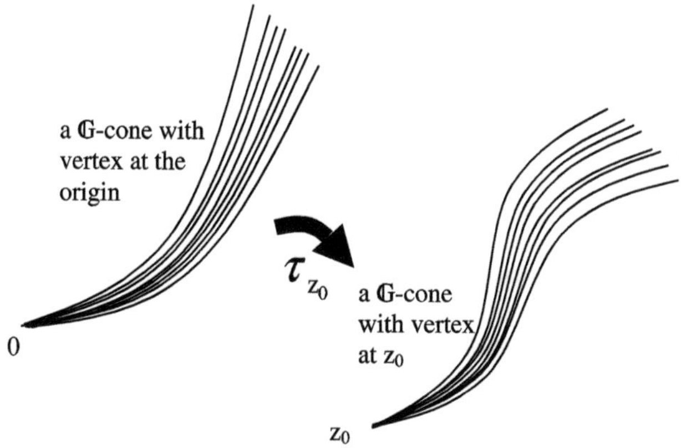

Fig. 10.3. $\mathbb{G}$-cones

Proposition 10.2.7. *Every open $\mathbb{G}$-cone is not a Q-set.*

Proof. It is enough to prove the assertion for an open $\mathbb{G}$-cone $\mathcal{C}$ with vertex at the origin. Let $x_0 \in \mathcal{C} \setminus \{0\}$ and $R_0 > 0$ be such that $B_d(x_0, R_0) \subseteq \mathcal{C}$. Then, since $\mathcal{C}$ is a $\mathbb{G}$-cone,

$$B_d(\delta_\lambda(x_0), \lambda R_0) \subseteq \mathcal{C} \quad \forall \lambda > 0.$$

The assertion follows from Proposition 10.2.5, since

$$z_\lambda := \delta_\lambda(x_0) \longrightarrow \infty \text{ as } \lambda \to \infty$$

and

$$\frac{\lambda R_0}{d(z_\lambda)} = \frac{R_0}{d(x_0)} \quad \forall \lambda > 0.$$

This completes the proof. $\square$

Corollary 10.2.8. *Every half-space of $\mathbb{G}$ is not a Q-set.*

Proof. Let

$$\Pi := \left\{ x = (x_1, \ldots, x_N) \in \mathbb{G} \, : \, \sum_{j=1}^{N} a_j x_j > \alpha \right\}$$

be an open half-space of $\mathbb{G}$ (here, $a_1, \ldots, a_N, \alpha \in \mathbb{R}$), and let us consider

$$\mathcal{C}_0 := \left\{ x = (x_1, \ldots, x_N) \in \Pi \, : \, a_j x_j \geq 0 \ \forall j = 1, \ldots, N \right\}.$$

$\mathcal{C}_0$ is a non-empty open subset of Π with the following property:

$$x \in \mathcal{C}_0, \lambda \geq 1 \quad \Longrightarrow \quad \delta_\lambda(x) \in \mathcal{C}_0.$$

Then, if we choose a point $x_0 \in \mathcal{C}_0$ and a real number $R_0 > 0$ such that $B_d(x_0, R_0) \subseteq \mathcal{C}_0$, we have $B_d(\delta_\lambda(x_0), \lambda R_0) \subseteq \mathcal{C}_0$ for every $\lambda \geq 1$. Since

$$z_\lambda := \delta_\lambda(x_0) \longrightarrow \infty \text{ as } \lambda \to \infty$$

and

$$\frac{\lambda R_0}{d(z_\lambda)} = \frac{R_0}{d(x_0)} \quad \forall \lambda \geq 1,$$

from Proposition 10.2.5 it follows that $\mathcal{C}_0$, and hence Π, is not a Q-set. □

10.3 The Maximum Principle on Unbounded Domains

We collect Theorems 10.1.3, 10.2.2 and 10.2.4, Proposition 10.2.5 and Corollaries 10.2.7 and 10.2.8 in the following theorem.

Theorem 10.3.1. *The open set $\Omega \subseteq \mathbb{G}$ is an MP set for $\mathcal{L}$ if one of the following (sufficient) conditions is satisfied:*

(1) $\mathbb{G} \setminus \Omega$ *is not $\mathcal{L}$-thin at infinity (this condition is also necessary),*
(2) $\mathbb{G} \setminus \Omega$ *is not a q-set for some $q > Q - 2$,*
(3) $\mathbb{G} \setminus \Omega$ *contains a sequence of d-balls $\{B_d(z_j, R_j)\}_{j \in \mathbb{N}}$ such that*

$$d(z_j) \to \infty, \qquad \liminf_{j \to \infty} \frac{R_j}{d(z_j)} > 0,$$

(4) $\mathbb{G} \setminus \Omega$ *contains an open $\mathbb{G}$-cone,*
(5) $\mathbb{G} \setminus \Omega$ *contains a half-space or, equivalently, Ω is contained in a half-space.*

We close this section by giving an application of the previous theorem.

Corollary 10.3.2 (A maximum principle on unbounded domains). *Let Ω be an open subset of $\mathbb{G}$ satisfying one of the five conditions of Theorem 10.3.1. Let $c : \Omega \to \mathbb{R}$, $c \leq 0$, and let u be a bounded from above function on Ω of class C^2 satisfying*

$$\begin{cases} \mathcal{L}u + c\,u \geq 0 & \text{in } \Omega, \\ \limsup_{x \to y} u(x) \leq 0 & \text{for every } y \in \partial\Omega. \end{cases} \tag{10.12}$$

Then $u \leq 0$ in Ω.

Proof. Define $v := \max\{u, 0\}$ and

$$\Omega_0 := \{x \in \Omega \mid u(x) > 0\}.$$

We have to show that $\Omega_0 = \emptyset$. Assume, by contradiction, $\Omega_0 \neq \emptyset$. Then, since $c \leq 0$ and $v = u$ in Ω_0, the function v is of class C^2 in Ω_0 and $\mathcal{L}v \geq 0$ in Ω_0. It follows that v is $\mathcal{L}$-subharmonic, hence $\mathcal{L}$-submean, in Ω_0. On the other hand, for every $x \in \Omega \setminus \Omega_0$ and $r > 0$ such that $\overline{B_d(x, r)} \subset \Omega$,

$$v(x) = 0 \leq \mathrm{M}_r(v)(x).$$

This shows that v is locally $\mathcal{L}$-submean in Ω. Thus $v \in \underline{S}_b(\Omega)$. From the boundary condition in (10.12) we also get $\limsup_{x \to y} v(x) \leq 0$ for every $y \in \partial\Omega$. Since Ω is an MP set (by Theorem 10.3.1), this implies $v \leq 0$ in Ω, i.e. $u \leq 0$ in Ω. Then $\Omega_0 = \emptyset$, and the proof is complete. □

10.4 Appendix to Chapter 10. The Proof of Lemma 10.2.3

The proof of Lemma 10.2.3 rests on the following deep lemma, an extension to the sub-Laplacian setting of a theorem by Cartan [Cart28] related to the classical Laplace operator (see also [HK76, pp. 131–134]).

In the proof of the lemma below, **c** denotes, as usual, the constant appearing in the pseudo-triangle inequality for the $\mathcal{L}$-gauge d (and $\Gamma = \beta_d\, d$ is the fundamental solution for $\mathcal{L}$).

Lemma 10.4.1 (Cartan-type covering lemma). *Let μ be a Radon measure on $\mathbb{G}$ with finite total mass $\mu_0 := \mu(\mathbb{G}) < \infty$. Let $q > Q-2$ (Q being the homogeneous dimension of $\mathbb{G}$). Then, if $h > 0$, the set*

$$\{x \in \mathbb{G} \,|\, (\Gamma * \mu)(x) \geq h\}$$

can be covered by a finite or countable family of closed d-balls $\overline{B_d(x_n, r_n)}$ satisfying the following condition

$$\sum_n {r_n}^q < A\,(\mu_0/h)^{q/(Q-2)}.$$

The constant A depends only on Q, q and **c**.

Proof. We fix $\nu \in \mathbb{N}$ and set

$$r_\nu := (\mu_0/h)^{1/(Q-2)}\, 2^{-\frac{2\nu}{q+Q-2}}.$$

Let

$$\mathcal{D}_\nu := \{\overline{D}_{k,\nu} = \overline{B_d(x_{k,\nu}, r_\nu/2)} \,|\, k = 1, \ldots, k_\nu\}$$

be a maximal family of disjoint closed balls such that

$$\mu(\overline{D}_{k,\nu}) \geq \frac{\mu_0}{2^\nu} \qquad \forall\, k = 1, \ldots, k_\nu.$$

Since $\mu_0 < \infty$ and the balls are disjoint, then $k_\nu \leq 2^\nu$. Define

$$F := \bigcup_{\nu\in\mathbb{N}} \bigcup_{k \leq k_\nu} \overline{B_d(x_{k,\nu}, \mathbf{c}\, r_\nu)}.$$

We observe that if $x \notin F$ then $\overline{B_d}(x, r_\nu/2)$ does not intersect any ball of the maximal family $\mathcal{D}_\nu$. Hence

$$\mu(\overline{B_d(x, r_\nu/2)}) < \frac{\mu_0}{2^\nu} \qquad \forall\, \nu \in \mathbb{N}. \tag{10.13}$$

In particular, we have

$$\mu(\{x\}) = \lim_{\nu\to\infty} \mu(\overline{B_d(x, r_\nu/2)}) = 0. \tag{10.14}$$

Hence, if $x \notin F$, by (10.14) we have

$$
\begin{aligned}
(\Gamma * \mu)(x) &= \int_{\mathbb{G}\setminus\{x\}} \Gamma(x^{-1} \circ y)\,\mathrm{d}\mu(y) \\
&= \left(\int_{d(x^{-1}\circ y) \geq \frac{r_1}{2}} + \sum_{\nu=1}^{\infty} \int_{\frac{r_{\nu+1}}{2} \leq d(x^{-1}\circ y) < \frac{r_\nu}{2}} \right) \Gamma(x^{-1} \circ y)\,\mathrm{d}\mu(y) \\
&\leq \beta_d\,(r_1/2)^{2-Q}\,\mu_0 + \sum_{\nu=1}^{\infty} \beta_d\,(r_{\nu+1}/2)^{2-Q} \int_{\frac{r_{\nu+1}}{2} \leq d(x^{-1}\circ y) < \frac{r_\nu}{2}} \mathrm{d}\mu(y) \\
&\leq \text{(by (10.13))} \quad \beta_d\,2^{Q-2}\,\mu_0 \sum_{\nu=1}^{\infty} 2^{1-\nu}\,{r_\nu}^{2-Q} \\
&= \beta_d\,2^{Q-1}\,h \sum_{\nu=1}^{\infty} 2^{\frac{Q-2-q}{Q-2+q}\,\nu} =: A_1\,h,
\end{aligned}
$$

where A_1 depends only on Q and q. Then

$$
\{x \in \mathbb{G} \,|\, (\Gamma * \mu)(x) > A_1\,h\} \subseteq \bigcup_{\nu\in\mathbb{N}} \bigcup_{k\leq k_\nu} \overline{B_d(x_{k,\nu}, \mathbf{c}\,r_\nu)}.
$$

Moreover, since $k_\nu \leq 2^\nu$, we have

$$
\sum_{\nu\in\mathbb{N}} \sum_{k\leq k_\nu} (\mathbf{c}\,r_\nu)^q \leq A_2 \left(\frac{\mu_0}{h} \right)^{q/(Q-2)},
$$

where A_2 depends only on q, Q and $\mathbf{c}$. This ends the proof. □

We are now ready to prove Lemma 10.2.3.

Proof (of Lemma 10.2.3). Let $\nu \in \mathbb{N}$. Define

$$
C_\nu := \{x \in \mathbb{G} \,|\, (2\,\mathbf{c})^\nu < d(x) \leq (2\mathbf{c})^{\nu+1}\},
$$

where $\mathbf{c}$ denotes the constant appearing in the pseudo-triangle inequality for d (see Proposition 5.1.7, page 231). By Theorem 9.6.1 (page 451), if μ is the $\mathcal{L}$-Riesz measure of u and $U := \sup_{\mathbb{G}} u$, we have

$$
U - u(x) = (\Gamma * \mu)(x) = I_1(x) + I_2(x) + I_3(x),
$$

where

$$
\begin{aligned}
I_1(x) &:= \int_{d(y)\leq(2\,\mathbf{c})^{\nu-1}} \beta_d\, d(x^{-1} \circ y)^{2-Q}\,\mathrm{d}\mu(y)\,; \\
I_2(x) &:= \int_{(2\,\mathbf{c})^{\nu-1} < d(y) < (2\,\mathbf{c})^{\nu+2}} \beta_d\, d(x^{-1} \circ y)^{2-Q}\,\mathrm{d}\mu(y)\,; \\
I_3(x) &:= \int_{d(y)\geq(2\,\mathbf{c})^{\nu+2}} \beta_d\, d(x^{-1} \circ y)^{2-Q}\,\mathrm{d}\mu(y).
\end{aligned}
$$

From now on, we shall denote by $C, C', \dots$ positive constants depending only on Q, $\mathbf{c}$ and β_d. We also let

$$n(t) := \mu(B_d(0,t)).$$

If $x \in C_\nu$, we have

$$\begin{aligned} I_1(x) &\le \beta_d\,(2\,\mathbf{c})^{(\nu-1)(2-Q)} \int_{d(y)\le(2\,\mathbf{c})^{\nu-1}} \mathrm{d}\mu(y) \le C\cdot\big((2\,\mathbf{c})^{\nu}\big)^{2-Q}\, n\big((2\,\mathbf{c})^{\nu}\big) \\ &\le (\text{because } n(t) \text{ is increasing})\quad C' \int_{(2\,\mathbf{c})^\nu}^{\infty} \frac{n(t)}{t^{Q-1}}\,\mathrm{d}t. \end{aligned}$$

Analogously,

$$\begin{aligned} I_3(x) &\le \beta_d \int_{d(y)\ge(2\,\mathbf{c})^{\nu+2}} \left(\frac{d(y)}{2\,\mathbf{c}}\right)^{2-Q} \mathrm{d}\mu(y) \le C \int_{(2\,\mathbf{c})^\nu}^{\infty} t^{2-Q}\,\mathrm{d}n(t) \\ &= C'\left[\frac{n(t)}{t^{Q-2}}\right]_{(2\,\mathbf{c})^\nu}^{\infty} + C'' \int_{(2\,\mathbf{c})^\nu}^{\infty} \frac{n(t)}{t^{Q-1}}\,\mathrm{d}t \le C''' \int_{(2\,\mathbf{c})^\nu}^{\infty} \frac{n(t)}{t^{Q-1}}\,\mathrm{d}t. \end{aligned}$$

The estimate of $I_2(x)$, $x \in C_\nu$, is the crucial step of the proof. Let $q > Q-2$ be fixed. Define

$$\begin{aligned} \mu_\nu &:= \mu\big(\{y \in \mathbb{G} \mid (2\,\mathbf{c})^{\nu-1} < d(y) < (2\,\mathbf{c})^{\nu+2}\}\big), \\ \eta_\nu &:= \mu_\nu\,(2\,\mathbf{c})^{(2-Q)\,\nu}, \\ \varepsilon_\nu &:= \eta_\nu^{1-(Q-2)/q}. \end{aligned}$$

Then

$$\sum_{\nu=1}^{\infty} \eta_\nu \le C \sum_{\nu=1}^{\infty} \int_{(2\,\mathbf{c})^{\nu-1}}^{(2\,\mathbf{c})^{\nu+2}} \frac{\mathrm{d}n(t)}{t^{Q-2}} \le C' \int_1^{\infty} \frac{\mathrm{d}n(t)}{t^{Q-2}} < \infty. \tag{10.15}$$

This last inequality follows from (9.29). On the other hand, by Lemma 10.4.1, there exists a countable family of closed balls $\big\{\overline{B_d(x_{k,\nu}, r_{k,\nu})}\big\}_{k\in J_\nu}$ such that

$$\{x \in C_\nu \mid I_2(x) < \varepsilon_\nu\} \supseteq C_\nu \setminus \bigcup_{k\in J_\nu} \overline{B_d(x_{k,\nu}, r_{k,\nu})}$$

and

$$\sum_{k\in J_\nu} (r_{k,\nu})^q < A\left(\frac{\mu_\nu}{\varepsilon_\nu}\right)^{q/(Q-2)}. \tag{10.16}$$

As a consequence,

$$(r_{k,\nu})^q \le A\left(\frac{\mu_\nu}{\varepsilon_\nu}\right)^{q/(Q-2)} = A\,\eta_\nu\,(2\mathbf{c})^{q\,\nu},$$

so that

$$r_{k,\nu} \le (A\,\eta_\nu)^{1/q}(2\mathbf{c})^{\nu}. \tag{10.17}$$

We may also suppose $\overline{B_d(x_{k,\nu}, r_{k,\nu})} \cap C_\nu \neq \emptyset$ for every $k \in J_\nu$. This implies $d(x_{k,\nu}) \geq (2\,\mathbf{c})^{\nu-2}$. Indeed,

$$\begin{aligned} d(x_{k,\nu}) \geq \text{(see (10.17))} \quad & \frac{1}{\mathbf{c}}(2\mathbf{c})^\nu - (A\,\eta_\nu)^{1/q}(2\mathbf{c})^\nu \\ &= (2\mathbf{c})^\nu\left(\frac{1}{\mathbf{c}} - (A\,\eta_\nu)^{1/q}\right), \end{aligned}$$

and, since $\eta_\nu \to 0$ as $\nu \to \infty$ (see (10.15)), the claim follows. As a consequence, from (10.15), (10.16), the choice of ε_ν and the bound for $d(x_{k,\nu})$ we get

$$\sum_{\nu=1}^{\infty}\sum_{k\in J_\nu}\left(\frac{r_{k,\nu}}{d(x_{k,\nu})}\right)^q \leq A\sum_{\nu=1}^{\infty}\frac{(2\,\mathbf{c})^{\nu q}}{(2\,\mathbf{c})^{(\nu-2)q}}\,\eta_\nu = C\sum_{\nu=1}^{\infty}\eta_\nu < \infty. \tag{10.18}$$

Collecting the estimates for I_1, I_2 and I_3, we finally obtain

$$U - u(x) < C\int_{(2\,\mathbf{c})^\nu}^{\infty}\frac{n(t)}{t^{Q-1}}\,\mathrm{d}t + \eta_\nu^{1-(Q-2)/q} \tag{10.19}$$

for every $x \in C_\nu \setminus \bigcup_{k\in J_\nu}\overline{B_d(x_{k,\nu}, r_{k,\nu})}$ and for every $\nu \in \mathbb{N}$. By (9.29) and the positivity of the exponent of η_ν, the right-hand side of (10.19) tends to zero, as $\nu \to \infty$. Together with (10.18), this proves that the set

$$F = \bigcup_{\nu=1}^{\infty}\bigcup_{k\in J_\nu}\overline{B_d(x_{k,\nu}, r_{k,\nu})}$$

is a q-set satisfying

$$\lim_{x\to\infty,\ x\notin F} u = \sup_{\mathbb{G}} u.$$

This ends the proof. □

Bibliographical Notes. In the recent literature, a special attention has been paid to the maximum principle on *unbounded domains*. Such a principle plays a key rôle in looking for *symmetry properties* of solutions to semilinear Poisson equations (see [BN91,BCN97,BHM00]). In order to provide our analogue in the Carnot group setting, we have been inspired by some ideas of J. Deny published in 1947 (see Deny's theorem [Den48, p. 142]; see also [HK76, Theorem 3.21]).

We would like to stress that maximum principles in half-spaces are crucial tools in looking for monotonicity and symmetry properties of solutions to the semilinear equation $\mathcal{L}u + f(u) = 0$ in $\mathbb{G}$. See [BHM00] and [BP99] for some recent noteworthy results in the cases $\mathcal{L} = \Delta$ and $\mathcal{L} = \Delta_{\mathbb{H}^N}$, respectively. See also [BL02,BL03,BP01, BP02].

Furthermore, we would like to note that Theorem 10.2.2 can be used as a starting point in studying the asymptotic behavior of $\mathcal{L}$-subharmonic functions *not* bounded from above in $\mathbb{G}$.

Some of the topics presented in this chapter also appear in [BL02].

10.5 Exercises of Chapter 10

1) Suppose that Ω is a bounded open set. Let $u : \Omega \to [-\infty, \infty[$ be u.s.c. Suppose that

$$\limsup_{x \to y} u(x) \le 0 \qquad \forall\, y \in \partial\Omega.$$

Then u is bounded from above. (*Hint*: Let $B \subseteq \Omega$ be an open set such that $\partial\Omega \subset \overline{B}$ and $u < 1$ on B. Then write $\Omega = A \cup B$ and observe that u is bounded from above on $\overline{A}$ which is a compact subset of Ω. Indeed, any u.s.c. function restricted to a compact set attains its maximum!)

2) Does there exist any $u \in \underline{S}_b(\Omega)$, non-constant and bounded from below? In case, provide examples. (*Hint*: See also Theorem 9.3.10.)

3) Let F_0 be a q-set. Then, $F \setminus F_0$ is not a q-set, whenever F is not.

4) Complete the proof of (10.11) arguing as in the last paragraph of Proposition 10.2.5.

5) Prove that every subset of a q-set is a q-set and every bounded set is a q-set (for every $q > 0$). Prove also that a finite union of q-sets is a q-set.

6) Consider the subset of $\mathbb{R}^3$

$$S = \{(x_1, x_2, x_3) \in \mathbb{R}^3 \,:\, x_1 \le -1,\ x_2 = 0,\ x_3 = 0\}.$$

Prove that S is not a q-set when $q = 1$ and the relevant d in Definition 10.2.1 is the Euclidean norm. (*Hint*: Show that if

$$\{B_j = B(c_j, r_j)\}_{j \in \mathbb{N}}$$

is a sequence of ordinary balls in $\mathbb{R}^3$, then there exists a constant $M > 0$ such that

$$\left| \int_H \frac{\mathrm{d}x}{x} \right| \le M \sum_{j \in \mathbb{N}} \frac{r_j}{|c_j|}, \tag{10.20}$$

where

$$H = \pi_1\Big(S \cap \bigcup_j B_j\Big)$$

and

$$\pi_1 : \mathbb{R}^3 \to \mathbb{R}, \quad \pi_1(x_1, x_2, x_3) = x_1.$$

Thus, if S were a 1-set and $\{B_j\}_j$ the relevant covering of S as in Definition 10.2.1, we would have $H =]-\infty, -1]$, so that the left-hand side of (10.20) would diverge, whereas the right-hand side is finite by the definition of q-set when $q = 1$.)

7) In this exercise, we show that the assertion of Lemma 10.2.3 may fail to be true for $q = Q - 2$.
For instance, let $\mathbb{G} = (\mathbb{R}^3, +)$ be the usual Euclidean group on $\mathbb{R}^3$, and let $\mathcal{L} = \Delta$ be the usual Laplace operator. Consider the function

$$u(x_1, x_2, x_3) = -\int_{-\infty}^{-1} \frac{((x_1 - y)^2 + x_2^2 + x_3^2)^{-1/2}}{1 - y}\, \mathrm{d}y.$$

Show that:

a) The function u coincides with the convolution

$$-\int_{\mathbb{G}} \Gamma(y^{-1} \circ x)\, \mathrm{d}\mu(y),$$

where μ is the measure on $\mathbb{R}^3$ supported on the set S of Ex. 6, and coinciding there with

$$\frac{\chi_{(-\infty,-1)}(y)}{1 - y}\, \mathrm{d}y.$$

Here, $\mathrm{d}y$ denotes the usual Hausdorff 1-dimensional measure on $\mathbb{R}^1$.

b) $u \in \underline{S}(\mathbb{R}^3)$ satisfies all the conditions in Theorem 9.6.1 (page 451). Indeed, with the notation in that theorem, we have

$$n(t) = \mu(B_d(0, t)) = \begin{cases} 0 & \text{if } t \in]0, 1] \\ \int_{-t}^{-1} \frac{1}{1-y}\, \mathrm{d}y = \log(1 + t) - \log 2 & \text{if } t > 1. \end{cases}$$

Thus

$$\int_1^{\infty} \frac{n(t)}{t^2}\, \mathrm{d}t < \infty.$$

c) Deduce that μ is the $\mathcal{L}$-Riesz measure of u in $\mathbb{R}^3$, $u(0) > -\infty$ and the least upper bound of u is 0. Moreover, $u = -\infty$ on S and is finite elsewhere.

Note that, if Lemma 10.2.3 were true for $q = Q - 2 = 1$, there would exist a 1-set $F \subseteq \mathbb{R}^3$ such that

$$\lim_{|x| \to \infty,\ x \notin F} u = \sup_{\mathbb{R}^3} u = 0.$$

Hence, F would definitely cover S, so that S would be a 1-set. But this is false, as shown by Exercise 6.

11

$\mathcal{L}$-capacity, $\mathcal{L}$-polar Sets and Applications

The aim of this chapter is to provide an ad hoc theory of capacity and of polar sets for a sub-Laplacian $\mathcal{L}$ on a Carnot group $\mathbb{G}$. The structure of the fundamental solution for $\mathcal{L}$ has a prominent rôle in the whole chapter.

Our starting points are a continuity principle for $\mathcal{L}$-potentials and a suitable version of the Maria–Frostman domination principle. Starting from these results, we develop a theory of capacity following two classical approaches: one based on the notion of $\mathcal{L}$-energy, see Section 11.4, and one based on the balayage, see Section 11.5. We compare these approaches to each other, showing that they lead to the same concept of capacity.

As an application of the above results we prove a Poisson–Jensen formula and the so-called "fundamental convergence theorem".

Notation. Throughout the chapter, we shall use the following notation and definitions. $\mathfrak{M}$ denotes the set of Radon measures μ on $\mathbb{G}$, i.e. of the Borel measures on $\mathbb{G}$ which are finite on compact sets. For $\mu \in \mathfrak{M}$, we denote by $\mathrm{supp}(\mu)$ the support of μ, i.e. the complement of the largest open set with μ-measure zero. We denote by $\mathfrak{M}_0$ the subset of $\mathfrak{M}$ of compactly supported Radon measures. Moreover, if E is any set, we denote by $\mathfrak{M}(E)$ (respectively, $\mathfrak{M}_0(E)$) the set of measures $\mu \in \mathfrak{M}$ (respectively, $\mu \in \mathfrak{M}_0$) such that $\mathrm{supp}(\mu) \subseteq E$. If $\mu \in \mathfrak{M}$ and A is any set, by $\mu|_A$ we mean the measure in $\mathfrak{M}$ defined by $\mu|_A(E) = \mu(E \cap A)$ for every Borel set E. Moreover, we write $A \Subset B$ whenever A is a compact subset of B. Finally, $\overline{S}^+(\Omega)$ denotes the set of the non-negative $\mathcal{L}$-superharmonic functions in Ω.

11.1 The Continuity Principle for $\mathcal{L}$-potentials

Theorem 11.1.1 (The continuity principle for $\mathcal{L}$-potentials). *Let $E \Subset \mathbb{G}$ be a compact set, and let μ be a Radon measure in $\mathbb{G}$ supported in E. Let $x_0 \in E$, and let the function $(\Gamma * \mu)|_E$ be finite and continuous at x_0 (as a function in E). Then $\Gamma * \mu$ is also continuous at x_0 (as a function in the whole $\mathbb{G}$).*

Proof. We can suppose that $x_0 \in \partial E$, since otherwise there is nothing to prove. Let us set for brevity $V_0 := (\Gamma * \mu)(x_0)$. We first observe that $\mu(\{x_0\}) = 0$ since

$$+\infty > V_0 \geq \int_{\{x_0\}} \Gamma(y^{-1} \circ x_0) \, \mathrm{d}\mu(y) = \Gamma(0) \cdot \mu(\{x_0\}) = \infty \cdot \mu(\{x_0\}).$$

If x_0 is an isolated point of E, then there exists an open neighborhood O of x_0 such that $O \cap E = \{x_0\}$. Since $\mu(\{x_0\}) = 0$ and $\operatorname{supp} \mu \subseteq E$, we get $\mu(O) = 0$. Then, by Theorem 9.3.5, page 433 (see also Corollary 9.3.3), $\Gamma * \mu$ is $\mathcal{L}$-harmonic in O and hence continuous. We can then suppose that $x_0 \in \partial E$ is a limit point of E. By hypothesis, we know that $\lim_{x \to x_0} (\Gamma * \mu)|_E(x) = V_0$, i.e. for fixed $\varepsilon > 0$, there exists $\rho_0 > 0$ such that

$$|(\Gamma * \mu)(\xi) - V_0| < \varepsilon \quad \forall \, \xi \in E_{\rho_0}. \tag{11.1}$$

Here and in the following we denote

$$E_r := E \cap \overline{B_d(x_0, r)}.$$

We have to prove that

$$\lim_{x \to x_0} (\Gamma * \mu)|_{(\mathbb{G} \setminus E)}(x) = V_0. \tag{11.2}$$

Since $\mu(\{x_0\}) = 0$ and $\int_{\mathbb{G}} \Gamma(y^{-1} \circ x_0) \, \mathrm{d}\mu(y) = V_0 \in \mathbb{R}$, we have

$$\int_{E_\rho} \Gamma(y^{-1} \circ x_0) \, \mathrm{d}\mu(y) \to 0 \quad \text{as } \rho \to 0^+.$$

Hence there exists $\rho_1 > 0$ ($\rho_1 \leq \rho_0$) such that

$$\int_{E_\rho} \Gamma(y^{-1} \circ x_0) \, \mathrm{d}\mu(y) < \varepsilon \qquad \forall \, \rho \leq \rho_1. \tag{11.3}$$

We now set

$$\mu_1 = \mu|_{E \setminus E_{\rho_1}},$$

and we consider $\Gamma * \mu_1$. Such a function is $\mathcal{L}$-harmonic outside the support of μ_1. In particular, it is continuous in a neighborhood of x_0. Therefore, the function

$$\int_{E_{\rho_1}} \Gamma(y^{-1} \circ \cdot) \, \mathrm{d}\mu(y) = \Gamma * \mu - \Gamma * \mu_1$$

is continuous at x_0 if restricted to E. In particular, recalling (11.3), there exists $\rho_2 > 0$ ($\rho_2 < \rho_1$) such that

$$\int_{E_{\rho_1}} \Gamma(y^{-1} \circ \xi) \, \mathrm{d}\mu(y) \leq \int_{E_{\rho_1}} \Gamma(y^{-1} \circ x_0) \, \mathrm{d}\mu(y) + \varepsilon < 2\varepsilon \qquad \forall \, \xi \in E_{\rho_2}. \tag{11.4}$$

By the continuity of $\Gamma * \mu_1$ at x_0, there also exists $\rho_3 > 0$ ($\rho_3 \leq \rho_2$) such that

$$|(\Gamma * \mu_1)(\eta) - (\Gamma * \mu_1)(x_0)| < \varepsilon \qquad \forall \, \eta \in \overline{B}_d(x_0, \rho_3). \tag{11.5}$$

We can now prove (11.2). Let $x \in \mathbb{G} \setminus E$ be such that $x \in B_d(x_0, (2\mathbf{c})^{-1}\rho_3)$, where $\mathbf{c} \geq 1$ is the constant in the pseudo-triangle inequality for d (see Proposition 5.1.8, page 231). Let $\overline{x} \in E_{\rho_1}$ be a point that minimizes the d-distance between x and E_{ρ_1}. We have

$$\begin{aligned}
&|(\Gamma * \mu)(x) - V_0| \\
&\quad \leq |(\Gamma * \mu)(x) - (\Gamma * \mu)(\overline{x})| + |(\Gamma * \mu)(\overline{x}) - V_0| \\
&\quad \leq |(\Gamma * \mu_1)(x) - (\Gamma * \mu_1)(x_0)| + |(\Gamma * \mu_1)(x_0) - (\Gamma * \mu_1)(\overline{x})| \\
&\qquad + \left| \int_{E_{\rho_1}} \Gamma(y^{-1} \circ \overline{x}) \, \mathrm{d}\mu(y) \right| + \left| \int_{E_{\rho_1}} \Gamma(y^{-1} \circ x) \, \mathrm{d}\mu(y) \right| + |(\Gamma * \mu)(\overline{x}) - V_0|.
\end{aligned}$$

Let us estimate the far right-hand side. The first two summands are smaller than ε by means of (11.5). Indeed, $x \in B_d(x_0, (2\mathbf{c})^{-1}\rho_3) \subset \overline{B}_d(x_0, \rho_3)$ and $\overline{x} \in B_d(x_0, \rho_3)$ since, recalling the choice of $\overline{x}$,

$$d(\overline{x}, x_0) \leq \mathbf{c}(d(\overline{x}, x) + d(x, x_0)) \leq 2\mathbf{c}\, d(x, x_0) < \rho_3.$$

The third summand is smaller than 2ε by means of (11.4) ($\overline{x} \in E_{\rho_2}$, as we have just seen that $d(\overline{x}, x_0) < \rho_3 \leq \rho_2$). Let us estimate the fourth summand. Again by the choice of $\overline{x}$, for every $y \in E_{\rho_1}$ we have

$$d(\overline{x}, y) \leq \mathbf{c}(d(\overline{x}, x) + d(x, y)) \leq 2\mathbf{c}\, d(x, y).$$

This immediately gives

$$\Gamma(y^{-1} \circ x) = d^{2-Q}(x, y) \leq (2\mathbf{c})^{Q-2} d^{2-Q}(\overline{x}, y) = (2\mathbf{c})^{Q-2} \Gamma(y^{-1} \circ \overline{x}),$$

and we thus derive

$$\int_{E_{\rho_1}} \Gamma(y^{-1} \circ x) \, \mathrm{d}\mu(y) \leq (2\mathbf{c})^{Q-2} \int_{E_{\rho_1}} \Gamma(y^{-1} \circ \overline{x}) \, \mathrm{d}\mu(y) < (2\mathbf{c})^{Q-2}\, 2\varepsilon$$

(here we have used the estimate of the third summand). Finally, the fifth summand is smaller than ε by means of (11.1). This proves (11.2) and completes the proof of the theorem. □

11.2 $\mathcal{L}$-polar Sets

Definition 11.2.1 ($\mathcal{L}$-polar set). *A set $E \subseteq \mathbb{G}$ is called* $\mathcal{L}$-polar *if there exist an open set $\Omega \supseteq E$ and a function $u \in \overline{S}(\Omega)$ such that $u \equiv \infty$ in E.*

Exercise 11.2.2. (i) Any singleton $\{x_0\}$ is $\mathcal{L}$-polar.
(ii) If $E_1 \subseteq E_2$ and E_2 is $\mathcal{L}$-polar, then also E_1 is $\mathcal{L}$-polar.
(iii) If E is $\mathcal{L}$-polar, then E is contained in an $\mathcal{L}$-polar set $\widetilde{E}$ of G_δ type (i.e. countable intersection of open sets).
(iv) If E is $\mathcal{L}$-polar, then E has Lebesgue measure zero.
(v) Any countable set is $\mathcal{L}$-polar.

Theorem 11.2.3 ($\mathcal{L}$-polar sets and $\mathcal{L}$-potentials. I). *Let $E \subseteq \mathbb{G}$ be an $\mathcal{L}$-polar set, and let $y \notin E$. Then there exists $\mu \in \mathfrak{M}$ such that $\Gamma * \mu \equiv \infty$ in E and $(\Gamma * \mu)(y) < \infty$. In particular,*[1] $\Gamma * \mu \in \overline{\mathcal{S}}(\mathbb{G})$.

Proof. By the definition of $\mathcal{L}$-polar set, there exists $u \in \overline{\mathcal{S}}(\Omega)$ (for some open set $\Omega \supseteq E$) such that $u \equiv \infty$ in E. Moreover, we can suppose that $u > 0$ in Ω (recall that the lower semicontinuity of u implies that $\{x \in \Omega \mid u(x) > 0\}$ is open). Let D_k be a sequence of open sets such that

$$D_k \subseteq \overline{D}_k \subseteq D_{k+1}, \qquad \bigcup_{k\in\mathbb{N}} D_k = \Omega \setminus \{y\}.$$

By the Riesz representation Theorem 9.4.4, page 442, there exists $h_k \in \mathcal{H}(D_k)$ such that

$$u(x) = h_k(x) + \int_{D_k} \Gamma(y^{-1} \circ x)\, \mathrm{d}\mu_u(y) \qquad \forall\, x \in D_k. \tag{11.6}$$

Let us set, for brevity, $v_k(x) := \int_{D_k} \Gamma(y^{-1} \circ x)\, \mathrm{d}\mu_u(y)$. We remark that $v_k \in \overline{\mathcal{S}}(\mathbb{G})$, by means of Corollary 9.3.3, page 433. Since v_k is $\mathcal{L}$-harmonic outside $\overline{D}_k$ (see Theorem 9.3.5, page 433), we have, in particular, $v_k(y) \in \mathbb{R}$ for every $k \in \mathbb{N}$. We then set

$$v(x) := \sum_{k\in\mathbb{N}} 2^{-k} v_k(x)/(v_k(y) + 1) \qquad \forall\, x \in \mathbb{G}.$$

Since $v(y) \le \sum_{k\in\mathbb{N}} 2^{-k} < \infty$, we have $v \in \overline{\mathcal{S}}(\mathbb{G})$ (see Corollary 8.2.8, page 403). Moreover, $v \ge 0$. Hence we can use Corollary 9.4.8, page 444, and obtain

$$v = \Gamma * \mu_v + \inf_{\mathbb{G}} v.$$

It is now easy to see that $\mu := \mu_v$ satisfies the assertion of the theorem. Indeed, $v(y) < \infty$ immediately gives $(\Gamma * \mu)(y) < \infty$. Moreover, for every $x_0 \in E$, there exists $k_0 \in \mathbb{N}$ such that $x_0 \in D_{k_0}$. Since $u \equiv \infty$ in E, (11.6) implies that $v_{k_0}(x_0) = \infty$. As a consequence, $(\Gamma * \mu)(x_0) = v(x_0) - \inf_{\mathbb{G}} v = \infty$. This ends the proof. □

Example 11.2.4. [There exist finite-valued discontinuous $\mathcal{L}$-superharmonic functions.] Indeed, let $E \subset \mathbb{G}$ be a non-closed $\mathcal{L}$-polar set (let, e.g. $E = \{x_n\}_{n\in\mathbb{N}}$ with $x_n \to 0$ and $0 \notin E$), and let x_0 be a limit point for E not belonging to E. By Theorem 11.2.3, there exists $v \in \overline{\mathcal{S}}(\mathbb{G})$ such that $v \equiv \infty$ on E and $v(x_0) < \infty$. Let now $u := \min\{v, v(x_0) + 1\}$. Then u clearly satisfies the requisites of the assertion. □

Corollary 11.2.5 ($\mathcal{L}$-polar sets and $\mathcal{L}$-potentials. II). *Let $E \subseteq \mathbb{G}$ be a* bounded *$\mathcal{L}$-polar set, and let $y \notin E$. Then there exists $\nu \in \mathfrak{M}_0$ such that $\Gamma * \nu \equiv \infty$ in E and $(\Gamma * \nu)(y) < \infty$.*

[1] By Theorem 9.3.2, page 432.

Proof. By Theorem 11.2.3, there exists $\mu \in \mathfrak{M}$ such that $\Gamma * \mu \equiv \infty$ in E and $(\Gamma * \mu)(y) < \infty$. We have

$$\overline{E \cup \{y\}} \Subset D(0, R) =: B,$$

for some $R > 0$. Setting $\nu := \mu|_{\overline{B}}$, we have $\nu \in \mathfrak{M}_0$. Moreover, by the Riesz representation Theorem 9.4.4, page 442 (see also Theorem 9.3.5, page 433), there exists $h \in \mathcal{H}(B)$ such that

$$\Gamma * \mu = h + \Gamma * \nu$$

in B. It is now immediate to recognize that $\Gamma * \nu$ has the required properties. This completes the proof. □

Corollary 11.2.6. *Let $\{E_j\}_{j\in\mathbb{N}}$ be a sequence of $\mathcal{L}$-polar sets. Then $\bigcup_{j\in\mathbb{N}} E_j$ is $\mathcal{L}$-polar.*

Proof. Using Theorem 11.2.3 and observing that $\mathcal{L}$-polar sets have Lebesgue measure zero (since $\mathcal{L}$-superharmonic functions are L^1_{loc}), we can find $y \in \mathbb{G}$ and $u_j \in \overline{\mathcal{S}}(\mathbb{G})$, $u_j \geq 0$, such that $u_j|_{E_j} \equiv \infty$ and $u_j(y) < \infty$ for every $j \in \mathbb{N}$. We then set

$$u(x) := \sum_{k\in\mathbb{N}} \frac{2^{-k} u_k(x)}{(u_k(y) + 1)} \qquad \forall\, x \in \mathbb{G}.$$

Since $u(y) \leq \sum_{k\in\mathbb{N}} 2^{-k} < \infty$, we have $u \in \overline{\mathcal{S}}(\mathbb{G})$ (see Corollary 8.2.8, page 403). Moreover, $u \equiv \infty$ in $\bigcup_{j\in\mathbb{N}} E_j$. □

We end the section with an improvement of the maximum principle for $\mathcal{L}$.

Theorem 11.2.7 (The extended maximum principle for $\mathcal{L}$). *Let $\Omega \subset \mathbb{G}$ be a bounded domain, and let $u \in \underline{\mathcal{S}}(\Omega)$ be bounded from above. If*

$$\limsup_{x\to\xi} u(x) \leq M \quad \textit{for every } \xi \in \partial\Omega$$

except for an $\mathcal{L}$-polar *set $E \subset \partial\Omega$, then $u \leq M$ on Ω.*

Proof. Let $x_0 \in \Omega$. By Theorem 11.2.3, there exists $\omega \in \underline{\mathcal{S}}(\mathbb{G})$, $\omega < 0$ and finite in x_0, such that $\omega|_E \equiv -\infty$. Then, for $\varepsilon > 0$, the function $u_\varepsilon = u + \varepsilon\,\omega$ is $\mathcal{L}$-subharmonic in Ω and has lim sup less than M on $\partial\Omega$. Therefore, by the weak maximum principle (Theorem 8.2.19, page 409), we get

$$u(x_0) + \varepsilon\,\omega(x_0) \leq M.$$

Letting $\varepsilon \to 0^+$ and recalling that $\omega(x_0) > -\infty$, we finally get $u(x_0) \leq M$. This ends the proof. □

11.3 The Maria–Frostman Domination Principle

Lemma 11.3.1. *Let $\Omega \subseteq \mathbb{G}$ be an open set, and let $f, g \in \underline{S}(\Omega)$. If $f \leq g$ a.e. in Ω, then $f \leq g$ in Ω. In particular, if $f = g$ a.e., then $f \equiv g$.*

Proof. By Theorem 8.2.11, page 405, we know that $f(x) = \lim_{r\to 0^+} \mathrm{M}_r(f)(x)$ and $g(x) = \lim_{r\to 0^+} \mathrm{M}_r(g)(x)$ for every $x \in \Omega$. On the other hand, $f \leq g$ a.e. implies $\mathrm{M}_r(f) \leq \mathrm{M}_r(g)$. Letting $r \to 0^+$, we get the assertion. □

The following result is a particular case of Exercise 24, Chapter 6. We provide the proof for the sake of completeness.

Lemma 11.3.2. *Let $\Omega_2 \subseteq \Omega_1 \subseteq \mathbb{G}$ be open sets, and let $s \in \underline{S}(\Omega_1)$, $u \in \underline{S}(\Omega_2)$ be such that $\limsup_{\Omega_2 \ni x \to y} u(x) \leq s(y)$ for every $y \in (\partial\Omega_2) \cap \Omega_1$. Then*

$$v : \Omega_1 \longrightarrow [-\infty, \infty[, \qquad v(x) := \begin{cases} \max\{s(x), u(x)\}, & x \in \Omega_2, \\ s(x), & x \in \Omega_1 \setminus \Omega_2, \end{cases}$$

is $\mathcal{L}$-subharmonic in Ω_1. (See Fig. 11.1.)

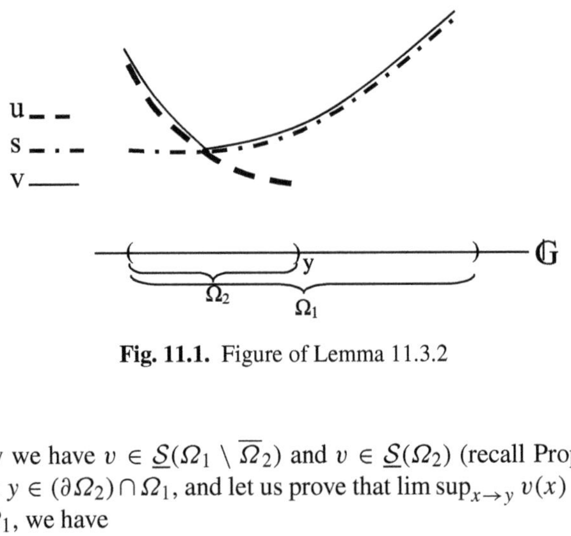

Fig. 11.1. Figure of Lemma 11.3.2

Proof. Clearly we have $v \in \underline{S}(\Omega_1 \setminus \overline{\Omega_2})$ and $v \in \underline{S}(\Omega_2)$ (recall Proposition 6.5.4, page 355). Let $y \in (\partial\Omega_2) \cap \Omega_1$, and let us prove that $\limsup_{x\to y} v(x) \leq v(y)$. Since s is u.s.c. in Ω_1, we have

$$\limsup_{\Omega_1 \setminus \Omega_2 \ni x \to y} v(x) = \limsup_{\Omega_1 \setminus \Omega_2 \ni x \to y} s(x) \leq s(y) = v(y),$$

$$\limsup_{\Omega_2 \ni x \to y} v(x) = \limsup_{\Omega_2 \ni x \to y} (\max\{s(x), u(x)\})$$

$$\leq \max\left\{ \limsup_{\Omega_2 \ni x \to y} s(x), \limsup_{\Omega_2 \ni x \to y} u(x) \right\} \leq s(y) = v(y).$$

This proves that v is u.s.c. in Ω_1. Moreover, v is finite in a dense subset of Ω_1, since $s \leq v$ has the same property. By Theorem 8.2.1, page 401, we only have to prove that

v is sub-mean. Let $y \in (\partial\Omega_2) \cap \Omega_1$, $\overline{B}_d(y,r) \subset \Omega_1$. Since $v \geq s$ and $s \in \underline{S}(\Omega_1)$, we have

$$\mathrm{M}_r(v)(y) \geq \mathrm{M}_r(s)(y) \geq s(y) = v(y).$$

Observing that $v \in \underline{S}(\Omega_1 \setminus \overline{\Omega_2})$ and $v \in \underline{S}(\Omega_2)$ imply that v is sub-mean in $\Omega_1 \setminus \partial\Omega_2$, this completes the proof. □

Lemma 11.3.3 (Lusin-type theorem for potentials). *Let $\mu \in \mathfrak{M}_0$ be such that $\Gamma * \mu < \infty$ in $K = \operatorname{supp}\mu \Subset \mathbb{G}$.*
Then, for every $\varepsilon > 0$, there exists a compact set $C \subseteq K$ such that:

(i) $\mu(K \setminus C) < \varepsilon$,
(ii) $\Gamma * (\mu|_C) \in C(\mathbb{G}, \mathbb{R})$.

Proof. We first observe that $\Gamma * \mu < \infty$ in $\mathbb{G}$, since $\Gamma * \mu < \infty$ in K by hypothesis and

$$\Gamma * \mu \in \mathcal{H}(\mathbb{G} \setminus \operatorname{supp}\mu)$$

by Theorem 9.3.5, page 433. Moreover, $\Gamma * \mu$ is l.s.c. Hence $\Gamma * \mu$ is measurable, and we can apply the Lusin theorem (see, e.g. [Rud87]): for every $\varepsilon > 0$, there exists a compact set $C \subseteq K$ such that $\mu(K \setminus C) < \varepsilon$ and $(\Gamma * \mu)|_C$ is continuous in C. Since $\Gamma * \mu < \infty$ in $\mathbb{G}$, we can write

$$\Gamma * (\mu|_C) = \Gamma * \mu - \Gamma * (\mu|_{K \setminus C}).$$

Since $\Gamma * (\mu|_C)$ is l.s.c., $-\Gamma * (\mu|_{K\setminus C})$ is u.s.c. and $(\Gamma * \mu)|_C$ is continuous in C, we obtain that $(\Gamma * (\mu|_C))|_C$ is continuous in C. We can now use the continuity principle for potentials (Theorem 11.1.1, page 489) and get that $\Gamma * (\mu|_C)$ is continuous in C. On the other hand, $\Gamma * (\mu|_C)$ is $\mathcal{L}$-harmonic and hence continuous in $\mathbb{G} \setminus \operatorname{supp}(\mu|_C)$. Therefore, $\Gamma * (\mu|_C)$ is continuous in $\mathbb{G}$. This ends the proof. □

Theorem 11.3.4 (The Maria–Frostman domination principle). *Let $\mu \in \mathfrak{M}$ be such that $\Gamma * \mu$ is finite in $\mathbb{G}$. Let P be an $\mathcal{L}$-polar set. If $u \in \overline{S}^{+}(\mathbb{G})$ and $u \geq \Gamma * \mu$ in $(\operatorname{supp}\mu) \setminus P$, then $u \geq \Gamma * \mu$ in $\mathbb{G}$.*

Proof. Since P is $\mathcal{L}$-polar, by Theorem 11.2.3 we can find $\nu \in \mathfrak{M}$ such that

$$\omega := \Gamma * \nu \in \overline{S}(\mathbb{G}), \quad \omega \equiv \infty \text{ in } P.$$

Let $\varepsilon > 0$. Clearly, we have

$$u + \varepsilon\,\omega \geq \Gamma * \mu \quad \text{in } \operatorname{supp}\mu. \tag{11.7}$$

Let $K_n \Subset \mathbb{G}$ be such that

$$K_n \subseteq K_{n+1} \text{ and } \textstyle\bigcup_{n\in\mathbb{N}} K_n = \mathbb{G}.$$

Let us set $\mu_n := \mu|_{K_n}$. Since we obviously have $\Gamma * \mu_n \leq \Gamma * \mu < \infty$ in $\mathbb{G}$, we can apply Lemma 11.3.3 and find

$$L_{n,\varepsilon} \Subset \operatorname{supp} \mu_n = K_n \cap \operatorname{supp} \mu$$

such that $\mu_n(K_n \setminus L_{n,\varepsilon}) < \varepsilon$ and $v_{n,\varepsilon} := \Gamma * (\mu_n|_{L_{n,\varepsilon}}) \in C(\mathbb{G})$. We explicitly remark that

$$v_{n,\varepsilon} \le \Gamma * \mu_n \le \Gamma * \mu < \infty \qquad \text{in } \mathbb{G}. \tag{11.8}$$

We now set

$$s(x) := \begin{cases} \max\{v_{n,\varepsilon}(x) - u(x) - \varepsilon\omega(x), 0\}, & x \in \mathbb{G} \setminus L_{n,\varepsilon}, \\ 0, & x \in L_{n,\varepsilon}. \end{cases}$$

Since $v_{n,\varepsilon}$ is $\mathcal{L}$-harmonic outside $L_{n,\varepsilon}$ (see Theorem 9.3.5, page 433), we have $v_{n,\varepsilon} - u - \varepsilon\omega \in \underline{S}(\mathbb{G} \setminus L_{n,\varepsilon})$. Moreover, for every $y \in \partial L_{n,\varepsilon}$, we have

$$\begin{aligned} &\limsup_{\mathbb{G}\setminus L_{n,\varepsilon} \ni x \to y} \ (v_{n,\varepsilon}(x) - u(x) - \varepsilon\omega(x)) \\ &\quad \le (v_{n,\varepsilon} \in C(\mathbb{G}),\, -u,\, -\varepsilon\omega \in \underline{S}(\mathbb{G})) \\ &\quad \le v_{n,\varepsilon}(y) - u(y) - \varepsilon\omega(y) \le 0, \end{aligned}$$

by recalling that

$$v_{n,\varepsilon} \le \Gamma * \mu \le u + \varepsilon\omega \qquad \text{in } \operatorname{supp}(\mu), \tag{11.9}$$

thanks to (11.8) and (11.7). We can then apply Lemma 11.3.2 with $\Omega_1 = \mathbb{G}$ and $\Omega_2 = \mathbb{G} \setminus L_{n,\varepsilon}$ and obtain that $s \in \underline{S}(\mathbb{G})$. Since

$$0 \le s \le v_{n,\varepsilon} \text{ in } \mathbb{G},\ s \in \underline{S}(\mathbb{G}),\ v_{n,\varepsilon} \in \overline{S}(\mathbb{G}),$$

by Theorem 6.6.1 (page 358) there exists $h \in \mathcal{H}(\mathbb{G})$ such that $0 \le s \le h \le v_{n,\varepsilon}$ in $\mathbb{G}$. We now use Theorem 9.3.7, page 434, and we obtain that $h \equiv 0$. This implies $s \equiv 0$ in $\mathbb{G}$. By the definition of s, this means that $v_{n,\varepsilon} - u - \varepsilon\omega \le 0$ in $\mathbb{G} \setminus L_{n,\varepsilon}$. Since also (11.9) holds, we obtain

$$v_{n,\varepsilon} \le u + \varepsilon\,\omega \quad \text{in } \mathbb{G}. \tag{11.10}$$

For every $x \in \mathbb{G} \setminus (K_n \cap \operatorname{supp} \mu)$, we have

$$\begin{aligned} &\int_{(K_n \cap \operatorname{supp} \mu)\setminus L_{n,\varepsilon}} \Gamma(y^{-1} \circ x)\, \mathrm{d}\mu_n(y) \\ &\quad \le \mu_n\big((K_n \cap \operatorname{supp} \mu) \setminus L_{n,\varepsilon}\big) \cdot \sup\big\{\Gamma(y^{-1} \circ x) \,|\, y \in (K_n \cap \operatorname{supp} \mu) \setminus L_{n,\varepsilon}\big\} \\ &\quad \le \varepsilon \sup\big\{\Gamma(y^{-1} \circ x) \,|\, y \in (K_n \cap \operatorname{supp} \mu)\big\} < \infty, \end{aligned}$$

and then, using (11.10), we obtain

$$\begin{aligned} (\Gamma * \mu_n)(x) &= v_{n,\varepsilon}(x) + \int_{(K_n \cap \operatorname{supp} \mu)\setminus L_{n,\varepsilon}} \Gamma(y^{-1} \circ x)\, \mathrm{d}\mu_n(y) \\ &\le u(x) + \varepsilon\omega(x) + \int_{(K_n \cap \operatorname{supp} \mu)\setminus L_{n,\varepsilon}} \Gamma(y^{-1} \circ x)\, \mathrm{d}\mu_n(y) \\ &\le u(x) + \varepsilon\omega(x) + \varepsilon \sup\big\{\Gamma(y^{-1} \circ x) \,|\, y \in (K_n \cap \operatorname{supp} \mu)\big\}. \end{aligned}$$

Since $\omega \in \overline{\mathcal{S}}(\mathbb{G})$, we have $\omega < \infty$ a.e. (see Corollary 8.2.4, page 402). Letting $\varepsilon \to 0$, we then obtain

$$\Gamma * \mu_n \le u \qquad \text{a.e. in } \mathbb{G} \setminus (K_n \cap \operatorname{supp} \mu).$$

On the other hand, we know by hypothesis that $\Gamma * \mu_n \le \Gamma * \mu \le u$ in $\operatorname{supp} \mu \setminus P$, hence a.e. in $\operatorname{supp} \mu$ (P is $\mathcal{L}$-polar and then has Lebesgue measure zero). Therefore, $\Gamma * \mu_n \le u$ a.e. in $\mathbb{G}$. Recalling that $\Gamma * \mu_n, u \in \overline{\mathcal{S}}(\mathbb{G})$, by Lemma 11.3.1, we obtain $\Gamma * \mu_n \le u$ in $\mathbb{G}$. Since $\Gamma * \mu_n \to \Gamma * \mu$ pointwise in $\mathbb{G}$, we finally get $\Gamma * \mu \le u$ in $\mathbb{G}$. □

11.4 $\mathcal{L}$-energy and $\mathcal{L}$-equilibrium Potentials

Definition 11.4.1 ($\mathcal{L}$-energy and $\mathcal{L}$-equilibrium value). *Let $E \Subset \mathbb{G}$ be a compact set. For any Radon measure μ on $\mathbb{G}$, supported in E, we set*

$$I(\mu) := \int_{\mathbb{G}} (\Gamma * \mu)(x)\, d\mu(x) = \int_E \int_E \Gamma(y^{-1} \circ x)\, d\mu(x)\, d\mu(y).$$

*$I(\mu)$ is called the $\mathcal{L}$-*energy *of μ.*

*We then define the $\mathcal{L}$-*equilibrium value *$V(E)$ of E as*

$$V(E) := \inf \big\{ I(\mu) \mid \mu \textit{ Radon measure}, \ \operatorname{supp} \mu \subseteq E, \ \mu(E) = 1 \big\}.$$

The next theorem states that the above infimum is actually a minimum.

Theorem 11.4.2 ($\mathcal{L}$-equilibrium potential). *Let $E \Subset \mathbb{G}$ be a compact set. Then there exists a Radon measure $\overline{\mu}$ supported in E such that $\overline{\mu}(E) = 1$ and*

$$I(\overline{\mu}) = V(E) = \min \big\{ I(\mu) \mid \mu \textit{ Radon measure}, \ \operatorname{supp} \mu \subseteq E, \ \mu(E) = 1 \big\}.$$

*Such a measure $\overline{\mu}$ will be called an $\mathcal{L}$-*equilibrium distribution *for E. The related potential $\Gamma * \overline{\mu}$ will be called an $\mathcal{L}$-*equilibrium potential *for E.*

In Section 11.5, we shall prove that the $\mathcal{L}$-equilibrium potential is unique if $V(E) < \infty$ (see Corollary 11.5.13).

Proof. We can assume that $V(E) < \infty$, otherwise there is nothing to prove. Let us take a minimizing sequence $\{\mu_n\}_{n\in\mathbb{N}}$ of Radon measures such that $\operatorname{supp} \mu_n \subseteq E$, $\mu_n(E) = 1$ and $V(E) = \lim_{n\to\infty} I(\mu_n)$. Then[2] there exists a subsequence $\{\mu_{n_k}\}_{k\in\mathbb{N}}$ and a Radon measure $\overline{\mu}$, $\operatorname{supp} \overline{\mu} \subseteq E$, such that $\mu_{n_k} \xrightarrow{w^*} \overline{\mu}$ as $k \to \infty$, i.e.

$$\lim_{k\to\infty} \int_E f \, d\mu_{n_k} = \int_E f \, d\overline{\mu} \quad \text{for every } f \in C(E).$$

[2] See, e.g. [Rud87].

In particular, taking $f \equiv 1$, we have $\overline{\mu}(E) = 1$. Then $V(E) \le I(\overline{\mu})$, by definition of $V(E)$. On the other hand, $\mu_{n_k} \xrightarrow{w^*} \overline{\mu}$ also implies that[3]

$$\mu_{n_k} \otimes \mu_{n_k} \xrightarrow{w^*} \overline{\mu} \otimes \overline{\mu} \quad \text{in } E \times E.$$

Observing that $(x, y) \mapsto \Gamma(y^{-1} \circ x)$ is l.s.c. in $E \times E$, we obtain

$$\begin{aligned} V(E) &= \lim_{k\to\infty} I(\mu_{n_k}) = \liminf_{k\to\infty} \left(\int_E \int_E \Gamma(y^{-1} \circ x) \, d\mu_{n_k}(x) \, d\mu_{n_k}(y) \right) \\ &\ge \int_E \int_E \Gamma(y^{-1} \circ x) \, d\overline{\mu}(x) \, d\overline{\mu}(y) = I(\overline{\mu}). \end{aligned}$$

We have here used a general property of weakly convergent measures.[4] Therefore $V(E) = I(\overline{\mu})$ and the proof is complete. □

The main goal of this section is to prove Theorem 11.4.5 below as a consequence of the Maria–Frostman domination principle proved in Section 11.3. To this end we need the following definition.

Definition 11.4.3 ($\mathcal{L}$-polar* set). *We shall say that a set $E \subset \mathbb{G}$ is* $\mathcal{L}$-polar* *if there exists a countable family of compact sets F_n such that $V(F_n) = \infty$ and $E \subseteq \bigcup_{n\in\mathbb{N}} F_n$.*

In Section 11.5 (Corollary 11.5.12) we shall prove that any $\mathcal{L}$-polar* set is an $\mathcal{L}$-polar set according to our previous Definition 11.2.1, page 491.

Lemma 11.4.4. *Let $E \Subset \mathbb{G}$ be a compact set and let ν be a Radon measure supported in E such that $\nu(E) > 0$ and $I(\nu) < \infty$. Then, for any $\mathcal{L}$-polar* set E_1, we have $\nu(E_1) = 0$.*

Proof. Let us prove the lemma in the case that E_1 is a compact set such that $V(E_1) = \infty$. The general case will follow immediately from the definition of $\mathcal{L}$-polar* set. If by contradiction $\nu(E_1) > 0$, we can consider $\widetilde{\mu} := \nu|_{E_1}/\nu(E_1)$. Obviously, $\operatorname{supp} \widetilde{\mu} \subseteq E_1$ and $\widetilde{\mu}(E_1) = 1$. Moreover,

$$I(\widetilde{\mu}) = \nu^{-2}(E_1) \int_{E_1} \int_{E_1} \Gamma(y^{-1} \circ x) \, d\nu(x) \, d\nu(y) \le \nu^{-2}(E_1) I(\nu) < \infty.$$

This contradicts $V(E_1) = \infty$ and completes the proof. □

Theorem 11.4.5 (The fundamental theorem on $\mathcal{L}$-equilibrium potentials). *Let $E \Subset \mathbb{G}$ be a compact set such that $V(E) < \infty$. Let μ be an $\mathcal{L}$-equilibrium distribution and $U_E = \Gamma * \mu$ an $\mathcal{L}$-equilibrium potential for E. Then we have*

$$\begin{cases} U_E(x) \le V(E) & \text{for all } x \in \mathbb{G}, \\ U_E(x) = V(E) & \text{for all } x \in E \setminus P \text{ for a suitable } \mathcal{L}\text{-polar* set } P. \end{cases} \tag{11.11}$$

[3] Ibidem.

[4] Ibidem.

Proof. We first observe that we only need to prove

$$U_E \le V(E) \quad \text{in } \operatorname{supp}(\mu), \tag{11.12}$$

$$U_E \ge V(E) \quad \text{in } E \setminus P \text{ for a suitable } \mathcal{L}\text{-polar}^* \text{ set } P. \tag{11.13}$$

Then the assertion of the theorem will follow from the Maria–Frostman domination principle (Theorem 11.3.4), observing that $U_E = \Gamma * \mu$ is $\mathcal{L}$-harmonic outside the support of μ and then it is finite in $\mathbb{G}$ if (11.12) also holds.

We now want to prove (11.13) by showing that

$$P := \{x \in E \mid U_E(x) < V(E)\}$$

is $\mathcal{L}$-polar*. We have $P = \bigcup_{n\in\mathbb{N}} A_n$, where

$$A_n := \{x \in E \mid U_E(x) \le V(E) - 1/n\}$$

are compact sets, since U_E is l.s.c. We have to show that $V(A_n) = \infty$. Let us assume by contradiction that $V(A_{n_0}) < \infty$ for some $n_0 \in \mathbb{N}$, i.e. that there exists a Radon measure $\overline{\sigma}$ such that $\operatorname{supp}\overline{\sigma} \subseteq A_{n_0}$, $\overline{\sigma}(A_{n_0}) = 1$ and $I(\overline{\sigma}) < \infty$. We set $\varepsilon := (2n_0)^{-1}$, and we observe that there exists $a_0^{(\varepsilon)} \in \operatorname{supp}\mu$ such that $U_E(a_0^{(\varepsilon)}) > V(E) - \varepsilon$. Indeed, $U_E \le V(E) - \varepsilon$ in $\operatorname{supp}\mu$ would give the contradiction

$$V(E) = \int U_E \,\mathrm{d}\mu \le (V(E) - \varepsilon)\,\mu(\operatorname{supp}\mu) = V(E) - \varepsilon.$$

Recalling again that U_E is l.s.c., we can find a radius $r^{(\varepsilon)} > 0$ such that

$$U_E > V(E) - \varepsilon \qquad \text{in } B_\varepsilon := B_d(a_0^{(\varepsilon)}, r^{(\varepsilon)})$$

and B_ε is at a positive distance from A_{n_0}. Moreover, $a_0^{(\varepsilon)} \in \operatorname{supp}\mu$ implies that $m^{(\varepsilon)} := \mu(B_\varepsilon) > 0$. We also have

$$m^{(\varepsilon)} \le \mu(\operatorname{supp}\mu) = 1.$$

We now set $\sigma^{(\varepsilon)} := m^{(\varepsilon)}\overline{\sigma}$, and we define a (signed) measure $\sigma_1^{(\varepsilon)}$ in E by

$$\sigma_1^{(\varepsilon)} = \begin{cases} \sigma^{(\varepsilon)} & \text{in } A_{n_0}, \\ -\mu & \text{in } B_\varepsilon, \\ 0 & \text{elsewhere.} \end{cases}$$

Then, for $0 < \eta < 1$, $\mu_1^{(\varepsilon)} := \mu + \eta\,\sigma_1^{(\varepsilon)}$ is a (positive) Radon measure supported in E and

$$\mu_1^{(\varepsilon)}(E) = \mu(E) + \eta\,(\sigma^{(\varepsilon)}(A_{n_0}) - \mu(B_\varepsilon)) = 1.$$

Moreover, it is easy to see that the integral

$$\widetilde{I}(\sigma_1^{(\varepsilon)}) := \int_E \int_E \Gamma(y^{-1} \circ x) \, \mathrm{d}\sigma_1^{(\varepsilon)}(x) \, \mathrm{d}\sigma_1^{(\varepsilon)}(y)$$

is convergent, since $I(\overline{\sigma}) < \infty$, $I(\mu) < \infty$ and B_ε is at a positive distance from A_{n_0}. Now, recalling that $\Gamma(x^{-1} \circ y) = \Gamma(y^{-1} \circ x)$, we have

$$\begin{aligned} I(\mu_1^{(\varepsilon)}) - I(\mu) &= \int_E \int_E \Gamma(y^{-1} \circ x) \, \mathrm{d}\mu(x) \, \eta \, \mathrm{d}\sigma_1^{(\varepsilon)}(y) \\ &\quad + \int_E \int_E \Gamma(y^{-1} \circ x) \eta \, \mathrm{d}\sigma_1^{(\varepsilon)}(x) \, \mathrm{d}\mu(y) + \eta^2 \widetilde{I}(\sigma_1^{(\varepsilon)}) \\ &= 2\eta \int_E \int_E \Gamma(y^{-1} \circ x) \, \mathrm{d}\mu(y) \, \mathrm{d}\sigma_1^{(\varepsilon)}(x) + \eta^2 \widetilde{I}(\sigma_1^{(\varepsilon)}) \\ &= 2\eta \int_E U_E \, \mathrm{d}\sigma_1^{(\varepsilon)} + \eta^2 \widetilde{I}(\sigma_1^{(\varepsilon)}) \\ &= 2\eta \left(\int_{A_{n_0}} U_E \, \mathrm{d}\sigma^{(\varepsilon)} - \int_{B_\varepsilon} U_E \, \mathrm{d}\mu \right) + \eta^2 \widetilde{I}(\sigma_1^{(\varepsilon)}) \\ &\le 2\eta \{ (V(E) - 2\varepsilon) \sigma^{(\varepsilon)}(A_{n_0}) - (V(E) - \varepsilon) \mu(B_\varepsilon) \} + \eta^2 \widetilde{I}(\sigma_1^{(\varepsilon)}) \\ &= 2\eta \, (\widetilde{I}(\sigma_1^{(\varepsilon)}) \eta/2 - \varepsilon m^{(\varepsilon)}) < 0 \end{aligned}$$

if η is small enough. This contradicts the minimality property of $I(\mu)$ and proves (11.13).

We now turn to the proof of (11.12). We argue by contradiction assuming that there exists $x_0 \in \operatorname{supp} \mu$ such that $U_E(x_0) > V(E)$. Since U_E is l.s.c., there exist $\varepsilon > 0$ and $\rho^{(\varepsilon)} > 0$ such that $U_E > V(E) + \varepsilon$ in $B_d(x_0, \rho^{(\varepsilon)})$. Moreover, $m_0^{(\varepsilon)} := \mu(B_d(x_0, \rho^{(\varepsilon)})) > 0$ since $x_0 \in \operatorname{supp} \mu$. We have already proved that the set

$$E_1 := \{ x \in \operatorname{supp} \mu \mid U_E(x) < V(E) \}$$

is $\mathcal{L}$-polar*. Then $\mu(E_1) = 0$ by Lemma 11.4.4. Therefore, setting

$$E_2 := (\operatorname{supp} \mu) \cap B_d(x_0, \rho^{(\varepsilon)}), \quad E_3 := (\operatorname{supp} \mu) \setminus (E_1 \cup E_2),$$

we have

$$\begin{aligned} V(E) &= \int U_E \, \mathrm{d}\mu = \int_{E_2} U_E \, \mathrm{d}\mu + \int_{E_3} U_E \, \mathrm{d}\mu \\ &\ge (V(E) + \varepsilon) \, \mu(E_2) + V(E) \, \mu(E_3) \\ &= (V(E) + \varepsilon) \, m_0^{(\varepsilon)} + V(E) \, (1 - m_0^{(\varepsilon)}) = V(E) + \varepsilon \, m_0^{(\varepsilon)} > V(E). \end{aligned}$$

This gives a contradiction and completes the proof of the theorem. □

11.5 $\mathcal{L}$-balayage and $\mathcal{L}$-capacity

In this section, we show some important properties of the reduced function and the balayage in the setting of the $\mathcal{L}$-harmonic spaces. We will start from the general abstract results proved in Section 6.11, page 375. Our main ingredients will be the characterizations of the $\mathcal{L}$-superharmonicity in terms of the average operators $\mathcal{M}_r$ and M_r, proved in Chapter 8.

To begin with, for reading convenience, we rewrite the definitions of reduced function and balayage in the $\mathcal{L}$-harmonic space.

Definition 11.5.1 ($\mathcal{L}$-reduced function, $\mathcal{L}$-balayage). *Given $A \subseteq \mathbb{G}$ and $u \in \overline{\mathcal{S}}^+(\mathbb{G})$, the* $\mathcal{L}$-reduced function (*or* $\mathcal{L}$-réduite) *of u relative to A is*

$$\mathbf{R}_A^u := \inf\{f \mid f \in \Phi_A^u\},$$

where

$$\Phi_A^u := \{f \in \overline{\mathcal{S}}^+(\mathbb{G}) \mid f \geq u \text{ in } A\}. \tag{11.14}$$

The $\mathcal{L}$-balayage *of u relative to A is the lower semicontinuous regularization $\widehat{\mathbf{R}}_A^u$ of $\mathbf{R}_A^u$, i.e. for any $x \in \mathbb{G}$,*

$$\widehat{\mathbf{R}}_A^u(x) := \liminf_{y \to x} \mathbf{R}_A^u(y).$$

Since all the results in Chapter 6 apply to the present setting, from Theorems 6.11.6 and 6.11.7 (page 377) we get a series of properties of the $\mathcal{L}$-reduced function and the $\mathcal{L}$-balayage that we list here for reading convenience.

(I) $\widehat{\mathbf{R}}_A^f \leq \mathbf{R}_A^f$ in $\mathbb{G}$, $\widehat{\mathbf{R}}_A^f = \mathbf{R}_A^f$ in $\mathbb{G} \setminus \partial A$, $\widehat{\mathbf{R}}_A^f = f$ in $\mathrm{Int}(A)$,
(II) $\widehat{\mathbf{R}}_A^f$ is $\mathcal{L}$-subharmonic in $\mathbb{G}$ and $\mathcal{L}$-harmonic in $\mathbb{G} \setminus \overline{A}$,
(III) if $0 \leq f \leq g$ then $\mathbf{R}_A^f \leq \mathbf{R}_A^g$, $\widehat{\mathbf{R}}_A^f \leq \widehat{\mathbf{R}}_A^g$,
(IV) if $A \subseteq B \subseteq \mathbb{G}$ then $\mathbf{R}_A^f \leq \mathbf{R}_B^f$, $\widehat{\mathbf{R}}_A^f \leq \widehat{\mathbf{R}}_B^f$.

From Theorem 9.5.6 (page 449) we also have the following crucial result

(V) $\widehat{\mathbf{R}}_A^f = \mathbf{R}_A^f$ almost everywhere in $\mathbb{G}$.

Moreover, by Corollary 9.5.8 (page 450),

(VI) $\widehat{\mathbf{R}}_A^f(x) = \lim_{r \to 0} \mathrm{M}_r(\mathbf{R}_A^f)(x)$ for every $x \in \mathbb{G}$,

where M_r is the mean value operator in (5.50f), page 259. Properties (V) and (VI) allow to give an easy proof of the following theorem.

Theorem 11.5.2 (Further properties of $\mathcal{L}$-reduction and $\mathcal{L}$-balayage). *Let $A \subseteq \mathbb{G}$, and let $u, v \in \overline{\mathcal{S}}^+(\mathbb{G})$. One has:*

(i) $\mathbf{R}_A^{u+v} \leq \mathbf{R}_A^u + \mathbf{R}_A^v$,
(ii) $\widehat{\mathbf{R}}_A^{u+v} \leq \widehat{\mathbf{R}}_A^u + \widehat{\mathbf{R}}_A^v$.

Proof. (i). If $f \in \Phi_A^u$ and $g \in \Phi_A^v$ then $f + g \in \Phi_A^{u+v}$. As a consequence,

$$\mathbf{R}_A^{u+v} \leq f + g.$$

By taking the infimum with respect to $f \in \Phi_A^u$ and $g \in \Phi_A^v$, we obtain (i).

(ii). By using property (VI), from (i) we immediately obtain

$$\begin{aligned}\widehat{\mathbf{R}}_A^{u+v}(x) &= \lim_{r\to 0} \mathrm{M}_r(\mathbf{R}_A^{u+v})(x) \le \lim_{r\to 0} \mathrm{M}_r(\mathbf{R}_A^{u})(x) + \lim_{r\to 0} \mathrm{M}_r(\mathbf{R}_A^{v})(x) \\ &= \widehat{\mathbf{R}}_A^u(x) + \widehat{\mathbf{R}}_A^v(x)\end{aligned}$$

at any point $x \in \mathbb{G}$. □

When $A \subseteq \mathbb{G}$ is compact, the balayage takes the following form.

Theorem 11.5.3. *Let $K \Subset \mathbb{G}$ be a compact set, and let $u \in \overline{\mathcal{S}}^+(\mathbb{G})$. Then there exists a Radon measure μ in $\mathbb{G}$ such that $\widehat{\mathbf{R}}_K^u = \Gamma * \mu$.*

Proof. Since $\widehat{\mathbf{R}}_K^u \in \overline{\mathcal{S}}^+(\mathbb{G})$ (see Theorem 6.11.6, page 377), by Corollary 9.4.8, page 444, we only need to prove that

$$\inf_{\mathbb{G}} \widehat{\mathbf{R}}_K^u = 0.$$

We first assume that u is bounded in K. Then there exists $\lambda > 0$ such that $u \le \lambda\,\Gamma$ in K, so that $\lambda\,\Gamma \in \Phi_K^u$. As a consequence,

$$0 \le \widehat{\mathbf{R}}_K^u \le \mathbf{R}_K^u \le \lambda\,\Gamma \quad \text{in } \mathbb{G},$$

which immediately gives $\inf_{\mathbb{G}} \widehat{\mathbf{R}}_K^u = 0$.

We now consider the general case. By the cited Corollary 9.4.8, we have $u = v + h$ in $\mathbb{G}$, where $v = \Gamma * \mu$ is an $\mathcal{L}$-potential and $h \equiv \inf_{\mathbb{G}} u$. Since $\inf_{\mathbb{G}} v = 0$ (see Theorem 9.3.7, page 434) and

$$0 \le \widehat{\mathbf{R}}_K^v \le \mathbf{R}_K^v \le v$$

we have $\inf_{\mathbb{G}} \widehat{\mathbf{R}}_K^v = 0$. Hence $\widehat{\mathbf{R}}_K^v$ is an $\mathcal{L}$-potential (again by Corollary 9.4.8). On the other hand, since h is bounded in K, from the first part of the proof it follows that also $\widehat{\mathbf{R}}_K^h$ is an $\mathcal{L}$-potential. Therefore, $\widehat{\mathbf{R}}_K^v + \widehat{\mathbf{R}}_K^h$ is an $\mathcal{L}$-potential and thus it has infimum $= 0$ (again by Theorem 9.3.7). We now use

$$\widehat{\mathbf{R}}_K^u \le \widehat{\mathbf{R}}_K^v + \widehat{\mathbf{R}}_K^h$$

(see (6.31b), page 377), and we obtain $\inf_{\mathbb{G}} \widehat{\mathbf{R}}_K^u = 0$. This ends the proof. □

We now give the definition of $\mathcal{L}$-capacity. In Theorem 11.5.8 below, we shall see that this $\mathcal{L}$-capacity turns out to be equal to the inverse of the $\mathcal{L}$-equilibrium value defined in Section 11.4.

Definition 11.5.4 ($\mathcal{L}$-capacity for a compact set). *Let $K \Subset \mathbb{G}$ be a compact set. We define*

$$W_K := \mathbf{R}_K^1, \quad V_K := \widehat{\mathbf{R}}_K^1.$$

By Theorem 11.5.3, *V_K is an $\mathcal{L}$-potential called $\mathcal{L}$-*capacitary potential *of K. Moreover, the $\mathcal{L}$-Riesz measure μ_K of V_K will be called the $\mathcal{L}$-*capacitary distribution *for K (so that $V_K = \Gamma * \mu_K$). We define the $\mathcal{L}$-*capacity *of K as*

$$\mathcal{C}(K) := \mu_K(K). \tag{11.15}$$

We shall use the following properties of the $\mathcal{L}$-capacitary potential:

$$0 \le V_K \le W_K \le 1 \text{ in } \mathbb{G}, \quad V_K = W_K \text{ in } \mathbb{G} \setminus \partial K, \quad W_K = 1 \text{ in } K, \tag{11.16a}$$

$$\operatorname{supp} \mu_K \subseteq \partial K, \quad V_K \in \mathcal{H}(\mathbb{G} \setminus \partial K), \quad \lim_{|x| \to \infty} V_K(x) = 0. \tag{11.16b}$$

The properties in (11.16a) and the inclusion $V_K \in \mathcal{H}(\mathbb{G} \setminus K)$ directly follow from properties (I) and (II) of the balayage listed in the previous pages. Let us complete the proof of (11.16b). Since the constant 1 is $\mathcal{L}$-harmonic and $V_K = 1$ in $\operatorname{Int}(K)$, we obtain that

$$V_K \in \mathcal{H}(\mathbb{G} \setminus \partial K).$$

Hence, the inclusion $\operatorname{supp} \mu_K \subseteq \partial K$ readily follows recalling that μ_K is the $\mathcal{L}$-Riesz measure of V_K. Finally, by the structure of the fundamental solution $\Gamma = d^{2-Q}$ (being d a suitable $\mathcal{L}$-gauge), we have

$$V_K(x) = \int_K \Gamma(y^{-1} \circ x) \, d\mu_K(y) \le d(x, K)^{2-Q} \mu_K(K) \to 0 \quad \text{as } |x| \to \infty.$$

Theorem 11.5.5 (On the $\mathcal{L}$-capacitary potential). *Let $K \Subset \mathbb{G}$ be a compact set and let $\mu \in \mathfrak{M}(K)$ be a Radon measure such that $\Gamma * \mu \le 1$ in $\mathbb{G}$. Then $\Gamma * \mu \le V_K$ in $\mathbb{G}$.*

Proof. Let $v \in \Phi_K^1$, i.e. $v \in \overline{\mathcal{S}}^+(\mathbb{G})$, $v \ge 1$ in K. Since

$$\Gamma * \mu \le 1 \le v \quad \text{in } \operatorname{supp} \mu,$$

we have $\Gamma * \mu \le v$ in $\mathbb{G}$, in force of the Maria–Frostman domination principle (Theorem 11.3.4, page 495). Therefore,

$$\Gamma * \mu \le \inf\{v \mid v \in \Phi_K^1\} = W_K.$$

Recalling that $\Gamma * \mu$ is l.s.c., we finally obtain $\Gamma * \mu \le V_K$ in $\mathbb{G}$. □

Theorem 11.5.6 (On the $\mathcal{L}$-capacity. I). *Let $K \Subset \mathbb{G}$ be a compact set. Then*

$$\mathcal{C}(K) = \max\left\{\mu(K) \mid \mu \in \mathfrak{M}(K), \Gamma * \mu \le 1 \text{ in } \mathbb{G}\right\}. \tag{11.17}$$

Proof. We first observe that the $\mathcal{L}$-capacitary distribution μ_K for K has the properties $\mu_K \in \mathfrak{M}(K)$, $\Gamma * \mu_K \le 1$ in $\mathbb{G}$, by (11.16a)-(11.16b). Let Ω be an open set such that $K \subset \Omega \subset \overline{\Omega} \Subset \mathbb{G}$, and let $\nu = \mu_{\overline{\Omega}}$ be the $\mathcal{L}$-capacitary distribution for $\overline{\Omega}$. Then

$$V_{\overline{\Omega}} = \Gamma * \nu \equiv 1 \quad \text{in } \Omega \supset K,$$

by (11.16a). Let now $\mu \in \mathfrak{M}(K)$ be such that $\Gamma * \mu \le 1$ in $\mathbb{G}$. By Theorem 11.5.5, we have

$$\Gamma * \mu \le V_K = \Gamma * \mu_K \quad \text{in } \mathbb{G}.$$

Recalling that $\Gamma(x) = \Gamma(x^{-1})$, we obtain

$$\begin{aligned}\mu(K) &= \int_K 1\,\mathrm{d}\mu = \int_K \Gamma * \nu\,\mathrm{d}\mu = \int_{\mathbb{G}} \Gamma * \nu\,\mathrm{d}\mu = \int_{\mathbb{G}} \Gamma * \mu\,\mathrm{d}\nu \le \int_{\mathbb{G}} \Gamma * \mu_K\,\mathrm{d}\nu \\ &= \int_{\mathbb{G}} \Gamma * \nu\,\mathrm{d}\mu_K = \int_K \Gamma * \nu\,\mathrm{d}\mu_K = \int_K 1\,\mathrm{d}\mu_K = \mu_K(K) = \mathcal{C}(K).\end{aligned}$$

This completes the proof. □

The $\mathcal{L}$-capacitary potentials have the following sub-additivity property.

Theorem 11.5.7 (Sub-additivity of the $\mathcal{L}$-capacitary potential). *Let K_1 and K_2 be compact subsets of $\mathbb{G}$. Then:*

(i) $W_{K_1\cup K_2} + W_{K_1\cap K_2} \le W_{K_1} + W_{K_2}$,
(ii) $V_{K_1\cup K_2} + V_{K_1\cap K_2} \le V_{K_1} + V_{K_2}$.

Proof. (i). The inequality trivially holds on $K_1 \cup K_2$. Indeed, if $y \in K_1 \cup K_2$, then $y \in K_1 \cap K_2$ or $y \in K_i \setminus K_j$ for $i \ne j$. In the first case,

$$W_{K_1\cup K_2}(y) = W_{K_1\cap K_2}(y) = W_{K_1}(y) = W_{K_2}(y) = 1,$$

while in the second case

$$W_{K_1\cup K_2}(y) = W_{K_2}(y) = 1, \qquad W_{K_1\cap K_2}(y) \le W_{K_2}(y).$$

To show that (i) holds in $\Omega := \mathbb{G} \setminus (K_1 \cup K_2)$, let us consider $u_i \in \Phi^1_{K_i}$, $i = 1, 2$. Since $W_K = V_K$ in $\mathbb{G} \setminus K$, for every compact K, and $V_K(x) \to 0$ as $|x| \to \infty$, we have

$$\begin{aligned}\liminf_{|x|\to\infty} w(x) &:= \liminf_{|x|\to\infty}(u_1(x) + u_2(x) - (W_{K_1\cup K_2}(x) + W_{K_1\cap K_2}(x))) \\ &\ge \liminf_{|x|\to\infty}(u_1(x) + u_2(x)) \ge 0.\end{aligned}$$

Moreover, if $y \in \partial\Omega = K_1 \cup K_2$, then

$$\liminf_{\Omega\ni x\to y} w(x) \ge u_1(y) + u_2(y) - \limsup_{\Omega\ni x\to y}(W_{K_1\cup K_2}(x) + W_{K_1\cap K_2}(x)) \ge 0.$$

Indeed, if $y \in K_1 \cap K_2$, we have

$$u_1(y) + u_2(y) \ge 2 \ge \sup_{\mathbb{G}}(W_{K_1\cup K_2} + W_{K_1\cap K_2}),$$

while, if $y \in K_i \setminus K_j$, for $i \ne j$,

$$u_i(y) \ge 1 \ge \sup_{\mathbb{G}} W_{K_1\cup K_2} \quad \text{and} \quad u_j(y) \ge W_{K_1\cap K_2}(y) = \lim_{x\to y} W_{K_1\cap K_2}(x),$$

since $W_{K_1\cap K_2}$ is $\mathcal{L}$-harmonic, hence continuous, in a neighborhood of y. Then, the minimum principle for $\mathcal{L}$-superharmonic functions (Theorem 8.2.19-(ii)) implies $w \ge 0$ in Ω.

We have thus proved that

$$u_1 + u_2 \geq W_{K_1 \cup K_2} + W_{K_1 \cap K_2} \quad \text{in } \mathbb{G},$$

for every $u_i \in \Phi^1_{K_i}$, $i = 1, 2$. From this inequality (i) immediately follows.

(ii). From (i) and property (VI) we have

$$V_{K_1 \cup K_2} + V_{K_1 \cap K_2} \leq V_{K_1} + V_{K_2} \quad \text{a.e. in } \mathbb{G}.$$

This inequality extends on $\mathbb{G}$ in force of Corollary 9.5.8, page 450. □

Theorem 11.5.8 (On the $\mathcal{L}$-capacity. II). *Let $K \Subset \mathbb{G}$ be a compact set, and let $\mathcal{C}(K)$ and $V(K)$ be the $\mathcal{L}$-capacity and the $\mathcal{L}$-equilibrium value of K. Then we have*

$$\mathcal{C}(K) = (V(K))^{-1}.$$

Moreover, if $\mathcal{C}(K) > 0$ and μ_K is the $\mathcal{L}$-capacitary distribution for K, then $\mathcal{C}(K)^{-1}\mu_K$ is an $\mathcal{L}$-equilibrium distribution for K.

Proof. We first observe that, by definition, it is always $\mathcal{C}(K) < \infty$ and $V(K) > 0$. If $\mathcal{C}(K) > 0$, setting $\nu = \mathcal{C}(K)^{-1}\mu_K$, we have $\nu(K) = 1$. Moreover, $\operatorname{supp} \nu \subseteq K$ (see (11.16b)). Hence, by Definition 11.4.1 of $V(K)$ and recalling that $\Gamma * \mu_K = V_K \leq 1$ (see (11.16a)), we have

$$V(K) \leq I(\nu) = \mathcal{C}(K)^{-2} \int (\Gamma * \mu_K)\, \mathrm{d}\mu_K \leq \mathcal{C}(K)^{-2} \mu_K(K) = \mathcal{C}(K)^{-1} < \infty.$$

On the other hand, if $V(K) < \infty$ and $\overline{\mu}$ is an $\mathcal{L}$-equilibrium distribution for K, then, setting $\mu = V(K)^{-1}\overline{\mu}$, we have

$$\Gamma * \mu = V(K)^{-1}(\Gamma * \overline{\mu}) \leq 1 \quad \text{in } \mathbb{G},$$

in force of the fundamental Theorem 11.4.5 on $\mathcal{L}$-equilibrium potentials. Therefore, from Theorem 11.5.6 it follows that

$$\mathcal{C}(K) \geq \mu(K) = V(K)^{-1}\overline{\mu}(K) = V(K)^{-1} > 0.$$

This proves that $\mathcal{C}(K) > 0$ iff $V(K) < \infty$ and that we have $\mathcal{C}(K) = V(K)^{-1}$. The last statement of the theorem is an immediate consequence of the above arguments. □

Lemma 11.5.9. *If $\{K_j\}_j$ is a decreasing sequence of compact sets, then $W_{K_j} \downarrow W_{\bigcap_j K_j}$ as $j \to \infty$.*

Proof. Let us set $K := \bigcap_j K_j$. From (6.31a) (page 377), it immediately follows that $W_{K_j} \downarrow$. We set

$$g := \lim_{j \to \infty} W_{K_j} = \inf_j W_{K_j} = \inf \left\{ v \,\middle|\, v \in \bigcup_{j \in \mathbb{N}} \Phi^1_{K_j} \right\}.$$

Using Proposition 6.5.4-(iii) (page 355), it is easily seen that the family $\bigcup_j \Phi^1_{K_j}$ is down directed. Moreover, it is bounded from below by zero. From Theorem 6.11.1, page 375, it follows that the lower semicontinuous regularization $\widehat{g}$ of g is $\mathcal{L}$-superharmonic in $\mathbb{G}$. By the Corollary 9.4.8 (page 444) we know that

$$\widehat{g} = \Gamma * \nu + \inf_{\mathbb{G}} \widehat{g} \quad \text{in } \mathbb{G},$$

where ν is the $\mathcal{L}$-Riesz measure of $\widehat{g}$. Since $0 \le g \le W_{K_j}$ in $\mathbb{G}$, we have

$$0 \le \widehat{g}(x) \le V_{K_j}(x) \to 0 \quad \text{at infinity}$$

(see (11.16b)). Hence $\inf_{\mathbb{G}} \widehat{g} = 0$ and

$$\widehat{g} = \Gamma * \nu \quad \text{in } \mathbb{G}.$$

For a fixed j_0, $(W_{K_j})_{j \ge j_0}$ is a decreasing sequence of $\mathcal{L}$-harmonic functions in $\mathbb{G} \setminus K_{j_0}$ (see (11.16a)–(11.16b)).

Hence, by Theorem 5.7.10 (page 268), also the limit g is $\mathcal{L}$-harmonic in $\mathbb{G} \setminus K_{j_0}$. Since j_0 is arbitrary, g is $\mathcal{L}$-harmonic in $\mathbb{G} \setminus K$ (in particular $\widehat{g} = g$ in $\mathbb{G} \setminus K$). Hence the support of ν is contained in K. Moreover, $W_{K_j} \le 1$ gives $g \le 1$ and then $\Gamma * \nu = \widehat{g} \le 1$ in $\mathbb{G}$. Thus, by Theorem 11.5.5, we have $\widehat{g} \le V_K$. On the other hand, since

$$W_{K_j} \ge W_K \ge V_K \quad \text{for every } j,$$

(see (6.31a)), it is $g \ge V_K$, and then $\widehat{g} \ge V_K$. Therefore $\widehat{g} = V_K$, and we obtain

$$\widehat{g} = g = \lim_{j\to\infty} W_{K_j} \ge W_K = V_K = \widehat{g} \qquad \text{in } \mathbb{G} \setminus K.$$

As a consequence, $\lim_{j\to\infty} W_{K_j} = W_K$ in $\mathbb{G} \setminus K$. Observing that $W_{K_j} = 1 = W_K$ in K, the proof is complete. □

Proposition 11.5.10. *Given a sequence $\{K_j\}_j$ of compact subsets of $\mathbb{G}$, the following properties of the $\mathcal{L}$-capacity hold:*

$$K_1 \subseteq K_2 \quad \Rightarrow \quad \mathcal{C}(K_1) \le \mathcal{C}(K_2), \tag{11.18}$$

$$K_j \downarrow \quad \Rightarrow \quad \mathcal{C}(K_j) \downarrow \mathcal{C}\Big(\bigcap_j K_j\Big), \tag{11.19}$$

$$\mathcal{C}(K_1 \cup K_2) + \mathcal{C}(K_1 \cap K_2) \le \mathcal{C}(K_1) + \mathcal{C}(K_2). \tag{11.20}$$

Proof. Fix any K_i involved in any of the above statements. Let Ω be an open set such that $K_i \subset \Omega \subset \overline{\Omega} \Subset \mathbb{G}$, and let ν be the $\mathcal{L}$-capacitary distribution for $\overline{\Omega}$. We have $V_{\overline{\Omega}} = \Gamma * \nu \equiv 1$ in Ω and $\operatorname{supp} \nu \subseteq \partial\Omega$ by (11.16a)–(11.16b). Let V_{K_i} and μ_{K_i} be, respectively, the $\mathcal{L}$-capacitary potential and the $\mathcal{L}$-capacitary distribution for K_i. Then $\operatorname{supp} \mu_{K_i} \subseteq K_i$ and $V_{K_i} \equiv W_{K_i}$ outside K_i, hence in $\partial\Omega$ (again by (11.16a)–(11.16b)).

(i). If $K_1 \subseteq K_2$, then $W_{K_1} \le W_{K_2}$, see (6.31a), page 377. Recalling that $\Gamma(x) = \Gamma(x^{-1})$, we obtain

$$\begin{aligned}\mathcal{C}(K_1) &= \int_{K_1} 1 \, \mathrm{d}\mu_{K_1} = \int_{K_1} \Gamma * \nu \, \mathrm{d}\mu_{K_1} = \int_{\mathbb{G}} \Gamma * \nu \, \mathrm{d}\mu_{K_1} = \int_{\mathbb{G}} \Gamma * \mu_{K_1} \, \mathrm{d}\nu \\ &= \int_{\partial\Omega} V_{K_1} \, \mathrm{d}\nu = \int_{\partial\Omega} W_{K_1} \, \mathrm{d}\nu \le \int_{\partial\Omega} W_{K_2} \, \mathrm{d}\nu = \int_{\partial\Omega} V_{K_2} \, \mathrm{d}\nu \\ &= \int_{\mathbb{G}} \Gamma * \mu_{K_2} \, \mathrm{d}\nu = \int_{\mathbb{G}} \Gamma * \nu \, \mathrm{d}\mu_{K_2} = \int_{K_2} 1 \, \mathrm{d}\mu_{K_2} = \mathcal{C}(K_2).\end{aligned}$$

(ii). If $K_j \downarrow$ and $K := \bigcap_j K_j$, then $W_{K_j} \downarrow W_K$ by Lemma 11.5.9. Recalling that $V_K \equiv W_K$ outside K, hence in $\partial\Omega$, and arguing as in (i), we obtain

$$\begin{aligned}\mathcal{C}(K_j) &= \int_{\mathbb{G}} \Gamma * \nu \, \mathrm{d}\mu_{K_j} = \int_{\mathbb{G}} \Gamma * \mu_{K_j} \, \mathrm{d}\nu = \int_{\partial\Omega} V_{K_j} \, \mathrm{d}\nu \\ &= \int_{\partial\Omega} W_{K_j} \, \mathrm{d}\nu \searrow \int_{\partial\Omega} W_K \, \mathrm{d}\nu = \int_{\partial\Omega} V_K \, \mathrm{d}\nu \\ &= \int_{\mathbb{G}} \Gamma * \mu_K \, \mathrm{d}\nu = \int_{\mathbb{G}} \Gamma * \nu \, \mathrm{d}\mu_K = \mathcal{C}(K).\end{aligned}$$

(iii). By (6.31c) (page 377), we have $W_{K_1 \cup K_2} + W_{K_1 \cap K_2} \le W_{K_1} + W_{K_2}$. The same arguments as in (i), (ii) give

$$\begin{aligned}&\mathcal{C}(K_1 \cup K_2) + \mathcal{C}(K_1 \cap K_2) \\ &= \int_{\mathbb{G}} \Gamma * \nu \, \mathrm{d}\mu_{K_1 \cup K_2} + \int_{\mathbb{G}} \Gamma * \nu \, \mathrm{d}\mu_{K_1 \cap K_2} \\ &= \int_{\mathbb{G}} \Gamma * \mu_{K_1 \cup K_2} \, \mathrm{d}\nu + \int_{\mathbb{G}} \Gamma * \mu_{K_1 \cap K_2} \, \mathrm{d}\nu \\ &= \int_{\partial\Omega} V_{K_1 \cup K_2} \, \mathrm{d}\nu + \int_{\partial\Omega} V_{K_1 \cap K_2} \, \mathrm{d}\nu \\ &= \int_{\partial\Omega} (W_{K_1 \cup K_2} + W_{K_1 \cap K_2}) \, \mathrm{d}\nu \\ &\le \int_{\partial\Omega} (W_{K_1} + W_{K_2}) \, \mathrm{d}\nu = \int_{\partial\Omega} (V_{K_1} + V_{K_2}) \, \mathrm{d}\nu \\ &= \int_{\mathbb{G}} (\Gamma * \mu_{K_1} + \Gamma * \mu_{K_2}) \, \mathrm{d}\nu = \int_{\mathbb{G}} \Gamma * \nu \, \mathrm{d}\mu_{K_1} + \int_{\mathbb{G}} \Gamma * \nu \, \mathrm{d}\mu_{K_2} \\ &= \mathcal{C}(K_1) + \mathcal{C}(K_2).\end{aligned}$$

This ends the proof. □

Theorem 11.5.11 (Characterization of $\mathcal{L}$-polarity for $K \Subset \mathbb{G}$). *Let $K \Subset \mathbb{G}$ be a compact set. Then K is $\mathcal{L}$-polar if and only if $\mathcal{C}(K) = 0$.*

Proof. First, suppose K is $\mathcal{L}$-polar. By Corollary 11.2.5, there exists $\mu \in \mathfrak{M}_0$ such that $\Gamma * \mu \equiv \infty$ in K. Recalling that $\Gamma(x) = \Gamma(x^{-1})$ and that $\Gamma * \mu_K = V_K \le 1$ (see (11.16a)), we obtain

$$\infty \cdot \mathcal{C}(K) = \infty \cdot \mu_K(K) = \int \Gamma * \mu \, \mathrm{d}\mu_K = \int \Gamma * \mu_K \, \mathrm{d}\mu \le \mu(\mathbb{G}) < \infty.$$

Therefore, it must be $\mathcal{C}(K) = 0$.

Suppose now $\mathcal{C}(K) = 0$. By (11.19), we can find a sequence of bounded open sets $\Omega_j \supset K$ such that

$$K_{j+1} := \overline{\Omega}_{j+1} \subset \Omega_j \quad \text{and } \mathcal{C}(K_j) < 1/j^2.$$

We now choose $x_0 \notin K_1$, and we set $m := \max_{y \in K_1} \Gamma(y^{-1} \circ x_0)$. We have

$$V_{K_j}(x_0) = \int_{K_j} \Gamma(y^{-1} \circ x_0) \, \mathrm{d}\mu_{K_j}(y) \le m \, \mu_{K_j}(K_j) = m \, \mathcal{C}(K_j) < m/j^2.$$

We now define $\omega = \sum_{j \in \mathbb{N}} V_{K_j}$. Then $\omega \in \overline{\mathcal{S}}(\mathbb{G})$, since $\omega(x_0) < \infty$ (see Corollary 8.2.8, page 403). Moreover, $\omega \equiv \infty$ in K, since $V_{K_j} \equiv 1$ in $\Omega_j \supset K$ (see (11.16a)). Therefore, K is $\mathcal{L}$-polar. □

Corollary 11.5.12 ($\mathcal{L}$-polar* ⇒ $\mathcal{L}$-polar). *Any $\mathcal{L}$-polar* set is $\mathcal{L}$-polar.*

Proof. Let $E \subset \mathbb{G}$ be an $\mathcal{L}$-polar* set, i.e. there exists a countable family of compact sets F_n such that $V(F_n) = \infty$ and $E \subseteq \bigcup_{n \in \mathbb{N}} F_n$. From Theorem 11.5.8 and Theorem 11.5.11 it follows that F_n is $\mathcal{L}$-polar for every $n \in \mathbb{N}$. We now only need to recall that any countable union of $\mathcal{L}$-polar sets is $\mathcal{L}$-polar, by means of Corollary 11.2.6. □

Corollary 11.5.13 (On the $\mathcal{L}$-equilibrium potential). *Let $K \Subset \mathbb{G}$ be a compact set with $\mathcal{L}$-equilibrium value $V(K) < \infty$ (or, equivalently, with $\mathcal{L}$-capacity $\mathcal{C}(K) > 0$).*

Then there exists a unique $\mathcal{L}$-equilibrium potential of K given by $U_K = \mathcal{C}(K)^{-1} V_K$, where V_K is the $\mathcal{L}$-capacitary potential of K.

Proof. Let $U_1 = \Gamma * \mu_1$, $U_2 = \Gamma * \mu_2$ be $\mathcal{L}$-equilibrium potentials of K. From the fundamental Theorem 11.4.5 on $\mathcal{L}$-equilibrium potentials it follows that

$$U_1, U_2 \le V(K) < \infty \quad \text{in } \mathbb{G}$$

and $U_1 = V(K) = U_2$ in $K \setminus P$, for a suitable $\mathcal{L}$-polar* set P. Observing that P is also $\mathcal{L}$-polar, by Corollary 11.5.12, and recalling that $\operatorname{supp} \mu_i \subseteq K$, we infer that $U_1 = U_2$ in $\mathbb{G}$, in force of the Maria–Frostman domination principle (Theorem 11.3.4). In order to complete the proof, we only need to observe that by Theorem 11.5.8, $U_K = \mathcal{C}(K)^{-1} V_K$ is an $\mathcal{L}$-equilibrium potential of K. □

We now aim to extend the notion of $\mathcal{L}$-capacity to a larger class of sets.

Definition 11.5.14 ($\mathcal{L}$-capacitable set and $\mathcal{L}$-capacity). *Given a non-empty set $E \subseteq \mathbb{G}$, we define the* interior $\mathcal{L}$-capacity *of E as*

$$\mathcal{C}_*(E) := \sup\{\mathcal{C}(K) \mid K \textit{ compact}, \, K \subseteq E\}.$$

We then define the exterior $\mathcal{L}$-capacity *of E as follows*

$$\mathcal{C}^*(E) := \inf\{\mathcal{C}_*(\Omega) \mid \Omega \text{ open}, \Omega \supseteq E\}.$$

We also set $\mathcal{C}_*(\emptyset), \mathcal{C}^*(\emptyset) := 0$. *A set* $E \subseteq \mathbb{G}$ *is called* $\mathcal{L}$-capacitable *if* $\mathcal{C}_*(E) = \mathcal{C}^*(E)$. *In this case, this common value is denoted by* $\mathcal{C}(E)$ *and it is called the* $\mathcal{L}$-capacity *of* E.

Proposition 11.5.15. *Any Borel set is $\mathcal{L}$-capacitable. Moreover, for compact sets the above definition of $\mathcal{L}$-capacity agrees with the one given in Definition 11.5.4.*

Proof. With Proposition 11.5.10 at hand, one can follow verbatim the classical capacitability theory as presented, e.g. in [Helm69] or [AG01]. We omit further details. □

Theorem 11.5.16 (Characterization of $\mathcal{L}$-polarity. I). *A set $E \subseteq \mathbb{G}$ is $\mathcal{L}$-polar if and only if E is capacitable and $\mathcal{C}(E) = 0$.*

Proof. First, suppose E is $\mathcal{L}$-polar. Then E is contained in a Borel $\mathcal{L}$-polar set $\widetilde{E}$ (see Ex. 11.2.2). Any compact subset K of $\widetilde{E}$ is $\mathcal{L}$-polar and then it has null $\mathcal{L}$-capacity by Theorem 11.5.11. Thus, we have $\mathcal{C}_*(\widetilde{E}) = 0$. Since $\widetilde{E}$ is a Borel set, it is capacitable by Proposition 11.5.15. Hence

$$0 = \mathcal{C}_*(\widetilde{E}) = \mathcal{C}^*(\widetilde{E}) \geq \mathcal{C}^*(E) \geq \mathcal{C}_*(E) \geq 0$$

(we remark that from (11.18) it readily follows the monotonicity of interior and exterior capacity). Therefore E is capacitable and $\mathcal{C}(E) = 0$.

We now want to prove the reverse implication. We first observe that we can assume that E is bounded. The general case follows by considering the sequence of bounded sets $E_n := E \cap D(0, n)$. From $\mathcal{C}(E) = 0$ we get

$$0 = \mathcal{C}^*(E) \geq \mathcal{C}^*(E_n) \geq \mathcal{C}_*(E_n) \geq 0,$$

and thus E_n is capacitable with $\mathcal{C}(E_n) = 0$, hence $\mathcal{L}$-polar; then, by Corollary 11.2.6, also $E = \bigcup_n E_n$ is $\mathcal{L}$-polar.

Suppose now that E is a bounded capacitable set and $\mathcal{C}(E) = 0$. Then, for every $j \in \mathbb{N}$, there exists an open set $\Omega_j \supseteq E$ such that $\mathcal{C}(\Omega_j) < 1/j^2$. Moreover, we can suppose that all the Ω_j's are contained in a fixed bounded set. In particular, there exists $x_0 \in \mathbb{G}$ such that $B_d(x_0, 1) \cap \Omega_j = \emptyset$ for every j, so that

$$\Gamma(\zeta^{-1} \circ x_0) \leq 1 \qquad \forall\, \zeta \in \bigcup_j \Omega_j. \tag{11.21}$$

For a fixed $j \in \mathbb{N}$, we consider a sequence $\{B_i\}_i$ of open sets such that

$$\overline{B}_i \subset B_{i+1}, \qquad \bigcup_i B_i = \Omega_j.$$

We set $K_i := \overline{B}_i$. From (11.21) we obtain

$$V_{K_i}(x_0) = \int_{K_i} \Gamma(\zeta^{-1} \circ x_0)\, \mathrm{d}\mu_{K_i}(\zeta) \le \int_{K_i} \mathrm{d}\mu_{K_i} = \mathcal{C}(K_i) \le \mathcal{C}(\Omega_j) \le j^{-2}.$$

From $K_i \subseteq K_{i+1}$ we infer $0 \le V_{K_i} \le V_{K_{i+1}} \le 1$. If we set

$$\omega_j = \lim_{i\to\infty} V_{K_i},$$

we have $\omega_j(x_0) \le j^{-2}$, $0 \le \omega_j \le 1$. Moreover, $\omega_j \in \overline{\mathcal{S}}(\mathbb{G})$ by Corollary 8.2.8, page 403. On the other hand, (11.16a) gives $1 \ge \omega_j \ge V_{K_i} = 1$ on B_i for every i, so that $\omega_j \equiv 1$ on $\bigcup_i B_i = \Omega_j \supseteq E$.

Finally, setting $\omega = \sum_{j\in\mathbb{N}} \omega_j$, we have $\omega(x_0) \in \mathbb{R}$, $\omega \equiv \infty$ on E and $\omega \in \overline{\mathcal{S}}(\mathbb{G})$ (again by Corollary 8.2.8). This implies that E is $\mathcal{L}$-polar. □

11.6 The Fundamental Convergence Theorem

The following theorem is sometimes referred to as *the fundamental convergence theorem of potential theory*. It generalizes to our sub-Laplacian setting a theorem due to Cartan [Cart45, Théorème 8.4] (see also [Helm69, Theorem 7.39] and [AG01, Theorem 5.7.1]).

Theorem 11.6.1 (The fundamental convergence theorem). *Let $\mathcal{F}$ be a family of $\mathcal{L}$-superharmonic functions uniformly bounded from below (for example, a family of $\mathcal{L}$-potentials).*

Then, $v = \inf_{u\in\mathcal{F}} u$ differs from its lower semicontinuous regularization $\widehat{v}$ at most on an $\mathcal{L}$-polar set.

Proof. Since $\mathbb{G}$ is a countable union of d-balls, and a countable union of $\mathcal{L}$-polar sets is $\mathcal{L}$-polar, it is sufficient to prove that $v = \widehat{v}$ in $B := B_d(x_0, r)$ except at most on an $\mathcal{L}$-polar subset of B. We can also assume that $u \ge 0$ for every $u \in \mathcal{F}$, since $\mathcal{F}$ is uniformly bounded from below. Moreover, replacing $\mathcal{F}$ by the collection of infimums of finitely many elements of $\mathcal{F}$, if necessary, we can assume that $\mathcal{F}$ is down directed.

Let us now consider the $\mathcal{L}$-balayage $\widehat{\mathbf{R}}^u_{\overline{B}}$ of $u \in \mathcal{F}$. By Theorem 6.11.6, page 377, we know that $\widehat{\mathbf{R}}^u_{\overline{B}}$ is $\mathcal{L}$-superharmonic and it is equal to u in B. Moreover, $\widehat{\mathbf{R}}^u_{\overline{B}}$ is the potential of a measure supported in $\overline{B}$, by Theorem 11.5.3 and Theorem 6.11.6. Observing also that the family

$$\{\widehat{\mathbf{R}}^u_{\overline{B}} : u \in \mathcal{F}\}$$

is still down directed, it is then not restrictive to assume that any $u \in \mathcal{F}$ is the potential of a measure supported in $\overline{B}$. Finally, recalling that we are working with a down directed family, we can obviously assume that there exists $u_0 \in \mathcal{F}$ such that

$$u \le u_0 \quad \text{for every } u \in \mathcal{F}.$$

Let now D be a d-ball such that $\overline{B} \subset D$, and let us consider the $\mathcal{L}$-capacitary potential

$$V_{\overline{D}} = \Gamma * \mu_{\overline{D}} \equiv 1 \text{ in } D.$$

For every $u = \Gamma * \mu \in \mathcal{F}$ ($\mu \in \mathfrak{M}(\overline{B})$), we have

$$\begin{aligned}\mu(\overline{B}) &= \int_{\overline{B}} \Gamma * \mu_{\overline{D}} \,\mathrm{d}\mu = \int_{\mathbb{G}} \Gamma * \mu_{\overline{D}} \,\mathrm{d}\mu = \int_{\mathbb{G}} \Gamma * \mu \,\mathrm{d}\mu_{\overline{D}} \\ &\leq \int_{\mathbb{G}} u_0 \,\mathrm{d}\mu_{\overline{D}} = \int_{\mathbb{G}} \Gamma * \mu_0 \,\mathrm{d}\mu_{\overline{D}} = \int_{\mathbb{G}} \Gamma * \mu_{\overline{D}} \,\mathrm{d}\mu_0 = \mu_0(\overline{B}),\end{aligned}$$

where $u_0 = \Gamma * \mu_0$, $\mu_0 \in \mathfrak{M}(\overline{B})$. Hence, the set

$$\left\{\mu(\overline{B}) \mid u \in \mathcal{F}, u = \Gamma * \mu\right\}$$

is bounded. The assertion of the theorem now follows from Lemma 11.6.2 below: we find $\nu \in \mathfrak{M}$ such that $\Gamma * \nu \leq v$ in B with equality except possibly on an $\mathcal{L}$-polar set P, and we deduce that $v = \Gamma * \nu \leq \widehat{v} \leq v$ in $B \setminus P$. □

Lemma 11.6.2. *Let $\{\mu_i\}_{i\in I}$ be a family of Radon measures on $\mathbb{G}$ with the following properties*:

(i) *there exists a compact set K such that* $\operatorname{supp}(\mu_i) \subseteq K$ *for every $i \in I$,*
(ii) *the family $\{\Gamma * \mu_i \mid i \in I\}$ is down directed,*
(iii) *there exists $m \in \mathbb{R}$ such that $\mu_i(\mathbb{G}) \leq m$ for every $i \in I$.*

Then there exists $\mu \in \mathfrak{M}$ such that

$$\Gamma * \mu \leq \inf_{i\in I} \Gamma * \mu_i,$$

and there exists an $\mathcal{L}$-polar set P such that

$$\Gamma * \mu = \inf_{i\in I} \Gamma * \mu_i \quad \text{in } \mathbb{G} \setminus P.$$

Proof. By Proposition 6.1.2, page 339, there exists a countable set $I_0 \subseteq I$ such that if g is l.s.c. on $\mathbb{G}$ and $g \leq \inf_{i\in I_0}(\Gamma * \mu_i)$ then $g \leq \inf_{i\in I}(\Gamma * \mu_i)$. Since the family

$$\{\Gamma * \mu_i \mid i \in I\}$$

is down directed, it is not restrictive to assume that $I_0 = \{i_n\}_{n\in\mathbb{N}}$ and $\{\Gamma * \mu_{i_n}\}_n$ is decreasing. Now,[5] up to extracting a subsequence, μ_{i_n} converges in the w^*-topology to some measure $\mu \in \mathfrak{M}(K)$, $\mu(\mathbb{G}) \leq m$, and

$$\begin{aligned}(\Gamma * \mu)(x) &= \int_{\mathbb{G}} \Gamma(y^{-1} \circ x) \,\mathrm{d}\mu(y) \leq \liminf_{n\to\infty} \int_{\mathbb{G}} \Gamma(y^{-1} \circ x) \,\mathrm{d}\mu_{i_n}(y) \\ &= \liminf_{n\to\infty} (\Gamma * \mu_{i_n})(x) \\ &= \inf_{i\in I_0} (\Gamma * \mu_i)(x).\end{aligned}$$

Recalling that $\Gamma * \mu$ is l.s.c. on $\mathbb{G}$, it follows that

[5] See, e.g. [Rud87].

$$\Gamma * \mu \leq \inf_{i\in I}(\Gamma * \mu_i).$$

Let us now consider the Borel set

$$E := \Big\{x \in \mathbb{G} \,|\, (\Gamma * \mu)(x) < \lim_{n\to\infty}(\Gamma * \mu_{i_n})(x)\Big\}.$$

Since

$$\Big\{x \in \mathbb{G} \,|\, (\Gamma * \mu)(x) < \inf_{i\in I}(\Gamma * \mu_i)(x)\Big\} \subseteq E,$$

it remains only to prove that E is $\mathcal{L}$-polar. By means of Theorem 11.5.16 and Proposition 11.5.15, it suffices to show that $\mathcal{C}_*(E) = 0$.

Assume by contradiction that there exists a compact set $C \subseteq E$ such that $\mathcal{C}(C) > 0$, and consider the $\mathcal{L}$-capacitary distribution μ_C. Then

$$\mu_C(C) = \mathcal{C}(C) > 0 \text{ and } \Gamma * \mu_C \leq 1.$$

By Lemma 11.3.3, we infer that there exists a compact set $C' \subseteq C$, $\mu_C(C') > 0$, such that setting $\nu = \mu_C|_{C'}$ we have $\Gamma * \nu \in C(\mathbb{G}, \mathbb{R})$. Thus

$$\int_{\mathbb{G}} \Gamma * \nu \, \mathrm{d}\mu_{i_n} \stackrel{n\to\infty}{\longrightarrow} \int_{\mathbb{G}} \Gamma * \nu \, \mathrm{d}\mu$$

and, by Fatou's lemma,

$$\begin{aligned}
\int_{\mathbb{G}} \lim_{n\to\infty} \Gamma * \mu_{i_n} \, \mathrm{d}\nu &\leq \liminf_{n\to\infty} \int_{\mathbb{G}} \Gamma * \mu_{i_n} \, \mathrm{d}\nu \\
&= \liminf_{n\to\infty} \int_{\mathbb{G}} \Gamma * \nu \, \mathrm{d}\mu_{i_n} \\
&= \int_{\mathbb{G}} \Gamma * \nu \, \mathrm{d}\mu = \int_{\mathbb{G}} \Gamma * \mu \, \mathrm{d}\nu < \infty
\end{aligned}$$

($\Gamma * \nu$ is continuous and μ is compactly supported). Therefore

$$\int_C (\Gamma * \mu - \lim_{n\to\infty} \Gamma * \mu_{i_n}) \, \mathrm{d}\nu \geq 0.$$

Since the integrand is strictly negative on C and $\nu(C) > 0$, we get a contradiction. This completes the proof. □

For the reader's convenience and for the future references, we collect some of the properties of the $\mathcal{L}$-balayage that we know so far.

Lemma 11.6.3. *Let* $E \subseteq \mathbb{G}$. *Let* $u \in \overline{\mathcal{S}}^+(\mathbb{G})$. *Then* $\widehat{\mathbf{R}}^u_E$ *is* $\mathcal{L}$*-superharmonic on* Ω. *Moreover, the following properties hold*:

(i) $u \geq \mathbf{R}^u_E \geq \widehat{\mathbf{R}}^u_E \geq 0$ *on* $\mathbb{G}$,
(ii) $u = \mathbf{R}^u_E$ *on* E,
(iii) $u = \mathbf{R}^u_E = \widehat{\mathbf{R}}^u_E$ *on the interior of* E,

(iv) $\mathbf{R}^u_E = \widehat{\mathbf{R}}^u_E$ *on* $\mathbb{G} \setminus \overline{E}$, *and they are* $\mathcal{L}$*-harmonic on* $\mathbb{G} \setminus \overline{E}$,
(v) $\mathbf{R}^u_E$ *differs from* $\widehat{\mathbf{R}}^u_E$ *on a (*$\mathcal{L}$*-polar) subset of* ∂E,
(vi) $E \subseteq F \Rightarrow \widehat{\mathbf{R}}^u_E \le \widehat{\mathbf{R}}^u_F$,
(vii) $v \in \overline{\mathcal{S}}^+(\mathbb{G})$ *and* $u \le v$ *on* $E \Rightarrow \widehat{\mathbf{R}}^u_E \le \widehat{\mathbf{R}}^v_E$,
(viii) $\lambda > 0 \Rightarrow \widehat{\mathbf{R}}^{\lambda u}_E = \lambda \widehat{\mathbf{R}}^u_E$,
(ix) $\widehat{\mathbf{R}}^{u+v}_E \le \widehat{\mathbf{R}}^u_E + \widehat{\mathbf{R}}^v_E$.

The following result holds.

Proposition 11.6.4 ($\mathbf{R}^u_E$ and $\widehat{\mathbf{R}}^u_E$. I). *Let E be any subset of $\mathbb{G}$ and $u \in \overline{\mathcal{S}}^+(\mathbb{G})$. Then the $\mathcal{L}$-reduced function $\mathbf{R}^u_E$ and the $\mathcal{L}$-balayage $\widehat{\mathbf{R}}^u_E$ coincide on $\mathbb{G} \setminus E$.*

Proof. To begin with, from Lemma 11.6.3 we know that $\mathbf{R}^u_E$ and $\widehat{\mathbf{R}}^u_E$ may differ only on a subset of ∂E. From Theorem 11.6.1 we derive that $\mathbf{R}^u_E \ne \widehat{\mathbf{R}}^u_E$ on an $\mathcal{L}$-polar set, say P. Let $Z := E \cap P$. Collecting the above remarks, we infer that Z is an $\mathcal{L}$-polar subset of $E \cap \partial E$.

Let $x_0 \notin E$. In particular, $x_0 \notin Z$, so that, by Theorem 11.2.3, there exists $w \in \overline{\mathcal{S}}^+(\mathbb{G})$ such that $w(x_0) \in \mathbb{R}$ and $w \equiv \infty$ on Z. Then, for every $\varepsilon > 0$, the function

$$v := \widehat{\mathbf{R}}^u_E + \varepsilon w$$

belongs to $\overline{\mathcal{S}}^+(\mathbb{G})$ and is $\ge u$ on E (indeed, if $z \in E \cap P = Z$ then $v(z) = \infty \ge u(z)$, whereas if $z \in E \setminus P$, by definition of P it holds $v(z) \ge \widehat{\mathbf{R}}^u_E(z) = \mathbf{R}^u_E(z) = u(z)$, see Lemma 11.6.3).

Then, by the very Definition 11.5.1 of $\mathbf{R}^u_E$, we have

$$\widehat{\mathbf{R}}^u_E + \varepsilon\, w \le \mathbf{R}^u_E \quad \text{on } \mathbb{G}.$$

In particular, $\widehat{\mathbf{R}}^u_E(x_0) + \varepsilon\, w(x_0) \le \mathbf{R}^u_E(x_0)$. Letting $\varepsilon \to 0^+$ (recall that $w(x_0) \in \mathbb{R}$), we infer $\widehat{\mathbf{R}}^u_E(x_0) \le \mathbf{R}^u_E(x_0)$, which together with Lemma 11.6.3-(i), proves $\widehat{\mathbf{R}}^u_E(x_0) = \mathbf{R}^u_E(x_0)$. Being $x_0 \notin E$ arbitrary, the assertion of the proposition follows. □

From Theorem 11.6.1 and Proposition 11.6.4, one immediately obtains the following remarkable result.

Theorem 11.6.5 ($\mathbf{R}^u_E$ and $\widehat{\mathbf{R}}^u_E$. II). *Let $E \subseteq \mathbb{G}$ and $u \in \overline{\mathcal{S}}^+(\mathbb{G})$. Then the $\mathcal{L}$-reduced function $\mathbf{R}^u_E$ differs from the $\mathcal{L}$-balayage $\widehat{\mathbf{R}}^u_E$ at most on an $\mathcal{L}$-polar subset of $E \cap \partial E$.*

In particular, we have the following assertion.

Corollary 11.6.6. *Let $\Omega \subseteq \mathbb{G}$ be open and $u \in \overline{\mathcal{S}}^+(\mathbb{G})$. Then $\mathbf{R}^u_\Omega \equiv \widehat{\mathbf{R}}^u_E$.*

Proof. Apply Theorem 11.6.5 and note that $\Omega \cap \partial\Omega = \emptyset$ since Ω is open. □

11.7 The Extended Poisson–Jensen Formula

The aim of this section is to prove Theorem 11.7.6, which extends the Poisson–Jensen formula of Theorem 9.5.1 (page 445) in the case when the open set Ω is not necessarily an $\mathcal{L}$-regular set. The notion of *extended* $\mathcal{L}$-Green function is also needed.

Throughout this section, we say that $\Omega \subseteq \mathbb{G}$ is a *domain* if it is an open and connected set.

Lemma 11.7.1. *Let $\Omega \subset \mathbb{G}$ be a bounded domain, and let $\zeta_0 \in \partial\Omega$. Suppose $\Omega \not\subseteq B_d(x_0, r)$ and set $E = \overline{B_d}(x_0, r) \setminus \Omega$. If*

$$U_E(\zeta_0) = V(E) < \infty$$

(being U_E the $\mathcal{L}$-equilibrium potential for E and $V(E)$ its $\mathcal{L}$-equilibrium value), then there exists an $\mathcal{L}$-barrier for Ω at ζ_0.

Proof. To begin with, we have

$$U_E \in \mathcal{H}(\mathbb{G} \setminus E), \qquad U_E \leq V(E) \quad \text{in } \mathbb{G}$$

(see Theorem 11.4.5). Since $U_E \to 0$ at infinity, from the strong maximum principle (Theorem 5.13.8, page 296) it follows that

$$U_E < V(E) \quad \text{in } \mathbb{G} \setminus E.$$

Moreover, U_E l.s.c. and $U_E(\zeta_0) = V(E)$ imply that there exists

$$\lim_{\mathbb{G}\setminus E \ni x \to \zeta_0} U_E(x) = V(E).$$

Hence, the function $w := V(E) - U_E$ is an $\mathcal{L}$-barrier for Ω at ζ_0. □

The following result is of unquestionable importance.

Theorem 11.7.2 (On the $\mathcal{L}$-polarity of $\mathcal{L}$-irregular points). *Let $\Omega \subset \mathbb{G}$ be a bounded domain. Then the subset of $\partial\Omega$ of the $\mathcal{L}$-irregular points is an $\mathcal{L}$-polar* set. In particular, it is $\mathcal{L}$-polar.*

Proof. Let $\{B_d(x_j, r_j)\}_{j \leq n}$ be a cover of $\partial\Omega$ such that $\Omega \not\subseteq B_d(x_j, r_j)$ for every $j \leq n$. We set $E_j = \overline{B_d}(x_j, r_j) \setminus \Omega$ and $F_j = \partial\Omega \cap \overline{B_d}(x_j, r_j)$. We write

$$\partial\Omega = \Bigg(\bigcup_{j:\, \mathcal{C}(F_j)=0} F_j \Bigg) \cup \Bigg(\bigcup_{j:\, \mathcal{C}(F_j)>0} F_j \Bigg) =: B_1 \cup B_2.$$

Let us recall once for all that $\mathcal{C}(K) = V(K)^{-1}$ for any compact set K, by Theorem 11.5.8. In particular, we have that B_1 is $\mathcal{L}$-polar*. Suppose next that $j \in \{1, \dots, n\}$ is such that $\mathcal{C}(F_j) > 0$. Since $F_j \subseteq E_j$, we also have $\mathcal{C}(E_j) > 0$ (see (11.18)). From Theorem 11.4.5 it follows that

$$U_{E_j}(x) = V(E_j) \text{ for all } x \in E_j \setminus P_j,$$

where $P_j \subseteq E_j$ is $\mathcal{L}$-polar*. Hence, for every $\zeta_0 \in F_j \setminus P_j$, we have $U_{E_j}(\zeta_0) = V(E_j) < \infty$. We can then apply Lemma 11.7.1 and obtain that there exists an $\mathcal{L}$-barrier for Ω at ζ_0. Thus ζ_0 is an $\mathcal{L}$-regular point, by Bouligand's theorem 6.10.4, page 371. We finally write

$$\begin{aligned}\partial\Omega &= B_1 \cup \Big(\bigcup_{j:\, \mathcal{C}(F_j)>0} (P_j \cap F_j) \Big) \cup \Big(\bigcup_{j:\, \mathcal{C}(F_j)>0} (F_j \setminus P_j) \Big) \\ &=: B_1 \cup B_2' \cup B_2'',\end{aligned}$$

and we remark that $B_1 \cup B_2'$ is $\mathcal{L}$-polar*, whereas all the points of B_2'' are $\mathcal{L}$-regular. This ends the proof. □

Theorem 11.7.3 (On the $\mathcal{L}$-harmonic measure and $\mathcal{L}$-polarity). *Let $\Omega \subset \mathbb{G}$ be a bounded domain. Let E be an $\mathcal{L}$-polar subset of $\partial\Omega$. Then, for every $x \in \Omega$, we have $\mu_x^\Omega(E) = 0$.*

Proof. Since μ_x^Ω is regular, it is enough to prove the theorem when E is compact. By Theorem 6.9.3 (page 367), we know that the characteristic function χ_E of E is $\mathcal{L}$-resolutive, and we have

$$\mu_x^\Omega(E) = \int_E \mathrm{d}\mu_x^\Omega(y) = \int_{\partial\Omega} \chi_E(y)\, \mathrm{d}\mu_x^\Omega(y) = H_{\chi_E}^\Omega(x).$$

In particular, $H_{\chi_E}^\Omega \in \mathcal{H}(\Omega)$ and $H_{\chi_E}^\Omega$ is bounded. We have to prove that $H_{\chi_E}^\Omega = 0$ in Ω. We start by showing that

$$\lim_{\Omega \ni x \to \zeta_0} H_{\chi_E}^\Omega(x) = 0 \quad \forall\, \zeta_0 \in \partial\Omega \setminus F, \tag{11.22}$$

where $F = E \cup E_0$ and E_0 is the set of the $\mathcal{L}$-irregular points of $\partial\Omega$. Consider a sequence $\delta_n \in C(\partial\Omega)$, $\delta_n \searrow \chi_E$, such that there exists n_0 with $\delta_{n_0}(\zeta_0) = 0$ (we can take $\delta_n(\zeta) := \max\{1 - n\,\mathrm{dist}(\zeta, E), 0\}$). We have

$$0 \le H_{\chi_E}^\Omega(x) \le H_{\delta_n}^\Omega(x) \quad \text{for every } x \in \Omega$$

(see Proposition 6.7.4-(i), page 360). Moreover, since ζ_0 is an $\mathcal{L}$-regular point,

$$\lim_{\Omega \ni x \to \zeta_0} H_{\delta_{n_0}}^\Omega(x) = \delta_{n_0}(\zeta_0) = 0.$$

Thus, (11.22) follows. On the other hand, E_0 is $\mathcal{L}$-polar* by means of Theorem 11.7.2, hence $\mathcal{L}$-polar by Corollary 11.5.12. Moreover, E is $\mathcal{L}$-polar by hypothesis. Thus also F is $\mathcal{L}$-polar (see Corollary 11.2.6). We can now use the extended maximum principle for $\mathcal{L}$ in Lemma 11.2.7, and we obtain $H_{\chi_E}^\Omega = 0$ in Ω. This ends the proof. □

Definition 11.7.4 (The extended $\mathcal{L}$-Green function). *Let $\Omega \subset \mathbb{G}$ be a bounded domain, and let $x \in \Omega$ be fixed. We define the* extended $\mathcal{L}$-Green function $\overline{G}_\Omega(x,\cdot)$ *for Ω by*

$$\overline{G}_\Omega(x,y) := \begin{cases} G_\Omega(x,y) & \text{if } y \in \Omega, \\ 0 & \text{if } y \in \mathbb{G}\setminus\overline{\Omega}, \\ \limsup_{\Omega\ni z\to y} G_\Omega(x,z) & \text{if } y \in \partial\Omega. \end{cases} \tag{11.23}$$

In particular, $\overline{G}_\Omega(x,y) = 0$ for every $\mathcal{L}$-regular point $y \in \partial\Omega$. We have the following remarkable result.

Theorem 11.7.5 (On the extended $\mathcal{L}$-Green function). *Let $\Omega \subset \mathbb{G}$ be a bounded domain, and let $x \in \Omega$ be fixed. Then the extended $\mathcal{L}$-Green function for Ω is $\mathcal{L}$-subharmonic in $\mathbb{G}\setminus\{x\}$. Moreover, for all $y \in \mathbb{G}$,*

$$\overline{G}_\Omega(x,y) = \Gamma(y^{-1}\circ x) - \int_{\partial\Omega} \Gamma\big(y^{-1}\circ\eta\big)\,\mathrm{d}\mu_x^\Omega(\eta). \tag{11.24}$$

Proof. Let $x \in \Omega$ be fixed. We define the function $\overline{g}_\Omega(x,\cdot)$ on $\mathbb{G}$ by

$$\overline{g}_\Omega(x,\cdot) := \Gamma(x^{-1}\circ\cdot) - \int_{\partial\Omega} \Gamma\big((\cdot)^{-1}\circ\eta\big)\,\mathrm{d}\mu_x^\Omega(\eta). \tag{11.25}$$

(It is easy to see that no indeterminate forms $\infty-\infty$ may occur.) From Theorem 9.3.2 (page 432) and the symmetry of Γ it follows that $\overline{g}_\Omega(x,\cdot) \in \underline{S}(\mathbb{G}\setminus\{x\})$. We are left to prove that $\overline{g}_\Omega = \overline{G}_\Omega$.

If $y \in \Omega$, then from the symmetry of G_Ω (see Proposition 9.2.10, page 431) it follows that

$$\overline{g}_\Omega(x,y) = G_\Omega(y,x) = G_\Omega(x,y).$$

Let now $y \in \mathbb{G}\setminus\overline{\Omega}$. Since $\Gamma(y^{-1}\circ\cdot) \in \mathcal{H}(\Omega)\cap C(\overline{\Omega})$, we have

$$\int_{\partial\Omega} \Gamma(y^{-1}\circ\eta)\,\mathrm{d}\mu_x^\Omega(\eta) = H^\Omega_{\Gamma(y^{-1}\circ\cdot)}(x) = \Gamma(y^{-1}\circ x)$$

(see Proposition 6.7.7, page 361). In this case, $\overline{g}_\Omega(x,y) = 0$.

We now show that $\overline{g}_\Omega(x,\cdot)$ is non-negative on $\mathbb{G}$. It suffices to prove it for $y \in \partial\Omega$. For $\eta \in \partial\Omega$, we set $f_n(\eta) := \min\{n, \Gamma(y^{-1}\circ\eta)\}$. Then $f_n \in C(\partial\Omega,\mathbb{R})$ and $f_n(\eta) \nearrow \Gamma(y^{-1}\circ\eta)$ as $n\to\infty$. Then, by monotone convergence,

$$\int_{\partial\Omega} \Gamma(y^{-1}\circ\eta)\,\mathrm{d}\mu_x^\Omega(\eta) = \lim_{n\to\infty} H^\Omega_{f_n}(x).$$

If we prove that $H^\Omega_{f_n}(x) \le \Gamma(y^{-1}\circ x)$ for any $x \in \Omega$, it follows $\overline{g}_\Omega(x,y) \ge 0$. By the extended maximum principle in Lemma 11.2.7, since the set $\partial_2\Omega$ of the $\mathcal{L}$-irregular points of $\partial\Omega$ is an $\mathcal{L}$-polar set (see Theorem 11.7.2 and Corollary 11.5.12), it is enough to show that

$$\limsup_{\Omega\ni x\to\zeta}\left(H^{\Omega}_{f_n}(x)-\Gamma(y^{-1}\circ x)\right)\le 0 \quad \text{for } \zeta\in\partial_1\Omega,$$

where $\partial_1\Omega$ is the set of the $\mathcal{L}$-regular points of $\partial\Omega$, and that $H^{\Omega}_{f_n}(x)-\Gamma(y^{-1}\circ x)$ is bounded from above. The latter fact follows from

$$H^{\Omega}_{f_n}(x)-\Gamma(y^{-1}\circ x)\le H^{\Omega}_{f_n}(x)\le \max_{\partial\Omega} f_n<\infty.$$

We then prove the former claim. Since each $\zeta\in\partial_1\Omega$ is an $\mathcal{L}$-regular point,

$$\limsup_{\Omega\ni x\to\zeta}\left(H^{\Omega}_{f_n}(x)-\Gamma(y^{-1}\circ x)\right)\le f_n(\zeta)+\limsup_{\Omega\ni x\to\zeta}(-\Gamma(y^{-1}\circ x))$$

which is ≤ 0 by definition of f_n. This proves that $\overline{g}_{\Omega}(x,y)\ge 0$.

In order to prove that $\overline{g}_{\Omega}$ coincides with $\overline{G}_{\Omega}$ defined in (11.23), it remains to show that, for any $y_0\in\partial\Omega$, it holds

$$\overline{g}_{\Omega}(x,y_0)=\limsup_{\Omega\ni y\to y_0} G_{\Omega}(x,y).$$

First, we suppose $y_0\in\partial_1\Omega$. Since $\partial_2\Omega$ is $\mathcal{L}$-polar, by Corollary 11.2.5 there exists $h\in\underline{S}(\mathbb{G})$, finite at y_0, $h<0$ such that $h|_{\partial_2\Omega}\equiv-\infty$. For any $\varepsilon>0$, let

$$g_{\varepsilon}(y):=\overline{g}_{\Omega}(x,y)+\varepsilon\, h(y), \qquad y\in\mathbb{G}\setminus\{x\}.$$

Since $\overline{g}_{\Omega}(x,y)=G_{\Omega}(x,y)$ for every $y\in\Omega$, $\overline{g}_{\Omega}(x,y)$ vanishes as y approaches $\partial_1\Omega$ from the inside of Ω and remains bounded as $y\in\Omega$ approaches $\partial_2\Omega$. Using also the fact that h is u.s.c., we have that, for any $y_1\in\partial\Omega$,

$$\limsup_{\Omega\ni y\to y_1} g_{\varepsilon}(y)<0.$$

Moreover,

$$\limsup_{\mathbb{G}\setminus\overline{\Omega}\ni y\to y_1} g_{\varepsilon}(y)<0,$$

since $\overline{g}_{\Omega}(x,y)=0$ for every $y\in\mathbb{G}\setminus\overline{\Omega}$. Then there exists a compact neighborhood N of $\partial\Omega$ such that $g_{\varepsilon}<0$ on $N\setminus\partial\Omega$. Recalling that

$$\overline{g}_{\Omega}(x,\cdot)\in\underline{S}(\mathbb{G}\setminus\{x\}),$$

from the maximum principle for $\mathcal{L}$-subharmonic functions (Theorem 8.2.19, page 409) we then obtain $g_{\varepsilon}\le 0$ in N. In particular, $g_{\varepsilon}(y_0)\le 0$ and, letting ε vanish, we get $\overline{g}_{\Omega}(x,y_0)\le 0$, which (jointly with $\overline{g}_{\Omega}(x,\cdot)\ge 0$) gives $\overline{g}_{\Omega}(x,y_0)=0$. We have thus proved that

$$\lim_{\Omega\ni y\to y_0} G_{\Omega}(x,y)=0=\overline{g}_{\Omega}(x,y_0)$$

if $y_0\in\partial_1\Omega$.

We now treat the case of $y_0\in\partial_2\Omega$. Let us set for brevity

$$E := \partial_2 \Omega, \quad A := \mathbb{G} \setminus \{x\}, \quad u := \overline{g}_\Omega(x, \cdot).$$

With this notation, we have E $\mathcal{L}$-polar and $u \in \underline{S}(A)$. We now set

$$\widetilde{u}(x) := \limsup_{A \setminus E \ni \xi \to x} u(\xi), \quad \text{for } x \in A \cap E, \qquad \widetilde{u}(x) := u(x), \quad \text{for } x \in A \setminus E.$$

Since we know that $u \geq 0$ is u.s.c., we immediately get $0 \leq \widetilde{u} \leq u$ in A. Moreover, since E has Lebesgue measure zero being $\mathcal{L}$-polar, the following equality of the solid means holds: $M_r(u) = M_r(\widetilde{u})$. As a consequence, $\widetilde{u}$ is sub-mean in A, being $u \in \underline{S}(A)$, hence sub-mean by Theorem 8.2.1 (page 401). Furthermore, it is easy to verify that $\widetilde{u}$ is u.s.c. Therefore, again from Theorem 8.2.1, we obtain $\widetilde{u} \in \underline{S}(A)$. We now use Lemma 11.3.1 and conclude that $\widetilde{u} \equiv u$ in A. In particular, we have

$$\begin{aligned}\overline{g}_\Omega(x, y_0) = u(y_0) = \widetilde{u}(y_0) &= \limsup_{\mathbb{G} \setminus \partial_2 \Omega \ni y \to y_0} \overline{g}_\Omega(x, y) \\ &= \limsup_{\Omega \ni y \to y_0} G_\Omega(x, y).\end{aligned}$$

In the last equality we used the following facts: $\overline{g}_\Omega(x, \cdot) \geq 0$, $\overline{g}_\Omega(x, \cdot) \equiv 0$ in $\mathbb{G} \setminus (\Omega \cup \partial_2 \Omega)$ and $\overline{g}_\Omega(x, \cdot) = G_\Omega(x, \cdot)$ in Ω. This completes the proof. □

Theorem 11.7.6 (The extended Poisson–Jensen formula). *Let Ω be a bounded domain of $\mathbb{G}$. Suppose u is $\mathcal{L}$-subharmonic on a neighborhood of $\overline{\Omega}$. Then, for all $x \in \Omega$, we have*

$$u(x) = \int_{\partial_1 \Omega} u(y) \, \mathrm{d}\mu_x^\Omega(y) - \int_{\Omega \cup \partial_2 \Omega} \overline{G}_\Omega(x, y) \, \mathrm{d}\mu_u(y), \qquad (11.26)$$

where $\partial_1 \Omega$, $\partial_2 \Omega$ denote, respectively, the subsets of $\partial \Omega$ of the regular and the irregular points for the Dirichlet problem related to $\mathcal{L}$, μ_u is the $\mathcal{L}$-Riesz measure of u, μ_x^Ω is the $\mathcal{L}$-harmonic measure for Ω at x, and $\overline{G}_\Omega$ is the extended $\mathcal{L}$-Green function for Ω.

Proof. Let O be a bounded domain such that $\overline{\Omega} \subset O$ and such that u is $\mathcal{L}$-subharmonic on a neighborhood of $\overline{O}$. By the Riesz representation Theorem 9.4.4 (page 442), there exists $h \in \mathcal{H}(O)$ such that

$$u(x) = h(x) - \int_{\overline{O}} \Gamma(y^{-1} \circ x) \, \mathrm{d}\mu_u(y) \qquad \forall\, x \in O.$$

We certainly have $h(x) = \int_{\partial \Omega} h(\eta) \, \mathrm{d}\mu_x^\Omega(\eta)$. We now let

$$v(x) := -\int_{\overline{O}} \Gamma(y^{-1} \circ x) \, \mathrm{d}\mu_u(y) \quad \text{for } x \in \mathbb{G}.$$

Recalling Theorem 9.3.5, page 433, $v \in \underline{S}(\mathbb{G})$ and μ_v coincides with μ_u on $\overline{O}$ and vanishes on $\mathbb{G} \setminus \overline{O}$. We claim that it is enough to prove (11.26) with u replaced by v.

Indeed, using the fact that $\mu_x^\Omega(\partial_2\Omega) = 0$ for every $x \in \Omega$ (which is an immediate consequence of Theorem 11.7.2 and Theorem 11.7.3), this would give

$$\begin{aligned} u(x) &= h(x) + v(x) \\ &= \int_{\partial\Omega} h(y)\,\mathrm{d}\mu_x^\Omega(y) + \int_{\partial_1\Omega} v(y)\,\mathrm{d}\mu_x^\Omega(y) - \int_{\partial_2\Omega\cup\Omega} \overline{G}_\Omega(x,y)\,\mathrm{d}\mu_v(y) \\ &= \int_{\partial_1\Omega} (h(y)+v(y))\,\mathrm{d}\mu_x^\Omega(y) - \int_{\partial_2\Omega\cup\Omega} \overline{G}_\Omega(x,y)\,\mathrm{d}\mu_u(y) \\ &= \int_{\partial_1\Omega} u(y)\,\mathrm{d}\mu_x^\Omega(y) - \int_{\partial_2\Omega\cup\Omega} \overline{G}_\Omega(x,y)\,\mathrm{d}\mu_u(y). \end{aligned}$$

We are then left to prove (11.26) for v. We recall that $\overline{G}_\Omega(x,\cdot) \equiv 0$ on $\partial_1\Omega$ and on $\mathbb{G}\setminus\overline{\Omega}$ (see the definition (11.23) of $\overline{G}_\Omega$). Using this fact, exploiting $\mu_x^\Omega(\partial_2\Omega) = 0$, and using (11.24), we obtain that, for every $x \in \Omega$,

$$\begin{aligned} &\int_{\partial_1\Omega} v(\eta)\,\mathrm{d}\mu_x^\Omega(\eta) - \int_{\partial_2\Omega\cup\Omega} \overline{G}_\Omega(x,y)\,\mathrm{d}\mu_v(y) \\ &= -\int_{\partial_1\Omega} \left(\int_{\overline{O}} \Gamma(y^{-1}\circ\eta)\,\mathrm{d}\mu_u(y)\right)\mathrm{d}\mu_x^\Omega(\eta) - \int_{\partial_2\Omega\cup\Omega} \overline{G}_\Omega(x,y)\,\mathrm{d}\mu_u(y) \\ &= -\int_{\overline{O}} \left(\int_{\partial\Omega} \Gamma(y^{-1}\circ\eta)\,\mathrm{d}\mu_x^\Omega(\eta)\right)\mathrm{d}\mu_u(y) - \int_{\overline{O}} \overline{G}_\Omega(x,y)\,\mathrm{d}\mu_u(y) \\ &= \int_{\overline{O}} \left(-\int_{\partial\Omega} \Gamma(y^{-1}\circ\eta)\,\mathrm{d}\mu_x^\Omega(\eta) - \overline{G}_\Omega(x,y)\right)\mathrm{d}\mu_u(y) \\ &= -\int_{\overline{O}} \Gamma(y^{-1}\circ x)\,\mathrm{d}\mu_u(y) = v(x). \end{aligned}$$

This completes the proof. □

11.8 Further Results. A Miscellanea

The aim of this section is to collect a miscellanea of results concerning with the $\mathcal{L}$-capacity. These assertions, besides completing the investigation of the previous sections, also provide useful results, some having an interest in its own.

Theorem 11.8.1 (The $\mathcal{L}$-balayage as a $\mathcal{L}$-potential). *Let $E \subset \mathbb{G}$ be bounded. Then the following assertions hold:*

(a) *For every $u \in \overline{\mathcal{S}}^+(\mathbb{G})$, the $\mathcal{L}$-balayage $\widehat{\mathbf{R}}_E^u$ is an $\mathcal{L}$-potential. In particular, if $u \equiv 1$, there exists a Radon measure ν_E such that $\widehat{\mathbf{R}}_E^1 = \Gamma * \nu_E$ (i.e. ν_E is the $\mathcal{L}$-Riesz measure of $\widehat{\mathbf{R}}_E^1$);*
(b) *With the above notation, we have*

$$C^*(E) = \nu_E(\mathbb{G}).$$

Proof. (a). Let $R \gg 1$ be such that $\overline{E} \subset B_d(0, R)$. Let

$$K_1 := \overline{B_d(0, R)}, \quad K_2 := \overline{B_d(0, R+1)}, \quad v := \widehat{\mathbf{R}}^u_{K_1}.$$

Since $v \in \mathcal{H}(\mathbb{G} \setminus K_1)$ (see Lemma 11.6.3-(iv)), there exists $a \gg 1$ such that $a\,\Gamma \geq v$ on ∂K_2. We set

$$w(x) := \begin{cases} v(x) & \text{if } x \in K_2, \\ \min\{a\,\Gamma(x), v(x)\} & \text{if } x \in \mathbb{G} \setminus K_2. \end{cases}$$

By Lemma 11.3.2, $w \in \overline{\mathcal{S}}^+(\mathbb{G})$. Moreover,[6] $w \geq u$ on E. Hence, by the definition of $\mathcal{L}$-balayage, $w \geq \mathbf{R}^u_E \geq \widehat{\mathbf{R}}^u_E$. Then, on $\mathbb{G} \setminus K_2$ we have

$$\widehat{\mathbf{R}}^u_E \leq w = \min\{a\,\Gamma, v\} \leq a\,\Gamma.$$

As a consequence,

$$\inf_{\mathbb{G}} \widehat{\mathbf{R}}^u_E \leq \inf_{\mathbb{G}\setminus K_2} \leq \inf_{\mathbb{G}\setminus K_2} a\,\Gamma = 0.$$

We are therefore in a position to apply Proposition 9.9.1 (see Ex. 5, Chapter 9, page 464) and derive that $\widehat{\mathbf{R}}^u_E$ is an $\mathcal{L}$-potential. This proves assertion (a).

(b). Let us now take $u = 1$ and apply what we have proved above: there exists a Radon measure ν_E such that $\widehat{\mathbf{R}}^1_E = \Gamma * \nu_E$. Our task is to prove that $\mathcal{C}^*(E) = \nu_E(\mathbb{G})$. The proof is split in two parts: first, we suppose that E is also open, then we consider the general case.

(b'). Let $E = \Omega$ be open and bounded. Let $\{K_n\}_n$ be a sequence of compact subsets of $\mathbb{G}$ such that

$$K_n \subset \mathrm{Int}(K_{n+1}) \subset K_{n+1} \subset \Omega \quad \text{and} \quad \bigcup_n K_n = \Omega.$$

By Ex. 15 at the end of the chapter, we infer (see also Corollary 11.6.6 and part (a) of the proof)

$$\lim_{n\to\infty} \widehat{\mathbf{R}}^1_{K_n} = \mathbf{R}^1_\Omega = \widehat{\mathbf{R}}^1_\Omega = \Gamma * \nu_\Omega. \tag{11.27}$$

As a consequence, we have

$$\mathcal{C}(\Omega) = \lim_{n\to\infty} \mathcal{C}(K_n) = \lim_{n\to\infty} \mu_{K_n}(\mathbb{G}) = \nu_\Omega(\mathbb{G}).$$

The first equality is a consequence of Ex. 16-(6) at the end of the chapter; the second is the definition of $\mathcal{L}$-capacity for a compact set; the third equality follows from Ex. 14 at the end of the chapter.[7]

[6] Indeed, since E is contained in the interior of K_1, we have $w = v = \widehat{\mathbf{R}}^u_{K_1} = \mathbf{R}^u_{K_1} = u$ on E, see Lemma 11.6.3.

[7] We apply Ex. 14 with $\nu = \nu_\Omega$, $\sigma_n = \mu_{K_n}$, $K = \overline{\Omega}$. Indeed, note that

$$\Gamma * \mu_{K_n} = V_{K_n} = \widehat{\mathbf{R}}^1_{K_n} \nearrow \Gamma * \nu_\Omega$$

by (11.27) and property (vi) of Lemma 11.6.3.

(b"). To begin with, we make some remarks on the definition of $\widehat{\mathbf{R}}^1_E$. We recall that

$$\widehat{\mathbf{R}}^1_E = \Big(\inf_{u\in\Phi^1_E} u\Big)^\frown$$

where

$$\Phi^1_E = \{u \in \overline{\mathcal{S}}^+(\mathbb{G}) \mid u \geq 1 \text{ on } E\}.$$

Since Φ^1_E is obviously down directed and $1 \in \Phi^1_E$, then[8]

$$\inf_{u\in\Phi^1_E} u = \inf_{u\in\Phi^1_E,\ u\leq 1} u.$$

Furthermore, by a very general result on the lower semicontinuous regularization of the infimum of an arbitrary family of functions,[9] there exists a sequence $\{u_n\}$ in Φ^1_E with $u_n \leq 1$ such that

$$\Big(\inf_n u_n\Big)^\frown = \Big(\inf_{u\in\Phi^1_E} u\Big)^\frown = \widehat{\mathbf{R}}^1_E.$$

We can also suppose that $u_1 = 1$. Let

$$v_n := \min\{u_1, u_2, \ldots, u_n\}.$$

We have $0 \leq v_n \leq 1$, $v_n \in \overline{\mathcal{S}}^+(\mathbb{G})$, $v_n \geq 1$ on E and $\{v_n\}_{n\in\mathbb{N}}$ is non-increasing. Also,

$$\lim_{n\to\infty} v_n = \inf_{n\in\mathbb{N}} u_n. \tag{11.28}$$

Let $\varepsilon \in (0, 1)$. Let Ω be an open set such that

$$E \subseteq \Omega \text{ and } \mathcal{C}(\Omega) < \mathcal{C}^*(E) + \varepsilon. \tag{11.29}$$

We can suppose that Ω is bounded, by replacing it by $\Omega \cap B_d(0, R)$ (with $R \gg 1$ such that $B_d(0, R) \supset E$; note that $\mathcal{C}(\Omega \cap B_d(0, R)) \leq \mathcal{C}(\Omega)$). We set

$$w_n := \mathbf{R}^{v_n}_\Omega \quad\text{and}\quad \Omega_n := \{x \in \Omega \mid w_n(x) > 1 - \varepsilon\}.$$

We collect some properties of w_n and Ω_n:

(a) $w_n := \widehat{\mathbf{R}}^{v_n}_\Omega$, for Ω is open (see Corollary 11.6.6);
(b) $\{w_n\}_n$ is decreasing (see Lemma 11.6.3-(vii));
(c) Ω_n is open, for w_n is l.s.c.;
(d) $\Omega_n \supseteq E$, for $w_n = \mathbf{R}^{v_n}_\Omega = v_n \geq 1 > 1 - \varepsilon$ on E;

---

[8] *If $\mathcal{F}$ is a down directed family of functions and $u_0 \in \mathcal{F}$, then* $\inf\{u \mid u \in \mathcal{F}\} = \inf\{u \mid u \in \mathcal{F}, u \leq u_0\}$. Indeed, obviously $\inf\{u \mid u \in \mathcal{F}\} \leq \inf\{u \mid u \in \mathcal{F}, u \leq u_0\}$. Moreover, for every $u \in \mathcal{F}$, there exists $v \in \mathcal{F}$ such that $v \leq \min\{u, u_0\} \leq u_0$. Hence $\inf\{u \mid u \in \mathcal{F}, u \leq u_0\} \leq v \leq u$. Taking the infimum over $\{u \mid u \in \mathcal{F}\}$ in the last inequality, the assertion follows.

[9] See Proposition 6.1.2, page 339.

(e) $w_n = \mathbf{R}_\Omega^{v_n} \in \mathcal{H}(\mathbb{G} \setminus \overline{\Omega_n})$ (see Lemma 11.6.3-(iv));
(f) $w_n \geq \widehat{\mathbf{R}}_E^1$ (indeed, $w_n \in \Phi_E^1$ by (d) above, whence $w_n \geq \mathbf{R}_E^1$, and take the lower semicontinuous regularization);
(g) there exists a Radon measure σ_n on $\mathbb{G}$ such that $w_n = \Gamma * \sigma_n$ (since $w_n = \widehat{\mathbf{R}}_\Omega^{v_n}$ and by the first part of the proof, being Ω bounded);
(h) $\mathbf{R}_{\Omega_n}^1 = \widehat{\mathbf{R}}_{\Omega_n}^1 = \Gamma * \nu_{\Omega_n}$ (by the first part of the proof);
(i) by (e) above, it follows that σ_n in (g) is supported in $\overline{\Omega_n}$ (hence compactly supported); the same is true for ν_{Ω_n}.

As a consequence of (f),

$$v_n \geq \mathbf{R}_\Omega^{v_n} = w_n \geq \widehat{\mathbf{R}}_E^1.$$

Letting $n \to \infty$ (see also (11.28)), we get

$$\inf_{n\in\mathbb{N}} u_n \geq W := \lim_{n\to\infty} w_n \geq \widehat{\mathbf{R}}_E^1 \quad \text{on } \mathbb{G}. \tag{11.30}$$

By (b) and (e) above, it follows that, on each component of $\mathbb{G} \setminus \overline{\Omega}$, W is $\mathcal{L}$-harmonic unless it is $-\infty$. From (11.30) we infer that the latter is impossible. Hence, $W = \widehat{W}$ on $\mathbb{G}\setminus\overline{\Omega}$ whence, taking lower semicontinuous regularization in (11.30), one obtains

$$\widehat{\mathbf{R}}_E^1 = \left(\inf_n u_n\right)^\wedge \geq \widehat{W} \geq \widehat{\widehat{\mathbf{R}}_E^1} = \widehat{\mathbf{R}}_E^1$$

on $\mathbb{G} \setminus \overline{\Omega}$. This proves $W = \widehat{\mathbf{R}}_E^1$ on $\mathbb{G} \setminus \overline{\Omega}$.

We claim that

$$w_n \geq (1-\varepsilon)\, \mathbf{R}_{\Omega_n}^1. \tag{11.31}$$

Indeed, the function $\widetilde{w}_n := w_n/(1-\varepsilon)$ belongs to $\overline{\mathcal{S}}^+(\mathbb{G})$ and, by the definition of Ω_n, $\widetilde{w}_n > 1$ on Ω_n. Thus $\widetilde{w}_n \in \Phi_{\Omega_n}^1$, so that

$$\mathbf{R}_{\Omega_n}^1 = \inf_{v\in\Phi_{\Omega_n}^1} v \leq \widetilde{w}_n = w_n/(1-\varepsilon),$$

which is the claimed (11.31).

By (g) and (h) above, (11.31) rewrites as follows

$$\Gamma * \sigma_n \geq (1-\varepsilon)\, \Gamma * \nu_{\Omega_n} = \Gamma * \big((1-\varepsilon)\, \nu_{\Omega_n}\big).$$

By Ex. 13 at the end of the chapter, the above inequality ensures that

$$\sigma_n(\mathbb{G}) \geq (1-\varepsilon)\, \nu_{\Omega_n}(\mathbb{G}). \tag{11.32}$$

The following facts hold:

(i) $\mathrm{supp}(\sigma_n) \subseteq \overline{\Omega} \Subset \mathbb{G}$ for every $n \in \mathbb{N}$,
(ii) $\Gamma * \sigma_n = w_n$ is decreasing,
(iii) on $\mathbb{G} \setminus \overline{\Omega}$, it holds $w_n \to W = \widehat{\mathbf{R}}_E^1 = \Gamma * \nu_E$ (by the definition of ν_E).

Hence, we are in the position to apply Ex. 14 at the end of the chapter to derive that $\nu_E(\mathbb{G}) = \lim_n \sigma_n(\mathbb{G})$. By (11.32), we get

$$\nu_E(\mathbb{G}) \geq (1-\varepsilon) \limsup_{n\to\infty} \nu_{\Omega_n}(\mathbb{G}). \tag{11.33}$$

Moreover, from part (a) of the proof, $E \subseteq \Omega$, Lemma 11.6.3-(vi) and the definition of ν_E, we get

$$\Gamma * \nu_\Omega = \widehat{\mathbf{R}}^1_\Omega \geq \widehat{\mathbf{R}}^1_E = \Gamma * \nu_E.$$

Hence (by another application of Ex. 13 at the end of the chapter) $\nu_\Omega(\mathbb{G}) \geq \nu_E(\mathbb{G})$. This inequality, together with (11.33) implies

$$\nu_\Omega(\mathbb{G}) \geq (1-\varepsilon) \limsup_{n\to\infty} \nu_{\Omega_n}(\mathbb{G}). \tag{11.34}$$

Finally, by (11.29), part (b') of the proof (for Ω), (11.34), again part (b') of the proof (this time for Ω_n) and $\Omega_n \supseteq E$ (whence $\mathcal{C}(\Omega_n) \geq \mathcal{C}^*(E)$), we obtain

$$\begin{aligned}\mathcal{C}^*(E) + \varepsilon > \mathcal{C}(\Omega) = \nu_\Omega(\mathbb{G}) \geq \nu_E(\mathbb{G}) &\geq (1-\varepsilon) \limsup_{n\to\infty} \nu_{\Omega_n}(\mathbb{G}) \\ &= (1-\varepsilon) \limsup_{n\to\infty} \mathcal{C}(\Omega_n) \geq (1-\varepsilon)\, \mathcal{C}^*(E).\end{aligned}$$

Passing to the limit as $\varepsilon \to 0^+$, we get $\mathcal{C}^*(E) = \nu_E(\mathbb{G})$, which we aimed to prove. □

Proposition 11.8.2 (On the inner and outer $\mathcal{L}$-capacity). *Let $E \subset \mathbb{G}$ be a bounded set. Then we have:*

(i) $\mathcal{C}_*(E) = \sup\{\mu(E) \mid \mu \in \mathfrak{M}(E),\ \Gamma * \mu \leq 1 \ \textit{on}\ \mathbb{G}\}$,
(ii) $\mathcal{C}^*(E) = \inf\{\mu(\mathbb{G}) \mid \mu \in \mathfrak{M}(\mathbb{G}),\ \Gamma * \mu \geq 1 \ \textit{on } E \textit{ except for a } \mathcal{L}\textit{-polar set}\}$.

Proof. (i). Let $\mu \in \mathfrak{M}(\mathbb{G})$ with $K := \mathrm{supp}(\mu) \subseteq E$ and $\Gamma * \mu \leq 1$ on $\mathbb{G}$. Let also $v \in \overline{\mathcal{S}}^+(\mathbb{G})$ be such that $v \geq 1$ on K. From Maria–Frostman Theorem 11.3.4, it follows $v \geq \Gamma * \mu$ on $\mathbb{G}$. Hence, being $v \in \Phi^1_K$ arbitrary, we get $\mathbf{R}^1_K \geq \Gamma * \mu$. By passing to the lower semicontinuous regularization, we infer (denoting the $\mathcal{L}$-capacitary distribution of K by ν_K)

$$\Gamma * \nu_K = V_K = \widehat{\mathbf{R}}^1_K \geq \Gamma * \mu \quad \text{on } \mathbb{G}.$$

By Ex. 13 at the end of the chapter, this yields $\mu(\mathbb{G}) \leq \nu_K(\mathbb{G})$, whence (being $\mu(\mathbb{G}) = \mu(E)$, $\nu_K(\mathbb{G}) = \mathcal{C}(K)$)

$$\mu(E) = \mu(K) \leq \nu_K(\mathbb{G}) \leq \sup\big\{\mathcal{C}(K) : K \subseteq E,\ K \text{ compact}\big\} = \mathcal{C}_*(E).$$

Passing to the supremum over the μ's as above, we get the "$\geq$" sign in assertion (i). On the other hand, for every compact subset K of E, it holds (see Theorem 11.5.6)

$$\begin{aligned}\mathcal{C}(K) &= \max\{\mu(K) = \mu(E) \mid \mu \in \mathfrak{M}(K),\ \Gamma * \mu \le 1 \text{ in } \mathbb{G}\}\\ &\le \sup\{\mu(E) \mid \mu \in \mathfrak{M}(E),\ \Gamma * \mu \le 1 \ \text{ on } \mathbb{G}\}.\end{aligned}$$

Finally, passing to the supremum over all possible compact subsets K of E, by definition of $\mathcal{C}_*(E)$, we get the remaining "$\le$" sign. Thus (i) is proved.

(ii). Let $\mu \in \mathfrak{M}$ be such that $\Gamma * \mu \ge 1$ on $E \setminus F$, where $F \subseteq E$ is an $\mathcal{L}$-polar set. By Corollary 11.2.5 (being $F \subseteq E$ and E bounded), there exists $\mu_0 \in \mathfrak{M}_0$ such that $\Gamma * \mu_0 \equiv \infty$ on F. Let $\varepsilon > 0$. Consider the measure

$$\mu_1 := \frac{\varepsilon}{2\,\mu_0(\mathbb{G})}\,\mu_0.$$

Then $\mu_1(\mathbb{G}) = \varepsilon/2$, $\mu_1 \in \mathfrak{M}_0$ and

$$\Gamma * \mu_1 = \frac{\varepsilon}{2\,\mu_0(\mathbb{G})}\,\Gamma * \mu_0 \equiv \infty \quad \text{on } F.$$

Obviously (by the minimum principle for $\overline{\mathcal{S}}(\mathbb{G})$), we have $\Gamma * \mu_1 > 0$ on $\mathbb{G}$ (otherwise $\mu_1 \equiv 0$, i.e. $\mu_0 \equiv 0$ contradicting $\Gamma * \mu_0 \equiv \infty$ on F). Let $\xi \in E$. Then we have

$$(\Gamma * \mu)(\xi) + (\Gamma * \mu_1)(\xi) \ge \begin{cases} 0 + \infty & \text{if } \xi \in F,\\ 1 + (\Gamma * \mu_1)(\xi) & \text{if } \xi \in E \setminus F, \end{cases} > 1.$$

Consequently, setting $\mu_2 := \mu + \mu_1$, we have $\Gamma * \mu_2 = \Gamma * \mu + \Gamma * \mu_1 > 1$ on E. Being $\Gamma * \mu_2$ l.s.c., for every $\xi \in E$ there exists an open neighborhood U_ξ of ξ such that $\Gamma * \mu_2 > 1$ on U_ξ. Then the set $U := \bigcup_{\xi \in E} U_\xi$ is an open neighborhood of E such that $\Gamma * \mu_2 > 1$ on U. It is also not restrictive to suppose that U is bounded (for E is). Then $\Gamma * \mu_2 \in \Phi_U^1$, so that

$$\Gamma * \mu_2 \ge \mathbf{R}_U^1 = \widehat{\mathbf{R}}_U^1 = \Gamma * \nu_U.$$

Here, in the first equality we used Corollary 11.6.6, whereas in the second equality we used Theorem 11.8.1-(a) and the notation therein. So, we have $\Gamma * \mu_2 \ge \Gamma * \nu_U$, whence (see Ex. 13 at the end of the chapter)

$$\mu_2(\mathbb{G}) \ge \nu_U(\mathbb{G}). \tag{11.35}$$

Now, from Theorem 11.8.1-(b) we have $\nu_U(\mathbb{G}) = \mathcal{C}^*(U) = \mathcal{C}(U)$ (since U is open), whereas (by the definition of μ_2 and μ_1) $\mu_2(\mathbb{G}) < \mu(\mathbb{G}) + \varepsilon$. Hence (11.35) yields

$$\mathcal{C}^*(E) \le \mathcal{C}(U) < \mu(\mathbb{G}) + \varepsilon.$$

Due to the arbitrariness of ε, we get $\mathcal{C}^*(E) \le \mu(\mathbb{G})$. Hence, we have proved the "$\le$" sign in assertion (ii).

On the other hand, since E is a bounded set, it obviously holds

$$\mathcal{C}^*(E) = \inf\{\mathcal{C}(\Omega) \mid \Omega \text{ open and bounded, } \Omega \supseteq E\}. \tag{11.36}$$

So, let Ω be open and bounded with $\Omega \supseteq E$. By Theorem 11.8.1, there exists $\nu_\Omega \in \mathfrak{M}$ such that $\widehat{\mathbf{R}}^1_\Omega = \Gamma * \nu_\Omega$ and $\mathcal{C}(\Omega) = \nu_\Omega(\mathbb{G})$. Hence (see Corollary 11.6.6) $\Gamma * \nu_\Omega = \mathbf{R}^1_\Omega = 1$ on $U \supseteq E$ (see Lemma 11.6.3-(iii)), so that

$$\begin{aligned}&\inf\{\mu(\mathbb{G}) \,|\, \mu \in \mathfrak{M}(\mathbb{G}),\ \Gamma * \mu \geq 1 \text{ on } E \text{ except for a } \mathcal{L}\text{-polar set}\}\\ &\leq \nu_\Omega(\mathbb{G}) = \mathcal{C}(\Omega).\end{aligned}$$

Passing to the infimum over the above Ω's, from (11.36) we get the "$\geq$" sign in assertion (ii). This ends the proof. □

Proposition 11.8.3. *Let $Z,\ E \subseteq \mathbb{G}$, and let Z be $\mathcal{L}$-polar. Let also $u \in \overline{\mathcal{S}}^+(\mathbb{G})$. Then $\widehat{\mathbf{R}}^u_E = \widehat{\mathbf{R}}^u_{E\setminus Z}$.*

Proof. From Lemma 11.6.3-(vi), we have $\widehat{\mathbf{R}}^u_E \geq \widehat{\mathbf{R}}^u_{E\setminus Z}$. We are thus left to prove the reverse inequality. Let $x_0 \notin Z$. By Theorem 11.2.3, there exists $v \in \overline{\mathcal{S}}^+(\mathbb{G})$ such that $v(x_0) \in \mathbb{R}$ and $v \equiv \infty$ on Z. Let $w \in \Phi^u_{E\setminus Z}$ (i.e. $w \in \overline{\mathcal{S}}^+(\mathbb{G})$ and $w \geq u$ on $E \setminus Z$). The function $w + \varepsilon\, v$ belongs to $\overline{\mathcal{S}}^+(\mathbb{G})$ and is clearly $\geq u$ on E. Then

$$\widehat{\mathbf{R}}^u_E \leq \mathbf{R}^u_E \leq w + \varepsilon\, v \quad \text{on } \mathbb{G}.$$

Evaluating this at x_0 and letting $\varepsilon \to 0^+$, one gets $\widehat{\mathbf{R}}^u_E(x_0) \leq w(x_0)$. Being $x_0 \in \mathbb{G} \setminus Z$ arbitrary, we infer $\widehat{\mathbf{R}}^u_E \leq w$ on $\mathbb{G} \setminus Z$, so that[10] we obtain $\widehat{\mathbf{R}}^u_E \leq w$ on $\mathbb{G}$. Taking the infimum over $w \in \Phi^u_{E\setminus Z}$, this gives $\widehat{\mathbf{R}}^u_E \leq \mathbf{R}^u_{E\setminus Z}$, then taking lower semicontinuous regularization, we get $\widehat{\mathbf{R}}^u_E \leq \widehat{\mathbf{R}}^u_{E\setminus Z}$. □

Proposition 11.8.4 (Characterization of $\mathcal{L}$-polarity. II). *Let $E \subseteq \mathbb{G}$ be any set. Then the following statements are equivalent:*

(a) *E is $\mathcal{L}$-polar,*
(b) *There exists $u \in \overline{\mathcal{S}}(\mathbb{G})$, u strictly positive on $\mathbb{G}$, such that $\widehat{\mathbf{R}}^u_E \equiv 0$,*
(c) *For every $u \in \overline{\mathcal{S}}^+(\mathbb{G})$, it holds $\widehat{\mathbf{R}}^u_E \equiv 0$.*

Proof. (a) $\Rightarrow$ (c). See Ex. 7 at the end of the chapter.

(c) $\Rightarrow$ (b). Take $u \equiv 1$ and apply directly (c).

(b) $\Rightarrow$ (a). Let u be as in (b) above. Since, by Theorem 11.6.5, $\mathbf{R}^u_E$ differs from $\widehat{\mathbf{R}}^u_E$ at most on an $\mathcal{L}$-polar set, they are equal almost everywhere on $\mathbb{G}$. Hence, by hypothesis (b), $\mathbf{R}^u_E = 0$ a.e. Let $x_0 \in \mathbb{G}$ be such that $\mathbf{R}^u_E(x_0) = 0$. By the very definition of $\mathbf{R}^u_E(x_0)$, there exists a sequence $\{v_n\}_{n\in\mathbb{N}}$ in $\overline{\mathcal{S}}^+(\mathbb{G})$ such that $v_n \geq u$ on E and $v_n(x_0) \leq 2^{-n}$. Set $v := \sum_{n\in\mathbb{N}} v_n$. Being $v(x_0) \leq 1$ and $v_n \geq u > 0$, we have $v \in \overline{\mathcal{S}}^+(\mathbb{G})$ (see Theorem 8.2.7, page 403). Moreover, $v \equiv \infty$ on E, since, for every $e \in E$, we have $v_n(e) \geq u(e) > 0$. Hence, by definition, E is $\mathcal{L}$-polar and (a) follows. □

---

[10] Recall that $\mathcal{L}$-polar sets have zero Lebesgue measure, and any $\mathcal{L}$-superharmonic function $f \in \overline{\mathcal{S}}(\Omega)$ satisfies $f(x) = \liminf_{\Omega \ni y \to x} f(y)$ for every $x \in \Omega$.

We have the following result, concerning with the "smallness" of $\mathcal{L}$-polar sets w.r.t. the $\mathcal{L}$-harmonic measures (in the classical context of Laplace's operator, see [AG01, Theorem 5.1.9]).

Proposition 11.8.5. *Let $O \subseteq \mathbb{G}$ be an open set, and let $u \in \overline{\mathcal{S}}(O)$ be locally bounded. If μ_u denotes the $\mathcal{L}$-Riesz measure of u on O, then $\mu_u(Z) = 0$ for every Borel $\mathcal{L}$-polar set $Z \subseteq O$.*

Proof. We only need to show that $\mu_u(Z \cap B) = 0$ for every d-ball B with $\overline{B} \subset O$.

First, we claim that there exists a compactly supported Radon measure ν on $\mathbb{G}$ such that $G_B * \nu = \infty$ on $Z \cap B$. Indeed, let $w \in \overline{\mathcal{S}}(\mathbb{G})$ be such that $w|_{Z\cap B} \equiv \infty$. The lower semicontinuity of w proves that w is bounded from below on B. Hence, by Riesz representation Theorem 9.4.7 (page 443), we have $w = G_B * \mu_w + h$ on B, for a suitable $h \in \mathcal{H}(B)$. Since h is finite-valued, this proves the claim.

Let $\alpha = \inf_B u$ ($\alpha \in \mathbb{R}$ since u is l.s.c.). Again from Riesz representation Theorem 9.4.7 applied to the function $w = u - \alpha$ (which is non-negative on B, hence endowed with an $\mathcal{L}$-harmonic minorant on B), we infer the existence of $h \in \mathcal{H}(B)$ such that

$$u(x) - \alpha = (G_B * \mu_u)(x) + h(x) \quad \text{for every } x \in B. \tag{11.37}$$

(We used $\mu_w = \mu_u$.) Since h is the greatest $\mathcal{L}$-harmonic minorant of $w \geq 0$ on B, it holds $h \geq 0$, so that (11.37) gives

$$(G_B * \mu_u)(x) \leq u(x) - \alpha \quad \text{for every } x \in B. \tag{11.38}$$

As a consequence, we have

$$\infty \cdot \mu_u(Z \cap B) \leq \int_B (G_B * \nu)(y)\, \mathrm{d}\mu_u(y) = \int_B (G_B * \mu_u)(y)\, \mathrm{d}\nu(y)$$
$$\text{(see (11.38))} \leq \Big(\sup_{\overline{B}} u - \alpha\Big) \cdot \nu(\mathbb{G}) < \infty \quad \text{(by the local boundedness of } u\text{)}.$$

(In the equality sign, we used Fubini–Tonelli's theorem jointly with the symmetry of G_B.) This gives $\mu_u(Z \cap B) = 0$, ending the proof. □

The following proposition shows some more "smallness" properties of the $\mathcal{L}$-polar sets.

Proposition 11.8.6. *The following assertions hold:*

(i) *If $A \subseteq \mathbb{G}$ is an open connected set and $E \subset A$ is $\mathcal{L}$-polar and relatively closed in A, then $A \setminus E$ is connected;*
(ii) *If $A \subseteq \mathbb{G}$ is open (and non-empty) and ∂A is $\mathcal{L}$-polar, then A is connected and dense in $\mathbb{G}$.*

Proof. (i). With the notation of the assertion, let A_0 be a connected component of $A \setminus E$. We are done if we show that $A_0 = A \setminus E$. Note that, since E is relatively closed in A, then $A \setminus E$ is open, whence A_0 is open too. Since E is $\mathcal{L}$-polar, there exists $u \in \overline{\mathcal{S}}(\mathbb{G})$ such that $u \equiv \infty$ on E. Set

$$v : A \to (-\infty, \infty], \quad v(x) = \begin{cases} u(x) & \text{if } x \in A_0, \\ \infty & \text{if } x \in A \setminus A_0. \end{cases}$$

Note that $\widetilde{A} := (A \setminus E) \setminus A_0$ is an open set since it coincides with the union of the connected components of the open set $A \setminus E$, except for A_0. It is easily seen that v is lower semicontinuous, not identically ∞ and $\mathcal{L}$-supermean on A, since $u|_{A_0}$ is $\mathcal{L}$-superharmonic. Then $v \in \overline{\mathcal{S}}(A)$. But $v \equiv \infty$ on the open set $\widetilde{A}$, and this can happen only if $\widetilde{A} = \emptyset$, for $\mathcal{L}$-superharmonic functions are locally integrable. This proves that A contains no connected components other than A_0, thus completing the proof of (i).

(ii). Since ∂A is closed in $\mathbb{G}$ (connected), by (i) we have $\mathbb{G} \setminus \partial A$ connected. If by contradiction $\mathbb{G} \setminus \overline{A} \neq \emptyset$, then $\mathbb{G} \setminus \partial A = A \cup (\mathbb{G} \setminus \overline{A})$ is the union of two open disjoint and non-empty sets, i.e. $\mathbb{G} \setminus \partial A$ is not connected (contradicting the above argument). This proves that $\overline{A} = \mathbb{G}$. Moreover, the above argument also gives

$$\mathbb{G} \setminus \partial A = A \cup (\mathbb{G} \setminus \overline{A}) = A,$$

so that A is connected, for $\mathbb{G} \setminus \partial A$ is. □

11.9 Further Reading and the Quasi-continuity Property

The aim of this section is to sketch a brief investigation of the so-called *quasi-continuity property* of $\mathcal{L}$-subharmonic functions.

As we have seen in Section 8.3 following [JLMS07], a function u which is subharmonic w.r.t. *every* sub-Laplacian on $\mathbb{G}$ is v-convex (even under the assumption of upper semicontinuity instead of continuity in the definition of v-convexity) and has fine regularity properties. On the other hand, if a function is superharmonic with respect to only *one* sub-Laplacian $\mathcal{L}$, it nonetheless possesses some (weaker) continuity property. Indeed, the following quasi-continuity property holds (see Theorem 11.9.1 below for the precise statement):

(Q-C) *Every $\mathcal{L}$-superharmonic function is continuous, if restricted to the complement of an open set with arbitrarily small $\mathcal{L}$-capacity.*

The quasi-continuity of superharmonic functions has important applications in the theory of PDE's, where it is often desirable to know the accurate pointwise behavior of the relevant super- and sub-solutions and their fine properties.

In the classical case of the Laplace operator on $\mathbb{G}$, the (Q-C) property was proved by H. Cartan [Cart28]. Proofs of the (Q-C) property for more general operators are nowadays available: see [HKM93, Theorem 10.9] for a class of non-linear elliptic operators; see [MV97b, Theorem 14.3] for a class of subelliptic operators generalizing the latter; see [TW02b, Theorem 4.1] for a wide class of quasilinear elliptic and subelliptic operators. The last two results also cover the case of sub-Laplacians.

Nonetheless, in the significant case of Carnot groups, a direct and simpler approach can be applied in proving (Q-C). See, for instance, [BC05]. In particular, the stratified context allows us to recover several fine results on the theory of the *energy* for $\mathcal{L}$, i.e. of the integral

$$\iint \Gamma(x,y)\,\mathrm{d}\mu(x)\mathrm{d}\mu(y) \quad \text{(where } \mu \text{ is a Radon measure on } \mathbb{G}\text{)}.$$

The proof of the (Q-C) property in [BC05] is in the spirit of this book and differs from those in [HKM93,MV97b,TW02b]. Indeed, the approach of these papers (due to the considered wider classes of operators) requires highly non-trivial techniques such as deep weak-convergence results, a suitable theory of the relevant Sobolev-type function spaces, some results from the theory of variational inequalities, a quantitative use of Caccioppoli-type and Harnack inequalities (both weak and strong), Hölder estimates for weak solutions, etc.

In the remainder of this section, we sketch a line of the proof of the following result, as approached in the cited paper [BC05].

Theorem 11.9.1 (Quasi-continuity of $\mathcal{L}$-superharmonic functions). *Let $u \in \overline{\mathcal{S}}(\Omega)$. Then, for every $\varepsilon > 0$, there exists an open set $O_\varepsilon \subseteq \Omega$ with $\mathcal{C}(O_\varepsilon) < \varepsilon$ and such that*

the restriction of u to $\Omega \setminus O_\varepsilon$ is continuous on $\Omega \setminus O_\varepsilon$.

Following Cartan's theory of energy and the related ideas in the classical potential theory, we give the following definition (we recall that $\mathfrak{M}$ denotes the set of the Radon measures on $\mathbb{G}$).

Definition 11.9.2 (Mutual $\mathcal{L}$-energy). *Let $\mu, \nu \in \mathfrak{M}$. The* mutual $\mathcal{L}$-energy *of μ and ν is*

$$\langle \mu, \nu \rangle = \int_{\mathbb{G}} \Gamma * \mu \,\mathrm{d}\nu.$$

The quantity $\| \mu \|^2 := \langle \mu, \mu \rangle$ will be called the $\mathcal{L}$-energy *of $\mu \in \mathfrak{M}$. Finally,*

$$\textit{we set} \quad \mathcal{E}^+ := \{\mu \in \mathfrak{M} : \|\mu\| < \infty\}.$$

We notice that $\|\mu\|^2 = I(\mu)$ if $\mu \in \mathfrak{M}_0$, where $I(\mu)$ is as in Definition 11.4.1 on page 497.

Proposition 11.9.3. *For every $x \in \mathbb{G}$ and $r > 0$, the measure $\lambda_{x,r}$ introduced in* (9.9) *(page 436) belongs to $\mathcal{E}^+$ and $\| \lambda_{x,r} \|^2 = r^{2-Q}$.*

Proof. Let $x \in \mathbb{G}$ and $r > 0$. Then, by (9.9) and (9.10a), we have

$$\begin{aligned}\|\lambda_{x,r}\|^2 &= \int_{\mathbb{G}\setminus\overline{B}_d(x,r)} \Gamma(z^{-1}\circ x)\,\mathrm{d}\lambda_{x,r}(z) + r^{2-Q}\int_{\overline{B}_d(x,r)} \mathrm{d}\lambda_{x,r}(z)\\ &= 0 + r^{2-Q}\int_{\partial B_d(x,r)} \mathbf{k}(x,z)\,\mathrm{d}H^{N-1}(z) = r^{2-Q}\,m_r[1](x) = r^{2-Q}.\end{aligned}$$

This completes the proof. □

The $\mathcal{L}$-polarity is related to the set $\mathcal{E}^+$, as assertion (i) of the following theorem states (see [BC05, Lemma 6.1 and Theorem 6.1]).

Theorem 11.9.4. *The following assertions hold*:

(i) *Let E be a Borel set. Then E is $\mathcal{L}$-polar iff $\mu(E) = 0$ for every $\mu \in \mathcal{E}^+$*;
(ii) (**$\mathcal{L}$-energy principle**) $\langle \mu, \nu \rangle \leq \|\mu\| \cdot \|\nu\|$ *for every* $\mu,\ \nu \in \mathfrak{M}$.

We next give a new definition. We denote by $\mathcal{E}$ the set of formal[11] differences $\mu - \nu$, $\mu,\ \nu \in \mathcal{E}^+$. By a slight abuse of notation, the mutual $\mathcal{L}$-energy and the $\mathcal{L}$-energy are defined as follows

$$\langle \mu_1 - \nu_1, \mu_2 - \nu_2 \rangle = \langle \mu_1, \mu_2 \rangle - \langle \mu_1, \nu_2 \rangle - \langle \nu_1, \mu_2 \rangle + \langle \nu_1, \nu_2 \rangle,$$
$$\|\mu - \nu\|^2 = \|\mu\|^2 - 2\langle \mu, \nu \rangle + \|\nu\|^2.$$

Thanks to the $\mathcal{L}$-energy principle in Theorem 11.9.4-(ii), we have

$$\big|\langle \mu, \nu \rangle\big| \leq \|\mu\| \cdot \|\nu\| \quad \text{for every } \mu,\ \nu \in \mathcal{E}.$$

It can be proved that $\langle \cdot, \cdot \rangle$ defines an inner product on $\mathcal{E}$. We need many preliminary results.

Proposition 11.9.5. *For every $x \in \mathbb{G}$ and every $r > 0$, $\tau_{x,r} := \lambda_{x,r/2} - \lambda_{x,r}$ belongs to $\mathcal{E}$ and $\|\tau_{x,r}\|^2 = (r/2)^{2-Q} - r^{2-Q}$.*

Proof. We fix $x \in \mathbb{G}$, $r > 0$. By Proposition 11.9.3, $\tau_{x,r} \in \mathcal{E}$. By (9.10a),

$$\langle \lambda_{x,r/2}, \lambda_{x,r} \rangle = \int_{\mathbb{G} \setminus \overline{B}_d(x,r/2)} \Gamma(z^{-1} \circ x)\, \mathrm{d}\lambda_{x,r}(z) + (r/2)^{2-Q} \int_{\overline{B}_d(x,r/2)} \mathrm{d}\lambda_{x,r}(z) = r^{2-Q},$$

so that $\|\lambda_{x,r/2} - \lambda_{x,r}\|^2$ equals

$$\|\lambda_{x,r/2}\|^2 + \|\lambda_{x,r}\|^2 - 2\langle \lambda_{x,r/2}, \lambda_{x,r} \rangle = (r/2)^{2-Q} - r^{2-Q}.$$

This ends the proof. $\square$

We now introduce a set of functions which play a prominent rôle in the theory of $\mathcal{L}$-energy. We set

$$\mathcal{T} = \{\mathrm{T}_{x,r} := \Gamma * \lambda_{x,r/2} - \Gamma * \lambda_{x,r} \mid x \in \mathbb{G},\ r > 0\}. \tag{11.39}$$

In the following definition, we agree to denote by $C_0^+(\mathbb{G})$ the set of non-negative compactly supported functions on $\mathbb{G}$.

Definition 11.9.6 (Total set in $C_0^+(\mathbb{G})$). *A set $T \subseteq C_0^+(\mathbb{G})$ is called* total *if it has the following property: for every $C \Subset \mathbb{G}$, for every open set $W \supset C$, for all $\varepsilon > 0$ and for all $v \in C_0^+(\mathbb{G})$ supported in C, there exist $\beta_1, \ldots, \beta_n > 0$, $f_1, \ldots, f_n \in T$ supported in W such that $\|v - \sum_{i=1}^n \beta_i f_i\|_\infty < \varepsilon$.*

[11] More precisely, $\mathcal{E}$ is the set of the couples $(\mu, \nu) \in \mathcal{E}^+ \times \mathcal{E}^+$ modulo the equivalence relation $(\mu_1, \nu_1) \sim (\mu_2, \nu_2)$ iff $\mu_1 + \nu_2 = \mu_2 + \nu_1$.

We have an important result, a $\mathbb{G}$-version of a theorem due to Cartan [Cart28]:

Theorem 11.9.7 (Bonfiglioli and Cinti, [BC05], Theorem 4.3). *Suppose $T \subseteq C_0^+(\mathbb{G})$ has the following properties*:

(1) *for every $f \in T$ and every $\alpha \in \mathbb{G}$, it holds $f(\alpha \circ \cdot) \in T$,*
(2) *for every $x \in \mathbb{G}$ and every $\delta > 0$, there exists $f \in T$, $f \not\equiv 0$, such that f vanishes outside $B_d(x, \delta)$.*

Then T is a total set (according to Definition 11.9.6).

From Theorem 11.9.7 and Ex. 10 at the end of this chapter we deduce the following remarkable assertion.

Corollary 11.9.8. *The set $\mathcal{T}$ in* (11.39) *is a total set of functions in $C_0^+(\mathbb{G})$.*

We can now prove the theorem below.

Theorem 11.9.9. *The square-root $\|\cdot\|$ of the $\mathcal{L}$-energy is a norm on $\mathcal{E}$, and $\mathcal{E}^+$ is a complete subset of the normed space $(\mathcal{E}, \|\cdot\|)$.*

Proof. We only prove the first part of the assertion. The second part is proved in [BC05, Theorem 6.3] by delicate results on the weak and weak* convergence in $\mathcal{E}$. We explicitly remark that $(\mathcal{E}, \|\cdot\|)$ is not complete even in the classical Euclidean case of Laplace's operator (see [Cart28, p. 87]).

We must show that $\mu - \nu \sim 0$ if $\mu - \nu$, $\mu, \nu \in \mathcal{E}^+$ satisfy $\|\mu - \nu\| = 0$. We have $|\langle\lambda, \mu - \nu\rangle| \leq \|\lambda\| \cdot \|\mu - \nu\| = 0$, i.e. $\langle\lambda, \mu\rangle = \langle\lambda, \nu\rangle$ for every $\lambda \in \mathcal{E}$. In particular, this holds for $\lambda = \lambda_{x,r/2} - \lambda_{x,r}$ (see Proposition 11.9.5). Thus, $\langle\lambda_{x,r/2}, \mu\rangle - \langle\lambda_{x,r}, \mu\rangle = \langle\lambda_{x,r/2}, \nu\rangle - \langle\lambda_{x,r}, \nu\rangle$, i.e.

$$\int_{\mathbb{G}} (\Gamma * \lambda_{x,r/2} - \Gamma * \lambda_{x,r})\, d\mu = \int_{\mathbb{G}} (\Gamma * \lambda_{x,r/2} - \Gamma * \lambda_{x,r})\, d\nu$$

for every $x \in \mathbb{G}$ and every $r > 0$. By Corollary 11.9.8, the functions

$$\Gamma * \lambda_{x,r/2} - \Gamma * \lambda_{x,r}$$

form a total set in $C_0^+(\mathbb{G})$, so that (invoking Ex. 11 at the end of this chapter) $\mu = \nu$, and the theorem is proved. □

In order to prove our (Q-C) property, we need the following lemma (whose proof follows by collecting together Lemmas 7.1, 7.2, 7.3 in [BC05]).

Lemma 11.9.10. *The following assertions hold*:

(i) *Let $\mu \in \mathcal{E}^+$. Then there exists a sequence $\mu_j \in \mathcal{E}^+$ converging to μ in the $\mathcal{L}$-energy norm with the following properties*: $\operatorname{supp}(\mu_j)$ *is compact, $\Gamma * \mu_j \leq \Gamma * \mu$ and $\Gamma * \mu_j \in C(\mathbb{G}, \mathbb{R})$ for every $j \in \mathbb{N}$.*
(ii) *Let $\mu, \nu \in \mathcal{E}^+$. Given $\varepsilon > 0$, we set $E = \{x \in \mathbb{G} \mid (\Gamma * \mu)(x) > (\Gamma * \nu)(x) + \varepsilon\}$. Then $C^*(E) \leq \varepsilon^{-2} \| \mu - \nu \|^2$.*

(iii) *Let* $\mu \in \mathfrak{M}$ *be such that* $\mu(\mathbb{G}) \le m \in \mathbb{R}$. *Then, for every* $\varepsilon > 0$, *the* $\mathcal{L}$-*capacity of the set* $A = \{x \in \mathbb{G} \mid (\Gamma * \mu)(x) > \varepsilon\}$ *does not exceed* m/ε.

The key-rôle in the proof of the (Q-C) property is played by the following lemma.

Lemma 11.9.11 (Quasi-continuity for $\boldsymbol{\Gamma}$-potentials). *Let* $\mu \in \mathfrak{M}$ *with* $\Gamma * \mu \not\equiv \infty$. *Then, for every* $\varepsilon > 0$, *there exists an open set* $O \subseteq \Omega$ *with* $\mathcal{C}(O) < \varepsilon$ *such that the restriction of* $\Gamma * \mu$ *to* $\Omega \setminus O$ *is continuous.*

Proof. We split the proof in three steps. Throughout the proof, $\varepsilon > 0$ is fixed.

(I) $\mu \in \mathcal{E}^+$. For any $n \in \mathbb{N}$, by Lemma 11.9.10-(i), there exists $\mu_n \in \mathcal{E}^+$ with compact support such that $\Gamma * \mu_n \le \Gamma * \mu$, $\Gamma * \mu_n \in C(\mathbb{G}, \mathbb{R})$ and $\| \mu_n - \mu \|^2 < \varepsilon/8^n$. We consider

$$E_n := \{x \in \mathbb{G} \mid (\Gamma * \mu)(x) > (\Gamma * \mu_n)(x) + 1/2^n\}.$$

E_n is an open set, hence $\mathcal{L}$-capacitable. By Lemma 11.9.10-(ii), we have $\mathcal{C}(E_n) \le 4^n \| \mu - \mu_n \|^2 < \varepsilon/2^n$. As a consequence, the $\mathcal{L}$-capacity of $E = \cup_n E_n$ does not exceed ε. Set $F = \mathbb{G} \setminus E$. It is easily seen that $\Gamma * \mu_n$ converges uniformly to $\Gamma * \mu$ on F. Since $\Gamma * \mu_n \in C(\mathbb{G}, \mathbb{R})$, this proves that $(\Gamma * \mu)|_F$ is continuous on F.

(II) $\mu \in \mathfrak{M}_0$. Let $p \in \mathbb{N}$ be such that $\mu(\mathbb{G})/p < \varepsilon/2$. Consider the function $\min\{\Gamma * \mu, p\}$. By Proposition 9.9.1-(b) (page 464), we have $\min\{\Gamma * \mu, p\} = \Gamma * \nu$ for some $\nu \in \mathfrak{M}$. Moreover, by applying Lemma 11.9.10-(iii) with $m = \mu(\mathbb{G})$, it follows $\mathcal{C}(A) < \varepsilon/2$, where $A = \{x \in \mathbb{G} \mid (\Gamma * \mu)(x) > p\}$. Obviously, $\Gamma * \nu \equiv \Gamma * \mu$ on $\mathbb{G} \setminus A$. Now, $\nu \in \mathcal{E}^+$. Indeed,

$$\|\nu\|^2 \le \int \Gamma * \mu \, \mathrm{d}\nu = \int \Gamma * \nu \, \mathrm{d}\mu \le p\, \mu(\mathbb{G}) \in \mathbb{R}.$$

Hence, by the part (I) of the proof, there exists an open set B with $\mathcal{C}(B) < \varepsilon/2$ such that $(\Gamma * \nu)|_{\mathbb{G}\setminus B}$ is continuous on $\mathbb{G} \setminus B$. Finally, the open set $O = A \cup B$ has $\mathcal{L}$-capacity less than ε and satisfies the assertion of the lemma. Indeed, $(\Gamma * \mu)|_F$ is continuous on $F = \mathbb{G} \setminus O$ since $\Gamma * \mu \equiv \Gamma * \nu$ on $\mathbb{G} \setminus A \supseteq F$, and since $(\Gamma * \nu)|_{\mathbb{G}\setminus B}$ is continuous on $\mathbb{G} \setminus B \supseteq F$.

(III) $\mu \in \mathfrak{M}$ with $\Gamma * \mu \not\equiv \infty$. For $j \in \mathbb{N}$, we denote

$$C_j = \{x \in \mathbb{G} \mid j - 1 \le d(x) < j\}$$

and $\mu_j = \mu|_{C_j}$. It then follows that $\Gamma * \mu = \sum_{j=1}^{\infty} \Gamma * \mu_j$. Since $\mathrm{supp}(\mu_j) \subseteq C_j \Subset \mathbb{G}$, we can apply the part (II) of the proof. Thus, let O_j be an open set with $\mathcal{C}(O_j) < \varepsilon/2^j$ and such that $(\Gamma * \mu_j)|_{F_j}$ is continuous on $F_j = \mathbb{G} \setminus O_j$. Let us prove that the set $O = \bigcup_{j=1}^{\infty} O_j$ satisfies the assertion of the lemma. Indeed, $\mathcal{C}(O) < \varepsilon$ and (for $F = \mathbb{G} \setminus O$)

$$(\Gamma * \mu)|_F \quad \text{is continuous on } F.$$

To see this, we fix $x_0 \in F$, $n_0 \in \mathbb{N}$ such that $x_0 \in C_{n_0}$ and an open neighborhood W of x_0 with $\overline{W} \Subset \{n_0 - 2 < d(x) < n_0\}$. Then

$$\Gamma * \mu = \sum_{j=1}^{n_0-2} \Gamma * \mu_j + \Gamma * \mu_{n_0-1} + \Gamma * \mu_{n_0} + \sum_{j=n_0+1}^{\infty} \Gamma * \mu_j.$$

Now, for $j = 1, \dots, n_0-2$, it holds $\Gamma*\mu_j \in \mathcal{H}(\{d(x) > n_0-2\})$, since $\mathrm{supp}(\mu_j) \subseteq \{d(x) \le n_0 - 2\}$. Moreover, for any $j \ge n_0 + 1$, we have

$$\mathrm{supp}(\mu_j) \subseteq \{d(x) \ge n_0\}, \quad \text{whence } \Gamma * \mu_j \in \mathcal{H}(\{d(x) < n_0\}).$$

As a consequence, the function $\sum_{j=n_0+1}^{\infty} \Gamma * \mu_j$ is $\mathcal{L}$-harmonic near x_0 since it is majorized by $\Gamma * \mu \not\equiv \infty$. Finally, $(\Gamma * \mu_j)|_F$ is continuous at x_0 for all $j \in \mathbb{N}$, since $F \subseteq F_j$ and $\Gamma * \mu_j \equiv (\Gamma * \mu_j)|_{F_j}$ on F. This ends the proof. □

With Lemma 11.9.11 at hand and a Riesz-representation argument, we can easily prove the quasi-continuity of arbitrary $\mathcal{L}$-superharmonic functions.

Proof (of Theorem 11.9.1). We consider a sequence $\{\Omega_j\}_{j\in\mathbb{N}}$ of open sets exhausting Ω with $\overline{\Omega}_j \Subset \Omega$, and we let $\mu_j = (\mu_u)|_{\Omega_j}$. Then $\mu_j \in \mathfrak{M}_0$. Thus, by Lemma 11.9.11, for given $\varepsilon > 0$, there exists an open set O_j such that

$$\mathcal{C}(O_j) < \varepsilon/2^j \quad \text{and} \quad (\Gamma * \mu_j)|_{F_j}$$

is continuous on $F_j = \mathbb{G} \setminus O_j$. Then $O = \Omega \cap \bigcup_j O_j$ is an open subset of Ω whose $\mathcal{L}$-capacity does not exceed ε. Finally, $u|_F$ is continuous on $F = \Omega \setminus O$. Indeed, if $x_0 \in F$, then $x_0 \in \Omega_{j_0}$ for a suitable $j_0 \in \mathbb{N}$. Now, by Riesz representation Theorem 9.4.4 on page 442, there exists $h \in \mathcal{H}(\Omega_{j_0})$ such that $u = h + \Gamma * \mu_{j_0}$ on Ω_{j_0}. Since $F \subseteq F_{j_0}$, this completes the proof. □

Bibliographical Notes. The topics developed in this chapter within the classical theory of Laplace's operator can be found in all the main monographs devoted to potential theory; see the references in the Bibliographical Notes of Chapter 8. We also mention the following references.

The paper by N.S. Trudinger and X.-J. Wang [TW02b] (and references therein) where several results of potential theory are established for a class of quasilinear subelliptic operators including sub-Laplacians; see also the series of papers by the same authors [TW97,TW99,TW02a] where fully non-linear *Hessian operators* are treated; see D. Labutin [Lab02] for the study of subharmonic functions related to this last class of operators; see the axiomatic non-linear potential theory contained in the monograph by J. Heinonen, T. Kilpelainen and O. Martio [HKM93] and in I.G. Markina and S.K. Vodopyanov [MV97a,MV97b], in a more general subelliptic context.

For recent results on capacity, see [AH96,CDG96,Cart45,DG98,Hei95b,Ne88, KR87].

Some of the topics presented in this chapter also appear in [BC05,BC04].

11.10 Exercises of Chapter 11

Ex. 1) Prove that $\mathcal{C}(\{x \in \mathbb{G} : \Gamma(x) \geq 1/r\}) = r$. (*Hint*: Let $K := \{x \in \mathbb{G} : \Gamma(x) \geq 1/r\}$. Compute $\Gamma * \mu(0) = \int_{\Gamma=1/r} \Gamma(y)\, d\mu(y)$, where μ is the $\mathcal{L}$-capacitary distribution of K and recall that μ is supported in ∂K.)

Ex. 2) Prove that for any compact set K, we have

$$\mathcal{C}(\delta_\lambda(K)) = \lambda^{Q-2}\,\mathcal{C}(K), \qquad \mathcal{C}(\alpha \circ K) = \mathcal{C}(K)$$

for every $\lambda > 0$ and every $\alpha \in \mathbb{G}$. (*Hint*: Use Theorem 11.5.6.)

Ex. 3) Prove Ex. 11.2.2, page 491.

Ex. 4) Show that

$$\mathcal{C}(K) = \sup\left\{\left(\int_K\int_K \Gamma(x,y)\, d\mu(x)\, d\mu(y)\right)^{1/(2-Q)} \,\middle|\, \mu \in \mathfrak{M},\ \mu(K) \geq 1\right\}$$

if $K \subset \mathbb{G}$ is compact.

Ex. 5) Re-derive Theorem 11.5.11 (page 507) as a consequence of Proposition 11.8.4.

Ex. 6) Show that the hypothesis "u bounded from above" cannot be removed in the extended maximum principle of Theorem 11.2.7. (*Hint*: Take $u = \Gamma$ in $\Omega = B_d(0,1) \setminus \{0\}$ and $E = \{0\}$.)

Ex. 7) Let $Z \subset \mathbb{G}$ be $\mathcal{L}$-polar. Let $u \in \overline{\mathcal{S}}^+(\mathbb{G})$. Show that

$$\mathbf{R}_Z^u = \begin{cases} u(x) & \text{if } x \in Z, \\ 0 & \text{otherwise.} \end{cases}$$

Derive $\widehat{\mathbf{R}}_Z^u \equiv 0$. (*Hint*: Let $y \notin Z$ and $v \in \overline{\mathcal{S}}^+(\mathbb{G})$ be such that $v(y) \in \mathbb{R}$ and $v|_Z \equiv \infty$. Then $v/n \in \Phi_Z^u$ for every $n \in \mathbb{N}$, so that... .)

Ex. 8) **(Reciprocity law.)** Prove that: *If Ω is an open subset of $\mathbb{G}$ with $\mathcal{L}$-Green function G_Ω and μ and ν are (positive) measures on Ω, then $\int_\Omega G_\Omega * \mu\, d\nu = \int_\Omega G_\Omega * \nu\, d\mu$.*
(*Hint*: Use Fubini–Tonelli's theorem and the symmetry of the G_Ω.)

Ex. 9) Following the notation in Definition 11.9.2, prove that:

- $\langle \mu, \nu \rangle$ is well defined;
- it holds $\langle \mu, \nu \rangle = \langle \nu, \mu \rangle$ (use the reciprocity law);
- if $\mu \in \mathfrak{M}_0$ and $\Gamma * \mu$ is bounded then $\mu \in \mathcal{E}^+$;
- if $\mu \in \mathcal{E}^+$ and $\lambda \in \mathfrak{M}$ satisfies $\Gamma * \lambda \leq \Gamma * \mu$, then $\lambda \in \mathcal{E}^+$ and $\|\lambda\| \leq \|\mu\|$.

Ex. 10) Following (11.39) (page 529) and using Theorem 9.3.10 (page 438), prove that:

- the generic element $\mathrm{T}_{x,r}(z)$ of $\mathcal{T}$ in (11.39) is a non-negative and *continuous* function of $z \in \mathbb{G}$;
- it vanishes outside $B_d(x,r)$, it equals $(r/2)^{2-Q} - r^{2-Q}$ on $\overline{B_d(x,r/2)}$ and it equals $\Gamma(z^{-1} \circ x) - r^{2-Q}$ in $B_d(x,r) \setminus \overline{B_d(x,r/2)}$;

- show that, for every $\alpha \in \mathbb{G}$, the function $z \mapsto \mathrm{T}_{x,r}(\alpha \circ z)$ still belongs to $\mathcal{T}$: precisely, it holds $\mathrm{T}_{x,r}(\alpha \circ z) = \mathrm{T}_{\alpha^{-1} \circ x,r}(z)$.

Ex. 11) Following Definition 11.9.6 (page 529), prove that if $\mu_1, \mu_2 \in \mathfrak{M}$ satisfy $\int_{\mathbb{G}} f \,\mathrm{d}\mu_1 = \int_{\mathbb{G}} f \,\mathrm{d}\mu_2$ for every f in a total set T, then $\mu_1 = \mu_2$.

Ex. 12) Let $u : B_d(0,1) \to \mathbb{R}$ be an $\mathcal{L}$-harmonic function. Let us denote by μ the $\mathcal{L}$-harmonic measure of $B_d(0,1)$ at $x_0 = 0$. Then the function

$$\lambda \mapsto \int_{\partial B_d(0,1)} |u(\delta_\lambda(x))|^p \mathrm{d}\mu(x), \qquad 1 \le p < \infty,$$

is monotone increasing for $0 < r < 1$. (*Hint*: $x \mapsto |u(x)|^p$ is $\mathcal{L}$-subharmonic. Then use the Poisson–Jensen formula.)

Ex. 13) *Let μ, ν be Radon measures in $\mathbb{G}$ and suppose that $\Gamma * \mu \le \Gamma * \nu$. Then $\mu(\mathbb{G}) \le \nu(\mathbb{G})$.*
Hint: It suffices to prove that $\mu(B(0,R)) \le \nu(\mathbb{G})$ for every $R > 0$. Let $K := \overline{B(0,R)}$. As usual, μ_K denotes the $\mathcal{L}$-capacitary distribution for K. Recall that $V_K = \Gamma * \mu_K = 1$ on the interior of K, i.e. on $B(0,R)$. Then we have

$$\begin{aligned}
\mu(B(0,R)) &= \int_{B(0,R)} 1 \mathrm{d}\mu = \int_{B(0,R)} \Gamma * \mu_K \,\mathrm{d}\mu \le \int_{\mathbb{G}} \Gamma * \mu_K \,\mathrm{d}\mu \\
&= \int_{\mathbb{G}} \Gamma * \mu \,\mathrm{d}\mu_K \le \int_{\mathbb{G}} \Gamma * \nu \,\mathrm{d}\mu_K = \int_{\mathbb{G}} \Gamma * \mu_K \,\mathrm{d}\nu \\
&= \int_{\mathbb{G}} V_K \,\mathrm{d}\nu \le \int_{\mathbb{G}} 1 \,\mathrm{d}\nu = \nu(\mathbb{G}).
\end{aligned}$$

Indeed, recall that $V_K \le 1$ on $\mathbb{G}$ (see (11.16a)).

Ex. 14) *Let ν, σ_n be Radon measures on $\mathbb{G}$ (for every $n \in \mathbb{N}$) with the following properties*:
(i) *There exists a compact set K such that* $\mathrm{supp}(\sigma_n) \subseteq K$, $\forall\, n \in \mathbb{N}$;
(ii) *$\Gamma * \sigma_n$ is decreasing*;
(iii) *on $\mathbb{G} \setminus K$, it holds $\Gamma * \sigma_n \longrightarrow \Gamma * \nu$.*
Then, $\nu(\mathbb{G}) = \lim_{n\to\infty} \sigma_n(\mathbb{G})$.
Hint: Let $A := \overline{B(0,r)}$ with $r \gg 1$ such that $K \subset B(0,r)$. Since $\Gamma * \sigma_n \in \mathcal{H}(\mathbb{G} \setminus K)$ and $\{\Gamma * \sigma_n\}_n$ is decreasing and non-negative, it holds $\Gamma * \nu \in \mathcal{H}(\mathbb{G} \setminus K)$, whence $\mathrm{supp}(\nu) \subseteq K$. Let μ_A denote the $\mathcal{L}$-capacitary distribution for A. Note that $\Gamma * \nu_A = 1$ on $B(0,r) \supset K$ and $\mathrm{supp}(\mu_A) \subseteq \partial A \subset \mathbb{G} \setminus K$. Thus Fubini's theorem and monotone (decreasing) convergence imply

$$\begin{aligned}
\sigma_n(\mathbb{G}) &= \int_K 1 \,\mathrm{d}\sigma_n = \int_K \Gamma * \mu_A \,\mathrm{d}\sigma_n = \int_{\mathbb{G}} \Gamma * \mu_A \,\mathrm{d}\sigma_n = \int_{\mathbb{G}} \Gamma * \sigma_n \,\mathrm{d}\mu_A \\
&= \int_{\mathrm{supp}(\mu_A)} \Gamma * \sigma_n \,\mathrm{d}\mu_A \to \int_{\mathrm{supp}(\mu_A)} \Gamma * \nu \,\mathrm{d}\mu_A = \int_{\mathbb{G}} \Gamma * \nu \,\mathrm{d}\mu_A \\
&= \int_{\mathbb{G}} \Gamma * \mu_A \,\mathrm{d}\nu = \int_K \Gamma * \mu_A \,\mathrm{d}\nu = \int_K 1 \,\mathrm{d}\nu = \nu(\mathbb{G}).
\end{aligned}$$

Ex. 15) *Let $u \in \overline{\mathcal{S}}^+(\mathbb{G})$. Let, for $n \in \mathbb{N}$, $K_n \subseteq K_{n+1}$ be compact subsets of $\mathbb{G}$. Suppose $\Omega := \bigcup_{n\in\mathbb{N}} K_n$ is open. Then*

$$\lim_{n\to\infty} \widehat{\mathbf{R}}^u_{K_n} = \mathbf{R}^u_{\Omega}.$$

(*Hint*: Provide details for the following arguments. $\{\widehat{\mathbf{R}}^u_{K_n}\}_n$ is a decreasing sequence in $\overline{\mathcal{S}}^+(\mathbb{G})$. Set $v := \lim_{n\to\infty} \widehat{\mathbf{R}}^u_{K_n}$. Then $v \in \overline{\mathcal{S}}^+(\mathbb{G})$ and $v \le u$ on $\mathbb{G}$. Moreover,

$$(\star) \qquad v \le \mathbf{R}^u_{K_n} \le \mathbf{R}^u_{\Omega} \quad \text{on } \mathbb{G}.$$

Being $\widehat{\mathbf{R}}^u_{K_n} = \mathbf{R}^u_{K_n}$ up to an $\mathcal{L}$-polar subset of ∂K_n, say P_n, we get $\widehat{\mathbf{R}}^u_{K_n}(x) = u(x)$ for every $x \in K_n \setminus P_n$. Let $P := \bigcup_{n\in\mathbb{N}} P_n$. If $x \in \Omega \setminus P$, then $x \notin P_n$ for every $n \in \mathbb{N}$, and there exists $n_0 \in \mathbb{N}$ such that $x \in K_{n_0} \subseteq K_n$ for every $n \ge n_0$. Hence

$$\widehat{\mathbf{R}}^u_{K_n}(x) = u(x) \quad \forall\, n \ge n_0.$$

Letting $n \to \infty$, we derive $v(x) = u(x)$ for every $x \in \Omega \setminus P$. Hence (recall that P is $\mathcal{L}$-polar) $v = u$ on Ω. Thus $v \in \Phi^u_{\Omega}$, so that $v \ge \mathbf{R}^u_{\Omega}$. By $(\star)$, this gives $v = \mathbf{R}^u_{\Omega}$.)

Ex. 16) In this exercise, we state the main properties of the exterior $\mathcal{L}$-capacity $\mathcal{C}^*$ defined in Definition 11.5.14. For the relevant proofs, one can follow verbatim the classical capacitability theory as presented, e.g. in [Helm69].

(1) Starting from the strong sub-additivity property in (11.20) of Proposition 11.5.10, one first shows that *if $\{K_i\}_{1\le i\le m}$ and $\{C_i\}_{1\le i\le m}$ are finite families of compact sets in $\mathbb{G}$ such that $C_i \subset K_i$ $(i \le m)$, then*

$$\mathcal{C}\left(\bigcup_{i=1}^m K_i\right) - \mathcal{C}\left(\bigcup_{i=1}^m C_i\right) \le \sum_{i=1}^m \big(\mathcal{C}(K_i) - \mathcal{C}(C_i)\big).$$

(2) The second step is to derive from the above that *if $\{U_i\}_{1\le i\le m}$ and $\{V_i\}_{1\le i\le m}$ are finite families of open sets in $\mathbb{G}$ such that $V_i \subset U_i$ $(i \le m)$, then*

$$\mathcal{C}_*\left(\bigcup_{i=1}^m U_i\right) + \sum_{i=1}^m \mathcal{C}_*(V_i) \le \sum_{i=1}^m \mathcal{C}_*(U_i) + \mathcal{C}_*\left(\bigcup_{i=1}^m V_i\right).$$

(3) From (1) and (2) it is possible to derive that *if $\{E_i\}_{1\le i\le m}$ and $\{F_i\}_{1\le i\le m}$ are arbitrary finite families of subsets of $\mathbb{G}$ such that $F_i \subset E_i$ $(i \le m)$, then*

$$\mathcal{C}^*\left(\bigcup_{i=1}^m E_i\right) + \sum_{i=1}^m \mathcal{C}^*(F_i) \le \mathcal{C}^*\left(\bigcup_{i=1}^m F_i\right) + \sum_{i=1}^m \mathcal{C}^*(E_i).$$

(4) From (3) it is possible to derive that $\mathcal{C}^*$ *is an exterior capacity.* We say that a non-negative real-valued function $\phi^*(\cdot)$ defined on the subsets of $\mathbb{G}$ is an *exterior capacity* if:

(i) $E \subset F \Rightarrow \phi^*(E) \le \phi^*(F)$;
(ii) if $\{E_j\}_j \subset \mathbb{G}$ is such that $E_j \subseteq E_{j+1}$, then $\phi^*(\bigcup_j E_j) = \lim_{j\to\infty} \phi^*(E_j)$;
(iii) if $\{K_j\}_j \Subset \mathbb{G}$ is such that $K_{j+1} \subseteq K_j$, then $\phi^*(\bigcap_j K_j) = \lim_{j\to\infty} \phi^*(K_j)$.

(5) From (4) we infer that $\mathcal{C}^*$ *is numerably sub-additive*. Indeed, if $\{E_j\}_j \subseteq \mathbb{G}$ is a sequence of arbitrary sets, we apply (ii) to $\{\widetilde{E}_j\}_j$, where $\widetilde{E}_j := \bigcup_{i=1}^j E_i$, so that

$$\mathcal{C}^*\Big(\bigcup_j E_j\Big) = \mathcal{C}^*\Big(\bigcup_j \widetilde{E}_j\Big) = \lim_j \mathcal{C}^*(\widetilde{E}_j) \le \sum_j \mathcal{C}^*(E_j),$$

for (apply (3) with empty F_i's)

$$\mathcal{C}^*(\widetilde{E}_j) = \mathcal{C}^*\Big(\bigcup_{i=1}^j E_i\Big) \le \sum_{i=1}^j \mathcal{C}^*(E_i) \le \sum_{i=1}^\infty \mathcal{C}^*(E_i).$$

(6) As a consequence of (4) and the $\mathcal{L}$-capacitability of open sets and compact sets, we infer that if $\Omega = \bigcup_j K_j$ with Ω open and $K_j \subseteq K_{j+1}$ compact sets, then

$$\mathcal{C}(\Omega) = \lim_{j\to\infty} \mathcal{C}(K_j).$$

12

$\mathcal{L}$-thinness and $\mathcal{L}$-fine Topology

In this chapter, we deal with the notion of fine topology generated by the superharmonic functions related to a sub-Laplacian $\mathcal{L}$. We mainly investigate the relationship between $\mathcal{L}$-thinness and $\mathcal{L}$-regularity of boundary points for the Dirichlet problem.

As usual, throughout the chapter $\mathbb{G} = (\mathbb{R}^N, \circ, \delta_\lambda)$ is a (homogeneous) Carnot group and $\mathcal{L}$ is a sub-Laplacian on $\mathbb{G}$. Moreover, $\Gamma = d^{2-Q}$ is the fundamental solution for $\mathcal{L}$.

12.1 The $\mathcal{L}$-fine Topology: A More Intrinsic Tool

Since $\mathcal{L}$-superharmonic functions play a central rôle in the potential theory for $\mathcal{L}$, it seems more natural to introduce a new topology—finer than the Euclidean one—with respect to which any $\mathcal{L}$-superharmonic function is continuous. In the sequel, if a function u takes on a ∞-value at a point x_0, we say that u is continuous (in the extended sense) at x_0, provided $\lim_{x\to x_0} u(x) = \infty$. Moreover, if X is a topological space, $A \subseteq X$ is any subset of X and $x \in X$, we say that x is a limit point of A if, for every open set U containing x, there exists $a \in A \cap U$ such that $a \neq x$. We denote by $\mathrm{Der}(A)$ (the *derived set* of A) the set of the limit points of A.

Definition 12.1.1 (The $\mathcal{L}$-fine topology). *The* $\mathcal{L}$-fine topology *on $\mathbb{G}$ is the smallest topology on $\mathbb{R}^N$ with respect to which all $\mathcal{L}$-superharmonic functions are continuous (in the extended sense).*

Throughout the chapter, all topological concepts (continuous, open, neighborhood, etc.) will be prefixed by "$\mathcal{L}$-fine" when related to the $\mathcal{L}$-fine topology. Instead, they will not be prefixed or prefixed by "Euclidean" or "metric", when related to the usual Euclidean topology of $\mathbb{R}^N$.

A few simple remarks are in order.

Remark 12.1.2. (1) Since $\mathcal{L}$-superharmonic functions are l.s.c., a basis for the $\mathcal{L}$-fine topology is given by all finite intersections of the sets

$$\{x \in \Omega : u(x) < \beta\} \tag{12.1}$$

as Ω ranges over all (Euclidean-)open subsets of $\mathbb{R}^N$, u is any function in $\overline{S}(\Omega)$, and β ranges over $]-\infty, \infty]$.

(2) The $\mathcal{L}$-fine topology is properly larger than the Euclidean one (see Example 11.2.4, page 492).

(3) In particular, if $\lim_{x\to x_0} u(x) = \lambda$ in the Euclidean sense, then the same is true in the $\mathcal{L}$-fine sense too, whereas the converse may be false. Vice versa, if $\{x_n\}_{n\in\mathbb{N}}$ is a sequence in $\mathbb{G}$ converging to $x_0 \in \mathbb{G}$ in the $\mathcal{L}$-fine sense, then $x_n \to x_0$ in the Euclidean sense too.

(4) Hence, the collection of $\mathcal{L}$-fine limit points of a set A is contained (possibly properly) in the set of usual limit points of A,

$$\mathrm{Der}_{\mathcal{L}\text{-fine}}(A) \subseteq \mathrm{Der}_{\mathrm{Eucl.}} A.$$

For example, if $x_0 \in \mathbb{G} \setminus \{0\}$, the set of the $\mathcal{L}$-fine limit points of $A = \{(1+\frac{1}{n})\, x_0\}_{n\in\mathbb{N}}$ is empty whereas x_0 is a limit point of A (see also Proposition 12.1.3 below).

Proposition 12.1.3. *Let $Z \subset \mathbb{G}$ be an $\mathcal{L}$-polar set. Then the set of the $\mathcal{L}$-fine limit points of Z is empty. In particular, the points of Z are $\mathcal{L}$-finely isolated points of Z.*

Proof. By contradiction, let x be an $\mathcal{L}$-fine limit point of Z. Then x is also a (Euclidean) limit point of Z. By Example 11.2.4 (page 492), there exists $u \in \overline{S}(\mathbb{G})$ such that

$$u(x) < \lim_{Z\ni y\to x} u(y) =: r < \infty.$$

Set $\varepsilon := \frac{1}{2}(r - u(x))$. Then there exists a neighborhood U of x such that $u(y) > r - \varepsilon$ for every $y \in (U \cap Z) \setminus \{x\}$. Since u is $\mathcal{L}$-fine continuous at x (being $\mathcal{L}$-superharmonic), there is an $\mathcal{L}$-fine neighborhood U^{fine} of x such that $u(y) < u(x)+\varepsilon$ for every $y \in U^{\mathrm{fine}}$. But $r - \varepsilon = \frac{1}{2}(r + u(x)) = u(x) + \varepsilon$, so that $(U \cap Z) \cap U^{\mathrm{fine}} = (U \cap U^{\mathrm{fine}}) \cap Z$ is contained in $\{x\}$. This means that $U \cap U^{\mathrm{fine}}$ is an $\mathcal{L}$-fine neighborhood of x having in common with Z at most x, whence x is not an $\mathcal{L}$-fine limit point of Z. This is a contradiction. □

12.2 $\mathcal{L}$-thinness at a Point

Consider a bounded open set $\Omega \subset \mathbb{G}$ and a point $x \in \partial\Omega$. We have seen in Proposition 7.1.5 (page 385) that if $\mathbb{G} \setminus \Omega$ contains a suitable non-characteristic ball, then x is regular for the Dirichlet problem related to $\mathcal{L}$. Hence, for $\mathcal{L}$-irregular boundary points of Ω, the complement of Ω must be quite *thin* (in a suitable sense, which also takes into account the "degeneracy" of the operator $\mathcal{L}$). The notion of $\mathcal{L}$-*thinness* will give a more precise sense to this.

Definition 12.2.1 ($\mathcal{L}$-thinness of a set at a point). *A set $E \subseteq \mathbb{G}$ is called* $\mathcal{L}$-thin at *$x \in \mathbb{G}$ if x is not an $\mathcal{L}$-fine limit point of E.*

In other words,

$$E \text{ is } \mathcal{L}\text{-thin at } x \quad \Leftrightarrow \quad x \notin \mathrm{Der}_{\mathcal{L}\text{-fine}}(E) \quad \Leftrightarrow \quad \{x\} \cap \mathrm{Der}_{\mathcal{L}\text{-fine}}(E) = \emptyset.$$

A few simple remarks are in order.

Remark 12.2.2. Obviously, E is not $\mathcal{L}$-thin at x iff x is an $\mathcal{L}$-fine limit point of E, i.e.

$$E \text{ is not } \mathcal{L}\text{-thin at } x \quad \Leftrightarrow \quad x \in \mathrm{Der}_{\mathcal{L}\text{-fine}}(E).$$

Hence, if E is not $\mathcal{L}$-thin at x, then x is a (Euclidean) limit point of E (indeed, $\mathrm{Der}_{\mathcal{L}\text{-fine}}(E) \subseteq \mathrm{Der}_{\mathrm{Eucl.}} E$). Equivalently, a sufficient condition that a set E is $\mathcal{L}$-thin at x is that x is not a metric limit point of E.

From Proposition 12.1.3 we immediately get the following result.

Corollary 12.2.3. *A $\mathcal{L}$-polar set is $\mathcal{L}$-thin at every point.*

The following result will be important in the sequel.

Theorem 12.2.4 (Characterization of $\mathcal{L}$-thinness at a point. I). *A set $E \subseteq \mathbb{G}$ is $\mathcal{L}$-thin at a (metric limit) point x of E if and only if there exists an $\mathcal{L}$-superharmonic function u on a (metric) neighborhood of x such that*

$$u(x) < \liminf_{E \ni y \to x} u(y). \tag{12.2}$$

Proof. To avoid heavy notation, it is not restrictive to suppose that $x \notin E$.

($\Rightarrow$): Suppose first that E is $\mathcal{L}$-thin at x. Then there exists an $\mathcal{L}$-fine neighborhood U^{fine} of x such that $E \cap U^{\mathrm{fine}} = \emptyset$. By taking finite intersections of sets as in (12.1), as a basis for $\mathcal{L}$-fine neighborhoods, we infer the existence of $\beta_1, \ldots, \beta_n \in]-\infty, \infty]$ and $u_1, \ldots, u_n \in \overline{\mathcal{S}}(W)$, W being an open neighborhood of x, such that $U^{\mathrm{fine}} = \bigcap_{i=1}^n \{y \in W : u_i(y) < \beta_i\}$. Since $x \in U^{\mathrm{fine}}$ (i.e. $u_i(x) < \beta_i$), by choosing $\varepsilon = \min\{\beta_i - u_i(x)\}$, we can replace U^{fine} by the set $\bigcap_{i=1}^n \{y \in W : u_i(y) < u_i(x) + \varepsilon\}$. Now let $u := \sum_{i=1}^n u_i$. We claim that u satisfies (12.2). By the lower semi-continuity of the u_i's at x, there exists a neighborhood $U \subset W$ of x such that

$$u_i(y) > u_i(x) - \varepsilon/n \quad \text{for every } y \in U \text{ and } i = 1, \ldots, n. \tag{12.3}$$

Now take any $y \in U \cap E$. Then, since $E \cap U^{\mathrm{fine}} = \emptyset$, there exists $i_0 \in \{1, \ldots, n\}$ such that

$$u_{i_0}(y) \geq u_{i_0}(x) + \varepsilon. \tag{12.4}$$

Collecting together (12.3) and (12.4), we get

$$u(y) = \sum_{i \neq i_0} u_i(y) + u_{i_0}(x) > \sum_{i \neq i_0} u_i(x) - \frac{n-1}{n}\varepsilon + u_{i_0}(x) + \varepsilon = u(x) + \frac{\varepsilon}{n}.$$

In short, $u(y) > u(x) + \frac{\varepsilon}{n}$ for every $y \in U \cap E$ which at once implies (12.2).

($\Leftarrow$): Suppose now there exists an $\mathcal{L}$-superharmonic function u on a neighborhood Ω of x such that $u(x) < \liminf_{E\ni y\to x} u(y) =: r$. We can assume that $r < \infty$ by replacing, if needed, u by $\min\{u, u(x)+1\}$. Set $\varepsilon = \frac{1}{2}(r - u(x))$. Consider $U^{\text{fine}} := \{y \in \Omega : u(y) < u(x) + \varepsilon\}$, which is an $\mathcal{L}$-fine neighborhood of x. From $\liminf_{E\ni y\to x} u(y) = r$ we infer that there exists a neighborhood U of x such that $\inf_{E\cap V} u > r - \varepsilon$. Now we argue *verbatim* as in the last part of the proof of Proposition 12.1.3. □

Since the value of an $\mathcal{L}$-superharmonic function at a point coincide with its lim inf there, (12.2) gives another equivalent definition of $\mathcal{L}$-thinness: *E is $\mathcal{L}$-thin at x if and only if*

$$\text{there exists an } \mathcal{L}\text{-superharmonic function } u \text{ on a neighborhood of } x \text{ such that } \liminf_{y\to x} u(y) < \liminf_{E\ni y\to x} u(y). \tag{12.5}$$

This last condition is the analogue of $\mathcal{L}$-thinness at infinity given in (10.2), page 474. Another remarkable result (making $\mathcal{L}$-thinness easier to handle with) is the following one.

Theorem 12.2.5 (Characterization of $\mathcal{L}$-thinness at a point. II). *A set $E \subseteq \mathbb{G}$ is $\mathcal{L}$-thin at a point $x \in \mathbb{G}$ (limit point of E) if and only if there exists an $\mathcal{L}$-superharmonic function w on $\mathbb{G}$, which is the $\mathcal{L}$-potential of a compactly supported Radon measure, such that*

$$w(x) < \lim_{E\ni y\to x} w(y) = \infty. \tag{12.6}$$

Proof. The "if" part follows from Theorem 12.2.4. To prove the converse, we assume that E is $\mathcal{L}$-thin at x (and, without affecting the generality, we may suppose $x \notin E$). It is enough[1] to prove the existence of an open neighborhood Ω of x and $u \in \overline{\mathcal{S}}(\Omega)$ such that (12.6) holds with w replaced by u.

To this end, let u be as in Theorem 12.2.4. We can assume that the right-hand side of (12.2) is finite (otherwise there is nothing to prove). Let $B = B_d(x, \varepsilon)$ with $\varepsilon > 0$ small enough. By arguing as in the (footnote of the) previous paragraph, we can suppose that $u = \Gamma * \nu$, where ν is a Radon measure supported in B. For n large enough, we have

[1] Indeed, in this case we can take any open set O with $x \in O \subset \overline{O} \Subset \Omega$ and apply Riesz representation Theorem 9.4.4 (page 442) to get

$$u(y) = (\Gamma * \nu)(y) + h(y) \quad \forall\, y \in O,$$

where $\nu = \mu_u|_{\overline{O}}$ and h is a suitable function in $\mathcal{H}(O)$. Hence, we have

$$\lim_{E\ni\to x} u(y) = \lim_{E\ni\to x} (\Gamma * \nu)(y) + h(x), \quad \text{whereas } u(x) = (\Gamma * \nu)(x) + h(x).$$

Consequently, (12.6) holds if and only if $(\Gamma * \nu)(x) < \lim_{E\ni\to x}(\Gamma * \nu)(y)$, and the choice $w := \Gamma * \nu$ fulfills the requirement of the assertion of the theorem.

$$\int_{B_d(x,1/n)} \Gamma(x,y)\,\mathrm{d}\nu(y) + \int_{B\setminus B_d(x,1/n)} \Gamma(x,y)\,\mathrm{d}\nu(y) = u(x) \in \mathbb{R},$$

and the second integral in the left-hand side increases[2] to $u(x)$ as $n \uparrow \infty$. As a consequence, setting $\nu_n := \nu|_{B_d(x,1/n)}$, we have $\lim_{n\to\infty}(\Gamma * \nu_n)(x) = 0$. This ensures the existence of $\varepsilon_n \downarrow 0$ such that $\sum_n u_n(x) < \infty$, where $u_n := \Gamma * \nu_{\varepsilon_n}$. Then the function $v := \sum_n u_n$ is finite at x and belongs to $\overline{\mathcal{S}}(B)$. Let $\delta := \liminf_{E\ni y\to x}(u(y) - u(x)) > 0$. By applying once again the Riesz representation theorem, we get $u = u_n + h_n$, where $h_n \in \mathcal{H}(B_d(x,\varepsilon_n))$. This gives

$$\begin{aligned}\liminf_{E\ni y\to x}(u_n(y) - u_n(x)) &\geq \liminf_{E\ni y\to x}(u_n(y) + h_n(y)) + \liminf_{E\ni y\to x}(-h_n(y) - u_n(x))\\ &= \liminf_{E\ni y\to x} u(y) - h_n(x) - u_n(x) = \liminf_{E\ni y\to x}(u(y) - u(x)) = \delta\end{aligned}$$

for every $n \in \mathbb{N}$, so that (by summing up)

$$\liminf_{E\ni y\to x}\left(\sum_{n=1}^{p} u_n(y) - \sum_{n=1}^{p} u_n(x)\right) \geq p\,\delta.$$

This gives $\liminf_{E\ni y\to x}\sum_{n=1}^{p} u_n(y) \geq \sum_{n=1}^{p} u_n(x) + p\,\delta \geq p\,\delta$. Since $v = \sum_n u_n \geq \sum_{n=1}^{p} u_n$, we derive (letting $p \to \infty$) $\liminf_{E\ni y\to x} v(y) = \infty$. □

From Theorem 12.2.5 one immediately obtains the following assertion.

Corollary 12.2.6. *Let $Z,\ E \subseteq \mathbb{G}$, $y \in \mathbb{G}$. Suppose Z is $\mathcal{L}$-polar. Then $E \setminus Z$ is $\mathcal{L}$-thin at y if and only if E is.*

Proof. Suppose first that E is $\mathcal{L}$-thin at y. Then, by Theorem 12.2.4, there exists an $\mathcal{L}$-superharmonic function u on a neighborhood of y such that $u(y) < \liminf_{E\ni z\to y} u(z)$. This gives

$$u(y) < \liminf_{E\ni z\to y} u(z) \leq \liminf_{E\setminus Z\ni z\to y} u(z),$$

so that $E \setminus Z$ is $\mathcal{L}$-thin at y, again by Theorem 12.2.4.

Vice versa, suppose that $E \setminus Z$ is $\mathcal{L}$-thin at y and (thanks to Theorem 12.2.5) consider an $\mathcal{L}$-superharmonic function u on $\mathbb{G}$ such that $u(y) < \lim_{E\setminus Z\ni z\to y} u(z) = \infty$. Since $Z \setminus \{y\}$ is $\mathcal{L}$-polar, there exists $w \in \overline{\mathcal{S}}^{+}(\mathbb{G})$ such that $w \equiv \infty$ on $Z \setminus \{y\}$ and $w(y) \in \mathbb{R}$. Consequently, we have

$$\lim_{E\ni z\to y}(u(z) + w(z)) = \infty > u(y) + w(y).$$

This proves that E is $\mathcal{L}$-thin at y by making use of Theorem 12.2.5. □

[2] This follows from $\nu(\{x\}) = 0$. Otherwise, we would have

$$\infty > u(x) \geq \int_{B_d(x,1/n)} \Gamma(x,y)\,\mathrm{d}\nu(y) \to \infty \cdot \nu(\{x\}) = \infty.$$

12.3 $\mathcal{L}$-thinness and $\mathcal{L}$-regularity

The aim of this section is to prove that the $\mathcal{L}$-regularity of a boundary point can be characterized in terms of $\mathcal{L}$-thinness (see Theorem 12.3.6 below). To this end, we provide some preliminary results. Some have an interest in their own and are collected in the following subsection.

12.3.1 Functions Peaking at a Point

We need a definition and some preliminary results.

Definition 12.3.1 (Function peaking at a point). *If u is an extended real-valued function defined on a neighborhood of x, we say that u* peaks at x *if $u(x) > \sup_{y \notin V} u(y)$ for every small neighborhood V of x.*

Lemma 12.3.2. *For every fixed $x \in \mathbb{G}$, there exists $w \in \overline{\mathcal{S}}^+(\mathbb{G}) \cap C(\mathbb{G})$ such that w has a strict absolute maximum at x (in particular, w peaks at x).*

Proof. Let $r > 0$. Then the function $w(z) := \mathrm{M}_r(\Gamma(z^{-1} \circ \cdot))(x)$ has the required properties (see Theorem 9.3.10, page 438; see also Fig. 12.1). □

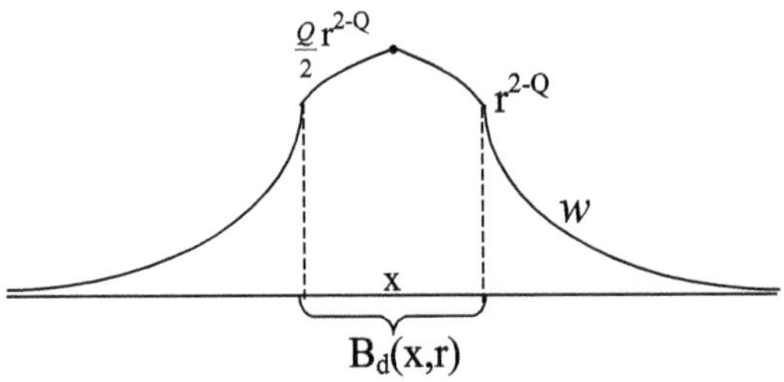

Fig. 12.1. The function w of Lemma 12.3.2

The relevance of functions peaking at a point within the theory of $\mathcal{L}$-thinness becomes apparent from the following result.

Lemma 12.3.3 (Characterization of $\mathcal{L}$-thinness. III). *Let $w \in \overline{\mathcal{S}}^+(\mathbb{G})$ be a function which peaks at x. Then a set E is $\mathcal{L}$-thin at x if and only if*

$$\widehat{\mathbf{R}}^w_E(x) \lneqq w(x).$$

Proof. Since E is $\mathcal{L}$-thin at x if and only if $E \setminus \{x\}$ is $\mathcal{L}$-thin at x (see, e.g. Corollary 12.2.6) and since $\widehat{\mathbf{R}}^w_E = \widehat{\mathbf{R}}^w_{E\setminus\{x\}}$ (see Proposition 11.8.3, page 525), we can suppose that $x \notin E$. As a consequence, from Theorem 11.6.5 (page 513) we derive

$$\widehat{\mathbf{R}}^w_E(x) = \mathbf{R}^w_E(x).$$

($\Leftarrow$): Let us suppose $\mathbf{R}_E^w(x) < w(x)$ and prove that E is $\mathcal{L}$-thin at x. We can suppose that x is a limit point of E (otherwise, nothing is left to prove). By the definition of $\mathcal{L}$-réduite function, there exists $v \in \overline{\mathcal{S}}^+(\mathbb{G})$ such that $v \geq w$ on E and $v(x) < w(x)$. This gives

$$\liminf_{E \ni y \to x} v(y) \geq \liminf_{E \ni y \to x} w(y) \geq \liminf_{y \to x} w(y) \geq w(x) > v(x),$$

whence E is $\mathcal{L}$-thin at x by Theorem 12.2.4.

($\Rightarrow$): Let now E be $\mathcal{L}$-thin at x. Suppose first that x is not a limit point of E, and let V be a neighborhood of x such that $V \cap E = \emptyset$. Then

$$w(x) > \sup_{\mathbb{G} \setminus V} w \geq \sup_E w \geq \mathbf{R}_E^w(x) = \widehat{\mathbf{R}}_E^w(x).$$

The third inequality follows from the fact that the constant function $\sup_E w$ belongs to Φ_E^w (see the notation at the beginning of Section 11.5, page 500). We are left with the case when x is a limit point of E. By Theorem 12.2.5, there exists $v \in \overline{\mathcal{S}}^+(\mathbb{G})$ such that $v(x) < \lim_{E \ni y \to x} v(y) = \infty$. Replacing v by $\widetilde{v} := \min\{v, v(x) + 1\}$, we can suppose that $v(x) < \lim_{E \ni y \to x} v(y) < \infty$ and v is bounded from above. Let α and β be such that

$$v(x) < \alpha < \beta < \lim_{E \ni y \to x} v(y). \tag{12.7}$$

For every $\lambda > 0$, we set

$$w_\lambda(\cdot) := w(x) + \lambda\,(v(\cdot) - \alpha).$$

First, we claim that *there exists a neighborhood $\overline{U}$ of x and $0 < \overline{\lambda} \ll 1$ such that $w_\lambda > 0$ on $\overline{U}$ for any $0 < \lambda \leq \overline{\lambda}$*. Indeed, from $w(x) > 0$ (w peaks at x) and the lower semi-continuity of w we get $\liminf_{y \to x} w(y) \geq w(x) > 0$. Hence, there exists a neighborhood $\overline{U}$ of x and $\varepsilon > 0$ such that $\inf_{\overline{U}} w \geq \varepsilon$. We choose $0 < \overline{\lambda} \ll 1$ such that $\overline{\lambda}\,|v(y) - \alpha| \leq \varepsilon/2$ (recall that v is bounded) for every $y \in \overline{U}$. This gives, for every $y \in \overline{U}$ and every $0 < \lambda \leq \overline{\lambda}$,

$$w_\lambda(y) = w(y) + \lambda(v(y) - \alpha) \geq \varepsilon - \overline{\lambda}\,|v(y) - \alpha| \geq \varepsilon/2.$$

The claim is proved. According to (12.7), there exists a neighborhood $U \subseteq \overline{U}$ of x such that $v(y) > \alpha$ for every $y \in E \cap U$. Then we have

$$w_\lambda(y) > w(x) \geq w(y) \qquad \forall\, y \in E \cap U. \tag{12.8}$$

Moreover, $w(x) - w(y) \geq \gamma > 0$ for every $y \notin U$ (again by the peaking property of w). Since v is bounded, there exists a small positive $\lambda_0 < \overline{\lambda}$ such that

$$\lambda_0\,|v(y) - \alpha| \leq \gamma \leq w(x) - w(y) \quad \forall\, y \notin U.$$

This proves that, for $y \notin U$, we have

$$\lambda_0\,|v(y) - \alpha| \geq -\lambda_0\,|v(y) - \alpha| \geq -(w(x) - w(y)) = w(y) - w(x),$$

so that
$$w_{\lambda_0}(y) = w(x) + \lambda_0\,(v(y) - \alpha) \geq w(y) \qquad \forall\, y \notin U.$$
This last inequality, together with (see also (12.8)) the inequality
$$w_{\lambda_0}(y) \geq w(y) \qquad \forall\, y \in E \cap U,$$
gives $w_{\lambda_0}(y) \geq w(y)$ for every $y \in E$. Moreover, from $w_{\lambda_0}(y) \geq 0$ on U (see the claim above) and $w_{\lambda_0} \geq w \geq 0$ on E we derive that $w_{\lambda_0}(y) \geq 0$. This proves that $w \in \overline{\mathcal{S}}^+(\mathbb{G})$ is such that $w_{\lambda_0} \geq w$ on E, whence, by the definition of $\mathcal{L}$-réduite,
$$\mathbf{R}_E^w(x) \leq w_{\lambda_0}(x) = w(x) + \lambda_0\,(v(x) - \alpha) < w(x).$$
This completes the proof. □

Theorem 12.3.4 ($\mathcal{L}$-thinness and $\mathcal{L}$-polarity. I). *Let $E \subseteq \mathbb{G}$ be any set. Then the subset of E where E is $\mathcal{L}$-thin is an $\mathcal{L}$-polar set. In other words, the set of the $\mathcal{L}$-finely isolated points of E is an $\mathcal{L}$-polar set.*

Proof. Let E be $\mathcal{L}$-thin at $x_0 \in E$. Let $w \in \overline{\mathcal{S}}^+(\mathbb{G}) \cap C(\mathbb{G})$ be such that w peaks at x_0 (see Lemma 12.3.2). By Lemma 12.3.3, we have $\widehat{\mathbf{R}}_E^w(x_0) \lneqq w(x_0) = \mathbf{R}_E^w(x_0)$ (for $x_0 \in E$). Then x_0 belongs to $P := \{x \in \mathbb{G} : \widehat{\mathbf{R}}_E^w(x_0) < \mathbf{R}_E^w(x_0)\}$ which is a $\mathcal{L}$-polar subset of $E \cap \partial E$ (see Theorem 11.6.5, page 513). From the arbitrariness of x_0, the assertion follows. □

Corollary 12.3.5 ($\mathcal{L}$-thinness and $\mathcal{L}$-polarity. II). *A set $Z \subset \mathbb{G}$ is $\mathcal{L}$-polar if and only if Z is $\mathcal{L}$-thin at any of its points.*

Proof. If Z is $\mathcal{L}$-polar, then Z is $\mathcal{L}$-thin at every point of $\mathbb{G}$ (see Proposition 12.1.3). The converse assertion follows immediately from Theorem 12.3.4. □

12.3.2 $\mathcal{L}$-thinness and $\mathcal{L}$-regularity

We are now in a position to state and prove the main results of the section.

Theorem 12.3.6 (Non-$\mathcal{L}$-thinness and $\mathcal{L}$-barriers). *Let $F \subseteq \mathbb{G}$ be a closed set. Let $x_0 \in \partial F$. Then F is not $\mathcal{L}$-thin at x_0 if and only if there exists a ($\mathcal{L}$-barrier) function b with the following properties*: $b \in \overline{\mathcal{S}}(\mathbb{G} \setminus F)$, $b \geq 0$ *on* $\mathbb{G} \setminus F$, $b > 0$ *on* $W := B_d(x_0, 1) \setminus F$ *and* $\lim_{W \ni y \to x_0} b(y) = 0$.

From Theorem 12.3.6 and Bouligand's Theorem 6.10.4 (page 371) we immediately derive the following $\mathcal{L}$-regularity criterion.

Theorem 12.3.7 ($\mathcal{L}$-regularity criterion in terms of $\mathcal{L}$-thinness). *Let $\Omega \subset \mathbb{G}$ be a bounded and connected open set. Let $x_0 \in \partial\Omega$. Then x_0 is an $\mathcal{L}$-regular point for Ω if and only if $\mathbb{G} \setminus \Omega$ is not $\mathcal{L}$-thin at x_0. Equivalently, x_0 is an $\mathcal{L}$-irregular point if and only if $\mathbb{G} \setminus \Omega$ is $\mathcal{L}$-thin at x_0.*

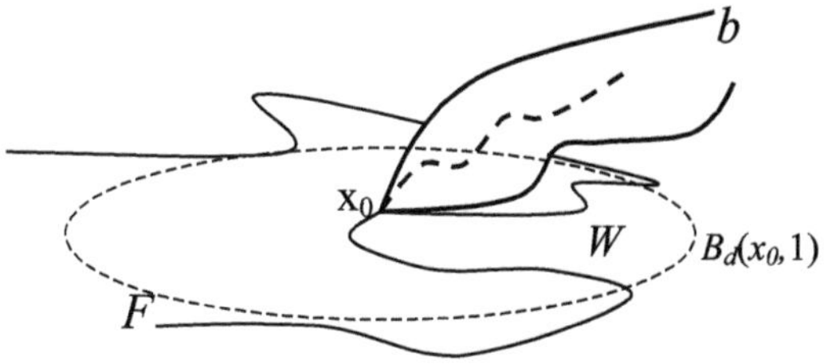

Fig. 12.2. The barrier function of Theorem 12.3.6

Proof (of Theorem 12.3.6). Let $F \subseteq \mathbb{G}$ be closed, and let $x_0 \in \partial F$ ($\subseteq F$).

($\Rightarrow$): Suppose F is not $\mathcal{L}$-thin at x_0. Let

$$w(z) := \mathrm{M}_1(\Gamma(z^{-1} \circ \cdot))(x_0).$$

Then (see Theorem 9.3.10, page 438) $w \in \overline{\mathcal{S}}(\mathbb{G}) \cap C(\mathbb{G})$, w peaks at x_0 and $w > 0$ on $\mathbb{G}$. Consider the $\mathcal{L}$-balayage $\widehat{\mathbf{R}}_F^w$ of w relative to F. Then

$$\widehat{\mathbf{R}}_F^w(x_0) = w(x_0).$$

Indeed, since F is not $\mathcal{L}$-thin at x_0, by Lemma 12.3.3 (and the general properties of the $\mathcal{L}$-balayage) we have $\widehat{\mathbf{R}}_F^w(x_0) \geq w(x_0) \geq \widehat{\mathbf{R}}_F^w(x_0)$. Hence,

$$\liminf_{\mathbb{G}\setminus F \ni y \to x_0} \widehat{\mathbf{R}}_F^w(y) \geq \liminf_{y \to x} \widehat{\mathbf{R}}_F^w(y) \geq \widehat{\mathbf{R}}_F^w(x_0) = w(x_0). \tag{12.9}$$

Let us now consider the function

$$b := w - \widehat{\mathbf{R}}_F^w.$$

The function b is non-negative[3] on $\mathbb{G}$ and $b \in \overline{\mathcal{S}}(\mathbb{G} \setminus F)$. This last fact follows from $w \in \overline{\mathcal{S}}(\mathbb{G})$, $\widehat{\mathbf{R}}_F^w \in \mathcal{H}(\mathbb{G} \setminus \overline{F})$ and $\overline{F} = F$. We now claim that $b > 0$ on $W := B_d(x_0, 1) \setminus F$. Indeed, it is enough to prove that $b > 0$ on every connected component C of W. Suppose, to the contrary, that b vanishes at some point of C. By the minimum principle,[4] this gives $b \equiv 0$ on C, whence $w \equiv \widehat{\mathbf{R}}_F^w$ on C, so that

$$\mathcal{L}w = 0 \quad \text{on } C, \tag{12.10}$$

for $\widehat{\mathbf{R}}_F^w$ is $\mathcal{L}$-harmonic on $\mathbb{G} \setminus F \supset W \supseteq C$. Now, a direct computation (see (9.10b) in Theorem 9.3.10, page 438) shows that, for every $z \in C$, we have[5]

$$\begin{aligned}\mathcal{L}w(z) &= \frac{2-Q}{2}\,\mathcal{L}(d^2(x_0, z)) \\ &= \frac{Q}{2-Q}\,\Gamma^{(2Q-2)/(2-Q)}(x_0^{-1} \circ z) \cdot |\nabla_{\mathcal{L}}\Gamma|^2(x_0^{-1} \circ z).\end{aligned}$$

---

[3] This follows from the general fact $w \geq \widehat{\mathbf{R}}_F^w$ if $w \in \overline{\mathcal{S}}(\mathbb{G})$.

[4] Note that $b \in \overline{\mathcal{S}}(W)$ since W is open, $W \subseteq \mathbb{G} \setminus F$ and $b \in \overline{\mathcal{S}}(\mathbb{G} \setminus F)$.

[5] Recall that $d = \Gamma^{1/(2-Q)}$, the general formula $\mathcal{L}(\alpha(u)) = \alpha''(u)\,|\nabla_{\mathcal{L}} u|^2 + \alpha'(u)\,\mathcal{L}u$ (where $u : \mathbb{R} \to \mathbb{R}$ is regular enough) and $\mathcal{L}\Gamma = 0$ outside the origin.

The far right-hand term cannot vanish on the open set C (otherwise $\nabla_{\mathcal{L}} \Gamma \equiv 0$ on C, i.e. $\Gamma \equiv 0$ on C, see Proposition 1.5.6, page 69) in contradiction with (12.10).

Moreover, we have

$$\begin{aligned} 0 \le \limsup_{W \ni y \to x_0} b(y) &= \limsup_{F \not\ni y \to x_0} b(y) = \limsup_{F \not\ni y \to x_0} (w(y) - \widehat{\mathbf{R}}^w_F(y)) \\ &\le w(x_0) - \liminf_{F \not\ni y \to x_0} \widehat{\mathbf{R}}^w_F(y) \le 0. \end{aligned}$$

Here, we exploited the continuity of w and (12.9). This finally proves that $\limsup_{W \ni y \to x_0} b(y) = 0$.

($\Leftarrow$): Let b be as in the assertion of Theorem 12.3.6 (see also Fig. 12.2). Let us suppose by contradiction that F is not $\mathcal{L}$-thin at $x_0 \in \partial F$. Then, by making use of Theorem 12.2.5, it is not difficult[6] to prove the existence of a bounded function $v \in \underline{\mathcal{S}}(\mathbb{G})$ and a neighborhood $U \subseteq B_d(x_0, 1)$ of x_0 such that $v(x_0) = 1$ and $v \le -1$ on $U \cap (F \setminus \{x_0\})$.

Let $B_1 := B_d(x_0, \rho)$ with $0 < \rho \ll 1$ such that $\overline{B_1} \subset U$. Then

$$v \le -1 \quad \text{on } F \cap \partial B_1 \ (\subseteq U \cap F \setminus \{x_0\}).$$

Since $F \cap \partial B_1$ is compact, this fact, together with the upper semi-continuity of v, implies the existence of an open neighborhood $\widetilde{W}$ of $F \cap \partial B_1$ such that $v \le -1/2$ on $\widetilde{W}$. Hence,[7] for every $\lambda > 0$

$$v - \lambda\, b \le 0 \quad \text{on } \widetilde{W} \cap (\partial B_1 \setminus F). \tag{12.11}$$

Being $b > 0$ on $\partial B_1 \setminus F$, we have

$$\inf_{\partial B_1 \setminus \widetilde{W}} b > 0, \tag{12.12}$$

for $\partial B_1 \setminus \widetilde{W}$ is a compact set which does not intersect a neighborhood of F and b is l.s.c. outside F. Hence, the boundedness of v and (12.12) ensure that there exists $\lambda_0 \gg 1$ such that

$$v - \lambda_0\, b \le 0 \quad \text{on } \partial B_1 \setminus \widetilde{W}. \tag{12.13}$$

Thus, for every $y \in \partial B_1 \setminus \widetilde{W}$ and every $\varepsilon > 0$, one has (recalling that $v - \lambda_0\, b$ is u.s.c. and using (12.13))

$$\limsup_{B_1 \setminus F \ni z \to y} \left(v(z) - \lambda_0\, b(z) - \varepsilon\, \Gamma(x_0^{-1} \circ z) \right) \le 0. \tag{12.14}$$

The same holds for $y = x_0$ (since $\Gamma(0) = \infty$) and for every $y \in \overline{B_1} \cap (\partial F \setminus \{x_0\})$, for this last set is contained in $U \cap (F \setminus \{x_0\})$ where $v \le -1$. Now, being $v - \lambda_0\, b - \varepsilon\, \Gamma(x_0^{-1} \circ \cdot) \in \underline{\mathcal{S}}(B_1 \setminus F)$, from the maximum principle one gets (also collecting (12.11), (12.14))

[6] See Ex. 5 at the end of the present Chapter.

[7] Indeed, $v \le -1/2$ on $\widetilde{W}$, $b > 0$ on $W = B_d(x_0, 1) \setminus F$ and $B_d(x_0, 1) \supseteq U \supset \overline{B_1} \supset \partial B_1$.

$$v - \lambda_0 b - \varepsilon \Gamma(x_0^{-1} \circ \cdot) \le 0 \quad \text{on } B_1 \setminus F.$$

Letting $\varepsilon \to 0^+$, we derive $v \le \lambda_0 b$ on $B_1 \setminus F$, so that

$$\limsup_{B_1 \setminus F \ni z \to x_0} v(z) \le \limsup_{B_1 \setminus F \ni z \to x_0} \lambda_0 b(z) = 0$$

by the definition of the barrier function b. Since

$$\limsup_{F \ni z \to x_0} v(z) \le -1$$

(for $v \le -1$ on $U \cap (F \setminus \{x_0\})$), we finally get

$$\limsup_{z \to x_0} v(z) \le 0.$$

But this is in contradiction to $\limsup_{z \to x_0} v(z) = v(x_0)$ (a consequence of the fact that v is $\mathcal{L}$-subharmonic in x_0) and $v(x_0) = 1$ (by construction). This contradiction proves the theorem. □

12.4 Wiener's Criterion for Sub-Laplacians

The aim of this section is to prove a test for $\mathcal{L}$-thinness similar to the well-known Wiener criterion for the classical Laplace operator (in this case, see, e.g. [AG01, Section 7.7]). For a generalization to more general operators, see also [NS87].

As usual, throughout the section, $\mathbb{G} = (\mathbb{R}^N, \circ, \delta_\lambda)$ is a Carnot group and $\mathcal{L}$ is a sub-Laplacian on $\mathbb{G}$. Moreover, $\Gamma = d^{2-Q}$ is the fundamental solution for $\mathcal{L}$. We shall also denote by $\beta \ge 1$ the constant appearing in the following inequalities:

$$d(a \circ b) \le \beta\, d(a) + d(b) \quad \forall\, a, b \in \mathbb{G}, \tag{12.15a}$$

$$d(a \circ b) \ge \frac{1}{\beta}\, |d(a) - d(b)| \quad \forall\, a, b \in \mathbb{G}. \tag{12.15b}$$

These are (subtle) improvements of the pseudo-triangle inequality for d. (12.15a) is proved in Proposition 5.14.1 on page 306. Then, by choosing $a = x \circ y$ and $b = y^{-1}$ in (12.15a) (and using the symmetry of d), we derive $d(x) - d(y) \le \beta\, d(x \circ y)$; then, replacing x and y in the latter with, respectively, y^{-1} and x^{-1} (and using twice the symmetry of d) we get $d(y) - d(x) = d(y^{-1}) - d(x^{-1}) \le \beta\, d(y^{-1} \circ x^{-1}) = \beta\, d((x \circ y)^{-1}) = d(x \circ y)$. This proves (12.15b).

12.4.1 A Technical Lemma

Throughout the section, we use the following notation: given a point $y \in \mathbb{G}$ and a constant $\alpha > 1$, for every $n \in \mathbb{N}$ we set

$$C_n := \big\{ x \in \mathbb{G} : \ \alpha^n \le \Gamma(y^{-1} \circ x) \le \alpha^{n+1} \big\}. \tag{12.16}$$

In other words, let

$$\underline{\alpha} := \left(\frac{1}{\alpha}\right)^{\frac{1}{Q-2}},$$

we have

$$C_n = \overline{B_d(y, \underline{\alpha}^n)} \setminus B_d(y, \underline{\alpha}^{n+1}).$$

We begin with a crucial lemma concerning with suitable estimates of the fundamental solution Γ for $\mathcal{L}$. We remark that the proof of Lemma 12.4.1 below is more delicate in our context than in the classical case (see [AG01, Lemma 7.7.1]).

We shall henceforth suppose that

$$\alpha > \beta^{Q-2}, \tag{12.17}$$

where β is the constant appearing in (12.15a) (note that in the Euclidean case, $\beta = 1$ and (12.17) reduces to $\alpha > 1$).

Lemma 12.4.1. *Let $\alpha > 1$ (be as in (12.17)). Then there exists $\mathbf{c} = \mathbf{c}(\alpha) > 0$ such that, for every $y \in \mathbb{G}$, for every Radon measure μ on $\overline{B_d(y, \frac{1}{2})}$, for every $n \in \mathbb{N}$, $n \geq 2$ and every $x \in C_n$ it holds*

$$\int_{\mathbb{G}\setminus(C_{n-1}\cup C_n \cup C_{n+1})} \Gamma(x^{-1} \circ z)\, \mathrm{d}\mu(z) \leq \mathbf{c} \int_{\mathbb{G}} \Gamma(y^{-1} \circ z)\, \mathrm{d}\mu(z). \tag{12.18}$$

Proof. Let $y \in \mathbb{G}$ and $\alpha > 1$ be fixed throughout. Let also $x \in C_n$, i.e.

$$\alpha^n \leq d^{2-Q}(y^{-1} \circ x) \leq \alpha^{n+1}. \tag{12.19}$$

We have

$$\begin{aligned}\mathbb{G} \setminus (C_{n-1} \cup C_n \cup C_{n+1}) = \{&z \in \mathbb{G} : \Gamma(y^{-1} \circ z) > \alpha^{n+2}\} \,\cup \\ &\cup \{z \in \mathbb{G} : \Gamma(y^{-1} \circ z) < \alpha^{n-1}\} =: I \cup II.\end{aligned}$$

We distinguish two cases:

$$\text{(i):}\quad z \in I, \text{ i.e.} \qquad d^{2-Q}(y^{-1} \circ z) > \alpha^{n+2}, \tag{12.20a}$$

$$\text{(ii):}\quad z \in II, \text{ i.e.} \qquad d^{2-Q}(y^{-1} \circ z) < \alpha^{n-1}. \tag{12.20b}$$

(i): Suppose first that $z \in I$. Then (see (12.15b), (12.19), (12.20a))

$$d(x^{-1} \circ z) \geq \frac{1}{\beta}\big(d(x^{-1} \circ y) - d(y^{-1} \circ z)\big) \geq \frac{1}{\beta}\, \alpha^{\frac{n+2}{2-Q}} (\alpha^{\frac{1}{Q-2}} - 1).$$

Equivalently (exploiting once again (12.20a) in the last inequality),

$$\Gamma(x^{-1} \circ z) \leq \left(\frac{\beta}{\alpha^{\frac{1}{Q-2}} - 1}\right)^{Q-2} \alpha^{n+2} =: \mathbf{c}(\alpha)\, \alpha^{n+2} < \mathbf{c}(\alpha)\, \Gamma(y^{-1} \circ z).$$

We have thus proved (indeed, we only used the second inequality in (12.19))

$$x \in C_n,\ z \in I \quad \Longrightarrow \quad \frac{\Gamma(x^{-1} \circ z)}{\Gamma(y^{-1} \circ z)} \le \mathbf{c}(\alpha). \tag{12.21}$$

(ii): Suppose now that $z \in II$. In this case, we also need an estimate similar to that in the far right-hand of (12.21), namely

$$x \in C_n,\ z \in II \quad \Longrightarrow \quad \frac{\Gamma(x^{-1} \circ z)}{\Gamma(y^{-1} \circ z)} \le \mathbf{c}(\alpha). \tag{12.22}$$

Since this estimate will intervene to estimate the integral in the left-hand side of (12.18), and the measure μ appearing there is supported in $\overline{B_d(y, \frac{1}{2})}$, we can suppose that

$$d(y^{-1} \circ z) \le 1/2. \tag{12.23}$$

Now, (12.22) follows if we, equivalently, prove that

$$\frac{d(y^{-1} \circ z)}{d(x^{-1} \circ z)} \le \widetilde{\mathbf{c}}(\alpha) \tag{12.24}$$

under the following[8] hypotheses on z (see (12.20b), (12.23))

$$\alpha^{\frac{n-1}{2-Q}} < d(y^{-1} \circ z) \le 1/2. \tag{12.25}$$

Suitable bounds for $d(x^{-1} \circ z)$ are given by the following inequalities:

$$d(x^{-1} \circ z) \le \beta\, d(x^{-1} \circ y) + d(y^{-1} \circ z) \le \beta\, \alpha^{\frac{n}{2-Q}} + d(y^{-1} \circ z),$$
$$d(x^{-1} \circ z) \ge d(y^{-1} \circ x \circ x^{-1} \circ z) - \beta\, d(y^{-1} \circ x) \ge d(y^{-1} \circ z) - \beta\, \alpha^{\frac{n}{2-Q}}.$$

The first estimate has been obtained by applying (12.15a) with $a = x^{-1} \circ y$, $b = y^{-1} \circ z$, jointly with the first inequality in (12.19); the second estimate has been obtained by applying (12.15a) with $a = y^{-1} \circ x$, $b = x^{-1} \circ z$, jointly with the first inequality in (12.19). Thus,

$$d(y^{-1} \circ z) - \beta\, \alpha^{\frac{n}{2-Q}} \le d(x^{-1} \circ z) \le d(y^{-1} \circ z) + \beta\, \alpha^{\frac{n}{2-Q}}. \tag{12.26}$$

Set $\lambda := d(y^{-1} \circ z)$, $\mu := d(x^{-1} \circ z)$, $A := \alpha^{\frac{n-1}{2-Q}}$ and $B := \beta\, \alpha^{\frac{n}{2-Q}}$. According to the hypotheses (12.25), (12.26) and the needed assertion (12.24), we have to prove

$$\begin{array}{c} A < \lambda \le 1/2 \\ \lambda - B \le \mu \le \lambda + B \end{array} \quad \Longrightarrow \quad \frac{\lambda}{\mu} \le \widetilde{\mathbf{c}}. \tag{12.27}$$

Now, (12.27) follows from a simple exercise of optimization (see Ex. 10 at the end of the present Chapter[9]) with $\widetilde{\mathbf{c}} = A/(A - B)$, i.e.

---

[8] We are tacitly supposing that n is large enough, so that $\alpha^{\frac{n-1}{2-Q}} < 1/2$, otherwise there is nothing to prove.

[9] Note that the hypothesis (12.17) on α ensures that $A \in]0, 1/2[$ and $B \in]0, A[$.

$$\widetilde{\mathbf{c}} = \frac{\alpha^{\frac{n-1}{2-Q}}}{\alpha^{\frac{n-1}{2-Q}} - \beta\,\alpha^{\frac{n}{2-Q}}} = \frac{1}{1+\beta\,\alpha^{1/(2-Q)}} = \mathbf{c}(\alpha).$$

(Note that $\mathbf{c}$ depends only on B/A and does not depend on n.)

Finally, (12.18) follows straightforwardly from (12.21) and (12.22). The lemma is completely proved. □

12.4.2 Wiener's Criterion for $\mathcal{L}$

Let $\alpha > 1$ be as in Lemma 12.4.1 at the beginning of the section. We also maintain the notation C_n given by (12.16). We are ready to state and prove our Wiener's test for sub-Laplacians.

Theorem 12.4.2 (Wiener's criterion for $\mathcal{L}$). *Let $\mathcal{C}^*$ denote the exterior $\mathcal{L}$-capacity relative to $\mathbb{G}$. Let also $E \subseteq \mathbb{G}$ and $y \in \mathbb{G}$. Then the following statements are equivalent:*

(i) *E is $\mathcal{L}$-thin at y,*
(ii) *It holds $\sum_{n=1}^{\infty} \alpha^n\, \mathcal{C}^*(E \cap C_n) < \infty$,*
(iii) *It holds $\sum_{n=1}^{\infty} \widehat{\mathbf{R}}^1_{E\cap C_n}(y) < \infty$,*
(iv) *It holds $\int_{\alpha}^{\infty} \mathcal{C}^*(\{x \in E : \Gamma(y^{-1} \circ x) \geq t\})\, \mathrm{d}t < \infty$.*

Proof. First, we remark that it is not restrictive to suppose that y is a limit point of E, otherwise (for suitable large n and t) $E \cap C_n$ and $E \cap B_d(y, t^{1/(2-Q)})$ would be contained in $\{y\}$ so that (i) through (iv) all hold (recall that a set is $\mathcal{L}$-thin in the complement of its limit points).

The proof is split into four steps: we prove one by one the equivalences (ii) $\Leftrightarrow$ (iii), (i) $\Leftrightarrow$ (iii), (ii) $\Leftrightarrow$ (iv).

(ii) $\Leftrightarrow$ (iii): Since $E \cap C_n$ is bounded, $\widehat{\mathbf{R}}^1_{E\cap C_n}$ is an $\mathcal{L}$-potential (see Theorem 11.8.1), say $\Gamma * \mu_n$ (where μ_n is supported in C_n for $\widehat{\mathbf{R}}^1_{E\cap C_n}$ is $\mathcal{L}$-harmonic in the complement of $\overline{E \cap C_n}$). Let us set

$$u(x) := \sum_{n\geq 1} \Gamma * \mu_n(x).$$

From Theorem 11.8.1 (page 519) we also infer $\mathcal{C}^*(E \cap C_n) = \mu_n(\mathbb{G})$. Let us assume (ii). Then (iii) holds too, for we have

$$\begin{aligned} u(y) &= \sum_{n\geq 1} \Gamma * \mu_n(y) = \sum_{n\geq 1} \int_{C_n} \Gamma(z^{-1} \circ y)\, \mathrm{d}\mu_n(z) \\ &\leq \sum_{n\geq 1} \alpha^{n+1}\, \mu_n(\mathbb{G}) = \sum_{n\geq 1} \alpha^{n+1}\, \mathcal{C}^*(E \cap C_n) < \infty. \end{aligned} \tag{12.28}$$

Vice versa, let us assume (iii). Then (ii) holds too, for we have

$$\sum_{n\geq 1} \alpha^{n+1}\, \mathcal{C}^*(E\cap C_n) = \sum_{n\geq 1} \alpha^{n+1}\, \mu_n(C_n)$$
$$\leq \sum_{n\geq 1}\int_{C_n} \Gamma(z^{-1}\circ y)\,\mathrm{d}\mu_n(z) = \sum_{n\geq 1} \Gamma * \mu_n(y) < \infty. \qquad (12.29)$$

(Proving the inequalities "$\leq$" in (12.28) and (12.29) we used the fact that, by definition, $\alpha^n \leq \Gamma(z^{-1}\circ y) \leq \alpha^{n+1}$ on C_n.)

(i) $\Leftrightarrow$ (iii): First, suppose that (iii) holds. Then[10] there exists a sequence b_n diverging to ∞ such that $\sum_n b_n\, \widehat{\mathbf{R}}^1_{E\cap C_n}(y) < \infty$. Set

$$v(x) := \sum_n b_n\, \widehat{\mathbf{R}}^1_{E\cap C_n}(x).$$

By Theorem 11.6.5 (page 513), there exists an $\mathcal{L}$-polar set $P_n \subseteq E\cap C_n$ such that

$$\widehat{\mathbf{R}}^1_{E\cap C_n}(x) = 1 \qquad \forall\, x\in (E\cap C_n)\setminus P_n. \qquad (12.30)$$

Set $A = \bigcup_n A_n$, $P = \bigcup_n P_n$. From (12.30) we infer $v(x) = \sum_n b_n = \infty$ for every $x \in (E\cap A)\setminus P$. As a consequence ($A$ being a neighborhood of y) $E\setminus P$ is $\mathcal{L}$-thin at y. This ensures that E is also $\mathcal{L}$-thin at y (see Corollary 12.2.6).

Now, suppose that (i) holds. Then, by Theorem 12.2.5, there exists $w \in \overline{\mathcal{S}}^+(\mathbb{G})$ such that

$$w(y) < \lim_{E\ni x\to y} w(x) = \infty. \qquad (12.31)$$

Set $E_0 := \bigcup_{k=k_0}^{\infty} E\cap C_{2k}$ where k_0 is chosen so that $C_{2k_0} \subset B_d(y, \frac{1}{2})$. Since E_0 is bounded, by Theorem 11.8.1 (page 519), there exists a measure ν_0 such that $\widehat{\mathbf{R}}^w_{E_0} = \Gamma * \nu_0$. By Theorem 11.6.5 (page 513), there exists an $\mathcal{L}$-polar set $F_0 \subseteq E_0\cap \partial E_0$ such that $\widehat{\mathbf{R}}^w_{E_0} = \mathbf{R}^w_{E_0}$ on $\mathbb{G}\setminus F_0$. This ensures that $\Gamma * \nu_0 = w$ on $E_0\setminus F_0$, whence (from (12.31) and $E_0\setminus F_0 \subseteq E$) we derive $\lim_{(E_0\setminus F_0)\ni x\to y} \Gamma * \nu_0(x) = \infty$, i.e.

$$\lim_{(E_0\setminus F_0)\ni x\to y} \int \Gamma(x^{-1}\circ z)\,\mathrm{d}\nu_0(z) = \infty. \qquad (12.32)$$

Furthermore, it holds

$$\int \Gamma(y^{-1}\circ z)\,\mathrm{d}\nu_0(z) = \Gamma * \nu_0(y) = \widehat{\mathbf{R}}^w_{E_0}(y) \leq w(y) < \infty. \qquad (12.33)$$

As a consequence, by applying Lemma 12.4.1 we get

$$\int_{\mathbb{G}\setminus(C_{n-1}\cup C_n\cup C_{n+1})} \Gamma(x^{-1}\circ z)\,\mathrm{d}\nu_0(z) \leq \mathbf{c}\, w(y) < \infty \quad \forall\, x\in C_n,\ n\geq 2. \quad (12.34)$$

[10] Here, we use the following simple analysis lemma: *If $\sum_n a_n < \infty$ with $a_n \geq 0$, then there exists a sequence $\{b_n\}$ such that $b_n \uparrow \infty$ and $\sum_n a_n b_n < \infty$.* Indeed, setting $A_n = \sum_{j\geq n} a_j$, we have $A_n \downarrow 0$. Choose $\{n_k\}_k$ such that $A_{n_k} \leq 4^{-k}$ and define $b_n = k$ whenever $n = n_k, n_k+1, \dots, n_{k+1}-1$.

Now, (12.32) and (12.34) may agree only if (for a suitable diverging sequence $\{\beta_n\}$)

$$\int_{C_{2n}} \Gamma(x^{-1} \circ z)\, \mathrm{d}\nu_0(z) \geq \beta_n \quad \forall\, x \in C_{2n} \cap (E_0 \setminus F_0),\ n \geq 1.$$

In particular, also setting $\mu_n := (\nu_0)|_{C_{2n}}$, this gives

$$(\Gamma * \mu_n)(x) \geq 1 \qquad \forall\, x \in (E \cap C_{2n}) \setminus F_0,\ n \gg 1.$$

By the definition of $\mathcal{L}$-balayage (being $\Gamma * \mu_n \in \overline{\mathcal{S}}^+(\mathbb{G})$), this proves

$$\Gamma * \mu_n \geq \widehat{\mathbf{R}}^1_{(E\cap C_{2n})\setminus F_0},$$

i.e. thanks to Proposition 11.8.3 (page 525),

$$\Gamma * (\nu_0|_{C_{2n}}) \geq \widehat{\mathbf{R}}^1_{E\cap C_{2n}}, \qquad n \gg 1. \tag{12.35}$$

By arguing analogously with

$$E_1 := \bigcup_{k\geq k_0} E \cap C_{2k+1}, \qquad \widehat{\mathbf{R}}^w_{E_1} = \Gamma * \nu_1,$$

we have

$$\Gamma * (\nu_1|_{C_{2n+1}}) \geq \widehat{\mathbf{R}}^1_{E\cap C_{2n+1}}, \qquad n \gg 1. \tag{12.36}$$

From (12.33), (12.35) and (12.36) we get (iii). Indeed, one has

$$\begin{aligned}\sum_{n\gg 1} \widehat{\mathbf{R}}^1_{C_n}(y) &= \sum_{n \text{ even}} + \sum_{n \text{ odd}} \\ &\leq \sum_{n\gg 1} \Gamma * (\nu_0|_{C_{2n}})(y) + \sum_{n\gg 1} \Gamma * (\nu_1|_{C_{2n+1}})(y) \\ &\leq (\Gamma * \nu_0)(y) + (\Gamma * \nu_1)(y) \leq 2\, w(y) < \infty.\end{aligned}$$

(ii) $\Leftrightarrow$ (iv): First, we show the implication "$\Rightarrow$",

$$\int_\alpha^\infty \mathcal{C}^*\big(\{x \in E : \Gamma(y^{-1} \circ x) \geq t\}\big)\, \mathrm{d}t = \sum_{n=1}^\infty \int_{\alpha^n}^{\alpha^{n+1}} [\cdots]$$

$$\leq \sum_{n=1}^\infty \mathcal{C}^*\big(\{x \in E : \Gamma(y^{-1} \circ x) \geq \alpha^n\}\big) \int_{\alpha^n}^{\alpha^{n+1}} \mathrm{d}t$$

(from $E \cap \{\Gamma(y^{-1} \circ x) \geq \alpha^n\} \subseteq \bigcup_{m=n}^\infty E \cap \{\alpha^m \leq \Gamma(y^{-1} \circ x) \leq \alpha^{m+1}\}$ and the sub-additivity of the exterior $\mathcal{L}$-capacity $\mathcal{C}^*$)

$$\leq \sum_{n=1}^\infty \sum_{m=n}^\infty \mathcal{C}^*(E \cap C_m)(\alpha^{n+1} - \alpha^n) = \sum_{m=1}^\infty \mathcal{C}^*(E \cap C_m) \sum_{n=1}^m (\alpha^{n+1} - \alpha^n)$$

$$= \sum_{m=1}^\infty \mathcal{C}^*(E \cap C_m)\, (\alpha^{m+1} - \alpha) \leq \sum_{m=1}^\infty \mathcal{C}^*(E \cap C_m)\, \alpha^{m+1} < \infty.$$

Finally, we show the implication "$\Leftarrow$",

$$
\begin{aligned}
\sum_{n=2}^{\infty} C^*(E \cap C_n)\,\alpha^n &\le \sum_{n=2}^{\infty} \alpha^n\, C^*\big(\{x \in E : \Gamma(y^{-1} \circ x) \ge \alpha^n\}\big) \\
&\quad \text{(use the monotonicity of } C^* \text{ and } \alpha^n = \tfrac{\alpha}{\alpha-1} \textstyle\int_{\alpha^{n-1}}^{\alpha^n} \mathrm{d}t) \\
&\le \frac{\alpha}{\alpha-1} \sum_{n=2}^{\infty} \int_{\alpha^{n-1}}^{\alpha^n} C^*\big(\{x \in E : \Gamma(y^{-1} \circ x) \ge t\}\big)\, \mathrm{d}t \\
&= \frac{\alpha}{\alpha-1} \int_{\alpha}^{\infty} C^*\big(\{x \in E : \Gamma(y^{-1} \circ x) \ge t\}\big)\, \mathrm{d}t < \infty.
\end{aligned}
$$

This completes the proof. □

Collecting together the $\mathcal{L}$-regularity criterion of Theorem 12.3.7 and Wiener's criterion for $\mathcal{L}$ of Theorem 12.4.2, we derive the following remarkable result (again, $\alpha > 1$ is a constant as in Lemma 12.4.1):

Theorem 12.4.3 (Wiener's regularity test for $\mathcal{L}$). *Let C^* denote the exterior $\mathcal{L}$-capacity relative to $\mathbb{G}$. Let Ω be an open and connected subset of $\mathbb{G}$ and $y \in \partial\Omega$. Then the following statements are equivalent:*

(i) *y is an $\mathcal{L}$-regular point for Ω;*
(ii) *$\mathbb{G} \setminus \Omega$ is not $\mathcal{L}$-thin at y;*
(iii) *It holds $\sum_{n=1}^{\infty} \alpha^n\, C^*(C_n \setminus \Omega) = \infty$;*
(iv) *It holds $\sum_{n=1}^{\infty} \widehat{\mathbf{R}}^1_{C_n \setminus \Omega}(y) = \infty$;*
(v) *It holds $\int_{\alpha}^{\infty} C^*(\{x \notin \Omega : \Gamma(y^{-1} \circ x) \ge t\})\, \mathrm{d}t = \infty$.*

Here, C_n is as in (12.16) at the beginning of the section.

Bibliographical Notes. For the topics presented in this chapter, we were inspired by the exposition of the same subjects in [Helm69, Chapter 10] and [AG01, Chapter 7].

The Wiener test in the context of the nilpotent Lie groups was proved by H. Hueber [Hu85] and then generalized for Hörmander vector fields by P. Negrini and V. Scornazzani [NS87].

For further results on the Wiener test, see, e.g. [HH87,Ne88] and on the fine topology see, e.g. [BR67,CC72].

12.5 Exercises of Chapter 12

Ex. 1) Prove that a set E is $\mathcal{L}$-thin at $x \in \mathbb{G}$ if and only if $E \setminus \{x\}$ is $\mathcal{L}$-thin at x (prove this by the very definition of $\mathcal{L}$-thinness).

Ex. 2) Prove that any non-empty $\mathcal{L}$-fine open set contains infinitely many distinct points. (*Hint*: If u is $\mathcal{L}$-superharmonic, then $u(x) = \liminf_{y\to x} u(y)$.)

Ex. 3) Prove that the $\mathcal{L}$-fine topology is not locally compact.
(*Hint*: Let F be the $\mathcal{L}$-fine closure of an $\mathcal{L}$-fine neighborhood of a point. Then, by Exercise 1, F contains an infinite set $Z = \{x_n\}_{n\in\mathbb{N}}$. By Proposition 12.1.3, no subsequence of Z can converge.)

Ex. 4) A set is $\mathcal{L}$-finely compact if and only if it is finite.
(*Hint*: See the hint for Exercise 2.)

Ex. 5) Prove that E is $\mathcal{L}$-thin at x if and only if there exists a bounded function $v \in \underline{S}(\mathbb{G})$ and a neighborhood U of x such that $v(x) = 1$ and $v \le -1$ on $U \cap (E \setminus \{x\})$. (*Hint*: By Theorem 12.2.5, let $u \in \overline{S}^{+}(\mathbb{G})$ be such that $u(x) < \lim_{E\ni y\to x} u(y) = \infty$. Replace u by $w = \min\{u, u(x)+1\}$ and take $U := \{y : u(y) > u(x) + 1/2\}$ and $v := 1 + 4(u(x) - u)$.)

Ex. 6) **(The exterior $\mathcal{L}$-cone regularity condition).** In the following statement, we say that C is a (truncated) $\mathcal{L}$-cone if

$$\delta_\lambda(x) \in C \qquad \forall\, x \in C,\ \ \forall\, \lambda \in]0,1[,$$

where $\{\delta_\lambda\}_{\lambda>0}$ is the family of dilations of the Carnot group $\mathbb{G} = (\mathbb{R}^N, \circ, \delta_\lambda)$. As usual, $\mathcal{L}$ is a sub-Laplacian on $\mathcal{L}$.
Let $\Omega \subseteq \mathbb{R}^N$ be a bounded open set, and let $y \in \partial\Omega$. Assume Ω has at y the property of the exterior $\mathcal{L}$-cone, i.e. there exists a bounded open (truncated) $\mathcal{L}$-cone C such that

$$\mathbb{R}^N \setminus \Omega \supseteq y \circ C.$$

Then y is an $\mathcal{L}$-regular point for Ω.
(*Hint*: We follow the notation of Theorem 12.4.3. Use the results of Exercise 2 at the end of Chapter 11 (page 533) to show that

$$C^*(C_n \setminus \Omega) \ge c_0\,(1/\alpha)^n.$$

This follows from the fact that

$$C_n \setminus \Omega \supseteq \delta_{r^n}\{\xi \in C : r \le d(\xi) \le 1\}, \quad \text{with } r = \alpha^{1/(2-Q)}.$$

Thus, $\sum_n \alpha^n C^*(C_n \setminus \Omega) \ge \sum_n c_0 \alpha^n \alpha^{-n} = \infty$, and the $\mathcal{L}$-regularity of y follows from Wiener's test in Theorem 12.4.3.)

Ex. 7) In the case of the classical Laplace operator, give a proof of the exterior $\mathcal{L}$-ball regularity condition in Ex. 9 at the end of Chapter 7 by making use of Wiener's criterion and geometrical arguments (instead of Bouligand's theorem).
Do the same in the case of the Heisenberg–Weyl group on $\mathbb{R}^3$.

Ex. 8) Let $\Omega \subseteq \mathbb{R}^N$ be a bounded open set, and let $y \in \partial\Omega$. Assume there exists $R > 0$ and $\gamma \in]0,1[$ such that

$$|B_d(y,r) \setminus \Omega| \ge \gamma\, r^Q \qquad \forall\, r \in]0,R[.$$

Then y is an $\mathcal{L}$-regular point for Ω. Here $|\cdot|$ stands for the Lebesgue measure.

Ex. 9) Let $\Omega \subseteq \mathbb{R}^N$ be a bounded open set, and let $y \in \partial\Omega$. For every $r > 0$, let us set

$$V_r := \widehat{\mathbf{R}}^1_{\Omega'_r(y)}, \quad \Omega'_r(y) = B_d(y,r) \setminus \Omega.$$

Then y is not an $\mathcal{L}$-regular point for Ω if and only if $V_r(y) \to 0$ as $r \to 0$.

Ex. 10) Let $A,\ B$ be fixed constants with $0 < B < A < 1/2$. Prove that

$$\begin{array}{c} A \leq \lambda \leq 1/2 \\ \lambda - B \leq \mu \leq \lambda + B \end{array} \quad \Longrightarrow \quad \frac{\lambda}{\mu} \leq \frac{A}{A-B}.$$

This proves (12.27) in the proof of Lemma 12.4.1.

Ex. 11) Prove that if A is $\mathcal{L}$-thin at x then, for every $B \subseteq A$, B is $\mathcal{L}$-thin at x. Vice versa, if A is not $\mathcal{L}$-thin at x then, for every $B \supseteq A$, B is not $\mathcal{L}$-thin at x. Derive the following facts:
Let Ω, $\widetilde{\Omega}$ be open, connected and bounded subsets of $\mathbb{G}$. Suppose also that $x_0 \in \partial\Omega \cap \partial\widetilde{\Omega}$. If x_0 is $\mathcal{L}$-regular for Ω and $\widetilde{\Omega} \subseteq \Omega$, then x_0 is $\mathcal{L}$-regular for $\widetilde{\Omega}$. Vice versa, if x_0 is $\mathcal{L}$-irregular for Ω and $\widetilde{\Omega} \supseteq \Omega$, then x_0 is $\mathcal{L}$-irregular for $\widetilde{\Omega}$.
Derive this both from Theorem 12.3.7 and from Bouligand's Theorem 6.10.4 (page 371).

Ex. 12) Let $0 < r_1 < r_2 < \infty$. Let $x_0 \in \mathbb{G}$ be fixed. Set $B_i := B_d(x_0, r_i)$ (for $i = 1, 2$) and $C := B_2 \setminus \overline{B_1}$. Prove that the annulus C is an $\mathcal{L}$-regular set by constructing explicit $\mathcal{L}$-barrier functions (see Definition 6.10.3, page 371). (*Hint*: $w_1 := r^{2-Q} - d^{2-Q}(x_0^{-1} \circ \cdot)$ is an $\mathcal{L}$-barrier function for the points in ∂B_1, whereas $w_2 := -r^{2-Q} + d^{2-Q}(x_0^{-1} \circ \cdot)$ is an $\mathcal{L}$-barrier function for the points in ∂B_2.)
Note that, by Theorem 12.3.7, the $\mathcal{L}$-regularity of a point $y \in \partial C$ implies that (actually, is equivalent to) $\mathbb{G} \setminus C = \overline{B_1} \cup (\mathbb{G} \setminus B_2)$ is not $\mathcal{L}$-thin at y. In turn, this is equivalent to

$$y \in \mathrm{Der}_{\mathcal{L}\text{-fine}}\big(\overline{B_1} \cup (\mathbb{G} \setminus B_2)\big).$$

Since $\mathrm{Der}(A \cup B) = \mathrm{Der}(A) \cup \mathrm{Der}(B)$ in any topological space (prove this), then the above argument shows that $\partial C \subseteq \mathrm{Der}_{\mathcal{L}\text{-fine}}(\overline{B_1}) \cup \mathrm{Der}_{\mathcal{L}\text{-fine}}(\mathbb{G} \setminus B_2)$. Recalling that $\mathrm{Der}_{\mathcal{L}\text{-fine}}(A) \subseteq \mathrm{Der}_{\mathrm{Eucl.}}(A)$, this easily proves that $\mathrm{Der}_{\mathcal{L}\text{-fine}}(\overline{B_1}) = \overline{B_1}$, $\mathrm{Der}_{\mathcal{L}\text{-fine}}(\mathbb{G} \setminus B_2) = \mathbb{G} \setminus B_2$. Is it true that we also have $\mathrm{Der}_{\mathcal{L}\text{-fine}}(B_1) = \overline{B_1}$ and $\mathrm{Der}_{\mathcal{L}\text{-fine}}(\mathbb{G} \setminus \overline{B_2}) = \mathbb{G} \setminus B_2$?

13

d-Hausdorff Measure and $\mathcal{L}$-capacity

The aim of this chapter is to provide some estimate of the "smallness" of $\mathcal{L}$-polar sets in terms of the Hausdorff measure naturally related to the quasi-distance defined by an $\mathcal{L}$-gauge d on $\mathbb{G}$.

To this end, we first need the relevant definitions of d-Hausdorff measure and d-Hausdorff dimension, besides several preliminary results. One of the main results is contained in Theorem 13.2.5.

Finally, in Section 13.3, we recall some results contained in the remarkable paper [BRSC03], in order to emphasize the new phenomena which can occur when the Hausdorff dimension is considered in the sub-elliptic context of Carnot groups (for instance, the Heisenberg–Weyl group $\mathbb{H}^1$).

As usual, throughout the section $\mathbb{G} = (\mathbb{R}^N, \circ, \delta_\lambda)$ is a homogeneous Carnot group and $\mathcal{L}$ is a sub-Laplacian on $\mathbb{G}$. Moreover, Γ is the fundamental solution for $\mathcal{L}$ and d is any $\mathcal{L}$-gauge function (i.e. a positive constant times $\Gamma^{1/(2-Q)}$). Q denotes, as usual, the homogeneous dimension of $\mathbb{G}$. We recall that d is a quasi-distance on $\mathbb{G}$ satisfying the pseudo-triangle inequality

$$d(x^{-1} \circ y) \le \mathbf{c}\,(d(x) + d(y)) \quad \text{for all } x,\ y \in \mathbb{G}. \tag{13.1}$$

The constant $\mathbf{c}\geq 1$ will always denote that appearing in (13.1). As usual, the d-balls related to d are denoted by $B_d(x, r)$.

13.1 d-Hausdorff Measure and Dimension

Definition 13.1.1 (The d-Hausdorff measure). *An increasing function $\phi : (0, \infty) \to (0, \infty]$ vanishing as $t \to 0+$ will be referred to as* measure function. *Given any set $E \subseteq \mathbb{G}$ and any $\rho > 0$, we set*

$$M_\phi^{(\rho)}(E) := \inf\Big\{ \sum_k \phi(r_k) \ :\ E \subseteq \bigcup_k B_d(x_k, r_k),\ r_k < \rho\ \forall\, k \in \mathbb{N} \Big\}, \tag{13.2}$$

*where the infimum is taken over all possible coverings E by a (finite or) countable family of d-balls $\{B_d(x_k, r_k)\}_{k\in\mathbb{N}}$ such that the rays r_k are strictly less than ρ (any such covering will be referred to as ρ-*covering of E).

Since the map $]0,\infty] \ni \rho \mapsto M_\phi^{(\rho)}$ is decreasing, it is well posed,

$$m_\phi(E) := \lim_{\rho\to 0^+} M_\phi^{(\rho)}(E) = \sup_{\rho>0} M_\phi^{(\rho)}(E). \tag{13.3}$$

In the sequel, $m_\phi(E)$ will be referred to as the d-Hausdorff ϕ-measure. When $\phi(t) = t^\alpha$ (for $\alpha > 0$), we simply write $M_{(\alpha)}^{(\rho)}$ instead of $M_{t^\alpha}^{(\rho)}$ and $m_{(\alpha)}$ instead of m_{t^α}. In the sequel, $m_{(\alpha)}(E)$ will be referred to as the d-Hausdorff (α)-measure of the set E.

Remark 13.1.2. With the above notation, it can be proved that $M_\phi^{(\rho)}$ and m_ϕ are outer measures on $\mathbb{G}$. Properly restricted to their related measurable sets, they are Borel regular measures. We leave the proof as an exercise.

We begin with some lemmas.

Lemma 13.1.3 (The d-Hausdorff dimension). *For every bounded set $E \subset \mathbb{G}$, there exists a number $\alpha(E) \in [0, Q]$ (where Q is the homogeneous dimension of $\mathbb{G}$) such that*

$$m_{(\alpha)}(E) = \infty \text{ for all } \alpha < \alpha(E), \qquad m_{(\alpha)}(E) = 0 \text{ for all } \alpha > \alpha(E). \tag{13.4}$$

We thus have

$$\alpha(E) = \inf\{\alpha > 0 : m_{(\alpha)}(E) = 0\}. \tag{13.5}$$

We call $\alpha(E)$ the d-Hausdorff dimension of E.

Proof. Let $\alpha > Q$ be fixed and take any $\varepsilon \in]0,1[$. Let Ω be any bounded open set containing E. Then there exists a countable collection of disjoint d-balls $B_k := B_d(x_k, \frac{r_k}{4})$ (with $\frac{r_k}{2} < \varepsilon$ for every k) and a set N such that $\Omega = N \cup \bigcup_k B_k$ and such that $\bigcup_k B_d(x_k, \mathbf{c}\, r_k) \supseteq \Omega$. This gives (for a suitable structural constant c_Q)

$$\begin{aligned} M_{(\alpha)}^{(\varepsilon)}(E) &\le \sum_k (\mathbf{c}\, r_k)^\alpha \le \varepsilon^{\alpha-Q} \sum_k (\mathbf{c}\, r_k)^Q = (4\mathbf{c})^Q \varepsilon^{\alpha-Q} \sum_k \left(\frac{r_k}{4}\right)^Q \\ &= c_Q \varepsilon^{\alpha-Q} \sum_k \mathrm{meas}(B_k) \le c_Q \varepsilon^{\alpha-Q}\, \mathrm{meas}(\Omega) \to 0 \quad \text{as } \varepsilon \to 0^+. \end{aligned}$$

This proves that $m_{(\alpha)}(E) = 0$ for every $\alpha > Q$. If we define $\alpha(E)$ as in (13.5), we thus get $\alpha(E) \in [0, Q]$.

Let now $\alpha > \alpha(E)$ and $\varepsilon > 0$. Hence, there exist a suitable $\beta \in]\alpha(E), \alpha[$ and an ε-covering $\{B_d(x_k, r_k)\}_k$ of E such that $\sum_k r_k^\beta < 1$. Consequently,

$$M_{(\alpha)}^{(\varepsilon)}(E) \le \sum_k r_k^\alpha = \sum_k r_k^{\alpha-\beta}\, r_k^\beta \le \varepsilon^{\alpha-\beta} \to 0 \quad \text{as } \varepsilon \to 0^+.$$

This proves the second part of (13.4). Let now $\alpha \in]0, \alpha(E)[$. By the definition of $\alpha(E)$, there exists $\beta \in]\alpha, \alpha(E)[$ with $m_{(\beta)}(E) > 0$. If $\{B_d(x_k, r_k)\}_k$ is an ε-covering of E, then $\sum_k r_k^\alpha = \sum_k r_k^{\alpha-\beta} r_k^\beta \geq \varepsilon^{\alpha-\beta} M_{(\beta)}^{(\varepsilon)}(E)$, whence

$$M_{(\alpha)}^{(\varepsilon)}(E) \geq \varepsilon^{\alpha-\beta} M_{(\beta)}^{(\varepsilon)}(E) \to \infty \quad \text{as } \varepsilon \to 0^+.$$

This proves the first part of (13.4), completing the proof of the lemma. □

Example 13.1.4. The d-Hausdorff dimension of $\mathbb{G}$ is Q. Indeed, there exists a positive constant c such that

$$m_{(Q)}(E) = c\, H^N(E) \qquad \text{for every } E \subseteq \mathbb{G},$$

where H^N stands for the usual (Euclidean) Hausdorff N-dimensional measure on $\mathbb{R}^N \equiv \mathbb{G}$. The proof is left as an exercise.

Remark 13.1.5. It is easy to see that $m_{(\alpha)}$ is invariant under left-translation and homogeneous of degree α with respect to the dilation of $\mathbb{G}$. In other words, for every $x \in \mathbb{G}$ and every $\lambda > 0$, we have

$$m_{(\alpha)}(x \circ E) = m_{(\alpha)}(E), \qquad m_{(\alpha)}(\delta_\lambda(E)) = \lambda^\alpha\, m_{(\alpha)}(E),$$

where E is any subset of $\mathbb{G}$. The proof is left as an exercise.

We next aim to prove Theorem 13.1.7 below. In order to do this, we need a lemma having an interest in its own.

Lemma 13.1.6. *Suppose $K \Subset \mathbb{G}$ is not an $\mathcal{L}$-polar set. Then there exists a Radon measure μ supported in K with $\mu(K) > 0$ such that $\Gamma * \mu \in C(\mathbb{G}, \mathbb{R})$.*

Proof. By Proposition 11.5.11 (page 507), since K is not $\mathcal{L}$-polar, we have

$$\mathcal{C}(K) = \mu_K(K) > 0.$$

Since $\Gamma * \mu_K \leq 1$, by Lemma 11.3.3 (page 495) there exists $C \Subset K$ such that

$$\mu_K(K \setminus C) < \mu_K(K)/2 \quad \text{and} \quad \Gamma * (\mu_K|_C) \in C(\mathbb{G}, \mathbb{R}).$$

We claim that $\mu := (\mu_K)|_C$ fulfills the assertion of the lemma. Indeed,

$$\mu_K(C) = \mu_K(K) - \mu_K(K \setminus C) > \mu_K(K)/2 > 0.$$

This ends the proof. □

We are now ready to give the proof of the following result.

Theorem 13.1.7 (d-Hausdorff measure and $\mathcal{L}$-polarity. I). *Let $E \subset \mathbb{G}$ be a bounded $\mathcal{L}$-capacitable set satisfying $m_{(Q-2)}(E) < \infty$ (as usual, Q is the homogeneous dimension of $\mathbb{G}$). Then E is $\mathcal{L}$-polar.*

For a partial converse of Theorem 13.1.7, see also Theorem 13.2.5.

Proof. We argue by contradiction. Let E be a non-$\mathcal{L}$-polar capacitable set, i.e. $\mathcal{C}(E) > 0$. This also gives $\mathcal{C}_*(E) > 0$, i.e. there exists $K \Subset E$ with $\mathcal{C}(K) > 0$. By the sub-additivity of the $\mathcal{L}$-capacity, there exists an open d-ball B_1 of radius 1 such that $\mathcal{C}(K \cap B_1) > 0$. Then, it is not restrictive to suppose that K is contained in such a d-ball B_1. From $\mathcal{C}(K) > 0$ and Lemma 13.1.6 we derive the existence of a Radon measure μ supported in K such that $\mu(K) > 0$ and $u := \Gamma * \mu \in C(\mathbb{G}, \mathbb{R})$. For any $0 < \rho < 1$ and $x \in \mathbb{G}$, we set

$$u_\rho(x) := \int_{\mathbb{G}} \Gamma(y^{-1} \circ x) \max\left\{0, 1 - \frac{d(y^{-1} \circ x)}{2\rho}\right\} \mathrm{d}\mu(y).$$

For any fixed $x_0 \in \mathbb{G}$, we have

$$\begin{aligned}
u_\rho(x_0) &\le \liminf_{x\to x_0} u_\rho(x) \le \limsup_{x\to x_0} u_\rho(x) \\
&\le \limsup_{x\to x_0} u(x) + \limsup_{x\to x_0}(u_\rho(x) - u(x)) \\
&= u(x_0) - \liminf_{x\to x_0}(u(x) - u_\rho(x)) \\
&= u(x_0) - \liminf_{x\to x_0} \int_{\mathbb{G}} \Gamma(y^{-1} \circ x)\left(1 - \max\left\{0, 1 - \frac{d(y^{-1} \circ x)}{2\rho}\right\}\right) \mathrm{d}\mu(y) \\
&\le u(x_0) - \int_{\mathbb{G}} \liminf_{x\to x_0}[\cdots] \\
&= u(x_0) - \int_{\mathbb{G}} \Gamma(y^{-1} \circ x_0)\left(1 - \max\left\{0, 1 - \frac{d(y^{-1} \circ x_0)}{2\rho}\right\}\right) \mathrm{d}\mu(y) \\
&= u_\rho(x_0).
\end{aligned}$$

In the first and the last inequalities, we used Fatou's lemma; in the first equality, we used the continuity of u. This proves that u_ρ is continuous on $\mathbb{G}$.

Being $\Gamma * \mu$ finite-valued, we have $\mu(\{x\}) = 0$ for every $x \in \mathbb{G}$. As a consequence, it holds (by dominated convergence)

$$\begin{aligned}
\lim_{\rho\to 0^+} u_\rho(x) &= \lim_{\rho\to 0^+} \int_{\mathbb{G}\setminus\{x\}} [\cdots]\,\mathrm{d}\mu \\
&= \int_{\mathbb{G}\setminus\{x\}} \Gamma(y^{-1} \circ x) \lim_{\rho\to 0^+} \max\left\{0, 1 - \frac{d(y^{-1} \circ x)}{2\rho}\right\} \mathrm{d}\mu(y) = 0.
\end{aligned}$$

The convergence is also monotone, for $u_{\rho'}(x) \le u_\rho(x)$ whenever $0 < \rho' < \rho$. By Dini's theorem, the convergence is uniform on compact sets. Consequently, there exists a sequence $\rho_n \downarrow 0$ such that $u_{\rho_n} < 2^{-n}$ on $\overline{B_1}$ for every $n \in \mathbb{N}$. Let us introduce the function $\phi : [0, \infty[\to \mathbb{R}$,

$$\phi(t) := \frac{1}{2} t^{2-Q} \sum_{n=1}^{\infty} \chi_{[0,\rho_n]}(t).$$

Since $\chi_{[0,\rho_n]}(d(y^{-1}\circ x)) = \chi_{\overline{B_d(x,\rho_n)}}(y)$ (and by monotone convergence), we have

$$\begin{aligned}
\int_K \phi(d(y^{-1}\circ x))\,\mathrm{d}\mu(y) &= \sum_{n=1}^{\infty}\frac{1}{2}\int_{\mathbb{G}} \Gamma(y^{-1}\circ x)\,\chi_{\overline{B_d(x,\rho_n)}}(y)\,\mathrm{d}\mu(y)\\
&\le \sum_{n=1}^{\infty}\frac{1}{2}\int_{\mathbb{G}} \Gamma(y^{-1}\circ x)\,\max\left\{0, 2-\frac{d(y^{-1}\circ x)}{\rho_n}\right\}\mathrm{d}\mu(y)\\
&= \sum_{n=1}^{\infty} u_{\rho_n}(x).
\end{aligned}$$

Hence,

$$\int_K \phi(d(y^{-1}\circ x))\,\mathrm{d}\mu(y) \le 1 \qquad \text{for every } x\in\overline{B_1}.$$

Let now n be large enough so that

$$\rho_n < \sup\{d(a^{-1}\circ b) : a\in K,\ b\in \mathbb{G}\setminus B_1\}. \tag{13.6}$$

Let us consider a ρ_n-covering of K, say $\{B_d(x_k,r_k)\}_k$, consisting of d-balls all intersecting K. By (13.6), it is easily seen that $x_k\in B_1$ for every $k\in\mathbb{N}$. This gives

$$\begin{aligned}
1 &\ge \int_K \phi(d(y^{-1}\circ x_k))\,\mathrm{d}\mu(y) \ge \int_{B_d(x_k,r_k)} [\cdots]\\
(\phi \text{ is decreasing}) &\ge \int_{B_d(x_k,r_k)} \phi(r_k)\,\mathrm{d}\mu(y) = \phi(r_k)\,\mu(B_d(x_k,r_k))\\
&\ge \frac{n}{2}(r_k)^{2-Q}\mu(B_d(x_k,r_k)).
\end{aligned}\tag{13.7}$$

The last inequality follows from $r_k<\rho_n$ and the very definition of ϕ. From (13.7) we derive the second inequality in the following chain of inequalities:

$$\mu(K)\le \sum_k \mu(B_d(x_k,r_k)) \le \frac{2}{n}\sum_k (r_k)^{Q-2}.$$

This gives (from $E\supseteq K$ and again $r_k<\rho_n$)

$$M^{(\rho_n)}_{(Q-2)}(E) \ge M^{(\rho_n)}_{(Q-2)}(K) \ge \sum_k (r_k)^{Q-2} \ge \frac{n}{2}\,\mu(K)\to\infty \quad \text{as } n\to\infty.$$

This gives $m_{(Q-2)}(E)=\infty$, contradicting the assertion of the theorem. □

13.2 d-Hausdorff Measure and $\mathcal{L}$-capacity

The aim of this section is to prove Theorem 13.2.5 (page 568). In order to do it, we need some lemmas. The first one is a simple result from the real analysis.

Lemma 13.2.1. *Let μ be a Radon measure on $\mathbb{R}^N$. Let $g : (0, R] \to [0, \infty)$ be a continuous decreasing function. Finally, let $f : \mathbb{R}^N \to [0, \infty)$ be a μ-measurable function. Set*

$$n(t) := \int_{f(y)<t} \mathrm{d}\mu(y). \tag{13.8}$$

Suppose $n(t)$ is finite-valued on $(0, \infty)$. Then g is integrable in the generalized Riemann–Stieltjes sense on $(0, R]$ with respect to $n(t)$ if and only if $g \circ f$ is μ-integrable on the set $\{y \in \mathbb{R}^N : 0 < f(y) < R\}$. Moreover,

$$\int_{0<f(y)<R} g(f(y))\,\mathrm{d}\mu(y) = \int_0^R g(t)\,\mathrm{d}n(t) \tag{13.9}$$

whenever one of the integrals is finite.

Proof. Since $t \mapsto n(t)$ is decreasing, it is of bounded variation on every compact sub-interval of $(0, R]$. Hence $g(t)$ (which is continuous) is $\mathrm{d}n(t)$-integrable on every compact sub-interval of $(0, R]$.

Suppose first that g is integrable in the generalized Riemann–Stieltjes sense on $(0, R]$ with respect to $n(t)$. If $\varepsilon > 0$, there exists $\delta = \delta(\varepsilon) > 0$ such that

$$\left| \int_0^R g(t)\,\mathrm{d}n(t) - \int_\delta^R g(t)\,\mathrm{d}n(t) \right| < \frac{\varepsilon}{2}.$$

We can assume $\delta(\varepsilon) < \varepsilon$. Moreover, there exists a partition of $[\delta, R]$, say $\delta = t_0 < t_1 < \cdots < t_p = R$, such that

$$\left| \int_\delta^R g(t)\,\mathrm{d}n(t) - \sum_{k=0}^{p-1} g(t_k)\,(n(t_{k+1}) - n(t_k)) \right| < \frac{\varepsilon}{2}. \tag{13.10}$$

We remark that $n(t_{k+1}) - n(t_k) = \int_{t_k \le f(y) < t_{k+1}} \mathrm{d}\mu(y)$. Moreover, from $t_k \le f(y) < t_{k+1}$ it follows $g(t_k) \ge g(f(y)) > g(t_{k+1})$, whence

$$\begin{aligned} \sum_{k=0}^{p-1} g(t_k) \int_{t_k \le f(y) < t_{k+1}} \mathrm{d}\mu(y) &\ge \sum_{k=0}^{p-1} \int_{t_k \le f(y) < t_{k+1}} g(f(y))\,\mathrm{d}\mu(y) \\ &= \int_{\delta(\varepsilon) \le f(y) < R} g(f(y))\,\mathrm{d}\mu(y). \end{aligned}$$

As a consequence, it holds

$$\begin{aligned} \int_0^R g(t)\,\mathrm{d}n(t) &> \int_{\delta(\varepsilon)}^R g(t)\,\mathrm{d}n(t) - \frac{\varepsilon}{2} \ge \sum_{k=0}^{p-1} g(t_k)\,(n(t_{k+1}) - n(t_k)) - \varepsilon \\ &\ge \int_{\delta(\varepsilon) \le f(y) < R} g(f(y))\,\mathrm{d}\mu(y) - \varepsilon. \end{aligned}$$

Making use of (13.10), we have

$$\left| \int_{\delta(\varepsilon) \le f(y) < R} g(f(y))\, \mathrm{d}\mu(y) - \int_0^R g(t)\, \mathrm{d}n(t) \right| < \varepsilon. \tag{13.11}$$

Being $g \ge 0$ and $\delta(\varepsilon) < \varepsilon$, from Beppo Levi's theorem we have

$$\int_{\delta(\varepsilon) \le f(y) < R} g(f(y))\, \mathrm{d}\mu(y) \nearrow \int_{0 < f(y) < R} g(f(y))\, \mathrm{d}\mu(y) \qquad \text{if } \varepsilon \to 0^+.$$

Hence (13.11) proves the assertion.

Let us suppose that $g(f(y))$ is μ-integrable on $\{y \in \mathbb{R}^N : 0 < f(y) < R\}$. Let $\delta \in]0, R[$ be fixed and $\varepsilon > 0$. Since g is continuous on $[\delta, R]$, there exists $\sigma = \sigma(\delta, \varepsilon)$ such that

$$|g(x') - g(x'')| < \varepsilon / n(R)$$

for every $x', x'' \in [\delta, R]$ with $|x' - x''| < \sigma$. Let us now fix any arbitrary partition $\delta = t_0 < t_1 < \cdots < t_p = R$ of $[\delta, R]$ such that $t_{k+1} - t_k < \sigma$, and for $k = 0, \ldots, p-1$, let us choose $\tau_k \in [t_k, t_{k+1}]$. We have

$$\begin{aligned}
&\sum_{k=0}^{p-1} g(\tau_k)(n(t_{k+1}) - n(t_k)) \\
&\quad \ge -\varepsilon/n(R) \cdot \sum_{k=0}^{p-1} (n(t_{k+1}) - n(t_k)) + \sum_{k=0}^{p-1} g(t_k)\,(n(t_{k+1}) - n(t_k)) \\
&\quad \ge -\varepsilon + \int_{\delta \le f(y) < R} g(f(y))\, \mathrm{d}\mu(y).
\end{aligned}$$

Analogously,

$$\sum_{k=0}^{p-1} g(\tau_k)(n(t_{k+1}) - n(t_k)) \le \varepsilon + \int_{\delta \le f(y) < R} g(f(y))\, \mathrm{d}\mu(y).$$

This proves that

$$\left| \int_{\delta \le f(y) < R} g(f(y))\, \mathrm{d}\mu(y) - \sum_{k=0}^{p-1} g(\tau_k)\,(n(t_{k+1}) - n(t_k)) \right| < \varepsilon$$

for every partition $\{t_k\}_k$ of $[\delta, R]$ with $t_{k+1} - t_k < \sigma$ and any $\tau_k \in [t_k, t_{k+1}]$. In other words, $g(t)$ is $\mathrm{d}n(t)$-integrable on $[\delta, R]$ and

$$\int_\delta^R g(t)\, \mathrm{d}n(t) = \int_{\delta \le f(y) < R} g(f(y))\, \mathrm{d}\mu(y).$$

Being δ arbitrary and taking into account the hypothesis

$$\lim_{\delta \to 0^+} \int_{\delta \le f(y) < R} g(f(y))\, \mathrm{d}\mu(y) = \int_{0 < f(y) < R} g(f(y))\, \mathrm{d}\mu(y),$$

we derive that g is integrable in the generalized sense of Riemann–Stieltjes on $(0, R]$ w.r.t. $n(t)$ and (13.9) holds. □

If we take $f(y) := d(y^{-1} \circ z)$ and $g(t) := t^{2-Q}$ in (13.9), we obtain

$$\int_{0<d(y^{-1}\circ z)<1} d^{2-Q}(y^{-1} \circ z)\,\mathrm{d}\mu(y) = \int_0^1 t^{2-Q}\,\mathrm{d}m_z(t), \tag{13.12}$$

where $m_z(t) = \mu(B_d(z,t))$.

Lemma 13.2.2. *Let μ be any compactly supported Radon measure on $\mathbb{G}$. Let also ϕ be a strictly positive measure function such that*

$$I(\phi, Q) := \int_0^1 \frac{\phi(t)}{t^{Q-1}}\,\mathrm{d}t < \infty. \tag{13.13}$$

Then there exists a constant $C > 0$ (precisely, it holds $C = (Q-2)\,I(\phi, Q)$) such that

$$(\Gamma * \mu)(z) \le \mu(\mathbb{G}) + C \sup_{0<r<1} \frac{\mu(B_d(z,r))}{\phi(r)} \quad \forall\, z \in \mathbb{G}. \tag{13.14}$$

Proof. We split the proof in three steps.

(I): *$z \in \mathbb{G}$ is such that $(\Gamma * \mu)(z) < \infty$*. This case ensures that $\mu(\{z\}) = 0$. Let us set

$$m_z(r) := \mu(B_d(z,r)), \quad r > 0.$$

First of all, we claim that

$$\lim_{\varepsilon \to 0^+} \varepsilon^{2-Q}\, m_z(\varepsilon) = 0. \tag{13.15}$$

Indeed, from $\mu(\{z\}) = 0$ we have $m_z(0^+) = 0$, so that

$$\varepsilon^{2-Q}\, m_z(\varepsilon) = \varepsilon^{2-Q}\,(m_z(\varepsilon) - m_z(0^+)) = \int_0^{\varepsilon} \varepsilon^{2-Q}\,\mathrm{d}m_z(t) \le \int_0^{\varepsilon} t^{2-Q}\,\mathrm{d}m_z(t),$$

where the far right-hand side vanishes as $\varepsilon \to 0$, for $\int_0^1 t^{2-Q}\,\mathrm{d}m_z(t) < \infty$. Indeed, making use of (13.12), we have

$$\int_0^1 t^{2-Q}\,\mathrm{d}m_z(t) = \int_{0<d(y^{-1}\circ z)<1} d^{2-Q}(y^{-1} \circ z)\,\mathrm{d}\mu(y) \le (\Gamma * \mu)(z),$$

where the far right-hand side is finite, by our assumption. Then

$$\begin{aligned}
(\Gamma * \mu)(z) &= \left(\int_{\{y=0\}} + \int_{0<d(y^{-1}\circ z)<1} + \int_{d(y^{-1}\circ z)\ge 1} \right) d^{2-Q}(y^{-1} \circ z)\,\mathrm{d}\mu(y) \\
&\le 0 + \int_{0<d(y^{-1}\circ z)<1} d^{2-Q}(y^{-1} \circ z)\,\mathrm{d}\mu(y) + \mu(\mathbb{G} \setminus B_d(z,1)) \\
\text{(see (13.12))} &= \int_0^1 t^{2-Q}\,\mathrm{d}m_z(t) + \mu(\mathbb{G} \setminus B_d(z,1))
\end{aligned}$$

$$\text{(by parts)} = \lim_{\varepsilon\to 0^+}\left(m_z(1) - \varepsilon^{2-Q}\, m_z(\varepsilon) - \int_\varepsilon^1 (2-Q)t^{1-Q}\, m_z(t)\,\mathrm{d}t \right.$$
$$\left. + \mu(\mathbb{G}\setminus B_d(z,1))\right)$$
$$\text{(see (13.15))} = m_z(1) + (Q-2)\int_0^1 t^{1-Q}\, m_z(t)\,\mathrm{d}t + \mu(\mathbb{G}\setminus B_d(z,1))$$
$$= \mu(\mathbb{G}) + (Q-2)\int_0^1 t^{1-Q}\,\frac{\mu(B_d(z,t))}{\phi(t)}\,\phi(t)\,\mathrm{d}t$$
$$\le \mu(\mathbb{G}) + (Q-2)\, I(\phi,Q)\sup_{0<t<1}\frac{\mu(B_d(z,t))}{\phi(t)},$$

where $I(\phi, Q)$ is as in (13.13). This proves (13.14).

(II): *$z \in \mathbb{G}$ is such that $(\Gamma * \mu)(z) = \infty$ and $\mu(\{z\}) = 0$.* Let $R \gg 1$ be such that $\operatorname{supp}(\mu) \subset B_d(0, R)$. We have (also using $\mu(\{z\}) = 0$)

$$\infty = (\Gamma * \mu)(z) = \int_{0<d(y^{-1}\circ z)<R} d^{2-Q}(y^{-1}\circ z)\,\mathrm{d}\mu(y) = \int_0^R t^{2-Q}\,\mathrm{d}m_z(t).$$

Since μ is a Radon measure, this also ensures that

$$\int_0^1 t^{2-Q}\,\mathrm{d}m_z(t) = \infty. \tag{13.16}$$

(13.14) is proved in case (II), if we show that

$$\sup_{0<r<1}\frac{\mu(B_d(z,r))}{\phi(r)} = \infty.$$

Suppose by contradiction that there exists $M > 0$ such that

$$m_z(r) = \mu(B_d(z,r)) \le M\,\phi(r) \quad \text{for all } r \in\,]0,1[. \tag{13.17}$$

This gives (see also (13.16) and (13.14))

$$\infty = \int_0^1 t^{2-Q}\,\mathrm{d}m_z(t) = \lim_{\varepsilon\to 0^+}\int_\varepsilon^1 t^{2-Q}\,\mathrm{d}m_z(t)$$
$$= \lim_{\varepsilon\to 0^+}\left(m_z(1) - \varepsilon^{2-Q}\, m_z(\varepsilon) + (Q-2)\int_\varepsilon^1 t^{1-Q}\, m_z(t)\,\mathrm{d}t\right)$$
$$\text{(see (13.17))} \le m_z(1) + (Q-2)\,M\int_0^1 t^{1-Q}\,\phi(t)\,\mathrm{d}t < \infty,$$

a contradiction.

(III): *$z \in \mathbb{G}$ is such that $(\Gamma * \mu)(z) = \infty$ and $\mu(\{z\}) > 0$.* From $\phi(0^+) = 0$ we derive

$$\sup_{0<r<1}\frac{\mu(B_d(z,r))}{\phi(r)} \ge \frac{\mu(B_d(z,1/n))}{\phi(1/n)} \ge \frac{\mu(\{z\})}{\phi(1/n)} \longrightarrow \infty$$

as $n \to \infty$, so that both sides of (13.14) equal ∞. □

We have the following general covering lemma.

Lemma 13.2.3 (Covering). *Let $\mathcal{D} = \{B_d(x_\alpha, r_\alpha)\}_{\alpha\in\mathcal{I}}$ be any collection of d-balls with $\sup_{\alpha\in\mathcal{I}} r_\alpha < \infty$. Let $E \subset \mathbb{G}$ be a bounded set covered by $\mathcal{D}$. Then there exists a disjoint at most countable sub-collection $\{B_d(x_{\alpha_k}, r_{\alpha_k})\}_{k\in\mathbb{N}}$ of $\mathcal{D}$ such that*

$$E \subseteq \bigcup_{k\in\mathbb{N}} B_d(x_{\alpha_k}, 3\mathbf{c}^2 r_{\alpha_k}).$$

Here, $\mathbf{c}$ *is the constant appearing in the pseudo-triangle inequality* (13.1).

Proof. We ignore the d-balls non-intersecting E, and we still denote by $\mathcal{I}$ the resulting sub-family of indices. Let also set $B_\alpha := B_d(x_\alpha, r_\alpha)$. Let $\alpha_1 \in \mathcal{I}$ be such that $r_{\alpha_1} \geq \frac{1}{2} \sup_{\alpha\in\mathcal{I}} r_\alpha$. Let $\alpha_2 \in \mathcal{I}$ (if it exists, otherwise the process ends) be such that

$$B_{\alpha_2} \cap B_{\alpha_1} = \emptyset, \quad r_{\alpha_2} \geq \frac{1}{2} \sup\left\{r_\alpha : B_\alpha \cap B_{\alpha_1} = \emptyset\right\}.$$

Proceeding by induction, chosen $\alpha_1, \ldots, \alpha_k$, we let $\alpha_{k+1} \in \mathcal{I}$ (if it exists, otherwise the process ends) be such that

$$\begin{aligned} &B_{\alpha_{k+1}} \cap B_{\alpha_j} = \emptyset \quad \text{for every } j = 1, \ldots, k, \\ &r_{\alpha_{k+1}} \geq \frac{1}{2} \sup\{r_\alpha : B_\alpha \cap B_{\alpha_j} = \emptyset \text{ for every } j = 1, \ldots, k\}. \end{aligned} \tag{13.18}$$

Set $\mathcal{A} := \{\alpha_k : k \in \mathbb{N}\}$. We split the proof in two steps.

(I): *The selection process* (13.18) *is finite.* Let $\alpha' \notin \mathcal{A}$. Since the selection process is finite, we have $B_{\alpha'} \cap B_{\alpha_j} \neq \emptyset$ for some j. Consequently, a simple application of the pseudo-triangle inequality (13.1) proves that $B_{\alpha'} \subseteq B_d(x_{\alpha_j}, 3\,\mathbf{c}^2\, r_{\alpha_j})$, and the proof is complete.

(II): *The selection process* (13.18) *is infinite.* Since $E \cap B_\alpha \neq \emptyset$ for every $\alpha \in \mathcal{I}$, the B_{α_k}'s are mutually disjoint and E is bounded, we claim that in this case

$$r_{\alpha_k} \longrightarrow 0 \quad \text{as } k \to \infty. \tag{13.19}$$

Suppose by contradiction that (13.19) is false. This means that there exist $\varepsilon_0 > 0$ and a sequence $\{k_j\}_j$ in $\mathbb{N}$ such that $r_{\alpha_{k_j}} > \varepsilon_0$ for every $j \in \mathbb{N}$. Consequently, we have

$$\infty = \sum_{j\in\mathbb{N}} \text{meas}(B_d(x_{\alpha_{k_j}}, \varepsilon_0)) \leq \sum_{j\in\mathbb{N}} \text{meas}(B_{\alpha_{k_j}}) = \text{meas}\Bigg(\bigcup_{j\in\mathbb{N}} B_{\alpha_{k_j}}\Bigg).$$

This is impossible since $\bigcup_k B_{\alpha_k}$ is bounded. This last fact is proved as follows. Since E is bounded, there exists $H > 0$ such that $E \subset B_d(0, H)$; if $\xi \in B_{\alpha_k} \cap E$, we have

$$\begin{aligned} d(z) &\leq \mathbf{c}(d(z,\xi) + d(\xi)) \leq \mathbf{c}^2 d(z, x_k) + \mathbf{c}^2 d(x_k, \xi) + \mathbf{c}\, d(\xi) \\ &\leq 2\mathbf{c}^2\, r_{\alpha_k} + \mathbf{c}\, H \leq 2\mathbf{c}^2 \sup_\alpha r_\alpha + \mathbf{c}\, H < \infty \end{aligned}$$

for every k and every $z \in B_{\alpha_k}$. The claimed (13.19) is proved.

If $\mathcal{I} = \mathcal{A}$, the proof is complete. Otherwise, let $\alpha' \notin \mathcal{A}$. If $r_{\alpha_{k+1}} \geq \frac{1}{2} r_{\alpha'}$ for every $k \in \mathbb{N}$, then (thanks to (13.19)) the selection process is finite, which is not the case. Hence, there exists k (necessarily $k \neq 0$, for $r_{\alpha_1} \geq \frac{1}{2} r_{\alpha'}$) such that $r_{\alpha_{k+1}} < \frac{1}{2} r_{\alpha'}$. Let k_0 be the least positive integer such that

$$r_{\alpha_{k_0+1}} < \frac{1}{2} r_{\alpha'}. \tag{13.20}$$

Then there exists $j \in \{1, \dots, k_0\}$ such that

$$B_{\alpha'} \cap B_{\alpha_j} \neq \emptyset \tag{13.21}$$

(indeed, if it were $B_{\alpha'} \cap B_{\alpha_j} = \emptyset$ for every $j = 1, \dots, k_0$, we would have $r_{\alpha_{k_0+1}} \geq \frac{1}{2} r_{\alpha'}$ by the very selection process (13.18), contradicting (13.20)). Then (13.21), together with $r_{\alpha_j} \geq \frac{1}{2} r_{\alpha'}$ (recall that k_0 is the least index satisfying (13.20)), proves

$$B_{\alpha'} \subseteq B_d(x_{\alpha_j}, 3\mathbf{c}^2 r_{\alpha_j})$$

by a simple application of the pseudo-triangle inequality (13.1). This ends the proof. $\square$

Lemma 13.2.4. *Let $E \subset \mathbb{G}$ be a bounded $\mathcal{L}$-polar set. Then, for every measure function ϕ such that*

$$\int_0^1 t^{1-Q} \phi(t) \, dt < \infty, \tag{13.22}$$

we have $m_\phi(E) = 0$.

Proof. From Corollary 11.2.5 (page 492), being E bounded, there exists $\mu \in \mathfrak{M}_0$ such that $\Gamma * \mu \equiv \infty$ on E. If ϕ is as in the assertion of the present lemma, we set

$$\phi_1(t) := \min \left\{ \phi(3\,\mathbf{c}^2 t), \, \phi(1/2) \right\}.$$

Here $\mathbf{c} \geq 1$ is the constant appearing in the pseudo-triangle inequality (13.1). It is easily seen that $\int_0^1 t^{1-Q} \phi_1(t) \, dt$ is finite, thanks to (13.22). Fix $a > 0$ such that

$$\frac{\mu(\mathbb{G})}{a} < \phi(1/2). \tag{13.23}$$

From Lemma 13.2.2 applied to the measure function ϕ_1 we get

$$(\Gamma * \mu)(x) \leq \mu(\mathbb{G}) + C(Q, \phi_1) \sup_{0<r<1} \frac{\mu(B_d(x,r))}{\phi_1(r)}$$

for any $x \in \mathbb{G}$. If, in particular, we take $x \in E$ (so that $\Gamma * \mu(x) = \infty$), being $\mu(\mathbb{G}) < \infty$, we derive the existence of $r(x) \in]0, 1[$ such that

$$\frac{\mu(B_d(x, r(x)))}{\phi_1(r(x))} > a. \tag{13.24}$$

By the covering Lemma 13.2.3 applied to the family

$$\{B_x := B_d(x, r(x)) : x \in E\},$$

there exists a sequence $x_k \in E$ such that the B_{x_k}'s are mutually disjoint and

$$E \subseteq \bigcup_k B_d(x_k, 3\mathbf{c}^2\, r(x_k)). \tag{13.25}$$

This gives (using (13.24) in the first inequality below)

$$\sum_k \phi_1(r(x_k)) < \frac{1}{a} \sum_k \mu(B_{x_k}) = \frac{1}{a}\, \mu\Big(\bigcup_k B_{x_k}\Big) \le \frac{1}{a}\, \mu(\mathbb{G}). \tag{13.26}$$

Taking into account the choice (13.23) of a, from (13.26) it follows that no summand $\phi_1(r(x_k))$ coincides with $\phi(1/2)$. By the very definition of ϕ_1, this gives

$$\phi_1(r(x_k)) = \phi(3\,\mathbf{c}^2\, r(x_k)) \quad \forall\, k \in \mathbb{N}. \tag{13.27}$$

This yields

$$M_\phi^{(3\mathbf{c}^2)}(E) \le \sum_k \phi(3\mathbf{c}^2\, r(x_k)) = \sum_k \phi_1(r(x_k)) \le \mu(\mathbb{G})/a. \tag{13.28}$$

In the first inequality we used (13.25), in the second we exploited (13.27), in the third we took into account (13.26).

Since we can take any $a \gg 1$ satisfying (13.23), we can let a tend to ∞ in (13.28) to derive that

$$M_\phi^{(3\mathbf{c}^2)}(E) = 0.$$

From the fact that ϕ is increasing and positive it easily[1] follows that $M_\phi^{(\rho)}(E) = 0$ for any $\rho > 0$, whence $m_\phi(E) = 0$. The proof is complete. □

As a consequence of Lemma 13.2.4, we have a partial converse of Theorem 13.1.7.

Theorem 13.2.5 (d-Hausdorff measure and $\mathcal{L}$-polarity. II). *Let $E \subset \mathbb{G}$ be a bounded $\mathcal{L}$-polar set. Then we have*

$$m_{(\alpha)}(E) = 0 \quad \textit{for every } \alpha > Q - 2. \tag{13.29}$$

Hence, the d-Hausdorff dimension of a bounded $\mathcal{L}$-polar set is $\le Q - 2$.

Proof. This follows from Lemma 13.2.4, by taking $\phi(t) = t^\alpha$ with $\alpha > Q - 2$ and the very definition of d-Hausdorff dimension (see Lemma 13.1.3). □

Collecting together Theorems 13.1.7 and 13.2.5, we have the following result.

---

[1] See Ex. 4 at the end of the chapter.

Remark 13.2.6. Let E be any bounded and $\mathcal{L}$-capacitable subset of $\mathbb{G}$, and let $\dim_H(E)$ be the d-Hausdorff dimension of E. We have:

(i) If $\dim_H(E) < Q-2$, then E is $\mathcal{L}$-polar,
(ii) If $\dim_H(E) > Q-2$, then E is not $\mathcal{L}$-polar,
(iii) If $\dim_H(E) = Q-2$ and $m_{(Q-2)}(E) \in [0, \infty[$, then E is $\mathcal{L}$-polar,
(iv) If $\dim_H(E) = Q-2$ and $m_{(Q-2)}(E) = \infty$, we cannot conclude anything about the $\mathcal{L}$-polarity of E.

13.3 New Phenomena Concerning the d-Hausdorff Dimension: The Case of the Heisenberg Group

The aim of this section is to provide a brief overview (without any proof) of some remarkable recent results on the Hausdorff dimension in the Heisenberg group, by Z.M. Balogh, M. Rickly and F. Serra Cassano [BRSC03].

Let $\mathbb{H}^1$ be the Heisenberg–Weyl group on $\mathbb{R}^3$. We denote its points by $z = (x, y, t)$. The relevant dilation is $\delta_\lambda(z) = (\lambda x, \lambda y, \lambda^2 t)$, and the composition law is given by

$$z \circ z' = \big(x + x', y + y', t + t' + 2(x'y - y'x)\big).$$

The homogeneous norm $d(z) = ((x^2 + y^2)^2 + t^2)^{1/4}$ is an $\mathcal{L}$-gauge, where $\mathcal{L}$ is the canonical sub-Laplacian on $\mathbb{H}^1$. We denote by e the usual Euclidean norm on $\mathbb{R}^3$, i.e. $e(x, y, t) = (x^2 + y^2 + t^2)^{1/2}$.

As it is expected, the Hausdorff measure and the Hausdorff dimension of a subset of $\mathbb{H}^1 \equiv \mathbb{R}^3$ w.r.t. d and e may be different. For example, the d-Hausdorff dimension of $\mathbb{H}^1$ is 4 (see Example 13.1.4), whereas its e-Hausdorff dimension is 3. Furthermore, the d-Hausdorff dimension of a regular surface is 3, and the d-Hausdorff dimension of a regular curve can be 1 or 2 (see [Gro96,Str92]).

In the sequel of this section, we use the notation H_e^α and H_d^β to denote, respectively, $m_{(\alpha)}$ when the relevant gauge is the Euclidean norm e and $m_{(\beta)}$ when the relevant gauge is the above d.

We fix some notation. We define the following functions:

$$f, \varphi, g, \gamma : [0, \infty) \to [0, \infty), \quad \text{where} \quad \begin{cases} f(\alpha) := \min\{2\alpha, \alpha + 1\}, \\ \varphi(\beta) := \max\{\frac{1}{2}\beta, \beta - 1\}, \\ g(\beta) := \min\{\beta, 1 + \beta/2\}, \\ \gamma(\alpha) := \max\{\alpha, 2\alpha - 2\}. \end{cases}$$

(Note that $f = \varphi^{-1}$ and $g = \gamma^{-1}$.) The following theorem holds.

Theorem 13.3.1 (Dimension jump theorem, [BRSC03]). *For every real numbers* $\alpha, \beta > 0$, *we have*

$$H_d^{f(\alpha)} \ll H_e^\alpha \ll H_d^{\gamma(\alpha)},$$
$$H_e^{g(\beta)} \ll H_d^\beta \ll H_e^{\varphi(\beta)}.$$

(Here, $\mu \ll \lambda$ means that the measure μ is absolutely continuous with respect to the measure λ.) For example, the above theorem (and the very definition of absolutely continuous measures) give

$$H_e^\alpha(E) = 0 \Longrightarrow H_d^{f(\alpha)}(E) = 0,$$
$$H_e^\alpha(E) > 0 \Longrightarrow H_d^{\gamma(\alpha)}(E) > 0,$$
$$H_d^\beta(E) = 0 \Longrightarrow H_e^{g(\beta)}(E) = 0,$$
$$H_d^\beta(E) > 0 \Longrightarrow H_e^{\varphi(\beta)} > 0.$$

Together with Theorem 13.3.1, Balogh, Rickly and Serra Cassano proved the following theorem.

Theorem 13.3.2 (Sharpness of the dimension jump, [BRSC03]). *The following assertions hold*:

(i) *Given* $\alpha \in (0, 3]$, *there exists a compact set* $E_\alpha \subset \mathbb{H}^1$ *such that*

$$H_e^\alpha(E_\alpha) < \infty \text{ and } H_d^{f(\alpha)}(E_\alpha) > 0$$

or, equivalently, given $\beta \in (0, 4]$, *there exists a compact set* $E_\beta \subset \mathbb{H}^1$ *such that*

$$H_e^{\varphi(\beta)}(E_\beta) < \infty \text{ and } H_d^\beta(E_\beta) > 0;$$

(ii)' *For* $\beta \in (0, 2) \cup \{4\}$, *there exists a compact set* $E_\beta \subset \mathbb{H}^1$ *such that*

$$H_d^\beta(E_\beta) < \infty \text{ and } H_e^{g(\beta)}(E_\beta) > 0$$

or, equivalently, given $\alpha \in (0, 2) \cup \{3\}$, *there exists a compact set* $E_\alpha \subset \mathbb{H}^1$ *such that*

$$H_d^{\gamma(\alpha)}(E_\alpha) < \infty \text{ and } H_e^\alpha(E_\alpha) > 0;$$

(ii)" *For* $\beta \in [2, 4)$ *and* $\varepsilon \in (0, 1)$, *there exists a compact set* $E_{\beta,\varepsilon} \subset \mathbb{H}^1$ *such that*

$$H_d^\beta(E_{\beta,\varepsilon}) = 0 \text{ and } H_e^{g(\beta)-\varepsilon}(E_{\beta,\varepsilon}) > 0.$$

We explicitly describe some of the sets E_α's in the previous theorem (see [BRSC03] for the details).

(i) Let $0 < s < 1$. We consider the Cantor[2] set $C_s \subset [0, 1]$ of (Euclidean) Hausdorff dimension s.

---

[2] We recall the construction of C_s. Let $s \in (0, 1)$ be fixed and set $\sigma := 2^{-1/s}$. Let $k \in \mathbb{N}\cup\{0\}$ and $j \in \{1, \dots, 2^k\}$. We define $I_{k,j}^s$ in the following way:

$$I_{0,1}^s = [0, 1],$$
$$I_{1,1}^s = [0, \sigma], \quad I_{1,2}^s = [1 - \sigma, 1],$$

(a) If $\alpha \in (0, 1]$, we let $E_\alpha = \{0\} \times \{0\} \times C_\alpha$ (here $C_1 := [0, 1]$). Obviously, $H_e^\alpha(E_\alpha)$ is positive and finite and the same is true of $H_d^{2\alpha}(E_\alpha) = H_d^{f(\alpha)}(E_\alpha)$ for, roughly speaking, the distance induced by the gauge d on the t-axis coincides with the $\frac{1}{2}$-power of the Euclidean distance. Note that the vertical segment $E_1 = \{0\} \times \{0\} \times [0, 1]$ has d-Hausdorff dimension 2, but e-Hausdorff dimension 1. Since it also has finite H_d^2-measure, by Theorem 13.1.7 we infer that E_1 is $\mathcal{L}$-polar, where $\mathcal{L}$ is the canonical sub-Laplacian of $\mathbb{H}^1$ (recall that here $Q - 2 = 2$).

(b) If $\alpha \in (1, 2)$, the set $E_\alpha = C_{\alpha-1} \times \{0\} \times [0, 1]$ (a Cantor set of vertical segments) has the requisites in Theorem 13.3.2-(i).

(c) If $\alpha = 2$, it suffices to take $E_2 = \{x^2 + y^2 + t^2 = 1\}$ (see also [Pan82]). Note that E_2 has positive and finite 2-dimensional Euclidean Hausdorff measure and positive 3-dimensional d-Hausdorff measure.

(d) If $\alpha \in (2, 3)$, then the choice

$$E_\alpha = \bigcup_{r \in C_{\alpha-2}} \partial B_d(0, r)$$

(a Cantor set of d-spheres) has the requisites in Theorem 13.3.2-(i).

(e) If $\alpha = 3$, take $E_3 = B_d(0, 1)$. Note that E_3 has positive and finite 3-dimensional Euclidean Hausdorff measure and the same is true of its 4-dimensional d-Hausdorff measure.

(ii) If $\beta \in (0, 1]$, the set $E_\beta = C_\beta \times \{0\} \times \{0\}$ (here C_1 stands for $[0, 1]$) has positive and finite H_d^β- and H_e^β-measures. If $\beta = 4$, it suffices to take $E_4 = B_d(0, 1)$, as already discussed. The case $\beta \in [2, 4)$ is more involved, and the reader is referred directly to [BRSC03].

Bibliographical Notes. In the case of the classical Laplace operator, the relationship between the Hausdorff measure and capacity can be found, e.g. in [AG01, Section 5.9] (whose exposition we followed here) and [HK76, Section 5.4]. See [BRSC03] for recent results on the Hausdorff measure in the context of the Heisenberg group. See also [Gro96,Str92].

$$I_{2,1}^s = [0, \sigma^2],\ I_{2,2}^s = [\sigma - \sigma^2, \sigma],\ I_{2,3}^s = [1 - \sigma, 1 - \sigma + \sigma^2],\ I_{1,2}^s = [1 - \sigma^2, 1],$$

$$\vdots$$

if we are given $I_{k,j}^s = [a_{k,j}^s, b_{k,j}^s]$ for $j = 1, \dots, 2^k$, then

$$I_{k+1,2j-1}^s = [a_{k,j}^s, a_{k,j}^s + \sigma^{k+1}], \quad I_{k+1,2j}^s = [b_{k,j}^s - \sigma^{k+1}, b_{k,j}^s].$$

Now, we define C_s as

$$C_s := \bigcap_{k=0}^{\infty} \bigcup_{j=1}^{2^k} I_{k,j} =: \bigcap_{k=0}^{\infty} C_k^s.$$

Note that $C_{k+1}^s \subset C_k^s$, so that $C_s = \lim_{k\to\infty} C_k^s$ with the obvious meaning. Moreover, C_k^s is the union of 2^k disjoint closed intervals of length σ^k. It can be proved that C_s is a compact subset of $[0, 1]$ with no interior points and such that $0 < H_e^s(C_s) < \infty$.

13.4 Exercises of Chapter 13

Ex. 1) Consider the Heisenberg–Weyl group $(\mathbb{H}^1, \circ)$ on $\mathbb{R}^3$. Let $z_0 = (x_0, y_0, t_0) \in \mathbb{H}^1$ be fixed, with $t_0 \neq 0$. Consider the following "segment of δ_λ-ray"

$$R := \{\delta_\lambda(z_0) \,:\, \lambda \in [a, b]\},$$

where $a < b$ are fixed positive real numbers. Finally, let $\Delta_{\mathbb{H}^1}$ be the canonical sub-Laplacian on $\mathbb{H}^1$. Prove the following facts:

- If Γ is the fundamental solution for $\Delta_{\mathbb{H}^1}$, the function
$$u(z) := \int_a^b \Gamma(\delta_\lambda(z_0^{-1}) \circ z)\,\mathrm{d}\lambda$$
belongs to $\overline{\mathcal{S}}(\mathbb{H}^1)$ (w.r.t. $\Delta_{\mathbb{H}^1}$) and $u(0) < \infty$;
- $u \equiv \infty$ on R;
- Derive that R is a bounded $\Delta_{\mathbb{H}^1}$-polar set. If d is a $\Delta_{\mathbb{H}^1}$-gauge, what is the d-Hausdorff dimension of R? (Use the remarks in Section 13.3.)

Hint: $u(0) = \Gamma(z_0) \int_a^b \lambda^{-2}\mathrm{d}\lambda$; $u \in \overline{\mathcal{S}}(\mathbb{H}^1)$ by Theorem 8.2.20, page 410; by the explicit form of $\circ$ and Γ, if $\lambda_0 \in [a, b]$ is fixed, it holds

$$u(\delta_{\lambda_0}(z_0)) = [\cdots] = \int_a^b \frac{1}{\{(\lambda_0 - \lambda)^4\,(x_0^2 + y_0^2)^2 + (\lambda_0^2 - \lambda^2)^2\,t_0^2\}^{1/2}}\,\mathrm{d}\lambda$$
$$\approx \int_{\lambda_0-\varepsilon}^{\lambda_0+\varepsilon} \frac{1}{|\lambda - \lambda_0|\,2\lambda_0\,|t_0|}\,\mathrm{d}\lambda = \infty.$$

Ex. 2) Let $\Omega \subset \mathbb{G}$ be any bounded open set, and let $\varepsilon > 0$ be any positive real number. Prove that there exists a countable collection of disjoint d-balls

$$B_k := B_d\left(x_k, \frac{r_k}{4}\right)$$

with $\frac{r_k}{2} < \varepsilon$ for every k and a set N such that

$$\begin{cases} \Omega = N \cup \bigcup_k B_k, \\ \bigcup_k B_d(x_k, \mathbf{c}\, r_k) \supseteq \Omega. \end{cases}$$

Ex. 3) Let μ be any compactly supported Radon measure on $\mathbb{G}$. Suppose $z \in \mathbb{G}$ is such that $(\Gamma * \mu)(z) < \infty$ and prove that $\mu(\{z\}) = 0$. (Compare to Exercise 22, Chapter 9.)

Ex. 4) Let $E \subset \mathbb{G}$ be any set and ϕ be a measure function. Suppose there exists $\rho_0 > 0$ such that

$$M_\phi^{(\rho_0)}(E) = 0.$$

Prove that $M_\phi^{(\rho)}(E) = 0$ for every $\rho > 0$, whence $m_\phi(E) = 0$.

(*Hint:* Since the map $]0,\infty] \ni \rho \mapsto M_\phi^{(\rho)}$ is non-negative and decreasing (prove this!), we have $M_\phi^{(\rho)}(E)=0$ for every $\rho \geq \rho_0$. Let now $\rho \in]0,\rho_0[$ be fixed. By the very definition of $M_\phi^{(\rho_0)}(E)=0$, we have

$$(\bigstar) \quad \forall\, \varepsilon>0 \quad \exists\, \{B_d(x_k(\varepsilon), r_k(\varepsilon))\}_{k\in\mathbb{N}} : \begin{cases} r_k(\varepsilon) < \rho_0, \\ E \subseteq \bigcup_k B_d(x_k(\varepsilon), r_k(\varepsilon)), \\ \sum_k \phi(r_k(\varepsilon)) < \varepsilon. \end{cases}$$

Since ϕ is positive on $]0,\infty[$, in $(\bigstar)$ we can take any $\varepsilon \in]0,\phi(\rho)[$. Hence, the third condition in the right-hand side of $(\bigstar)$ gives

$$\phi(r_k(\varepsilon)) \leq \sum_k \phi(r_k(\varepsilon)) < \varepsilon < \phi(\rho) \quad \forall\, k \in \mathbb{N}.$$

Since ϕ is increasing, this gives $r_k(\varepsilon)<\rho$, so that $M_\phi^{(\rho)}(E)=0$.)

Ex. 5) Prove the assertions of Example 13.1.4 and Remark 13.1.5. (*Hint:* The following facts may help. For every $x \in \mathbb{G}$ and $r>0$, we have $r^Q = c\, H^N(B_d(x,r))$, where $c = H^N(B_d(0,1))$. If $E \subseteq \bigcup_k B_d(x_k, r_k)$, then

$$x \circ E \subseteq \bigcup_k B_d(x\circ x_k, r_k), \qquad \delta_\lambda(E) \subseteq \bigcup_k B_d(\delta_\lambda(x_k), \lambda r_k),$$

whence $H_{(\alpha)}^{(\rho)}(x\circ E) = H_{(\alpha)}^{(\rho)}(E)$ and $H_{(\alpha)}^{(\lambda\rho)}(\delta_\lambda(E)) = \lambda^\alpha H_{(\alpha)}^{(\rho)}(E)$.)

Ex. 6) Prove the assertion in Remark 13.1.2. More precisely, we have the following results. For any $E\subseteq\mathbb{G}$, set

$$\mathrm{diam}_d(E) := \sup\{d(x^{-1}\circ y) : x,y\in E\}, \quad \mathrm{diam}_d(\emptyset) := 0.$$

If ϕ is a measure function as in Definition 13.1.1, we let

$$H_\phi^{(\rho)}(E) := \inf\Big\{ \sum_k \phi(\mathrm{diam}_d(E_k)) : E \subseteq \bigcup_k E_k,\ \mathrm{diam}_d(E_k) < \rho\ \forall\, k \Big\},$$

$$M_\phi^{(\rho)}(E) := \inf\Big\{ \sum_k \phi(r_k) : E \subseteq \bigcup_k B_d(x_k, r_k),\ r_k < \rho\ \forall\, k \Big\}.$$

- Prove that $H_\phi^{(\rho)}$, $M_\phi^{(\rho)}$, h_ϕ, m_ϕ are outer measures on $\mathbb{G}$, where

$$h_\phi(E) := \sup_{\rho>0} H_\phi^{(\rho)}(E), \quad m_\phi(E) := \sup_{\rho>0} M_\phi^{(\rho)}(E).$$

- Let A, B be arbitrary subsets of $\mathbb{G}$. Prove that

$$\mathrm{dist}_d(A,B) > 0 \quad \Longrightarrow \quad \begin{cases} h_\phi(A\cup B) = h_\phi(A) + h_\phi(B), \\ m_\phi(A\cup B) = m_\phi(A) + m_\phi(B). \end{cases} \tag{13.30}$$

Here, we have set

$$\mathrm{dist}_d(A, B) := \inf\{d(a^{-1} \circ b) \,:\, a \in A,\ b \in B\}.$$

Note that $\mathrm{dist}_d(A, B) > 0$ iff the usual Euclidean distance between A and B is positive.[3] Hence, derive from (13.30) and from very general results of measure theory, that every (open and then every) Borel set in $\mathbb{G}$ is both h_ϕ and m_ϕ measurable. Hence, if restricted to their related measurable sets, h_ϕ and m_ϕ are Borel measures.
Now, we take $\phi(t) = t^\alpha$ with $\alpha > 0$, and write $h_{(\alpha)}$ and $m_{(\alpha)}$ instead of h_{t^α} and m_{t^α}.
• Prove that

$$2^{-\alpha} m_{(\alpha)}(E) \le h_{(\alpha)}(E) \le (2\mathbf{c})^\alpha\, m_{(\alpha)}(E) \quad \text{for every } E \subseteq \mathbb{G}, \qquad (13.31)$$

where $\mathbf{c}$ is the constant in the pseudo-triangle inequality for d. The following facts might help: $r \le \mathrm{diam}\big(B_d(x, r)\big) \le 2\mathbf{c}\, r$; if $e \in E$, then $E \subseteq \overline{B_d}(e, \mathrm{diam}(E)) \subset B_d(e, 2\mathrm{diam}(E))$;

$$H^{\varepsilon}_{t^\alpha}(E) \le (2\mathbf{c})^\alpha H^{\varepsilon/(2\mathbf{c})}_{t^\alpha}(E); \qquad M^{2\varepsilon}_{t^\alpha}(E) \le 2^\alpha H^{\varepsilon}_{t^\alpha}(E).$$

• Derive from (13.31) that, for every bounded set $E \subset \mathbb{G}$, there exists a number $\alpha(E) \in [0, Q]$ (Q is the homogeneous dimension of $\mathbb{G}$) such that

$$h_{(\alpha)}(E) = m_{(\alpha)}(E) = \infty \text{ for all } \alpha < \alpha(E),$$
$$h_{(\alpha)}(E) = m_{(\alpha)}(E) = 0 \text{ for all } \alpha > \alpha(E).$$

Thus, we have

$$\alpha(E) = \inf\{\alpha > 0 : h_{(\alpha)}(E) = 0\} = \inf\{\alpha > 0 : m_{(\alpha)}(E) = 0\}.$$

[3] Indeed, if $a_j \in A$, $b_j \in B$, then observe that $a_j^{-1} \circ b_j \to 0$ in the Euclidean metric iff $d(a_j^{-1} \circ b_j) \to 0$.

Part III

Further Topics on Carnot Groups

14

Some Remarks on Free Lie Algebras

The aim of this chapter is to investigate a few properties of the *free nilpotent Lie algebras*. After recalling the definition of the free Lie algebra $\mathfrak{f}_{m,r}$ with m generators and nilpotent of step r, we provide an algorithm (due to M. Hall) to construct a basis for $\mathfrak{f}_{m,r}$, jointly with several examples. Furthermore, we describe another useful algorithm concerning with free Lie algebras in an analytic context, an algorithm due to M. Grayson and R. Grossman.

In Section 14.2, we furnish a canonical way to construct free Carnot groups (i.e. Carnot groups whose Lie algebras is free and nilpotent) by means of the Campbell–Hausdorff formula.

As we shall see in Chapter 16, free Carnot groups are the appropriate generalizations of the usual Euclidean group, when we are concerned with the "equivalence" of all sub-Laplacians.

14.1 Free Lie Algebras and Free Lie Groups

We first recall the definition of free Lie algebra with m generators and nilpotent of step r (see, e.g. [VSC92, p. 45]; see also [Var84, p. 174]).

Definition 14.1.1 (The free Lie algebra $\mathfrak{f}_{m,r}$). *Let $m \geq 2$ and $r \geq 1$ be fixed integers. We say that $\mathfrak{f}_{m,r}$ is the* free Lie algebra with m generators $x_1, \dots, x_m$ and nilpotent of step r *if:*

(i) $\mathfrak{f}_{m,r}$ *is a Lie algebra generated by its elements $x_1, \dots, x_m$, i.e.*

$$\mathfrak{f}_{m,r} = \mathrm{Lie}\{x_1, \dots, x_m\};$$

(ii) $\mathfrak{f}_{m,r}$ *is nilpotent of step r;*

(iii) *for every Lie algebra $\mathfrak{n}$ nilpotent of step r and for every map φ from the set $\{x_1, \dots, x_m\}$ to $\mathfrak{n}$, there exists a (unique) homomorphism of Lie algebras $\widetilde{\varphi}$ from $\mathfrak{f}_{m,r}$ to $\mathfrak{n}$ which extends φ.*

Remark 14.1.2. It is easy to see that, fixed m and r, the Lie algebra $\mathfrak{f}_{m,r}$ is unique up to isomorphism: it is enough to compare two algebras $\mathfrak{f}$, $\mathfrak{f}'$ satisfying the above definition and recognize that the morphism as in (iii) which takes the generators of $\mathfrak{f}$ into the generators of $\mathfrak{f}'$ is the inverse of the morphism taking the generators of $\mathfrak{f}'$ into those of $\mathfrak{f}$. Moreover, if $\mathfrak{f}_{m,r}$ is isomorphic to $\mathfrak{f}_{m',r'}$, then it necessarily holds $m = m'$ and $r = r'$.

We explicitly remark that it is not trivial to prove the existence of such an algebra $\mathfrak{f}_{m,r}$. We refer the reader to [Var84] for this topic.

Definition 14.1.3 (Free Carnot group). *We say that a Carnot group $\mathbb{G}$ is a* free Carnot group *if its Lie algebra $\mathfrak{g}$ is isomorphic to $\mathfrak{f}_{m,r}$ for some m and r.*

The following lemma concerning with free Lie algebras $\mathfrak{f}_{m,r}$ will be useful in the sequel.

Lemma 14.1.4. *Let $\mathfrak{f}_{m,r}$ be the free Lie algebra with m generators $x_1, \dots, x_m$, nilpotent of step r. Let*

$$\varphi : \operatorname{span}\{x_1, \dots, x_m\} \longrightarrow \operatorname{span}\{x_1, \dots, x_m\}$$

be a linear bijective map. Then there exists one (and only one) isomorphism of Lie algebras

$$\widetilde{\varphi} : \mathfrak{f}_{m,r} \longrightarrow \mathfrak{f}_{m,r}$$

extending φ.

Proof. We consider the maps

$$\varphi,\ \varphi^{-1} : \operatorname{span}\{x_1, \dots, x_m\} \longrightarrow \operatorname{span}\{x_1, \dots, x_m\}.$$

By the very definition of $\mathfrak{f}_{m,r}$, there exist two homomorphisms of Lie algebras $\widetilde{\varphi}$, $\widetilde{(\varphi^{-1})}$ defined on $\mathfrak{f}_{m,r}$ into itself which extend φ and φ^{-1}, respectively. We consider the map

$$\psi := \widetilde{(\varphi^{-1})} \circ \widetilde{\varphi}.$$

The proof is complete if we show that ψ is the identity map. To this end, we note that, for every $i = 1, \dots, m$, one has

$$\psi(x_i) = \widetilde{(\varphi^{-1})}(\varphi(x_i)) = \varphi^{-1}(\varphi(x_i)) = x_i,$$

since $\varphi(x_i) \in \operatorname{span}\{x_1, \dots, x_m\}$. By the definition of $\mathfrak{f}_{m,r}$, there exists a *unique* homomorphism of the Lie algebra $\mathfrak{f}_{m,r}$ into itself which extends

$$\psi|_{\operatorname{span}\{x_1,\dots,x_m\}}.$$

Since both ψ and the identity map satisfy this requirement, the assertion immediately follows. □

We next give a model for $\mathfrak{f}_{m,r}$. It is known that, setting

$$H(m,r) := \dim(\mathfrak{f}_{m,r}),$$

$\mathfrak{f}_{m,r}$ may be represented by upper triangular matrices of order $H(m,r)$ (see [Var84]). There exists another remarkable representation of $\mathfrak{f}_{m,r}$ via vector fields, useful in the applications and easy to be dealt with, due to M. Grayson and R. Grossman (see [GG90]), which we now briefly recall. This representation makes use of the so-called *Hall basis* for $\mathfrak{f}_{m,r}$.

Definition 14.1.5 (The Hall basis for $\mathfrak{f}_{m,r}$). *Let $\mathfrak{f}_{m,r}$ be the free Lie algebra with m generators $x_1, \dots, x_m$, nilpotent of step r. We define the elements of the* Hall basis for $\mathfrak{f}_{m,r}$ *in the following way:*

- *$x_1, \dots, x_m$ are the first m elements of the Hall basis and they are referred to as the* standard monomials of height 1;
- *We now let $2 \le n \le r$, and we define the* standard monomials of height n: *suppose we have already defined the standard monomials of height $1, \dots, n-1$, and suppose we have ordered them in such a way that u precedes v (we write $u < v$) if the height of u is strictly lower than that of v; then the Lie bracket $[u,v]$ is a* standard monomial of height n *if the sum of the heights of u and v equals n and if, moreover, the following conditions are satisfied:*
  1. *u and v are standard monomials such that $u > v$;*
  2. *if $u = [x,y]$ is the form of the standard monomial u, then $v \ge y$.*

The collection of all the standard monomials of height $1, \dots, r$ is referred to as the Hall basis for $\mathfrak{f}_{m,r}$.

The following result holds.

Theorem 14.1.6 (Hall, [Hal50]). *The Hall basis for $\mathfrak{f}_{m,r}$ is a basis (in the sense of vector fields) for $\mathfrak{f}_{m,r}$.*

Proof. The assertion follows from the arguments in the paper by M. Hall [Hal50, Theorem 3.1], in which a general basis for free Lie *rings* is constructed exactly as in Definition 14.1.5 (but with no restriction on r, i.e. $r = \infty$). Since $\mathfrak{f}_{m,r}$ is nilpotent of step r, it is enough to consider the elements of the basis constructed in [Hal50] which are brackets of height r at most. □

Example 14.1.7. We next give some examples of Hall bases:

- The Hall basis for $\mathfrak{f}_{3,4}$ is given by:

$$\begin{array}{ll}
\text{height 1:} & \qquad x_1, \quad x_2, \quad x_3, \\
\text{height 2:} & \quad [x_2,x_1], \ [x_3,x_1], \ [x_3,x_2], \\
\text{height 3:} & \left\{ \begin{array}{l} [[x_2,x_1],x_1], \ [[x_3,x_1],x_1], \\ [[x_2,x_1],x_2], \ [[x_3,x_1],x_2], \ [[x_3,x_2],x_2], \\ [[x_2,x_1],x_3], \ [[x_3,x_1],x_3], \ [[x_3,x_2],x_3], \end{array} \right.
\end{array}$$

$$\text{height 4:} \begin{cases} [[[x_2,x_1]x_1]x_1], \ [[[x_3,x_1]x_1]x_1], \\ [[[x_2,x_1]x_1]x_2], \ [[[x_3,x_1]x_1]x_2], \\ [[[x_2,x_1]x_2]x_2], \ [[[x_3,x_1]x_2]x_2], \ [[[x_3,x_2]x_2]x_2], \\ [[[x_2,x_1]x_1]x_3], \ [[[x_3,x_1]x_1]x_3], \\ [[[x_2,x_1]x_2]x_3], \ [[[x_3,x_1]x_2]x_3], \ [[[x_3,x_2]x_2]x_3], \\ [[[x_2,x_1]x_3]x_3], \ [[[x_3,x_1]x_3]x_3], \ [[[x_3,x_2]x_3]x_3], \\ [[x_3,x_1],[x_2,x_1]], \ [[x_3,x_2],[x_2,x_1]], \\ [[x_3,x_2],[x_3,x_1]]. \end{cases}$$

- The Hall basis for $\mathfrak{f}_{4,2}$ is given by:

$$\begin{aligned} &\text{height 1:} \quad x_1, \quad x_2, \quad x_3, \quad x_4, \\ &\text{height 2:} \begin{cases} [x_2,x_1], [x_3,x_1], [x_4,x_1], \\ \qquad\quad\ [x_3,x_2], [x_4,x_2], \\ \qquad\qquad\qquad\ [x_4,x_3]. \end{cases} \end{aligned}$$

- The Hall basis for $\mathfrak{f}_{m,2}$ is given by:

$$\begin{aligned} &\text{height 1:} \quad x_1, \quad x_2, \quad x_3, \quad \dots, \quad x_m, \\ &\text{height 2:} \begin{cases} [x_2,x_1], [x_3,x_1], \cdots [x_m,x_1], \\ \qquad\quad\ [x_3,x_2], \cdots [x_m,x_2], \\ \qquad\qquad\qquad \ddots \quad \vdots \\ \qquad\qquad\qquad\ [x_m,x_{m-1}]. \end{cases} \end{aligned}$$

Remark 14.1.8. We explicitly remark that

$$H(m,2) = \dim(\mathfrak{f}_{m,2}) = \sum_{j=1}^{m} j = \frac{m(m+1)}{2}.$$

In particular, this implies, the following result.

Remark 14.1.9. The Heisenberg group $\mathbb{H}^N$ is a free Carnot group if and only if $N = 1$.

Indeed, since the Lie algebra $\mathfrak{h}_N$ of $\mathbb{H}^N$ has step 2 and $2N$ generators, a necessary condition for $\mathbb{H}^N$ to be free is that $2N+1 = \dim\mathfrak{h}_N = H(2N,2) = 2N(2N+1)/2$, i.e. that $N = 1$. On the other hand, it is immediate to recognize that $\mathfrak{h}_1$ is isomorphic to $\mathfrak{f}_{2,2}$. $\square$

As we shall see more closely in Chapter 17, this shows that, even if the vector fields related to the Heisenberg group $\mathbb{H}^N$ are naturally defined on $\mathbb{R}^{2N+1}$, the *free* (*lifted*) counterpart of such vector fields acts on $\mathbb{R}^{2N(2N+1)/2}$.

We finally turn our attention to the cited model for $\mathfrak{f}_{m,r}$ by Grayson–Grossman, which we now describe. We order the elements of the Hall basis for $\mathfrak{f}_{m,r}$ according[1] to Definition 14.1.5. We set $H := \dim \mathfrak{f}_{m,r}$, and we denote

$$E_1, \ldots, E_H$$

the elements of the Hall basis with such a fixed ordering. We fix an element E_i. According to the construction of the Hall basis in Definition 14.1.5, it certainly has the following form (if $i > m$)

$$E_i = [E_{j_1}, E_{k_1}], \quad \text{with } j_1 > k_1.$$

Taking E_{k_1} fixed (whatever its height is) and arguing in the same way for E_{j_1}, one has

$$E_i = [[E_{j_2}, E_{k_2}]; E_{k_1}], \quad \text{with } j_2 > k_2.$$

After finitely many steps, we have obtained the expression

$$\begin{aligned} E_i &= [E_{j_1}, E_{k_1}] \\ &= [[E_{j_2}, E_{k_2}]; E_{k_1}] \\ &= [[[E_{j_3}, E_{k_3}], E_{k_2}]; E_{k_1}] \\ &= [[[[E_{j_4}, E_{k_4}], E_{k_3}], E_{k_2}]; E_{k_1}] \\ &= \cdots \\ &= [[\cdots[[E_{j_n}, E_{k_n}], E_{k_{n-1}}], \ldots, E_{k_2}]; E_{k_1}], \end{aligned}$$

with

$$1 \le k_n < j_n \le m \quad \text{and} \quad k_{l+1} \le k_l \quad \text{for } 1 \le l \le n-1.$$

This maximal expression for E_i involves n commutations (or, equivalently, there are n indices of the type k's), and then we set, by definition,

$$d(i) := n$$

with the convention

$$d(1) = d(2) = \cdots = d(m) := 0.$$

---

[1] A somewhat "canonical" way to order the elements of the Hall basis is the following one:

1. $x_1, x_2, \ldots, x_m$ is a fixed ordering of the standard monomials of height 1.
2. If $2 \le n \le r$, we now show how to canonically order the standard monomials of height n. Suppose the standard monomials of heights $1, \ldots, n-1$ have already been ordered. Let u and v be two standard monomials of height n in the form

$$u = [x_1, y_1], \qquad v = [x_2, y_2].$$

We then follow the rules:

- if $y_1 \ne y_2$, then u precedes v if and only if y_1 precedes y_2;
- if $y_1 = y_2$, then u precedes v if and only if x_1 precedes x_2.

This process associates in a natural way a multi-index $I(i) = (a_1, a_2, \dots, a_H)$ to E_i, where

$$a_s := \text{cardinality of the set } \{t \mid k_t = s\}.$$

In other words, for every $s = 1, \dots, H$, a_s is the number of times the element E_s appears within

$$E_{k_n}, E_{k_{n-1}}, \dots, E_{k_2}; E_{k_1}.$$

By definition, we set

$$I(1) = I(2) = \cdots = I(m) := (0, \dots, 0).$$

Moreover, we say that E_i is a *direct descendant* of all the elements of the type E_{j_l}

$$E_{j_1}, E_{j_2}, \dots, E_{j_{n-1}}, E_{j_n},$$

and we write

$$j_l \prec i.$$

We remark that $\prec$ is a partial ordering, and that if $a \prec b$, then every entry of $I(a)$ is $\leq$ of the corresponding entry of $I(b)$. For every pair a, b satisfying $a \prec b$, we define the monomial

$$a \prec b \quad \Longrightarrow \quad P_{a,b} := \frac{(-1)^{d(b)-d(a)}}{(I(b) - I(a))!}\, x^{I(b)-I(a)}.$$

For example, if

$$\begin{aligned} E_i &= [[[[[[[E_2, E_1], E_1], E_1], E_2], E_4], E_4]; E_7] \quad \text{and} \\ E_k &= [[[E_2, E_1], E_1]; E_1], \end{aligned}$$

then $d(i) = 7$, $d(k) = 3$, $d(2) = 0$ and

$$\begin{aligned} I(i) &= (3, 1, 0, 2, 0, 0, 1, 0, \dots, 0), \\ I(k) &= (3, 0, 0, 0, 0, 0, 0, 0, \dots, 0), \end{aligned}$$

whereas

$$P_{2,i} = -\frac{x_1^3 x_2 x_4^2 x_7}{3!\,2!}, \qquad P_{k,i} = \frac{x_2 x_4^2 x_7}{2!}.$$

The following remarkable result holds (see [GG90, Theorem 2.1]).

Theorem 14.1.10 (Grayson–Grossman, [GG90]). *Let $r \geq 1$, $m \geq 2$, and let H be the dimension of $\mathfrak{f}_{m,r}$. Then the vector fields*

$$\begin{aligned} E_1 &:= \frac{\partial}{\partial x_1}, \\ E_2 &:= \frac{\partial}{\partial x_2} + \sum_{j \succ 2} P_{2,j} \frac{\partial}{\partial x_j}, \\ &\vdots \\ E_m &:= \frac{\partial}{\partial x_m} + \sum_{j \succ m} P_{m,j} \frac{\partial}{\partial x_j} \end{aligned} \tag{14.1}$$

with polynomial coefficients on $\mathbb{R}^H$ have the following properties:

(1) *$E_1, \dots, E_m$ are homogeneous of degree one with respect to the dilation canonically induced on $\mathbb{R}^H$ by the stratification*

$$\mathbb{R}^H \equiv \mathfrak{f}_{m,r} = V_1 \oplus \cdots \oplus V_r,$$

where V_i is the vector space spanned by the standard monomials of the Hall basis with height i;

(2) *The elements of the Hall basis generated by the vector fields $E_1, \dots, E_m$ satisfy*

$$E_i(0) = \frac{\partial}{\partial x_i}, \quad i = 1, \dots, H;$$

(3) *The Lie algebra of vector fields on $\mathbb{R}^H$ generated by $E_1, \dots, E_m$ is isomorphic to $\mathfrak{f}_{m,r}$.*

Remark 14.1.11. We explicitly remark that the vector fields in (14.1) satisfy conditions **(H0)**–**(H1)**–**(H2)** in Section 4.2, page 191. Then they naturally define, as described in Section 4.2 (see, in particular, Theorem 4.2.10), a Carnot group $\mathbb{G}$ which is *free* (of step r and with m generators), according to Definition 14.1.3: indeed, by (3) in the above Theorem 14.1.10, we immediately infer that the Lie algebra of $\mathbb{G}$ is isomorphic to $\mathfrak{f}_{m,r}$. Moreover, by (2) in the above Theorem 14.1.10, the canonical sub-Laplacian $\Delta_{\mathbb{G}}$ is simply given by $\Delta_{\mathbb{G}} = \sum_{j=1}^m E_j^2$. □

Example 14.1.12. To end this section, we give an example of a model for $\mathfrak{f}_{4,2}$. Let E_1, E_2, E_3, E_4 be the generators of $\mathfrak{f}_{4,2}$. We set

$$\begin{aligned} &E_5 := [E_2, E_1], \quad E_6 := [E_3, E_1], \quad E_7 := [E_4, E_1], \\ &E_8 := [E_3, E_2], \quad E_9 := [E_4, E_2], \quad E_{10} := [E_4, E_3]. \end{aligned}$$

With the above notation, one has

$$\begin{aligned} &d(1) = d(2) = d(3) = d(4) = 0, \\ &d(5) = d(6) = d(7) = d(8) = d(9) = d(10) = 1, \\ &I(1) = I(2) = I(3) = I(4) = (0,0,0,0,0,0,0,0,0,0), \\ &I(5) = I(6) = I(7) = (1,0,0,0,0,0,0,0,0,0), \\ &I(8) = I(9) = (0,1,0,0,0,0,0,0,0,0), \\ &I(10) = (0,0,1,0,0,0,0,0,0,0), \\ &2 \prec 5, \quad 3 \prec 6, \quad 4 \prec 7, \quad 3 \prec 8, \quad 4 \prec 9, \quad 4 \prec 10, \\ &P_{2,5} = P_{3,6} = P_{4,7} = -x_1, \\ &P_{3,8} = P_{4,9} = -x_2, \\ &P_{4,10} = -x_3. \end{aligned}$$

Then, according to Theorem 14.1.10, a model for $\mathfrak{f}_{4,2}$ is given by the Lie algebra of vector fields generated by the following vector fields on $\mathbb{R}^{10}$:

$$E_1 := \frac{\partial}{\partial x_1},$$
$$E_2 := \frac{\partial}{\partial x_2} - x_1 \frac{\partial}{\partial x_5},$$
$$E_3 := \frac{\partial}{\partial x_3} - x_1 \frac{\partial}{\partial x_6} - x_2 \frac{\partial}{\partial x_8},$$
$$E_4 := \frac{\partial}{\partial x_4} - x_1 \frac{\partial}{\partial x_7} - x_2 \frac{\partial}{\partial x_9} - x_3 \frac{\partial}{\partial x_{10}}$$

with the commutator identities

$$E_5 = [E_2, E_1] = \frac{\partial}{\partial x_5},$$
$$E_6 = [E_3, E_1] = \frac{\partial}{\partial x_6},$$
$$E_7 = [E_4, E_1] = \frac{\partial}{\partial x_7},$$
$$E_8 = [E_3, E_2] = \frac{\partial}{\partial x_8},$$
$$E_9 = [E_4, E_2] = \frac{\partial}{\partial x_9},$$
$$E_{10} = [E_4, E_3] = \frac{\partial}{\partial x_{10}}.$$

As observed in Remark 14.1.11, the following second order differential operator on $\mathbb{R}^{10}$

$$\begin{aligned} &E_1^2 + E_2^2 + E_3^2 + E_4^2 \\ &\quad = (\partial_{x_1})^2 + (\partial_{x_2} - x_1\,\partial_{x_5})^2 \\ &\qquad + (\partial_{x_3} - x_1\,\partial_{x_6} - x_2\,\partial_{x_8})^2 + (\partial_{x_4} - x_1\,\partial_{x_7} - x_2\,\partial_{x_9} - x_3\,\partial_{x_{10}})^2 \end{aligned}$$

is then the canonical sub-Laplacian related to a suitable free homogeneous Carnot group on $\mathbb{R}^{10}$, with 4 generators and nilpotent of step 2. Finally, we explicitly remark that the above is not the only possible model for $\mathfrak{f}_{4,2}$ (see, e.g. Example 14.2.5 in Section 14.2).

14.2 A Canonical Way to Construct Free Carnot Groups

14.2.1 The Campbell–Hausdorff Composition ⋄

First of all, we recall some references on the so-called Campbell–Hausdorff formula. Roughly speaking, if X and Y are two non-commuting indeterminates, the (Baker–)Campbell–(Dynkin–)Hausdorff formula states that the formal expression

$$\text{“}\log(\exp(X)\exp(Y))\text{”}$$

can be expressed in terms of an infinite sum of iterated commutators of X and Y. This statement can be made precise in many contexts such as for formal power series, for matrix algebras, for general normed Banach algebras, for finite-dimensional Lie groups, for solutions of differential equations, etc. (See Bibliographical Notes at the end of the chapter.)

Since we are mainly interested in the setting of Lie groups and Lie algebras, we briefly recall how the Campbell–Hausdorff formula naturally arises in this context (see also Section 2.2, page 121, where we discussed this topic; for the reader's convenience, the main results we need here are reproduced). Let $(\mathfrak{n}, [\cdot,\cdot])$ be an abstract nilpotent Lie algebra. For $X, Y \in \mathfrak{n}$, we set[2]

$$\begin{aligned} X \diamond Y &:= \sum_{n\geq 1} \frac{(-1)^{n+1}}{n} \sum_{\substack{p_i+q_i\geq 1\\ 1\leq i\leq n}} \frac{(\operatorname{ad} \mathrm{X})^{p_1}(\operatorname{ad} \mathrm{Y})^{q_1}\cdots(\operatorname{ad} \mathrm{X})^{p_n}(\operatorname{ad} \mathrm{Y})^{q_n-1} Y}{(\sum_{j=1}^n (p_j+q_j))\, p_1! q_1! \cdots p_n! q_n!} \\ &= X + Y + \frac{1}{2}[X,Y] \\ &\quad + \frac{1}{12}[X,[X,Y]] - \frac{1}{12}[Y,[X,Y]] \\ &\quad - \frac{1}{48}[Y,[X,[X,Y]]] - \frac{1}{48}[X,[Y,[X,Y]]] \\ &\quad + \{\text{brackets of height} \geq 5\}. \end{aligned} \tag{14.2}$$

Since $\mathfrak{n}$ is nilpotent, (14.2) is a finite sum and $\diamond$ determines a binary operation on $\mathfrak{n}$, which is defined *in a universal way* by a sum of Lie monomials with rational coefficients.

Now, let $(\mathbb{F}, *)$ be a Lie group with the Lie algebra $\mathfrak{f}$. Suppose the exponential map $\operatorname{Exp} : \mathfrak{f} \to \mathbb{F}$ has an inverse function Log globally defined. Then the operation

$$\mathfrak{f} \times \mathfrak{f} \ni (X, Y) \mapsto \operatorname{Log}\big(\operatorname{Exp}(X) * \operatorname{Exp}(Y)\big) \in \mathfrak{f}$$

is well posed on $\mathfrak{f}$. With this last composition law, $\mathfrak{f}$ is obviously a Lie group isomorphic to $(\mathbb{F}, *)$ via Exp. The following result states that this last composition law is (under suitable hypotheses on $\mathbb{F}$) precisely the universal one defined in (14.2), i.e. the following *Campbell–Hausdorff formula* holds:

$$\operatorname{Log}\big(\operatorname{Exp}(X) * \operatorname{Exp}(Y)\big) = X \diamond Y \qquad \text{for all } X, Y \in \mathfrak{f}. \tag{14.3}$$

---

[2] Here, we use the notation $(\operatorname{ad} A)B = [A, B]$. Moreover, if $q_n = 0$, the term in the sum (14.2) is, by convention,

$$(\operatorname{ad} \mathrm{X})^{p_1}(\operatorname{ad} \mathrm{Y})^{q_1}\cdots(\operatorname{ad} \mathrm{X})^{p_{n-1}}(\operatorname{ad} \mathrm{Y})^{q_{n-1}}(\operatorname{ad} \mathrm{X})^{p_n-1} X.$$

Clearly, if $q_n > 1$, or $q_n = 0$ and $p_n > 1$, the term is zero.

Theorem 14.2.1 (Corwin and Greenleaf, [CG90], Theorem 1.2.1). *Let* $(\mathbb{F}, *)$ *be a connected and simply connected Lie group. Suppose that the Lie algebra* $\mathfrak{f}$ *of* $\mathbb{F}$ *is nilpotent. Let* $\diamond$ *be the operation defined in* (14.2). *Then* $\diamond$ *defines a Lie group structure on* $\mathfrak{f}$ *and* $\operatorname{Exp} : (\mathfrak{f}, \diamond) \to (\mathbb{F}, *)$ *is a Lie group isomorphism. In particular, if* Log *is the inverse function of* Exp, *then* $\operatorname{Log} : (\mathbb{F}, *) \to (\mathfrak{f}, \diamond)$ *is a Lie group isomorphism and* (14.3) *holds.*

We then recall the third fundamental theorem of Lie, which is another deep result in the theory of Lie groups.

Theorem 14.2.2 (Varadarajan, [Var84], Theorem 3.15.1). *Let* $\mathfrak{f}$ *be a finite-dimensional Lie algebra. Then there exists a connected and simply connected Lie group whose Lie algebra is isomorphic to* $\mathfrak{f}$.

From the above two theorems we obtain the following result (for its proof, see Corollary 2.2.15, page 130).

Theorem 14.2.3. *If* $\mathfrak{f}$ *is a finite-dimensional nilpotent Lie algebra, then the operation* $\diamond$ *introduced in* (14.2) *defines a Lie group structure on* $\mathfrak{f}$. *We call* $\diamond$ *the* Campbell–Hausdorff operation *on* $\mathfrak{f}$.

14.2.2 A Canonical Way to Construct Free Carnot Groups

We now apply the results of the previous section in order to show a standard way to construct Carnot groups on $\mathbb{R}^N$. This construction makes also use of $\mathfrak{f}_{m,r}$, the free nilpotent Lie algebra of step r with m generators, which we treated in Section 14.1.

Let $r \geq 1$ and $m \geq 2$ be given, and set $H := \dim(\mathfrak{f}_{m,r})$. Let $E_1, \dots, E_m$ be a system of generators for $\mathfrak{f}_{m,r}$. Suppose that $\{E_i\}_{i \leq H}$ is an enumeration of the Hall basis for $\mathfrak{f}_{m,r}$ which preserves the natural ordering $\leq$ given in Definition 14.1.5. We define a family of dilations $\{D_\lambda\}_{\lambda>0}$ on $\mathfrak{f}_{m,r}$ by

$$D_\lambda\left(\sum_{i=1}^{H} x_i E_i\right) = \sum_{i=1}^{H} \lambda^{\alpha_i} x_i E_i,$$

where α_i is the height of E_i. Finally, we identify $\mathfrak{f}_{m,r}$ with $\mathbb{R}^H$ in the natural way by introducing the linear isomorphism

$$\pi : \mathbb{R}^H \to \mathfrak{f}_{m,r} \quad \text{such that } x \mapsto \sum_{i=1}^{H} x_i\, E_i.$$

It is now not difficult to prove the following result (we explicitly observe that, by Theorem 14.2.3, $(\mathfrak{f}_{m,r}, \diamond)$ is a Lie group).

Theorem 14.2.4 (Free homogeneous Carnot groups). *With the above notation, for every* $x, y \in \mathbb{R}^H$ *and* $\lambda > 0$, *we set*

$$x \circ y := \pi^{-1}\big(\pi(x) \diamond \pi(y)\big), \qquad \delta_\lambda(x) := \pi^{-1}\big(D_\lambda(\pi(x))\big).$$

Then $\mathbb{G} := (\mathbb{R}^H, \circ, \delta_\lambda)$ *is a free homogeneous Carnot group.*

Example 14.2.5. We give a simple descriptive example of this construction for a free homogeneous Carnot group $\mathbb{G} = (\mathbb{R}^{10}, \circ, \delta_\lambda)$ of step 2 and with 4 generators. Suppose $\mathfrak{f}_{4,2}$ is generated by $E_1, \dots, E_4$. The Hall basis $E_1, \dots, E_{10}$ for $\mathfrak{f}_{4,2}$ is given by

$$E_1,\ E_2,\ E_3,\ E_4;\quad [E_2, E_1],\ [E_3, E_1],\ [E_4, E_1],\ [E_3, E_2],\ [E_4, E_2],\ [E_4, E_3].$$

The group of dilations on $\mathbb{G}$ is given by

$$\delta_\lambda(x) = (\lambda x_1, \dots, \lambda x_4, \lambda^2 x_5, \dots, \lambda^2 x_{10}). \tag{14.4}$$

The Campbell–Hausdorff formula on a Lie algebra nilpotent of step 2 is given by

$$X \diamond Y = X + Y + \frac{1}{2}[X, Y].$$

The group law $x \circ y$ is simply obtained by explicitly writing

$$\Big(\sum_{i \le 10} x_i E_i\Big) \diamond \Big(\sum_{i \le 10} y_i E_i\Big)$$

with respect to the Hall basis, making use of bilinearity, skew-symmetry and the Jacobi identity

$$\begin{aligned}
&\sum_{i \le 10} x_i E_i \diamond \sum_{i \le 10} y_i E_i \\
&\quad = \sum_{i \le 10} x_i E_i + \sum_{i \le 10} y_i E_i + \frac{1}{2} \sum_{i,j \le 10} x_i y_j [E_i, E_j] = \cdots \\
&\quad = \sum_{i \le 4} (x_i + y_i) E_i + \Big(x_5 + y_5 + \frac{1}{2}(x_2 y_1 - x_1 y_2)\Big) E_5 + \cdots \\
&\quad\quad + \Big(x_{10} + y_{10} + \frac{1}{2}(x_4 y_3 - x_3 y_4)\Big) E_{10}.
\end{aligned}$$

This yields, for every $x,\ y \in \mathbb{G}$,

$$x \circ y = \begin{pmatrix}
x_1 + y_1 \\
x_2 + y_2 \\
x_3 + y_3 \\
x_4 + y_4 \\
x_5 + y_5 + \frac{1}{2}(x_2\, y_1 - x_1\, y_2) \\
x_6 + y_6 + \frac{1}{2}(x_3\, y_1 - x_1\, y_3) \\
x_7 + y_7 + \frac{1}{2}(x_4\, y_1 - x_1\, y_4) \\
x_8 + y_8 + \frac{1}{2}(x_3\, y_2 - x_2\, y_3) \\
x_9 + y_9 + \frac{1}{2}(x_4\, y_2 - x_2\, y_4) \\
x_{10} + y_{10} + \frac{1}{2}(x_4\, y_3 - x_3\, y_4)
\end{pmatrix}.$$

With the above composition law $\circ$ and the dilation δ_λ defined in (14.4), $\mathbb{G} = (\mathbb{R}^{10}, \circ, \delta_\lambda)$ is a free homogeneous Carnot group of step 2 and with 4 generators.

Now, by making use of (1.34) (page 19), it is very simple to find the first 4 elements $Z_1, \dots, Z_4$ of the Jacobian basis of the algebra of $\mathbb{G}$: given $\varphi \in C^\infty(\mathbb{R}^{10})$, we have

$$\begin{aligned}
(Z_1\varphi)(x) &= (\partial/\partial y_1)|_{y=0}\varphi(x \circ y) \\
&= \partial_{x_1}\varphi(x) + \frac{1}{2}x_2\,\partial_{x_5}\varphi(x) + \frac{1}{2}x_3\,\partial_{x_6}\varphi(x) + \frac{1}{2}x_4\,\partial_{x_7}\varphi(x), \\
(Z_2\varphi)(x) &= (\partial/\partial y_2)|_{y=0}\varphi(x \circ y) \\
&= \partial_{x_2}\varphi(x) - \frac{1}{2}x_1\,\partial_{x_5}\varphi(x) + \frac{1}{2}x_3\,\partial_{x_8}\varphi(x) + \frac{1}{2}x_4\,\partial_{x_9}\varphi(x), \\
(Z_3\varphi)(x) &= (\partial/\partial y_3)|_{y=0}\varphi(x \circ y) \\
&= \partial_{x_3}\varphi(x) - \frac{1}{2}x_1\,\partial_{x_6}\varphi(x) - \frac{1}{2}x_2\,\partial_{x_8}\varphi(x) + \frac{1}{2}x_4\,\partial_{x_{10}}\varphi(x), \\
(Z_4\varphi)(x) &= (\partial/\partial y_4)|_{y=0}\varphi(x \circ y) \\
&= \partial_{x_4}\varphi(x) - \frac{1}{2}x_1\,\partial_{x_7}\varphi(x) - \frac{1}{2}x_2\,\partial_{x_9}\varphi(x) - \frac{1}{2}x_3\,\partial_{x_{10}}\varphi(x).
\end{aligned}$$

As a consequence, the following second order differential operator on $\mathbb{R}^{10}$

$$\begin{aligned}
&Z_1^2 + Z_2^2 + Z_3^2 + Z_4^2 \\
&\quad = \left(\partial_{x_1} + \frac{1}{2}x_2\,\partial_{x_5} + \frac{1}{2}x_3\,\partial_{x_6} + \frac{1}{2}x_4\,\partial_{x_7}\right)^2 \\
&\qquad + \left(\partial_{x_2} - \frac{1}{2}x_1\,\partial_{x_5} + \frac{1}{2}x_3\,\partial_{x_8} + \frac{1}{2}x_4\,\partial_{x_9}\right)^2 \\
&\qquad + \left(\partial_{x_3} - \frac{1}{2}x_1\,\partial_{x_6} - \frac{1}{2}x_2\,\partial_{x_8} + \frac{1}{2}x_4\,\partial_{x_{10}}\right)^2 \\
&\qquad + \left(\partial_{x_4} - \frac{1}{2}x_1\,\partial_{x_7} - \frac{1}{2}x_2\,\partial_{x_9} - \frac{1}{2}x_3\,\partial_{x_{10}}\right)^2
\end{aligned}$$

is the canonical sub-Laplacian related to a free homogeneous Carnot group on $\mathbb{R}^{10}$, with 4 generators and nilpotent of step 2 (for another example, see Example 14.1.12).

Bibliographical Notes. The motivation for this brief investigation on free Lie algebras is primarily due to the Rothschild and Stein lifting theorem, according to which any sum of squares of Hörmander vector fields can be "approximated as close as we want" (in a suitable sense; see [RS76]) by a sub-Laplacian on a (larger) manifold, which is a stratified group whose Lie algebra is *free*.

14.3 Exercises of Chapter 14

Ex. 1) Prove that $\mathfrak{f}_{m,r}$ is isomorphic to $\mathfrak{f}_{m',r'}$ if and only if $m = m'$ and $r = r'$.

Ex. 2) Write down the Hall bases for $\mathfrak{f}_{2,3}$, $\mathfrak{f}_{2,4}$ and $\mathfrak{f}_{4,3}$.

Ex. 3) This exercise concerns with the canonical way of constructing Carnot groups via the Campbell–Hausdorff formula described in Section 14.2. Suppose it is given an algebra generated by the two elements X_1, X_2 with the following commutator relations (indeed, it should be noted that these relations are consistent with the properties of the bracket operation: bilinearity, skew-symmetry and the Jacobi identity; see Definition 2.1.39): X_1, X_2, $[X_1, X_2]$ and $[X_1, [X_1, X_2]]$ are linearly independent, whereas there exists $\alpha \in \mathbb{R}$ such that

$$[X_2, [X_1, X_2]] = \alpha\, [X_1, [X_1, X_2]]$$

and all commutators of height ≥ 4 vanish.
Consider the Campbell–Hausdorff operation for an algebra nilpotent of step 3

$$X \diamond Y = X + Y + \frac{1}{2}\,[X, Y] + \frac{1}{12}\,[X, [X, Y]] - \frac{1}{12}\,[Y, [X, Y]].$$

Write down explicitly the Campbell–Hausdorff operation

$$\begin{aligned}&\big(x_1 X_1 + x_2 X_2 + x_3\,[X_1, X_2] + x_4\,[X_1, [X_1, X_2]]\big)\\ &\qquad \diamond \big(y_1 X_1 + y_2 X_2 + y_3\,[X_1, X_2] + y_4\,[X_1, [X_1, X_2]]\big)\end{aligned}$$

and write the result with respect to the basis X_1, X_2, $[X_1, X_2]$ and $[X_1, [X_1, X_2]]$. Denote the 4-tuple of the coordinates thus obtained as $(x_1, x_2, x_3, x_4) \circ (y_1, y_2, y_3, y_4)$. Then, arguing as in Theorem 14.2.4, $\circ$ defines on $\mathbb{R}^4$ a homogeneous Carnot group. Verify that $\circ$ is given by

$$\begin{pmatrix} x_1 + y_1 \\ x_2 + y_2 \\ x_3 + y_3 + \frac{1}{2}(x_1\, y_2 - x_2\, y_1) \\ x_4 + y_4 + \frac{1}{2}(x_1\, y_3 - x_3\, y_1) + \frac{\alpha}{2}(x_2\, y_3 - x_3\, y_2) \\ \quad + \frac{1}{12}(x_1 - y_1)\,(x_1\, y_2 - x_2\, y_1) + \frac{\alpha}{12}(x_2 - y_2)\,(x_1\, y_2 - x_2\, y_1) \end{pmatrix}.$$

Ex. 4) We follow the ideas in Ex. 1 to re-derive the Heisenberg–Weyl group on $\mathbb{R}^3$: the Campbell–Hausdorff operation for a Lie algebra $\mathfrak{h}$ of step 2 has the form

$$X \diamond Y = X + Y + \frac{1}{2}[X, Y].$$

We consider an algebra of step two with the basis Z_1, Z_2 and $Z_3 := [Z_1, Z_2]$. Then we have

$$\begin{aligned}&\big(\xi_1\, Z_1 + \xi_2\, Z_2 + \xi_3\, Z_3\big) \diamond \big(\eta_1\, Z_1 + \eta_2\, Z_2 + \eta_3\, Z_3\big)\\ &\quad = \big(\xi_1\, Z_1 + \xi_2\, Z_2 + \xi_3\, Z_3\big) + \big(\eta_1\, Z_1 + \eta_2\, Z_2 + \eta_3\, Z_3\big)\end{aligned}$$

$$+\frac{1}{2}[\xi_1 Z_1 + \xi_2 Z_2 + \xi_3 Z_3, \eta_1 Z_1 + \eta_2 Z_2 + \eta_3 Z_3]$$

$\big([Z_1, Z_2] = Z_3, [Z_2, Z_1] = -Z_3$ and commutators of height > 2 vanish$\big)$

$$= (\xi_1 + \eta_1) Z_1 + (\xi_2 + \eta_2) Z_2 + \left(\xi_3 + \eta_3 + \frac{1}{2}\xi_1 \eta_2 - \frac{1}{2}\eta_1 \xi_2\right) Z_3.$$

By the natural identification

$$\mathfrak{h} \ni \xi_1 Z_1 + \xi_2 Z_2 + \xi_3 Z_3 \longleftrightarrow (\xi_1, \xi_2, \xi_3) \in \mathbb{R}^3,$$

we have re-obtained the Heisenberg–Weyl group $\mathbb{H}^1$ on $\mathbb{R}^3$ whose composition law is

$$\xi \circ \eta = \left(\xi_1 + \eta_1, \xi_2 + \eta_2, \xi_3 + \eta_3 + \frac{1}{2}\xi_1 \eta_2 - \frac{1}{2}\eta_1 \xi_2\right).$$

Carry out a similar exercise with an algebra of step two whose basis is

$$X_1, \dots, X_N; \quad Y_1, \dots, Y_N; \qquad [X_1, Y_1]$$

with the following commutator identities:

a) $[X_j, Y_j] = [X_1, Y_1]$ for all $j = 1, \dots, N$,
b) $[X_i, X_j] = 0 = [Y_i, Y_j]$ for all $i, j = 1, \dots, N$,
c) $[X_i, Y_j] = 0$ for all $i, j = 1, \dots, N$ such that $i \neq j$.

Recognize the Lie group in (4.16), page 201.

Ex. 5) Give a detailed proof of Theorem 14.2.4.

Ex. 6) By means of Example 14.1.7 and the ideas in Section 14.2, find a Carnot group whose Lie algebra is isomorphic to $\mathfrak{f}_{4,2}$. Finally, consider the general case of $\mathfrak{f}_{m,2}$.

Ex. 7) **a)** Use Grayson–Grossman's Theorem 14.1.10 to find vector fields generating $\mathfrak{f}_{5,2}$.

b) Show that they satisfy conditions **(H0)**–**(H1)**–**(H2)** in Section 4.2 (page 191) and construct the relevant homogeneous Carnot group as described in that section.

c) Construct a Carnot group whose Lie algebra is isomorphic to $\mathfrak{f}_{5,2}$ following the ideas in Section 14.2. Compare to point **(b)**.

Ex. 8) Let $m, r \in \mathbb{N}$ ($m \geq 2$) and consider $\mathfrak{f}_{m,r}$. Set $H(m, r) = \dim(\mathfrak{f}_{m,r})$. Finally, let $\mathfrak{f}$ be any finite-dimensional (real) Lie algebra and set $H := \dim(\mathfrak{f})$. What of the following conditions are sufficient in order to have $\mathfrak{f}_{m,r}$ and $\mathfrak{f}$ isomorphic as Lie algebras?

a) $H(m, r) = H$. (*Hint*: Consider $\mathfrak{f}_{2,2}$ and $\mathfrak{f}_{3,1}$ or, equivalently, the Lie algebra of the Heisenberg–Weyl group $\mathbb{H}^1$ on $\mathbb{R}^3$ and the Lie algebra of the Euclidean group $(\mathbb{R}^3, +)$.)

b) $H(m, r) = H$ and $\mathfrak{f}_{m,r}$ and $\mathfrak{f}$ are nilpotent of step r. (*Hint*: Consider $\mathfrak{f}_{3,2}$ and the Lie algebra $\mathfrak{f}$, nilpotent of step 2, whose basis is $\{X, Y, A, B, C, [X, Y]\}$ such that A, B, C commute with each other and with X and Y; roughly speaking, $\mathfrak{f}$ is the Lie algebra of the sum of $\mathbb{H}^1$ and $(\mathbb{R}^3, +)$ (see Section 4.1.5, page 190).)

c) $\mathfrak{f}_{m,r}$ and $\mathfrak{f}$ are both nilpotent of step r and Lie-generated by m of their elements. (*Hint*: Consider $\mathfrak{f}_{3,2}$ and the Lie algebra $\mathfrak{f}$, nilpotent of step 2, whose basis is $\{X, Y, A, [X, Y]\}$ such that A commutes with X and Y; roughly speaking, $\mathfrak{f}$ is the Lie algebra of the sum of $\mathbb{H}^1$ and $(\mathbb{R}^1, +)$.)

d) $H(m, r) = H$ and $\mathfrak{f}_{m,r}$ and $\mathfrak{f}$ are Lie-generated by m of their elements. (*Hint*: Consider $\mathfrak{f}_{3,2}$ and the Lie algebra $\mathfrak{f}$, nilpotent of step 3, whose basis is

$$\big\{X_1, X_2, A, X_3 := [X_1, X_2],\\ X_4 := [X_1, [X_1, X_2]], X_5 := [X_1, [X_1, [X_1, X_2]]]\big\},$$

such that A commutes with X_i $(i = 1, \dots, 5)$, $[X_1, X_5] = 0$, $[X_i, X_j] = 0$ for every $1 < i < j \le 5$. Roughly speaking, $\mathfrak{f}$ is obtained from the filiform Lie algebra $\mathfrak{h}$ in Example 4.3.5 (page 208) when $i = 5$, by adding an extra generator A commuting with every other generator.)

e) $H(m, r) = H$ and $\mathfrak{f}_{m,r}$ and $\mathfrak{f}$ are nilpotent of step r and both are Lie-generated by m of their elements.

Hint: Condition (**e**) is actually sufficient. Indeed, suppose

$$\mathfrak{f} = \text{Lie}\{F_1, \dots, F_m\}$$

(here $F_1, \dots, F_m \in \mathfrak{f}$) is nilpotent of step r. Let also $x_1, \dots, x_m$ denote the generators for $\mathfrak{f}_{m,r}$, as in Definition 14.1.1.

According to this very definition (replacing $\mathfrak{n}$ with $\mathfrak{f}$, what is possible *since* $\mathfrak{f}$ *is nilpotent of step* r), if we consider the map φ sending x_i in F_i (for every $i = 1, \dots, m$), there exists a homomorphism of Lie algebras $\widetilde{\varphi}$ from $\mathfrak{f}_{m,r}$ to $\mathfrak{f}$ which extends φ.

We are done if we show that $\widetilde{\varphi}$ is a vector space isomorphism. *Since* $\mathfrak{f}_{m,r}$ *and* $\mathfrak{f}$ *have the same dimension*, it suffices to prove that $\widetilde{\varphi}$ is onto. This follows by the argument below

$$\begin{aligned}\mathfrak{f} &= \text{Lie}\{F_1, \dots, F_m\} = \text{Lie}\{\varphi(x_1), \dots, \varphi(x_m)\}\\ &= \text{Lie}\{\widetilde{\varphi}(x_1), \dots, \widetilde{\varphi}(x_m)\} = \widetilde{\varphi}(\text{Lie}\{x_1, \dots, x_m\}) = \widetilde{\varphi}(\mathfrak{f}_{m,r}).\end{aligned}$$

Provide the proofs of the above equalities.

15

More on the Campbell–Hausdorff Formula

The main aim of this chapter is to sketch a proof of Lemma 4.2.4 in Section 4.2 (page 194). This lemma played a crucial rôle in the construction of Carnot groups. We could refer to this lemma as a sort of *Campbell–Hausdorff formula for stratified vector fields*.

Our proof here is articulated in the following way. In Section 15.1, we show *in details* how to derive Lemma 4.2.4 from a general version of the Campbell–Hausdorff formula *for formal power series*. Then, in Sections 15.2 and 15.3, we provide two proofs of this formula for formal power series: the former is due to D. Djoković [Djo75], the latter is due to M. Eichler [Eic68].

The first proof makes use of results from the theory of formal power series; the second one rests on some properties of Lie polynomials. For the sake of brevity, we shall leave to the reader the details of the proofs of these properties, which are actually very intuitive. As a result, the proofs of the Campbell–Hausdorff formula are surprisingly simple and short, as in the spirit of the cited authors.

Finally, Section 15.4 provides the version of the Campbell–Hausdorff formula mostly cited in analytic contexts. Once again, it is unavoidable to make use of the formula for formal power series. Furthermore, the main task is to show how to control and give precise estimates of the remainder terms in the relevant series expansions.

15.1 A Proof of the Campbell–Hausdorff Formula for Stratified Vector Fields

We recall the notation of Section 4.2: $X_1, \ldots, X_m$ is a given set of smooth vector fields on $\mathbb{R}^N$ satisfying hypotheses **(H0)**–**(H1)**–**(H2)** of Section 4.2, i.e. the following conditions are fulfilled:

(H0) $X_1, \ldots, X_m$ are linearly independent and δ_λ-homogeneous of degree one with respect to a suitable family of dilations $\{\delta_\lambda\}_{\lambda>0}$ of the following type

$$\delta_\lambda : \mathbb{R}^N \to \mathbb{R}^N, \quad \delta_\lambda(x) = \delta_\lambda(x^{(1)}, \ldots, x^{(r)}) := (\lambda x^{(1)}, \ldots, \lambda^r x^{(r)}),$$

where $r \geq 1$ is an integer, $x^{(i)} \in \mathbb{R}^{N_i}$ for $i = 1, \ldots, r$, $N_1 = m$ and $N_1 + \cdots + N_r = N$.

(H1) We set $W^{(k)} = \text{span}\{X_J \mid J \in \{1, \ldots, m\}^k\}$, where

$$X_{(j_1,\ldots,j_k)} = [X_{j_1}, \ldots [X_{j_{k-1}}, X_{j_k}] \ldots]$$

if $J = (j_1, \ldots, j_k)$. Then, $\dim(W^{(k)}) = \dim\{XI(0) : X \in W^{(k)}\}$ for every $k = 1, \ldots, r$.

(H2) $\dim(\text{Lie}\{X_1, \ldots, X_m\}I(0)) = N$.

Let

$$\mathfrak{a} = \text{Lie}\{X_1, \ldots, X_m\}.$$

For every $k = 1, \ldots, r$, $Z_1^{(k)}, \ldots, Z_{N_k}^{(k)}$ will be a fixed basis for $W^{(k)}$. We know that

$$\{Z_1, \ldots, Z_N\} := \{Z_1^{(1)}, \ldots, Z_{N_1}^{(1)}, \ldots, Z_1^{(r)}, \ldots Z_{N_r}^{(r)}\}$$

is a basis for $\mathfrak{a}$. For every $\xi = (\xi_1, \ldots, \xi_N) = (\xi^{(1)}, \ldots, \xi^{(r)})$ we set

$$\xi \cdot Z = \sum_{j=1}^{N} \xi_j Z_j = \sum_{k=1}^{r} \sum_{j=1}^{N_k} \xi_j^{(k)} Z_j^{(k)}.$$

We know that[1] the map $(x, t) \mapsto \exp(t\xi \cdot Z)(x)$ is well defined for every $x \in \mathbb{R}^N$ and $t \in \mathbb{R}$. Moreover,

$$\text{Exp} : \mathbb{R}^N \longrightarrow \mathbb{R}^N, \quad \text{Exp}\,(\xi) := \exp(\xi \cdot Z)(0)$$

is a global diffeomorphism.[2] Its inverse function is denoted by Log. In Section 4.2, we defined a composition law on $\mathbb{R}^N$ by means of

$$x,\ y \in \mathbb{R}^N, \quad x \circ y := \exp(\text{Log}\,(y) \cdot Z)(x).$$

Our aim there was to show that $\mathbb{G} := (\mathbb{R}^N, \circ, \delta_\lambda)$ is a Carnot group whose Lie algebra $\mathfrak{g}$ coincides with $\mathfrak{a}$. The main task of the proof definitely was to show that $\circ$ *is associative*. To this end, we needed the cited Lemma 4.2.4, which stated that

For every $X, Y \in \mathfrak{a}$, there exists a unique $V \in \mathfrak{a}$ such that

$$\exp(Y)\big(\exp(X)(x)\big) = \exp(V)(x) \tag{15.1}$$

for every $x \in \mathbb{R}^N$. In this section, we at last prove (15.1) as a consequence of the following theorem.

---

[1] We recall that, given a smooth vector field X on $\mathbb{R}^N$, $\exp(tX)(x) := \gamma(t, x)$ denotes the solution $\gamma(\cdot) = \gamma(\cdot, x)$ to the ordinary differential system

$$\begin{cases} \dot{\gamma} = XI(\gamma), \\ \gamma(0) = x. \end{cases}$$

[2] We explicitly remark that we are using the notation Exp to denote a map on $\mathbb{R}^N$ instead of a map on an algebra of vector fields.

Theorem 15.1.1 (The Campbell–Hausdorff formula for stratified vector fields). *Let $X, Y \in \mathfrak{a}$ be fixed. Let V be the differential operator defined by the formal expansion*

$$\sum_{j=1}^{r} \frac{(-1)^{j+1}}{j} \left(\sum_{k_1+k_2=1}^{r} \frac{X^{k_2}\, Y^{k_1}}{k_2!\, k_1!} \right)^j$$
$$= V + \left\{ \text{summands of the type } Y^{y_n} X^{x_n} \cdots Y^{y_1} X^{x_1} \text{ with } \sum_{i=1}^{n} (y_i + x_i) > r \right\}. \tag{15.2}$$

Then V turns out to be a vector field belonging to Lie$\{X, Y\}$ *(hence belonging to $\mathfrak{a}$) such that*

$$\exp(Y)\big(\exp(X)(x)\big) = \exp(V)(x) \quad \forall\, x \in \mathbb{R}^N. \tag{15.3}$$

We explicitly remark that (15.3) proves (15.1), i.e. it proves Lemma 4.2.4. For example, a direct computation shows that

$$V = X \diamond Y := X + Y + \frac{1}{2}[X, Y] + \frac{1}{12}[X, [X, Y]] - \frac{1}{12}[Y, [X, Y]]$$
$$- \frac{1}{48}[Y, [X, [X, Y]]] - \frac{1}{48}[X, [Y, [X, Y]]] + \cdots.$$

We recognize here the Campbell–Hausdorff composition law $X \diamond Y$ defined in (14.2), page 585.

Before proceeding with the proof of Theorem 15.1.1, we give an example of how the formal expansion in (15.2) behaves. For instance, taking $r = 3$, we have

$$\sum_{j=1}^{3} \frac{(-1)^{j+1}}{j} \left(\sum_{k_1+k_2=1}^{3} \frac{X^{k_1}\, Y^{k_2}}{k_1!\, k_2!} \right)^j$$
$$= \sum_{j=1}^{3} \frac{(-1)^{j+1}}{j} \left(\frac{X^0 Y^1}{0!\,1!} + \frac{X^1 Y^0}{1!\,0!} + \frac{X^0 Y^2}{0!\,2!} + \frac{X^1 Y^1}{1!\,1!} + \frac{X^2 Y^0}{2!\,0!} \right.$$
$$\left. + \frac{X^0 Y^3}{0!\,3!} + \frac{X^1 Y^2}{1!\,2!} + \frac{X^2 Y^1}{2!\,1!} + \frac{X^3 Y^0}{3!\,0!} \right)^j$$
$$= \sum_{j=1}^{3} \frac{(-1)^{j+1}}{j} \left(Y + X + \frac{Y^2}{2} + XY + \frac{X^2}{2} + \frac{Y^3}{6} + \frac{XY^2}{2} + \frac{X^2Y}{2} + \frac{X^3}{6} \right)^j$$
$$= \left(Y + X + \frac{Y^2}{2} + XY + \frac{X^2}{2} + \frac{Y^3}{6} + \frac{XY^2}{2} + \frac{X^2Y}{2} + \frac{X^3}{6} \right)^1$$
$$- \frac{1}{2}\left(Y + X + \frac{Y^2}{2} + XY + \frac{X^2}{2} + \frac{Y^3}{6} + \frac{XY^2}{2} + \frac{X^2Y}{2} + \frac{X^3}{6} \right)^2$$
$$+ \frac{1}{3}\left(Y + X + \frac{Y^2}{2} + XY + \frac{X^2}{2} + \frac{Y^3}{6} + \frac{XY^2}{2} + \frac{X^2Y}{2} + \frac{X^3}{6} \right)^3$$

$$
\begin{aligned}
&= Y + X + \frac{Y^2}{2} + XY + \frac{X^2}{2} + \frac{Y^3}{6} + \frac{XY^2}{2} + \frac{X^2Y}{2} + \frac{X^3}{6} \\
&\quad - \frac{1}{2}\Big(Y^2 + YX + \frac{Y^3}{2} + YXY + \frac{YX^2}{2} + XY + X^2 + \frac{XY^2}{2} + X^2Y \\
&\quad + \frac{X^3}{2} + \frac{Y^3}{2} + \frac{Y^2X}{2} + XY^2 + XYX + \frac{X^2Y}{2} + \frac{X^3}{2}\Big) \\
&\quad + \frac{1}{3}(Y^3 + Y^2X + YXY + YX^2 + XY^2 + XYX + X^2Y + X^3) \\
&\quad + \{\text{summands of height} \geq 4\} \\
&= X + Y + X^2\Big(\frac{1}{2} - \frac{1}{2}\Big) + Y^2\Big(\frac{1}{2} - \frac{1}{2}\Big) + XY\Big(1 - \frac{1}{2}\Big) + YX\Big(-\frac{1}{2}\Big) \\
&\quad + X^3\Big(\frac{1}{6} - \frac{1}{4} - \frac{1}{4} + \frac{1}{3}\Big) + Y^3\Big(\frac{1}{6} - \frac{1}{4} - \frac{1}{4} + \frac{1}{3}\Big) \\
&\quad + X^2Y\Big(\frac{1}{2} - \frac{1}{2} - \frac{1}{4} + \frac{1}{3}\Big) \\
&\quad + Y^2X\Big(-\frac{1}{4} + \frac{1}{3}\Big) + XY^2\Big(\frac{1}{2} - \frac{1}{4} - \frac{1}{2} + \frac{1}{3}\Big) + YX^2\Big(-\frac{1}{4} + \frac{1}{3}\Big) \\
&\quad + YXY\Big(-\frac{1}{2} + \frac{1}{3}\Big) + XYX\Big(-\frac{1}{2} + \frac{1}{3}\Big) \\
&\quad + \{\text{summands of height} \geq 4\} \\
&= X + Y + \frac{1}{2}XY - \frac{1}{2}YX + \frac{1}{12}X^2Y + \frac{1}{12}Y^2X + \frac{1}{12}XY^2 \\
&\quad + \frac{1}{12}YX^2 - \frac{1}{6}YXY - \frac{1}{6}XYX \\
&\quad + \{\text{summands of height} \geq 4\} \\
&= X + Y + \frac{1}{2}[X, Y] + \frac{1}{12}[X, [X, Y]] - \frac{1}{12}[Y, [X, Y]] \\
&\quad + \{\text{summands of height} \geq 4\}.
\end{aligned}
$$

Proof (of Theorem 15.1.1). Throughout the proof, $X, Y \in \mathfrak{a}$ are fixed. We begin by noticing that (see the arguments in Remark 1.1.3, page 9) the map $\mathbb{R} \times \mathbb{R}^N \ni (t, x) \mapsto \exp(tX)(x)$ has polynomial component functions. Moreover, we have

$$
\sum_{i=1}^{n} (y_i + x_i) > r \quad \Rightarrow \quad Y^{y_n} X^{x_n} \cdots Y^{y_1} X^{x_1} I \equiv 0. \tag{15.4}
$$

This follows by recalling that any field in $\mathfrak{a}$ is a sum of vector fields δ_λ-homogeneous of degree at least 1 and by observing that the component functions of the identity map I are δ_λ-homogeneous monomials of degree at most r. Since $\exp(tX)(x)$ is an analytic function of t, (1.7) (page 7) gives

$$
\exp(X) \equiv \sum_{k=0}^{r} \frac{1}{k!} X^k I.
$$

We henceforth fix $x \in \mathbb{R}^N$ and set

$$\Phi(t_1, t_2) := \exp(t_1 Y)\big(\exp(t_2 X)(x)\big), \quad t_1,\ t_2 \in \mathbb{R}.$$

Clearly, any component function of Φ is a polynomial in $t_1,\ t_2$. From (1.15) (page 10), for every smooth vector field Z on $\mathbb{R}^N$, $f \in C^\infty(\mathbb{R}^N, \mathbb{R}^N)$, $k \in \mathbb{N}$ and every $z \in \mathbb{R}^N$, we have

$$\left(\frac{\mathrm{d}^k}{\mathrm{d}t^k}\right)_{t=0} \big(f(\exp(tZ)(z))\big) = (Z^k f)(z). \tag{15.5}$$

This gives, for every $k_1,\ k_2 \in \mathbb{N}$,

$$\begin{aligned}
&\left.\left(\frac{\partial^{k_1+k_2}}{\partial t_1^{k_1} \partial t_2^{k_2}}\right)\right|_{(t_1,t_2)=(0,0)} \Phi(t_1, t_2) \\
&= \left(\frac{\partial^{k_2}}{\partial t_2^{k_2}}\right)_{t_2=0} \left(\frac{\partial^{k_1}}{\partial t_1^{k_1}}\right)_{t_1=0} I\big(\exp(t_1 Y)\big(\exp(t_2 X)(x)\big)\big) \\
&= \left(\frac{\partial^{k_2}}{\partial t_2^{k_2}}\right)_{t_2=0} (Y^{k_1} I)\big(\exp(t_2 X)(x)\big) \\
&= (X^{k_2} Y^{k_1} I)(x).
\end{aligned} \tag{15.6}$$

Indeed, in the second equality we used (15.5) with $f = I$, $k = k_1$, $Z = Y$, $z = \exp(t_2 X)(x)$, whereas in the third equality we used (15.5) with $f = Y^{k_1} I$, $k = k_2$, $Z = X$, $z = x$.

We now use Taylor's formula for Φ. Since Φ is a polynomial, exploiting (15.4), we derive

$$\exp(Y)\big(\exp(X)(x)\big) = \Phi(1, 1) = \sum_{k_1+k_2=0}^{r} \frac{1}{k_1!\,k_2!} (X^{k_2} Y^{k_1} I)(x). \tag{15.7}$$

We now introduce the following higher order differential operator

$$W(t, X, Y) := \sum_{j=1}^{r} \frac{(-1)^{j+1}}{j} \left(\sum_{k_1+k_2=1}^{r} \frac{t^{k_1+k_2}}{k_1!\,k_2!} X^{k_2} Y^{k_1} \right)^j.$$

We formally expand $W(t, X, Y)$, and we order it as a polynomial in t setting

$$\begin{aligned}
W(t, X, Y) &= \sum_{k=1}^{r} t^k Z_k(X, Y) + \sum_{k=r+1}^{r^2} t^k Z_k(X, Y) \\
&=: V(t, X, Y) + R(t, X, Y).
\end{aligned}$$

We explicitly remark that the differential operator V appearing in (15.2) in the statement of the theorem is simply given by $V(t, X, Y)$ with $t = 1$. It is easy to recognize that, by (15.4), any power of $R(t, X, Y)$ annihilates the identity map, i.e.

$$(R(t, X, Y))^k I \equiv 0 \quad \text{for every } k \geq 0. \tag{15.8}$$

We claim that

$$Z_k(X, Y) \in \mathrm{Lie}\{X, Y\} \quad \text{for every } k \geq r. \tag{15.9}$$

Before proving the claimed (15.9), we first show that (15.9) implies the assertion of the theorem. Indeed, if we set

$$V := V(1, X, Y),$$

then from the definition of $V(t, X, Y)$ and (15.9) we derive that

$$V \in \mathrm{Lie}\{X, Y\}.$$

Finally, we have to prove that (15.3) is satisfied with the above choice of V. Indeed, we have

$$\begin{aligned}
\exp(V)(x) &= \sum_{k=0}^{r} \frac{1}{k!} V^k I(x) \\
&= \sum_{k=0}^{r} \frac{1}{k!} \big(V(1, X, Y) + R(1, X, Y)\big)^k I(x) \\
&= \sum_{k=0}^{r} \frac{1}{k!} \big(W(1, X, Y)\big)^k I(x) \\
&= \sum_{k=0}^{r} \frac{1}{k!} \left\{ \sum_{j=1}^{r} \frac{(-1)^{j+1}}{j} \left(\sum_{k_1+k_2=1}^{r} \frac{1}{k_1!\,k_2!} X^{k_2} Y^{k_1} \right)^j \right\}^k I(x) \\
&= I(x) + \sum_{k_1+k_2=1}^{r} \frac{1}{k_1!\,k_2!} X^{k_2} Y^{k_1} I(x) = \exp(Y)\big(\exp(X)(x)\big).
\end{aligned}$$

We now motivate every equality appearing above:

1) the first equality follows from the fact that $V \in \mathfrak{a}$ and from (1.7);
2) the second one follows from (15.8) and simple homogeneity arguments;
3) the third one is the very definition of $W(1, X, Y)$;
4) the fourth equality is a consequence of the formal power series expansion of the identity $1 + x = \exp(\log(1 + x))$, jointly with (15.4): more precisely, the fourth equality is a consequence of the identity

$$\sum_{k=0}^{r} \frac{1}{k!} \left\{ \sum_{j=1}^{r} \frac{(-1)^{j+1}}{j} H^j \right\}^k = I + H + \sum_{j=r+1}^{r^2} c_j H^j \qquad \forall H \in \mathfrak{a}$$

   and of (15.4);
5) the last equality follows from (15.7).

Finally, we are left with the proof of the claimed (15.9). A proof of it can be derived from a general result by Djoković [Djo75], which we present in Section 15.2. Indeed, in [Djo75] an assertion analogous to (15.9) is derived in the more general context[3] of the *formal power series in two non-commuting indeterminates X and Y*. This completes the proof of the theorem. □

Remark 15.1.2. An odd fact, arising from the proof of Theorem 15.1.1, has to be remarked: from (15.7) and the fact that $\exp(X) = \sum_{k=0}^{r} \frac{1}{k!} X^k I$ (see (1.7)) we derived

$$\sum_{k_1=0}^{r} \frac{Y^{k_1}}{k_1!} I \left(\sum_{k_2=0}^{r} \frac{X^{k_2}}{k_2!} I(x) \right) = \sum_{k_1+k_2=0}^{r} \frac{1}{k_2!\,k_1!} X^{k_2} Y^{k_1} I(x).$$

The order of composition of X and Y seems reversed, on the face of it. Thus, formal computations such as

$$\text{“} \quad \sum_{k_1} \frac{Y^{k_1}}{k_1!} \sum_{k_2} \frac{X^{k_2}}{k_2!} = \sum_{k_1+k_2} \frac{1}{k_1!\,k_2!} Y^{k_1} X^{k_2} \quad \text{”}$$

are erroneous in our context of composition of exponential maps. This is another reason why we proved in detail our Campbell–Hausdorff-type Theorem 15.1.1 rather than reducing it, with no further justification, to formal power series arguments.

15.2 A Proof of the Campbell–Hausdorff Formula for Formal Power Series

The aim of this section is to present the arguments (with few more details) in [Djo75] for a proof of the Campbell–Hausdorff formula for formal power series. Here X and Y will denote general *non-commuting indeterminates in a multiplicative algebra over the field* $\mathbb{Q}$ *of the rational numbers* (for instance, X and Y may be vector fields on $\mathbb{R}^N$, as in the previous section). We shall perform computations in the algebra of *formal power series* in X and Y.

We first fix a very natural[4] notation:

$$\mathbf{e}^X := \sum_{k \geq 0} \frac{X^k}{k!} \tag{15.10}$$

[3] At this point, a simple remark has to be given, in order to justify the fact that we can replace general calculus between two non-commuting indeterminates X and Y with our vector fields $X, Y \in \mathfrak{a}$. To this end, we explicitly remark that the identities between formal power series used in [Djo75] in order to prove (15.9) readily reduce, in our context, to identities between *finite sums* of vector fields: it will be enough to make use of arguments such as (15.4).

[4] As Remark 15.1.2 has shown, though this is a very natural definition, when X is a vector field on $\mathbb{R}^N$, $\mathbf{e}^X$ should not be confused with the exponential $\exp(X)$ of a vector field as the integral curve to the system $\dot{\gamma} = X(\gamma)$.

denotes the formal expression of the exponential of X. The Campbell–Hausdorff formula simply states that there exists Z which is a formal infinite sum of summands in $\mathrm{Lie}\{X, Y\}$ such that $\mathbf{e}^Y \cdot \mathbf{e}^X = \mathbf{e}^Z$. More precisely, see Theorem 15.2.2 below.

In the next result and for the future references, we need the definition of Lie-polynomial.

Definition 15.2.1 (Lie polynomial). *If $\mathbb{K}$ is any field, we denote by $\mathbb{K}[X, Y]$ the vector space (over $\mathbb{K}$) of the polynomials in two non-commuting indeterminates X, Y, i.e. $\mathbb{K}[X, Y]$ is $\mathbb{K}$-spanned by the formal expressions*

$$X^{\alpha_1} Y^{\beta_1} \dots X^{\alpha_i} Y^{\beta_i}$$

(called, a monomial of degree $\alpha_1 + \beta_1 + \dots + \alpha_i + \beta_i$), where $i \in \mathbb{N}$ and $\alpha_1, \beta_1, \cdots, \alpha_i, \beta_i \in \mathbb{N} \cup \{0\}$. The degree of a polynomial is the maximum degree of its monomials.

Let X and Y be two non-commuting indeterminates in an associative algebra. Set $[X, Y] = XY - YX$ (which we call a Lie-bracket). *Given $k \in \mathbb{N}$, $k \geq 2$, and $Z_1, \dots, Z_k \in \{X, Y\}$, we say that the Lie-bracket*

$$[Z_1[Z_2[Z_3[\cdots[Z_{k-1}, Z_k]\cdots]]]],$$

is a Lie(-bracket) monomial of length k in X and Y. We also say that X and Y are Lie-monomials of length 1.

A linear combination of Lie monomials of length k in X and Y is called a homogeneous Lie-polynomial of length k. Any finite sum of Lie-bracket monomials is called a Lie-polynomial. Analogous definitions are given in the setting of Lie algebras.

Obviously, $\mathbb{K}[X, Y]$ can be equipped with an associative algebra structure in an obvious way.[5] We are now ready to state and prove the following theorem.

Theorem 15.2.2 (The Campbell–Hausdorff formula for formal power series). *For every positive integer k, there exists a homogeneous Lie-bracket-polynomial Z_k of length k (with rational coefficients) in two non-commuting indeterminates such that, using the notation $Z_k(\cdot, \cdot)$ to express the dependence of Z_k on the two indeterminates, we have:*

$$\text{if } Z(X, Y) = \sum_{k=1}^{\infty} Z_k(X, Y) \quad \text{then } \mathbf{e}^Y \cdot \mathbf{e}^X = \mathbf{e}^Z \tag{15.11}$$

for every couple of non-commuting indeterminates X, Y. For example,

$$Z_1(X, Y) = X + Y,$$
$$Z_2(X, Y) = \frac{1}{2}[Y, X],$$

[5] For example, $XY^2 \cdot Y^3X^4 = XY^5X^4$ and $XY^2 \cdot X^4 = XY^2X^4$, $XY^2X^5 \cdot XY^3X^4Y = XY^2X^6Y^3X^4Y$, etc.

$$Z_3(X,Y) = \frac{1}{12}[Y,[Y,X]] - \frac{1}{12}[X,[Y,X]],$$
$$Z_4(X,Y) = -\frac{1}{48}[X,[Y,[Y,X]]] - \frac{1}{48}[Y,[X,[Y,X]]]. \tag{15.12}$$

Proof. For every $t \in \mathbb{R}$, we have

$$\begin{aligned}\mathbf{e}^{t\,Y}\cdot\mathbf{e}^{t\,X} &= \sum_{k_2\geq 0}\frac{(t\,Y)^{k_2}}{k_2!}\cdot\sum_{k_1\geq 0}\frac{(t\,X)^{k_1}}{k_1!} = \sum_{k_1,k_2\geq 0}\frac{t^{k_1+k_2}}{k_1!k_2!}Y^{k_2}X^{k_1}\\ &= I + \sum_{k_1+k_2\geq 1}\frac{t^{k_1+k_2}}{k_1!k_2!}Y^{k_2}X^{k_1} =: I + H_{X,Y}(t).\end{aligned}$$

If $H_{X,Y}(t)$ is as above, we set

$$Z_{X,Y}(t) := \ln(I + H_{X,Y}(t)) = \sum_{j\geq 1}\frac{(-1)^{j+1}}{j}\left(\sum_{k_1+k_2\geq 1}\frac{t^{k_1+k_2}}{k_1!k_2!}Y^{k_2}X^{k_1}\right)^j. \tag{15.13a}$$

From the formal power series expansion of the identity $e^{\ln(1+x)} = 1 + x$, it then follows that

$$\mathbf{e}^{Z_{X,Y}(t)} = I + H_{X,Y}(t) = \mathbf{e}^{t\,Y}\cdot\mathbf{e}^{t\,X}. \tag{15.13b}$$

Now, (15.11) is formally fulfilled with the natural choice

$$Z(X,Y) := Z_{X,Y}(1). \tag{15.13c}$$

The theorem is then proved if we show that $Z_{X,Y}(1)$ can be put in the form $Z(X,Y) = \sum_{k=1}^{\infty} Z_k(X,Y)$ with the Z_k's as in the assertion of the theorem. To this end, it suffices to prove the following power series expansion w.r.t. t:

$$Z_{X,Y}(t) = \sum_{k=1}^{\infty} t^k\cdot Z_k(X,Y), \tag{15.13d}$$

where

$$Z_k(X,Y) \in \mathrm{Lie}\{X,Y\} \quad \text{for every } k \in \mathbb{N} \tag{15.13e}$$

and $Z_k(X,Y)$ is homogeneous of length k.

To begin with, we observe that, differentiating the far right-hand side of (15.13b) w.r.t. t (and writing henceforth $Z(t)$ instead of $Z_{X,Y}(t)$), we have

$$\begin{aligned}\partial_t\left(\mathbf{e}^{Z(t)}\right) &= \partial_t\left(\mathbf{e}^{t\,Y}\cdot\mathbf{e}^{t\,X}\right) = \left(\partial_t\mathbf{e}^{t\,Y}\right)\cdot\mathbf{e}^{t\,X} + \mathbf{e}^{t\,Y}\cdot\left(\partial_t\mathbf{e}^{t\,X}\right)\\ &= Y\cdot\mathbf{e}^{t\,Y}\cdot\mathbf{e}^{t\,X} + \mathbf{e}^{t\,Y}\cdot\mathbf{e}^{t\,X}\cdot X = Y\cdot\mathbf{e}^{Z(t)} + \mathbf{e}^{Z(t)}\cdot X.\end{aligned}$$

Hence, we obtain

$$\partial_t\left(\mathbf{e}^{Z(t)}\right)\cdot\mathbf{e}^{-Z(t)} = \mathbf{e}^{Z(t)}\cdot X\cdot\mathbf{e}^{-Z(t)} + Y. \tag{15.13f}$$

As a general fact, it holds

$$\mathbf{e}^Z \cdot X \cdot \mathbf{e}^{-Z} = \sum_{m\geq 0} \frac{Z^m}{m!} \cdot X \cdot \sum_{k\geq 0} (-1)^k \frac{Z^k}{k!} = \sum_{m,k\geq 0} (-1)^k \frac{Z^m \cdot X \cdot Z^k}{m!\,k!}.$$

Then, setting $m+k =: n$, the expression for $\mathbf{e}^Z \cdot X \cdot \mathbf{e}^{-Z}$ becomes[6]

$$\mathbf{e}^Z \cdot X \cdot \mathbf{e}^{-Z} = \sum_{n\geq 0} \frac{1}{n!} \sum_{k=0}^{n} (-1)^k \binom{n}{k} Z^{n-k} \cdot X \cdot Z^k. \tag{15.13g}$$

Throughout the sequel, we use the following classical notation for the left and right multiplication:

$$\begin{aligned} &\mathrm{L}_Z(X) := Z \cdot X, \quad \mathrm{R}_Z(X) := X \cdot Z, \\ &(\mathrm{Ad}\, Z)(X) := \big(\mathrm{L}_Z - \mathrm{R}_Z\big)(X) = Z \cdot X - X \cdot Z. \end{aligned}$$

Obviously, L_Z and R_Z commute.[7] As a consequence, we can apply Newton's binomial expansion to get

$$\begin{aligned} \big(\mathrm{Ad}\, Z\big)^n(X) = \big(\mathrm{L}_Z - \mathrm{R}_Z\big)^n(X) &= \sum_{k=0}^{n} (-1)^k \binom{n}{k} \mathrm{L}_Z^{n-k} \cdot \mathrm{R}_Z^k\,(X) \\ &= \sum_{k=0}^{n} (-1)^k \binom{n}{k} Z^{n-k} \cdot X \cdot Z^k. \end{aligned} \tag{15.13h}$$

Taking into account (15.13h), the expression (15.13g) turns out to be

$$\mathbf{e}^Z \cdot X \cdot \mathbf{e}^{-Z} = \sum_{n\geq 0} \frac{1}{n!} \big(\mathrm{Ad}\, Z\big)^n(X) =: \mathbf{e}^{\mathrm{Ad}\, Z}(X).$$

This gives (see (15.13f))

$$\partial_t\big(\mathbf{e}^{Z(t)}\big) \cdot \mathbf{e}^{-Z(t)} = Y + \mathbf{e}^{\mathrm{Ad}\, Z(t)}(X). \tag{15.13i}$$

On the other hand, a direct computation yields

$$\begin{aligned} \partial_t\big(\mathbf{e}^{Z(t)}\big) \cdot \mathbf{e}^{-Z(t)} &= \Bigg(\sum_{m\geq 1} \frac{1}{m!} \partial_t(Z(t)^m)\Bigg) \cdot \Bigg(\sum_{k\geq 0} \frac{(-1)^k}{k!} Z(t)^k\Bigg) \\ (k =: n-m) &= \sum_{n\geq 1} \frac{1}{n!} \sum_{m=1}^{n} (-1)^{n-m} \binom{n}{m} \partial_t(Z(t)^m)\, Z(t)^{n-m} =: (\star). \end{aligned}$$

---

[6] Here we wrote

$$\frac{1}{m!\,k!} = \binom{n}{k} \frac{1}{n!}.$$

[7] We obviously have $\mathrm{L}_Z(\mathrm{R}_Z(X)) = Z \cdot X \cdot Z = \mathrm{R}_Z(\mathrm{L}_Z(X))$.

Thus, since one has

$$\partial_t(Z(t)^m) = \sum_{k=0}^{m-1} Z(t)^k \cdot (\partial_t Z(t)) \cdot Z(t)^{m-k-1},$$

we derive that

$$(\star) = \sum_{n\geq 1} \frac{1}{n!} \sum_{m=1}^{n} \sum_{k=0}^{m-1} (-1)^{n-m} \binom{n}{m} Z(t)^k \cdot (\partial_t Z(t)) \cdot Z(t)^{n-k-1}$$

$$= \sum_{n\geq 1} \frac{1}{n!} \sum_{k=0}^{n-1} \left(\sum_{m=k+1}^{n} (-1)^{n-m} \binom{n}{m} \right) Z(t)^k \cdot (\partial_t Z(t)) \cdot Z(t)^{n-k-1}$$

(we set $r := n - m$ and use properties of the binomial coefficients)

$$= \sum_{n\geq 1} \frac{1}{n!} \sum_{k=0}^{n-1} \left(\sum_{r=0}^{n-k-1} (-1)^r \binom{n}{r} \right) Z(t)^k \cdot (\partial_t Z(t)) \cdot Z(t)^{n-k-1} =: (\star\star).$$

An inductive argument shows that

$$\sum_{r=0}^{m} (-1)^r \binom{n}{r} = (-1)^m \binom{n-1}{m}, \qquad 0 \leq m \leq n-1.$$

Consequently, we infer

$$(\star\star) = \sum_{n\geq 1} \frac{1}{n!} \sum_{k=0}^{n-1} \left((-1)^{n-k-1} \binom{n-1}{k} \right) Z(t)^k \cdot (\partial_t Z(t)) \cdot Z(t)^{n-k-1}$$

$$= \sum_{n\geq 1} \frac{1}{n!} \sum_{j=0}^{n-1} \left((-1)^j \binom{n-1}{j} \right) Z(t)^{n-1-j} \cdot (\partial_t Z(t)) \cdot Z(t)^j$$

(see (15.13h)) $$= \sum_{n\geq 1} \frac{1}{n!} (\mathrm{Ad}\, Z(t))^{n-1} \big(\partial_t Z(t)\big).$$

Finally, comparing to (15.13i), we have

$$Y + \mathbf{e}^{\mathrm{Ad}\, Z(t)}(X) = \sum_{n\geq 1} \frac{1}{n!} (\mathrm{Ad}\, Z(t))^{n-1} \big(\partial_t Z(t)\big),$$

i.e. by the definition of the exponential of $\mathrm{Ad}\, Z(t)$,

$$Y + \sum_{n\geq 0} \frac{1}{n!} (\mathrm{Ad}\, Z(t))^n\, (X) = \sum_{n\geq 1} \frac{1}{n!} (\mathrm{Ad}\, Z(t))^{n-1} \big(\partial_t Z(t)\big). \tag{15.13j}$$

Now, the power series $\sum_{n\geq 1} x^{n-1}/n!$ furnishes the formal expansion of $\varphi(x) := (e^x - 1)/x$. Then, since $\varphi(0) = 1 \neq 0$, there exists a reciprocal series expansion of $\sum_{n\geq 1} x^{n-1}/n!$, say $g(x) := \sum_{n\geq 0} a_n\, x^n$. More precisely, the following formal identity holds

$$\Big(\sum_{n\geq 0} a_n\, x^n\Big) \cdot \Big(\sum_{n\geq 1} x^{n-1}/n!\Big) = 1.$$

As a consequence, multiplying both sides of (15.13j) times $g(\operatorname{Ad} Z(t))$, we obtain

$$\begin{aligned} & g\big(\operatorname{Ad} Z(t)\big)\Big(Y + \sum_{n\geq 0} \frac{1}{n!}(\operatorname{Ad} Z(t))^n\,(X)\Big) \\ &= \sum_{n\geq 0} a_n (\operatorname{Ad} Z(t))^n \cdot \Big(\sum_{n\geq 1} \frac{1}{n!}(\operatorname{Ad} Z(t))^{n-1}\big(\partial_t\, Z(t)\big)\Big) \\ &= \partial_t\, Z(t). \end{aligned}$$

This proves that $Z(t)$ satisfies the *differential equation*

$$\partial_t\, Z(t) = g(\operatorname{Ad} Z(t))\Big(Y + \sum_{n\geq 0} \frac{1}{n!}(\operatorname{Ad} Z(t))^n\,(X)\Big). \tag{15.13k}$$

By reordering the sum in (15.13a) which defines $Z(t)$, we obtain

$$Z(t) = \sum_{j\geq 1} \frac{(-1)^{j+1}}{j}\Big(\sum_{k=1}^{\infty}\Big(\sum_{k_1+k_2=k} \frac{1}{k_1!\,k_2!}\, Y^{k_2}\, X^{k_1}\Big) t^k\Big)^j.$$

If we set

$$b_k := \sum_{k_1+k_2=k} \frac{1}{k_1!\,k_2!}\, Y^{k_2}\, X^{k_1}, \qquad k \geq 1,$$

by means of the formula for the j-th power of a power series, we derive

$$\begin{aligned} Z(t) &= \sum_{j\geq 1} \frac{(-1)^{j+1}}{j} \sum_{k=1}^{\infty} t^k \Big(\sum_{\substack{n_1+\cdots+n_j=k \\ n_1,\ldots,n_j\geq 1}} b_{n_1}\cdots b_{n_j}\Big) \\ &= \sum_{k=1}^{\infty} t^k \Big(\sum_{j=1}^{\infty} \frac{(-1)^{j+1}}{j} \sum_{\substack{n_1+\cdots+n_j=k \\ n_1,\ldots,n_j\geq 1}} b_{n_1}\cdots b_{n_j}\Big) \\ &= \sum_{k=1}^{\infty} t^k \Big(\sum_{j=1}^{\infty} \frac{(-1)^{j+1}}{j} \sum_{\substack{(r_1+s_1)+\cdots+(r_j+s_j)=k \\ r_1+s_1\geq 1,\ldots,r_j+s_j\geq 1}} \frac{Y^{r_1} X^{s_1} \ldots Y^{r_j} X^{s_j}}{r_1!s_1!\cdots r_j!s_j!}\Big) \\ &=: \sum_{k=1}^{\infty} t^k \cdot Z_k(X,Y). \end{aligned}$$

Following this notation, the left-hand side in (15.13k) (i.e. the derivative of $Z(t)$) equals

$$\sum_{k=0}^{\infty} t^k \cdot (k+1)\, Z_{k+1}(X, Y).$$

Following again the notation $Z(t) = \sum_{k \geq 1} t^k\, Z_k$, recalling that $g(x) = \sum_{n \geq 0} a_n\, x^n$, and by using the formula for the power of a series in t, the right-hand side of (15.13k) equals

$$\begin{aligned} X + Y &+ \sum_{k=1}^{\infty} t^k \sum_{n=1}^{\infty} \frac{1}{n!} \sum_{m_1+\cdots+m_n=k} \operatorname{Ad} Z_{m_1} \cdots \operatorname{Ad} Z_{m_n}(X) \\ &+ \sum_{k=1}^{\infty} t^k \sum_{n=1}^{\infty} a_n \sum_{m_1+\cdots+m_n=k} \operatorname{Ad} Z_{m_1} \cdots \operatorname{Ad} Z_{m_n} \Bigg(X + Y \\ &+ \sum_{l=1}^{\infty} t^l \sum_{j=1}^{\infty} \frac{1}{j!} \sum_{h_1+\cdots+h_j=l} \operatorname{Ad} Z_{h_1} \cdots \operatorname{Ad} Z_{h_j}(X) \Bigg). \end{aligned}$$

It has to be noticed that the coefficient of t^k in this last expression is a linear combination of terms of the type

$$\operatorname{Ad} Z_{m_1} \cdots \operatorname{Ad} Z_{m_n}(X) \quad \text{and} \quad \operatorname{Ad} Z_{m_1} \cdots \operatorname{Ad} Z_{m_n}(Y)$$

with $m_1 + \cdots + m_n = k$. In particular, by equating the coefficients of t^k from both sides of (15.13k), one has

$$Z_1(X, Y) = X + Y,$$

whereas $Z_2(X, Y)$ is a linear combination of commutators of height 2 of $X + Y$ with X (and Y); moreover $Z_3(X, Y)$ is a linear combination of commutators of height 3 of $Z_2(X, Y)$ (or $X + Y$) and X (or Y) and so on. An inductive argument now proves that $Z_k(X, Y)$ is a linear combination of commutators of height k of X and Y. In particular, $Z_k(X, Y) \in \operatorname{Lie}\{X, Y\}$ for every $k \in \mathbb{N}$. This actually demonstrates (15.13e), and the proof is complete. □

15.3 Another Proof of the Campbell–Hausdorff Formula for Formal Power Series

In this section, we provide another proof of the Campbell–Hausdorff formula for formal power series, as given by M. Eichler in [Eic68]. This proof essentially rests on some properties of Lie polynomials. (In due course of the proof, we shall leave to the reader the details for these properties, which are indeed very intuitive.)

In the rest of the section, we consider the associative algebra of formal power series in two non-commuting indeterminates A, B over the field of the rational numbers $\mathbb{Q}$. In particular, we shall use the definition of the exponential as in (15.10) and the notion of Lie-polynomial (see Definition 15.2.1). We are now ready to give another proof of Theorem 15.2.2.

Proof (of Theorem 15.2.2). We aim to prove that

$$\mathbf{e}^{A}\cdot\mathbf{e}^{B}=\mathbf{e}^{\sum_{n=1}^{\infty}F_n(A,B)}, \tag{15.14}$$

where, for every $n\in\mathbb{N}$, $F_n(A,B)$ is a *Lie-polynomial* in A, B which is homogeneous of length n.

First of all, we remark that it certainly exist unique homogeneous polynomials F_n's in A, B satisfying (15.14) (note that we are not yet asking for *Lie*-polynomials). Indeed, there exists a unique formal power series Z in A, B such that $\mathbf{e}^{A}\cdot\mathbf{e}^{B}=\mathbf{e}^{Z}$. It suffices to take

$$Z:=\mathbf{log}(\mathbf{e}^{A}\cdot\mathbf{e}^{B})=\mathbf{log}\Big(1+\sum_{n+m\geq 1}A^n/n!\,B^m/m!\Big)$$

or, more precisely,

$$Z=\sum_{j\geq 1}\frac{(-1)^{j+1}}{j}\Big(\sum_{n+m\geq 1}A^n/n!\,B^m/m!\Big)^j.$$

Then we reorder the summands defining Z in such a (unique) way that they are homogeneous polynomials in A, B. Our main task is to prove that the resulting F_n's are Lie-polynomials in A, B.

We find F_1 and F_2. F_1 results from the choice $(j=1, n=1, m=0)$ or from $(j=1, n=0, m=1)$ thus giving

$$F_1(A,B)=A+B.$$

F_2 results from the choice

$$(j,n,m)\in\{(1,1,1),\ (1,2,0),\ (1,0,2),\ (2,1,0),\ (2,0,1)\},$$

so that

$$\begin{aligned}F_2(A,B)&=AB+\frac{A^2}{2}+\frac{B^2}{2}-\frac{1}{2}(A+B)^2\\&=AB+\frac{A^2}{2}+\frac{B^2}{2}-\frac{1}{2}(A^2+AB+BA+B^2)\\&=\frac{1}{2}AB-\frac{1}{2}BA=\frac{1}{2}[A,B].\end{aligned}$$

Hence, F_1 and F_2 are homogeneous Lie-polynomials of the desired degree. We now argue by induction. Let $n\geq 3$ and suppose that $F_1,\dots,F_{n-1}$ are Lie-polynomials of degree $1,\dots,n-1$, respectively. We are done if we show that F_n is a Lie-polynomial (its degree is obviously n for it is the sum of polynomials of degree n).

The crucial device is to take *three* non-commuting indeterminates A, B, C and to use associativity to equal the expansions of

$$(\mathbf{e}^A \cdot \mathbf{e}^B) \cdot \mathbf{e}^C = \mathbf{e}^A \cdot (\mathbf{e}^B \cdot \mathbf{e}^C). \tag{15.15}$$

By (using twice) (15.14), the left-hand side of (15.15) is

$$\mathbf{e}^{\sum_{j=1}^{\infty} F_j(A,B)} \cdot \mathbf{e}^C = \mathbf{e}^{\sum_{i=1}^{\infty} F_i\left(\sum_{j=1}^{\infty} F_j(A,B),C\right)}.$$

The reader should appreciate that we are here using the universality of the F_j's as functions of two "arbitrary" non-commuting indeterminates.[8] Analogously, the right-hand side of (15.15) is

$$\mathbf{e}^A \cdot \mathbf{e}^{\sum_{j=1}^{\infty} F_j(B,C)} = \mathbf{e}^{\sum_{i=1}^{\infty} F_i\left(A,\sum_{j=1}^{\infty} F_j(B,C)\right)}.$$

We thus have

$$W := \sum_{i=1}^{\infty} F_i\left(\sum_{j=1}^{\infty} F_j(A,B), C\right) = \sum_{i=1}^{\infty} F_i\left(A, \sum_{j=1}^{\infty} F_j(B,C)\right). \tag{15.16}$$

It is easy to verify that whenever F_i and F_j are Lie-polynomials in two indeterminates, then $F_i(F_j(A,B),C)$ and $F_i(A,F_j(B,C))$ are Lie-polynomials in A, B, C. Moreover, it is also not difficult to verify that the *homogeneous* polynomials which are the summands into which a Lie-polynomial splits up are homogeneous Lie-polynomials.

We now compare homogeneous terms of degree n (in A, B, C) on both sides of (15.16). Such terms appear in summands of the type

$$F_i(F_j(A,B),C), \qquad F_i(A,F_j(B,C))$$

if and only if $i = 1, j = n$; $i = n, j = 1$; or $i + j = n$ (in the case $i > 1$, $j > 1$). Thus, with the remarks of the preceding paragraph at hand, by our inductive assumption, we see that all homogeneous terms of degree n in both expressions for W are Lie-polynomials, except at most for

$$F_n(A,B) + F_n(A+B,C)$$

in the left-hand side of (15.16) and except for

$$F_n(A,B+C) + F_n(B,C)$$

in its right-hand side.

For the sake of brevity, we introduce the equivalence relation $\sim$ between polynomials p_1, p_2 in A, B, C by saying that $p_1 \sim p_2$ whenever $p_1 - p_2$ is a Lie-polynomial. The assertion made at the end of the preceding paragraph thus rewrites as

$$F_n(A,B) + F_n(A+B,C) \sim F_n(A,B+C) + F_n(B,C). \tag{15.17}$$

[8] After all, the arbitrary nature of the non-commuting indeterminates lies in the very concept of "indeterminates".

We now make some trivial remarks on the $\sim$ relation. If $\alpha, \beta \in \mathbb{Q}$, then (since $\alpha\, A$ and $\beta\, A$ commute) we have

$$\mathbf{e}^{\alpha\, A} \cdot \mathbf{e}^{\beta\, A} = \mathbf{e}^{\alpha\, A + \beta\, A} = \mathbf{e}^{F_1(\alpha\, A, \beta\, A)}.$$

Consequently, for every $n \geq 2$, $F_n(\alpha\, A, \beta\, A) = 0$, i.e.

$$F_n(\alpha\, A, \beta\, A) \sim 0. \tag{15.18}$$

In particular, this holds for $\alpha = 1$ and $\beta = 0$.

Now, we show that (15.17) and (15.18) imply

$$F_n(A, B) \sim 0,$$

which is our goal. First, the choice $C := -B$ in (15.17) yields

$$F_n(A, B) \sim -F_n(A + B, -B) + F_n(A, 0) + F_n(B, -B) \sim -F_n(A + B, -B),$$

i.e.

$$F_n(A, B) \sim -F_n(A + B, -B). \tag{15.19}$$

Analogously, the choice $A := -B$ in (15.17) yields

$$F_n(B, C) \sim F_n(-B, B) + F_n(0, C) - F_n(-B, B + C) \sim -F_n(-B, B + C),$$

i.e. by renaming $B \mapsto A$, $C \mapsto B$,

$$F_n(A, B) \sim -F_n(-A, A + B). \tag{15.20}$$

Applying (15.20), (15.19) and again (15.20), we get

$$\begin{aligned} F_n(A, B) &\sim -F_n(-A, A + B) \sim -\big(- F_n(-A + A + B, -A - B)\big) \\ &= F_n(B, -A - B) \sim -F_n(-B, B - A - B) = -F_n(-B, -A) \\ &= -(-1)^n\, F_n(B, A). \end{aligned}$$

The last equality is a consequence of the length-homogeneity n of F_n as a polynomial. This gives

$$F_n(A, B) \sim (-1)^{n+1}\, F_n(B, A). \tag{15.21}$$

Now, we insert $C = -\frac{1}{2}\, B$ in (15.17). This gives

$$\begin{aligned} F_n(A, B) &\sim -F_n\Big(A + B, -\frac{1}{2}\, B\Big) + F_n\Big(A, B - \frac{1}{2}\, B\Big) + F_n\Big(B, -\frac{1}{2}\, B\Big) \\ &\sim -F_n\Big(A + B, -\frac{1}{2}\, B\Big) + F_n\Big(A, \frac{1}{2}\, B\Big), \end{aligned}$$

i.e.

$$F_n(A, B) \sim F_n\Big(A, \frac{1}{2}\, B\Big) - F_n\Big(A + B, -\frac{1}{2}\, B\Big). \tag{15.22}$$

Analogously, taking $A := -\frac{1}{2}B$ in (15.17) and renaming $B \mapsto A$, $C \mapsto B$, we get

$$F_n(A,B) \sim F_n\Big(\frac{1}{2}A, B\Big) - F_n\Big(-\frac{1}{2}A, A+B\Big). \tag{15.23}$$

An application of (15.22) to both summands in the right-hand side of (15.23) yields

$$\begin{aligned} F_n(A,B) \sim [\cdots] = {} & F_n\Big(\frac{1}{2}A, \frac{1}{2}B\Big) - F_n\Big(\frac{1}{2}A+B, -\frac{1}{2}B\Big) \\ & - F_n\Big(-\frac{1}{2}A, \frac{1}{2}A + \frac{1}{2}B\Big) + F_n\Big(\frac{1}{2}A+B, -\frac{1}{2}A - \frac{1}{2}B\Big). \end{aligned}$$

Now, we apply (15.20) to the third summand in the far right-hand of the above expression and (15.19) to the second and fourth summands. We thus get

$$\begin{aligned} F_n(A,B) \sim {} & F_n\Big(\frac{1}{2}A, \frac{1}{2}B\Big) + F_n\Big(\frac{1}{2}A + \frac{1}{2}B, \frac{1}{2}B\Big) \\ & + F_n\Big(\frac{1}{2}A, \frac{1}{2}B\Big) - F_n\Big(\frac{1}{2}B, \frac{1}{2}A + \frac{1}{2}B\Big) \\ = {} & 2^{1-n}\, F_n(A,B) + 2^{-n}\, F_n(A+B,B) - 2^{-n}\, F_n(B, A+B) \\ \sim {} & 2^{1-n}\, F_n(A,B) + 2^{-n}(1+(-1)^n)\, F_n(A+B,B). \end{aligned}$$

In the equality we used the degree-homogeneity n of F_n, and we used (15.21) for the last $\sim$ sign. This gives

$$(1-2^{1-n})\, F_n(A,B) \sim 2^{-n}(1+(-1)^n)\, F_n(A+B,B), \tag{15.24}$$

which proves that $F_n(A,B) \sim 0$ whenever $n \neq 1$ in odd. As for the case when n is even, we argue as follows. We write $A-B$ instead of A in (15.24)

$$(1-2^{1-n})\, F_n(A-B,B) \sim 2^{-n}(1+(-1)^n)\, F_n(A,B),$$

and we use (15.19) in the above left-hand side getting (after multiplication times $(1-2^{1-n})^{-1}$)

$$F_n(A,-B) \sim -(1-2^{1-n})^{-1}\, 2^{-n}(1+(-1)^n)\, F_n(A,B). \tag{15.25}$$

We write $-B$ instead of B in (15.25), and we finally apply (15.25) once again obtaining

$$F_n(A,B) \sim (1-2^{1-n})^{-2}\, 2^{-2n}(1+(-1)^n)^2\, F_n(A,B).$$

When n is even and $n \neq 2$, this is possible only if $F_n(A,B) \sim 0$. This ends the proof. □

15.4 The Campbell–Hausdorff Formula for General Smooth Vector Fields

At this point of the exposition, the Campbell–Hausdorff formula has been exhaustively proved in the abstract setting of formal power series and in the more concrete setting of the vector fields in the algebra of a Carnot group. As a matter of fact, it is easy to recognize that our proof of Theorem 15.1.1 can be applied for any couple of vector fields X, Y which are δ_λ-homogeneous of positive degree with respect to any dilation $\delta_\lambda : \mathbb{R}^N \to \mathbb{R}^N$ of the form

$$\delta_\lambda(x_1, \dots, x_N) = (\lambda^{\alpha_1} x_1, \dots, \lambda^{\alpha_N} x_N)$$

with positive α_i's. The reason for this fact is that the cited δ_λ-homogeneity ensures that $\exp(X)$, $\exp(Y)$ are indeed given by finite sums and there exists a large $m \in \mathbb{N}$ such that $X^m I,\ Y^m I \equiv 0$. As a consequence, all the computations with formal power series in two indeterminates make precise sense, and the given proof of the Campbell–Hausdorff formula for formal power series can be straightforwardly adapted.

We have so far provided the proof of the Campbell–Hausdorff formula for many interesting cases in literature, but we are now concerned in giving the most useful version of such an important formula. Theorem 15.4.1 below is the version which analysts mostly refer to. For example, it is, almost verbatim, the version needed in the very recent paper [Mor00, Proposition 2.3] or a restatement of Proposition 4.3 in the pioneering paper [NSW85] by Nagel, Stein and Wainger or, furthermore, an analogue of the statement needed for the celebrated lifting by Rothschild and Stein (see [RS76, Section 10]). The aim of this section is to provide its proof.

As usual, given a smooth vector field X on an open set $\Omega \subseteq \mathbb{R}^N$ and fixed $x \in \Omega$, we denote by

$$t \mapsto \gamma(t) = \exp(tX)(x)$$

the solution to the first order differential system in $\mathbb{R}^N$

$$\dot\gamma = XI(\gamma),\ \gamma(0) = x. \qquad (*)$$

Equivalently (see for example, Ex. 7 at the end of Chapter 1), if $\mathcal{D}(X, x)$ is the maximal domain of existence of the solution of $(*)$, then the solution of the system

$$\dot\nu(s) = (tX)I\big(\nu(s)\big),\ \nu(0) = x$$

exists at $s = 1$ for every $t \in \mathcal{D}(X, x)$, and it holds

$$\nu(1) = \gamma(t).$$

Theorem 15.4.1 (The Campbell–Hausdorff formula for smooth vector fields). *Let X and Y be smooth vector fields on the open set $\Omega \subseteq \mathbb{R}^N$. Then the following formal equality holds:*

$$\exp(sY)\circ\exp(tX) = \exp\left(sY+tX+\frac{st}{2}[X,Y]+\sum_{k+j>2} s^k t^j\, C_{k,j}(X,Y)\right). \quad (15.26)$$

Here $C_{k,j}(X,Y)$ denotes a finite linear combination of commutators of X and Y. Any summand in $C_{k,j}(X,Y)$ contains k times the field X and j times the field Y. The precise meaning of (15.26) *is the following*:

For any fixed compact set $K \subset \Omega$ and any given couple of integers k_0, j_0, there exists a real number $r_0 > 0$ (depending on K, k_0, j_0, X, Y) such that if $|s|$, $|t| < r_0$, then we have (see also Figure 15.1)

$$\begin{aligned}\exp(sY)\big(\exp(tX)(x)\big) &= \exp\Bigg(sY+tX+\frac{st}{2}[X,Y] \\ &\quad + \sum_{\substack{1\le k\le k_0,\ 1\le j\le j_0\\ k+j>2}} s^k t^j\, C_{k,j}(X,Y)\Bigg)(x) \\ &\quad + \mathcal{O}(s^{k_0+1}) + \mathcal{O}(t^{j_0+1}) \end{aligned} \quad (15.27)$$

for every $x \in K$, where the above $\mathcal{O}$'s mean $|\mathcal{O}(r^m)| \le c\, r^m$ for every $|r| \le r_0$ and uniformly in $x \in K$ (the constant c depends on r_0, K, k_0, j_0, X, Y).

Finally, the $C_{k,j}$'s are "universal" Lie-polynomials in two non-commuting indeterminates, as given in Theorem 15.2.2*. For example,*

$$\begin{aligned} C_{1,1}(X,Y) &= \frac{1}{2}[X,Y], \\ C_{1,2}(X,Y) &= \frac{1}{12}[X,[X,Y]], \quad C_{2,1}(X,Y) = -\frac{1}{12}[Y,[X,Y]], \\ C_{2,2}(X,Y) &= -\frac{1}{48}[Y,[X,[X,Y]]] - \frac{1}{48}[X,[Y,[X,Y]]], \\ C_{1,3}(X,Y) &= C_{3,1}(X,Y) = 0. \end{aligned}$$

Proof. Consider the notation in the statement of the Campbell–Hausdorff formula for formal power series (Theorem 15.2.2). Then replace X and Y in that statement with, respectively, $s\mathbf{Y}$ and $t\mathbf{X}$, where $\mathbf{X}, \mathbf{Y}$ are still two non-commuting indeterminates and s, t are real numbers. Within the context of formal power series, formulas from (15.11) to (15.13d) then yield

$$\sum_{j\ge 0}\frac{(t\mathbf{X})^j}{j!}\cdot\sum_{k\ge 0}\frac{(s\mathbf{Y})^k}{k!} = \sum_{n\ge 0}\frac{\left(Z_{s\mathbf{Y},t\mathbf{X}}\right)^n}{n!}, \quad (15.28)$$

where $Z_{s\mathbf{Y},t\mathbf{X}} = \sum_{n=1}^{\infty} Z_n(s\mathbf{Y}, t\mathbf{X})$, and $Z_n(\cdot,\cdot)$ is a "universal" homogeneous Lie-bracket-polynomial of length n in two indeterminates. Now, for every fixed $k, j \in \mathbb{N}\cup\{0\}$, we group together the terms containing $s^k t^j$ in the $Z_n(s\mathbf{Y}, t\mathbf{X})$'s. Denote the resulting group by $C_{k,j}(\mathbf{X}, \mathbf{Y})$. Thus, $Z_{s\mathbf{Y},t\mathbf{X}}$ rewrites as

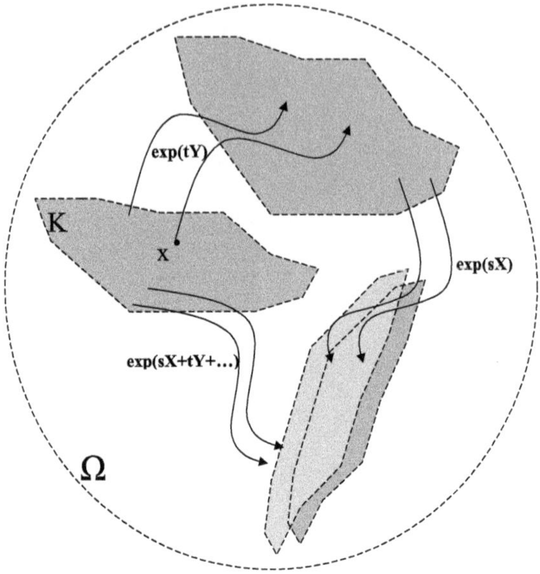

Fig. 15.1. The Campbell–Hausdorff flow

$$Z_{s\mathbf{Y},t\mathbf{X}} = s\mathbf{Y} + t\mathbf{X} + \frac{st}{2}[\mathbf{X}, \mathbf{Y}] + \sum_{k+j>2} s^k t^j \, C_{k,j}(\mathbf{X}, \mathbf{Y}). \tag{15.29}$$

For the future references, we see that (15.28) and (15.29) group together to give

$$\sum_{j\geq 0} \frac{(t\mathbf{X})^j}{j!} \cdot \sum_{k\geq 0} \frac{(s\mathbf{Y})^k}{k!} = \sum_{n\geq 0} \frac{1}{n!} \Big(s\mathbf{Y} + t\mathbf{X} + \sum_{k+j\geq 2} s^k t^j \, C_{k,j}(\mathbf{X}, \mathbf{Y}) \Big)^n . \tag{15.30}$$

The $C_{k,j}(X, Y)$'s in the assertion of the present theorem are defined by making use of the above "universal" $C_{k,j}$'s. Note that, obviously, any summand in $C_{k,j}(X, Y)$ contains k times the field X and j times the field Y.

Throughout the proof, we fix a compact set $K \subset \Omega$ and a couple of integers k_0, j_0. By general results on differential systems of ODE's (see, e.g. [Har82]), there exists a positive $r_0 = r_0(X, Y, K)$ such that the set

$$\big\{ \exp(sY)\big(\exp(tX)(x)\big) \,\big|\, x \in K, \ |s|, \ |t| \leq r_0 \big\}$$

makes sense and is a compact subset of Ω. Analogously, using general results on the dependence on the initial data for ODE's, it is possible to prove the existence of $r_0 > 0$ (we may assume it is the same as above) such that the integral curve of the vector field

$$V(s,t) := sY + tX + \frac{st}{2}[X,Y] + \sum_{\substack{1\le k\le k_0,\ 1\le j\le j_0 \\ k+j>2}} s^k t^j\, C_{k,j}(X,Y) \tag{15.31}$$

exists at unit time, whenever $x \in K$ and $|s|,\ |t| \le r_0$. (The $C_{k,j}$'s have been defined in the first part of the proof.)

Thus, the functions

$$u(s,t) := \exp(sY)\big(\exp(tX)(x)\big), \quad v(s,t) := \exp(V(s,t))(x),$$

are well-posed on $Q_0 := (-r_0, r_0) \times (-r_0, r_0)$ for every fixed $x \in K$. Our aim is to prove that the maps $(s,t) \mapsto u(s,t),\ v(s,t)$ have the same Taylor expansion at $(s,t) = (0,0)$, up to order k_0 w.r.t. s and up to order j_0 w.r.t. t.

To begin with, we exhibit the expansion of u. Recalling that, for every function $f \in C^\infty(\Omega, \mathbb{R}^N)$, every smooth vector field Z on Ω, every $z \in \Omega$ and every $n \in \mathbb{N} \cup \{0\}$, it holds

$$\left(\frac{\mathrm{d}}{\mathrm{d}r}\right)^n (f(\exp(rZ)(z))) = (Z^n f)(\exp(rZ)(z)), \tag{15.32}$$

we infer

$$\begin{aligned}
\left(\frac{\partial^{k+j}}{\partial s^k\,\partial t^j}\right) u(s,t) &= \left(\frac{\partial^j}{\partial t^j}\right)\left(\frac{\partial^k}{\partial s^k}\right)\big\{\exp(sY)\big(\exp(tX)(x)\big)\big\} \\
&\quad \text{(use (15.32) with } r = s, n = k, f = I, Z = Y, z = \exp(tX)(x)) \\
&= \left(\frac{\partial^j}{\partial t^j}\right)(Y^k I)\big(\exp(sY)\big(\exp(tX)(x)\big)\big) \\
&\quad \text{(use (15.32) with } r = t, n = j, f = Y^k I(\exp(sY)(\cdot)), Z = X, z = x) \\
&= X^j\big(Y^k I\big(\exp(sY)(\cdot)\big)\big)(\exp(tX)(x)).
\end{aligned}$$

Thus, we have proved that, for every $j,\ k \in \mathbb{N} \cup \{0\}$,

$$\left(\frac{\partial^{k+j}}{\partial s^k\,\partial t^j}\right) u(s,t) = X^j\big(Y^k I\big(\exp(sY)(\cdot)\big)\big)(\exp(tX)(x)). \tag{15.33}$$

Now, when $s = 0 = t$, the map $\exp(sY)(\cdot)$ is the identity map, so that the above computation reduces to (15.6), i.e.

$$\left(\frac{\partial^{k+j}}{\partial s^k\,\partial t^j}\right) u(0,0) = (X^j Y^k I)(x).$$

As a consequence, the Taylor polynomial of $u(s,t)$ at $(0,0)$ up to order k_0 w.r.t. s and up to order j_0 w.r.t. t is given by

$$\begin{aligned}
&I(x) + sYI(x) + tXI(x) + \sum_{\substack{1\le k\le k_0 \\ 1\le j\le j_0}} \frac{s^k t^j}{k!\,j!}(X^j Y^k I)(x) \\
&= \left(\sum_{j=0}^{j_0} \frac{t^j}{j!} X^j\right)\left(\sum_{k=0}^{k_0} \frac{s^k}{k!} Y^k\right) I(x).
\end{aligned}$$

We now estimate the remainder. From the Taylor formula with the Lagrange-remainder up to height $j_0 + k_0$ applied to the i-th component of u, say u_i, for every $(s,t) \in Q_0$ we infer the existence of $(\sigma_i, \tau_i) \in Q_0$ such that

$$u_i(s,t) = \sum_{\substack{0 \le k \le k_0 \\ 0 \le j \le j_0}} \frac{s^k t^j}{k!\, j!} (X^j Y^k I_i)(x) + \sum_{\substack{k > k_0 \text{ vel } j > j_0 \\ k+j \le k_0 + j_0}} \frac{s^k t^j}{k!\, j!} (X^j Y^k I_i)(x)$$

$$+ \sum_{k+j = k_0 + j_0 + 1} \frac{s^k t^j}{k!\, j!} X^j \big(Y^k I_i \big(\exp(\sigma_i Y)(\cdot)\big)\big)(\exp(\tau_i X)(x)).$$

Clearly, the second sum in the above right-hand side is given by

$$\mathcal{O}(s^{k_0+1}) + \mathcal{O}(t^{j_0+1}),$$

uniformly in $x \in K$ (due to the smoothness of X, Y and the compactness of K). The same estimate is true of the third sum, for (see (15.33))

$$|X^j (Y^k I_i (\exp(\sigma_i Y)(\cdot)))(\exp(\tau_i X)(x))| \le M(Q_0, K, X, Y)$$

uniformly in $x \in K$. All these facts prove that

$$u(s,t) = \left(\sum_{j=0}^{j_0} \frac{t^j}{j!} X^j \right) \left(\sum_{k=0}^{k_0} \frac{s^k}{k!} Y^k \right) I(x) + \mathcal{O}(s^{k_0+1}) + \mathcal{O}(t^{j_0+1}). \tag{15.34}$$

We now turn to the expansion of $v(s,t)$. By the very definition $v(s,t) = \exp(V(s,t))(x)$, we have $v(s,t) = \gamma_{s,t}(1)$, where $r \mapsto \gamma_{s,t}(r)$ is the solution to

$$\dot\gamma_{s,t}(r) = \big(V(s,t) I\big)(\gamma_{s,t}(r)), \quad \gamma_{s,t}(0) = x.$$

A simple computation (see the note to Ex. 5) shows that

$$\left(\frac{\mathrm{d}}{\mathrm{d}r}\right)^n \gamma_{s,t}(r) = \big(V(s,t)^n I\big)(\gamma_{s,t}(r)),$$

so that, by Taylor's expansion with an integral remainder, we have

$$v(s,t) = \gamma_{s,t}(1) = \sum_{n=0}^{N} \frac{1}{n!} \left(\frac{\mathrm{d}}{\mathrm{d}r}\right)^n \gamma_{s,t}(0) + \frac{1}{N!} \int_0^1 (1-r)^N \left(\frac{\mathrm{d}}{\mathrm{d}r}\right)^{N+1} \gamma_{s,t}(r)\, \mathrm{d}r$$

$$= \sum_{n=0}^{N} \frac{1}{n!} \big(V(s,t)^n I\big)(x) + \frac{1}{N!} \int_0^1 (1-r)^N \big(V(s,t)^{N+1} I\big)(\gamma_{s,t}(r))\, \mathrm{d}r.$$

Let us remark that, for $r \in [0,1]$, $(s,t) \in Q_0$ and $x \in K$, all the points $\gamma_{s,t}(r)$ belong to a fixed compact set, say K_0. Hence, if we choose $N = k_0 + j_0$, we have (see also (15.31))

$$
\begin{aligned}
&\big|\big(V(s,t)^{k_0+j_0+1}I\big)(\gamma_{s,t}(r))\big| \\
&= \Bigg|\Bigg(sY+tX+\sum_{\substack{1\le k\le k_0,\,1\le j\le j_0\\ k+j\ge 2}} s^k t^j\, C_{k,j}(X,Y)\Bigg)^{k_0+j_0+1} I(\gamma_{s,t}(r))\Bigg| \\
&\le \mathbf{c}_0\,(s^{k_0+1}+t^{j_0+1}) \\
&\qquad \times \max\Big\{\sup_{\xi\in K_O} |HI(\xi)| \,:\, H\in\mathbb{R}[X,Y] \text{ is a monomial with degree} \\
&\qquad\qquad \le (k_0+j_0)(k_0+j_0+1)\Big\} \\
&= \mathcal{O}(s^{k_0+1})+\mathcal{O}(t^{j_0+1})
\end{aligned}
$$

uniformly in $x\in K$. Here we have denoted by $\mathbb{R}[X,Y]$ the set of the polynomials in the (non-commuting) indeterminates X, Y (see Definition 15.2.1). This shows that

$$
v(s,t)=\sum_{n=0}^{k_0+j_0}\frac{1}{n!}V(s,t)^n I(x)+\mathcal{O}(s^{k_0+1})+\mathcal{O}(t^{j_0+1}), \tag{15.35}
$$

uniformly in $x\in K$. Collecting together (15.34), (15.35) and the definition (15.31) of $V(s,t)$, we see that the asserted (15.27) will follow if we prove that

$$
\begin{aligned}
&\Bigg(\sum_{j=0}^{j_0}\frac{t^j}{j!}X^j\Bigg)\Bigg(\sum_{k=0}^{k_0}\frac{s^k}{k!}Y^k\Bigg)I(x) \\
&\quad=\sum_{n=0}^{k_0+j_0}\frac{1}{n!}\Bigg(sY+tX+\sum_{\substack{1\le k\le k_0,\,1\le j\le j_0\\ k+j\ge 2}} s^k t^j\, C_{k,j}(X,Y)\Bigg)^n I(x) \\
&\qquad+\mathcal{O}(s^{k_0+1})+\mathcal{O}(t^{j_0+1}),
\end{aligned}
\tag{15.36}
$$

again uniformly in $x\in K$.

Now, let us return to identity (15.30), an identity in the context of formal power series in two non-commuting indeterminates $\mathbf{X}, \mathbf{Y}$. By the very definition of formal power series, we derive from (15.30) an identity between terms involving $\mathbf{X}$ up to j_0 times and $\mathbf{Y}$ up to k_0 times, namely

$$
\begin{aligned}
&\sum_{j=0}^{j_0}\frac{(t\mathbf{X})^j}{j!}\cdot\sum_{k=0}^{k_0}\frac{(s\mathbf{Y})^k}{k!} \\
&\quad\equiv\sum_{n=0}^{k_0+j_0}\frac{1}{n!}\Bigg(s\mathbf{Y}+t\mathbf{X}+\sum_{\substack{1\le k\le k_0,\,1\le j\le j_0\\ k+j\ge 2}} s^k t^j\, C_{k,j}(\mathbf{X},\mathbf{Y})\Bigg)^n \\
&\qquad\{\textit{modulo} \text{ terms with } \mathbf{X} \text{ more than } j_0 \text{ times or } \mathbf{Y} \text{ more than } k_0 \text{ times}\}.
\end{aligned}
\tag{15.37}
$$

In particular, (15.37) *holds true if* $\mathbf{X}, \mathbf{Y}$ *are our smooth vector fields in* Ω*, as differential operators in the* (*non-commutative*) *algebra* $T(\Omega, \mathbb{R}^N)$ *of the smooth vector fields on* Ω.

As a consequence, (15.36) will follow from (15.37) if we prove the following fact. After carrying out the n-th power in the second line of (15.36), those terms where X appears more than j_0 times or where Y appears more than k_0 times, when applied to I and evaluated at x, all together give $\mathcal{O}(s^{k_0+1}) + \mathcal{O}(t^{j_0+1})$, uniformly in $x \in K$.

This is straightforwardly seen. Indeed, in the second line of (15.36) we have only a finite number of summands (depending on j_0, k_0) and any summand

$$((tX)^{\alpha_1}(sY)^{\beta_1} \cdots (tX)^{\alpha_i}(sY)^{\beta_i})I(x)$$

with $\alpha_1 + \cdots + \alpha_i > j_0$ or $\beta_1 + \cdots + \beta_i > k_0$ is bounded by

$$t^{\alpha_1+\cdots+\alpha_i} s^{\beta_1+\cdots+\beta_i} \sup_{x\in K} \left|\left(X^{\alpha_1} Y^{\beta_1} \cdots X^{\alpha_i} Y^{\beta_i}\right) I(x)\right|$$
$$\leq \begin{cases} \mathbf{c}_0\, t^{j_0+1} & \text{if } \alpha_1 + \cdots + \alpha_i > j_0, \\ \mathbf{c}_0\, s^{k_0+1} & \text{if } \beta_1 + \cdots + \beta_i > k_0. \end{cases}$$

Here $\mathbf{c}_0$ depends only on Q_0, k_0, j_0, K and on

$$\max\left\{ \sup_{x\in K} |HI(x)| \ : \ H \in \mathbb{R}[X, Y] \text{ monomial with degree} \leq (k_0 + j_0)^2 \right\}.$$

This completes the proof. □

Bibliographical Notes. Classical references on the Campbell–Hausdorff formula are Bourbaki [Bou89], M. Hausner and J.T. Scwartz [HS68], G. Hochschild [Hoc68], N. Jacobson [Jac62], V.S. Varadarajan [Var84].

Some applications of this tool to analysis are discussed, for example, in [Hor67, RS76,VSC92], see also the paper [GG90], already mentioned in Section 14.1.

We would also like to cite [DST91,Egg93,Oki95,Ote91,Str87,Tho82] for other remarkable applications, referring the reader to the references therein for further details.

15.5 Exercises of Chapter 15

Ex. 1) Given a smooth vector field X on $\mathbb{R}^N$ and a point $x \in \mathbb{R}^N$, we recall that by $\exp(X)(x)$ we mean the point reached at unit time (if this point exists) by the integral curve of X starting at x. In (1.7) of Chapter 1 (page 7), we found the exponential-type expansion

$$\exp(X)(x) = I(x) + XI(x) + \frac{1}{2!}X^2 I(x) + \cdots . \qquad (15.38)$$

Suppose that there exists a dilation δ_λ on $\mathbb{R}^N$ of the form $\delta_\lambda(x) = \delta_\lambda(x_1, \ldots, x_N) = (\lambda^{\sigma_1} x_1, \ldots, \lambda^{\sigma_N} x_N)$ with $0 < \sigma_1 \le \cdots \le \sigma_N$ such that X is δ_λ-homogeneous of positive degree. Prove that any integral curve of X exists on the whole $\mathbb{R}$ (that is to say that X is *complete*) and there exists $n \in \mathbb{N}$ such that $X^n I \equiv 0$ (here X^n denotes the n-th power of X as a differential operator). Deduce that formula (15.38) makes a precise sense, for the sum is finite.

Ex. 2) By means of the above exercise, find the integral curves of the vector fields on $\mathbb{R}^2$ given by ∂_x, $x\,\partial_y$ and $(x - \frac{1}{2})\,\partial_y + \partial_x$ both by solving a suitable system of ODE's and by the above formula (15.38). Then check the Campbell–Hausdorff formula (15.3), which, in this case, gives

$$\exp(\partial_x)\big(\exp(x\,\partial_y)(x,y)\big) = \exp\left(x\,\partial_y + \partial_x + \frac{1}{2}[x\,\partial_y, \partial_x]\right)(x,y).$$

The reader is once again invited to appreciate the interchange of the order of the two vector fields from one side to the other of the formula

$$\exp(X)\big(\exp(Y)(x,y)\big) = \exp(Y \diamond X)(x,y) = \exp\left(Y + X + \frac{1}{2}[Y,X]\right)(x,y).$$

Ex. 3) Taking into account the actual expressions (15.12) (page 601) of the first four terms in the Campbell–Hausdorff formula, verify directly for $n = 1, 2, 3, 4$ the alternating skew-symmetry relation

$$Z_n(A,B) = (-1)^{n+1} Z_n(B,A).$$

Compare to (15.21) (page 608).

Ex. 4) Find the Campbell–Hausdorff composition $\diamond$ up to step five.

Ex. 5) Prove Theorem 15.4.1 when $k_0 = 1 = j_0$, with direct arguments and without the aid of the Campbell–Hausdorff formula. The assertion to be proved is the following one.

Let X, Y be smooth vector fields on the open set $\Omega \subseteq \mathbb{R}^N$, and let $K \subset \Omega$ be any fixed compact set. Then there exists a real number $r_0 > 0$ (depending on K, X, Y) such that if $|s|$, $|t| < r_0$, then, for every $x \in K$, we have

$$\exp(sY)\big(\exp(tX)(x)\big) = \exp\left(sY + tX + \frac{st}{2}[X,Y]\right)(x) + \mathcal{O}(s^2) + \mathcal{O}(t^2), \qquad (15.39)$$

where the above $\mathcal{O}$'s mean $|\mathcal{O}(r^2)| \le c\,r^2$ for every $|r| \le r_0$ and uniformly in $x \in K$.

Hint: The Taylor expansion w.r.t. (s, t) up to order 2 of the left-hand side of (15.39) is[9] given by

$$I(x) + s\,YI(x) + t\,XI(x) + \frac{st}{2}\,XYI(x) + \mathcal{O}(s^2) + \mathcal{O}(t^2).$$

Prove by a simple computation that this is also the expansion of

$$\exp\Big(sY + tX + \frac{st}{2}\,[X,Y]\Big)(x).$$

This can be derived by recalling that, by the Taylor formula with an integral remainder, we have

$$\begin{aligned}
&\exp\Big(sY + tX + \frac{st}{2}\,[X,Y]\Big)(x)\\
&= \exp\Big(r\Big(sY + tX + \frac{st}{2}\,[X,Y]\Big)\Big)(x)\Big|_{r=1}\\
&= I(x) + \Big(sY + tX + \frac{st}{2}\,[X,Y]\Big)I(x)\\
&\quad + \frac{1}{2}\Big(sY + tX + \frac{st}{2}\,[X,Y]\Big)^2 I(x)\\
&\quad + \frac{1}{2}\int_0^1 (1-r)^2\Big(sY + tX + \frac{st}{2}\,[X,Y]\Big)^3 I(\gamma_{s,t,x}(r))\,\mathrm{d}r,
\end{aligned}$$

where $r \mapsto \gamma_{s,t,x}(r)$ is the integral curve of the vector field $sY + tX + \frac{st}{2}\,[X,Y]$ starting at x. Notice that the integral remainder in the above far right-hand side is $\mathcal{O}(s^2) + \mathcal{O}(t^2)$, whereas the third line of the above expression equals, modulo $\mathcal{O}(s^2) + \mathcal{O}(t^2)$,

---

[9] See (15.6); see also Ex. 8 at the end of Chapter 1 where it is proved that if X is a smooth vector field on an open set $\Omega \subseteq \mathbb{R}^N$, $f \in C^\infty(\Omega, \mathbb{R}^N)$ and $k \in \mathbb{N}$, it holds

$$\Big(\frac{\mathrm{d}}{\mathrm{d}t}\Big)^k \Big(f(\gamma(t))\Big) = (X^k f)(\gamma(t)),$$

whenever γ satisfies $\dot\gamma(t) = XI(\gamma(t))$. Indeed, when $k = 1$, we have

$$\frac{\mathrm{d}}{\mathrm{d}t}\Big(f(\gamma(t))\Big) = \langle(\nabla f)(\gamma(t)), \dot\gamma(t)\rangle = \langle(\nabla f)(\gamma(t)), XI(\gamma(t))\rangle = (Xf)(\gamma(t)).$$

Moreover, arguing by induction and by using the inductive step and the case $n = 1$ *with f replaced by $X^k f$*, we have

$$\begin{aligned}
\Big(\frac{\mathrm{d}}{\mathrm{d}t}\Big)^{1+k}\Big(f(\gamma(t))\Big) &= \Big(\frac{\mathrm{d}}{\mathrm{d}t}\Big)\Big(\frac{\mathrm{d}}{\mathrm{d}t}\Big)^k\Big(f(\gamma(t))\Big) = \Big(\frac{\mathrm{d}}{\mathrm{d}t}\Big)(X^k f)(\gamma(t))\\
&= \Big(X(X^k f)\Big)(\gamma(t)) = (X^{1+k} f)(\gamma(t)).
\end{aligned}$$

$$\begin{aligned}
&I(x)+sYI(x)+tXI(x)+\frac{st}{2}\,(XYI(x)-YXI(x))+\frac{st}{2}\,YXI(x)\\
&=I(x)+s\,YI(x)+t\,XI(x)+\frac{st}{2}\,XYI(x).
\end{aligned}$$

This ends the proof.

16

Families of Diffeomorphic Sub-Laplacians

Let $\mathbb{G}$ be a homogeneous Carnot group, and let $\Delta_{\mathbb{G}} = \sum_{i=1}^{m} X_i^2$ be the canonical sub-Laplacian on $\mathbb{G}$. Suppose it is given a positive-definite symmetric matrix $A = (a_{i,j})_{1 \le i,j \le m}$. We consider the second order operator modeled on the matrix A and the vector fields X_i's

$$\mathcal{L}_A := \sum_{i,j=1}^{m} a_{i,j}\, X_i X_j. \tag{16.1}$$

For example, in the classical case $\mathbb{G} = (\mathbb{R}^N, +)$,

$$\Delta_A = \sum_{i,j=1}^{N} a_{i,j}\, \partial_{x_i} \partial_{x_j}$$

is a constant coefficient second order operator of elliptic type, whereas $\Delta = \sum_{i=1}^{N} (\partial_{x_i})^2$ is the usual Laplace operator. It is well known that a *linear change of coordinates* in $\mathbb{R}^N$ transforms the above operator Δ_A into Δ. Thus, the operator Δ_A is "equivalent" to the operator Δ (in a new system of coordinates).

Following the above naive idea, in order to study the operator $\mathcal{L}_A$ in (16.1), it is natural to ask the following question. *Does there exist a diffeomorphism $T = T_A : \mathbb{G} \to \mathbb{G}$ such that, in the new coordinate system defined by T, the operator* (16.1) *is turned*[1] *into $\Delta_{\mathbb{G}}$?*

Unfortunately, when the X_i's have not constant coefficients, classical changes of basis may fail to apply, and simple examples (provided in this chapter) show that T *may not exist* at all. Then, to face the above question, it seems natural to make an additional assumption on the group $\mathbb{G}$. Indeed, we shall prove the existence of such a diffeomorphism T *whenever $\mathbb{G}$ is a free Carnot group.*

Finally, in Section 16.4, we present an example of application to PDE's of the topics developed in this chapter.

---

[1] We say that T turns $\mathcal{L}_A$ into $\Delta_{\mathbb{G}}$ if $\mathcal{L}_A(u \circ T) = (\Delta_{\mathbb{G}} u) \circ T$ for every $u \in C^\infty(\mathbb{G})$.

16.1 An Isomorphism Turning the Operator $\sum_{i,j} a_{i,j}\, X_i X_j$ into the Canonical Sub-Laplacian $\Delta_{\mathbb{G}}$

Let $\mathbb{G}$ be a homogeneous Carnot group, and let

$$\Delta_{\mathbb{G}} = \sum_{i=1}^{m} X_i^2$$

be the canonical[2] sub-Laplacian on $\mathbb{G}$. Suppose it is given a positive-definite symmetric matrix $A = (a_{i,j})_{1\le i,j\le m}$. We then consider the following new second order differential operator modeled on the matrix A and the vector fields X_i's

$$\mathcal{L}_A := \sum_{i,j=1}^{m} a_{i,j}\, X_i X_j. \tag{16.2}$$

This is exactly what is done in the classical case $\mathbb{G} = (\mathbb{R}^N, +)$ when, starting from the classical Laplace operator

$$\Delta := \sum_{i=1}^{N} \left(\frac{\partial}{\partial x_i}\right)^2,$$

one considers the following constant coefficient second order operator of elliptic type

$$\Delta_A = \sum_{i,j=1}^{N} a_{i,j}\, \frac{\partial^2}{\partial x_i\, \partial x_j}.$$

It is well known and easy to prove that a *linear change of coordinates* in $\mathbb{R}^N$ (naturally related to the matrix $A^{-1/2}$) transforms the above operator Δ_A into Δ. Indeed, consider the new coordinates in $\mathbb{R}^N$ given by

$$y = T(x) := C\,x, \quad \text{where } C \text{ is a symmetric matrix such that } C^2 = A^{-1},$$

i.e. $C = B^{-1}$ is the inverse of the symmetric (square root of A) matrix B such that $B^2 = A$. With this choice, for every $u \in C^\infty(\mathbb{R}^N)$, $u = u(y)$, we have

[2] We recall that the *Jacobian basis* of $\mathfrak{g}$ (=algebra of $\mathbb{G}$) is the basis of vector fields in $\mathfrak{g}$ agreeing at the origin with the coordinate partial derivatives of $\mathbb{R}^N \equiv \mathbb{G}$. Then, the canonical sub-Laplacian on $\mathbb{G}$ is the second order differential operator $\Delta_{\mathbb{G}} = \sum_{i=1}^m X_i^2$, where $X_1, \ldots, X_m$ are the first m vector fields of the Jacobian basis for $\mathfrak{g}$. Here $m = N_1$ is the dimension of the first layer of the stratification of $\mathfrak{g}$ or, equivalently, $m = N_1$ is the dimension of the layer of coordinates where the dilation group δ_λ of $\mathbb{G}$

$$\delta_\lambda(x) = \delta_\lambda(x^{(1)}, x^{(2)}, \ldots, x^{(r)}) = (\lambda x^{(1)}, \lambda^2 x^{(2)}, \ldots, \lambda^r x^{(r)}), \qquad x^{(i)} \in \mathbb{R}^{N_i},$$

acts like the multiplication times λ.

$$\begin{aligned}\Delta_A\big(u(T(x))\big) &= \sum_{i,j=1}^{N} a_{i,j}\, \partial_{x_i}\left(\sum_{h=1}^{N} c_{h,j}\, (\partial_{y_h} u)(T(x))\right)\\ &= \sum_{h,k=1}^{N}\left(\sum_{i,j=1}^{N} a_{i,j}\, c_{h,j}\, c_{k,i}\right)(\partial^2_{y_k,y_h} u)(T(x))\\ &= \sum_{h,k=1}^{N} ({}^tC\cdot A\cdot C)_{k,h}\, (\partial^2_{y_k,y_h} u)(T(x))\\ &= \sum_{h=1}^{N} (\partial^2_{y_h,y_h} u)(T(x)) = (\Delta u)(T(x)),\end{aligned}$$

since C is symmetric and $C \cdot A \cdot C = B^{-1} \cdot B^2 \cdot B^{-1} = \mathbb{I}_N$, the identity matrix of order N. Thus, the operator Δ_A is "equivalent" to the operator Δ (in a new system of coordinates).

Following the above naive idea, in order to study the operator $\mathcal{L}_A$ in (16.2), it is natural to ask the following question.

Does there exist a diffeomorphism $T = T_A : \mathbb{G} \to \mathbb{G}$ such that, in the new coordinate system defined by T, the operator (16.2) *is turned into $\Delta_{\mathbb{G}}$?*

Here, we said that T turns $\mathcal{L}_A$ into $\Delta_{\mathbb{G}}$ if

$$\mathcal{L}_A(u \circ T) = (\Delta_{\mathbb{G}} u) \circ T$$

for every $u \in C^\infty(\mathbb{G})$. A further justification for this natural question is given by the following assertion.

Proposition 16.1.1 ($\{\mathcal{L}_A : A\}$ and $\{\mathcal{L}\}$). *Any sub-Laplacian on $\mathbb{G}$ is of the form* (16.2) *for a suitable symmetric positive definite matrix A. Vice versa, any operator $\mathcal{L}_A$ in that form is a sub-Laplacian*

$$\mathcal{L}_A = \sum_{k=1}^{m} Y_k^2 \quad \textit{with } Y_k = \textstyle\sum_{j=1}^{m} (A^{1/2})_{k,j}\, X_j. \tag{16.3}$$

Proof. First, suppose that $\mathcal{L} = \sum_{k=1}^{m} Y_k^2$ is a sub-Laplacian on $\mathbb{G}$. By the definition given in Section 1.5 (page 62), this means that $\{Y_1, \dots, Y_m\}$ and $\{X_1, \dots, X_m\}$ are both bases for the first layer of the stratification of the algebra of $\mathbb{G}$. In particular, there exists a non-singular $m \times m$ matrix $D = (d_{k,j})_{k,j}$ such that

$$Y_k = \sum_{j=1}^{m} d_{k,j}\, X_j \quad \text{for every } k = 1, \dots, m.$$

Then we have

$$\begin{aligned}\sum_{k=1}^{m} Y_k^2 &= \sum_{k=1}^{m}\left(\sum_{i=1}^{m} d_{k,i} X_i\right)\left(\sum_{j=1}^{m} d_{k,j} X_j\right)\\ &= \sum_{i,j=1}^{m}\left(\sum_{k=1}^{m} d_{k,i}\, d_{k,j}\right) X_i X_j =: \sum_{i,j=1}^{m} a_{i,j}\, X_i X_j = \mathcal{L}_A,\end{aligned}$$

where we have set

$$A = (a_{i,j})_{i,j} := \left(\sum_{k=1}^{m} d_{k,i}\, d_{k,j} \right)_{i,j} = {}^t D \cdot D$$

(clearly, A is a symmetric positive-definite matrix).

Vice versa, if A is a symmetric positive-definite matrix and $\mathcal{L}_A$ is the operator (16.2), denoting by $B = (b_{i,j})_{1 \le i,j \le m}$ the symmetric positive-definite matrix such that $A = B^2$ (i.e. $B = A^{1/2}$), one has

$$\begin{aligned} \mathcal{L}_A &= \sum_{i,j=1}^{m} a_{i,j}\, X_i X_j = \sum_{i,j=1}^{m} \sum_{k=1}^{m} b_{i,k} b_{k,j}\, X_i X_j \\ &= \sum_{k=1}^{m} \left(\sum_{i=1}^{m} b_{k,i} X_i \right) \left(\sum_{j=1}^{m} b_{k,j} X_j \right) = \sum_{k=1}^{m} Y_k^2, \end{aligned}$$

where $Y_k := \sum_{j=1}^{m} b_{k,j} X_j$ for every $k = 1, \dots, m$. This proves that $\mathcal{L}_A$ is a sub-Laplacian. □

Proposition 16.1.1 shows that, looking for a change of coordinates T turning $\mathcal{L}_A$ into $\Delta_{\mathbb{G}}$, it is natural to look for a T turning

$$Y_k = \sum_{j=1}^{m} (A^{1/2})_{k,j}\, X_j$$

into X_k, for $k = 1, \dots, m$. In the simple case when $X_i = \partial/\partial x_i$, we showed at the beginning of the section that the problem always has a solution via a *linear* change of coordinates in $\mathbb{R}^N$. However, when the X_i's have not constant coefficients, classical changes of basis may fail to apply. Moreover, and this is the most striking fact, simple examples (see Remark 16.2.2 and Remark 16.2.3 in the next section) show that T *may not exist* in the general case! Some further assumptions either on the coefficient matrix A or on the structure of the group $\mathbb{G}$ must be made.

To face this question, rather than restricting the form of the matrix A, it seems more natural to make an additional assumption on the group $\mathbb{G}$. Indeed, we shall prove the existence of such a diffeomorphism T, *whenever* $\mathbb{G}$ *is a free Carnot group* (we recall below the relevant definition; see also Chapter 14, page 577). *The free group setting turns out to be the most natural one in order to generalize the classical Euclidean case.* Though, we stress that T is not necessarily a linear application, in the general case (see Remark 16.2.1 in the next section).

We now recall the definition of $\mathfrak{f}_{m,r}$, the free nilpotent Lie algebra of step r with m (≥ 2) generators $x_1, \dots, x_m$. By definition, $\mathfrak{f}_{m,r}$ is the unique (up to isomorphism) nilpotent Lie algebra of step r generated by m of its elements $x_1, \dots, x_m$ such that, for every nilpotent Lie algebra $\mathfrak{n}$ of step r and for every map φ from $\{x_1, \dots, x_m\}$ to $\mathfrak{n}$, there exists a (unique) Lie algebra morphism $\widetilde{\varphi}$ from $\mathfrak{f}_{m,r}$ to $\mathfrak{n}$ extending φ. (See Chapter 14 for more details.) We say that the Carnot group $\mathbb{G}$ is a *free Carnot group* if its Lie algebra $\mathfrak{g}$ is isomorphic to $\mathfrak{f}_{m,r}$, for some m and r. $\mathbb{R}^N$ equipped with the ordinary Abelian structure is an example of free Carnot group. The Heisenberg group $\mathbb{H}^1$ is also a free Carnot group, while $\mathbb{H}^n$ is not free for any $n \geq 2$, as can be seen by a dimensional argument.

We are now in the position to prove the main result of this chapter.

Theorem 16.1.2 (Equivalence of sub-Laplacians on free groups). *Let $\mathbb{G}$ be a* free *homogeneous Carnot group, and let A be a given positive-definite symmetric matrix. Let $X_1, \dots, X_m$ denote the first vector fields in the Jacobian basis related to $\mathbb{G}$, i.e.*

$$\Delta_{\mathbb{G}} = \sum_{k=1}^{m} X_k^2$$

is the canonical sub-Laplacian on $\mathbb{G}$. Finally, let the Y_k's be defined by

$$Y_k = \sum_{j=1}^{m} (A^{1/2})_{k,j} \, X_j, \quad k = 1, \dots, m. \tag{16.4}$$

Consider the related (non-canonical) sub-Laplacian

$$\mathcal{L}_A = \sum_{k=1}^{m} Y_k^2 = \sum_{i,j=1}^{m} a_{i,j} \, X_i X_j.$$

Then there exists a Lie group automorphism T_A of $\mathbb{G}$ such that

$$Y_k(u \circ T_A) = (X_k u) \circ T_A, \quad k = 1, \dots, m, \tag{16.5a}$$
$$\mathcal{L}_A(u \circ T_A) = (\Delta_{\mathbb{G}} u) \circ T_A \tag{16.5b}$$

for every smooth function $u : \mathbb{G} \to \mathbb{R}$. Moreover, T_A has polynomial component functions and commutes with the dilations of $\mathbb{G}$.

Proof. First of all, we remark that (16.5b) is an immediate consequence of (16.5a) since $\mathcal{L}_A = \sum_{k=1}^{m} Y_k^2$ whereas $\Delta_{\mathbb{G}} = \sum_{k=1}^{m} X_k^2$.

On the other hand, by the very definition (16.4) of Y_k, (16.5a) reads

$$\left(\sum_{j=1}^{m} (A^{1/2})_{k,j} \, X_j \right) (u \circ T_A) = (X_k u) \circ T_A, \quad k = 1, \dots, m,$$

which, in turn, is trivially equivalent to

$$X_k(u \circ T_A) = \sum_{j=1}^{m} (A^{-1/2})_{k,j} \, (X_j u) \circ T_A, \quad k = 1, \dots, m. \tag{16.6}$$

Moreover, if we already knew that T_A (denoted henceforth simply by T) is a Lie group morphism, then it would be sufficient to prove that (16.6) holds at the origin. Indeed, suppose that

$$X_k(u \circ T)(0) = \sum_{j=1}^{m} b_{k,j}\,(X_j u)(0) \qquad \forall\, u \in C^\infty(\mathbb{R}^N),\ k = 1, \ldots, m, \qquad (16.7)$$

where we have set for brevity $B = (b_{k,j})_{k,j} = A^{-1/2}$. We fix $y \in \mathbb{G}$, $v \in C^\infty(\mathbb{R}^N)$ and apply (16.7) to

$$u(x) = v(T(y) \circ x).$$

Since T is a Lie group morphism and X_i is left-invariant, we have

$$\begin{aligned} X_k\big(v \circ T\big)(y) &= X_k\big(v\big(T(y \circ \cdot)\big)\big)(0) = X_k\big(v\big(T(y) \circ T(\cdot)\big)\big)(0) \\ &= \sum_{j=1}^{m} b_{k,j}\, X_j\big(v(T(y) \circ \cdot)\big)(0) = \sum_{j=1}^{m} b_{k,j}\,(X_j v)(T(y)), \end{aligned}$$

i.e. (16.6) holds.

We now turn to show (16.7). We first observe that, by the chain rule and since B is symmetric, (16.7) is equivalent to

$$\mathcal{J}_T(0)\,(XI)(0) = (XI)(0)\,B. \qquad (16.8)$$

Here we have denoted by I the identity map on $\mathbb{G}$, by $XI = \big(X_1 I \cdots X_m I\big)$ the $(N \times m)$-matrix whose i-th column is $X_i I$ and by $\mathcal{J}_T$ the Jacobian matrix of T.

We thus aim to construct a *Lie group automorphism* T of $\mathbb{G}$ satisfying (16.8). This will complete the proof of the first statement of the theorem. Consider the linear map defined as follows

$$\varphi : \operatorname{span}\{X_1, \ldots, X_m\} \to \operatorname{span}\{X_1, \ldots, X_m\}, \quad X_i \mapsto \sum_{j=1}^{m} b_{i,j}\, X_j.$$

Since $X_1, \ldots, X_m$ are linearly independent, φ is well posed. Moreover, B being a non-singular matrix, φ is a bijective linear map. Since $\mathbb{G}$ is a free Carnot group and its Lie algebra $\mathfrak{g}$ is nilpotent of step r and generated by $\{X_1, \ldots, X_m\}$, clearly $\mathfrak{g}$ is isomorphic to $\mathfrak{f}_{m,r}$. Then, by Lemma 14.1.4 (page 578), there exists a unique Lie algebra automorphism $\widetilde{\varphi} : \mathfrak{g} \to \mathfrak{g}$ extending φ. For the simplicity of notation, we also denote $\widetilde{\varphi}$ by φ. We are now in the position to define T. We set

$$T : \mathbb{G} \xrightarrow{\mathrm{Log}} \mathfrak{g} \xrightarrow{\varphi} \mathfrak{g} \xrightarrow{\mathrm{Exp}} \mathbb{G}, \quad T := \mathrm{Exp} \circ \varphi \circ \mathrm{Log}, \qquad (16.9)$$

where Exp denotes the exponential map and Log its inverse function. Clearly, T is bijective,

$$T^{-1} = \mathrm{Exp} \circ \varphi^{-1} \circ \mathrm{Log},$$

whence both T and T^{-1} are C^∞ maps. Moreover, T is a Lie group automorphism of $\mathbb{G}$: indeed, when $\mathfrak{g}$ is equipped with the Campbell–Hausdorff[3] group law $\diamond$, then Exp, Log and any Lie algebra morphism of $\mathfrak{g}$ are Lie group morphisms. Indeed, the

---

[3] See Section 14.2, page 584, for the definition and the properties of $\diamond$.

fact that Exp and Log are Lie-algebra morphisms between $(\mathfrak{g}, \diamond)$ and $(\mathbb{G}, \circ)$ is a consequence of Theorem 14.2.1 (page 586). The fact that any Lie-algebra morphism of $(\mathfrak{g}, [\cdot, \cdot])$ into itself is actually a Lie-group morphism of $(\mathfrak{g}, \diamond)$ into itself is easily seen.[4]

Let us prove the matrix equality (16.8). We recall that, by definition,

$$(XI)(0) = \begin{pmatrix} \mathbb{I}_m \\ 0 \end{pmatrix}$$

where $\mathbb{I}_m$ is the identity matrix of order m and 0 is the null matrix of order $(N-m) \times m$. If $\mathfrak{g}$ is equipped with the well-known Jacobian basis, we have (see, for instance, the arguments in Remark 1.3.30, page 50)

$$\mathcal{J}_{\mathrm{Log}}(0) = (\mathcal{J}_{\mathrm{Exp}}(0))^{-1} = \mathbb{I}_N,$$

whence

$$\mathcal{J}_T(0) = \mathcal{J}_{\mathrm{Exp}}(0)\, \mathcal{J}_\varphi(0)\, \mathcal{J}_{\mathrm{Log}}(0) = \mathcal{J}_\varphi(0).$$

Thus, in order to prove (16.8), it is enough to prove that the $(N \times m)$-matrix of the first m-columns of $\mathcal{J}_\varphi(0)$ is equal to

$$\begin{pmatrix} \mathbb{I}_m \\ 0 \end{pmatrix} B = \begin{pmatrix} B \\ 0 \end{pmatrix},$$

which straightforwardly follows from the definition of φ and from the fact that $X_1, \dots, X_m$ are the first m vectors of the Jacobian basis.

We now turn to the proof of the last statement of the theorem. We recall that Exp and Log have polynomial components (see Theorem 1.3.28, 49). Hence T has polynomial component functions since φ is linear. Finally, we prove that T commutes with δ_λ, the dilations on $\mathbb{G}$. Recalling that δ_λ has the form

$$\delta_\lambda(x) = \delta_\lambda(x^{(1)}, \dots, x^{(r)}) = (\lambda x^{(1)}, \dots, \lambda^r x^{(r)}), \qquad x^{(i)} \in \mathbb{R}^{N_i},$$

let us accordingly denote by

$$Z_1^{(1)}, \dots, Z_{N_1}^{(1)}, \dots, Z_1^{(r)}, \dots, Z_{N_r}^{(r)}$$

the Jacobian basis of $\mathfrak{g}$, and let δ_λ also denote the map

---

4 Indeed, let $\varphi : \mathfrak{g} \to \mathfrak{g}$ be a Lie-algebra morphism, i.e. $\varphi([X, Y]) = [\varphi(X), \varphi(Y)]$ for every $X, Y \in \mathfrak{g}$. Then, comparing to the definition (14.2) of the $\diamond$ operation, we have

$$\begin{aligned} \varphi(X \diamond Y) &= \varphi\left(X + Y + \frac{1}{2}[X, Y] + \frac{1}{12}[X, [X, Y]] + \cdots \right) \\ &= \varphi(X) + \varphi(Y) + \frac{1}{2}[\varphi(X), \varphi(Y)] + \frac{1}{12}[\varphi(X), [\varphi(X), \varphi(Y)]] + \cdots \\ &= \varphi(X) \diamond \varphi(Y). \end{aligned}$$

This amounts to say that φ is a Lie-group morphism of $(\mathfrak{g}, \diamond)$ into itself.

$$\mathfrak{g} \ni \sum_{i=1}^{r} \sum_{j=1}^{N_i} \xi_j^{(i)} Z_j^{(i)} \mapsto \sum_{i=1}^{r} \sum_{j=1}^{N_i} \lambda^i \xi_j^{(i)} Z_j^{(i)} \in \mathfrak{g}.$$

With this notation, both Exp and Log commute with δ_λ (see (1.71), page 49, in Theorem 1.3.28). Thus, we only have to show

$$\varphi \circ \delta_\lambda = \delta_\lambda \circ \varphi.$$

Now, $Z_j^{(k)}$ is a δ_λ-homogeneous vector field of degree k, hence in particular $Z_j^{(k)}$ can be expressed as a homogeneous Lie polynomial of degree k in the generators $X_1, \ldots, X_m$. By the definition of the Lie algebra morphism φ, $\varphi(Z_j^{(k)})$ is a homogeneous Lie polynomial of degree k in $X_1, \ldots, X_m$ as well. Hence, there exist scalars $c_{i,j}^{(k)}$ such that

$$\varphi(Z_j^{(k)}) = \sum_{i=1}^{N_k} c_{i,j}^{(k)} Z_i^{(k)}.$$

Consequently, for every $k = 1, \ldots, r$ and $j = 1, \ldots, N_k$, we have

$$\begin{aligned}(\varphi \circ \delta_\lambda)(Z_j^{(k)}) &= \varphi(\lambda^k Z_j^{(k)}) = \lambda^k \varphi(Z_j^{(k)}) = \lambda^k \sum_{i=1}^{N_k} c_{i,j}^{(k)} Z_j^{(k)} \\ &= \sum_{i=1}^{N_k} c_{i,j}^{(k)} \lambda^k Z_j^{(k)} = \sum_{i=1}^{N_k} c_{i,j}^{(k)} \delta_\lambda(Z_j^{(k)}) = \delta_\lambda \left(\sum_{i=1}^{N_k} c_{i,j}^{(k)} Z_j^{(k)} \right) \\ &= (\delta_\lambda \circ \varphi)(Z_j^{(k)}).\end{aligned}$$

This completes the proof. □

We explicitly remark that, in the proof of Theorem 16.1.2, we have proved that the diffeomorphism T has the following form

$$T : \mathbb{G} \xrightarrow{\text{Log}} \mathfrak{g} \xrightarrow{\varphi} \mathfrak{g} \xrightarrow{\text{Exp}} \mathbb{G}, \qquad T := \text{Exp} \circ \varphi \circ \text{Log},$$

where φ is a linear map which, w.r.t. the Jacobian basis, is related to a block-diagonal matrix of the type

$$\begin{pmatrix} C^{(1)} & 0 & \cdots & 0 \\ 0 & C^{(2)} & \cdots & 0 \\ \vdots & \vdots & \ddots & \vdots \\ 0 & 0 & \cdots & C^{(r)} \end{pmatrix}, \quad \text{where} \quad C^{(k)} = \left(c_{i,j}^{(k)}\right)_{1 \le i,j \le N_k}.$$

16.2 Examples and Counter-examples

In this section, we give some examples and counter-examples concerning with the topics of the previous section. We follow all the notation of Section 16.1.

Remark 16.2.1. The diffeomorphism T_A constructed in Theorem 16.1.2 may fail to be linear.

Proof. We consider the free homogeneous Carnot group $\mathbb{G} = (\mathbb{R}^3, \circ, \delta_\lambda)$, where

$$x \circ y = (x_1 + y_1, x_2 + y_2, x_3 + y_3 + x_1\, y_2)$$

and

$$\delta_\lambda(x) = (\lambda\, x_1, \lambda\, x_2, \lambda^2\, x_3).$$

The Jacobian basis for $\mathfrak{g}$ is given by

$$X_1 = \partial_1, \;\; X_2 = \partial_2 + x_1\, \partial_3, \;\; X_3 = [X_1, X_2] = \partial_3.$$

Moreover, we have

$$\mathrm{Exp}\,(\xi_1\, X_1 + \xi_2\, X_2 + \xi_3\, X_3) = \left(\xi_1, \xi_2, \xi_3 + \frac{1}{2}\,\xi_1\,\xi_2\right),$$

$$\mathrm{Log}\,(x) = x_1\, X_1 + x_2\, X_2 + \left(x_3 - \frac{1}{2}\, x_1\, x_2\right) X_3.$$

Let now $B = (b_{i,j})_{i,j \le 2}$ be an assigned symmetric matrix. Since $\mathbb{G}$ is free, there exists a unique Lie algebra morphism φ from $\mathfrak{g}$ into itself which maps the generators X_1, X_2 respectively in

$$b_{1,1}\, X_1 + b_{1,2}\, X_2, \qquad b_{1,2}\, X_1 + b_{2,2}\, X_2.$$

In Jacobian coordinates, φ is represented by the block-diagonal matrix

$$\begin{pmatrix} b_{1,1} & b_{1,2} & 0 \\ b_{1,2} & b_{2,2} & 0 \\ 0 & 0 & \det(B) \end{pmatrix}.$$

We now set

$$T = \mathrm{Exp} \circ \varphi \circ \mathrm{Log}\,.$$

A direct computation gives

$$T(x) = \begin{pmatrix} b_{1,1}\, x_1 + b_{1,2}\, x_2 \\ b_{1,2}\, x_1 + b_{2,2}\, x_2 \\ \det(B)\, x_3 + \frac{1}{2}\, b_{1,1}\, b_{1,2}\, x_1^2 + \frac{1}{2}\, b_{1,2}\, b_{2,2}\, x_2^2 + b_{1,2}^2\, x_1\, x_2 \end{pmatrix}.$$

In particular, if we choose

$$b_{1,1} = 2, b_{1,2} = b_{2,1} = 1, b_{2,2} = 4,$$

we obtain

$$T(x) = (2\, x_1 + x_2, x_1 + 4\, x_2, 7\, x_3 + x_1^2 + 2\, x_2^2 + x_1\, x_2).$$

With the notation in the proof of Theorem 16.1.2, we see that, choosing

$$A = B^{-2} = \begin{pmatrix} +\frac{17}{49} & -\frac{6}{49} \\ -\frac{6}{49} & +\frac{5}{49} \end{pmatrix},$$

the diffeomorphism $T = T_A$ turns

$$Y_1 = \tfrac{4}{7}\,X_1 - \tfrac{1}{7}\,X_2 \quad \text{and} \quad Y_2 = -\tfrac{1}{7}\,X_1 + \tfrac{2}{7}\,X_2$$

respectively into X_1 and X_2. As a consequence, T_A turns the sub-Laplacian

$$\mathcal{L}_A = \frac{17}{49}\,X_1^2 - \frac{6}{49}\,X_1X_2 - \frac{6}{49}\,X_2X_1 + \frac{5}{49}\,X_2^2$$

into the canonical sub-Laplacian

$$\Delta_{\mathbb{G}} = X_1^2 + X_2^2.$$

We remark that T_A is not linear. □

We observe that the group $\mathbb{G}$ in the above remark is isomorphic to $\mathbb{H}^1$. However, on $\mathbb{H}^1$, T_A is always linear. This follows from the fact that, in $\mathbb{H}^1$, the map Exp is linear (with the obvious choices of basis in $\mathbb{H}^1$ and in its Lie algebra).

Remark 16.2.2. If $\mathbb{G}$ is not free, a diffeomorphism T_A satisfying (16.5a) of Theorem 16.1.2 may not exist.

Proof. We consider the group $\mathbb{G} = \mathbb{H}^2$, the Heisenberg group on $\mathbb{R}^5$. If the points of $\mathbb{H}^2$ are denoted by (a, b, c), with $a,\ b \in \mathbb{R}^2$, $c \in \mathbb{R}$, we have

$$(a, b, c) \circ (\alpha, \beta, \gamma) = (a + \alpha, b + \beta, c + \gamma + 2\langle b, \alpha\rangle - 2\langle a, \beta\rangle)$$

and

$$\delta_\lambda(a, b, c) = (\lambda\, a, \lambda\, b, \lambda^2 c),$$

and the canonical sub-Laplacian on $\mathbb{H}^2$ is given by

$$\sum_{j=1}^{2} (A_j^2 + B_j^2),$$

where

$$A_j = \partial_{a_j} + 2\, b_j\, \partial_c, \quad B_j = \partial_{b_j} - 2\, a_j\, \partial_c, \quad j = 1,\ 2.$$

$(\mathbb{H}^2, \circ, \delta_\lambda)$ is then a homogeneous Carnot group of step 2 with 4 generators. The dimension of the Lie algebra of $\mathbb{H}^2$ is 5, whereas

$$\dim(\mathfrak{f}_{4,2}) = 10,$$

hence $\mathbb{H}^2$ is *not* free. We observe that *there does not exist any diffeomorphism on $\mathbb{H}^2$ turning the set of vector fields*

$$\mathcal{Y} = \{A_1 + B_2, A_2, B_1, B_2\}$$

into the set

$$\mathcal{X} = \{A_1, A_2, B_1, B_2\}.$$

Indeed, each vector field in $\mathcal{X}$ has exactly one non-vanishing commutator with any other vector field in $\mathcal{X}$; on the contrary, $A_1 + B_2 \in \mathcal{Y}$ has two non-vanishing commutators with the other vector fields in $\mathcal{Y}$. □

According to Remark 16.2.2 (we follow the notation therein), we may expect that the two sub-Laplacians on $\mathbb{H}^2$

$$(A_1 + B_2)^2 + A_2^2 + B_1^2 + B_2^2 \quad \text{and} \quad A_1^2 + A_2^2 + B_1^2 + B_2^2$$

cannot be turned into each other, thus furnishing an example of two sub-Laplacians on the same group which are not "equivalent". However, since (16.5a) is only sufficient but not necessary to deduce (16.5b), further comments must be made: these are given in the next theorem.

Theorem 16.2.3. *If $\mathbb{G}$ is not free, a diffeomorphism T_A satisfying* (16.5b) *of Theorem* 16.1.2 *may not exist.*

Proof. The proof of this result makes use of some Liouville-type theorems for sub-Laplacians proved in Section 5.8, page 269. We shall exhibit a rather elaborated counter-example.

First of all, if $A = (a_{i,j})_{i,j \le m}$ is a fixed symmetric matrix and the X_i's are smooth vector fields, we provide a lemma characterizing the maps T turning the operator $\sum_{i,j} a_{i,j} X_i X_j$ into the operator $\sum_i X_i^2$.

Lemma 16.2.4. *Suppose $A = (a_{i,j})_{i,j \le m}$ is a fixed symmetric matrix and*

$$X_i = \sum_{k=1}^{N} \alpha_k^{(i)}(x)\, \partial_k, \qquad i = 1, \dots, m,$$

are smooth vector fields on $\mathbb{R}^N$. Then a map $T \in C^2(\mathbb{R}^N, \mathbb{R}^N)$ satisfies

$$\sum_{i,j=1}^{m} a_{i,j}\, X_i X_j (u \circ T) = \left(\sum_{i=1}^{m} X_i^2 u \right) \circ T \qquad \forall\, u \in C^\infty(\mathbb{R}^N) \tag{16.10}$$

(i.e. T turns the operator $\sum_{i,j} a_{i,j} X_i X_j$ into the operator $\sum_i X_i^2$) if and only if the following system of quasi-linear PDE's is satisfied

$$\begin{cases} \displaystyle\sum_{i,j=1}^{m} a_{i,j}\, X_i X_j T_k = \sum_{i=1}^{m} (X_i \alpha_k^{(i)}) \circ T & \forall\, k = 1, \dots, N, \\[2ex] \displaystyle\sum_{i,j=1}^{m} a_{i,j}\, X_i T_l\, X_j T_k = \sum_{i=1}^{m} (\alpha_k^{(i)}\, \alpha_l^{(i)}) \circ T & \forall\, k, l = 1, \dots, N. \end{cases} \tag{16.11}$$

Proof. The left-hand side of (16.10) is equal to

$$\sum_{k,l=1}^{N}\Big(\sum_{i,j} a_{i,j}\, X_iT_l\, X_jT_k\Big)\cdot(\partial_{k,l}u)\circ T+\sum_{k=1}^{N}\Big(\sum_{i,j} a_{i,j}\, X_iX_jT_k\Big)\cdot(\partial_k u)\circ T.$$

With the position $X_i=\sum_{k=1}^{N}\alpha_k^{(i)}(x)\,\partial_k$, the right-hand side of (16.10) equals

$$\sum_{k,l=1}^{N}\left(\sum_{i=1}^{m}(\alpha_k^{(i)}\,\alpha_l^{(i)})\circ T\right)\cdot(\partial_{k,l}u)\circ T+\sum_{k=1}^{N}\left(\sum_{i=1}^{m}(X_i\alpha_k^{(i)})\circ T\right)\cdot(\partial_k u)\circ T.$$

Now, if we put $u(x)=x_k$ in (16.10) and equate the left-hand and right-hand sides of (16.10) expressed above, we obtain

$$\sum_{i,j=1}^{m} a_{i,j}\, X_iX_jT_k=\sum_{i=1}^{m}(X_i\alpha_k^{(i)})\circ T \qquad \forall\, k=1,\dots,N. \tag{16.12}$$

Moreover, if we put $u(x)=x_k\,x_l$ in (16.10), and (16.12) holds, we get (recall that $a_{i,j}=a_{j,i}$)

$$\sum_{i,j=1}^{m} a_{i,j}\, X_iT_l\, X_jT_k=\sum_{i=1}^{m}(\alpha_k^{(i)}\,\alpha_l^{(i)})\circ T \qquad \forall\, k,\, l=1,\dots,N. \tag{16.13}$$

This proves that (16.10) is equivalent to (16.11). □

We explicitly remark that, if we set

$$\mathcal{J}_T^X:=(X_jT_i)_{1\le i\le N,\ 1\le j\le m} \qquad \text{(the “}X\text{-Jacobian” matrix of the map } T\text{)}$$

and if we consider the $N\times m$ matrix X whose j-th column is given by the N-tuple of the coefficient functions of X_jI, i.e.

$$X:=(X_1I\cdots X_mI)=\big(\alpha_i^{(j)}\big)_{1\le i\le N,\ 1\le j\le m}, \tag{16.14}$$

then the second equation in (16.11) turns out to be equivalent to

$$\mathcal{J}_T^X\cdot A\cdot {}^t\mathcal{J}_T^X=\big(X\cdot {}^tX\big)\circ T. \tag{16.15}$$

We now return to the proof of Theorem 16.2.3. We consider the special case of the Heisenberg group $\mathbb{H}^2$ (whose points are denoted by $(x_1,\dots,x_5)$), with the natural choice of the fields $X_1,\dots,X_4$, i.e.

$$X_1:=\partial_1+2\,x_3\,\partial_5, \qquad X_2:=\partial_2+2\,x_4\,\partial_5,$$
$$X_3:=\partial_3-2\,x_1\,\partial_5, \qquad X_4:=\partial_4-2\,x_2\,\partial_5.$$

We aim to prove that there does not exist any map $T\in C^2(\mathbb{H}^2,\mathbb{H}^2)$ satisfying (16.10), for a suitable matrix A to be specified in the sequel (henceforth, we write

$\mathcal{L}_A := \sum_{i,j=1}^m a_{i,j} X_i X_j$). To this end, we suppose by contradiction that there exists such a T.

Obviously, it is not restrictive to suppose that $T(0) = 0$ (since $X_1, \ldots, X_4$ are left-translation invariant). With the above choice of $X_1, \ldots, X_4$ (and following the notation in (16.14)), one has

$$X \cdot {}^t X = \begin{pmatrix} 1 & 0 & 0 & 0 & 2x_3 \\ 0 & 1 & 0 & 0 & 2x_4 \\ 0 & 0 & 1 & 0 & -2x_1 \\ 0 & 0 & 0 & 1 & -2x_2 \\ 2x_3 & 2x_4 & -2x_1 & -2x_2 & 4(x_1^2 + x_2^2 + x_3^2 + x_4^2) \end{pmatrix}. \tag{16.16}$$

Hence, the first equation in (16.11) becomes

$$\mathcal{L}_A T_k = \sum_{i,j=1}^4 a_{i,j} X_i X_j T_k = 0 \qquad \forall\, k = 1, \ldots, 5. \tag{16.17}$$

The second equation in (16.11) gives, in particular,

$$\sum_{i,j=1}^4 a_{i,j}\, X_i T_h\, X_j T_h = 1 \qquad \forall\, h = 1, \ldots, 4, \tag{16.18a}$$

$$\sum_{i,j=1}^4 a_{i,j}\, X_i T_5\, X_j T_5 = 4\,(T_1^2 + T_2^2 + T_3^2 + T_4^2). \tag{16.18b}$$

Moreover, from (16.17) and (16.18a) we easily obtain

$$\mathcal{L}_A\big((T_k)^2\big) = 2\,T_k\,\mathcal{L}_A T_k + 2\sum_{i,j=1}^4 a_{i,j}\, X_i T_k\, X_j T_k = 2. \tag{16.19}$$

Suppose now that $A = (a_{i,j})_{i,j\le 4}$ is a symmetric *positive-definite* matrix. Then, by Proposition 16.1.1, $\mathcal{L}_A$ is a sub-Laplacian on $\mathbb{H}^2$. In particular, we are in the position to apply the Liouville-type theorems for sub-Laplacians in Section 5.8 (page 269), Chapter 5.

By the hypoellipticity of $\mathcal{L}_A$, from (16.17) it follows that $T_k \in C^\infty(\mathbb{R}^N)$ for every $k = 1, \ldots, 5$. Let now $k \in \{1, \ldots, 4\}$ be fixed. We note that $(T_k)^2$ is a non-negative function such that

$$\mathcal{L}_A((T_k)^2)$$

is a polynomial of $\mathbb{H}^2$-degree[5] zero (see (16.19)). As a consequence, by Liouville Theorem 5.8.4 (page 270), $(T_k)^2$ is a polynomial function of $\mathbb{H}^2$-degree at most 2 for every $k = 1, \ldots, 4$. In particular, this proves that T_k is bounded from below by a polynomial function of $\mathbb{H}^2$-degree at most 1. Hence, by Theorem 5.8.2, page 270

[5] For the definition of $\mathbb{G}$-degree, see Definition 1.3.3, page 33.

(being $\mathcal{L}_A T_k = 0$), it follows that, for any $k = 1, \dots, 4$, T_k is a polynomial of $\mathbb{H}^2$-degree at most 1. Having supposed that $T(0) = 0$, there exists a constant matrix $B = (b_{i,j})_{i,j \le 4}$ such that

$$\begin{pmatrix} T_1 \\ T_2 \\ T_3 \\ T_4 \end{pmatrix} = B \cdot \begin{pmatrix} x_1 \\ x_2 \\ x_3 \\ x_4 \end{pmatrix}. \tag{16.20}$$

From (16.17) and (16.18b) we obtain

$$\mathcal{L}_A\big((T_5)^2\big) = 2\, T_5\, \mathcal{L}_A T_5 + 2 \sum_{i,j=1}^{4} a_{i,j}\, X_i T_5\, X_j T_5 = 8\,(T_1^2 + T_2^2 + T_3^2 + T_4^2).$$

From (16.20) we then infer that $\mathcal{L}_A((T_5)^2)$ is a polynomial function of $\mathbb{H}^2$-degree at most 2. Arguing as above, we derive that T_5 is a polynomial function of $\mathbb{H}^2$-degree at most 2. Then there exist $c_5 \in \mathbb{R}$, a 1×4 constant matrix $D = (d_i)_{i \le 4}$ and a *symmetric* 4×4 constant matrix $C = (c_{i,j})_{i,j \le 4}$ such that (setting $\overline{x} = {}^t(x_1, x_2, x_3, x_4)$)

$$T_5(x) = c_5\, x_5 + D \cdot \overline{x} + {}^t\overline{x} \cdot C \cdot \overline{x}.$$

First, we show that the entries of D are zero. From (16.20) we obtain

$$X_i T_l = \partial_i T_l = b_{l,i}$$

for every $l = 1, \dots, 4$ and $i = 1, \dots, 4$. Hence, if we take $1 \le l \le 4$ and $1 \le k \le 4$ in the second equation of (16.11), then we derive (see also (16.16))

$$B \cdot A \cdot {}^t B = \mathbb{I}_4 \tag{16.21}$$

($\mathbb{I}_n$ denoting henceforth the $m \times n$ identity matrix). In particular, we notice that B is non-singular. If in the second equation of (16.11) we take $l \in \{1, \dots, 4\}$ and $k = 5$, we obtain

$$B \cdot A \cdot \begin{pmatrix} X_1 T_5 \\ X_2 T_5 \\ X_3 T_5 \\ X_4 T_5 \end{pmatrix} = \begin{pmatrix} 2\, T_3 \\ 2\, T_4 \\ -2\, T_1 \\ -2\, T_2 \end{pmatrix}.$$

By evaluating both sides of this last identity at 0 (and recalling the form of T_5), one has

$$(X_i T_5)(0) = (\partial_i T_5)(0) = d_i,$$

hence $B \cdot A \cdot D = 0$. Since A and B are non-singular, it follows $D = 0$.

Consequently, T_5 is a homogeneous polynomial of the form

$$T_5(x) = c_5\, x_5 + {}^t\overline{x} \cdot C \cdot \overline{x}, \qquad \text{where } {}^t C = C, \quad \overline{x} = {}^t(x_1, x_2, x_3, x_4). \tag{16.22}$$

By writing (16.15) in a more compact form and making use of (16.20) and (16.22), one gets

$$\mathcal{J}_T^X = \begin{pmatrix} B \\ 2\,{}^t\overline{x}\cdot{}^t\widehat{C} \end{pmatrix},$$

where we have set

$$\widehat{C} := C + c_5\,\widehat{I} \qquad \text{and} \quad \widehat{I} := \begin{pmatrix} 0 & \mathbb{I}_2 \\ -\mathbb{I}_2 & 0 \end{pmatrix}.$$

Thus, with a block notation, we infer

$$\mathcal{J}_T^X \cdot A \cdot {}^t\mathcal{J}_T^X = \begin{pmatrix} B\cdot A\cdot{}^tB & 2\,B\cdot A\cdot\widehat{C}\cdot\overline{x} \\ 2\,{}^t\overline{x}\cdot{}^t\widehat{C}\cdot A\cdot{}^tB & 4\,{}^t\overline{x}\cdot{}^t\widehat{C}\cdot A\cdot\widehat{C}\cdot\overline{x} \end{pmatrix}.$$

Again from (16.20) and (16.16), we also recognize that

$$(X\cdot{}^tX)\circ T = \begin{pmatrix} \mathbb{I}_4 & 2\,\widehat{I}\cdot B\cdot\overline{x} \\ 2\,{}^t\overline{x}\cdot{}^tB\cdot{}^t\widehat{I} & 4\,{}^t\overline{x}\cdot{}^tB\cdot{}^t\widehat{I}\cdot\widehat{I}\cdot B\cdot\overline{x} \end{pmatrix}.$$

Collecting together all the above facts, we have proved that (16.15) is equivalent to

$$\begin{cases} B\cdot A\cdot{}^tB = \mathbb{I}_4, \\ 2\,B\cdot A\cdot\widehat{C}\cdot\overline{x} = 2\,\widehat{I}\cdot B\cdot\overline{x}, \\ 2\,{}^t\overline{x}\cdot{}^t\widehat{C}\cdot A\cdot{}^tB = 2\,{}^t\overline{x}\cdot{}^tB\cdot{}^t\widehat{I}, \\ 4\,{}^t\overline{x}\cdot{}^t\widehat{C}\cdot A\cdot\widehat{C}\cdot\overline{x} = 4\,{}^t\overline{x}\cdot{}^tB\cdot{}^t\widehat{I}\cdot\widehat{I}\cdot B\cdot\overline{x}. \end{cases}$$

The third equation follows from the second one by transposing, whereas the fourth one follows from the first and the second equations. As a consequence, (16.15) turns out to be equivalent to

$$\begin{cases} B\cdot A\cdot{}^tB = \mathbb{I}_4, \\ 2\,B\cdot A\cdot\widehat{C} = 2\,\widehat{I}\cdot B, \end{cases}$$

or equivalently,

$$\begin{cases} B\cdot A\cdot{}^tB = \mathbb{I}_4, \\ \widehat{C} = {}^tB\cdot\widehat{I}\cdot B. \end{cases} \tag{16.23}$$

We now explicitly observe that $\widehat{I}$ is skew-symmetric, then the same is true of ${}^tB\cdot\widehat{I}\cdot B$. From the second equation in (16.23) we infer that the matrix

$$\widehat{C} = \begin{pmatrix} c_{11} & c_{12} & c_{13}+c_5 & c_{14} \\ c_{12} & c_{22} & c_{23} & c_{24}+c_5 \\ c_{13}-c_5 & c_{23} & c_{33} & c_{34} \\ c_{14} & c_{24}-c_5 & c_{34} & c_{44} \end{pmatrix}$$

must be skew-symmetric. This is possible if and only if $C = 0$. Thus, (16.23) is equivalent to

$$\begin{cases} B\cdot A\cdot{}^tB = \mathbb{I}_4, \\ c_5\,\widehat{I} = {}^tB\cdot\widehat{I}\cdot B. \end{cases} \tag{16.24}$$

From the second equation of (16.24) it follows that $c_5 \neq 0$ (thus, T is bijective!).

Finally, we choose the symmetric positive-definite matrix A of the following form: $A = {}^tS \cdot S$, where S is a suitable non-singular matrix. From the first equation in (16.24) we then have

$$(B \cdot S) \cdot {}^t(B \cdot S) = \mathbb{I}_4,$$

i.e. $B = O \cdot S^{-1}$, with an orthogonal matrix O. In particular, the second equation of (16.24) becomes

$${}^tS \cdot \widehat{I} \cdot S = \frac{1}{c_5}\,{}^tO \cdot \widehat{I} \cdot O, \quad \text{where } O \text{ is an orthogonal matrix.} \tag{16.25}$$

We choose

$$S = \begin{pmatrix} 1 & 0 & 0 & 1 \\ 0 & 1 & 0 & 0 \\ 0 & 0 & 1 & 0 \\ 0 & 0 & 0 & 1 \end{pmatrix}.$$

Thus, (16.25) is equivalent to

$$\begin{pmatrix} 0 & 0 & 1 & 0 \\ 0 & 0 & 0 & 1 \\ -1 & 0 & 0 & -1 \\ 0 & -1 & 1 & 0 \end{pmatrix} = \frac{1}{c_5}\,{}^tO \cdot \widehat{I} \cdot O, \quad \text{where } O \text{ is an orthogonal matrix.} \tag{16.26}$$

In particular, the eigenvalues of these matrices should be equal. But the eigenvalues of $\frac{1}{c_5}\,{}^tO \cdot \widehat{I} \cdot O$ (being O orthogonal) are $\pm\frac{1}{c_5}\,i$, whereas the eigenvalues of the matrix in the left-hand side of (16.26) are $\pm i\sqrt{\frac{3\pm\sqrt{5}}{2}}$. This gives a contradiction, proving that, with the above choice of S and consequently of A, *it does not exist any map* $T \in C^2(\mathbb{H}^2, \mathbb{H}^2)$ *satisfying* (16.10). This ends the proof of Theorem 16.2.3. □

Remark 16.2.5. In the proof of Theorem 16.2.3, we explicitly proved that the operator $\mathcal{L}_A = \sum_{i,j} a_{i,j} X_i X_j$ on $\mathbb{H}^2$, where

$$A = {}^tS \cdot S = \begin{pmatrix} 1 & 0 & 0 & 0 \\ 0 & 1 & 0 & 0 \\ 0 & 0 & 1 & 0 \\ 1 & 0 & 0 & 1 \end{pmatrix} \cdot \begin{pmatrix} 1 & 0 & 0 & 1 \\ 0 & 1 & 0 & 0 \\ 0 & 0 & 1 & 0 \\ 0 & 0 & 0 & 1 \end{pmatrix} = \begin{pmatrix} 1 & 0 & 0 & 1 \\ 0 & 1 & 0 & 0 \\ 0 & 0 & 1 & 0 \\ 1 & 0 & 0 & 2 \end{pmatrix},$$

cannot be turned into the canonical sub-Laplacian $\Delta_{\mathbb{H}^2} = \sum_i X_i^2$ by *any* C^2-map T. However, by Proposition 16.1.1, we know that $\mathcal{L}_A$ is (exactly as $\Delta_{\mathbb{H}^2}$) a sub-Laplacian on $\mathbb{H}^2$. Since the symmetric square root B of A is

$$B = \begin{pmatrix} \frac{2}{\sqrt{5}} & 0 & 0 & \frac{1}{\sqrt{5}} \\ 0 & 1 & 0 & 0 \\ 0 & 0 & 1 & 0 \\ \frac{1}{\sqrt{5}} & 0 & 0 & \frac{3}{\sqrt{5}} \end{pmatrix},$$

again from the result in Proposition 16.1.1 we know that $\mathcal{L}_A$ is equal to the following sub-Laplacian

$$\mathcal{L}_A = \sum_{k=1}^{m} Y_k^2, \quad \text{with } Y_k = \sum_{j=1}^{m} (A^{1/2})_{k,j}\, X_j,$$

i.e.

$$Y_1 = \frac{2}{\sqrt{5}} X_1 + \frac{1}{\sqrt{5}} X_4, \quad Y_2 = X_2, \quad Y_3 = X_3, \quad Y_4 = \frac{1}{\sqrt{5}} X_1 + \frac{3}{\sqrt{5}} X_4.$$

This proves that the following two sub-Laplacians on $\mathbb{H}^2$ *cannot be turned into each other by any C^2-map*:

$$\begin{aligned}
\Delta_{\mathbb{H}^2} &= (\partial_1 + 2\,x_3\,\partial_5)^2 + (\partial_2 + 2\,x_4\,\partial_5)^2 + (\partial_3 - 2\,x_1\,\partial_5)^2 + (\partial_4 - 2\,x_2\,\partial_5)^2,\\
\mathcal{L}_A &= \left(\frac{2}{\sqrt{5}}\,\partial_1 + \frac{1}{\sqrt{5}}\,\partial_4 + \left(\frac{4}{\sqrt{5}}\,x_3 - \frac{2}{\sqrt{5}}\,x_2\right)\partial_5\right)^2\\
&\quad + (\partial_2 + 2\,x_4\,\partial_5)^2 + (\partial_3 - 2\,x_1\,\partial_5)^2\\
&\quad + \left(\frac{1}{\sqrt{5}}\,\partial_1 + \frac{3}{\sqrt{5}}\,\partial_4 + \left(\frac{2}{\sqrt{5}}\,x_3 - \frac{6}{\sqrt{5}}\,x_2\right)\partial_5\right)^2\\
&= X_1^2 + X_2^2 + X_3^2 + 2\,X_4^2 + X_1\,X_4 + X_4\,X_1\\
&= (\partial_1 + 2\,x_3\,\partial_5)^2 + (\partial_2 + 2\,x_4\,\partial_5)^2 + (\partial_3 - 2\,x_1\,\partial_5)^2 + 2\,(\partial_4 - 2\,x_2\,\partial_5)^2\\
&\quad + 2\,(\partial_1 + 2\,x_3\,\partial_5)\,(\partial_4 - 2\,x_2\,\partial_5).
\end{aligned}$$

However, another natural question still arises: is it possible to create a *new* Carnot group $\mathbb{G}$ on $\mathbb{R}^5$, (different but) diffeomorphic to $\mathbb{H}^2$, such that its *canonical* sub-Laplacian $\Delta_{\mathbb{G}}$ is turned into (i.e. equivalent to) the *non-canonical* $\mathcal{L}_A$? We answer to this question in the next section.

16.3 Canonical or Non-canonical?

In this section, we deal with the following problem. Let $\mathbb{G} = (\mathbb{R}^N, \circ, \delta_\lambda)$ be a homogeneous Carnot group with an assigned (not necessarily canonical) sub-Laplacian $\mathcal{L}$. We know that there may not exist any C^2-map *on* $\mathbb{G}$ *itself* turning $\mathcal{L}$ into the canonical sub-Laplacian $\Delta_{\mathbb{G}}$ of $\mathbb{G}$ (see, for instance, Theorem 16.2.3). We aim to show that *there exists a different homogeneous Carnot group* $\mathbb{H} = (\mathbb{R}^N, *, \delta_\lambda)$ (with the same underlying manifold $\mathbb{R}^N$ and the same group of dilations δ_λ) *and a Lie-group isomorphism from* $\mathbb{G}$ *to* $\mathbb{H}$ *turning* $\mathcal{L}$ *into the canonical sub-Laplacian* $\Delta_{\mathbb{H}}$. Roughly speaking, even if we are given a *non-canonical* sub-Laplacian $\mathcal{L}$, up to an isomorphism of Lie groups, we can consider $\mathcal{L}$ to be equivalent to a *canonical* sub-Laplacian, modulo a change of the Lie group.

Let us denote by $\mathcal{L} = \sum_{j=1}^m Y_j^2$ a given sub-Laplacian on $\mathbb{G}$ and by $\Delta_{\mathbb{G}} = \sum_{j=1}^m X_j^2$ the canonical sub-Laplacian on $\mathbb{G}$. By definition, there exists a non-singular matrix

$$B = (b_{i,j})_{i,j \le m}$$

such that

$$Y_j = \sum_{i=1}^{m} b_{i,j}\, X_i \quad \text{for every } j = 1, \ldots, m. \tag{16.27}$$

With reference to the particular form

$$\delta_\lambda(x) = \delta_\lambda(x^{(1)}, x^{(2)}, \ldots, x^{(r)}) = (\lambda x^{(1)}, \lambda^2 x^{(1)}, \ldots, \lambda^r x^{(r)}), \qquad x^{(i)} \in \mathbb{R}^{N_i},$$

of the dilations on $\mathbb{G}$, we introduce a block-diagonal matrix

$$C = \begin{pmatrix} C^{(1)} & 0 & \cdots & 0 \\ 0 & C^{(2)} & \ddots & \vdots \\ \vdots & \ddots & \ddots & 0 \\ 0 & \cdots & 0 & C^{(r)} \end{pmatrix}, \tag{16.28}$$

where $C^{(i)}$ is a matrix of order $N_i \times N_i$, for every $i = 1, \ldots, r$, and we choose $C^{(i)}$ in such a way that

$$C^{(1)} = B^{-1} \quad \text{and} \quad C^{(2)}, \ldots, C^{(r)} \text{ are non-singular.} \tag{16.29}$$

In the sequel, we also denote by C the linear map

$$C : \mathbb{R}^N \to \mathbb{R}^N, \qquad x \mapsto y = C \cdot x.$$

We now consider a new Lie group $\mathbb{H} = (\mathbb{R}^N, *)$, where $*$ is defined by

$$\mathbb{H} \times \mathbb{H} \ni (x, y) \mapsto x * y := C\big(C^{-1}(x) \circ C^{-1}(y)\big) \in \mathbb{H}. \tag{16.30}$$

We claim that:

1) $C : (\mathbb{G}, \circ) \to (\mathbb{H}, *)$ is a Lie group isomorphism;
2) $\mathbb{H} = (\mathbb{R}^N, *, \delta_\lambda)$ is a *homogeneous* Carnot group (this is not obvious, since we know that the homogeneity is not an invariant of equivalence classes of abstract Carnot groups);
3) C turns $\mathcal{L}$ on $\mathbb{G}$ into the canonical sub-Laplacian $\Delta_{\mathbb{H}}$ of $\mathbb{H}$.

Now, 1) is trivial (see also Example 2.1.48, page 112): for every $\xi, \eta \in \mathbb{G}$ we have, by (16.30),

$$C(\xi) * C(\eta) = C\big(C^{-1}(C(\xi)) \circ C^{-1}(C(\eta))\big) = C(\xi \circ \eta).$$

We now prove 2). We must show that:

2') the first m vector fields of the Jacobian basis of $\mathbb{H}$ are Lie-generators of the Lie algebra $\mathfrak{h}$ of $\mathbb{H}$; and
2") δ_λ is a Lie group morphism of $(\mathbb{H}, *)$.

First of all, we show that

$$C \text{ turns } Y_j \text{ into the } j\text{-th vector field of the Jacobian basis of } \mathbb{H}. \tag{16.31}$$

We denote by $H_1, \dots, H_N$ the Jacobian basis of $\mathfrak{h}$, the Lie algebra of $\mathbb{H}$. In order to prove (16.31), we have to show that

$$C \text{ turns } Y_j \text{ into } H_j \text{ for every } j = 1, \dots, m.$$

Now, from Example 2.1.48 (page 112) we know that the map C turns the vector field Y_j into a $*$-left-invariant vector field $\widetilde{Y}_j$ on $\mathbb{H}$ such that (see (2.17a) in Example 2.1.25, page 101)

$$\widetilde{Y}_j I(y) = \mathcal{J}_C(C^{-1}(y)) \cdot (Y_j I)(C^{-1}(y)).$$

In particular, we have

$$\begin{aligned}
\widetilde{Y}_j I(0) &= \mathcal{J}_C(C^{-1}(0)) \cdot (Y_j I)(C^{-1}(0)) = C \cdot (Y_j I)(0) \\
\text{(see (16.27))} &= C \cdot \left(\sum_{i=1}^m b_{i,j}\, X_i I(0) \right) \\
&= C \cdot (b_{1,j}, \dots, b_{m,j}, 0, \dots, 0)^T \\
&= (C^{(1)} \cdot b_j^{(1)}, 0^{(2)}, \dots, 0^{(r)}),
\end{aligned}$$

where $b_j^{(1)}$ denotes the j-th column of B. By the choice (16.29) of $C^{(1)}$, this proves that $\widetilde{Y}_j I(0)$ is the j-th vector of the canonical basis of $\mathbb{R}^N$. Since the same is true of $H_j I(0)$ (by the definition of Jacobian basis) and the left-invariant vector fields are determined by their value at the origin, this proves (16.31).

From Theorem 2.1.50 (page 114) we know that $\widetilde{Y}_j = \mathrm{d}C(Y_j)$. Hence, the above computation proves that

$$\mathrm{d}C(Y_j) = H_j. \tag{16.32}$$

Since $X_1, \dots, X_m$ is a system of Lie-generators for the Lie-algebra $\mathfrak{g}$ of $\mathbb{G}$ and the X_j's are linear combinations of the Y_j's, then every element of $\mathfrak{g}$ is a linear combination of commutators of the type

$$U = [U_1[U_2[U_3[\cdots[U_{n-1}, U_n]\cdots]]]],$$

with $U_1, \dots, U_n \in \{Y_1, \dots, Y_m\}$. Now, take any $H \in \mathfrak{h}$; since $\mathrm{d}C : \mathfrak{g} \to \mathfrak{h}$ is a Lie-algebra isomorphism (see Theorem 2.1.50, page 114), then $(\mathrm{d}C)^{-1}(H) \in \mathfrak{g}$ is a linear combination of brackets as the above U. Consequently, H is a linear combination of elements in $\mathfrak{h}$ of the type

$$\begin{aligned}
(\mathrm{d}C)(U) &= (\mathrm{d}C)([U_1[U_2[U_3[\cdots[U_{n-1}, U_n]\cdots]]]]) \\
&= [(\mathrm{d}C)U_1[(\mathrm{d}C)U_2[(\mathrm{d}C)U_3[\cdots[(\mathrm{d}C)U_{n-1}, (\mathrm{d}C)U_n]\cdots]]]] \\
&= [V_1[V_2[V_3[\cdots[V_{n-1}, V_n]\cdots]]]],
\end{aligned}$$

where $V_1, \dots, V_n \in \{(dC)Y_1, \dots, (dC)Y_m\} = \{H_1, \dots, H_m\}$ (see (16.32)). This proves (2').

In order to prove (2"), we first remark that δ_λ commutes with C, i.e.

$$\delta_\lambda(C(x)) = C(\delta_\lambda(x)) \quad \forall x \in \mathbb{R}^N, \; \forall \lambda > 0. \tag{16.33}$$

Indeed, following our usual notation, we have

$$C(\delta_\lambda(x)) = \begin{pmatrix} C^{(1)} & 0 & \cdots & 0 \\ 0 & C^{(2)} & \ddots & \vdots \\ \vdots & \ddots & \ddots & 0 \\ 0 & \cdots & 0 & C^{(r)} \end{pmatrix} \cdot \begin{pmatrix} \lambda x^{(1)} \\ \lambda^2 x^{(2)} \\ \vdots \\ \lambda^r x^{(r)} \end{pmatrix} = \begin{pmatrix} C^{(1)} \lambda x^{(1)} \\ C^{(2)} \lambda^2 x^{(2)} \\ \vdots \\ C^{(r)} \lambda^r x^{(r)} \end{pmatrix}$$

$$= \delta_\lambda \begin{pmatrix} C^{(1)} x^{(1)} \\ C^{(2)} x^{(2)} \\ \vdots \\ C^{(r)} x^{(r)} \end{pmatrix} = \delta_\lambda(C(x)).$$

Obviously, (16.33) implies that δ_λ commutes with C^{-1} too. Now, (2") is equivalent to

$$\delta_\lambda(x * y) = (\delta_\lambda(x)) * (\delta_\lambda(y)) \quad \forall x, y \in \mathbb{H}, \; \forall \lambda > 0,$$

which follows from the simple argument below

$$\begin{aligned}
(\delta_\lambda(x)) * (\delta_\lambda(y)) &= \text{(see (16.30))} \quad C\big(C^{-1}(\delta_\lambda(x)) \circ C^{-1}(\delta_\lambda(y))\big) \\
&= \text{(see (16.33))} \quad C\big((\delta_\lambda(C^{-1}(x))) \circ (\delta_\lambda(C^{-1}(y)))\big) \\
&\quad (\delta_\lambda \text{ is a Lie group morphism of } (\mathbb{G}, \circ)) \\
&= C\big(\delta_\lambda\big((C^{-1}(x)) \circ (C^{-1}(y))\big)\big) \\
&= \text{(see (16.33))} \quad \delta_\lambda\big(C\big((C^{-1}(x)) \circ (C^{-1}(y))\big)\big) \\
&= \text{(see (16.30))} \quad \delta_\lambda(x * y).
\end{aligned}$$

Finally, (3) is a straightforward consequence of (16.31). □

In the next subsection, we shall see an explicit example of non-canonical sub-Laplacian turning into a canonical one.

Remark 16.3.1. The above computations also prove the following fact. Let $(\mathbb{R}^N, \circ, \delta_\lambda)$ be a homogeneous Carnot group. As usual, we denote the points of $\mathbb{G}$ by

$$x = (x^{(1)}, \dots, x^{(r)}) \qquad \text{with } x^{(i)} \in \mathbb{R}^{N_i}$$

and the dilation group by $\delta_\lambda(x) = (\lambda x^{(1)}, \dots, \lambda^r x^{(r)})$. Let

$$C^{(1)}, \dots, C^{(r)}$$

be fixed non-singular matrices, with $C^{(i)}$ of dimension $N_i \times N_i$ for every $i = 1, \dots, r$. We denote by C the matrix as in (16.28). Then C defines a linear change of coordinates on $\mathbb{R}^N$. The expression of $\circ$ in the new coordinates defined by $\xi = C(x)$

is precisely the composition $*$ defined in (16.30). Then, $\mathbb{H} = (\mathbb{R}^N, *, \delta_\lambda)$ is a homogeneous Carnot group isomorphic to $\mathbb{G} = (\mathbb{R}^N, \circ, \delta_\lambda)$, and C turns a left-invariant vector field X on $(\mathbb{G}, \circ)$ into the left-invariant vector field $\mathrm{d}C(X)$ on $\mathbb{H} = (\mathbb{R}^N, *)$; in particular, $\mathrm{d}C(X)$ coincides with $C \cdot XI(0)$ at the origin, i.e.

$$\big(\mathrm{d}C(X)\big)I(0) = C \cdot XI(0). \tag{16.34}$$

This last fact can be proved as follows. From Example 2.1.48 (page 112) we have

$$\big(\mathrm{d}C(X)\big)I(y) = \mathcal{J}_C(C^{-1}(y)) \cdot (XI)(C^{-1}(y)) = C \cdot (XI)(C^{-1}(y)).$$

When $y = 0$, we have (16.34). □

16.3.1 Example of a Non-canonical Sub-Laplacian Turning into a Canonical One

In Theorem 16.2.3, we proved that the following sub-Laplacian on $\mathbb{H}^2$

$$\begin{aligned}\mathcal{L} = &\left(\frac{2}{\sqrt{5}}\,\partial_1 + \frac{1}{\sqrt{5}}\,\partial_4 + \left(\frac{4}{\sqrt{5}}\,x_3 - \frac{2}{\sqrt{5}}\,x_2\right)\partial_5\right)^2 \\ &+ (\partial_2 + 2\,x_4\,\partial_5)^2 + (\partial_3 - 2\,x_1\,\partial_5)^2 \\ &+ \left(\frac{1}{\sqrt{5}}\,\partial_1 + \frac{3}{\sqrt{5}}\,\partial_4 + \left(\frac{2}{\sqrt{5}}\,x_3 - \frac{6}{\sqrt{5}}\,x_2\right)\partial_5\right)^2\end{aligned}$$

cannot be turned into the canonical sub-Laplacian $\Delta_{\mathbb{H}^2}$ of $\mathbb{H}^2$ itself by any C^2-map. Nonetheless, we have just showed a constructive method which allows us to turn the above $\mathcal{L}$ into the canonical sub-Laplacian on another homogeneous Carnot group.

Indeed (following the notation in Section 16.3), we have $\mathbb{G} = (\mathbb{R}^5, \circ, \delta_\lambda)$, where

$$x \circ y = \begin{pmatrix} x_1 + y_1 \\ x_2 + y_2 \\ x_3 + y_3 \\ x_4 + y_4 \\ x_5 + y_5 + 2(x_3\,y_1 + x_4\,y_2 - x_1\,y_3 - x_2\,y_4) \end{pmatrix}$$

and

$$\delta_\lambda(x_1, x_2, x_3, x_4, x_5) = (\lambda x_1, \lambda x_2, \lambda x_3, \lambda x_4, \lambda^2 x_5).$$

Moreover,

$$\mathcal{L} = \sum_{j=1}^4 Y_j^2, \quad \text{where } Y_j = \sum_{i=1}^4 b_{i,j}\,X_i$$

with

$$B = (b_{i,j})_{i,j\le 4} = \begin{pmatrix} \frac{2}{\sqrt{5}} & 0 & 0 & \frac{1}{\sqrt{5}} \\ 0 & 1 & 0 & 0 \\ 0 & 0 & 1 & 0 \\ \frac{1}{\sqrt{5}} & 0 & 0 & \frac{3}{\sqrt{5}} \end{pmatrix}.$$

More explicitly, since

$$X_1 = \partial_1 + 2\,x_3\,\partial_5, \qquad X_2 = \partial_2 + 2\,x_4\,\partial_5,$$
$$X_3 = \partial_3 - 2\,x_1\,\partial_5, \qquad X_4 = \partial_4 - 2\,x_2\,\partial_5,$$

the Y_j's are given by

$$Y_1 = \frac{2}{\sqrt{5}}\,X_1 + \frac{1}{\sqrt{5}}\,X_4, \quad Y_2 = X_2, \quad Y_3 = X_3, \quad Y_4 = \frac{1}{\sqrt{5}}\,X_1 + \frac{3}{\sqrt{5}}\,X_4.$$

We have

$$C^{(1)} = B^{-1} = \begin{pmatrix} \frac{3}{\sqrt{5}} & 0 & 0 & -\frac{1}{\sqrt{5}} \\ 0 & 1 & 0 & 0 \\ 0 & 0 & 1 & 0 \\ -\frac{1}{\sqrt{5}} & 0 & 0 & \frac{2}{\sqrt{5}} \end{pmatrix}.$$

Consequently, we define

$$C := \begin{pmatrix} C^{(1)} & 0 \\ 0 & 1 \end{pmatrix} = \begin{pmatrix} \frac{3}{\sqrt{5}} & 0 & 0 & -\frac{1}{\sqrt{5}} & 0 \\ 0 & 1 & 0 & 0 & 0 \\ 0 & 0 & 1 & 0 & 0 \\ -\frac{1}{\sqrt{5}} & 0 & 0 & \frac{2}{\sqrt{5}} & 0 \\ 0 & 0 & 0 & 0 & 1 \end{pmatrix}$$

and the relevant linear map

$$C : \mathbb{R}^5 \to \mathbb{R}^5, \quad x \mapsto y = C\,x = \begin{pmatrix} \frac{3}{\sqrt{5}}\,x_1 - \frac{1}{\sqrt{5}}\,x_4 \\ x_2 \\ x_3 \\ -\frac{1}{\sqrt{5}}\,x_1 + \frac{2}{\sqrt{5}}\,x_4 \\ x_5 \end{pmatrix}.$$

Following (16.30), we equip $\mathbb{R}^5$ with a new Lie group law $*$ defined by

$$\begin{aligned} y * \eta &:= C\big(C^{-1}(y) \circ C^{-1}(\eta)\big) \\ &= \Big(y_1 + \eta_1,\ y_2 + \eta_2,\ y_3 + \eta_3,\ y_4 + \eta_4, \\ &\qquad y_5 + \eta_5 + \frac{2}{\sqrt{5}}\big(y_3(2\eta_1 + \eta_4) + (y_1 + 3y_4)\eta_2 - (2y_1 + y_4)\eta_3 \\ &\qquad - y_2(\eta_1 + 3\eta_4)\big)\Big). \end{aligned}$$

We explicitly remark that this Lie group composition law can be written as in (3.6) (page 158) of Section 3.2, i.e.

$$\begin{aligned} y * \eta = (&y_1 + \eta_1, y_2 + \eta_2, y_3 + \eta_3, y_4 + \eta_4, \\ &y_5 + \eta_5 + \langle H(y_1, y_2, y_3, y_4), (\eta_1, \eta_2, \eta_3, \eta_4)\rangle), \end{aligned}$$

where

$$H = \frac{1}{\sqrt{5}} \begin{pmatrix} 0 & 1 & -2 & 0 \\ -1 & 0 & 0 & -3 \\ 2 & 0 & 0 & 1 \\ 0 & 3 & -1 & 0 \end{pmatrix}.$$

Finally, we observe that the first four vector fields of the Jacobian basis of $(\mathbb{R}^5, *)$ are given by[6]

$$\begin{aligned} H_1 &= \partial_{y_1} + \left(\frac{4}{\sqrt{5}} y_3 - \frac{2}{\sqrt{5}} y_2 \right) \partial_{y_5}, \\ H_2 &= \partial_{y_2} + \left(\frac{2}{\sqrt{5}} y_1 + \frac{6}{\sqrt{5}} y_4 \right) \partial_{y_5}, \\ H_3 &= \partial_{y_3} + \left(-\frac{4}{\sqrt{5}} y_1 - \frac{2}{\sqrt{5}} y_4 \right) \partial_{y_5}, \\ H_4 &= \partial_{y_4} + \left(\frac{2}{\sqrt{5}} y_3 - \frac{6}{\sqrt{5}} y_2 \right) \partial_{y_5}. \end{aligned}$$

We obviously recognize that the H_j's are the expression of the Y_j's w.r.t. the new system of coordinates $y = C\,x$, i.e.

$$(H_j f)(C(x)) = Y_j(f \circ C)(x) \qquad \forall\, f \in C^\infty(\mathbb{R}^N, \mathbb{R}),\ \forall\, x \in \mathbb{R}^N.$$

As a consequence, the sub-Laplacian $\mathcal{L}$ on $\mathbb{H}^2$ is turned by C into the *canonical* sub-Laplacian of $\mathbb{H}^* = (\mathbb{R}^5, *)$

$$\begin{aligned} \Delta_{\mathbb{H}^*} &= \sum_{j=1}^{4} H_j^2 \\ &= \left(\partial_{y_1} + \left(\frac{4}{\sqrt{5}} y_3 - \frac{2}{\sqrt{5}} y_2 \right) \partial_{y_5} \right)^2 \\ &\quad + \left(\partial_{y_2} + \left(\frac{2}{\sqrt{5}} y_1 + \frac{6}{\sqrt{5}} y_4 \right) \partial_{y_5} \right)^2 \\ &\quad + \left(\partial_{y_3} + \left(-\frac{4}{\sqrt{5}} y_1 - \frac{2}{\sqrt{5}} y_4 \right) \partial_{y_5} \right)^2 \\ &\quad + \left(\partial_{y_4} + \left(\frac{2}{\sqrt{5}} y_3 - \frac{6}{\sqrt{5}} y_2 \right) \partial_{y_5} \right)^2. \end{aligned}$$

---

[6] Here, we use the formula

$$(H_j \varphi)(y) = (\partial/\partial \eta_j)|_{\eta=0} \varphi(y \circ \eta).$$

16.4 Further Reading: An Example of Application to PDE's

We close the chapter, by giving an example of application to PDE's of the topics treated in Section 16.1. To this end, let us introduce the parabolic-type operator in *non-divergence* form

$$\sum_{i,j} a_{i,j}(x)\, X_i X_j - \partial_t, \tag{16.35}$$

being $\{X_i\}_i$ the generators of the Lie algebra of a Carnot group $\mathbb{G}$ and $a_{i,j}$ Hölder continuous functions.

The fundamental solution for (16.35) can be constructed via the Levi-parametrix method[7], provided suitable uniform estimates of the fundamental solutions for the *frozen* operators are established (see [BLU03]). By the frozen operators, we mean the *constant coefficient* operators

$$\mathcal{H}_A = \mathcal{L}_A - \partial_t = \sum_{i,j=1}^m a_{i,j}\, X_i X_j - \partial_t, \quad A = (a_{i,j})_{i,j} \in \mathcal{M}_\Lambda,$$

where $\mathcal{M}_\Lambda$ is the set of $m \times m$ symmetric matrices A such that

$$\Lambda^{-1}|\xi|^2 \le \langle A\xi, \xi\rangle \le \Lambda|\xi|^2 \qquad \forall \xi \in \mathbb{R}^m$$

($\Lambda \ge 1$ being a fixed constant). The Levi-parametrix method for the operator (16.35) requires the following Gaussian *uniform* estimates

$$\begin{aligned} &\left|X_{i_1} \cdots X_{i_p} (\partial_t)^q\, \Gamma_A(x,t) - X_{i_1} \cdots X_{i_p} (\partial_t)^q\, \Gamma_B(x,t)\right| \\ &\le c_{\Lambda,p,q}\, \|A - B\|^{1/r}\, t^{-(Q+p+2q)/2} \exp\left(-\frac{d_{\mathbb{G}}^2(x)}{\mathbf{c}_\Lambda\, t}\right) \\ &\text{for every } x \in \mathbb{G},\ t > 0,\ A, B \in \mathcal{M}_\Lambda, \end{aligned} \tag{16.36}$$

where we have denoted by Γ_A the fundamental solution for $\mathcal{H}_A$ and by $d_{\mathbb{G}}$ a *fixed* symmetric homogeneous norm[8] on $\mathbb{G}$.

A natural question in approaching (16.36) (whose proof can be found in [BLU02]) is to ask whether the sub-Laplacians $\mathcal{L}_A$'s are all diffeomorphic to the canonical operator $\Delta_{\mathbb{G}}$, via a change of variables. This naive idea stands at the basis of the Levi-parametrix method in the classical case. As we showed in Section 16.1, in the general case of a Carnot group, this problem is not trivial.

Theorem 16.1.2 allows us to obtain the fundamental solution Γ_A for $\mathcal{H}_A$ as the composition of T_A with the fundamental solution $\Gamma_{\mathbb{G}}$ for $\mathcal{H}_{\mathbb{G}}$, where

$$\mathcal{H}_{\mathbb{G}} = \Delta_{\mathbb{G}} - \partial_t.$$

[7] See, e.g. [Kal92] for a reference to the Levi-parametrix method in the classical case.

[8] I.e. a continuous function $d_{\mathbb{G}} : \mathbb{R}^N \to [0, \infty[$, smooth away from the origin, such that $d_{\mathbb{G}}(\delta_\lambda(x)) = \lambda\, d_{\mathbb{G}}(x)$, $d_{\mathbb{G}}(x^{-1}) = d_{\mathbb{G}}(x)$ and $d_{\mathbb{G}}(x) = 0$ iff $x = 0$; see Section 5.1, page 229.

Indeed, if $\mathbb{G}$ is free, it turns out that

$$\Gamma_A(x,t;\xi,\tau) = |\det \mathcal{J}_{T_A}(x)|\ \Gamma_{\mathbb{G}}(T_A(x),t;T_A(\xi),\tau) \tag{16.37}$$

for every $x,\ \xi \in \mathbb{R}^N$ and $t,\ \tau \in \mathbb{R}$. A crucial step in order to obtain the uniform estimates in (16.36) is then to establish *ad hoc* uniform estimates for T_A (these can be found in [BU04b, Theorem 2.7]).

In order to handle the case of an arbitrary Carnot group $\mathbb{G}$, a possible approach is to *lift* $\mathbb{G}$ to a free Carnot group $\widetilde{\mathbb{G}}$ in such a way that

$$\Delta_{\mathbb{G}} \text{ is lifted to } \Delta_{\widetilde{\mathbb{G}}}$$

(the lifting technique is treated in Chapter 17). Indeed, the following result holds (see Theorem 17.1.5).

There exists a free homogeneous Carnot group $\widetilde{\mathbb{G}}$ on $\mathbb{R}^H$ (with $H \geq N$) such that, denoting by $\pi : \mathbb{R}^H \to \mathbb{R}^N$ the projection on the first N coordinates, we have

$$\widetilde{X}_i(u \circ \pi) = (X_i u) \circ \pi \quad \forall u \in C^\infty(\mathbb{R}^N),$$

where $\sum_{i=1}^m X_i^2$ and $\sum_{i=1}^m \widetilde{X}_i^2$ are the canonical sub-Laplacians $\Delta_{\mathbb{G}}$ and $\Delta_{\widetilde{\mathbb{G}}}$, respectively.

This result, together with Theorem 16.1.2, allows to prove (16.36) in the general case. We briefly describe how (we refer to [BLU02] for the complete proofs).

A family of sub-Laplacians $\{\mathcal{L}_A\}$ on an arbitrary Carnot group $\mathbb{G}$ is lifted to a family of sub-Laplacians $\{\widetilde{\mathcal{L}}_A\}$ on a free $\widetilde{\mathbb{G}}$. The related family $\{\widetilde{\Gamma}_A\}$ of fundamental solutions fulfills identity (16.37). By means of the uniform estimates of T_A, this allows to derive (16.36) for $\{\widetilde{\Gamma}_A\}$. Finally, the fundamental solutions Γ_A are explicitly represented by integrating $\widetilde{\Gamma}_A$ over the added variables. As a consequence, estimate (16.36) can be obtained on $\mathbb{G}$.

Bibliographical Notes. For the details of the topics presented in Section 16.4, we refer the reader to [BLU02,BLU03]. Some of the topics presented in this chapter also appear in [BU04b].

16.5 Exercises of Chapter 16

Ex. 1) Prove that $\mathbb{R}^N$ equipped with the ordinary Abelian structure is a free Carnot group. Then prove that the Heisenberg group $\mathbb{H}^1$ is also a free Carnot group.

Ex. 2) Consider the Heisenberg group $\mathbb{H}^1$. Let B be a 2×2 symmetric non-singular matrix. Arguing as in Theorem 16.1.2, write down explicitly a diffeomorphism T which turns the sub-Laplacian $\mathcal{L}_{B^{-2}}$ into $\Delta_{\mathbb{H}^1}$. Recognize that T so constructed turns out to be linear. Compare with Remark 16.2.1.

Ex. 3) Consider the following sub-Laplacian on the Heisenberg group $\mathbb{H}^2$

$$\mathcal{L} = X_1^2 + X_2^2 + X_4^2 + (X_2 + X_3)^2,$$

where $\sum_{j=1}^{4} X_j^2$ is the canonical sub-Laplacian of $\mathbb{H}^2$. Following the arguments of Section 16.3, find a new group law $*$ on $\mathbb{R}^5$ such that $\mathbb{H}^* := (\mathbb{R}^5, *)$ is a homogeneous Carnot group isomorphic to $\mathbb{H}^2$, via a Lie group isomorphism which turns $\mathcal{L}$ into the canonical sub-Laplacian $\Delta_{\mathbb{H}^*}$ of $\mathbb{H}^*$. Write down explicitly $\Delta_{\mathbb{H}^*}$.

Ex. 4) Let $(\mathbb{G}, \circ)$ be a free homogeneous Carnot group of step r, and let $d_{\mathbb{G}}$ be a symmetric homogeneous norm on $\mathbb{G}$. Given a fixed constant $\Lambda \geq 1$, let $\mathcal{M}_\Lambda$ denote the set of symmetric $m \times m$ constant matrices A such that

$$\Lambda^{-1}|\xi|^2 \leq \langle A\xi, \xi\rangle \leq \Lambda|\xi|^2 \qquad \forall\, \xi \in \mathbb{R}^m.$$

In what follows we shall denote by $\mathbf{c}_\Lambda$ any positive constant depending only on Λ and the structure of $\mathbb{G}$. Finally, $\|A\|$ will stand for the matrix norm $\max_{|\xi|=1} |A\xi|$.

Consider the diffeomorphism T_A constructed in Theorem 16.1.2 and set, for $x \in \mathbb{G}$,

$$J_A(x) = |\det \mathcal{J}_{T_A}(x)|.$$

Prove that the following properties hold for any $A,\ A_1,\ A_2 \in \mathcal{M}_\Lambda$ and $x \in \mathbb{G}$:

a) J_A is constant in x,
b) $(\mathbf{c}_\Lambda)^{-1} \leq J_A \leq \mathbf{c}_\Lambda$,
c) $|J_{A_1} - J_{A_2}| \leq \mathbf{c}_\Lambda\, \|A_1 - A_2\|$,
d) $(\mathbf{c}_\Lambda)^{-1}\, d_{\mathbb{G}}(x) \leq d_{\mathbb{G}}(T_A(x)) \leq \mathbf{c}_\Lambda\, d_{\mathbb{G}}(x)$,
e) $d_{\mathbb{G}}((T_{A_2}(x))^{-1} \circ T_{A_1}(x)) \leq \mathbf{c}_\Lambda\, \|A_1 - A_2\|^{1/r}\, d_{\mathbb{G}}(x)$.

Hint: First prove Ex. 5 below. Then recall that T_A is defined in (16.9), where $\varphi = \varphi_A$ is a linear map represented in Jacobian coordinates by a block-diagonal matrix whose entries are polynomials in the entries of $A^{-1/2}$. Use the following facts

$$\det \mathcal{J}_{\mathrm{Exp}} = \det \mathcal{J}_{\mathrm{Log}} \equiv 1$$

to prove **(a)**. Then use **(b)** of Ex. 5 to prove **(b)**. Now, observing that J_A is in the form

$$J_A = |\Psi(A^{-1/2})|,$$

being Ψ a polynomial function in the entries of $A^{-1/2}$, estimate $|J_{A_1} - J_{A_2}|$ by the mean value theorem. Use then **(b)** and **(c)** of Exercise 5 to prove **(c)**. Exploiting the fact that T_A commutes with the dilations of $\mathbb{G}$ (see Theorem 16.1.2) and that $d_{\mathbb{G}}$ is δ_λ-homogeneous of degree 1, reduce the proofs of **(d)** and **(e)** to the case

$$x \in S_{\mathbb{G}} := \{\xi \in \mathbb{G} \,|\, d_{\mathbb{G}}(\xi) = 1\}.$$

Then prove that the map $\mathcal{M}_\Lambda \times S_{\mathbb{G}} \ni (A, \xi) \mapsto T_A(\xi)$ is continuous, observing that $T_A(\xi)$ is a polynomial function both in ξ and in the entries of $A^{-1/2}$, and using **(c)** of Ex. 5. Observe also that $d_{\mathbb{G}}(T_A(\xi)) > 0$ for all $\xi \in S_{\mathbb{G}}$, and then prove **(d)**. Finally, apply (5.5) to $d_{\mathbb{G}}$ and $K := \{T_A(\xi) |\, A \in \mathcal{M}_\Lambda,\ \xi \in S_{\mathbb{G}}\}$ in order to obtain **(e)**.

Ex. 5) With the notation of Ex. 4, prove that:

a) If $A \in \mathcal{M}_\Lambda$, then $A^{-1} \in \mathcal{M}_\Lambda$;

b) If $A \in \mathcal{M}_\Lambda$, then

$$\Lambda^{-1} \le \|A\|, \|A^{-1}\| \le \Lambda \quad \text{and} \quad \Lambda^{-1/2} \le \|A^{1/2}\|, \|A^{-1/2}\| \le \Lambda^{1/2};$$

c) For every $A,\ B \in \mathcal{M}_\Lambda$, we have

$$\|A^{1/2} - B^{1/2}\|, \|A^{-1} - B^{-1}\|, \|A^{-1/2} - B^{-1/2}\| \le \mathbf{c}_\Lambda \, \|A - B\|.$$

Ex. 6) In the setting and with the notation of Ex. 4 (and using the very results of Ex. 4, besides Theorem 16.1.2), prove the following facts:

a) For every $A \in \mathcal{M}_\Lambda$, we have

$$\Gamma_A(x) = J_A \cdot \Gamma_{\mathbb{G}}(T_A(x)), \qquad x \in \mathbb{R}^N,$$

where $\Gamma_{\mathbb{G}}$ and Γ_A denote, respectively, the fundamental solutions for the canonical sub-Laplacian

$$\Delta_{\mathbb{G}} = \sum_{k=1}^{m} X_k^2$$

and for the sub-Laplacian

$$\mathcal{L}_A = \sum_{i,j=1}^{m} a_{i,j} \, X_i X_j;$$

b) The fundamental solution Γ_A satisfy the following uniform estimates

$$\mathbf{c}_\Lambda^{-1} \, (d_{\mathbb{G}}(x))^{2-Q} \le \Gamma_A(x) \le \mathbf{c}_\Lambda \, (d_{\mathbb{G}}(x))^{2-Q}, \quad x \in \mathbb{R}^N \setminus \{0\},$$

for every $A \in \mathcal{M}_\Lambda$ (as usual, Q denotes the homogeneous dimension of $\mathbb{G}$);

c) The derivatives of the fundamental solution Γ_A satisfy the following uniform estimates

$$\left| X_{i_1} \cdots X_{i_p} \, \Gamma_A(x) \right| \le \mathbf{c}_{\Lambda,p} \, (d_{\mathbb{G}}(x))^{2-Q-p}, \quad x \in \mathbb{R}^N \setminus \{0\},$$

for every $A \in \mathcal{M}_\Lambda$ and for every $i_1, \ldots, i_p \in \{1, \ldots, m\}$.

17

Lifting of Carnot Groups

As we discussed in the Preface of this book, L.P. Rothschild and E.M. Stein [RS76] obtained sharp regularity results for the sum of squares of Hörmander vector fields $\sum_{j=1}^{m} X_j^2$ by using analysis on nilpotent Lie groups. A crucial step in [RS76] is the construction of new vector fields $\widetilde{X}_1, \dots, \widetilde{X}_m$ (on a larger manifold) which "*lift*" the X_j's and which can be locally approximated by left-invariant vector fields on a stratified group.

The aim of this chapter is to study the lifting procedure in the special case when the X_j's generate the Lie algebra of a Carnot group $\mathbb{G}$. In this case, we prove that $\mathbb{G}$ can be directly lifted to a free group $\widetilde{\mathbb{G}}$ which preserves the homogeneous structure of $\mathbb{G}$, besides being itself a homogeneous Carnot group.

We also give an example of application of the lifting technique to PDE's. Indeed, in Section 17.3 we write an explicit formula for the fundamental solutions for *all* the sub-Laplacians on Carnot groups of *step two*. This formula is given in terms only of the fundamental solution for the canonical sub-Laplacian on a fixed free Carnot group.

17.1 Lifting to Free Carnot Groups

We first recall the basic notation for Carnot groups that we are going to use throughout the chapter. We denote, as usual, by $\mathbb{G} = (\mathbb{R}^N, \circ, \delta_\lambda)$ a homogeneous Carnot group. The dilations on $\mathbb{G}$ have the usual form

$$\delta_\lambda(x) = \delta_\lambda(x^{(1)}, \dots, x^{(r)}) = (\lambda x^{(1)}, \dots, \lambda^r x^{(r)}). \qquad (17.1)$$

Here $x^{(i)} \in \mathbb{R}^{N_i}$ for $i = 1, \dots, r$ and $N_1 + \cdots + N_r = N$. We denote by $\mathfrak{g}$ the Lie algebra of $\mathbb{G}$. For $i = 1, \dots, N$, let Z_i be the vector field in $\mathfrak{g}$ that agrees at the origin with $\partial/\partial x_i$, i.e. the Z_i's form the Jacobian basis for $\mathfrak{g}$.

We know that the crucial hypothesis on $\mathbb{G}$ is that *the Lie algebra generated by* $Z_1, \dots, Z_{N_1}$ *is the whole* $\mathfrak{g}$. We also say that $\mathbb{G}$ is of step r and has $m := N_1$ generators. Let us also explicitly recall that $\mathfrak{g}$ is thus an N-dimensional nilpotent Lie algebra of step r generated by $Z_1, \dots, Z_m$. Moreover, if $Z_j^{(i)} \in \mathfrak{g}$ agrees at the origin with $\partial/\partial x_j^{(i)}$, then $Z_j^{(i)}$ is δ_λ-homogeneous of degree i (see Remark 1.4.5, page 58).

As usual, we denote by $Q = \sum_{j=1}^r j\, N_j$ the homogeneous dimension of $\mathbb{G}$. The canonical sub-Laplacian on $\mathbb{G}$ is the second order differential operator

$$\Delta_{\mathbb{G}} = \sum_{i=1}^m Z_i^2.$$

If $Y_1, \dots, Y_m$ is any basis for $\operatorname{span}\{Z_1, \dots, Z_m\}$, the second order differential operator

$$\mathcal{L} = \sum_{i=1}^m Y_i^2$$

is called a sub-Laplacian on $\mathbb{G}$.

We now recall the definition of $\mathfrak{f}_{m,r}$, the free nilpotent Lie algebra of step r with m (≥ 2) generators $x_1, \dots, x_m$ (see also Chapter 14). By definition, $\mathfrak{f}_{m,r}$ is the unique (up to isomorphism) nilpotent Lie algebra of step r generated by m of its elements $F_1, \dots, F_m$ such that, for every nilpotent Lie algebra $\mathfrak{n}$ of step r and for every map φ from $\{F_1, \dots, F_m\}$ to $\mathfrak{n}$, there exists a (unique) Lie algebra morphism $\widetilde{\varphi}$ from $\mathfrak{f}_{m,r}$ to $\mathfrak{n}$ extending φ. We say that the Carnot group $\mathbb{G}$ is a *free Carnot group* if its Lie algebra $\mathfrak{g}$ is isomorphic to $\mathfrak{f}_{m,r}$ for some m and r.

Following the above notation, by the definition of $\mathfrak{f}_{m,r}$, there exists a unique Lie algebra morphism Π from $\mathfrak{f}_{m,r}$ to $\mathfrak{g}$ such that

$$\Pi(F_i) = Z_i \quad \text{for every } i = 1, \dots, m.$$

Clearly, Π is surjective, whence

$$\dim \mathfrak{f}_{m,r} = \dim \ker(\Pi) + \dim \mathfrak{g}.$$

We set $H = \dim \mathfrak{f}_{m,r}$. The following result will be relevant in the sequel.

Proposition 17.1.1 (A lifting basis $\mathcal{F}$). *There exists a basis*

$$\mathcal{F} = \{F_1, \dots, F_H\}$$

for $\mathfrak{f}_{m,r}$ such that

$$\begin{cases} \Pi(F_j) = Z_j & \text{for every } j = 1, \dots, N, \\ \Pi(F_j) = 0 & \text{for every } j = N+1, \dots, H, \end{cases} \tag{17.2}$$

and each F_j ($j = 1, \dots, H$) is a homogeneous Lie polynomial in $F_1, \dots, F_m$.

Proof. Let α_j denote the δ_λ-homogeneity degree of Z_j. In particular, $\alpha_1 = \dots = \alpha_m = 1$. For a multi-index $I = (i_1, \dots, i_k)$ with $i_1, \dots, i_k \in \{1, \dots, m\}$, we set $|I| = k$ (the *height* of I) and define

$$Z_I := [Z_{i_1}, [Z_{i_2} \dots [Z_{i_{k-1}}, Z_{i_k}] \dots]].$$

Then, by simple homogeneity arguments, we have (see Proposition 1.3.9, page 36, and Proposition 1.1.7, page 12)

$$Z_j = \sum_{I \in \mathcal{I}_j} c_I^{(j)} Z_I, \quad c_I^{(j)} \in \mathbb{R}, \quad \text{for every } j = 1, \dots, N,$$

where $\mathcal{I}_j$ is a set of multi-indices all with height α_j. If we analogously set

$$F_I := [F_{i_1}, [F_{i_2} \dots [F_{i_{k-1}}, F_{i_k}] \dots]],$$

the first N elements of the basis $\mathcal{F}$ can be chosen as

$$F_j = \sum_{I \in \mathcal{I}_j} c_I^{(j)} F_I, \qquad j = 1, \dots, N.$$

Indeed $\Pi(F_j) = Z_j$ and $F_1, \dots, F_N$ are linearly independent homogeneous Lie polynomials in the generators $F_1, \dots, F_m$. Let now $\widehat{F}_{N+1}, \dots, \widehat{F}_H$ be a basis for $\ker(\Pi)$. We can write

$$\widehat{F}_j = \sum_{I \in \mathcal{A}_j} q_I^{(j)} F_I, \quad j = N+1, \dots, H,$$

for a certain set of multi-indices $\mathcal{A}_j$ and scalars $q_I^{(j)}$'s. For every $k = 1, \dots, r$, we set $\mathcal{A}_j^{(k)} := \{I \in \mathcal{A}_j \ : \ |I| = k\}$. We evidently have

$$\mathcal{A}_j = \mathcal{A}_j^{(1)} \cup \dots \cup \mathcal{A}_j^{(r)}.$$

Then

$$\widehat{F}_j = \sum_{I \in \mathcal{A}_j^{(1)}} q_I^{(j)} F_I + \dots + \sum_{I \in \mathcal{A}_j^{(r)}} q_I^{(j)} F_I =: \widehat{F}_j^{(1)} + \dots + \widehat{F}_j^{(r)}.$$

The system of vectors

$$\widehat{\mathcal{F}} := \{\widehat{F}_j^{(k)} : j = N+1, \dots, H, \ k = 1, \dots, r\}$$

has the following properties:

1) each of its vectors is a homogeneous Lie polynomial in the generators $F_1, \dots, F_m$;
2) the system spans $\ker(\Pi)$;
3) for every $j = N+1, \dots, H$ and $k = 1, \dots, r$, we have $\widehat{F}_j^{(k)} \in \ker(\Pi)$.

Let us prove this last assertion: for $j = N+1, \dots, H$, we have

$$0 = \Pi(\widehat{F}_j) = \Pi(\widehat{F}_j^{(1)}) + \dots + \Pi(\widehat{F}_j^{(r)}).$$

We notice that $\Pi(\widehat{F}_j^{(k)})$ is a δ_λ-homogeneous vector field of degree k, unless it vanishes. Consequently, since the cited δ_λ-degrees are all *different*, each $\Pi(\widehat{F}_j^{(k)})$ must vanish identically (see Proposition 1.3.9, page 36).

Finally, the proposition is proved by extracting from the system of vectors $\widehat{\mathcal{F}}$ a basis $F_{N+1}, \ldots, F_H$ for $\ker(\Pi)$. $\square$

In the sequel, we shall consider some abstract finite-dimensional algebras; we hence introduce a useful notation, which will allow us to deal with ordinary $\mathbb{R}^n$ spaces.

Let $\mathfrak{h}$ be a finite-dimensional nilpotent real Lie algebra. It is possible to prove (see Theorem 14.2.3, page 586) that $\mathfrak{h}$ is equipped with a Lie group structure by the so-called Campbell–Hausdorff composition law defined by (14.2), page 585. For X, $Y \in \mathfrak{h}$, the first few terms in the sum (14.2) (which is a finite sum since $\mathfrak{h}$ is nilpotent) are given by

$$X \diamond Y = X + Y + \frac{1}{2}[X, Y] + \frac{1}{12}[X, [X, Y]] - \frac{1}{12}[Y, [X, Y]] + \cdots .$$

We explicitly remark that $\diamond$ is defined in a *universal way* (independent of $\mathfrak{h}$) as a Lie polynomial in X and Y. We now fix a basis

$$\mathcal{E} = (E_1, \ldots, E_N)$$

for $\mathfrak{h}$ ($N := \dim \mathfrak{h}$), and we identify $\mathfrak{h}$ with $\mathbb{R}^N$ via the map

$$\pi_{\mathcal{E}} : \mathfrak{h} \to \mathbb{R}^N, \qquad \sum_{i=1}^N \xi_i E_i \mapsto (\xi_1, \ldots, \xi_N).$$

The group law $\diamond$ is then turned into a group law $\diamond_{\mathcal{E}}$ on $\mathbb{R}^N$ in the natural way

$$a \diamond_{\mathcal{E}} b := \pi_{\mathcal{E}}\big(\pi_{\mathcal{E}}^{-1}(a) \diamond \pi_{\mathcal{E}}^{-1}(b)\big), \qquad a,\, b \in \mathbb{R}^N.$$

In other words, if $a,\ b \in \mathbb{R}^N$, then $a \diamond_{\mathcal{E}} b$ is the only element of $\mathbb{R}^N$ such that

$$\left(\sum_{i=1}^N a_i E_i\right) \diamond \left(\sum_{i=1}^N b_i E_i\right) =: \sum_{i=1}^N (a \diamond_{\mathcal{E}} b)_i E_i .$$

The Lie groups $(\mathbb{R}^N, \diamond_{\mathcal{E}})$ and $(\mathfrak{h}, \diamond)$ are clearly isomorphic[1] via $\pi_{\mathcal{E}}$. We stress that the Lie group morphism $\pi_{\mathcal{E}}$ is also a linear map.

[1] It is evident that the particular form of the operation $\diamond_{\mathcal{E}}$ on $\mathbb{R}^N$ does depend on the choice of the basis $\mathcal{E}$ on $\mathfrak{h}$. However, if $\mathcal{E}_1$ and $\mathcal{E}_2$ are two different bases of $\mathfrak{h}$, obviously the two Lie groups $(\mathbb{R}^N, \diamond_{\mathcal{E}_1})$ and $(\mathbb{R}^N, \diamond_{\mathcal{E}_2})$ are canonically Lie-isomorphic via the natural linear change of basis from $\mathcal{E}_1$ to $\mathcal{E}_2$:

$$\begin{array}{ccc} (\mathfrak{h}, \diamond) & = \!=\!=\!=\!=\!= & (\mathfrak{h}, \diamond) \\ \big\downarrow {\scriptstyle \pi_{\mathcal{E}_1}} & & \big\downarrow {\scriptstyle \pi_{\mathcal{E}_2}} \\ (\mathbb{R}^N, \diamond_{\mathcal{E}_1}) & \xrightarrow{\text{change of basis}} & (\mathbb{R}^N, \diamond_{\mathcal{E}_2}). \end{array}$$

With the above notation, we now consider the Lie groups $(\mathbb{R}^N, \diamond_{\mathcal{Z}})$ and $(\mathbb{R}^H, \diamond_{\mathcal{F}})$, where $\mathcal{Z}$ is the Jacobian basis of $\mathfrak{g}$, whereas $\mathcal{F}$ is the basis of $\mathfrak{f}_{m,r}$ introduced in Proposition 17.1.1. First of all, we remark that the map

$$\pi := \pi_{\mathcal{Z}} \circ \Pi \circ \pi_{\mathcal{F}}^{-1} : (\mathbb{R}^H, \diamond_{\mathcal{F}}) \to (\mathbb{R}^N, \diamond_{\mathcal{Z}})$$

is a surjective Lie group morphism coinciding with the usual projection of $\mathbb{R}^H$ onto $\mathbb{R}^N$

$$\mathbb{R}^H \ni (\xi_1, \ldots, \xi_N, \xi_{N+1}, \ldots, \xi_H) \mapsto (\xi_1, \ldots, \xi_N) \in \mathbb{R}^N.$$

The following diagram may be considered

$$\begin{array}{ccc}
(\mathfrak{f}_{m,r}, \diamond) & \xrightarrow{\quad \Pi \quad} & (\mathfrak{g}, \diamond) \\
\Big\uparrow {\scriptstyle \pi_{\mathcal{F}}^{-1}} & & \Big\downarrow {\scriptstyle \pi_{\mathcal{Z}}} \\
(\mathbb{R}^H, \diamond_{\mathcal{F}}) & \xrightarrow{\text{canonical projection}} & (\mathbb{R}^N, \diamond_{\mathcal{Z}}).
\end{array}$$

Indeed, if $\xi \in \mathbb{R}^H$, we have

$$\begin{aligned}
(\pi_{\mathcal{Z}} \circ \Pi \circ \pi_{\mathcal{F}}^{-1})(\xi) &= (\pi_{\mathcal{Z}} \circ \Pi)\left(\sum_{i=1}^{H} \xi_i \, F_i \right) \\
&= \pi_{\mathcal{Z}}\left(\sum_{i=1}^{H} \xi_i \, \Pi(F_i) \right) \\
\text{(see (17.2))} &= \pi_{\mathcal{Z}}\left(\sum_{i=1}^{N} \xi_i \, Z_i \right) = (\xi_1, \ldots, \xi_N).
\end{aligned}$$

We have thus proved a first kind of *lifting* result (we notice that the Lie algebra of vector fields $\mathfrak{g}$ may be replaced by any finite-dimensional nilpotent Lie algebra $\mathfrak{h}$ generated by m of its elements).

Proposition 17.1.2 (Lifting via a lifting basis $\mathcal{F}$). *Let $\mathfrak{h}$ be a finite-dimensional Lie algebra generated by m of its elements, nilpotent of step r. Let $N := \dim(\mathfrak{h})$. Let $\mathcal{E}$ be a fixed arbitrary linear basis of $\mathfrak{h}$. Then there exists a basis $\mathcal{F}$ of $\mathfrak{f}_{m,r}$ such that the projection*

$$\begin{array}{cccc}
\pi : & (\mathbb{R}^H, \diamond_{\mathcal{F}}) & \longrightarrow & (\mathbb{R}^N, \diamond_{\mathcal{E}}), \\
& (\xi_1, \ldots, \xi_N, \xi_{N+1}, \ldots, \xi_H) & \mapsto & (\xi_1, \ldots, \xi_N)
\end{array}$$

is a surjective Lie group homomorphism.

We now turn to the Lie algebras of the Lie groups $(\mathbb{R}^N, \diamond_{\mathcal{Z}})$ and $(\mathbb{R}^H, \diamond_{\mathcal{F}})$ considered above: let $\mathfrak{r}^N$ and $\mathfrak{r}^H$ denote these Lie algebras, respectively. Since π is a Lie group morphism, then its differential

$$d\pi : \mathfrak{r}^H \to \mathfrak{r}^N$$

is a Lie algebra morphism (see Theorem 2.1.50, page 114) with the following property

$$\big(\mathrm{d}\pi(E)\big)_{\pi(\xi)} = \mathrm{d}\pi(E_\xi) \qquad \forall\, E \in \mathfrak{r}^H, \quad \forall \xi \in \mathbb{R}^H. \tag{17.3}$$

Roughly speaking, the differential of a Lie group morphism which coincides with a projection gives a *lifting of vector fields.*

More precisely, we have the following lemma.

Lemma 17.1.3. *With the above notation, if $E \in \mathfrak{r}^H$, then E is a* lifting *of* $\mathrm{d}\pi(E)$ *in the following sense*:

if $f \in C^\infty(\mathbb{R}^H)$ depends only on $\xi_1, \ldots, \xi_N$, i.e. there exists $g \in C^\infty(\mathbb{R}^N)$ such that

$$f(\xi_1, \ldots, \xi_H) = g(\xi_1, \ldots, \xi_N)$$

for every $\xi \in \mathbb{R}^H$, then we have

$$\big(Ef\big)(\xi_1, \ldots, \xi_H) = \big(\mathrm{d}\pi(E)g\big)(\xi_1, \ldots, \xi_N) \quad \forall \xi \in \mathbb{R}^H.$$

Proof. Let $g \in C^\infty(\mathbb{R}^N)$ and $f \in C^\infty(\mathbb{R}^H)$ be such that $f = g \circ \pi$. We have to prove

$$E_\xi(f) = \big(\mathrm{d}\pi(E)\big)_{\pi(\xi)}(g) \quad \text{for every } \xi \in \mathbb{R}^H.$$

From (17.3) we immediately obtain

$$\big(\mathrm{d}\pi(E)\big)_{\pi(\xi)}(g) = \big(\mathrm{d}\pi(E_\xi)\big)(g) = E_\xi(g \circ \pi) = E_\xi(f).$$

This ends the proof. □

The rest of the lifting method consists in the transferring of this result to the group $\mathbb{G}$, after a suitable definition of the larger group which projects onto $\mathbb{G}$. We refer the reader to the following diagram of Lie group morphisms:

$$\begin{array}{ccccccc}
 & & (\mathfrak{f}_{m,r}, \diamond) & \stackrel{\Pi}{\longrightarrow} & (\mathfrak{g}, \diamond) & \stackrel{\mathrm{Exp}}{\longleftrightarrow} & (\mathbb{G}, \circ) \\
 & & \pi_{\mathcal{F}} \updownarrow & & \updownarrow \pi_{\mathcal{Z}} & & \\
(\mathbb{G} \times \mathbb{R}^{H-N}, \bullet) & \stackrel{\Phi}{\longleftrightarrow} & (\mathbb{R}^H, \diamond_{\mathcal{F}}) & \stackrel{\pi}{\longrightarrow} & (\mathbb{R}^N, \diamond_{\mathcal{Z}}). & &
\end{array}$$

We recall that $(\mathbb{R}^N, \diamond_{\mathcal{Z}})$ is isomorphic to $(\mathfrak{g}, \diamond)$ via $\pi_{\mathcal{Z}}$. On the other hand, the exponential map

$$\mathrm{Exp} : (\mathfrak{g}, \diamond) \to (\mathbb{G}, \circ)$$

is a Lie group isomorphism (see Theorem 2.2.13, page 129). As a consequence, the map

$$\Psi := \mathrm{Exp} \circ \pi_{\mathcal{Z}}^{-1}$$

is a Lie group isomorphism from $(\mathbb{R}^N, \diamond_{\mathcal{Z}})$ to $(\mathbb{G}, \circ)$.

We then look for a suitable group structure on $\mathbb{G} \times \mathbb{R}^{H-N}$ and a Lie group isomorphism

$$\Phi : (\mathbb{R}^H, \diamond_{\mathcal{F}}) \to (\mathbb{G} \times \mathbb{R}^{H-N}, \bullet)$$

such that

$$\vartheta := \Psi \circ \pi \circ \Phi^{-1} = \operatorname{Exp} \circ \pi_{\mathcal{Z}}^{-1} \circ \pi \circ \Phi^{-1}$$

is the projection of $\mathbb{G} \times \mathbb{R}^{H-N}$ onto $\mathbb{G}$. To this end, we set

$$\Phi(\xi_1, \dots, \xi_H) := (\Psi(\xi_1, \dots, \xi_N), \xi_{N+1}, \dots, \xi_H).$$

Clearly, Φ is an invertible map of class C^∞ and the same is true of its inverse map. We then define on $\mathbb{G} \times \mathbb{R}^{H-N}$ the composition law "induced" by Φ

$$(g_1, a_1) \bullet (g_2, a_2) := \Phi\big(\Phi^{-1}(g_1, a_1) \diamond_{\mathcal{F}} \Phi^{-1}(g_2, a_2)\big),$$

so that Φ becomes a Lie group isomorphism between the groups $(\mathbb{R}^H, \diamond_{\mathcal{F}})$ and $(\mathbb{G} \times \mathbb{R}^{H-N}, \bullet)$. Finally, the Lie group morphism

$$\vartheta := \Psi \circ \pi \circ \Phi^{-1} : (\mathbb{G} \times \mathbb{R}^{H-N}, \bullet) \to (\mathbb{G}, \circ)$$

is, by construction, the natural projection. Indeed,

$$\begin{aligned}\vartheta(x_1, \dots, x_H) &= (\Psi \circ \pi)(\Psi^{-1}(x_1, \dots, x_N), x_{N+1}, \dots, x_H) \\ &= \Psi(\Psi^{-1}(x_1, \dots, x_N)) = (x_1, \dots, x_N).\end{aligned}$$

We denote by $\mathfrak{g}\,\mathfrak{r}^{H-N}$ the Lie algebra of $(\mathbb{G} \times \mathbb{R}^{H-N}, \bullet)$.

The proof of the following result is simply a restatement of the proof of Lemma 17.1.3.

Lemma 17.1.4. *If $W \in \mathfrak{g}\,\mathfrak{r}^{H-N}$, then W is a lifting of $\mathrm{d}\vartheta(W)$ in the following sense: if $f \in C^\infty(\mathbb{G} \times \mathbb{R}^{H-N})$ depends only on $x_1, \dots, x_N$, i.e. there exists $g \in C^\infty(\mathbb{G})$ such that*

$$f(x_1, \dots, x_H) = g(x_1, \dots, x_N) \quad \textit{for every } x \in \mathbb{G} \times \mathbb{R}^{H-N},$$

then we have

$$\big(Wf\big)(x_1, \dots, x_H) = \big(\mathrm{d}\vartheta(W)g\big)(x_1, \dots, x_N) \quad \forall x \in \mathbb{G} \times \mathbb{R}^{H-N}.$$

Proof. Let $g \in C^\infty(\mathbb{G})$ and $f \in C^\infty(\mathbb{G} \times \mathbb{R}^{H-N})$ be such that $f = g \circ \vartheta$, i.e.

$$f(x_1, \dots, x_H) = g(x_1, \dots, x_N) \qquad \forall x \in \mathbb{G} \times \mathbb{R}^{H-N}.$$

We have to prove that

$$W_x(f) = (\mathrm{d}\vartheta(W))_{\vartheta(x)}(g) \qquad \forall x \in \mathbb{G} \times \mathbb{R}^{H-N}.$$

To this end, it suffices to notice that

$$(\mathrm{d}\vartheta(W))_{\vartheta(x)}(g) = (\mathrm{d}\vartheta(W_x))(g) = W_x(g \circ \vartheta) = W_x(f).$$

This ends the proof. $\square$

We claim that the first N vector fields of the Jacobian basis $W_1, \dots, W_H$ for $\mathfrak{g}\,\mathfrak{r}^{H-N}$ lift orderly the Jacobian basis $Z_1, \dots, Z_N$ for $\mathfrak{g}$. Indeed, by Lemma 17.1.4, it suffices to show that

$$\mathrm{d}\vartheta(W_k) = Z_k \qquad \text{for every } k = 1, \dots, N.$$

To this end, we remark that, for all $f \in C^\infty(\mathbb{G})$, we have

$$\big(\mathrm{d}\vartheta(W_k)\big)_0(f) = (W_k)_0(f \circ \vartheta) = (\partial_{x_k} f)(0) = \big(Z_k\big)_0 f,$$

which proves the assertion, since two left-invariant vector fields are equal if and only if they coincide at the origin.

We are now in the position to prove the main result of this section.

Theorem 17.1.5 (Lifting). *Let $\mathbb{G}$ be a homogeneous Carnot group on $\mathbb{R}^N$ of step r and m $(= N_1)$ generators. Then there exists a free homogeneous Carnot group $\widetilde{\mathbb{G}}$ on $\mathbb{R}^H$ $(H = \dim \mathfrak{f}_{m,r})$ with the properties* (i) *and* (ii) *stated below.*

We fix the following notation:

$$\delta_\lambda(x) = \delta_\lambda(x^{(1)}, x^{(2)}, \dots, x^{(r)}) = (\lambda x^{(1)}, \lambda^2 x^{(2)}, \dots, \lambda^r x^{(r)})$$

and

$$\widetilde{\delta}_\lambda(\widetilde{x}) = \widetilde{\delta}_\lambda(\widetilde{x}^{(1)}, \widetilde{x}^{(2)}, \dots, \widetilde{x}^{(r)}) = (\lambda \widetilde{x}^{(1)}, \lambda^2 \widetilde{x}^{(2)}, \dots, \lambda^r \widetilde{x}^{(r)})$$

denote the dilations on $\mathbb{G}$ and $\widetilde{\mathbb{G}}$, respectively, with the usual notation

$$x^{(i)} \in \mathbb{R}^{N_i}, \qquad i = 1, \dots, r, \quad N_1 + \dots + N_r = N,$$

and analogously

$$\widetilde{x}^{(i)} \in \mathbb{R}^{\widetilde{N}_i}, \qquad i = 1, \dots, r, \quad \widetilde{N}_1 + \dots + \widetilde{N}_r = H;$$

the vector fields

$$Z_j^{(i)}, \qquad i = 1, \dots, r, \quad j = 1, \dots, N_i,$$

denote the Jacobian basis of the Lie algebra $\mathfrak{g}$ of $\mathbb{G}$, and analogously

$$\widetilde{Z}_j^{(i)}, \qquad i = 1, \dots, r, \quad j = 1, \dots, \widetilde{N}_i,$$

denote the Jacobian basis of the Lie algebra $\widetilde{\mathfrak{g}}$ of $\widetilde{\mathbb{G}}$. Then:

(i) *$\widetilde{\mathbb{G}}$ has step r and m generators, and its Lie algebra is isomorphic to $\mathfrak{f}_{m,r}$.*
(ii) *For a certain $i_0 \in \{1, \dots, r\}$, we have*

$$\widetilde{N}_i = N_i, \quad i = 1, \dots, i_0 \quad \text{and} \quad \widetilde{N}_i > N_i, \quad i = i_0 + 1, \dots, r;$$

moreover, if $\varrho^{(i)} : \mathbb{R}^{\widetilde{N}_i} \to \mathbb{R}^{N_i}$ denotes the projection on the first N_i coordinates and $\varrho : \mathbb{R}^H \to \mathbb{R}^N$ is defined by

$$\varrho(\widetilde{x}) = (\varrho^{(1)}(\widetilde{x}^{(1)}), \dots, \varrho^{(r)}(\widetilde{x}^{(r)})),$$

then

$$\widetilde{Z}_j^{(i)}(u \circ \varrho) = \big(Z_j^{(i)} u\big) \circ \varrho \qquad \forall\, u \in C^\infty(\mathbb{R}^N), \quad i \le r, \ j \le N_i, \tag{17.4}$$

i.e. $\widetilde{Z}_j^{(i)}$ lifts $Z_j^{(i)}$. Moreover, ϱ is a Lie group morphism.

Proof. Let $(\mathbb{G} \times \mathbb{R}^{H-N}, \bullet)$ be the previously defined Lie group on $\mathbb{R}^H$. We show that $\widetilde{\mathbb{G}}$ can be constructed from $\mathbb{G} \times \mathbb{R}^{H-N}$ by a permutation of the coordinates. Let $\mathcal{F} = (F_1, \dots, F_H)$ be the basis for $\mathfrak{f}_{m,r}$ as in Proposition 17.1.1. Then F_j is a homogeneous Lie polynomial of degree α_j in the generators $F_1, \dots, F_m$. We stress that, by Proposition 17.1.1, α_j is also the δ_λ-homogeneity degree of Z_j for $j = 1, \dots, N$, i.e. the dilation δ_λ on $\mathbb{R}^N$ can be written as

$$\delta_\lambda(x_1, \dots, x_N) = (\lambda^{\alpha_1} x_1, \dots, \lambda^{\alpha_N} x_N).$$

We also observe that only $F_1, \dots, F_m$ have degree 1. We define the dilations on $\mathfrak{f}_{m,r}$ as follows

$$\widehat{\delta}_\lambda \left(\sum_{i=1}^{H} \xi_i F_i \right) := \sum_{i=1}^{H} \lambda^{\alpha_i} \xi_i F_i.$$

First, we prove that $\widehat{\delta}_\lambda$ is a Lie algebra automorphism of $\mathfrak{f}_{m,r}$. Indeed, for all $i, j \in \{1, \dots, H\}$, one has

$$\widehat{\delta}_\lambda\big([F_i, F_j]\big) = \lambda^{\alpha_i + \alpha_j} [F_i, F_j] = [\lambda^{\alpha_i} F_i, \lambda^{\alpha_j} F_j] = [\widehat{\delta}_\lambda(F_i), \widehat{\delta}_\lambda(F_j)].$$

The first equality holds since $[F_i, F_j]$ is a homogeneous Lie polynomial in $F_1, \dots, F_m$ of degree $\alpha_i + \alpha_j$. Then $\widehat{\delta}_\lambda$ is also a Lie group automorphism of $(\mathfrak{f}_{m,r}, \diamond)$ by Remark 2.2.16, page 130. As a consequence, the map

$$\delta^*_\lambda := \pi_{\mathcal{F}} \circ \widehat{\delta}_\lambda \circ \pi_{\mathcal{F}}^{-1}$$

is a Lie group automorphism of $(\mathbb{R}^H, \diamond_{\mathcal{F}})$. Analogously, the map

$$\widetilde{\delta}_\lambda := \Phi \circ \delta^*_\lambda \circ \Phi^{-1}$$

is a Lie group automorphism of $(\mathbb{G} \times \mathbb{R}^{H-N}, \bullet)$. If $x \in \mathbb{G} \times \mathbb{R}^{H-N}$, we have

$$\begin{aligned}\widetilde{\delta}_\lambda(x) &= ((\Psi \circ \delta_\lambda \circ \Psi^{-1})(x_1, \dots, x_N), \lambda^{\alpha_{N+1}} x_{N+1}, \dots, \lambda^{\alpha_H} x_H) \\ &= (\delta_\lambda(x_1, \dots, x_N), \lambda^{\alpha_{N+1}} x_{N+1}, \dots, \lambda^{\alpha_H} x_H).\end{aligned}$$

Indeed, Ψ commutes with the dilations of $\mathbb{G}$ (see Theorem 1.3.28, page 49).

We then reorder the coordinates of $\mathbb{G} \times \mathbb{R}^{H-N}$ in the following way. Let $(x, y) \in \mathbb{G} \times \mathbb{R}^{H-N}$, where

$$x = (x^{(1)}, \dots, x^{(r)}) \in \mathbb{G} \quad \text{and} \quad y = (y_{N+1}, \dots, y_H) \in \mathbb{R}^{H-N}.$$

We can suppose that the coordinates of y are ordered in such a way that $\alpha_{N+1} \le \dots \le \alpha_H$. Setting $i_0 := \alpha_{N+1} - 1$, we can write

$$y = (y^{(i_0+1)}, \dots, y^{(r)}),$$

where to each coordinate y_s of $y^{(k)}$ $(i_0 + 1 \le k \le r)$ corresponds a degree of homogeneity α_s equal to k. We now set

$$P : \mathbb{G} \times \mathbb{R}^{H-N} \to \mathbb{R}^H,$$
$$(x, y) \mapsto (x^{(1)}; \dots; x^{(i_0)}; x^{(i_0+1)}, y^{(i_0+1)}; \dots; x^{(r)}, y^{(r)}).$$

We define $\widetilde{\mathbb{G}}$ as $\mathbb{R}^H$ with the group structure naturally induced by P from $(\mathbb{G} \times \mathbb{R}^{H-N}, \bullet)$.

With respect to the coordinates of $\widetilde{\mathbb{G}}$, $\widetilde{\delta}_\lambda$ assumes the usual form

$$\big(\lambda x^{(1)}; \dots; \lambda^{i_0} x^{(i_0)}; \lambda^{i_0+1} x^{(i_0+1)}, \lambda^{i_0+1} y^{(i_0+1)}; \dots; \lambda^r x^{(r)}, \lambda^r y^{(r)}\big).$$

We now set $\varrho := \vartheta \circ P^{-1}$. It is then easy to recognize that (17.4) follows from Lemma 17.1.4.

In order to complete the proof, we have to show that $\widetilde{\mathbb{G}}$ is a homogeneous Carnot group with Lie algebra $\widetilde{\mathfrak{g}}$ isomorphic to $\mathfrak{f}_{m,r}$. To this purpose, let $(\mathbb{G}_{m,r}, *)$ be a free homogeneous Carnot group[2] on $\mathbb{R}^H$ whose Lie algebra is isomorphic to $\mathfrak{f}_{m,r}$. It is not restrictive to consider $\mathfrak{f}_{m,r}$ to be the Lie algebra of $\mathbb{G}_{m,r}$ itself. If we denote by Exp_* the exponential map from $(\mathfrak{f}_{m,r}, \diamond)$ to $(\mathbb{G}_{m,r}, *)$, then the map

$$\operatorname{Exp}_* \circ \pi_{\mathcal{F}}^{-1} : (\mathbb{R}^H, \diamond_{\mathcal{F}}) \to (\mathbb{G}_{m,r}, *)$$

is a Lie group isomorphism (see Theorem 2.2.13, page 129). As a consequence, by Theorem 2.1.50, page 114, the differential

$$\mathrm{d}(\operatorname{Exp}_* \circ \pi_{\mathcal{F}}^{-1})$$

is a Lie algebra isomorphism from $\mathfrak{r}^H$ to $\mathfrak{f}_{m,r}$. Let $E_1, \dots, E_H$ be the Jacobian basis for $\mathfrak{r}^H$. We now prove that

$$\mathrm{d}(\operatorname{Exp}_* \circ \pi_{\mathcal{F}}^{-1})(E_i) = F_i, \quad i = 1, \dots, m. \tag{17.5}$$

In particular, since $F_1, \dots, F_m$ are generators for $\mathfrak{f}_{m,r}$, this will prove that $E_1, \dots, E_m$ are generators for $\mathfrak{r}^H$. If $f \in C^\infty(\mathbb{G}_{m,r})$, we have

$$\begin{aligned}
&\big(\mathrm{d}\,(\operatorname{Exp}_* \circ \pi_{\mathcal{F}}^{-1})(E_i)\big)_0(f) = (E_i)_0\big(f \circ \operatorname{Exp}_* \circ \pi_{\mathcal{F}}^{-1}\big) \\
&= (\partial_{\xi_i}|_{\xi=0}) f\left(\operatorname{Exp}_*\left(\sum_{j=1}^H \xi_j F_j\right)\right) = \left.\frac{\mathrm{d}}{\mathrm{d}t}\right|_{t=0} f\big(\operatorname{Exp}_*(t\,F_i)\big) \\
&= \left.\frac{\mathrm{d}}{\mathrm{d}t}\right|_{t=0} f\big(\exp_{F_i}(t)\big) = (F_i)_0(f).
\end{aligned}$$

Since a left invariant vector field is determined by its value at the origin, this proves (17.5). An analogous argument shows that

$$\mathrm{d}\Phi : \mathfrak{r}^H \to \mathfrak{g}\,\mathfrak{r}^{H-N}$$

[2] The existence of such a group can be easily derived by the third fundamental theorem of Lie (see Theorem 2.2.14, page 130) and by Theorem 2.2.18, page 131.

maps the first m fields of the Jacobian basis for $\mathfrak{r}^H$ into the first m fields of the Jacobian basis for $\mathfrak{g}\,\mathfrak{r}^{H-N}$. Finally, since $\widetilde{\mathbb{G}}$ is obtained from $\mathbb{G} \times \mathbb{R}^{H-N}$ by the Lie group isomorphism P (which permutes the coordinates, leaving the first m coordinates unaltered), we can assert that $\widetilde{\mathbb{G}}$ is a homogeneous Carnot group. Moreover, $\widetilde{\mathfrak{g}}$ is isomorphic to $\mathfrak{f}_{m,r}$ via the Lie algebra isomorphism

$$\mathrm{d}(\mathrm{Exp}_* \circ \pi_{\mathcal{F}}^{-1} \circ \Phi^{-1} \circ P^{-1}).$$

This ends the proof. □

In Section 17.2 below, we give an example of the lifting method just described. The reader is referred to this example for a straightforward comprehension of the topics presented so far.

17.2 An Example of Lifting

We here give an example of a homogeneous Carnot group, and we lift it to a free homogeneous Carnot group. We shall follow the notation introduced in Section 17.1. We consider the homogeneous Carnot group $\mathbb{G}$ on $\mathbb{R}^5$ with the group law and dilation defined as follows:

$$x \circ y = (x_1 + y_1, x_2 + y_2, x_3 + y_3, x_4 + y_4, x_5 + y_5 + x_1\, y_3 + x_2\, y_4)$$

and

$$\delta_\lambda(x) = (\lambda\, x_1, \lambda\, x_2, \lambda\, x_3, \lambda\, x_4, \lambda^2 x_5).$$

The Jacobian basis $\mathcal{Z}$ for $\mathfrak{g}$ is given by

$$Z_1 = \partial_1,\ Z_2 = \partial_2,\ Z_3 = \partial_3 + x_1\, \partial_5,\ Z_4 = \partial_4 + x_2\, \partial_5,\ Z_5 = \partial_5.$$

The Campbell–Hausdorff formula on a Lie algebra nilpotent of step 2 is

$$X \diamond Y = X + Y + \frac{1}{2}[X, Y].$$

Hence the group $(\mathbb{R}^5, \diamond_{\mathcal{Z}})$ has the composition law given by

$$\xi \diamond_{\mathcal{Z}} \eta = \Big(\xi_1 + \eta_1, \xi_2 + \eta_2, \xi_3 + \eta_3, \xi_4 + \eta_4,$$
$$\xi_5 + \eta_5 + \frac{1}{2}(\xi_1\, \eta_3 - \xi_3\, \eta_1 + \xi_2\, \eta_4 - \xi_4\, \eta_2)\Big).$$

With few modifications, we recognize that $\diamond_{\mathcal{Z}}$ is the usual group law on $\mathbb{H}^2$, the Heisenberg group on $\mathbb{R}^5$. The Jacobian basis $\mathcal{X}$ for $\mathfrak{r}^5$ is given by

$$X_1 = \partial_1 - \frac{\xi_3}{2}\,\partial_5, \quad X_2 = \partial_2 - \frac{\xi_4}{2}\,\partial_5,$$
$$X_3 = \partial_3 + \frac{\xi_1}{2}\,\partial_5, \quad X_4 = \partial_4 + \frac{\xi_2}{2}\,\partial_5, \quad X_5 = \partial_5.$$

We now turn to consider $\mathfrak{f}_{4,2}$. We have $\dim \mathfrak{f}_{4,2} = 10$. Let $\Pi : \mathfrak{f}_{4,2} \to \mathfrak{g}$ be the Lie algebra morphism such that $\Pi(F_i) = Z_i$ for every $i = 1, \ldots, 4$. The following is a basis $\mathcal{F}$ for $\mathfrak{f}_{4,2}$ as in Proposition 17.1.1:

$$F_1,\ F_2,\ F_3,\ F_4,\ [F_1, F_3];$$
$$[F_1, F_2],\ [F_1, F_4],\ [F_2, F_3],\ [F_3, F_4],\ [F_2, F_4] - [F_1, F_3].$$

One can easily recognize that the last 5 vectors in $\mathcal{F}$ form a basis for $\ker(\Pi)$. The Lie group $(\mathbb{R}^{10}, \diamond_{\mathcal{F}})$ has the composition law given by

$$\xi \diamond_{\mathcal{F}} \eta = \Big(\xi_1 + \eta_1, \xi_2 + \eta_2, \xi_3 + \eta_3, \xi_4 + \eta_4,$$
$$\xi_5 + \eta_5 + \frac{1}{2}(\xi_1\,\eta_3 - \xi_3\,\eta_1 + \xi_2\,\eta_4 - \xi_4\,\eta_2),$$
$$\xi_6 + \eta_6 + \frac{1}{2}(\xi_1\,\eta_2 - \xi_2\,\eta_1),$$
$$\xi_7 + \eta_7 + \frac{1}{2}(\xi_1\,\eta_4 - \xi_4\,\eta_1), \xi_8 + \eta_8 + \frac{1}{2}(\xi_2\,\eta_3 - \xi_3\,\eta_2),$$
$$\xi_9 + \eta_9 + \frac{1}{2}(\xi_3\,\eta_4 - \xi_4\,\eta_3), \xi_{10} + \eta_{10} + \frac{1}{2}(\xi_2\,\eta_4 - \xi_4\,\eta_2)\Big).$$

The first 5 vector fields of the Jacobian basis for $\mathfrak{r}^{10}$ are then given by

$$E_1 = \partial_1 - \frac{\xi_3}{2}\,\partial_5 - \frac{\xi_2}{2}\,\partial_6 - \frac{\xi_4}{2}\,\partial_7,$$
$$E_2 = \partial_2 - \frac{\xi_4}{2}\,\partial_5 + \frac{\xi_1}{2}\,\partial_6 - \frac{\xi_3}{2}\,\partial_8 - \frac{\xi_4}{2}\,\partial_{10},$$
$$E_3 = \partial_3 + \frac{\xi_1}{2}\,\partial_5 + \frac{\xi_2}{2}\,\partial_8 - \frac{\xi_4}{2}\,\partial_9,$$
$$E_4 = \partial_4 + \frac{\xi_2}{2}\,\partial_5 + \frac{\xi_1}{2}\,\partial_7 + \frac{\xi_3}{2}\,\partial_9 + \frac{\xi_2}{2}\,\partial_{10},$$
$$E_5 = \partial_5.$$

It is evident that E_i lifts $d\pi(E_i) = X_i$ for every $i = 1, \ldots, 5$, as stated in Lemma 17.1.3. By a straightforward calculation of the exponential map from $\mathfrak{g}$ to $\mathbb{G}$, we have

$$\Psi = \operatorname{Exp} \circ \pi_{\mathcal{Z}}^{-1} : (\mathbb{R}^5, \diamond_{\mathcal{Z}}) \to (\mathbb{G}, \circ),$$
$$\Psi(\xi) = \Big(\xi_1, \xi_2, \xi_3, \xi_4, \xi_5 + \frac{1}{2}(\xi_1\,\xi_3 + \xi_2\,\xi_4) \Big).$$

In particular, the Lie group isomorphism $\Phi : (\mathbb{R}^{10}, \diamond_{\mathcal{F}}) \to (\mathbb{G} \times \mathbb{R}^5, \bullet)$ is

$$\Phi(\xi) = \Big(\xi_1, \xi_2, \xi_3, \xi_4, \xi_5 + \frac{1}{2}(\xi_1\,\xi_3 + \xi_2\,\xi_4); \xi_6, \xi_7, \xi_8, \xi_9, \xi_{10}\Big).$$

As a consequence, the group law on $\mathbb{G} \times \mathbb{R}^5$ is given by

$$\begin{aligned} x \bullet y = \Big(& x_1 + y_1, x_2 + y_2, x_3 + y_3, x_4 + y_4, \\ & x_5 + y_5 + x_1\,y_3 + x_2\,y_4, x_6 + y_6 + \frac{1}{2}(x_1\,y_2 - x_2\,y_1), \\ & x_7 + y_7 + \frac{1}{2}(x_1\,y_4 - x_4\,y_1), x_8 + y_8 + \frac{1}{2}(x_2\,y_3 - x_3\,y_2), \\ & x_9 + y_9 + \frac{1}{2}(x_3\,y_4 - x_4\,y_3), x_{10} + y_{10} + \frac{1}{2}(x_2\,y_4 - x_4\,y_2)\Big). \end{aligned}$$

The first 5 vector fields of the Jacobian basis for $\mathfrak{g}\,\mathfrak{r}^5$ are then given by

$$\begin{aligned} W_1 &= \partial_1 - \frac{x_2}{2}\,\partial_6 - \frac{x_4}{2}\,\partial_7, \\ W_2 &= \partial_2 + \frac{x_1}{2}\,\partial_6 - \frac{x_3}{2}\,\partial_8 - \frac{x_4}{2}\,\partial_{10}, \\ W_3 &= \partial_3 + x_1\,\partial_5 + \frac{x_2}{2}\,\partial_8 - \frac{x_4}{2}\,\partial_9, \\ W_4 &= \partial_4 + x_2\,\partial_5 + \frac{x_1}{2}\,\partial_7 + \frac{x_3}{2}\,\partial_9 + \frac{x_2}{2}\,\partial_{10}, \\ W_5 &= \partial_5. \end{aligned}$$

As stated in Lemma 17.1.4, W_i lifts $\mathrm{d}\vartheta(W_i) = Z_i$ for every $i = 1, \dots, 5$. Finally, in this case, the free homogeneous Carnot group $\widetilde{\mathbb{G}}$ which lifts $\mathbb{G}$ has the same group law as $(\mathbb{G} \times \mathbb{R}^5, \bullet)$ since the permutation of the coordinates P can be chosen as the identity map. We remark that the dilations of $\widetilde{\mathbb{G}}$ are given by

$$\delta_\lambda(x) = \big(\lambda\,x_1, \lambda\,x_2, \lambda\,x_3, \lambda\,x_4, \lambda^2 x_5, \lambda^2 x_6, \lambda^2 x_7, \lambda^2 x_8, \lambda^2 x_9, \lambda^2 x_{10}\big).$$

17.3 An Example of Application to PDE's

In this section, we focus our attention on Carnot groups of step two. We recall a characterization of the composition law on such groups, and we explicitly exhibit their lifting. As a consequence, we derive a direct formula for the fundamental solutions for all the sub-Laplacians on groups of step two. This formula is given in terms only of the fundamental solution for the canonical sub-Laplacian on a fixed free Carnot group. The latter fundamental solution can be written in a somewhat explicit form by means of a result by Beals, Gaveau and Greiner [BGG96].

Let us start by recalling the result below which easily follows from Theorem 1.3.15, page 39 (see also Theorem 3.2.2, page 160).

Remark 17.3.1. The N-dimensional homogeneous Carnot groups of step two and m generators are characterized by being $(\mathbb{R}^N, \circ)$ with the following Lie group law $(N = m + n,\, x \in \mathbb{R}^m,\, t \in \mathbb{R}^n)$

$$(x,t)\circ(\xi,\tau) = \begin{pmatrix} x_j + \xi_j, \quad j = 1,\dots,m \\ t_j + \tau_j + \frac{1}{2}\langle x, B^{(j)}\xi\rangle, \quad j = 1,\dots,n \end{pmatrix},$$

where the $B^{(j)}$'s are $m \times m$ matrices whose skew-symmetric parts

$$\frac{1}{2}(B^{(j)} - (B^{(j)})^T)$$

are linearly independent.

We fix $\mathbb{G} = (\mathbb{R}^N, \circ)$, a Carnot group of step two as above, i.e. we fix matrices $B^{(1)},\dots,B^{(n)}$ as in Remark 17.3.1. Let $\mathcal{L} = \sum_{j=1}^m Y_j^2$ be a fixed sub-Laplacian on $\mathbb{G}$. Then there exists a non-singular $m \times m$ matrix $A = (a_{k,j})_{k,j}$ such that

$$Y_j = \sum_{k=1}^m a_{k,j}\, Z_k,$$

where $Z_1,\dots,Z_m$ denote the first m vector fields of the Jacobian basis of $\mathfrak{g}$.

Our aim here is to write the fundamental solution for $\mathcal{L}$ in terms only of the matrices

$$A, \quad B^{(1)},\dots,B^{(n)}$$

and the fundamental solution for the canonical sub-Laplacian of the prototype free Carnot group $(\mathbb{F}_{m,2}, \star)$, which we introduced in Section 3.3, page 163.

For the reader's convenience, we recall the notation. We set

$$N = m(m+1)/2,$$

and we denote the points of $\mathbb{R}^N = \mathbb{R}^m \times \mathbb{R}^{m(m-1)/2}$ by (x,γ), where $x \in \mathbb{R}^m$ and $\gamma \in \mathbb{R}^{m(m-1)/2}$, and we agree to write the coordinates of γ by $\gamma_{i,j}$, where (i,j) varies in the set

$$\mathcal{I} = \{(i,j) \mid 1 \le j < i \le m\}.$$

Then the composition law $\star$ on $\mathbb{R}^N$ is given by

$$(x,\gamma)\star(x',\gamma') = \begin{pmatrix} x_h + x'_h, \quad h = 1,\dots,m \\ \gamma_{i,j} + \gamma'_{i,j} + \frac{1}{2}(x_i\, x'_j - x_j\, x'_i), \quad (i,j) \in \mathcal{I} \end{pmatrix}.$$

It is easily proved that the Lie algebra of $(\mathbb{F}_{m,2}, \star)$ is (isomorphic to) $\mathfrak{f}_{m,2}$. Then $\mathbb{F}_{m,2}$ is a free homogeneous Carnot group of step two on $\mathbb{R}^N$, with m generators.

Since, by the assumption, the skew-symmetric parts of $B^{(1)},\dots,B^{(n)}$ are linearly independent and the matrix A is non-singular, then the matrices

$$A^T\, \frac{B^{(r)} - (B^{(r)})^T}{2}\, A, \qquad r = 1,\dots,n,$$

are also linearly independent. Hence, there exist indices $(i_1, j_1), \dots, (i_n, j_n) \in \mathcal{I}$ such that the following $n \times n$ matrix

$$\left(\left(A^T \frac{B^{(r)} - (B^{(r)})^T}{2} A\right)_{i_s,\, j_s}\right)_{1\le r,s\le n}$$

is non-singular. We denote by $K = (k_{r,s})_{r,s}$ the inverse of the above matrix. We also define the subset of indices

$$\mathcal{C} = \mathcal{I} \setminus \{(i_1, j_1), \dots, (i_n, j_n)\}.$$

With the above notation, we shall also denote the points of $\mathbb{F}_{m,2}$ by (x, t, β), where $x \in \mathbb{R}^m$, $t \in \mathbb{R}^n$, and

$$\beta = (\beta_{h,k})_{(h,k)\in\mathcal{C}} \in \mathbb{R}^{m(m-1)/2-n}.$$

We are now in the position to state the main result of this section.

Proposition 17.3.2. *With the above notation, let $\Gamma_{\mathcal{L}}$ and $\Gamma_{m,2}$ denote respectively the fundamental solutions of $\mathcal{L}$ and of the canonical sub-Laplacian $\Delta_{\mathbb{F}_{m,2}}$. Then the following formula holds*

$$\begin{aligned} \Gamma_{\mathcal{L}}(\mathbf{x}, \mathbf{t}) = \frac{|\det K|}{|\det A|} \int_{\beta\in\mathbb{R}^{m(m-1)/2-n}} & \mathrm{d}\beta \\ \times\, \Gamma_{m,2}\Bigg(A^{-1}\mathbf{x}, \Bigg[\sum_{s\le n} k_{r,s} \Bigg\{ & \mathbf{t}_s - \frac{1}{4}\langle \mathbf{x}, B^{(s)}\mathbf{x}\rangle \\ - \sum_{(h,k)\in\mathcal{C}} \beta_{h,k} & \left(A^T \frac{B^{(s)} - (B^{(s)})^T}{2} A\right)_{h,k}\Bigg\}\Bigg]_{r\le n}, \beta\Bigg). \end{aligned}$$

The above result can be applied in order to manage, in a uniform way, families of sub-Laplacians (letting A vary) and families of step two Carnot groups with the same number of generators (letting $B^{(1)}, \dots, B^{(n)}$ vary). Indeed, an example of application is given in [BLU02,BLU03] where uniform estimates for some families of sub-Laplacians are derived and, as a consequence, the fundamental solutions for non-divergence form operators with Hölder-continuous coefficients are constructed.

Remark 17.3.3. We explicitly remark that a rather explicit formula is given by Beals, Gaveau and Greiner in [BGG96] for the fundamental solution of the canonical sub-Laplacian on any step two Carnot group $\mathbb{G}$:

$$\Gamma_{\Delta_{\mathbb{G}}}(x, t) = \int_{\mathbb{R}^n} V(\rho)\, f(x, t, \rho)^{(2-Q)/2}\, \mathrm{d}\rho,$$

where f is the action associated to a complex Hamiltonian problem and V solves a transport equation. Collecting together this formula in the case of the free group $\mathbb{F}_{m,2}$ and our Proposition 17.3.2, we obtain the following formula

$$\Gamma_{\mathcal{L}}(\mathbf{x},\mathbf{t}) = \frac{|\det K|}{|\det A|}\int_{\beta\in\mathbb{R}^{m(m-1)/2-n}} \mathrm{d}\beta \int_{\rho\in\mathbb{R}^{m(m-1)/2}} \mathrm{d}\rho$$
$$\times\left\{\det\left(\frac{S(\rho)}{\sinh S(\rho)}\right)\right\}^{1/2}\left\{\frac{1}{2}\Big\langle S(\rho)\coth S(\rho)\,A^{-1}\mathbf{x},\,A^{-1}\mathbf{x}\Big\rangle\right.$$
$$-\iota\left(\sum_{r=1}^{n}\rho_{i_r,j_r}\left[\sum_{s\le n}k_{r,s}\left(\mathbf{t}_s-\frac{1}{4}\langle\mathbf{x},\,B^{(s)}\mathbf{x}\rangle\right.\right.\right.$$
$$\left.\left.\left.\left.-\sum_{(h,k)\in\mathcal{C}}\beta_{h,k}\left(A^T\frac{B^{(s)}-(B^{(s)})^T}{2}A\right)_{h,k}\right)\right]+\sum_{i,j\in\mathcal{C}}\rho_{i,j}\,\beta_{i,j}\right)\right\}^{(2-m^2)/2}.$$

Here, $S(\rho)$ is the matrix $S(\rho)=\iota\sum_{i,j\in\mathcal{I}}\rho_{i,j}\,S^{(i,j)}$, where the matrices $S^{(i,j)}$ have been introduced in Section 3.3, page 163, and ι is the imaginary unit. So, we also have an even more explicit representation

$$\Gamma_{\mathcal{L}}(\mathbf{x},\mathbf{t}) = \frac{|\det K|}{|\det A|}\int_{\beta\in\mathbb{R}^{m(m-1)/2-n}} \mathrm{d}\beta \int_{\rho\in\mathbb{R}^{m(m-1)/2}} \mathrm{d}\rho$$
$$\times\left\{\det\left[\sum_{p=0}^{\infty}\frac{(-1)^p}{(2p+1)!}\left(-\sum_{i,j\in\mathcal{I}}\rho_{i,j}\,S^{(i,j)}\right)^{2p}\right]\right\}^{-1/2}$$
$$\times\left\{\frac{1}{2}\left\langle\left[\sum_{p=0}^{\infty}\frac{(-1)^p2^{(2p)}B_{2p}}{(2p)!}\left(-\sum_{i,j\in\mathcal{I}}\rho_{i,j}\,S^{(i,j)}\right)^{2p}\right]A^{-1}\mathbf{x},\,A^{-1}\mathbf{x}\right\rangle\right.$$
$$-\iota\left(\sum_{r=1}^{n}\rho_{i_r,j_r}\left[\sum_{s\le n}k_{r,s}\left(\mathbf{t}_s-\frac{1}{4}\langle\mathbf{x},\,B^{(s)}\mathbf{x}\rangle\right.\right.\right.$$
$$\left.\left.-\sum_{(h,k)\in\mathcal{C}}\beta_{h,k}\left(A^T\frac{B^{(s)}-(B^{(s)})^T}{2}A\right)_{h,k}\right)\right]$$
$$\left.\left.+\sum_{i,j\in\mathcal{C}}\rho_{i,j}\,\beta_{i,j}\right)\right\}^{(2-m-2n)/2}.$$

As an example, if $m=3$, $\Gamma_{3,2}$ has been explicitly written in [BGG96, p. 322]. Hence, the above formula writes

$$\Gamma_{\mathcal{L}}(\mathbf{x},\mathbf{t}) = -\frac{\mathrm{Gamma}(7/2)}{(2\pi)^{7/2}}\frac{|\det K|}{|\det A|}\int_{\beta\in\mathbb{R}^{3-n}}\mathrm{d}\beta\int_{\rho\in\mathbb{R}^3}\mathrm{d}\rho\,\frac{2|\rho|}{\sinh(2|\rho|)}$$
$$\times\left\{\frac{\langle A^{-1}\mathbf{x},\,\rho\rangle^2}{2|\rho|^2}+|\rho|\coth(2|\rho|)\left[|A^{-1}\mathbf{x}|^2-\frac{\langle A^{-1}\mathbf{x},\,\rho\rangle^2}{|\rho|^2}\right.\right.$$
$$-\iota\left(\sum_{r=1}^{n}\rho_r\left\{\sum_{s\le n}k_{r,s}\left(\mathbf{t}_s-\frac{1}{4}\langle\mathbf{x},\,B^{(s)}\mathbf{x}\rangle\right.\right.\right.$$

$$- \sum_{(h,k)\in C} \beta_{h,k} \left(A^T \frac{B^{(s)} - (B^{(s)})^T}{2} A \right)_{h,k} \Bigg) \Bigg\}$$

$$+ \sum_{r=n+1}^{3} \rho_r \, \beta_{r-n} \Bigg) \Bigg] \Bigg\}^{-7/2}.$$

Proof (of Proposition 17.3.2). As the first step, we turn the arbitrary sub-Laplacian $\mathcal{L}$ on $\mathbb{G}$ into the canonical sub-Laplacian on a new group $\mathbb{H}$. This can be done by the isomorphism of Lie groups

$$\mathbb{H} \ni (\xi, \tau) \mapsto (A\xi, \tau) \in \mathbb{G}$$

which turns the composition $\circ$ on $\mathbb{G}$ into

$$(\xi, \tau) \bullet (\xi', \tau') = \begin{pmatrix} \xi_j + \xi'_j, \quad j = 1, \dots, m, \\ \tau_j + \tau'_j + \frac{1}{2} \langle \xi, A^T B^{(j)} A\xi' \rangle, \quad j = 1, \dots, n \end{pmatrix}.$$

Then, it is easy to see that we have the relation

$$\Gamma_{\mathcal{L}}(x, t) = |\det A|^{-1} \Gamma_{\Delta_{\mathbb{H}}}(A^{-1}x, t). \tag{17.6}$$

As the second step, we turn the composition law $\bullet$ into a composition law $*$ whose associated matrices (according to Remark 17.3.1) are skew-symmetric. This can be done by identifying $\mathbb{H}$ with its Lie algebra via the exponential map. Then, the isomorphism of Lie groups

$$\mathbb{P} \ni (\mathbf{x}, \mathbf{t}) \mapsto \left(\mathbf{x}, \mathbf{t}_j + \frac{1}{4} \langle \mathbf{x}, A^T B^{(j)} A\mathbf{x} \rangle \right) \in \mathbb{H}$$

turns the composition $\bullet$ on $\mathbb{H}$ into

$$(\mathbf{x}, \mathbf{t}) * (\mathbf{x}', \mathbf{t}') = \begin{pmatrix} \mathbf{x}_j + \mathbf{x}'_j, \quad j = 1, \dots, m, \\ \mathbf{t}_j + \mathbf{t}'_j + \frac{1}{2} \langle \mathbf{x}, A^T (\frac{B^{(j)} - (B^{(j)})^T}{2}) A\mathbf{x}' \rangle, \quad j = 1, \dots, n \end{pmatrix},$$

where the associated matrices are skew-symmetric. It is easy to see that we have the relation

$$\Gamma_{\Delta_{\mathbb{H}}}(\xi, \tau) = \Gamma_{\Delta_{\mathbb{P}}}\big(\xi, \tau_j - \frac{1}{4} \langle \xi, A^T B^{(j)} A\xi \rangle\big). \tag{17.7}$$

As the third step, we lift $\mathbb{P}$ to a free Carnot group $\mathbb{V}$, as in Theorem 17.1.5. The fundamental solution of the canonical sub-Laplacian on $\mathbb{P}$ can be obtained by integrating the fundamental solution of the canonical sub-Laplacian on $\mathbb{V}$ with respect to the added variables, i.e.

$$\Gamma_{\Delta_{\mathbb{P}}}(\mathbf{x}, \mathbf{t}) = \int_{\beta \in \mathbb{R}^{m(m-1)/2-n}} \Gamma_{\Delta_{\mathbb{V}}}(\mathbf{x}, \mathbf{t}, \beta) \, d\beta. \tag{17.8}$$

We need the explicit form of the lifting for an arbitrary group of step two (with associated skew-symmetric matrices). This is given in the following lemma, which can be proved by retracing the proof of Theorem 17.1.5.

Lemma 17.3.4. *Let* $(\mathbb{R}^N, \circ)$ *be a Carnot group of step two and* m *generators as in Remark 17.3.1, where its associated matrices* $B^{(j)}$*'s are linearly independent and skew-symmetric. Then* $(\mathbb{R}^N, \circ)$ *is lifted to the group*

$$(\mathbb{R}^N \times \mathbb{R}^{m(m-1)/2-n}, \widetilde{\circ})$$

(according to Theorem 17.1.5*), where (using the notation* $x \in \mathbb{R}^m$, $t \in \mathbb{R}^n$, $\beta \in \mathbb{R}^{m(m-1)/2-n}$*)*

$$(x, t, \beta) \circ (x', t', \beta') = \begin{pmatrix} x_j + x'_j, \quad j = 1, \dots, m \\ t_j + t'_j + \frac{1}{2} \langle x, B^{(j)} x' \rangle, \quad j = 1, \dots, n \\ \beta_{i,j} + \beta'_{i,j} + \frac{1}{2}(x_i x'_j - x_j x'_i), \quad (i, j) \in \mathcal{C} \end{pmatrix}.$$

The next step is to explicitly write a Lie group isomorphism between $\mathbb{V}$ and $\mathbb{F}_{m,2}$. This can be obtained by the composition

$$\mathrm{Exp} \circ \varphi \circ \mathrm{Log} =: T,$$

where Exp is the exponential map on $\mathbb{F}_{m,2}$, Log is the logarithmic map on $\mathbb{V}$, and φ is the unique Lie algebra morphism mapping the first m vector fields of the Jacobian basis related to $\mathbb{V}$ into the first m vector fields of the Jacobian basis related to $\mathbb{F}_{m,2}$. A laborious computation shows that the map

$$T : (\mathbb{V}, \widetilde{*}) \to (\mathbb{F}_{m,2}, \star)$$

is defined by

$$T(x, t, \beta) := \left(x, \left[\sum_{s=1}^{n} k_{r,s} \left(t_s - \sum_{(h,k)\in\mathcal{C}} \beta_{h,k} \left(A^T \frac{B^{(s)} - (B^{(s)})^T}{2} A \right)_{h,k} \right) \right]_{r \le n}, \beta \right).$$

It is easy to see that we have the relation

$$\Gamma_{\Delta_{\mathbb{V}}}(\mathbf{x}, \mathbf{t}, \beta) = |\det K| \, \Gamma_{m,2}(T(\mathbf{x}, \mathbf{t}, \beta)). \tag{17.9}$$

Finally, collecting together equations (17.6) to (17.9), we obtain the desired formula in Proposition 17.3.2. □

17.4 Folland's Lifting of Homogeneous Vector Fields

The aim of this section is to provide another lifting theorem for vector fields more general than those considered in Section 17.1. An example will clarify the new situation. Let us consider on $\mathbb{R}^2$ the vector fields $X = \partial_x$ and $Y = x\,\partial_y$. As we saw in Example 1.2.14 (page 18), they are not left-invariant with respect to any group law in $\mathbb{R}^2$, so that the lifting Theorem 17.1.5 cannot be applied. In this section, we shall provide a lifting result which *can* be applied to X, Y. It will give a constructive way to build new vector fields (generating a homogeneous Carnot group!) lifting X and Y (namely, $\widetilde{X} = \partial_x$ and $\widetilde{Y} = \partial_z + x\,\partial_y$ on $\mathbb{R}^3$, whose points are denoted by (x, z, y)). This lifting result (due to Folland, see [Fol77]) only relies on the homogeneity properties of X and Y (namely, they are homogeneous of degree one, w.r.t. the dilation $\delta_\lambda(x, y) = (\lambda x, \lambda^2 y)$ on $\mathbb{R}^2$).

Before entering into the details of this new lifting result, it is useful to focus on the example above and to see how this result works.

Example 17.4.1. Let us denote the points of $\mathbb{R}^2$ by (x_1, x_2), and let us consider the following vector fields on $\mathbb{R}^2$: $X_1 = \partial_{x_1}$ and $X_2 = x_1\,\partial_{x_2}$. They are δ_λ-homogeneous of degree one w.r.t. the dilation $\delta_\lambda(x_1, x_2) = (\lambda x_1, \lambda^2 x_2)$ on $\mathbb{R}^2$. As a consequence, the Lie algebra $\mathfrak{a}$ generated by them in $T(\mathbb{R}^2)$ (i.e. the set of smooth vector fields on $\mathbb{R}^2$ equipped with the usual bracket operation) is nilpotent (of step two), namely, since $[X_1, X_2] = \partial_{x_2}$ and all other commutators vanish, we have

$$\mathfrak{a} = \mathrm{Lie}\{X_1, X_2\} = \mathrm{span}\{X_1, X_2, [X_1, X_2]\}, \quad \text{whence} \quad \dim(\mathfrak{a}) = 3.$$

As we saw in Theorem 2.2.13 (page 129; see also the relevant Definition 2.2.11), the Campbell–Hausdorff operation $\diamond$ (truncated of step two)

$$X \diamond Y = X + Y + \frac{1}{2}[X, Y]$$

defines on $\mathfrak{a}$ a Lie group structure $(\mathfrak{a}, \diamond)$. We first realize this group as a Lie group on $\mathbb{R}^3$ in the usual way. We fix the basis $\mathcal{X} = (X_1, X_2, [X_1, X_2])$ for $\mathfrak{a}$, and we identify $\mathfrak{a}$ with $\mathbb{R}^3$ via the map

$$\phi : \mathbb{R}^3 \to \mathfrak{a}, \qquad (\xi_1, \xi_2, \xi_3) \mapsto \xi_1 X_1 + \xi_2 X_2 + \xi_3 [X_1, X_2].$$

The group law $\diamond$ is then turned into a group law[3] $\circ$ on $\mathbb{R}^3$ in the natural way

$$a \circ b := \phi^{-1}\big(\phi(a) \diamond \phi(b)\big), \qquad a,\, b \in \mathbb{R}^3.$$

The Lie group $\mathbb{G} = (\mathbb{R}^3, \circ)$ is obviously isomorphic to $(\mathfrak{a}, \diamond)$, and the Lie algebra $\mathfrak{g}$ of $\mathbb{G}$ is isomorphic (as a Lie algebra) to the Lie algebra $\mathfrak{a}$ (see Ex. 3 of Chapter 2, page 148). A direct computation shows that in this case we have

$$a \circ b = \left(a_1 + b_1, a_2 + b_2, a_3 + b_3 + \frac{1}{2}(a_1 b_2 - a_2 b_1)\right).$$

Denoting the points of $\mathbb{R}^3$ by $\xi = (\xi_1, \xi_2, \xi_3)$, we have $\mathfrak{g} = \mathrm{span}\{Z_1, Z_2, Z_3\}$, where

$$Z_1 = \partial_{\xi_1} - \frac{1}{2}\xi_2\,\partial_{\xi_3}, \;\; Z_2 := \partial_{\xi_2} + \frac{1}{2}\xi_1\,\partial_{\xi_3}, \;\; Z_3 := [Z_1, Z_2] = \partial_{\xi_3}.$$

The Lie algebra isomorphism between $\mathfrak{g}$ and $\mathfrak{a}$ is the linear map $\alpha : \mathfrak{g} \to \mathfrak{a}$ such that

---

[3] In the previous sections, we denoted it by $\diamond_{\mathcal{X}}$, and we also wrote $\phi^{-1} = \pi_{\mathcal{X}}$.

$$\alpha(Z_1) = X_1, \quad \alpha(Z_2) = X_2, \quad \alpha(Z_3) = [X_1, X_2].$$

Let us denote by Log the logarithmic map related to the Lie group $(\mathbb{G}, \circ)$. Moreover, we write I for the identity map on $\mathbb{R}^2$. We then introduce the following map

$$\begin{aligned} &\pi : \mathbb{R}^3 \longrightarrow \mathbb{R}^2, \\ &\pi(\xi) := e^{\alpha(\mathrm{Log}\,(\xi))} I(0,0) := \sum_{k\geq 0} \frac{\big(\alpha(\mathrm{Log}\,(\xi))\big)^k}{k!} I(0,0). \end{aligned} \tag{17.10}$$

We explicitly remark that the map π is well defined. Indeed, if $\xi \in \mathbb{R}^3 \equiv \mathbb{G}$, then $\mathrm{Log}\,(\xi) \in \mathfrak{g}$, so that $\alpha(\mathrm{Log}\,\xi) \in \mathfrak{a}$. As a consequence, *by the properties of δ_λ-homogeneity of the vector fields in* $\mathfrak{a}$, any power $\big(\alpha(\mathrm{Log}\,(\xi))\big)^k$ with $k \geq 3$ annihilates the identity map I, so that the sum in (17.10) is finite. We compute π explicitly,

$$\begin{aligned} \pi(\xi) &= \left(I + (\alpha(\xi_1 Z_1 + \xi_2 Z_2 + \xi_3 Z_3))I + \frac{1}{2}(\alpha(\xi_1 Z_1 + \xi_2 Z_2 + \xi_3 Z_3))^2 I\right)\Big|_{x=0} (0) \\ &= \left(I + (\xi_1 \partial_{x_1} + \xi_2 x_1 \partial_{x_2} + \xi_3 \partial_{x_2})I + \frac{1}{2}(\xi_1 \partial_{x_1} + \xi_2 x_1 \partial_{x_2} + \xi_3 \partial_{x_2})^2 I\right)\Big|_{x=0} (0) \\ &= \left(I + \left(\xi_1 \partial_{x_1} + \xi_3 \partial_{x_2} + \frac{1}{2}\xi_1 \xi_2 \partial_{x_2}\right) I\right)\Big|_{x=0} (0) \\ &= \left(\xi_1, \xi_3 + \frac{1}{2}\xi_1 \xi_2\right). \end{aligned}$$

We now verify that Z_1 and Z_2 (the vector fields in $\mathfrak{g}$ corresponding to X_1 and X_2 in $\mathfrak{a}$ via α) *lift* X_1 and X_2 via π in the sense that

$$X_i|_{\pi(\xi)} = \mathrm{d}_\xi \pi(Z_i) \quad \text{for all } \xi \in \mathbb{R}^3 \text{ and } i = 1, 2.$$

Indeed, since the component functions of $X_i|_{\pi(\xi)}$ are given by $X_i I(\pi(\xi))$ whereas those of $\mathrm{d}_\xi \pi(Z_i)$ are given by $\mathcal{J}_\pi(\xi) \cdot Z_i I(\xi)$, we only have to check the matrix identity

$$\begin{pmatrix} 1 & 0 \\ 0 & \xi_1 \end{pmatrix} = \begin{pmatrix} 1 & 0 & 0 \\ \frac{1}{2}\xi_2 & \frac{1}{2}\xi_1 & 1 \end{pmatrix} \cdot \begin{pmatrix} 1 & 0 \\ 0 & 1 \\ -\frac{1}{2}\xi_2 & \frac{1}{2}\xi_1 \end{pmatrix},$$

which actually holds.

The ideas presented so far appear in the paper [Fol77] by G.B. Folland. We now aim to add a new feature: we are interested in a "change of coordinates" on $\mathbb{R}^3$ turning Z_i into *a vector field on* $\mathbb{R}^3$ lifting X_i. To this aim, we define the following map

$$T : \mathbb{R}^3 \to \mathbb{R}^3, \quad T(\xi) = (\pi(\xi), \xi_2) = \left(\xi_1, \xi_3 + \frac{1}{2}\xi_1\xi_2, \xi_2\right).$$

We obtained T by completing π with the only coordinate function (i.e. ξ_2) not appearing in the first order expansion of $\pi(\xi)$. We remark that T is a smooth diffeomorphism of $\mathbb{R}^3$ onto itself (with polynomial coordinates and the same holds for T^{-1}). We perform a change of coordinates on $\mathbb{R}^3$ by introducing new variables $x = (x_1, x_2, x_3)$ as follows

$$x = T(\xi), \quad x, \xi \in \mathbb{R}^3.$$

We express Z_1 and Z_2 (the vector fields in $\mathfrak{g}$ corresponding to X_1 and X_2 in $\mathfrak{a}$ via α) in these new coordinates and obtain $\widetilde{Z}_1$ and $\widetilde{Z}_2$. *We claim that $\widetilde{Z}_1$ and $\widetilde{Z}_2$ lift X_1 and X_2*. Indeed, for every $x \in \mathbb{R}^3$ and $i = 1, 2$, we have

$$\widetilde{Z}_i = \langle \widetilde{Z}_i I(x), \nabla_x \rangle, \quad \text{where} \quad \widetilde{Z}_i I(x) = \mathcal{J}_T(T^{-1}(x)) \cdot (Z_i I)(T^{-1}(x)),$$

so that (after a simple computation)

$$\widetilde{Z}_1 = \partial_{x_1} \quad \text{and} \quad \widetilde{Z}_2 = \partial_{x_3} + x_1\, \partial_{x_2}.$$

It is now immediately seen that $\widetilde{Z}_1$ and $\widetilde{Z}_2$ lift X_1 and X_2, respectively. □

The aim of the rest of the section is to generalize the previous example to sets of vector fields satisfying suitable hypotheses.

17.4.1 The Hypotheses on the Vector Fields

Assume that we are given on $\mathbb{R}^n$ (whose points will be denoted by $x = (x_1, \dots, x_n)$) a set of m smooth vector fields $X_1, \dots, X_m$ satisfying the following conditions:

(F1) $X_1, \dots, X_m$ are linearly independent and d_λ-homogeneous of degree one with respect to a suitable family of dilations $\{d_\lambda\}_{\lambda>0}$ of the following type

$$d_\lambda : \mathbb{R}^n \to \mathbb{R}^n, \quad d_\lambda(x) = (\lambda^{\sigma_1} x_1, \dots, \lambda^{\sigma_n} x_n),$$

where $1 = \sigma_1 \le \cdots \le \sigma_n$.

(F2) $\dim(\mathrm{Lie}\{X_1, \dots, X_m\} I(0)) = n$.

The reader should notice that these conditions are resemblant to conditions **(H0)** and **(H2)** on page 191, but condition **(H1)** on that page has no analogue. For example, the vector fields in Example 17.4.1 fulfill hypotheses **(F1)** and **(F2)** but not **(H1)**.

We henceforth use the following notation: $\mathfrak{a}$ denotes the Lie algebra generated by the X_j's (sub-algebra of $T(\mathbb{R}^n)$, the set of the smooth vector fields on $\mathbb{R}^n$ equipped with the usual bracket of differential operators, not necessarily left-invariant on a Lie-group), i.e.

$$\mathfrak{a} := \mathrm{Lie}\{X_1, \dots, X_m\}. \tag{17.11}$$

The first task of the section is to prove the following theorem (see also [Fol77, Theorem 1]).

Theorem 17.4.2 (Folland's lifting, [Fol77]). *Let $X_1, \dots, X_m$ satisfy the above conditions* **(F1)** *and* **(F2)**. *Then there exist a homogeneous Carnot group $\mathbb{G}$ on $\mathbb{R}^N$ with m generators and nilpotent of step r (where $N \ge m$ is the dimension of $\mathfrak{a}$ and r the step of nilpotence of $\mathfrak{a}$), a polynomial surjective map*

$$\pi : \mathbb{R}^N \to \mathbb{R}^n,$$

and a system of Lie-generators $Z_1, \ldots, Z_m$ *for the algebra of* $\mathbb{G}$ *such that* Z_i *lift* X_i *via* π, *i.e.*

$$\mathrm{d}_\xi \pi(Z_i) = (X_i)_{\pi(\xi)} \qquad \text{for every } \xi \in \mathbb{R}^N. \tag{17.12}$$

More precisely, Z_i *is the* i*-th vector of the Jacobian basis for* $\mathbb{G}$.

We give a constructive proof of Theorem 17.4.2 consisting of six steps.

STEP 1 (**The properties of** $\mathfrak{a}$**).** We leave as an exercise the verification of the following results. If $X_1, \ldots, X_m$ satisfy the above conditions **(F1)** and **(F2)**, then the following facts hold:

(i) Each X_j has the following form

$$X_j = \sum_{i=1}^{n} a_{i,j}(x)\, \partial_{x_i},$$

where $a_{i,j}$ is a d_λ-homogeneous polynomial of degree $\sigma_i - 1$ (so that $a_{i,j}$ depends only on the x_k's such that $\sigma_k \le \sigma_i - 1$).

(ii) $\mathfrak{a}$ is nilpotent (of step at most σ_n, say r). As a consequence, since $\mathfrak{a}$ is spanned by high-order brackets of the form

$$[X_{i_k} \cdots [X_{i_1}, X_{i_2}] \cdots], \quad i_1, \ldots, i_k \in \{1, \ldots, m\}, \quad k \in \mathbb{N},$$

and only a finite number of them are not vanishing (due to the nilpotence of $\mathfrak{a}$), then $\mathfrak{a}$ *is finite-dimensional*. We set

$$N := \dim(\mathfrak{a}). \tag{17.13}$$

(iii) The Lie algebra $\mathfrak{a}$ is stratified, i.e. (see also Definition 2.2.3, page 122)

$$\mathfrak{a} = \mathfrak{a}_1 \oplus \mathfrak{a}_2 \oplus \cdots \oplus \mathfrak{a}_r \quad \text{with} \quad \begin{cases} [\mathfrak{a}_1, \mathfrak{a}_{i-1}] = \mathfrak{a}_i & \text{if } 2 \le i \le r, \\ [\mathfrak{a}_1, \mathfrak{a}_r] = \{0\}. \end{cases} \tag{17.14}$$

Here $\mathfrak{a}_1 = \mathrm{span}\{X_1, \ldots, X_m\}$ and r is the step of nilpotence of $\mathfrak{a}$. We explicitly remark that

$$\textit{any operator in } \mathfrak{a}_i \textit{ is } d_\lambda\textit{-homogeneous of degree } i, \tag{17.15}$$

for it is a linear combination of elements of the type

$$X_{j_1} X_{j_2} \cdots X_{j_s} \quad \text{with } j_1 + j_2 + \cdots + j_s = i$$

(and any X_j is d_λ-homogeneous of degree 1 by hypothesis **(F1)**). The above decomposition also allows us to define a group of dilations $\{\delta_\lambda\}_{\lambda>0}$ on $\mathfrak{a}$ in the following way:

$$\delta_\lambda : \mathfrak{a} \to \mathfrak{a}, \quad \delta_\lambda\left(\sum_{i=1}^{r} A_i\right) := \sum_{i=1}^{r} \lambda^i A_i, \quad \text{where } A_i \in \mathfrak{a}_i \text{ for } i = 1, \ldots, r. \tag{17.16}$$

STEP 2 (**The choice of a basis for $\mathfrak{a}$**). Recalling hypothesis (**F1**) and (17.13), we can complete $X_1, \dots, X_m$ to a basis of $\mathfrak{a}$

$$\mathcal{X} = (X_1, \dots, X_m, X_{m+1}, \dots, X_N)$$

satisfying the following two conditions:

(1) we have (by making use of hypothesis (**F2**))

$$\operatorname{span}\{X_1 I(0), \dots, X_m I(0), X_{m+1} I(0), \dots, X_N I(0)\} = \mathbb{R}^n, \tag{17.17}$$

(2) $\mathcal{X}$ is adapted to the stratification of $\mathfrak{a}$, i.e. we have

$$\mathcal{X} = \big(X_1^{(1)}, \dots, X_{m_1}^{(1)};\ X_1^{(2)}, \dots, X_{m_2}^{(2)};\ \cdots;\ X_1^{(r)}, \dots, X_{m_r}^{(r)}\ \big), \tag{17.18}$$

where $m_1 = m$, $X_i^{(1)} = X_i$ for every $1 \le i \le m$ and, with reference to (17.14),

$$\mathfrak{a}_i = \operatorname{span}\{X_1^{(i)}, \dots, X_{m_i}^{(i)}\} \qquad \text{for every } i = 2, \dots, r.$$

STEP 3 (**The group operations**). It is well known that (see Theorem 2.2.13, page 129; see also the relevant Definition 2.2.11), the Campbell–Hausdorff operation $\diamond$ defines on $\mathfrak{a}$ a Lie group structure $(\mathfrak{a}, \diamond)$. We can realize this group as a Lie group on $\mathbb{R}^N$ in the usual way. Fixed a basis $\mathcal{X}$ for $\mathfrak{a}$ as in Step 2, we identify $\mathfrak{a}$ with $\mathbb{R}^N$ via the map

$$\phi : \mathbb{R}^N \to \mathfrak{a}, \qquad \xi = (\xi_1, \dots, \xi_N) \mapsto \sum_{j=1}^{N} \xi_j X_j.$$

The group law $\diamond$ is then turned into a group law[4] $*$ on $\mathbb{R}^N$ in the natural way

$$a * b := \phi^{-1}\big(\phi(a) \diamond \phi(b)\big), \qquad a, b \in \mathbb{R}^N.$$

The Lie group $\mathbb{G} := (\mathbb{R}^N, *)$ is obviously isomorphic to $(\mathfrak{a}, \diamond)$, for the map $\phi : (\mathbb{G}, *) \to (\mathfrak{a}, \diamond)$ is a Lie group isomorphism (which is also linear, when $\mathfrak{a}$ and $\mathbb{G} \equiv \mathbb{R}^N$ are equipped with their vector space structures).

Analogously, we transfer the dilations $\{\delta_\lambda\}_{\lambda>0}$ on $\mathfrak{a}$ defined in (17.16) to a group of dilations $\{D_\lambda\}_{\lambda>0}$ on $\mathbb{G}$ by setting

$$D_\lambda : \mathbb{G} \to \mathbb{G}, \quad D_\lambda(\xi) := \phi^{-1}\big(\delta_\lambda(\phi(\xi))\big). \tag{17.19}$$

Since δ_λ is an automorphism of the Lie group $(\mathfrak{a}, \diamond)$ (see Ex. 5 at the end of the Chapter), it immediately follows that D_λ is an automorphism of the Lie group $(\mathbb{G}, *)$.

We claim that $\mathbb{G} = (\mathbb{R}^N, *, D_\lambda)$ *is a homogeneous Carnot group.* Let us prove this fact. With reference to the choice of the basis $\mathcal{X}$ for $\mathfrak{a}$ in (17.18), for any $\xi \in \mathbb{R}^N$, we write

$$\xi = (\xi^{(1)}, \dots, \xi^{(r)}), \quad \xi^{(i)} \in \mathbb{R}^{m_i}, \quad i = 1, \dots, r.$$

---

[4] In the previous sections, we denoted it by $\diamond_{\mathcal{X}}$, and we also wrote $\phi^{-1} = \pi_{\mathcal{X}}$.

First, we aim to show that, with this notation,

$$D_\lambda(\xi^{(1)}, \xi^{(2)}, \dots, \xi^{(r)}) = (\lambda \xi^{(1)}, \lambda^2 \xi^{(2)}, \dots, \lambda^r \xi^{(r)}). \tag{17.20}$$

We have

$$\begin{aligned} D_\lambda(\xi) &= \phi^{-1}(\delta_\lambda(\phi(\xi))) = \phi^{-1}\left(\delta_\lambda\left(\sum_{j=1}^{N} \xi_j X_j\right)\right) \\ &= \phi^{-1}\left(\delta_\lambda\left(\sum_{i=1}^{r}\sum_{j=1}^{m_i} \xi_j^{(i)} X_j^{(i)}\right)\right) = \phi^{-1}\left(\sum_{i=1}^{r}\sum_{j=1}^{m_i} \xi_j^{(i)} \lambda^i X_j^{(i)}\right) \\ &= (\lambda \xi^{(1)}, \lambda^2 \xi^{(2)}, \dots, \lambda^r \xi^{(r)}). \end{aligned}$$

Denote by $Y_1^{(1)}, \dots, Y_m^{(1)}$ the first m vector fields of the Jacobian basis of $\mathfrak{g}$. In order to prove that $\mathbb{G}$ is a homogeneous Carnot group, we have to show

$$\mathrm{Lie}\{Y_1^{(1)}, \dots, Y_m^{(1)}\} = \mathfrak{g}. \tag{17.21}$$

Let us prove this fact. Let $Y_j^{(i)}$ denote the vector field in $\mathfrak{g}$ coinciding at 0 with

$$\partial_{\xi_j^{(i)}}|_0$$

(i.e. $Y_j^{(i)}$ is the vector field in the Jacobian basis of $\mathfrak{g}$ relative to the coordinate $\xi_j^{(i)}$). Then, $Y_j^{(i)}$ is D_λ-homogeneous of degree i: indeed, it holds

$$\begin{aligned} Y_j^{(i)}|_\xi(u \circ D_\lambda) &= \mathrm{d}_0\tau_\xi(Y_j^{(i)}|_0)(u \circ D_\lambda) = (Y_j^{(i)}|_0)(u \circ D_\lambda \circ \tau_\xi) \\ &= (Y_j^{(i)}|_0)(u \circ \tau_{D_\lambda(\xi)} \circ D_\lambda) \\ &= (\partial_{x_j^{(i)}}\big|_0)(u \circ \tau_{D_\lambda(\xi)})(\lambda x^{(1)}, \dots, \lambda^i x^{(i)}, \dots, \lambda^r x^{(r)}) \\ &= \lambda^i (\partial_{y_j^{(i)}}\big|_0)(u \circ \tau_{D_\lambda(\xi)})(y^{(1)}, \dots, y^{(i)}, \dots, y^{(r)}) \\ &= \lambda^i\, (Y_j^{(i)}|_0)(u \circ \tau_{D_\lambda(\xi)}) = \lambda^i Y_j^{(i)}\big|_{D_\lambda(\xi)} u. \end{aligned}$$

Hence, $\mathfrak{g}$ is split into the direct sum $\mathfrak{g} = H^{(1)} \oplus \dots \oplus H^{(r)}$, where $H^{(i)}$ is the set of the left-invariant vector fields, D_λ-homogeneous of degree i, i.e. $H^{(i)} = \mathrm{span}\{H_1^{(i)}, \dots, H_{m_i}^{(i)}\}$. In order to prove (17.21), it is sufficient to find another set $\{Z_1, \dots, Z_m\}$ of independent vector fields in $\mathfrak{g}$ which are D_λ-homogeneous of degree 1 and such that $\mathrm{Lie}\{Z_1, \dots, Z_m\} = \mathfrak{g}$. This is done in the next step.

STEP 4 (**Relations between $\mathfrak{g}$ and $\mathfrak{a}$**). Consider the following commutative (see Theorem 2.1.59, page 119) diagram:

$$\begin{array}{ccc} (\mathbb{G}, *) & \xrightarrow{\quad \phi \quad} & (\mathfrak{a}, \diamond) \\ \uparrow{\scriptstyle \mathrm{Exp}_{\mathbb{G}}} & & \uparrow{\scriptstyle \mathrm{Exp}_{\mathfrak{a}}} \\ \mathfrak{g} & \xrightarrow{\quad \mathrm{d}\phi \quad} & \mathrm{Lie}(\mathfrak{a}). \end{array}$$

It is proved in Ex. 4 at the end of the chapter that $\mathrm{Exp}_{\mathfrak{a}}$ *is linear and it is a Lie algebra isomorphism*, when $\mathfrak{a}$ is equipped with its former structure of Lie algebra (and $\mathrm{Lie}(\mathfrak{a})$ with its obvious one). Hence (recall that $\mathrm{d}\phi$ is a Lie algebra isomorphism, see Theorem 2.1.50, page 114)

$$\alpha : \mathfrak{g} \to \mathfrak{a}, \quad \alpha := \phi \circ \mathrm{Exp}_{\mathbb{G}} = \mathrm{Exp}_{\mathfrak{a}} \circ \mathrm{d}\phi \tag{17.22}$$

is a Lie algebra isomorphism. The reader should notice that the elements of $\mathfrak{a}$, which are vector fields on $\mathbb{R}^n$, have now been put into a bijective correspondence with vector fields on $\mathbb{R}^N$ (the elements of $\mathfrak{g}$). We consider the vector fields $Z_1, \ldots, Z_m$ in $\mathfrak{g}$ corresponding to $X_1, \ldots, X_m$, respectively, i.e.

$$Z_i \in \mathfrak{g} : \quad \alpha(Z_i) = X_i \qquad \text{for every } i = 1, \ldots, m. \tag{17.23}$$

We claim that Z_i *is the i-th vector of the Jacobian basis for* $\mathfrak{g}$. Indeed,

$$\begin{aligned} Z_i|_0 u &= (\alpha^{-1}(X_i))|_0 u = (\mathrm{d}\phi^{-1}(\mathrm{Log}_{\mathfrak{a}}(X_i)))_0 u = (\mathrm{d}_0\phi^{-1}((\mathrm{Log}_{\mathfrak{a}}(X_i))_0))u \\ &= (\mathrm{Log}_{\mathfrak{a}}(X_i))_0(u \circ \phi^{-1}) = \frac{\mathrm{d}}{\mathrm{d}t}\Big|_{t=0} (u \circ \phi^{-1})(tX_i) \\ &= \frac{\mathrm{d}}{\mathrm{d}t}\Big|_{t=0} u(te_i) = \frac{\partial}{\partial \xi_i}\Big|_{\xi=0} (u(\xi)). \end{aligned}$$

For the fifth equality, we used the results of Ex. 6 at the end of the Chapter (recall that $(tX) \diamond (sX) = (t+s)X$ for every $X \in \mathfrak{a}$ and $s, t \in \mathbb{R}$). Finally, we notice that (17.23), the fact that α is a Lie algebra isomorphism, and $\mathrm{Lie}\{X_1, \ldots, X_m\} = \mathfrak{a}$ imply $\mathrm{Lie}\{Z_1, \ldots, Z_m\} = \mathfrak{g}$. This completes the proof of Step 3.

The aim of the following step is to find a "lifting" map $\pi : \mathbb{R}^N \to \mathbb{R}^n$ relating the Z_i's to the X_i's as in (17.12).

STEP 5 (**The lifting map π**). Let us denote by Log the logarithmic map related to the Lie group $(\mathbb{G}, \circ)$. Moreover, we write I for the identity map on $\mathbb{R}^n$. We then introduce the following map

$$\begin{aligned} &\pi : \mathbb{R}^N \longrightarrow \mathbb{R}^n, \\ &\pi(\xi) := e^{\alpha(\mathrm{Log}\,(\xi))} I(0) := \sum_{k \geq 0} \frac{\big(\alpha(\mathrm{Log}\,(\xi))\big)^k}{k!} I(0). \end{aligned} \tag{17.24}$$

We explicitly remark that the map π is well defined. Indeed, if $\xi \in \mathbb{R}^N \equiv \mathbb{G}$, then $\mathrm{Log}\,(\xi) \in \mathfrak{g}$, so that $\alpha(\mathrm{Log}\,\xi) \in \mathfrak{a}$. As a consequence, *by the properties of δ_λ-homogeneity of the vector fields in* $\mathfrak{a}$ (see hypothesis **(F1)**), any power $(\alpha(\mathrm{Log}\,(\xi)))^k$ with $k \geq n+1$ annihilates the identity map I of $\mathbb{R}^n$, so that the sum in (17.24) is finite. By the very definition of α in (17.22), we have $\alpha(\mathrm{Log}\,(\xi)) = \phi(\xi) = \sum_{j=1}^N \xi_j X_j$, so that

$$\pi(\xi) := \sum_{k \geq 0} \frac{\big(\sum_{j=1}^N \xi_j X_j\big)^k}{k!} I(0) \quad \text{for all } \xi \in \mathbb{R}^N. \tag{17.25}$$

STEP 6 (**The properties of π**). The map π has the following properties:

1. (HOMOGENEITY PROPERTY.) We have

$$\pi(D_\lambda(\xi)) = d_\lambda(\pi(\xi)) \qquad \forall\, \xi \in \mathbb{R}^N,\ \lambda > 0. \tag{17.26}$$

2. (SURJECTIVE PROPERTY.) The map $\pi : \mathbb{R}^N \to \mathbb{R}^n$ is onto.
3. (EXPONENTIAL PROPERTY.) For every $X \in \mathfrak{a}$, we have

$$\pi(\phi^{-1}(X)) = \exp(X)(0), \tag{17.27}$$

where the right-hand side is meant as a formal exponential.[5]
4. (LIFTING PROPERTY.) It holds

$$\mathrm{d}_\xi \pi(Z_i) = (X_i)_{\pi(\xi)} \qquad \text{for every } \xi \in \mathbb{R}^N. \tag{17.28}$$

Let us begin with the homogeneity property. If we shortly write

$$D_\lambda(\xi) = (\lambda^{\beta_1}\xi_1, \ldots, \lambda^{\beta_N}\xi_N), \tag{17.29}$$

then from (17.25) we have

$$\pi(D_\lambda(\xi)) := \sum_{k\geq 0} \frac{\left(\sum_{j=1}^N \xi_j\, \lambda^{\beta_j}\, X_j\right)^k}{k!} I(0).$$

Now, let us fix $i_0 \in \{1, \ldots, n\}$. The claimed (17.26) will follow if we show that

$$\sum_{k\geq 0} \frac{\left(\sum_{j=1}^N \xi_j\, \lambda^{\beta_j}\, X_j\right)^k}{k!} I_{i_0}(0) = \lambda^{\sigma_{i_0}} \sum_{k\geq 0} \frac{\left(\sum_{j=1}^N \xi_j\, X_j\right)^k}{k!} I_{i_0}(0). \tag{17.30}$$

Let us prove (17.30). From (17.15), the choice (17.18) of the basis $\mathcal{X}$, the very definition (17.16) of δ_λ, (17.20) and the notation in (17.29) it follows that X_j is d_λ-homogeneous of degree β_j. Moreover, any non-vanishing summand in the left-hand side of (17.30) has the following form (for some $c \in \mathbb{R}$, $i_1, \ldots, i_s \in \{1, \ldots, N\}$)

$$c\, \lambda^{\beta_{i_1}+\cdots+\beta_{i_s}}\, \xi_{i_1} \cdots \xi_{i_s}\, X_{i_1} \cdots X_{i_s} I_{i_0}(0) \qquad \text{with } \beta_{i_1} + \cdots + \beta_{i_s} = \sigma_{i_0},$$

because the function I_{i_0} is d_λ-homogeneous of degree σ_{i_0}. This proves (17.30).

Let us turn to the surjective property. From (17.25) it is evident that

$$\pi(\xi) := \sum_{j=1}^N \xi_j\, X_j I(0) + \mathcal{O}(|\xi|^2), \quad \text{as } \xi \to 0.$$

[5] Id est, the formal exponential of a high-order differential operator on $\mathbb{R}^n$

$$\exp(X)(0) := \sum_{k\geq 0} \frac{X^k}{k!} I(0),$$

which is well defined for the sum is finite (recall the d_λ-homogeneity properties of the operators in $\mathfrak{a}$).

This gives

$$\mathcal{J}_\pi(0) = \big(X_1 I(0) \cdots X_N I(0)\big),$$

so that, recalling (17.17),

$$\operatorname{rank}\big(\mathcal{J}_\pi(0)\big) = n.$$

By the rank theorem from elementary calculus, this proves the existence of an open neighborhood $W \subseteq \mathbb{R}^n$ of $\pi(0) = 0$ such that $\pi : \mathbb{R}^N \to W$ is surjective. We next demonstrate that π is also onto $\mathbb{R}^n$. Indeed, if $x \in \mathbb{R}^n$, there exists $\lambda = \lambda(x) > 0$ such that $d_\lambda(x) \in W$. Let $\xi \in \mathbb{R}^N$ be such that $\pi(\xi) = d_\lambda(x)$. Then, thanks to (17.26),

$$\pi(D_{1/\lambda}(\xi)) = d_{1/\lambda}(\pi(\xi)) = d_{1/\lambda}(d_\lambda(x)) = x.$$

This shows that π is surjective. Obviously, π has polynomial entries, for the sum in (17.25) is finite.

Let us prove (17.27). This is just the definition (17.24) of π jointly with $(\alpha \circ \operatorname{Log} \circ \phi^{-1})(X) = (\phi \circ \phi)(X) = X$.

Finally, Theorem 17.4.2 will be completely proved if we show that (17.28) holds (the lifting property of π). Let us fix $\xi \in \mathbb{R}^N$. Since in (17.12) we are testing the equality of two first order differential operators on $\mathbb{R}^n$, (17.12) follows if we demonstrate that

$$\mathrm{d}_\xi \pi(Z_i) I_s = (X_i)_{\pi(\xi)} I_s \qquad \text{for every } s = 1, \ldots, n, \tag{17.31}$$

where $I_s(x) = x_s$ is the s-th coordinate projection of $\mathbb{R}^n$. Recalling that Z_i is the i-th vector of the Jacobian basis for $\mathfrak{g}$, we have

$$\begin{aligned}
\mathrm{d}_\xi \pi(Z_i) I_s &= Z_i|_\xi (I_s \circ \pi) = \frac{\mathrm{d}}{\mathrm{d}t}\Big|_{t=0} \big((I_s \circ \pi)(\xi * (te_i))\big) \\
&= \frac{\mathrm{d}}{\mathrm{d}t}\Big|_{t=0} (I_s \circ \pi)\big(\phi^{-1}(\phi(\xi) \diamond \phi(te_i))\big) \\
&= \frac{\mathrm{d}}{\mathrm{d}t}\Big|_{t=0} (I_s \circ \pi \circ \phi^{-1})\left(\left(\sum_{j=1}^N \xi_j X_j\right) \diamond (tX_i)\right) \\
&= \frac{\mathrm{d}}{\mathrm{d}t}\Big|_{t=0} I_s \circ \exp\left(\left(\sum_{j=1}^N \xi_j X_j\right) \diamond (tX_i)\right)(0) \\
&= \frac{\mathrm{d}}{\mathrm{d}t}\Big|_{t=0} I_s \circ \left(\exp(tX_i)\left(\exp\left(\sum_{j=1}^N \xi_j X_j\right)(0)\right)\right)_g \\
&= (X_i I_s)\left(\exp\left(\sum_{j=1}^N \xi_j X_j\right)(0)\right) = (X_i I_s)(\pi(\xi)),
\end{aligned}$$

which gives (17.31). Here we used the following facts: The first equality is the definition of the differential; the second equality follows from the left-invariance of Z_i; the third equality is the definition of $*$; the fourth equality is the definition of ϕ; the fifth equality follows from (17.27); the sixth equality follows from the Campbell–Hausdorff formula (applied to the vector fields of $\mathfrak{a}$ which have d_λ-homogeneity properties; see (15.3) page 595); the seventh equality is trivial; the last equality follows again from (17.27). This ends the proof. □

Bibliographical Notes. Different proofs of the lifting procedure have been provided. Besides the pioneering paper by L.P. Rothschild and E.M. Stein [RS76], we refer the reader to L. Hörmander and A. Melin [HM78], G.B. Folland [Fol77], R.W. Goodman [Goo78]. In [Fol77], the case described in Section 17.4 is considered. Some of the topics presented in this chapter also appear in [BU05a].

17.5 Exercises of Chapter 17

Ex. 1) Let $\alpha \in \mathbb{R}$ be fixed. Lift the Carnot group on $\mathbb{R}^4$ with the composition law

$$\begin{pmatrix} x_1 + y_1, \\ x_2 + y_2, \\ x_3 + y_3 + \frac{1}{2}(x_1\, y_2 - x_2\, y_1), \\ x_4 + y_4 + \frac{1}{2}(x_1\, y_3 - x_3\, y_1) + \frac{\alpha}{2}(x_2\, y_3 - x_3\, y_2) \\ +\frac{1}{12}(x_1 - y_1)\,(x_1\, y_2 - x_2\, y_1) + \frac{\alpha}{12}(x_2 - y_2)\,(x_1\, y_2 - x_2\, y_1) \end{pmatrix}.$$

Verify that, following the notation and definitions in Section 17.1, the following facts hold:

a) The Jacobian basis $\mathcal{Z}$ for $\mathfrak{g}$ is

$$Z_1 = \partial_1 - \frac{1}{2}x_2\,\partial_3 - \left(\frac{1}{2}x_3 + \frac{1}{12}x_2(x_1 + \alpha\, x_2)\right)\partial_4,$$
$$Z_2 = \partial_2 + \frac{1}{2}x_1\,\partial_3 + \left(-\frac{1}{2}\alpha\, x_3 + \frac{1}{12}x_1(x_1 + \alpha\, x_2)\right)\partial_4,$$
$$Z_3 = \partial_3 + \left(\frac{1}{2}x_1 + \frac{1}{2}\alpha\, x_2\right)\partial_4,$$
$$Z_4 = \partial_4.$$

b) The Campbell–Hausdorff law related to $\mathcal{Z}$ on $\mathbb{R}^4$ coincides with $\circ$. Hence, the Jacobian basis $\mathcal{X}$ for $\mathfrak{r}^4$ is given by

$$X_1 = \partial_1 - \frac{1}{2}\xi_2\,\partial_3 - \left(\frac{1}{2}\xi_3 + \frac{1}{12}\xi_2(\xi_1 + \alpha\,\xi_2)\right)\partial_4,$$
$$X_2 = \partial_2 + \frac{1}{2}\xi_1\,\partial_3 + \left(-\frac{1}{2}\alpha\,\xi_3 + \frac{1}{12}\xi_1(\xi_1 + \alpha\,\xi_2)\right)\partial_4,$$

$$X_3 = \partial_3 + \left(\frac{1}{2}\xi_1 + \frac{1}{2}\alpha\,\xi_2\right)\partial_4,$$
$$X_4 = \partial_4.$$

c) Consider $\mathfrak{f}_{2,3}$. We have $\dim \mathfrak{f}_{2,3} = 5$. Let $\Pi : \mathfrak{f}_{2,3} \to \mathfrak{g}$ be the Lie algebra morphism such that $\Pi(F_i) = Z_i$ for every $i = 1, 2$. The Hall basis for $\mathfrak{f}_{2,3}$ is given by

$$F_1, \quad F_2, \quad F_3 = [F_2, F_1], \quad F_4 = [[F_2, F_1], F_1],$$
$$F_5 = [[F_2, F_1], F_2].$$

The following is a basis $\mathcal{F}$ for $\mathfrak{f}_{2,3}$, as in Proposition 17.1.1,

$$F_1, \quad F_2, \quad -F_3, \quad F_4; \qquad F_5 - \alpha\,F_4.$$

d) The Lie group $(\mathbb{R}^5, \diamond_{\mathcal{F}})$ is such that $\xi \diamond_{\mathcal{F}} \eta$ equals

$$\begin{pmatrix} \xi_1 + \eta_1, \\ \xi_2 + \eta_2, \\ \xi_3 + \eta_3 + \frac{1}{2}(\xi_1\,\eta_2 - \xi_2\,\eta_1), \\ \xi_4 + \eta_4 + \frac{1}{2}(\xi_1\,\eta_3 - \xi_3\,\eta_1) + \frac{\alpha}{2}(\xi_2\,\eta_3 - \xi_3\,\eta_2) \\ +\frac{1}{12}(\xi_1 - \eta_1)(\xi_1\,\eta_2 - \xi_2\,\eta_1) + \frac{\alpha}{12}(\xi_2 - \eta_2)(\xi_1\,\eta_2 - \xi_2\,\eta_1), \\ \xi_5 + \eta_5 + \frac{1}{2}(\xi_2\,\eta_3 - \xi_3\,\eta_2) + \frac{1}{12}(\xi_2 - \eta_2)(\xi_1\,\eta_2 - \xi_2\,\eta_1) \end{pmatrix}.$$

Verify that the first 4 vector fields of the Jacobian basis for $\mathfrak{r}^5$ lift X_i, for every $i = 1, \dots, 4$, as stated in Lemma 17.1.3.

e) Verify that

$$\psi = \mathrm{Exp} \circ \pi_{\mathcal{Z}}^{-1} : (\mathbb{R}^4, \diamond_{\mathcal{Z}}) \longrightarrow (\mathbb{G}, \circ)$$

satisfies $\psi(\xi_1, \xi_2, \xi_3, \xi_4) = (\xi_1, \xi_2, \xi_3, \xi_4) \in \mathbb{G}$. In particular, the Lie group isomorphism $\Phi : (\mathbb{R}^5, \diamond_{\mathcal{F}}) \to (\mathbb{G} \times \mathbb{R}, \bullet)$ is defined by

$$\Phi(\xi_1, \xi_2, \xi_3, \xi_4, \xi_5) = (\xi_1, \xi_2, \xi_3, \xi_4; \xi_5) \in \mathbb{G} \times \mathbb{R}.$$

As a consequence, for the group law $\bullet$ on $\mathbb{G} \times \mathbb{R}$, we have that $x \bullet y$ equals

$$\begin{pmatrix} x_1 + y_1, \\ x_2 + y_2, \\ x_3 + y_3 + \frac{1}{2}(x_1\,y_2 - x_2\,y_1), \\ x_4 + y_4 + \frac{1}{2}(x_1\,y_3 - x_3\,y_1) + \frac{\alpha}{2}(x_2\,y_3 - x_3\,y_2) \\ +\frac{1}{12}(x_1 - y_1)(x_1\,y_2 - x_2\,y_1) + \frac{\alpha}{12}(x_2 - y_2)(x_1\,y_2 - x_2\,y_1), \\ x_5 + y_5 + \frac{1}{2}(x_2\,y_3 - x_3\,y_2) + \frac{1}{12}(x_2 - y_2)(x_1\,y_2 - x_2\,y_1) \end{pmatrix}.$$

Find the Jacobian basis $W_1, \dots, W_5$ for $\mathfrak{g}\,\mathfrak{r}^1$. Verify that, in particular, it holds

$$W_1 = \partial_1 - \frac{1}{2}x_2\,\partial_3 - \left(\frac{1}{2}x_3 + \frac{1}{12}x_2(x_1 + \alpha\,x_2)\right)\partial_4 - \frac{1}{12}x_2^2\,\partial_5,$$
$$W_2 = \partial_2 + \frac{1}{2}x_1\,\partial_3 + \left(-\frac{1}{2}\alpha\,x_3 + \frac{1}{12}x_1(x_1 + \alpha\,x_2)\right)\partial_4$$
$$+\left(-\frac{1}{2}x_3 + \frac{1}{12}x_1x_2\right)\partial_5.$$

Verify that, as stated in Lemma 17.1.4, W_i lifts $\mathrm{d}\vartheta(W_i) = Z_i$, for every $i = 1, \ldots, 4$.

Ex. 2) Give a detailed proof of Lemma 17.3.4, page 666.

Ex. 3) Prove properties (i)–(ii)–(iii) in Step 1, page 670.

Ex. 4) Prove the following result.

Proposition 17.5.1. *Let $\mathfrak{a}$ be a (finite-dimensional) nilpotent Lie algebra. Consider the Lie group $(\mathfrak{a}, \diamond)$, where $\diamond$ is the Campbell–Hausdorff operation*[6] *on $\mathfrak{a}$. Denote by* $\mathrm{Lie}(\mathfrak{a})$ *the Lie algebra of the Lie group $(\mathfrak{a}, \diamond)$ and by* $\mathrm{Exp}_{\mathfrak{a}} : \mathrm{Lie}(\mathfrak{a}) \to \mathfrak{a}$ *the relevant exponential map. Then,* $\mathrm{Exp}_{\mathfrak{a}}$ *is linear and it is a Lie algebra isomorphism, when $\mathfrak{a}$ is equipped with its former structure of Lie algebra.*

(*Hint*: By the third fundamental theorem of Lie (see Theorem 2.2.14, page 130) there exists a connected, simply connected Lie group $(G, \cdot)$ such that $\mathfrak{g} := \mathrm{Lie}(G)$ is isomorphic to $\mathfrak{a}$ as Lie algebras. Let $\varphi : \mathfrak{a} \to \mathfrak{g}$ be the relevant Lie algebra isomorphism. Then $\varphi : (\mathfrak{a}, \diamond) \to (\mathfrak{g}, \diamond)$ is a Lie group isomorphism (see Remark 2.2.16, page 130). Consider the following commutative diagram

$$\begin{array}{ccccc}
(\mathfrak{a}, \diamond) & \xrightarrow{\ \varphi\ } & (\mathfrak{g}, \diamond) & \xrightarrow{\ \mathrm{Exp}_G\ } & (G, \cdot) \\
\uparrow{\scriptstyle \mathrm{Exp}_{\mathfrak{a}}} & & \uparrow & & \uparrow{\scriptstyle \mathrm{Exp}_G A} \\
\mathrm{Lie}(\mathfrak{a}) & \xrightarrow{\ \mathrm{d}\varphi\ } & \mathrm{Lie}(\mathfrak{g}) & \xrightarrow{\ \mathrm{d}\,\mathrm{Exp}_G\ } & \mathfrak{g}.
\end{array}$$

We have

$$\mathrm{Exp}_{\mathfrak{a}} = \varphi^{-1} \circ (\mathrm{Exp}_G)^{-1} \circ \mathrm{Exp}_G \circ \mathrm{dExp}_G \circ \mathrm{d}\varphi = \varphi^{-1} \circ \mathrm{dExp}_G \circ \mathrm{d}\varphi,$$

which is evidently a Lie algebra isomorphism.)

Ex. 5) Let $\mathfrak{a}$ be a (finite-dimensional) nilpotent Lie algebra. Consider the Lie group $(\mathfrak{a}, \diamond)$, where $\diamond$ is the Campbell–Hausdorff operation on $\mathfrak{a}$. Suppose $\mathfrak{a}$ is stratified, i.e. it admits a decomposition of the form

$$\mathfrak{a} = \mathfrak{a}_1 \oplus \mathfrak{a}_2 \oplus \cdots \oplus \mathfrak{a}_r \quad \text{with} \quad \begin{cases} [\mathfrak{a}_1, \mathfrak{a}_{i-1}] = \mathfrak{a}_i & \text{if } 2 \le i \le r, \\ [\mathfrak{a}_1, \mathfrak{a}_r] = \{0\}. \end{cases}$$

Define a group of dilations $\{\delta_\lambda\}_{\lambda>0}$ on $\mathfrak{a}$ in the following way:

[6] See Theorem 2.2.13, page 129; see also the relevant Definition 2.2.11.

$$\delta_\lambda : \mathfrak{a} \to \mathfrak{a}, \quad \delta_\lambda\Big(\sum_{i=1}^r A_i\Big) := \sum_{i=1}^r \lambda^i A_i, \quad \text{where } A_i \in \mathfrak{a}_i \text{ for } i = 1, \dots, r.$$

Prove that δ_λ is an automorphism of the Lie group $(\mathfrak{a}, \diamond)$. (*Hint*: First show that if $X \in \mathfrak{a}_i$ and $Y \in \mathfrak{a}_j$, then $[X, Y] \in \mathfrak{a}_{i+j}$ (where $\mathfrak{a}_i := \{0\}$ for $i > r$). Then prove that $\delta_\lambda([X, Y]) = [\delta_\lambda(X), \delta_\lambda(Y)]$ for every $X, Y \in \mathfrak{a}$. End by using Remark 2.2.16.)

Ex. 6) Let V be a real (finite-dimensional) vector space equipped with a Lie group structure by the operation $*$. Denote by $\mathfrak{v}$ the relevant Lie algebra, by Exp and Log the relevant exponential and logarithmic maps. Furthermore, suppose that

$$(\bigstar) \qquad (tv) * (sv) = (t + s)v \quad \text{for every } v \in V \text{ and } s, t \in \mathbb{R}.$$

Let us introduce the following notation: for every $v \in V$, $\Xi(v)$ denotes the tangent vector at $0 \in V$ such that

$$\Xi(v)f = \frac{\mathrm{d}}{\mathrm{d}t}\Big|_{t=0} \big(f(tv)\big) \qquad \forall\, V \in C^\infty(V, \mathbb{R}).$$

Prove the following facts:

a) Show that $\Xi(v)$ is indeed an element of $T_0(V)$;

b) Show that
$$T_0(V) = \{\Xi(v) \,:\, v \in V\},$$
whence the natural identification between $T_0(V)$ and V;

c) Denote by $\alpha : \mathfrak{v} \to T_0(V)$ the usual identification between $\mathfrak{v}$ and $T_0(V)$, i.e.
$$\alpha(X) = X_0 \quad \text{for every } X \in \mathfrak{v}.$$
Prove that the flow of $X \in \mathfrak{v}$ is $t \mapsto t\, v$, where $v = \Xi^{-1}(X_0)$.

d) Derive that $\mathrm{Exp}\,(X) = \Xi^{-1}(X_0)$, i.e.
$$\big(\mathrm{Exp} \circ \alpha^{-1} \circ \Xi\big)(v) = v \quad \text{for every } v \in V,$$
or analogously, $\alpha \circ \mathrm{Log} = \Xi$, that is, for every $v \in V$, it holds
$$(\mathrm{Log}\, v)_0 f = \frac{\mathrm{d}}{\mathrm{d}t}\Big|_{t=0} f(tv) \qquad \text{for every } f \in C^\infty(V).$$

(*Hint*: The main task is to prove (c). We must show that $\gamma : \mathbb{R} \to V$, $\gamma(t) = tv$ solves the Cauchy problem $\dot\gamma(t) = X_{\gamma(t)}$, $\gamma(0) = 0$. Note that
$$X_{\gamma(t)} = \mathrm{d}_0\tau_{\gamma(t)}(X_0) = \mathrm{d}_0\tau_{\gamma(t)}(\Xi(v)).$$
We thus have to prove that, for every $f \in C^\infty(V)$,
$$\frac{\mathrm{d}}{\mathrm{d}t}\Big|_t \big(f(t\,v)\big) \overset{?}{=} \mathrm{d}_0\tau_{\gamma(t)}(\Xi(v))f = [\cdots] = \frac{\mathrm{d}}{\mathrm{d}s}\Big|_{s=0} f((tv) * (sv)),$$
which follows from $(\bigstar)$.)

Ex. 7) Prove that the vector fields on $\mathbb{R}^3$ (whose points are denoted by (x, y, t)) defined by

$$X := \partial_x + y^2\,\partial_t, \qquad Y := \partial_y$$

are d_λ-homogeneous of degree 1, where

$$d_\lambda : \mathbb{R}^3 \longrightarrow \mathbb{R}^3, \qquad d_\lambda(x, y, t) := (\lambda x, \lambda y, \lambda^3 t),$$

and satisfy hypotheses **(F1)**–**(F2)** of page 669 (these are the fields considered in Section 4.4.2, page 212, as an example of fields not satisfying hypothesis **(H1)** of page 191). Lift them by the method illustrated in Section 17.4.

18

Groups of Heisenberg Type

In this chapter, we treat the so-called Heisenberg-type groups (also referred to as H-type groups), a significant class of Carnot groups of step two, generalizing the classical Heisenberg–Weyl groups.

In Section 18.2, we give a direct characterization of Heisenberg-type groups, via a suitable choice of a coordinate system, with respect to which the canonical sub-Laplacian has a simple and suggestive representation. We compare the definition of (abstract) H-type group given in the present chapter to our previous definition of *prototype* H-type group (see Definition 3.6.1): roughly speaking, any abstract H-type group is naturally isomorphic to a prototype H-type group.

In Section 18.3, we prove in a direct way the existence of a fundamental solution for some sub-Laplacians on Heisenberg-type groups, following the proof by A. Kaplan in [Kap80]. In Sections 18.4 and 18.5 we define an inversion and a Kelvin-transform on H-type groups, generalizing those from the classical theory of Laplace's operator. Unfortunately, this Kelvin transform has remarkable properties only on a sub-class of the H-type groups: the so-called *Iwasawa-type groups* (the relevant definition is given in Section 18.4). Indeed given a H-type group $\mathbb{H}$, the following result holds: *The H-Kelvin transform of a $\Delta_{\mathbb{H}}$-harmonic function is $\Delta_{\mathbb{H}}$-harmonic if and only if $\mathbb{H}$ is of Iwasawa-type.*

18.1 Heisenberg-type Groups

Let us begin with the definition of Heisenberg-type algebra and Heisenberg-type group.

Definition 18.1.1 (H-type algebra, H-type group). *A* Heisenberg-type *algebra* (H-type *algebra, in short) is a finite-dimensional real Lie algebra $\mathfrak{g}$ which can be endowed with an inner product $\langle\ ,\ \rangle$ such that*

$$[\mathfrak{z}^{\perp}, \mathfrak{z}^{\perp}] = \mathfrak{z},$$

where $\mathfrak{z}$ is the center of $\mathfrak{g}$ and moreover, for every fixed $z \in \mathfrak{z}$, the map

$$J_z : \mathfrak{z}^\perp \to \mathfrak{z}^\perp$$

defined by[1]

$$\langle J_z(v), w\rangle = \langle z, [v, w]\rangle \qquad \forall\, w \in \mathfrak{z}^\perp \tag{18.1}$$

is an orthogonal map whenever $\langle z, z\rangle = 1$.

A Heisenberg-type *group* (H-type *group, in short*) *is a connected and simply connected Lie group whose Lie algebra is an H-type algebra.*

We recall that the center of $\mathfrak{g}$ is, by definition,

$$\mathfrak{z} = \{z \in \mathfrak{g} \mid [z, g] = 0\ \forall\, g \in \mathfrak{g}\}.$$

In the following we shall always suppose that $\mathfrak{z}$ is not the null subspace. We shall often use the notation

$$\mathfrak{b} := \mathfrak{z}^\perp.$$

Let us consider some examples.

Example 18.1.2 (The Heisenberg–Weyl group is an H-type group). The classical Heisenberg–Weyl group $\mathbb{H}^N$ is an H-type group. As usual, we shall denote by (x, y, t) the points of $\mathbb{H}^N \equiv \mathbb{R}^{2N+1}$, $x, y \in \mathbb{R}^N$, $t \in \mathbb{R}$. Consider the vector fields

$$X_j := \partial_{x_j} + 2y_j\, \partial_t, \qquad Y_j := \partial_{y_j} - 2x_j\, \partial_t, \quad j = 1, \ldots, N.$$

Then we have

$$\mathfrak{g} = \mathrm{Lie}\{X_j, Y_j \mid j = 1, \ldots, N\}, \qquad \mathfrak{z} = \mathrm{span}\{\partial_t\}.$$

Let us now fix on $\mathfrak{g}$ the coordinate system associated to the basis

$$\{X_1, \ldots, X_N, Y_1, \ldots, Y_N, -4\,\partial_t\},$$

and let us consider on $\mathfrak{g}$ the standard inner product generated by this system. In other words, we identify $\mathfrak{g}$ with $\mathbb{R}^{2N+1}$ in the following way

$$\mathfrak{g} \ni \sum_{j=1}^N (a_j\, X_j + b_j\, Y_j) - 4c\, \partial_t \longleftrightarrow (a, b, c) \in \mathbb{R}^{2N+1},$$

and we set

$$\big\langle (a, b, c), (a', b', c')\big\rangle := \sum_{j=1}^N (a_j\, a'_j + b_j\, b'_j) + c\, c'.$$

In this coordinate system, the Lie bracket has the following expression

$$[(a, b, c), (a', b', c')] = \left(0, 0, \sum_{j=1}^N (a_j\, b'_j - b_j\, a'_j)\right).$$

[1] J_z is well-posed, see (18.3) below.

Then we have

$$\mathfrak{b} := \mathfrak{z}^{\perp} = \mathrm{span}\{X_j,\ Y_j \mid j = 1, \ldots, N\},$$

which gives $[\mathfrak{b}, \mathfrak{b}] = \mathfrak{z}$. Let now $z \in \mathfrak{z}$ be such that $|z| = 1$, i.e.

$$z = \mp 4\, \partial_t \equiv (0, 0, \pm 1).$$

With the notation of Definition 18.1.1, we have

$$J_z\left(\sum_{j=1}^{N}(a_j\, X_j + b_j\, Y_j)\right) = \pm \sum_{j=1}^{N}(-b_j\, X_j + a_j\, Y_j),$$

and, clearly, J_z is an orthogonal endomorphism of $\mathfrak{b}$. □

Example 18.1.3. It can be proved (and it will be proved in Section 18.2) that the following group law on $\mathbb{R}^6$ defines an H-type group:

$$x \circ y := \left(x^{(1)} + y^{(1)},\ x_1^{(2)} + y_1^{(2)} + \frac{1}{2}\langle P_1\, x^{(1)}, y^{(1)}\rangle,\ x_2^{(2)} + y_2^{(2)} + \frac{1}{2}\langle P_2\, x^{(1)}, y^{(1)}\rangle\right),$$

where $x = (x^{(1)}, x_1^{(2)}, x_2^{(2)}) \in \mathbb{R}^6$, $x^{(1)} \in \mathbb{R}^4$, $x_1^{(2)},\ x_2^{(2)} \in \mathbb{R}$ and

$$P_1 := \frac{\sqrt{2}}{2}\begin{pmatrix} 0 & 1 & 1 & 0 \\ -1 & 0 & 0 & -1 \\ -1 & 0 & 0 & 1 \\ 0 & 1 & -1 & 0 \end{pmatrix}, \quad P_2 := \frac{\sqrt{2}}{2}\begin{pmatrix} 0 & -1 & 1 & 0 \\ 1 & 0 & 0 & -1 \\ -1 & 0 & 0 & -1 \\ 0 & 1 & 1 & 0 \end{pmatrix}.$$

Example 18.1.4. The Lie group obtained as the direct product of the group $(\mathbb{R}, +)$ with the Heisenberg group $(\mathbb{H}^1, \circ)$ is *not* an H-type group. More precisely, if we consider the algebra $\mathfrak{g}$ generated by the vector fields on $\mathbb{R}^4$

$$\partial_{x_1}, \quad \partial_{x_2} + 2x_3\partial_{x_4}, \quad \partial_{x_3} - 2x_2\partial_{x_4},$$

we observe that the center $\mathfrak{z}$ of $\mathfrak{g}$ is given by $\mathrm{span}\{\partial_{x_1}, \partial_{x_4}\}$, but ∂_{x_1} cannot be obtained as linear combinations of brackets in $\mathfrak{g}$. Thus the hypothesis $[\mathfrak{b}, \mathfrak{b}] = \mathfrak{z}$ in Definition 18.1.1 cannot be satisfied by *any* $\mathfrak{b} \subseteq \mathfrak{g}$. □

Remark 18.1.5. If a Lie algebra $\mathfrak{g}$ has a center $\mathfrak{z}$ such that $\dim(\mathfrak{g}) - \dim(\mathfrak{z})$ is *odd*, then $\mathfrak{g}$ is not an H-type algebra. This fact will be proved in Section 18.2 (see also Remark 18.1.6 below). □

Remark 18.1.6. We have the following general result (see A. Kaplan [Kap80, Corollary 1]). Let $N_1, N_2 \in \mathbb{N} \setminus \{0\}$. Then there exists an H-type algebra of dimension $N_1 + N_2$ whose center has dimension N_2 (thus N_1 is the dimension of the orthogonal completion of the center) *if and only if* $N_2 < \rho(N_1)$, where ρ is the so-called *Hurwitz–Radon function*, i.e.

$$\rho : \mathbb{N} \to \mathbb{N}, \qquad \rho(n) := 8p + q, \quad \text{where} \quad n = (\text{odd}) \cdot 2^{4p+q},\ 0 \le q \le 3.$$

We observe that $\rho(N_1) = 0$ if N_1 is odd. Thus there cannot be H-type algebras where the dimension of the orthogonal completion of the center is odd.

We explicitly remark that the relation

$$\langle J_z(v), v' \rangle = \langle z, [v, v'] \rangle \qquad \forall\, v' \in \mathfrak{v} \tag{18.2}$$

in Definition 18.1.1 actually defines an endomorphism of $\mathfrak{v}$. Indeed, for fixed $v \in \mathfrak{v}$ and $z \in \mathfrak{z}$, the map

$$\Psi : \mathfrak{v} \to \mathbb{R}, \quad v' \mapsto \Psi(v') := \langle z, [v, v'] \rangle$$

is linear. Hence, there exists exactly one $w \in \mathfrak{v}$ (depending only on v and z) such that $\Psi(v') = \langle w, v' \rangle$ for every $v' \in \mathfrak{v}$. Then we set $J_z(v) := w$. We finally show that, for fixed $z \in \mathfrak{z}$, $J_z(\cdot)$ is linear: if $u,\ v \in \mathfrak{v}$, $\alpha,\ \beta \in \mathbb{R}$, we have

$$\begin{aligned}
\langle J_z(\alpha u + \beta v), v' \rangle &= \langle z, [\alpha u + \beta v, v'] \rangle = \alpha \langle z, [u, v'] \rangle + \beta \langle z, [v, v'] \rangle \\
&= \alpha \langle J_z(u), v' \rangle + \beta \langle J_z(v), v' \rangle \\
&= \langle \alpha J_z(u) + \beta J_z(v), v' \rangle \qquad \forall\, v' \in \mathfrak{v}.
\end{aligned}$$

Then we have

$$J_z(\alpha u + \beta v) - \alpha J_z(u) - \beta J_z(v) \in \mathfrak{v}^{\perp} \cap \mathfrak{v} = \{0\}.$$

Moreover, for fixed $v \in \mathfrak{v}$, the map $J_{(\cdot)}(v) : \mathfrak{z} \to \mathfrak{v}$, $z \mapsto J_z(v)$, is a linear map. Indeed, if $z,\ z' \in \mathfrak{z}$ and $\alpha,\ \beta \in \mathbb{R}$, we have

$$\begin{aligned}
\langle J_{(\alpha z + \beta z')}(v), v' \rangle &= \langle \alpha z + \beta z', [v, v'] \rangle = \alpha \langle z, [v, v'] \rangle + \beta \langle z', [v, v'] \rangle \\
&= \alpha \langle J_z(v), v' \rangle + \beta \langle J_{z'}(v), v' \rangle \\
&= \langle \alpha J_z(v) + \beta J_{z'}(v), v' \rangle \qquad \forall\, v' \in \mathfrak{v}.
\end{aligned}$$

Thus $J_{(\alpha z + \beta z')}(v) = \alpha J_z(v) + \beta J_{z'}(v)$ as above. This shows that the map

$$J_{(\cdot)} : \mathfrak{z} \to \mathrm{End}(\mathfrak{v}), \quad z \mapsto J_z, \quad \text{defined by (18.2) is well-posed and linear.} \tag{18.3}$$

Let us make some other simple but useful remarks on H-type algebras.

Remark 18.1.7. Let $\mathfrak{g}$ be an H-type algebra. With the above notation, we have:

1) is a nilpotent Lie algebra of step two;
2) $\mathfrak{g} = \mathfrak{v} \oplus \mathfrak{z}$, $[\mathfrak{v}, \mathfrak{v}] = \mathfrak{z}$, $[\mathfrak{v}, \mathfrak{z}] = \{0\}$; hence, if $\mathbb{G}$ is an H-type group, then $\mathbb{G}$ is a step two Carnot group;
3) the notion of H-type algebra (respectively, of H-type group) is invariant under Lie algebra (respectively, Lie group) isomorphisms.

Proof. 1) Since $\mathfrak{v}$ is the orthogonal completion of $\mathfrak{z} \neq \{0\}$ with respect to $\langle\,,\,\rangle$, we have $\mathfrak{g} = \mathfrak{v} \oplus \mathfrak{z}$. Let now $g_1,\ g_2,\ g_3 \in \mathfrak{g}$. We have $g_i = v_i + z_i$, with $v_i \in \mathfrak{v}$ and $z_i \in \mathfrak{z}$, for every $i = 1, 2, 3$. Since $\mathfrak{z}$ is the center of $\mathfrak{g}$, every z_i commutes with any element of $\mathfrak{g}$. As a consequence,

$$[g_1, [g_2, g_3]] = [v_1 + z_1, [v_2 + z_2, v_3 + z_3]] = [v_1, [v_2, v_3]] = 0,$$

since we have $[v_2, v_3] \in [\mathfrak{b}, \mathfrak{b}] = \mathfrak{z}$.

2) It directly follows from the above argument.

3) Let $\mathfrak{g}$ be an H-type algebra, let $\mathfrak{g}^*$ be a Lie algebra isomorphic to $\mathfrak{g}$ and let $\varphi : \mathfrak{g}^* \to \mathfrak{g}$ be a Lie algebra isomorphism. Since φ is a vector space isomorphism, the bilinear form

$$\langle g_1^*, g_2^* \rangle^* := \langle \varphi(g_1^*), \varphi(g_2^*) \rangle, \quad g_1^*, g_2^* \in \mathfrak{g}^*,$$

is an inner product in $\mathfrak{g}^*$. Let $\mathfrak{z}^* := \varphi^{-1}(\mathfrak{z})$. It is immediate to recognize that $\mathfrak{z}^*$ is the center of $\mathfrak{g}^*$. Moreover, setting $\mathfrak{b}^* := (\mathfrak{z}^*)^\perp$ (orthogonal completion of $\mathfrak{z}^*$ with respect to $\langle\, , \,\rangle^*$), we have $\mathfrak{b}^* = \varphi^{-1}(\mathfrak{z}^\perp) = \varphi^{-1}(\mathfrak{b})$. It follows that

$$[\mathfrak{b}^*, \mathfrak{b}^*] = [\varphi^{-1}(\mathfrak{b}), \varphi^{-1}(\mathfrak{b})] = \varphi^{-1}([\mathfrak{b}, \mathfrak{b}]) = \varphi^{-1}(\mathfrak{z}) = \mathfrak{z}^*.$$

For $z^* \in \mathfrak{z}^*$ fixed, we consider the linear map $J_{z^*}^* : \mathfrak{b}^* \to \mathfrak{b}^*$ defined as follows: if $v^* \in \mathfrak{b}$, $J_{z^*}^*(v^*)$ is defined by the identity

$$\langle J_{z^*}^*(v^*), (v')^* \rangle^* = \langle z^*, [v^*, (v')^*] \rangle^* \qquad \forall\, (v')^* \in \mathfrak{b}^*.$$

It is immediate to verify that

$$J_{z^*}^* = \varphi^{-1} \circ J_{\varphi(z^*)} \circ \varphi.$$

Let now $z^* \in \mathfrak{z}^*$ be such that $\langle\, z^*, z^* \,\rangle^* = 1$. Then $\langle\, \varphi(z^*), \varphi(z^*) \,\rangle = 1$. Thus $J_{z^*}^*$ is an orthogonal map. □

Proposition 18.1.8. *Let $\mathfrak{g}$ be an H-type algebra. With the above notation,*

$$\langle J_z(v), v \rangle = 0, \tag{18.4a}$$
$$\langle J_z(v), v' \rangle = -\langle v, J_z(v') \rangle, \tag{18.4b}$$
$$|J_z(v)| = |z| \cdot |v|, \tag{18.4c}$$
$$\langle J_z(v), J_{z'}(v) \rangle = \langle z, z' \rangle \cdot |v|^2, \tag{18.4d}$$
$$[v, J_z(v)] = |v|^2 \cdot z \tag{18.4e}$$

for every $z,\ z' \in \mathfrak{z}$ and for every $v,\ v' \in \mathfrak{b}$.

Proof. Let $z,\ z' \in \mathfrak{z}$ and $v,\ v' \in \mathfrak{b}$ be fixed. From (18.2) we have

$$\langle J_z(v), v \rangle = \langle z, [v, v] \rangle = 0,$$

which gives (18.4a). It follows that

$$\begin{aligned} 0 &= \langle J_z(v + v'), v + v' \rangle = \langle J_z(v), v \rangle + \langle J_z(v), v' \rangle + \langle J_z(v'), v \rangle + \langle J_z(v'), v' \rangle \\ &= \langle J_z(v), v' \rangle + \langle J_z(v'), v \rangle, \end{aligned}$$

which gives (18.4b). If $z = 0$, we have $J_z \equiv 0$ and (18.4c) trivially follows. If $z \neq 0$, from (18.3) it follows that $J_z = |z|\, J_{z/|z|}$. In this case, by the definition of H-type algebra, we have that $J_{z/|z|}$ is an orthogonal map. As a consequence,

$$|J_z(v)| = |z| \cdot |J_{z/|z|}(v)| = |z| \cdot |v|.$$

This proves (18.4c). Using (18.3), it is immediate to verify that we have

$$\langle J_{z+z'}(v), J_{z+z'}(v)\rangle = \langle J_z(v), J_z(v)\rangle + \langle J_{z'}(v), J_{z'}(v)\rangle + 2\,\langle J_z(v), J_{z'}(v)\rangle.$$

Thus, from (18.4c) it follows that

$$\begin{aligned}\langle J_z(v), J_{z'}(v)\rangle &= \frac{1}{2}(\langle J_{z+z'}(v), J_{z+z'}(v)\rangle - \langle J_z(v), J_z(v)\rangle - \langle J_{z'}(v), J_{z'}(v)\rangle)\\ &= \frac{1}{2}(|z+z'|^2 \cdot |v|^2 - |z|^2 \cdot |v|^2 - |z'|^2 \cdot |v|^2) = \langle z, z'\rangle \cdot |v|^2,\end{aligned}$$

which gives (18.4d). Finally, we have

$$\langle z', [v, J_z(v)]\rangle = \langle J_{z'}(v), J_z(v)\rangle = \langle z', z\rangle \cdot |v|^2 = \langle z', |v|^2 \cdot z\rangle.$$

The first equality follows from (18.2), the second equality follows from (18.4d). As a consequence, we have

$$\langle z', [v, J_z(v)] - |v|^2 \cdot z\rangle = 0 \qquad \forall\, z' \in \mathfrak{z}.$$

Since $|v|^2 \cdot z \in \mathfrak{z}$ and $[v, J_z(v)] \in [\mathfrak{b}, \mathfrak{b}] = \mathfrak{z}$, we immediately get (18.4e). □

18.2 A Direct Characterization of H-type Groups

The main aim of this section is to prove that (under a suitable system of coordinates) the group law of any H-type group has a somewhat explicit form (see (18.6) below). Roughly speaking, following our definitions from Section 3.6 (page 169), any *abstract* H-type group is naturally isomorphic to a *prototype* H-type group. Moreover, we furnish several useful details on H-type groups.

By means of the natural identification of a Carnot group with its Lie algebra (see Section 2.2 on page 121 for details), we already proved that (see Theorem 3.2.2, page 160, for the precise statement) any Carnot group of step two is canonically isomorphic to $\mathbb{R}^N$ equipped with the following group law ($N = m + n$, $x^{(1)} \in \mathbb{R}^m$, $x^{(2)} \in \mathbb{R}^n$)

$$(x^{(1)}, x^{(2)}) \circ (y^{(1)}, y^{(2)}) = \begin{pmatrix} x_j^{(1)} + y_j^{(1)}, \quad j = 1, \dots, m \\ x_j^{(2)} + y_j^{(2)} + \frac{1}{2}\langle B^{(j)} x^{(1)}, y^{(1)}\rangle, \quad j = 1, \dots, n \end{pmatrix}, \tag{18.5}$$

where the $B^{(j)}$'s are $m \times m$ linearly independent skew-symmetric matrices. Let now $\mathbb{G}$ be a general H-type group according to Definition 18.1.1. We set

$$m := \dim(\mathfrak{z}^\perp) \text{ and } n := \dim(\mathfrak{z}).$$

Since $\mathbb{G}$ has step two and since the stratification of the Lie algebra $\mathfrak{g}$ is evidently $\mathfrak{z}^\perp \oplus \mathfrak{z}$ (see Remark 18.1.7), in the sequel we shall fix on $\mathbb{G}$ a system of coordinates

(x, t), and we shall suppose that the dilations on $\mathbb{G}$ are $\delta_\lambda(x, t) = (\lambda x, \lambda^2 t)$ and that the group law has the form

$$(x, t) \circ (\xi, \tau) = \begin{pmatrix} x_j + \xi_j, \quad j = 1, \dots, m \\ t_j + \tau_j + \frac{1}{2} \langle U^{(j)} x, \xi \rangle, \quad j = 1, \dots, n \end{pmatrix},$$
$$x, \xi \in \mathbb{R}^m, \ t, \tau \in \mathbb{R}^n, \tag{18.6}$$

for suitable skew-symmetric matrices $U^{(j)}$'s. Our main aim is to give a necessary and sufficient condition on the $U^{(j)}$'s such that the composition law (18.6) defines on $\mathbb{R}^{m+n}$ an H-type group. A complete answer is given by the following result.

Theorem 18.2.1 (Characterization). *$\mathbb{G}$ is an H-type group if and only if $\mathbb{G}$ is (isomorphic to) $\mathbb{R}^{m+n}$ with the group law* (18.6) *and the matrices $U^{(1)}, \dots, U^{(n)}$ have the following properties*:

1) *$U^{(j)}$ is an $m \times m$ skew-symmetric and orthogonal matrix for every $j \le n$;*
2) *$U^{(i)} U^{(j)} + U^{(j)} U^{(i)} = 0$ for every $i, j \in \{1, \dots, n\}$ with $i \neq j$.*

Remark 18.2.2. $\mathbb{R}^{m+n}$ equipped with the group law (18.6) (the matrices $U^{(j)}$'s being as in Theorem 18.2.1) will be referred to as a *prototype* H-type group. Theorem 18.2.1 states that *any H-type group is naturally isomorphic to a prototype H-type group.*

Remark 18.2.3. Conditions 1) and 2) imply that $U^{(1)}, \dots, U^{(n)}$ are linearly independent. Namely, as it will appear from the proof, if 1) and 2) hold, then $\sum_{s=1}^n z_s U^{(s)}$ is the product of $|z|$ times an orthogonal matrix.

For example, the following three matrices

$$U^{(1)} = \begin{pmatrix} 0 & -1 & 0 & 0 \\ 1 & 0 & 0 & 0 \\ 0 & 0 & 0 & -1 \\ 0 & 0 & 1 & 0 \end{pmatrix}, \qquad U^{(2)} = \begin{pmatrix} 0 & 0 & 1 & 0 \\ 0 & 0 & 0 & -1 \\ -1 & 0 & 0 & 0 \\ 0 & 1 & 0 & 0 \end{pmatrix},$$
$$U^{(3)} = \begin{pmatrix} 0 & 0 & 0 & 1 \\ 0 & 0 & 1 & 0 \\ 0 & -1 & 0 & 0 \\ -1 & 0 & 0 & 0 \end{pmatrix}$$

satisfy conditions 1)–2) and, together with (18.6), define on $\mathbb{R}^7$ an H-type group whose center has dimension 3.

Proof (of Theorem 18.2.1). First of all, we prove the "only if" part of the theorem. Let $\mathbb{G}$ be an H-type group: we make use of all the notation in Definition 18.1.1. Let $B_1, \dots, B_m$ and $Z_1, \dots, Z_n$ be orthonormal bases for $\mathfrak{b} := \mathfrak{z}^\perp$ and $\mathfrak{z}$, respectively. The hypothesis $[\mathfrak{b}, \mathfrak{b}] = \mathfrak{z}$ ensures that there exist scalars $U_{i,j}^{(s)}$ such that

$$[B_i, B_j] = \sum_{s=1}^n U_{j,i}^{(s)} Z_s \quad \forall\, i, j \in \{1, \dots, m\}. \tag{18.7}$$

With the notation in (18.7), we define the following square matrices

$$U^{(s)} := \big(U^{(s)}_{i,j}\big)_{i,j\le m}, \qquad s = 1, \dots, n. \tag{18.8}$$

Since $[B_i, B_j] = -[B_j, B_i]$, and recalling that $Z_1, \dots, Z_n$ is a basis of $\mathfrak{z}$, $U^{(s)}$ is skew-symmetric. We now recall that the group $(\mathbb{G}, \circ)$ is canonically isomorphic (via the exponential map) to its Lie algebra $\mathfrak{g}$ (endowed with the Campbell–Hausdorff operation $(X, Y) \mapsto X + Y + \frac{1}{2}[X, Y]$). As a consequence, by means of the fixed basis $B_1, \dots, B_m, Z_1, \dots, Z_n$ on $\mathfrak{g}$, we can identify $\mathbb{G}$ to $\mathfrak{g} \equiv \mathbb{R}^{m+n}$ with the group law (18.6), where the matrices $U^{(j)}$'s are as in (18.8).

We fix $v = \sum_{i=1}^m v_i B_i$ and $z = \sum_{i=1}^n z_i Z_i$. We look for $x_1, \dots, x_m \in \mathbb{R}$ such that $J_z(v) = \sum_{i=1}^m x_i B_i$ satisfies condition (18.1) of Definition 18.1.1 for every $w = \sum_{i=1}^m w_i B_i$. Precisely,

$$\begin{aligned}
\sum_{j=1}^m x_j w_j &= \Big\langle \sum_{i=1}^m x_i B_i, \sum_{i=1}^m w_i B_i \Big\rangle = \langle J_z(v), w\rangle = \langle z, [v, w]\rangle \\
&= \Big\langle \sum_{i=1}^n z_i Z_i, \Big[\sum_{i=1}^m v_i B_i, \sum_{j=1}^m w_j B_j\Big]\Big\rangle \\
&= \Big\langle \sum_{i=1}^n z_i Z_i, \sum_{i,j=1}^m v_i w_j [B_i, B_j]\Big\rangle \\
&= \Big\langle \sum_{i=1}^n z_i Z_i, \sum_{i,j=1}^m v_i w_j \sum_{s=1}^n U^{(s)}_{j,i} Z_s\Big\rangle \\
&= \sum_{j=1}^m w_j \Big(\sum_{r,s=1}^n \sum_{i=1}^m v_i\, U^{(r)}_{j,i}\, z_s \langle Z_r, Z_s\rangle \Big) \\
&= \sum_{j=1}^m w_j \Big(\sum_{s=1}^n \sum_{i=1}^m v_i U^{(s)}_{j,i} z_s \Big).
\end{aligned} \tag{18.9}$$

Since w is arbitrary, this gives

$$x_j = \sum_{s\le n} \sum_{i\le m} v_i U^{(s)}_{j,i} z_s,$$

whence

$$J_z : \mathfrak{b} \to \mathfrak{b}, \qquad \sum_{j=1}^m v_j B_j \mapsto \sum_{j=1}^m \Big(\sum_{s\le n} \sum_{i\le m} v_i U^{(s)}_{j,i} z_s \Big) B_j.$$

As a consequence, w.r.t. the orthonormal basis $B_1, \dots, B_m$ for $\mathfrak{b}$, the endomorphism J_z is represented by the following matrix

$$\sum_{s=1}^{n} z_s \begin{pmatrix} U^{(s)}_{1,1} & \cdots & U^{(s)}_{1,m} \\ \vdots & \ddots & \vdots \\ U^{(s)}_{m,1} & \cdots & U^{(s)}_{m,m} \end{pmatrix} = \sum_{s=1}^{n} z_s U^{(s)}. \tag{18.10}$$

By Definition 18.1.1, this matrix has to be orthogonal, whenever $\sum_{s=1}^{n}(z_s)^2 = 1$. In particular, this implies that every $U^{(s)}$ is orthogonal (whence property 1) of the assertion is proved). We explicitly remark that since $U^{(s)}$ is both skew-symmetric and orthogonal, $U^{(s)}$ has no real eigenvalues, whence m is necessarily even. Moreover,

$$-U^{(s)}U^{(s)} = \mathbb{I}_m$$

($\mathbb{I}_m$ denotes the unit matrix of order m). If $\sum_{s=1}^{n}(z_s)^2 = 1$, the matrix $\sum_{s=1}^{n} z_s U^{(s)}$ is orthogonal if and only if

$$\begin{aligned} \mathbb{I}_m &= \left(\sum_{s=1}^{n} z_s U^{(s)}\right)\cdot\left(\sum_{s=1}^{n} z_s U^{(s)}\right)^T = -\sum_{r,s\le n} z_r z_s\, U^{(r)}U^{(s)} \\ &= -\sum_{r\le n} z_r^2\, U^{(r)}U^{(r)} - \sum_{r,s\le n,\, r\ne s} z_r z_s\, U^{(r)}U^{(s)} \\ &= \mathbb{I}_m - \sum_{r,s\le n,\, r\ne s} z_r z_s\, U^{(r)}U^{(s)}. \end{aligned}$$

We have here used the fact that $(U^{(r)})^T = -U^{(r)}$ and

$$-(U^{(r)})^2 = U^{(r)}\cdot(-U^{(r)}) = U^{(r)}\cdot(U^{(r)})^T = \mathbb{I}_m,$$

since $U^{(r)}$ is skew-symmetric and orthogonal. Therefore, we have

$$\sum_{r,s=1,\, r\ne s}^{n} z_r z_s\, U^{(r)}U^{(s)} = 0 \qquad \forall\, z_1,\dots,z_n:\ \sum_{s=1}^{n} z_s^2 = 1. \tag{18.11}$$

If in (18.11) we take $z = (0,\dots,1/\sqrt{2},\dots,1/\sqrt{2},\dots,0)$, we obtain

$$U^{(i)}U^{(j)} + U^{(j)}U^{(i)} = 0 \qquad \text{for every } i,\ j\in\{1,\dots,n\} \text{ with } i\ne j, \tag{18.12}$$

which is property 2) of the assertion. Since we also have

$$\sum_{r,s\le n,\, r\ne s} z_r z_s\, U^{(r)}U^{(s)} = \sum_{r,s\le n,\, r<s} z_r z_s\, (U^{(r)}U^{(s)} + U^{(s)}U^{(r)}),$$

(18.11) turns out to be equivalent to (18.12).

We now prove the "if" part of the theorem. Let $U^{(1)},\dots,U^{(n)}$ be matrices having properties 1)–2) of the assertion. Suppose $\mathbb{R}^{m+n}$ is endowed with the composition law (18.6). It is immediately verified that $\circ$ defines a Lie group, nilpotent of step at most two, in which the identity is the origin and the inverse of (x,t) is $(-x,-t)$. Moreover, $\delta_\lambda(x,t) = (\lambda x, \lambda^2 t)$ is a group of automorphisms. An easy computation shows that the vector field in the algebra $\mathfrak{g}$ of $\mathbb{G} = (\mathbb{R}^{m+n}, \circ)$ that agrees at the origin with $\partial/\partial x_j$ ($j = 1, \dots, m$) is given by

$$X_j = (\partial/\partial x_j) + \frac{1}{2}\sum_{s=1}^{n}\left(\sum_{i=1}^{m} U_{j,i}^{(s)}\, x_i\right)(\partial/\partial t_s), \tag{18.13}$$

and that $\mathfrak{g}$ is spanned by

$$X_1, \dots, X_m, \quad \partial/\partial t_1, \dots, \partial/\partial t_n.$$

From (18.13) and the skew-symmetry of $U^{(s)}$ we obtain

$$[X_i, X_j] = \sum_{s=1}^{n} U_{j,i}^{(s)}\,(\partial/\partial t_s)$$

for every $i,\ j \in \{1, \dots, m\}$. Now, since $U^{(1)}, \dots, U^{(n)}$ are linearly independent (see Remark 18.2.3), the dimension of the vector space spanned by $(U_{j,i}^{(1)}, \dots, U_{j,i}^{(n)})$ as $i, j \in \{1, \dots, m\}$ equals n. As a consequence, $\mathbb{G}$ is a homogeneous Carnot group.

For every $s = 1, \dots, n$, we set $Z_s = \partial/\partial t_s$. We claim that $\mathfrak{z}$, the center of $\mathfrak{g}$, is spanned by $Z_1, \dots, Z_n$. Indeed, we suppose by contradiction that (for suitable scalars α_i's) $\sum_{i=1}^{m} \alpha_i X_i \in \mathfrak{z}$, i.e.

$$0 = \left[\sum_{i=1}^{m} \alpha_i X_i, X_j\right] = \sum_{s=1}^{n}\sum_{i=1}^{m} \alpha_i U_{j,i}^{(s)}\, Z_s \quad \text{for every } j \in \{1, \dots, m\}.$$

Since the Z_s's are linearly independent, this means that

$$\sum_{i=1}^{m} \alpha_i U_{j,i}^{(s)} = 0$$

for every $s \le n$ and every $j \le m$, i.e. $\alpha = (\alpha_1, \dots, \alpha_m)$ belongs to the kernel of the transpose matrix of $U^{(s)}$ for every $s \le n$. Since every $U^{(s)}$ is orthogonal, this is possible only if $\alpha = 0$, which proves the claim.

Finally, *let* $\langle\, ,\rangle$ *be the standard inner product on* $\mathfrak{g}$ *w.r.t. the basis*

$$X_1, \dots, X_m, \;\; Z_1, \dots, Z_n.$$

For what has been proved above, we have $\mathfrak{z}^{\perp} = \operatorname{span}\{X_1, \dots, X_m\}$ and

$$[\mathfrak{z}^{\perp}, \mathfrak{z}^{\perp}] = \operatorname{span}\{Z_1, \dots, Z_n\} = \mathfrak{z}.$$

By means of a computation analogous to (18.9), it is easy to recognize that *properties* 1) *and* 2) *of the assertion ensure that, with the above choice of* $\langle\, ,\rangle$, $\mathbb{G}$ *is an H-type group according to Definition* 18.1.1. Indeed, for a fixed $z = \sum_{j=1}^{n} z_j Z_j$, setting

$$v := \sum_{i=1}^{m} v_i\, X_i, \qquad w := \sum_{i=1}^{m} w_i\, X_i,$$

we have

$$\begin{aligned}
\langle z, [v, w]\rangle &= \left\langle \sum_{i=1}^{n} z_i\, Z_i, \left[\sum_{i=1}^{m} v_i\, X_i, \sum_{j=1}^{m} w_j\, X_j\right]\right\rangle \\
&= \left\langle \sum_{i=1}^{n} z_i\, Z_i, \sum_{i,j=1}^{m} v_i\, w_j [X_i, X_j]\right\rangle \\
&= \left\langle \sum_{i=1}^{n} z_i\, Z_i, \sum_{i,j=1}^{m} v_i\, w_j \sum_{s=1}^{n} U_{j,i}^{(s)} Z_s\right\rangle \\
&= \sum_{j=1}^{m} w_j \left(\sum_{r,s=1}^{n} \sum_{i=1}^{m} v_i\, U_{j,i}^{(s)}\, z_r\, \langle Z_r, Z_s\rangle \right) \\
(\langle Z_r, Z_s\rangle = \delta_{r,s}) &= \sum_{j=1}^{m} w_j \left(\sum_{r=1}^{n} \sum_{i=1}^{m} v_i\, U_{j,i}^{(r)}\, z_r \right).
\end{aligned}$$

As a consequence, by definition of $\langle , \rangle$, setting

$$J_z(v) := \sum_{j=1}^{m} x_j\, X_j$$

with

$$x_j := \sum_{r=1}^{n} \sum_{i=1}^{m} v_i\, U_{j,i}^{(r)}\, z_r,$$

we have

$$\langle J_z(v), w\rangle = \langle z, [v, w]\rangle$$

for every $w = \sum_{i=1}^{m} w_i\, X_i$. Repeating exactly the same computations previously done, we see that if $U^{(1)}, \ldots, U^{(n)}$ are orthogonal matrices satisfying

$$U^{(r)} \cdot U^{(s)} + U^{(s)} \cdot U^{(r)} = 0$$

for every $r,\ s \in \{1, \ldots, n\}$ with $r \neq s$, then the map

$$\sum_{j=1}^{m} v_j\, X_j \mapsto J_z(v)$$

defines an orthogonal endomorphism on $\operatorname{span}\{X_1, \ldots, X_m\}$ for every choice of $z = \sum_{j=1}^{n} z_j\, Z_j$ such that $\sum_{j=1}^{n} |z_j|^2 = 1$. This completes the proof. □

From the explicitness of the operation (18.6) we can derive some more properties of H-type groups. We first give the explicit form of the canonical sub-Laplacian.

Proposition 18.2.4. *With the notation in the proof of Theorem* 18.2.1, *the canonical sub-Laplacian on the H-type group* $\mathbb{G}$ *is given by*

$$\Delta_{\mathbb{G}} = \Delta_x + \frac{1}{4}|x|^2 \Delta_t + \sum_{s=1}^{n} \langle U^{(s)} x, \nabla_x \rangle \frac{\partial}{\partial t_s}, \tag{18.14}$$

where the $U^{(s)}$ *are as in Theorem* 18.2.1. *Here we used the notation*

$$\Delta_x = \sum_{j=1}^{m} \left(\frac{\partial}{\partial x_j}\right)^2, \quad \Delta_t = \sum_{s=1}^{n} \left(\frac{\partial}{\partial t_s}\right)^2, \quad \nabla_x = \left(\frac{\partial}{\partial x_1}, \ldots, \frac{\partial}{\partial x_m}\right).$$

Moreover, on functions $u(x,t) = \widetilde{u}(|x|, t)$, $\Delta_{\mathbb{G}}$ *has the form*

$$\Delta_{\mathbb{G}} = \Delta_x + \frac{1}{4}|x|^2 \Delta_t = \left(\frac{\partial}{\partial r}\right)^2 + \frac{m-1}{r}\frac{\partial}{\partial r} + \frac{1}{4} r^2 \Delta_t, \quad r = |x| \neq 0. \tag{18.15}$$

Proof. Since the canonical sub-Laplacian is $\Delta_{\mathbb{G}} = \sum_{j=1}^{m} X_j^2$, where X_j is given by (18.13), one has

$$\begin{aligned}
\Delta_{\mathbb{G}} &= \Delta_x + \frac{1}{4}\sum_{j=1}^{m}\sum_{s=1}^{n}\left(\sum_{i=1}^{m} x_i U_{j,i}^{(s)}\right)^2 \left(\frac{\partial}{\partial t_s}\right)^2 + \frac{1}{2}\sum_{j=1}^{m}\sum_{s=1}^{n} U_{j,j}^{(s)} \frac{\partial}{\partial t_s} \\
&\quad + \frac{1}{4}\sum_{j,h,k=1}^{m}\ \sum_{r,s=1,\, r\neq s}^{n} x_h x_k\, U_{j,h}^{(r)} U_{j,k}^{(s)} \frac{\partial^2}{\partial t_r \partial t_s} \\
&\quad + \sum_{i,j=1}^{m}\sum_{s=1}^{n} x_i U_{j,i}^{(s)} \frac{\partial^2}{\partial x_j \partial t_s} \\
&= \Delta_x + \frac{1}{4}\sum_{s=1}^{n} |U^{(s)}x|^2 \left(\frac{\partial}{\partial t_s}\right)^2 + \sum_{i,j=1}^{m}\sum_{s=1}^{n} x_i U_{j,i}^{(s)} \frac{\partial^2}{\partial x_j \partial t_s} \\
&\quad + \frac{1}{2}\sum_{r,s=1,\, r<s}^{n} \langle U^{(r)}x, U^{(s)}x\rangle \frac{\partial^2}{\partial t_r \partial t_s} \\
&= \Delta_x + \frac{1}{4}|x|^2 \Delta_t + \sum_{i,j=1}^{m}\sum_{s=1}^{n} x_i U_{j,i}^{(s)} \frac{\partial^2}{\partial x_j \partial t_s} \\
&\quad + \frac{1}{2}\sum_{r,s=1,\, r<s}^{n} \langle U^{(r)}U^{(s)}x, x\rangle \frac{\partial^2}{\partial t_r \partial t_s} \\
&= \Delta_x + \frac{1}{4}|x|^2 \Delta_t + \sum_{i,j=1}^{m}\sum_{s=1}^{n} x_i U_{j,i}^{(s)} \frac{\partial^2}{\partial x_j \partial t_s}.
\end{aligned}$$

Here we used the following facts: $U_{j,j}^{(s)} = 0$ since $U^{(s)}$ is skew-symmetric; $|U^{(s)}x| = |x|$ since $U^{(s)}$ is orthogonal; from (18.12) we have

$$\langle U^{(r)}x, U^{(s)}x\rangle = \langle -U^{(s)}U^{(r)}x, x\rangle = \langle U^{(r)}U^{(s)}\,x, x\rangle$$

for every $r \neq s$. Again from (18.12) it follows that $U^{(r)}U^{(s)}$ is skew-symmetric (for $r \neq s$) since

$$(U^{(r)}U^{(s)})^T = (-U^{(s)})(-U^{(r)}) = U^{(s)}U^{(r)} = -U^{(r)}U^{(s)},$$

whence $\langle U^{(r)}U^{(s)}x, x\rangle = 0$ for every $r \neq s$, since for every skew-symmetric matrix A we have $\langle A\,x, x\rangle = 0$. This proves the first part of Proposition 18.2.4.

We now prove the second part. The third differential summand in the right-hand side of (18.14) vanishes on functions $u(|x|, t)$. Indeed, we have

$$\begin{aligned}&\left(\sum_{i,j=1}^{m}\sum_{s=1}^{n} x_i U_{j,i}^{(s)} \frac{\partial^2}{\partial x_j \partial t_s}\right)u(|x|, t)\\ &\quad = \frac{1}{r}\sum_{s=1}^{n}\left(\sum_{i,j=1}^{m} x_i x_j U_{j,i}^{(s)}\right)\left(\frac{\partial^2 u}{\partial r \partial t_s}\right)(|x|, t) = 0,\end{aligned}$$

since for every $s \leq n$, $\sum_{i,j=1}^{m} x_i x_j U_{j,i}^{(s)} = \langle U^{(s)}\,x, x\rangle = 0$, being $U^{(s)}$ skew-symmetric. □

Remark 18.2.5. Starting from (18.13), another direct computation shows

$$\begin{aligned}|\nabla_{\mathbb{G}} u|^2 &= |\nabla_x u|^2 + \frac{1}{4}\,|x|^2 \cdot |\nabla_t u|^2 + \sum_{s=1}^{n}\left(\sum_{i,j=1}^{m} x_i\, U_{j,i}^{(s)} \frac{\partial u}{\partial x_j}\right)\frac{\partial u}{\partial t_s}\\ &= |\nabla_x u|^2 + \frac{1}{4}\,|x|^2 \cdot |\nabla_t u|^2 + \sum_{s=1}^{n}\langle U^{(s)}x, \nabla_x u\rangle\frac{\partial u}{\partial t_s}.\end{aligned}$$

Remark 18.2.6. Let $(\mathbb{G}, \circ)$ be an H-type group. Suppose that the center of the algebra of $\mathbb{G}$ has dimension 1. Then $\mathbb{G}$ is isomorphic to a Heisenberg group $\mathbb{H}^k$ (here $k = \frac{1}{2}(\dim(\mathbb{G}) - 1)$).

Indeed, by Theorem 18.2.1 (and by the hypothesis $n = 1$) it is not restrictive to suppose that $\mathbb{G}$ is $\mathbb{R}^{m+1}$ equipped with the group law

$$(x, t) \circ (\xi, \tau) = \left(x + \xi, t + \tau + \frac{1}{2}\langle U^{(1)}\,x, \xi\rangle\right).$$

Let M be an $m \times m$ non-singular matrix and consider the following bijection

$$\widetilde{M} : \mathbb{G} \to \mathbb{R}^{m+1}, \qquad (x, t) \mapsto (M\,x, t).$$

Clearly, if $\widetilde{\mathbb{G}}$ is $\mathbb{R}^{m+1}$ equipped with the composition

$$(M\,x, t) * (M\,\xi, \tau) := \widetilde{M}((x, t) \circ (\xi, \tau)),$$

then $\widetilde{M} : (\mathbb{G}, \circ) \to (\widetilde{\mathbb{G}}, *)$ is a Lie group isomorphism. It is then sufficient to show that there exists M such that $(\widetilde{\mathbb{G}}, *)$ is isomorphic to a Heisenberg group. In order to prove it, we notice that

$$\begin{aligned}(\widetilde{x},\widetilde{t}) * (\widetilde{\xi},\widetilde{\tau}) &= \widetilde{M}\big((M^{-1}\widetilde{x},\widetilde{t}) \circ (M^{-1}\widetilde{\xi},\widetilde{\tau})\big)\\ &= \widetilde{M}\left(M^{-1}\widetilde{x} + M^{-1}\widetilde{\xi}, \widetilde{t} + \widetilde{\tau} + \frac{1}{2}\langle U^{(1)} M^{-1}\widetilde{x}, M^{-1}\widetilde{\xi}\rangle\right)\\ &= \left(\widetilde{x} + \widetilde{\xi}, \widetilde{t} + \widetilde{\tau} + \frac{1}{2}\langle M\, U^{(1)} M^{-1}\widetilde{x}, \widetilde{\xi}\rangle\right)\end{aligned}$$

if $(\widetilde{x},\widetilde{t})$, $(\widetilde{\xi},\widetilde{\tau}) \in \widetilde{\mathbb{G}}$. It is known that (see, e.g. [HJ85, Corollary 2.5.14]) every skew-symmetric orthogonal matrix is congruent to a block diagonal matrix of the type

$$J = \mathrm{diag}\left\{ \begin{pmatrix} 0 & -1 \\ 1 & 0 \end{pmatrix}, \dots, \begin{pmatrix} 0 & -1 \\ 1 & 0 \end{pmatrix} \right\}.$$

Hence, if P is a non-singular matrix such that $P^T\, U^{(1)}\, P = J$ and if we take $M = P^{-1}$, Remark 18.2.6 is proved. □

Remark 18.2.7. We end this section with the following natural question. Let $\mathbb{R}^{m+n}$ be equipped with a homogeneous Carnot group structure $\mathbb{G} = (\mathbb{R}^{m+n}, \delta_\lambda, \circ)$ by the usual dilations $\delta_\lambda(x,t) = (\lambda x, \lambda^2 t)$ and the usual operation

$$\begin{aligned}&(x,t)\circ(\xi,\tau) = \begin{pmatrix} x_j + \xi_j, \quad j = 1,\dots,m \\ t_j + \tau_j + \frac{1}{2}\langle B^{(j)}x, \xi\rangle, \quad j = 1,\dots,n \end{pmatrix},\\ &x,\xi \in \mathbb{R}^m, \ t,\tau \in \mathbb{R}^n,\end{aligned} \tag{18.16}$$

for suitable skew-symmetric linearly independent matrices $B^{(j)}$.

We aim to answer to the following question. *When is* $\mathbb{G}$ *an H-type group?* Obviously, we cannot say that this happens *if and only if*[2] the $B^{(j)}$'s satisfy conditions 1) and 2) of Theorem 18.2.1, for this is only the characterization of *prototype* H-type groups, but not the characterization of *all* H-type groups. (Consider the Heisenberg–Weyl group as a counterexample.)

The proof of Theorem 18.2.1 helps us finding the answer to this question. Indeed, let $\mathbb{G}$ be an H-type group. Suppose

$$\mathcal{B} = \{B_1,\dots,B_m, Z_1,\dots,Z_n\}$$

is any fixed basis for the algebra $\mathfrak{g}$ of $\mathbb{G}$, which is an orthonormal basis w.r.t. an inner product endowing $\mathfrak{g}$ with a structure of H-type algebra (see Definition 18.1.1, page 681). Moreover, identify $\mathbb{G}$ to $\mathfrak{g}$ (via the exponential map) and then identify $\mathfrak{g}$ to $\mathbb{R}^{m+n}$ (via coordinates w.r.t. $\mathcal{B}$). Denote by $\mathbb{H}$ this last group, which is obviously isomorphic to $\mathbb{G}$. We demonstrated in due course of the proof of Theorem 18.2.1 that $\mathbb{H}$ *is a prototype H-type group.*

[2] The "if" part is true, the "only if" is not.

Roughly speaking, if $\mathbb{G}$ is an H-type group, then a suitable choice of a new basis of $\mathfrak{g}$ turns $\mathbb{G}$ into a prototype H-type group. As we already studied the action of a change of basis in the algebra of a homogeneous Carnot group in Remark 2.2.20 (precisely, see page 153), we can infer that[3] there exist two non-singular matrices U (of order $m \times m$) and V (of order $n \times n$) such that the matrices

$$U^{(i)} := U^T \cdot \left(\sum_{j=1}^{n} w_{i,j}\, B^{(j)} \right) \cdot U,$$

where $V^{-1} = (w_{i,j})_{i,j \le n}$, fulfill the requirements 1)–2) of Theorem 18.2.1. We thus have proved the following result.

Corollary 18.2.8. *The homogeneous Carnot group $\mathbb{G} = (\mathbb{R}^{m+n}, \delta_\lambda, \circ)$ with $\circ$ as in* (18.16) *is an H-type group if and only if there exist two non-singular matrices U (of order $m \times m$) and V (of order $n \times n$) such that the matrices*

$$U^{(i)} := U^T \cdot \left(\sum_{j=1}^{n} w_{i,j}\, B^{(j)} \right) \cdot U,$$

where $V^{-1} = (w_{i,j})_{i,j \le n}$ fulfill the requirements 1)–2) *of Theorem* 18.2.1.

In particular, if the second layer of the stratification of $\mathbb{G}$ is one-dimensional (i.e. if $n = 1$), then $\mathbb{G}$ is an H-type group if and only if there exists $w \in \mathbb{R}$ such that the matrices $w\, B^{(j)}$'s are simultaneously congruent to a set of matrices $U^{(j)}$'s as in 1)–2) *of Theorem* 18.2.1.

Unfortunately, this result does not seem very operative in order to answer to the proposed question.

18.3 The Fundamental Solution for Sub-Laplacians on H-type Groups

Throughout this section, $(\mathbb{G}, \circ)$ will denote a fixed H-type group with the Lie algebra $\mathfrak{g}$. Moreover, we shall denote by $\langle\, ,\, \rangle$ the given inner product on $\mathfrak{g}$ as in Definition 18.1.1. It is not restrictive to suppose that, if $\mathfrak{z}$ is the center of $\mathfrak{g}$, $\mathfrak{b}$ is the orthogonal completion of $\mathfrak{z}$ and $N_1 := \dim(\mathfrak{b})$, $N_2 := \dim(\mathfrak{z})$, then $\mathbb{G}$ is a homogeneous Carnot group on $\mathbb{R}^{N_1+N_2}$ with the dilations

$$\delta_\lambda(x) = \delta_\lambda(x^{(1)}, x^{(2)}) = (\lambda x^{(1)}, \lambda^2 x^{(2)}), \qquad x^{(1)} \in \mathbb{R}^{N_1},\ x^{(2)} \in \mathbb{R}^{N_2}.$$

We finally set $N := N_1 + N_2$, and we observe that $Q = N_1 + 2N_2$ is the homogeneous dimension of $\mathbb{G}$. Under our hypotheses, we have $N_2 \ge 1$ and thus $Q \ge 4$.

[3] This fact is proved in details at the cited page 153.

We fix an orthonormal basis $X_1, \dots, X_{N_1}$ of $\mathfrak{b}$ and an orthonormal basis $Z_1, \dots, Z_{N_2}$ of $\mathfrak{z}$. Then $X_1, \dots, X_{N_1}, Z_1, \dots, Z_{N_2}$ is an orthonormal basis of $\mathfrak{g}$, and we have

$$v = \sum_{j=1}^{N_1} \langle v, X_j \rangle X_j, \quad |v|^2 = \sum_{j=1}^{N_1} \langle v, X_j \rangle^2 \qquad \forall\, v \in \mathfrak{b}, \tag{18.17a}$$

$$z = \sum_{j=1}^{N_2} \langle z, Z_j \rangle Z_j, \quad |z|^2 = \sum_{j=1}^{N_2} \langle z, Z_j \rangle^2 \qquad \forall\, z \in \mathfrak{z}, \tag{18.17b}$$

$$|v + z|^2 = |v|^2 + |z|^2 \qquad \forall\, v \in \mathfrak{b}, \quad \forall\, z \in \mathfrak{z}, \tag{18.17c}$$

$$\sum_{j=1}^{N_1} |X_j|^2 = N_1, \quad \sum_{j=1}^{N_2} |Z_j|^2 = N_2. \tag{18.17d}$$

The aim of this section is to find the fundamental solution for the sub-Laplacian

$$\mathcal{L} := \sum_{j=1}^{N_1} X_j^2.$$

To this end, we start by introducing the following functions on $\mathbb{G}$:

$$v : \mathbb{G} \to \mathfrak{b}, \qquad v(x) := \sum_{j=1}^{N_1} \langle \mathrm{Log}\,(x), X_j \rangle X_j,$$

$$z : \mathbb{G} \to \mathfrak{z}, \qquad z(x) := \sum_{j=1}^{N_2} \langle \mathrm{Log}\,(x), Z_j \rangle Z_j.$$

From the definition it immediately follows that v and z are characterized by the following property:

$$\forall\, x \in \mathbb{G}, \quad x = \mathrm{Exp}\,(v(x) + z(x)), \qquad v(x) \in \mathfrak{b}, \quad z(x) \in \mathfrak{z}. \tag{18.18}$$

We want to prove the following result, due to A. Kaplan [Kap80].

Theorem 18.3.1 (A. Kaplan [Kap80]). *Let $X_1, \dots, X_{N_1}$ be an orthonormal basis of $\mathfrak{b}$, and let $\mathcal{L}$ be the sub-Laplacian*

$$\mathcal{L} = \sum_{j=1}^{N_1} X_j^2.$$

Then, with the above notation, there exists a positive constant c such that the function

$$\Phi(x) := c \cdot \left\{ |v(x)|^4 + 16 |z(x)|^2 \right\}^{(2-Q)/4}$$

is the fundamental solution for $\mathcal{L}$.

Remark 18.3.2. We stress that Theorem 18.3.1 gives a fundamental solution for $\mathcal{L} = \sum_{j=1}^{N_1} X_j^2$ only when $X_1, \dots, X_{N_1}$ is an orthonormal basis of $\mathfrak{b}$ and not for any sub-Laplacian $\mathcal{L}$ on the H-type group $\mathbb{G}$.

We observe that

$$d(x) := (|v(x)|^4 + 16|z(x)|^2)^{1/4}$$

is a symmetric homogeneous norm on $\mathbb{G}$. Indeed, from the definition of $v(x)$ and $z(x)$ it immediately follows $d \in C^\infty(\mathbb{G} \setminus \{0\}) \cap C(\mathbb{G})$ and

$$v(\delta_\lambda(x)) = \sum_{j=1}^{N_1} \langle \mathrm{Log}\,(\delta_\lambda(x)), X_j\rangle X_j = \sum_{j=1}^{N_1} \langle \lambda \mathrm{Log}\,(x), X_j\rangle X_j = \lambda v(x),$$

$$z(\delta_\lambda(x)) = \sum_{j=1}^{N_2} \langle \mathrm{Log}\,(\delta_\lambda(x)), Z_j\rangle Z_j = \sum_{j=1}^{N_2} \langle \lambda^2 \mathrm{Log}\,(x), Z_j\rangle Z_j = \lambda^2 z(x).$$

Therefore, $d(\delta_\lambda(x)) = \lambda d(x)$ for every $\lambda > 0$. Arguing analogously, we prove that $v(x^{-1}) = -v(x)$ and $z(x^{-1}) = -z(x)$, which give $d(x^{-1}) = d(x)$.

Remark 18.3.3. When $\mathbb{G}$ is a prototype H-type group as in Theorem 18.2.1, from Theorem 18.3.1 we get the following more explicit formula for the fundamental solution Φ of the canonical sub-Laplacian $\Delta_{\mathbb{G}}$ in (18.14). Namely, with the notation of Theorem 18.2.1, the fundamental solution of the operator

$$\Delta_{\mathbb{G}} = \Delta_x + \frac{1}{4}\,|x|^2\,\Delta_t + \sum_{s=1}^{n} U^{(s)}\langle x, \nabla_x\rangle \frac{\partial}{\partial t_s},$$

is given by

$$\Phi(x) := c \cdot (|x|^4 + 16|t|^2)^{(2-Q)/4},$$

for some positive constant c.

Proof (of Remark 18.3.3). Let $X_1, \dots, X_m$ be defined as in (18.13), so that $\Delta_{\mathbb{G}} = \sum_{j=1}^{m} X_j^2$, and let $Z_i = \partial_{t_i}$, $i = 1, \dots, n$. Then $\{X_1, \dots, X_m\}$ is an orthonormal basis of $\mathfrak{b}$ and $\{X_1, \dots, X_m, Z_1, \dots, Z_n\}$ is an orthonormal basis of $\mathfrak{g}$ (see the proof of Theorem 18.2.1). By means of Theorem 18.3.1, we are only left to prove that

$$|v(x,t)| = |x|, \qquad |z(x,t)| = |t|. \tag{18.19}$$

To this end, we need to investigate the exponential map $\mathrm{Exp} : \mathfrak{g} \to \mathbb{G}$. For a fixed $(a, b) \in \mathbb{R}^{m+n}$, we have (by the definition of Exp)

$$\mathrm{Exp}\left(\sum_{j=1}^{m} a_j X_j + \sum_{i=1}^{n} b_i Z_i\right) = \gamma(1),$$

where γ solves

$$\begin{cases} \dot{\gamma}(s) = \sum_{j=1}^m a_j X_j(\gamma(s)) + \sum_{i=1}^n b_i Z_i(\gamma(s)), \\ \gamma(0) = 0. \end{cases}$$

Setting $\gamma(s) = (x(s), t(s))$, we get

$$\begin{cases} \dot{x}_j(s) = a_j, \\ \dot{t}_i(s) = b_i + \frac{1}{2}\langle U^{(i)} x(s), a\rangle, \\ x(0) = 0, \quad t(0) = 0, \end{cases}$$

so that $x_j(s) = s\, a_j$ and

$$\dot{t}_i(s) = b_i + \frac{s}{2}\langle U^{(i)} a, a\rangle = b_i,$$

since $U^{(i)}$ is skew-symmetric. Therefore, $t_i(s) = s\, b_i$, and we finally get

$$\operatorname{Exp}\left(\sum_{j=1}^m a_j X_j + \sum_{i=1}^n b_i Z_i\right) = \gamma(1) = (x(1), t(1)) = (a, b).$$

As a consequence, we get

$$\operatorname{Log} : \mathbb{G} \to \mathfrak{g}, \quad (x, t) \mapsto \sum_{j=1}^m x_j X_j + \sum_{i=1}^n t_i Z_i$$

and then

$$v(x, t) = \sum_{j=1}^m x_j X_j, \quad z(x, t) = \sum_{i=1}^n t_i Z_i$$

which finally give (18.19). □

Remark 18.3.4. We refer to Example 5.4.7 on page 250 for another, more direct proof of Remark 18.3.3 above.

Proof (of Theorem 18.3.1). We introduce a family of regular functions Φ_ε approximating Φ, and we compute $\mathcal{L}(\Phi_\varepsilon)$. If $\varepsilon > 0$ and $x \in \mathbb{G}$, we set

$$\Phi_\varepsilon(x) := c \cdot \left\{(|v(x)|^2 + \varepsilon^2)^2 + 16|z(x)|^2\right\}^{(2-Q)/4}.$$

It is clear (being Exp and Log analytic functions) that Φ_ε is an analytic function for every fixed $\varepsilon > 0$, and Φ is analytic on $\mathbb{G}\setminus\{0\}$. From the definition of the exponential map we have

$$\mathcal{L}(\Phi_\varepsilon)(x) = \sum_{j=1}^{N_1} X_j^2(\Phi_\varepsilon)(x) = \sum_{j=1}^{N_1} \left(\frac{\mathrm{d}}{\mathrm{d}t}\right)^2\Bigg|_{t=0} \Phi_\varepsilon(x \circ \operatorname{Exp}(tX_j)).$$

Let us fix $\varepsilon > 0$ and $x \in \mathbb{G}$. In order to compute $\mathcal{L}(\Phi_\varepsilon)(x)$, we need to compute the first and the second derivatives of

$$\phi_j(t) := \left(|v(x \circ \mathrm{Exp}\,(tX_j))|^2 + \varepsilon^2\right)^2 + 16|z(x \circ \mathrm{Exp}\,(tX_j))|^2, \qquad j = 1, \dots, N_1.$$

We observe that

$$\phi_j(0) = (|v(x)|^2 + \varepsilon^2)^2 + 16|z(x)|^2$$

does not depend on j; then we set $\phi(0) := \phi_j(0)$. In the above notation, setting also $k := (Q-2)/4$, we then have

$$\begin{aligned}
\mathcal{L}(\Phi_\varepsilon)(x) &= \sum_{j=1}^{N_1} \left(\frac{\mathrm{d}}{\mathrm{d}t}\right)^2\bigg|_{t=0} c\,(\phi_j(t))^{-k} = c \sum_{j=1}^{N_1} \frac{\mathrm{d}}{\mathrm{d}t}\bigg|_{t=0} \left(-k\phi_j(t)^{-k-1}\phi_j'(t)\right) \\
&= c\,k(k+1) \sum_{j=1}^{N_1} \phi_j(0)^{-k-2}(\phi_j'(0))^2 - c\,k \sum_{j=1}^{N_1} \phi_j(0)^{-k-1}\phi_j''(0) \\
&= c\,k\,\phi(0)^{-k-2}\left((k+1)\sum_{j=1}^{N_1}(\phi_j'(0))^2 - \phi(0)\sum_{j=1}^{N_1}\phi_j''(0)\right).
\end{aligned}$$

We now fix $j = 1, \dots, N_1$ and $t \in \mathbb{R}$. We look for simple expressions for

$$v(x \circ \mathrm{Exp}\,(tX_j)), \quad z(x \circ \mathrm{Exp}\,(tX_j)).$$

We recall that, since $\mathbb{G}$ is a step two nilpotent group, the following Campbell–Hausdorff formula holds:

$$\mathrm{Exp}\,(A) \circ \mathrm{Exp}\,(B) = \mathrm{Exp}\left(A + B + \frac{1}{2}[A, B]\right) \qquad \forall\, A,\, B \in \mathfrak{g}.$$

Thus we have

$$\begin{aligned}
&\mathrm{Exp}\left(v(x \circ \mathrm{Exp}\,(tX_j)) + z(x \circ \mathrm{Exp}\,(tX_j))\right) \quad \text{(by (18.18))} \\
&\quad = x \circ \mathrm{Exp}\,(tX_j) \\
&\quad = \mathrm{Exp}\,(\mathrm{Log}\,(x)) \circ \mathrm{Exp}\,(tX_j) \quad \text{(by Campbell–Hausdorff)} \\
&\quad = \mathrm{Exp}\left(\mathrm{Log}\,(x) + tX_j + \frac{t}{2}[\mathrm{Log}\,(x), X_j]\right) \quad \text{(by (18.18))} \\
&\quad = \mathrm{Exp}\left(v(x) + z(x) + tX_j + \frac{t}{2}[v(x) + z(x), X_j]\right) \quad (z(x) \in \mathfrak{z},\ \forall x \in \mathbb{G}) \\
&\quad = \mathrm{Exp}\left(v(x) + tX_j + z(x) + \frac{t}{2}[v(x), X_j]\right).
\end{aligned}$$

Comparing the first term of this equality to the last one and observing that

$$v(x) + tX_j \in \mathfrak{v}, \qquad z(x) + \frac{t}{2}[v(x), X_j] \in \mathfrak{z},$$

from (18.18) we get

$$v(x \circ \operatorname{Exp}(tX_j)) = v(x) + tX_j, \quad z(x \circ \operatorname{Exp}(tX_j)) = z(x) + \frac{t}{2}[v(x), X_j]. \quad (18.20a)$$

Hence we have

$$\begin{aligned}
\phi_j(t) &= \left(|v(x) + tX_j|^2 + \varepsilon^2\right)^2 + 16|z(x) + \frac{t}{2}[v(x), X_j]|^2 \\
&= \left(|v(x)|^2 + t^2 + 2t\,\langle v(x), X_j\rangle + \varepsilon^2\right)^2 \\
&\quad + 16\left(|z(x)|^2 + \frac{t^2}{4}|[v(x), X_j]|^2 + t\,\langle z(x), [v(x), X_j]\rangle\right).
\end{aligned}$$

This gives

$$\begin{aligned}
\phi_j'(t) &= 2\left(|v(x)|^2 + t^2 + 2t\,\langle v(x), X_j\rangle + \varepsilon^2\right) \cdot (2t + 2\,\langle v(x), X_j\rangle) \\
&\quad + 16\left(\frac{t}{2}|[v(x), X_j]|^2 + \langle z(x), [v(x), X_j]\rangle\right); \\
\phi_j''(0) &= 8\langle v(x), X_j\rangle^2 + 4(|v(x)|^2 + \varepsilon^2) + 8|[v(x), X_j]|^2; \\
\phi_j'(0) &= 4(|v(x)|^2 + \varepsilon^2) \cdot \langle v(x), X_j\rangle + 16\,\langle z(x), [v(x), X_j]\rangle \\
&= 4(|v(x)|^2 + \varepsilon^2) \cdot \langle v(x), X_j\rangle + 16\,\langle J_{z(x)}(v(x)), X_j\rangle \\
&= 4\langle(|v(x)|^2 + \varepsilon^2) \cdot v(x) + 4\,J_{z(x)}(v(x)), X_j\rangle.
\end{aligned}$$

In particular, we obtain

$$\begin{aligned}
\sum_{j=1}^{N_1} (\phi_j'(0))^2 &= 16 \sum_{j=1}^{N_1} \langle(|v(x)|^2 + \varepsilon^2) \cdot v(x) + 4\,J_{z(x)}(v(x)), X_j\rangle^2 \\
\text{(by (18.17a))} \quad &= 16\,|(|v(x)|^2 + \varepsilon^2) \cdot v(x) + 4\,J_{z(x)}(v(x))|^2 \\
\text{(by (18.4a))} \quad &= 16\,(|v(x)|^2 + \varepsilon^2)^2 \cdot |v(x)|^2 + 16^2\,|J_{z(x)}(v(x))|^2 \\
\text{(by (18.4c))} \quad &= 16\,|v(x)|^2\,\{(|v(x)|^2 + \varepsilon^2)^2 + 16\,|z(x)|^2\} \\
&= 16\,|v(x)|^2\,\phi(0); \\
\sum_{j=1}^{N_1} \phi_j''(0) &= 8 \sum_{j=1}^{N_1} \langle v(x), X_j\rangle^2 + 4\,N_1\,(|v(x)|^2 + \varepsilon^2) + 8 \sum_{j=1}^{N_1} |[v(x), X_j]|^2 \\
\text{(by (18.17a))} \quad &= 8\,|v(x)|^2 + 4\,N_1\,(|v(x)|^2 + \varepsilon^2) + 8 \sum_{j=1}^{N_1} |[v(x), X_j]|^2.
\end{aligned}$$

Now $[v(x), X_j] \in [\mathfrak{b}, \mathfrak{b}] = \mathfrak{z}$ and thus, from (18.17b) we have

$$\begin{aligned}
\sum_{j=1}^{N_1} \left|[v(x), X_j]\right|^2 &= \sum_{j=1}^{N_1} \sum_{i=1}^{N_2} \langle Z_i, [v(x), X_j]\rangle^2 = \sum_{j=1}^{N_1} \sum_{i=1}^{N_2} \langle J_{Z_i}(v(x)), X_j\rangle^2 \\
&= \sum_{i=1}^{N_2} \sum_{j=1}^{N_1} \langle J_{Z_i}(v(x)), X_j\rangle^2 = \sum_{i=1}^{N_2} |Z_i|^2\,|v(x)|^2 = N_2\,|v(x)|^2.
\end{aligned}$$

(In the fourth equality, we used (18.17a) and (18.4c).) As a consequence,

$$\sum_{j=1}^{N_1} \phi_j''(0) = 8\,|v(x)|^2 + 4\,N_1\,(|v(x)|^2 + \varepsilon^2) + 8\,N_2\,|v(x)|^2$$
$$= 16(k+1)\,|v(x)|^2 + 4\,N_1\,\varepsilon^2,$$

and then

$$\begin{aligned}\mathcal{L}(\Phi_\varepsilon)(x) &= c\,k\,\phi(0)^{-k-2}\big(16(k+1)\,|v(x)|^2\,\phi(0) \\ &\qquad - 16(k+1)\,|v(x)|^2\,\phi(0) - 4\,N_1\,\varepsilon^2\,\phi(0)\big) \\ &= -4\,N_1\,c\,k\,\varepsilon^2\,\phi(0)^{-k-1} \\ &= -4\,N_1\,c\,k\,\varepsilon^2\,\big((|v(x)|^2+\varepsilon^2)^2 + 16|z(x)|^2\big)^{(-2-Q)/4}. \qquad (18.20\text{b})\end{aligned}$$

For $\varepsilon = 0$, we clearly have

$$\mathcal{L}\Phi(x) = 0 \qquad \forall\, x \neq 0. \qquad (18.20\text{c})$$

Moreover, from $\Phi = c\,d^{2-Q}$ we immediately get

$$\Phi \in L^1_{\text{loc}}(\mathbb{G}). \qquad (18.20\text{d})$$

In order to prove Theorem 18.3.1, we are only left to show that $\mathcal{L}\Phi = -\text{Dirac}_0$ (Dirac_0 denotes the Dirac mass at the origin) in the sense of distributions. Let us first observe that, since v and z are δ_λ-homogeneous of degree one and two, respectively, we immediately get

$$\mathcal{L}(\Phi_\varepsilon)(\delta_\varepsilon(x)) = \varepsilon^{-Q}\mathcal{L}(\Phi_1)(x). \qquad (18.20\text{e})$$

We now observe that

$$\mathcal{L}(\Phi_1)(x) = -c\,(Q-2)\,N_1\,\big((|v(x)|^2+1)^2 + 16|z(x)|^2\big)^{(-2-Q)/4}$$

is a C^∞ function on $\mathbb{G}$ whose norm is bounded at infinity by a function δ_λ-homogeneous of degree $-2-Q$. Thus $\mathcal{L}(\Phi_1)(x)$ has finite integral on $\mathbb{G}$, and we can choose $c > 0$ such that

$$\int_{\mathbb{G}} \mathcal{L}(\Phi_1)(x)\,\mathrm{d}x = -1. \qquad (18.20\text{f})$$

Let us fix $f \in C_0^\infty(\mathbb{G})$. We want to prove that

$$\int_{\mathbb{G}} \Phi(x)\,\mathcal{L}f(x)\,\mathrm{d}x = -f(0). \qquad (18.20\text{g})$$

By dominated convergence, since $|\Phi_\varepsilon| \le \Phi \in L^1_{\text{loc}}$, we immediately get

$$\lim_{\varepsilon\to 0^+} \int_{\mathbb{G}} \Phi_\varepsilon(x)\,\mathcal{L}f(x)\,\mathrm{d}x = \int_{\mathbb{G}} \Phi(x)\,\mathcal{L}f(x)\,\mathrm{d}x.$$

On the other hand, since $\Phi_\varepsilon \in C^\infty(\mathbb{G})$, f has compact support and $\mathcal{L}$ is self-adjoint, we have

$$\begin{aligned}\int_{\mathbb{G}} \Phi_\varepsilon(x)\,\mathcal{L}f(x)\,\mathrm{d}x &= \int_{\mathbb{G}} f(x)\,\mathcal{L}\Phi_\varepsilon(x)\,\mathrm{d}x \\ &\quad \text{(by (18.20e) with the change of variable } x=\delta_\varepsilon(z)) \\ &= \int_{\mathbb{G}} f(\delta_\varepsilon z)\,\mathcal{L}\Phi_1(z)\,\mathrm{d}z \\ &\overset{\varepsilon\to 0^+}{\longrightarrow} \int_{\mathbb{G}} f(0)\,\mathcal{L}\Phi_1(z)\,\mathrm{d}z = -f(0) \quad \text{(by (18.20f))}.\end{aligned}$$

This proves (18.20g) and completes the proof of Theorem 18.3.1. □

Remark 18.3.5. In the above proof (see (18.20b)), we have incidentally proved that, for every $\varepsilon > 0$, the function

$$x \mapsto \big((|v(x)|^2+\varepsilon^2)^2+16|z(x)|^2\big)^{(2-Q)/4}$$

is, up to a multiplicative constant, a solution of the semilinear differential equation

$$-\mathcal{L}u = u^{(Q+2)/(Q-2)}.$$

18.4 H-type Groups of Iwasawa-type

We begin with the following definition.

Definition 18.4.1 (H-type Iwasawa group). *Let $\mathbb{H}$ be an H-type group, and let, as usual, $\mathfrak{h} = \mathfrak{b} \oplus \mathfrak{z}$ be the decomposition of its algebra, as in Definition* 18.1.1 *(page* 681*). For any $Z \in \mathfrak{z}$, let J_Z be the endomorphism of $\mathfrak{b}$ defined in* (18.1)*. Then $\mathbb{H}$ is called an* H-type Iwasawa group *(in the sequel, an* Iwasawa group*) if, for every $B \in \mathfrak{b}$ and for every $Z, Z' \in \mathfrak{z}$ with $Z \perp Z'$, there exists $Z'' \in \mathfrak{z}$ (depending on B, Z, Z') such that*

$$J_Z(J_{Z'}(B)) = J_{Z''}(B). \tag{18.21}$$

In literature, condition (18.21) is referred to as the "J^2-condition" (see, e.g. [CDKR91,CK84]).

Remark 18.4.2 ($\mathbb{H}^n$ is an Iwasawa group). Any Heisenberg–Weyl group $\mathbb{H}^k$ is an H-type group of Iwasawa-type. Indeed, we know that (see Remark 3.6.5, page 172) the classical Heisenberg–Weyl group $\mathbb{H}^k$ on $\mathbb{R}^{2k+1}$ is canonically isomorphic to the (prototype) H-type group $\mathbb{H}$ corresponding to the case $m = 2k$, $n = 1$ and

$$B^{(1)} = \begin{pmatrix} 0 & -\mathbb{I}_k \\ \mathbb{I}_k & 0 \end{pmatrix}.$$

In turn, $\mathbb{H}$ is obviously an Iwasawa group since its center $\mathfrak{z}$ is one-dimensional, and if $Z, Z' \in \mathfrak{z}$ are orthogonal, then at least one of them is the null vector, so that (for every $B \in \mathfrak{b}$) we have

$$J_Z(J_{Z'}(B)) = 0$$

and (18.21) is satisfied by choosing $Z'' = 0$. This proves that *any H-type group with one-dimensional center is an Iwasawa group.* $\square$

We now explicitly write the J^2-condition (when $\mathbb{H}$ is a prototype H-type group; see Theorem 18.2.1 and Remark 18.2.2 for all the details) in terms of the usual matrices $U^{(r)}$'s defining the composition law in $\mathbb{H}$. First, we fix orthonormal bases $\{B_1, \dots, B_m\}$, $\{Z_1, \dots, Z_n\}$ of $\mathfrak{b}$ and $\mathfrak{z}$, respectively, and we recall that, if $Z = \sum_{r=1}^n z_r Z_r \in \mathfrak{z}$, then we have

$$\begin{aligned} &J_Z : \mathfrak{b} \to \mathfrak{b}, \\ &J_Z\left(\sum_{j=1}^m v_j B_j\right) = \sum_{j=1}^m \left(\sum_{r=1}^n \sum_{i=1}^m v_i U_{j,i}^{(r)} z_r\right) B_j. \end{aligned} \tag{18.22}$$

More explicitly, J_Z is represented (w.r.t. the basis $\{B_1, \dots, B_m\}$ of $\mathfrak{b}$) by the $m \times m$ matrix

$$\sum_{r=1}^n z_r U^{(r)}.$$

Now, if we write

$$B = \sum_{j=1}^m v_j B_j, \quad Z = \sum_{r=1}^n z_r Z_r, \quad Z' = \sum_{r=1}^n z'_r Z_r, \quad Z'' = \sum_{r=1}^n z''_r Z_r,$$

then it is easy[4] to see that (18.21) writes as

$$\begin{aligned} &\left\{ \forall\, v \in \mathbb{R}^m, \quad \forall\, z, z' \in \mathbb{R}^n : \sum_{r=1}^n z_r z'_r = 0 \right\} \\ &\Longrightarrow \left\{ \exists\, z'' \in \mathbb{R}^n : \left(\sum_{1 \le s < r \le n} (z_s z'_r - z_r z'_s) U^{(s)} U^{(r)} \right) v = \left(\sum_{r=1}^n z''_r U^{(r)} \right) v \right\}. \end{aligned} \tag{18.23}$$

Example 18.4.3 (An H-type non-Iwasawa group). Let us consider the prototype H-type group $\mathbb{H}$ on $\mathbb{R}^6$ defined by taking $m = 4$, $n = 2$ and

$$U^{(1)} = \begin{pmatrix} 0 & 1 & 0 & 0 \\ -1 & 0 & 0 & 0 \\ 0 & 0 & 0 & 1 \\ 0 & 0 & -1 & 0 \end{pmatrix}, \quad U^{(2)} = \begin{pmatrix} 0 & 0 & -1 & 0 \\ 0 & 0 & 0 & 1 \\ 1 & 0 & 0 & 0 \\ 0 & -1 & 0 & 0 \end{pmatrix}.$$

[4] The reader is invited to verify (18.23) by making use of the well-known properties of the matrices $U^{(r)}$'s

$$(U^{(r)})^2 = -\mathbb{I}_m, \quad U^{(r)} U^{(s)} = -U^{(s)} U^{(r)} \ (r \neq s)$$

and (since $Z \perp Z'$) $\sum_{r=1}^n z_r z'_r = 0$.

More explicitly, $\mathbb{H}$ has the composition law

$$\begin{pmatrix} x_1 \\ x_2 \\ x_3 \\ x_4 \\ t_1 \\ t_2 \end{pmatrix} \circ \begin{pmatrix} \xi_1 \\ \xi_2 \\ \xi_3 \\ \xi_4 \\ \tau_1 \\ \tau_2 \end{pmatrix} = \begin{pmatrix} x_1 + \xi_1 \\ x_2 + \xi_2 \\ x_3 + \xi_3 \\ x_4 + \xi_4 \\ t_1 + \tau_1 + \frac{1}{2}(x_2\xi_1 - x_1\xi_2 + x_4\xi_3 - x_3\xi_4) \\ t_2 + \tau_2 + \frac{1}{2}(-x_3\xi_1 + x_4\xi_2 + x_1\xi_3 - x_2\xi_4) \end{pmatrix}.$$

Then it is easy to see that (18.23) has no solution $z'' \in \mathbb{R}^2$ if we choose

$$v = (0, 0, 1, 0) \in \mathbb{R}^4, \quad Z = (1, 0) \in \mathbb{R}^2, \quad Z' = (0, 1).$$

Hence $\mathbb{H}$ is an H-type group which is not an Iwasawa group. □

Note 18.4.4. If $\mathfrak{g}$ is a simple Lie algebra of rank one and $\mathfrak{n}$ is the Lie algebra in the so-called Iwasawa decomposition

$$\mathfrak{k} \oplus \mathfrak{a} \oplus \mathfrak{n}$$

of $\mathfrak{g}$ (also referred to as the KAN decomposition), then (equipped with a suitable inner product) $\mathfrak{n}$ is an H-type algebra satisfying the "J^2-condition" (18.21), and vice versa. More precisely, by [CDKR91, Theorem 1.1], the Iwasawa $\mathfrak{n}$-components of simple Lie groups of real rank one are *exactly* the H-type algebras which satisfy the "J^2-condition". This motivates the name *of Iwasawa-type*.

The importance of the Iwasawa-type groups appears naturally in dealing with Kelvin-type transforms, which we shall briefly treat in the next section (see [CDKR91]).

18.5 The Inversion and the Kelvin Transform on H-type Groups

In the classical theory of harmonic functions, the map

$$\mathbb{R}^N \setminus \{0\} \ni x \mapsto \sigma(x) := -x/|x|^2 \in \mathbb{R}^N \setminus \{0\}$$

is the so-called *inversion map* (with respect to the unit sphere centered at the origin). This map is used to define the *Kelvin-transform* of a function u (defined, for simplicity, on $\mathbb{R}^N$ with $N \geq 3$), namely

$$u^* : \mathbb{R}^N \setminus \{0\} \to \mathbb{R}, \quad u^*(x) := |x|^{2-N}\, u(\sigma(x)).$$

As it is well known (a simple but tedious computation is enough to the purpose), if u is a solution to the classical Laplace equation on $\mathbb{R}^N$, then the same is true of u^* on $\mathbb{R}^N \setminus \{0\}$.

Unfortunately (at present), a well-behaved analogue of the Kelvin transform for an arbitrary Carnot group is not known, except for the case of H-type groups of

Iwasawa-type. Indeed, in this section we explicitly write a suitable inversion map for a general H-type group: it turns out that the relevant Kelvin transform is well-behaved (namely, preserves the $\mathcal{L}$-harmonicity) *only* for H-type groups of Iwasawa-type (hence, in particular, for Heisenberg–Weyl groups).

We fix the notation: throughout the remaining of this section, $\mathbb{H} = (\mathbb{R}^{m+n}, \circ, \delta_\lambda)$ is a fixed prototype H-type group (any general H-type group is naturally isomorphic to a prototype one, as we proved in Theorem 18.2.1) with usual coordinates (x,t), $x = (x_1, \dots, x_m) \in \mathbb{R}^m$, $t = (t_1, \dots, t_n) \in \mathbb{R}^n$, and the composition map

$$(x,t)\circ(\xi,\tau) = \left(x+\xi, t_1+\tau_1+\frac{1}{2}\langle U^{(1)}x,\xi\rangle, \dots, t_n+\tau_n+\frac{1}{2}\langle U^{(n)}x,\xi\rangle\right)$$

where $U^{(1)}, \dots, U^{(n)}$ are fixed $m \times m$ matrices with the following properties:

(H1) $U^{(r)}$ is an $m \times m$ skew-symmetric and orthogonal matrix for every $r \le n$;
(H2) $U^{(r)}U^{(s)} = -U^{(s)}U^{(r)}$ for every $r, s \in \{1, \dots, n\}$ with $r \neq s$.

Moreover, the dilation group is given by

$$\delta_\lambda(x,t) = (\lambda\, x, \lambda^2 t).$$

As usual, $\Delta_{\mathbb{H}}$ denotes the canonical sub-Laplacian for $\mathbb{H}$. We recall that (see Theorem 18.3.1 and Remark 18.3.3) the fundamental solution for $\Delta_{\mathbb{H}}$ has the form $\Gamma = c\, d^{2-m-2n}$ for a suitable positive constant c and the following gauge on $\mathbb{H}$

$$d(x,t) = (|x|^4 + 16\,|t|^2)^{1/4}.$$

It is known that d defines an actual distance on $\mathbb{H}$, i.e. d satisfies the pseudo-triangle inequality (see Proposition 5.1.7, page 231) with the constant $\mathbf{c} = 1$ (see, e.g. [Cyg81]). We are now ready to explicitly write the inversion map and the Kelvin transform.

Definition 18.5.1 (H-inversion and H-Kelvin transform). *Let $\mathbb{H}$ be an (prototype) H-type group. Following all the above notation, we set*

$$\begin{aligned} &\sigma : \mathbb{H}\setminus\{0\} \to \mathbb{H}\setminus\{0\}, \\ &\sigma(x,t) := \left(-\frac{|x|^2 x - 4\sum_{k=1}^n t_k U^{(k)}x}{|x|^4+16|t|^2}, -\frac{t}{|x|^4+16|t|^2}\right). \end{aligned} \tag{18.24}$$

If σ is as above, for any function $u : \mathbb{H} \to \mathbb{R}$ we let

$$\begin{aligned} &u^* : \mathbb{H}\setminus\{0\} \to \mathbb{R}, \\ &u^*(x,t) := d(x,t)^{2-m-2n}\, u(\sigma(x,t)). \end{aligned} \tag{18.25}$$

We call σ the H-inversion map *on $\mathbb{H}$ and u^* the* H-Kelvin transform of u.

The remarkable rôle played by the H-Kelvin transform on Iwasawa-type groups is comparable to that of the classical Kelvin transform.

The definitions of the inversion map and the Kelvin transform in the abstract setting of an (general) H-type group (not necessarily written in the exponential coordinates) are slightly more complicated. Our (non-restrictive) choice of *prototype* coordinates makes it much simpler to give the very explicit definitions in (18.24) and (18.25).

We collect some easy properties of the H-inversion map in the following proposition.

Proposition 18.5.2. *Let σ be the H-inversion map defined in* (18.24). *We also write $\sigma(x,t) = (\sigma^{(1)}(x,t), \sigma^{(2)}(x,t))$ with the usual stratified splitting of the coordinates. Then, for every $(x,t) \in \mathbb{H} \setminus \{0\}$, we have:*

1) *if $|\cdot|$ denotes the Euclidean norm (in $\mathbb{R}^m$ or $\mathbb{R}^n$, accordingly), it holds*

$$|\sigma^{(1)}(x,t)| = \frac{|x|}{d^2(x,t)}, \qquad |\sigma^{(2)}(x,t)| = \frac{|t|}{d^4(x,t)}; \tag{18.26}$$

2) *for every $\lambda > 0$, we have*

$$\sigma(\delta_\lambda(x,t)) = \delta_{1/\lambda}\big(\sigma(x,t)\big);$$

3) *the inversion is involutive, i.e. $\sigma(\sigma(x,t)) = (x,t)$; in particular, σ is a bijection of $\mathbb{H} \setminus \{0\}$ onto itself;*

4) *it holds*

$$d(\sigma(x,t)) = \frac{1}{d(x,t)}, \tag{18.27}$$

whence σ maps the punctured d-ball

$$B_d(0,1) \setminus \{0\} = \{(x,t) \in \mathbb{H} : 0 < d(x,t) < 1\}$$

onto $\mathbb{H} \setminus \overline{B_d(0,1)}$, and vice versa, and maps the d-sphere $\partial B_d(0,1)$ onto itself.

Proof. Throughout the proof, (x,t) is a fixed point of $\mathbb{H} \setminus \{0\}$ and (for the sake of brevity) we write d instead of $d(x,t)$.

1) From the very definition of σ we have

$$\begin{aligned}
&|\sigma^{(1)}(x,t)| \\
&= \left| \frac{|x|^2 x - 4\sum_{k=1}^n t_k U^{(k)} x}{|x|^4 + 16|t|^2} \right| \\
&= d^{-4} \left| |x|^2 x - 4 \sum_{k=1}^n t_k U^{(k)} x \right| \\
&= d^{-4} \left\langle \left(|x|^2\, \mathbb{I}_m - 4 \sum_{k=1}^n t_k U^{(k)} \right) x, \left(|x|^2\, \mathbb{I}_m - 4 \sum_{k=1}^n t_k U^{(k)} \right) x \right\rangle^{1/2}
\end{aligned}$$

$$= d^{-4}\left\langle \left(|x|^2\,\mathbb{I}_m - 4\sum_{k=1}^n t_k U^{(k)}\right)^T \cdot \left(|x|^2\,\mathbb{I}_m - 4\sum_{k=1}^n t_k U^{(k)}\right)x, x\right\rangle^{1/2}$$
$$= d^{-4}\left\langle \left(|x|^2\,\mathbb{I}_m + 4\sum_{k=1}^n t_k U^{(k)}\right) \cdot \left(|x|^2\,\mathbb{I}_m - 4\sum_{k=1}^n t_k U^{(k)}\right)x, x\right\rangle^{1/2}$$
$$= d^{-4}\left\langle \left(|x|^4\,\mathbb{I}_m + 4\sum_{k=1}^n t_k U^{(k)} - 4\sum_{k=1}^n t_k U^{(k)}\right.\right.$$
$$\left.\left. - 16\sum_{j,k=1}^n t_j\, t_k\, U^{(k)} U^{(j)}\right)x, x\right\rangle^{1/2}$$
$$= d^{-4}\langle(|x|^4\,\mathbb{I}_m + 16\,|t|^2\,\mathbb{I}_m)x, x\rangle^{1/2} = d^{-4}\,d^2\,|x| = |x|d^{-2},$$

thus proving the first identity in (18.26). In the fifth equality, we used the skew-symmetry of the $U^{(r)}$'s; in the seventh equality, we used the following fact

$$\sum_{j,k=1}^n t_j\, t_k\, U^{(k)} U^{(j)}$$
$$= \left(\sum_{1\le j=k\le n} + \sum_{1\le j<k\le n} + \sum_{1\le k<j\le n}\right) t_j\, t_k\, U^{(k)} U^{(j)}$$
$$= \sum_{j=1}^n t_j^2 (U^{(j)})^2 + \sum_{1\le r<s\le n} t_r\, t_s\, U^{(s)} U^{(r)} + \sum_{1\le r<s\le n} t_s\, t_r\, U^{(r)} U^{(s)}$$
$$= -\sum_{j=1}^n t_j^2 \mathbb{I}_m + \sum_{1\le r<s\le n} t_r\, t_s\,(U^{(s)} U^{(r)} + U^{(r)} U^{(s)}) = -|t|^2\,\mathbb{I}_m, \qquad (18.28)$$

by the assumptions (H1)–(H2) on the matrices $U^{(r)}$'s defining an H-type group. As for the second identity in (18.26), we immediately have

$$|\sigma^{(2)}(x,t)| = \left|\frac{t}{|x|^4 + 16|t|^2}\right| = |t|\,d^{-4}.$$

This proves 1).

2) A simple computation.

3) From the results in (18.26), we infer

$$\sigma^{(1)}(\sigma(x,t)) = -\frac{|\sigma^{(1)}(x,t)|^2\,\sigma^{(1)}(x,t) - 4\sum_{k=1}^n \sigma^{(2)}(x,t)_k\, U^{(k)}\sigma^{(1)}(x,t)}{|\sigma^{(1)}(x,t)|^4 + 16|\sigma^{(2)}(x,t)|^2}$$
$$= -\frac{|x|^2\,d^{-4}\,\sigma^{(1)}(x,t) + 4\sum_{k=1}^n t_k\, d^{-4}\,U^{(k)}\sigma^{(1)}(x,t)}{|x|^4\,d^{-8} + 16|t|^2\,d^{-8}}$$
$$= -\left(|x|^2\sigma^{(1)}(x,t) + 4\sum_{k=1}^n t_k\, U^{(k)}\sigma^{(1)}(x,t)\right)$$

$$= d^{-4}\left(|x|^2\left(|x|^2 x - 4\sum_{j=1}^{n} t_j U^{(j)} x\right)\right.$$
$$\left. + 4\sum_{k=1}^{n} t_k\, U^{(k)}\left(|x|^2 x - 4\sum_{j=1}^{n} t_j U^{(j)} x\right)\right)$$
$$= d^{-4}\left(|x|^4 x - 16\left\{\sum_{j,k=1}^{n} t_j\, t_k\, U^{(k)} U^{(j)}\right\} x\right)$$
$$= d^{-4}\,(|x|^4 x + 16\,|t|^2 x) = d^{-4}\, d^4\, x = x.$$

To derive the sixth equality, we used the computations in (18.28). Moreover, it holds

$$\sigma^{(2)}(\sigma(x,t)) = -\frac{\sigma^{(2)}(x,t)}{|\sigma^{(1)}(x,t)|^4 + 16|\sigma^{(2)}(x,t)|^2}$$
$$= -\frac{-t\, d^{-4}}{|x|^4\, d^{-8} + 16|t|^2\, d^{-8}} = t.$$

This proves 3).

4) From (18.26) we have

$$d(\sigma(x,t)) = \left(|\sigma^{(1)}(x,t)|^4 + 16\,|\sigma^{(2)}(x,t)|^2\right)^{1/4}$$
$$= \left(|x|^4\, d^{-8}(x,t) + 16\,|t|^2\, d^{-8}(x,t)\right)^{1/4}$$
$$= d^{-2}(x,t)\,(|x|^4 + 16\,|t|^2)^{1/4} = d^{-2}(x,t)\, d(x,t) = d^{-1}(x,t),$$

which proves (18.27). The second part of what asserted in 4) immediately follows from (18.27) and the results in 3). This ends the proof. □

We end the section by stating the following result on the H-Kelvin transform in the setting of Iwasawa-type groups. We denote by $\Delta_{\mathbb{H}}$ the canonical sub-Laplacian of the H-type group $\mathbb{H}$ fixed at the beginning of the section.

Theorem 18.5.3 (Cowling et al., [CDKR91], Theorem 4.2). *The H-Kelvin transform of every $\Delta_{\mathbb{H}}$-harmonic function is always $\Delta_{\mathbb{H}}$-harmonic if and only if $\mathbb{H}$ is of Iwasawa-type. In this case, the following formula holds*

$$\Delta_{\mathbb{H}}(u^*(x,t)) = d^{-2-m-2n}(x,t)\,(\Delta_{\mathbb{H}} u)(\sigma(x,t)) \tag{18.29}$$

for every $u \in C^2(\mathbb{H} \setminus \{0\})$ and every $(x,t) \in \mathbb{H} \setminus \{0\}$.

Remark 18.5.4. We explicitly remark that formula (18.29) and the "if" part of Theorem 18.5.3 could also be proved by a direct computation, since both the canonical sub-Laplacian $\Delta_{\mathbb{H}}$ and the H-Kelvin transform are explicitly written for our prototype groups (see (18.14), page 692, for the expression of $\Delta_{\mathbb{H}}$).

Bibliographical Notes. H-type groups have been introduced by A. Kaplan in 1980 (see [Kap80]). The definition we have given in this chapter is not exactly the original one given by Kaplan, but it is the one that is usually adopted in the most recent literature. It is not difficult to prove the equivalence of the two definitions.

The inversion map and the Kelvin transform were first introduced on Heisenberg–Weyl groups by A. Korányi [Kor82] and then generalized to the H-type groups, see [CK84,CDKR91]. It is far from our scopes here to provide complete results and an exhaustive list of references from the existing literature on H-type groups and the related topics, which is nowadays still growing fast.

The result asserting that the H-Kelvin transform of a $\Delta_{\mathbb{H}}$-harmonic function is $\Delta_{\mathbb{H}}$-harmonic if and only if $\mathbb{H}$ is of Iwasawa-type (which is not proved in this chapter) is demonstrated in the paper by M. Cowling, A. H. Dooley, A. Korányi and F. Ricci [CDKR91].

Some of the topics presented in this chapter also appear in [BU04a].

18.6 Exercises of Chapter 18

Ex. 1) Prove that (18.21) writes as (18.23) when $\mathbb{H}$ is an (prototype) H-type group.

Ex. 2) Prove in details the assertions made in Example 18.4.3, page 703.

Ex. 3) Verify that the group introduced in Example 18.4.3 is indeed an H-type group and that an orthonormal basis for its algebra is the Jacobian basis

$$
\begin{aligned}
X_1 &= \partial_{x_1} + \frac{1}{2}(x_2 \partial_{t_1} - x_3 \partial_{t_2}),\\
X_2 &= \partial_{x_2} + \frac{1}{2}(-x_1 \partial_{t_1} + x_4 \partial_{t_2}),\\
X_3 &= \partial_{x_3} + \frac{1}{2}(x_4 \partial_{t_1} + x_1 \partial_{t_2}),\\
X_4 &= \partial_{x_4} + \frac{1}{2}(-x_3 \partial_{t_1} - x_2 \partial_{t_2}),\\
T_1 &= \partial_{t_1}, \qquad T_2 = \partial_{t_2}.
\end{aligned}
$$

Ex. 4) Verify that the H-inversion map σ for the group introduced in Example 18.4.3 is given by

$$
\sigma(x,t) = -(|x|^4 + 16\,|t|^2)^{-1} \begin{pmatrix} |x|^2\,x_1 + 4(-t_1\,x_2 + t_2\,x_3) \\ |x|^2\,x_2 + 4(t_1\,x_1 - t_2\,x_4) \\ |x|^2\,x_3 + 4(-t_1\,x_4 - t_2\,x_1) \\ |x|^2\,x_4 + 4(t_1\,x_3 + t_2\,x_2) \\ t_1 \\ t_2 \end{pmatrix}.
$$

Ex. 5) Write the explicit expression of the H-inversion map σ for the Heisenberg–Weyl group $\mathbb{H}^1$.

Ex. 6) Prove formula (18.29) for the Heisenberg–Weyl group $\mathbb{H}^1$ (using the calculus formulas given at the end of Ex. 6, Chapter 1, page 77).

Ex. 7) Let $X := \{X_1, \dots, X_m\}$ be a given set of smooth vector fields on $\mathbb{R}^N$. If $\Omega \subseteq \mathbb{G}$ is an open set with smooth boundary, we recall that the *characteristic set* (*related to* X) of Ω is the set

$$X\text{-Char}(\Omega) := \{x \in \partial\Omega \mid X_i(x) \in \mathrm{T}_x(\partial\Omega),\ i = 1, \dots, m\},$$

$\mathrm{T}_x(\partial\Omega)$ being the tangent space to $\partial\Omega$ at the point x.
Provide the details for the following result.
Let $\mathbb{G}$ *be an H-type group. Choose coordinates on* $\mathbb{G}$ *as in Theorem* 18.2.1. *Let* $\Delta_{\mathbb{G}} = \sum_{j=1}^m X_j^2$ *be the canonical sub-Laplacian of* $\mathbb{G}$, *as in Proposition* 18.2.4. *Finally, let* $X = \{X_1, \dots, X_m\}$. *Consider the general half-space* Π *of* $\mathbb{G}$

$$\Pi = \{(x, t) \in \mathbb{G} \mid \langle a, x\rangle + \langle b, t\rangle > c\},$$

where $a \in \mathbb{R}^m$, $b \in \mathbb{R}^n$ *and* $c \in \mathbb{R}$ *are fixed. Then* Π *possesses* X-*characteristic points if and only if* $b \neq 0$. *Moreover, any half-space with characteristic points can be left-translated into a half-space of the type*

$$\{(x, t) \in \mathbb{G} \mid \langle b, t\rangle > 0\}.$$

Indeed, a point $(x, t) \in \partial\Pi$ is characteristic if and only if (why? use also (18.13))

$$a_j + \frac{1}{2}\sum_{s=1}^n \left(\sum_{i=1}^m x_i\, U_{i,j}^{(s)}\right) b_s = 0 \text{ for every } j \leq m,$$

or equivalently,

$$\text{(Ch)} \qquad \begin{cases} \langle a, x\rangle + \langle b, t\rangle = c, \\ \left(\displaystyle\sum_{s=1}^n b_s U^{(s)}\right)^T x = -2a. \end{cases}$$

If $b = 0$, then $a \neq 0$, and the second equation in (Ch) has clearly no solution (why?). If $b \neq 0$, then (why?) the matrix $(\sum_{s=1}^n b_s/|b| U^{(s)})^T$ is non-singular. As a consequence, the second equation in (Ch) admits a solution $\overline{x} \in \mathbb{R}^m$. The characteristic set for Π is then given by

$$\{(x, t) \in \mathbb{G} \mid x = \overline{x},\ \ \langle b, t\rangle = c - \langle a, \overline{x}\rangle\}.$$

Finally, let $b \neq 0$. If $(\xi, \tau) \in \mathbb{G}$ is fixed, we set $\Pi = (\xi, \tau) \circ \widetilde{\Pi}$, i.e. $\widetilde{\Pi}$ is the set of all points $(\widetilde{x}, \widetilde{t}) \in \mathbb{G}$ such that (why?)

$$\sum_{j=1}^n b_j(\widetilde{t}_j + \tau_j) + \sum_{j=1}^m a_j \xi_j + \sum_{j=1}^m \widetilde{x}_j \left(a_j + \frac{1}{2}\sum_{s=1}^n b_s \sum_{i=1}^m \xi_i U_{i,j}^{(s)}\right) > c.$$

There exists (why?) $\overline{\xi} \in \mathbb{R}^m$ such that

$$a_j + \frac{1}{2}\sum_{s=1}^{n} b_s \sum_{i=1}^{m} \overline{\xi}_i U_{i,j}^{(s)} = 0$$

for every $j \le m$. Then, there exists (why?) $\overline{\tau} \in \mathbb{R}^n$ such that

$$\sum_{j=1}^{n} b_j \overline{\tau_j} = c - \sum_{j=1}^{m} a_j \overline{\xi}_j.$$

If $\widetilde{\Pi} = (\overline{\xi}, \overline{\tau})^{-1} \circ \Pi$, we have $\widetilde{\Pi} = \{(\widetilde{x}, \widetilde{t}) \in \mathbb{G} \,|\, \langle b, \widetilde{t}\rangle > 0\}$.

Ex. 8) Let $(\mathbb{G}, \circ)$ be a H-type group. Fix $z_0 \in \mathbb{G}$. We denote by Z^{z_0} the following vector field on $\mathbb{G}$

$$Z^{z_0} I(z) = \frac{\mathrm{d}}{\mathrm{d}h}\bigg|_{h=0} ((hz_0) \circ z).$$

Prove that, if $z_0 = (x^0, t^0)$, then the above vector field $Z^{z_0} I(x, t)$ is given by (we follow our usual notation on $\circ$)

$$Z^{z_0} I(x, t) = \begin{pmatrix} x_j^0, \quad j = 1, \dots, m \\ t_j^0 + \frac{1}{2}\langle x^0, U^{(j)} x\rangle, \quad j = 1, \dots, n \end{pmatrix}.$$

Ex. 9) Denote by $\xi = (z, t) = (x + iy, t) \equiv (x, y, t)$ the points of the Heisenberg group $\mathbb{H}^n = \mathbb{C}^n \times \mathbb{R} \equiv \mathbb{R}^{2n+1}$. The group law on $\mathbb{H}^n$ is given by

$$\xi \circ \xi' = (z + z', t + t' + 2\,\mathrm{Im}(z\bar{z}')).$$

Define

$$F : \mathbb{H}^n \to \mathbb{S}^{2n+1} \setminus \{(0, \dots, 0, -1)\},$$
$$F(\xi) = F(z, t) = \left(\frac{2iz}{t + i(1 + |z|^2)}, \frac{-t + i(1 - |z|^2)}{t + i(1 + |z|^2)}\right).$$

Here $\mathbb{S}^{2n+1}$ denotes the unit sphere of $\mathbb{C}^{n+1}$. The map F is a CR equivalence between the Heisenberg group and the sphere minus the south pole, which is the analogue in CR geometry of the stereographic projection. Define now

$$\mathcal{K} : \mathbb{H}^n \setminus \{0\} \to \mathbb{H}^n \setminus \{0\}$$

to be the map conjugate to the inversion in $\mathbb{C}^{n+1}$ of $\mathbb{S}^{2n+1}$ through the equivalence F, i.e.

$$\mathcal{K} := F \circ (-\mathrm{Id}_{\mathbb{C}^{n+1}}) \circ F^{-1}.$$

Prove that $\mathcal{K}$ has the following expression

$$\mathcal{K}(z, t) = \left(\frac{-iz}{t + i|z|^2}, -\frac{t}{d^4}\right), \qquad \text{where } d := (|z|^4 + t^2)^{1/4}.$$

Compare with the H-inversion map defined in (18.24). Finally compute the Jacobian determinant of $\mathcal{K}$ and recognize that

$$\det(\mathcal{J}_{\mathcal{K}}(z,t)) = \frac{1}{d^{2Q}},$$

where $Q = 2n + 2$ is the homogeneous dimension of $\mathbb{H}^n$.

Ex. 10) Let us consider the Heisenberg group $\mathbb{H}^n$ with the usual notation. An application $\rho : \mathbb{H}^n \to \mathbb{H}^n$ is called a *rotation around the t-axis* if there exists a complex unitary transformation $F : \mathbb{C}^n \to \mathbb{C}^n$ such that

$$\rho(z,t) = (F(z), t).$$

Let Ω be a half-space of $\mathbb{H}^n$, and let us define

$$\Omega_1 = \{(x, y, t) \in \mathbb{R}^{2n+1} \mid x_1 > 0\}$$

and

$$\Omega_t = \{(x, y, t) \in \mathbb{R}^{2n+1} \mid t > 0\}.$$

Prove that one of the two following cases always occurs:
(i) there exists $\xi_0 \in \mathbb{H}^n$ such that either $\Omega = \tau_{\xi_0}(\Omega_t)$ or $\Omega = \tau_{\xi_0}(-\Omega_t)$,
(ii) there exist $\xi_0 \in \mathbb{H}^n$ and a rotation ρ around the t-axis such that $\Omega = \rho(\tau_{\xi_0}(\Omega_1))$.

Prove, in particular, that the first case occurs when $\partial\Omega$ and the t-axis intersect, the second one when they are parallel. Finally prove that $\Delta_{\mathbb{H}^n}$ is invariant with respect to rotations around the t-axis, i.e.

$$\Delta_{\mathbb{H}^n}(u \circ \rho) = (\Delta_{\mathbb{H}^n} u) \circ \rho$$

for every rotation ρ around the t-axis.

Ex. 11) Let us consider the Heisenberg group $\mathbb{H}^n$ with the usual notation, and let $\Omega = \{\xi = (x, y, t) \in \mathbb{R}^{2n+1} \mid x_1 > 0\}$. For every $\xi \in \Omega$, set

$$r(\xi) = \frac{x_1}{2}, \quad B_\xi = B_d(\xi, r(\xi)),$$

$$K : \Omega \times \Omega \to \mathbb{R}, \quad K(\xi, \xi') = m_d \, \frac{\psi(\xi, \xi')}{r(\xi)^Q} \, \chi_{B_\xi}(\xi'),$$

where χ_{B_ξ} is the characteristic function of B_ξ and

$$\psi(\xi, \xi') = \frac{|z - z'|^2}{d(\xi, \xi')^2}.$$

For $u \in L^1_{\mathrm{loc}}(\Omega)$, define

$$Tu : \Omega \to \mathbb{R}, \quad (Tu)(\xi) = (M_{r(\xi)} u)(\xi) = \int_\Omega K(\xi, \xi')\, u(\xi')\, d\xi'.$$

(Compare to (5.50f) on page 259 where m_d was also introduced.)

a) Observe that $Tu = u$ for any $\Delta_{\mathbb{H}^n}$-harmonic function u and deduce that

$$\int_\Omega K(\xi, \xi')\, d\xi' = 1 \quad \forall\, \xi \in \Omega.$$

b) Prove that the integration of $K(\xi, \xi')$ with respect to the variable ξ also yields a constant value, i.e. there exists a constant β such that

$$\int_\Omega K(\xi, \xi_0)\, d\xi = \beta$$

for every $\xi_0 \in \Omega$.

c) Recognize now that T is a bounded linear operator in $L^p(\Omega)$ for every $p \in [1, +\infty]$. More precisely,

$$\|Tu\|_\infty \le \|u\|_\infty \quad \forall\, u \in L^\infty(\Omega),$$
$$\|Tu\|_p^p \le \beta\, \|u\|_p^p \quad \forall\, u \in L^p(\Omega),\ \forall p \in [1, +\infty[.$$

Moreover,

$$\int\limits_\Omega Tu = \beta \int\limits_\Omega u \quad \forall\, u \in L^1(\Omega).$$

d) Prove that $\beta > 1$.

e) Prove the following Liouville type theorem. *If u is a non-negative $\Delta_{\mathbb{H}^n}$-superharmonic function such that*

$$u \in L^1(\Omega),$$

then $u \equiv 0$ in Ω.

f) Using Ex. 10, prove that the Liouville-type theorem stated in (e) holds for any half-space Ω of $\mathbb{H}^n$ whose boundary does not intersect the center of $\mathbb{H}^n$.

(*Hint*: In order to prove (b), use the invariance of Ω (and of r) under the left translations "parallel" to the boundary. To prove (d), test the solid mean value formula (5.51), page 259, with a suitable $\Delta_{\mathbb{H}^n}$-subharmonic function v (for instance, the function d^{-Q}, suitably left translated) in order to obtain that, chosen a compact subset K of Ω with $|K| > 0$, there exists $\varepsilon > 0$ such that $Tv \ge v + \varepsilon$ in K. Use then (c) and deduce (d). To prove (e), observe that $u \ge Tu$ gives

$$\int_\Omega u \ge \int_\Omega Tu = \beta \int_\Omega u,$$

and then use (d). All the details of this exercise can be found in [Ugu99].)

Ex. 12) a) Prove that the Liouville-type theorem stated in Ex. 11 no longer holds on every $\mathbb{H}^n$ if we replace the hypothesis $u \in L^1(\Omega)$ with $u \in L^p(\Omega)$.

b) Prove that the function

$$-\partial_t (d(\xi)^{2-Q}) = n\, t\, (|z|^4 + t^2)^{-1-\frac{n}{2}}$$

is $\Delta_{\mathbb{H}^n}$-harmonic away from the origin, and in the half-space $\{t > 1\}$ it is positive and belongs to L^p for all $p > 1$ and also to the weak L^1.

(*Hint*: To prove (a), take $p \in]1, +\infty]$, $n > \frac{1}{p-1}$ and consider the fundamental solution of $-\Delta_{\mathbb{H}^n}$ with pole outside Ω.)

19

The Carathéodory–Chow–Rashevsky Theorem

The aim of this chapter is to furnish the proof of the so-called Carathéodory–Chow–Rashevsky theorem, i.e. the result asserting the connectedness of $\mathbb{R}^N$ with respect to a family X of Hörmander vector fields. We shall restrict to the particular but significant case when the given vector fields are the generators of the first layer of a homogeneous Carnot group $\mathbb{G}$.

To this aim, we shall make use of two crucial tools. First, it will intervene the Carnot–Carathéodory distance related to X (already introduced in Section 5.2, page 232). Second, we shall exploit the deep properties of the Campbell–Hausdorff operation $\diamond$ recalled in Section 14.2.1 (page 584). In particular, we shall make use of a result asserting that, roughly speaking, we can approximate any nested commutator

$$[X_q, \cdots [X_2, X_1] \cdots]$$

of the vector fields X_j's in the algebra of $\mathbb{G}$ by means of a $\diamond$-composition of the vector fields $\pm X_1, \ldots, \pm X_q$ (up to commutators of length $\geq q+1$).

As an application of the connectivity theorem, we shall prove that any Carnot group $\mathbb{G}$ admits the decomposition

$$\mathbb{G} = \operatorname{Exp}(\mathfrak{g}_1) * \cdots * \operatorname{Exp}(\mathfrak{g}_1)$$

into a precise number of factors which are the images under the exponential map of the first layer $\mathfrak{g}_1$ of the stratification of $\mathfrak{g}$ (see Theorem 19.2.1). We shall make a crucial use of this result again in Chapter 20, where a suitable version of Lagrange mean value theorem in the stratified setting will be provided.

19.1 The Carathéodory–Chow–Rashevsky Theorem for Stratified Vector Fields

In this section, we give a proof of the Carathéodory–Chow–Rashevsky connectivity theorem in the case of stratified vector fields. First of all, we recall the definitions of X-subunit path and of X-connectedness, given in Section 5.2, page 232.

Definition 19.1.1 (X-subunit path). *Let $X = \{X_1, \dots, X_m\}$ be any family of vector fields in $\mathbb{R}^N$. A piecewise regular path $\gamma : [0, T] \to \mathbb{R}^N$ is said to be* X-subunit *if*

$$\langle \dot\gamma(t), \xi\rangle^2 \le \sum_{j=1}^{m} \langle X_j I(\gamma(t)), \xi\rangle^2 \quad \forall \xi \in \mathbb{R}^N,$$

for almost every $t \in [0, T]$.

We denote by $\mathcal{S}(X)$ the set of all X-subunit paths, and we put

$$l(\gamma) = T$$

if $[0, T]$ is the domain of $\gamma \in \mathcal{S}(X)$.

We explicitly remark that every integral curve of $\pm X_j$, $j \in \{1, \dots, m\}$ is X-subunit.

Definition 19.1.2 (X-connectedness. Carnot–Carathéodory distance). *We say that $\mathbb{R}^N$ is* X-connected *if and only if, for every $x, y \in \mathbb{R}^N$, there exists*

$$\gamma \in \mathcal{S}(X), \quad \gamma : [0, T] \to \mathbb{R}^N \quad \textit{such that } \gamma(0) = x \textit{ and } \gamma(T) = y.$$

If $\mathbb{R}^N$ is X-connected, the following definition makes sense

$$d_X(x, y) := \inf\left\{T > 0 \mid \exists\, \gamma : [0, T] \to \mathbb{R}^N,\ \gamma \in \mathcal{S}(X),\ \gamma(0) = x,\ \gamma(T) = y\right\}$$

for every $x, y \in \mathbb{R}^N$. Moreover, the function $(x, y) \mapsto d_X(x, y)$ is a metric[1] *on $\mathbb{R}^N$, called the* X-control distance *or the* Carnot–Carathéodory distance *related to X.*

In what follows, we shall simply write d instead of d_X.

We now specialize to the case of stratified vector fields. From now on, throughout this section, we always denote by $\mathbb{G} = (\mathbb{R}^N, \circ, \delta_\lambda)$ a fixed homogeneous Carnot group with the Lie algebra $\mathfrak{g}$. The dilations $\{\delta_\lambda\}_{\lambda>0}$ of $\mathbb{G}$ are given by

$$\delta_\lambda(x) = \delta_\lambda(x^{(1)}, \dots, x^{(r)}) = (\lambda x^{(1)}, \dots, \lambda^r x^{(r)}), \quad x^{(i)} \in \mathbb{R}^{N_i}, \quad 1 \le i \le r.$$

We shall denote by $m := N_1$ the number of generators of $\mathbb{G}$ and by

$$Q = N_1 + 2\,N_2 + \dots + r\,N_r$$

the homogeneous dimension of $\mathbb{G}$.

The following theorem then holds.

Theorem 19.1.3 (Carathéodory–Chow–Rashevsky theorem). *Let the preceding hypotheses and notation hold. Let us suppose that $Z_1, \dots, Z_m$ are vector fields in $\mathfrak{g}$ such that*

$$\mathfrak{g} = \mathrm{Lie}\{Z_1, \dots, Z_m\}$$

[1] See Proposition 5.2.3, page 232.

(for example, we can choose $Z_1, \dots, Z_m$ to be the first m vector fields of the Jacobian basis for $\mathfrak{g}$, i.e. the left invariant vector fields on $\mathbb{G}$ such that $Z_j(0) = (\partial/\partial x_j)|_0$, $j = 1, \dots, m$). Then, setting $Z = \{Z_1, \dots, Z_m\}$, we have that

$$\mathbb{G} \text{ is } Z\text{-connected.}$$

Moreover, the Z-connectedness is given by paths that are piecewise integral curves of the vector fields $Z_1, \dots, Z_m$. In particular, the Z-control distance d is well-defined. Finally, setting

$$d_0 := d(\cdot, 0),$$

d_0 turns out to be a homogeneous norm[2] *on $\mathbb{G}$.*

(*Note.* We explicitly remark that X-connectivity also holds when X is a system of vector fields in $\mathbb{R}^N$ satisfying Hörmander's rank condition. See the Bibliographical Notes at the end of this chapter for references.)

A main step in the proof of Theorem 19.1.3 is Lemma 19.1.4 below which, roughly speaking, states that we can approximate any nested commutator

$$[X_q, \cdots [X_2, X_1] \cdots]$$

of vector fields X_j's in $\mathfrak{g}$ by means of a composition of the vector fields $\pm X_1, \dots,$ $\pm X_q$ (up to commutators of length $\geq q+1$). The composition is meant in the sense of iterated application of the Campbell–Hausdorff operation $\diamond$ (see Definition 2.2.11 on page 128; see also Section 14.2.1, page 584). More precisely, we have

$$[X_q, \cdots [X_2, X_1] \cdots] = X_{j_1} \diamond \cdots \diamond X_{j_{c(q)}} + R(X_1, \dots, X_q),$$

where $c(q)$ is an integer depending only on q (namely, $c(q) = 3 \cdot 2^{q-1} - 2$) *and not depending on the X_j's*, and

$$X_{j_1}, \dots, X_{j_{c(q)}} \in \{\pm X_1, \dots, \pm X_q\}$$

can be chosen in a "universal" way, again *not depending on the X_j's*. Finally, the remainder $R(X_1, \dots, X_q)$ is a linear combination of brackets of height $\geq q+1$ of the vector fields $X_1, \dots, X_q$, and the form of R can, once more, be chosen *independently of the X_j's*.

In order to be more precise, we need to introduce some more notation and definitions from algebra. We leave the details to the interested reader. (See also Definition 15.2.1, page 600, and Section 1.1, page 3.)

Let $q \in \mathbb{N}$, let $x_1, \dots, x_q$ be non-commuting letters and let us denote by $\mathfrak{a}$ the set of the formal words in the letters $x_1, \dots, x_q$. For example, an element of $\mathfrak{a}$ is $x_1^3 x_2 x_3 x_1^4 x_3$. Let also $\mathfrak{a}$ be equipped with its natural structure of associative algebra[3] $(\mathfrak{a}, \cdot)$ over $\mathbb{R}$ and then of a Lie algebra $(\mathfrak{a}, [\cdot, \cdot])$ by setting $[x, y] := xy - yx$ for every $x, y \in \mathfrak{a}$.

---

[2] See Section 5.1, page 229, for the definition of homogeneous norm on $\mathbb{G}$.

[3] For example, $x_1 x_3^2 x_4 \cdot x_4^3 x_1 x_7 = x_1 x_3^2 x_4^4 x_1 x_7$ and $x_1 x_3^2 \cdot x_7^4 = x_1 x_3^2 x_7^4$, etc.

For every $k \in \mathbb{N}$ and any given multi-index $J = (j_1, \dots, j_k) \in \{1, \dots, q\}^k$, we set

$$x_J := [x_{j_1}, \cdots [x_{j_{k-1}}, x_{j_k}] \cdots].$$

We say that x_J is a commutator of length (or height) k of $x_1, \dots, x_q$. If $J = j_1$, we also say that $x_J := x_{j_1}$ is a commutator of length 1 of $x_1, \dots, x_q$.

For any $q, k \in \mathbb{N}$, we shall use the notation

$$\mathcal{P}_k[x_1, \dots, x_q] := \operatorname{span}\big\{x_J = [x_{j_1}, \dots [x_{j_{k-1}}, x_{j_k}] \dots] \ : J \in \{1, \dots, q\}^k\big\}$$

to denote the vector space generated by the symbols x_J, $J \in \{1, \dots, q\}^k$, i.e. the set of linear combinations with real coefficients of the formal objects x_J. We shall refer to $\mathcal{P}_k[x_1, \dots, x_q]$ as the set of formal Lie-polynomials homogeneous of length k on the (non-commuting) indeterminates $x_1, \dots, x_q$, and x_J will be called a formal (nested) Lie-monomial (homogeneous of length k on $x_1, \dots, x_q$). For a fixed $q \in \mathbb{N}$, any sum of elements in $\mathcal{P}_k[x_1, \dots, x_q]$ with a finite number of k's will be referred to as a formal Lie-polynomial on the indeterminates $x_1, \dots, x_q$.

Let us now denote by $\mathfrak{h}$ any (abstract) nilpotent Lie algebra. For example, $\mathfrak{h}$ will be soon equal to $\mathfrak{g}$, the Lie algebra of a homogeneous Carnot group $\mathbb{G}$. For any $q, k \in \mathbb{N}$ and every $P \in \mathcal{P}_k[x_1, \dots, x_q]$, we wish to give a precise meaning to $P(X_1, \dots, X_q)$, whenever $X_1, \dots, X_q \in \mathfrak{h}$. Let $P \in \mathcal{P}_k[x_1, \dots, x_q]$. There exist a subset $\mathcal{I}$ of $\{1, \dots, q\}^k$ and real numbers a_J's (for every $J \in \mathcal{I}$) such that

$$P = \sum_{J \in \mathcal{I}} a_J x_J.$$

We set

$$P(X_1, \dots, X_q) := \sum_{J \in \mathcal{I}} a_J X_J$$

if $X_1, \dots, X_q \in \mathfrak{g}$, where

$$(\star) \qquad X_J := [X_{j_1}, \cdots [X_{j_{k-1}}, X_{j_k}] \cdots]$$

if $x_J = [x_{j_1}, \cdots [x_{j_{k-1}}, x_{j_k}] \cdots]$. Note that the signs $[\cdot, \cdot]$ in $(\star)$ denote the *Lie brackets in* $\mathfrak{h}$.

For example, if $\mathfrak{h} = \mathfrak{g}$, then, given

$$P = [x_1, [x_2, x_3]] - [x_3, [x_4, x_1]] + 2[x_2, [x_3, x_3]] \in \mathcal{P}_3[x_1, \dots, x_4]$$

and given vector fields $X_1, \dots, X_4 \in \mathfrak{g}$, we have

$$P(X_1, \dots, X_4) = [X_1, [X_2, X_3]] - [X_3, [X_4, X_1]] \in \mathfrak{g}.$$

If r is the step of nilpotency of $\mathfrak{h}$, and s is any fixed integer in $\{1, \dots, r\}$, we denote by

$$\bigoplus_{k=s}^{r} \mathcal{P}_k[x_1, \dots, x_q]$$

the set of formal sums $\sum_{k=s}^{r} P^{(k)}$, where $P^{(k)} \in \mathcal{P}_k[x_1, \dots, x_q]$. If $\mathfrak{h} = \mathfrak{g}$, $X_1, \dots, X_q \in \mathfrak{g}$ and $R_s \in \bigoplus_{k=s}^{r} \mathcal{P}_k[x_1, \dots, x_q]$, then

$$R_s(X_1, \dots, X_q)$$

will denote a sort of "remainder", a linear combination of brackets of height $\geq s$ of the vector fields $X_1, \dots, X_q$. Vice versa, recalling Proposition 1.1.7, page 12, any linear combination of brackets of height $\geq s$ of the vector fields $X_1, \dots, X_q \in \mathfrak{g}$ can be written as $R_s(X_1, \dots, X_q)$ for some $R_s \in \bigoplus_{k=s}^{r} \mathcal{P}_k[x_1, \dots, x_q]$.

With the above notation at hand, the Campbell–Hausdorff operation $\diamond$ in *every* nilpotent Lie algebra $\mathfrak{h}$ can be written as

$$\begin{aligned} X \diamond Y &= X + Y + H_2(X, Y) \\ &= X + Y + \frac{1}{2}[X, Y] + H_3(X, Y) \qquad \forall\, X, Y \in \mathfrak{h}, \end{aligned} \tag{19.1}$$

where

$$H_2 \in \bigoplus_{k=2}^{r} \mathcal{P}_k[x_1, x_2], \qquad H_3 \in \bigoplus_{k=3}^{r} \mathcal{P}_k[x_1, x_2]$$

depend[4] only on the step of nilpotency r of $\mathfrak{h}$.

Let us now henceforth fix $\mathfrak{h} = \mathfrak{g}$, the Lie algebra of the homogeneous Carnot group $\mathbb{G}$, nilpotent of step r. We want to show some notable algebraic properties of the $\diamond$ operation on $\mathfrak{g}$ in the simple cases $q = 2$ and $q = 3$.

• For $q = 2$, we claim that there exists

$$R_3 \in \bigoplus_{k=3}^{r} \mathcal{P}_k[x_1, x_2]$$

such that

$$[Y, X] = Y \diamond X \diamond (-Y) \diamond (-X) + R_3(X, Y) \qquad \forall\, X, Y \in \mathfrak{g} \tag{19.2}$$

(we recall that $\diamond$ is associative!). Indeed, (19.1) gives

$$\begin{aligned} &(Y \diamond X) \diamond ((-Y) \diamond (-X)) \\ &= Y \diamond X + (-Y) \diamond (-X) + \frac{1}{2}[Y \diamond X, (-Y) \diamond (-X)] \\ &\quad + H_3(Y \diamond X, (-Y) \diamond (-X)) \\ &= \left(Y + X + \frac{1}{2}[Y, X] + H_3(Y, X)\right) + \left(-Y - X + \frac{1}{2}[Y, X] + H_3(-Y, -X)\right) \\ &\quad + \frac{1}{2}[Y + X + H_2(Y, X), -(Y + X) + H_2(-Y, -X)] + P_3(X, Y) \end{aligned}$$

[4] The "universality" of H_2, H_3 follows from the "universality" of the Campbell–Hausdorff operation (14.2), page 585.

$$
\begin{aligned}
&= [Y, X] + H_3(Y, X) + H_3(-Y, -X) \\
&\quad + \frac{1}{2}([Y + X, -(Y + X)] + P_3'(X, Y)) + P_3(X, Y) \\
&= [Y, X] + P_3''(X, Y)
\end{aligned}
$$

for some $P_3, P_3', P_3'' \in \bigoplus_{k=3}^{r} \mathcal{P}_k[x_1, x_2]$ depending only on the step r of $\mathbb{G}$.

• For $q = 3$, we claim that there exists

$$R_4 \in \bigoplus_{k=4}^{r} \mathcal{P}_k[x_1, x_2, x_3]$$

such that

$$
\begin{aligned}
[Z, [Y, X]] = {} & Z \diamond Y \diamond X \diamond (-Y) \diamond (-X) \diamond (-Z) \diamond X \diamond Y \diamond (-X) \diamond (-Y) \\
& + R_4(X, Y, Z) \qquad \text{for every } X, Y, Z \in \mathfrak{g}.
\end{aligned} \tag{19.3}
$$

In order to prove (19.3), we start by recalling that $-W$ is the $\diamond$-inverse of W (for any $W \in \mathfrak{g}$), i.e. $-W$ is the unique element of $\mathfrak{g}$ such that

$$(-W) \diamond W = 0 = W \diamond (-W).$$

As a consequence, we get

$$X \diamond Y \diamond (-X) \diamond (-Y) = -(Y \diamond X \diamond (-Y) \diamond (-X)), \tag{19.4}$$

since

$$
\begin{aligned}
&(X \diamond Y \diamond (-X) \diamond (-Y)) \diamond (Y \diamond X \diamond (-Y) \diamond (-X)) \\
&\quad = (X \diamond Y \diamond (-X)) \diamond ((-Y) \diamond Y) \diamond (X \diamond (-Y) \diamond (-X)) \\
&\quad = (X \diamond Y) \diamond ((-X) \diamond X) \diamond ((-Y) \diamond (-X)) \\
&\quad = \cdots = 0,
\end{aligned}
$$

and analogously

$$(Y \diamond X \diamond (-Y) \diamond (-X)) \diamond (X \diamond Y \diamond (-X) \diamond (-Y)) = 0.$$

We now introduce the notation

$$\alpha(X, Y) := Y \diamond X \diamond (-Y) \diamond (-X) \qquad \forall\, X, Y \in \mathfrak{g}, \tag{19.5}$$

so that (19.2) reads

$$[Y, X] = \alpha(X, Y) + R_3(X, Y) \qquad \forall\, X, Y \in \mathfrak{g}. \tag{19.6}$$

From (19.4) and (19.6) it follows that

$$\begin{aligned}
&Z \diamond (Y \diamond X \diamond (-Y) \diamond (-X)) \diamond (-Z) \diamond (X \diamond Y \diamond (-X) \diamond (-Y)) \\
&\quad = Z \diamond \alpha(X,Y) \diamond (-Z) \diamond (-\alpha(X,Y)) \\
&\quad = \alpha(\alpha(X,Y), Z) \\
&\quad = [Z, \alpha(X,Y)] - R_3(\alpha(X,Y), Z) \\
&\quad = [Z, [Y,X] - R_3(X,Y)] - R_3([Y,X] - R_3(X,Y), Z) \\
&\quad = [Z, [Y,X]] + P_4(X,Y,Z),
\end{aligned}$$

for some $P_4 \in \bigoplus_{k=4}^r \mathcal{P}_k[x_1, x_2, x_3]$ depending only on the step r of nilpotency of $\mathfrak{g}$. This proves (19.3). □

Let us now consider the case of general q. Further (in particular, in (19.7) below) we use the following notation: if $X_1, \dots, X_q$ is a family of vector fields, we agree to let

$$X_{-i} := -X_i \qquad \forall\, i \in \{1, \dots, q\}.$$

Lemma 19.1.4. *Let the notation in the preceding paragraphs be fixed. Let $q \in \mathbb{N}$, $q \geq 2$, and set $c(q) = 3 \cdot 2^{q-1} - 2$. Recall also that r denotes the step of the (fixed) Carnot group $\mathbb{G}$. Then there exists an "indexing map"*

$$j_q : \{1, \dots, c(q)\} \to \{-q, \dots, -1, 1, \dots, q\}$$

and there exists

$$R_{q+1} \in \bigoplus_{k=q+1}^{r} \mathcal{P}_k[x_1, \dots, x_q],$$

$$R_{q+1} = \sum_{k=q+1}^{r} R_k^{(q)}, \qquad R_k^{(q)} \in \mathcal{P}_k[x_1, \dots, x_q]$$

(we agree to let $R_{q+1} = 0$ if $q \geq r$) such that

$$[X_q, \cdots [X_2, X_1] \cdots] = X_{j_q(1)} \diamond \cdots \diamond X_{j_q(c(q))} + R_{q+1}(X_1, \dots, X_q) \quad (19.7)$$

for every $X_1, \dots, X_q \in \mathfrak{g}$.

For example:

- when $q = 2$, we have (see (19.2)) $c(2) = 4$ and

$$\begin{aligned}
&j_2 : \{1,2,3,4\} \to \{-2,-1,1,2\}, \\
&j_2(1) = 2,\ j_2(2) = 1,\ j_2(3) = -2,\ j_2(4) = -1,
\end{aligned}$$

so that

$$\begin{aligned}
&X_{j_2(1)} \diamond X_{j_2(2)} \diamond X_{j_2(3)} \diamond X_{j_2(4)} \\
&\quad = X_2 \diamond X_1 \diamond X_{-2} \diamond X_{-1} \\
&\quad = X_2 \diamond X_1 \diamond (-X_2) \diamond (-X_1);
\end{aligned}$$

- when $q = 3$, we have $c(3) = 10$ and (see (19.3))

$$j_3 : \{1, \dots, 10\} \to \{-3, -2, -1, 1, 2, 3\},$$
$$j_3(1) = 3,\ j_3(2) = 2,\ j_3(3) = 1,\ j_3(4) = -2,\ j_3(5) = -1,$$
$$j_3(6) = -3,\ j_3(7) = 1,\ j_3(8) = 2,\ j_3(9) = -1,\ j_3(10) = -2.$$

Proof (of Lemma 19.1.4). (We explicitly remark that the proof makes a crucial use of the "universality" of the Campbell–Hausdorff operation $\diamond$.)

We argue by induction on $q \geq 2$. The case $q = 2$ follows from (19.2). Let us now prove the statement for $q + 1$ assuming it to be true for q. First of all, let us observe that, arguing as in (19.4), we get

$$(-X_{j_q(c(q))}) \diamond \cdots \diamond (-X_{j_q(1)}) = -(X_{j_q(1)} \diamond \ldots \diamond X_{j_q(c(q))}).$$

Hence, recalling (19.5) and using (19.6), we have

$$\begin{aligned}
&X_{q+1} \diamond (X_{j_q(1)} \diamond \cdots \diamond X_{j_q(c(q))}) \diamond (-X_{q+1}) \diamond ((-X_{j_q(c(q))}) \diamond \cdots \diamond (-X_{j_q(1)}))\\
&= \alpha(X_{j_q(1)} \diamond \cdots \diamond X_{j_q(c(q))}, X_{q+1})\\
&= [X_{q+1}, X_{j_q(1)} \diamond \cdots \diamond X_{j_q(c(q))}] - R_3(X_{j_q(1)} \diamond \cdots \diamond X_{j_q(c(q))}, X_{q+1})\\
&= \big[X_{q+1}, [X_q, \cdots [X_2, X_1] \cdots] - R_{q+1}(X_1, \ldots, X_q)\big] + P_{q+2}(X_1, \ldots, X_{q+1})\\
&= [X_{q+1}, [X_q, \cdots [X_2, X_1] \cdots]] + P'_{q+2}(X_1, \ldots, X_{q+1}),
\end{aligned}$$

for some $P_{q+2}, P'_{q+2} \in \bigoplus_{k=q+2}^{r} \mathcal{P}_k[x_1, \ldots, x_{q+1}]$ depending only on the step r of $\mathbb{G}$. Since

$$2c(q) + 2 = 2(3 \cdot 2^{q-1} - 2) + 2 = 3 \cdot 2^q - 2 = c(q+1),$$

the proof is complete. □

We are now ready to prove Theorem 19.1.3.

Proof (of Theorem 19.1.3). Using $\mathfrak{g} = \text{Lie}\{Z_1, \ldots, Z_m\}$ and recalling again Proposition 1.1.7 on page 12, we conclude that there exist multi-indices

$$J_1 \in \{1, \ldots, m\}^{q_1}, \ldots, J_N \in \{1, \ldots, m\}^{q_N}, \qquad 1 \leq q_i \leq r,$$

such that

$$\mathfrak{g} = \text{span}\{Z_{J_1}, \ldots, Z_{J_N}\}. \tag{19.8a}$$

For every $i \in \{1, \ldots, N\}$, we also consider the multi-index

$$J_{-i} := (J_i(1), \ldots, J_i(q_i - 2), J_i(q_i), J_i(q_i - 1))$$

obtained by interchanging the last two entries of J_i, so that

$$Z_{J_{-i}} = -Z_{J_i}. \tag{19.8b}$$

By means of Lemma 19.1.4, for every $i \in \{-N, \dots, -1, 1, \dots, N\}$ there exists an indexing map

$$h^i : \{1, \dots, c(q_i)\} \to \{-m, \dots, -1, 1, \dots, m\}$$

$(q_{-i} := q_i)$ and there exists

$$W_k^i \in \mathcal{P}_k[x_1, \dots, x_m] \qquad \text{for every } k \in \{q_i + 1, \dots, r\}$$

such that, setting

$$\begin{aligned} W^i(t) &:= \sum_{k=q_i+1}^{r} R_k^{(q_i)}(t^{1/q_i} Z_{J_i(q_i)}, \dots, t^{1/q_i} Z_{J_i(1)}) \\ &= \sum_{k=q_i+1}^{r} t^{k/q_i} W_k^i(Z_1, \dots, Z_m) \qquad \forall\, t \geq 0 \end{aligned} \tag{19.8c}$$

(we agree to let $W^i \equiv 0$ if $q_i \geq r$), we have

$$\begin{aligned} t\, Z_{J_i} &= [t^{1/q_i} Z_{J_i(1)}, \dots [t^{1/q_i} Z_{J_i(q_i-1)}, t^{1/q_i} Z_{J_i(q_i)}] \dots] \\ &= (t^{1/q_i} Z_{h^i(1)}) \diamond \dots \diamond (t^{1/q_i} Z_{h^i(c(q_i))}) \\ &\quad + R_{q_i+1}(t^{1/q_i} Z_{J_i(q_i)}, \dots, t^{1/q_i} Z_{J_i(1)}) \\ &= (t^{1/q_i} Z_{h^i(1)}) \diamond \dots \diamond (t^{1/q_i} Z_{h^i(c(q_i))}) \\ &\quad + \sum_{k=q_i+1}^{r} R_k^{(q_i)}(t^{1/q_i} Z_{J_i(q_i)}, \dots, t^{1/q_i} Z_{J_i(1)}) \\ &= (t^{1/q_i} Z_{h^i(1)}) \diamond \dots \diamond (t^{1/q_i} Z_{h^i(c(q_i))}) \\ &\quad + \sum_{k=q_i+1}^{r} t^{k/q_i} R_k^{(q_i)}(Z_{J_i(q_i)}, \dots, Z_{J_i(1)}) \\ &= (t^{1/q_i} Z_{h^i(1)}) \diamond \dots \diamond (t^{1/q_i} Z_{h^i(c(q_i))}) \\ &\quad + \sum_{k=q_i+1}^{r} t^{k/q_i} W_k^i(Z_1, \dots, Z_m) \\ &= (t^{1/q_i} Z_{h^i(1)}) \diamond \dots \diamond (t^{1/q_i} Z_{h^i(c(q_i))}) + W^i(t) \end{aligned} \tag{19.8d}$$

for every $t \geq 0$. As usual, we agree to let $Z_{-j} := -Z_j$ for every $j \in \{1, \dots, m\}$.

For every $i \in \{-N, \dots, -1, 1, \dots, N\}$, we now define

$$\begin{aligned} &E_i : \mathbb{G} \times [0, +\infty[\to \mathbb{G}, \\ &E_i(x, t) := (\exp(t^{1/q_i} Z_{h^i(c(q_i))}) \circ \dots \circ \exp(t^{1/q_i} Z_{h^i(1)}))(x). \end{aligned} \tag{19.8e}$$

(Here $\circ$ denotes the usual composition of maps.) Let us recall that, by means of Theorem 15.1.1, we have

$$\exp(Y)\big(\exp(X)(x)\big) = \exp(X \diamond Y)(x) \quad \forall\, X, Y \in \mathfrak{g},\ \forall\, x \in \mathbb{G}. \tag{19.8f}$$

As a consequence, we can write

$$\begin{aligned} E_i(x,t) &= \exp\Big(\big(t^{1/q_i} Z_{h^i(1)}\big) \diamond \cdots \diamond \big(t^{1/q_i} Z_{h^i(c(q_i))}\big)\Big)(x) \\ &= \exp\big(t Z_{J_i} - W^i(t)\big)(x). \end{aligned}$$

We have used here (19.8d). We now recall[5] formula (1.9), page 8,

$$\exp(X)(x) = \sum_{k=0}^{r} \frac{1}{k!} X^k I(x) \qquad \forall\, X \in \mathfrak{g},\ \forall\, x \in \mathbb{G}. \tag{19.8g}$$

Taking $X = tZ_{J_i} - W^i(t)$, we obtain

$$\begin{aligned} E_i(x,t) &= x + \big(tZ_{J_i} - W^i(t)\big) I(x) + \sum_{k=2}^{r} \frac{1}{k!} \big(tZ_{J_i} - W^i(t)\big)^k I(x) \\ &= x + tZ_{J_i} I(x) + \omega_i(x,t), \end{aligned} \tag{19.8h}$$

for some remainder function $\omega_i : \mathbb{G} \times [0, +\infty[\to \mathbb{G}$. From the very definition (19.8c) of W^i it follows that such remainder function has the following regularity properties:

$$\begin{aligned} &\omega_i \in C^1\big(\mathbb{G} \times [0, +\infty[, \mathbb{G}\big), \\ &\omega_i(x,t) = t^{1+1/q_i}\, \widetilde{\omega}_i(x,t), \qquad \widetilde{\omega}_i \in C\big(\mathbb{G} \times [0, +\infty[, \mathbb{G}\big), \\ &(\partial_t \omega_i)(x, 0) = 0. \end{aligned} \tag{19.8i}$$

Roughly speaking, formula (19.8h) above states that a *translation* along a commutator Z_J of arbitrary length can be approximated by a composition of a finite number of *elementary translations* along $\pm Z_1, \ldots, \pm Z_m$, at least for small positive times. We now want to take into account negative times too (see formula (19.8l) below). For every $i \in \{1, \ldots, N\}$, we set

$$r_i : \mathbb{G} \times \mathbb{R} \to \mathbb{G}, \qquad r_i(x,t) := \begin{cases} \omega_i(x,t) & \text{if } t \geq 0, \\ \omega_{-i}(x,-t) & \text{if } t < 0. \end{cases}$$

From (19.8i) it follows that r_i has the following regularity properties:

$$\begin{aligned} &r_i \in C^1(\mathbb{G} \times \mathbb{R}, \mathbb{G}), \\ &r_i(x,t) = |t|^{1+1/(2q_i)}\, \widetilde{r}_i(x,t), \qquad \widetilde{r}_i \in C(\mathbb{G} \times \mathbb{R}, \mathbb{G}), \\ &(\partial_t r_i)(x,0) = 0. \end{aligned} \tag{19.8j}$$

[5] We explicitly remark that (19.8g) is a particular case of (1.9), page 8, because $X^j I \equiv 0$ for every $j > r$, since (for those j's) X^j is a sum of differential operators δ_λ-homogeneous of degree $> r$, whereas the component functions of the identity map I are δ_λ-homogeneous of degree at most r.

For every $i \in \{1, \dots, N\}$, we now define

$$B_i : \mathbb{G} \times \mathbb{R} \to \mathbb{G}, \qquad B_i(x,t) := \begin{cases} E_i(x,t) & \text{if } t \geq 0, \\ E_{-i}(x,-t) & \text{if } t < 0. \end{cases} \tag{19.8k}$$

Recalling (19.8h) and (19.8b), we can write

$$B_i(x,t) = x + t Z_{J_i} I(x) + r_i(x,t) \qquad \forall\, (x,t) \in \mathbb{G} \times \mathbb{R}. \tag{19.8l}$$

As a consequence, it turns out that

$$B_i \in C^1(\mathbb{G} \times \mathbb{R}, \mathbb{G}).$$

Moreover, we have

$$\begin{aligned} &B_i(x,0) = x, \\ &\mathcal{J}_{B_i}(x,0) = (\mathbb{I}_N, Z_{J_i} I(x)). \end{aligned} \tag{19.8m}$$

We now fix $x_0 \in \mathbb{G}$, and we define

$$\begin{aligned} &F : \mathbb{R}^N \to \mathbb{R}^N \equiv \mathbb{G}, \\ &F(t) := B_N\big(\dots B_2(B_1(x_0,t_1),t_2)\dots,t_N\big). \end{aligned} \tag{19.8n}$$

We have

$$F \in C^1(\mathbb{R}^N, \mathbb{R}^N),$$

$$\begin{aligned} \mathcal{J}_F(t) = {} & \mathcal{J}_{B_N}(B_{N-1}(\dots B_2(B_1(x_0,t_1),t_2)\dots,t_{N-1})) \\ & \cdot \operatorname{diag}[\dots \operatorname{diag}[\mathcal{J}_{B_2}(B_1(x_0,t_1),t_2) \operatorname{diag}[\partial_{t_1} B_1(x_0,t_1),1],1]\dots,1], \end{aligned}$$

where, for any $p \times q$ matrix A, we have denoted by $\operatorname{diag}[A,1]$ the $(p+1) \times (q+1)$ matrix

$$\operatorname{diag}[A,1] := \begin{pmatrix} A & 0 \\ 0 & 1 \end{pmatrix}.$$

Using (19.8m), we obtain

$$\begin{aligned} \mathcal{J}_F(0) &= \mathcal{J}_{B_N}(x_0,0) \operatorname{diag}[\dots \operatorname{diag}[\mathcal{J}_{B_2}(x_0,0) \operatorname{diag}[Z_{J_1} I(x_0),1],1]\dots,1] \\ &= (\mathbb{I}_N, Z_{J_N} I(x_0)) \\ &\quad \cdot \operatorname{diag}[\dots \operatorname{diag}[(\mathbb{I}_N, Z_{J_2} I(x_0)) \operatorname{diag}[Z_{J_1} I(x_0),1],1]\dots,1] \\ &= (Z_{J_1} I(x_0), \dots, Z_{J_N} I(x_0)). \end{aligned}$$

As a consequence, recalling the choice of $J_1, \dots, J_N$ (see (19.8a)), we get

$$\det \mathcal{J}_F(0) \neq 0,$$

and F is a diffeomorphism from a neighborhood V of the origin to a neighborhood W of $x_0 = F(0)$.

In particular, for every $y \in W$, there exists $t \in V$ such that $y = F(t)$. We now observe that (by the definition of F, see (19.8n), (19.8k), (19.8e)) $F(t)$ is the final point of a $(Z_1, \dots, Z_m)$-subunit path (recall Definition 19.1.1 of subunit path) starting from x_0. Furthermore, such a path is piecewise an integral curve of the vector fields $Z_1, \dots, Z_m$. As a consequence, *there exists a neighborhood of x_0 whose points are $(Z_1, \dots, Z_m)$-connected to x_0 along paths that are piecewise integral curves of the vector fields $Z_1, \dots, Z_m$*.

A standard argument now allows to pass from local to global connectivity. We set $Z = (Z_1, \dots, Z_m)$ and, once $x_0 \in \mathbb{R}^N$ is fixed, we consider the set E of points Z-connected to x_0 along paths that are piecewise integral curves of the vector fields of Z. Such a set E is non-empty, since $x_0 \in E$. Moreover, E is open. Indeed, if $y \in E$, from the above arguments it follows that there exists an open neighborhood of y whose points are Z-connected to y along paths that are piecewise integral curves of the vector fields of Z.

Therefore, x_0 is Z-connected to any point of such a neighborhood of y. Finally, E is closed. In fact, let y_n be a sequence in E such that $y_n \to y$. By the above argument, there exists an open neighborhood W of y whose points are Z-connected to y along paths that are piecewise integral curves of the vector fields of Z.

Moreover, there exists $\bar{n} \in \mathbb{N}$ such that $y_{\bar{n}} \in W$. As a consequence, x_0 is Z-connected to y (passing through $y_{\bar{n}}$) along paths that are piecewise integral curves of the vector fields of Z, and then $y \in E$. This completes the proof of the $(Z_1, \dots, Z_m)$-connectivity of $\mathbb{G}$.

We now want to prove the second statement of Theorem 19.1.3, i.e. that $d_0 = d(\cdot, 0)$ is a homogeneous norm on $\mathbb{G}$. By means of Proposition 5.2.6, page 234, we only need to prove that d_0 is continuous. For every $i \in \{-N, \dots, -1, 1, \dots, N\}$, we have

$$d(x, E_i(x,t)) \le c(q_i)\, t^{1/q_i}. \tag{19.8o}$$

Indeed, for every fixed $t \ge 0$, the point $E_i(x,t)$ is connected to x by a $(Z_1, \dots, Z_m)$-subunit path, which is piecewise an integral curve of one of the vector fields $Z_1, \dots, Z_m$. Each piece of this path connects two points which have d-distance $\le t^{1/q_i}$. Hence, from the triangle inequality for d we get

$$d(x, E_i(x,t)) \le \sum_{j=1}^{c(q_i)} t^{1/q_i} = c(q_i)\, t^{1/q_i}.$$

Recalling the definition of F (see (19.8n), (19.8k)) and setting

$$x_1 := B_1(x_0, t_1), \quad x_2 := B_2(x_1, t_2), \quad \dots, \quad x_N := B_N(x_{N-1}, t_N) = F(t),$$

from the triangle inequality we also get (for $|t| \le 1$)

$$\begin{aligned} d(x_0, F(t)) &\le d(x_0, x_1) + d(x_1, x_2) + \cdots + d(x_{N-1}, x_N) \\ &\le c(q_1)\, |t_1|^{1/q_1} + c(q_2)\, |t_2|^{1/q_2} + \cdots + c(q_N)\, |t_N|^{1/q_N} \\ &\le c\, |t|^{1/r}. \end{aligned}$$

Therefore,

$$
\begin{aligned}
|d_0(x) - d_0(x_0)| &= |d(x,0) - d(x_0,0)| \\
&\leq d(x_0,x) = d(x_0, F(F^{-1}(x))) \leq c|F^{-1}(x)|^{1/r} \\
&\longrightarrow c|F^{-1}(x_0)|^{1/r} = 0 \qquad \text{as } x \to x_0.
\end{aligned}
$$

This completes the proof. □

19.2 An Application of Carathéodory–Chow–Rashevsky Theorem

As an application of Theorem 19.1.3, we have the following result, which has an interest in its own.

Theorem 19.2.1. *Let* $(\mathbb{G}, *)$ *be a homogeneous Carnot group, let* $\mathfrak{g}_1$ *be the first layer of a stratification of the algebra* $\mathfrak{g}$ *of* $\mathbb{G}$ *and let* $\mathcal{Z} = \{Z_1, \dots, Z_m\}$ *be a basis for* $\mathfrak{g}_1$*. Finally, fix a homogeneous norm* d *on* $\mathbb{G}$*.*

Then there exists an absolute constant $\mathbf{M} \in \mathbb{N}$ *(only depending on* $\mathbb{G}$*) such that*

$$\mathbb{G} = \underbrace{\operatorname{Exp}(\mathfrak{g}_1) * \cdots * \operatorname{Exp}(\mathfrak{g}_1)}_{\mathbf{M}\ \textit{times}}.$$

More precisely, for every $x \in \mathbb{G}$*, there exist* $x_1, \dots, x_{\mathbf{M}} \in \operatorname{Exp}(\mathfrak{g}_1)$ *with the following properties*:

$$x = x_1 * \cdots * x_{\mathbf{M}}, \tag{19.9a}$$

$$d(x_j) \leq \mathbf{c}_0\, d(x) \quad \textit{for all } j = 1, \dots, \mathbf{M}, \tag{19.9b}$$

$$\begin{aligned} &\textit{for every } j = 1, \dots, \mathbf{M}, \textit{ there exist } t_j \in \mathbb{R} \textit{ and } i_j \in \{1, \dots, m\} \\ &\quad \textit{such that} \quad x_j = \operatorname{Exp}(t_j Z_{i_j}). \end{aligned} \tag{19.9c}$$

Here $\mathbf{c}_0 > 0$ *is a constant depending on* $\mathbb{G}$*,* d *and* $\mathcal{Z}$*, but not depending on* x *and of the* x_j*'s.*

Before proving Theorem 19.2.1, we need a simple preliminary remark (used also in the proof of Theorem 20.3.1, page 746).

Remark 19.2.2. Let us recall Definition 2.2.6, page 125. Let $\mathfrak{g} = \mathfrak{g}_1 \oplus \cdots \oplus \mathfrak{g}_r$ be a stratification of the algebra $\mathfrak{g}$ of a homogeneous Carnot group $\mathbb{G}$. For every $k = 1, \dots, r$, let us fix a basis for $\mathfrak{g}_k$, say $X_1^{(k)}, \dots, X_{N_k}^{(k)}$ (where $N_k = \dim(\mathfrak{g}_k)$). As usual, we say that

$$\{X_1, \dots, X_N\} := \{X_1^{(1)}, \cdots, X_{N_1}^{(1)}, \cdots, X_1^{(r)}, \cdots, X_{N_r}^{(r)}\}$$

is a basis of $\mathfrak{g}$ adapted to the stratification. For every $\xi = (\xi_1, \dots, \xi_N) = (\xi^{(1)}, \dots, \xi^{(r)})$, we set

$$\xi \cdot X = \sum_{j=1}^{N} \xi_j X_j = \sum_{k=1}^{r} \sum_{j=1}^{N_k} \xi_j^{(k)} X_j^{(k)}. \tag{19.10a}$$

If the dilations on $\mathbb{G}$ are indifferently denoted by

$$\delta_\lambda(x) = (\lambda^{\sigma_1} x_1, \dots, \lambda^{\sigma_N} x_N) = (\lambda x^{(1)}, \dots, \lambda^r x^{(r)}),$$

we can define a dilation on $\mathfrak{g}$ (see Definition 1.3.24, page 46) still denoted by δ_λ, extending by linearity the map such that $\delta_\lambda(X_j) := \lambda^{\sigma_j} X_j$ for every $j = 1, \dots, N$. In other words, we have

$$\delta_\lambda\left(\sum_{j=1}^{N} \xi_j X_j\right) = \sum_{j=1}^{N} \xi_j \lambda^{\sigma_j} X_j = \sum_{k=1}^{r} \sum_{j=1}^{N_k} \xi_j^{(k)} \lambda^k X_j^{(k)}.$$

Following (19.10a), this can be rewritten as

$$\delta_\lambda(\xi \cdot X) = \delta_\lambda(\xi) \cdot X. \tag{19.10b}$$

Now, in (4.12) (page 196), we have proved that, for every $x \in \mathbb{G}$, $\xi \in \mathbb{R}^N$ and $\lambda > 0$, we have

$$\delta_\lambda\big(\exp(\xi \cdot X)(x)\big) = \exp((\delta_\lambda \xi) \cdot X)(\delta_\lambda(x)).$$

Following (19.10b), this also reads as

$$\delta_\lambda\big(\exp(\xi \cdot X)(x)\big) = \exp\big(\delta_\lambda(\xi \cdot X)\big)(\delta_\lambda(x)). \tag{19.10c}$$

In particular, if we take as ξ the i-th coordinate vector of the standard basis for $\mathbb{R}^N$, this gives

$$\delta_\lambda\big(\exp(X_i)(x)\big) = \exp\big(\lambda^{\sigma_i} X_i\big)(\delta_\lambda(x)). \tag{19.10d}$$

Here $\lambda > 0$, $x \in \mathbb{G}$ and (due to the arbitrariness of the adapted basis) X_i can be any left-invariant (non identically vanishing) vector field in $\mathfrak{g}$ which is δ_λ-homogeneous of degree σ_i. As a consequence, if $\lambda = t^{1/\sigma_i}$, we have

$$\delta_{t^{1/\sigma_i}}\big(\exp(X_i)(x)\big) = \exp\big(t X_i\big)(\delta_{t^{1/\sigma_i}}(x)) \quad \text{for all } t > 0 \text{ and } x \in \mathbb{G},$$

so that, if d is any homogeneous norm on $\mathbb{G}$, we have

$$d\big(\exp(t X_i)(\delta_{t^{1/\sigma_i}}(x))\big) = t^{1/\sigma_i}\, d\big((\exp(X_i)(x))\big). \tag{19.10e}$$

Proof (of Theorem 19.2.1). We claim that the proof of this theorem is contained in the proof of the Carathéodory–Chow–Rashevsky Theorem 19.1.3. We closely analyze that proof. We know that the following map (take $x_0 = 0$ in (19.8n))

$$F : \mathbb{R}^N \to \mathbb{R}^N \equiv \mathbb{G},$$
$$F(t) := B_N\big(\dots B_2(B_1(0, t_1), t_2) \dots, t_N\big)$$

is a diffeomorphism from a neighborhood V of the origin to a neighborhood W of the origin. It is not restrictive to replace V by

$$V_1 := V \cap \{t \in \mathbb{G} : |t_1|^{1/q_1} + \cdots + t_N^{1/q_N} < 1\}$$

and take $W_1 := F(V_1)$. (We notice that the q_i's and the map F itself depend only on the structure of $\mathbb{G}$.)

We have (see (19.8k) and (19.8e)): for every $1 \le i \le N$, $B_i(x, t_i)$ is the composition of $c(q_i) = 3 \cdot 2^{q_i - 1} - 2$ exponential maps (starting from x) along one of the following vector fields

$$\pm t_i^{1/q_i} Z_1, \ldots, \pm t_i^{1/q_i} Z_m.$$

But since $1 \le q_i \le r$ for every $1 \le i \le N$, we can infer that $F(t)$ is the composition of at most

$$\mathbf{M} := N(3 \cdot 2^{r-1} - 2) \ge \sum_{i=1}^{N} (3 \cdot 2^{q_i - 1} - 2) = \sum_{i=1}^{N} c(q_i)$$

vector fields of the above type.

Thus (setting for brevity $\tau_i := 0$ if needed) any $\zeta \in W_1$ can be written as

$$\zeta = \exp(\tau_{\mathbf{M}} Z_{i_{\mathbf{M}}})(\cdots \exp(\tau_2 Z_{i_2})(\exp(\tau_1 Z_{i_1})(0)) \cdots), \tag{19.11a}$$

where $i_1, \ldots, i_{\mathbf{M}} \in \{1, \ldots, m\}$ and $|\tau_i| < 1$ are suitably chosen. We set

$$\zeta_1 := \exp(\tau_1 Z_{i_1})(0), \qquad \zeta_k := \exp(\tau_k Z_{i_k})(\zeta_{k-1}), \quad k = 2, \ldots, \mathbf{M}. \tag{19.11b}$$

By the well-known formula $\exp(Z)(x) = x * \mathrm{Exp}\,(Z)$, from (19.11b) we have $\zeta_k := \zeta_{k-1} * \mathrm{Exp}\,(\tau_k Z_{i_k})$ (here $\zeta_0 := 0$), so that from (19.11b) we also derive

$$\begin{aligned}\zeta &= \zeta_{\mathbf{M}} = \zeta_{\mathbf{M}-1} * \mathrm{Exp}\,(\tau_{\mathbf{M}} Z_{i_{\mathbf{M}}}) \\ &= \zeta_{\mathbf{M}-2} * \mathrm{Exp}\,(\tau_{\mathbf{M}-1} Z_{i_{\mathbf{M}-1}}) * \mathrm{Exp}\,(\tau_{\mathbf{M}} Z_{i_{\mathbf{M}}}) = [\cdots] \\ &= \mathrm{Exp}\,(\tau_1 Z_{i_1}) * \cdots * \mathrm{Exp}\,(\tau_{\mathbf{M}-1} Z_{i_{\mathbf{M}-1}}) * \mathrm{Exp}\,(\tau_{\mathbf{M}} Z_{i_{\mathbf{M}}}).\end{aligned}$$

As a consequence, we have the decomposition

$$\zeta = \xi_1 * \cdots * \xi_{\mathbf{M}}, \qquad \text{where } \xi_j = \mathrm{Exp}\,(\tau_j Z_{i_j}). \tag{19.11c}$$

We explicitly remark that, since the Z_{i_j}'s all belong to the first layer of $\mathfrak{g}$, from (19.10e) we get

$$d(\xi_j) = d\big(\mathrm{Exp}\,(\tau_j Z_{i_j})\big) = |\tau_j|\, d\big(\mathrm{Exp}\,(\pm Z_{i_j})\big).$$

Recalling that $|\tau_j| < 1$ and setting

$$\alpha_d := \max\big\{d\big(\mathrm{Exp}\,(\pm Z_1)\big), \ldots, d\big(\mathrm{Exp}\,(\pm Z_m)\big)\big\},$$

we have proved the estimate

$$d(\xi_j) \le \alpha_d \quad \text{for every } j = 1, \ldots, \mathbf{M}. \tag{19.11d}$$

Thus, (19.11c) and (19.11d) demonstrate (19.9a)–(19.9c) for points in W_1.

Finally, take any $x \in \mathbb{G}$. We shall use a simple δ_λ-homogeneity argument. Since W_1 is a neighborhood of the origin, there exists $r > 0$ such that $\overline{B_d(0,r)} \subset W_1$. Consequently, we have $\zeta := \delta_{r/d(x)}(x) \in W_1$. Thus, from what we proved above, one gets (setting $\lambda := d(x)/r$)

$$x = \delta_\lambda(\zeta) = \delta_\lambda(\xi_1 * \cdots * \xi_{\mathbf{M}}) = \delta_\lambda(\xi_1) * \cdots * \delta_\lambda(\xi_{\mathbf{M}}) =: x_1 * \cdots * x_{\mathbf{M}}.$$

We notice that (by using (19.10d))

$$x_j = \delta_\lambda(\xi_j) = \delta_\lambda\big(\operatorname{Exp}(\tau_j Z_{i_j})\big) = \operatorname{Exp}(\lambda \tau_j Z_{i_j}),$$

whence (19.9c) follows by setting $t_j = \lambda \tau_j$. Moreover, it holds

$$d(x_j) = d(\delta_\lambda(\xi_j)) = \lambda\, d(\xi_j) = \frac{d(x)}{r}\, d(\xi_j) \le \frac{\alpha_d}{r}\, d(x)$$

(using (19.11d)), so that (19.9b) follows as well by taking $\mathbf{c}_0 := \alpha_d/r$. This ends the proof of the theorem. □

Bibliographical Notes. The connectivity Theorem 19.1.3 (indeed, a more general version of it) is due to W.L. Chow [Cho39] and P.K. Rashevsky [Ras38] (the version of this theorem in the $\mathbb{R}^3$ case and for two Hörmander vector fields is also known as *Carathéodory's theorem*, see [Car09]). For modern proofs, see also [Bel96,Gro96, Her68,NSW85,VSC92]. For applications, see e.g., P. Hajłasz and P. Koskela [HK00]. We explicitly remark that X-connectivity also holds when X is a system of vector fields in $\mathbb{R}^N$ satisfying Hörmander's rank condition (see the references above).

For a proof of Theorem 19.2.1, see also G.B. Folland and E.M. Stein [FS82, Lemma 1.40] and P. Pansu [Pan89, Corollary]. Theorem 19.2.1 is called "generating property" by V. Magnani [Mag06, Proposition 3.12] and is applied in [Mag06] to derive properties of convex functions in Carnot groups.

19.3 Exercises of Chapter 19

Ex. 1) Let $\Omega \subseteq \mathbb{R}^N$ be an open and connected set, and let $X = \{X_1, \ldots, X_m\}$ be any family of locally Lipschitz-continuous vector fields defined in Ω. Obviously, the Definition 5.2.1 of X-subunit path $\gamma \in \mathcal{S}(X)$ makes sense (see page 232). Given $x, y \in \Omega$, we let, by definition, $d_X(x,y) = \infty$ iff there does not exist any $\gamma \in \mathcal{S}(X)$, $\gamma : [0,T] \to \Omega$, such that $\gamma(0) = x$ and $\gamma(T) = y$. Otherwise, $d_X(x,y)$ is defined as in Definition 5.2.2, page 232.

Prove that Ω splits into a (possibly infinite and not denumerable) family of sets $\Omega = \bigcup_i \Omega_i$ such that:

a) $x, y \in \Omega_i$ iff there exists $\gamma \in \mathcal{S}(X)$ connecting x and y;
b) for every i, $(\Omega_i, d_X|_{\Omega_i \times \Omega_i})$ is a metric space;
c) if $x \in \Omega_i$ and $y \in \Omega_j$ with $i \neq j$, then $d_X(x, y) = \infty$.

Are the Ω_i's open sets?
With the above notation, let $\Omega = \mathbb{R}^2$, $X = \{\partial_{x_1}\}$. Prove that $d_X(x, y) < \infty$ iff x and y both lie on the same line parallel to the x_1-axis. Find the relevant Ω_i's.

Ex. 2) Write down and prove a simplified version of Lemma 19.1.4 in the case of step two Carnot groups.

Ex. 3) Write down explicitly the indexing map j_q of Lemma 19.1.4 in the cases $q = 4$ and $q = 5$.

20

Taylor Formula on Homogeneous Carnot Groups

The aim of this chapter is to prove the Lagrange mean value theorem and, consequently, several versions of the Taylor formula for smooth functions on homogeneous Carnot groups.

We begin with a brief sketch of polynomial functions and derivatives of a homogeneous Carnot group. Then, we introduce and investigate the Taylor polynomials and we prove suitable versions of the Lagrange mean value theorem and of the Taylor formula.

The latter will be derived from the former by an adaptation of standard arguments, whereas the proof of the Lagrange mean value theorem will deeply rely on the Carathéodory–Chow–Rashevsky connectivity theorem in Chapter 19.

Due to the importance of the Lagrange mean value theorem on stratified groups, we provide two proofs of it both relying on the cited Theorem 19.1.3. The first one makes also use of the Carnot–Carathéodory distance d_Z, the second one does not involve d_Z, but it makes use of some precise estimates in the connectivity theorem.

Finally, Taylor formulas with remainder of Lagrange-type, of Peano-type and in integral form will be given.

Throughout the chapter, we fix the following notation. $\mathbb{G} = (\mathbb{R}^N, \delta_\lambda, *)$ is a homogeneous Carnot group nilpotent of step r, whose points are denoted indifferently by

$$x = (x_1, \dots, x_N) = (x^{(1)}, \dots, x^{(r)}), \tag{20.1}$$

where $x^{(i)} \in \mathbb{R}^{N_i}$. Following our usual notation for the dilations $\delta_\lambda : \mathbb{R}^N \to \mathbb{R}^N$,

$$\delta_\lambda(x) = (\lambda^{\sigma_1} x_1, \dots, \lambda^{\sigma_N} x_N) = (\lambda\, x^{(1)}, \dots, \lambda^r x^{(r)}). \tag{20.2}$$

α will always denote a multi-index with N entries, i.e. $\alpha = (\alpha_1, \dots, \alpha_N)$ with $\alpha_1, \dots, \alpha_N \in \mathbb{N} \cup \{0\}$. As usual, if α is a multi-index, we set

$$x^\alpha = x_1^{\alpha_1} \cdots x_N^{\alpha_N}, \qquad (D_x)^\alpha = \partial_{x_1}^{\alpha_1} \cdots \partial_{x_N}^{\alpha_N}, \quad \text{and} \quad \alpha! = \alpha_1! \cdots \alpha_N!.$$

We shall also use the following notation

$$|\alpha| = \alpha_1 + \cdots + \alpha_N, \quad |\alpha|_{\mathbb{G}} = \alpha_1\, \sigma_1 + \cdots + \alpha_N\, \sigma_N \tag{20.3}$$

to denote, respectively, the Euclidean length (also said isotropic length) of α and the homogeneous length (also said δ_λ-length or $\mathbb{G}$-length) of α.

Furthermore, we denote by $\{Z_1, \dots, Z_N\}$ the Jacobian basis for the algebra $\mathfrak{g}$ of $\mathbb{G}$, i.e. for every $j \in \{1, \dots, N\}$, Z_j is the left invariant vector field on $\mathbb{G}$ such that

$$Z_j|_0 = \partial_{x_j}|_0.$$

When $\xi = (\xi_1, \dots, \xi_N) \in \mathbb{R}^N$, we use the useful notation $\xi \cdot Z$ to denote $\sum_{j=1}^N \xi_j Z_j$, the vector field in $\mathfrak{g}$ whose Jacobian coordinates are given by $(\xi_1, \dots, \xi_N)$. Finally, we denote by

$$\operatorname{Exp} : \mathfrak{g} \to \mathbb{G}, \qquad \operatorname{Log} : \mathbb{G} \to \mathfrak{g},$$

respectively, the usual exponential and logarithmic maps related to $(\mathbb{G}, *)$.

20.1 Polynomials and Derivatives on Homogeneous Carnot Groups

20.1.1 Polynomial Functions on $\mathbb{G}$

We begin with a very natural definition; although, despite this naturalness, we shall abandon it very shortly (after some due remarks).

Definition 20.1.1 ($\mathbb{G}$-polynomial). *A function $P : \mathbb{G} \to \mathbb{R}$ is called a* $\mathbb{G}$-polynomial *if $p := P \circ \operatorname{Exp}$ is a polynomial function on the vector space $\mathfrak{g}$, i.e. p is a polynomial function when expressed in coordinates w.r.t. any (or equivalently, w.r.t. at least one) basis for $\mathfrak{g}$.*

Since $\{Z_1, \dots, Z_N\}$ is a basis for $\mathfrak{g}$, the set of $\mathbb{G}$-polynomials is given by

$$\big\{P = p \circ \operatorname{Log} \mid \widetilde{p}(\xi) := p(\xi \cdot Z) \text{ is a polynomial in } \xi\big\}. \tag{20.4}$$

We now recall formulas (1.75a) and (1.75b) (page 50), where we proved that Exp and Log have the following coordinate form

$$\operatorname{Exp}(\xi \cdot Z) = \begin{pmatrix} \xi_1 \\ \xi_2 + B_2(\xi_1) \\ \vdots \\ \xi_N + B_N(\xi_1, \dots, \xi_{N-1}) \end{pmatrix} \qquad \forall\, \xi \in \mathbb{R}^N,$$

$$\operatorname{Log}(x) = \begin{pmatrix} x_1 \\ x_2 + C_2(x_1) \\ \vdots \\ x_N + C_N(x_1, \dots, x_{N-1}) \end{pmatrix} \cdot Z \qquad \forall\, x \in \mathbb{G}, \tag{20.5}$$

where the B_i's and C_i's are polynomial functions on $\mathbb{R}^N$ (δ_λ-homogeneous of degree σ_i) completely determined by the composition law on $\mathbb{G}$. It is then immediately seen that *the set of $\mathbb{G}$-polynomials coincides with the set of polynomial functions with respect to the fixed coordinate system $(x_1, \dots, x_N)$ on $\mathbb{G}$.*

Indeed, if Q is any polynomial function on $\mathbb{G}$, then the function $q := Q \circ \operatorname{Exp}$ is a polynomial on $\mathfrak{g}$, for (thanks to (20.5))

$$q(\xi \cdot Z) = Q(\operatorname{Exp}(\xi \cdot Z)) = Q(\xi_1, \xi_2 + B_2(\xi_1), \dots, \xi_N + B_N(\xi_1, \dots, \xi_{N-1}))$$

is evidently a polynomial function in ξ. Vice versa, let P be a $\mathbb{G}$-polynomial. Then (setting $p = P \circ \operatorname{Exp}$ and following the notation in (20.4))

$$\begin{aligned} P(x) &= (P \circ \operatorname{Exp} \circ \operatorname{Log})(x) = p((x_1, \dots, x_N + C_N(x_1, \dots, x_{N-1})) \cdot Z) \\ &= \widetilde{p}(x_1, x_2 + C_2(x_1), \dots, x_N + C_N(x_1, \dots, x_{N-1})) \end{aligned}$$

is a polynomial function in x, since $\widetilde{p}$ and the C_i's are polynomials.

If $p = p(x)$ is any polynomial function in the fixed coordinates $(x_1, \dots, x_N)$ on $\mathbb{G}$, we set (also following notation (20.3))

$$\deg(p) := \max\left\{ |\alpha| \ : \ p(x) = \sum_\alpha c_\alpha \, x^\alpha \ \text{ with } c_\alpha \neq 0 \text{ for every } \alpha \right\},$$

$$\deg_{\mathbb{G}}(p) := \max\left\{ |\alpha|_{\mathbb{G}} \ : \ p(x) = \sum_\alpha c_\alpha \, x^\alpha \ \text{ with } c_\alpha \neq 0 \text{ for every } \alpha \right\}$$

to denote, respectively, the Euclidean degree (also said isotropic degree) of p and the homogeneous degree (also said δ_λ-degree or $\mathbb{G}$-degree) of p. For any $n \in \mathbb{N} \cup \{0\}$, we give the following definition

$$\mathcal{P}_n := \{p \text{ polynomial on } \mathbb{G} \ : \ \deg_{\mathbb{G}}(p) \leq n\}. \tag{20.6}$$

Obviously, $\mathcal{P}_n$ is a vector space over the field $\mathbb{R}$. We denote its dimension by

$$\mu_n := \dim_{\mathbb{R}}(\mathcal{P}_n). \tag{20.7}$$

For every $n \in \mathbb{N} \cup \{0\}$, we introduce the set of multi-indices

$$\mathcal{I}_n := \{\alpha \text{ multi-index in } (\mathbb{N} \cup \{0\})^N \ : \ |\alpha|_{\mathbb{G}} \leq n\}. \tag{20.8}$$

With the above definitions, we have the following proposition, whose simple proof is left as an exercise.

Proposition 20.1.2 (On the vector space $\mathcal{P}_n$). *For every $n \in \mathbb{N} \cup \{0\}$, it holds $\mathcal{P}_n = \operatorname{span}\{x^\alpha : \alpha \in \mathcal{I}_n\}$, so that μ_n equals the cardinality of $\mathcal{I}_n$. Moreover, if $\mathcal{I}_n$ has been ordered in some fixed way, then the map*

$$\mathcal{P}_n \to \mathbb{R}^{\mu_n}, \qquad p \mapsto ((D_x)^\alpha p(0))_{\alpha \in \mathcal{I}_n}$$

is a vector space isomorphism. More explicitly, if $\beta \in \mathcal{I}_n$, the above map sends x^β into $(0, \dots, \beta!, \dots, 0)$, where the only non-vanishing entry appears at the same place where β appears in the fixed ordering of $\mathcal{I}_n$.

Example 20.1.3. If $\mathbb{G} = \mathbb{H}^1$ is the Heisenberg–Weyl group on $\mathbb{R}^3$ (whose points are denoted by (x, y, t)) with dilations $\delta_\lambda(x, y, t) = (\lambda x, \lambda y, \lambda^2 t)$, we have

$$\begin{aligned}
&\mathcal{I}_0 = \{(0, 0, 0)\},\\
&\mathcal{I}_1 = \{(0, 0, 0), (1, 0, 0), (0, 1, 0)\},\\
&\mathcal{I}_2 = \mathcal{I}_1 \cup \{(0, 0, 1), (2, 0, 0), (0, 2, 0), (1, 1, 0)\},\\
&\mathcal{I}_3 = \mathcal{I}_2 \cup \{(1, 0, 1), (0, 1, 1), (3, 0, 0), (0, 3, 0), (2, 1, 0), (1, 2, 0)\},\\
&\mathcal{P}_0 = \operatorname{span}\{1\}, \qquad \mu_0 = 1,\\
&\mathcal{P}_1 = \operatorname{span}\{1, x, y\}, \qquad \mu_1 = 3,\\
&\mathcal{P}_2 = \operatorname{span}\{1, x, y; t, x^2, y^2, xy\}, \qquad \mu_1 = 7,\\
&\mathcal{P}_3 = \operatorname{span}\{1, x, y; t, x^2, y^2, xy; xt, yt, x^3, y^3, x^2 y, xy^2\}, \qquad \mu_1 = 13.
\end{aligned}$$

20.1.2 Derivatives on $\mathbb{G}$

We now relate the usual Euclidean derivatives on $\mathbb{R}^N \equiv \mathbb{G}$ to the derivatives along the Jacobian basis related to $\mathbb{G}$. We recall formulas (1.37) and (1.38) (page 22) where we showed that, for any smooth function u on $\mathbb{G}$ and every $x \in \mathbb{G}$, it holds

$$\begin{aligned}
\big(Z_1 u(x), \ldots, Z_N u(x)\big) &= \big(\partial_{x_1} u(x), \ldots, \partial_{x_N} u(x)\big) \cdot \mathcal{J}_{\tau_x}(0),\\
\big(\partial_{x_1} u(x), \ldots, \partial_{x_N} u(x)\big) &= \big(Z_1 u(x), \ldots, Z_N u(x)\big) \cdot \mathcal{J}_{\tau_{x^{-1}}}(0),
\end{aligned} \tag{20.9}$$

where

$$\mathcal{J}_{\tau_x}(0) = \begin{pmatrix} 1 & 0 & \cdots & 0\\ a_{2,1}(x) & 1 & \ddots & \vdots\\ \vdots & \ddots & \ddots & 0\\ a_{N,1}(x) & \cdots & a_{N,N-1}(x) & 1 \end{pmatrix}$$

is the Jacobian matrix of the left translation by x on $\mathbb{G}$ (i.e. $\tau_x(y) = x * y$) and the $a_{i,j}$'s are polynomial functions on $\mathbb{G}$. More explicitly, the usual partial derivatives w.r.t. the fixed coordinates x_i's on $\mathbb{G}$ and the vector fields of the Jacobian basis are related by the following "triangle-shaped" formulas:

$$\begin{aligned}
Z_1 &= \partial_{x_1} + a_{2,1}(x)\,\partial_{x_2} + a_{3,1}(x)\,\partial_{x_3} + \cdots + a_{N,1}(x)\,\partial_{x_N},\\
Z_2 &= \qquad\qquad\quad \partial_{x_2} + a_{3,2}(x)\,\partial_{x_3} + \cdots + a_{N,2}(x)\,\partial_{x_N},\\
&\ \vdots \qquad\qquad\qquad\qquad\qquad\qquad\qquad\qquad \vdots\\
Z_N &= \qquad\qquad\qquad\qquad\qquad\qquad\qquad\qquad\quad \partial_{x_N}
\end{aligned} \tag{20.10}$$

and, setting $b_{i,j}(x) := a_{i,j}(x^{-1})$,

$$\begin{aligned}
\partial_{x_1} &= Z_1 + b_{2,1}(x)\,Z_2 + b_{3,1}(x)\,Z_3 + \cdots + b_{N,1}(x)\,Z_N,\\
\partial_{x_2} &= \qquad\quad Z_2 + b_{3,2}(x)\,Z_3 + \cdots + b_{N,2}(x)\,Z_N,\\
&\ \vdots \qquad\qquad\qquad\qquad\qquad\qquad\qquad \vdots\\
\partial_{x_N} &= \qquad\qquad\qquad\qquad\qquad\qquad\qquad\quad Z_N.
\end{aligned} \tag{20.11}$$

Taking into account the δ_λ-degrees of the Z_h's and of the ∂_{x_h}'s we have the formulas

$$
\begin{aligned}
Z_h &= \partial_{x_h} + \sum_{j:\,\sigma_j>\sigma_h} a_{j,h}(x)\,\partial_{x_j}, \\
\partial_{x_h} &= Z_h + \sum_{j:\,\sigma_j>\sigma_h} b_{j,h}(x)\,Z_j.
\end{aligned}
\tag{20.12}
$$

For example, if $\mathbb{G}$ is the Heisenberg–Weyl group on $\mathbb{R}^3$ (see Example 20.1.3),

$$Z_1 = \partial_x + 2y\partial_t, \quad Z_2 = \partial_y - 2x\partial_t, \quad Z_3 = \partial_t$$

is the Jacobian basis. It is immediately seen that

$$
\begin{aligned}
Z_1 &= \partial_x + 2y\,\partial_t, \\
Z_2 &= \partial_y - 2x\,\partial_t, \\
Z_3 &= \partial_t, \\
\\
\partial_x &= Z_1 - 2y\,Z_3, \\
\partial_y &= Z_2 + 2x\,Z_3, \\
\partial_t &= Z_3.
\end{aligned}
$$

For any multi-index α, we introduce the higher order Z-derivative setting

$$Z^\alpha := Z_1^{\alpha_1} \cdots Z_N^{\alpha_N}. \tag{20.13}$$

Obviously, Z^α is a left invariant differential operator (not necessarily a vector field) and, as a differential operator, it is δ_λ-homogeneous of degree

$$\alpha_1\sigma_1 + \cdots + \alpha_N\sigma_N = |\alpha|_{\mathbb{G}}$$

(for Z_j is δ_λ-homogeneous of degree σ_j).

We explicitly observe that, a priori, Z_2Z_1 is not necessarily contained in the set of the Z^α's. Nonetheless, we have the following proposition.

Proposition 20.1.4. *For every $k \in \mathbb{N}$, every $i_1, \ldots, i_k \in \{1, \ldots, N\}$, and every $\beta_1, \ldots, \beta_k \in \mathbb{N} \cup \{0\}$, we have*

$$(Z_{i_1})^{\beta_1} \cdots (Z_{i_k})^{\beta_k} \in \operatorname{span}\{(Z_1 \cdots Z_N)^\alpha : \alpha \text{ multi-index}\}.$$

More precisely, the multi-indices in the right-hand side run over the set of the α's such that[1] $|\alpha|_{\mathbb{G}} = \sigma_{i_1}\beta_1 + \cdots + \sigma_{i_k}\beta_k$.

Proof. First, we write

$$(Z_{i_1})^{\beta_1} \cdots (Z_{i_k})^{\beta_k} = \underbrace{Z_{i_1} \cdots Z_{i_1}}_{\beta_1 \text{ times}} \cdots \underbrace{Z_{i_k} \cdots Z_{i_k}}_{\beta_k \text{ times}} = Z_{i_1} \cdots (Z_i Z_j) \cdots Z_{i_k},$$

[1] Here, the σ_{i_j}'s are the exponents in (20.2).

where by i and j we have indicated an arbitrary couple of contiguous indices. Now, since $Z_i Z_j = [Z_i, Z_j] + Z_j Z_i$, we have

$$Z_{i_1} \cdots (Z_i Z_j) \cdots Z_{i_k} = Z_{i_1} \cdots [Z_i, Z_j] \cdots Z_{i_k} + Z_{i_1} \cdots (Z_j Z_i) \cdots Z_{i_k}. \quad (20.14)$$

Since $[Z_i, Z_j]$ is a left invariant vector field, it can be written as a linear combination of the Z_k's, say $[Z_i, Z_j] = \sum_r c_r Z_r$, so that the first summand in the right-hand side of (20.14) equals $\sum_r c_r Z_{i_1} \cdots Z_r \cdots Z_{i_k}$. Moreover, the Z_r's can be chosen so that $\sigma_r = \sigma_i + \sigma_j$.

As a consequence, we have written the left-hand side of (20.14) as the sum of a term with Z_i and Z_j interchanged plus a linear combination of summands each with $\beta_1 + \cdots + \beta_k - 1$ factors of the type Z_h's. Also, these two terms are δ_λ-homogeneous operators of the same degree as the left-hand side (namely, $\sigma_{i_1}\beta_1 + \cdots + \sigma_{i_k}\beta_k$). Now, this argument can be iterated and, arguing by induction, the assertion can be easily proved. $\square$

For example, if $\mathbb{G}$ is the Heisenberg–Weyl group on $\mathbb{R}^3$ (see Example 20.1.3) and $Z_1 = \partial_x + 2y\partial_t$, $Z_2 = \partial_y - 2x\partial_t$, $Z_3 = \partial_t$ is the Jacobian basis, then $Z_2 Z_1$ does not belong to the set of the Z_α's in (20.13). Indeed, it holds

$$\begin{aligned}
Z_1^2 &= \partial_{x,x} + 4y\,\partial_{x,t} + 4y^2\,\partial_{t,t},\\
Z_2^2 &= \partial_{y,y} - 4x\,\partial_{y,t} + 4x^2\,\partial_{t,t},\\
Z_1 Z_2 &= \partial_{x,y} - 2\,\partial_t - 2x\,\partial_{x,t} + 2y\,\partial_{y,t} - 4xy\,\partial_{t,t},\\
Z_2 Z_1 &= \partial_{x,y} + 2\,\partial_t + 2y\,\partial_{y,t} - 2x\,\partial_{x,t} - 4xy\,\partial_{t,t}.
\end{aligned}$$

Though, as asserted in the above proposition, it holds

$$Z_2 Z_1 = Z_1 Z_2 + 4\,Z_3 = (Z_1 Z_2 Z_3)^{(1,1,0)} + 4\,(Z_1 Z_2 Z_3)^{(0,0,1)}.$$

We now provide a higher-order version of formula (20.12).

Proposition 20.1.5 (Z^α's and $(D_x)^\alpha$'s). *For every multi-index α, we have*

$$Z^\alpha = (D_x)^\alpha + \sum_{\beta:\,\beta\neq\alpha,\ |\beta|\le|\alpha|,\ |\beta|_{\mathbb{G}}\ge|\alpha|_{\mathbb{G}}} a_{\beta,\alpha}(x)\,(D_x)^\beta, \quad (20.15)$$

where the $a_{\beta,\alpha}$'s are polynomials, δ_λ-homogeneous of degree $|\beta|_{\mathbb{G}} - |\alpha|_{\mathbb{G}}$.

Example 20.1.6. If $\mathbb{G}$ is the Heisenberg–Weyl group on $\mathbb{R}^3$ (see Example 20.1.3) and $Z_1 = \partial_x + 2y\partial_t$, $Z_2 = \partial_y - 2x\partial_t$, $Z_3 = \partial_t$ is the Jacobian basis, we have

$$\begin{aligned}
(Z_1 Z_2 Z_3)^{(1,1,0)} &= \partial_{x,y} - 2\partial_t - 2x\partial_{x,t} + 2y\partial_{y,t} - 4xy\partial_{t,t}\\
&= (D_{x,y,t})^{(1,1,0)} - 2(D_{x,y,t})^{(0,0,1)} - 2x(D_{x,y,t})^{(1,0,1)}\\
&\quad + 2y(D_{x,y,t})^{(0,1,1)} - 4xy(D_{x,y,t})^{(0,0,2)}.
\end{aligned}$$

We compare this to (20.15). We notice that the far right-hand side contains $(D_x)^\alpha$ plus 4 summands having $(D_x)^\beta$'s terms with $\beta \neq \alpha$: the first term satisfies $|\beta| < |\alpha|$ and $|\beta|_{\mathbb{G}} = |\alpha|_{\mathbb{G}}$, the second, third and fourth terms satisfy $|\beta| = |\alpha|$ and $|\beta|_{\mathbb{G}} > |\alpha|_{\mathbb{G}}$.

Proof (of Proposition 20.1.5). We argue by induction on the (isotropic) length of α. If $|\alpha| = 1$, then $\alpha = e_h$ (the h-th vector of the standard basis of $\mathbb{R}^N$) and $Z^\alpha = Z_h$. In this case, (20.15) immediately reduces to (20.12) (see also Proposition 1.3.5, page 34).

Suppose now that (20.15) has been proved for every multi-index α with $|\alpha| \le k$. Let us prove it for a multi-index α with $|\alpha| = k+1$. We have

$$\alpha = (0, \dots, 0, \alpha_h, \dots, \alpha_N), \qquad \text{with } \alpha_h \neq 0.$$

In particular, it holds $Z^\alpha = Z_h\, Z^{\widehat{\alpha}}$, where $\widehat{\alpha} = \alpha - e_h$, so that $|\widehat{\alpha}| = k$. Moreover, we have $|\widehat{\alpha}|_{\mathbb{G}} = |\alpha|_{\mathbb{G}} - \sigma_h$. As a consequence, by the inductive hypothesis, we have

$$\begin{aligned} Z^\alpha &= Z_h \Bigg(\sum_{\beta:\, |\beta| \le |\widehat{\alpha}|,\, |\beta|_{\mathbb{G}} \ge |\widehat{\alpha}|_{\mathbb{G}}} a_{\beta,\widehat{\alpha}}(x)\, (D_x)^\beta \Bigg) \\ &= \Bigg(\sum_{\sigma_j \ge \sigma_h} a_j\, \partial_{x_j} \Bigg) \Bigg(\sum_{|\beta| \le k,\, |\beta|_{\mathbb{G}} \ge |\widehat{\alpha}|_{\mathbb{G}}} \widehat{a}_\beta\, (D_x)^\beta \Bigg) \\ &= \sum_{\sigma_j \ge \sigma_h,\, |\beta| \le k,\, |\beta|_{\mathbb{G}} \ge |\widehat{\alpha}|_{\mathbb{G}}} a_j (\partial_{x_j} \widehat{a}_\beta)\, (D_x)^\beta \\ &\quad + \sum_{\sigma_j \ge \sigma_h,\, |\beta| \le k,\, |\beta|_{\mathbb{G}} \ge |\widehat{\alpha}|_{\mathbb{G}}} a_j\, \widehat{a}_\beta\, (D_x)^{\beta + e_j}. \end{aligned}$$

Now we treat each of the two sums appearing in the above far right-hand side. First, we notice that $\widehat{a}_\beta$ is δ_λ-homogeneous of degree $|\beta|_{\mathbb{G}} - |\widehat{\alpha}|_{\mathbb{G}} = |\beta|_{\mathbb{G}} - |\alpha|_{\mathbb{G}} + \sigma_h$ and that a_j is δ_λ-homogeneous of degree $\sigma_j - \sigma_h$. Then the $\mathbb{G}$-degree of $a_j(\partial_{x_j}\widehat{a}_\beta)$ is $\sigma_j - \sigma_h + |\beta|_{\mathbb{G}} - |\alpha|_{\mathbb{G}} + \sigma_h - \sigma_j = |\beta|_{\mathbb{G}} - |\alpha|_{\mathbb{G}}$. Moreover, $\partial_{x_j}\widehat{a}_\beta \equiv 0$ whenever the $\mathbb{G}$-degree of $\widehat{a}_\beta$ is $< \sigma_j$, i.e. whenever $|\beta|_{\mathbb{G}} - |\alpha|_{\mathbb{G}} < \sigma_j - \sigma_h$. So we can suppose that $|\beta|_{\mathbb{G}} - |\alpha|_{\mathbb{G}} \ge \sigma_j - \sigma_h \ge 0$.

Finally, as for the second sum, the $\mathbb{G}$-degree of $a_j\,\widehat{a}_\beta$ is $\sigma_j - \sigma_h + |\beta|_{\mathbb{G}} - |\alpha|_{\mathbb{G}} + \sigma_h = \sigma_j + |\beta|_{\mathbb{G}} - |\alpha|_{\mathbb{G}} = |\beta + e_j|_{\mathbb{G}} - |\alpha|_{\mathbb{G}}$. Moreover, $|\beta + e_j|_{\mathbb{G}} = |\beta|_{\mathbb{G}} + \sigma_j \ge |\widehat{\alpha}|_{\mathbb{G}} + \sigma_h = |\alpha|_{\mathbb{G}}$.

We now isolate the coefficient of $(D_x)^\alpha$. In the first sum, this term does not appear, since the sum runs over the β's with $|\beta| \le k$, whereas $|\alpha| = k+1$. In the second sum, this term appears only when j and β (satisfying also $\sigma_j \ge \sigma_h$, $|\beta| \le k$ and $|\beta|_{\mathbb{G}} \ge |\alpha|_{\mathbb{G}} - \sigma_h$) are such that $\beta + e_j = \alpha$. In particular, $\beta = \alpha - e_j$, so that $|\beta|_{\mathbb{G}} = |\alpha|_{\mathbb{G}} - \sigma_j$, which is $\ge |\alpha|_{\mathbb{G}} - \sigma_h$ only if $\sigma_j \le \sigma_h$. But since the sum runs over the j's such that $\sigma_j \ge \sigma_h$, it must be $\sigma_j = \sigma_h$. Now, from the first equality in (20.12), we see that the only summand with $\sigma_j = \sigma_h$ is the one with $j = h$, and in this case $a_h = 1$. This gives $\beta = \alpha - e_h = \widehat{\alpha}$, so that

$$a_j\, \widehat{a}_\beta = a_j\, a_{\beta,\widehat{\alpha}} = a_j\, a_{\widehat{\alpha},\widehat{\alpha}} = 1,$$

also using the inductive hypothesis $a_{\widehat{\alpha},\widehat{\alpha}} = 1$. This completes the proof. □

A completely analogous proof (this time making use of the second formula in (20.12)) demonstrates the following result.

Proposition 20.1.7 (**$(D_x)^\alpha$'s and Z^α's**). *For every multi-index α, we have*

$$(D_x)^\alpha = Z^\alpha + \sum_{\beta:\,\beta\neq\alpha,\ |\beta|\le|\alpha|,\ |\beta|_{\mathbb{G}}\ge|\alpha|_{\mathbb{G}}} b_{\beta,\alpha}(x)\, Z^\beta, \tag{20.16}$$

where the $b_{\beta,\alpha}$'s are polynomials, δ_λ-homogeneous of degree $|\beta|_{\mathbb{G}} - |\alpha|_{\mathbb{G}}$.

From Propositions 20.1.5 and 20.1.7 we derive the following corollary.

Corollary 20.1.8. *Let u be a smooth real-valued function on $\mathbb{G}$. Then u is a polynomial if and only if there exists $k \in \mathbb{N}$ such that $Z^\alpha u \equiv 0$ for every multi-index α with $|\alpha|_{\mathbb{G}} \ge k$ (or with $|\alpha| \ge k$).*

The same assertion holds if there exists $k \in \mathbb{N}$ such that $Z^\alpha u \equiv 0$ for every multi-index α with $|\alpha|_{\mathbb{G}} = k$.

Proof. First, we remark that (see (20.3))

$$\sigma_1\, |\alpha| \le |\alpha|_{\mathbb{G}} \le \sigma_N\, |\alpha|. \tag{20.17}$$

As a consequence, the "only if" part follows from (20.15). Vice versa, if $Z^\beta u \equiv 0$ for every multi-index β with $|\beta|_{\mathbb{G}} \ge k$, then (20.16) proves that $(D_x)^\alpha u \equiv 0$ whenever $|\alpha|_{\mathbb{G}} \ge k$. Thanks to (20.17), if $|\alpha| \ge k/\sigma_1$, then $|\alpha|_{\mathbb{G}} \ge k$, so that $(D_x)^\alpha u \equiv 0$. Hence u is a polynomial.

As for the last statement of the corollary, we only need to prove that if there exists $k \in \mathbb{N}$ such that $Z^\alpha u \equiv 0$ whenever $|\alpha|_{\mathbb{G}} = k$, then there exists $k^* \in \mathbb{N}$ such that $Z^\alpha u \equiv 0$ whenever $|\alpha|_{\mathbb{G}} \ge k^*$. Now, it suffices to take k^* large enough depending on k and $\sigma_1, \dots, \sigma_N$. Indeed[2], we argue by induction noticing that

$$Z^\alpha = Z_{i_0}^{\alpha_{i_0}}\, Z_{1+i_0}^{\alpha_{1+i_0}} \cdots Z_N^{\alpha_N} = Z_{i_0}^{\alpha_{i_0}-1}\big(Z_{i_0}^{\alpha_{i_0}-1}\, Z_{1+i_0}^{\alpha_{1+i_0}} \cdots Z_N^{\alpha_N}\big)$$

(here i_0 is the first index such that $\alpha_{i_0} \neq 0$) and

$$\big|(0, \dots, 0, \alpha_{i_0} - 1, \alpha_{1+i_0}, \dots, \alpha_N)\big|_{\mathbb{G}} = |\alpha|_{\mathbb{G}} - \sigma_{i_0}.$$

This ends the proof. □

We now recall that, since $\mathbb{G}$ is a Carnot group, then the elements of the Jacobian basis laying in the first layer of the stratification of $\mathfrak{g}$ Lie-generate the whole $\mathfrak{g}$. Setting $m := N_1$ in (20.1), we denote these vectors by $Z_1, \dots, Z_m$. We thus have

$$\mathrm{Lie}\{Z_1, \dots, Z_m\} = \mathrm{span}\{Z_1, \dots, Z_N\} = \mathfrak{g}. \tag{20.18}$$

The following result holds.

---

[2] Notice that the choice $k^* = k$ will not do. For instance, in the Heisenberg group on $\mathbb{R}^3$ of Example 20.1.6, if we know that $Z^\alpha u \equiv 0$ whenever $|\alpha|_{\mathbb{G}} \ge 5$, then we cannot immediately deduce by induction that $Z^\alpha u \equiv 0$ whenever $|\alpha|_{\mathbb{G}} = 6$. For example, $Z_3^3 = Z^{(0,0,3)}$ and $|(0,0,3)|_{\mathbb{G}} = 6$, but $Z_3^3 = Z_3\, Z_3^2 = Z_3(Z^{(0,0,2)})$ and $|(0,0,2)|_{\mathbb{G}} = 4$.

Proposition 20.1.9. *It holds*

$$\begin{aligned}&\operatorname{span}\{Z_{i_1}\cdots Z_{i_k} : i_1,\dots,i_k\in\{1,\dots,m\},\ k\in\mathbb{N}\}\\&\quad=\operatorname{span}\{(Z_1\cdots Z_N)^\alpha : \alpha \textit{ multi-index}\}.\end{aligned}$$

Proof. The inclusion $\subseteq$ follows from Proposition 20.1.4. In order to prove the opposite inclusion, it suffices to prove that $Z_1,\dots,Z_N$ can all be written as linear combinations of terms of the type $Z_{i_1}\cdots Z_{i_k}$. This fact now easily follows from (20.18), since

$$[\cdots[[Z_{i_1},Z_{i_2}],Z_{i_3}]\cdots Z_{i_k}]\in\operatorname{span}\{Z_{i_1}\cdots Z_{i_k} : i_j\in\{1,\dots,m\},\ k\in\mathbb{N}\}.$$

This ends the proof. □

With the aid of Lemma 20.1.11 below (whose proof is left as an exercise), it can immediately be proved the following result (compare it to Corollary 1.5.5, page 68).

Corollary 20.1.10. *Let u be a smooth real-valued function on $\mathbb{G}$. Then u is a polynomial if and only if there exists $k\in\mathbb{N}$ such that $Z_{i_1}\cdots Z_{i_k}u\equiv 0$ for every $i_1,\dots,i_k\in\{1,\dots,m\}$.*

Proof. It follows by collecting together Corollary 20.1.8, Proposition 20.1.9 and Lemma 20.1.11 below. □

Lemma 20.1.11. *If $H_1,\dots,H_n$ are (not identically vanishing) differential operators on $\mathbb{R}^N\equiv\mathbb{G}$ which are δ_λ-homogeneous of degrees $d_1,\dots,d_n$, respectively, with $d_i\neq d_j$ for every $i\neq j$, then they are linearly independent.*

Example 20.1.12. Let $\mathbb{G}$ be the Heisenberg–Weyl group on $\mathbb{R}^3$ (see Example 20.1.6). Then, for example, we know that u is a polynomial provided that

$$Z_1^2u,\ Z_1Z_2u,\ Z_2^2u,\quad Z_3u$$

all vanish identically: this is stated in the last part of Corollary 20.1.8. Also, we know that u is a polynomial function provided that

$$Z_1^2u,\ Z_1Z_2u,\ Z_2Z_1u,\ Z_2^2u$$

all vanish identically: this is stated in Corollary 20.1.10. These conditions are equivalent, though different in appearance.

20.2 Taylor Polynomials on Homogeneous Carnot Groups

We begin with another result on the derivatives on $\mathbb{G}$, helpful for our Taylor formula. We follow the notation in the previous sections (see in particular (20.6)–(20.8) and (20.13)).

Theorem 20.2.1. *Let $n \in \mathbb{N} \cup \{0\}$ and consider the space $\mathcal{P}_n$ of the polynomials δ_λ-homogeneous of degree at most n. Moreover, let $\mathcal{I}_n$ be ordered in some fixed way. Then the map*

$$\mathcal{P}_n \to \mathbb{R}^{\mu_n}, \qquad p \mapsto ((Z^\alpha p)(0))_{\alpha \in \mathcal{I}_n}$$

is a vector space isomorphism.

Proof. We consider (20.15) of Proposition 20.1.5. Since $a_{\beta,\alpha}(x)$ is δ_λ-homogeneous of degree $|\beta|_{\mathbb{G}} - |\alpha|_{\mathbb{G}}$, we have $a_{\beta,\alpha}(0) = 0$ whenever $|\beta|_{\mathbb{G}} > |\alpha|_{\mathbb{G}}$ and $a_{\beta,\alpha}(x)$ =constant (say, $a_{\beta,\alpha}$) whenever $|\beta|_{\mathbb{G}} = |\alpha|_{\mathbb{G}}$. Setting $a_{\alpha,\alpha} = 1$, from (20.15) we thus derive

$$Z^\alpha|_{x=0} = \sum_{\beta:\, |\beta| \le |\alpha|,\, |\beta|_{\mathbb{G}} = |\alpha|_{\mathbb{G}}} a_{\beta,\alpha}\, (D_x)^\beta|_{x=0}. \tag{20.19}$$

Now, we recall the dual formula

$$(D_x)^\alpha|_{x=0} = \sum_{\beta:\, |\beta| \le |\alpha|,\, |\beta|_{\mathbb{G}} = |\alpha|_{\mathbb{G}}} b_{\beta,\alpha}\, Z^\beta|_{x=0} \tag{20.20}$$

implied by (20.16) of Proposition 20.1.7. These facts prove the following result: If we indicate by $\Phi : \mathcal{P}_n \to \mathbb{R}^{\mu_n}$ the map in Proposition 20.1.2 and by $\Psi : \mathcal{P}_n \to \mathbb{R}^{\mu_n}$ the map in Theorem 20.2.1, then (20.19) shows that there exists a linear map $\phi : \mathbb{R}^{\mu_n} \to \mathbb{R}^{\mu_n}$ such that

$$\Psi(p) = \phi(\Phi(p)) \quad \text{for every } p \in \mathcal{P}_n;$$

analogously, (20.20) shows that there exists a linear map $\psi : \mathbb{R}^{\mu_n} \to \mathbb{R}^{\mu_n}$ such that

$$\Phi(p) = \psi(\Psi(p)) \quad \text{for every } p \in \mathcal{P}_n.$$

These facts together demonstrate that ψ and ϕ are inverse to each other, so that $\Psi = \phi \circ \Phi$ is a linear isomorphism since Φ is (see Proposition 20.1.2). □

Theorem 20.2.1 has the following important corollary.

Corollary 20.2.2. *Let $n \in \mathbb{N} \cup \{0\}$. Let, as usual, $\mathcal{P}_n$ be the space of polynomials, δ_λ-homogeneous of degree at most n, and let μ_n be the dimension of $\mathcal{P}_n$. Then, for any μ_n-tuple of real numbers z, there exists one and only one polynomial p in $\mathcal{P}_n$ such that*

$$((Z^\alpha p)(0))_{\alpha \in \mathcal{I}_n} = z. \tag{20.21}$$

Here, as usual, we have fixed an ordering on $\mathcal{I}_n$, the set of multi-indices α such that $|\alpha|_{\mathbb{G}} \le n$.

Corollary 20.2.2 proves that the following definition is well-posed.

Definition 20.2.3 (Z-Mac Laurin polynomial). *Let $f \in C^\infty(\mathbb{G}, \mathbb{R})$. Let, as usual, $Z_1, \ldots, Z_N$ denote the Jacobian basis related to $\mathbb{G}$. Then, for every $n \in \mathbb{N} \cup \{0\}$, there exists one and only one polynomial p, δ_λ-homogeneous of degree at most n, such that*

$$(Z_1 \cdots Z_N)^\alpha(p)(0) = (Z_1 \cdots Z_N)^\alpha(f)(0) \qquad \textit{for every } \alpha \in \mathcal{I}_n. \tag{20.22}$$

We say that $p(x) = P_n(f,0)(x)$ *is the* Z-Mac Laurin polynomial *of* δ_λ*-degree* n*, related to* f.

Indeed, it suffices to apply Corollary 20.2.2 with

$$z = ((Z^\alpha f)(0))_{\alpha \in \mathcal{I}_n}.$$

Remark 20.2.4. In (20.56) of Section 20.3.2, we shall provide an explicit formula for $P_n(f,0)$.

An example is in order.

Example 20.2.5. Let us consider $\mathbb{G} = \mathbb{H}^1$, the Heisenberg–Weyl group on $\mathbb{R}^3$ (see also Example 20.1.3), with its Jacobian basis $Z_1 = \partial_x + 2y\partial_t$, $Z_2 = \partial_y - 2x\partial_t$, $Z_3 = \partial_t$. Let $f \in C^\infty(\mathbb{H}^1, \mathbb{R})$. Let us find $p = P_1(f,0)$. Since $p \in \mathcal{P}_1$, there exist $a, b, c \in \mathbb{R}$ such that $p(x,y,t) = a + b\,x + c\,y$. Then (20.22) gives

$$\begin{aligned} a &= p(0) = f(0), \\ b &= Z_1 p(0) = Z_1 f(0), \\ c &= Z_2 p(0) = Z_2 f(0), \end{aligned}$$

so that

$$P_1(f,0)(\mathbf{x},\mathbf{y},\mathbf{t}) = f(0) + Z_1 f(0)\,\mathbf{x} + Z_2 f(0)\,\mathbf{y}.$$

Let us now find $q = P_2(f,0)$. Since $q \in \mathcal{P}_2$, there exist $a,b,c,d,e,f,g \in \mathbb{R}$ such that $q(x,y,t) = a + b\,x + c\,y + d\,t + e\,x^2 + f\,y^2 + g\,xy$. Then (20.22) gives

$$\begin{aligned} a &= q(0) = f(0), \\ b &= Z_1 q(0) = Z_1 f(0), \\ c &= Z_2 q(0) = Z_2 f(0), \\ d &= Z_3 q(0) = Z_3 f(0), \\ 2e &= (Z_1)^2 q(0) = (Z_1)^2 f(0), \\ 2f &= (Z_2)^2 q(0) = (Z_2)^2 f(0), \\ g - 2d &= Z_1 Z_2 q(0) = Z_1 Z_2 f(0), \end{aligned}$$

so that

$$\begin{aligned} P_2(f,0)(\mathbf{x},\mathbf{y},\mathbf{t}) = {} & f(0) + Z_1 f(0)\,\mathbf{x} + Z_2 f(0)\,\mathbf{y} + Z_3 f(0)\,\mathbf{t} \\ & + \frac{1}{2}(Z_1)^2 f(0)\,\mathbf{x^2} + \frac{1}{2}(Z_2)^2 f(0)\,\mathbf{y^2} + (2Z_3 f(0) \\ & + Z_1 Z_2 f(0))\,\mathbf{xy}. \end{aligned}$$

Since in this case $Z_3 = -\frac{1}{4}(Z_1 Z_2 - Z_2 Z_1)$, this formula can be rewritten as

$$\begin{aligned}
&P_2(f,0)(\mathbf{x},\mathbf{y},\mathbf{t})\\
&\quad = f(0) + Z_1 f(0)\,\mathbf{x} + Z_2 f(0)\,\mathbf{y} - \frac{1}{4}(Z_1Z_2 - Z_2Z_1)f(0)\,\mathbf{t}\\
&\qquad + \frac{1}{2}(Z_1)^2 f(0)\,\mathbf{x}^2 + \frac{1}{2}(Z_2)^2 f(0)\,\mathbf{y}^2 + \frac{1}{2}(Z_1Z_2 + Z_2Z_1)f(0)\mathbf{xy}. \quad (20.23)
\end{aligned}$$

Notice that, though unique, the Z-Mac Laurin polynomial can be written in many different symbolic ways.

For instance, observe that (20.23) can be rewritten in the suggesting form

$$\begin{aligned}
P_2(f,0)(\mathbf{x},\mathbf{y},\mathbf{t}) &= f(0) + \nabla_{\mathbb{G}} f(0)\cdot\begin{pmatrix}\mathbf{x}\\ \mathbf{y}\end{pmatrix}\\
&\quad + Z_3 f(0)\,\mathbf{t} + \frac{1}{2}\,(\mathbf{x},\mathbf{y})\cdot \mathrm{Hess}_{\mathrm{sym}} f(0)\cdot\begin{pmatrix}\mathbf{x}\\ \mathbf{y}\end{pmatrix}, \quad (20.24)
\end{aligned}$$

where $\nabla_{\mathbb{G}} f(0) = (Z_1 f(0), Z_2 f(0))$ is the *horizontal gradient* of f at 0 and

$$\mathrm{Hess}_{\mathrm{sym}} f(0) := \begin{pmatrix} Z_1^2 f(0) & \frac{1}{2}(Z_1Z_2 f(0) + Z_2Z_1 f(0))\\ \frac{1}{2}(Z_1Z_2 f(0) + Z_2Z_1 f(0)) & Z_2^2 f(0)\end{pmatrix}$$

is the so-called *symmetrized horizontal Hessian* of f at 0. See Ex. 4 for the computation of $P_3(f,0)$ in the present case. See Ex. 6 for a suitable generalization of (20.24) to other Carnot groups and see Section 20.3.2 for a general formula for $P_n(f,0)$ for every n and every Carnot group. □

We leave as an exercise (see Ex. 7 at the end of the chapter) to prove the following assertion.

Remark 20.2.6. Let $\mathcal{P}_n$ be the vector space of the polynomials with $\mathbb{G}$-degree at most n. Let also $\mathcal{P}$ be the set of polynomial functions on $\mathbb{G}$. Since the set of all multi-indices $\mathcal{I} = (\mathbb{N}\cup\{0\})^N$ can be decomposed as $\mathcal{I} = \bigcup_{n\geq 0}\mathcal{I}_n$, the following map is well-defined

$$\pi_n : \mathcal{P} \longrightarrow \mathcal{P}_n, \qquad p \mapsto p_n,$$

where

$$p(x) = \sum_{\alpha} c_\alpha\, x^\alpha = \sum_{\alpha\in\mathcal{I}_n} c_\alpha\, x^\alpha + \sum_{\alpha\notin\mathcal{I}_n} c_\alpha\, x^\alpha =: p_n(x) + p_n^*(x).$$

Then, if $Q_n(f,0)(x)$ denotes the usual Euclidean Mac Laurin polynomial of f, it is easy to see that

$$P_n(f,0)(x) = \pi_n\big(Q_n(f,0)(x)\big).$$

In other words, *the Z-Mac Laurin polynomial of f coincides with the sum of the terms, in the usual Euclidean Mac Laurin polynomial of f, which are δ_λ-homogeneous of degree at most n.*

The following result is not so evident, a priori.

Proposition 20.2.7 (Nested property of the Z-Mac Laurin polynomials). *Let $n \in \mathbb{N}$ and $f \in C^\infty(\mathbb{G}, \mathbb{R})$. Then $P_n(f, 0) - P_{n-1}(f, 0)$ is a δ_λ-homogeneous polynomial of degree n. In other words,*

$$P_n(f, 0) = P_{n-1}(f, 0) + \{a \text{ polynomial of } \mathbb{G}\text{-degree} = n\}.$$

Proof. Let us write $P_n(f, 0) = p + q$, where p has $\mathbb{G}$-degree $\leq n - 1$, so that q has $\mathbb{G}$-degree $= n$ (unless it vanishes identically). We have to prove that $p = P_{n-1}(f, 0)$. By Definition 20.2.3, this follows if we show that

$$Z^\alpha(p)(0) = Z^\alpha(f)(0) \qquad \text{for every } \alpha \in \mathcal{I}_{n-1}. \tag{20.25}$$

If $\alpha \in \mathcal{I}_{n-1}$, then $\alpha \in \mathcal{I}_n$, so that by the very definition of $P_n(f, 0)$, we have

$$Z^\alpha(f)(0) = Z^\alpha(P_n(f, 0))(0) = Z^\alpha(p + q)(0) = Z^\alpha p(0) + Z^\alpha q(0).$$

Then, (20.25) will follow if we show that $Z^\alpha q(0) = 0$. This fact is a consequence of the fact that $Z^\alpha q$ is a polynomial, δ_λ-homogeneous of positive degree, precisely $n - |\alpha|_{\mathbb{G}}$ (recall that q has $\mathbb{G}$-degree $= n$ and $|\alpha|_{\mathbb{G}} \leq n - 1$, since $\alpha \in \mathcal{I}_{n-1}$). This ends the proof. □

The following definition is definitely not unexpected.

Definition 20.2.8 (Z-Taylor polynomial). *Let $f \in C^\infty(\mathbb{G}, \mathbb{R})$, $x_0 \in \mathbb{G}$ and $n \in \mathbb{N} \cup \{0\}$ be fixed. Let us consider the Z-Mac Laurin polynomial of the function*

$$x \mapsto f(x_0 * x),$$

i.e. the polynomial $P_n(f \circ \tau_{x_0}, 0)$. We say that the polynomial

$$P_n(f, x_0)(x) := P_n(f \circ \tau_{x_0}, 0)(x_0^{-1} * x) \tag{20.26}$$

is the Z-Taylor polynomial *of $\mathbb{G}$-degree n related to f and centered at x_0.*

Remark 20.2.9. In (20.56) of Section 20.3.2, we shall provide an explicit formula for $P_n(f, x_0)$.

The following characterization of $P_n(f, x_0)$ holds.

Proposition 20.2.10 (Characterization). *With the notation of the above definition, $P_n(f, x_0)$ is characterized by being the only polynomial p, δ_λ-homogeneous of degree at most n, such that*

$$(Z_1 \cdots Z_N)^\alpha(p)(x_0) = (Z_1 \cdots Z_N)^\alpha(f)(x_0) \qquad \textit{for every } \alpha \in \mathcal{I}_n. \tag{20.27}$$

Proof. An immediate consequence of Definitions 20.2.3, 20.2.8 and the left invariance of $Z_1, \ldots, Z_N$. □

With reference to Example 20.2.5, setting $z_0 = (x_0, y_0, t_0)$, we have

$$
\begin{aligned}
P_2(f \circ \tau_{z_0}, 0) = f(z_0) + Z_1 f(z_0)\, \mathbf{x} + Z_2 f(z_0)\, \mathbf{y} + Z_3 f(z_0)\, \mathbf{t} \\
+ \frac{1}{2}(Z_1)^2 f(z_0)\, \mathbf{x^2} + \frac{1}{2}(Z_2)^2 f(z_0)\, \mathbf{y^2} \\
+ \frac{1}{2}(Z_1 Z_2 + Z_2 Z_1) f(z_0)\, \mathbf{xy},
\end{aligned}
$$

so that

$$
\begin{aligned}
P_2(f, z_0)(\mathbf{x}, \mathbf{y}, \mathbf{t}) &= P_2(f \circ \tau_{z_0}, 0)(z_0^{-1} * (\mathbf{x}, \mathbf{y}, \mathbf{t})) \\
&= f(z_0) + Z_1 f(z_0)\, (\mathbf{x} - x_0) + Z_2 f(z_0)\, (\mathbf{y} - y_0) \\
&\quad + Z_3 f(z_0)\, (\mathbf{t} - t_0 - 2y_0\mathbf{x} + 2x_0\mathbf{y}) + \frac{1}{2}(Z_1)^2 f(z_0)\, (\mathbf{x} - x_0)^2 \\
&\quad + \frac{1}{2}(Z_2)^2 f(z_0)\, (\mathbf{y} - y_0)^2 \\
&\quad + \frac{1}{2}(Z_1 Z_2 + Z_2 Z_1) f(z_0)\, (\mathbf{x} - x_0)(\mathbf{y} - y_0).
\end{aligned}
$$

20.3 Taylor Formula on Homogeneous Carnot Groups

The aim of this section is to provide our $\mathbb{G}$-version of Taylor formula. In order to do this, we use a suitable version of the mean value theorem, using paths along vector fields of the first layer of the stratification of $\mathbb{G}$. For this reason, we shall call it the *stratified Lagrange mean value theorem*. To derive this theorem, we need some important prerequisites. Indeed, we shall make a crucial use of the results in Chapter 19 (and in particular, of Lemma 19.1.4, a subtle consequence of the Campbell–Hausdorff formula).

Throughout the section, $(\mathbb{G}, *)$ is a fixed homogeneous Carnot group with m generators. We take $Z_1, \ldots, Z_m$ (the first m vectors of the Jacobian basis) as generators for the algebra of $\mathbb{G}$. Moreover, we shall denote by $\mathrm{Exp} : \mathfrak{g} \to \mathbb{G}$ the usual exponential map related to $(\mathbb{G}, *)$ and by d any homogeneous norm on $\mathbb{G}$.

Theorem 20.3.1 (The stratified Lagrange mean value theorem). *There exist absolute constants $\mathbf{c}_1, \mathbf{b} > 0$, depending only on the Carnot group $\mathbb{G}$ and on the homogeneous norm d, such that*

$$
|f(x * h) - f(x)| \le \mathbf{c}_1\, d(h) \sup_{z:\, d(z) \le \mathbf{b}\, d(h)} |(Z_1 f(x * z), \ldots, Z_m f(x * z))|, \tag{20.28}
$$

for all $f \in C^1(\mathbb{G}, \mathbb{R})$ and every $x, h \in \mathbb{G}$.

Proof. Due to the importance of this result, we provide two proofs of it, both relying on the Carathéodory–Chow–Rashevsky connectivity Theorem 19.1.3 in Chapter 19.

The first one is very simple, but it makes also use of the Carnot–Carathéodory distance d_Z related to the set of vector fields $Z = \{Z_1, \ldots, Z_m\}$ (see Section 5.2, Chapter 5, page 232). The second one does not involve d_Z but it makes use of the precise estimate in Theorem 19.2.1, page 727.

First proof. Since all homogeneous norms are equivalent (see Proposition 5.1.4, page 230) we can replace d by the Carnot–Carathéodory distance d_Z related to $Z = \{Z_1, \ldots, Z_m\}$. By the very definition of d_Z (see (5.6), page 232), there exists a sequence $T_n \downarrow d_Z(x * h, x) = d_Z(h, 0)$ and, for every $n \in \mathbb{N}$, there exists an absolutely continuous curve $\gamma_n : [0, T_n] \to \mathbb{G}$ connecting x to $x * h$ (i.e. $\gamma_n(0) = x$ and $\gamma_n(T_n) = x * h$) which is $\{Z_1, \ldots, Z_m\}$-subunit, i.e.

$$\langle \dot{\gamma}_n(t), \xi \rangle^2 \le \sum_{j=1}^{m} \langle X_j I(\gamma_n(t)), \xi \rangle^2 \quad \forall \xi \in \mathbb{R}^N$$

almost everywhere in $[0, T]$.

Then we have (also setting $Zf(y) = (Z_1 f(y), \ldots, Z_m f(y))$)

$$\begin{aligned} |f(x*h) - f(x)| &= \left| \int_0^{T_n} \frac{\mathrm{d}}{\mathrm{d}s} (f(\gamma_n(s)))\,\mathrm{d}s \right| = \left| \int_0^{T_n} \langle (\nabla f)(\gamma_n(s)), \dot{\gamma}_n(s) \rangle\,\mathrm{d}s \right| \\ &\le \int_0^{T_n} \left(\sum_{j=1}^{m} \langle Z_j I(\gamma_n(s)), (\nabla f)(\gamma_n(s)) \rangle^2 \right)^{1/2} \mathrm{d}s \\ &= \int_0^{T_n} \big|(Zf)(\gamma_n(s))\big|\,\mathrm{d}s \le T_n \sup_{s \in [0, T_n]} |Zf(\gamma_n(s))|. \end{aligned} \tag{20.29}$$

Now, by the very definition of d_Z, for every $s \in [0, T_n]$ we obviously have

$$d_Z(x, \gamma_n(s)) \le d_Z(x, \gamma_n(T_n)) = d_Z(x, x*h) = d_Z(h, 0),$$

so (20.29) yields

$$|f(x*h) - f(x)| \le T_n \sup_{y:\, d_Z(x,y) \le d_Z(h,0)} |Zf(y)|. \tag{20.30}$$

Let now $n \to \infty$ in (20.30), and recall that $T_n \downarrow d_Z(h, 0)$. This gives[3]

$$|f(x*h) - f(x)| \le d_Z(h, 0) \sup_{y:\, d_Z(x,y) \le d_Z(h,0)} |Zf(y)|. \tag{20.31}$$

Finally, (20.31) implies (20.28) by setting $y = x * z$ and by taking $\mathbf{b} = \mathbf{c}^2$ and $\mathbf{c}_1 = \mathbf{c}$, where $\mathbf{c}$ is the constant such that $\mathbf{c}^{-1} d \le d_Z \le \mathbf{c}\, d$ on $\mathbb{G} \times \mathbb{G}$.

Second proof. First, suppose that $h \in \operatorname{Exp}(\operatorname{span}\{Z_j\})$ for some $j \in \{1, \ldots, m\}$. We can suppose that $h = \operatorname{Exp}(t Z_j)$ for some $t > 0$ (the case $t < 0$ follows by replacing Z_j by $-Z_j$). Then (19.10e) gives $d(h) = t\, d(\operatorname{Exp}(Z_j))$ (recall that $\{Z_1, \ldots, Z_m\}$ span the first layer of the stratification, i.e. $\sigma_j = 1$ for $1 \le j \le m$). Let us set

[3] (20.31) shows that the stratified mean-value theorem in (20.28) holds with constants $\mathbf{c}_1 = 1$ and $\mathbf{b} = 1$ when d is the Carnot–Carathéodory distance.

$$\mathbf{c} := \max\left\{d^{-1}(\operatorname{Exp}(Z_1)), \dots, d^{-1}(\operatorname{Exp}(Z_m))\right\}.$$

With this position, we have

$$t = d(h)\, d^{-1}(\operatorname{Exp}(Z_j)) \le \mathbf{c}\, d(h).$$

Thus, for every $x \in \mathbb{G}$, it holds

$$\begin{aligned}
&|f(x * h) - f(x)| \\
&\quad = \left|f\big(x * \operatorname{Exp}(tZ_j)\big) - f(x)\right| = \left|f\big(\exp(tZ_j)(x)\big) - f(x)\right| \\
&\quad (\text{set } \gamma(t) := \exp(tZ_j)(x)) \\
&\quad = \left|\int_0^t \frac{\mathrm{d}}{\mathrm{d}s}(f(\gamma(s)))\,\mathrm{d}s\right| = \left|\int_0^t (Z_j f)(\gamma(s))\,\mathrm{d}s\right| \\
&\quad \le t \sup_{s\in[0,t]} |Z_j f(x * \operatorname{Exp}(sZ_j))| \le \mathbf{c}\, d(h) \sup_{z:\, d(z)\le d(h)} |Zf(x * z)|. \qquad (20.32)
\end{aligned}$$

Here we have set $Zf(y) = (Z_1 f(y), \dots, Z_m f(y))$, and we have used again the fact that, for every $s \in [0, t]$, one has

$$d(\operatorname{Exp}(sZ_j)) = s\, d(\operatorname{Exp}(Z_j)) \le t\, d(\operatorname{Exp}(Z_j)) = d(\operatorname{Exp}(tZ_j)) = d(h).$$

In the case of an arbitrary $h \in \mathbb{G}$, we first remark that by Theorem 19.2.1 (page 727), we have $h = h_1 * \cdots * h_{\mathbf{M}}$, where any h_j's belong to $\operatorname{Exp}(\operatorname{span}\{Z_i\})$ for some $i \in \{1, \dots, m\}$ and $d(h_j) \le \mathbf{c}_0\, d(h)$ for all $j = 1, \dots, \mathbf{M}$. Thus, using (20.32) repeatedly, one gets (also setting $h_0 := 0$)

$$\begin{aligned}
|f(x * h) - f(x)| &= \left|f\big(x * h_1 * \cdots * h_{\mathbf{M}}\big) - f(x)\right| \\
&\le \sum_{j=1}^{\mathbf{M}} \left|f\big(x * h_1 * \cdots * h_j\big) - f(x * h_1 * \cdots * h_{j-1})\right| \\
(\text{by } (20.32)) &\le \sum_{j=1}^{\mathbf{M}} \mathbf{c}\, d(h_j) \sup_{z:\, d(z)\le d(h_j)} |Zf(x * h_1 * \cdots * h_{j-1} * z)|. \qquad (20.33)
\end{aligned}$$

By the pseudo-triangle inequality for d, $d(x * y) \le c\,(d(x) + d(y))$, we have

$$\begin{aligned}
d(h_1 * \cdots * h_{j-1} * z) &\le c^{j-1}\,(d(h_1) + \cdots + d(h_{j-1}) + d(z)) \\
(\text{since } d(z) \le d(h_j)) &\le c^{j-1}\,(d(h_1) + \cdots + d(h_j)) \\
(\text{since } d(h_j) \le \mathbf{c}_0\, d(h)) &\le c^{j-1}\, j\, \mathbf{c}_0\, d(h) \le c^{j-1}\, \mathbf{M}\, \mathbf{c}_0\, d(h) =: \mathbf{b}\, d(h).
\end{aligned}$$

As a consequence, (20.33) proves that

$$|f(x * h) - f(x)| \le \mathbf{M}\, \mathbf{c}\, \mathbf{c}_0\, d(h) \sup_{\zeta:\, d(\zeta)\le \mathbf{b}\, d(h)} |Zf(x * \zeta)|,$$

so that the position $\mathbf{c}_1 := \mathbf{M}\, \mathbf{c}\, \mathbf{c}_0$ finally gives (20.28). □

With Theorem 20.3.1 at hand, we can prove the following theorem. Note that here and in the proof below, α is a multi-index with m entries (not N, as in the previous paragraphs). Note that the Euclidean length $|\alpha|$ coincides with

$$|(\alpha, 0, \dots, 0)|_{\mathbb{G}},$$

the 0's occurring $N - m$ times.

Theorem 20.3.2 (Stratified Taylor inequality). *Let $(\mathbb{G}, *)$ be a homogeneous Carnot group and d a homogeneous norm on $\mathbb{G}$. For every $n \in \mathbb{N} \cup \{0\}$, there exists a constant $\mathbf{c}_n > 0$ (depending only on n, $\mathbb{G}$ and d) such that*

$$\begin{aligned} &|f(x*h) - P_n(f,x)(x*h)| \le \mathbf{c}_n\, d^n(h) \\ &\quad\times \sup_{d(z)\le \mathbf{b}^n d(h)} \Big\{ \big|(Z_1,\dots,Z_m)^\alpha f(x*z) - (Z_1,\dots,Z_m)^\alpha f(x)\big| \,:\, |\alpha| = n \Big\} \end{aligned} \tag{20.34}$$

for all $f \in C^n(\mathbb{G}, \mathbb{R})$ and every $x, h \in \mathbb{G}$. Here $\mathbf{b}$ is as in Theorem 20.3.1.

Proof. We fix $x \in \mathbb{G}$ and $n \in \mathbb{N} \cup \{0\}$, and we let

$$\Phi(h) := f(x*h) - P_n(f,x)(x*h).$$

By Proposition 20.2.10, it holds $Z^\beta(f - P_n(f,x))(x) = 0$ whenever β is a N-dimensional multi-index with $|\beta|_{\mathbb{G}} \le n$, i.e. (by using the left-invariance of the Z^β's)

$$(Z_1,\dots,Z_N)^\beta \Phi(0) = 0 \qquad \forall\, \beta \in \big(\mathbb{N}\cup\{0\}\big)^N, \quad |\beta|_{\mathbb{G}} \le n.$$

By taking a multi-index β with 0 entries from the $(m+1)$-th one, we get

$$(Z_1,\dots,Z_m)^\alpha \Phi(0) = 0 \qquad \forall\, \alpha : |\alpha| \le n. \tag{20.35}$$

Let us prove by induction on $j = 0, 1, \dots, n$ that if $|\alpha| = n - j$ then

$$\begin{aligned} &\big|(Z_1,\dots,Z_m)^\alpha \Phi(h)\big| \le \mathbf{c}_j\, d^j(h) \\ &\quad\times \sup_{d(z)\le \mathbf{b}^j d(h)} \Big\{ \big|(Z_1,\dots,Z_m)^\alpha f(x*z) - (Z_1,\dots,Z_m)^\alpha f(x)\big| \,:\, |\alpha| = n \Big\}. \end{aligned} \tag{20.36}$$

When $j = n$ (whence $|\alpha| = 0$, so that $(Z_1,\dots,Z_m)^\alpha\Phi(h) = \Phi(h)$) this will obviously prove (20.34).

Step zero: For the sake of brevity, we set $\widehat{Z} = (Z_1,\dots,Z_m)$ (to distinguish it from the former $Z = (Z_1,\dots,Z_N)$). If $j = 0$, i.e. $|\alpha| = n$, then $\widehat{Z}^\alpha P_n(f,x)(\cdot)$ has $\mathbb{G}$-degree $= 0$, i.e. it is constant, so that

$$\big(\widehat{Z}^\alpha P_n(f,x)\big)(x*h) = \big(\widehat{Z}^\alpha P_n(f,x)\big)(x) = (\widehat{Z}^\alpha f)(x),$$

whence the far left-hand side of (20.36) equals

$$\left|\widehat{Z}^\alpha f(x*h) - \widehat{Z}^\alpha f(x)\right|.$$

In this case (20.36) is trivially true (for the left-hand side is one of the elements in the curly brackets when $z = h$).

Step of induction: Suppose that (20.36) holds true for indices $0 \le \cdots \le j-1$ and let us prove it for j. Let α be a multi-index with $|\alpha| = n - j$. We apply the stratified mean-value Theorem 20.3.1 to the function $f = \widehat{Z}^\alpha \Phi$ and the points $x = 0, h \in \mathbb{G}$ (so that $f(x) = (\widehat{Z}^\alpha \Phi)(0) = 0$ thanks to (20.35)). Then from (20.28) we infer

$$|\widehat{Z}^\alpha \Phi(h)| \le \mathbf{c}_1\, d(h) \sup_{z:\, d(z) \le \mathbf{b}\, d(h)} |(Z_1 \widehat{Z}^\alpha \Phi(z), \dots, Z_m \widehat{Z}^\alpha \Phi(z))| =: (\bigstar). \quad (20.37)$$

But $Z_1 \widehat{Z}^\alpha, \dots, Z_m \widehat{Z}^\alpha$ are linear combinations of the $\widehat{Z}^\beta$'s with $|\beta| = |\alpha| + 1 = n - (j-1)$ (see the proof of Proposition 20.1.4 and apply, if needed, a simple homogeneity argument). As a consequence, we can apply the inductive hypothesis to estimate the right-hand side of (20.37). We get (here $c(Z)$ denotes a structural constant depending only on the algebraic structure properties of the set of vector fields $\{Z_1, \dots, Z_N\}$; see the proof of Proposition 20.1.4)

$$\begin{aligned}
(\bigstar) &\le \mathbf{c}_1\, d(h) \sup_{z:\, d(z) \le \mathbf{b}\, d(h)} c(Z)\, \mathbf{c}_{j-1}\, d^{j-1}(z) \\
&\quad \times \sup_{\substack{\zeta:\, d(\zeta) \le \mathbf{b}^{j-1} d(z) \\ |\alpha| = n}} \{|(Z_1, \dots, Z_m)^\alpha f(x*\zeta) - (Z_1, \dots, Z_m)^\alpha f(x)|\} \\
&\le \mathbf{c}_1\, c(Z)\, \mathbf{c}_{j-1}\, \mathbf{b}^{j-1}\, d^j(h) \\
&\quad \times \sup_{\substack{\zeta:\, d(\zeta) \le \mathbf{b}^{j} d(h) \\ |\alpha| = n}} \{|(Z_1, \dots, Z_m)^\alpha f(x*\zeta) - (Z_1, \dots, Z_m)^\alpha f(x)|\}.
\end{aligned}$$

This is precisely (20.36). If we set

$$\mathbf{c}_j := \mathbf{c}_1\, c(Z)\, \mathbf{c}_{j-1}\, \mathbf{b}^{j-1},$$

the proof is complete. □

Starting from Theorems 20.3.1 and 20.3.2 it is a standard matter to prove the following stratified Taylor formula with remainder. Here, α is a multi-index with m entries (not N, as in the paragraphs preceding Theorem 20.3.2).

Theorem 20.3.3 (Stratified Taylor formula). *Let $(\mathbb{G}, *)$ be a homogeneous Carnot group. For every $n \in \mathbb{N} \cup \{0\}$, there exists a constant $\mathbf{c}_n > 0$ (depending only on n and $\mathbb{G}$) such that*

$$\begin{aligned}
&|f(x*h) - P_n(f,x)(x*h)| \le \mathbf{c}_n\, d^{n+1}(h) \\
&\quad \times \sup_{d(z) \le \mathbf{b}^{n+1} d(h)} \{|(Z_1, \dots, Z_m)^\alpha f(x*z)| \,:\, |\alpha| = n+1\} \qquad (20.38)
\end{aligned}$$

for all $f \in C^{n+1}(\mathbb{G}, \mathbb{R})$ and every $x, h \in \mathbb{G}$. Here $\mathbf{b}$ is as in Theorem 20.3.1.

Proof. First, apply the stratified Taylor inequality in Theorem 20.3.2, and then estimate the right-hand side of (20.34) by means of the stratified mean value Theorem 20.3.1. This ends the proof. □

Example 20.3.4. Consider the Heisenberg–Weyl group $\mathbb{H}^1$ on $\mathbb{R}^3$. The relevant Jacobian basis is

$$Z_1 = \partial_{x_1} + 2x_2\, \partial_{x_3}, \quad Z_2 = \partial_{x_2} - 2x_1\, \partial_{x_3}, \quad Z_3 = \partial_{x_3},$$

and a homogeneous norm on $\mathbb{H}^1$ is, for example,

$$d(x) = |x_1| + |x_2| + |x_3|^{1/2}.$$

Hence formula (20.38) in Theorem 20.3.3 gives the estimate

$$\begin{aligned} &\big|u(x) - \big(u(0) + x_1\, Z_1 u(0) + x_2\, Z_2 u(0)\big)\big| \\ &\quad \le \mathbf{c}_1 \left(|x_1|^2 + |x_2|^2 + |x_3|\right) \times \sup_{d(z) \le b^2\, d(x)} \{|Z_1^2 u|(z),\ |Z_2^2 u|(z),\ |Z_1 Z_2 u|(z)\}. \end{aligned}$$

As an application, if $u \in C^2(\mathbb{G}, \mathbb{R})$ is such that

$$Z_1^2 u,\ Z_1 Z_2 u,\ Z_2^2 u \equiv 0 \quad \text{in } \mathbb{H}^1,$$

then

$$u(x) = u(0) + x_1\, Z_1 u(0) + x_2\, Z_2 u(0).$$

In particular, $Z_3 u \equiv 0$ whence also $Z_2 Z_1 u = Z_1 Z_2 u + 4 Z_3 u \equiv 0$. This is quite surprising, since we have derived an information on $\{Z_3, Z_2 Z_1\}$ starting from an information only on $\{Z_1^2, Z_1 Z_2, Z_2^2\}$, but

$$Z_3,\ Z_2 Z_1 \notin \operatorname{span}\{Z_1^2, Z_1 Z_2, Z_2^2\}.$$

See Ex. 7 for a generalization of the above comments.

20.3.1 Stratified Taylor Formula with Peano Remainder

From Theorem 20.3.3 we immediately derive the following corollary.

Corollary 20.3.5 (Stratified Taylor formula–Peano remainder). *Let $(\mathbb{G}, *)$ be a homogeneous Carnot group. For every $f \in C^{n+1}(\mathbb{G}, \mathbb{R})$, $x_0 \in \mathbb{G}$ and $n \in \mathbb{N} \cup \{0\}$, we have*

$$f(x) = P_n(f, x_0)(x) + \mathcal{O}_{x \to x_0}\big(d^{n+1}(x_0^{-1} * x)\big). \tag{20.39}$$

There is another way to prove the stratified Taylor formula with Peano remainder in (20.39), by starting from the usual Euclidean Taylor formula.

This alternative proof, which we now describe, is much simpler than the proof given in Section 20.3. Nonetheless, it does not provide any estimate of the remainder (in terms of the sole Z^α's, $|\alpha|_{\mathbb{G}} = n + 1$) as in Theorem 20.3.3. We begin with a simple lemma.

Lemma 20.3.6. *Let f be a smooth function on $\mathbb{R}^N$ such that*

$$f(x) = \mathcal{O}_{x\to 0}(|x|^{n+1}),$$

where $|\cdot|$ is the Euclidean norm on $\mathbb{R}^N$. Then $Z^\alpha f(0) = 0$ for every multi-index α with $|\alpha|_{\mathbb{G}} \le n$.

Proof. It is known[4] that the assertion holds in the Euclidean setting, i.e. when $Z^\alpha = (D_x)^\alpha$. In the present setting, note that, by (20.15),

$$Z^\alpha|_0 = (D_x)^\alpha|_0 + \sum_{\beta:\,\beta\neq\alpha,\ |\beta|\le|\alpha|,\ |\beta|_{\mathbb{G}}=|\alpha|_{\mathbb{G}}} a_{\beta,\alpha}(0)\,(D_x)^\beta|_0.$$

Hence, $Z^\alpha f(0)$ is a linear combination of terms $(D_x)^\beta f(0)$ with $|\beta| \le |\alpha|$. Note that if $|\alpha|_{\mathbb{G}} \le n$, then $|\alpha| \le n$, for (in general) it holds $|\alpha| \le |\alpha|_{\mathbb{G}}$. Consequently, $Z^\alpha f(0)$ is a linear combination of terms $(D_x)^\beta f(0)$ with $|\beta| \le n$, and the proof is complete, thanks to the cited Euclidean version of the lemma. □

By the classical Mac Laurin formula, for a given $f \in C^\infty(\mathbb{R}^N, \mathbb{R})$, we have

$$f(x) = Q_n(f,0)(x) + R(x), \quad \text{where } R(x) = \mathcal{O}_{x\to 0}(|x|^{n+1}). \tag{20.40}$$

Here $Q_n(f,0)(x) =: q(x)$ is the (only) polynomial in x of ordinary degree n such that

$$(D_x)^\alpha q(0) = (D_x)^\alpha f(0) \quad \text{for all } |\alpha| \le n.$$

Now, we decompose (in a unique way) q as q_1+q_2, where q_1 contains only monomials which are δ_λ-homogeneous of degree $\le n$. Note that q_2 contains only monomials of δ_λ-degree $\ge n+1$. Consequently,

$$f = q_1 + (q_2 + R). \tag{20.41}$$

Next, we take any multi-index α with $|\alpha|_{\mathbb{G}} \le n$. Then

$$Z^\alpha f = Z^\alpha q_1 + Z^\alpha q_2 + Z^\alpha R,$$

so that

$$Z^\alpha f(0) = Z^\alpha q_1(0) + Z^\alpha q_2(0) + Z^\alpha R(0). \tag{20.42}$$

Now, $Z^\alpha q_2$ is a polynomial of δ_λ-degree $\ge n+1-|\alpha|_G \ge 1$, so that $Z^\alpha q_2(0) = 0$. Moreover, since $R(x) = \mathcal{O}_{x\to 0}(|x|^{n+1})$, by means of Lemma 20.3.6, it follows that $Z^\alpha R(0) = 0$, whence (20.42) yields

$$Z^\alpha f(0) = Z^\alpha q_1(0) \quad \text{whenever } |\alpha|_{\mathbb{G}} \le n$$

i.e. by definition, q_1 is the Z-Mac Laurin polynomial of f.

---

[4] One of the simplest way to see this is to apply the Euclidean Taylor formula. Note that when $N = 1$, a Cauchy-type theorem is needed.

Consider now the following δ_λ-homogeneous norm on $\mathbb{G}$

$$\widetilde{d}(x) = \widetilde{d}(x^{(1)}, x^{(2)}, \dots, x^{(r)}) = |x^{(1)}| + |x^{(2)}|^{1/2} + \dots + |x^{(r)}|^{1/r}$$

(here $|\cdot|$ denotes the Euclidean norm on any $\mathbb{R}^{N_i}$, $i = 1, \dots, r$). We have

$$q_2(x),\ \ R(x) = \mathcal{O}_{x\to 0}(\widetilde{d}(x)^{n+1}). \tag{20.43}$$

Indeed, the estimate of q_2 follows by recalling that q_2 contains only monomials of δ_λ-degree $\geq n+1$. Moreover, noticing that (for every $i = 1, \dots, r$) it holds $|x^{(i)}| \leq |x^{(i)}|^{1/i}$ near the origin, we can use (20.40) to estimate R.

By collecting together (20.41) and (20.43), we have proved the formula

$$f(x) = q_1(x) + \mathcal{O}_{x\to 0}(\widetilde{d}^{n+1}), \tag{20.44}$$

where q_1 is the Z-Mac Laurin polynomial of f or, *equivalently, the "projection" on $\mathcal{P}_n$ of the usual Mac Laurin polynomial of f*, $\mathcal{P}_n$ being the vector space of the polynomials of δ_λ-degree $\leq n$.

Finally, if d is any homogeneous norm on $\mathbb{G}$, then d is equivalent to $\widetilde{d}$, so that (20.44) gives back (20.39). (Incidentally, we have also demonstrated the assertion in Remark 20.2.6.)

The following remark shows that there is only one polynomial of δ_λ-degree at most n which plays the role of $P_n(f, x_0)$ in (20.39).

Remark 20.3.7. Suppose $f \in C^{n+1}(\mathbb{G}, \mathbb{R})$, $x_0 \in \mathbb{G}$ and $n \in \mathbb{N} \cup \{0\}$. If $p(x)$ is a polynomial of δ_λ-degree at most n such that

$$(\star) \qquad f(x) = p(x) + \mathcal{O}_{x\to x_0}\big(d^{n+1}(x_0^{-1} * x)\big),$$

then p coincides with $P_n(f, x_0)$. Indeed, $(\star)$ and (20.39) give

$$q(x) := P_n(f, x_0)(x) - p(x) = \mathcal{O}_{x\to x_0}\big(d^{n+1}(x_0^{-1} * x)\big).$$

We remark that q is a polynomial of δ_λ-degree at most n. Setting $\widetilde{q}(y) := q(x_0 * y)$, the above is equivalent to

$$\widetilde{q}(y) = \mathcal{O}_{y\to 0}\big(d^{n+1}(y)\big). \tag{20.45}$$

Now, since q is a polynomial of δ_λ-degree at most n and the i-th component function of $x_0 * y$ is a polynomial function in y of δ_λ-degree[5] at most σ_i (see Theorem 1.3.15-(3), page 39), then $\widetilde{q}(y) = q(x_0 * y)$ is a polynomial function in y of δ_λ-degree at most n. Hence, (20.45) can hold if and only if $\widetilde{q} \equiv 0$, i.e. $q(x_0 * y) = 0$ for every $y \in \mathbb{G}$. Recalling the definition of q, this is in turn equivalent to

$$p(x_0 * y) = P_n(f, x_0)(x_0 * y) \quad \forall\, y \in \mathbb{G}.$$

Obviously, this means $p = P_n(f, x_0)$.

---

[5] Here, we are writing, as usual, the dilation δ_λ on $\mathbb{G}$ as

$$\delta_\lambda(x) = (\lambda^{\sigma_1} x_1, \dots, \lambda^{\sigma_N} x_N).$$

20.3.2 Stratified Taylor Formula with Integral Remainder

In this section, we derive a stratified Taylor formula with integral remainder which will furnish yet another proof of the Taylor formulas in the previous sections. The resulting Taylor polynomials will be apparently different from the $P_n(f, x_0)$'s introduced in Section 20.2, but they will prove to be equivalent, thanks to the uniqueness result in Remark 20.3.7. Thus, we shall obtain another way to represent the Z-Taylor polynomials.

As usual, $\mathbb{G} = (\mathbb{R}^N, *, \delta_\lambda)$ is a homogeneous Carnot group of step r and m generators and $Z_1, \ldots, Z_N$ is the relevant Jacobian basis of $\mathfrak{g}$, the Lie algebra of $\mathbb{G}$. We denote by Exp and Log the exponential and logarithmic maps related to $\mathbb{G}$. Furthermore, we denote, as usual, the dilation δ_λ on $\mathbb{G}$ by

$$\delta_\lambda(x) = (\lambda^{\sigma_1} x_1, \ldots, \lambda^{\sigma_N} x_N).$$

We begin by recalling a known result. Let $X \in \mathfrak{g}$ be given, and suppose that $\gamma(t)$ is an arbitrary integral curve of X, i.e. $\dot{\gamma}(t) = XI(\gamma(t))$ for every $t \in \mathbb{R}$. Then, for every $u \in C^{n+1}(\mathbb{G}, \mathbb{R})$, we have (see Ex. 8 in Chapter 1)

$$\begin{aligned} u(\gamma(t)) &= \sum_{k=0}^{n} \frac{t^k}{k!} (X^k u)(\gamma(0)) \\ &\quad + \frac{1}{n!} \int_0^t (t-s)^n \left(X^{n+1} u\right)(\gamma(s))\, \mathrm{d}s. \end{aligned} \tag{20.46}$$

Indeed, it suffices to apply the ordinary Taylor formula with integral remainder to $t \mapsto u(\gamma(t))$, by noticing that

$$\frac{\mathrm{d}^k}{\mathrm{d}t^k}\big(u(\gamma(t))\big) = (X^k u)(\gamma(t)).$$

We thus immediately get the following assertion.

Lemma 20.3.8. *Let $x, h \in \mathbb{G}$. Let also $u \in C^{n+1}(\mathbb{G}, \mathbb{R})$. Then we have*

$$\begin{aligned} u(x * h) &= \sum_{k=0}^{n} \frac{1}{k!} ((\operatorname{Log} h)^k u)(x) \\ &\quad + \frac{1}{n!} \int_0^1 (1-s)^n ((\operatorname{Log} h)^{n+1} u)(x * \operatorname{Exp}(s \operatorname{Log} h))\, \mathrm{d}s. \end{aligned} \tag{20.47}$$

Proof. We know that (see Corollary 1.2.24, page 24) the integral curve γ of $X := \operatorname{Log} h \in \mathfrak{g}$ starting at x is given by

$$s \mapsto \gamma(s) = \exp\big(s \operatorname{Log} h\big)(x) = x * \operatorname{Exp}\big(s \operatorname{Log} h\big).$$

Note that $\gamma(0) = x$, $\gamma(1) = x * \operatorname{Exp}\big(\operatorname{Log} h\big) = x * h$, so that, by applying (20.46) with $t = 1$, we get (20.47). $\square$

A restatement of the lemma above gives the following corollary.

Corollary 20.3.9. *Let $x, h \in \mathbb{G}$ and $u \in C^{n+1}(\mathbb{G}, \mathbb{R})$. Let $(X_1, \dots, X_N)$ be any basis of $\mathfrak{g}$. Then we have*

$$u(x * h) = u(x) + \sum_{k=1}^{n} \sum_{i_1,\dots,i_k=1}^{N} \frac{\zeta_{i_1}(h) \cdots \zeta_{i_k}(h)}{k!} X_{i_1} \cdots X_{i_k} u(x)$$
$$+ \sum_{i_1,\dots,i_{n+1}=1}^{N} \frac{\zeta_{i_1}(h) \cdots \zeta_{i_{n+1}}(h)}{n!}$$
$$\times \int_0^1 (X_{i_1} \cdots X_{i_{n+1}} u)\left(x * \mathrm{Exp}\left(\sum_{i=1}^{N} s\, \zeta_i(h)\, X_i\right)\right) (1-s)^n \, \mathrm{d}s. \quad (20.48)$$

Here, we have used the following notation

$$\mathrm{Log}\, h = \zeta_1(h)\, X_1 + \cdots + \zeta_N(h)\, X_N \quad \forall\, h \in \mathbb{G}, \qquad (20.49)$$

i.e. $\zeta_1(h), \dots, \zeta_N(h)$ are the components of Log h *w.r.t. $(X_1, \dots, X_N)$.*

Proof. Obviously, there exist (polynomial) functions

$$\mathbb{G} \ni h \mapsto \zeta_i(h) \in \mathbb{R}, \quad i = 1, \dots, N,$$

such that (20.49) holds. For every $k \in \mathbb{N}$, we thus obtain

$$(\mathrm{Log}\, h)^k = \left(\sum_{i=1}^{N} \zeta_i(h)\, X_i\right)^k = \sum_{i_1,\dots,i_k=1}^{N} \zeta_{i_1}(h) \cdots \zeta_{i_k}(h)\, X_{i_1} \cdots X_{i_k}.$$

Now, (20.48) follows directly from (20.47). □

Next, we recall that the Lie algebra of $\mathbb{G}$ is equipped with the stratification

$$\mathfrak{g} = V^{(1)} \oplus \cdots \oplus V^{(r)} \quad \text{with} \quad [V^{(1)}, V^{(i)}] = \begin{cases} V^{(i+1)} & \text{if } i = 1, \dots, r-1, \\ \{0\} & \text{if } i = r. \end{cases}$$

We say that a basis of $\mathfrak{g}$ is adapted to the stratification if it has the form

$$X = (X_1^{(1)}, \dots, X_{N_1}^{(1)} \; ; \dots ; \; X_1^{(r)}, \dots, X_{N_r}^{(r)}), \qquad (20.50)$$

where, for every $i = 1, \dots, r$,

$$X_1^{(i)}, \dots, X_{N_i}^{(i)} \quad \text{is a basis of } V^{(i)}.$$

If X is a basis for $\mathfrak{g}$ adapted to the stratification as in (20.50), then we can define on $\mathfrak{g}$ a group of dilations, still denoted by $\{\delta_\lambda\}_{\lambda>0}$, such that

$$\delta_\lambda(X_j^{(i)}) = \lambda^i\, X_j^{(i)} \quad \text{for every } i = 1, \dots, r \text{ and every } j = 1, \dots, N_i.$$

We then know that (argue as in Theorem 1.3.28 on page 49)

$$\delta_\lambda\big(\mathrm{Log}\,(x)\big) = \mathrm{Log}\,\big(\delta_\lambda(x)\big), \quad \delta_\lambda\big(\mathrm{Exp}\,(X)\big) = \mathrm{Exp}\,\big(\delta_\lambda(X)\big) \quad \forall\, x \in \mathbb{G},\ X \in \mathfrak{g}. \tag{20.51}$$

In the proof of Proposition 20.3.11 below, we shall need the following lemma having an interest in its own.

Lemma 20.3.10. *Let d be any homogeneous norm on $\mathbb{G}$. Then there exists a constant $\mathbf{c} = \mathbf{c}(d, \mathbb{G})$ such that, for every $x \in \mathbb{G}$ and every $s \in [0, 1]$, we have $d(\gamma(s)) \le \mathbf{c}\, d(x)$, where γ is the integral curve of* Log (x) *starting at the origin. More explicitly,*

$$d\big(\mathrm{Exp}\,(s\mathrm{Log}\,(x))\big) \le \mathbf{c}\, d(x) \quad \text{for every } x \in \mathbb{G} \text{ and every } s \in [0, 1]. \tag{20.52}$$

Proof. We know that the integral curve γ of Log (x) starting at the origin is $s \mapsto \mathrm{Exp}\,(s\mathrm{Log}\,(x))$, hence we have to prove (20.52). It is not restrictive to suppose $x \neq 0$, otherwise (20.52) holds for any $\mathbf{c}$.

Let X be any basis for $\mathfrak{g}$ adapted to the stratification (for example, the Jacobian basis), and let δ_λ denote both the dilation on $\mathfrak{g}$ defined above and the usual dilation on $\mathbb{G}$. Then, thanks to (20.51), the linearity of δ_λ and the fact that d is δ_λ-homogeneous of degree 1, we have

$$\begin{aligned}\frac{d(\mathrm{Exp}\,(s\mathrm{Log}\,(x)))}{d(x)} &= d(\delta_{1/d(x)}\mathrm{Exp}\,(s\mathrm{Log}\,(x))) = d(\mathrm{Exp}\,(\delta_{1/d(x)}(s\mathrm{Log}\,(x))))\\ &= d(\mathrm{Exp}\,(s\delta_{1/d(x)}(\mathrm{Log}\,(x)))) = d(\mathrm{Exp}\,(s\mathrm{Log}\,(\delta_{1/d(x)}(x)))).\end{aligned}$$

As a consequence, being $d(\delta_{1/d(x)}(x)) = 1$, we have

$$\frac{d(\mathrm{Exp}\,(s\mathrm{Log}\,(x)))}{d(x)} \le \max_{\substack{\xi \in \mathbb{G}:\, d(\xi)=1\\ s \in [0,1]}} d(\mathrm{Exp}\,(s\mathrm{Log}\,(\xi))) =: \mathbf{c}.$$

Note that $\mathbf{c}$ is finite, for $\{\xi : d(\xi) = 1\}$ is compact and d, Exp, Log are continuous functions. □

We are in a position to prove the following result.

Proposition 20.3.11 (Stratified Mac Laurin formula—integral remainder). *Let $n \in \mathbb{N} \cup \{0\}$, and let $u \in C^{n+1}(\mathbb{G}, \mathbb{R})$. Suppose $\mathcal{X} = (X_1, \dots, X_N)$ is any basis of $\mathfrak{g}$ adapted to the stratification of $\mathfrak{g}$. Then, following the notation in* (20.49)*, we have*

$$u(\mathbf{x}) = u(0) + \sum_{h=1}^{n} \sum_{\substack{k=1,\dots,n\\ i_1,\dots,i_k \le N\\ \sigma_{i_1}+\dots+\sigma_{i_k}=h}} \frac{\zeta_{i_1}(\mathbf{x}) \cdots \zeta_{i_k}(\mathbf{x})}{k!} X_{i_1} \cdots X_{i_k} u(0) + R_n(\mathbf{x}), \tag{20.53}$$

where

$$R_n(\mathbf{x}) = \Bigg\{ \sum_{k=1}^{n} \sum_{\substack{i_1,\dots,i_k \le N \\ \sigma_{i_1}+\dots+\sigma_{i_k} \ge n+1}} \frac{\zeta_{i_1}(\mathbf{x}) \cdots \zeta_{i_k}(\mathbf{x})}{k!} X_{i_1} \cdots X_{i_k} u(0)$$

$$+ \sum_{i_1,\dots,i_{n+1}=1}^{N} \frac{\zeta_{i_1}(\mathbf{x}) \cdots \zeta_{i_{n+1}}(\mathbf{x})}{n!}$$

$$\times \int_0^1 (X_{i_1} \cdots X_{i_{n+1}} u) \left(\mathrm{Exp} \left(\sum_{i=1}^{N} s\, \zeta_i(\mathbf{x})\, X_i \right) \right) (1-s)^n \, \mathrm{d}s \Bigg\}. \quad (20.54)$$

Moreover, for every fixed homogeneous norm d on $\mathbb{G}$ and every $n \in \mathbb{N} \cup \{0\}$, there exists $\mathbf{c}_n > 0$ (depending on n, $\mathbb{G}$, d and the basis $\mathcal{X}$) such that, for $\mathbf{x}$ near 0,

$$|R_n(\mathbf{x})| \le \mathbf{c}_n\, d^{n+1}(\mathbf{x}) \times \sup_{d(z) \le \mathbf{c}\, d(\mathbf{x})} \{ |X_{i_1} \cdots X_{i_M} u(z)| :$$
$$n+1 \le M \le r(n+1),\ i_1, \dots, i_M \in \{1, \dots, m\} \}. \quad (20.55)$$

Finally, when $\mathcal{X}$ is the Jacobian basis related to $\mathbb{G}$, the polynomial function

$$q_n(\mathbf{x}) = u(0) + \sum_{h=1}^{n} \sum_{\substack{k=1,\dots,n \\ i_1,\dots,i_k \le N \\ \sigma_{i_1}+\dots+\sigma_{i_k}=h}} \frac{\zeta_{i_1}(\mathbf{x}) \cdots \zeta_{i_k}(\mathbf{x})}{k!} X_{i_1} \cdots X_{i_k} u(0) \quad (20.56)$$

coincides with the Z-Mac Laurin polynomial of δ_λ-degree n related to u.

Proof. Let us begin by taking $x = 0$ and $h = \mathbf{x}$ in (20.48). We get

$$u(\mathbf{x}) = u(0) + \sum_{k=1}^{n} \sum_{i_1,\dots,i_k=1}^{N} \frac{\zeta_{i_1}(\mathbf{x}) \cdots \zeta_{i_k}(\mathbf{x})}{k!} X_{i_1} \cdots X_{i_k} u(0)$$

$$+ \sum_{i_1,\dots,i_{n+1}=1}^{N} \frac{\zeta_{i_1}(\mathbf{x}) \cdots \zeta_{i_{n+1}}(\mathbf{x})}{n!}$$

$$\times \int_0^1 (X_{i_1} \cdots X_{i_{n+1}} u) \left(\mathrm{Exp} \left(\sum_{i=1}^{N} s\, \zeta_i(\mathbf{x})\, X_i \right) \right) (1-s)^n \, \mathrm{d}s. \quad (20.57)$$

If $Z = (Z_1, \dots, Z_N)$ is the Jacobian basis of $\mathbb{G}$, then it is easily seen that (with the obvious notation), for every $i = 1, \dots, r$,

$$Z_1^{(i)}, \dots, Z_{N_i}^{(i)} \quad \text{is a basis of } V^{(i)}.$$

Hence, there exist r non-singular square matrices $M^{(1)}, \dots, M^{(r)}$ such that, for $i = 1, \dots, r$, $M^{(i)}$ has order N_i and such that the matrix

$$\begin{pmatrix} M^{(1)} & \dots & 0 \\ \vdots & \ddots & \vdots \\ 0 & \dots & M^{(r)} \end{pmatrix}$$

represents the transformation of the coordinates between the bases $\mathcal{X}$ and Z. Hence, by the corresponding results for the Jacobian basis in Theorem 1.3.28 (page 49), we can infer that, following the notation in (20.49), $\zeta_i(\mathbf{x})$ is a polynomial function in $\mathbf{x}$, δ_λ-homogeneous of degree σ_i. In particular, if d is any homogeneous norm on $\mathbb{G}$, there exists $\mathbf{c} = \mathbf{c}(d, \mathrm{Log}) > 0$ such that

$$\mathbf{c}^{-1} d^{\sigma_i}(\mathbf{x}) \le |\zeta_i(\mathbf{x})| \le \mathbf{c}\, d^{\sigma_i}(\mathbf{x}) \quad \forall\, \mathbf{x} \in \mathbb{G},\ i \le N. \tag{20.58}$$

As a consequence, for every $k \in \mathbb{N} \cup \{0\}$, $\zeta_{i_1}(\mathbf{x}) \cdots \zeta_{i_k}(\mathbf{x})$ is a polynomial function in $\mathbf{x}$, δ_λ-homogeneous of degree

$$\sigma_{i_1} + \cdots + \sigma_{i_k}.$$

Observe that this is also the δ_λ-degree of $X_{i_1} \cdots X_{i_k}$ as a differential operator. In particular, $\zeta_{i_1}(\mathbf{x}) \cdots \zeta_{i_{n+1}}(\mathbf{x})$ is δ_λ-homogeneous of degree $\ge n+1$ and, in the integral summands, there appear only higher order derivatives of u with δ_λ-height $\ge n+1$.

We rewrite (20.57) pointing out, in the right-hand side, the polynomial of δ_λ-degree $\le n$

$$\begin{aligned} u(\mathbf{x}) &= u(0) + \sum_{h=1}^{n} \sum_{\substack{k=1,\dots,n \\ i_1,\dots,i_k \le N \\ \sigma_{i_1}+\cdots+\sigma_{i_k}=h}} \frac{\zeta_{i_1}(\mathbf{x}) \cdots \zeta_{i_k}(\mathbf{x})}{k!} X_{i_1} \cdots X_{i_k} u(0) \\ &\quad + \Bigg\{ \sum_{k=1}^{n} \sum_{\substack{i_1,\dots,i_k \le N \\ \sigma_{i_1}+\cdots+\sigma_{i_k} \ge n+1}} \frac{\zeta_{i_1}(\mathbf{x}) \cdots \zeta_{i_k}(\mathbf{x})}{k!} X_{i_1} \cdots X_{i_k} u(0) \\ &\quad + \sum_{i_1,\dots,i_{n+1}=1}^{N} \frac{\zeta_{i_1}(\mathbf{x}) \cdots \zeta_{i_{n+1}}(\mathbf{x})}{n!} \\ &\quad \times \int_0^1 (X_{i_1} \cdots X_{i_{n+1}} u)\left(\mathrm{Exp}\left(\sum_{i=1}^{N} s\, \zeta_i(\mathbf{x})\, X_i \right)\right) (1-s)^n \, ds \Bigg\} \\ &=: q_n(\mathbf{x}) + \{R_n(\mathbf{x})\}. \end{aligned} \tag{20.59}$$

As we remarked above, the function

$$q_n(\mathbf{x}) = \sum_{h=0}^{n} q_n^{(h)}(\mathbf{x}),$$

where

$$q_n^{(h)}(\mathbf{x}) = \begin{cases} u(0) & \text{if } h = 0, \\ \displaystyle\sum_{\substack{k=1,\dots,n \\ i_1,\dots,i_k \le N \\ \sigma_{i_1}+\cdots+\sigma_{i_k}=h}} \frac{\zeta_{i_1}(\mathbf{x}) \cdots \zeta_{i_k}(\mathbf{x})}{k!} X_{i_1} \cdots X_{i_k} u(0) & \text{if } h = 1, \dots, n, \end{cases} \tag{20.60}$$

is a polynomial function of δ_λ-degree $\le n$ and each $q_n^{(h)}(\mathbf{x})$ is a homogeneous polynomial of δ_λ-degree $= h$.

We now turn our attention to $R_n(\mathbf{x})$. The first sum in braces (ranging over $k = 1, \dots, n$) is a polynomial function of δ_λ-degree $\ge n + 1$. The remaining summand can be estimated as follows (for $\mathbf{x}$ in a neighborhood of 0)

$$\begin{aligned} &\Bigg| \sum_{i_1,\dots,i_{n+1}=1}^{N} \frac{\zeta_{i_1}(\mathbf{x}) \cdots \zeta_{i_{n+1}}(\mathbf{x})}{n!} \\ &\quad \times \int_0^1 (X_{i_1} \cdots X_{i_{n+1}} u)\left(\operatorname{Exp}\left(\sum_{i=1}^N s\, \zeta_i(\mathbf{x})\, X_i \right)\right) (1-s)^n \, \mathrm{d}s \Bigg| \\ &\le \sum_{i_1,\dots,i_{n+1}=1}^{N} \frac{|\zeta_{i_1}(\mathbf{x})| \cdots |\zeta_{i_{n+1}}(\mathbf{x})|}{(n+1)!} \\ &\quad \times \sup_{d(z) \le \mathbf{c}\, d(\mathbf{x})} \{ |X_{i_1} \cdots X_{i_{n+1}} u(z)| \, : i_1, \dots, i_{n+1} \in \{1, \dots, N\} \}. \end{aligned}$$

Indeed, we recall that, by definition, it holds

$$\sum_{i=1}^N s\, \zeta_i(\mathbf{x})\, X_i = s \operatorname{Log}(\mathbf{x}).$$

Then, we can apply Lemma 20.3.10 to derive that

$$\operatorname{Exp}\left(\sum_{i=1}^N s\, \zeta_i(\mathbf{x})\, X_i \right) \in B_d(0, \mathbf{c}\, d(\mathbf{x})),$$

so that

$$\begin{aligned} &\left| (X_{i_1} \cdots X_{i_{n+1}} u)\left(\operatorname{Exp}\left(\sum_{i=1}^N s\, \zeta_i(\mathbf{x})\, X_i \right)\right) \right| \\ &\quad \le \sup_{d(z) \le \mathbf{c}\, d(\mathbf{x})} \{ |X_{i_1} \cdots X_{i_{n+1}} u(z)| \, : i_1, \dots, i_{n+1} \in \{1, \dots, N\} \}. \end{aligned}$$

As a consequence, $R_n(\mathbf{x})$ can be estimated as follows (for $\mathbf{x}$ near 0).

$$
\begin{aligned}
|R_n(\mathbf{x})| &\le \sum_{k=1}^{n} \sum_{\substack{i_1,\dots,i_k\le N \\ \sigma_{i_1}+\dots+\sigma_{i_k}\ge n+1}} \frac{|\zeta_{i_1}(\mathbf{x})|\cdots|\zeta_{i_k}(\mathbf{x})|}{k!} \big|X_{i_1}\cdots X_{i_k}u(0)\big| \\
&\quad + \sum_{i_1,\dots,i_{n+1}=1}^{N} \frac{|\zeta_{i_1}(\mathbf{x})|\cdots|\zeta_{i_{n+1}}(\mathbf{x})|}{(n+1)!} \\
&\quad \times \sup_{d(z)\le \mathbf{c}\,d(\mathbf{x})} \{|X_{i_1}\cdots X_{i_{n+1}}u(z)| : i_1,\dots,i_{n+1}\in\{1,\dots,N\}\} \\
&\le \mathbf{c}_n\, d^{n+1}(\mathbf{x}) \times \sup_{d(z)\le \mathbf{c}\,d(\mathbf{x})} \{|X_{i_1}\cdots X_{i_k}u(z)| : \\
&\qquad k\le n+1,\ i_1,\dots,i_k\le N,\ \sigma_{i_1}+\dots+\sigma_{i_k}\ge n+1\} \\
&\le \widetilde{\mathbf{c}}_n\, d^{n+1}(\mathbf{x}) \times \sup_{d(z)\le \mathbf{c}\,d(\mathbf{x})} \{|X_{i_1}\cdots X_{i_M}u(z)| : \\
&\qquad n+1\le M\le r(n+1),\ i_1,\dots,i_M\in\{1,\dots,m\}\}.
\end{aligned}
$$

Here, we have used the fact that (thanks to the stratification of $\mathfrak{g}$) any X_j (for $j \in \{1,\dots,N\}$) is a precise linear combination (with scalars fixed together with $\mathfrak{g}$ and the basis $(X_1,\dots,X_N)$) of terms like

$$X_{i_1}\cdots X_{i_h}$$

with $1 \le h \le r$ (r being the step of nilpotency of $\mathfrak{g}$) and $i_1,\dots,i_h \in \{1,\dots,m\}$.

Since the above estimate of R_n gives

$$R_n(\mathbf{x}) = \mathcal{O}(d^{n+1}(\mathbf{x})) \quad \text{as } \mathbf{x}\to 0,$$

the decomposition $u(\mathbf{x}) = q_n(\mathbf{x}) + R_n(\mathbf{x})$ in (20.59) rewrites as

$$u(\mathbf{x}) = q_n(\mathbf{x}) + \mathcal{O}(d^{n+1}(\mathbf{x})) \quad \text{as } \mathbf{x}\to 0.$$

Thus (being q_n a polynomial of δ_λ-degree $\le n$) via an application of what has been proved in Remark 20.3.7, we infer that $q_n(\mathbf{x})$ is necessarily the Z-Mac Laurin polynomial of δ_λ-degree n related to u. This ends the proof. □

For an improvement of (20.55) for large $\mathbf{x}$'s, see Exercise 9 at the end of the chapter. For the future references, we state the explicit formula for $P_n(u,0)$ proved above in the following corollary.

Corollary 20.3.12 (Explicit form of the Z-Mac Laurin polynomial). *Let $n \in \mathbb{N}\cup\{0\}$, and let $u \in C^\infty(\mathbb{G},\mathbb{R})$. Let $Z = (Z_1,\dots,Z_N)$ be the Jacobian basis related to $\mathbb{G}$. We set*

$$\mathrm{Log}\,(\mathbf{x}) = \zeta_1(\mathbf{x})\,Z_1 + \dots + \zeta_N(\mathbf{x})\,Z_N \quad \forall\,\mathbf{x}\in\mathbb{G},$$

where Log *is the logarithmic map, i.e. $\zeta_1(\mathbf{x}),\dots,\zeta_N(\mathbf{x})$ are the components of* Log h *w.r.t. Z or, equivalently,*[6]

[6] See Remark 1.2.20 on page 21.

$$\zeta_j(\mathbf{x}) = \operatorname{Log}(\mathbf{x}) I_j(0) \quad \forall\, j \in \{1, \dots, N\}.$$

Then the Z-Mac Laurin polynomial of δ_λ-degree n related to u, i.e. the only polynomial p of δ_λ-degree $\le n$ such that

$$(Z_1 \cdots Z_N)^\alpha (p)(0) = (Z_1 \cdots Z_N)^\alpha (u)(0),$$

for every multi-index α with $|\alpha|_{\mathbb{G}} \le n$, is given by the following formula

$$P_n(u,0)(\mathbf{x}) = u(0) + \sum_{h=1}^{n} \sum_{\substack{k=1,\dots,n \\ i_1,\dots,i_k \le N \\ \sigma_{i_1}+\dots+\sigma_{i_k}=h}} \frac{\zeta_{i_1}(\mathbf{x}) \cdots \zeta_{i_k}(\mathbf{x})}{k!} Z_{i_1} \cdots Z_{i_k} u(0).$$

For example, if $n = 2$, the above formula for $P_2(u,0)$ rewrites as

$$\begin{aligned} P_2(u,0)(\mathbf{x}) &= u(0) + \sum_{i:\sigma_i=1} \zeta_i(\mathbf{x})\,(Z_i u)(0) + \sum_{i:\sigma_i=2} \zeta_i(\mathbf{x})\,(Z_i u)(0) \\ &\quad + \sum_{i,j:\ \sigma_i=\sigma_j=1} \frac{\zeta_i(\mathbf{x})\,\zeta_j(\mathbf{x})}{2} (Z_i Z_j u)(0). \end{aligned} \tag{20.61}$$

Example 20.3.13. Consider the Heisenberg–Weyl group $\mathbb{H}^1$ on $\mathbb{R}^3$. In order to obtain the expansion of a smooth u on $\mathbb{H}^1$ by Proposition 20.3.11, we must take $n = 1$. Hence, if we consider the Jacobian basis

$$Z_1 = \partial_{x_1} + 2x_2\,\partial_{x_3}, \quad Z_2 = \partial_{x_2} - 2x_1\,\partial_{x_3}, \quad Z_3 = \partial_{x_3},$$

then we have

$$\operatorname{Exp}(\xi_1 Z_1 + \xi_2 Z_2 + \xi_3 Z_3) = (\xi_1, \xi_2, \xi_3), \ \operatorname{Log}(x_1, x_2, x_3) = x_1 Z_1 + x_2 Z_2 + x_3 Z_3),$$

so that formula (20.53) gives

$$\begin{aligned} u(\mathbf{x}) &= u(0) + \mathbf{x}_1\, Z_1 u(0) + \mathbf{x}_2\, Z_2 u(0) \\ &\quad + \left\{ \mathbf{x}_3\, Z_3 u(0) + \sum_{i,j=1}^{3} \mathbf{x}_i \mathbf{x}_j \int_0^1 (X_i X_j u)(s\mathbf{x})\,(1-s)\,\mathrm{d}s \right\}. \end{aligned} \tag{20.62}$$

This shows that formula (20.53), despite its very simple proof, is not as "optimal" as formula (20.38) in Theorem 20.3.3. Indeed, note that the remainder in braces in (20.62) contains also derivatives of u of order 4 (w.r.t. Z_1, Z_2) such as

$$Z_3 Z_3 u = \frac{1}{16}\big(Z_2 Z_1 Z_2 Z_1 - Z_2 Z_1^2 Z_2 - Z_1 Z_2^2 Z_1 + Z_1 Z_2 Z_1 Z_2\big) u.$$

Instead, formula (20.38) gives the estimate

$$\begin{aligned} &\big|u(\mathbf{x}) - \big(u(0) + \mathbf{x}_1\, Z_1 u(0) + \mathbf{x}_2\, Z_2 u(0)\big)\big| \\ &\le \mathbf{c}_1 \left(|\mathbf{x}_1|^2 + |\mathbf{x}_2|^2 + |\mathbf{x}_3|\right) \times \sup_{d(z) \le \mathbf{b}^2 d(\mathbf{x})} \{|Z_1^2 u|(z),\ |Z_2^2 u|(z),\ |Z_1 Z_2 u|(z)\}. \end{aligned}$$

If $x_0 \in \mathbb{G}$ is any given point, by replacing u and $\mathbf{x}$ in the above results respectively with $u \circ \tau_{x_0}$ and $x_0^{-1} * \mathbf{x}$, we obtain the corresponding results for the stratified Taylor formula. We state them without further comment, recalling only that

$$P_n(u, x_0) = P_n(u \circ \tau_{x_0}, 0) \circ \tau_{x_0^{-1}}.$$

Proposition 20.3.14 (Stratified Taylor formula-integral remainder). *Let $n \in \mathbb{N} \cup \{0\}$, $u \in C^{n+1}(\mathbb{G}, \mathbb{R})$, $x_0 \in \mathbb{G}$. Suppose $\mathcal{X} = (X_1, \dots, X_N)$ is any basis of $\mathfrak{g}$ adapted to the stratification of $\mathfrak{g}$. Then, following the notation in (20.49), we have*

$$u(\mathbf{x}) = u(x_0) + \sum_{h=1}^{n} \sum_{\substack{k=1,\dots,n \\ i_1,\dots,i_k \le N \\ \sigma_{i_1}+\cdots+\sigma_{i_k}=h}} \frac{\zeta_{i_1}(x_0^{-1} * \mathbf{x}) \cdots \zeta_{i_k}(x_0^{-1} * \mathbf{x})}{k!} \times \big(X_{i_1} \cdots X_{i_k} u\big)(x_0) + R_n(\mathbf{x}, x_0), \tag{20.63}$$

where

$$R_n(\mathbf{x}, x_0) = \Bigg\{ \sum_{k=1}^{n} \sum_{\substack{i_1,\dots,i_k \le N \\ \sigma_{i_1}+\cdots+\sigma_{i_k} \ge n+1}} \frac{\zeta_{i_1}(x_0^{-1} * \mathbf{x}) \cdots \zeta_{i_k}(x_0^{-1} * \mathbf{x})}{k!} \times \big(X_{i_1} \cdots X_{i_k} u\big)(x_0) + \sum_{i_1,\dots,i_{n+1}=1}^{N} \frac{\zeta_{i_1}(x_0^{-1} * \mathbf{x}) \cdots \zeta_{i_{n+1}}(x_0^{-1} * \mathbf{x})}{n!} \times \int_0^1 (X_{i_1} \cdots X_{i_{n+1}} u)\left(x_0 * \mathrm{Exp}\left(\sum_{i=1}^{N} s\, \zeta_i(x_0^{-1} * \mathbf{x})\, X_i \right)\right) (1-s)^n \, ds \Bigg\}. \tag{20.64}$$

Moreover, for every fixed homogeneous norm d on $\mathbb{G}$ and every $n \in \mathbb{N} \cup \{0\}$, there exists $\mathbf{c}_n > 0$ (depending on n, $\mathbb{G}$, d and the basis $\mathcal{X}$) such that (for $\mathbf{x}$ near x_0)

$$|R_n(\mathbf{x}, x_0)| \le \mathbf{c}_n\, d^{n+1}(x_0^{-1} * \mathbf{x}) \times \sup_{d(z) \le \mathbf{c}\, d(x_0^{-1} * \mathbf{x})} \{ |(X_{i_1} \cdots X_{i_M} u)(x_0 * z)| : n+1 \le M \le r(n+1),\ i_1, \dots, i_M \in \{1, \dots, m\} \}. \tag{20.65}$$

Finally, when $\mathcal{X}$ is the Jacobian basis related to $\mathbb{G}$, the polynomial function

$$q_n(\mathbf{x}, x_0) = u(x_0) + \sum_{h=1}^{n} \sum_{\substack{k=1,\dots,n \\ i_1,\dots,i_k \le N \\ \sigma_{i_1}+\cdots+\sigma_{i_k}=h}} \frac{\zeta_{i_1}(x_0^{-1} * \mathbf{x}) \cdots \zeta_{i_k}(x_0^{-1} * \mathbf{x})}{k!} \times (X_{i_1} \cdots X_{i_k} u)(x_0) \tag{20.66}$$

coincides with the Z-Taylor polynomial of δ_λ-degree n related to u and x_0.

For an improvement of (20.65) for large $\mathbf{x}$'s, see Exercise 9 at the end of the chapter.

Corollary 20.3.15 (Explicit form of the Z-Taylor polynomial). *Let $n \in \mathbb{N} \cup \{0\}$, $u \in C^\infty(\mathbb{G}, \mathbb{R})$ and $x_0 \in \mathbb{G}$. Let $Z = (Z_1, \dots, Z_N)$ be the Jacobian basis related to $\mathbb{G}$. If the ζ_i's are as in Corollary 20.3.12, then the Z-Taylor polynomial of δ_λ-degree n related to u and x_0, i.e. the only polynomial p of δ_λ-degree $\leq n$ such that*

$$(Z_1 \cdots Z_N)^\alpha(p)(x_0) = (Z_1 \cdots Z_N)^\alpha(u)(x_0),$$

for every multi-index α with $|\alpha|_{\mathbb{G}} \leq n$, is given by the following formula

$$P_n(u, x_0)(\mathbf{x}) = u(x_0) + \sum_{h=1}^{n} \sum_{\substack{k=1,\dots,n \\ i_1,\dots,i_k \leq N \\ \sigma_{i_1}+\cdots+\sigma_{i_k}=h}} \frac{\zeta_{i_1}(x_0 * \mathbf{x}) \cdots \zeta_{i_k}(x_0 * \mathbf{x})}{k!} \times (Z_{i_1} \cdots Z_{i_k} u)(x_0).$$

By suitably choosing the "increment" h in Corollary 20.3.9 to belong to the "first layer" of $\mathbb{G}$, we obtain the following *"horizontal" Taylor formula.*

Corollary 20.3.16 (Horizontal Taylor formula). *Let $x \in \mathbb{G}$, and let $u \in C^{n+1}(\mathbb{G}, \mathbb{R})$. Let $(X_1, \dots, X_m)$ be any basis of the first layer of the stratification of $\mathfrak{g}$. Then, for every $\xi_1, \dots, \xi_m \in \mathbb{R}$, we have*

$$\begin{aligned} &u\left(x * \operatorname{Exp}\left(\sum_{j=1}^m \xi_j X_j\right)\right) \\ &= u(x) + \sum_{k=1}^{n} \sum_{i_1,\dots,i_k=1}^{m} \frac{\xi_{i_1} \cdots \xi_{i_k}}{k!} X_{i_1} \cdots X_{i_k} u(x) \\ &\quad + \sum_{i_1,\dots,i_{n+1}=1}^{m} \frac{\xi_{i_1} \cdots \xi_{i_{n+1}}}{n!} \\ &\quad \times \int_0^1 (X_{i_1} \cdots X_{i_{n+1}} u)\left(x * \delta_s\left(\operatorname{Exp}\left(\sum_{j=1}^m \xi_j X_j\right)\right)\right)(1-s)^n \, \mathrm{d}s. \end{aligned} \qquad (20.67)$$

Proof. It suffices to apply (20.67) with

$$h = \operatorname{Exp}\left(\sum_{j=1}^m \xi_j X_j\right)$$

and to observe that, in this case, $\operatorname{Log}(h) = \sum_{j=1}^m \xi_j X_j$, so that (with the notation in (20.67))

$$\zeta_j(h) = \begin{cases} \xi_j & \text{if } j \in \{1, \dots, m\}, \\ 0 & \text{if } j \in \{m+1, \dots, N\}. \end{cases}$$

Finally, notice that (see, e.g. (19.10d) on page 728)

$$\mathrm{Exp}\left(\sum_{j=1}^{m} s\,\xi_j\,X_j\right) = \mathrm{Exp}\left(\delta_s\left(\sum_{j=1}^{m} \xi_j\,X_j\right)\right) = \delta_s\left(\mathrm{Exp}\left(\sum_{j=1}^{m} \xi_j\,X_j\right)\right).$$

This ends the proof. □

For example, if we take $n = 1$ in the above corollary, we get the *horizontal Taylor formula with the integral remainder of order two*

$$\begin{aligned} u&\left(x * \mathrm{Exp}\left(\sum_{j=1}^{m} \xi_j X_j\right)\right) \\ &= u(x) + \sum_{j=1}^{m} \xi_j\, X_j u(x) \\ &\quad + \int_0^1 \sum_{h,k=1}^{m} \xi_h \xi_k (X_h X_k u)\left(x * \delta_s\left(\mathrm{Exp}\left(\sum_{j=1}^{m} \xi_j\,X_j\right)\right)\right)(1-s)\,\mathrm{d}s \qquad (20.68) \end{aligned}$$

for every $\xi_1, \ldots, \xi_m \in \mathbb{R}$. Since, for a given arbitrary $m \times m$ matrix $A = (a_{h,k})_{h,k}$, it holds

$$\sum_{h,k=1}^{m} \xi_h \xi_k\, a_{h,k} = \sum_{h,k=1}^{m} \xi_h \xi_k\, \frac{a_{h,k} + a_{k,h}}{2} = \left\langle \frac{A + A^T}{2}\,\xi, \xi \right\rangle,$$

(20.68) also rewrites as

$$\begin{aligned} u(x * \mathrm{Exp}\,(\xi \cdot X)) &= u(x) + (\xi \cdot X) u(x) \\ &\quad + \int_0^1 \langle \mathrm{Hess}_{\mathrm{sym}} u(x * \delta_s(\mathrm{Exp}\,(\xi \cdot X)))\,\xi, \xi\rangle\,(1-s)\,\mathrm{d}s, \end{aligned} \qquad (20.69)$$

where we have set

$$\xi \cdot X = \sum_{j=1}^{m} \xi_j X_j$$

and

$$\mathrm{Hess}_{\mathrm{sym}} u = \left(\frac{X_h X_k u + X_k X_h u}{2}\right)_{h,k=1,\ldots,m},$$

the so-called *symmetrized horizontal Hessian* of u.

As an application of Taylor's formula, we give the following result.

Proposition 20.3.17 ($\mathcal{L}$-harmonicity of Taylor polynomials). *Let $(G, *)$ be a homogeneous Carnot group, and let $Z = (Z_1, \ldots, Z_N)$ be the Jacobian basis related to $\mathbb{G}$. Let also $\mathcal{L}$ be a sub-Laplacian on $\mathbb{G}$. Suppose that $u \in C^\infty(\mathbb{G}, \mathbb{R})$ is an $\mathcal{L}$-harmonic function.*

Then, for every $n \in \mathbb{N} \cup \{0\}$ and every $x_0 \in \mathbb{G}$, the function $x \mapsto P_n(u, x_0)(x)$ is $\mathcal{L}$-harmonic in $\mathbb{G}$.

Proof. Suppose we have proved the assertion for $x_0 = 0$. Then, by applying the assertion for $u \circ \tau_{x_0}$ (which is $\mathcal{L}$-harmonic on $\mathbb{G}$ since u is) we infer that $y \mapsto P_n(u \circ \tau_{x_0}, 0)(y)$ is $\mathcal{L}$-harmonic on $\mathbb{G}$. As a consequence, since (20.26) gives

$$P_n(u, x_0) = P_n(u \circ \tau_{x_0}, 0) \circ \tau_{x_0^{-1}},$$

we infer that $x \mapsto P_n(u, x_0)(x)$ is $\mathcal{L}$-harmonic in $\mathbb{G}$.

The above argument shows that it suffices to prove the assertion when $x_0 = 0$. By Lemma 20.3.18 below, we know that, for every $n \geq 2$,

$$\mathcal{L}(P_n(u, 0)) = P_{n-2}(\mathcal{L}u, 0) = P_{n-2}(0, 0) = 0.$$

Since $\mathcal{L}(P_n(u, 0)) = 0$ also for $n = 0, 1$ (since P_0 and P_1 are polynomials of $\mathbb{G}$-degree ≤ 1), this completes the proof. □

Lemma 20.3.18. *Let $u \in C^\infty(\mathbb{G}, \mathbb{R})$. Then we have*

$$\mathcal{L}(P_n(u, 0)) = P_{n-2}(\mathcal{L}u, 0) \quad \text{for every } n \geq 2. \tag{20.70}$$

Proof. By Proposition 20.3.11, we have $u(x) = P_n(u, 0)(x) + R_n(x)$, so that

$$(\star) \qquad \mathcal{L}u(x) = \mathcal{L}(P_n(u, 0))(x) + \mathcal{L}R_n(x).$$

If we show that, for every $n \geq 2$, one has

$$(\star\star) \qquad \mathcal{L}R_n(x) = \mathcal{O}_{x \to 0}(d^{n-1}),$$

then, by Remark 20.3.7, $(\star)$ will give (20.70). We are then left with the proof of $(\star\star)$. Denote the expression of R_n in (20.54) as follows

$$R_n(x) = \sum_{k=1}^{n} Q_k(x) + \sum_{I=(i_1,\dots,i_{n+1})} A_I(x) \times B_I(x),$$

where B_I is the integral in the far right-hand side of (20.54).

Clearly, each Q_k is a δ_λ-homogeneous polynomial function of degree $\geq n + 1$, so that $\mathcal{L}Q_k$ is a δ_λ-homogeneous polynomial function of degree $\geq n - 1$, whence $\mathcal{L}Q_k(x) = \mathcal{O}_{x \to 0}(d^{n-1})$. Dropping the subscript I in A_I and B_I, we have (see Ex. 6, Chapter 1)

$$\mathcal{L}(A\,B) = B\,\mathcal{L}A + A\,\mathcal{L}B + 2\,\langle \nabla_{\mathcal{L}} A, \nabla_{\mathcal{L}} B \rangle.$$

Since $A(x) = \zeta_{i_1}(x) \cdots \zeta_{i_{n+1}}(x)/n!$, it clearly holds

$$\mathcal{L}A = \mathcal{O}_{x \to 0}(d^{n-1}), \quad A = \mathcal{O}_{x \to 0}(d^{n+1}(x)), \quad |\nabla_{\mathcal{L}} A| = \mathcal{O}_{x \to 0}(d^n).$$

This proves that $\mathcal{L}(A\,B) = \mathcal{O}_{x \to 0}(d^{n-1})$, since B, $\mathcal{L}B$ and $\nabla_{\mathcal{L}} B$ are bounded. This completes the proof. □

Bibliographical Notes. For the topics presented in Section 20.2, we are much indebted to the presentation of the same subject in [FS82] by G.B. Folland and E.M. Stein. For results concerning Taylor's formula on Carnot groups of step two, see G. Arena, A. Caruso, A. Causa [ACC06]. For other results of calculus on Carnot groups, see [CM06,Hei95a].

20.4 Exercises of Chapter 20

Ex. 1) Prove Proposition 20.1.2, page 735. (*Hint*: Show that the map introduced therein is linear and sends a basis into a basis.)

Ex. 2) Retracing the proof of Proposition 20.1.5, prove in details Proposition 20.1.7.

Ex. 3) Prove Lemma 20.1.11, page 741. (*Hint*: If $\sum_j c_j H_j \equiv 0$, then $0 = (\sum_j c_j H_j) \times (u(\delta_\lambda(x))) = \sum_j c_j \lambda^{d_j} (H_j u)(\delta_\lambda(x))$. Now take as u any function of the type x^α and complete the argument.)

Ex. 4) With reference to Example 20.2.5, let us now find $r = P_3(f, 0)$. Since $r \in \mathcal{P}_3$, there exist real numbers, $a, b, \ldots, m$ such that $r(x, y, t) = a + b\,\mathbf{x} + c\,\mathbf{y} + d\,\mathbf{t} + e\,\mathbf{x}^2 + f\,\mathbf{y}^2 + g\,\mathbf{xy} + h\,\mathbf{xt} + i\,\mathbf{yt} + j\,\mathbf{x}^3 + k\,\mathbf{y}^3 + l\,\mathbf{x}^2\mathbf{y} + m\,\mathbf{xy}^2$. Then (20.22) gives

$$
\begin{aligned}
a &= r(0) = f(0),\\
b &= Z_1 r(0) = Z_1 f(0),\\
c &= Z_2 r(0) = Z_2 f(0),\\
d &= Z_3 r(0) = Z_3 f(0),\\
2e &= (Z_1)^2 r(0) = (Z_1)^2 f(0),\\
2f &= (Z_2)^2 r(0) = (Z_2)^2 f(0),\\
g - 2d &= Z_1 Z_2 r(0) = Z_1 Z_2 f(0),\\
h &= Z_1 Z_3 r(0) = Z_1 Z_3 f(0),\\
i &= Z_2 Z_3 r(0) = Z_2 Z_3 f(0),\\
3!j &= (Z_1)^3 r(0) = (Z_1)^3 f(0),\\
3!k &= (Z_2)^3 r(0) = (Z_2)^3 f(0),\\
2l - 4h &= (Z_1)^2 Z_2 r(0) = (Z_1)^2 Z_2 f(0),\\
2m - 4i &= Z_1 (Z_2)^2 r(0) = Z_1 (Z_2)^2 f(0),
\end{aligned}
$$

so that

$$
\begin{aligned}
P_3(f, 0)(\mathbf{x}, \mathbf{y}, \mathbf{t}) = {} & f(0) + Z_1 f(0)\,\mathbf{x} + Z_2 f(0)\,\mathbf{y} + Z_3 f(0)\,\mathbf{t}\\
& + \frac{1}{2}(Z_1)^2 f(0)\,\mathbf{x}^2 + \frac{1}{2}(Z_2)^2 f(0)\,\mathbf{y}^2 + (2Z_3 f(0)
\end{aligned}
$$

$$+ Z_1 Z_2 f(0))\,\mathbf{xy} + Z_1 Z_3 f(0)\,\mathbf{xt} + Z_2 Z_3 f(0)\,\mathbf{yt}$$
$$+ \frac{1}{3!}(Z_1)^3 f(0)\,\mathbf{x^3} + \frac{1}{3!}(Z_2)^3\,\mathbf{y^3} + (2 Z_1 Z_3 f(0)$$
$$+ \frac{1}{2} Z_1^2 Z_2 f(0))\,\mathbf{x^2 y} + (2 Z_2 Z_3 f(0) + \frac{1}{2} Z_1 Z_2^2 f(0))\,\mathbf{xy^2}.$$

Ex. 5) Consider the homogeneous Carnot group $\mathbb{G}$ on $\mathbb{R}^3$ with the composition law

$$x \circ y = (x_1 + y_1, x_2 + y_2, x_3 + y_3 + x_1\, y_2).$$

Prove that the related Jacobian basis is given by

$$Z_1 = \partial_1, \quad Z_2 = \partial_2 + x_1\, \partial_3, \quad Z_3 = [Z_1, Z_2] = \partial_3.$$

Show that

$$\begin{aligned} P_2(f,0)(\mathbf{x_1}, \mathbf{x_2}, \mathbf{x_3}) &= f(0) + Z_1 f(0)\,\mathbf{x_1} + Z_2 f(0)\,\mathbf{x_2} + Z_3 f(0)\,\mathbf{x_3} \\ &\quad + \frac{1}{2}(Z_1)^2 f(0)\,\mathbf{x_1^2} + \frac{1}{2}(Z_2)^2 f(0)\,\mathbf{x_2^2} \\ &\quad + (Z_2 Z_1) f(0)\mathbf{x_1}\,\mathbf{x_2} \end{aligned} \tag{20.71}$$

is the Z-Mac Laurin polynomial of $\mathbb{G}$-degree 2.
Deduce that formula (20.24) *does not* apply in this case.
(Indeed, this would hold true iff $\frac{1}{2}(Z_1 Z_2 + Z_2 Z_1) = Z_2 Z_1$, i.e. $[Z_1, Z_2] = 0$ which is false.) This shows that (20.24) strongly depends on the properties of the fixed coordinate system on $\mathbb{G}$ or, equivalently, on the properties of the composition law.
Notice that the inverse on $\mathbb{G}$ differs from $-x$ (and the coordinate system on $\mathbb{G}$ is not the logarithmic one). See also Ex. 6 below.

Finally, compare to (20.61). Notice that here we have

$$\mathrm{Exp}\,(\xi_1 Z_1 + \xi_2 Z_2 + \xi_3 Z_3) = (\xi_1, \xi_2, \xi_3 + \xi_1\, \xi_2/2),$$

so that

$$\mathrm{Log}\,(x_1, x_2, x_3) = x_1 Z_1 + x_2 Z_2 + (x_3 - x_1 x_2/2) Z_3.$$

With the notation of Corollary 20.3.12, this means that

$$\zeta_1(\mathbf{x}) = \mathbf{x}_1, \;\; \zeta_2(\mathbf{x}) = \mathbf{x}_2, \;\; \zeta_3(\mathbf{x}) = \mathbf{x}_3 - \mathbf{x}_1 \mathbf{x}_2/2,$$

whence (20.61) gives

$$\begin{aligned} &P_2(f,0)(\mathbf{x_1}, \mathbf{x_2}, \mathbf{x_3}) \\ &\quad = f(0) + Z_1 f(0)\,\mathbf{x_1} + Z_2 f(0)\,\mathbf{x_2} \\ &\qquad + Z_3 f(0)\,(\mathbf{x_3} - \mathbf{x}_1 \mathbf{x}_2/2) + \frac{1}{2}(Z_1)^2 f(0)\,\mathbf{x_1^2} + \frac{1}{2}(Z_2)^2 f(0)\,\mathbf{x_2^2} \end{aligned}$$

$$+\frac{1}{2}(Z_1Z_2)f(0)\mathbf{x_1}\,\mathbf{x_2}+\frac{1}{2}(Z_2Z_1)f(0)\mathbf{x_2}\,\mathbf{x_1}$$
$$=f(0)+Z_1f(0)\,\mathbf{x_1}+Z_2f(0)\,\mathbf{x_2}+Z_3f(0)\,\mathbf{x_3}$$
$$+\frac{1}{2}(Z_1)^2f(0)\,\mathbf{x_1^2}+\frac{1}{2}(Z_2)^2f(0)\,\mathbf{x_2^2}$$
$$+\frac{1}{2}(Z_1Z_2+Z_2Z_1-Z_3)f(0)\mathbf{x_1}\,\mathbf{x_2},$$

which is (20.71), for

$$Z_1Z_2+Z_2Z_1-Z_3=Z_1Z_2+Z_2Z_1-[Z_1,Z_2]=2Z_2Z_1.$$

Ex. 6) (Symmetrized Z-Mac Laurin formula of δ_λ-degree two). *Suppose that $\mathbb{G}$ is a homogeneous Carnot group (of step $r\geq 2$) for which the inversion equals $-x$. Then*[7] *the Z-Mac Laurin polynomial of δ_λ-degree two is given by*

$$P_2(f,0)(\mathbf{x^{(1)}},\mathbf{x^{(2)}},\mathbf{x^{(3)}},\dots,\mathbf{x^{(r)}})=f(0)+\nabla^{(1)}f(0)\cdot\mathbf{x^{(1)}}$$
$$+\nabla^{(2)}f(0)\cdot\mathbf{x^{(2)}}+\frac{1}{2}\langle\mathbf{x^{(1)}},\mathrm{Hess}_{\mathrm{sym}}f(0)\cdot\mathbf{x^{(1)}}\rangle. \tag{20.72}$$

Here

$$\mathrm{Hess}_{\mathrm{sym}}f(0):=\left(\frac{1}{2}\{Z_h^{(1)}Z_k^{(1)}f(0)+Z_k^{(1)}Z_h^{(1)}f(0)\}\right)_{h,k=1,\dots,N_1}$$

is the so-called *symmetrized horizontal Hessian* of f at 0.
For example, a sufficient condition for $x^{-1}=-x$ to hold (and hence, for (20.72) to hold) is that $\mathbb{G}$ is equipped with logarithmic coordinates. Note that (20.72) holds for the Heisenberg–Weyl group $\mathbb{H}^n$ since the inversion on $\mathbb{H}^n$ is $-x$ (even if $\mathbb{H}^n$ is not generally equipped with logarithmic coordinates). Compare also to the previous Ex. to show that (20.72) does not hold in general unless suitable hypotheses on the coordinates are made. *Hint*: In order to prove (20.72), by Definition 20.2.3, it suffices to show that the polynomial (say q) in the right-hand side of (20.72) has δ_λ-degree two (obvious) and the following facts hold:

$$
\begin{aligned}
&\text{(i)} && q(0)=f(0),\\
&\text{(ii)} && Z_h^{(1)}q(0)=Z_h^{(1)}f(0)\quad \text{for } h=1,\dots,N_1,\\
&\text{(iii)} && Z_h^{(2)}q(0)=Z_i^{(2)}f(0)\quad \text{for } h=1,\dots,N_2,\\
&\text{(iv)} && (Z_h^{(1)}Z_k^{(1)})q(0)=(Z_h^{(1)}Z_k^{(1)})f(0)\quad \text{for } 1\leq h\leq k\leq N_1.
\end{aligned}
$$

---

[7] Here we used the following notation. The usual notation (1.79a) (page 56) for the "stratified" coordinates on a homogeneous group; moreover, for $i=1,2$,

$$\nabla^{(i)}f(0)=\left(Z_1^{(i)}f(0),\dots,Z_{N_i}^{(i)}f(0)\right),$$

i.e. $Z_j^{(i)}$ is the left-invariant vector field coinciding with $\partial_{x_j^{(i)}}$ at 0 or, equivalently, $Z_1^{(i)},\dots,Z_{N_i}^{(i)}$ are the vector fields of the Jacobian basis which are δ_λ-homogeneous of degree i. Notice that $\nabla^{(1)}f(0)$ is the *horizontal gradient* of f at 0.

Now, (i) is trivial, whereas (ii) and (iii) are simple consequences of the fact that $Z_h^{(i)}|_0 = (\partial/\partial x_h^{(i)})|_0$. In order to show (iv), recall (20.10), so that

$$Z_h^{(1)} = \partial_{x_h^{(1)}} + \sum_{j:\, \sigma_j>1} a_{j,h}(x)\, \partial_{x_j}, \qquad h = 1, \ldots, N_1,$$

where

$$\left(a_{j,h}\right)_{j,h=1,\ldots,N} = \begin{pmatrix} 1 & 0 & \cdots & 0 \\ a_{2,1}(x) & 1 & \ddots & \vdots \\ \vdots & \ddots & \ddots & 0 \\ a_{N,1}(x) & \cdots & a_{N,N-1}(x) & 1 \end{pmatrix} = \mathcal{J}_{\tau_x}(0)$$

is the Jacobian matrix of the left translation by x on $\mathbb{G}$ (i.e. $\tau_x(y) = x * y$) and the $a_{j,h}$'s are polynomial functions, δ_λ-homogeneous of degree $\sigma_j - \sigma_h$. As a consequence, $Z_h^{(1)} Z_k^{(1)}$ coincides with

$$\frac{\partial^2}{\partial x_h^{(1)} x_k^{(1)}} + \sum_{j:\, \sigma_j=2} \frac{\partial a_{j,k}}{\partial x_h^{(1)}}(0)\, \partial_{x_j} = \frac{\partial^2}{\partial x_h^{(1)} x_k^{(1)}} + \sum_{j=1}^{N_2} \frac{\partial a_{j,k}^{(2)}}{\partial x_h^{(1)}}(0)\, \partial_{x_j^{(2)}}$$

at 0. Here, we used a "stratified" notation for $\mathcal{J}_{\tau_x}(0)$, namely

$$\mathcal{J}_{\tau_x}(0) = \begin{pmatrix} \mathbb{I}_{N_1} & 0 & \cdots & 0 \\ A_1^{(2)}(x) & \mathbb{I}_{N_2} & \ddots & \vdots \\ \vdots & \ddots & \ddots & 0 \\ A_1^{(r)}(x) & \cdots & A_{r-1}^{(r)}(x) & \mathbb{I}_{N_r} \end{pmatrix}.$$

Notice that

$$(a_{j,k}^{(2)}(x))_{1\le j\le N_2,\, 1\le k\le N_1} = A_1^{(2)}(x).$$

This shows that (iv) is equivalent to

$$\frac{1}{2}\{Z_h^{(1)} Z_k^{(1)} f(0) + Z_k^{(1)} Z_h^{(1)} f(0)\} + \sum_{j=1}^{N_2} \frac{\partial a_{j,k}^{(2)}}{\partial x_h^{(1)}}(0)\, Z_j^{(2)} f(0)$$
$$\stackrel{?}{=} \frac{\partial^2 f(0)}{\partial x_h^{(1)} x_k^{(1)}} + \sum_{j=1}^{N_2} \frac{\partial a_{j,k}^{(2)}}{\partial x_h^{(1)}}(0)\, \partial_{x_j^{(2)}} f(0).$$

Arguing as above, this is in turn equivalent to

$$\frac{\partial^2 f(0)}{\partial x_h^{(1)} x_k^{(1)}} + \sum_{j=1}^{N_2} \frac{\partial a_{j,k}^{(2)}}{\partial x_h^{(1)}}(0)\, \partial_{x_j^{(2)}} f(0)$$

$$+\frac{1}{2}\sum_{j=1}^{N_2}\left\{\frac{\partial\, a^{(2)}_{j,k}}{\partial\, x^{(1)}_h}(0)+\frac{\partial\, a^{(2)}_{j,h}}{\partial\, x^{(1)}_k}(0)\right\}\partial_{x^{(2)}_j} f(0)$$

$$\overset{?}{=}\frac{\partial^2 f(0)}{\partial\, x^{(1)}_h x^{(1)}_k}+\sum_{j=1}^{N_2}\frac{\partial\, a^{(2)}_{j,k}}{\partial\, x^{(1)}_h}(0)\,\partial_{x^{(2)}_j} f(0).$$

Thus, we are left to prove that, for all $j\le N_2$ and all $h,k\le N_1$, it holds

$$\frac{\partial\, a^{(2)}_{j,k}}{\partial\, x^{(1)}_h}(0)+\frac{\partial\, a^{(2)}_{j,h}}{\partial\, x^{(1)}_k}(0)\overset{?}{=}0.$$

Now, the matrix $\big(a^{(2)}_{j,k}\big)=A^{(2)}_1$ involves the (derivatives of the) components of the left translation along the second layer. We know that the left translation is given by

$$\tau_x(y)=x*y=\big(x^{(1)}+y^{(1)},x^{(2)}+y^{(2)}+Q^{(2)}(x,y),\dots\big)$$

where $Q^{(2)}(x,y)$ is a polynomial in $x^{(1)}$, $y^{(1)}$ of (ordinary) degree 2, mixed in x,y. Hence, for every $j\in\{1,\dots,N_2\}$, there exists a square matrix B^j of order N_1 such that

$$(\tau_x(y))^{(2)}_j=x^{(2)}_j+y^{(2)}_j+\langle B^j\cdot x^{(1)},y^{(1)}\rangle.$$

We now use the hypothesis $x^{-1}=-x$ to derive that B^j is skew-symmetric. Consequently, this gives

$$a^{(2)}_{j,k}(x)=\frac{\partial}{\partial\, y^{(1)}_k}\bigg|_0\big(\tau_x(y)\big)^{(2)}_j=\sum_{i=1}^{N_1}x^{(1)}_i\,B^j_{i,k}.$$

Hence,

$$\frac{\partial\, a^{(2)}_{j,k}}{\partial\, x^{(1)}_h}(0)+\frac{\partial\, a^{(2)}_{j,h}}{\partial\, x^{(1)}_k}(0)=B^j_{h,k}+B^j_{k,h}=0,$$

for B^j is skew-symmetric. The assertion is proved. □

Ex. 7) Suppose $f,g\in C^{n+1}(\mathbb{G},\mathbb{R})$ are such that

$$(Z_1,\dots,Z_m)^\alpha f\equiv(Z_1,\dots,Z_m)^\alpha g\quad\forall\alpha\in(\mathbb{N}\cup\{0\})^m:\ |\alpha|=n+1.$$

Show that there exists a polynomial function p, δ_λ-homogeneous of degree at most n, such that $f=g+p$ on $\mathbb{G}$. (*Hint*: Apply Theorem 20.3.3 to $f-g$ and derive that $f(h)-g(h)=P_n(f-g,0)(h)$ for every $h\in\mathbb{G}$.) Derive that

$$(Z_{i_1},\dots,Z_{i_m})^\alpha f\equiv(Z_{i_1},\dots,Z_{i_m})^\alpha g$$

for every $\alpha\in(\mathbb{N}\cup\{0\})^m$ such that $|\alpha|=n+1$ and for every $i_1,\dots,i_m\in\{1,\dots,m\}$.
This proves that *the $(Z_1,\dots,Z_m)^\alpha f$'s with $|\alpha|=n+1$ determine all the $(Z_{i_1},\dots,Z_{i_m})^\alpha f$'s with $|\alpha|=n+1$.*

Ex. 8) Provide a detailed proof of Theorem 20.3.3.

Ex. 9) Prove the following improvement of (20.55) and (20.65). For *every* $\mathbf{x} \in \mathbb{G}$, we have

$$|R_n(\mathbf{x}, x_0)| \le \sum_{k=1}^{n+1} \frac{C^k}{k!} \times \sum_{\substack{i_1,\dots,i_k \le N, \\ \sigma_{i_1}+\dots+\sigma_{i_k} \ge n+1}} d(x_0^{-1} \circ \mathbf{x})^{\sigma_{i_1}+\dots+\sigma_{i_k}} \times \sup_{d(z) \le C\, d(x_0^{-1} \circ \mathbf{x})} |Z_{i_1} \cdots Z_{i_k} u(x_0 \circ z)|,$$

where C is a constant depending only on $\mathbb{G}$, d and the basis $\mathcal{X}$. Here, the σ_i's are the same as in (20.2).

References

[AH96] Adams, D.R., Hedberg, L.I.: Function Spaces and Potential Theory. Grundlehren der Mathematischen Wissenschaften, **314**, Springer-Verlag, Berlin, Heidelberg (1996)

[AE96] Aikawa, H., Essén, M.: Potential Theory—Selected Topics. Lecture Notes in Mathematics, **1633**, Springer-Verlag, Berlin (1996)

[Ale92] Alexopoulos, G.: An application of homogenization theory to harmonic analysis: Harnack inequalities and Riesz transforms on Lie groups of polynomial growth. Canad. J. Math., **44**, 691–727 (1992)

[Ale02] Alexopoulos, G.: Sub-Laplacians with drift on Lie groups of polynomial volume growth. Mem. Amer. Math. Soc., **155** (2002)

[Alt99] Altafini, C.: A matrix Lie group of Carnot type for filiform sub-Riemannian structures and its application to control systems in chained form. In: Breda, Azevedo, et al. (eds.), Proceedings of the Summer School on Differential Geometry, **1**, Coimbra Portugal (1999)

[Ama79] Amano, K.: Maximum principles for degenerate elliptic–parabolic operators. Indiana Univ. Math. J., **29**, 545–557 (1979)

[AM03] Ambrosio, L., Magnani, V.: Weak differentiability of BV functions on stratified groups. Math. Z., **245**, 123–153 (2003)

[ACC06] Arena, G., Caruso, A., Causa, A.: Taylor formula for Carnot groups and applications. Matematiche (Catania), **60**, 375–383 (2005)

[AG01] Armitage, D.H., Gardiner, S.J.: Classical Potential Theory, Springer-Verlag, London (2001)

[ABR92] Axler, S., Bourdon, P., Ramey, W.: Harmonic Function Theory. Graduate Texts in Mathematics, Springer-Verlag, New York (1992)

[Bah04] Bahri, A.: Recent results in contact form geometry. Partial differential equations and applications. Discrete Contin. Dyn. Syst., **10**, 21–30 (2004)

[Bah03] Bahri, A.: Recent progress in conformal geometry (with an emphasis on the use of configuration spaces). Lectures on Partial Differential Equations. New Stud. Adv. Math., **2**, Int. Press, Somerville, MA, 1–34 (2003)

[BR02] Balogh, Z.M., Rickly, M.: Regularity of convex functions on Heisenberg groups. Ann. Sc. Norm. Super. Pisa Cl. Sci. (5), **2**, 847–868 (2003)

[BRSC03] Balogh, Z.M., Rickly, M., Serra Cassano, F.: Comparison of Hausdorff measures with respect to the Euclidean and the Heisenberg metric. Publ. Mat., **47**, 237–259 (2003)

[BT02] Balogh, Z.M., Tyson, J.T.: Polar coordinates in Carnot group. Math. Z., **241**, 697–730 (2002)

[BT03] Balogh, Z.M., Tyson, J.T.: Potential theory in Carnot groups. Beckner, William, et al. (ed.), Harmonic Analysis at Mount Holyoke. Proceedings of Research Conference. Contemp. Math., **320**, American Mathematical Society, Providence, RI, 15–27 (2003)

[BG72] Baouendi, M.S., Goulaouic, C.: Nonanalytic-hypoellipticity for some degenerate elliptic operators. Bull. Amer. Math. Soc., **78**, 483–486 (1972)

[BD01] Bardi, M., Da Lio, F.: Propagation of maxima and strong maximum principle for viscosity solutions of degenerate elliptic equations. I: convex operators. Nonlinear Analysis, **44**, 991–1006 (2001)

[Bar03] Barletta, E.: Hörmander systems and harmonic morphisms. Ann. Sc. Norm. Super. Pisa Cl. Sci. (5), **2**, 379–394 (2003)

[BD04] Barletta, E., Dragomir, S.: On the regularity of weak subelliptic F-harmonic maps. Tsukuba J. Math., **28**, 417–436 (2004)

[BDU01] Barletta, E., Dragomir, S., Urakawa, H.: Pseudoharmonic maps from nondegenerate CR manifolds to Riemannian manifolds. Indiana Univ. Math. J., **50**, 719–746 (2001)

[Bau66] Bauer, H.: Harmonische Räume und ihre Potentialtheorie. Lecture Notes in Mathematics, **22**, Springer-Verlag, Berlin, New York (1966)

[BGG96] Beals, R., Gaveau, B., Greiner, P.: The Green function of model step two hypoelliptic operators and the analysis of certain tangential Cauchy–Riemann complexes. Adv. Math., **121**, 288–345 (1996)

[BM95] Bell, D.R., Mohammed, S.A.: An extension of Hörmander's theorem for infinitely degenerate second-order operators. Duke Math. J., **78**, 453–475 (1995)

[Bel96] Bellaïche, A.: The tangent space in sub-Riemannian geometry. Sub-Riemannian Geometry. Progress in Mathematics, **144**, Birkhäuser Verlag, Basel, 1–78 (1996)

[BR96] Bellaïche, A., Risler, J.-J. (eds.), Sub-Riemannian Geometry. Progress in Mathematics, **144**, Birkhäuser Verlag, Basel (1996)

[BAKS84] Ben Arous, G., Kusuoka, S., Stroock, D.W.: The Poisson kernel for certain degenerate elliptic operators. J. Funct. Anal., **56**, 171–209 (1984)

[BCN97] Berestycki, H., Caffarelli, L., Nirenberg, L.: Monotonicity for elliptic equations in unbounded Lipschitz domains. Comm. Pure Appl. Math., **50**, 1089–1111 (1997)

[BHM00] Berestycki, H., Hamel, F., Monneau, R.: One-dimensional symmetry of bounded entire solutions of some elliptic equations. Duke Math. J., **103**, 375–396 (2000)

[BN91] Berestycki, H., Nirenberg, L.: On the method of moving planes and the sliding method. Bol. Soc. Braseleira Mat., **22**, 1–37 (1991)

[BCC97] Birindelli, I., Capuzzo Dolcetta, I., Cutrì, A.: Liouville theorems for semilinear equations on the Heisenberg group. Ann. Inst. H. Poincaré Anal. Non Linéaire, **14**, 295–308 (1997)

[BCC98] Birindelli, I., Capuzzo Dolcetta, I., Cutrì, A.: Indefinite semi-linear equations on the Heisenberg group: a priori bounds and existence. Comm. Partial Differential Equations, **23**, 1123–1157 (1998)

[BC95] Birindelli, I., Cutrì, A.: A semi-linear problem for the Heisenberg Laplacian. Rend. Sem. Mat. Univ. Padova, **94**, 137–153 (1995)

[BL02] Birindelli, I., Lanconelli, E.: A note on one dimensional symmetry in Carnot groups. Atti Accad. Naz. Lincei Cl. Sci. Fis. Mat. Natur. Rend. Lincei (9) Mat. Appl., **13**, 17–22 (2002)

[BL03] Birindelli, I., Lanconelli, E.: A negative answer to a one-dimensional symmetry problem in the Heisenberg group. Calc. Var. Partial Differential Equations, **18**, 357–372 (2003)

[BP99] Birindelli, I., Prajapat, J.: Nonlinear Liouville theorems in the Heisenberg group via the moving plane method. Comm. Partial Differential Equations, **24**, 1875–1890 (1999)

[BP01] Birindelli, I., Prajapat, J.: One-dimensional symmetry in the Heisenberg group. Ann. Scuola Norm. Sup. Pisa Cl. Sci. (4), **30**, 269–284 (2001)

[BP02] Birindelli, I., Prajapat, J.: Monotonicity and symmetry results for degenerate elliptic equations on nilpotent Lie groups. Pacific J. Math., **204**, 1–17 (2002)

[Bir95] Biroli, M. (ed.), Potential theory and degenerate partial differential operators. Proceedings. Potential Anal., **4**, 309–467 (1995)

[BM95] Biroli, M., Mosco, U.: Sobolev inequalities on homogeneous spaces. Potential Anal., **4**, 311–324 (1995)

[Bon04] Bonfiglioli, A.: Homogeneous Carnot groups related to sets of vector fields. Boll. Unione Mat. Ital. Sez. B Artic. Ric. Mat. (8), **7**, 79–107 (2004)

[BC04] Bonfiglioli, A., Cinti, C.: A Poisson–Jensen type representation formula for subharmonic functions on stratified Lie groups. Potential Anal., **22**, 151–169 (2005)

[BC05] Bonfiglioli, A., Cinti, C.: The theory of energy for sub-Laplacians with an application to quasi-continuity. Manuscripta Math., **118**, 283–309 (2005)

[BL01] Bonfiglioli, A., Lanconelli, E.: Liouville-type theorems for real sub-Laplacians. Manuscripta Math., **105**, 111–124 (2001)

[BL02] Bonfiglioli, A., Lanconelli, E.: Maximum Principle on unbounded domains for sub-Laplacians: a potential theory approach. Proc. Amer. Math. Soc., **130**, 2295–2304 (2002)

[BL03] Bonfiglioli, A., Lanconelli, E.: Subharmonic functions on Carnot groups. Math. Ann., **325**, 97–122 (2003)

[BL07] Bonfiglioli, A., Lanconelli, E.: Gauge functions, Eikonal equations and Bôcher's theorem on stratified Lie groups. to appear in: Calc. Var. Partial Differential Equations (2007)

[BLU02] Bonfiglioli, A., Lanconelli, E., Uguzzoni, F.: Uniform Gaussian estimates of the fundamental solutions for heat operators on Carnot groups. Adv. Differential Equations, **7**, 1153–1192 (2002)

[BLU03] Bonfiglioli, A., Lanconelli, E., Uguzzoni, F.: Fundamental solutions for non-divergence form operators on stratified groups. Trans. Amer. Math. Soc., **356**, 2709–2737 (2003)

[BU04a] Bonfiglioli, A., Uguzzoni, F.: Nonlinear Liouville theorems for some critical problems on H-type groups. J. Funct. Anal., **207**, 161–215 (2004)

[BU04b] Bonfiglioli, A., Uguzzoni, F.: Families of diffeomorphic sub-Laplacians and free Carnot groups. Forum Math., **16**, 403–415 (2004)

[BU05a] Bonfiglioli, A., Uguzzoni, F.: A note on lifting of Carnot groups. Rev. Mat. Iberoamericana, **21**, 1–23 (2005)

[BU05b] Bonfiglioli, A., Uguzzoni, F.: Harnack inequality for non-divergence form operators on stratified groups. Trans. Amer. Math. Soc., **359**, 2463–2481 (2007)

[Bon69] Bony, J.-M.: Principe du maximum, inégalité de Harnack et unicité du problème de Cauchy pour les opérateurs elliptiques dégénérés. Ann. Inst. Fourier (Grenoble), **19**, 277–304 (1969)

[Bon70] Bony, J.-M.: Opérateurs elliptiques dégénérés associés aux axiomatiques de la théorie du potentiel. Potential Theory (C.I.M.E., I Ciclo, Stresa, 1969), Edizioni Cremonese, Rome, 69–119 (1970)

[Bou89] Bourbaki, N.: Lie Groups and Lie Algebras. Chapters 1–3, Elements of Mathematics, Springer-Verlag, Berlin (1989)

[BRS98] Brandolini, L., Rigoli, M., Setti, A.G.: Positive solutions of Yamabe-type equations on the Heisenberg group. Duke Math. J., **91**, 41–296 (1998)

[Bra74] Bratzlavsky, F.: Classification des algèbres de Lie nilpotentes de dimension n, de classe $n-1$, dont l'idéal dérivé est commutatif. Bull. Cl. Sci., V. Sér., Acad. R. Belg., **60**, 858–865 (1974)

[BR67] Brelot, M.: Recherches sur la topologie fine et ses applications: Théorie du potentiel. Ann. Inst. Fourier (Grenoble), **17**, 395–423 (1967)

[Bus55] Busemann, H.: The Geometry of Geodesics. Academic Press, New York (1955)

[CC95] Caffarelli, L.A., Cabré, X.: Fully Nonlinear Elliptic Equations. Amer. Math. Soc. Colloquium Publications, **43**, Amer. Math. Soc., Providence, RI (1995)

[Cap99] Capogna, L.: Regularity for quasilinear equations and 1-quasiconformal maps in Carnot groups. Math. Ann., **313**, 263–295 (1999)

[CC69] Capogna, L., Cowling, M.: Conformality and Q-harmonicity in Carnot groups. Duke Math. J., **135**, 455–479 (2006)

[CDG96] Capogna, L., Danielli, D., Garofalo, N.: Capacitary estimates and the local behavior of solutions of nonlinear subelliptic equations. Amer. J. Math., **118**, 1153–1196 (1986)

[CDG97] Capogna, L., Danielli, D., Garofalo, N.: Subelliptic mollifiers and a basic pointwise estimate of Poincaré type. Math. Z., **226**, 147–154 (1997)

[CG98] Capogna, L., Garofalo, N.: Boundary behavior of nonnegative solutions of subelliptic equations in NTA domains for Carnot–Carathéodory metrics. J. Fourier Anal. Appl., **4**, 403–432 (1998)

[CG03] Capogna, L., Garofalo, N.: Regularity of minimizers of the calculus of variations in Carnot groups via hypoellipticity of systems of Hörmander type. J. Eur. Math. Soc., **5**, 1–40 (2003)

[CG06] Capogna, L., Garofalo, N.: Ahlfors type estimates for perimeter measures in Carnot–Carathéodory spaces. J. Geom. Anal., **16**, 455–497 (2006)

[CGN00] Capogna, L., Garofalo, N., Nhieu, D.-M.: Examples of uniform and NTA domains in Carnot groups. Proceedings on Analysis and Geometry (Novosibirsk Akademgorodok, 1999), Izdat. Ross. Akad. Nauk Sib. Otd. Inst. Mat., Novosibirsk, 103–121 (2000)

[CGN02] Capogna, L., Garofalo, N., Nhieu, D.-M.: Properties of harmonic measures in the Dirichlet problem for nilpotent Lie groups of Heisenberg type. Amer. J. Math., **124**, 273–306 (2002)

[CD98] Capuzzo Dolcetta, I.: Liouville theorems and a priori estimates for elliptic semilinear equations. Rend. Sem. Mat. Fis. Milano, **68**, 1–18 (1998)

[Car09] Carathéodory, C.: Untersuchungen über die Grundlagen der Thermodynamik. Math. Ann., **67**, 355–386 (1909)

[Cart28] Cartan, H.: Sur le systèmes de fonctions holomorphes á variétés linéaires lacunaires et leurs applications. Ann. Sci. École Norm. Sup., **45**, 255–346 (1928)

[Cart45] Cartan, H.: Théorie du potentiel newtonien, énergie, capacité, suites de potentiels. Bull. Soc. Math. France, **73**, 73–106 (1945)

[Cha43] Chandresekhar, S.: Stochastic problems in physics and astronomy. Rev. Modern Phys., **15**, 1–89 (1943)

[Cho39] Chow, W.L.: Über Systeme non linearen partiellen Differentialgleichungen erster Ordung. Math. Ann., **117**, 98–105 (1939)

[Chr91] Christ, M.: Certain sums of squares of vector fields fail to be analytic hypoelliptic. Comm. Partial Differential Equations, **16**, 1695–1707 (1991)

[Chr93] Christ, M.: Analytic hypoellipticity, representations of nilpotent groups, and a nonlinear eigenvalue problem. Duke Math. J., **72**, 595–639 (1993)

[Chr95] Christ, M.: Nonexistence of invariant analytic hypoelliptic differential operators on nilpotent groups of step greater than two. Essays on Fourier Analysis in Honor of Elias M. Stein (Princeton, NJ, 1991). Princeton Math. Ser., **42**, Princeton Univ. Press, Princeton, NJ, 127–145 (1995)

[Cit98] Citti, G.: C^{∞} regularity of solutions of the Levi equation. Ann. Inst. H. Poincaré Anal. Non Linéaire, **15**, 517–534 (1998)

[CGL93] Citti, G., Garofalo, N., Lanconelli, E.: Harnack's inequality for sum of squares of vector fields plus a potential. Amer. J. Math., **115**, 699–734 (1993)

[CLM02] Citti, G., Lanconelli, E., Montanari, A.: Smoothness of Lipschitz continuous graphs with nonvanishing Levi curvature. Acta Math., **188**, 87–128 (2002)

[CM06] Citti, G., Manfredini, M.: Implicit function theorem in Carnot–Carathéodory spaces. Commun. Contemp. Math., **8**, 657–680 (2006)

[CMS04] Citti, G., Manfredini, M., Sarti, A.: Neuronal oscillations in the visual cortex: Γ-convergence to the Riemannian Mumford–Shah functional. SIAM J. Math. Anal., **35**, 1394–1419 (2004)

[CM00] Citti, G., Montanari, A.: Strong solutions for the Levi curvature equation. Adv. Differential Equations, **5**, 323–342 (2000)

[CPP01] Citti, G., Pascucci, A., Polidoro, S.: On the regularity of solutions to a nonlinear ultraparabolic equation arising in mathematical finance. Differential Integral Equations, **14**, 701–738 (2001)

[CS06] Citti, G., Sarti, A.: A cortical based model of perceptual completion in the roto-translation space. J. Math. Imaging Vision, **24**, 307–326 (2006)

[CT04] Citti, G., Tomassini, G.: Levi equation for almost complex structures. Rev. Mat. Iberoamericana, **20**, 151–182 (2004)

[CC72] Constantinescu, C., Cornea, A.: Potential Theory on Harmonic Spaces. Springer-Verlag, Berlin, (1972)

[CG90] Corwin, L.J., Greenleaf, F.P.: Representations of Nilpotent Lie Groups and their Applications (Part I: Basic Theory and Examples). Cambridge Studies in Advanced Mathematics, **18**, Cambridge University Press, Cambridge (1990)

[CDKR02] Cowling, M., De Mari, F., Korányi, A., Reimann, H.M.: Contact and conformal maps on Iwasawa N groups. Harmonic analysis on complex homogeneous domains and Lie groups. Atti Accad. Naz. Lincei Cl. Sci. Fis. Mat. Natur. Rend. Lincei (9) Mat. Appl., **13**, 219–232 (2002)

[CDKR91] Cowling, M., Dooley, A.H., Korányi, A., Ricci, F.: H-type groups and Iwasawa decompositions. Adv. Math., **87**, 1–41 (1991)

[CK84] Cowling, M., Korányi, A.: Harmonic analysis on Heisenberg type groups from a geometric viewpoint. Lie Group Representation III, Proceedings, University of Maryland, 1982–1983. Lect. Notes in Math., **1077**, Springer-Verlag, 60–100 (1984)

[CR03] Cowling, M., Reimann, H.M.: Quasiconformal mappings on Carnot groups: three examples. Harmonic Analysis at Mount Holyoke (South Hadley, MA, 2001). Contemp. Math., **320**, Amer. Math. Soc., Providence, RI, 111–118 (2003)

[Cyg81] Cygan, J.: Subadditivity of homogeneous norms on certain nilpotent Lie groups. Proc. Amer. Math. Soc., **83**, 69–70 (1981)

[Dai00] Dairbekov, N.S.: Mappings with bounded distortion of two-step Carnot groups. Vodop'yanov, S.K. (ed.), Proceedings on Analysis and Geometry. International Conference, Novosibirsk, Russia, 1999, Izdatel'stvo Instituta Matematiki Im. S. L. Soboleva SO RAN, Novosibirsk, 122–155 (2000)

[DG98] Danielli, D., Garofalo, N.: Geometric properties of solutions to subelliptic equations in nilpotent Lie groups. Reaction–Diffusion Systems (Trieste, 1995). Lecture Notes in Pure and Appl. Math., **194**, Dekker, New York, 89–105 (1998)

[DGN03] Danielli, D., Garofalo, N., Nhieu, D.-M.: Notions of convexity in Carnot groups. Comm. Anal. Geom., **11**, 263–341 (2003)

[DGNT04] Danielli, D., Garofalo, N., Nhieu, D.-M., Tournier, F.: The theorem of Busemann–Feller–Alexandrov in Carnot groups. Comm. Anal. Geom., **12**, 853–886 (2004)

[DGS03] Danielli, D., Garofalo, N., Salsa, S.: Variational inequalities with lack of ellipticity. I. Optimal interior regularity and non-degeneracy of the free boundary. Indiana Univ. Math. J., **52**, 361–398 (2003)

[Den48] Deny, J.: Un théorème sur les ensembles effilés. Annales Univ. Grenoble, Sect. Sci. Math. Phys., **23**, 139–142 (1948)

[Der71] Derridj, M.: Un problème aux limites pour une classe d'opérateurs du second ordre hypoelliptiques. Ann. Inst. Fourier (Grenoble), **21**, 99–148 (1971)

[DST91] Day, J., So, W., Thompson, R.C.: Some properties of the Campbell–Baker–Hausdorff series. Linear Multilinear Algebra, **29**, 207–224 (1991)

[DFGL05] Di Fazio, G., Gutiérrez, C.E., Lanconelli, E.: Covering theorems, inequalities on metric spaces and applications to PDE's. Preprint (2006)

[Djo75] Djoković, D.: An elementary proof of the Baker–Campbell–Hausdorff–Dynkin formula. Math. Z., **143**, 209–211 (1975)

[Doo01] Doob, J.L.: Classical Potential Theory and Its Probabilistic Counterpart. Classics in Mathematics, Springer-Verlag, Berlin (2001)

[Dra01] Dragomir, S. (ed.), Selected Topics in Cauchy–Riemann Geometry. Quaderni di Matematica (Mathematics Series), **9**, Department of Mathematics, Seconda Università di Napoli, Caserta (2001)

[duP70] du Plessis, N.: An Introduction to Potential Theory. Oliver and Boyd, Edinburgh (1970)

[Egg93] Eggert, A.: Extending the Campbell–Hausdorff multiplication. Geom. Dedicata, **46**, 35–45 (1993)

[Eic68] Eichler, M.: A new proof of the Baker–Campbell–Hausdorff formula. J. Math. Soc. Japan, **20**, 23–25 (1968)

[Fed70] Fediĭ, V.S.: Estimates in $H_{(s)}$ norms and hypoellipticity. Dokl. Akad. Nauk SSSR, **193**, 301–303 (1970)

[Fic56a] Fichera, G.: Sulle equazioni differenziali lineari ellittico–paraboliche del secondo ordine. Atti Accad. Naz. Lincei. Mem. Cl. Sci. Fis. Mat. Nat. Sez. I. (8), **5**, 1–30 (1956)

[Fic56b] Fichera, G.: Sulle equazioni alle derivate parziali del secondo ordine ellittico–paraboliche. Univ. e Politec. Torino. Rend. Sem. Mat., **15**, 27–47 (1956)

[Fok14] Fokker, A.D.: Ann. Phys. (Leipzig), **43**, 812 (1914)

[Fol75] Folland, G.B.: Subelliptic estimates and function spaces on nilpotent Lie groups. Ark. Mat., **13**, 161–207 (1975)

[Fol77] Folland, G.B.: On the Rothschild–Stein lifting theorem. Comm. Partial Differential Equations, **2**, 165–191 (1977)

[FS74] Folland, G.B., Stein, E.M.: Estimates for the $\overline{\partial_b}$ complex and analysis on the Heisenberg group. Comm. Pure Appl. Math., **27**, 429–522 (1974)

[FS82] Folland, G.B., Stein, E.M.: Hardy Spaces on Homogeneous Groups. Mathematical Notes, **28**, Princeton University Press, University of Tokyo Press, Princeton, NJ, Tokyo (1982)

[FGvN05] Franchi, B., Gutiérrez, C.E., van Nguyen, T.: Homogenization and convergence of correctors in Carnot groups. Comm. Partial Differential Equations, **30**, 1817–1841 (2005)

[FGW94] Franchi, B., Gutiérrez, C.E., Wheeden, R.L.: Weighted Sobolev–Poincaré inequalities for Grushin type operators. Comm. Partial Differential Equations, **19**, 523–604 (1994)

[FL82] Franchi, B., Lanconelli, E.: Une métrique associée à une classe d'opérateurs elliptiques dégénérés. Conference on Linear Partial and Pseudodifferential Operators (1982), Rend. Sem. Mat. Univ. Politec. Torino (1983), Special Issue, 105–114 (1984)

[FL83] Franchi, B., Lanconelli, E.: Hölder regularity theorem for a class of linear nonuniformly elliptic operators with measurable coefficients. Ann. Scuola Norm. Sup. Pisa Cl. Sci. (4), **10**, 523–541 (1983)

[FSS03a] Franchi, B., Serapioni, R., Serra Cassano, F.: Regular hypersurfaces, intrinsic perimeter and implicit function theorem in Carnot groups. Comm. Anal. Geom., **11**, 909–944 (2003)

[FSS03b] Franchi, B., Serapioni, R., Serra Cassano, F.: On the structure of finite perimeter sets in step 2 Carnot groups. J. Geom. Anal., **13**, 421–466 (2003)

[FSS01] Franchi, B., Serapioni, R., Serra Cassano, F.: Rectifiability and perimeter in the Heisenberg group. Math. Ann., **321**, 479–531 (2001)

[FSS97] Franchi, B., Serapioni, R., Serra Cassano, F.: Approximation and imbedding theorems for weighted Sobolev spaces associated with Lipschitz continuous vector fields. Boll. Unione Mat. Ital. Sez. B Artic. Ric. Mat., **11**, 83–117 (1997)

[Fri58] Friedman, A.: Remarks on the maximum principle for parabolic equations and its applications. Pacific J. Math., **8**, 201–211 (1958)

[Fri44] Friedrichs, K.: The identity of weak and strong extensions of differential operators. Trans. Amer. Math. Soc., **55**, 132–151 (1944)

[Gal82] Gallardo, L.: Capacités, mouvement Brownien et problème de l'épine de Lebesgue sur les groupes de Lie nilpotents. Proc. VII Oberwolfach Conference on Probability Measures on Groups. Lectures Notes in Math., **928**, Springer-Verlag, Berlin, New York (1982)

[Gam01] Gamara, N.: The CR Yamabe conjecture: the case $n = 1$. J. Eur. Math. Soc., **3**, 105–137 (2001)

[GY01] Gamara, N., Yacoub, R., C.R. Yamabe conjecture—the conformally flat case. Pacific J. Math., **201**, 121–175 (2001)

[GL89] Garofalo, N., Lanconelli, E.: Asymptotic behavior of fundamental solutions and potential theory of parabolic operators with variable coefficients. Math. Ann., **283**, 211–239 (1989)

[GL92] Garofalo, N., Lanconelli, E.: Existence and nonexistence results for semilinear equations on the Heisenberg group. Indiana Univ. Math. J., **41**, 71–98 (1992)

[GN98] Garofalo, N., Nhieu, D.-M.: Lipschitz continuity, global smooth approximations and extension theorems for Sobolev functions in Carnot–Carathéodory spaces. J. Anal. Math., **74**, 67–97 (1998)

[GT06] Garofalo, N., Tournier, F.: New properties of convex functions in the Heisenberg group. Trans. Amer. Math. Soc., **358**, 2011–2055 (2006)

[GV00] Garofalo, N., Vassilev, D.: Regularity near the characteristic set in the non-linear Dirichlet problem and conformal geometry of sub-Laplacians on Carnot groups. Math. Ann., **318**, 453–516 (2000)

[Gav77] Gaveau, L.: Principe de moindre action, propagation de la chaleur et estimées sous elliptiques sur certains groupes nilpotents. Acta Math., **139**, 95–153 (1977)

[Gel83] Geller, D.: Liouville's theorem for homogeneous groups. Comm. Partial Differential Equations, **8**, 1665–1677 (1983)

[GT77] Gilbarg, D., Trudinger, N.S.: Elliptic Partial Differential Equations of Second Order. Grundlehren der Mathematischen Wissenschaften **224**, Springer-Verlag, New York (1977)

[GK95] Golé, C., Karidi, R.: A note on Carnot geodesics in nilpotent Lie groups. J. Dyn. Control Syst., **1**, 535–549 (1995)

[Goo78] Goodman, R.W.: Lifting vector fields to nilpotent Lie groups. J. Math. Pures Appl., **57**, 77–85 (1978)

[GG90] Grayson, M., Grossman, R.: Models for free nilpotent Lie algebras. J. Algebra, **135**, 177–191 (1990)

[GS85] Grigis, A., Sjöstrand, J.: Front d'onde analytique et sommes des carrés de champs de vecteurs. Duke Math. J., **52**, 35–51 (1985)

[Gro96] Gromov, M.: Carnot–Carathéodory spaces seen from within. In: Bellaïche, A., Risler, J.-J. (eds.), Sub-Riemannian Geometry. Progress in Mathematics, **144**, Birkhäuser, Basel, 79–323 (1996)

[GL03] Gutiérrez, C.E., Lanconelli, E.: Maximum principle, nonhomogeneous Harnack inequality, and Liouville theorems for X-elliptic operators. Comm. Partial Differential Equations, **28**, 1833–1862 (2003)

[GM04a] Gutiérrez, C.E., Montanari, A.: Maximum and comparison principles for convex functions on the Heisenberg group. Comm. Partial Differential Equations, **29**, 1305–1334 (2004)

[GM04b] Gutiérrez, C.E., Montanari, A.: On the second order derivatives of convex functions on the Heisenberg group. Ann. Sc. Norm. Super. Pisa Cl. Sci. (5), **3**, 349–366 (2004)

[HK00] Hajłasz, P., Koskela, P.: Sobolev met Poincaré. Mem. Amer. Math. Soc., **145** (2000)

[Hal50] Hall, M.: A basis for free Lie rings and higher commutators in free groups. Proc. Amer. Math. Soc., **1**, 575–581 (1950)

[HH91] Hanges, N., Himonas, A.A.: Singular solutions for sums of squares of vector fields. Comm. Partial Differential Equations, **16**, 1503–1511 (1991)

[HH87] Hansen, W., Hueber, H.: The Dirichlet problem for sub-Laplacians on nilpotent Lie groups–geometric criteria for regularity. Math. Ann., **276**, 537–547 (1987)

[Har82] Hartman, P.: Ordinary Differential Equations. 2nd edition, Birkhäuser, Boston, Basel, Stuttgart (1982)

[HS68] Hausner, M., Schwartz, J.T.: Lie Groups. Lie Algebras. Notes on Mathematics and Its Applications, Gordon and Breach, New York, London, Paris (1968)

[HK76] Hayman, W.K., Kennedy, P.B.: Sub-Harmonic Functions, Volume I. Academic Press, London (1976)

[Hed72] Hedberg, L.I.: On certain convolution inequalities. Proc. Amer. Math. Soc., **36**, 505–510 (1972)

[Hei95a] Heinonen, J.: Calculus on Carnot groups. Kilpeläinen, T. (ed.), Fall School in Analysis, Jyväskylä, Finland, October 3–7, 1994. Univ. Jyväskylä. Bericht, **68**, University of Jyväskylä, Jyväskylä, 1–31 (1995)

[Hei95b] Heinonen, J.: A capacity estimate on Carnot groups. Bull. Sci. Math., **119**, 475–484 (1995)

[HH97] Heinonen, J., Holopainen, I.: Quasiregular maps on Carnot groups. J. Geom. Anal., **7**, 109–148 (1997)

[HKM93] Heinonen, J., Kilpelainen, T., Martio, O.: Nonlinear Potential Theory of Degenerate Elliptic Equations, Oxford Univ. Press (1993)

[HK98] Heinonen, J., Koskela, P.: Quasiconformal maps in metric spaces with controlled geometry. Acta Math., **181**, 1–61 (1998)

[Hel82] Helffer, B.: Conditions nécessaires d'hypoanalyticité pour des opérateurs invariants à gauche homogènes sur un groupe nilpotent gradué. J. Diff. Eq., **44**, 460–481 (1982)

[HN79] Helffer, B., Nourrigat, J.: Caractérisation des opérateurs hypoelliptiques homogènes invariants à gauche sur un groupe de Lie nilpotent gradué. Comm. Partial Differential Equations, **4**, 899–958 (1979)

[Helm69] Helms, L.L.: Introduction to Potential Theory. Pure and Applied Mathematics, **22**, Wiley–Interscience (1969)

[Her68] Herman, R.: Differential Geometry and the Calculus of Variations. Academic Press (1968)

[Hil70] Hill, C.D.: A sharp maximum principle for degenerate elliptic-parabolic equations. Indiana Univ. Math. J., **20**, 213–229 (1970)

[Him98] Himonas, A.A.: Analytic hypoellipticity for sums of squares of vector fields. Complex Analysis and Applications (Varsaw, 1997). Ann. Polon. Math., **70**, 117–129 (1998)

[Hoc68] Hochschild, G.: La Structure de Groupes de Lie. Monographies universitaires de mathématique, **27**, Dunod, Paris (1968)

[Hop52] Hopf, E.: A remark on linear elliptic differential equations of second order. Proc. Amer. Math. Soc., **3**, 791–793 (1952)

[Hor67] Hörmander, L.: Hypoelliptic second order differential equations. Acta Math., **119**, 147–171 (1967)

[Hor69] Hörmander, L.: Linear Partial Differential Equations. Springer-Verlag, Heidelberg, New York (1969)

[Hor94] Hörmander, L.: Notions of Convexity. Progress in Mathematics, **127**, Birkhäuser, Boston (1994)

[HM78] Hörmander, L., Melin, A.: Free systems of vector fields. Ark. Mat., **16**, 83–88 (1978)

[How80] Howe, R.: On the role of the Heisenberg group in harmonic analysis. Bull. Amer. Math. Soc., **3**, 821–843 (1980)

[Hu85] Hueber, H.: Wiener's criterion in potential theory with applications to nilpotent Lie groups. Math. Z., **190**, 527–542 (1985)

[HK99] Huisken, G., Klingenberg, W.: Flow of real hypersurfaces by the trace of the Levi form. Math. Res. Lett., **6**, 645–661 (1999)

[HJ85] Horn, R.A., Johnson, C.A.: Matrix Analysis. Cambridge University Press, Cambridge (1985)

[Ish03] Ishiwata, S.: A central limit theorem on a covering graph with a transformation group of polynomial growth. J. Math. Soc. Japan, **55**, 837–853 (2003)

[Jac62] Jacobson, N.: Lie Algebras. Interscience Tracts in Pure and Applied Mathematics, **10**, Wiley, New York, London (1962)

[JL87] Jerison, D., Lee, J.M.: The Yamabe problem on CR manifolds. J. Differential Geom., **25**, 167–197 (1987)

[JL88] Jerison, D., Lee, J.M.: Extremals for the Sobolev inequality on the Heisenberg group and the CR Yamabe problem. J. Amer. Math. Soc., **1**, 1–13 (1988)

[JL89] Jerison, D., Lee, J.M.: Intrinsic CR normal coordinates and the CR Yamabe problem. J. Differential Geom., **29**, 303–343 (1989)

[JSC87] Jerison, D., Sanchez-Calle, A.: Subelliptic, second order differential operators. In: Complex Analysis, III (College Park, Md., 1985–86). Lecture Notes in Math., **1277**, Springer-Verlag, 46–77 (1987)

[JLMS07] Juutinen, P., Lu, G., Manfredi, J.J., Stroffolini, B.: Convex functions on Carnot groups, to appear in: Rev. Mat. Iberoamericana (2007)

[Kal92] Kalf, H.: On E.E. Levi's method of constructing a fundamental solution for second-order elliptic equations. Rend. Circ. Mat. Palermo, **41**, 251–294 (1992)

[Kap80] Kaplan A.: Fundamental solutions for a class of hypoelliptic PDE generated by composition of quadratic forms. Trans. Amer. Math. Soc., **258**, 147–153 (1980)

[KK98] Kawohl, B., Kutev, N.: Strong maximum principle for semicontinuous viscosity solutions of nonlinear partial differential equations. Arch. Math. (Basel), **70**, 470–478 (1998)

[Kil94] Kilpeläinen, T.: Nonlinear potential theory and PDEs. Bertin, Emile, ICPT '91. Proceedings from the International Conference on Potential Theory, Amersfoort, The Netherlands, August 18–24, 1991. Kluwer Academic Publishers, Dordrecht, 107–118 (1994)

[Koh73] Kohn, J.J.: Pseudo-differential operators and hypoellipticity. Proc. Sympos. Pure Math., **23**, Amer. Math. Soc., 61–70 (1973)

[Kol34] Kolmogorov, A.N.: Zufällige Bewegungen. Ann. of Math., **35**, 116–117 (1934)

[Kor82] Korányi, A.: Kelvin transforms and harmonic polynomials on the Heisenberg group. J. Funct. Anal., **49**, 177–185 (1982)

[KR90] Korányi, A., Reimann, H.M.: Quasiconformal mappings on CR manifolds. Complex Geometry and Analysis (Pisa, 1988). Lecture Notes in Math., **1422**, Springer-Verlag, Berlin, 59–75 (1990)

[KR95] Korányi, A., Reimann, H.M.: Foundations for the theory of quasiconformal mappings on the Heisenberg group. Adv. Math., **111**, 1–87 (1995)

[KR87] Korányi, A., Reimann, H.M.: Horizontal normal vectors and conformal capacity of spherical rings in the Heisenberg group. Bull. Sci. Math. (2), **111**, 3–21 (1987)

[KoSt85] Korányi, A., Stanton, N.: Liouville type theorems for some complex hypoelliptic operators. J. Funct. Anal., **60**, 370–377 (1985)

[KuSt85] Kusuoka, S., Stroock, D.: Applications of the Malliavin calculus II, J. Fac. Sci. Univ. Tokyo, Sect. IA, Math., **32**, 1–76 (1985)

[Lab02] Labutin, D.: Potential estimates for a class of fully nonlinear elliptic equations. Duke Math. J., **111**, 1–49 (2002)

[LM97a] Lancia, M.R., Marchi, M.V.: Harnack inequalities for nonsymmetric operators of Hörmander type with discontinuous coefficients. Adv. Math. Sci. Appl., **7**, 833–857 (1997)

[LM97a] Lancia, M.R., Marchi, M.V.: Liouville theorems for Fuchsian-type operators on the Heisenberg group. Z. Anal. Anwendungen, **16**, 653–668 (1997)

[Lan83] Lanconelli, E.: Stime sub-ellittiche e metriche Riemanniane singolari. In: Seminario di Analisi Matematica. Istituto Matematico dell'Universitá di Bologna. A.A. 1982–83 (1983)

[Lan90] Lanconelli, E.: Formule di media per operatori subellittici del secondo ordine. In: Seminario di Analisi Matematica del Dipartimento di Matematica dell'Universitá di Bologna, A.A. 1988–89, XI.1–XI.22 (1990)

[Lan03] Lanconelli, E.: Nonlinear equations on Carnot groups and curvature problems for CR manifolds. Atti Accad. Naz. Lincei, Cl. Sci. Fis. Mat. Nat., IX. Ser., Rend. Lincei, Mat. Appl., **14**, 227–238 (2003)

[LPP02] Lanconelli, E., Pascucci, A., Polidoro, S.: Linear and nonlinear ultraparabolic equations of Kolmogorov type arising in diffusion theory and in finance. Nonlinear Problems in Mathematical Physics and Related Topics, II. Int. Math. Ser. (N. Y.), **2**, Kluwer/Plenum, New York, 243–265 (2002)

[LK00] Lanconelli, E., Kogoj, A.E.: X-elliptic operators and X-control distances. Ricerche Mat., **49**, 223–243 (2000)

[LP94] Lanconelli, E., Polidoro, S.: On a class of hypoelliptic evolution operators. Rend. Sem. Mat. Univ. Politec. Torino, **52**, 29–63 (1994)

[LU00] Lanconelli, E., Uguzzoni, F.: Non-existence results for semilinear Kohn-Laplace equations in unbounded domains. Comm. Partial Differential Equations, **25**, 1703–1739 (2000)

[Lan72] Landkof, N.S.: Foundations of Modern Potential Theory. Grundlehren der Mathematischen Wissenschaften, **180**, Springer-Verlag, Berlin, Heidelberg, New York (1972)

[LMS04] Lu, G., Manfredi, J.J., Stroffolini, B.: Convex functions on the Heisenberg group. Calc. Var. Partial Differential Equations, **19**, 1–22 (2004)

[LW97] Lu, G., Wei, J.: On positive entire solutions to the Yamabe-type problem on the Heisenberg and stratified groups. Electron. Res. Announc. Amer. Math. Soc., **3**, 83–89 (1997)

[Luo97] Luo, X.: Liouville's theorem for homogeneous differential operators. Comm. Partial Differential Equations, **22**, 1837–1848 (1997)

[Mag06] Magnani, V.: Lipschitz continuity, Aleksandrov theorem and characterizations for H-convex functions. Math. Ann., **334**, 199–233 (2006)

[MU02] Malchiodi, A., Uguzzoni, F.: A perturbation result for the Webster scalar curvature problem on the CR sphere. J. Math. Pures Appl., **81**, 983–997 (2002)

[MS02] Manfredi, J.J., Stroffolini, B.: A version of the Hopf–Lax formula in the Heisenberg group. Comm. Partial Differential Equations, **27**, 1139–1159 (2002)

[MV97a] Markina, I.G., Vodopyanov, S.K.: Fundamentals of the nonlinear potential theory for subelliptic equations, I. Siberian Adv. Math., **7**, 32–62 (1997)

[MV97b] Markina, I.G., Vodopyanov, S.K.: Fundamentals of the nonlinear potential theory for subelliptic equations, II. Siberian Adv. Math., **7**, 18–63 (1997)

[Met80] Métivier, G.: Hypoellipticité analytique sur des groupes nilpotents de rang 2. Duke Math. J., **47**, 195–213 (1980)

[Met81] Métivier, G.: Analytic hypoellipticity for operators with multiple characteristics. Comm. Partial Differential Equations, **6**, 1–90 (1981)

[Mil88] Miller, K.G.: Microlocal regularity on step two nilpotent Lie groups. Proc. Edinburgh Math. Soc., **31**, 25–39 (1988)

[Mit85] Mitchell, J.: On Carnot–Carathéodory metrics. Journal of Differential Geom., **21**, 35–45 (1985)

[Mo01] Montanari, A.: Real hypersurfaces evolving by Levi curvature: smooth regularity of solutions to the parabolic Levi equation. Comm. Partial Differential Equations, **26**, 1633–1664 (2001)

[ML04] Montanari, A., Lanconelli, E.: Pseudoconvex fully nonlinear partial differential operators: strong comparison theorems. J. Differential Equations, **202**, 306–331 (2004)

[ML01] Montanari, A., Lascialfari, F.: The Levi Monge-Ampére equation: smooth regularity of strictly Levi convex solutions. J. Geom. Anal., **14**, 331–353 (2004)

[Mon02] Montgomery, R.: A Tour of Subriemannian Geometries, Their Geodesics and Applications. Mathematical Surveys and Monographs, **91**, American Mathematical Society, Providence, RI (2002)

[MSS97] Montgomery, R., Shapiro, M., Stolin, A.: A nonintegrable sub-Riemannian geodesic flow on a Carnot group. J. Dyn. Control Syst., **3**, 519–530 (1997)

[MM05] Monti, R., Morbidelli, D.: Regular domains in homogeneous groups. Trans. Amer. Math. Soc., **357**, 2975–3011 (2005)

[MSC01] Monti, R., Serra Cassano, F.: Surface measures in Carnot–Carathéodory spaces. Calc. Var. Partial Differential Equations, **13**, 339–376 (2001)

[MR05] Monti, R., Rickly, M.: Geodetically convex sets in the Heisenberg group. J. Convex Anal., **12**, 187–196 (2005)

[Mor00] Morbidelli, D.: Fractional Sobolev norms and structure of Carnot–Carathèodory balls for Hörmander vector fields. Studia Math., **139**, 213–242 (2000)

[NSW85] Nagel, A., Stein, E.M., Wainger, S.: Balls and metrics defined by vector fields I: basic properties. Acta Math., **155**, 103–147 (1985)

[Ne88] Negrini, P.: Some remarks on capacity and the Wiener test for degenerate elliptic operators. Boll. Unione Mat. Ital. Sez. B Artic. Ric. Mat. (7) **2**-B, 555–570 (1988)

[NS87] Negrini, P., Scornazzani, V.: Wiener criterion for a class of degenerate elliptic operators. J. Differential Equations, **66**, 151–164 (1987)

[NV94] Netuka, I., Veselý, J.: Mean value property and harmonic functions. In: Classical and Modern Potential Theory and Applications (Chateau de Bonas, 1993). NATO Adv. Sci. Inst. Ser. C Math. Phys. Sci., **430**, Kluwer Acad. Publ., Dordrecht, 359–398 (1994)

[Nir53] Nirenberg, L.: A strong maximum principle for parabolic equations. Comm. Pure Appl. Math., **6**, 167–177 (1953)

[Oki95] Okikiolu, K.: The Campbell–Hausdorff theorem for elliptic operators and a related trace formula. Duke Math. J., **79**, 687–722 (1995)

[Ole52] Oleĭnik, O.A.: On properties of solutions of certain boundary problems for equations of elliptic type. Mat. Sbornik N.S., **30**, 695–702 (1952)

[OR72] Oleĭnik, O.A., Radkevič, E.V.: On the analyticity of solutions of linear differential equations and systems. Dokl. Akad. Nauk SSSR, **207**, 1614–1618 (1972)

[OR73] Oleĭnik, O.A., Radkevič, E.V.: Second Order Equations with Nonnegative Characteristic Form. Amer. Math. Soc. Providence, RI (1973)

[OV94] Onishchik, A.L., Vinberg, E.B. (eds.), Lie Groups and Lie Algebras III (Structure of Lie Groups and Lie algebras). Encyclopaedia of Mathematical Sciences, **41**, Springer-Verlag, Berlin, Heidelberg (1994)

[Ote91] Oteo, J.A.: The Baker–Campbell–Hausdorff formula and nested commutator identities. J. Math. Phys., **32**, 419–424 (1991)

[Pan82] Pansu, P.: Une inégalité isopérimétrique sur le groupe d'Heisenberg. C. R. Acad. Sci. Paris, **295**, 127–131 (1982)

[Pan89] Pansu, P.: Métriques de Carnot–Carathéodory et quasi-isométries des espaces symétriques de rang un. Ann. of Math., **129**, 1–60 (1989)

[PT99] Petitot, J., Tondut, Y.: Vers une neurogéométrie. Fibrations corticales, structures de contact et contours subjectifs modaux. Math. Inform. Sci. Humaines, **145**, 5–101 (1999)

[PR80] Pham The Lai, Robert, D.: Sur un problème aux valeurs propres non linéaire. Israel J. Math., **36**, 169–186 (1980)

[Pic13] Picone, M.: Teoremi di unicita nei problemi dei valori al contorno per le equazioni ellittiche e paraboliche. Born. Acc. L. Rend., **22**, 275–282 (1913)

[Pic27] Picone, M.: Maggiorazione degli integrali delle equazioni lineari ellittico–paraboliche alle derivate parziali del second'ordine. Rendiconti Accad. d. L. Roma (6), **5**, 138–143 (1927)

[Pla17] Planck, M.: Sitzungsber. Preuss. Akad. Wiss. Phys. Math. Kl., **324** (1917)

[PW67] Protter, M.H., Weinberger, H.F.: Maximum Principles in Differential Equations. Prentice-Hall, Englewood Cliffs, NJ (1967)

[Ras38] Rashevsky, P.K.: Any two points of a totally nonholonomic space may be connected by an admissible line. Uch. Zap. Ped. Inst. im. Liebknechta, Ser. Phys. Math., **2** 83–94 (1938)

[Red71] Redheffer, R.M.: The sharp maximum principle for nonlinear inequalities. Indiana Univ. Math. J., **21**, 227–248 (1971)

[Rei01a] Reimann, H.M.: Rigidity of H-type groups. Math. Z., **237**, 697–725 (2001)

[Rei01b] Reimann, H.M.: H-type groups and Clifford modules. Adv. Appl. Clifford Algebras, **11**, 277–287 (2001)

[Ric06] Rickly, M.: First-order regularity of convex functions on Carnot groups. J. Geom. Anal., **16**, 679–702 (2006)

[Rot83] Rothschild, L.P.: A remark on hypoellipticity of homogeneous invariant differential operators on nilpotent Lie groups. Comm. Partial Differential Equations, **8**, 1679–1682 (1983)

[Rot84] Rothschild, L.P.: Analyticity of solutions of partial differential equations on nilpotent Lie groups. Lie Group Representation III, Proceedings (University of Maryland, 1982–1983). Lect. Notes in Math., **1077**, Springer-Verlag, 389–395 (1984)

[RS76] Rothschild, L.P., Stein, E.M.: Hypoelliptic differential operators and nilpotent groups. Acta Math., **137**, 247–320 (1976)

[Rud87] Rudin, W.: Real and Complex Analysis. McGraw–Hill, New York (1987)

[SaC90] Saloff-Coste, L.: Analyse sur les groupes de Lie à croissance polynômiale, Ark. Mat., **28**, 315–331 (1990)

[SCS91] Saloff-Coste, L., Stroock, D.W.: Opérateurs uniformément sous-elliptiques sur les groupes de Lie. J. Funct. Anal., **98**, 97–121 (1991)

[ST91] Slodkowski, Z., Tomassini, G.: Weak solutions for the Levi equation and envelope of holomorphy. J. Funct. Anal., **101**, 392–407 (1991)

[Spe81] Sperb, R.P.: Maximum Principles and Their Applications. Mathematics in Science and Engineering, **157**, Academic Press Inc., New York (1981)

[Ste81] Stein, E.M.: Harmonic Analysis: Real-variable Methods, Orthogonality, and Oscillatory Integrals. Princeton Mathematical Series, **43**, Princeton University Press, Princeton, NJ (1993)

[Str87] Strichartz, R.S.: The Campbell–Baker–Hausdorff–Dynkin formula and solutions of differential equations. J. Funct. Anal., **72**, 320–345 (1987)

[Str92] Strichartz, R.: Self-similarity on nilpotent Lie groups. Contemp. Math., **140**, 123–157 (1992)

[SV72] Stroock, D.W., Varadhan, S.R.S.: On degenerate elliptic-parabolic operators of second order and their associated diffusions. Comm. Pure Appl. Math., **25**, 651–713 (1972)

[SY06] Sun, M., Yang, X.: Lipschitz continuity for H-convex functions in Carnot groups. Commun. Contemp. Math., **8**, 1–8 (2006)

[Tai88] Taira, K.: Diffusion Processes and Partial Differential Equations. Academic Press Inc., Boston, MA (1988)

[Tar80] Tartakoff, D.S.: On the local real analyticity of solutions to $\Box_b$ and $\overline{\partial}$-Neumann problem. Acta Math., **145**, 117–204 (1980)

[Tha98] Thangavelu, S.: Harmonic Analysis on the Heisenberg Group. Progress in Mathematics, **159**, Birkhäuser (1998)

[Tha94] Thangavelu, S.: A Paley–Wiener theorem for step two nilpotent Lie groups. Rev. Mat. Iberoamericana, **10**, 177–187 (1994)

[Tho82] Thompson, R.C.: Cyclic relations and the Goldberg coefficients in the Campbell–Baker–Hausdorff formula. Proc. Am. Math. Soc., **86**, 12–14 (1982)

[Tre67] Trèves, F.: Topological Vector Spaces, Distributions and Kernels. Academic Press, New York, London (1967)

[Tre78] Trèves, F.: Analytic hypo-ellipticity of a class of pseudodifferential operators with double characteristics and applications to the $\overline{\partial}$-Neumann problem. Comm. Partial Differential Equations, **3**, 475–642 (1978)

[TW97] Trudinger, N.S., Wang, X.-J.: Hessian measures. I. Topol. Methods Nonlinear Anal., **10**, 225–239 (1997)

[TW99] Trudinger, N.S., Wang, X.-J.: Hessian measures. II. Ann. of Math., **150**, 579–604 (1999)

[TW02a] Trudinger, N.S., Wang, X.-J.: Hessian measures. III. J. Funct. Anal., **193**, 1–23 (2002)

[TW02b] Trudinger, N.S., Wang, X.-J.: On the weak continuity of elliptic operators and applications to potential theory. Amer. J. Math., **124**, 369–410 (2002)

[Ugu99] Uguzzoni, F.: A Liouville-type Theorem on halfspaces for the Kohn Laplacian. Proc. Amer. Math. Soc., **127**, 117–123 (1999)

[Ugu00] Uguzzoni, F.: A note on Yamabe-type equations on the Heisenberg group. Hiroshima Math. J., **30**, 179–189 (2000)

[UL97] Uguzzoni, F., Lanconelli, E.: On the Poisson kernel for the Kohn Laplacian. Rend. Mat. Appl., **17**, 659–677 (1997)

[UL02] Uguzzoni, F., Lanconelli, E.: Degree theory for VMO maps on metric spaces and applications to Hörmander operators. Ann. Scuola Norm. Sup. Pisa Cl. Sci. (V), **1**, 569–601 (2002)

[Var84] Varadarajan, V.S.: Lie Groups, Lie Algebras and Their Representations. Graduate Texts in Mathematics, Springer-Verlag, New York, (1984)

[Varo87] Varopoulos, N.T.: Fonctions harmoniques sur les groups de Lie. C. R. Acad. Sci. Paris, **304**, 519–521 (1987)

[Varo00] Varopoulos, N.T.: Geometric and potential theoretic results on Lie groups. Canad. J. Math., **52**, 412–437 (2000)

[VSC92] Varopoulos, N.T., Saloff-Coste, L., Coulhon, T.: Analysis and Geometry on Groups. Cambridge Tracts in Mathematics, **100**, Cambridge University Press, Cambridge (1992)

[VM96] Vodopýanov, S.K., Markina, I.G.: Foundations of the nonlinear potential theory of subelliptic equations. Reshetnyak, Y.G., et al. (ed.), Sobolev Spaces and Adjacent Problems of Analysis. Tr. Inst. Mat. Im. S. L. Soboleva SO RAN, **31**, Izdatel'stvo Instituta Matematiki SO RAN, Novosibirsk, 100–160 (1996)

[Wan05] Wang, C.: Viscosity convex functions on Carnot groups. Proc. Amer. Math. Soc., **133**, 1247–1253 (2005)

[War83] Warner, F.W.: Foundations of Differentiable Manifolds and Lie Groups. Graduate Texts in Mathematics, **94**, Springer-Verlag, New York (1983)

[Weyl31] Weyl, H.: The Theory of Groups and Quantum Mechanics. Methuen, London (1931)

Index of the Basic Notation[1]

[1] The page number refers to the first time the symbol is introduced within the text. In case of multiple numbering, this means that the symbol is employed in multiple contexts.

Index

Springer Monographs in Mathematics

This series publishes advanced monographs giving well-written presentations of the "state-of-the-art" in fields of mathematical research that have acquired the maturity needed for such a treatment. They are sufficiently self-contained to be accessible to more than just the intimate specialists of the subject, and sufficiently comprehensive to remain valuable references for many years. Besides the current state of knowledge in its field, an SMM volume should also describe its relevance to and interaction with neighbouring fields of mathematics, and give pointers to future directions of research.

Abhyankar, S.S. **Resolution of Singularities of Embedded Algebraic Surfaces** 2nd enlarged ed. 1998
Alexandrov, A.D. **Convex Polyhedra** 2005
Andrievskii, V.V.; Blatt, H.-P. **Discrepancy of Signed Measures and Polynomial Approximation** 2002
Angell, T.S.; Kirsch, A. **Optimization Methods in Electromagnetic Radiation** 2004
Ara, P.; Mathieu, M. **Local Multipliers of C*-Algebras** 2003
Armitage, D.H.; Gardiner, S.J. **Classical Potential Theory** 2001
Arnold, L. **Random Dynamical Systems** corr. 2nd printing 2003 (1st ed. 1998)
Arveson, W. **Noncommutative Dynamics and E-Semigroups** 2003
Aubin, T. **Some Nonlinear Problems in Riemannian Geometry** 1998
Auslender, A.; Teboulle, M. **Asymptotic Cones and Functions in Optimization and Variational Inequalities** 2003
Banagl, M. **Topological Invariants of Stratified Spaces** 2006
Banasiak, J.; Arlotti, L. **Perturbations of Positive Semigroups with Applications** 2006
Bang-Jensen, J.; Gutin, G. **Digraphs** 2001
Baues, H.-J. **Combinatorial Foundation of Homology and Homotopy** 1999
Bonfiglioli, A., Lanconelli, E., Uguzzoni, F. **Stratified Lie Groups and Potential Theory for their Sub-Laplacians** 2007
Böttcher, A.; Silbermann, B. **Analysis of Toeplitz Operators** 2006
Brown, K.S. **Buildings** 3rd printing 2000 (1st ed. 1998)
Chang, K.-C. **Methods in Nonlinear Analysis** 2005
Cherry, W.; Ye, Z. **Nevanlinna's Theory of Value Distribution** 2001
Ching, W.K. **Iterative Methods for Queuing and Manufacturing Systems** 2001
Chudinovich, I. **Variational and Potential Methods for a Class of Linear Hyperbolic Evolutionary Processes** 2005
Coates, J.; Sujatha, R. **Cyclotomic Fields and Zeta Values** 2006
Crabb, M.C.; James, I.M. **Fibrewise Homotopy Theory** 1998
Dineen, S. **Complex Analysis on Infinite Dimensional Spaces** 1999
Dugundji, J.; Granas, A. **Fixed Point Theory** 2003
Ebbinghaus, H.-D.; Flum, J. **Finite Model Theory** 2006
Edmunds, D.E.; Evans, W.D. **Hardy Operators, Function Spaces and Embeddings** 2004
Elstrodt, J.; Grunewald, F.; Mennicke, J. **Groups Acting on Hyperbolic Space** 1998
Engler, A.J.; Prestel, A. **Valued Fields** 2005
Fadell, E.R.; Husseini, S.Y. **Geometry and Topology of Configuration Spaces** 2001
Fedorov, Yu. N.; Kozlov, V.V. **A Memoir on Integrable Systems** 2001
Flenner, H.; O'Carroll, L.; Vogel, W. **Joins and Intersections** 1999
Gelfand, S.I.; Manin, Y.I. **Methods of Homological Algebra** 2nd ed. 2003 (1st ed. 1996)
Griess, R.L. Jr. **Twelve Sporadic Groups** 1998
Gras, G. **Class Field Theory** corr. 2nd printing 2005
Greuel, G.-M.; Lossen, C.; Shustin, E. **Introduction to Singularities and Deformations** 2007
Hida, H. ***p*-Adic Automorphic Forms on Shimura Varieties** 2004
Ischebeck, E; Rao, R.A. **Ideals and Reality** 2005
Ivrii, V. **Microlocal Analysis and Precise Spectral Asymptotics** 1998
Jakimovski, A.; Sharma, A.; Szabados, J. **Walsh Equiconvergence of Complex Interpolating Polynomials** 2006
Jech, T. **Set Theory** (3rd revised edition 2002)
Jorgenson, J.; Lang, S. **Spherical Inversion on SLn (R)** 2001

Kanamori, A. **The Higher Infinite** corr. 2nd printing 2005 (2nd ed. 2003)
Kanovei, V. **Nonstandard Analysis, Axiomatically** 2005
Khoshnevisan, D. **Multiparameter Processes** 2002
Koch, H. **Galois Theory of *p*-Extensions** 2002
Komornik, V. **Fourier Series in Control Theory** 2005
Kozlov, V.; Maz'ya, V. **Differential Equations with Operator Coefficients** 1999
Lam, T.Y. **Serre's Problem on Projective Modules** 2006
Landsman, N.P. **Mathematical Topics between Classical & Quantum Mechanics** 1998
Leach, J.A.; Needham, D.J. **Matched Asymptotic Expansions in Reaction-Diffusion Theory** 2004
Lebedev, L.P.; Vorovich, I.I. **Functional Analysis in Mechanics** 2002
Lemmermeyer, F. **Reciprocity Laws: From Euler to Eisenstein** 2000
Malle, G.; Matzat, B.H. **Inverse Galois Theory** 1999
Mardesic, S. **Strong Shape and Homology** 2000
Margulis, G.A. **On Some Aspects of the Theory of Anosov Systems** 2004
Miyake, T. **Modular Forms** 2006
Murdock, J. **Normal Forms and Unfoldings for Local Dynamical Systems** 2002
Narkiewicz, W. **Elementary and Analytic Theory of Algebraic Numbers** 3rd ed. 2004
Narkiewicz, W. **The Development of Prime Number Theory** 2000
Neittaanmaki, P.; Sprekels, J.; Tiba, D. **Optimization of Elliptic Systems. Theory and Applications** 2006
Onishchik, A.L. **Projective and Cayley–Klein Geometries** 2006
Parker, C.; Rowley, P. **Symplectic Amalgams** 2002
Peller, V. (Ed.) **Hankel Operators and Their Applications** 2003
Prestel, A.; Delzell, C.N. **Positive Polynomials** 2001
Puig, L. **Blocks of Finite Groups** 2002
Ranicki, A. **High-dimensional Knot Theory** 1998
Ribenboim, P. **The Theory of Classical Valuations** 1999
Rowe, E.G.P. **Geometrical Physics in Minkowski Spacetime** 2001
Rudyak, Y.B. **On Thorn Spectra, Orientability and Cobordism** 1998
Ryan, R.A. **Introduction to Tensor Products of Banach Spaces** 2002
Saranen, J.; Vainikko, G. **Periodic Integral and Pseudodifferential Equations with Numerical Approximation** 2002
Schneider, P. **Nonarchimedean Functional Analysis** 2002
Serre, J-P. **Complex Semisimple Lie Algebras** 2001 (reprint of first ed. 1987)
Serre, J-P. **Galois Cohomology** corr. 2nd printing 2002 (1st ed. 1997)
Serre, J-P. **Local Algebra** 2000
Serre, J-P. **Trees** corr. 2nd printing 2003 (1st ed. 1980)
Smirnov, E. **Hausdorff Spectra in Functional Analysis** 2002
Springer, T.A.; Veldkamp, F.D. **Octonions, Jordan Algebras, and Exceptional Groups** 2000
Székelyhidi, L. **Discrete Spectral Synthesis and Its Applications** 2006
Sznitman, A.-S. **Brownian Motion, Obstacles and Random Media** 1998
Taira, K. **Semigroups, Boundary Value Problems and Markov Processes** 2003
Talagrand, M. **The Generic Chaining** 2005
Tauvel, P.; Yu, R.W.T. **Lie Algebras and Algebraic Groups** 2005
Tits, J.; Weiss, R.M. **Moufang Polygons** 2002
Uchiyama, A. **Hardy Spaces on the Euclidean Space** 2001
Üstünel, A.-S.; Zakai, M. **Transformation of Measure on Wiener Space** 2000
Vasconcelos, W. **Integral Closure. Rees Algebras, Multiplicities, Algorithms** 2005
Yang, Y. **Solitons in Field Theory and Nonlinear Analysis** 2001
Zieschang, P.-H. **Theory of Association Schemes** 2005